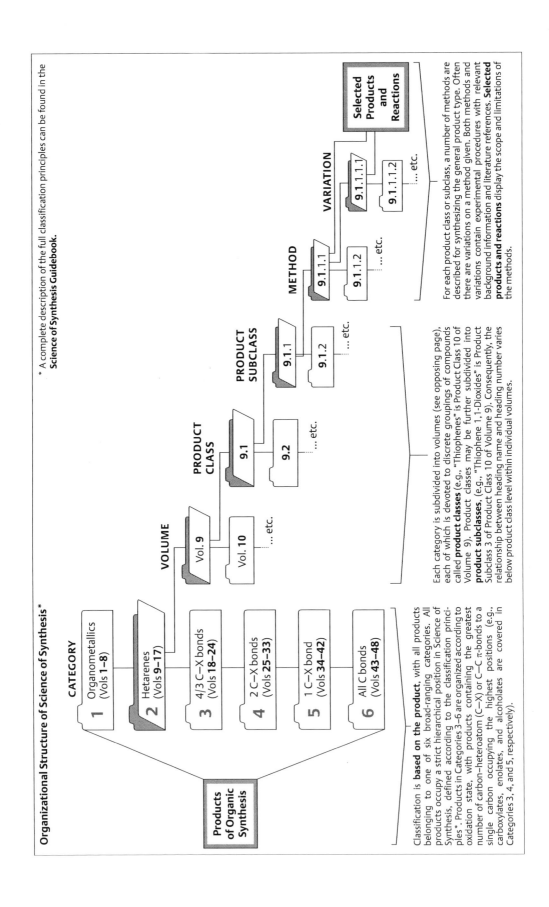

Science of Synthesis Reference Library

The **Science of Synthesis Reference Library** comprises volumes covering special topics of organic chemistry in a modular fashion, with six main classifications: (1) Classical, (2) Advances, (3) Transformations, (4) Applications, (5) Structures, and (6) Techniques. Volumes in the **Science of Synthesis Reference Library** focus on subjects of particular current interest with content that is evaluated by experts in their field. **Science of Synthesis**, including the **Knowledge Updates** and the **Reference Library**, is the complete information source for the modern synthetic chemist.

Science of Synthesis

Science of Synthesis is the authoritative and comprehensive reference work for the entire field of organic and organometallic synthesis.

Science of Synthesis presents the important synthetic methods for all classes of compounds and includes:
- Methods critically evaluated by leading scientists
- Background information and detailed experimental procedures
- Schemes and tables which illustrate the reaction scope

 **Science of Synthesis**

Editorial Board	E. M. Carreira	E. Schaumann
	C. P. Decicco	M. Shibasaki
	A. Fuerstner	E. J. Thomas
	G. Koch	B. M. Trost
	G. A. Molander	
Managing Editor	M. F. Shortt de Hernandez	
Senior Scientific Editors	K. M. Muirhead-Hofmann	
	T. B. Reeve	
	A. G. Russell	
Scientific Editors	E. L. Hughes	M. Weston
	J. S. O'Donnell	F. Wuggenig
	E. Smeaton	
Scientific Consultant	J. P. Richmond	

Georg Thieme Verlag KG
Stuttgart · New York

Science of Synthesis

Metal-Catalyzed Cyclization Reactions 2

Volume Editors	**S. Ma**
	S. Gao
Authors	U. Bora
	G. Domínguez
	H. Du
	L. K. B. Garve
	M. Harmata
	W. Hu
	D. E. Jones
	D. Lee
	X. Li
	M. Mondal
	J. Pérez-Castells
	V. Reddy Sabbasani
	Y. Shibata
	K. Tanaka
	W. Tang
	D. B. Werz
	F. Xia
	X. Xu
	S. Ye

2017
Georg Thieme Verlag KG
Stuttgart · New York

© 2017 Georg Thieme Verlag KG
Rüdigerstrasse 14
D-70469 Stuttgart

Printed in Germany

Typesetting: Ziegler + Müller, Kirchentellinsfurt
Printing and Binding: AZ Druck und Datentechnik GmbH, Kempten

Bibliographic Information published by
Die Deutsche Bibliothek

Die Deutsche Bibliothek lists this publication in the Deutsche Nationalbibliografie; detailed bibliographic data is available on the internet at <http://dnb.ddb.de>

Library of Congress Card No.: applied for

British Library Cataloguing in Publication Data

A catalogue record for this book is available from the British Library

ISBN 978-3-13-199821-7

Date of publication: December 14, 2016

Copyright and all related rights reserved, especially the right of copying and distribution, multiplication and reproduction, as well as of translation. No part of this publication may be reproduced by any process, whether by photostat or microfilm or any other procedure, without previous written consent by the publisher. This also includes the use of electronic media of data processing or reproduction of any kind.

This reference work mentions numerous commercial and proprietary trade names, registered trademarks and the like (not necessarily marked as such), patents, production and manufacturing procedures, registered designs, and designations. The editors and publishers wish to point out very clearly that the present legal situation in respect of these names or designations or trademarks must be carefully examined before making any commercial use of the same. Industrially produced apparatus and equipment are included to a necessarily restricted extent only and any exclusion of products not mentioned in this reference work does not imply that any such selection of exclusion has been based on quality criteria or quality considerations.

Warning! Read carefully the following: Although this reference work has been written by experts, the user must be advised that the handling of chemicals, microorganisms, and chemical apparatus carries potentially life-threatening risks. For example, serious dangers could occur through quantities being incorrectly given. The authors took the utmost care that the quantities and experimental details described herein reflected the current state of the art of science when the work was published. However, the authors, editors, and publishers take no responsibility as to the correctness of the content. Further, scientific knowledge is constantly changing. As new information becomes available, the user must consult it. Although the authors, publishers, and editors took great care in publishing this work, it is possible that typographical errors exist, including errors in the formulas given herein. Therefore, **it is imperative that and the responsibility of every user to carefully check whether quantities, experimental details, or other information given herein are correct based on the user's own understanding as a scientist.** Scale-up of experimental procedures published in **Science of Synthesis** carries additional risks. In cases of doubt, the user is strongly advised to seek the opinion of an expert in the field, the publishers, the editors, or the authors. When using the information described herein, the user is ultimately responsible for his or her own actions, as well as the actions of subordinates and assistants, and the consequences arising therefrom.

Preface

As the pace and breadth of research intensifies, organic synthesis is playing an increasingly central role in the discovery process within all imaginable areas of science: from pharmaceuticals, agrochemicals, and materials science to areas of biology and physics, the most impactful investigations are becoming more and more molecular. As an enabling science, synthetic organic chemistry is uniquely poised to provide access to compounds with exciting and valuable new properties. Organic molecules of extreme complexity can, given expert knowledge, be prepared with exquisite efficiency and selectivity, allowing virtually any phenomenon to be probed at levels never before imagined. With ready access to materials of remarkable structural diversity, critical studies can be conducted that reveal the intimate workings of chemical, biological, or physical processes with stunning detail.

The sheer variety of chemical structural space required for these investigations and the design elements necessary to assemble molecular targets of increasing intricacy place extraordinary demands on the individual synthetic methods used. They must be robust and provide reliably high yields on both small and large scales, have broad applicability, and exhibit high selectivity. Increasingly, synthetic approaches to organic molecules must take into account environmental sustainability. Thus, atom economy and the overall environmental impact of the transformations are taking on increased importance.

The need to provide a dependable source of information on evaluated synthetic methods in organic chemistry embracing these characteristics was first acknowledged over 100 years ago, when the highly regarded reference source **Houben–Weyl Methoden der Organischen Chemie** was first introduced. Recognizing the necessity to provide a modernized, comprehensive, and critical assessment of synthetic organic chemistry, in 2000 Thieme launched **Science of Synthesis, Houben–Weyl Methods of Molecular Transformations**. This effort, assembled by almost 1000 leading experts from both industry and academia, provides a balanced and critical analysis of the entire literature from the early 1800s until the year of publication. The accompanying online version of **Science of Synthesis** provides text, structure, substructure, and reaction searching capabilities by a powerful, yet easy-to-use, intuitive interface.

From 2010 onward, **Science of Synthesis** is being updated quarterly with high-quality content via **Science of Synthesis Knowledge Updates**. The goal of the **Science of Synthesis Knowledge Updates** is to provide a continuous review of the field of synthetic organic chemistry, with an eye toward evaluating and analyzing significant new developments in synthetic methods. A list of stringent criteria for inclusion of each synthetic transformation ensures that only the best and most reliable synthetic methods are incorporated. These efforts guarantee that **Science of Synthesis** will continue to be the most up-to-date electronic database available for the documentation of validated synthetic methods.

Also from 2010, **Science of Synthesis** includes the **Science of Synthesis Reference Library**, comprising volumes covering special topics of organic chemistry in a modular fashion, with six main classifications: (1) Classical, (2) Advances, (3) Transformations, (4) Applications, (5) Structures, and (6) Techniques. Titles will include *Stereoselective Synthesis*, *Water in Organic Synthesis*, and *Asymmetric Organocatalysis*, among others. With expert-evaluated content focusing on subjects of particular current interest, the **Science of Synthesis Reference Library** complements the **Science of Synthesis Knowledge Updates**, to make **Science of Synthesis** the complete information source for the modern synthetic chemist.

The overarching goal of the **Science of Synthesis** Editorial Board is to make the suite of **Science of Synthesis** resources the first and foremost focal point for critically evaluated information on chemical transformations for those individuals involved in the design and construction of organic molecules.

Throughout the years, the chemical community has benefited tremendously from the outstanding contribution of hundreds of highly dedicated expert authors who have devoted their energies and intellectual capital to these projects. We thank all of these individuals for the heroic efforts they have made throughout the entire publication process to make **Science of Synthesis** a reference work of the highest integrity and quality.

The Editorial Board July 2010

E. M. Carreira (Zurich, Switzerland)
C. P. Decicco (Princeton, USA)
A. Fuerstner (Muelheim, Germany)
G. A. Molander (Philadelphia, USA)
P. J. Reider (Princeton, USA)

E. Schaumann (Clausthal-Zellerfeld, Germany)
M. Shibasaki (Tokyo, Japan)
E. J. Thomas (Manchester, UK)
B. M. Trost (Stanford, USA)

Science of Synthesis Reference Library

Metal-Catalyzed Cyclization Reactions (2 Vols.)
Applications of Domino Transformations in Organic Synthesis (2 Vols.)
Catalytic Transformations via C—H Activation (2 Vols.)
Biocatalysis in Organic Synthesis (3 Vols.)
C-1 Building Blocks in Organic Synthesis (2 Vols.)
Multicomponent Reactions (2 Vols.)
Cross Coupling and Heck-Type Reactions (3 Vols.)
Water in Organic Synthesis
Asymmetric Organocatalysis (2 Vols.)
Stereoselective Synthesis (3 Vols.)

Volume Editors' Preface

Metal-catalyzed reactions, especially cyclizations, remain the most useful methods for the efficient construction of cyclic compounds and thus have continuously attracted attention. These methodologies have also been comprehensively applied as the key steps in various novel strategies for the synthesis of natural products and drug molecules. Many new discoveries and advances have been reported, which provide new tools for synthetic organic chemists, medicinal chemists, and even materials chemists. A timely summary is now required of the well-established advances that will shape the future development of this field and its application. On the basis of these considerations, the Editorial Board of *Science of Synthesis* planned two volumes in the Reference Library that focus on metal-catalyzed cyclization reactions. After a careful selection made by the Volume Editors, some of the most significant and practical metal-catalyzed reactions for modern organic synthesis are presented. The organization of these two volumes is based on the types of reaction, which mainly include metal-catalyzed C—C, C—O, and C—N bond formations, as well as epoxidation, aziridination, cyclopropanation, Pauson–Khand reactions, cycloadditions, radical reactions, and metathesis. We hope that these volumes can serve as a reference work for chemists in related areas to inspire future research and the development of new applications.

We would like to take this opportunity to express our sincere thanks for the support and contributions from all of the outstanding authors; without their dedication and professionalism, this project would not have been possible. We are also grateful to Alex Russell and Joe P. Richmond, scientific editors at *Science of Synthesis*, who solved a lot of problems to enable the project to proceed smoothly over the past two and a half years. It has been a very pleasant journey to work with the editorial team at Thieme, especially Michaela Frey and Guido F. Herrmann. We appreciate the professional spirit and passion of all members of this team, which have brought the manuscripts together into a two-volume set.

Volume Editors

Shengming Ma and Shuanhu Gao   Shanghai, May 2016

Abstracts

2.1 Epoxidation and Aziridination Reactions
F. Xia and S. Ye

p 1

Due to the strain associated with three-membered rings, epoxides and aziridines react with various nucleophiles to give 1,2-difunctionalized products. This makes them valuable intermediates in organic synthesis, including in the synthesis of bioactive natural and nonnatural compounds. This chapter summarizes recent advances in metal-catalyzed epoxidation and aziridination reactions of alkenes, as well as their synthetic applications.

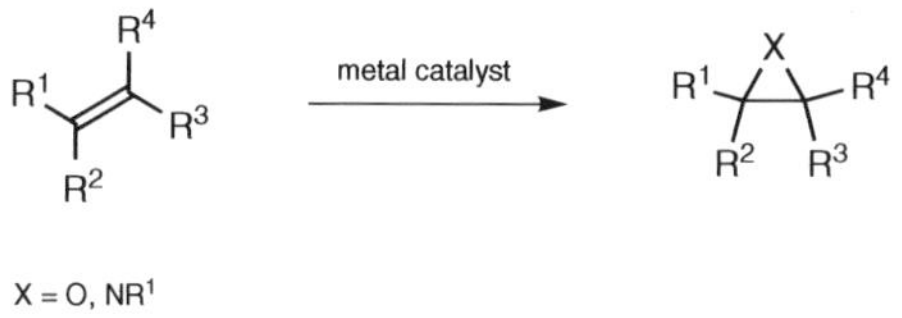

$X = O, NR^1$

Keywords: epoxidation · aziridination · expoxides · aziridines · alkenes · metal catalysis · cyclization · enantioselectivity · diastereoselectivity · asymmetric synthesis

2.2 Metal-Catalyzed Cyclopropanation
L. K. B. Garve and D. B. Werz

p 49

This chapter describes the most important metal-catalyzed methods to generate cyclopropanes, the smallest class of cycloalkanes. In the past fifty years, the use of metals in combination with chiral ligands for diastereo- and enantiodiscrimination in cyclopropane synthesis has been intensively studied. Two main approaches have emerged. Utilizing carbenes in the form of metal–carbene complexes has led to a renaissance of three-membered-rings in organic synthesis. In another approach, metal cations such as gold(I) and platinum(II) interact with alkynes, forming novel cyclopropane motifs. Finally, further metal-catalyzed cyclopropanations are mentioned.

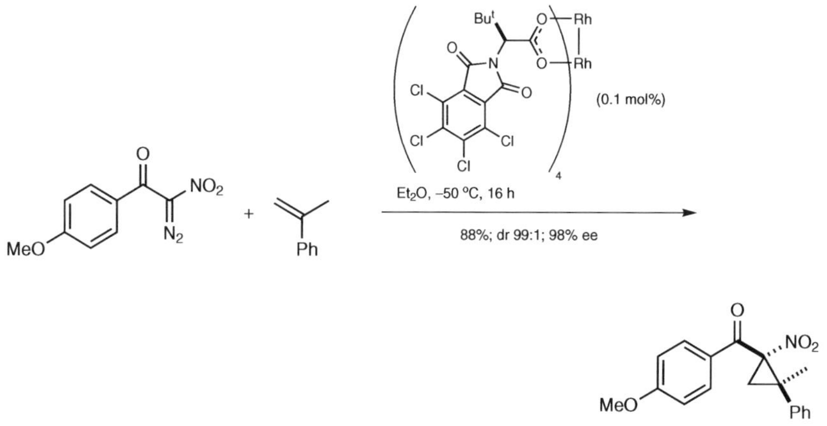

Keywords: cyclopropanes · three-membered rings · metal-catalyzed cyclopropanation · asymmetric cyclopropanation · carbenes · metal–carbene complexes · Kulinkovich cyclopropanation · enynes · copper · rhodium · cobalt · iridium · ruthenium · iron · palladium · gold · platinum · titanium · bicyclic molecules

─── p 99 ───

2.3 Pauson–Khand Reactions
G. Domínguez and J. Pérez-Castells

The Pauson–Khand (2+2+1) cycloaddition (PKR) is one of the best ways to construct a cyclopentenone. In this review, standard procedures to perform both stoichiometric and catalytic Pauson–Khand reactions are presented. These include the use of homo- and heterogeneous catalysts as well as different promoters. Asymmetric versions are described, as are cascade processes where the Pauson–Khand reaction is combined with other transformations. The chapter ends with a summary of representative total syntheses of natural products where a Pauson–Khand reaction is used as the key step.

Keywords: Pauson–Khand · cycloaddition · transition-metal catalysis · stereoselectivity · asymmetric synthesis · cobalt · rhodium · iridium · heterogeneous catalysis · cascade synthesis · total synthesis

─── p 167 ───

2.4 1,3-Dipolar Cycloadditions Involving Carbonyl or Azomethine Ylides
X. Xu and W. Hu

Carbonyl ylides, which behave as active 1,3-dipole species, have found numerous applications in organic synthesis, especially in the formation of five-membered heterocycles. Among the versatile transformations of carbonyl ylides, 1,3-dipolar cycloadditions with π-bonds, including (3+2)-cycloaddition reactions with carbon–carbon π-bonds, aldehydes, and imines, are ubiquitous and important reactions. This chapter focuses on recent advances in these catalytic (3+2)-cycloaddition reactions and the examples presented emphasize the chemo-, diastereo-, and enantiocontrol that can be achieved. Also described are selected examples of cycloaddition reactions with azomethine ylides, which behave with similar reactivity to carbonyl ylides, to give the corresponding N-heterocycles. In addition, some selected applications of these (3+2)-cycloaddition reactions in natural product synthesis are highlighted.

$X = O, NR^5; Y=Z$ = aldehyde, ketone, alkene, alkyne, imine, etc.

Keywords: (3+2) cycloaddition · carbonyl ylides · azomethine ylides · 1,3-dipoles · diazo compounds · chiral dirhodium carboxylates · heterocycles · five-membered rings · natural product synthesis · asymmetric catalysis

2.5 Metal-Catalyzed Asymmetric Diels–Alder and Hetero-Diels–Alder Reactions
H. Du

p 205

This chapter focuses on asymmetric Diels–Alder, oxa-Diels–Alder, and aza-Diels–Alder reactions promoted by various chiral, Lewis acidic metal catalysts. As well as describing the different approaches and catalysts, selected examples of these reactions in the synthesis of natural products and drug molecules are presented.

$X = CHR^6$, O, NR^6

Keywords: asymmetric synthesis · aza-Diels–Alder reactions · Diels–Alder reactions · Lewis acid catalysis · metal-catalyzed reactions · oxa-Diels–Alder reactions

2.6 Metal-Catalyzed (2+2+2) Cycloadditions
K. Tanaka and Y. Shibata

p 265

The transition-metal-catalyzed (2+2+2) cycloaddition is a useful and atom-economical method for the synthesis of substituted six-membered-ring molecules. In this chapter, standard procedures for the transition-metal-catalyzed (2+2+2) cycloaddition involving not only alkynes but also nitriles, heterocumulenes, alkenes, and carbonyl compounds are presented. Applications in the synthesis of chiral aromatic molecules (biaryls, cyclophanes, and helicenes) and biologically active molecules are also described.

Keywords: alkenes · alkynes · asymmetric catalysis · benzenes · biaryls · carbocycles · carbodiimides · carbon dioxide · carbonyl compounds · cycloaddition · cyclophanes · helicenes · heterocumulenes · heterocycles · isocyanates · metallacycles · nitriles · pyridines · pyridones · transition metals

2.7 Metal-Catalyzed (4+3) Cycloadditions Involving Allylic Cations
D. E. Jones and M. Harmata

p 319

The (4+3)-cycloaddition reaction is a reaction between an allylic cation and diene. It is a powerful process and one of the few reactions that directly generates seven-membered ring systems. This reaction exhibits high levels of diastereoselectivity and regioselectivity, which makes it attractive for the rapid construction of complex targets. Several select, reliable, and reproducible methods for the generation and capture of allylic cations by 1,3-dienes using metal catalysis or mediation are presented in this review along with some representative examples of the use of (4+3) cycloadditions in the synthesis of natural products.

Keywords: (4+3) cycloaddition · intramolecular · intermolecular · allylic cations · oxyallylic cations · chiral cations · 1,3-dienes · regioselective · diastereoselective · stereoselective · asymmetric · enantioselective · metal-catalyzed · metal-mediated

p 349

2.8 Metal-Catalyzed (5+1), (5+2), and (5+2+1) Cycloadditions
X. Li and W. Tang

Metal-catalyzed (5+n) cycloaddition is a powerful strategy for the synthesis of six-, seven-, and eight-membered carbocycles and heterocycles. These cycloadditions usually involve oxidative cyclization to a metallacycle; insertion into the C–M bond (e.g., by carbon monoxide, an alkene, alkyne, or allene, or a combination thereof); and reductive elimination. Vinylcyclopropanes and 3-acyloxy-1,4-enynes are the most common five-carbon synthons. Recent advances in transition-metal-catalyzed (5+1), (5+2), and (5+2+1) cycloadditions including their development, mechanistic studies, and applications are reviewed in this chapter.

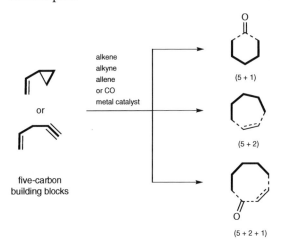

Keywords: cycloaddition · transition metals · catalysis · vinylcyclopropanes · propargylic esters · (5+1) cycloaddition · (5+2) cycloaddition · (5+2+1) cycloaddition · carbocycles · heterocycles · seven-membered rings · alkenes · alkynes · allenes · imines · cyclopropanes · epoxides · rhodium · ruthenium · nickel · iron · dictamnol · aphanamol · allocyathin B_2 · tremulenediol A · tremulenolide A · pseudolaric acid B · frondosin A · rameswaralide · hirsutene · desoxyhypnophilin · asteriscanolide

2.9 Intramolecular Free-Radical Cyclization Reactions
M. Mondal and U. Bora

p 383

Over the past five decades, the metal-mediated intramolecular free-radical cyclization strategy has developed into a widely applicable synthetic methodology. With properties including wide functional-group tolerance, selectivity, and spatial retention, radical chemistry has attracted considerable attention from chemists dealing with the synthesis and functionalization of diverse organic compounds. The coverage of this chapter is limited to the best methods available for the metal-mediated synthesis of cyclic organic and biomolecular species for practical use in both an academic setting and in industry.

Keywords: intramolecular · free-radical · cyclization · manganese(III) acetate · titanocenes · samarium(II) iodide · copper · tandem · cascade · pyrroles · alkylidenes · indoles · azaindolines · indolinones · oxindoles · lactams · quinolines · furans · lactones · phthalides · pyrans · benzopyrans · piperidines · cycloalkanes · cycloalkanols · allyl alcohols · amino alcohols · cyclic ketones · amino ketones · phenanthridines · benzothiazoles · thiophenes · total synthesis

2.10 Ring-Closing Metathesis
D. Lee and V. Reddy Sabbasani

p 543

Ring-closing metathesis (RCM) has emerged as a powerful synthetic tool. Depending on the unsaturated functional groups involved, ring-closing-metathesis reactions are classified into three categories: diene ring-closing metathesis, enyne ring-closing metathesis, and diyne ring-closing metathesis. These are mediated/catalyzed by metal alkylidenes or alkylidynes to form cyclic alkenes or alkynes, with ring sizes ranging from small to large, and including both carbocycles and heterocycles. Mechanistically, diene and diyne ring-closing metathesis involves an exchange reaction between the participating alkenes or alkynes, whereas enyne ring-closing metathesis involves a formal addition reaction between an alkene and an alkyne. This chapter summarizes the distinctive features of these different ring-closing metathesis processes in terms of the advancement of mechanistic understanding and the development of effective catalyst systems and their application to natural product synthesis.

Keywords: diene metathesis · diyne metathesis · enyne metathesis · ring-closing metathesis · ring rearrangement · ruthenium alkylidenes · metallotropic shift · molybdenum alkylidenes · molybdenum alkylidynes

Metal-Catalyzed Cyclization Reactions 2

	Preface	V
	Volume Editors' Preface	IX
	Abstracts	XI
	Table of Contents	XIX
2.1	**Epoxidation and Aziridination Reactions** F. Xia and S. Ye	1
2.2	**Metal-Catalyzed Cyclopropanation** L. K. B. Garve and D. B. Werz	49
2.3	**Pauson–Khand Reactions** G. Domínguez and J. Pérez-Castells	99
2.4	**1,3-Dipolar Cycloadditions Involving Carbonyl or Azomethine Ylides** X. Xu and W. Hu	167
2.5	**Metal-Catalyzed Asymmetric Diels–Alder and Hetero-Diels–Alder Reactions** H. Du	205
2.6	**Metal-Catalyzed (2 + 2 + 2) Cycloadditions** K. Tanaka and Y. Shibata	265
2.7	**Metal-Catalyzed (4 + 3) Cycloadditions Involving Allylic Cations** D. E. Jones and M. Harmata	319
2.8	**Metal-Catalyzed (5 + 1), (5 + 2), and (5 + 2 + 1) Cycloadditions** X. Li and W. Tang	349
2.9	**Intramolecular Free-Radical Cyclization Reactions** M. Mondal and U. Bora	383
2.10	**Ring-Closing Metathesis** D. Lee and V. Reddy Sabbasani	543
	Keyword Index	679
	Author Index	705
	Abbreviations	729

Table of Contents

2.1	**Epoxidation and Aziridination Reactions** F. Xia and S. Ye	
2.1	**Epoxidation and Aziridination Reactions**	1
2.1.1	Metal-Catalyzed Epoxidation Reactions	1
2.1.1.1	Titanium-Catalyzed Epoxidation	1
2.1.1.2	Vanadium-Catalyzed Epoxidation	3
2.1.1.3	Manganese-Catalyzed Epoxidation	7
2.1.1.4	Iron-Catalyzed Epoxidation	11
2.1.1.5	Ruthenium-Catalyzed Epoxidation	14
2.1.1.6	Molybdenum-Catalyzed Epoxidation	16
2.1.1.7	Tungsten-Catalyzed Epoxidation	17
2.1.1.8	Other Metal-Catalyzed Epoxidations	20
2.1.2	Metal-Catalyzed Aziridination Reactions	24
2.1.2.1	Copper-Catalyzed Aziridination	24
2.1.2.2	Rhodium-Catalyzed Aziridination	28
2.1.2.3	Ruthenium-Catalyzed Aziridination	32
2.1.2.4	Iron-Catalyzed Aziridination	34
2.1.2.5	Cobalt-Catalyzed Aziridination	35
2.1.2.6	Silver-Catalyzed Aziridination	39
2.1.3	Applications in the Syntheses of Natural Products and Drug Molecules	41
2.1.4	Conclusions and Future Perspectives	46
2.2	**Metal-Catalyzed Cyclopropanation** L. K. B. Garve and D. B. Werz	
2.2	**Metal-Catalyzed Cyclopropanation**	49
2.2.1	Copper-Catalyzed Cyclopropanation	50
2.2.1.1	Copper-Catalyzed Asymmetric Cyclopropanation Using a Bis(4,5-dihydrooxazole) Ligand	51
2.2.1.2	Copper-Catalyzed Cyclopropanation of Electron-Rich Double Bonds in Heterocyclic Systems	52
2.2.1.3	Copper-Catalyzed Asymmetric Cyclopropanation Leading to 1-Nitrocyclopropane-1-carboxylates	53

2.2.2	Rhodium-Catalyzed Cyclopropanation	54
2.2.2.1	Rhodium(II) Acetate Dimer Catalyzed Cyclopropanation Using Alkyl Diazoacetates	55
2.2.2.2	Rhodium-Catalyzed Enantio- and Diastereoselective Cyclopropanation Starting with Decomposition of 1-Sulfonyl-1,2,3-triazoles	56
2.2.2.3	Rhodium-Catalyzed Enantio- and Diastereoselective Cyclopropanation Leading to Cyclopropanes Substituted by Nitro and Cyano Groups	57
2.2.2.4	Rhodium-Catalyzed Enantioselective Cyclopropanation of Electron-Deficient Alkenes	59
2.2.3	Cobalt-Catalyzed Cyclopropanation	60
2.2.3.1	Cobalt(II)-Catalyzed Cyclopropanation Using Diazo Sulfones	60
2.2.3.2	Stereoselective Cyclopropanation of Alkenes with 2-Cyano-2-diazoacetates	61
2.2.4	Iridium-Catalyzed Cyclopropanation	62
2.2.4.1	*cis*-Selective Cyclopropanation of Conjugated and Nonconjugated Alkenes Using Aryliridium–Bis(salicylidene)ethylenediamine (Salen) Complexes	62
2.2.5	Ruthenium-Catalyzed Cyclopropanation	63
2.2.5.1	Enantioselective Cyclopropanation Using Ruthenium(II)–Diphenyl-4,5-dihydrooxazole Complexes	64
2.2.5.2	Ruthenium-Catalyzed Asymmetric Construction of 3-Azabicyclo[3.1.0]hexanes	65
2.2.6	Iron-Catalyzed Cyclopropanation	66
2.2.6.1	Iron-Catalyzed Cyclopropanation of Styrenes with In Situ Generation of Diazomethane	67
2.2.6.2	Iron-Catalyzed Cyclopropanation with In Situ Generated 2-Diazo-1,1,1-trifluoroethane	68
2.2.6.3	Iron-Catalyzed Cyclopropanation with Ethyl Glycinate Hydrochloride in Water	69
2.2.6.4	Enantioselective Iron-Catalyzed Intramolecular Cyclopropanation Reactions	69
2.2.7	Palladium-Catalyzed Cyclopropanation	70
2.2.7.1	Palladium(II) Acetate Catalyzed Cyclopropanation Using Diazomethane	71
2.2.7.2	Palladium-Catalyzed Cyclopropanation of Benzyl Bromides via C—H Activation	71
2.2.7.3	Palladium-Catalyzed Cyclopropanation of Acyclic Amides with Substituted Allyl Carbonates	73
2.2.7.4	Palladium-Catalyzed Cross-Coupling Reaction of Styrenes with Aryl Methyl Ketones	74
2.2.8	Gold-Catalyzed Cyclopropanation	75
2.2.8.1	Intramolecular Gold(I)-Catalyzed Cyclopropanation Forming Bicyclo[3.1.0]hexene Derivatives	76
2.2.8.2	Gold(I)-Catalyzed Asymmetric Cycloisomerization of Heteroatom-Tethered 1,6-Enynes	76
2.2.8.3	Intermolecular Gold-Catalyzed Cyclopropanation	78
2.2.8.4	Gold(I)-Catalyzed Intramolecular Biscyclopropanation of Dienynes	81

2.2.8.5	Gold(I)-Catalyzed Intermolecular Cyclopropanation To Give Functionalized Benzonorcaradienes	82
2.2.8.6	Cyclopropanation with Gold(I) Carbenes by Retro-Buchner Reaction from Cycloheptatrienes	83
2.2.9	Platinum-Catalyzed Cyclopropanation	85
2.2.9.1	Platinum(II)-Catalyzed Cyclopropanation of Dienes	85
2.2.9.2	Platinum-Catalyzed Enyne Cyclization and Acid-Catalyzed Ring-Opening Reaction	86
2.2.9.3	Platinum(II)-Catalyzed Formation of Bicyclic Alkylidenecyclopropanes	87
2.2.9.4	Platinum(II) Chloride Catalyzed Cycloisomerization of 5-En-1-yn-3-ol Precursors	88
2.2.10	Titanium-Catalyzed Cyclopropanation	88
2.2.11	Applications in the Syntheses of Natural Products and Drug Molecules	90
2.2.11.1	Gibberellin GA_{103} by Copper-Catalyzed Cyclopropanation	91
2.2.11.2	Echinopines A and B by Rhodium-Catalyzed Cyclopropanation	91
2.2.11.3	Antidepressive Agent GSK 1360707 by Gold-Catalyzed Cyclopropanation	92
2.2.11.4	Sabina Ketone by Platinum-Catalyzed Cyclopropanation	93
2.2.11.5	(S)-Cleonin by Titanium-Catalyzed Cyclopropanation	94
2.2.12	Conclusions and Future Perspectives	94

2.3 Pauson–Khand Reactions
G. Domínguez and J. Pérez-Castells

2.3	**Pauson–Khand Reactions**	99
2.3.1	General Overview of the Pauson–Khand Reaction	99
2.3.1.1	Origins, History, and Versions	99
2.3.1.2	Regioselectivity Issues in the Intermolecular Pauson–Khand Reaction	100
2.3.1.3	Promotion of the Reaction	102
2.3.1.4	Scope and Limitations	106
2.3.2	The Reaction Pathway	112
2.3.2.1	Stoichiometric Reaction	112
2.3.2.2	Catalytic Reaction	114
2.3.3	The Catalytic Pauson–Khand Reaction and Pauson–Khand-Type Reaction	115
2.3.3.1	Cobalt-Catalyzed Reactions	115
2.3.3.2	Rhodium-Catalyzed Reactions	120
2.3.3.2.1	Dienes as Alkene Partners: The Diene Effect	123
2.3.3.2.2	Allenes as Alkene Partners	125
2.3.3.2.3	Reaction in the Absence of Carbon Monoxide	127

2.3.3.3	Iridium-Catalyzed Reactions	129
2.3.3.4	Other Metal-Catalyzed Pauson–Khand-Type Reactions	130
2.3.3.5	Heterogeneous Catalysts	135
2.3.4	Inducing Asymmetry in the Pauson–Khand Reaction	138
2.3.4.1	Transferring Chirality from the Substrate	138
2.3.4.2	Using Chiral Auxiliaries	139
2.3.4.3	Using Chiral Metal Complexes	144
2.3.4.4	Using Chiral Ligands	145
2.3.4.5	Using Chiral Promoters	147
2.3.5	Cascade Reactions Involving Pauson–Khand Reactions	147
2.3.5.1	Cascade Synthesis of Enynes/Pauson–Khand Reaction	149
2.3.5.2	Tandem Carbocyclizations Involving (2+2+1) Reactions	151
2.3.5.3	Tandem Pauson–Khand Reactions	153
2.3.5.4	Reactions Occurring After the Pauson–Khand Process	155
2.3.6	Hetero-Pauson–Khand Reaction	156
2.3.7	Total Synthesis Using the Pauson–Khand Reaction as the Key Step	159
2.3.8	Conclusions	161

2.4	**1,3-Dipolar Cycloadditions Involving Carbonyl or Azomethine Ylides** X. Xu and W. Hu	
2.4	**1,3-Dipolar Cycloadditions Involving Carbonyl or Azomethine Ylides**	167
2.4.1	(3+2)-Cycloaddition Reactions of Carbonyl Ylides with Carbon–Carbon π-Bonds	170
2.4.1.1	Diastereoselective (3+2)-Cycloaddition Reactions of Carbonyl Ylides with Carbon–Carbon π-Bonds	170
2.4.1.2	Enantioselective (3+2)-Cycloaddition Reactions of Carbonyl Ylides with Carbon–Carbon π-Bonds	175
2.4.2	(3+2)-Cycloaddition Reactions of Carbonyl Ylides with Aldehydes	184
2.4.2.1	Diastereoselective (3+2)-Cycloaddition Reactions of Carbonyl Ylides with Aldehydes	184
2.4.2.2	Enantioselective (3+2)-Cycloaddition Reactions of Carbonyl Ylides with Aldehydes	187
2.4.3	Diastereoselective (3+2)-Cycloaddition Reactions of Carbonyl Ylides with Imines	188
2.4.4	1,3-Dipolar (3+2)-Cycloaddition Reactions of Azomethine Ylides	192
2.4.5	(3+2)-Cycloaddition Reactions of Carbonyl Ylides in Natural Product Synthesis	200
2.4.6	Conclusions	202

2.5	**Metal-Catalyzed Asymmetric Diels–Alder and Hetero-Diels–Alder Reactions**	
	H. Du	
2.5	**Metal-Catalyzed Asymmetric Diels–Alder and Hetero-Diels–Alder Reactions**	205
2.5.1	Diels–Alder Reactions	206
2.5.1.1	Reaction Using Chiral Boron Complexes	206
2.5.1.1.1	(Acyloxy)borane Catalysts	206
2.5.1.1.2	Superacidic Lewis Acid Catalysts	207
2.5.1.1.3	Brønsted Acid Activated Borane Catalysts	208
2.5.1.1.4	Lewis Acid Activated Borane Catalysts	211
2.5.1.2	Reaction Using Chiral Aluminum Complexes	212
2.5.1.3	Reaction Using Chiral Indium Complexes	213
2.5.1.4	Reaction Using Chiral Copper Complexes	215
2.5.1.4.1	Bis(oxazoline)–Copper Catalysts	215
2.5.1.4.2	Bis(sulfinyl)imidoamidine–Copper Catalysts	216
2.5.1.5	Reaction Using Chiral Titanium Complexes	217
2.5.1.6	Reaction Using Chiral Chromium Complexes	219
2.5.1.7	Reaction Using Chiral Cobalt Complexes	220
2.5.1.8	Reaction Using Chiral Ruthenium Complexes	221
2.5.1.9	Reaction Using Chiral Rare Earth Complexes	222
2.5.2	Oxa-Diels–Alder Reactions	224
2.5.2.1	Reaction Using Chiral Aluminum Complexes	225
2.5.2.2	Reaction Using Chiral Indium Complexes	226
2.5.2.3	Reaction Using Chiral Titanium Complexes	228
2.5.2.3.1	1,1′-Binaphthalene-2,2′-diol/Titanium (1:1) Catalysts	228
2.5.2.3.2	1,1′-Binaphthalene-2,2′-diol/Titanium (2:1) Catalysts	229
2.5.2.3.3	Chiral Schiff Base/Titanium Catalysts	230
2.5.2.4	Reaction Using Chiral Zirconium Complexes	232
2.5.2.5	Reaction Using Chiral Chromium Complexes	233
2.5.2.5.1	Chromium–Salen Catalysts	233
2.5.2.5.2	Tridentate Chromium Catalysts	234
2.5.2.6	Reaction Using Chiral Copper Complexes	236
2.5.2.6.1	Normal-Electron-Demand Oxa-Diels–Alder Reactions	236
2.5.2.6.2	Inverse-Electron-Demand Oxa-Diels–Alder Reactions	237
2.5.2.7	Reaction Using Chiral Zinc Complexes	238
2.5.2.8	Reaction Using Chiral Rhodium Complexes	241

2.5.2.9	Reaction Using Chiral Palladium Complexes	242
2.5.2.10	Reaction Using Chiral Rare Earth Metal Complexes	244
2.5.3	Aza-Diels–Alder Reactions	245
2.5.3.1	Reaction Using Chiral Copper Complexes	245
2.5.3.1.1	Normal-Electron-Demand Aza-Diels–Alder Reactions	245
2.5.3.1.2	Inverse-Electron-Demand Aza-Diels–Alder Reactions	246
2.5.3.2	Reaction Using Chiral Zinc Complexes	247
2.5.3.3	Reaction Using Chiral Silver Complexes	249
2.5.3.4	Reaction Using Chiral Zirconium Complexes	250
2.5.3.5	Reaction Using Chiral Niobium Complexes	251
2.5.3.6	Reaction Using Chiral Nickel Complexes	252
2.5.3.7	Reaction Using Chiral Rhodium Complexes	253
2.5.3.8	Reaction Using Chiral Rare Earth Metal Complexes	254
2.5.3.8.1	Normal-Electron-Demand Aza-Diels–Alder Reactions	254
2.5.3.8.2	Inverse-Electron-Demand Aza-Diels–Alder Reactions	255
2.5.4	Applications in the Syntheses of Natural Products and Drug Molecules	257
2.5.5	Conclusions and Future Perspectives	261

2.6 Metal-Catalyzed (2+2+2) Cycloadditions
K. Tanaka and Y. Shibata

2.6	**Metal-Catalyzed (2+2+2) Cycloadditions**	265
2.6.1	(2+2+2) Cycloadditions of Alkynes	266
2.6.1.1	Intermolecular Reactions	266
2.6.1.2	Intramolecular Reactions	281
2.6.2	(2+2+2) Cycloadditions of Alkynes with Nitriles	282
2.6.2.1	Intermolecular Reactions	283
2.6.2.2	Intramolecular Reactions	291
2.6.3	(2+2+2) Cycloadditions Involving Heterocumulenes	292
2.6.3.1	Isocyanates	292
2.6.3.2	Carbodiimides and Carbon Dioxide	299
2.6.4	(2+2+2) Cycloadditions Involving $C(sp^2)$ Multiple Bonds	301
2.6.4.1	Alkenes	301
2.6.4.2	Carbonyl Compounds	309
2.6.5	Applications in the Syntheses of Natural Products and Drug Molecules	312
2.6.6	Conclusions and Future Perspectives	315

2.7	**Metal-Catalyzed (4 + 3) Cycloadditions Involving Allylic Cations**	
	D. E. Jones and M. Harmata	
2.7	**Metal-Catalyzed (4 + 3) Cycloadditions Involving Allylic Cations**	319
2.7.1	Carbon-Substituted Allylic Cations in Cycloaddition Reactions	321
2.7.1.1	Reduction of α,α′-Dihalo Ketones	321
2.7.1.1.1	Reductive Cycloaddition with Copper Bronze	321
2.7.1.1.2	Reductive Cycloaddition with Zinc/Copper Couple	322
2.7.1.1.3	Reductive Cycloaddition with Iron Carbonyls	323
2.7.1.1.4	Reduction with Diethylzinc	324
2.7.1.2	Ionization of α-Halo Enol Ethers and Related Species	325
2.7.1.2.1	Silver-Mediated Ionization of Enol Ethers	325
2.7.1.2.2	Silver-Mediated Ionization of α-Chloroenamines	326
2.7.1.3	Other Approaches to Carbon-Substituted Allylic Cations	327
2.7.1.3.1	Furfuryl Alcohols as Allylic Cation Precursors	327
2.7.1.3.2	Dienones as Allylic Cation Precursors	328
2.7.1.3.3	Alkoxy Allylic Sulfones as Allylic Cation Precursors	329
2.7.1.3.4	Allenes as Allylic Cation Precursors	330
2.7.1.3.5	Decomposition of Vinyl Diazoacetates	332
2.7.2	Heteroatom-Substituted Allylic Cations in Cycloaddition Reactions	333
2.7.2.1	Nitrogen-Stabilized Allylic Cations	333
2.7.2.1.1	Allenamides as Vyniliminium Ion Precursors	334
2.7.2.1.2	Indole-3-methanols as Allylic Cation Precursors	338
2.7.2.2	Oxygen-Stabilized Allylic Cations	339
2.7.2.2.1	2-Siloxyacroleins as Vinyloxocarbenium Ion Precursors	339
2.7.2.2.2	Allylic Acetals as Vinyloxocarbenium Ion Precursors	341
2.7.2.3	Sulfur-Stabilized Allylic Cations	342
2.7.2.3.1	Sulfur-Substituted Allylic Acetals as Vinylthionium Ion Precursors	342
2.7.2.3.2	Sulfur-Substituted Allylic Sulfones as Vinylthionium Ion Precursors	342
2.7.3	Applications of Metal-Catalyzed (4 + 3) Cycloadditions to the Synthesis of Natural Products	343
2.7.3.1	(±)-Frondosin B	343
2.7.3.2	(−)-5-*epi*-Vibsanin E	344
2.7.3.3	(±)-Widdrol	345
2.7.3.4	(±)-Urechitol A	345
2.7.4	Conclusions and Future Perspectives	346

2.8	**Metal-Catalyzed (5 + 1), (5 + 2), and (5 + 2 + 1) Cycloadditions**
	X. Li and W. Tang

2.8	**Metal-Catalyzed (5 + 1), (5 + 2), and (5 + 2 + 1) Cycloadditions**	349
2.8.1	Metal-Catalyzed (5 + 1) Cycloadditions	349
2.8.1.1	Iron-Catalyzed (5 + 1) Cycloaddition	349
2.8.1.2	Chromium- or Molybdenum-Catalyzed (5 + 1) Cycloaddition	350
2.8.1.3	Cobalt-Catalyzed (5 + 1) Cycloaddition	351
2.8.1.4	Iridium-Catalyzed (5 + 1) Cycloaddition	353
2.8.1.5	Ruthenium-Catalyzed (5 + 1) Cycloaddition	353
2.8.1.6	Rhodium-Catalyzed (5 + 1) Cycloaddition	354
2.8.1.7	Applications of Metal-Catalyzed (5 + 1) Cycloaddition in Natural Product Synthesis	358
2.8.2	Metal-Catalyzed (5 + 2) Cycloadditions	359
2.8.2.1	Rhodium-Catalyzed (5 + 2) Cycloaddition	359
2.8.2.1.1	Rhodium-Catalyzed Cycloaddition with Vinylcyclopropanes	359
2.8.2.1.2	Rhodium-Catalyzed Cycloaddition with 3-Acyloxy-1,4-enynes	368
2.8.2.1.3	Mechanisms of the Two Types of Rhodium-Catalyzed (5 + 2) Cycloaddition	370
2.8.2.2	Ruthenium-Catalyzed (5 + 2) Cycloaddition with Vinylcyclopropanes	370
2.8.2.3	Nickel-Catalyzed (5 + 2) Cycloaddition with Vinylcyclopropanes	371
2.8.2.4	Iron-Catalyzed (5 + 2) Cycloaddition with Vinylcyclopropanes	372
2.8.2.5	Applications of Metal-Catalyzed (5 + 2) Cycloaddition in Natural Product Synthesis	373
2.8.3	Metal-Catalyzed (5 + 2 + 1) Cycloadditions	376
2.8.3.1	Rhodium-Catalyzed (5 + 2 + 1) Cycloaddition with Vinylcyclopropanes	376
2.8.3.2	Applications of Metal-Catalyzed (5 + 2 + 1) Cycloaddition in Natural Product Synthesis	379
2.8.4	Conclusions	380

2.9	**Intramolecular Free-Radical Cyclization Reactions**
	M. Mondal and U. Bora

2.9	**Intramolecular Free-Radical Cyclization Reactions**	383
2.9.1	Introduction to Radical Cyclization	383
2.9.1.1	Manganese(III) Acetate Based Radical Reactions	384
2.9.1.2	Titanocene(III)-Based Radical Reactions	387
2.9.1.3	Samarium(II) Iodide Based Radical Reactions	392
2.9.2	Intramolecular Free-Radical Cyclization Routes to N-Heterocycles	394
2.9.2.1	Synthesis of Pyrrole-Containing Moieties	394

2.9.2.1.1	Synthesis of 2-Arylpyrrole Derivatives	395
2.9.2.1.2	Synthesis of Fused Pyrrole Derivatives	397
2.9.2.1.3	Synthesis of Pyrrolidine Derivatives	400
2.9.2.1.3.1	Manganese(III) Acetate Mediated Cyclization of N-Substituted Internal Alkyne Esters	401
2.9.2.1.3.2	Copper(I) Trifluoromethanesulfonate–Benzene Complex Mediated Synthesis of 2,5-Disubstituted Pyrrolidines	401
2.9.2.1.3.3	Bis(η^5-cyclopentadienyl)dimethyltitanium(IV) (Petasis Reagent) Mediated Intramolecular Hydroamination/Cyclization of Alkynamines	403
2.9.2.1.4	Synthesis of Pyrrolidinone Derivatives	404
2.9.2.1.4.1	Manganese(III)-Mediated Synthesis of exo-Alkylidene Heterocycles	404
2.9.2.1.4.2	Tandem Visible-Light-Mediated Radical Cyclization/Rearrangement to Tricyclic Pyrrolidinones	406
2.9.2.2	Synthesis of the Indole Moiety	410
2.9.2.2.1	Synthesis of Indole Derivatives	410
2.9.2.2.1.1	Synthesis of N-Methylindole Derivatives	410
2.9.2.2.1.2	Photoredox Cyclization of Indole Substrates	411
2.9.2.2.1.3	Synthesis of Indole-3-carbaldehyde Derivatives	412
2.9.2.2.2	Synthesis of Indolines and Azaindolines	415
2.9.2.2.3	Synthesis of Oxindole Derivatives	418
2.9.2.2.3.1	Oxindoles and Indole-2,3-diones	418
2.9.2.2.3.2	Synthesis of 3,3-Disubstituted Oxindoles	420
2.9.2.3	Synthesis of Lactam Derivatives	422
2.9.2.3.1	Manganese(III)-Mediated Diastereoselective 4-exo-trig Cyclization of Enamides	422
2.9.2.3.2	Synthesis of γ-Lactams	423
2.9.2.3.2.1	Manganese(III)-Mediated Spirolactam Synthesis	423
2.9.2.3.2.2	γ-Lactams by Reverse Atom-Transfer Radical Cyclization of α-Polychloro-N-allylamides	423
2.9.2.4	Synthesis of Piperidines	425
2.9.2.4.1	Synthesis of 3-Chloropiperidine Derivatives	425
2.9.2.4.2	Synthesis of exo-Alkylidene Piperidinones	427
2.9.2.5	Synthesis of Quinoline Derivatives	427
2.9.2.5.1	Manganese(III)-Mediated Oxidative 6-endo-trig Cyclization	427
2.9.2.5.2	Manganese(III)-Mediated Oxidative 6-exo-trig Cyclization	429
2.9.3	Intramolecular Free-Radical Cyclization Routes to O-Heterocycles	431
2.9.3.1	Synthesis of Furan Derivatives	431
2.9.3.2	Synthesis of Tetrahydrofuran Derivatives	433

2.9.3.2.1	Titanocene(III)-Mediated Synthesis of Trisubstituted Tetrahydrofurans	433
2.9.3.2.2	Titanocene(III)-Mediated Synthesis of Multifunctional Tetrahydrofurans from Alkenyl Iodo Ethers	434
2.9.3.2.3	(2 + 2) Cycloadditions by Oxidative Visible Light Tris(2,2′-bipyridyl)-ruthenium(II) Bis(hexafluorophosphate) Mediated Photocatalysis	435
2.9.3.3	Synthesis of Lactone Derivatives	439
2.9.3.3.1	Synthesis of γ-Substituted Phthalides by Benzyl Radical Cyclization in Water	439
2.9.3.3.2	Chemoselective Synthesis of δ-Lactones through Benzyl Radical Cyclization Using Potassium Persulfate/Copper(II) Chloride	440
2.9.3.3.3	Atom-Transfer Radical Cyclization Reactions of Various Trichloroacetates to Macrolactones	440
2.9.3.3.4	Manganese(III)-Mediated Synthesis of Densely Functionalized and Sterically Crowded Lactones	442
2.9.3.3.4.1	Synthesis of Fused Tricyclic γ-Lactones	442
2.9.3.3.4.2	Cyclization of 2-Alkenylmalonates for the Synthesis of Bicyclo[3.3.0] γ-Lactones	444
2.9.3.3.4.3	Cyclization of γ-Lactones to Tricyclo[5.2.1.0^{1,5}] Bis(lactones)	447
2.9.3.3.4.4	Synthesis of Densely Functionalized and Sterically Crowded Cyclopentane-Fused Lactones	447
2.9.3.3.5	Synthesis of Tricyclic γ-Lactones	449
2.9.3.4	Synthesis of Pyrans and Derivatives	452
2.9.3.4.1	Synthesis of Polycyclic Dihydropyrans	452
2.9.3.4.2	Synthesis of Polysubstituted Tetrahydropyrans	454
2.9.3.4.3	Synthesis of Benzopyran Derivatives	458
2.9.3.4.4	Synthesis of Hexahydro-2H-1-benzopyran Derivatives	460
2.9.3.5	Synthesis of Oxepin Derivatives by an Intramolecular Ring Expansion Approach	461
2.9.4	Intramolecular Free-Radical Cyclization Routes to Carbocycles	463
2.9.4.1	Synthesis of Cycloalkane Derivatives	463
2.9.4.1.1	Synthesis of Monosubstituted and 1,1-Disubstituted Cyclopropane Derivatives	463
2.9.4.1.2	Photoredox-Mediated Radical Cyclization to Five- and Six-Membered Rings	465
2.9.4.1.3	5-exo Cyclization of Unsaturated Epoxides to Cyclic Derivatives	467
2.9.4.1.4	Photocatalyzed (3 + 2) Cycloadditions of Unsaturated Aryl Cyclopropyl Ketones To Give Cyclopentanes	469
2.9.4.2	Synthesis of Cycloalkanol Derivatives	470
2.9.4.2.1	Synthesis of anti-Cyclopropanol Derivatives	470
2.9.4.2.2	4-exo-trig Cyclizations of Unsaturated Aldehydes to Functionalized Cyclobutanols	471
2.9.4.2.3	Enantioselective Reductive Cyclization of Ketonitriles to Cycloalkanol Derivatives: Synthesis of Cyclic α-Hydroxy Ketones	473

2.9.4.2.4	Reductive Annulations of Ketones Bearing a Distal Vinyl Epoxide Moiety: Synthesis of Allyl Alcohols	476
2.9.4.2.5	Reductive Cyclization of Unactivated Alkenes: Synthesis of Five- and Six-Membered Cycloalkanols	478
2.9.4.2.6	Small-Ring 3-*exo* and 4-*exo* Cyclizations	480
2.9.4.2.7	Asymmetric Pinacol-Type Ketone *tert*-Butylsulfinyl Imine Reductive Coupling: Synthesis of *trans*-1,2-Vicinal Amino Alcohols	482
2.9.4.3	Synthesis of Substituted Cyclooctanols by an 8-*endo*-Radical Cyclization Process	484
2.9.4.4	Synthesis of Cyclopentanone Derivatives	487
2.9.4.4.1	Titanium(III)-Catalyzed Synthesis of Cyclic Amino Ketones	487
2.9.4.4.2	Synthesis of Bicyclo[4.3.0]nonan-8-one and Bicyclo[3.3.0]octan-3-one Derivatives from Bis(α,β-unsaturated esters) by Samarium(II) Iodide Induced Tandem Reductive Coupling/Dieckmann Condensation Reaction	488
2.9.4.4.3	Intramolecular Cyclization of Dicarbonyls to 3-Heterobicyclo[3.1.0]hexan-2-ones	490
2.9.4.5	Synthesis of 6-(Trifluoromethyl)phenanthridine Derivatives	491
2.9.5	Intramolecular Free-Radical Cyclization Routes to S-Heterocycles	495
2.9.5.1	Synthesis of Benzothiazole Derivatives	495
2.9.5.1.1	Synthesis of 2-Arylbenzothiazoles	495
2.9.5.1.2	Synthesis of 2-Substituted Benzothiazoles	496
2.9.5.2	Synthesis of Substituted 2,3-Dihydrothiophenes	497
2.9.6	Synthesis of the Core Frameworks of Biologically Active Molecules by Metal-Catalyzed Intramolecular Free-Radical Cyclization Reactions	498
2.9.6.1	Manganese(III)-Based Total Synthesis of Biologically Active Molecules	498
2.9.6.1.1	Synthesis of ($\pm$)-Garcibracteatone	498
2.9.6.1.2	Synthesis of a Precursor to 7,11-Cyclobotryococca-5,12,26-triene	499
2.9.6.1.3	Total Syntheses of Bakkenolides I, J, and S	500
2.9.6.1.4	Synthesis of the Core Framework of the Welwitindolinone Alkaloids	501
2.9.6.1.5	Synthesis of Ageliferins and Palau'amine	503
2.9.6.1.6	Synthesis of ($\pm$)-Ialibinones A and B	504
2.9.6.1.7	Synthesis of the ABC Ring System of Zoanthenol	505
2.9.6.1.8	Synthesis of the Tetracyclic Core of Tronocarpine	506
2.9.6.1.9	Synthesis of the ABC Ring of Hexacyclinic Acid	507
2.9.6.1.10	Synthesis of the Tetracyclic Framework of Azadiradione	508
2.9.6.1.11	Biomimetic Total Synthesis of ($\pm$)-Yezo'otogirin A	510
2.9.6.2	Titanium-Catalyzed Total Synthesis of Biologically Active Molecules	510
2.9.6.2.1	Synthesis of Magnofargesin and 7'-Epimagnofargesin	511
2.9.6.2.2	Short and Stereoselective Total Synthesis of ($\pm$)-Sesamin	512

2.9.6.2.3	Total Synthesis of (±)-Dihydroprotolichesterinic Acid and Formal Synthesis of (±)-Roccellaric Acid	513
2.9.6.2.4	Stereoselective Total Synthesis of Furano and Furofuran Lignans	514
2.9.6.2.5	Formal Total Synthesis of (±)-Fragranol by Template-Catalyzed 4-*exo* Cyclization	517
2.9.6.2.6	Total Synthesis of Entecavir	518
2.9.6.2.7	Enantioselective Synthesis of α-Ambrinol	519
2.9.6.2.8	Total Synthesis of (±)-Platencin	521
2.9.6.2.9	Total Synthesis of (±)-Smenospondiol	521
2.9.6.2.10	Total Synthesis of Fomitellic Acid B	522
2.9.6.2.11	Synthesis of the BCDE Molecular Fragment of Azadiradione	523
2.9.6.2.12	Approach to Bis(lactone) Skeletons: Total Synthesis of (±)-Penifulvin A	524
2.9.6.2.13	Synthesis of Eudesmanolides	525
2.9.6.3	Samarium(II)-Based Total Synthesis of Biologically Active Molecules	526
2.9.6.3.1	Total Synthesis of (±)-Lundurines A and B	527
2.9.6.3.2	Synthesis of Trehazolin from D-Glucose	528
2.9.6.3.3	Synthesis of (±)-Cryptotanshinone	529
2.9.6.3.4	Total Synthesis of Pradimicinone	530
2.9.7	Conclusions and Future Perspectives	531

2.10	**Ring-Closing Metathesis** D. Lee and V. Reddy Sabbasani	
2.10	**Ring-Closing Metathesis**	543
2.10.1	Brief Historical Background	544
2.10.2	Diene Ring-Closing Metathesis	546
2.10.2.1	Catalysts and Mechanism	546
2.10.2.2	Selectivity and Ring Size	548
2.10.2.2.1	*E*/*Z* Selectivity for Small and Medium Rings	548
2.10.2.2.2	*E*/*Z* Selectivity for Macrocycles	550
2.10.2.2.2.1	*E*-Selective Ring-Closing Metathesis	551
2.10.2.2.2.2	*Z*-Selective Ring-Closing Metathesis	551
2.10.2.2.3	Chemoselectivity with Multiple Double Bonds	554
2.10.2.2.3.1	Control Based on Ring Size	554
2.10.2.2.3.2	Stereochemistry-Based Control	559
2.10.2.2.3.3	Relay Metathesis Based Control	560
2.10.2.2.4	Diastereo- and Enantioselective Ring-Closing Metathesis	562
2.10.2.3	Applications of Diene Ring-Closing Metathesis to Natural Product Synthesis	567

2.10.2.3.1	Ring-Closing Metathesis with 1,n-Dienes	567
2.10.2.3.2	Ring-Closing Metathesis with Multiple Sequences (Ring Rearrangement)	602
2.10.2.4	Scope and Limitations	611
2.10.3	Enyne Ring-Closing Metathesis	611
2.10.3.1	Catalysts and Mechanism	611
2.10.3.2	Selectivity, Ring Size, and Substrates	616
2.10.3.2.1	Chemoselectivity with Multiple Double and Triple Bonds	617
2.10.3.2.1.1	Substituent-Based Control	618
2.10.3.2.1.2	Control Based on Ring Size	620
2.10.3.2.1.3	Relay Metathesis Based Control	622
2.10.3.2.2	*endo/exo* Mode and *E/Z* Selectivity	623
2.10.3.2.2.1	Small and Medium Rings	623
2.10.3.2.2.2	Macrocycles	625
2.10.3.2.3	Regioselectivity in Enyne Ring-Closing Metathesis–Metallotropic [1,3]-Shift	628
2.10.3.3	Applications of Enyne Ring-Closing Metathesis to Natural Product Synthesis	631
2.10.3.3.1	Ring-Closing Metathesis with 1,n-Enynes	631
2.10.3.3.2	Double Ring-Closing Metathesis with Dienynes	635
2.10.3.4	Diverse Applications of Enyne Ring-Closing Metathesis in Synthesis	642
2.10.3.4.1	Ring-Rearrangement by Enyne Ring-Closing Metathesis	642
2.10.3.4.2	Multiple Combinations of Metathesis	646
2.10.3.5	Scope and Limitations	649
2.10.4	Diyne Ring-Closing Metathesis	649
2.10.4.1	Catalyst and Mechanism	650
2.10.4.2	Ring Size and Substrates	652
2.10.4.2.1	1,n-Diynes	652
2.10.4.2.2	1,n-Bis-1,3-diynes	654
2.10.4.3	Applications of Diyne Ring-Closing Metathesis to Natural Product Synthesis	655
2.10.4.3.1	Ring-Closing Metathesis–Semireduction for the Construction of *Z*-Alkenes	655
2.10.4.3.2	Ring-Closing Metathesis–Semireduction for the Construction of *E*-Alkenes	661
2.10.4.3.3	Ring-Closing Alkyne Metathesis–Semireduction for the Synthesis of Cyclic Conjugated 1,3-Dienes	664
2.10.4.3.4	Ring-Closing Alkyne Metathesis for the Synthesis of Cyclic Conjugated 1,3-Diynes	668
2.10.4.4	Applications of Ring-Closing Alkyne Metathesis to the Synthesis of Cyclic Oligomers	668
2.10.4.5	Scope and Limitations	671
2.10.5	Conclusions	671

Keyword Index 679

Author Index 705

Abbreviations 729

2.1 Epoxidation and Aziridination Reactions

F. Xia and S. Ye

General Introduction

Owing to the strain associated with three-membered rings, epoxides and aziridines react with various nucleophiles to give 1,2-difunctionalized products, thereby establishing the stereochemistry of the two vicinal carbon atoms. Thus, epoxides and aziridines are valuable reagents for the synthesis of bioactive natural and unnatural compounds. The development of synthetic methods to access epoxides and aziridines has been recognized as an important goal in modern organic chemistry and many kinds of methods have been developed over the years. This chapter summarizes the advances in metal-catalyzed epoxidation[1-6] and aziridination[7-10] reactions, particularly those in recent decades.

2.1.1 Metal-Catalyzed Epoxidation Reactions

2.1.1.1 Titanium-Catalyzed Epoxidation

In 1980, Katsuki and Sharpless reported the titanium–tartrate complex catalyzed asymmetric epoxidation of allylic alcohols using *tert*-butyl hydroperoxide as the terminal oxidant.[11] The titanium–tartrate catalyst, prepared from a dialkyl tartrate and titanium(IV) isopropoxide, exhibits high enantioselectivity (up to 90% ee) for a wide range of substrates. In further studies, titanium–salalen complex **1** and titanium–salan complex **2** have been shown to catalyze the asymmetric epoxidation of alkenes using aqueous hydrogen peroxide as the oxidant.[12,13] Catalyst **1** can be applied to nonactivated alkenes to give the products with up to 99% ee. Catalyst **2** is easy to synthesize and tune but is less robust and selective than catalyst **1** (Scheme 1).[14]

Scheme 1 Asymmetric Epoxidation of Alkenes with Titanium–Salalen and Titanium–Salan Complexes[14]

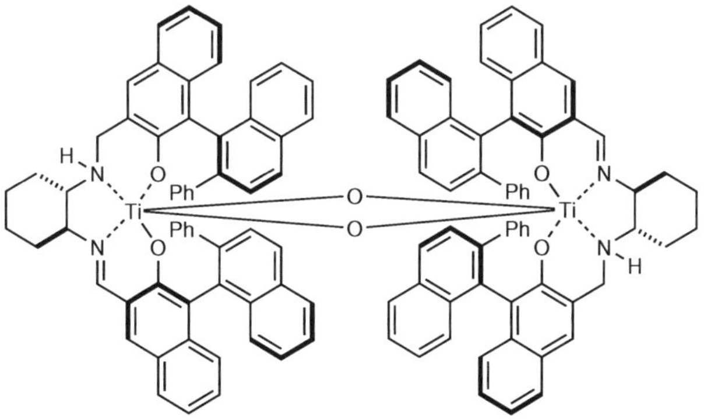

for references see p 47

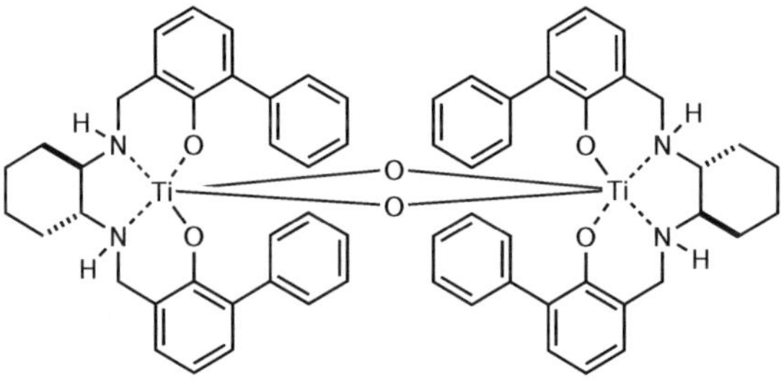

2

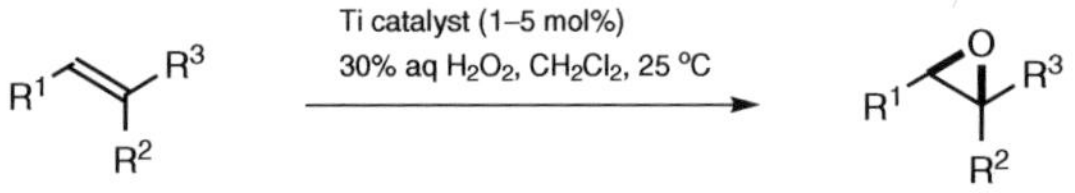

R¹	R²	R³	Catalyst	ee (%)	Yield (%)	Ref
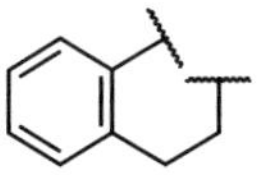 (tetralinyl)		H	1	99	99	[14]
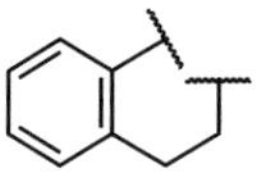 (tetralinyl)		H	2	96[a]	87	[14]
Ph	H	H	1	93	90	[14]
Ph	H	H	2	82[a]	47	[14]
H	(CH₂)₅Me	H	1	82	70	[14]
H	(CH₂)₅Me	H	2	55[a]	25	[14]

[a] Opposite enantiomer to that shown in scheme.

Katsuki and co-workers reported that titanium–salalen complex **1** is also an effective catalyst for the asymmetric epoxidation of (Z)-alkenylsilanes **3**.[15] cis-Epoxysilanes **4** are obtained with complete enantioselectivity using 0.5–2 mol% of catalyst **1** (Scheme 2).

Scheme 2 Asymmetric Epoxidation of (Z)-Alkenylsilanes with a Titanium–Salalen Complex[15]

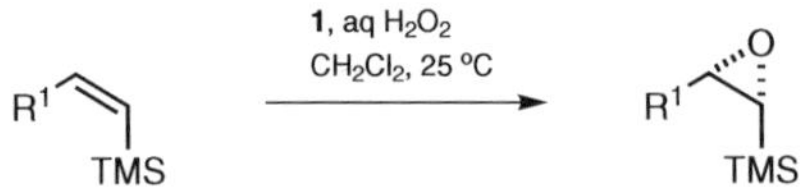

R¹	ee (%)	Yield (%)	Ref
Ph	99	87	[15]
2-MeOC₆H₄	99	95	[15]
3-BrC₆H₄	99	98	[15]
4-PhC₆H₄	99	99	[15]
1-naphthyl	99	92	[15]
2-naphthyl	99	96	[15]

This group has also developed novel C_1-asymmetric titanium–salalen complex **5** derived from proline and has used it in the highly enantioselective catalytic epoxidation of styrenes **6** using aqueous hydrogen peroxide as the oxidant.[16] Synthetically valuable styrene oxides **7** are obtained with 96–98% ee (Scheme 3).

Scheme 3 Asymmetric Epoxidation of Styrenes with a Titanium–Salan Complex[16]

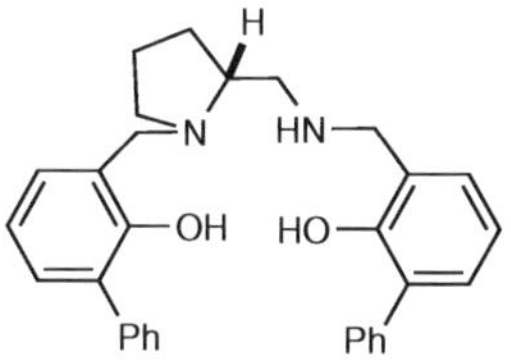

Ar^1	ee (%)	Yield (%)	Ref
Ph	98	60	[16]
2-Tol	97	64	[16]
3-Tol	98	70	[16]
4-Tol	98	74	[16]
2-ClC$_6$H$_4$	96	14	[16]
3-ClC$_6$H$_4$	98	55	[16]
4-ClC$_6$H$_4$	98	58	[16]

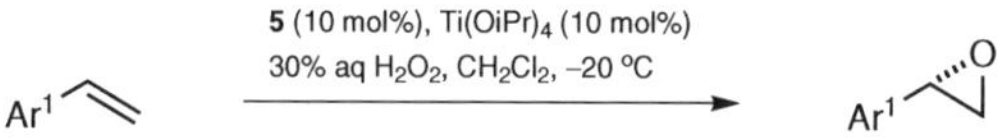

2-Aryl-3-(trimethylsilyl)oxiranes 4; General Procedure:[15]

Ti(salalen) complex **1** (4.5–18 mg, 0.5–2.0 mol%) and (Z)-alkenylsilane **3** (0.50 mmol) were dissolved in CH$_2$Cl$_2$ (0.50 mL). Then, 30–35% aq H$_2$O$_2$ (85 µL) was added, and the mixture was stirred at 25 °C. Upon completion of the reaction, the mixture was purified by chromatography (silica gel, pentane or hexane).

2-Aryloxiranes 7; General Procedure:[16]

Ligand **5** (46.5 mg, 10 mol%) was dissolved in 0.1 M Ti(OiPr)$_4$ in CH$_2$Cl$_2$ (1.0 mL, 10 mol%). After the mixture had been stirred at rt for 30 min, styrene derivative **6** (1.0 mmol) was added. The resulting soln was cooled to −20 °C before brine (0.5 mL) was added. After the addition of 30% aq H$_2$O$_2$ (170 µL, 1.5 mmol), the mixture was stirred at −20 °C for 48 h. The mixture was extracted with CH$_2$Cl$_2$ and the extracts were dried (Na$_2$SO$_4$), filtered, and concentrated under reduced pressure. The resulting residue was purified by column chromatography (silica gel, pentane/Et$_2$O 20:1 or pentane).

2.1.1.2 Vanadium-Catalyzed Epoxidation

The asymmetric epoxidation of allylic alcohols using vanadium and an optically active hydroxamic acid was first reported by Sharpless.[17,18] In 2000, Yamamoto and co-workers discovered that the chiral vanadium complex prepared from triisopropoxy(oxo)vanadium(V) and α-amino acid based hydroxamic acid **8** is an efficient catalyst for the epoxidation of mono- or disubstituted allylic alcohols **9** to provide products **10** in high yields with high enantioselectivities (Scheme 4).[19]

Scheme 4 Asymmetric Epoxidation of Mono- or Disubstituted Allylic Alcohols with Triisopropoxy(oxo)vanadium(V) and a Hydroxamic Acid[19]

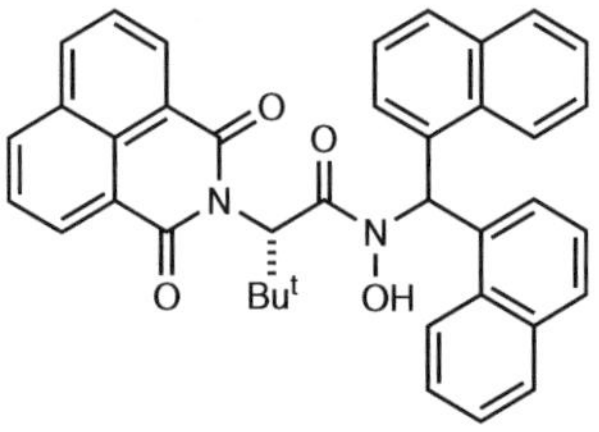

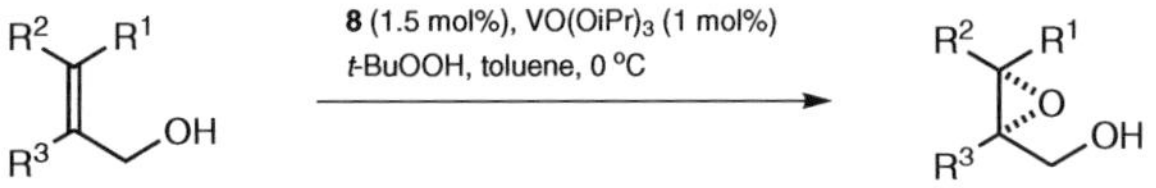

R¹	R²	R³	ee (%)	Yield (%)	Ref
H	Ph	Ph	96	93	[19]
H	Ph	Me	95	97	[19]
H	Ph	H	87	58	[19]
H	(CH₂)₄		93	82	[19]
Me	(CH₂)₂CH=CMe₂	H	81	95	[19]
H	(CH₂)₄Me	H	83	94	[19]

In 2005, Yamamoto and co-workers reported that vanadium complex **11**, prepared from a C_2-symmetric bishydroxamic acid, catalyzes the epoxidation of allylic alcohols **12** to give products **13** with up to 97% ee (Scheme 5).[20] The catalytic system has many advantages, including high enantioselectivity for a wide scope of allylic alcohols, low catalyst loading, mild reaction conditions, and tolerance of aqueous peroxide oxidants. Notably, the enantioselectivity of the reaction can be best explained by proposed reaction intermediate **14** (Scheme 6).

Scheme 5 Asymmetric Epoxidation of Allylic Alcohols with a Vanadium–Bishydroxamic Acid Complex[20]

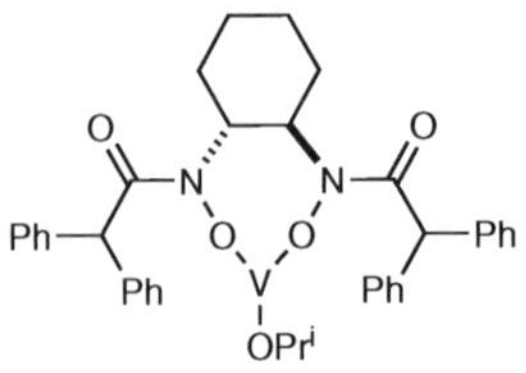

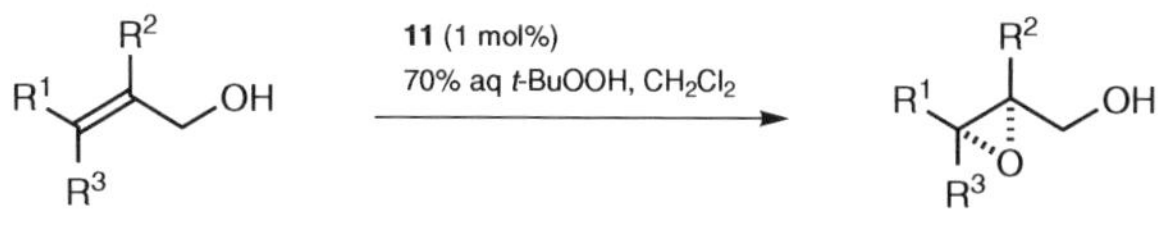

2.1.1 Metal-Catalyzed Epoxidation Reactions

R¹	R²	R³	ee (%)	Yield (%)	Ref
Ph	Ph	H	97	91	[20]
Ph	Me	H	97	84	[20]
Ph	H	H	97	53	[20]
Pr	H	H	95	56	[20]
(CH$_2$)$_4$Me	Ph	H	95	72	[20]
H	Ph	H	95	73	[20]
(CH$_2$)$_4$		H	95	79	[20]
(CH$_2$)$_2$CH=CMe$_2$	H	Me	95	68	[20]

Scheme 6 Postulated Model for the Epoxidation of Allylic Alcohols Catalyzed by a Vanadium Complex[20]

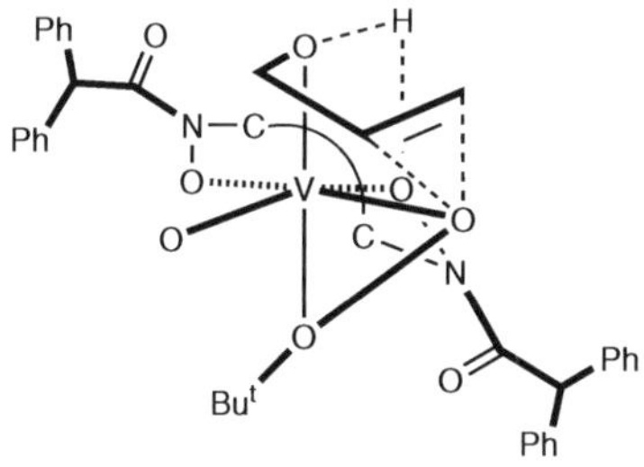

In 2007, Yamamoto and Zhang designed chiral bishydroxamic acid ligand **15**. This ligand has been shown to give excellent enantioselectivities for the vanadium-catalyzed asymmetric epoxidation of homoallylic alcohols (Scheme 7).[21] Both *trans*- and *cis*-substituted epoxides **16** can be obtained in satisfactory yields with virtually complete enantioselectivity.

Scheme 7 Asymmetric Epoxidation of Homoallylic Alcohols with Triisopropoxy(oxo)vanadium(V) and a Chiral Bishydroxamic Acid Ligand[21]

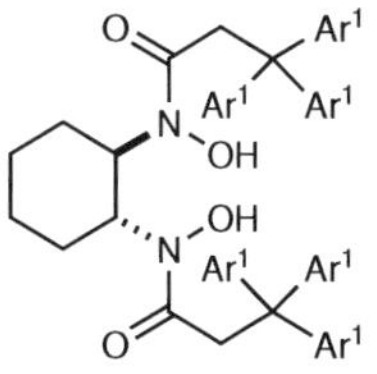

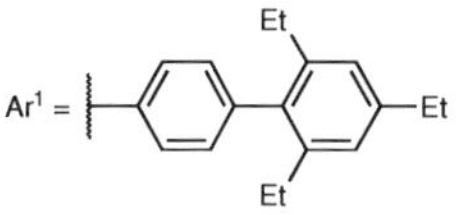

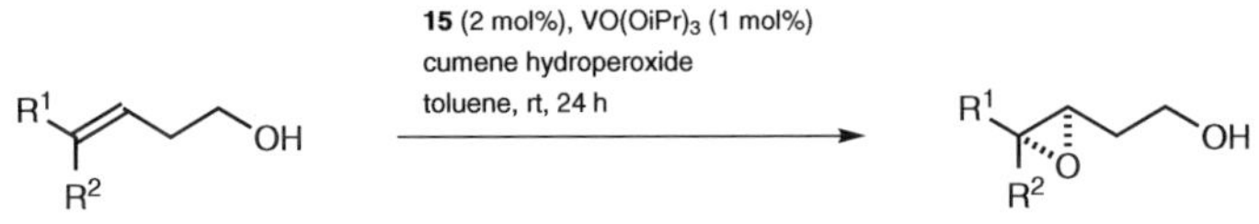

R¹	R²	ee (%)	Yield (%)	Ref
Et	H	93	85	[21]
$(CH_2)_4Me$	H	96	89	[21]
$(CH_2)_5Me$	H	98	92	[21]
H	Et	95	92	[21]
H	Pr	97	90	[21]
H	Bu	99	91	[21]

In 2008, Yamamoto and co-workers reported a vanadium/bishydroxamic acid catalyst system for the highly enantioselective desymmetrization of *meso* secondary allylic alcohols and homoallylic alcohols.[22] By using ligand **17**, typical *meso* secondary allylic alcohols **18** such as *E*- and *Z*-1,1-disubstituted and unsubstituted divinyl carbinols can be desymmetrized to give corresponding epoxy allylic alcohols **19** in good yields with high stereoselectivities (Scheme 8).

Scheme 8 Desymmetrization of *meso* Secondary Allylic Alcohols with Triisopropoxy(oxo)vanadium(V) and a Bishydroxamic Acid Ligand[22]

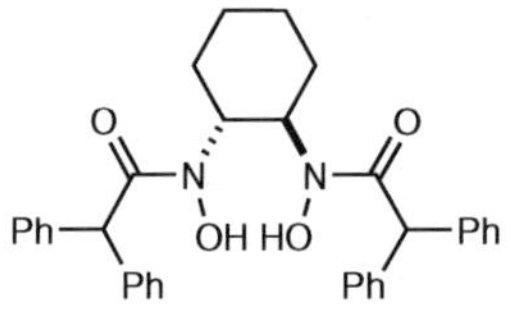

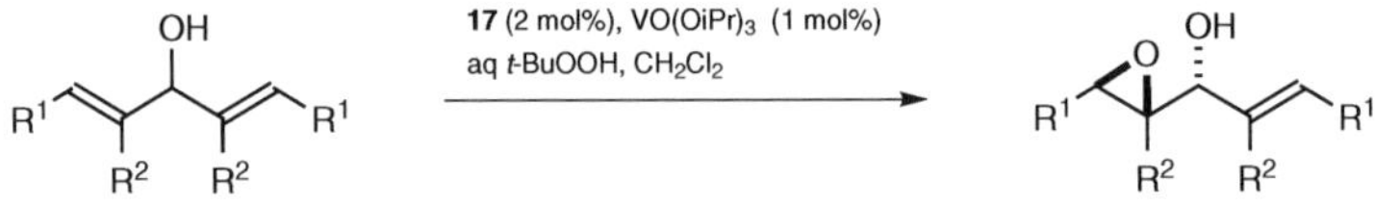

R¹	R²	ee (%)	Yield (%)	Ref
Ph	H	95	52	[22]
H	Ph	95	60	[22]
H	Me	95	62	[22]

[(2R,3R)-2,3-Diphenyloxiran-2-yl]methanol (13, R¹ = R² = Ph; R³ = H); Typical Procedure:[20]
VO(OiPr)₃ (0.0025 mL, 0.0104 mmol) was added to a soln of the bishydroxamic acid (11.2 mg, 0.0210 mmol) in CH₂Cl₂ or toluene (1 mL), and the mixture was stirred for 1 h at rt. The resulting soln of complex **11** was cooled to 0 °C, and 70% aq *t*-BuOOH (0.22 mL, 1.59 mmol) and (*E*)-2,3-diphenylprop-2-en-1-ol (220 mg, 1.05 mmol) were added. The mixture was stirred at the same temperature for 12 h. The progress of the epoxidation was monitored by TLC. Sat. aq Na₂SO₃ was added, and the mixture was stirred for 1 h at 0 °C. The mixture was then warmed to rt and extracted with Et₂O. The extracts were dried

(Na_2SO_4), filtered, and concentrated under reduced pressure. The remaining residue was purified by flash column chromatography (silica gel) to provide the title compound; yield: 91%; 97% ee.

2-[(2R,3R)-3-Ethyloxiran-2-yl]ethan-1-ol (16, R^1 = Et; R^2 = H); Typical Procedure:[21]
VO(OiPr)$_3$ (0.0025 mL, 0.0104 mmol) was added to a soln of bishydroxamic acid **15** (35.2 mg, 0.0210 mmol) in toluene (0.25 mL), and the mixture was stirred at rt for 8 h. 88% cumene hydroperoxide (0.25 mL, 1.50 mmol) and (E)-hex-3-en-1-ol (1.05 mmol) were then added and stirring was continued at the same temperature for 24 h. The progress of the epoxidation was monitored by TLC. The mixture was then purified by flash column chromatography (silica gel) to afford the title compound; yield: 85%; 93% ee.

(1R,2E)-3-Phenyl-1-[(2R,3R)-3-phenyloxiran-2-yl]prop-2-en-1-ol (19, R^1 = Ph; R^2 = H); Typical Procedure:[22]
VO(OiPr)$_3$ (2.5 µL, 0.0104 mmol) was added to a soln of bishydroxamic acid **17** (11.2 mg, 0.0210 mmol) in CH_2Cl_2 (0.5 mL), and the mixture was stirred for 1 h at rt. The resulting soln was cooled to 0 °C. (1E,4E)-1,5-Diphenylpenta-1,4-dien-3-ol (250 mg, 1.06 mmol) and 70% aq t-BuOOH (0.17 mL, 1.23 mmol) were sequentially added, and the mixture was stirred at the same temperature for 10 h (the progress of the epoxidation was monitored by TLC). Sat. aq Na_2SO_3 was then added, and the mixture was stirred for 1 h at 0 °C before it was warmed to rt and extracted with Et_2O. The extracts were dried (Na_2SO_4), filtered, and concentrated under reduced pressure. The resulting residue was purified by flash column chromatography (silica gel) to provide the title compound; yield: 52%; 95% ee.

2.1.1.3 Manganese-Catalyzed Epoxidation

The groups of Jacobsen and Katsuki pioneered the manganese–salen catalyzed asymmetric epoxidation of alkenes.[23,24] Katsuki and co-workers reported that hydrogen peroxide can serve as an effective terminal oxidant for the manganese–salen catalyzed asymmetric epoxidation of benzopyran (chromene) derivatives.[25] The same group has also disclosed that manganese–salen complex **20** bearing a nucleophilic substituent at the diimine unit serves as an efficient catalyst for the epoxidation of conjugated Z-alkenes **21** using 30% aqueous hydrogen peroxide to afford epoxides **22**.[26] The epoxidation of benzopyran derivatives and other cyclic and electron-rich alkenes proceeds in excellent yields with high enantioselectivities (Scheme 9).

Scheme 9 Asymmetric Epoxidation of Benzopyran Derivatives, 6,7-Dihydro-5H-benzo[7]annulene, and Prop-1-enylbenzene Catalyzed by a Manganese–Salen Complex[26]

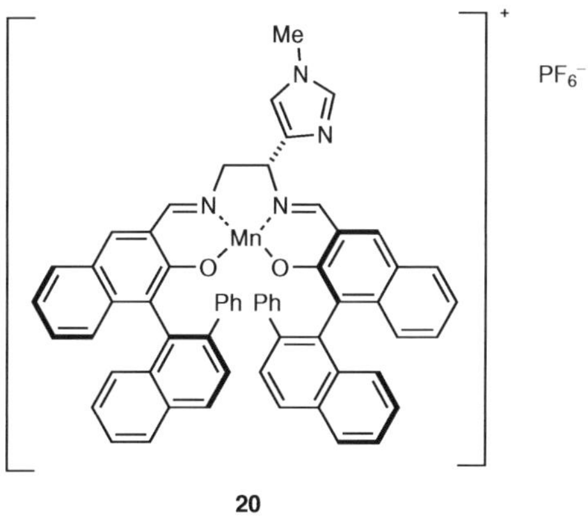

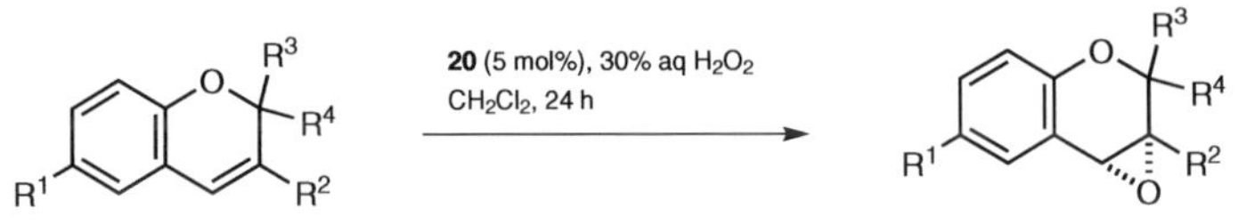
20 (5 mol%), 30% aq H_2O_2
CH_2Cl_2, 24 h
R1
R2
R3
R4
O

21 22

R^1	R^2	R^3	R^4	ee (%)	Yield (%)	Ref
H	H	Me	Me	98	80	[26]
Br	H	Me	Me	98	98	[26]
H	H	$(CH_2)_5$		98	84	[26]
H	Me	Me	Me	97	84	[26]

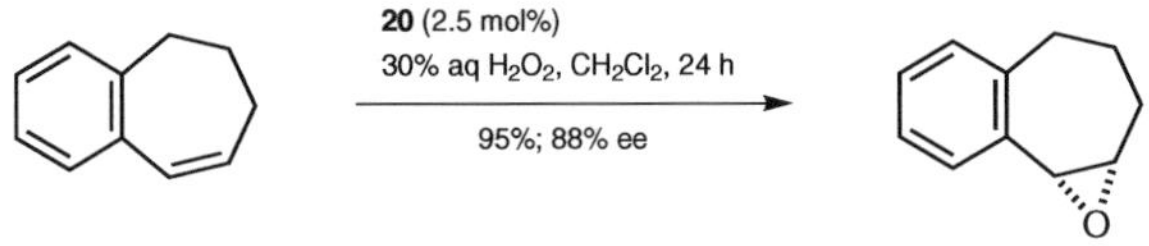
20 (2.5 mol%)
30% aq H_2O_2, CH_2Cl_2, 24 h
95%; 88% ee
O

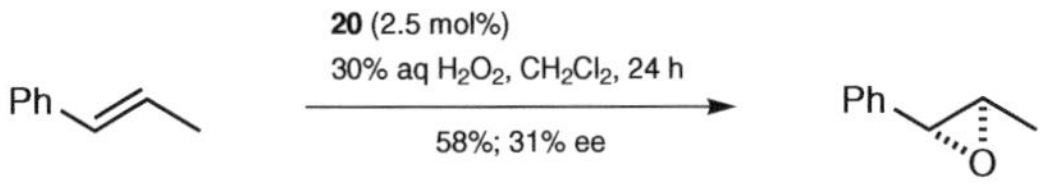
20 (2.5 mol%)
30% aq H_2O_2, CH_2Cl_2, 24 h
58%; 31% ee
Ph
Ph
O

For many years, the ligands of choice for the asymmetric epoxidation of alkenes were chiral salens and their derivatives. A parallel line of research has been developed involving the use of manganese complexes of tetradentate nitrogen-containing ligands. Stack and co-workers reported that a manganese complex of N_4 ligands in combination with peracetic acid acts as a very fast and efficient epoxidation catalyst.[27] The catalytic system can rapidly epoxidize a wide range of alkenes at room temperature with a low catalyst loading, but no enantioselectivity is reported.

In 2009, Xia, Sun, and co-workers synthesized a novel family of chiral manganese complexes **23** of N_4 ligands that promote the highly enantioselective epoxidation of various enones **24** with hydrogen peroxide.[28] In the presence of acetic acid, nice results for epoxides **25** are obtained with a catalyst loading of 1 mol%. The enantioselective epoxidation of α,β-unsaturated ketones proceeds with nearly full conversion and with enantiomeric excess values up to 89% (Scheme 10).

Scheme 10 Asymmetric Epoxidation of α,β-Unsaturated Ketones Catalyzed by a Chiral Manganese Complex[28]

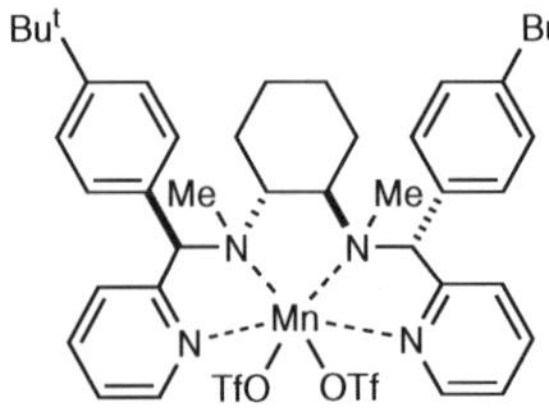
Bu^t
Bu^t
Me
Me
N
N
Mn
N
N
TfO
OTf

23

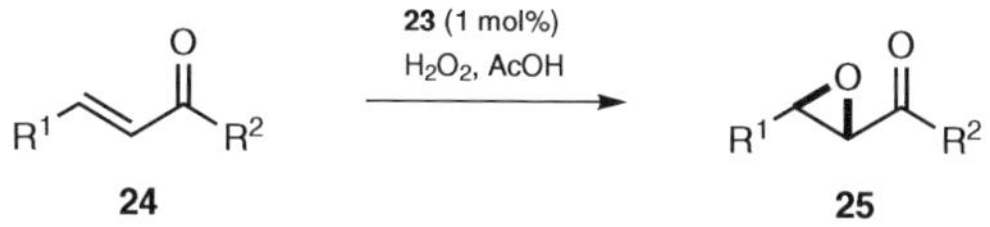

2.1.1 Metal-Catalyzed Epoxidation Reactions

R¹	R²	ee (%)	Yield (%)	Ref
Ph	Ph	78	91	[28]
Ph	4-MeOC$_6$H$_4$	76	63	[28]
Ph	4-O$_2$NC$_6$H$_4$	86	82	[28]
Ph	4-ClC$_6$H$_4$	72	89	[28]
4-ClC$_6$H$_4$	Ph	85	89	[28]
2-BrC$_6$H$_4$	Ph	89	87	[28]

In 2013, Gao and co-workers reported a complex formed from manganese(II) trifluoromethanesulfonate and porphyrin-inspired chiral ligand **26** that is capable of catalyzing the asymmetric epoxidation of alkenes using hydrogen peroxide as the terminal oxidant.[29] The system allows the general epoxidation of a wide variety of alkenes **27** and provides epoxide products **28** in excellent yields with enantiomeric excess values up to 99% (Scheme 11).

Scheme 11 Asymmetric Epoxidation of Alkenes Catalyzed by Manganese(II) Trifluoromethanesulfonate and a Porphyrin-Inspired Chiral Ligand[29]

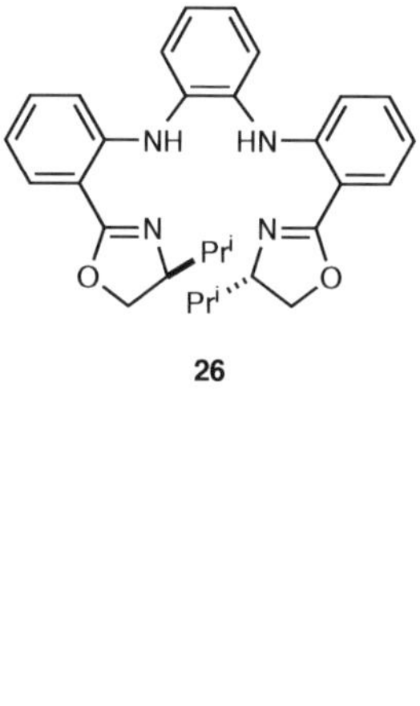

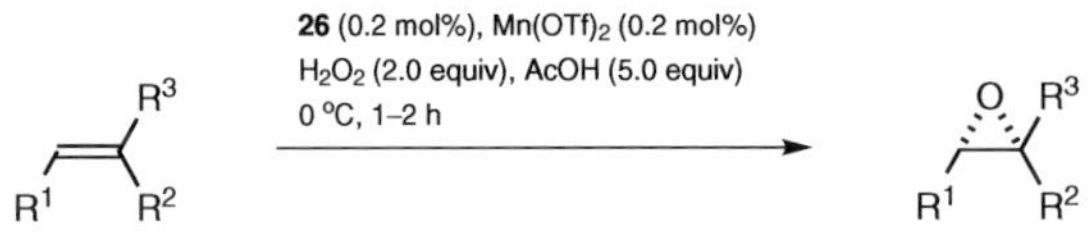

26 (0.2 mol%), Mn(OTf)$_2$ (0.2 mol%)
H$_2$O$_2$ (2.0 equiv), AcOH (5.0 equiv)
0 °C, 1–2 h

27 → **28**

R^1	R^2	R^3	ee (%)	Yield (%)	Ref
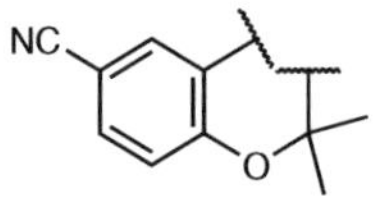 NC-benzofuran-spiro		H	95	95	[29]
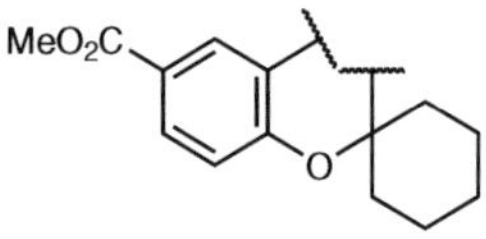 MeO$_2$C-benzofuran-spirocyclohexane		H	90	97	[29]
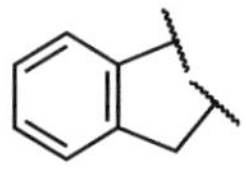 indanyl		H	84	93	[29]
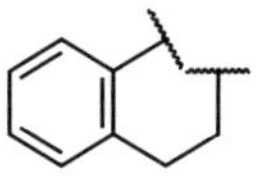 tetrahydronaphthyl		H	96	99	[29]
Ph	H	Ph	92	95	[29]

Talsi and co-workers have proposed a probable reaction mechanism (Scheme 12).[30] The starting manganese(II) complex is converted into (hydroxoperoxo)manganese(III) complex **29**, which exchanges its solvent ligand (S) with the carboxylic acid to give complex **30**. Elimination of water from complex **30** via seven-membered cyclic intermediate **31**, which undergoes O–O bond heterolysis, affords reactive oxometal(V) complex **32**, which reacts with the alkene to give the final epoxide.

Scheme 12 Proposed Mechanism for Manganese-Catalyzed Asymmetric Epoxidation[30]

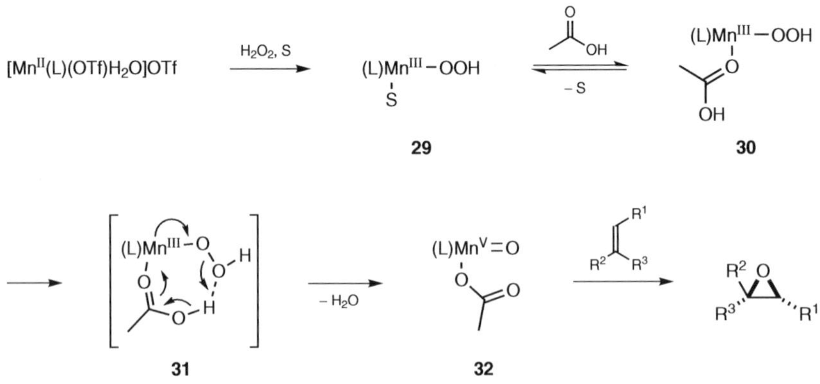

[Mn^{II}(L)(OTf)H$_2$O]OTf $\xrightarrow{H_2O_2, S}$ (L)Mn^{III}–OOH·S **29** $\rightleftharpoons$ (L)Mn^{III}–OOH with AcOH chelate **30**

→ [(L)Mn^{III}–O···carboxylate cyclic intermediate] **31** $\xrightarrow{-H_2O}$ (L)Mn^V=O (carboxylate) **32** → epoxide product

S = MeCN, H$_2$O

2.1.1 Metal-Catalyzed Epoxidation Reactions

(1aR,7bR)-2,2-Dimethyl-1a,7b-dihydro-2H-oxireno[c][1]benzopyran (22, $R^1 = R^2 = H$; $R^3 = R^4 = Me$); Typical Procedure:[26]

Mn(salen) complex **20** (5.4 mg, 5 µmol) and 2,2-dimethyl-1-benzopyran (0.1 mmol) were dissolved in CH_2Cl_2 (1 mL). After the addition of 30% aq H_2O_2 (0.3 mmol) at 0 °C, the resultant mixture was stirred for 24 h. The solvent was removed under reduced pressure, and the residue was purified by chromatography (silica gel, pentane/Et_2O 20:1) to give the title compound; yield: 14 mg (80%); 98% ee.

[(2S,3R)-3-(2-Bromophenyl)oxiran-2-yl](phenyl)methanone (25, $R^1 = 2\text{-}BrC_6H_4$; $R^2 = Ph$); Typical Procedure:[28]

Chiral complex **23** (2.5 µmol), (E)-3-(2-bromophenyl)-1-phenylprop-2-en-1-one (71 mg, 0.25 mmol), and AcOH (5 equiv) were added to MeCN (0.75 mL) under an argon atmosphere, and the soln was stirred for 5 min. Then, 50% H_2O_2 (1.5 mmol, 6 equiv) diluted with MeCN (0.75 mL) was added dropwise over 3 min, and the mixture was stirred at rt for 60–90 min. The crude product was purified by chromatography (silica gel, EtOAc/petroleum ether 40:1) to afford the title compound; yield: 66 mg (87%); 89% ee.

Epoxides 28; General Procedure:[29]

A 0.00168 M soln of $Mn(OTf)_2$ in MeCN (0.5 mL, 0.84 µmol) was added to a 0.00168 M soln of ligand **26** in MeCN (0.5 mL, 0.84 µmol) at rt. The mixture was stirred at rt for 1 h. Substrate **27** (0.42 mmol) and AcOH (2.1 mmol) were added to the soln of the manganese complex. Then, the temperature was decreased to 0 °C and 50% H_2O_2 (0.84 mmol) diluted with MeCN (1 mL) was added dropwise by syringe pump over 1 h. After being stirred at 0 °C for 1 h, the reaction was quenched by adding a mixture of 10% aq $Na_2S_2O_3$ and sat. aq $NaHCO_3$, and the mixture was diluted with CH_2Cl_2. The organic layer was separated and washed with brine, dried ($MgSO_4$), and concentrated under reduced pressure. The residue was purified by column chromatography (silica gel).

2.1.1.4 Iron-Catalyzed Epoxidation

The use of iron complexes for alkene epoxidation is similar to that of manganese catalysts. In 1983, the first asymmetric epoxidation catalyzed by a chiral iron porphyrin was reported by Groves and Myers[31] to give styrene oxide with 48% ee. After that, many chiral iron porphyrins were developed by Groves[32] and others[33–37] for epoxidation. After studies pursuing non-heme iron catalysts that could be easily prepared and modified, Beller and co-workers reported that the best results in terms of enantioselectivity are obtained using iron-catalyzed asymmetric epoxidation of stilbene derivatives.[38,39] However, the high selectivity is obtained for only one specific substrate with a catalyst loading of 10 mol%.

Subsequently, the iron-catalyzed asymmetric epoxidation of acyclic β,β-disubstituted enones **34** was developed by Yamamoto and Nishikawa in 2011 (Scheme 13).[40] Essential for the success of this reaction is the use of the iron complex derived from iron(II) trifluoromethanesulfonate and 2 equivalents of phenanthroline ligand **33**. The reaction provides highly enantioenriched β,β-disubstituted epoxy ketones **35** (up to 92% ee), which can be further converted into functionalized β-ketoaldehydes with an all-carbon quaternary center.

Scheme 13 Asymmetric Epoxidation of α,β-Unsaturated Ketones Catalyzed by Iron(II) Trifluoromethanesulfonate with a Phenanthroline Ligand[40]

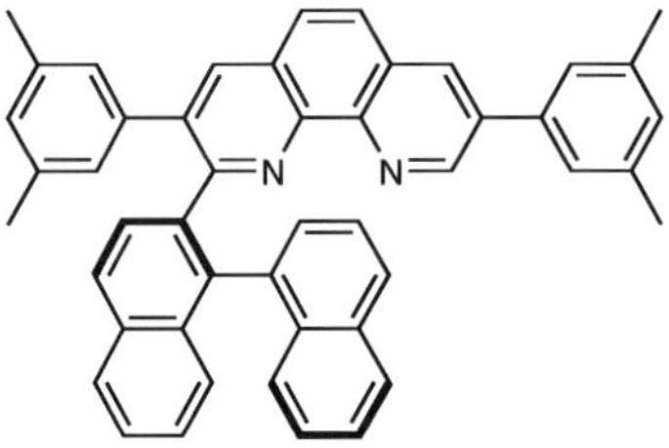

R¹	R²	R³	ee (%)	Yield (%)	Ref
Ph	Ph	Me	91	80	[40]
4-MeOC$_6$H$_4$	Ph	Me	90	78	[40]
4-Tol	Ph	Me	92	77	[40]
4-FC$_6$H$_4$	Ph	Me	92	78	[40]
4-F$_3$CC$_6$H$_4$	Ph	Me	89	70	[40]
2-naphthyl	Ph	Me	90	88	[40]
Ph	4-ClC$_6$H$_4$	Me	92	88	[40]
Ph	Ph	Et	92	72	[40]

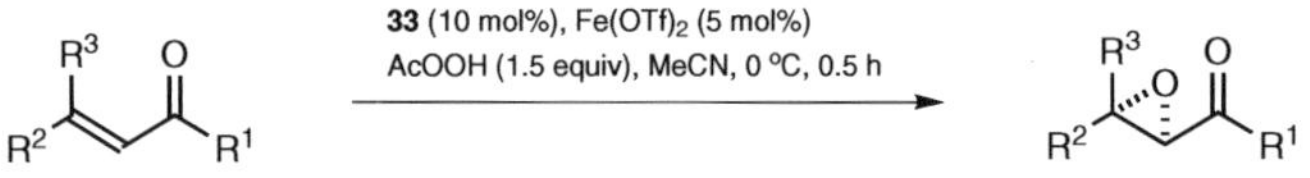

In 2012, Nakada and co-workers reported that iron(III) complex **36** containing a carbazole-based tridentate ligand catalyzes the highly enantioselective asymmetric epoxidation of *E*-alkenes at room temperature in the presence of chloro[1,3-bis(2,6-diisopropylphenyl)-imidazol-2-ylidene]silver(I) (**37**, SIPrAgCl) (Scheme 14).[41] The system uses iodosylbenzene as the oxidant in the presence of sodium tetrakis[3,5-bis(trifluoromethyl)phenyl]borate (NaBARF) and provides epoxides **38** with high enantioselectivities up to 97% ee. The carbazole-based tridentate ligand has been shown to be a good surrogate for traditional porphyrins, which allows flexibility in the design as well as simple and straightforward preparation.

Scheme 14 Asymmetric Epoxidation of Alkenes Catalyzed by an Iron(III) Complex[41]

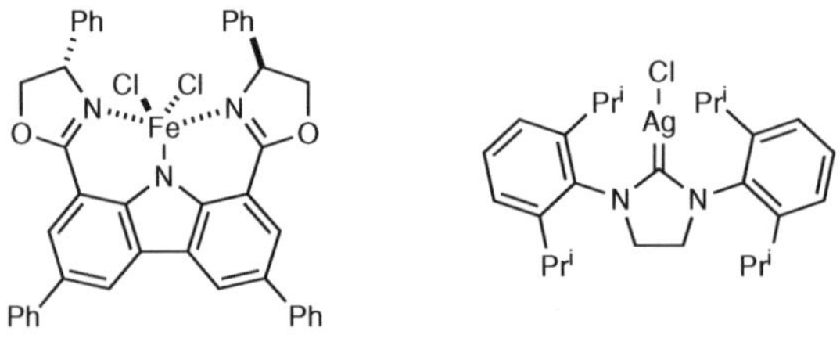

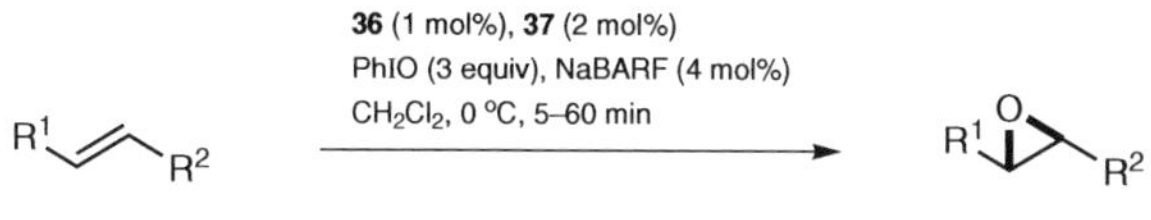

2.1.1 Metal-Catalyzed Epoxidation Reactions

R¹R²-alkene → **38** (epoxide)

Conditions: **36** (1 mol%), **37** (2 mol%), PhIO (3 equiv), NaBARF (4 mol%), CH₂Cl₂, 0 °C, 5–60 min

R¹	R²	ee (%)	Yield (%)	Ref
Ph	Ph	89	55	[41]
Ph	2-naphthyl	93	45	[41]
Ph	CH₂OMOM	76	90	[41]

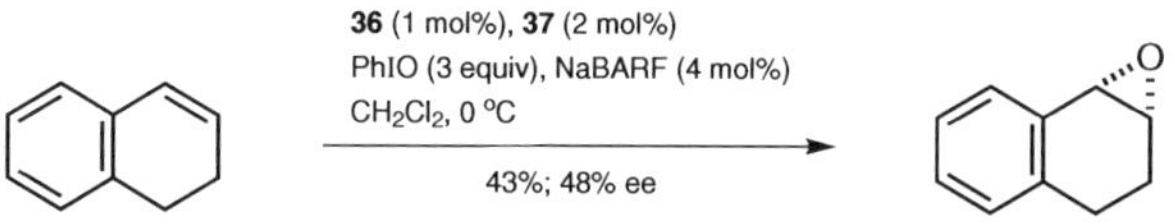

Dihydronaphthalene → epoxide

Conditions: **36** (1 mol%), **37** (2 mol%), PhIO (3 equiv), NaBARF (4 mol%), CH₂Cl₂, 0 °C

43%; 48% ee

The catalytic cycle proposed by the group is depicted in Scheme 15. Oxidation of iron(III) complex **39** by iodosylbenzene generates iron(IV)–oxo complex **40** bearing a π-cation radical, which works as the key intermediate to oxidize the alkenes to the epoxides.

Scheme 15 Proposed Reaction Mechanism of the Iron(III)-Catalyzed Asymmetric Epoxidation[41]

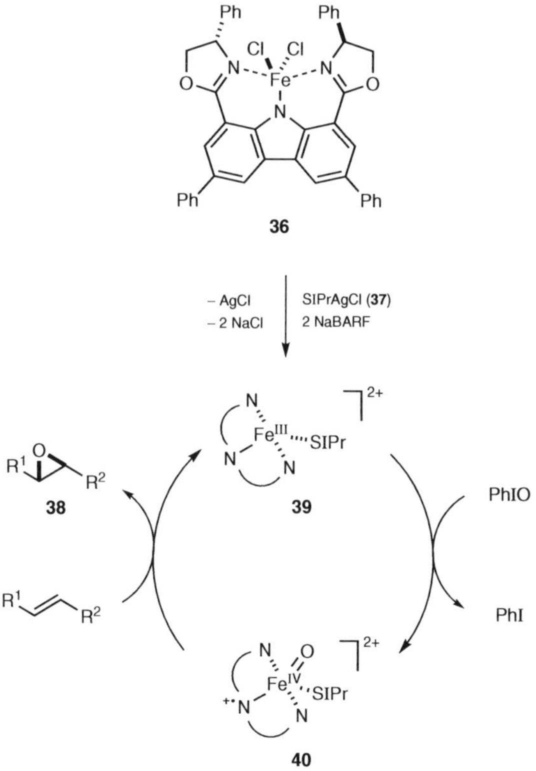

**[(2R,3S)-3-Methyl-3-phenyloxiran-2-yl](phenyl)methanone (35, $R^1 = R^2 = Ph$; $R^3 = Me$);
Typical Procedure:**[40]
A 0.025 M soln of Fe(OTf)$_2$ in MeCN (0.31 mL, 7.8 µmol) was added to ligand **33** (10 mg, 15.6 µmol) under an atmosphere of N$_2$ at rt. The mixture was rinsed with MeCN (0.31 mL), and then stirred at rt for 3 h. A soln of (E)-1,3-diphenylbut-2-en-1-one (35 mg, 0.156 mmol) in MeCN (0.1 mL) was added to the soln of the iron complex, and the mixture was cooled in an ice bath. A soln of 32 wt% AcOOH in AcOH (50 µL, 0.234 mmol) was rapidly added to the mixture. After stirring in an ice bath for 30 min, the reaction was quenched by adding a mixture of 10% aq Na$_2$S$_2$O$_3$ and sat. aq NaHCO$_3$, and the mixture was diluted with EtOAc. The organic layer was separated and washed with brine, dried (MgSO$_4$), and concentrated under reduced pressure. The residue was purified by column chromatography (silica gel and NH-silica gel, EtOAc/hexane 0:100 to 15:85) to afford the title compound; yield: 80%; 91% ee.

(2S,3S)-2,3-Diphenyloxirane (38, $R^1 = R^2 = Ph$); Typical Procedure:[41]
A mixture of iron complex **36** (0.0020 mmol, 0.010 equiv), NaBARF (7.1 mg, 0.0080 mmol, 0.040 equiv), and complex **37** (2.1 mg, 0.0040 mmol, 0.020 equiv) was suspended in CH$_2$Cl$_2$ (0.2 mL) and stirred for 5 min at rt. The soln was added to a mixture of (E)-stilbene (0.20 mmol, 1.0 equiv) and PhIO (132 mg, 0.6 mmol, 3.0 equiv) in CH$_2$Cl$_2$ (1.8 mL) at 0°C, and the mixture was stirred for 60 min at the same temperature. Sat. aq NaHCO$_3$ (2.0 mL) was added, the aqueous layer was extracted with CH$_2$Cl$_2$ (2 × 2.0 mL), and the combined organic layer was washed with brine (1 × 3.0 mL). The extract was dried (Na$_2$SO$_4$), filtered, and concentrated under reduced pressure. The residue was purified by preparative TLC (hexane/EtOAc 10:1) to afford the title compound as a white solid; yield: 21 mg (55%); 89% ee.

2.1.1.5 Ruthenium-Catalyzed Epoxidation

High-valent ruthenium oxides are powerful reagents for the oxidization of alkenes, and their use mostly results in cleavage of the double bond. The use of less-reactive, low-valent ruthenium complexes in combination with various oxidants for the preparation of epoxides from simple alkenes has been described.[42–44]

The first asymmetric ruthenium-catalyzed epoxidation with the use of hydrogen peroxide as the primary oxidant was reported in 1999 by Stoop and Mezzetti.[45] Up to 40% ee was obtained in the asymmetric epoxidation of styrene and other unfunctionalized alkenes. In 2003, Berkessel and co-workers reported that a ruthenium–porphyrin complex efficiently catalyzes the enantioselective epoxidation of arylalkenes.[46] Enantioselectivities up to 83% ee are obtained with this catalyst in the epoxidation of 1,2-dihydronaphthalene, whereas simple alkenes such as oct-1-ene react poorly and give epoxides with low enantioselectivity. In 2004, a more general procedure for the ruthenium-catalyzed asymmetric epoxidation of alkenes using hydrogen peroxide was developed by Beller and co-workers.[47] An important factor for the success of the reaction is the introduction of new pyboxazine-type ligands, as in catalyst **41**. In this case, epoxides **42** are obtained with enantioselective excesses up to 84% for various arylalkenes (Scheme 16).

2.1.1 Metal-Catalyzed Epoxidation Reactions

Scheme 16 Asymmetric Epoxidation of Alkenes Catalyzed by a Ruthenium–Pyboxazine Complex[47]

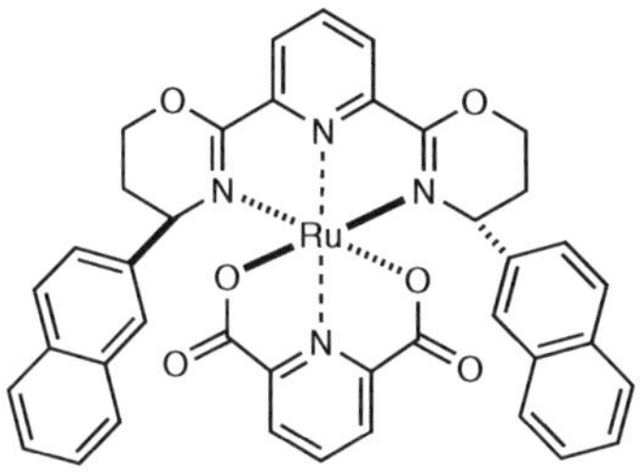

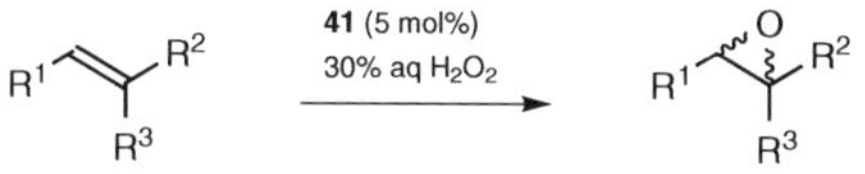

R¹	R²	R³	ee (%)	Yield (%)	Ref
Ph	H	H	48	59	[47]
4-ClC$_6$H$_4$	H	H	54	76	[47]
4-FC$_6$H$_4$	H	H	60	82	[47]
4-Tol	H	H	58	80	[47]
Ph	Ph	H	54	100	[47]
Ph	Me	H	72	95	[47]
Ph	CH$_2$OAc	H	48	83	[47]
Ph	CH$_2$Cl	H	28	68	[47]
Ph	Me	Me	84	91	[47]

In 2012, Katsuki and co-workers achieved the highly enantioselective epoxidation of conjugated alkenes using durable ruthenium–aqua–salen complex **43** as the catalyst.[48] At 25 °C in air or at 0 °C under an atmosphere of oxygen, epoxides **45** are obtained in excellent yields (up to 92%) with good enantioselectivities (up to 95% ee) from conjugated alkenes **44** (Scheme 17).

Scheme 17 Asymmetric Epoxidation of Alkenes Catalyzed by a Ruthenium–Aqua–Salen Complex[48]

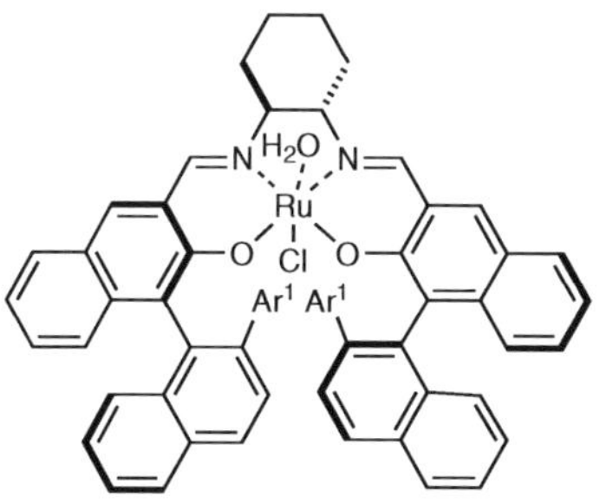

Ar¹ = 3,5-Cl$_2$C$_6$H$_3$

Scheme/Reaction:

R¹R²C=CR³R⁴ (**44**) → (with **43** (5 mol%), chlorobenzene) → epoxide **45**

R¹	R²	R³	R⁴	Conditions	ee (%)	Yield (%)	Ref
H	H	Me	Ph	air, 25 °C	93	62	[48]
Me	Me	H	Ph	O_2, 0 °C	84	89	[48]
H	Me	Me	Ph	O_2, 0 °C	80	44	[48]
H	Me	H	4-Tol	O_2, 0 °C	83	80	[48]
H	Me	H	3-Tol	air, 25 °C	92	92	[48]
H	Me	H	2-Tol	O_2, 0 °C	53	76	[48]

(R)-2,2-Dimethyl-3-phenyloxirane (42, R¹ = Ph; R² = R³ = Me); Typical Procedure:[47]

In a 25-mL Schlenk tube, catalyst **41** (19 mg, 0.025 mmol) was stirred at rt in 2-methylbutan-2-ol (9 mL) for 10 min. (2-Methylprop-1-enyl)benzene (66 mg, 0.5 mmol) and dodecane (GC internal standard; 100 µL) were added. A soln of 30% aq H_2O_2 (170 µL, 1.5 mmol) in 2-methylbutan-2-ol (830 µL) was added to the mixture over a period of 12 h using a syringe pump. After the addition, aliquots were taken from the mixture and analyzed by GC to determine the yield and conversion. The reaction was then quenched with a soln of Na_2SO_3 (10 mL), the mixture was extracted with CH_2Cl_2 (2 × 10 mL), and the extracts were washed with H_2O (20 mL). The combined organic layer was dried ($MgSO_4$) and concentrated to give the title compound; yield: 91%; 84% ee.

Epoxides 45; General Procedure:[48]

Alkene **44** (0.5 mmol), chlorobenzene (2.5 mL), and brine (2.5 mL) were placed in a test tube in air. Complex **43** (27.9 mg, 25 µmol) was added to the soln, and the mixture was stirred at 25 °C for 48 h. The aqueous layer was separated and extracted with Et_2O (2.2 mL). The combined organic layer was dried (Na_2SO_4) and concentrated under reduced pressure. The residue was subjected to chromatography (silica gel, pentane/Et_2O 1:0 to 20:1).

2.1.1.6 Molybdenum-Catalyzed Epoxidation

The use of molybdenum-based systems for the epoxidation of alkenes has been extensively explored over the last 40 years. Numerous attempts have also been undertaken to synthesize chiral molybdenum complexes for the enantioselective epoxidation of alkenes. Early reports of asymmetric epoxidation with a chiral molybdenum complex include the use of a stoichiometric molybdenum–(S)-lactic acid piperidineamide system by Schurig, Kagan, and co-workers.[49] Consequently, considerable effort has been directed toward the development of enantioselective epoxidation protocols in which chiral molybdenum catalysts are used.

In 2006, Yamamoto and co-workers found that a molybdenum complex derived from bishydroxamic acid **46** catalyzes the asymmetric epoxidation of alkenes.[50] Epoxides **47** are obtained in high yields up to 98% with enantiomeric excess values up to 96% under mild conditions (Scheme 18).

2.1.1 Metal-Catalyzed Epoxidation Reactions

Scheme 18 Asymmetric Epoxidation of Alkenes Catalyzed by Bis(acetylacetonato)-dioxomolybdenum(VI) and a Chiral Bishydroxamic Acid Ligand[50]

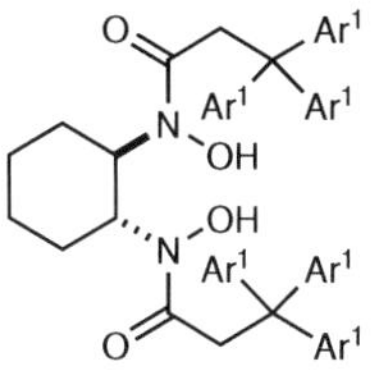

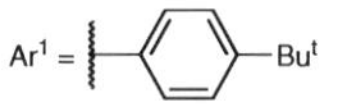

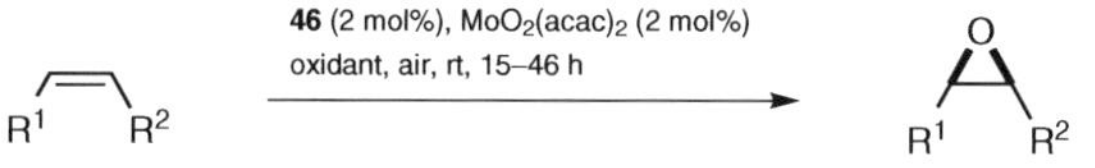

R^1	R^2	Oxidant	ee (%)	Yield (%)	Ref
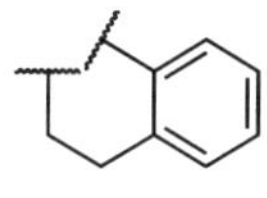 (tetrahydronaphthalenyl)		TrOOH	95	98	[50]
Me	Ph	TrOOH	90	60	[50]
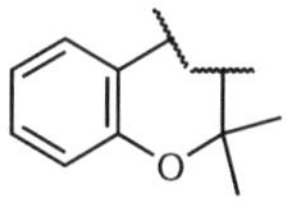 (chromanyl)		t-BuOOH	96	92	[50]
	CH=CH(CH$_2$)$_2$	Me$_2$C(Ph)OOH	64	77	[50]
Ph	H	Me$_2$C(Ph)OOH	85	95	[50]
Cy	H	Me$_2$C(Ph)OOH	85	95	[50]

Epoxides 47; General Procedure:[50]

MoO$_2$(acac)$_2$ (0.02 mmol) was added to a soln of ligand **46** (0.022 mmol) in CH$_2$Cl$_2$ (1 mL), and the mixture was stirred for 1 h at rt. The alkene (1.0 mmol) and alkyl hydroperoxide (1.5 mmol) were added to the resulting soln, and stirring was continued at the same temperature for 18 h. The progress of the oxidation was monitored by TLC. Sat. aq Na$_2$SO$_3$ was then added, and the mixture was stirred for 30 min at rt. The mixture was extracted with CH$_2$Cl$_2$ and the extracts were dried (Na$_2$SO$_4$), filtered, and concentrated under reduced pressure. The remaining residue was purified by flash column chromatography (silica gel).

2.1.1.7 Tungsten-Catalyzed Epoxidation

Tungstate-based epoxidation systems with hydrogen peroxide as the oxidant have attracted much attention because of their high reactivities and inherent poor activity for the decomposition of hydrogen peroxide. In 1959, Payne and Williams reported the epoxidation of α,β-unsaturated acids with a sodium tungstate catalyst in combination with aqueous hydrogen peroxide as the oxidant.[51] The key to the success of this reaction is careful control of the pH (4–5.5) of the reaction media. A practical method for the epoxidation of aliphatic α-alkenes with 30% hydrogen peroxide under halide-free conditions has been developed by Noyori and co-workers.[52] In this reaction, sodium tungstate (2 mol%), (aminomethyl)phosphonic acid (1 mol%), and methyltrioctylammonium hydrogen sulfate (1 mol%) are employed as catalysts.

In 1999, Jacobs and co-workers reported the tungsten-catalyzed epoxidation of α-pinene using hydrogen peroxide as the oxidant.[53] The epoxides of α-pinene, 1-phenylcyclohexene, and indene are obtained with high levels of conversion and good selectivities. In 2003, Mizuno and co-workers found that the tetrabutylammonium salt of a Keggin-type silicodecatungstate $\{[\gamma\text{-SiW}_{10}O_{34}(H_2O)_2]^{4-}\}$ catalyzes the epoxidation of various alkene substrates with aqueous hydrogen peroxide as the terminal oxidant.[54] The effectiveness of this catalyst is evidenced by 99% selectivity to the epoxide. Furthermore, the catalyst can be recovered and recycled up to five times without any loss of activity or selectivity (Scheme 19).

Scheme 19 Epoxidation of Alkenes Catalyzed by the Tetrabutylammonium Salt of a Keggin-Type Silicodecatungstate[54]

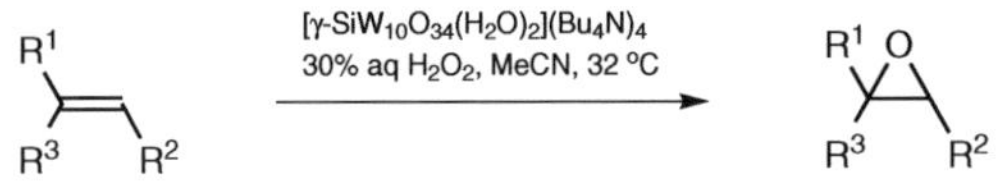

R^1	R^2	R^3	Yield (%)	Ref
Me	H	H	90	[54]
Et	H	H	88	[54]
CH=CH$_2$	H	H	91	[54]
H	(CH$_2$)$_4$Me	Me	>99	[54]
Me	(CH$_2$)$_4$Me	H	91	[54]
H	(CH$_2$)$_4$		84	[54]
H			>99	[54]
H	(CH$_2$)$_6$		99	[54]
	(CH$_2$)$_{10}$	H	97	[54]

In 2009, Mizuno and co-workers reported the efficient hydrogen-bond-assisted epoxidation of homoallylic and allylic alcohols. A novel selenium-containing dinuclear peroxotungstate complex $\{[SeO_4\{WO(O_2)_2\}_2](Bu_4N)_2\}$ shows high catalytic activity for the epoxidation of homoallylic and allylic alcohols with 1 equivalent of hydrogen peroxide (Scheme 20).[55]

Scheme 20 Epoxidation of Homoallylic and Allylic Alcohols Catalyzed by a Selenium-Containing Dinuclear Peroxotungstate Complex[55]

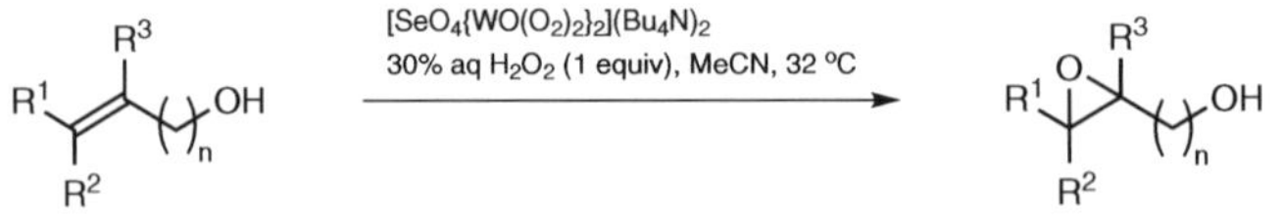

n = 1, 2

2.1.1 Metal-Catalyzed Epoxidation Reactions

R^1	R^2	R^3	n	Yield (%)	Ref
H	H	Me	2	76	[55]
H	(CH$_2$)$_4$Me	H	2	75	[55]
Et	H	H	2	66	[55]
Me	H	H	1	93	[55]
Me	Me	H	1	94	[55]
H	H	Me	1	76	[55]
Pr	H	H	1	89	[55]

Recently, Yamamoto and co-workers found that a tungsten complex formed from bishydroxamic acid ligand **48** catalyzes the asymmetric epoxidation of primary, secondary, and tertiary allylic alcohols **49** in addition to homoallylic alcohols with aqueous 30% hydrogen peroxide as the oxidant to give epoxides **50** (Scheme 21).[56]

Scheme 21 Asymmetric Epoxidation of Allylic Alcohols Catalyzed by a Tungsten–Bishydroxamic Acid Complex[56]

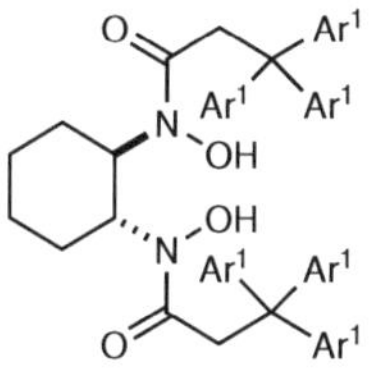

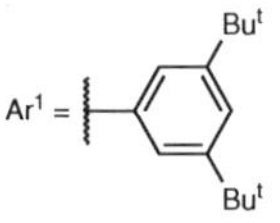

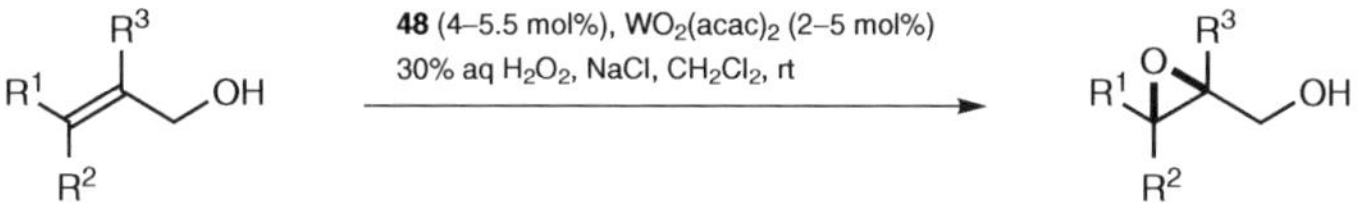

R^1	R^2	R^3	ee (%)	Yield (%)	Ref
H	Pr	H	95	92	[56]
H	(CH$_2$)$_5$Me	H	96	94	[56]
Pr	H	H	93	90	[56]
t-Bu	H	H	98	88	[56]
Ph	H	Ph	86	79	[56]
Ph	H	Me	84	87	[56]

2-(Hydroxymethyl)oxiranes 50; General Procedure:[56]
30% aq H$_2$O$_2$ (102 µL, 1.0 mmol) and allylic alcohol **49** (0.50 mmol) were added to a stirred soln of bishydroxamic acid **48** (0.012 or 0.0275 mmol, 2.4 or 5.4 mol%), WO$_2$(acac)$_2$ (0.01 or 0.025 mmol, 2 or 5 mol%), and NaCl (14.6 mg, 0.25 mmol) in CH$_2$Cl$_2$ (5.0 mL). After stirring for 24 h, the solvent was removed under reduced pressure, and the residue was purified by flash chromatography (silica gel).

2.1.1.8 Other Metal-Catalyzed Epoxidations

In 2008, Katsuki and Egami reported a novel dimeric oxo-bridged niobium complex formed with salan ligand **51** as a catalyst for the asymmetric epoxidation of allylic alcohols using hydrogen peroxide as the oxidant with good enantioselectivity.[57] Subsequent studies have indicated that the oxo complex dissociates into a monomeric species prior to epoxidation.[58] This monomeric complex catalyzes the epoxidation of allylic alcohols with high enantioselectivities ranging from 74 to 95% ee. In some cases, the resulting epoxy alcohols are partially oxidized under the experimental conditions to the corresponding epoxy aldehydes (Scheme 22).

Scheme 22 Asymmetric Epoxidation of Allylic Alcohols with a Niobium–Salan Complex[58]

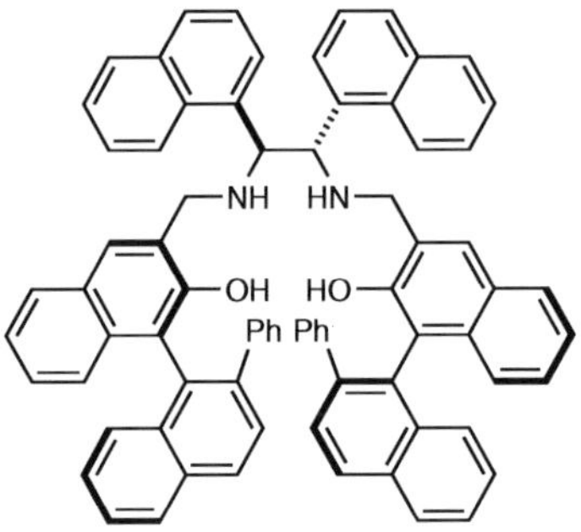

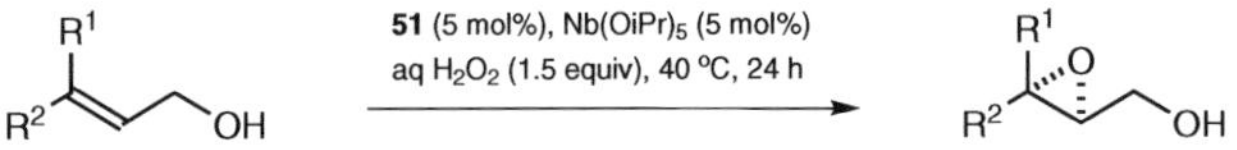

R^1	R^2	ee (%)	Yield (%)	Ref
H	Pr	91	57	[58]
H	$(CH_2)_4Me$	93	79	[58]
H	Cy	93	82	[58]
H	t-Bu	95	52	[58]
H	Ph	74	61	[58]

The group has proposed that precoordination of the allylic alcohol to catalyst **52** is essential for the epoxidation (Scheme 23).

Scheme 23 Plausible Mechanism for the Niobium-Catalyzed Epoxidation[58]

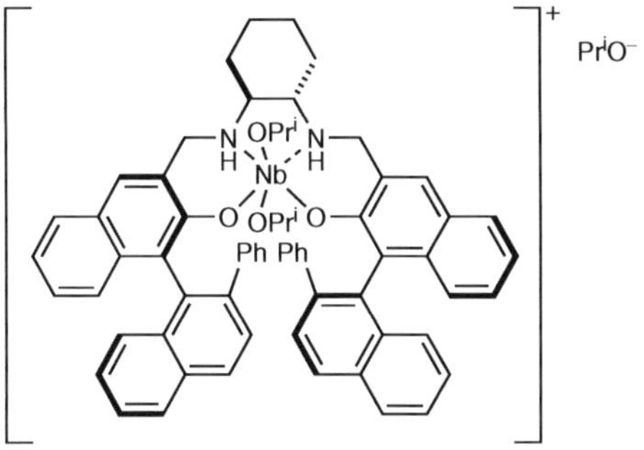
PriO−

52

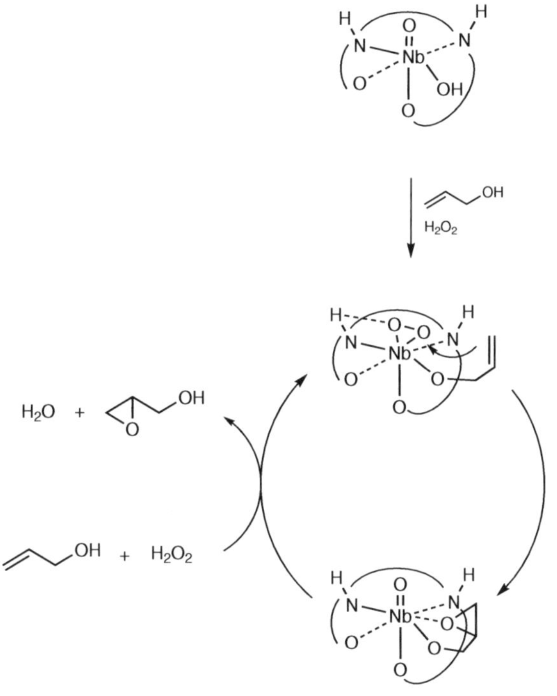
H2O2
H2O +
OH
+ H2O2
O-NH-NH-O = salan ligand

In 2012, a simple and hydrogen peroxide efficient asymmetric epoxidation of α,β-unsaturated carbonyl compounds was accomplished by Feng and co-workers, who used a scandium(III) complex derived from chiral *N*,*N*-dioxide **53** as the catalyst.[59] A number of optically active epoxides **55** can be obtained from corresponding α,β-unsaturated ketones **54** under additive-free conditions. The catalytic system also features good water and air tolerance and provides the products in excellent yields with excellent enantioselectivities (Scheme 24)

Scheme 24 Asymmetric Epoxidation of α,β-Unsaturated Ketones Using a Chiral N,N-Dioxide–Scandium(III) Complex[59]

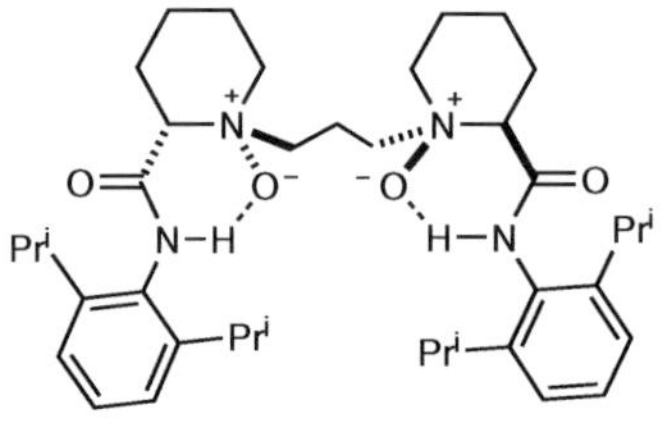

R¹	R²	ee (%)	Yield (%)	Ref
Ph	Ph	98	99	[59]
Ph	4-MeOC$_6$H$_4$	98	95	[59]
Ph	4-BrC$_6$H$_4$	97	99	[59]
3-Tol	Ph	96	99	[59]
2-naphthyl	Ph	98	99	[59]
t-Bu	Ph	96	99	[59]

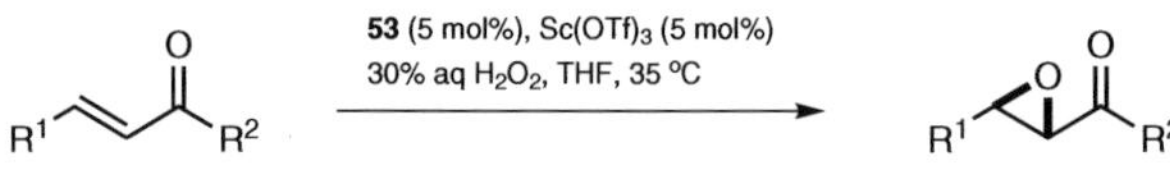

Yamamoto and co-workers have reported zirconium- and hafnium-catalyzed asymmetric epoxidation using C_2-symmetric chiral bishydroxamic acid **17**. The route is efficient for the epoxidation of homoallylic alcohols and bishomoallylic alcohols (Scheme 25).[60]

Scheme 25 Asymmetric Epoxidation of Homoallylic Alcohols Catalyzed by Zirconium–Bishydroxamic Acid and Hafnium–Bishydroxamic Acid Complexes[60]

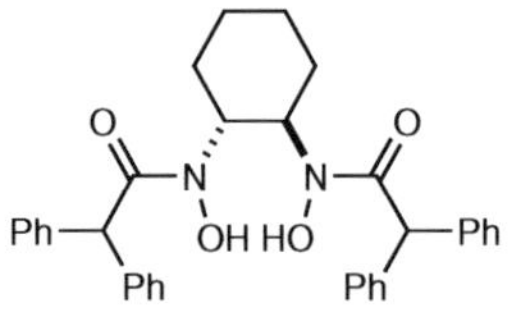

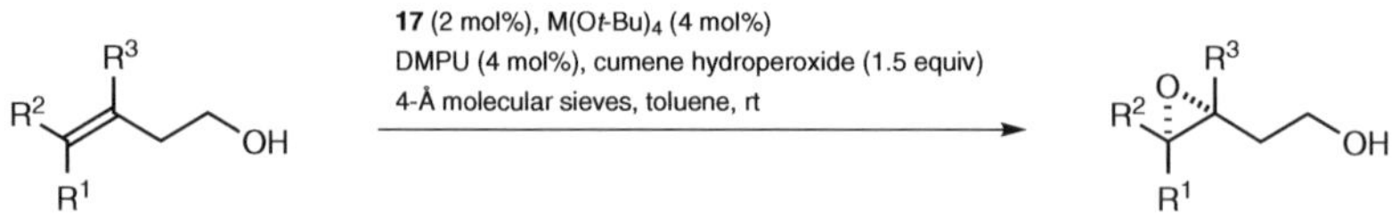

R¹	R²	R³	M	ee (%)	Yield (%)	Ref
H	H	Me	Zr	91	61	[60]
H	H	Me	Hf	97	81	[60]
H	H	H	Hf	63	37	[60]
Et	H	H	Zr	93[a]	45	[60]
Et	H	H	Hf	94[a]	82	[60]

2.1.1 Metal-Catalyzed Epoxidation Reactions

R¹	R²	R³	M	ee (%)	Yield (%)	Ref
Bu	H	H	Zr	92[a]	80	[60]
Bu	H	H	Hf	96[a]	83	[60]
H	H	Ph	Zr	92[b]	67	[60]
H	H	Ph	Hf	98[b]	69	[60]
H	(CH₂)₄		Zr	76[b]	72	[60]
H	(CH₂)₄		Hf	89[b]	81	[60]

[a] Opposite enantiomer to that shown in scheme.
[b] Absolute stereochemistry not specified.

Recently, Yamamoto and co-workers have reported that a hafnium(IV)–bishydroxamic acid complex catalyzes the enantioselective epoxidation of N-alkenyl sulfonamides **56** and N-tosylimines.[61] A number of optically active epoxides **57** are obtained in excellent yields (up to 97% yield) with excellent enantioselectivities (up to 92% ee) (Scheme 26).

Scheme 26 Asymmetric Epoxidation of N-Allyl Sulfonamides Catalyzed by a Hafnium(IV)–Bishydroxamic Acid Complex[61]

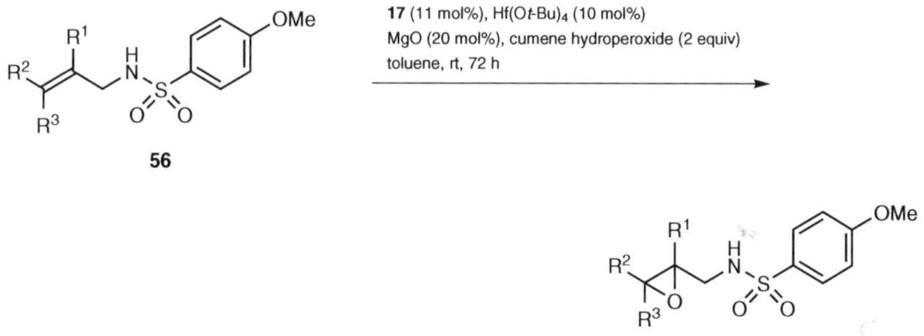

Reagents: **17** (11 mol%), Hf(O*t*-Bu)₄ (10 mol%), MgO (20 mol%), cumene hydroperoxide (2 equiv), toluene, rt, 72 h

R¹	R²	R³	ee[a] (%)	Yield (%)	Ref
H	H	H	93[b]	49	[61]
H	H	Me	91	92	[61]
H	Me	H	20	48	[61]
H	Me	Me	92	97	[61]
Me	Me	H	62	84	[61]
Me	H	H	42	92	[61]

[a] Unless otherwise specified, the absolute configuration was not reported.
[b] The R-product was obtained.

2-Benzoyloxiranes 55; General Procedure:[59]

α,β-Unsaturated ketone **54** (0.1 mmol), Sc(OTf)₃ (2.5 mg, 5 mol%), and ligand **53** (3.3 mg, 5 mol%) were weighed into a reaction tube, and then THF (0.5 mL) was added. The mixture was stirred at 35 °C for 30 min, and then 30% aq H₂O₂ (28.2 µL, 0.3 mmol) was added. The mixture was then stirred at 35 °C. The residue was purified by flash chromatography (silica gel).

2-{[(4-Methoxyphenyl)sulfonamido]methyl}oxiranes 57; General Procedure:[61]
Hf(Ot-Bu)$_4$ (16 mg, 0.04 mmol) was added to a soln of bishydroxamic acid **17** (23 mg, 0.044 mmol) in toluene. The catalyst mixture was stirred for 3 h at rt. MgO (3.2 mg, 0.08 mmol), N-alkenyl sulfonamide **56** (0.4 mmol), and 80% cumene hydroperoxide (147 μL, 0.8 mmol) were then added sequentially. The resulting mixture was efficiently stirred at rt for 72 h. The mixture was purified by flash column chromatography (silica gel, hexanes/EtOAc 2:1).

2.1.2 Metal-Catalyzed Aziridination Reactions

2.1.2.1 Copper-Catalyzed Aziridination

The most commonly employed chiral catalytic systems that have been developed to date for enantioselective aziridination by nitrene transfer to alkenes are based on copper complexes. The aziridination of alkenes using copper–bisoxazoline [bis(dihydrooxazole)] complexes was pioneered by Evans and co-workers.[62] In 1993, Jacobsen and co-workers discovered that chiral diimines can serve as excellent ligands for the copper-catalyzed asymmetric aziridination of alkenes.[63] Xu and co-workers have applied some bisoxazoline ligands for the asymmetric copper-mediated aziridination of chalcones **59**. The reactions are performed with copper(I) trifluoromethanesulfonate as the catalyst and [N-(4-toluenesulfonyl)imino]phenyliodinane (PhINTs) as the nitrogen source. The use of 1,8-bis(dihydrooxazolyl)anthracene **58** as the ligand furnishes aziridines **60** in good yields with up to 99% ee (Scheme 27).[64]

Scheme 27 Asymmetric Aziridination of Chalcones Catalyzed by a Copper–1,8-Bis(dihydrooxazolyl)anthracene Complex[64]

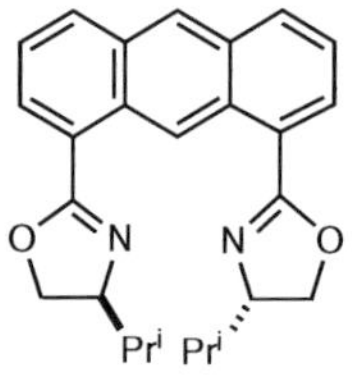

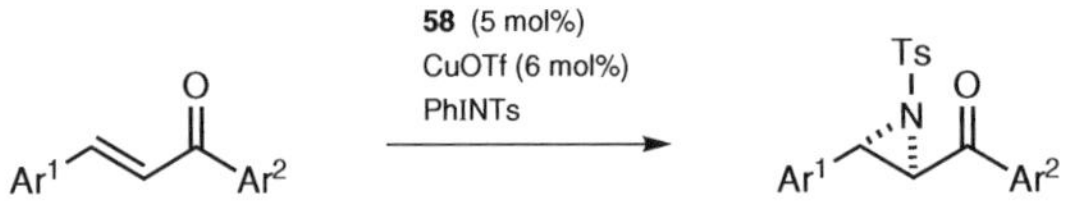

Ar^1	Ar^2	ee (%)	Yield (%)	Ref
Ph	Ph	96	80	[64]
4-Tol	Ph	98	86	[64]
3-ClC$_6$H$_4$	Ph	84	76	[64]
2-ClC$_6$H$_4$	Ph	79	91	[64]
Ph	4-Tol	>99	92	[64]
4-Tol	4-Tol	>99	59	[64]

In 2006, Ding and co-workers reported that novel chiral C_2-symmetric 1,4-diimine **61** derived from D-mannitol is an effective chiral ligand for the copper-catalyzed asymmetric

2.1.2 Metal-Catalyzed Aziridination Reactions

aziridination of alkene derivatives **62** with [N-(4-toluenesulfonyl)imino]phenyliodinane as the nitrene source to afford corresponding N-sulfonylaziridine derivatives **63** in good to excellent yields with up to 99% ee (Scheme 28).[65]

Scheme 28 Asymmetric Aziridination of Acrylates Catalyzed by a Copper–1,4-Diimine Complex[65]

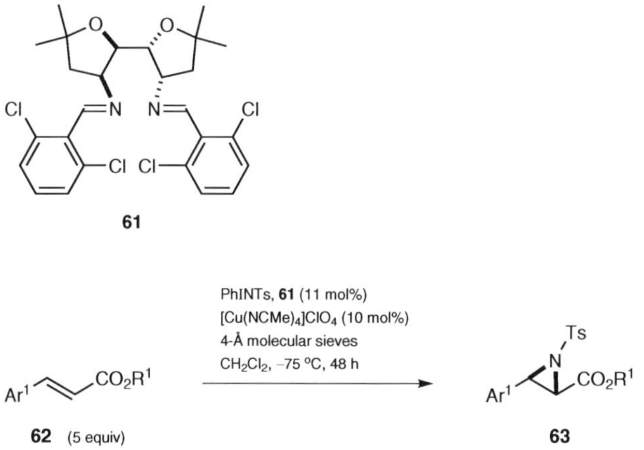

Ar[1]	R[1]	ee (%)	Yield (%)	Ref
Ph	Me	88	97	[65]
Ph	Ph	87	96	[65]
Ph	t-Bu	>99	99	[65]
4-FC$_6$H$_4$	t-Bu	98	99	[65]
4-ClC$_6$H$_4$	t-Bu	98	97	[65]
4-BrC$_6$H$_4$	t-Bu	98	97	[65]
4-Tol	t-Bu	94	86	[65]
4-MeOC$_6$H$_4$	t-Bu	80[a]	95	[65]
2-O$_2$NC$_6$H$_4$	t-Bu	99[a]	63	[65]

[a] Opposite enantiomer to that shown in scheme.

In 2007, Lebel and co-workers found that a copper–pyridine complex effectively catalyzes the intermolecular aziridination of styrenes using 2,2,2-trichloroethyl (tosyloxy)carbamate (**64**) as the nitrene source to provide trichloroethyl derivatives **65** (Scheme 29).[66]

Scheme 29 Intermolecular Aziridination of Various Styrenes with a Copper–Pyridine Complex[66]

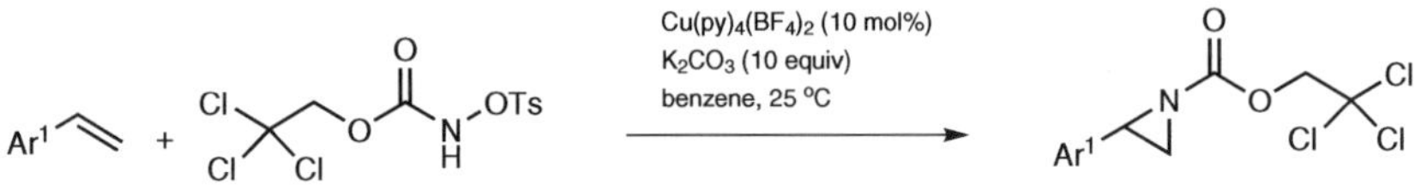

Ar[1]	Yield (%)	Ref
4-Tol	68	[66]
4-FC$_6$H$_4$	55	[66]
4-ClC$_6$H$_4$	62	[66]
4-BrC$_6$H$_4$	56	[66]
3-MeOC$_6$H$_4$	62	[66]
4-O$_2$NC$_6$H$_4$	51	[66]

In 2008, aziridination of aliphatic alkenes catalyzed by N-heterocyclic carbene–copper complex **66** was described by Appella and Xu. Upon using iodosylbenzene as the oxidant, a wide variety of aliphatic alkenes can be converted into aziridines **67** in moderate to high yields (Scheme 30).[67]

Scheme 30 Aziridination of Aliphatic Alkenes Catalyzed by a Copper–N-Heterocyclic Carbene Complex[67]

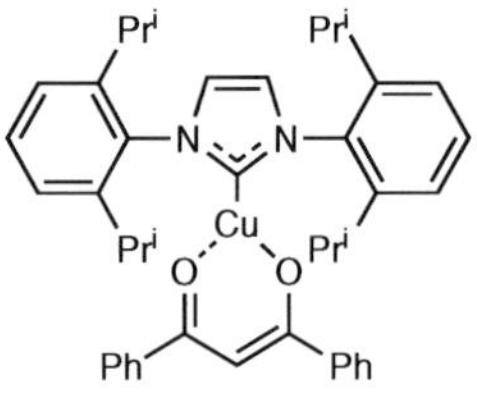

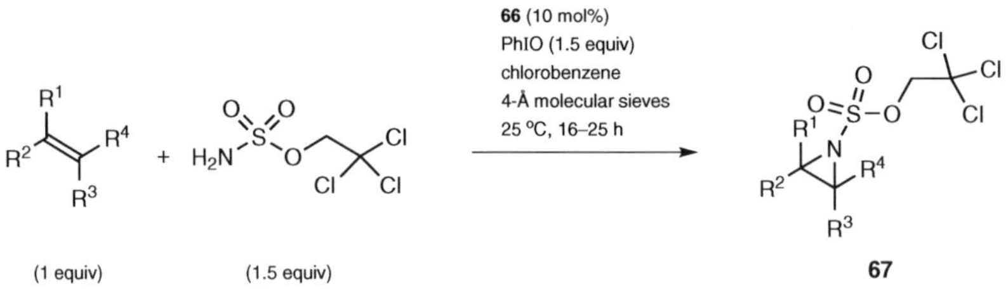

R[1]	R[2]	R[3]	R[4]	Yield (%)	Ref
Bu	H	H	H	75	[67]
t-Bu	H	H	H	51	[67]
Me	Pr	H	H	61	[67]

2.1.2 Metal-Catalyzed Aziridination Reactions

R^1	R^2	R^3	R^4	Yield (%)	Ref
H	Pr	H	Et	35^a	[67]
H	(CH$_2$)$_3$	H		57	[67]
Me	CO$_2$Me	H	H	82	[67]

^{a} Ratio (*trans/cis*) 8:1.

In 2012, Chan and co-workers reported that copper(II) trifluoromethanesulfonate catalyzes the aziridination of 2-alkyl-substituted 1,3-dicarbonyl compounds **68** with [*N*-(4-toluenesulfonyl)imino]phenyliodinane as the nitrene source.[68] In the presence of 1,10-phenanthroline, divergence in product selectivity is possible through slight modification of the reaction conditions. Increasing the amount of the iodinane from 1.2 to 2–3 equivalents affords corresponding aziridine products **69** in yields of 61–99% (Scheme 31).

Scheme 31 Aziridination of 2-Alkyl-Substituted 1,3-Dicarbonyl Compounds Catalyzed by Copper(II) Trifluoromethanesulfonate/1,10-Phenanthroline[68]

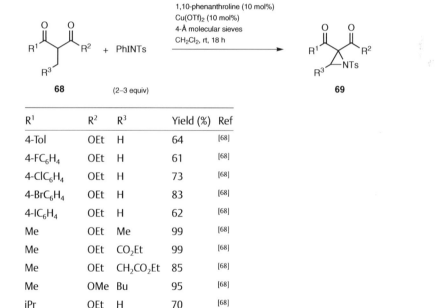

R^1	R^2	R^3	Yield (%)	Ref
4-Tol	OEt	H	64	[68]
4-FC$_6$H$_4$	OEt	H	61	[68]
4-ClC$_6$H$_4$	OEt	H	73	[68]
4-BrC$_6$H$_4$	OEt	H	83	[68]
4-IC$_6$H$_4$	OEt	H	62	[68]
Me	OEt	Me	99	[68]
Me	OEt	CO$_2$Et	99	[68]
Me	OEt	CH$_2$CO$_2$Et	85	[68]
Me	OMe	Bu	95	[68]
iPr	OEt	H	70	[68]
iPr	OEt	Me	76	[68]
cyclopropyl	OEt	Me	96	[68]

2-Aryl-3-benzoyl-1-tosylaziridines 60; General Procedure:[64]

A 25-mL three-necked flask was charged with chalcone **59** (1.50 mmol), 1,8-bisoxazolidinylanthracene (**58**; 24 mg, 0.06 mmol), and CuOTf (13 mg, 0.05 mmol) under a N$_2$ atmosphere. CH$_2$Cl$_2$ (8 mL) was added by syringe, and the resulting mixture was stirred for 1 h at 24 °C. PhINTs (373 mg, 1.00 mmol) was added portionwise to the mixture over 2 h. After the addition, the mixture was stirred for another 3 h. Flash chromatography [silica gel, petroleum ether (60–90 °C)/EtOAc 6:1] afforded the title compounds.

3-Aryl-1-tosylaziridine-2-carboxylates 63; General Procedure:[65]

Anhyd CH$_2$Cl$_2$ (3 mL) was added to a Schlenk tube containing the copper salt (0.0125 mmol) and chiral diimine ligand **61** (0.014 mmol) under an atmosphere of argon. The mixture was then stirred at rt for 1 h. After this time, the mixture was cooled to −75 °C and alkene **62** (0.625 mmol) and PhINTs (47 mg, 0.125 mmol) were added sequentially to the stirred soln against a slow positive flow of argon. The progress of the reaction was monitored by TLC. Upon completion of the reaction, the mixture was concentrated to dryness, and the residue was purified by flash chromatography (silica gel, EtOAc/hexane 1:4 to 1:6).

2,2,2-Trichloroethyl 2-Arylaziridine-1-carboxylates 65; General Procedure:[66]

$Cu(py)_4(BF_4)_2$ (28 mg, 0.050 mmol) was suspended in benzene (5 mL) (**CAUTION**: *carcinogen*) in a 20-mL scintillation vial in air. K_2CO_3 (692 mg, 5.00 mmol) and the styrene derivative (2.50 mmol) were successively added. The heterogeneous soln was stirred for 15 min at 25°C and carbamate **64** (182 mg, 0.50 mmol) was added in one portion. The soln was stirred at 25°C overnight. Et_2O (15 mL) was added, and the resulting mixture was filtered. The solid was washed with Et_2O (4 × 10 mL). The combined filtrate was concentrated under reduced pressure. The residue was purified by flash chromatography (silica gel, Et_2O/hexanes).

1-[(2,2,2-Trichloroethoxy)sulfonyl]aziridines 67; General Procedure:[67]

4-Å molecular sieves (600 mg) were added to a 10-mL, oven-dried, round-bottomed flask containing copper complex **66** (34 mg, 0.05 mmol), iodosylbenzene (166 mg, 0.75 mmol), and 2,2,2-trichloroethyl sulfamate (171 mg, 0.75 mmol). The flask was sealed with a septum and was purged with N_2 for 2 min while stirring. A soln of the alkene (0.5 mmol) in chlorobenzene (1.25 mL) was added. After the addition, the N_2 line was removed and the septum was sealed with grease. The green suspension was vigorously stirred at rt for 20–25 h. The mixture was diluted with CH_2Cl_2 (2 mL) and filtered through Celite, which was washed with CH_2Cl_2 (3 mL). The solvents were evaporated under reduced pressure, and the residue was purified by flash chromatography.

1-Tosylaziridine-2-carboxylates 69; General Procedure:[68]

CH_2Cl_2 (2 mL) was added to a mixture of $Cu(OTf)_2$ (18.1 mg, 0.05 mmol), 1,10-phenanthroline (9.9 mg, 0.05 mmol), and powdered 4-Å molecular sieves (400 mg). After stirring for 1 h, PhINTs (373 mg, 1.0 mmol or 560 mg, 1.5 mmol) and dicarbonyl compound **68** (0.5 mmol) were added. The mixture was stirred for another 18 h at rt, after which the mixture was filtered through Celite, washed with EtOAc (50 mL), concentrated to dryness, and purified by flash column chromatography (hexanes/EtOAc 4:1).

2.1.2.2 Rhodium-Catalyzed Aziridination

Rhodium catalysts are complementary in scope to copper catalysts in carbene-transfer reactions. Du Bois and Guthikonda report that tetrakis(μ-trifluoroacetamidato)dirhodium(II) catalyzes the aziridination of alkenes using 2,2,2-trichloroethyl sulfamate as a nitrogen source.[69] (Diacetoxyiodo)benzene is utilized as the oxidant to promote N-atom transfer reactions, and aziridine formation proceeds in good yields with a range of alkenes. Importantly, aziridination is stereospecific with *E*- and *Z*-prop-1-enylbenzene and *E*- and *Z*-dec-2-ene substrates (Scheme 32).

Scheme 32 Aziridination of Alkenes Catalyzed by Tetrakis(μ-trifluoroacetamidato)dirhodium(II)[69]

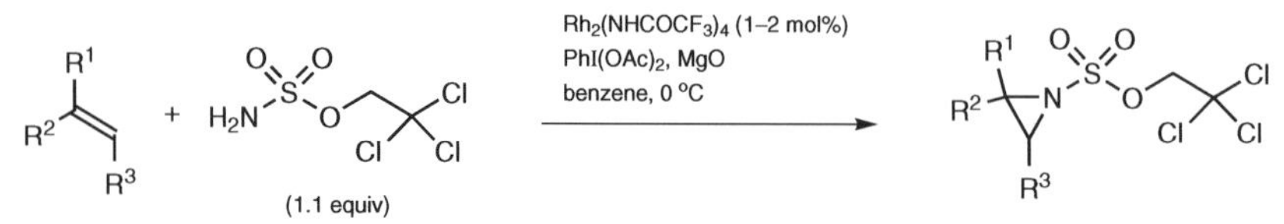

2.1.2 Metal-Catalyzed Aziridination Reactions

R¹	R²	R³	Yield (%)	Ref
Ph	H	Me	85	[69]
H	Ph	Me	85	[69]
2-BrC$_6$H$_4$	H	H	79	[69]
3-O$_2$NC$_6$H$_4$	H	H	95	[69]
4-Tol	H	H	91	[69]
4-ClC$_6$H$_4$	H	H	88	[69]
H	(CH$_2$)$_6$		82	[69]
Me	Pr	H	72	[69]

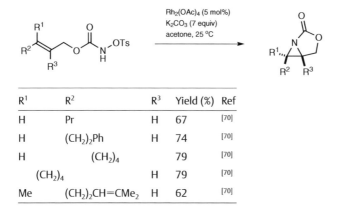

Lebel and co-workers reported that N-(tosyloxy)carbamates can be used as a source of nitrenes in rhodium-catalyzed intramolecular aziridination reactions.[70] In the presence of rhodium(II) acetate (5 mol%) and an excess of potassium carbonate, the reaction proceeds equally well with disubstituted and trisubstituted alkenes (Scheme 33).

Scheme 33 Intramolecular Aziridination of O-Allyl N-(Tosyloxy)carbamates Catalyzed by Rhodium(II) Acetate[70]

R¹	R²	R³	Yield (%)	Ref
H	Pr	H	67	[70]
H	(CH$_2$)$_2$Ph	H	74	[70]
H	(CH$_2$)$_4$		79	[70]
(CH$_2$)$_4$		H	79	[70]
Me	(CH$_2$)$_2$CH=CMe$_2$	H	62	[70]

Doyle and co-workers report that dirhodium(II) caprolactamate catalyzes the aziridination of alkenes.[71] In the presence of 4-toluenesulfonamide, N-bromosuccinimide, and potassium carbonate, the reaction provides the corresponding aziridines in yields up to 95% under mild conditions (Scheme 34).

Scheme 34 Intermolecular Aziridination of Alkenes Catalyzed by a Dirhodium Caprolactamate Complex[71]

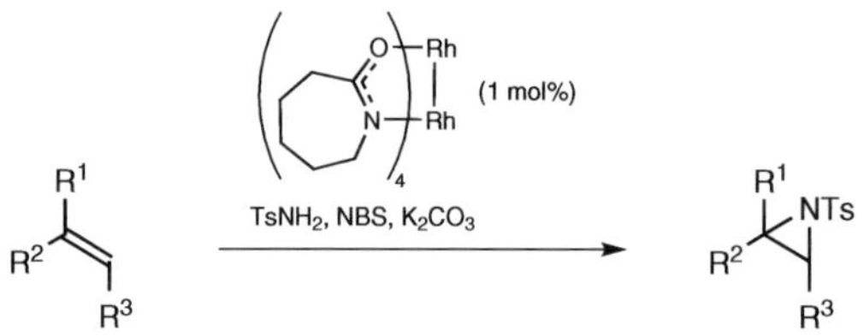

R¹	R²	R³	Yield (%)	Ref
Ph	H	H	77	[71]
Ph	H	Me	69	[71]
H	Ph	Me	77	[71]
2-naphthyl	H	H	65	[71]
H	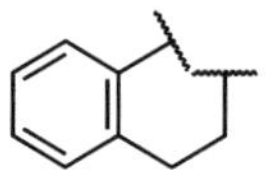 (tetralin)		95	[71]
H	$(CH_2)_6$		74	[71]
H	$(CH_2)_4$		60	[71]
Bu	H	H	77	[71]

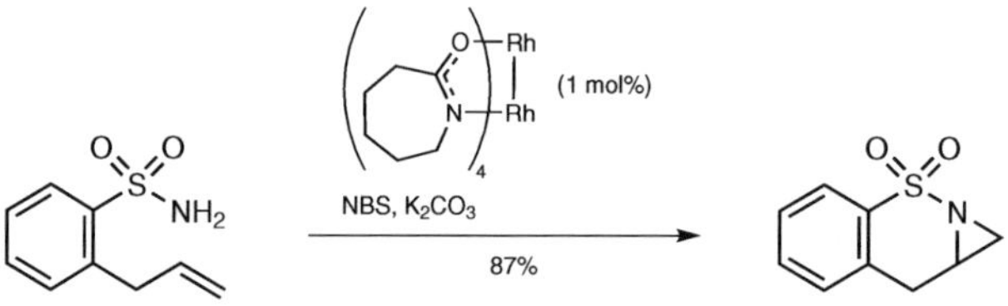

In 2011, Lebel and co-workers disclosed the rhodium-catalyzed intermolecular aziridination of alkenes using chiral N-(tosyloxy)carbamate **71** as the nitrene source in the presence of catalyst **70**.[72] Good to high yields and stereoselectivities of products **72** are achieved using this readily available chiral N-(tosyloxy)carbamate and a stoichiometric amount of the alkene substrate (Scheme 35).

Scheme 35 Intermolecular Aziridination of Styrenes Catalyzed by a Dirhodium Carboxylate Complex[72]

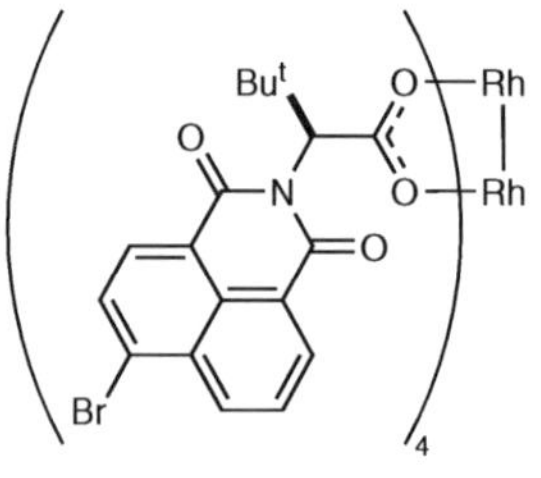

70

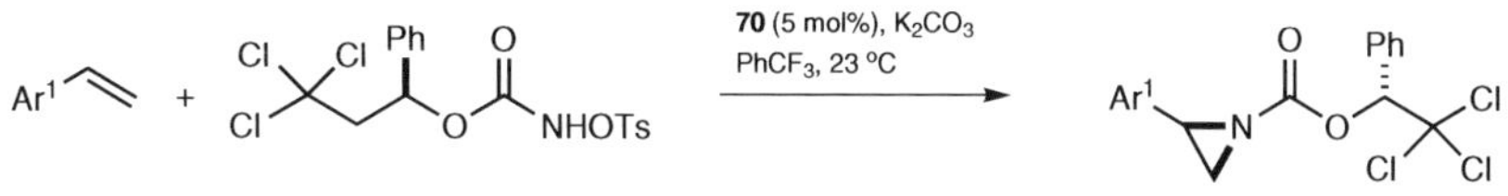

2.1.2 Metal-Catalyzed Aziridination Reactions

Ar^1	dr (%)	Yield (%)	Ref
2-ClC$_6$H$_4$	30:1	84	[72]
2-BrC$_6$H$_4$	33:1	78	[72]
2-Tol	6:1	72	[72]
4-O$_2$NC$_6$H$_4$	>50:1	63	[72]
4-F$_3$CC$_6$H$_4$	8:1	74	[72]
2-Br-4-MeC$_6$H$_3$	25:1	79	[72]

Recently, a useful new catalytic method to convert unactivated alkenes into aziridines was reported by Kürti, Falck, Ess, and co-workers that uses O-(2,4-dinitrophenyl)hydroxylamines **74** as the aminating agents and dirhodium complex **73** as the catalyst. This method is operationally simple, scalable, and fast at ambient temperature, and it furnishes N—H aziridines **75** in good to excellent yields (Scheme 36).[73]

Scheme 36 Aziridination of Alkenes Catalyzed by a Dirhodium Carboxylate Complex[73]

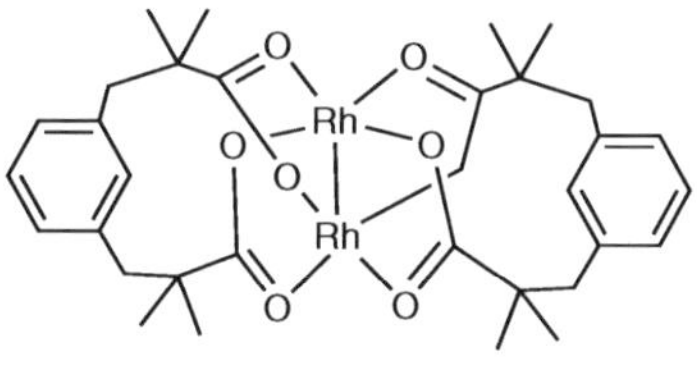

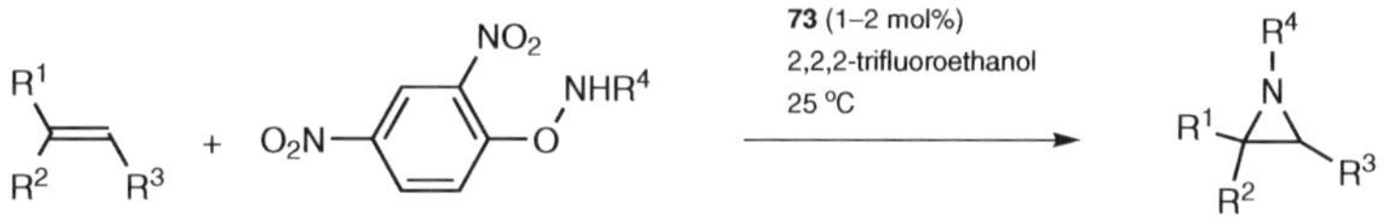

R^1	R^2	R^3	R^4	Yield (%)	Ref
(CH$_2$)$_4$OH	H	H	H	59	[73]
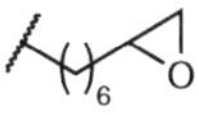 (cyclohexyl epoxide)$_6$	H	H	H	77	[73]
H	(CH$_2$)$_8$OTBDPS	CH$_2$OH	H	64	[73]
(CH$_2$)$_7$Me	H	(CH$_2$)$_7$CO$_2$Me	H	91	[73]
Et	H	(CH$_2$)$_2$CO$_2$Me	H	83	[73]
Me	Me	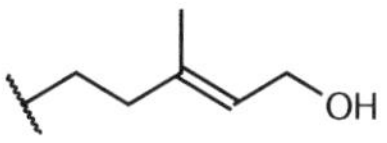 (prenyl-OH chain)	H	71	[73]

R¹	R²	R³	R⁴	Yield (%)	Ref
H	(CH$_2$)$_4$		H	71	[73]
Me			H	72	[73]
Me	Me	Ph	Me	81	[73]

2,2,2-Trichloro-1-phenylethyl 2-Phenylaziridine-1-carboxylates 72; General Procedure:[72]
Catalyst **70** (22 mg, 5 mol%) and the styrene derivative (0.250 mmol) were added to a soln of (1R)-2,2,2-trichloro-1-phenylethyl N-(tosyloxy)carbamate (**71**; 132 mg, 0.300 mmol) in (trifluoromethyl)benzene (1.0 mL). The mixture was stirred for 5 min and sat. aq K$_2$CO$_3$ (47 μL, 0.38 mmol) was added. The heterogeneous soln was stirred at rt for 90 min. The reaction was then quenched with pyridine (2 drops) and filtered through Celite with Et$_2$O. The filtrate was concentrated under reduced pressure to give the crude product. Filtration through silica gel (pretreated with 5% Et$_3$N/hexanes) to remove the rhodium catalyst was performed prior to HPLC analysis (no separation of the diastereomers).

Aziridines 75; General Procedure:[73]
A round-bottomed flask equipped with a magnetic stirrer was charged with the alkene (0.5 mmol, 1.0 equiv) and 2,2,2-trifluoroethanol (5 mL). Catalyst **73** (3.8 mg, 5 μmol, 1 mol%) and aminating agent **74** (0.119 g, 0.6 mmol, 1.2 equiv) were then added to this soln at 25 °C. The mixture was stirred at this temperature, and the progress of the reaction was monitored by TLC. More catalyst and aminating agent were added if required. Upon completion of the reaction, the mixture was diluted with CH$_2$Cl$_2$ (10 mL) and washed with 5% aq NaHCO$_3$ (1 × 5 mL). The aqueous layer was extracted with CH$_2$Cl$_2$ (2 × 10 mL), and the combined organic portion was washed with brine (1 × 5 mL), dried (Na$_2$SO$_4$), and concentrated under reduced pressure. The residue was purified on a prepacked silica gel column using a CombiFlash chromatograph.

2.1.2.3 Ruthenium-Catalyzed Aziridination

Katsuki and co-workers reported that chiral ruthenium–salen complex **76** catalyzes the asymmetric aziridination of a wide range of alkenes using azide compounds as the nitrene precursors.[74] Chiral complex **76** is effective in the aziridination of styrene and provides corresponding aziridines in high yields with enantioselectivities up to 99% ee (Scheme 37). The highly enantioselective aziridination of vinyl ketones using chiral ruthenium–salen complex **76** has also been reported.[75]

2.1.2 Metal-Catalyzed Aziridination Reactions

Scheme 37 Aziridination of Alkenes Catalyzed by a Ruthenium–Salen Complex[75]

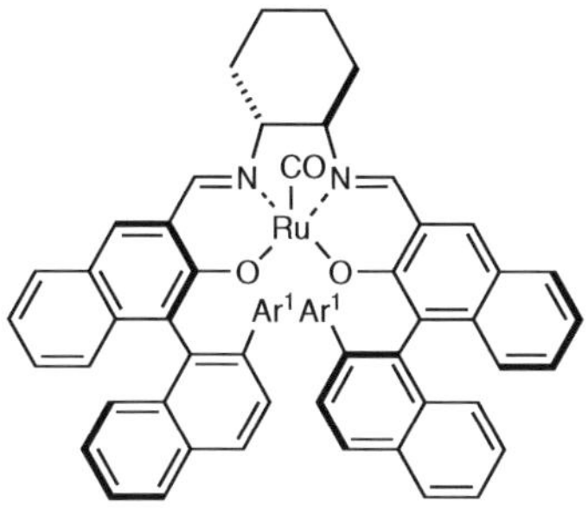

Ar^1 = 3,5-Cl_2-4-TMSC$_6$H$_2$

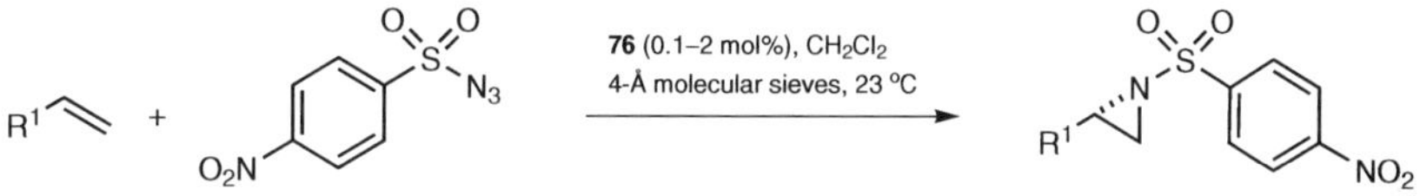

R^1	ee (%)	Yield (%)	Ref
Ph	87	90	[75]
4-BrC$_6$H$_4$	83	93	[75]
C≡CPh	98	98	[75]
(CH$_2$)$_5$Me	56	32	[75]

In 2012, ruthenium–salen complex **77** was found to be an efficient catalyst for the highly enantioselective aziridination of aromatic and aliphatic terminal alkenes in the presence of 2-(trimethylsilyl)ethane-1-sulfonyl azide as a nitrene source.[76] Aziridines **78** are obtained in high yields with excellent enantioselectivities (Scheme 38).

Scheme 38 Asymmetric Aziridination of Alkenes Catalyzed by a Ruthenium–Salen Complex[76]

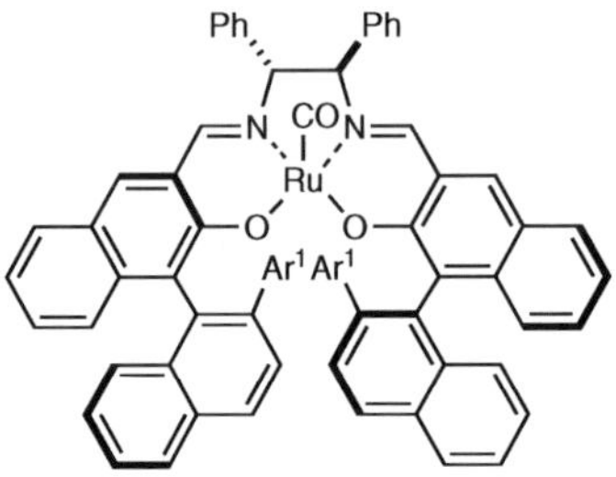

$Ar^1 = 3,5-(CF_3)_2C_6H_3$

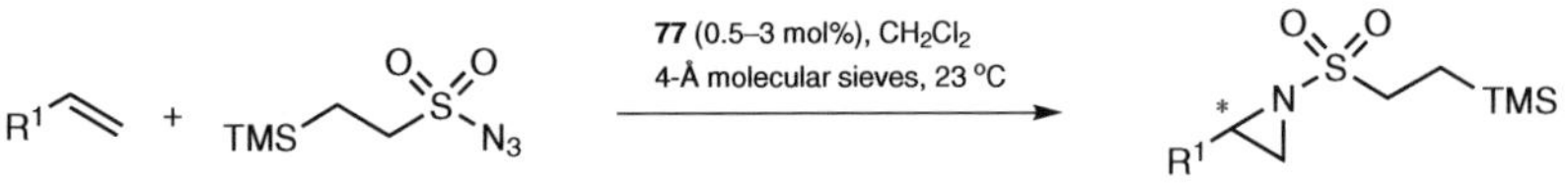

R^1	ee[a] (%)	Yield (%)	Ref
Bu	99	74	[76]
(E)-CH$_2$CH=CHMe	91	58	[76]
Bn	90	91	[76]
2-Tol	97	95	[76]
4-Tol	89[b]	95	[76]
4-ClC$_6$H$_4$	90	96	[76]

[a] Unless otherwise specified, the absolute configuration was not reported.
[b] The R-product was obtained.

Aziridines 78; General Procedure:[76]
A dried Schlenk tube was charged with 4-Å molecular sieves (50 mg) and then additionally dried with a heat gun for 10 min. The Schlenk tube was then evacuated, backfilled with N$_2$, and equipped with a magnetic stirrer bar. Complex **77** (0.5–3 mol%) and solvent (0.4 mL) were added, followed by the alkene (0.36–0.9 mmol) and the azide (0.3 mmol) at rt. After being stirred for another 6–24 h, the mixture was filtered through a pad of Celite. Concentration of the resulting soln and chromatographic separation (silica gel, hexane/EtOAc 10:1 to 5:1) gave the title compounds.

2.1.2.4 Iron-Catalyzed Aziridination

In recent years, there has been a surge of interest in developing iron catalysts for aziridination reactions. Relative to other transition-metal catalysts, iron complexes are inexpensive and biocompatible. Among the iron complexes reported for this endeavor, iron porphyrins are commercially available, air and moisture stable, and can be easily prepared and modified. In 2008, Che and co-workers reported the use of terpyridine–iron complex **79** as an efficient catalyst for the inter- and intramolecular aziridination of alkenes using sulfonamides.[77] In the presence of [N-(4-toluenesulfonyl)imino]phenyliodinane, alkenes afford corresponding aziridines **80** (Scheme 39).

2.1.2 Metal-Catalyzed Aziridination Reactions

Scheme 39 Aziridination of Alkenes Catalyzed by a Terpyridine–Iron Complex[77]

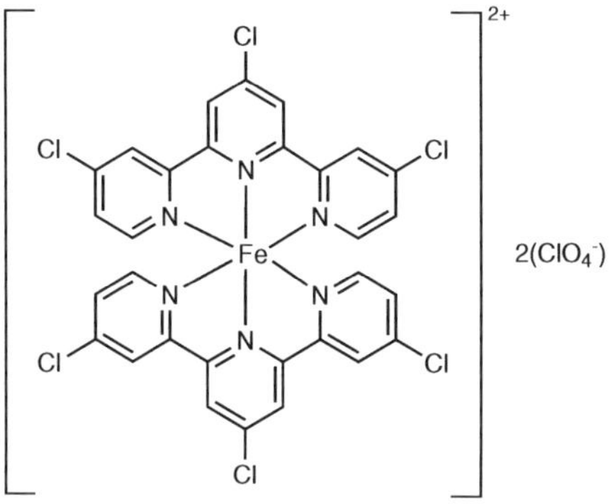

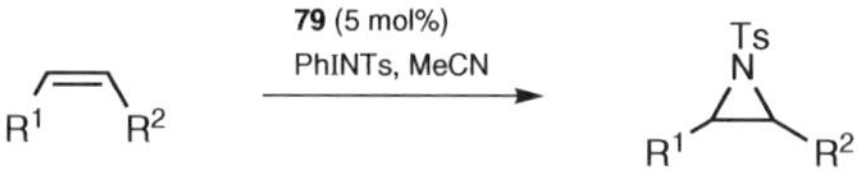

R^1	R^2	Yield (%)	Ref
Ph	H	95	[77]
4-F$_3$CC$_6$H$_4$	H	86	[77]
(CH$_2$)$_4$		68	[77]
(CH$_2$)$_7$Me	H	78	[77]

1-Tosylaziridines 80; General Procedure:[77]
PhINTs (1.5 equiv) was added to a soln of the substrate (0.2 mmol, 1 equiv) and catalyst **79** (5 mol%) in anhyd MeCN (3 mL) at 40 °C under an argon atmosphere. The soln was stirred for 12 h, diluted with CH$_2$Cl$_2$, and washed with H$_2$O. The combined organic layer was dried (Na$_2$SO$_4$), filtered, and concentrated to dryness under reduced pressure. The crude product was purified by flash column chromatography.

2.1.2.5 Cobalt-Catalyzed Aziridination

In 2005, Zhang and co-workers reported that cobalt porphyrin **81** is capable of catalyzing the aziridination of alkenes with bromamine-T as the nitrene source.[78] This catalytic system is suitable for many different types of alkene substrates, such as aromatic and conjugated alkenes and acyclic and cyclic aliphatic alkenes, and forms the desired N-sulfonylated aziridine derivatives in high to excellent yields (Scheme 40).

Scheme 40 Aziridination of Alkenes Catalyzed by a Cobalt–Porphyrin Complex[78]

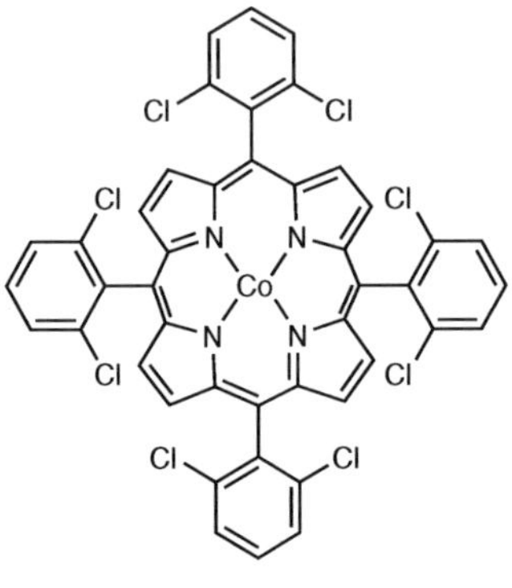

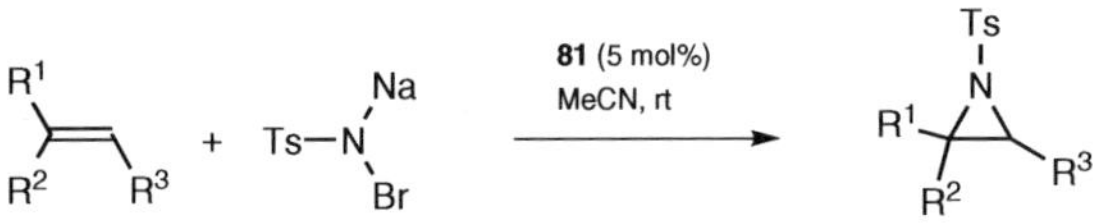

R¹	R²	R³	Yield (%)	Ref
4-Tol	H	H	76	[78]
4-F₃CC₆H₄	H	H	90	[78]
4-FC₆H₄	H	H	86	[78]
Ph	Me	H	70	[78]
2-naphthyl	H	H	53	[78]
H	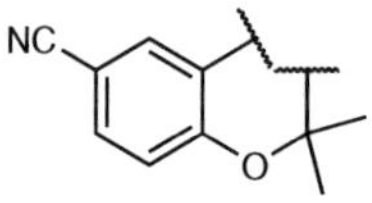 \<NC-aryl-dihydrofuran fused\>		66	[78]
H	Ph	Ph	92	[78]
H	(CH₂)₃		61	[78]
H	(CH₂)₄		66	[78]
H	(CH₂)₅		79	[78]
	(CH₂)₅	H	67	[78]

The same group has developed diphenyl phosphorazidate as a new nitrene source for catalytic aziridination using cobalt(II)–tetraphenylporphyrin **82**.[79] This catalyst system allows the direct synthesis of N-phosphorus-substituted aziridines from alkenes with nitrogen gas as the byproduct (Scheme 41).

2.1.2 Metal-Catalyzed Aziridination Reactions

Scheme 41 Aziridination of Alkenes Catalyzed by a Cobalt–Tetraphenylporphyrin Complex[79]

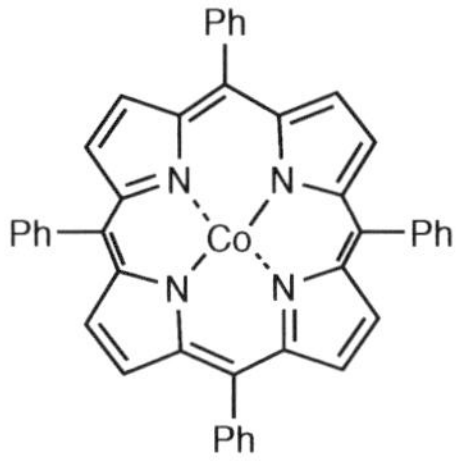

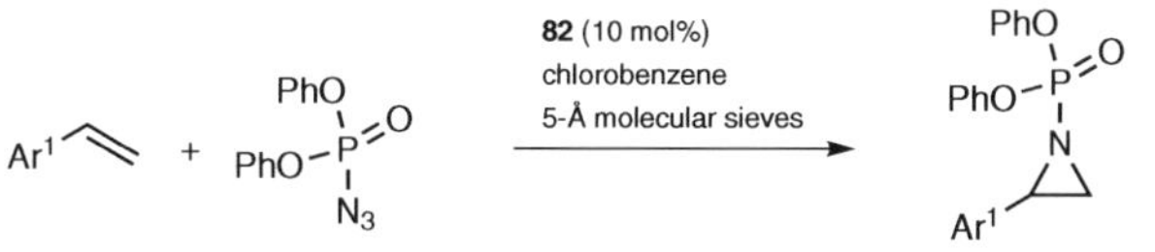

Ar[1]	Yield (%)	Ref
3-Tol	64	[79]
4-t-BuC$_6$H$_4$	43	[79]
4-BrC$_6$H$_4$	54	[79]
4-ClC$_6$H$_4$	52	[79]
4-F$_3$CC$_6$H$_4$	60	[79]
3-O$_2$NC$_6$H$_4$	68	[79]

In 2009, Zhang and co-workers reported that cobalt(II) complex **83** of a D_2-symmetric chiral porphyrin is able to catalyze a highly enantioselective aziridination process using 2,2,2-trichloroethyl sulfurazidate (**84**) as a new nitrene source.[80] The reaction provides products **85** in moderate to good yields with high enantioselectivity (er >90:10) for aromatic and aliphatic alkenes. Furthermore, the catalytic system can be conveniently recycled and reused multiple times through a simple precipitation/filtration protocol without significant loss of reactivity and selectivity (Scheme 42).

Scheme 42 Aziridination of Alkenes Catalyzed by a Cobalt–Porphyrin Complex[80]

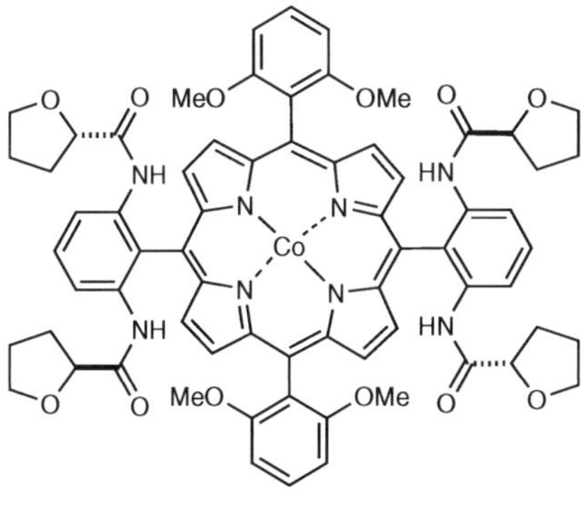

for references see p 47

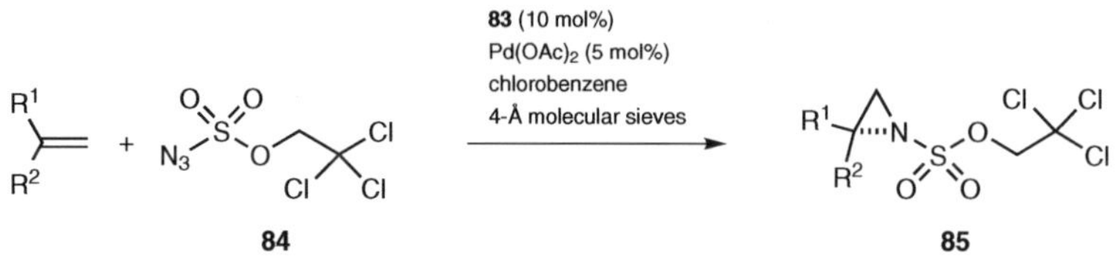

R¹	R²	Temp (°C)	ee (%)	Yield (%)	Ref
H	Ph	0	94	91	[80]
H	4-Tol	0	90	89	[80]
H	2-Tol	rt	84	86	[80]
Me	Ph	rt	80	48	[80]
Bu	H	40	91[a]	42	[80]
CMe=CH₂	Me	rt	87[a]	53	[80]

[a] Opposite enantiomer to that shown in scheme.

Recently, Zhang and co-workers found that cobalt–porphyrin **86** can efficiently catalyze the asymmetric aziridination of alkenes using fluoroaryl azides as the nitrene source.[81] Styrene reacts with several fluoroaryl azides in fluorobenzene as the solvent under mild conditions to furnish enantioenriched N-(fluoroaryl)aziridines with good levels of enantioselectivity (Scheme 43).

Scheme 43 Aziridination of Styrenes Catalyzed by a Cobalt–Porphyrin Complex[81]

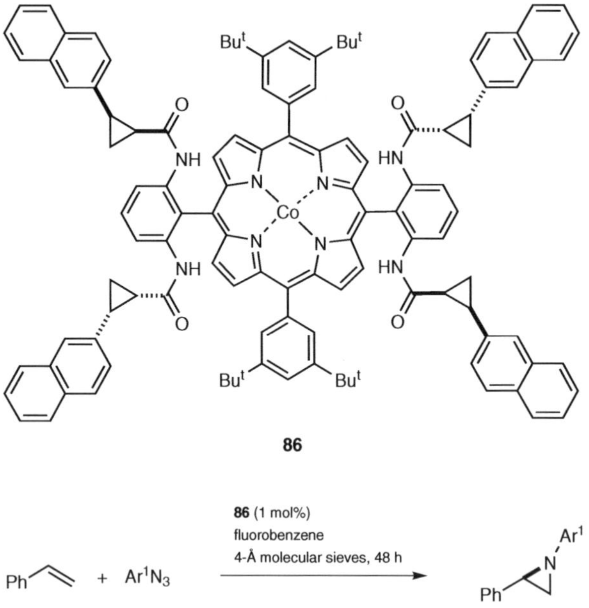

Ar¹	ee (%)	Yield (%)	Ref
2-FC₆H₄	75	52	[81]
2,4,5-F₃C₆H₂	80	64	[81]
2,6-F₂C₆H₃	96	80	[81]
2,3,5,6-F₄C₆H	89	95	[81]

2.1.2 Metal-Catalyzed Aziridination Reactions

Ar¹	ee (%)	Yield (%)	Ref
C₆F₅	92	99	[81]
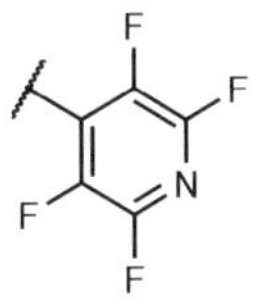 (tetrafluoropyridyl)	90	96	[81]
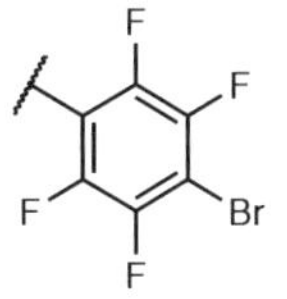 (tetrafluoro-bromophenyl)	80	95	[81]
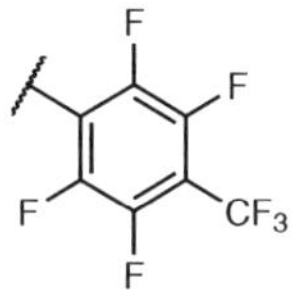 (tetrafluoro-trifluoromethylphenyl)	68	99	[81]

1-[(2,2,2-Trichloroethoxy)sulfonyl]aziridines 85; General Procedure:[80]
An oven-dried Schlenk tube, previously evacuated and backfilled with N_2, was charged with catalyst **83** (0.05 mmol), Pd(OAc)$_2$ (0.05 mmol), and 4-Å molecular sieves (100 mg). The Schlenk tube was then evacuated and backfilled with N_2 again. The Teflon screw cap was replaced with a rubber septum, and solvent (0.5 mL) was added at rt followed by the styrene derivative (0.5 mmol), another portion of solvent at 0 °C, then the azide **84** (0.1 mmol), and the remaining solvent (total 1 mL). The Schlenk tube was then purged with N_2 for 1 min, and the rubber septum was replaced with the Teflon screw cap. The Schlenk tube was then left at rt, 0 °C, or 40 °C for 24–48 h. Following completion of the reaction, the mixture was purified by flash chromatography.

2.1.2.6 Silver-Catalyzed Aziridination

In 2003, He and Cui discovered the efficient aziridination of alkenes using a novel disilver(I) compound as the catalyst and [N-(4-toluenesulfonyl)imino]phenyliodinane as the nitrene source in acetonitrile at ambient temperature.[82] The disilver(I) compound is prepared from tridentate 4,4′,4″-tri-*tert*-butyl-2,2′:6,2″-terpyridine (**87**) and silver(I) nitrate. This silver-catalyzed aziridination reaction furnishes the products **88** in good to excellent yields for a range of alkene substrates, and terminal alkene substrates can be converted into aziridines in good yields (Scheme 44).

Scheme 44 Aziridination of Alkenes Catalyzed by a Silver–Terpyridine Complex[82]

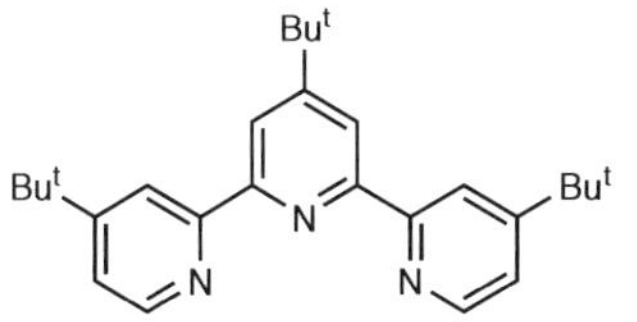

87

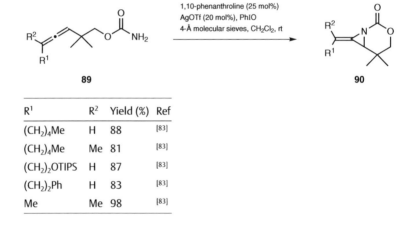

R¹	R²	Yield (%)	Ref
Pr	H	66	[82]
(CH$_2$)$_4$		81	[82]
(CH$_2$)$_6$		88	[82]
4-ClC$_6$H$_4$	H	74	[82]
Ph	Me	90	[82]
Ph	Ph	86	[82]

In 2013, Schomaker and co-workers reported the use of 1,10-phenanthroline as the ligand and iodosylbenzene as the oxidant in the silver-catalyzed aziridination of allene carbamates **89**.[83] This catalytic system improves the scope, yield, and chemoselectivity of the process to form bicyclic methylene aziridines **90** by allene aziridination (Scheme 45).

Scheme 45 Aziridination of Allene Carbamates Catalyzed by Silver(I) Trifluoromethanesulfonate/1,10-Phenanthroline[83]

R¹	R²	Yield (%)	Ref
(CH$_2$)$_4$Me	H	88	[83]
(CH$_2$)$_4$Me	Me	81	[83]
(CH$_2$)$_2$OTIPS	H	87	[83]
(CH$_2$)$_2$Ph	H	83	[83]
Me	Me	98	[83]

1-Tosylaziridines 88; General Procedure:[82]
In a dry 10-mL Schlenk tube, a mixture of AgNO$_3$ (1.7 mg, 0.01 mmol) and terpyridine **87** (4.0 mg, 0.01 mmol) in MeCN (2 mL) was stirred for 5–10 min. The resulting slightly yellow mixture was cooled to 0 °C and then PhINTs (186.5 mg, 0.5 mmol) was added together with activated 4-Å molecular sieves (0.5 g). The suspension quickly turned deep brown. The alkene (2.5 mmol, 5 equiv) was added, and the soln was then stirred at 0 °C for 0.5 h, warmed to rt, and stirred for another 8–20 h, during which time the color of the soln faded to pale yellow. The mixture was diluted with CH$_2$Cl$_2$ (10 mL) and filtered through Celite. The filter cake was washed thoroughly with CH$_2$Cl$_2$ (2 × 5 mL) and the combined filtrate was concentrated under reduced pressure. The isolated material was purified by chromatography (silica gel).

2.1.3 Applications in the Syntheses of Natural Products and Drug Molecules

5,5-Dimethyl-7-methylene-3-oxa-1-azabicyclo[4.1.0]heptan-2-ones 90;
General Procedure:[83]
A predried reaction flask was charged with AgOTf (26 mg, 0.1 mmol, 0.2 equiv) and 1,10-phenanthroline (23 mg, 0.125 mmol, 0.25 equiv). CH_2Cl_2 (1 mL) was then added. The mixture was stirred vigorously for 30 min, and then a soln of homoallenic carbamate **89** (0.5 mmol, 1 equiv) in CH_2Cl_2 (1 mL) was added, followed by 4-Å molecular sieves (1 mmol substrate/g of sieves or 0.25 mmol substrate/g of sieves). PhIO (220 mg, 1 mmol, 2 equiv) was then added in one portion, and the mixture was stirred at rt until TLC indicated complete consumption of the starting material (~14 h). The mixture was then filtered through a glass frit, and the filtrate was concentrated under reduced pressure. The crude product was then purified by column chromatography (silica gel).

2.1.3 Applications in the Syntheses of Natural Products and Drug Molecules

Owing to the remarkable reactivity associated with three-membered rings, epoxides and aziridines play important roles in the synthesis of natural products and pharmaceutically useful molecules.

Jacobsen epoxidation with salen–manganese(II) catalysts is one of the most commonly used approaches for the synthesis of chiral epoxides. The enantioselective epoxidation of indene catalyzed by a salen–manganese(II) complex results in the corresponding epoxide, which can be transformed into the HIV-protease inhibitor indinavir.[84] Recently, Fuchs and co-workers reported Jacobsen epoxidation of substrate **92** using catalyst **91** for the large-scale (up to 300–400 g) preparation of epoxy vinyl sulfonate **93** using hydrogen peroxide as the oxidant. This protocol has been successfully used in the synthesis of advanced intermediate **94**, which may lead to (+)-pretazettine (Scheme 46).[85]

Scheme 46 Synthesis of (+)-Pretazettine[85]

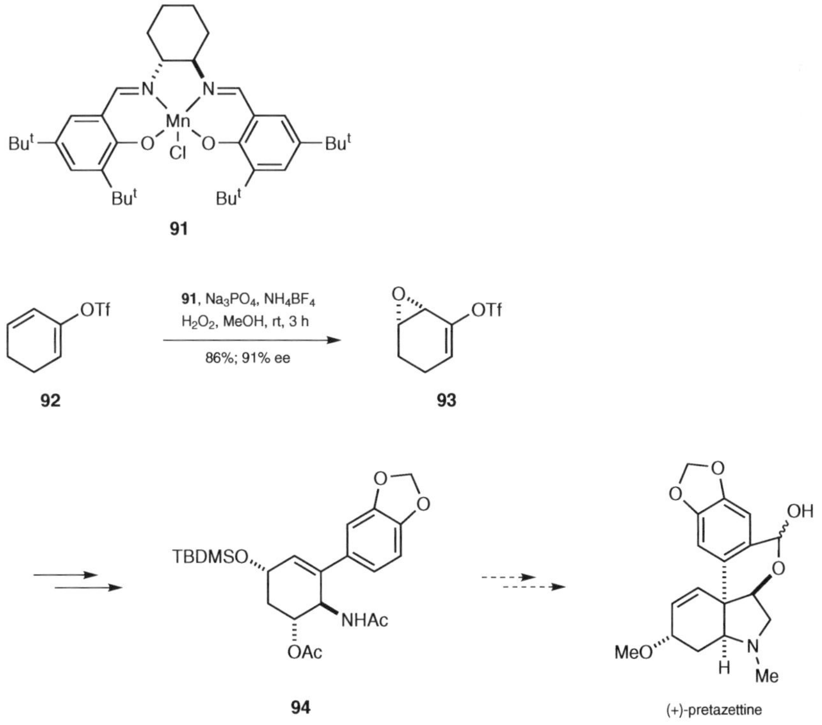

for references see p 47

In 2003, the Yamamoto group developed a total synthesis of (−)-α-bisabolol. The key step of the reaction involves the preparation of intermediate **97** through the asymmetric epoxidation of homoallylic alcohol **96** using a chiral vanadium catalyst prepared from triisopropoxy(oxo)vanadium(V) and ligand **95** (Scheme 47).[86]

Scheme 47 Synthesis of (−)-α-Bisabolol[86]

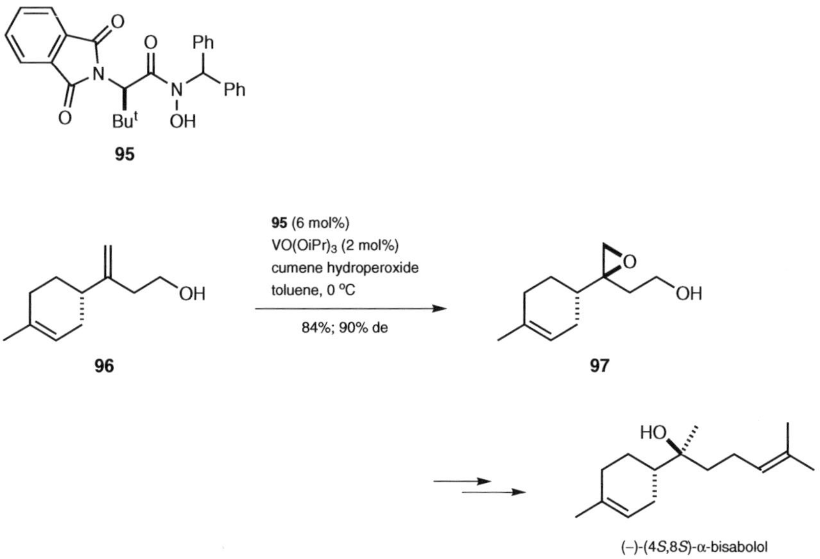

In 2013, Gao reported that a manganese complex catalyzes the asymmetric epoxidation of benzopyran **98** using hydrogen peroxide as the oxidant to provide corresponding epoxide **99** in 92% yield with 94% ee. This chiral epoxide can be further converted into the potential drug (S)-levcromakalim in 61% yield with up to 97% ee (Scheme 48).[29]

Scheme 48 Synthesis of (S)-Levcromakalim[29]

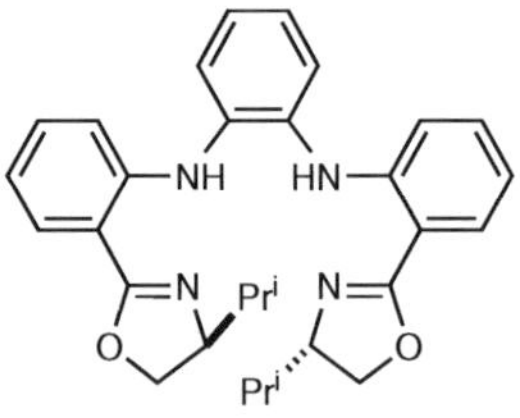

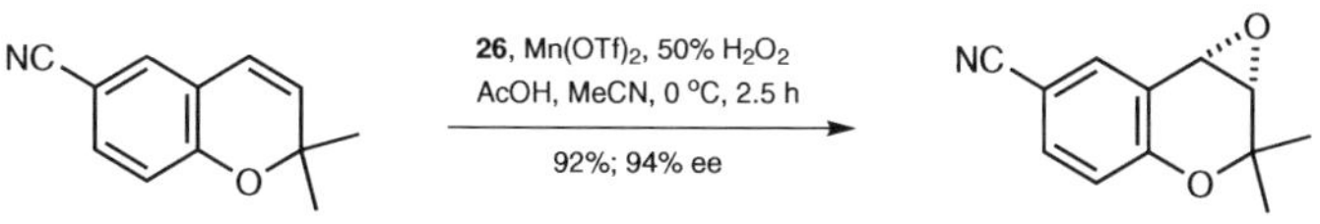

2.1.3 Applications in the Syntheses of Natural Products and Drug Molecules

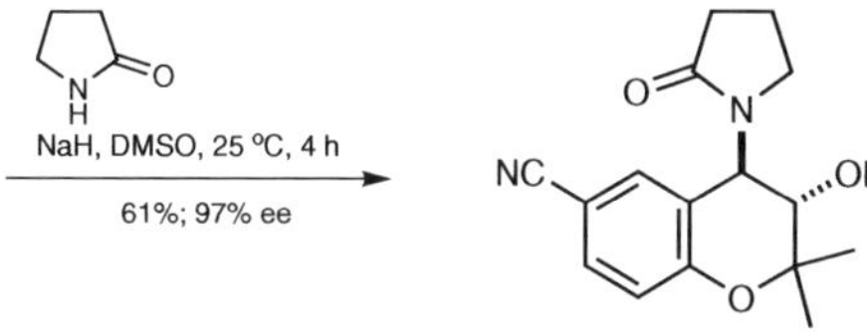

In 2006, Trost and Dong developed a total synthesis of (+)-agelastatin A with aziridine **102** as the key intermediate. In the presence of [N-(4-toluenesulfonyl)imino]phenyliodinane as the nitrene source, copper–N-heterocyclic carbene complex **100** catalyzes the aziridination of piperazinone **101** to give aziridine **102** in 52% yield, and this aziridine can be further converted into (+)-agelastatin A in four steps (Scheme 49).[87]

Scheme 49 Synthesis of (+)-Agelastatin A[87]

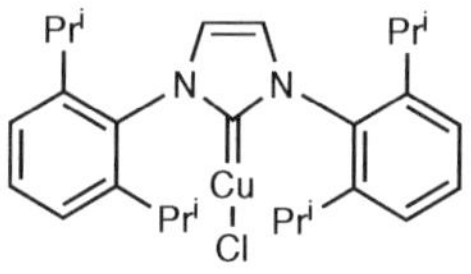

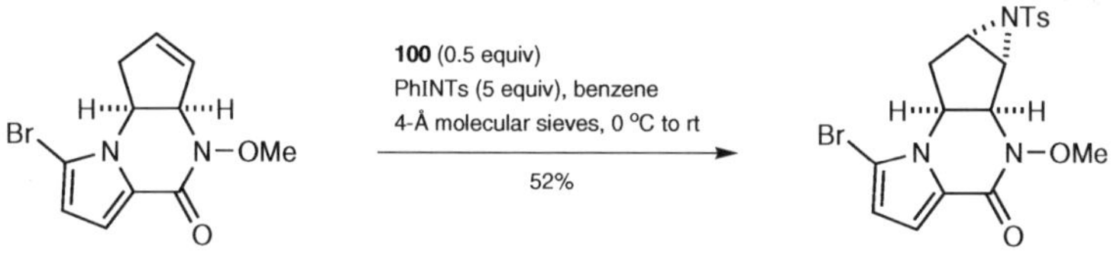

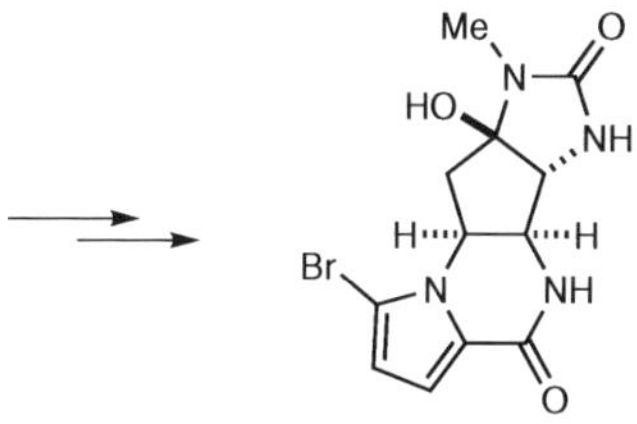

The nonproteinogenic amino acid (2R,3R)-β-methoxytyrosine is a constituent of several cyclic depsipeptide natural products such as callipeltin A and papuamides A and B. All of these related natural products have been shown to possess potent anti-HIV activity and broad-spectrum cytotoxicity. In 2007, Cranfill and Lipton reported the use of Evans' bis-oxazoline ligand **103** for the asymmetric aziridination of cinnamate ester **104**, followed by a ring-opening reaction with methanol to give α-amino-β-methoxy ester **105** in 89% yield with an enantiomeric ratio of >28:1 and a diastereomeric ratio of 19:1 (Scheme 50).[88]

Scheme 50 Synthesis of (R,R)-β-Methoxytyrosine[88]

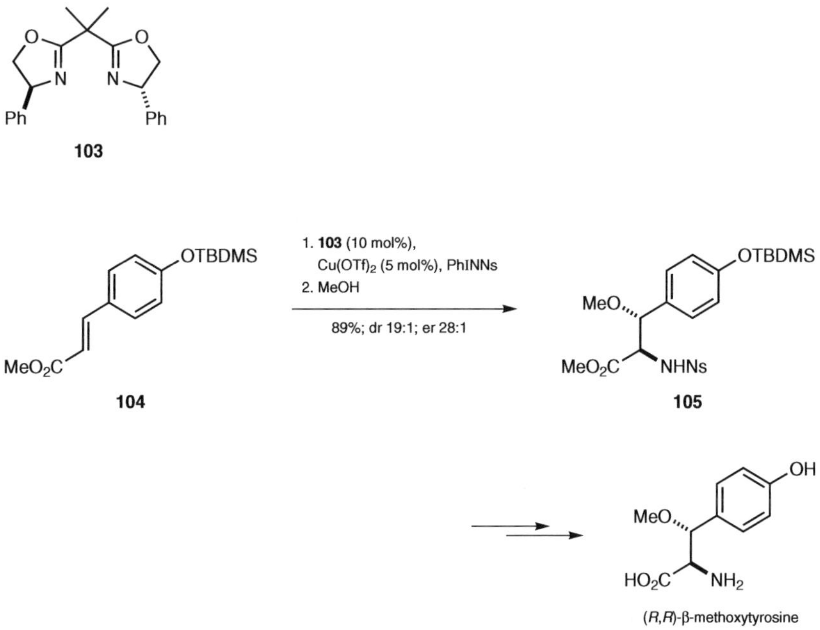

Ns = 4-nitrophenylsulfonyl

In 2008, Trost and Zhang developed the total synthesis of (−)-oseltamivir on the basis of the asymmetric aziridination of chiral diene **106**. This reaction is performed in the presence of 2-(trimethylsilyl)ethane-1-sulfonamide as the nitrene source and dirhodium species **73** as the catalyst. The corresponding aziridine **107** is isolated in 86% yield with exclusive diastereoselectivity, and it can be further converted into the desired (−)-oseltamivir (Scheme 51).[89]

Scheme 51 Synthesis of (−)-Oseltamivir[89]

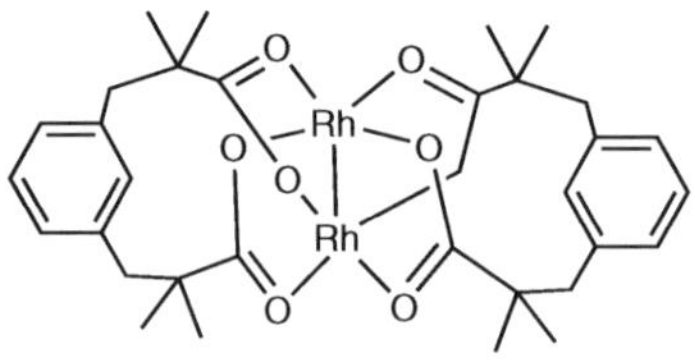

2.1.3 Applications in the Syntheses of Natural Products and Drug Molecules

(1S,6S)-7-Oxabicyclo[4.1.0]hept-2-en-2-yl Trifluoromethanesulfonate (93); Typical Procedure:[85]

A soln of cyclohexa-1,3-dien-2-yl trifluoromethanesulfonate (92; 1.24 g, 5.43 mmol) in MeOH (30 mL) was cooled to 0 °C and then NH_4BF_4 (557 mg, 5.43 mmol) was added, followed by Na_3PO_4 (462 mg, 2.81 mmol). The mixture was stirred for 5 min and then catalyst 91 (87 mg, 0.137 mmol) was added. After stirring for another 5 min, 30% H_2O_2 (1.5–2.5 mL, 3–5 equiv) was added and the soln was stirred for an additional 2 h. The mixture was diluted with Et_2O and washed with brine (2 ×). The organic layer was dried (Na_2SO_4) and concentrated under reduced pressure. Flash column chromatography (neutral alumina, Et_2O/hexanes 2.5:97.5) gave the title compound; yield: 1.14 g (86%). The neat oily compound starts to decompose after 30 min at rt; thus, the material must be stored in the freezer as a soln in benzene (**CAUTION**: *carcinogen*) to extend its lifetime.

2-{(2S)-2-[(1S)-4-Methylcyclohex-3-en-1-yl]oxiran-2-yl}ethan-1-ol (97); Typical Procedure:[86]

CAUTION: *Trimethyl phosphite is flammable and has a powerful, obnoxious odor. Induces headache. Severe skin and eye irritant. Corrosive and irritating to the respiratory tract.*

A mixture of $VO(OiPr)_3$ (5 mL, 20 µmol) and hydroxamic acid 95 (26.6 mg, 60 µmol) in toluene (1 mL) was stirred for 1 h and then cooled to 0 °C. Cumene hydroperoxide (275 µL, 1.5 mmol) and (S)-3-(4-methylcyclohex-3-enyl)but-3-en-1-ol (96; 166 mg, 1.0 mmol) were added at 0 °C. The mixture was stirred for 10 h, and then $(MeO)_3P$ (177 µL, 1.5 mmol) was added at that temperature. The mixture was allowed to reach rt and was then extracted with EtOAc. The extracts were dried (Na_2SO_4) and concentrated and the crude product was purified by column chromatography (silica gel, EtOAc/hexane 1:1); yield: 84%; 90% de.

(1aS,7bS)-2,2-Dimethyl-1a,7b-dihydro-2H-oxireno[c][1]benzopyran-6-carbonitrile (99); Typical Procedure:[29]

A 0.00168 M soln of $Mn(OTf)_2$ in MeCN (0.5 mL, 0.84 µmol) was added to a 0.00168 M soln of ligand 26 in MeCN (0.5 mL, 0.84 µmol) at rt. The mixture was stirred at rt for 1 h. Nitrile 98 (0.42 mmol) and AcOH (2.1 mmol) were added to the soln of the manganese complex. Then, the temperature was decreased to 0 °C and 50% H_2O_2 (0.84 mmol, diluted with 1 mL MeCN) was added dropwise by syringe pump over 1 h. The mixture was stirred at 0 °C for 1 h, and the reaction was quenched by adding a mixture of 10% aq $Na_2S_2O_3$ and sat. aq $NaHCO_3$ diluted with CH_2Cl_2. The organic layer was separated, washed with brine, dried ($MgSO_4$), and concentrated under reduced pressure. The residue was purified by column chromatography (silica gel); yield: 92%; 94% ee.

(5aS,5bR,6aS,7aS)-1-Bromo-5-methoxy-6-tosyl-5a,5b,6,6a,7,7a-hexahydroazirino-[2′,3′:4,5]cyclopenta[1,2-e]pyrrolo[1,2-a]pyrazin-4(5H)-one (102); Typical Procedure:[87]
Benzene (1 mL) (**CAUTION**: *carcinogen*) was added to a mixture of 4-Å molecular sieves (75 mg), compound **101** (20 mg, 0.071 mmol), PhINTs (132 mg, 0.355 mmol), and catalyst **100** (17.3 mg, 0.0355 mmol) under an atmosphere of N_2. The resulting soln was stirred at rt for 4 h and then filtered through a silica gel cake; the cake was rinsed with EtOAc. After removing the solvent under reduced pressure, the crude product was purified by flash column chromatography (neutral alumina, activity 3; petroleum ether/EtOAc 8:1, 4:1, then 3:1) to give a colorless foam; yield: 16.6 mg (52%).

Methyl (2R,3R)-3-[4-(*tert*-Butyldimethylsiloxy)phenyl]-3-methoxy-2-[(4-nitrophenyl)sulfonamido]propanoate (105); Typical Procedure:[88]
Bis(dihydrooxazole) ligand **103** (18 mg, 0.054 mmol), 4-Å molecular sieves, and CH_2Cl_2 (2 mL) were added to $Cu(OTf)_2$ (10 mg, 0.027 mmol). The mixture was stirred under an atmosphere of N_2 for 15 min. A soln of silyl ether **104** (557 mg, 1.90 mmol) in CH_2Cl_2 (2 mL) was added to the mixture, followed by [N-(4-nitrobenzenesulfonyl)imino]phenyliodinane (PhINNs; 220 mg, 0.54 mmol), and the reaction vessel was flushed with N_2. The reaction was allowed to proceed until PhINNs had completely dissolved (typically 1.5 h), at which point the reaction was stopped by filtration through a small pad of silica gel. The filtrate was concentrated under reduced pressure, and the residue was redissolved in MeOH (6 mL). The mixture was stirred for 1.5 h, after which it was concentrated under reduced pressure. The residue was purified by flash chromatography (EtOAc/hexanes 3:7) to afford a pale yellow oil; yield: 255 mg (89%).

Ethyl (1S,5S,6R)-5-Phthalimido-7-{[2-(trimethylsilyl)ethyl]sulfonyl}-7-azabicyclo-[4.1.0]hept-2-ene-3-carboxylate (107); Typical Procedure:[89]
Under a positive pressure of N_2, diene **106** (149 mg, 0.5 mmol), MgO (46 mg, 1.15 mmol), and dirhodium complex **73** (7.6 mg, 0.01 mmol) were added sequentially to a soln of 2-(trimethylsilyl)ethane-1-sulfonamide in chlorobenzene (1 mL). The mixture was cooled to 0 °C. $PhI(O_2C\textit{t}\text{-Bu})_2$ (264 mg, 0.65 mmol) was added in one portion. The mixture was slowly warmed to rt and was then stirred for 4 h while monitoring the progress of the reaction by TLC. The solid was filtered off and washed thoroughly with CH_2Cl_2. The combined filtrate was concentrated under reduced pressure with gentle heating. Chromatography (silica gel, petroleum ether/EtOAc 10:1) delivered the title compound as a white foam; yield: 206 mg (86%).

2.1.4 Conclusions and Future Perspectives

Metal-catalyzed epoxidation of alkenes has witnessed exponential growth during the last few decades, and significant contributions from various research groups have outlined remarkable results. Systems based on titanium, vanadium, manganese, iron, ruthenium, tungsten, and scandium have been shown to be effective catalysts for the asymmetric epoxidation of alkenes with a broad substrate scope. However, discovering highly enantioselective metal catalysts and environmentally friendly oxidants has proven to be challenging. The development of less expensive and environmentally more benign catalysts and oxidant systems is a major goal for organic synthesis.

Significant progress has also been made in the metal-catalyzed aziridination of alkenes using catalysts based on metals such as copper, rhodium, ruthenium, iron, cobalt, and silver. Good yields and high enantioselectivities have been obtained in various cases. However, regioselectivity and stereoselectivity are still problematic, which makes the stereoselective synthesis of aziridines a field in need of more attention.

References

[1] Shi, Y., *Acc. Chem. Res.*, (2004) **37**, 488.
[2] Yang, D., *Acc. Chem. Res.*, (2004) **37**, 497.
[3] Aggarwal, V. K.; Winn, C. L., *Acc. Chem. Res.*, (2004) **37**, 611.
[4] Adolfsson, H.; Balan, D., In *Aziridines and Epoxides in Organic Synthesis*, Yudin, A. K., Ed.; Wiley-VCH: Weinheim, Germany, (2006); Chapter 6.
[5] Faveri, G. D.; Ilyashenko, G.; Watkinson, M., *Chem. Soc. Rev.*, (2011) **40**, 1722.
[6] Srour, H.; Le Maux, P.; Chevance, S.; Simonneaux, G., *Coord. Chem. Rev.*, (2013) **257**, 3030.
[7] Müller, P.; Fruit, C., *Chem. Rev.*, (2003) **103**, 2905.
[8] Sweeney, J. B., In *Aziridines and Epoxides in Organic Synthesis*, Yudin, A. K., Ed.; Wiley-VCH: Weinheim, Germany, (2006); Chapter 4.
[9] Pellissier, H., *Tetrahedron*, (2010) **66**, 1509.
[10] Degennaro, L.; Trinchera, P.; Luisi, R., *Chem. Rev.*, (2014) **114**, 7881.
[11] Katsuki, T.; Sharpless, K. B., *J. Am. Chem. Soc.*, (1980) **102**, 5974.
[12] Matsumoto, K.; Sawada, Y.; Saito, B.; Sakai, K.; Katsuki, T., *Angew. Chem. Int. Ed.*, (2005) **44**, 4935.
[13] Sawada, Y.; Matsumoto, K.; Kondo, S.; Watanabe, H.; Ozawa, T.; Suzuki, K.; Saito, B.; Katsuki, T., *Angew. Chem. Int. Ed.*, (2006) **45**, 3478.
[14] Arends, I. W. C. E., *Angew. Chem. Int. Ed.*, (2006) **45**, 6250.
[15] Matsumoto, K.; Kubo, T.; Katsuki, T., *Chem.–Eur. J.*, (2009) **15**, 6573.
[16] Matsumoto, K.; Oguma, T.; Katsuki, T., *Angew. Chem. Int. Ed.*, (2009) **48**, 7432.
[17] Michaelson, R. C.; Palermo, R. E.; Sharpless, K. B., *J. Am. Chem. Soc.*, (1977) **99**, 1990.
[18] Berrisford, D. J.; Bolm, C.; Sharpless, K. B., *Angew. Chem. Int. Ed.*, (1995) **34**, 1059.
[19] Hoshino, Y.; Yamamoto, H., *J. Am. Chem. Soc.*, (2000) **122**, 10452.
[20] Zhang, W.; Basak, A.; Kosugi, Y.; Hoshino, Y.; Yamamoto, H., *Angew. Chem. Int. Ed.*, (2005) **44**, 4389.
[21] Zhang, W.; Yamamoto, H., *J. Am. Chem. Soc.*, (2007) **129**, 286.
[22] Li, Z.; Zhang, W.; Yamamoto, H., *Angew. Chem. Int. Ed.*, (2008) **47**, 7520.
[23] Irie, R.; Noda, K.; Ito, Y.; Matsumoto, N.; Katsuki, T., *Tetrahedron Lett.*, (1990) **31**, 7345.
[24] Zhang, W.; Loebach, J. L.; Wilson, S. R.; Jacobsen, E. N., *J. Am. Chem. Soc.*, (1990) **112**, 2801.
[25] Irie, R.; Hosoya, N.; Katsuki, T., *Synlett*, (1994), 255.
[26] Shitama, H.; Katsuki, T., *Tetrahedron Lett.*, (2006) **47**, 3203.
[27] Murphy, A.; Dubois, G.; Stack, T. D. P., *J. Am. Chem. Soc.*, (2003) **125**, 5250.
[28] Wu, M.; Wang, B.; Wang, S.; Xia, C.; Sun, W., *Org. Lett.*, (2009) **11**, 3622.
[29] Dai, W.; Li, J.; Li, G.; Yang, H.; Wang, L.; Gao, S., *Org. Lett.*, (2013) **15**, 4138.
[30] Lyakin, O. Y.; Ottenbacher, R. V.; Bryliakov, K. P.; Talsi, E. P., *ACS Catal.*, (2012) **2**, 1196.
[31] Groves, J. T.; Myers, R. S., *J. Am. Chem. Soc.*, (1983) **105**, 5791.
[32] Groves, J. T.; Viski, P., *J. Org. Chem.*, (1990) **55**, 3628.
[33] Ferrand, Y.; Daviaud, R.; Maux, P. L.; Simonneaux, G., *Tetrahedron: Asymmetry*, (2006) **17**, 952.
[34] Rose, E.; Andrioletti, B.; Zrig, S.; Quelquejeu-Ethève, M., *Chem. Soc. Rev.*, (2005) **34**, 573.
[35] Collman, J. P.; Zhang, X.; Lee, V. J.; Uffelman, E. S.; Brauman, J. I., *Science (Washington, D. C.)*, (1993) **261**, 1404.
[36] Gross, Z.; Ini, S., *J. Org. Chem.*, (1997) **62**, 5514.
[37] Naruta, Y.; Tani, F.; Ishihara, N.; Maruyama, K., *J. Am. Chem. Soc.*, (1991) **113**, 6865.
[38] Gelalcha, F. G.; Bitterlich, B.; Anilkumar, G.; Tse, M. K.; Beller, M., *Angew. Chem.*, (2007) **119**, 7431; *Angew. Chem. Int. Ed.*, (2007) **46**, 7293.
[39] Gelalcha, F. G.; Anilkumar, G.; Tse, M. K.; Brückner, A.; Beller, M., *Chem.–Eur. J.*, (2008) **14**, 7687.
[40] Nishikawa, Y.; Yamamoto, H., *J. Am. Chem. Soc.*, (2011) **133**, 8432.
[41] Niwa, T.; Nakada, M., *J. Am. Chem. Soc.*, (2012) **134**, 13538.
[42] Groves, J. T.; Bonchio, M.; Carofiglio, T.; Shalyaev, K., *J. Am. Chem. Soc.*, (1996) **118**, 8961.
[43] Higuchi, T.; Ohtake, H.; Hirobe, M., *Tetrahedron Lett.*, (1989) **30**, 6545.
[44] Ohtake, H.; Higuchi, T.; Hirobe, M., *Tetrahedron Lett.*, (1992) **33**, 2521.
[45] Stoop, R. M.; Mezzetti, A., *Green Chem.*, (1999) **1**, 39.
[46] Berkessel, A.; Kaiser, P.; Lex, J., *Chem.–Eur. J.*, (2003) **9**, 4746.
[47] Tse, M. K.; Döbler, C.; Bhor, S.; Klawonn, M.; Mägerlein, W.; Hugl, H.; Beller, M., *Angew. Chem. Int. Ed.*, (2004) **43**, 5255.
[48] Koya, S.; Nishioka, Y.; Mizoguchi, H.; Uchida, T.; Katsuki, T., *Angew. Chem. Int. Ed.*, (2012) **51**, 8243.

[49] Schurig, V.; Hintzer, K.; Leyrer, U.; Mark, C.; Pitchen, P.; Kagan, H. B., *J. Organomet. Chem.*, (1989) **370**, 81.
[50] Barlan, A. U.; Basak, A.; Yamamoto, H., *Angew. Chem. Int. Ed.*, (2006) **45**, 5849.
[51] Payne, G. B.; Williams, P. H., *J. Org. Chem.*, (1959) **24**, 54.
[52] Sato, K.; Aoki, M.; Noyori, R., *Science (Washington, D. C.)*, (1998) **281**, 1646.
[53] Villa, A. L.; Sels, B. F.; De Vos, D. E.; Jacobs, P. A., *J. Org. Chem.*, (1999) **64**, 7267.
[54] Kamata, K.; Yonehara, K.; Sumida, Y.; Yamaguchi, K.; Hikichi, S.; Mizuno, N., *Science (Washington, D. C.)*, (2003) **300**, 964.
[55] Kamata, K.; Hirano, T.; Kuzuya, S.; Mizuno, N., *J. Am. Chem. Soc.*, (2009) **131**, 6997.
[56] Wang, C.; Yamamoto, H., *J. Am. Chem. Soc.*, (2014) **136**, 1222.
[57] Egami, H.; Katsuki, T., *Angew. Chem. Int. Ed.*, (2008) **47**, 5171.
[58] Egami, H.; Oguma, T.; Katsuki, T., *J. Am. Chem. Soc.*, (2010) **132**, 5886.
[59] Chu, Y.; Liu, X.; Li, W.; Hu, X.; Lin, L.; Feng, X., *Chem. Sci.*, (2012) **3**, 1996.
[60] Li, Z.; Yamamoto, H., *J. Am. Chem. Soc.*, (2010) **132**, 7878.
[61] Olivares-Romero, J. L.; Li, Z.; Yamamoto, H., *J. Am. Chem. Soc.*, (2012) **134**, 5440.
[62] Evans, D. A.; Woerpel, K. A.; Hinman, M. M.; Faul, M. M., *J. Am. Chem. Soc.*, (1991) **113**, 726.
[63] Li, Z.; Conser, K. R.; Jacobsen, E. N., *J. Am. Chem. Soc.*, (1993) **115**, 5326.
[64] Xu, J.; Ma, L.; Jiao, P., *Chem. Commun. (Cambridge)*, (2004), 1616.
[65] Wang, X.; Ding, K., *Chem.–Eur. J.*, (2006) **12**, 4568.
[66] Lebel, H.; Lectard, S.; Parmentier, M., *Org. Lett.*, (2007) **9**, 4797.
[67] Xu, Q.; Appella, D. H., *Org. Lett.*, (2008) **10**, 1497.
[68] Ton, T. M. U.; Tejo, C.; Tiong, D. L. Y.; Chan, P. W. H., *J. Am. Chem. Soc.*, (2012) **134**, 7344.
[69] Guthikonda, K.; Du Bois, J., *J. Am. Chem. Soc.*, (2002) **124**, 13672.
[70] Lebel, H.; Huard, K.; Lectard, S., *J. Am. Chem. Soc.*, (2005) **127**, 14198.
[71] Catino, A. J.; Nichols, J. M.; Forslund, R. E.; Doyle, M. P., *Org. Lett.*, (2005) **7**, 2787.
[72] Lebel, H.; Spitz, C.; Leogane, O.; Trudel, C.; Parmentier, M., *Org. Lett.*, (2011) **13**, 5460.
[73] Jat, J. L.; Paudyal, M. P.; Gao, H.; Xu, Q.-L.; Yousufuddin, M.; Devarajan, D.; Ess, D. H.; Kürti, L.; Falck, J. R., *Science (Washington, D. C.)*, (2014) **343**, 61.
[74] Kawabata, H.; Omura, K.; Katsuki, T., *Tetrahedron Lett.*, (2006) **47**, 1571.
[75] Fukunag, Y.; Uchida, T.; Ito, Y.; Matsumoto, K.; Katsuki, T., *Org. Lett.*, (2012) **14**, 4658.
[76] Kim, C.; Uchida, T.; Katsuki, T., *Chem. Commun. (Cambridge)*, (2012) **48**, 7188.
[77] Liu, P.; Wong, E. L.-M.; Yuen, A. W.-H.; Che, C.-M., *Org. Lett.*, (2008) **10**, 3275.
[78] Gao, G.-Y.; Harden, J. D.; Zhang, X. P., *Org. Lett.*, (2005) **7**, 3191.
[79] Gao, G.-Y.; Jones, J. E.; Vyas, R.; Harden, J. D.; Zhang, X. P., *J. Org. Chem.*, (2006) **71**, 6655.
[80] Subbarayan, V.; Ruppel, J. V.; Zhu, S.; Perman, J. A.; Zhang, X. P., *Chem. Commun. (Cambridge)*, (2009), 4266.
[81] Jin, L.-M.; Lu, H.; Cui, X.; Wojtas, L.; Zhang, X. P., *Angew. Chem.*, (2013) **125**, 5417; *Angew. Chem. Int. Ed.*, (2013) **52**, 5309.
[82] Cui, Y.; He, C., *J. Am. Chem. Soc.*, (2003) **125**, 16202.
[83] Rigoli, J. W.; Weatherly, C. D.; Vo, B. T.; Neale, S.; Meis, A. R.; Schomaker, J. M., *Org. Lett.*, (2013) **15**, 290.
[84] Vacca, J. P.; Dorsey, B. D.; Schleif, W. A.; Levin, R. B.; McDaniel, S. L.; Darke, P. L.; Zugay, J.; Quintero, J. C.; Blahy, O. M.; Roth, E.; Sardana, V.; Schlabach, A. J.; Graham, P. I.; Condra, J. H.; Gotlib, L.; Holloway, M. K.; Lin, J.; Chen, I.-W.; Vastag, K.; Ostovic, D.; Anderson, P. S.; Emini, E. A.; Huff, J. R., *Proc. Natl. Acad. Sci. U. S. A.*, (1994) **91**, 4096.
[85] Ebrahimian, G. R.; du Jourdin, X. M.; Fuchs, P. L., *Org. Lett.*, (2012) **14**, 2630.
[86] Makita, N.; Hoshino, Y.; Yamamoto, H., *Angew. Chem. Int. Ed.*, (2003) **42**, 941.
[87] Trost, B. M.; Dong, G., *J. Am. Chem. Soc.*, (2006) **128**, 6054.
[88] Cranfill, D. C.; Lipton, M. A., *Org. Lett.*, (2007) **9**, 3511.
[89] Trost, B. M.; Zhang, T., *Angew. Chem. Int. Ed.*, (2008) **47**, 3759.

2.2   Metal-Catalyzed Cyclopropanation

L. K. B. Garve and D. B. Werz

General Introduction

Cyclopropanes, the smallest cycloalkanes, have fascinated chemists for more than 100 years. Despite the high ring strain and the respective thermodynamic destabilization, the large majority of these three-membered-ring hydrocarbons are kinetically stable.[1] Only if strong polarization effects are added to these systems is unique reactivity observed (so-called donor–acceptor cyclopropanes).[2,3] Cyclopropane moieties are even found in natural products as rather rigid scaffolds, and the pharmaceutical industry often uses cyclopropyl residues as a steric compromise between an ethyl and an isopropyl group. Numerous methods for their preparation have been developed since the first synthesis in 1882 by August Freund.[4] These transformations include ring-contraction reactions by elimination of small molecules such as nitrogen or carbon monoxide,[5] ring-closing reactions starting from three-carbon substrates that form one new C—C bond, and (2 + 1)-cycloaddition reactions. Particularly for the latter type of transformation, catalytic methods have come into play to overcome the inherent limitations of the reactive carbene species that are commonly employed. It is important to emphasize that this review will not cover cyclopropane synthesis in a comprehensive approach. Such information can be found in the *Houben–Weyl* series (Vols. 4/3 and E 17 a-c), in *The Chemistry of the Cyclopropyl Group*,[6,7] in the *Science of Synthesis* contribution on cyclopropanes [see Vol. 48 (Section 48.2)], and in several review articles.[8–10] In this contribution, the focus is solely on reliable metal-catalyzed transformations to access cyclopropane units.

Regarding the mechanism, two different pathways are discussed (Scheme 1).[11,12] Both pathways start with the assumption that a diazo compound **1** binds to a metal center. Subsequently, nitrogen is liberated and a stable metal–carbene complex **2** is formed. The reaction of an alkene and the carbene species leads to two possible pathways: a concerted route (A) and a stepwise route (B). Route A proceeds via a concerted (2 + 1) cycloaddition with the alkene to give the corresponding cyclopropane. In the stepwise route B, a formal (2 + 2) cycloaddition occurs leading to a metallacyclobutane intermediate **3**. After reductive elimination, the metal center is recovered and the desired cyclopropane is formed. Route B is apparently limited to certain metals such as palladium; route A is the most common mechanism for reactions involving copper, iron, rhodium, and ruthenium carbenes.

Scheme 1 Proposed Mechanism for Alkene Cyclopropanation Using Diazo Compounds[11,12]

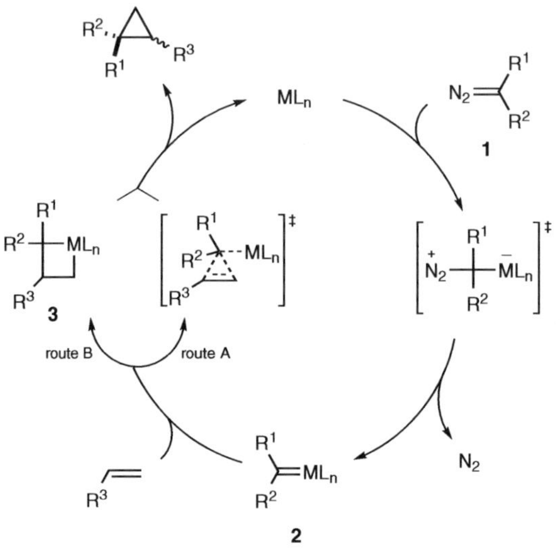

2.2.1 Copper-Catalyzed Cyclopropanation

The very first reports on metal-catalyzed cyclopropanation are associated with the use of copper. Its complexes have become the most inexpensive catalysts for both racemic and asymmetric cyclopropanation reactions; however, in recent years, other metals have come into play for more specialized solutions.

In the 1960s, catalytic decomposition of diazoalkanes by copper chelates was observed and investigated with respect to insertion of the carbenoid into alkenes.[13–15] Very quickly, a first asymmetric version was realized by reacting styrene with ethyl 2-diazoacetate in the presence of a chiral chelating ligand.[16] For this system, only 6% ee was obtained. However, this breakthrough paved the way for a variety of asymmetric copper complexes which show excellent enantiodiscrimination. Milestones in this field are the semi-corrin ligand **4**[17] and the box-type ligands such as **5** and **6** (Scheme 2),[18] but also other ligand scaffolds are successfully utilized such as diamine ligands (e.g., **7**).[19] Intermolecular asymmetric copper-catalyzed cyclopropanation reactions commonly work much better than their intramolecular counterparts. For the latter type of reactions, rhodium has become the metal of choice.

2.2.1 Copper-Catalyzed Cyclopropanation

Scheme 2 Ligands for Copper-Catalyzed Asymmetric Cyclopropanation[17–19]

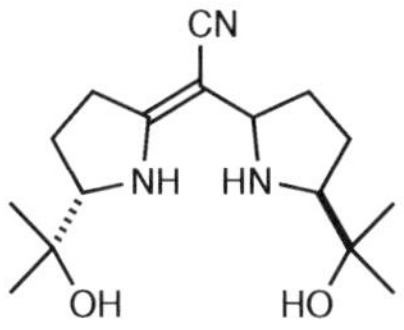

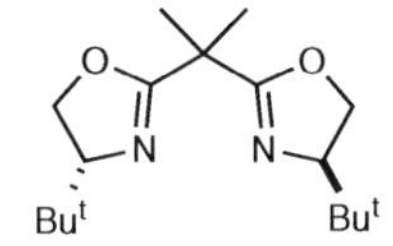

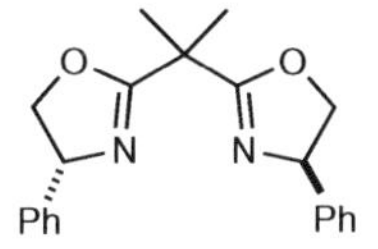

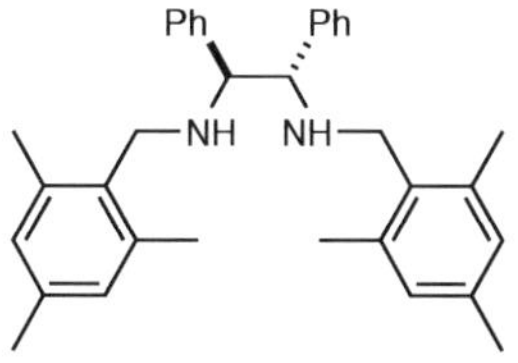

2.2.1.1 Copper-Catalyzed Asymmetric Cyclopropanation Using a Bis(4,5-dihydrooxazole) Ligand

Complexes of chiral bis(4,5-dihydrooxazole) [bis(oxazoline)] ligands with copper(I) trifluoromethanesulfonate are highly effective catalysts for a cyclopropanation that proceeds rapidly at room temperature.[18] Respective *cis/trans*-mixtures of the cyclopropyl esters **9** are obtained from styrene and 2-diazoacetates **8** (Scheme 3). Copper(II) salts do not catalyze the reaction except when temperatures higher than 65 °C are applied, or if they are pretreated with phenylhydrazine to convert the copper(II) into a copper(I) species. Copper(I) salts other than trifluoromethanesulfonate show only very little catalytic activity. The choice of the ester residue at the diazo compound proves to be crucial for the diastereo- and enantiodiscrimination.[18] Commonly, sterically encumbered ester groups lead to better ratios and even higher yields.

Scheme 3 Copper-Catalyzed Asymmetric Cyclopropanation of Styrene Using a Bis(4,5-dihydrooxazole) Copper Complex[18]

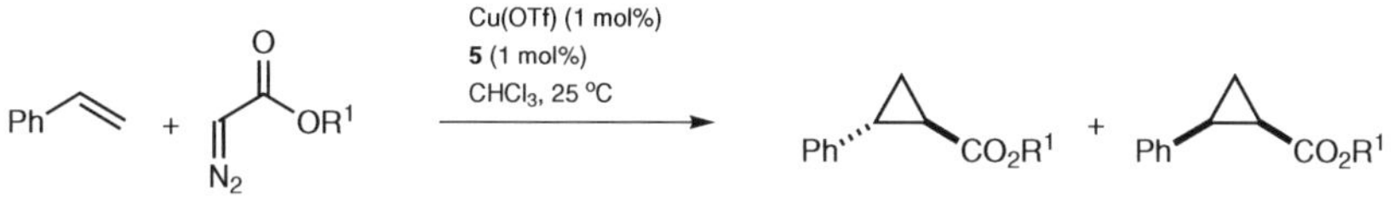

R¹	Ratio (9A/9B)	ee (%) 9A	ee (%) 9B	Yield^a (%)	Ref
Et	73:27	99	97	77	[18]
t-Bu	81:19	96	93	75	[18]
2,6-Me$_2$C$_6$H$_3$	86:14	97	96	68	[18]
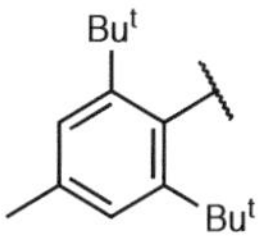 2,6-(t-Bu)$_2$-4-Me-C$_6$H$_2$	94:6	99	n.d.^{b}	85	[18]

^{a} Combined yield.
^{b} n.d. = not determined.

Ethyl (1R,2R)-2-Phenylcyclopropane-1-carboxylate (9A, R¹ = Et) and Ethyl (1R,2S)-2-Phenyl-cyclopropane-1-carboxylate (9B, R¹ = Et); General Procedure:[18]

To a suspension of Cu(OTf) (5.7 mg, 0.027 mmol, 1 mol%) in anhyd $CHCl_3$ (25 mL) was added a soln of ligand **5** (8 mg, 0.027 mmol, 1 mol%) in $CHCl_3$ (9 mL). After 1 h, the resulting blue-green soln was filtered through glass wool under N_2 into a cooled (0 °C) 500-mL, three-necked, round-bottomed flask previously charged with styrene (1.45 g, 1.6 mL, 14 mmol) and $CHCl_3$ (20 mL). A soln of ethyl 2-diazoacetate (315 mg, 0.29 mL, 2.73 mmol) in $CHCl_3$ (100 mL) was added to the cooled soln over 5 h. The mixture was allowed to warm to 25 °C over 14 h. Volatiles were removed by rotary evaporation and the crude product was purified by column chromatography (silica gel); yield: 77%; ratio (**9A/9B**) 73:27; 99% ee for **9A** and 97% ee for **9B**.

2.2.1.2 Copper-Catalyzed Cyclopropanation of Electron-Rich Double Bonds in Heterocyclic Systems

Copper(I) trifluoromethanesulfonate serves also as a catalyst for the cyclopropanation of heterocyclic systems (Scheme 4).[20,21] Again, the most common method is to generate the copper(I) species in situ from copper(II) trifluoromethanesulfonate by the action of phenylhydrazine. If two C=C bonds strongly differing in their electronic properties are present, the more electron-rich one is more rapidly cyclopropanated with ethyl diazoacetate, as shown for methyl furan-2-carboxylate (**10**) which gives the monocyclopropanated product **11**.[22] In many cases, enantiodiscrimination can again be achieved by adding a bis(4,5-dihydrooxazole) ligand. Whereas furan (**12**, X = O) and 1-(*tert*-butoxycarbonyl)-1*H*-pyrrole (**12**, X = NBoc) can be converted into the doubly cyclopropanated products **13**,[20,21,23,24] the more aromatic thiophene (**12**, X = S) needs a respective rhodium catalyst for the (mono)cyclopropanation.[25] With α-substituted diazoacetates, a rhodium catalyst gives better yields.[24] In rare cases, elemental copper in the form of a powder can be also used for the cyclopropanation of enol ethers. However, substoichiometric amounts (30 mol%) and elevated temperatures (e.g., toluene, 100 °C) are commonly required.[21,26]

Scheme 4 Copper-Catalyzed Cyclopropanation of Heterocyclic Systems with Electron-Rich Double Bonds[21–23]

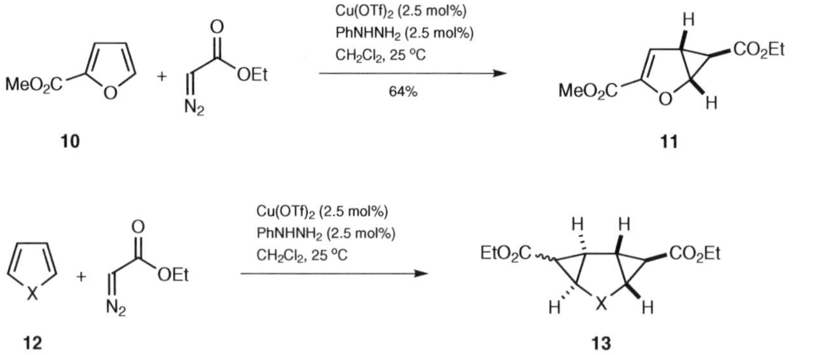

X	Yield (%)	Ref
O	77	[21]
NBoc	63[a]	[23]

[a] 1 mol% catalyst was used.

3,7-Diethyl (1R*,2R*,3R*,4R*,6R*,7R*)-5-(*tert*-Butoxycarbonyl)-5-azatricyclo[4.1.0.0²,⁴]heptane-3,7-dicarboxylate (13, X = NBoc); Typical Procedure:[23]

To a soln of 1-(*tert*-butoxycarbonyl)pyrrole (**12**, X = NBoc; 5.00 g, 29.9 mmol, 1.0 equiv), Cu(OTf)$_2$ (108 mg, 299 µmol, 1.0 mol%), and PhNHNH$_2$ (29 µL, 32 mg, 299 µmol, 1.0 mol%) in hexane (50 mL) was added a soln of ethyl diazoacetate (11.0 g, 96.3 mmol, 3.2 equiv) in CH$_2$Cl$_2$ (500 mL) via a dropping funnel over a period of 8 h, and the resulting mixture was stirred at rt overnight. The solvent was removed and the crude product was purified by column chromatography (silica gel, pentane/EtOAc 8:1 to 5:1) to give a slightly yellow oil; yield: 6.38 g (63%).

2.2.1.3 Copper-Catalyzed Asymmetric Cyclopropanation Leading to 1-Nitrocyclopropane-1-carboxylates

Copper catalysis with chiral bis(oxazoline) ligands (e.g., **6**) allows the stereoselective formation of 1-nitrocyclopropane-1-carboxylates **16**. Styrene derivatives **14** and methyl 2-nitroacetate (**15**) are employed as starting materials (Scheme 5).[27] The reactive species is an iodonium ylide, which is formed under oxidative conditions and serves as surrogate for the commonly used diazo compound. Good *cis*/*trans* and enantiomeric ratios are obtained. With more electron-rich double bonds a highly polarized three-membered ring is formed leading to an immediate ring enlargement of the emerging donor–acceptor cyclopropane.[28]

Scheme 5 Copper-Catalyzed Asymmetric Cyclopropanation Leading to Nitrocyclopropane Carboxylates Using In Situ Generated Iodonium Ylides[27]

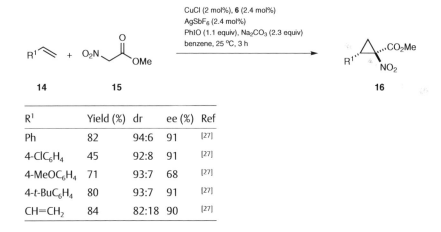

R¹	Yield (%)	dr	ee (%)	Ref
Ph	82	94:6	91	[27]
4-ClC$_6$H$_4$	45	92:8	91	[27]
4-MeOC$_6$H$_4$	71	93:7	68	[27]
4-*t*-BuC$_6$H$_4$	80	93:7	91	[27]
CH=CH$_2$	84	82:18	90	[27]

Methyl 2-Aryl-1-nitrocyclopropane-1-carboxylates 16; General Procedure:[27]

CuCl (1.0 mg, 0.01 mmol, 2 mol%), AgSbF$_6$ (4.1 mg, 0.012 mmol, 2.4 mol%), and bis(4,5-dihydrooxazole) ligand **6** (0.012 mmol, 2.4 mol%) were weighed in a glovebox and mixed in a flask. The flask was removed from the glovebox and benzene (5 mL) (**CAUTION:** *carcinogen*) was added. After the mixture had been stirred for 1 h at rt, the styrene **14** (2.5 mmol, 5.0 equiv) was added. In a separate vial, PhIO (121 mg, 0.55 mmol, 1.10 equiv), Na$_2$CO$_3$ (122 mg, 1.15 mmol, 2.30 equiv), and molecular sieves (ca. 100 mg) were charged and the vial was purged with argon for 10 min. The solids were then quickly transferred to the catalyst soln, followed by the addition of methyl 2-nitroacetate (**15**; 59.5 mg, 0.50 mmol). After the mixture had been stirred at rt for 3 h, the reaction was quenched with H$_2$O (5 mL) and the aqueous layer was extracted with EtOAc (2 × 10 mL). The combined organic layer was dried (Na$_2$SO$_4$), filtered, and concentrated under reduced pressure. The crude oil residue was purified by flash column chromatography to give the pure products.

2.2.2 Rhodium-Catalyzed Cyclopropanation

Beside copper-based systems (Section 2.2.1), rhodium catalysts have become the most prominent class of catalysts for cyclopropanation reactions. In particular, rhodium(II) dimers have come under the focus of synthetic organic chemists. The parent catalyst rhodium(II) acetate dimer [{Rh(OAc)$_2$}$_2$] can easily be modified by using other carboxylates or carboxamides instead of acetates. As a result, catalysts with different activities and selectivities are obtained (Scheme 6).[29,30] As a rule of thumb, one might state that strongly electron-withdrawing carboxylate ligands, such as fluorinated ones, lead to increased reactivity but lower selectivity compared to alkyl carboxylate ligands. In contrast, more electron-releasing ligands, such as amides and lactams, build up a less reactive, but often more selective catalyst.

Scheme 6 Overview of the Reactivity and Selectivity Profile of Rhodium Dimers as Cyclopropanation Catalysts[29]

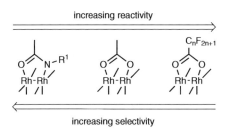

In early examples of rhodium-catalyzed cyclopropanation chemistry, diazoacetates that release an electron-poor carbenoid were commonly used.[31] Transformations using such species often suffered from low diastereoselectivities although the enantiodiscrimination was good. Progress in the area of donor–acceptor rhodium carbenes has led to cyclopropanations that also exhibit exceptional diastereoselectivities.[32,33] With these systems (at least in cases where an allylic hydrogen is present), however, C–H activation processes occurring next to the double bond that should be cyclopropanated are also possible.[32] Monosubstituted, 1,1-disubstituted, and cis-1,2-disubstituted alkenes undergo almost exclusive formation of the three-membered ring.[34,35] In contrast, C–H activation is the more favored process with trans-1,2-disubstituted and sterically encumbered alkenes. Numerous ligand systems have been developed to complex the rhodium dimer, as shown in Scheme 7 for the rhodium complexes **17–21**.[29] Lactam-based systems such as **17** favor the formation of cis-substituted cyclopropanes, with very good enantioselectivity.[36] However, better cis selectivities are reached today by the use of iridium-based systems (see Section 2.2.4). The high costs of this element are definitely a general drawback of rhodium catalysis; in some cases, copper-based systems perform analogous cyclopropanation reactions, but commonly in lower yield.

2.2.2 Rhodium-Catalyzed Cyclopropanation

Scheme 7 Chiral Rhodium Complexes Used as Catalyst in Cyclopropanation Reactions[29,36]

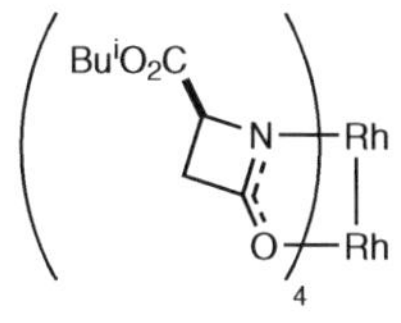

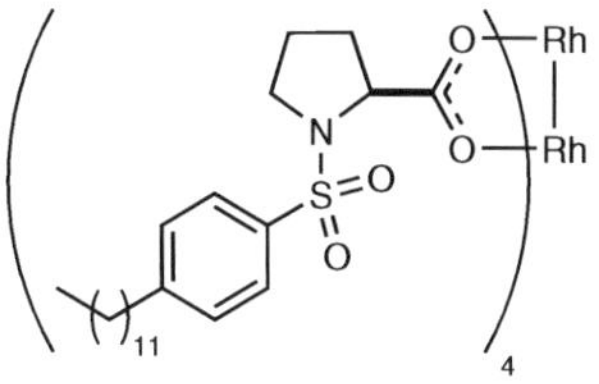

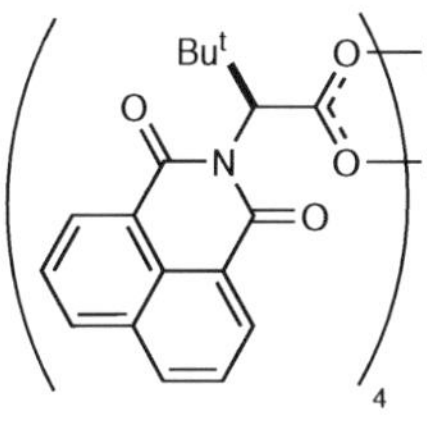

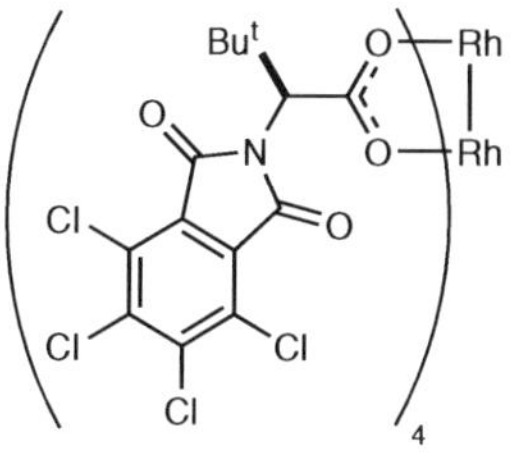

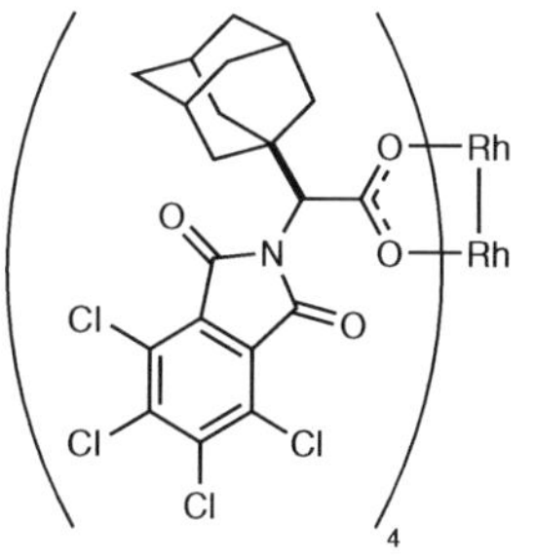

2.2.2.1 Rhodium(II) Acetate Dimer Catalyzed Cyclopropanation Using Alkyl Diazoacetates

The use of rhodium catalysis for cyclopropanation dates back to 1976 when Hubert decomposed alkyl diazoacetates in the presence of alkenes.[31] The yields of the formed cyclopropanes **24** strongly depend on the type of alkyl diazoacetate **23** (Table 1).[37] Larger residues at the ester (e.g., Bu) commonly lead to better yields than small residues such as methyl (compare entries 2 and 3). This fact might be traced back to the increased solubility and stabilization of the intermediately formed carbenoid species. Nevertheless, ethyl 2-diazoacetate (**23**, R^3 = Et) is often used in such transformations, representing a good compromise between size and solubility (entries 4 and 5). Beside electron-neutral alkenes **22**, electron-rich double bonds in enol ethers and even furans and pyrroles can easily be cyclopropanated.[23,24] In cases where there is a choice of two double bonds, the more electron-rich one gets transformed whereas the other one is unaffected (see Section 2.2.1.2).

Table 1 Rhodium(II) Acetate Dimer Catalyzed Cyclopropanation Reaction Using Alkyl Diazoacetates[37]

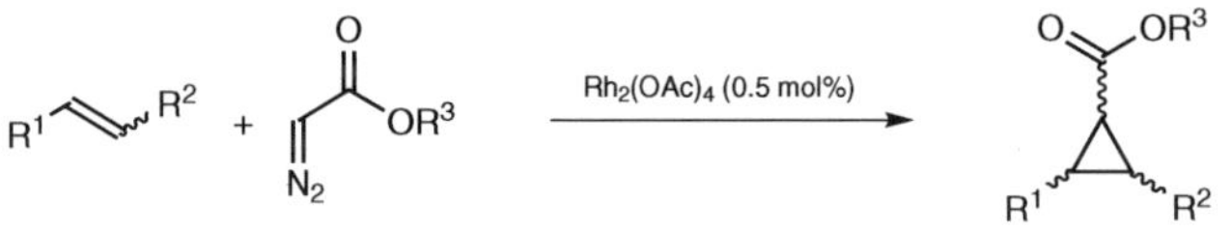

Entry	Substrates		Product	Yield (%)	Ref
	Alkene	Diazoacetate			
1	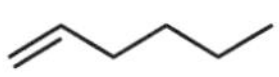 (propene/propyl alkene)	N_2CHCO_2Me	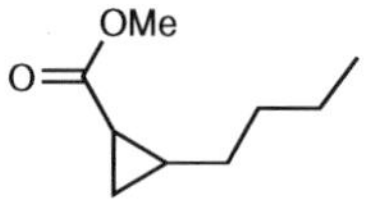 (cyclopropane with CO$_2$Me and propyl)	86	[37]
2	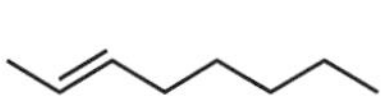 (internal alkene)	N_2CHCO_2Me	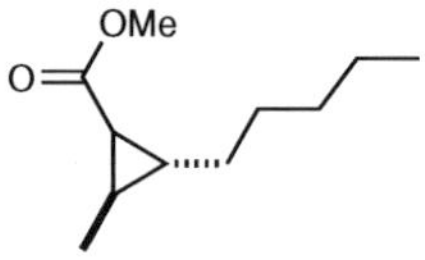 (cyclopropane with CO$_2$Me)	24	[37]
3	 (internal alkene)	N_2CHCO_2Bu	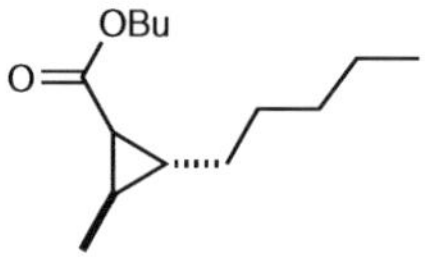 (cyclopropane with CO$_2$Bu)	70	[37]
4	 cyclohexene	N_2CHCO_2Et	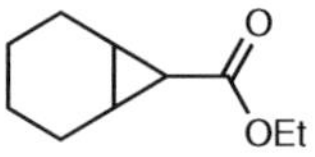 bicyclic cyclopropane with CO$_2$Et	88	[37]
5	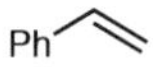 Ph-CH=CH$_2$	N_2CHCO_2Et	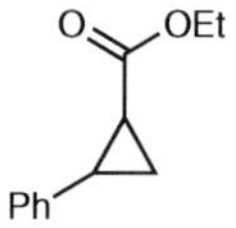 Ph-cyclopropane with CO$_2$Et	90	[37]

Cyclopropanecarboxylates 24; General Procedure:[37]
The alkyl diazoacetate **23** (13 mmol) was added slowly over a period of 10 h (using a syringe pump) to a soln or suspension of Rh$_2$(OAc)$_4$ (0.5 mol%) in the neat alkene **22** at 20 °C with stirring under N$_2$. The resulting mixture was filtered and the filtrate was distilled to give the cyclopropanation product in about 95% purity. Further purification, i.e. removal of dialkyl maleate and fumarate, was achieved by column chromatography (alumina).

2.2.2.2 Rhodium-Catalyzed Enantio- and Diastereoselective Cyclopropanation Starting with Decomposition of 1-Sulfonyl-1,2,3-triazoles

1-Sulfonyl-1,2,3-triazoles **25** have become versatile precursors for azavinyl carbenes in recent years.[38] Rhodium catalysis decomposes the five-membered heterocyclic system with loss of nitrogen. The emerging reactive intermediate, the azavinyl carbene, can be regarded as a synthetic equivalent of formyl carbene in which both amine and aldehyde functionalities can be accessed by simple transformations. The respective 1,2,3-triazoles **25** are easily obtained by copper-catalyzed Huisgen cycloaddition reaction of sulfonyl azides with alkynes. The cyclopropanation of styrene derivatives **26** with dirhodium complex **19**

2.2.2 Rhodium-Catalyzed Cyclopropanation

gives cyclopropane-1-carbaldehydes **27** in good diastereoselectivity, very good yields, and in many cases excellent enantioselectivities except when Z-alkenes are employed (Scheme 8).[39]

Scheme 8 Rhodium-Catalyzed Enantio- and Diastereoselective Cyclopropanation Reaction Starting with the Decomposition of 1-Sulfonyl-1,2,3-triazoles[39]

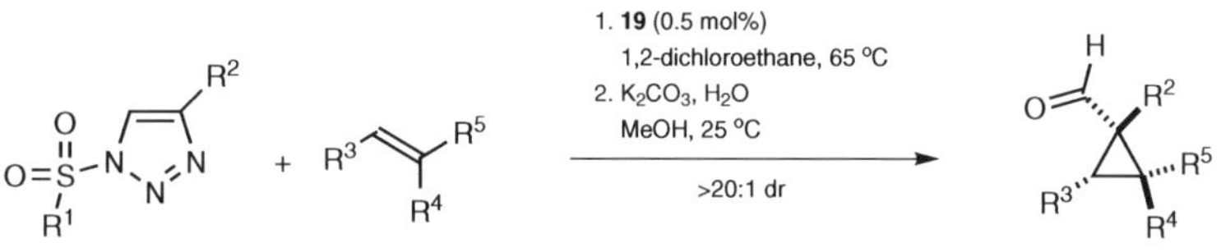

R^1	R^2	R^3	R^4	R^5	Yield (%)	ee (%)	Ref
4-Tol	Ph	H	Ph	H	92	96	[39]
4-Tol	Ph	H	4-F$_3$CC$_6$H$_4$	H	83	98	[39]
4-Tol	Ph	H	2-ClC$_6$H$_4$	H	88	96	[39]
4-Tol	Ph	H	Bu	H	72	96	[39]
4-Tol	Ph	Me	H	Ph	62	98	[39]
4-Tol	Ph	Me	Ph	H	59	8	[39]
Me	4-F$_3$CC$_6$H$_4$	H	Ph	H	80	97	[39]
Me	3-thienyl	H	Ph	H	57	93	[39]
Me	cyclohex-1-enyl	H	Ph	H	57	98	[39]

***trans*-1-Phenylcyclopropane-1-carbaldehydes 27 (R^2 = Ph); General Procedure:**[39]
A 4-phenyl-1-sulfonyl-1*H*-1,2,3-triazole **25** (R^2 = Ph; 0.5 mmol) and the Rh catalyst **19** (0.5 mol%) were added under ambient atmosphere to a 4-mL screw-cap vial, followed by anhyd 1,2-dichloroethane (1.25 mL) and an alkene **26** (0.6 mmol). The mixture was stirred in the capped vial at 65 °C for 4–5 h until the triazole **25** had been completely consumed as judged by HPLC or TLC analysis. An equal volume of MeOH, a few drops of H$_2$O, and anhyd K$_2$CO$_3$ (138 mg, 1.0 mmol) were added to the mixture, and the resulting suspension was vigorously stirred for 0.5–1 h until hydrolysis of the imine was complete. The solvents were removed under reduced pressure, and the residue was resuspended in CH$_2$Cl$_2$ (10 mL). The mixture was dried (Na$_2$SO$_4$), filtered, and analyzed by ^{1}H NMR spectroscopy to determine the ratio of diastereomers. Flash chromatography (silica gel) of the crude residue afforded the product as a white solid or colorless oil.

2.2.2.3 Rhodium-Catalyzed Enantio- and Diastereoselective Cyclopropanation Leading to Cyclopropanes Substituted by Nitro and Cyano Groups

The enantio- and diastereoselective rhodium-catalyzed cyclopropanation of alkenes using diazo reagents (e.g., **28**) bearing two acceptor groups (nitro and keto group) has only recently become a synthetically useful method (Scheme 9).[40,41] Although the transformation has its limitations when using styrene derivatives, a variety of highly useful building blocks **30** for pharmaceutical sciences are easily obtained. Careful selection of the residue at the ketone is of utmost importance. For the nitro-substituted derivatives, the 4-methoxyphenyl group is essential.[40]

for references see p 95

Scheme 9 Enantio- and Diastereoselective Rhodium-Catalyzed Cyclopropanation Leading to Nitro-Substituted Cyclopropanes[40]

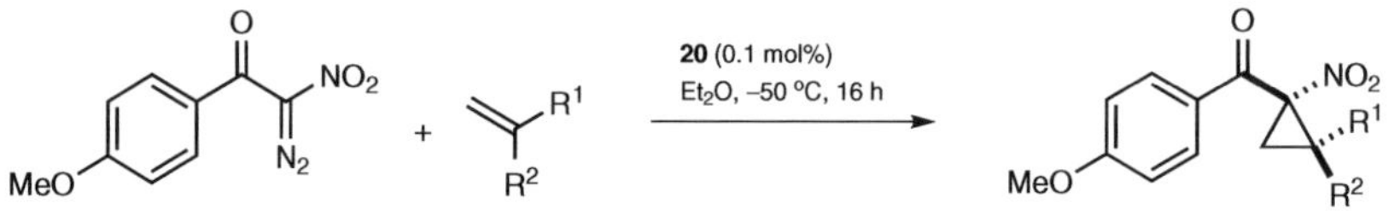

R¹	R²	Yield (%)	dr	ee (%)	Ref
Ph	H	81	98:2	93	[40]
4-Tol	H	76	98:2	93	[40]
4-FC$_6$H$_4$	H	88	98:2	92	[40]
4-O$_2$NC$_6$H$_4$	H	74	99:1	95	[40]
2-FC$_6$H$_4$	H	54	97:3	93	[40]
Ph	Me	88	99:1	98	[40]
Bu	H	5	54:46	42[a]	[40]

[a] ee of the major (*cis*) product.

With α-cyano acetamides (e.g., **31**), the addition of trifluoromethanesulfonamide as a hydrogen-bond-donor additive is crucial to enhance the selectivity of the rhodium-catalyzed reaction of alkenes **32** to give cyclopropane-1-carbonitriles **33** (Scheme 10).[41]

Scheme 10 Enantio- and Diastereoselective Rhodium-Catalyzed Cyclopropanation Leading to Cyano-Substituted Cyclopropanes[41]

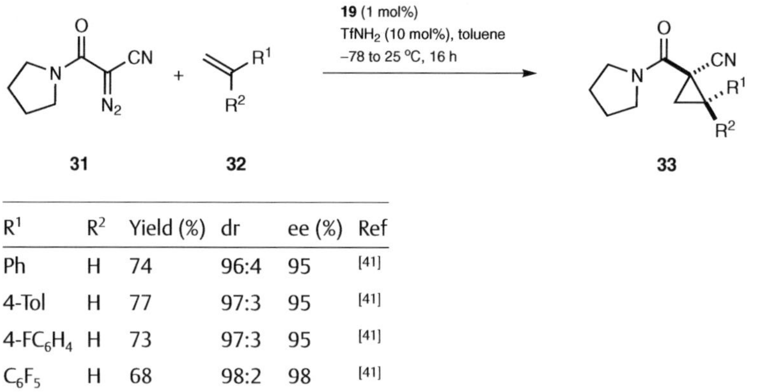

R¹	R²	Yield (%)	dr	ee (%)	Ref
Ph	H	74	96:4	95	[41]
4-Tol	H	77	97:3	95	[41]
4-FC$_6$H$_4$	H	73	97:3	95	[41]
C$_6$F$_5$	H	68	98:2	98	[41]
2-FC$_6$H$_4$	H	70	98:2	96	[41]
Ph	Me	63	85:15	91	[41]
Bu	H	68	93:7	90	[41]

(2-Aryl-1-nitrocyclopropyl)(4-methoxyphenyl)methanones 30; General Procedure:[40]
A flame-dried, round-bottomed flask capped with a rubber septum was charged with a 0.2 mM soln of rhodium catalyst **20** in anhyd Et$_2$O (1 mL, 0.2 µmol, 0.1 mol%), anhyd Et$_2$O (2 mL), and an alkene **29** (1 mmol, 5 equiv) at rt. The resulting green homogeneous soln was cooled to –50 °C and stirred for 15 min at that temperature before the diazo compound **28** (44.2 mg, 0.20 mmol, 1 equiv) was added as a solid under air. The flask was sealed again with the rubber septum and the resulting heterogeneous yellow mixture

was stirred for a further 16 h at −50 °C, keeping an argon line on the flask to minimize water condensation during the reaction. After 16 h, the resulting milky heterogeneous mixture was allowed to slowly warm to rt over a period of 2 h, diluted with CH_2Cl_2 (2 mL), and concentrated to dryness. The resulting residue was purified by column chromatography (silica gel, CH_2Cl_2/hexanes) to afford the product as a mixture of diastereomers.

2-Aryl-1-(pyrrolidine-1-carbonyl)cyclopropane-1-carbonitriles 33; General Procedure:[41]
A 25-mL, round-bottomed flask was charged with rhodium catalyst **19** (5.4 mg, 4 µmol, 1 mol%) and $TfNH_2$ (5.8 mg, 40 µmol, 10 mol%). The flask was purged with argon. Toluene (4 mL) and the alkene (2.00 mmol, 5.00 equiv) were added. The mixture was cooled to −78 °C and stirred for 15 min. A soln of 2-diazo-3-oxo-3-(pyrrolidin-1-yl)propanenitrile (**31**; 66.0 mg, 0.40 mmol, 1.0 equiv) in toluene (2 mL) was added to the mixture over a period of 2 h, using a syringe pump. Following complete addition, the resulting mixture was stirred for an additional 4 h at −78 °C. The reaction was warmed to rt over a period of 16 h, and then the solvent was removed and the residue was purified by chromatography (silica gel, hexane/Et_2O 100:0 to 0:100) to afford the corresponding cyclopropane. In several cases, the rhodium dimer binds to the product, forming catalyst–product complexes. In these cases, the green mixture was dissolved in CH_2Cl_2, and poly(4-vinylpyridine) (20 mg) was added. The mixture changed from green to red and was then filtered through Celite to afford a rhodium-free product following concentration under reduced pressure.

2.2.2.4 Rhodium-Catalyzed Enantioselective Cyclopropanation of Electron-Deficient Alkenes

For a long time, only electron-rich and electron-neutral alkenes could be cyclopropanated by rhodium catalysts; however, the ligand system **21** based on adamantylglycine allows highly stereoselective cyclopropanation reactions of electron-deficient alkenes **35** with substituted aryl and vinyl diazoacetates **34** to give *trans*-cyclopropane-1,2-dicarboxylates **36** (Scheme 11).[42]

Scheme 11 Enantioselective Rhodium-Catalyzed Cyclopropanation of Electron-Deficient Alkenes[42]

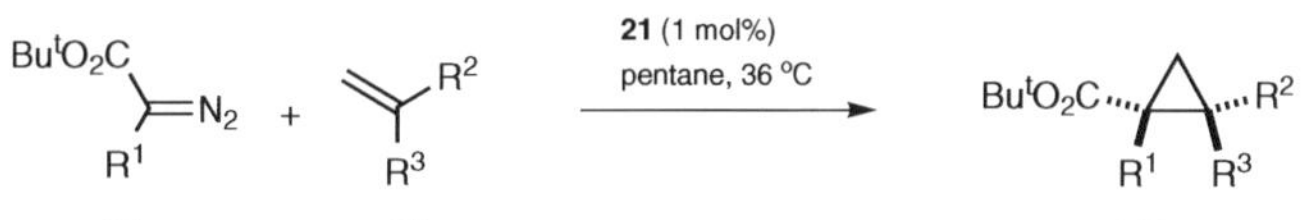

R^1	R^2	R^3	Yield (%)	dr	ee (%)	Ref
Ph	H	CO_2Et	78	>97:3	91	[42]
4-MeOC$_6$H$_4$	H	CO_2Et	83	>97:3	88	[42]
4-F$_3$CC$_6$H$_4$	H	CO_2Et	64	90:10	81	[42]
4-O$_2$NC$_6$H$_4$	H	CO_2Et	22	92:8	91	[42]
4-BrC$_6$H$_4$	H	CO_2Et	89	96:4	93	[42]
4-BrC$_6$H$_4$	H	CO_2Ph	74	>97:3	92	[42]
4-BrC$_6$H$_4$	H	$CONMe_2$	51	>97:3	94	[42]
4-BrC$_6$H$_4$	Me	CO_2Me	55	84:16	77	[42]
(*E*)-CH=CHPh	H	CO_2Et	89	>97:3	97	[42]
(*E*)-4-MeOC$_6$H$_4$CH=CH	H	CO_2Et	77	>97:3	97	[42]
(*E*)-4-F$_3$CC$_6$H$_4$CH=CH	H	CO_2Et	86	>97:3	95	[42]

***trans*-Cyclopropane-1,2-dicarboxylates 36; General Procedure:**[42]
Under argon, to a soln of an alkene **35** (5 equiv) and rhodium catalyst **21** (1 mol%) in pentane (3–5 mL) at reflux was added a soln of a diazo compound **34** (0.5 mmol, 1 equiv) in pentane (3–5 mL) over 2 h. The resulting mixture was stirred at reflux for another 15 min to 1 h before it was cooled to rt. The volatiles were removed under reduced pressure and the crude mixture was purified by flash column chromatography (silica gel, hexane/EtOAc).

2.2.3 Cobalt-Catalyzed Cyclopropanation

Cobalt-catalyzed procedures are well established for effective and stereoselective cyclopropanation reactions. Chiral cobalt(II) porphyrin complexes, especially, are widely used and show both diastereo- and enantioselectivity in cyclopropanation reactions using commonly accessible diazo compounds. These catalysts are superior to the corresponding ruthenium(II)–bis(salicylidene)ethylenediamine (salen) complexes.[43] In contrast to chiral copper and rhodium catalysts, chiral cobalt(II) porphyrin catalysts give access to the three-membered ring with the use of stoichiometric or nearly stoichiometric amounts of the alkene, thereby avoiding carbene dimer formation. Cobalt-catalyzed cyclopropanation protocols tolerate α,β-unsaturated carbonyl compounds and nitriles as substrates.[43]

2.2.3.1 Cobalt(II)-Catalyzed Cyclopropanation Using Diazo Sulfones

An asymmetric cyclopropanation of alkenes **38** with diazo sulfones (e.g., **39**) is achieved using the bulky chiral porphyrin cobalt(II) catalyst **37** and gives cyclopropyl sulfones **40** (Scheme 12). The introduction of a sulfone group leads to full diastereoselectivity and very high enantiomeric excess.[44] Beside aryl moieties, ester, ketone, and cyano groups can also be employed as substituents of the alkene. The yields are usually quite high.

Scheme 12 Cobalt(II)-Catalyzed Cyclopropanation Using Diazo Sulfones[44]

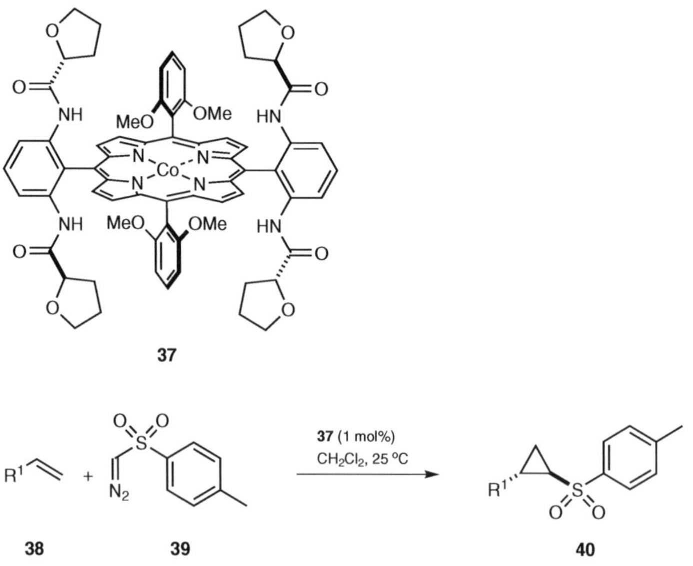

2.2.3 Cobalt-Catalyzed Cyclopropanation

R^1	Yield (%)	ee (%)	Ref
Ph	99	92	[44]
4-MeOC$_6$H$_4$	72	95	[44]
4-t-BuC$_6$H$_4$	57	94	[44]
3-O$_2$NC$_6$H$_4$	77	96	[44]
2-naphthyl	81	93	[44]
CO$_2$Me	96	89	[44]
Ac	64	97	[44]
CN	81	61	[44]

Cyclopropyl Sulfones 40; General Procedure:[44]
Cobalt catalyst **37** (1 mol%) was placed in an oven-dried Schlenk tube. The tube was capped with a Teflon screw cap, evacuated, and backfilled with N$_2$. The screw cap was replaced with a rubber septum, and a styrene **38** (0.25 mmol) in CH$_2$Cl$_2$ (0.5 mL) was added using a syringe, followed by diazo sulfone **39** (0.30 mmol) and CH$_2$Cl$_2$ (0.5 mL). The tube was purged with N$_2$ for 1 min and its contents were stirred at rt. After the reaction finished, the resulting mixture was concentrated and the residue was purified by flash column chromatography (silica gel).

2.2.3.2 Stereoselective Cyclopropanation of Alkenes with 2-Cyano-2-diazoacetates

An asymmetric cyclopropanation of alkenes **42** with 2-cyano-2-diazoacetates (e.g., **43**) is achieved using cobalt(II) catalyst **41**. Complete diastereoselectivity and high enantiomeric excesses are obtained (Scheme 13).[45] The formed 1-cyanocyclopropane-1-carboxylate products **44** open various possibilities for further transformations. Ester groups can be converted into primary alcohols without affecting the nitrile, while nitriles can be reduced to amines without affecting the ester group.

Scheme 13 Stereoselective Cyclopropanation of Alkenes with a 2-Cyano-2-diazoacetate[45]

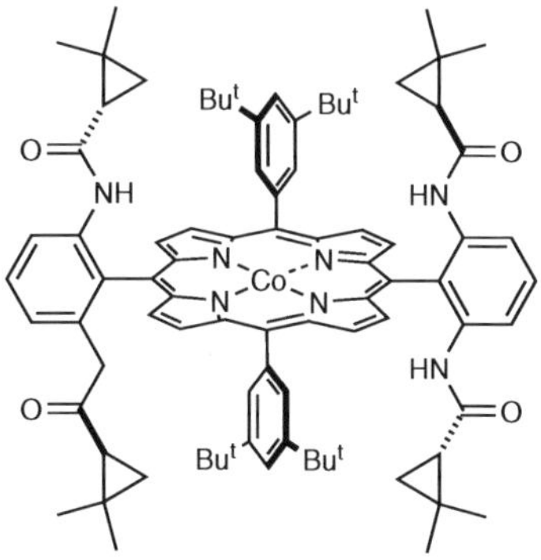

R^1 ══ + NC-C(N₂)-CO₂Bu^t →[**41** (1 mol%), hexane, −20 °C, 24 h] cyclopropane with R^1, CN, CO₂Bu^t

42   **43**   **44**

R¹	Yield (%)	ee (%)	Ref
Ph	96	98	[45]
4-MeOC₆H₄	88	99	[45]
4-F₃CC₆H₄	81	98	[45]
3-O₂NC₆H₄	90	98	[45]
C₆F₅	73	99	[45]
CO₂Me	90	88	[45]
Ac	81	92	[45]
CN	99	82	[45]
OBz	80	88	[45]
Bu	88	96	[45]

1-Cyanocyclopropane-1-carboxylates 44; General Procedure:[45]
Catalyst **41** (1 mol%) was placed in an oven-dried Schlenk tube under N₂. A soln of a styrene **42** (0.25 mmol) in hexane (1.0 mL) was added using a syringe. After the mixture had been cooled to −20 °C, tert-butyl 2-cyano-2-diazoacetate (**43**; 0.30 mmol) was added and the mixture was stirred for 24 h. The resulting mixture was concentrated and the residue was purified by flash column chromatography (silica gel).

2.2.4 Iridium-Catalyzed Cyclopropanation

Copper, rhodium, and cobalt are most often used to synthesize chiral complexes for asymmetric cyclopropanation reactions. Commonly, the *trans*-isomers are formed as major products. In 2007, an asymmetric iridium–bis(salicylidene)ethylenediamine (salen) complex was reported to cyclopropanate a wide range of alkenes with *tert*-butyl 2-diazoacetate.[46] The main advantage is the complete *cis* selectivity of the generated cyclopropanes. But to obtain high yields, a large excess of the corresponding alkene (up to 10 equiv) is necessary.

2.2.4.1 *cis*-Selective Cyclopropanation of Conjugated and Nonconjugated Alkenes Using Aryliridium–Bis(salicylidene)ethylenediamine (Salen) Complexes

Iridium–salen complexes such as **45** are efficient catalysts for *cis*-selective asymmetric cyclopropanation reactions with *tert*-butyl 2-diazoacetate (**47**) (Scheme 14).[47] The reaction is carried out at −78 °C in tetrahydrofuran to achieve complete *cis* selectivity for the produced *tert*-butyl cyclopropane-1-carboxylates **48**. This methodology can be applied to conjugated, polysubstituted, and nonactivated alkenes (e.g., **46**), both high *cis* selectivity and enantioselectivity is observed. The presence of functional groups such as ethers or esters is tolerated.

2.2.5 Ruthenium-Catalyzed Cyclopropanation

Scheme 14 *cis*-Selective Cyclopropanation Using Aryliridium–Salen Complexes[47]

R¹	Yield (%)	ee (%)	Ref
Ph	99	>99	[47]
4-MeOC$_6$H$_4$	99	97	[47]
4-ClC$_6$H$_4$	99	98	[47]
4-F$_3$CC$_6$H$_4$	73	97	[47]
2-naphthyl	97	97	[47]
C≡CPh	95	84	[47]
(CH$_2$)$_5$Me	62	98	[47]
OBz	87	94	[47]

2-Substituted *tert*-Butyl *cis*-Cyclopropane-1-carboxylates 48; General Procedure:[47]
In a 5-mL Schlenk tube, *tert*-butyl 2-diazoacetate (**47**; 0.1 mmol) was dissolved in THF (0.24 mL) under N$_2$. The alkene **46** (1.0 mmol) was added, and the mixture was cooled to −78 °C and stirred for 10 min. The iridium–salen complex **45** (1.1 mg, 1.0 µmol) was added, and the mixture was stirred at −78 °C for 24 h, allowed to warm to rt, passed through a pad of silica gel, and concentrated on a rotary evaporator. The residue was purified by chromatography (silica gel, hexane/iPr$_2$O 1:0 to 4:1). The ee of the product was determined by HPLC analysis.

2.2.5 Ruthenium-Catalyzed Cyclopropanation

Ruthenium catalysts are useful reagents for carbene-transfer to unsaturated moieties. In comparison to rhodium, ruthenium is roughly only one tenth of the price and has more possible oxidation states.[48] However, the main disadvantage of ruthenium-catalyzed cyclopropanation reactions is the low electrophilicity of the metal–carbene species, which hinders its application to highly substituted double bonds. Moreover, ruthenium carbenes easily undergo undesired metathesis reactions.[49]

Groundbreaking ruthenium-catalyzed cyclopropanation reactions with diazoacetates made use of a ruthenium–2,6-bis(4,5-dihydrooxazol-2-yl)pyridine (pybox) complex (for asymmetric cyclopropanation),[50] porphyrin ligands (high enantioselectivities),[51] and bis(salicylidene)ethylenediamine ligands.[52] Despite that, copper-, rhodium-, and cobalt-catalyzed cyclopropanation reactions generally show better conversion, higher diastereoselectivity, and higher group tolerance. For that reason, the above mentioned examples are not discussed further in this chapter.

2.2.5.1 Enantioselective Cyclopropanation Using Ruthenium(II)–Diphenyl-4,5-dihydrooxazole Complexes

The ruthenium(II)–diphenyl-4,5-dihydrooxazole complex **49** is an efficient catalyst for the cyclopropanation reaction of monosubstituted alkenes **50** with 2,5-dioxopyrrolidin-1-yl 2-diazoacetate (**51**) (Scheme 15).[53] The desired cyclopropanes are obtained in over 90% yield with excellent diastereoselectivity (*trans/cis* >99:1). Subsequently, the succinimidyl carboxylates are reduced to give the primary alcohols **52** with enantiomeric excesses between 91 and 99%.

Scheme 15 Ruthenium(II)-Catalyzed Enantioselective Cyclopropanation Leading to Cyclopropylmethanols[53]

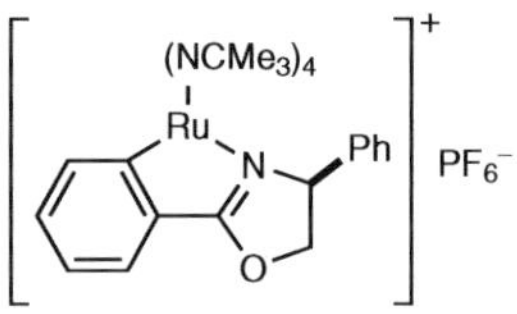

R^1	Yield (%)	ee (%)	Ref
Ph	78	95	[53]
4-Tol	78	99	[53]
3-Tol	84	91	[53]
4-MeOC$_6$H$_4$	79	92	[53]
4-ClC$_6$H$_4$	75	99	[53]
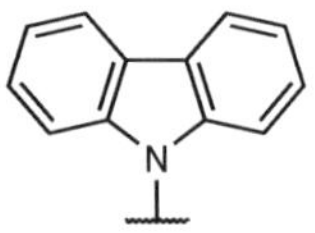 carbazolyl	89	98	[53]
OBu	66	99	[53]

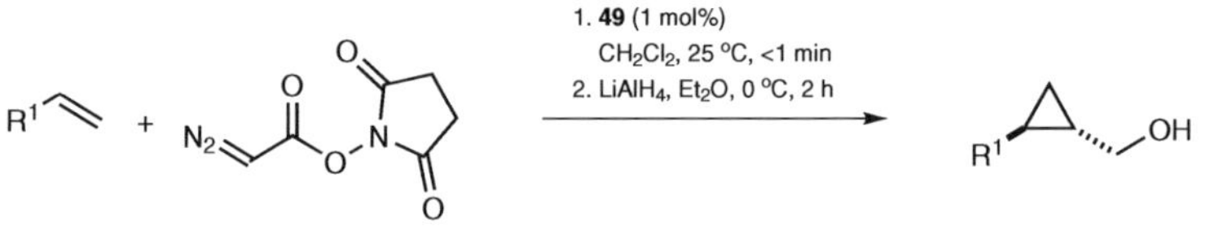

[(1S,2S)-2-Phenylcyclopropyl]methanol (52, R^1 = Ph); Typical Procedure:[53]

CAUTION: *Solid lithium aluminum hydride reacts vigorously with a variety of substances, and can ignite on rubbing or vigorous grinding.*

To a soln of 2,5-dioxopyrrolidin-1-yl 2-diazoacetate (**51**; 36.6 mg, 0.2 mmol) and styrene (**50**, R^1 = Ph; 114.6 µL, 1 mmol) in CH$_2$Cl$_2$ (2.0 mL) was added ruthenium catalyst **49**

2.2.5 Ruthenium-Catalyzed Cyclopropanation

(2 µmol) at rt. After 1 min, the solvent was removed and the residue was purified by column chromatography (silica gel, hexane/EtOAc 3:1).

To a soln of the 2,5-dioxopyrrolidin-1-yl 2-phenylcyclopropane-1-carboxylate derivative (0.2 mmol) in Et_2O (3.0 mL) was added $LiAlH_4$ (15.2 mg, 0.4 mmol) at 0°C. After the mixture had been stirred for 2 h at the same temperature, H_2O (0.5 mL) was added. The mixture was filtered and the filtrate was concentrated under reduced pressure. The residue was purified by flash column chromatography (silica gel, hexane/EtOAc 2:1); yield: 78%; 95% ee.

2.2.5.2 Ruthenium-Catalyzed Asymmetric Construction of 3-Azabicyclo[3.1.0]hexanes

Like gold and platinum catalysts, ruthenium also catalyzes the cycloisomerization of 1,6-enynes. In this case, 1,6-enynes **54** with a racemic propargylic alcohol moiety are treated with cyclopentadienyl ruthenium catalyst **53** containing a tethered chiral sulfoxide to give 3-azabicyclo[3.1.0]hexanes **55** with high enantioselectivity (Scheme 16).[54] This reaction represents the first asymmetric ruthenium-catalyzed redox cycloisomerization process.

Scheme 16 Ruthenium-Catalyzed Asymmetric Construction of 3-Azabicyclo[3.1.0]hexanes[54]

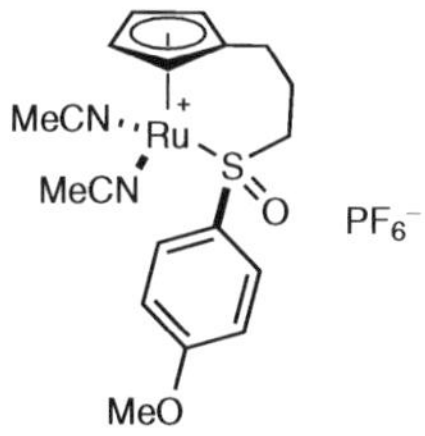

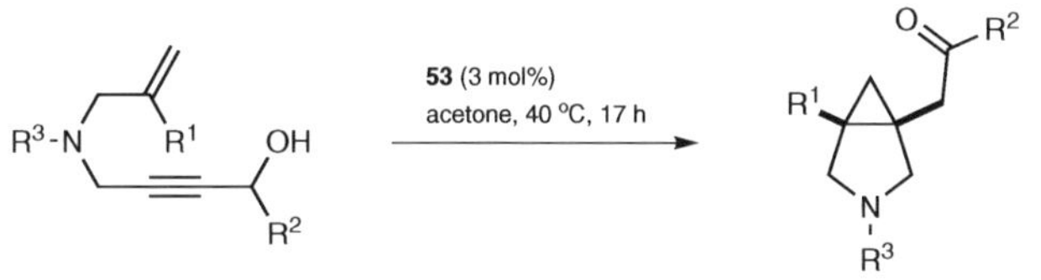

R^1	R^2	R^3	Yield (%)	ee (%)	Ref
Me	Me	2,4,6-iPr$_3$C$_6$H$_2$SO$_2$	85	81	[54]
Bn	Me	Ts	84	90	[54]
(CH$_2$)$_8$CH=CH$_2$	Me	2,4,6-iPr$_3$C$_6$H$_2$SO$_2$	57	87	[54]
Cy	cyclopropyl	2,4,6-iPr$_3$C$_6$H$_2$SO$_2$	75	80	[54]
Me	Me	Ts	81	88	[54]
Me	Me	P(O)(OPh)$_2$	84	96	[54]
iPr	cyclopentyl	P(O)(OPh)$_2$	64	92	[54]

A plausible mechanistic scenario (Scheme 17) starts with the coordination of the alcohol to the ruthenium catalyst. In the coordination sphere of the ruthenium, a redox isomerization takes place, followed by a (2 + 2) cycloaddition leading to a ruthenacyclobutane (e.g., **56**). The formation of the cyclopropane occurs through a reductive elimination step.

Scheme 17 Plausible Mechanism for the Ruthenium-Catalyzed Asymmetric Construction of 3-Azabicyclo[3.1.0]hexanes[54]

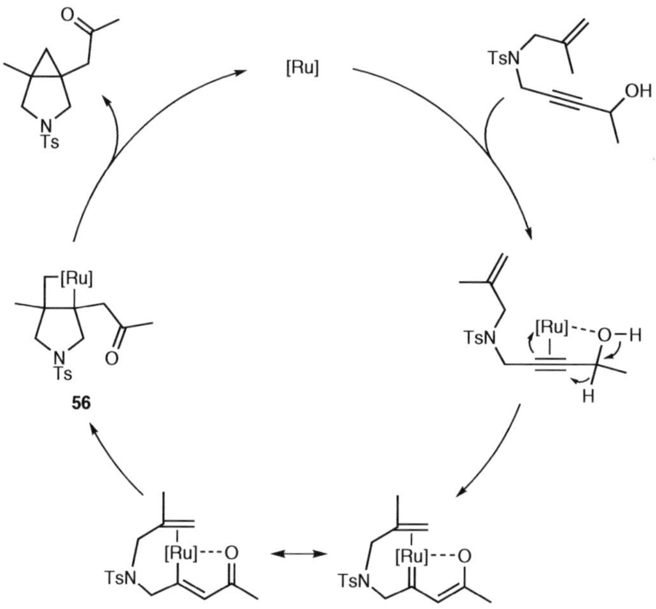

3-Azabicyclo[3.1.0]hexanes 55; General Procedure:[54]

An oven-dried microwave vial was charged with Ru catalyst **53** (0.03 equiv), fitted with a septum, and placed under argon. A 0.25 M soln of a propargyl alcohol **54** (1 equiv) in acetone was added using a syringe and the reaction was heated to 40 °C in an oil bath for 17 h. The mixture was cooled to rt and concentrated under reduced pressure. The crude residue was purified by chromatography (silica gel) to afford the pure product.

2.2.6 Iron-Catalyzed Cyclopropanation

Iron-catalyzed cyclopropanation reactions have only recently become synthetically useful, with the report of a method that starts from the respective amines in aqueous media.[55–57] Earlier procedures made use of diazo compounds, but could not compete with the simple and successful rhodium- and copper-catalyzed variants because either half-sandwich complexes or iron–porphyrin catalysts were needed.[58,59] With the advent of the in situ generation of the diazo compounds the situation changed, although iron–porphyrin complexes are still required. Nowadays, a variety of alkenes including terminal alkenes, dienes, and enynes can be converted into cyclopropanes in good yields. Biphasic systems arising from the alkenic substrate and water are a prerequisite for these processes. A major advantage is the in situ generation of the toxic and explosive diazo compounds from the corresponding amines or ammonium salts, thus avoiding the aforementioned hazards.

2.2.6.1 Iron-Catalyzed Cyclopropanation of Styrenes with In Situ Generation of Diazomethane

Styrenes **58** undergo smooth cyclopropanation with iron–porphyrin catalyst **57** under strong alkaline conditions, which are used to generate diazomethane from the water-soluble reagent sodium 3-(*N*-methyl-*N*-nitrososulfamoyl)benzoate (**59**) (Scheme 18).[60] Dienes and enynes are cyclopropanated solely at the terminal double bond. The method is limited to substrates that can withstand the strong basic reaction medium.

Scheme 18 Diazomethane Generation Followed by Iron-Catalyzed Cyclopropanation[60]

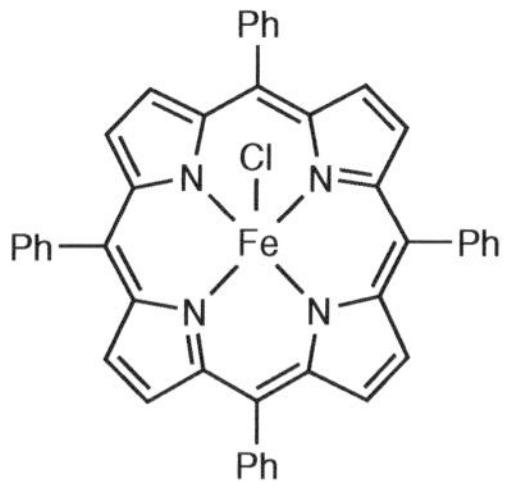

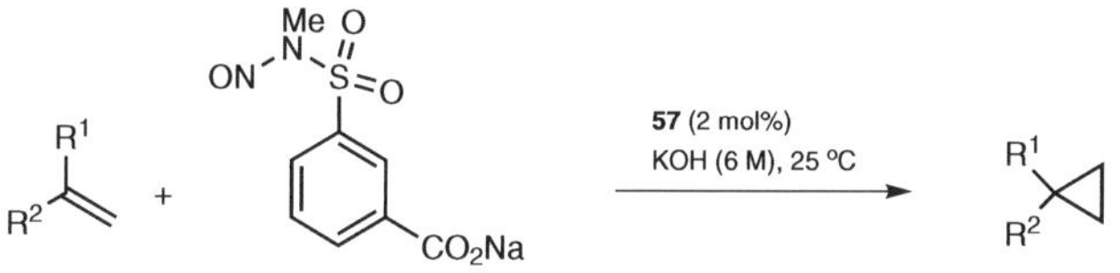

R¹	R²	Yield (%)	Ref
H	2-MeOC$_6$H$_4$	80	[60]
H	4-MeOC$_6$H$_4$	89	[60]
Me	4-ClC$_6$H$_4$	74	[60]
H	3-O$_2$NC$_6$H$_4$	64	[60]
H	(*E*)-2-MeOC$_6$H$_4$CH=CH	78	[60]
H	C≡CPh	74	[60]

1-Cyclopropyl-4-methoxybenzene (60, R¹ = H; R² = 4-MeOC$_6$H$_4$); Typical Procedure:[60]
Iron–porphyrin catalyst **57** (3.1 mg, 4.4 µmol) was dissolved in 4-methoxystyrene (**58**, R¹ = H; R² = 4-MeOC$_6$H$_4$; 0.22 mmol) and 6 M aq KOH (2 mL) was added in an open vial. To this vigorously and steadily stirred soln (>1000 rpm) was added a 0.13 M aqueous soln of sodium 3-(*N*-methyl-*N*-nitrososulfamoyl)benzoate (**59**; 5 mL, 0.66 mmol) at rt over 4 h, using a syringe pump. After a further 30 min, the soln was diluted with H$_2$O and extracted with CH$_2$Cl$_2$ (3 ×). The combined organic phases were dried (MgSO$_4$) and concentrated under reduced pressure. After NMR analysis of the crude mixture, the pure product was obtained by flash column chromatography (pentane/Et$_2$O) as a colorless oil; yield: 29 mg (89%).

2.2.6.2 Iron-Catalyzed Cyclopropanation with In Situ Generated 2-Diazo-1,1,1-trifluoroethane

Trifluoromethyl-substituted cyclopropanes **63** can be obtained using 2,2,2-trifluoroethylamine hydrochloride (**62**) under diazotization conditions followed by carbene generation and cyclopropanation in the presence of iron–porphyrin catalyst **57** (Scheme 19).[55,56] The great advantage is that handling of the gaseous 2-diazo-1,1,1-trifluoroethane can be avoided. The transformation with styrene derivatives and alkenes **61** strongly favors the *trans*-diastereomers (commonly dr >95:5). The respective products are obtained in good to excellent yield; however, with nonactivated alkenes only low yields are observed.

Scheme 19 Cyclopropanation with In Situ Generated 2-Diazo-1,1,1-trifluoroethane[55,56]

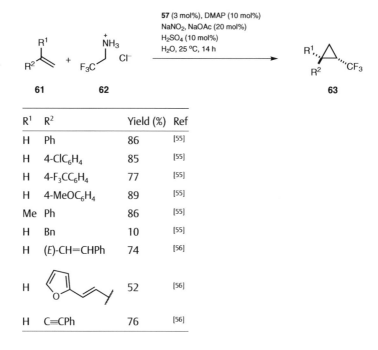

R¹	R²	Yield (%)	Ref
H	Ph	86	[55]
H	4-ClC$_6$H$_4$	85	[55]
H	4-F$_3$CC$_6$H$_4$	77	[55]
H	4-MeOC$_6$H$_4$	89	[55]
Me	Ph	86	[55]
H	Bn	10	[55]
H	(*E*)-CH=CHPh	74	[56]
H	(furyl-CH=CH)	52	[56]
H	C≡CPh	76	[56]

***trans*-1-Aryl-2-(trifluoromethyl)cyclopropanes 63; General Procedure:**[55,56]
Iron–porphyrin catalyst **57** (4.6 mg, 6.6 µmol), DMAP (2.6 mg, 0.022 mmol), and NaOAc (3.6 mg, 0.044 mmol) were dissolved in degassed, distilled H$_2$O (0.8 mL). 2,2,2-Trifluoroethylamine hydrochloride (**62**; 45 mg, 0.33 mmol) and concd H$_2$SO$_4$ (1.2 µL, 0.022 mmol) were added, and the soln was degassed for 1 min by sparging with argon. The alkene **61** (0.22 mmol) was subsequently added, and a soln of NaNO$_2$ (27 mg) in H$_2$O (0.5 mL) was added over 10 h, using a syringe pump. After 4 h, CH$_2$Cl$_2$ and H$_2$O were added, and the aqueous phase was extracted with CH$_2$Cl$_2$ (3 ×). The combined organic phases were dried (MgSO$_4$) and concentrated under reduced pressure. After NMR analysis of the crude mixture to determine the diastereoselectivity, the product was purified by chromatography (silica gel, pentane/Et$_2$O).

2.2.6.3 Iron-Catalyzed Cyclopropanation with Ethyl Glycinate Hydrochloride in Water

A one-pot process of diazotization/iron-catalyzed cyclopropanation of styrenes **64** can also be conducted with ethyl glycinate hydrochloride (**65**) in water with iron–porphyrin catalyst **57** (Scheme 20).[57] The respective ester-substituted cyclopropane derivatives **66** are accessed in good yields and diastereoselectivity. The highest yields are obtained for electron-rich styrenes, whereas electron-deficient ones and diene systems give poorer yields.

Scheme 20 Cyclopropanation with In Situ Generated Ethyl 2-Diazoacetate[57]

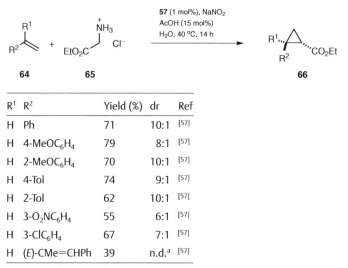

R^1	R^2	Yield (%)	dr	Ref
H	Ph	71	10:1	[57]
H	4-MeOC$_6$H$_4$	79	8:1	[57]
H	2-MeOC$_6$H$_4$	70	10:1	[57]
H	4-Tol	74	9:1	[57]
H	2-Tol	62	10:1	[57]
H	3-O$_2$NC$_6$H$_4$	55	6:1	[57]
H	3-ClC$_6$H$_4$	67	7:1	[57]
H	(E)-CMe=CHPh	39	n.d.^{a}	[57]

^{a} n.d. = not determined.

Ethyl *trans*-2-Arylcyclopropane-1-carboxylates 66; General Procedure:[57]
Iron–porphyrin catalyst **57** (7.0 mg, 0.01 mmol, 1 mol%) was dissolved in an alkene **64** (1.0 mmol) in an open vial under air. Ethyl glycinate hydrochloride (**65**; 279 mg, 2.0 mmol, 2.0 equiv), H$_2$O (5.0 mL), and AcOH (10.0 mg, 0.15 mmol, 0.15 equiv) were added. To the vigorously stirred heterogeneous mixture was added NaNO$_2$ (166 mg, 2.4 mmol, 2.4 equiv) in one portion at 40 °C. After 14 h, the mixture was diluted with H$_2$O and extracted with CH$_2$Cl$_2$ (3 ×). The combined extracts were dried (MgSO$_4$) and concentrated under reduced pressure. The crude product was then analyzed by ^{1}H NMR spectroscopy to determine the conversion and the dr. Purification by flash column chromatography afforded the pure product.

2.2.6.4 Enantioselective Iron-Catalyzed Intramolecular Cyclopropanation Reactions

Chiral spiro-bis(4,5-dihydrooxazole) ligands such as **67**, in the presence of iron salts and sodium tetrakis[3,5-bis(trifluoromethyl)phenyl]borate (NaBARF), can be utilized to realize asymmetric intramolecular cyclopropanation reactions of diazo compounds **68** (Scheme 21).[61] This method yields synthetically useful chiral [3.1.0]bicycloalkanes **69** in very good yield and with moderate to high enantioselectivity. Iron(II) salts exhibit higher reactivity than iron(III) salts. Electron-rich alkenes are preferentially transformed, suggesting that electrophilic iron carbenoid species have to be taken into account.

for references see p 95

Scheme 21 Enantioselective Iron-Catalyzed Intramolecular Cyclopropanation[61]

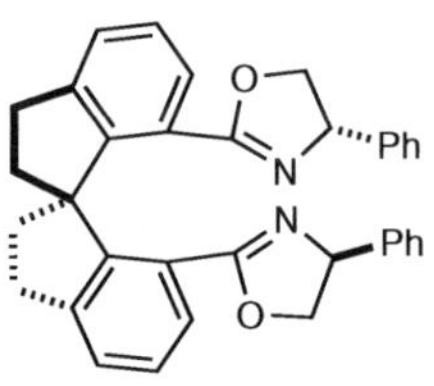

Ar¹	R¹	R²	R³	Time (h)	Yield (%)	ee (%)	Ref
Ph	Me	H	H	7	94	92	[61]
2-FC$_6$H$_4$	Me	H	H	7	93	93	[61]
3-F$_3$CC$_6$H$_4$	Me	H	H	7	89	96	[61]
3-Tol	Me	H	H	20	86	90	[61]
4-MeOC$_6$H$_4$	Me	H	H	7	93	44	[61]
Ph	Ph	H	H	10	87	97	[61]
Ph	Et	H	H	30	82	81	[61]
Ph	H	H	H	30	52	6	[61]
Ph	H	Me	Me	7	85	33	[61]

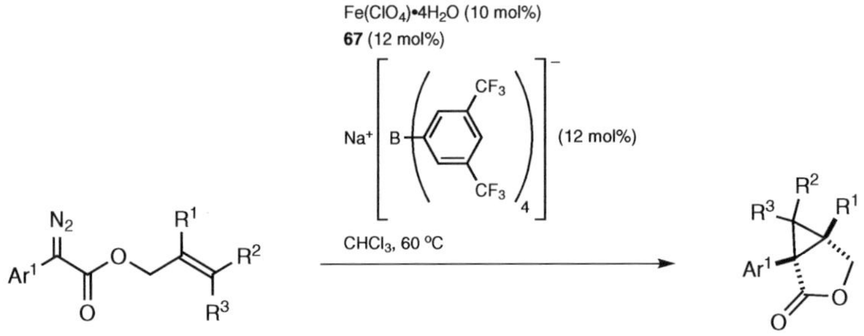

3-Oxabicyclo[3.1.0]hexan-2-ones 69; General Procedure:[61]
A mixture of Fe(ClO$_4$)$_2$·4H$_2$O (10.0 mg, 0.03 mmol, 10 mol%), ligand **67** (18.4 mg, 0.036 mmol, 12 mol%), sodium tetrakis[3,5-bis(trifluoromethyl)phenyl]borate (32 mg, 0.036 mmol, 12 mol%), and CHCl$_3$ (4 mL) was stirred under argon at rt for 4 h. Then, a diazo compound **68** (0.3 mmol) was injected into the mixture and it was stirred at 60 °C for the specified time. The solvent was removed under reduced pressure and the residue was purified by flash column chromatography (silica gel, petroleum ether/EtOAc 5:1).

2.2.7 Palladium-Catalyzed Cyclopropanation

Palladium is a very versatile catalyst for cyclopropanation reactions. On one hand, it catalyzes the reaction of unsaturated carbon–carbon bonds with diazomethane. On the other hand, it acts as catalyst for cross-coupling reactions, C—H activation reactions, or activation of double bonds, which also lead to the cyclopropyl motif.[62] In this chapter, a cross section of methods over different reaction modes of this metal with respect to cyclopropane formation is provided.

2.2.7 Palladium-Catalyzed Cyclopropanation

2.2.7.1 Palladium(II) Acetate Catalyzed Cyclopropanation Using Diazomethane

Terminal and electron-deficient double bonds can be efficiently cyclopropanated in the presence of palladium and diazomethane to provide, for example, cyclopropanes **71** (Scheme 22).[63–66] The great advantage is the chemoselective cyclopropanation of electron-deficient alkenes (e.g., **70**), whereas electron-rich double bonds stay untouched. Asymmetric palladium-catalyzed cyclopropanation reactions using diazomethane have not been achieved.

Scheme 22 Palladium(II) Acetate Catalyzed Cyclopropanation Using Diazomethane[63–66]

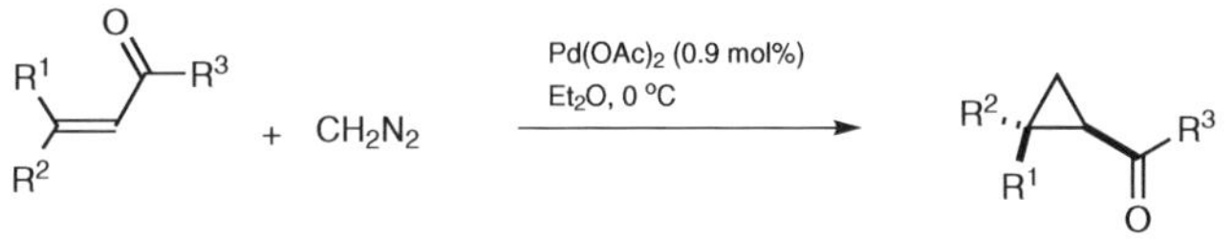

R¹	R²	R³	Yield (%)	Ref
H	Ph	Ph	98	[63]
H	Ph	OEt	90	[63]
Ph	H	OEt	85	[63]
H	(CH$_2$)$_3$CO$_2$Me	H	80	[64]
H	(CH$_2$)$_{18}$Me	OMe	99	[65]
H	MeOC$_6$H$_4$	OMe	95	[66]

trans-Phenyl(2-phenylcyclopropyl)methanone (**71**, R¹ = H; R² = R³ = Ph);
Typical Procedure:[63]

CAUTION: *Diazomethane is explosive by shock, friction, or heat, and is highly toxic by inhalation. Special glassware must be employed.*

To (*E*)-1,3-diphenylprop-2-en-1-one (**70**, R¹ = H; R² = R³ = Ph; 1.0 g, 4.8 mmol) and Pd(OAc)$_2$ (10 mg) in Et$_2$O (15 mL) an ethereal soln of CH$_2$N$_2$ [20 mL, prepared from *N*-methyl-*N*-nitrosourea (2 g, 19.4 mmol)] was added dropwise at 0 °C with continuous stirring during 10 min. After evaporation of the solvent, the residue was purified by chromatography (silica gel, hexane); yield: 98%.

2.2.7.2 Palladium-Catalyzed Cyclopropanation of Benzyl Bromides via C–H Activation

Benzyl bromides **72** and norbornene derivatives **73** can be transformed via a palladium(0)-catalyzed cyclopropanation into cyclopropane-fused tri- and tetracyclic compounds **74** (Scheme 23).[67]

Scheme 23 Cyclopropanation via Domino Heck-Type Coupling/C(sp³)–H Bond Activation of Benzyl Bromides with Norborne Derivatives[67]

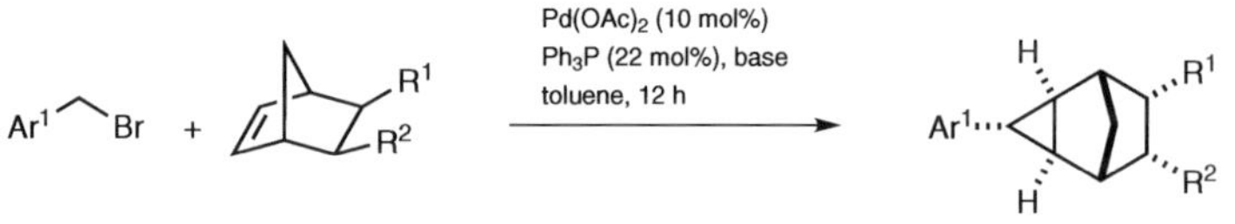

Ar¹	R¹	R²	Base (Equiv)	Temp (°C)	Yield (%)	Ref
Ph	H	H	K_2CO_3 (0.5)	120	89	[67]
3-MeOC$_6$H$_4$	H	H	K_2CO_3 (0.5)	120	93	[67]
Ph	CH=CHCH$_2$		K_2CO_3 (0.5)	100	90	[67]
4-FC$_6$H$_4$	CH=CHCH$_2$		K_2CO_3 (0.5)	100	86	[67]
1-naphthyl	C(O)N(4-Tol)C(O)		NaOMe (3)	100	85	[67]
4-Tol	C(O)N(iBu)C(O)		NaOMe (3)	100	90	[67]

The process involves a domino Heck-type coupling/C(sp³)–H bond activation (Scheme 24).[67] This cyclopropanation is impressive as it has simple reaction conditions [Pd(OAc)$_2$, Ph$_3$P, base, toluene] and gives high yields.

Scheme 24 Possible Mechanism of the Palladium(0)-Catalyzed Cyclopropanation via Domino Heck-Type Coupling/C(sp³)–H Bond Activation[67]

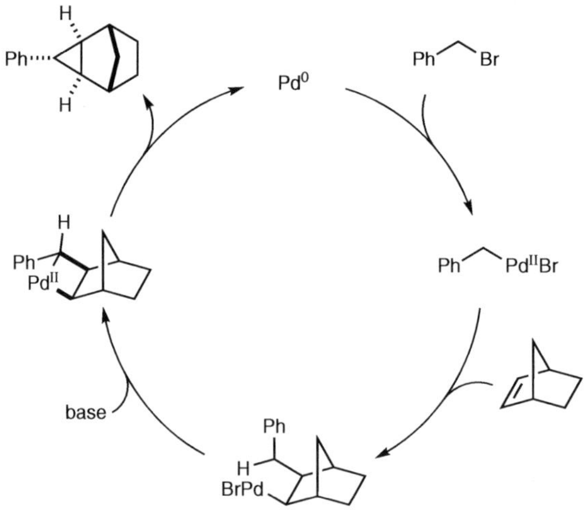

Cyclopropane-Fused Tri- and Tetracyclic Compounds 74; General Procedure:[67]
A benzyl bromide **72** (0.55 mmol, 1.1 equiv), a norbornene derivative **73** (0.5 mmol, 1.0 equiv), Pd(OAc)$_2$ (11.2 mg, 0.05 mmol, 10 mol%), Ph$_3$P (28.9 mg, 0.11 mmol, 22 mol%), K$_2$CO$_3$ (34.5 mg, 0.25 mmol, 0.5 equiv), and toluene (2 mL) were added sequentially to a flame-dried Teflon-screw-capped tube equipped with a magnetic stirrer bar under N$_2$. The tube was sealed with the Teflon-lined cap, and the mixture was stirred at 100 °C or 120 °C for 12 h. After completion of the reaction, the resulting mixture was cooled to rt, diluted with CH$_2$Cl$_2$ (10 mL), and filtered through a short pad of silica gel (EtOAc, 30 mL). The filtrate was concentrated under reduced pressure and the residue was purified by column chromatography (silica gel).

2.2.7.3 Palladium-Catalyzed Cyclopropanation of Acyclic Amides with Substituted Allyl Carbonates

Acyclic amides **76** and monosubstituted allyl carbonates **77** serve as substrates in a palladium-catalyzed asymmetric cyclopropanation reaction with chiral ferrocene ligand **75** (Scheme 25).[68] The cyclopropane derivatives **78** bearing three contiguous stereocenters are obtained with very high diastereo- and enantioselectivities.

Scheme 25 Palladium-Catalyzed Cyclopropanation of Acyclic Amides with Substituted Allyl Carbonates[68]

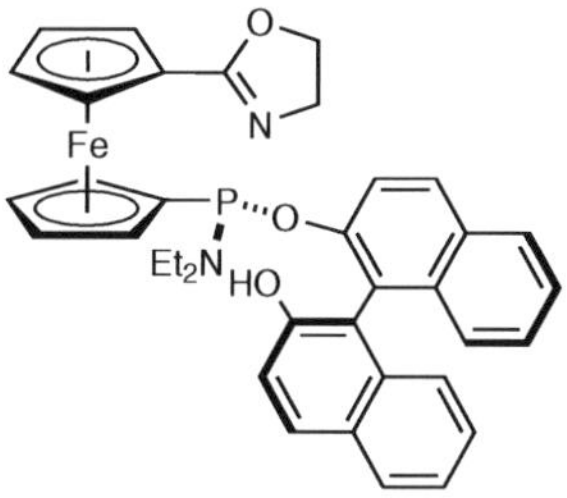

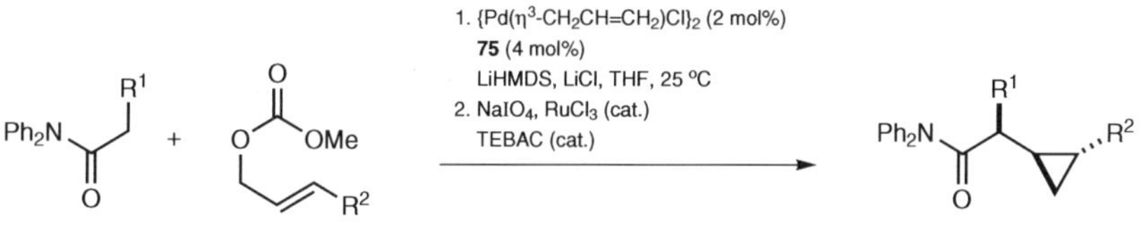

R^1	R^2	Yield (%)	dr	ee (%)	Ref
Me	Ph	73	12:1	95	[68]
Et	Ph	72	12:1	97	[68]
Me	4-Tol	69	15:1	94	[68]
Me	4-MeOC$_6$H$_4$	68	23:1	96	[68]
Me	4-FC$_6$H$_4$	67	11:1	84	[68]
iPr	4-BrC$_6$H$_4$	74	5:1	92	[68]
Me	1-naphthyl	68	8:1	93	[68]

Coordination of the palladium to the double bond results in a π-allyl complex **79**, with liberation of the carbonate. A subsequent attack of the nucleophile at the C2-position of the π-allyl complex forms the cyclopropane accompanied by reductive elimination of the palladium (Scheme 26).[68]

Scheme 26 Mechanism of the Palladium-Catalyzed Cyclopropanation of Acyclic Amides with Substituted Allyl Carbonates[68]

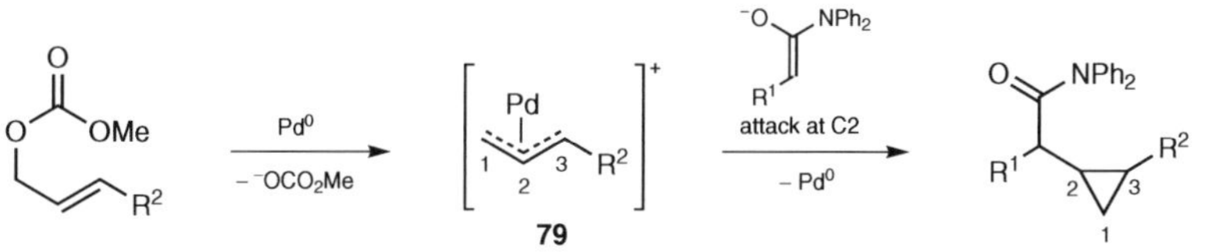

N,N-Diphenyl-2-cyclopropylacetamides 78; General Procedure:[68]
A dry Schlenk tube containing LiCl (12.7 mg, 0.30 mmol) was flame-dried and flushed with argon. Amide **76** (0.30 mmol) and THF (2.0 mL) were added. A 1.0 M soln of LiHMDS in THF (0.30 mL, 0.30 mmol) was added dropwise with stirring for 20 min at 0 °C. In a separate flask, {Pd(η^3-CH$_2$CH=CH$_2$)Cl}$_2$ (2.2 mg, 6 µmol) and ligand **75** (7.7 mg, 0.012 mmol) were added followed by THF (1.0 mL), and the mixture was stirred at rt for 30 min. Then, the catalyst soln was added to the enolate soln. The allylic carbonate **77** (0.36 mmol) was added and the mixture was stirred at rt for 4 h. When the reaction was complete, H$_2$O (5 mL) was added and the mixture was extracted with Et$_2$O (3 × 10 mL). The combined organic layers were dried (Na$_2$SO$_4$), filtered, and concentrated under reduced pressure to afford a crude oil. The diastereoselectivity was determined by GC. The crude oil was dissolved in EtOAc (2 mL) and transferred to a tube. RuCl$_3$ (1 mg, 5 µmol), TEBAC (10 mg, 0.044 mmol), and NaIO$_4$ (214 mg, 1 mmol) in H$_2$O (3 mL) were added slowly at rt. The resulting soln was stirred for 1 h. EtOAc (8 mL) was added, and the organic layer was separated and washed with H$_2$O. The aqueous phase was extracted with Et$_2$O (3 × 10 mL), and the combined organic layers were dried (Na$_2$SO$_4$), filtered, and concentrated under reduced pressure. The residue was purified by column chromatography (silica gel).

2.2.7.4 Palladium-Catalyzed Cross-Coupling Reaction of Styrenes with Aryl Methyl Ketones

A cyclopropanation reaction of aryl methyl ketones **80** with styrenes **81** is accomplished using palladium(II) acetate and copper(II) acetate in the presence of dioxygen (Scheme 27).[69] The use of readily available starting materials and the avoidance of special carbene species are the main advantages of this protocol. The yields of the aryl(arylcyclopropyl)-methanones **82** are high, and the products are always obtained with a *trans/cis* ratio higher than 98:2.

Scheme 27 Palladium-Catalyzed Cross-Coupling Reaction of Styrenes with Aryl Methyl Ketones[69]

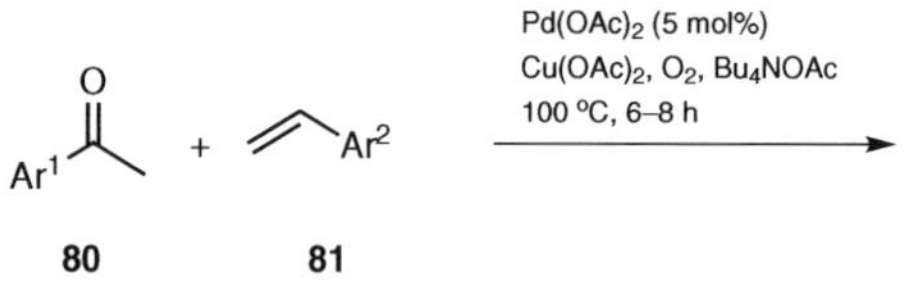

Ar^1	Ar^2	Yield (%)	Ref
Ph	Ph	95	[69]
4-MeOC$_6$H$_4$	4-F$_3$CC$_6$H$_4$	77	[69]
2-Tol	Ph	93	[69]
2-ClC$_6$H$_4$	Ph	92	[69]
Ph	4-*t*-BuC$_6$H$_4$	89	[69]
4-FC$_6$H$_4$	4-ClC$_6$H$_4$	85	[69]
Ph	2-naphthyl	50	[69]

Aryl(2-arylcyclopropyl)methanones 82; General Procedure:[69]
Bu$_4$NOAc (0.2 g), methyl ketone **80** (0.25 mmol), styrene derivative **81** (0.25 mmol), Pd(OAc)$_2$ (5 mol%), and Cu(OAc)$_2$ (0.25 mmol) were added to a 5-mL vial equipped with a side arm connected to an O$_2$ line, a screw cap, and a magnetic stirrer bar. The mixture was stirred and heated at 100 °C for 6 to 8 h, washed with aq HCl, and extracted with EtOAc (3 × 3 mL). The combined organic layers were dried (MgSO$_4$). The solvent was re-

2.2.8 Gold-Catalyzed Cyclopropanation

moved under reduced pressure and the crude mixture was purified by chromatography on a short pad of silica gel (hexane/EtOAc).

2.2.8 Gold-Catalyzed Cyclopropanation

Alkynes can be easily activated by π-acidic carbophilic gold(I) catalysts.[70] In particular, 1,n-enynes undergo several skeletal rearrangements in the presence of gold. After side-on coordination of the gold catalyst, the alkene attacks the activated alkyne unit leading to an *endo-* or *exo-dig* cyclization. If 1,5-enynes are used, a 5-*endo-dig* mode will lead to bicyclic compounds bearing a cyclopropane.[70] The cationic intermediate can be represented by many structures ranging from a carbocation to a gold carbenoid (e.g., **83**) (Scheme 28). This gold–carbene intermediate can undergo further transformation, e.g. a 1,2-hydride shift, a second cyclopropanation, or it can be trapped by a nucleophile. The 1,2-hydride shift leads to bicyclo[3.1.0]hexene derivatives.

Scheme 28 Intramolecular 5-*endo-dig* Gold(I)-Catalyzed Cyclopropanation of 1,5-Enynes[70]

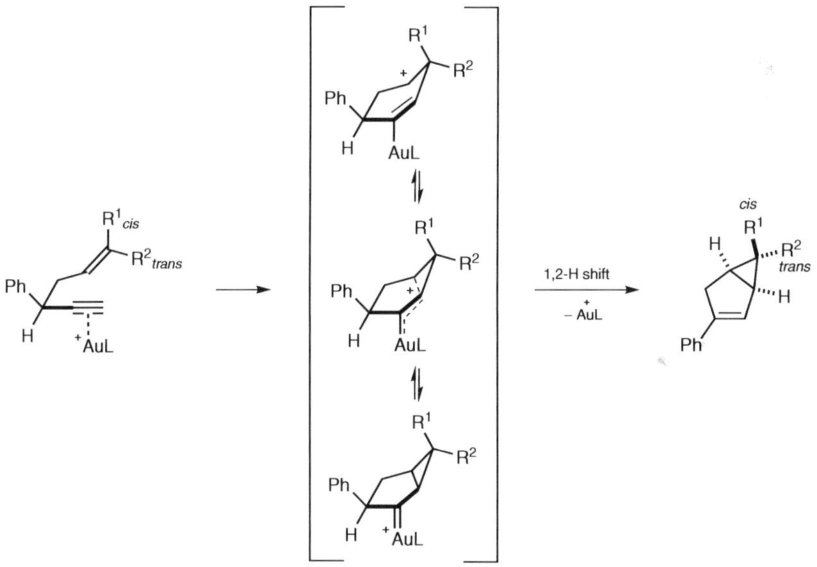

The synthesis of bicyclo[4.1.0]hept-4-enes is accomplished by an *endo* cyclization of 1,6-enynes (Scheme 29).[71] In the absence of nucleophiles the heptane derivatives are accessed by proton loss and protodemetalation.

Scheme 29 Intramolecular Cycloisomerizations of 1,6-Enynes[71]

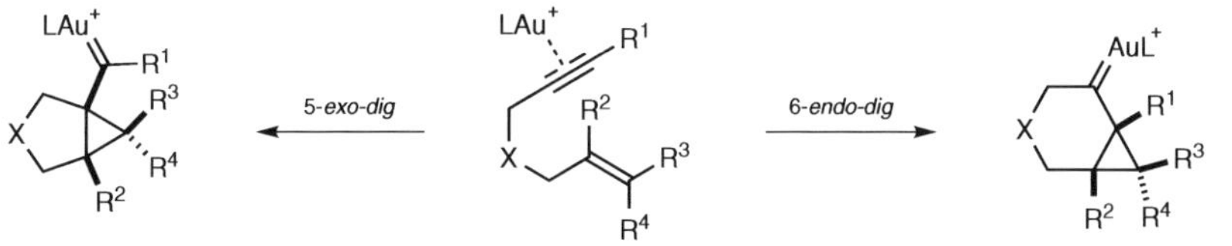

2.2.8.1 Intramolecular Gold(I)-Catalyzed Cyclopropanation Forming Bicyclo[3.1.0]hexene Derivatives

The synthesis of bicyclo[3.1.0]hexenes **85** is achieved by a (triphenylphosphine)gold(I)-catalyzed rearrangement of 1,5-enynes **84** (Scheme 30).[72] The propargylic position can be either unsubstituted or substituted with aryl or alkyl residues. The introduction of alkyl groups at the allylic position (R^3) is also tolerated, with a high diastereomeric excess obtained. Internal alkenes show high diastereomeric ratios and are transformed in good to excellent yields.

Scheme 30 Gold(I)-Catalyzed Cyclopropanation of 1,5-Enynes[72]

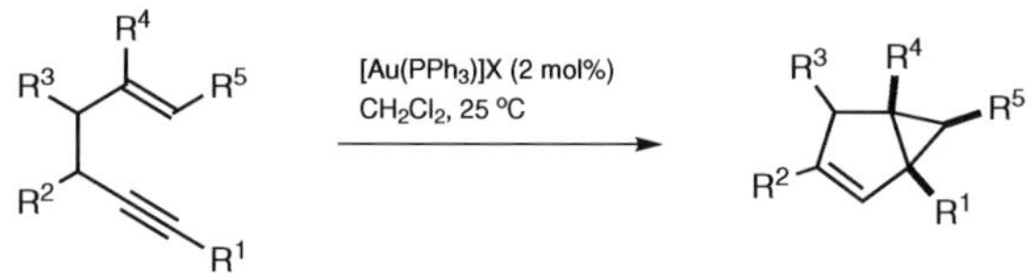

R^1	R^2	R^3	R^4	R^5	X	Yield (%)	dr	Ref
H	Ph	H	H	H	PF_6	99	–	[72]
Ph	Ph	H	H	H	SbF_6	94	–	[72]
H	Ph	H	CH_2OAc	H	SbF_6	96	–	[72]
H	H	Bn	H	H	SbF_6	82	10:1	[72]
H	H	H	H	CH_2OTIPS	SbF_6	61	>99:1	[72]
H	Bn	H	H	Pr	SbF_6	98	>99:1	[72]

Bicyclo[3.1.0]hexenes 85; General Procedure:[72]
To a 1-dram vial with a threaded cap containing a magnetic stirrer bar and a 0.5 M soln of a 1,5-enyne **84** (100 mg, 1 equiv) in CH_2Cl_2 (0.5 M) were added $AgSbF_6$ (1–5 mol%) and $AuCl(PPh_3)$ (2 mol%) sequentially. A cloudy white mixture was formed during the course of the reaction. The mixture was stirred at rt and monitored by TLC analysis. Upon completion, the mixture was filtered through a short plug of silica gel and eluted with CH_2Cl_2. The solvent was evaporated and the residue was purified by column chromatography (silica gel, hexane/Et_2O).

2.2.8.2 Gold(I)-Catalyzed Asymmetric Cycloisomerization of Heteroatom-Tethered 1,6-Enynes

An asymmetric enyne cycloisomerization reaction of nitrogen-tethered 1,6-enynes **87** gives 3-azabicyclo[4.1.0]hept-4-enes **88** in the presence of gold–phosphoramidite catalyst **86** (Scheme 31).[73] The use of phosphoramidites binding to an α,α,α′,α′-tetraaryl-1,3-dioxolane-4,5-dimethanol (TADDOL) related motif with an acyclic backbone gives great enantioselectivities as well as good to excellent yields. The transformations are independent of the N-protecting group. However, the substituent on the alkyne (R^2) has to be an aryl group for high enantiodiscrimination to be achieved.

2.2.8 Gold-Catalyzed Cyclopropanation

Scheme 31 Gold(I)-Catalyzed Cyclopropanation of Nitrogen-Tethered Enynes[73]

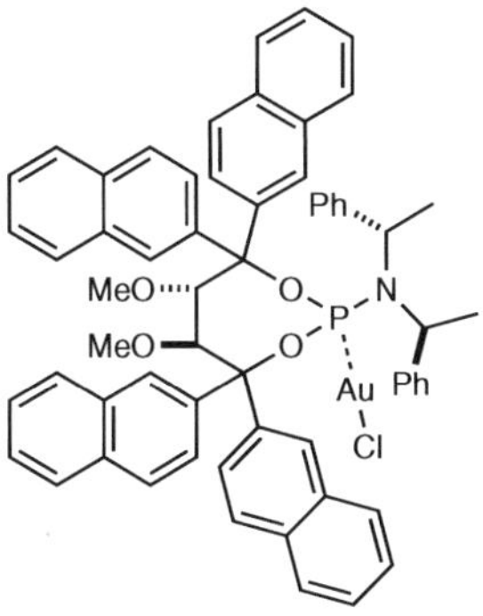

86

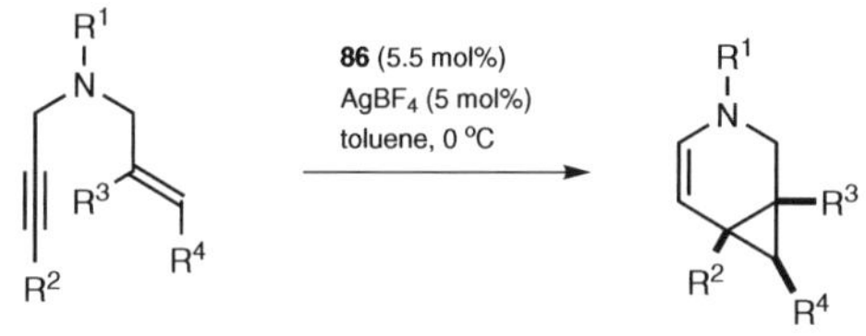

87 → **88**

Conditions: **86** (5.5 mol%), AgBF$_4$ (5 mol%), toluene, 0 °C

R^1	R^2	R^3	R^4	Yield (%)	ee (%)	Ref
Cbz	Ph	H	H	72	98	[73]
Cbz	Ph	Me	Ph	89	95	[73]
Cbz	Ph	H	H	53	96	[73]
Ts	Ph	Me	H	92	96	[73]
Ms	Ph	H	H	87	93	[73]
Ms	Ph	Me	H	92	89	[73]
CO$_2$Me	Ph	Me	H	74	93	[73]
Ms	4-BrC$_6$H$_4$	H	H	94	95	[73]
Cbz	Me	H	H	73	38	[73]

A variety of different oxygen-tethered 1,6-enynes **90** can be cyclized to form the corresponding oxabicyclo[4.1.0]hept-4-ene scaffold **91** bearing a cyclopropane ring (Scheme 32).[73] High enantiomeric excess is achieved using the chiral gold–phosphoramidite complex **89**. A disadvantage is the requirement of an aryl substituent attached to the triple bond; otherwise, the enantioselectivity decreases dramatically.

Scheme 32 Gold(I)-Catalyzed Cyclopropanation of Oxygen-Tethered Enynes[73]

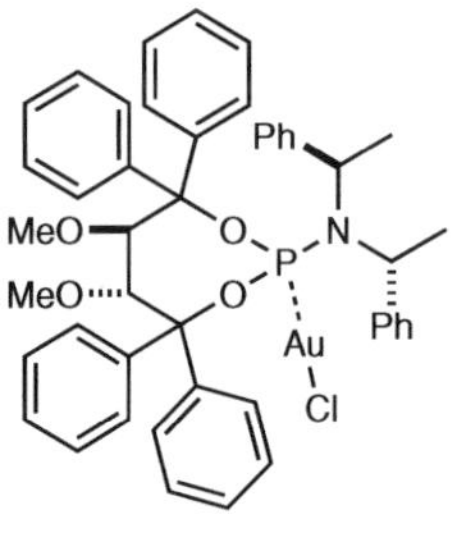

Ar¹	R¹	R²	R³	Yield (%)	ee (%)	Ref
Ph	H	H	Ph	72	83	[73]
Ph	Ph	H	Ph	64	93	[73]
Ph	Me	O(CH$_2$)$_3$		67	94	[73]
Ph	Ph	O(CH$_2$)$_3$		77	99	[73]
4-Tol	Ph	O(CH$_2$)$_3$		74	99	[73]
4-MeOC$_6$H$_4$	Ph	O(CH$_2$)$_3$		74	99	[73]
4-BrC$_6$H$_4$	Ph	O(CH$_2$)$_3$		69	99	[73]

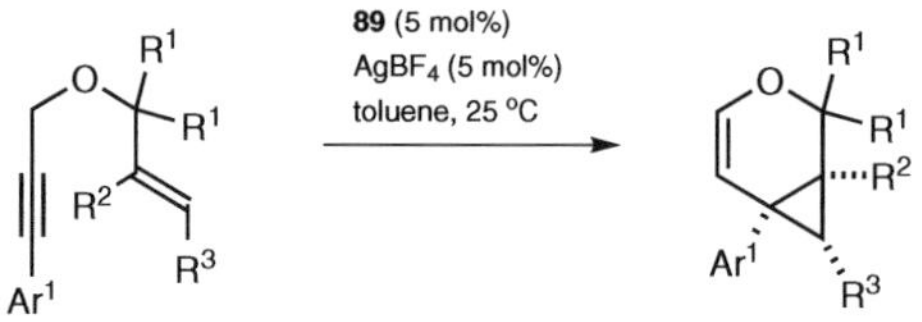

Methyl (1R,6R)-1-Methyl-6-phenyl-3-azabicyclo[4.1.0]hept-4-ene-3-carboxylate (88, R¹ = CO$_2$Me; R² = Ph; R³ = Me; R⁴ = H); Typical Procedure:[73]

A mixture containing the gold(I)–phosphoramidite complex **86** (6.3 mg, 5.5 μmol, 5.5 mol%) and AgBF$_4$ (1.0 mg, 5.0 μmol, 5.0 mol%) in toluene (1 mL) was stirred for 10 min at rt and 5 min at 0 °C in a capped vial before it was transferred to a cold (0 °C) soln of methyl (2-methylallyl)(3-phenylprop-2-ynyl)carbamate (**87**, R¹ = CO$_2$Me; R² = Ph; R³ = Me; R⁴ = H; 24.3 mg, 0.10 mmol) in toluene (1 mL) via a cannula equipped with a filter to retain the precipitates. The resulting colorless soln was stirred until GC or HPLC analysis showed complete conversion of the substrate. At this point, the soln was loaded on top of a silica gel column and the product was purified by flash column chromatography (hexane/ t-BuOMe 19:1 to 9:1) to give the product as a colorless liquid; yield: 18 mg (74%); 93% ee.

2.2.8.3 Intermolecular Gold-Catalyzed Cyclopropanation

Propargylic esters **92** and alkenes **93** can be transformed into the corresponding vinylcyclopropanes **94** in good yield and high diastereoselectivity by a (triphenylphosphine)-gold(I)-catalyzed cyclopropanation (Table 2).[74]

2.2.8 Gold-Catalyzed Cyclopropanation

Table 2 Intermolecular Gold(I)-Catalyzed Cyclopropanation of Alkenes and Propargylic Esters[74]

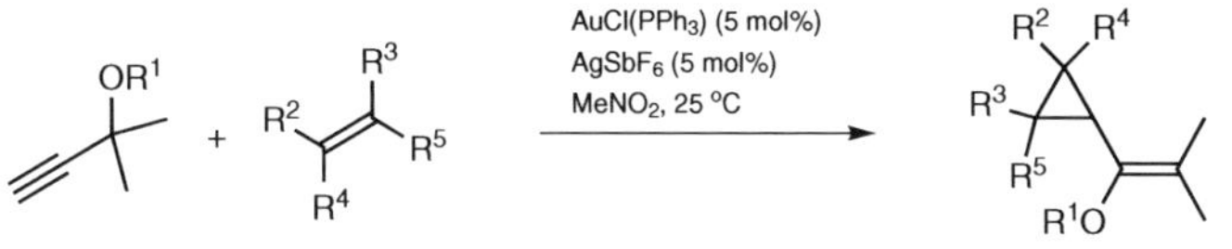

Substrates		Product	Yield (%)	Ratio (cis/trans)	Ref
Alkyne	Alkene				
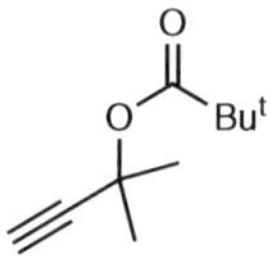 *tert*-butyl carbonate propargyl ester	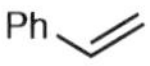 Ph–CH=CH₂	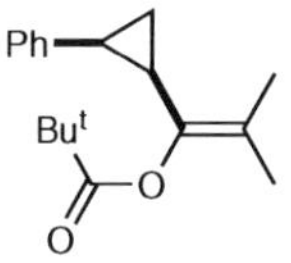 cyclopropane product	74	6:1	[74]
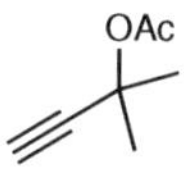 OAc propargyl ester	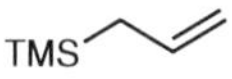 TMS-allyl	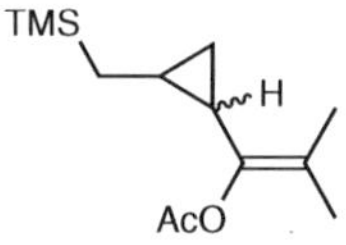 TMS cyclopropane product	62	1.3:1	[74]
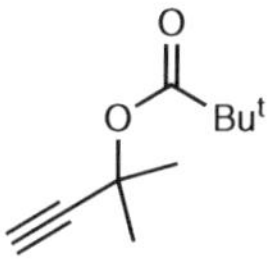 *tert*-butyl carbonate propargyl ester	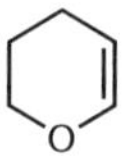 3,4-dihydro-2H-pyran	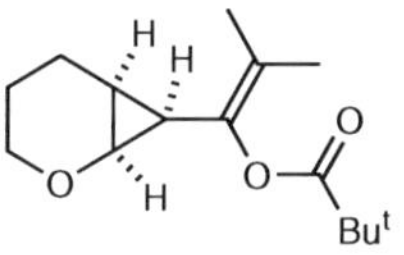 bicyclic product	61	>20:1	[74]
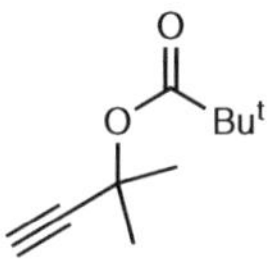 *tert*-butyl carbonate propargyl ester	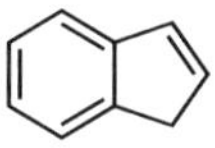 indene	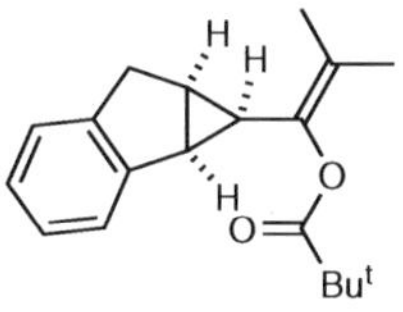 fused bicyclic product	68	>20:1	[74]
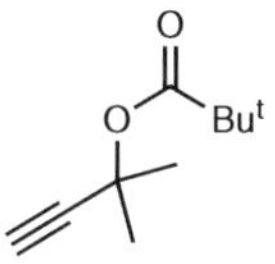 *tert*-butyl carbonate propargyl ester	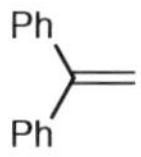 1,1-diphenylethylene	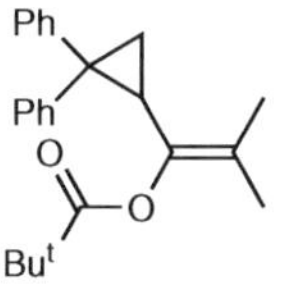 Ph,Ph cyclopropane product	73	–	[74]

Table 2 (cont.)

Substrates		Product	Yield (%)	Ratio (cis/trans)	Ref
Alkyne	Alkene				
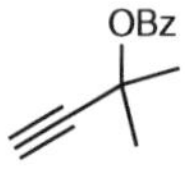 (OBz propargyl)	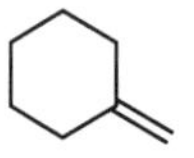 methylenecyclohexane	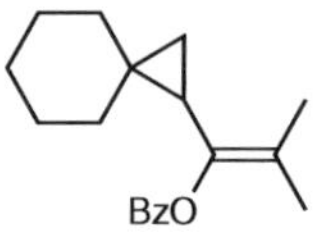 spiro product (BzO)	73	–	[74]
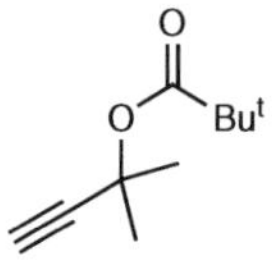 (pivaloyl propargyl, But)	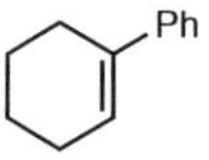 phenylcyclohexene (Ph)	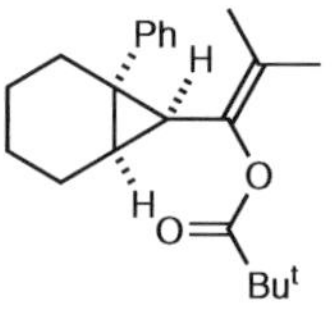 bicyclic cyclopropane (Ph, H, But)	84	5:1	[74]
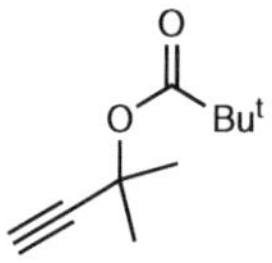 (pivaloyl propargyl, But)	 isobutylene	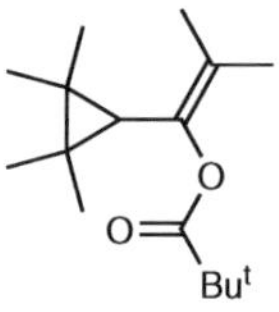 vinylcyclopropane (But)	67	–	[74]

Mechanistically, one assumes the formation of a gold carbenoid **95** concomitant to a 1,2-migration of the OR¹ residue. The *cis* selectivity observed in this reaction is in accordance with the stereochemical model of the transition state **96** proposed for this cyclopropanation reaction involving carbene transfer from the gold carbenoid. The phenyl moiety of the styrene evades the gold bearing the bulky ligand (Scheme 33).

Scheme 33 Mechanistic Hypothesis for the Intermolecular Gold(I)-Catalyzed Cyclopropanation of Alkenes and Propargylic Esters[74]

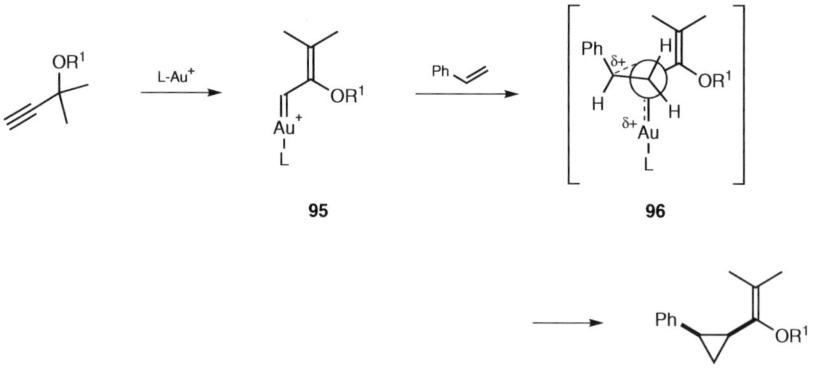

Vinylcyclopropanes 94; General Procedure:[74]

CAUTION: *Nitromethane is flammable, a shock- and heat-sensitive explosive, and an eye, skin, and respiratory tract irritant.*

AgSbF$_6$ (5 mol%), AuCl(PPh$_3$) (5 mol%), and MeNO$_2$ (0.1 M based on the propargyl ester) were added to a 1-dram vial with a threaded cap. After allowing the mixture to sit without stirring for 10 min, the alkene **93** (4 equiv) was added, followed by a soln of propargyl ester **92** (1 equiv) in MeNO$_2$ (0.1 M). The reaction was monitored by TLC until all starting material had been consumed (20–40 min). The mixture was concentrated and the residue

2.2.8.4 Gold(I)-Catalyzed Intramolecular Biscyclopropanation of Dienynes

Alkynophilic gold(I) complexes catalyze the biscyclopropanation of dienynes **97** to form complex tetracyclic skeletons **98** (Table 3).[75] Platinum(II) promotes similar biscyclopropanation reactions, but much less efficiently.[76] Starting from dienynes **97**, the tetracyclization leads to a single diastereomer. These tetracycles possess the same carbon skeleton as myliol, an unusual sesquiterpene isolated from the liverwort *Mylia taylorii*.[77,78]

Table 3 Gold(I)-Catalyzed Intramolecular Biscyclopropanation of Dienynes[75]

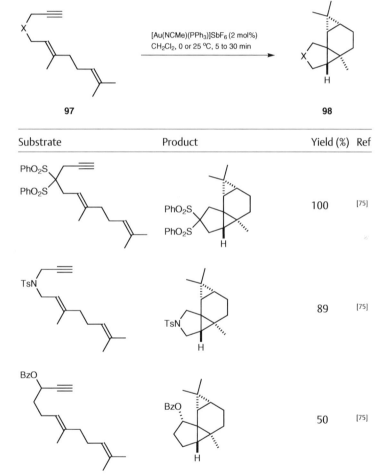

Table 3 (cont.)

Substrate	Product	Yield (%)	Ref
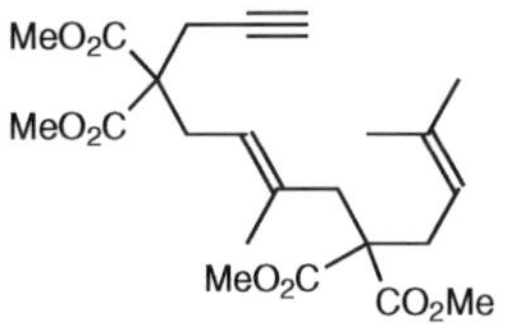 [dienyne structure with MeO$_2$C, MeO$_2$C, CO$_2$Me groups]	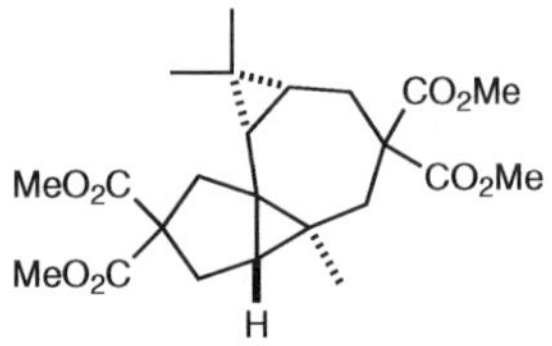 [tetracyclic product with MeO$_2$C, MeO$_2$C, CO$_2$Me, CO$_2$Me]	88	[75]
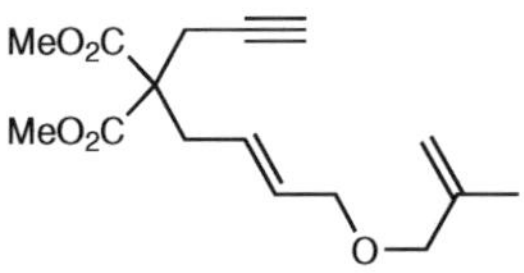 [dienyne structure with MeO$_2$C, MeO$_2$C, allyl ether]	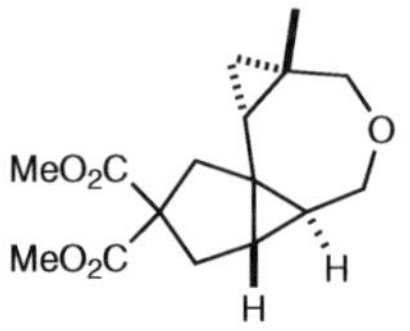 [tetracyclic product with MeO$_2$C, MeO$_2$C, O]	56	[75]

Tetracycles 98; General Procedure:[75]
A dienyne **97** (0.10–0.50 mmol) in CH$_2$Cl$_2$ (1 mL) was added to a mixture of [Au(NCMe)(PPh$_3$)]SbF$_6$ (2 mol%) in CH$_2$Cl$_2$ (2 mL) and the mixture was stirred for 5–30 min at 0 °C or rt. The resulting mixture was filtered through silica gel and the solvent was evaporated to give the corresponding product, which was purified by column chromatography (EtOAc/hexane mixtures).

2.2.8.5 Gold(I)-Catalyzed Intermolecular Cyclopropanation To Give Functionalized Benzonorcaradienes

The intermolecular transformation of diynyl ester **99** and vinylarenes **100** leads to functionalized benzonorcaradienes **101** (Scheme 34).[79] This cationic-gold(I)-catalyzed domino cyclopropanation/hydroarylation reaction produces formal (4 + 3)-annulation products with excellent regiocontrol.

Scheme 34 Gold(I)-Catalyzed Intermolecular Cyclopropanation To Give Functionalized Benzonorcaradienes[79]

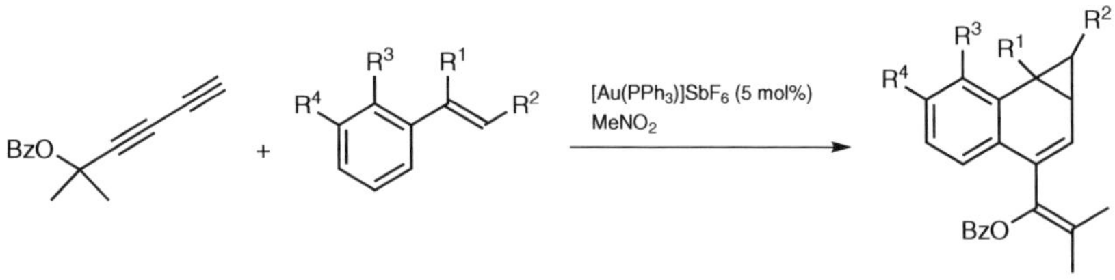

[Scheme showing BzO-substituted diyne **99** + vinylarene **100** with R^1, R^2, R^3, R^4 substituents, reacting with [Au(PPh$_3$)]SbF$_6$ (5 mol%) in MeNO$_2$ to give benzonorcaradiene **101**]

R^1	R^2	R^3	R^4	Temp (°C)	Yield (%)	Ref
H	H	Me	H	25	82	[79]
H	H	(CH=CH)$_2$		25	74	[79]
Ph	H	H	H	50	71	[79]
(CH$_2$)$_4$		H	H	25	60	[79]

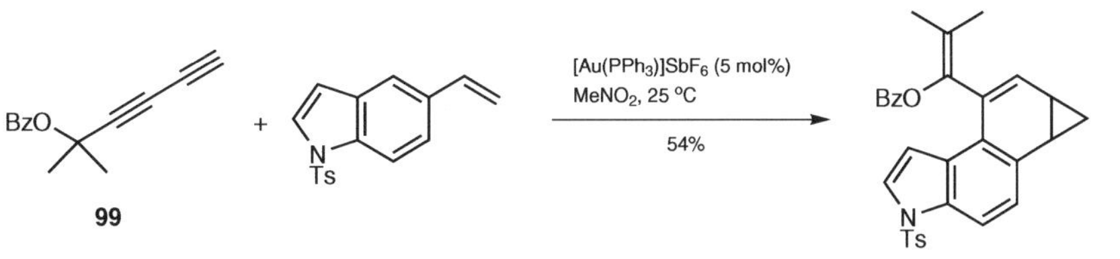

2.2.8 Gold-Catalyzed Cyclopropanation

Benzonorcaradienes 101; General Procedure:[79]

> **CAUTION:** *Nitromethane is flammable, a shock- and heat-sensitive explosive, and an eye, skin, and respiratory tract irritant.*

The gold catalyst was generated in a 1-dram vial with a threaded cap by addition of AgSbF$_6$ (5 mol%), AuCl(PPh$_3$) (5 mol%), and MeNO$_2$ (0.4 M based on the diynyl ester **99**). After allowing the catalyst mixture to sit without stirring for 10 min, a soln of the diynyl ester **99** (1 equiv) and the alkene **100** (4 equiv) in MeNO$_2$ (0.4 M) was added. The resulting mixture (0.2 M) was left overnight. Analysis by TLC generally indicated complete consumption of starting material within minutes, and subsequent conversion of the enynecyclopropane into the benzonorcaradiene **101** over several hours. The mixture was loaded directly onto a column (silica gel), and the product was eluted with hexanes/EtOAc.

2.2.8.6 Cyclopropanation with Gold(I) Carbenes by Retro-Buchner Reaction from Cycloheptatrienes

Cationic gold(I) catalysts (e.g., **104**) offer an elegant access to 1,2,3-trisubstituted cyclopropanes **105** from 1,2-disubstituted alkenes **103** and 7-substituted 1,3,5-cycloheptatrienes **102** (Scheme 35).[80] This catalytic process is a safe alternative to the use of explosive diazo compounds as the carbene precursor. 7-Substituted 1,3,5-cycloheptatrienes are easily prepared by a one-step procedure using an organometallic reagent and commercially available tropylium tetrafluoroborate.

Scheme 35 Cyclopropanation with Gold(I) Carbenes by Retro-Buchner Reaction from Cycloheptatrienes[80]

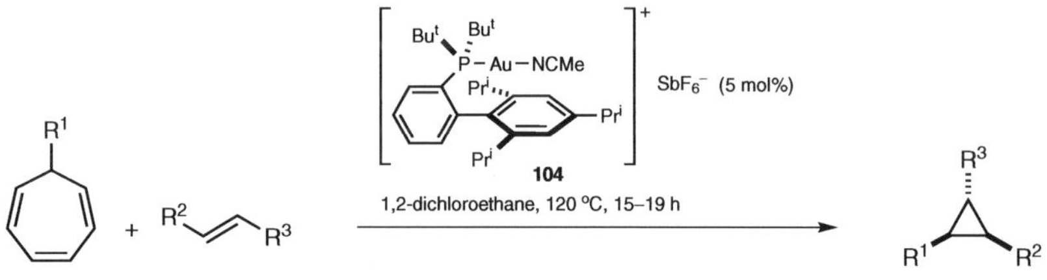

R¹	R²	R³	Yield (%)	dr	Ref
Ph	Ph	Ph	84	–	[80]
4-ClC$_6$H$_4$	Ph	Ph	91	–	[80]
4-BrC$_6$H$_4$	Ph	Ph	82	–	[80]
4-MeOC$_6$H$_4$	Ph	Ph	67	–	[80]
2-naphthyl	Ph	Ph	44[a]	–	[80]
4-ClC$_6$H$_4$	H	4-Tol	70	9:1	[80]
4-BrC$_6$H$_4$	H	3-Tol	81	9:1	[80]
4-FC$_6$H$_4$	H	3-O$_2$NC$_6$H$_4$	69	11:1	[80]
4-MeOC$_6$H$_4$	H	3-O$_2$NC$_6$H$_4$	77	4.5:1	[80]
4-FC$_6$H$_4$	Ph	Me	62	4:1	[80]
(E)-CH=CHPh	4-FC$_6$H$_4$	Ph	62	1.3:1	[80]

[a] 94% based on recovered starting material.

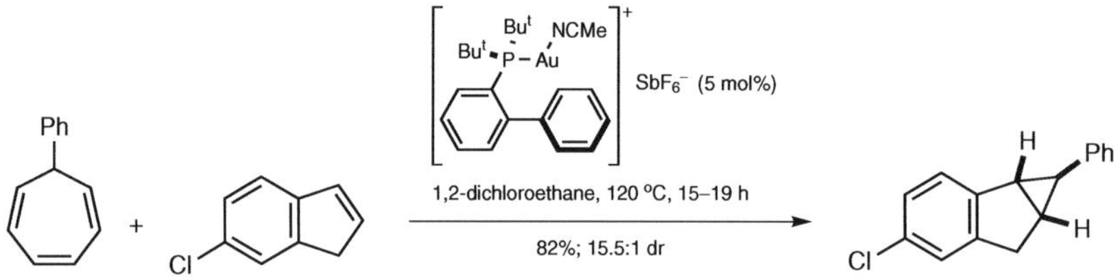

Mechanistically, the cationic gold(I) catalyst promotes the retro-Buchner reaction of the 7-substituted 1,3,5-cycloheptatriene, leading to a gold(I) carbene (Scheme 36).[80] Two C—C bonds of a norcaradiene, which are in equilibrium with the cycloheptatrienes, are cleaved by the gold(I) catalyst.

Scheme 36 Mechanism of the Cyclopropanation with Gold(I) Carbenes by Retro-Buchner Reaction[80]

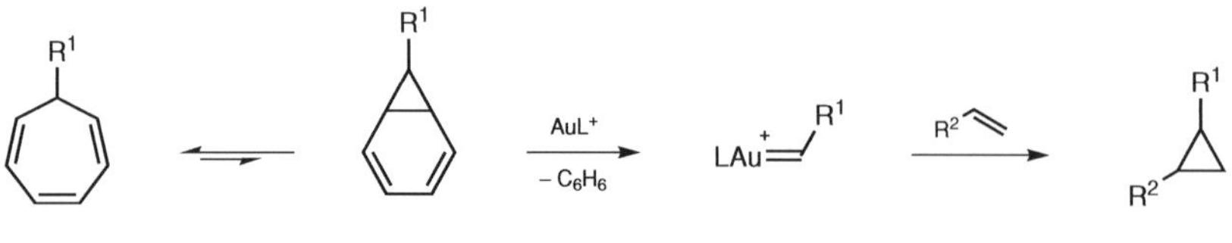

2.2.9 Platinum-Catalyzed Cyclopropanation

Di- or Trisubstituted Cyclopropanes 105; General Procedure:[80]

CAUTION: *These reactions are carried out in a sealed tube at a temperature higher than the boiling point of 1,2-dichloroethane. For larger-scale reactions, the appropriate safety precautions should be undertaken.*

An oven-dried Schlenk tube possessing a Teflon screw valve was charged with cycloheptatriene **102** (1.0 equiv), alkene **103** (1.5–2.0 equiv), and catalyst **104** (5 mol%). 1,2-Dichloroethane (0.5 M) was added using a syringe, and the Schlenk tube was sealed. The mixture was heated to 120 °C until the cycloheptatriene **102** had been completely consumed (15–19 h). The mixture was then allowed to cool to rt and filtered through a thin pad of Celite. The filtrate was concentrated under reduced pressure and the crude material was purified by flash column chromatography (silica gel) to yield the pure product.

2.2.9 Platinum-Catalyzed Cyclopropanation

The soft noble metal cation platinum(II) activates alkynes and alkenes for attack by (tethered) nucleophiles including arenes, alkenes, or heteroatom moieties. The similarities between gold and platinum catalysts with respect to cyclopropanation reactions are huge.[81] However, in contrast to its neighbor in the periodic table, platinum catalyzes the cyclizations of dienes in an *exo-* or *endo-trig* fashion.

2.2.9.1 Platinum(II)-Catalyzed Cyclopropanation of Dienes

In contrast to gold-catalyzed cycloisomerization reactions, platinum(II) catalysts cyclize dienes (e.g., **106**) to afford bicyclic targets (e.g., **107**) (Scheme 37).[82] To date, the exact mechanism of the cyclization to the bicyclo[3.1.0]alkanes has not been clarified. Both, 5-*exo-* and 6-*endo-trig* pathways are possible. The transformation of β-citronellene leads to α-thujone in 72% yield and with a diastereomeric ratio of 57:1.

Scheme 37 Cycloisomerization of Dienes with [Bis(diphenylphosphino)methane](trimethylphosphine)platinum(II) Tetrafluoroborate[82]

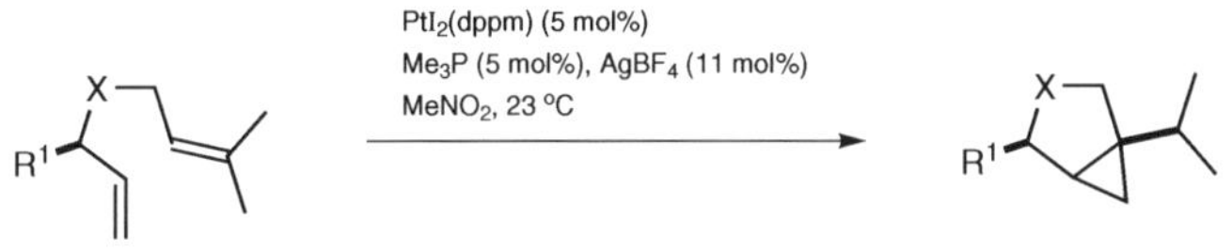

X	R^1	Time (h)	Yield (%)	dr	Ref
CH$_2$	Me	1.5	72	57:1	[82]
CH$_2$	H	3	70	–	[82]
(CH$_2$)$_2$	H	19	62	–	[82]
C(CO$_2$Me)$_2$	H	3	58	4.5:1	[82]
C(SO$_2$Ph)$_2$	H	1	66	5.8:1	[82]
C(O)	H	45	64	–	[82]

Bicyclo[3.1.0]hexanes 107; General Procedure:[82]

CAUTION: *Nitromethane is flammable, a shock- and heat-sensitive explosive, and an eye, skin, and respiratory tract irritant.*

CAUTION: *Trimethylphosphine is pyrophoric and has a very unpleasant odor.*

Me$_3$P (1 equiv) was added to a suspension of PtI$_2$(dppm) (13 µmol) in MeNO$_2$ (0.06 M) in a glass scintillation vial. The suspension was stirred until the PtI$_2$(dppm) completely dissolved. Diene **106** (20 equiv) was added, followed by AgBF$_4$ (5.6 mg, 29 µmol). The soln was stirred in the dark for several hours, and then the reaction was quenched by addition of MeNO$_2$. The product was extracted with Et$_2$O. The organic layer was washed with H$_2$O several times to remove MeNO$_2$, dried (MgSO$_4$), filtered, and concentrated under reduced pressure. The residue was purified by column chromatography (silver-doted silica gel).

2.2.9.2 Platinum-Catalyzed Enyne Cyclization and Acid-Catalyzed Ring-Opening Reaction

1,6-Enynes react with platinum(II) chloride as a catalyst in nonpolar solvents to give bicyclo[4.1.0]heptene derivatives (Scheme 38).[83] The cyclopropanation likely proceeds via a platinum carbene as intermediate. The use of enol ethers as the alkene unit (e.g., **108**) leads to bicyclic compounds **109** bearing an alkoxy substituent next to R^2. The reaction runs with total regiocontrol by 6-*endo*-dig cyclization and complete stereoselectivity. Treatment of substrates **109** with acid leads to 3,4-dihydro-2*H*-1-benzopyrans (**110**, n = 2) or 2,3-dihydrobenzo[*b*]furans **110**, n = 1).

Scheme 38 Platinum-Catalyzed Enyne Cyclization and Acid-Catalyzed Ring-Opening Reaction[83]

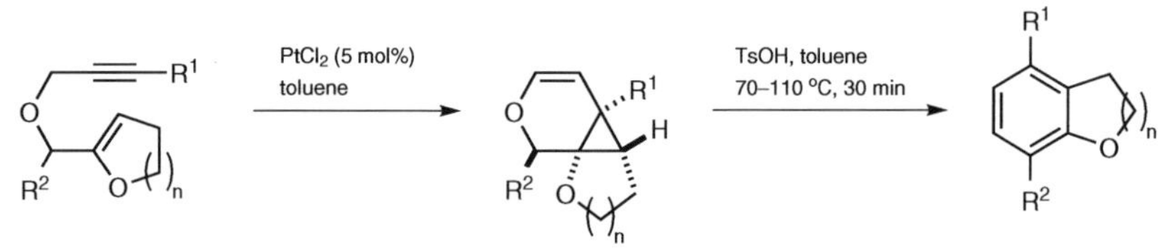

n	R^1	R^2	Temp (°C)	Time (h)	Yield (%) of **109**	Ref
2	Me	Ph	80	6	84	[83]
2	Me	1-naphthyl	50	2	97	[83]
2	Ph	Ph	80	12	72	[83]
2	Me	CH=CHPh	0	17	61	[83]
2	Me	iPr	80	4	74	[83]
1	Me	Ph	50	24	58	[83]
1	Et	4-Tol	80	1	70	[83]
1	Me	*t*-Bu	70	1.5	71	[83]

2,3,3a,3b-Tetrahydro-7*H*-furo[2′,3′:1,3]cyclopropa[1,2-*c*]pyrans 109 (n = 1); General Procedure:[83]

A mixture of an enyne **108** (n = 1; 0.5 mmol, 1 equiv) and PtCl$_2$ (0.05 equiv) was dissolved in toluene (2.5 mL). The soln was warmed under the stated conditions and then filtered through a short pad of Celite. The solvent was removed under reduced pressure and the

2.2.9.3 Platinum(II)-Catalyzed Formation of Bicyclic Alkylidenecyclopropanes

Complexes of platinum(II) with secondary phosphine oxides (e.g., **113**) catalyze the cyclopropanation between norbornene derivatives **111** and phenylacetylene (**112**) to provide bicyclic alkylidenecyclopropanes **114** (Scheme 39).[84] Diaza- and oxazanorbornene substrates can also be used which are very attractive synthetic building blocks. After cyclopropanation, reductive cleavage of the N—N or N—O bond offers a fast access to functionalized bicyclo[3.1.0]hexanes.

Scheme 39 Platinum(II)-Catalyzed Synthesis of Bicyclic Alkylidenecyclopropanes[84]

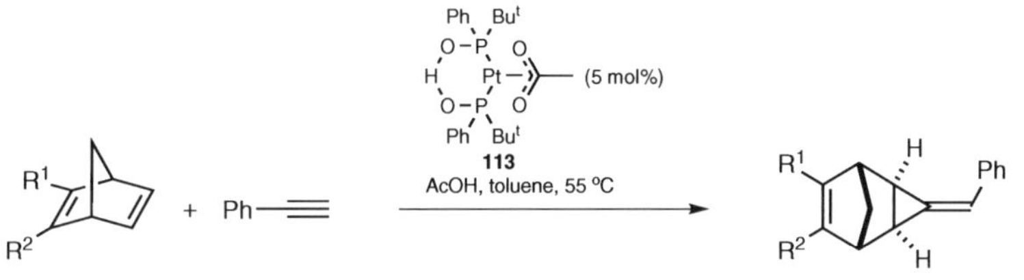

R^1	R^2	Yield (%)	Ref
H	H	81	[84]
(CH=CH)$_2$		68	[84]
CO$_2$Me	CO$_2$Me	82	[84]
CH$_2$OCO$_2$Me	CH$_2$OCO$_2$Me	85	[84]

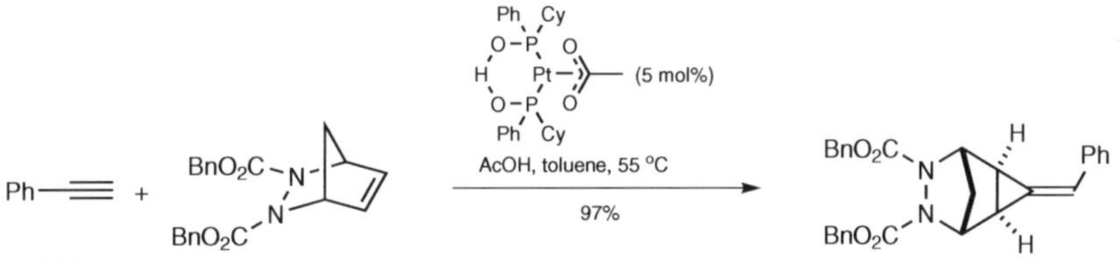

Bicyclic Alkylidenecyclopropanes 114; General Procedure:[84]
Pt complex **113** (0.025 mmol, 5 mol%) was dissolved in anhyd and degassed toluene (1 mL) in a 10-mL flame-dried Schlenk tube under argon. A soln of phenylacetylene (**112**; 1.0 mmol) and a norbornene derivative **111** (0.5 mmol) in anhyd and degassed toluene

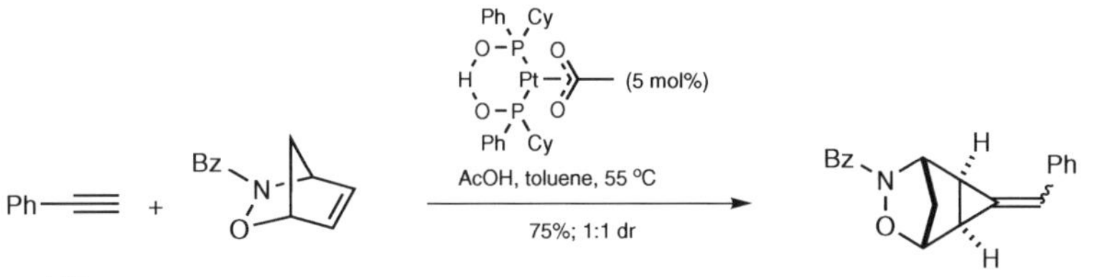

(1.5 mL) and AcOH (0.5 mmol) were added successively. The resulting mixture was stirred at 55 °C for 14–60 h. The solvent was evaporated under reduced pressure and the residue was purified by flash column chromatography.

2.2.9.4 Platinum(II) Chloride Catalyzed Cycloisomerization of 5-En-1-yn-3-ol Precursors

Acylated 5-en-1-yn-3-ols (e.g., **115**) are highly versatile substrates for platinum(II) chloride catalyzed cycloisomerization reactions (Scheme 40).[85] After electrophilic activation of the alkyne moiety, an O-acyl migration takes place leading to bicyclic unsaturated esters **116**. The cycloisomerization products can be converted into bicyclo[3.1.0]hexan-2-ones **117** by treatment with sodium hydroxide.

Scheme 40 Platinum(II) Chloride Catalyzed Cycloisomerizations of 5-En-1-yn-3-ol Precursors[85]

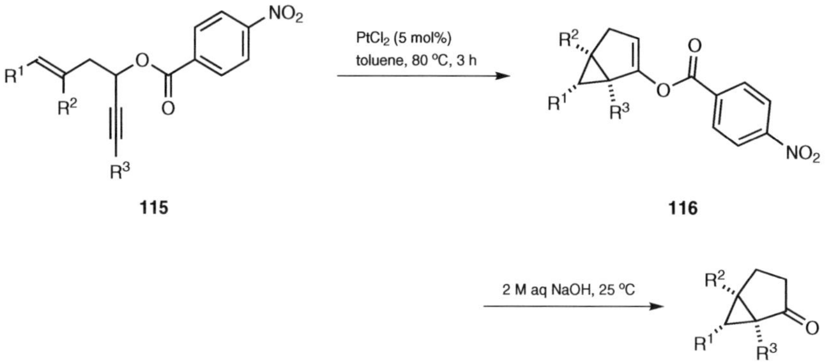

R^1	R^2	R^3	Yield (%) of **116**	Ref
H	H	H	85	[85]
Ph	H	H	86	[85]
H	Me	CO_2Me	50	[85]

Bicyclo[3.1.0]hex-2-en-2-yl 4-Nitrobenzoates 116; General Procedure:[85]
A 0.025 M soln of an enyne **115** in toluene was submitted to argon bubbling for 15 min. $PtCl_2$ (5 mol%) was then added and the mixture was heated at 80 °C for 3 h. The solvent was evaporated and the crude product was purified by flash column chromatography (silica gel, hexane/EtOAc).

2.2.10 Titanium-Catalyzed Cyclopropanation

Kulinkovich and co-workers have developed a titanium-catalyzed synthesis of 1-alkylcyclopropan-1-ols that proceeds in good to excellent yields.[86] Treatment of a carboxylic acid ester **118** with a catalytic amount of titanium(IV) isopropoxide and an excess of ethylmagnesium bromide leads to the desired cyclopropanols **119** (Scheme 41). The reaction tolerates alkenyl, cycloalkyl, and aryl carboxylates as well as carboxylates bearing halogens, acetals, or phosphoryl groups (Scheme 41).[87–96] Diesters and even triesters can be converted into bis- and triscyclopropanols, respectively. The reaction runs at ambient temperature under mild conditions.

2.2.10 Titanium-Catalyzed Cyclopropanation

There are several important modifications of this procedure, such as the Kulinkovich–de Meijere reaction,[87] which leads to cyclopropylamines from amides, and the Kulinkovich–Szymoniak reaction,[97] which leads to cyclopropylamines from nitriles. In both of these cases, best results are obtained using a stoichiometric amount of titanium(IV) isopropoxide. For that reason, these two cyclopropylamine syntheses are not discussed in detail.

Scheme 41 Titanium-Catalyzed Synthesis of Cyclopropanols[87–96]

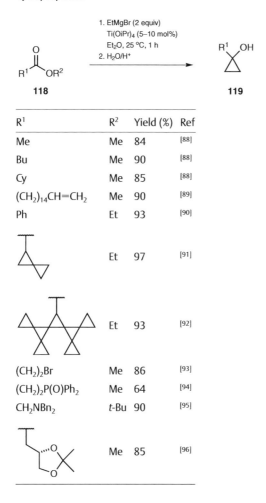

R^1	R^2	Yield (%)	Ref
Me	Me	84	[88]
Bu	Me	90	[88]
Cy	Me	85	[88]
$(CH_2)_{14}CH=CH_2$	Me	90	[89]
Ph	Et	93	[90]
(cyclopropyl-substituted)	Et	97	[91]
(bis-cyclopropyl-substituted)	Et	93	[92]
$(CH_2)_2Br$	Me	86	[93]
$(CH_2)_2P(O)Ph_2$	Me	64	[94]
CH_2NBn_2	t-Bu	90	[95]
(acetal-substituted)	Me	85	[96]

The mechanism of the Kulinkovich reaction begins with the dialkylation of titanium(IV) isopropoxide with ethylmagnesium bromide to form a diethyltitanium intermediate **120**, which liberates ethane to give the titanacyclopropane **121** (Scheme 42).[98] The subsequent addition of a carboxylic ester and ethylmagnesium bromide leads to the formation of the first C—C bond and builds up an oxatitanacyclopentane "ate"-complex **122**. The second C—C bond formation results in a titanium cyclopropoxide intermediate **123**. Alkylation at the metal center by ethylmagnesium bromide regenerates the active titanium catalyst and forms the magnesium salt of the desired product. Aqueous workup affords the desired cyclopropanol.

Scheme 42 Mechanism of the Titanium-Catalyzed Kulinkovich Reaction[98]

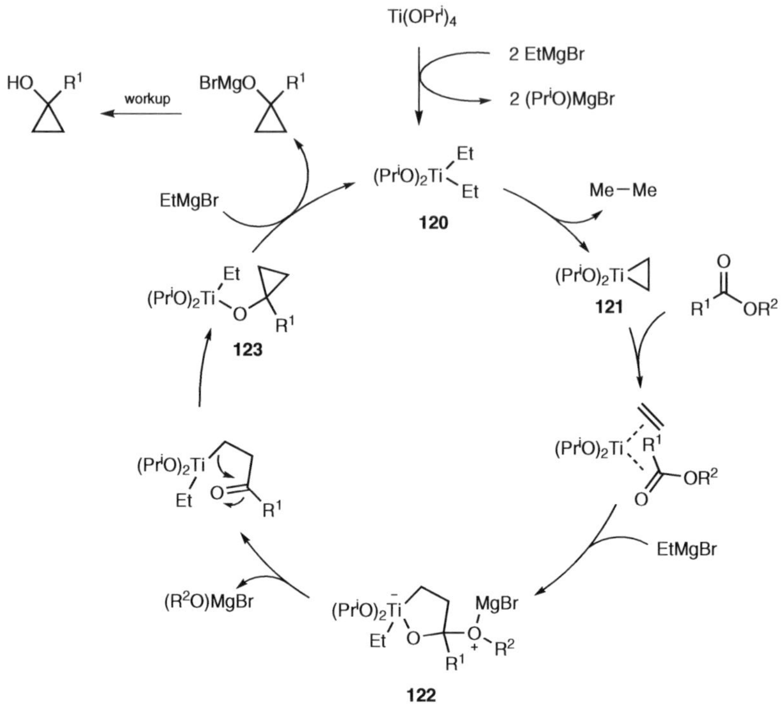

Cyclopropanols 119; General Procedure:[88]
To a stirred soln of the ester **118** (25 mmol) and Ti(OiPr)$_4$ (710 mg, 0.74 mL, 2.5 mmol, 10 mol%) in Et$_2$O (80 mL) was added a soln of EtMgBr (53 mmol) in Et$_2$O (60 mL) slowly over a period of 1 h at rt, and stirring was continued for 10 min. The mixture was poured into cooled (5 °C) 10% aq H$_2$SO$_4$ (250 mL) and extracted with Et$_2$O (3 × 50 mL). The combined Et$_2$O extracts were washed with H$_2$O (50 mL) and dried (Na$_2$SO$_4$), and the solvent was removed. The product was obtained by distillation or column chromatography (silica gel).

2.2.11 Applications in the Syntheses of Natural Products and Drug Molecules

A variety of natural products possess a cyclopropane ring as part of their complex framework. As with synthetic methods, the cyclopropane ring can be derived biosynthetically in different ways. One pathway involves methylation of alkenes using S-adenosylmethionine, followed by cyclization and proton loss. The latter two steps can also be initiated differently, without the use of S-adenosylmethionine (e.g., in the biosynthesis of chrysanthemic acid). Another possibility makes use of Wagner–Meerwein rearrangements of carbocationic intermediates (e.g., leading to terpenes of the thujane type). However, biomimetic syntheses of cyclopropane-derived natural products are almost unknown because the three-membered-ring formation event is associated with unfavorable thermodynamics and requires very specific stabilizing interactions in a carefully chosen active site of an enzyme. Therefore, metal-catalyzed methods are of crucial importance, as the examples described in the following sections demonstrate.

2.2.11 Applications in the Syntheses of Natural Products and Drug Molecules

2.2.11.1 Gibberellin GA$_{103}$ by Copper-Catalyzed Cyclopropanation

Gibberellins, which have been isolated in trace amounts from developing apple seeds, consist of a complex oligocyclic framework. A cleverly designed synthetic route reported in the 1990s by Mander and co-workers capitalized on the dearomatization of an electron-rich benzene ring by carbene insertion to form norcaradiene **125** (Scheme 43).[99] Thus, treatment of diazo ketone **124** with copper(II) acetate affords a copper carbenoid intermediate; this highly reactive species is set up for three-membered-ring formation. Further key steps on the way to the final product were a Diels–Alder reaction to afford **126** and a six- to five-membered-ring contraction by Wolff rearrangement.

Scheme 43 Copper-Catalyzed Cyclopropanation of an Electron-Rich Benzene Ring as Key Step in the Total Synthesis of Gibberellin GA$_{103}$[99]

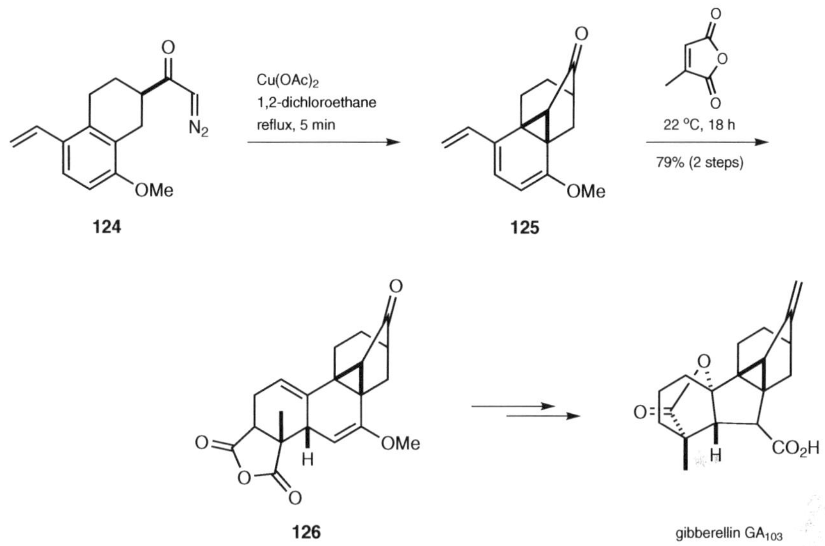

2.2.11.2 Echinopines A and B by Rhodium-Catalyzed Cyclopropanation

A complex tetracyclic sesquiterpene derived from the roots of *Echinops spinosus* has a unique [3.5.5.7] carbon framework onto which a methylene group and a carboxymethyl group are appended. An enantioselective total synthesis of this challenging target molecule starts from *tert*-butyldimethylsilyl-protected allylic cyclopentene alcohol **127** (Scheme 44).[100] Several steps lead to an α-diazo β-oxo ester **128** with an exocyclic double bond in the vicinity. An intramolecular rhodium-catalyzed cyclopropanation furnishes the tricycle **129**. Several other steps, including a samarium(II) iodide mediated ring closure as a key transformation, pave the way to the natural products echinopines A (**130**, R^1 = H) and B (**130**, R^1 = Me).

Scheme 44 Rhodium-Catalyzed Cyclopropanation as Key Step in the Total Synthesis of Echinopines A and B[100]

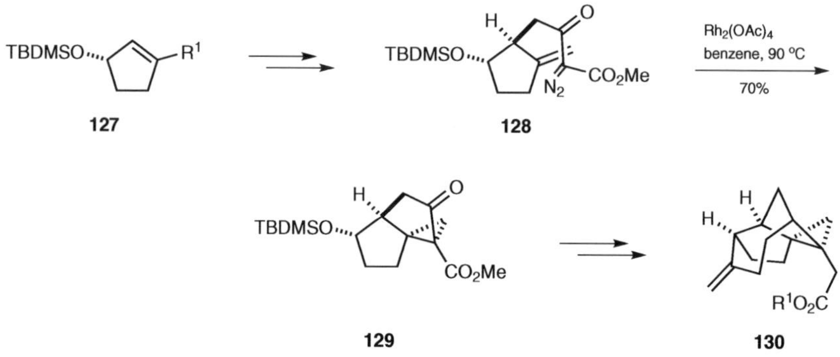

2.2.11.3 Antidepressive Agent GSK 1360707 by Gold-Catalyzed Cyclopropanation

The preparation of azabicyclo[4.1.0]hept-4-enes described in Section 2.2.8.2 has been utilized as part of an important synthesis of GSK 1 360 707 (**133**).[101] This inhibitor is currently in clinical development as a drug for the treatment of severe depression. The enantioselective synthesis of the antidepressive agent starts with a commercially available inexpensive propargylic amine. After three steps, the cycloisomerization precursor **131** is obtained with a three-step yield of 86%.[73] The use of cationic gold complex **86** and silver(I) tetrafluoroborate in toluene at 0 °C furnishes the desired product **132** in 88% yield and with an enantiomeric excess of 95% (Scheme 45). Just two more high-yielding steps lead to the drug candidate **133**.

2.2.11 Applications in the Syntheses of Natural Products and Drug Molecules

Scheme 45 Enantioselective Synthesis of the Antidepressive Agent GSK 1360707[73]

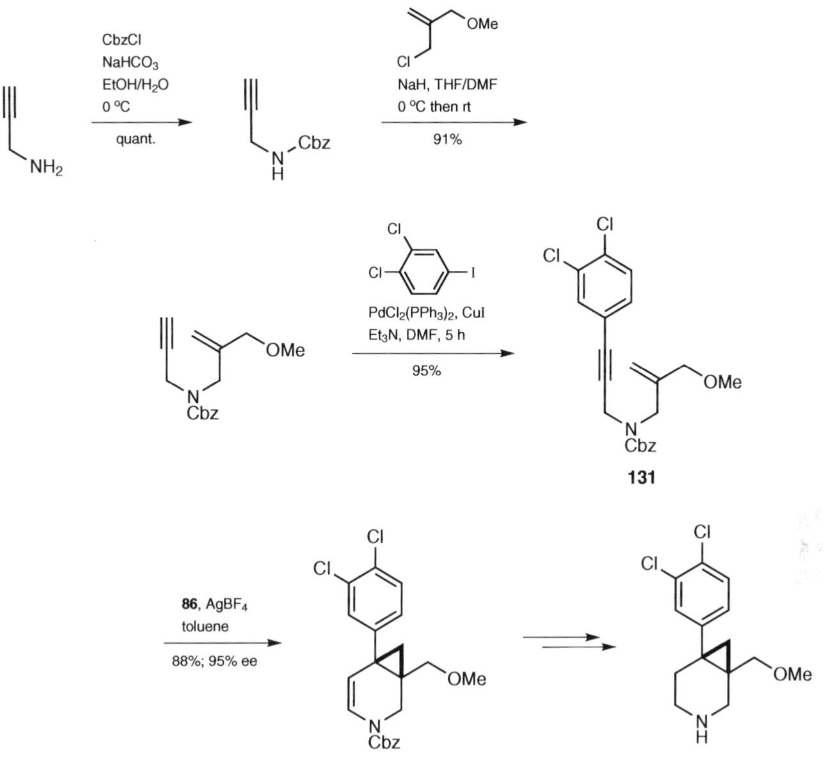

2.2.11.4 Sabina Ketone by Platinum-Catalyzed Cyclopropanation

The natural product sabina ketone (**136**) is an important intermediate for the synthesis of monoterpenes such as sabinene and sabinene hydrates.[102,103] The key step in the total synthesis of this natural product is the formation of the bicycle by a platinum-catalyzed cycloisomerization reaction of enyne **134** (Scheme 46; see also Section 2.2.9.4).[85] Cleavage of the protecting group in the resulting bicyclo[3.1.0]hex-2-en-2-ol ester **135** affords the desired product **136**.

Scheme 46 Platinum-Catalyzed Cycloisomerization To Afford the Bicyclic Skeleton of Sabinenes[85]

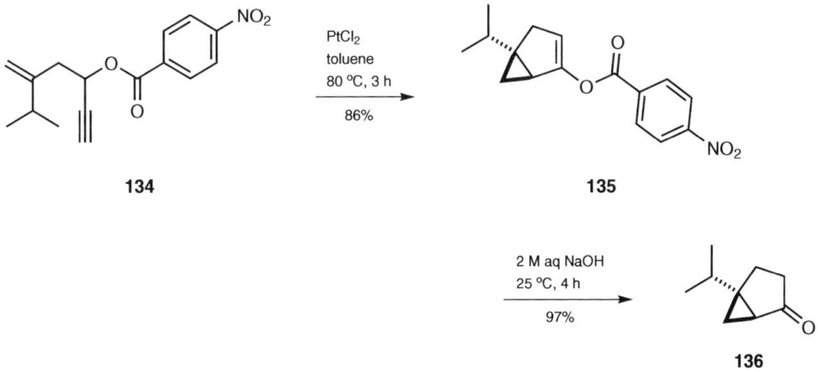

for references see p 95

2.2.11.5 (S)-Cleonin by Titanium-Catalyzed Cyclopropanation

(S)-Cleonin (**140**), an amino acid with a cyclopropanol substructure, is the key component of the very complex antitumor agent cleomycin. By making use of the Kulinkovich reaction (see Section 2.2.10) it is facile to prepare this important cyclopropane derivative from (R)-serine (**137**). Cyclopropanol **139** is formed from the methyl ester of N-benzyloxycarbonyl serine acetonide **138** under common Kulinkovich conditions employing a Grignard reagent and substoichiometric amounts of titanium(IV) isopropoxide (Scheme 47).[104] Acetonide removal, followed by oxidation of the primary alcohol, and benzyloxycarbonyl cleavage furnish the non-proteinogenic amino acid (S)-cleonin (**140**).

Scheme 47 Synthesis of (S)-Cleonin by the Kulinkovich Reaction[104]

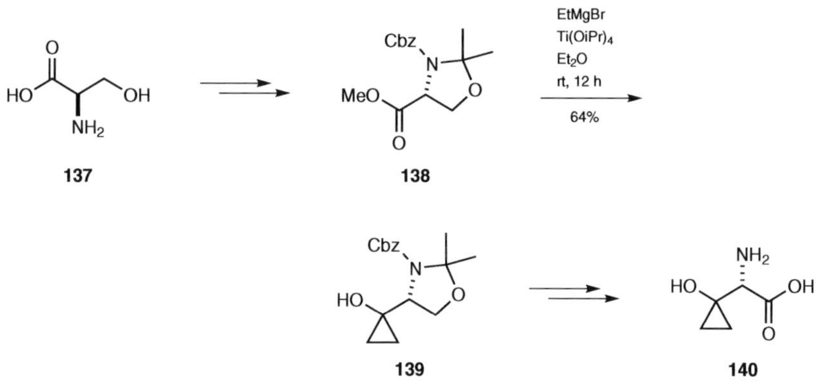

2.2.12 Conclusions and Future Perspectives

Numerous metal-catalyzed methods to prepare cyclopropanes have been developed over the past fifty years. Notably, the taming of carbenes in the form of metal–carbene complexes has catapulted three-membered-ring syntheses back to the forefront of organic synthesis over the past decades. With the advent of the use of metals in cyclopropane synthesis, chiral ligands for diastereo- and enantiodiscrimination have also come under focus and this strategy has opened new avenues to access these three-membered rings. Apart from the use of metals to catalyze the decomposition of diazo compounds, the action of carbophilic gold(I) cations interacting with alkynes, in particular, has led to amazing syntheses of the cyclopropane moiety. However, challenges in this area still remain. Many metal–carbene complexes do not react with tetrasubstituted double bonds; thus, the development of asymmetric methods for the preparation of hexasubstituted cyclopropanes would be highly desirable. Distinct deconstruction reactions of such three-membered rings could pave the way to highly substituted heterocyclic compounds or open-chain moieties. For many metals, diazo compounds are still needed to generate the reactive complex. Synthetic routes that circumvent these potentially hazardous compounds would be a great advance. An ideal synthesis of cyclopropanes, a dream reaction, would be the ring closure of alkane subunits or the respective transformation of alkenes with alkanes, both under oxidative conditions with the release of water. Of course, these concepts suffer from unfavorable thermodynamics, but this does not mean that they are generally impossible. The smallest ring, the cyclopropane, has not yet disclosed all of its synthetic secrets.

References

[1] de Meijere, A., *Angew. Chem. Int. Ed. Engl.*, (1979) **18**, 809.
[2] Reissig, H.-U.; Zimmer, R., *Chem. Rev.*, (2003) **103**, 1151.
[3] Schneider, T. F.; Kaschel, J.; Werz, D. B., *Angew. Chem. Int. Ed.*, (2014) **53**, 5504.
[4] Freund, A., *J. Prakt. Chem.*, (1882) **26**, 367.
[5] Maier, G., *Angew. Chem. Int. Ed. Engl.*, (1988) **27**, 309.
[6] *The Chemistry of the Cyclopropyl Group*, Rappoport, Z., Ed.; Wiley: Chichester, UK, (1987); Vol. 1.
[7] *The Chemistry of the Cyclopropyl Group*, Rappoport, Z., Ed.; Wiley: Chichester, UK, (1995); Vol. 2.
[8] de Meijere, A.; Kozhushkov, S. I., *Chem. Rev.*, (2000) **100**, 93.
[9] de Meijere, A.; Kozhushkov, S. I.; Schill, H., *Chem. Rev.*, (2006) **106**, 4926.
[10] Bartoli, G.; Bencivenni, G.; Dalpozzo, R., *Synthesis*, (2014) **46**, 979.
[11] Brookhart, M.; Studabaker, B., *Chem. Rev.*, (1987) **87**, 411.
[12] Fraile, J. M.; García, J. I.; Martínez-Merino, V.; Mayoral, J. A.; Salvatella, L., *J. Am. Chem. Soc.*, (2001) **123**, 7616.
[13] Cowan, D. O.; Couch, M. M.; Kopecky, K. R.; Hammond, G. S., *J. Org. Chem.*, (1964) **29**, 1922.
[14] Huisgen, R.; Binsch, G.; Ghosez, L., *Chem. Ber.*, (1964) **97**, 2628.
[15] Nozaki, H.; Moriuti, S.; Yamabe, M.; Noyori, R., *Tetrahedron Lett.*, (1966), 59.
[16] Nozaki, H.; Moriuti, S.; Takaya, H.; Noyori, R., *Tetrahedron Lett.*, (1966), 5239.
[17] Fritschi, H.; Leutenegger, U.; Pfaltz, A., *Angew. Chem. Int. Ed. Engl.*, (1986) **25**, 1005.
[18] Evans, D. A.; Woerpel, K. A.; Hinman, M. M.; Faul, M. M., *J. Am. Chem. Soc.*, (1991) **113**, 726.
[19] Kanemasa, S.; Hamura, S.; Harada, E.; Yamamoto, H., *Tetrahedron Lett.*, (1994) **35**, 7985.
[20] Schneider, T. F.; Kaschel, J.; Dittrich, B.; Werz, D. B., *Org. Lett.*, (2009) **11**, 2317.
[21] Schneider, T. F.; Kaschel, J.; Awan, S. I.; Dittrich, B.; Werz, D. B., *Chem.–Eur. J.*, (2010) **16**, 11 276.
[22] Böhm, C.; Schinnerl, M.; Bubert, C.; Zabel, M.; Labahn, T.; Parisini, E.; Reiser, O., *Eur. J. Org. Chem.*, (2000), 2955.
[23] Kaschel, J.; Schneider, T. F.; Schirmer, P.; Maaß, C.; Stalke, D.; Werz, D. B., *Eur. J. Org. Chem.*, (2013), 4539.
[24] Kaschel, J.; Schneider, T. F.; Kratzert, D.; Stalke, D.; Werz, D. B., *Org. Biomol. Chem.*, (2013) **11**, 3494.
[25] Waser, M.; Moher, E. D.; Borders, S. S. K.; Hansen, M. H.; Hoard, D. W.; Laurila, M. E.; LeTourneau, M. E.; Miller, R. D.; Phillips, M. L.; Sullivan, K. A.; Ward, J. A.; Xie, C.; Bye, C. A.; Leitner, T.; Herzog-Krimbacher, B.; Kordian, M.; Müllner, M., *Org. Process Res. Dev.*, (2011) **15**, 1266.
[26] Brand, C.; Rauch, G.; Zanoni, M.; Dittrich, B.; Werz, D. B., *J. Org. Chem.*, (2009) **74**, 8779.
[27] Moreau, B.; Alberico, D.; Lindsay, V. N. G.; Charette, A. B., *Tetrahedron*, (2012) **68**, 3487.
[28] Schmidt, C. D.; Kaschel, J.; Schneider, T. F.; Kratzert, D.; Stalke, D.; Werz, D. B., *Org. Lett.*, (2013) **15**, 6098.
[29] Doyle, M. P.; Forbes, D. C., *Chem. Rev.*, (1998) **98**, 911.
[30] Doyle, M. P.; Bagheri, V.; Wandless, T. J.; Harn, N. K.; Brinker, D. A.; Eagle, C. T.; Loh, K.-L., *J. Am. Chem. Soc.*, (1990) **112**, 1906.
[31] Hubert, A. J.; Noels, A. F.; Anciaux, A. J.; Teyssié, P., *Synthesis*, (1976), 600.
[32] Davies, H. M. L.; Morton, D., *Chem. Soc. Rev.*, (2011) **40**, 1857.
[33] Davies, H. M. L.; Denton, J. R., *Chem. Soc. Rev.*, (2009) **38**, 3061.
[34] Ventura, D. L.; Li, Z.; Coleman, M. G.; Davies, H. M. L., *Tetrahedron*, (2009) **65**, 3052.
[35] Davies, H. M. L.; Coleman, M. G.; Ventura, D. L., *Org. Lett.*, (2007) **9**, 4971.
[36] Doyle, M. P.; Zhou, Q.-L.; Simonsen, S. H.; Lynch, V., *Synlett*, (1996), 697.
[37] Anciaux, A. J.; Hubert, A. J.; Noels, A. F.; Petiniot, N.; Teyssié, P., *J. Org. Chem.*, (1980) **45**, 695.
[38] Horneff, T.; Chuprakov, S.; Chernyak, N.; Gevorgyan, V.; Fokin, V. V., *J. Am. Chem. Soc.*, (2008) **130**, 14 972.
[39] Chuprakov, S.; Kwok, S. W.; Zhang, L.; Lercher, L.; Fokin, V. V., *J. Am. Chem. Soc.*, (2009) **131**, 18 034.
[40] Lindsay, V. N. G.; Nicolas, C.; Charette, A. B., *J. Am. Chem. Soc.*, (2011) **133**, 8972.
[41] Marcoux, D.; Azzi, S.; Charette, A. B., *J. Am. Chem. Soc.*, (2009) **131**, 6970.
[42] Wang, H.; Guptill, D. M.; Varela-Alvarez, A.; Musaev, D. G.; Davies, H. M. L., *Chem. Sci.*, (2013) **4**, 2844.
[43] Doyle, M. P., *Angew. Chem. Int. Ed.*, (2009) **48**, 850.
[44] Zhu, S.; Ruppel, J. V.; Lu, H.; Wojtas, L.; Zhang, X. P., *J. Am. Chem. Soc.*, (2008) **130**, 5042.

[45] Zhu, S.; Xu, X.; Perman, J. A.; Zhang, X. P., *J. Am. Chem. Soc.*, (2010) **132**, 12796.
[46] Kanchiku, S.; Suematsu, H.; Matsumoto, K.; Uchida, T.; Katsuki, T., *Angew. Chem. Int. Ed.*, (2007) **46**, 3889.
[47] Suematsu, H.; Kanchiku, S.; Uchida, T.; Katsuki, T., *J. Am. Chem. Soc.*, (2008) **130**, 10327.
[48] Maas, G., *Chem. Soc. Rev.*, (2004) **33**, 183.
[49] Pellissier, H., *Tetrahedron*, (2008) **64**, 7041.
[50] Nishiyama, H.; Itoh, Y.; Matsumoto, H.; Park, S.-B.; Itoh, K., *J. Am. Chem. Soc.*, (1994) **116**, 2223.
[51] Che, C.-M.; Huang, J.-S.; Lee, F.-W.; Li, Y.; Lai, T.-S.; Kwong, H.-L.; Teng, P.-F.; Lee, W.-S.; Lo, W.-C.; Peng, S.-M.; Zhou, Z.-Y., *J. Am. Chem. Soc.*, (2001) **123**, 4119.
[52] Miller, J. A.; Hennessy, E. J.; Marshall, W. J.; Scialdone, M. A.; Nguyen, S. T., *J. Org. Chem.*, (2003) **68**, 7884.
[53] Chanthamath, S.; Phomkeona, K.; Shibatomi, K.; Iwasa, S., *Chem. Commun. (Cambridge)*, (2012) **48**, 7750.
[54] Trost, B. M.; Ryan, M. C.; Rao, M.; Markovic, T. Z., *J. Am. Chem. Soc.*, (2014) **136**, 17422.
[55] Morandi, B.; Carreira, E. M., *Angew. Chem. Int. Ed.*, (2010) **49**, 938.
[56] Morandi, B.; Cheang, J.; Carreira, E. M., *Org. Lett.*, (2011) **13**, 3080.
[57] Morandi, B.; Dolva, A.; Carreira, E. M., *Org. Lett.*, (2012) **14**, 2162.
[58] Seitz, W. J.; Saha, A. K.; Casper, D.; Hossain, M. M., *Tetrahedron Lett.*, (1992) **33**, 7755.
[59] Wolf, J. R.; Hamaker, C. G.; Djukic, J.-P.; Kodadek, T.; Woo, L. K., *J. Am. Chem. Soc.*, (1995) **117**, 9194.
[60] Morandi, B.; Carreira, E. M., *Science (Washington, D. C.)*, (2012) **335**, 1471.
[61] Shen, J.-J.; Zhu, S.-F.; Cai, Y.; Xu, H.; Xie, X.-L.; Zhou, Q.-L., *Angew. Chem. Int. Ed.*, (2014) **53**, 13188.
[62] Reiser, O., In *Handbook of Organopalladium Chemistry for Organic Synthesis*, Negishi, E.-i., Ed.; Wiley: New York, (2002); pp 1561–1577.
[63] Mende, U.; Radüchel, B.; Skuballa, V.; Vorbrüggen, H., *Tetrahedron Lett.*, (1975), 629.
[64] Spur, B.; Crea, A.; Peters, W., *Z. Naturforsch., B*, (1984) **39**, 125.
[65] Gangadhar, A.; Subbarao, R.; Lakshminarayana, G., *J. Am. Oil Chem. Soc.*, (1988) **65**, 601.
[66] Arvidsson, L.-E.; Johansson, A. M.; Hacksell, U.; Nilsson, J. L. G.; Svensson, K.; Hjorth, S.; Magnusson, T.; Carlsson, A.; Lindberg, P.; Andersson, B.; Sanchez, D.; Wikström, H.; Sundell, S., *J. Med. Chem.*, (1988) **31**, 92.
[67] Mao, J.; Zhang, S.-Q.; Shi, B.-F.; Bao, W., *Chem. Commun. (Cambridge)*, (2014) **50**, 3692.
[68] Liu, W.; Chen, D.; Zhu, X.-Z.; Wan, X.-L.; Hou, X.-L., *J. Am. Chem. Soc.*, (2009) **131**, 8734.
[69] Cotugno, P.; Monopoli, A.; Ciminale, F.; Milella, A.; Nacci, A., *Angew. Chem. Int. Ed.*, (2014) **53**, 13563.
[70] Qian, D.; Zhang, J., *Chem. Soc. Rev.*, (2015) **44**, 677.
[71] Jiménez-Núñez, E.; Echavarren, A. M., *Chem. Rev.*, (2008) **108**, 3326.
[72] Luzung, M. R.; Markham, J. P.; Toste, F. D., *J. Am. Chem. Soc.*, (2004) **126**, 10858.
[73] Teller, H.; Corbet, M.; Mantilli, L.; Gopakumar, G.; Goddard, R.; Thiel, W.; Fürstner, A., *J. Am. Chem. Soc.*, (2012) **134**, 15331.
[74] Johansson, M. J.; Gorin, D. J.; Staben, S. T.; Toste, F. D., *J. Am. Chem. Soc.*, (2005) **127**, 18002.
[75] Nieto-Oberhuber, C.; López, S.; Muñoz, M. P.; Jiménez-Núñez, E.; Buñuel, E.; Cárdenas, D. J.; Echavarren, A. M., *Chem.–Eur. J.*, (2006) **12**, 1694.
[76] Mainetti, E.; Mouriès, V.; Fensterbank, L.; Malacria, M.; Marco-Contelles, J., *Angew. Chem. Int. Ed.*, (2002) **41**, 2132.
[77] Nozaki, H., *J. Chem. Soc., Perkin Trans. 2*, (1979), 514.
[78] Matsuo, A.; Nozaki, H.; Nakayama, M.; Kushi, Y.; Hayashi, S.; Kamijo, N.; Benesova, V.; Herout, V., *J. Chem. Soc., Chem. Commun.*, (1976), 1006.
[79] Gorin, D. J.; Dubé, P.; Toste, F. D., *J. Am. Chem. Soc.*, (2006) **128**, 14480.
[80] Solorio-Alvarado, C. R.; Wang, Y.; Echavarren, A. M., *J. Am. Chem. Soc.*, (2011) **133**, 11952.
[81] Zhang, L.; Sun, J.; Kozmin, S. A., *Adv. Synth. Catal.*, (2006) **348**, 2271.
[82] Feducia, J. A.; Campbell, A. N.; Doherty, M. Q.; Gagné, M. R., *J. Am. Chem. Soc.*, (2006) **128**, 13290.
[83] Nevado, C.; Ferrer, C.; Echavarren, A. M., *Org. Lett.*, (2004) **6**, 3191.
[84] Bigeault, J.; Giordano, L.; de Riggi, I.; Gimbert, Y.; Buono, G., *Org. Lett.*, (2007) **9**, 3567.
[85] Harrak, Y.; Blaszykowski, C.; Bernard, M.; Cariou, K.; Mainetti, E.; Mouriès, V.; Dhimane, A.-L.; Fensterbank, L.; Malacria, M., *J. Am. Chem. Soc.*, (2004) **126**, 8656.
[86] Kulinkovich, O. G.; de Meijere, A., *Chem. Rev.*, (2000) **100**, 2789.
[87] de Meijere, A.; Kozhushkov, S. I.; Savchenko, A. I., *J. Organomet. Chem.*, (2004) **689**, 2033.
[88] Kulinkovich, O. G.; Vasilevski, D. A.; Savchenko, A. I.; Sviridov, S. V., *Russ. J. Org. Chem. (Engl. Transl.)*, (1991) **27**, 1249.

[89] Dolgopalets, V. I.; Volkov, S. M.; Kisel, M. A.; Kozhevko, A. N.; Kulinkovich, O. G., *Russ. J. Org. Chem. (Engl. Transl.)*, (1999) **35**, 1436.
[90] Kulinkovich, O. G.; Sviridov, S. V.; Vasilevski, D. A.; Pritytskaya, T. S., *Russ. J. Org. Chem. (Engl. Transl.)*, (1989) **25**, 2027.
[91] de Meijere, A.; Kozhushkov, S. I.; Spaeth, T.; Zefirov, N. S., *J. Org. Chem.*, (1993) **58**, 502.
[92] Kozhushkov, S. I.; Haumann, T.; Boese, R.; de Meijere, A., *Angew. Chem. Int. Ed. Engl.*, (1993) **32**, 401.
[93] Sviridov, S. V.; Vasilevskii, D. A.; Kulinkovich, O. G., *Russ. J. Org. Chem. (Engl. Transl.)*, (1991) **27**, 1251.
[94] Winsel, H.; Gazizova, V.; Kulinkovich, O.; Pavlov, V.; de Meijere, A., *Synlett*, (1999), 1999.
[95] Shevchuk, T. A.; Kulinkovich, O. G., *Russ. J. Org. Chem. (Engl. Transl.)*, (2000) **36**, 491.
[96] Achmatowiz, B.; Jankowski, P.; Wicha, J., *Tetrahedron Lett.*, (1996) **37**, 5589.
[97] Bertus, P.; Szymoniak, J., *Synlett*, (2007), 1346.
[98] Eisch, J. J.; Adeosun, A. A.; Gitua, J. N., *Eur. J. Org. Chem.*, (2003), 4721.
[99] King, G. R.; Mander, L. N.; Monck, N. J. T.; Morris, J. C.; Zhang, H., *J. Am. Chem. Soc.*, (1997) **119**, 3828.
[100] Nicolaou, K. C.; Ding, H.; Richard, J.-A.; Chen, D. Y.-K., *J. Am. Chem. Soc.*, (2010) **132**, 3815.
[101] Micheli, F.; Cavanni, P.; Andreotti, D.; Arban, R.; Benedetti, R.; Bertani, B.; Bettati, M.; Bettelini, L.; Bonanomi, G.; Braggio, S.; Carletti, R.; Checchia, A.; Corsi, M.; Fazzolari, E.; Fontana, S.; Marchioro, C.; Merlo-Pich, E.; Negri, M.; Oliosi, B.; Ratti, E.; Rea, K. D.; Roscic, M.; Sartori, I.; Spada, S.; Tedesco, G.; Tarsi, L.; Terreni, S.; Visentini, F.; Zocchi, A.; Zonzini, L.; Di Fabio, R., *J. Med. Chem.*, (2010) **53**, 4989.
[102] Barberis, M.; Pérez-Prieto, J., *Tetrahedron Lett.*, (2003) **44**, 6683.
[103] Galopin, C. C., *Tetrahedron Lett.*, (2001) **42**, 5589.
[104] Esposito, A.; Piras, P. P.; Ramazzotti, D.; Taddei, M., *Org. Lett.*, (2001) **3**, 3273.

2.3 Pauson–Khand Reactions

G. Domínguez and J. Pérez-Castells

General Introduction

The Pauson–Khand reaction is a rare example of a transformation that in one step produces a great increase in molecular complexity;[1–4] only a few other transformations, such as the Diels–Alder reaction or the cyclotrimerization of alkynes, can compete. The Pauson–Khand reaction occurs by the (2+2+1) cycloaddition of a triple bond, a double bond, and carbon monoxide to form a cyclopentenone (Scheme 1). Cyclopentenones are useful synthetic intermediates that can be converted into other structures, e.g. those present in natural products, or subjected to further transformation.

Scheme 1 Connectivity of the Pauson–Khand Reaction

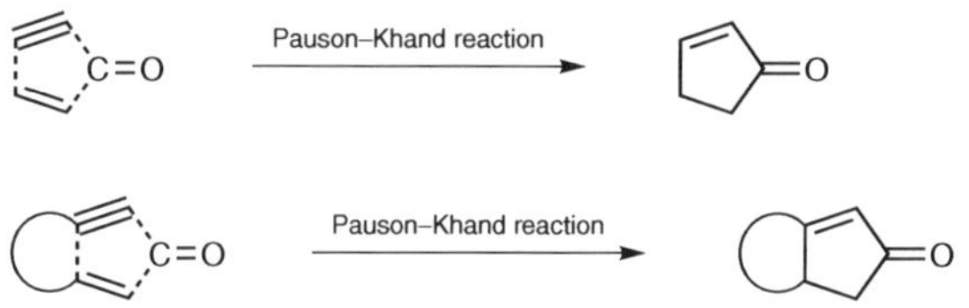

For a previous discussion of the Pauson–Khand reaction, see *Science of Synthesis*, Vol. 1 (Section 1.4.4.14.3).

2.3.1 General Overview of the Pauson–Khand Reaction

2.3.1.1 Origins, History, and Versions

The Pauson–Khand reaction was first reported in the early 1970s as an unexpected result when trying to obtain new organometallic cobalt complexes.[5] Initially it was performed by heating the alkyne–cobalt cluster with an alkene; this resulted in reactions with generally low efficiency. Until the mid-1990s, octacarbonyldicobalt(0) was the only reagent used to mediate the Pauson–Khand reaction. The large number of byproducts, low yields, and narrow scope delayed the development of this reaction for more than 20 years.

The main versions of the Pauson–Khand reaction are the stoichiometric and the catalytic methods. In addition to cobalt, the stoichiometric reaction has been performed with zirconium, nickel, iron, titanium, tungsten, and molybdenum derivatives; the catalytic version is performed with cobalt, titanium, ruthenium, iridium, and rhodium complexes. Many heterogeneous catalytic systems are available that are generally based on cobalt or combinations of cobalt and palladium or cobalt and ruthenium, either supported on clays or charcoal or as nanoparticles in suspension.[6]

In addition, the Pauson–Khand reaction can be classified into inter- or intramolecular versions. Intermolecular Pauson–Khand reactions were developed first, and in this version unstrained alkenes generally react poorly and there is a lack of general catalytic procedures. On the other hand, the intramolecular Pauson–Khand reaction, although developed later, avoids regioselectivity problems and is successful with more types of double bond. The first intramolecular Pauson–Khand reaction was reported by Schore and allowed the formation of 5,5- and 5,6-fused bicycles.[7] In general, good conversions are achieved only with *gem*-disubstituted enynes. Some 5,7-bicycles are now available by the intramolecular Pauson–Khand reaction, and the alkene source has been expanded to include allenes, carbonyls, imines, and heterocumulenes.

2.3.1.2 Regioselectivity Issues in the Intermolecular Pauson–Khand Reaction

Regioselectivity in the alkyne insertion is driven by the steric bulk of the substituents. Thus, for unsymmetrically substituted alkynes, the larger substituent in the alkyne is located adjacent to the carbonyl in the cyclopentenone product. Internal alkynes sometimes give mixtures of regioisomers and studies using unsymmetrical alkynes **1** disubstituted by electronically different aromatic rings[8–10] showed that if the electronic properties of the substituents are markedly different, then electronic issues can also affect the regioselectivity. In general, if the substrate is an alkyne with one electron-withdrawing-group-substituted aryl, hence an electron-deficient moiety, this group will generally occupy the β-position relative to the carbonyl, giving selectivity for **2** over **3** (R^2 = EWG) (Scheme 2).[9]

Scheme 2 Regioselectivity in the Intermolecular Pauson–Khand Reaction of Unsymmetrical Alkynes: Electronic Effects[9]

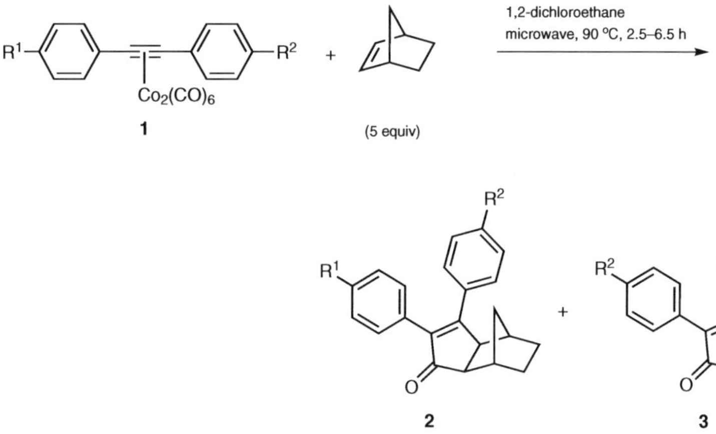

R^1	R^2	Ratio (2/3)	Combined Yield (%)	Ref
H	F	1:1	80	[9]
H	Ac	1.25:1	100	[9]
H	CF$_3$	1.3:1	95	[9]
Me	CO$_2$Et	1.3:1	100	[9]
OMe	CO$_2$Et	1.5:1	73	[9]
OMe	CF$_3$	1.5:1	91	[9]

An indirect way to reverse the regiochemistry from the alkyne side uses trifluoromethyl as a removable steering group. Thus, the Pauson–Khand reaction of norbornadiene with cobalt complexes derived from (trifluoromethyl)acetylenes provides indenone products with excellent regioselectivity for the α-(trifluoromethyl)cyclopentenone; treatment of these adducts with 1,8-diazabicyclo[5.4.0]undec-7-ene results in the removal of the trifluoromethyl moiety. For example, β-substituted Pauson–Khand adducts are synthesized by reaction of the alkyne–cobalt complex **4** with norbornadiene to give β-aryl-α-(trifluoromethyl)cyclopentenone **5**, which then undergoes detrifluoromethylation to give the β-substituted Pauson–Khand adduct (Scheme 3).[11]

Scheme 3 Pauson–Khand Reaction with a (Trifluoromethyl)acetylene: Synthesis of β-Substituted Pauson–Khand Adducts[11]

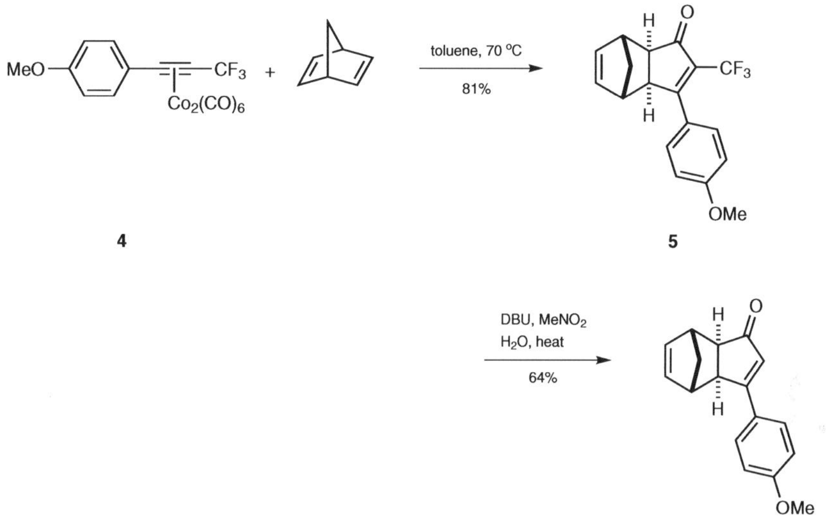

Unsymmetrical alkenes generally give mixtures of regioisomers, and substitution at the double bond is restricted since disubstituted alkenes generally react poorly. If disubstituted alkynes are used, the regioselectivity of the alkene insertion is enhanced in favor of the α-substituted cyclopentenone. A route to achieve good regioselectivity is by the so-called directed Pauson–Khand reaction, first developed by Krafft.[12] This variation consists of situating a coordinating atom at a certain distance from the alkene moiety; coordination of this directing group enables high regioselectivity. An extension of this strategy uses allylphosphonates and phosphonates tethered through an ester moiety to direct the Pauson–Khand reaction with unsymmetrical alkynes. For example, the reaction of 4-tolylacetylene–cobalt complex **6** with phosphonate **7** gives a mixture of **8** and **9** in a 12:1 ratio in favor of the α,α′-substituted product **8** (Scheme 4).[13]

Scheme 4 The Directed Pauson–Khand Reaction[13]

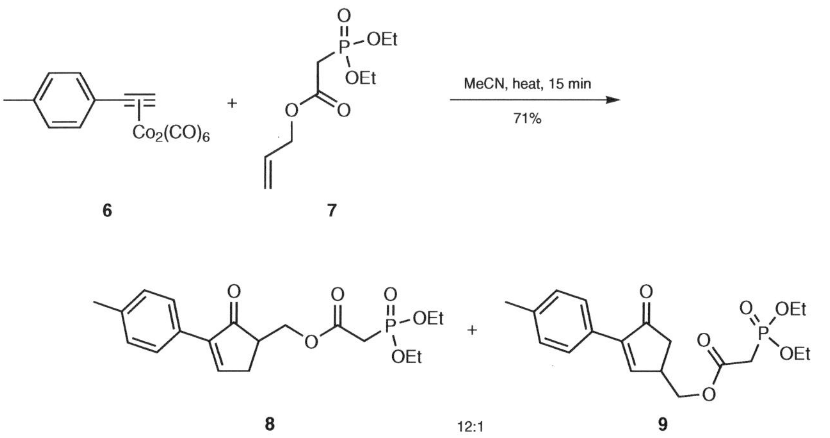

for references see p 162

2.3.1.3 Promotion of the Reaction

There are several methods to accelerate the Pauson–Khand reaction; conversions are generally only moderate in the thermally promoted version. The aim is to design procedures that give a clean reaction under mild conditions and which give the Pauson–Khand adduct in high yields and free from metal-containing byproducts.

The first significant method used to promote the Pauson–Khand reaction consisted of adsorbing the reagents on to solid supports (dry state adsorption conditions, DSAC) (Table 1, entries 1 and 2). Under these conditions, the conversion improves, although side products are frequently isolated.[14]

The promotion of the Pauson–Khand reaction generally involves the use of chemical additives and the first examples were described by Schreiber[15] and then by Jeong;[16] they both utilized amine N-oxides as the additive and these have since become the most widely applied conditions. Amine N-oxide additives act by oxidizing one carbon monoxide (carbonyl or CO) ligand into carbon dioxide, thus forming a vacancy in the cobalt cluster. As discussed in Section 2.3.2.1, this is the rate-determining step in the Pauson-Khand reaction pathway. Thus, it is possible to use milder conditions in the Pauson-Khand reaction by utilizing amine N-oxides as additives (Table 1, entries 3, 4, and 6–8). The most generally used N-oxides are trimethylamine N-oxide (Table 1, entries 3 and 6–8)[16–18] and 4-methylmorpholine N-oxide (Table 1, entry 4).[19] Aqueous conditions using hexadecyl(trimethyl)-ammonium bromide as a phase-transfer catalyst are also successful (Table 1, entry 5).[20] Reactive alkenes, such as norbornadiene or cyclopropene, react under very mild conditions (Table 1, entries 3–5). Further improvements are achieved by combining amine N-oxides with high intensity sonication (Table 1, entry 6).[17] Molecular sieves have a positive effect on both the catalytic and stoichiometric versions of the Pauson–Khand reaction. It is thought that these zeolites retain carbon monoxide molecules, thus increasing the conversion, even with unfavorable substrates such as substituted alkenes. For example, the use of N-amine oxide promotion alone (Table 1, entry 7) gives the Pauson–Khand adduct in 35% yield, whereas using zeolites combined with amine N-oxide promotion gives the same product in 90% yield (Table 1, entry 8).[18]

Table 1 Chemical Promotion of the Intermolecular Pauson–Khand Reaction[14–20]

Entry	Starting Materials		Conditions	Product	Yield (%)	Ref
	Alkyne	Alkene				
1	Ph-≡-, Co$_2$(CO)$_6$	norbornene	silica gel, dry state adsorption conditions, 55 °C, 2 h	Ph-cyclopentenone adduct	86	[14]
2	≡, Co$_2$(CO)$_6$	methylenecyclopropane	alumina, dry state adsorption conditions, 50 °C, 2 h	two products (5:1)	77	[14]

2.3.1 General Overview of the Pauson–Khand Reaction

Table 1 (cont.)

Entry	Starting Materials		Conditions	Product	Yield (%)	Ref
	Alkyne	Alkene				
3	Ph—≡—[Co$_2$(CO)$_6$]	norbornene	TMANO, CH$_2$Cl$_2$, 0 °C, 2 h	Ph-cyclopentenone fused	80	[16]
4	Bu^t—≡—[Co$_2$(CO)$_6$]	cyclopropene	NMO, CH$_2$Cl$_2$, –35 °C, 5 min	Bu^t-product	93	[19]
5	Ph—≡—[Co$_2$(CO)$_6$]	norbornene	Me(CH$_2$)$_{15}$NMe$_3$Br, H$_2$O, 70 °C, 18 h	Ph-product	62	[20]
6	Ph—≡—[Co$_2$(CO)$_6$]	cyclopentene	TMANO, toluene, ultrasound, 25 °C, 10 min	Ph-product **10**	97	[17]
7	Et—≡—Et [Co$_2$(CO)$_6$]	norbornene	TMANO, toluene, 0 °C, 16 h	Et-product	35	[18]
8	Et—≡—Et [Co$_2$(CO)$_6$]	norbornene	TMANO, 4-Å molecular sieves, toluene, 0 °C, 16 h	Et-product	90	[18]

The addition of coordinating molecules also has favorable effects. A study on the role of Lewis base promoters by density functional theory (DFT) calculations shows that these promoters do not accelerate the departure of a carbon monoxide ligand from the acetylene–dicobalt complex to create the requisite site for alkene coordination, but rather stabilize the alkene-insertion product so that product formation is essentially irreversible.[21] Some examples of Lewis base promoted Pauson–Khand reactions are given in Table 2. Phosphine oxides were initially used as the Lewis base (Table 2, entry 1),[22] and they were followed by cyclohexylamine, which is a very efficient promoter, albeit with a limited scope (Table 2, entry 2).[23] This lack of generality led to the introduction of sulfides as promoters and they are now widely employed; the conditions used are mild and they give high yields with numerous examples (e.g., Table 2, entry 3).[24] However, low-molecular-weight sulfides have unpleasant odors; this problem can be avoided by using dodecyl methyl sulfide as a good alternative (Table 2, entries 4–6).[25]

Table 2 Intermolecular Pauson–Khand Reactions Promoted by Lewis Bases[22–25]

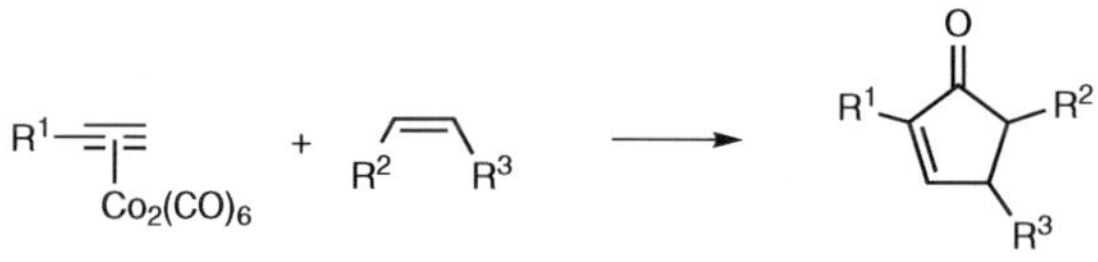

Entry	Starting Materials		Additive	Conditions	Product	Yield (%)	Ref
	Alkyne	Alkene					
1	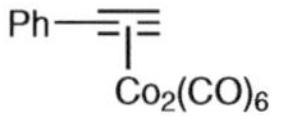 Ph—≡— Co$_2$(CO)$_6$	 cyclopentene	Bu$_3$PO	hexane, 69 °C, 36 h	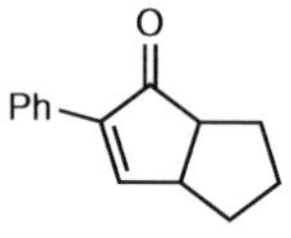 Ph-bicyclic enone	70	[22]
2	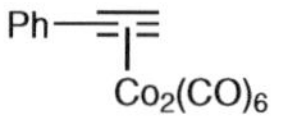 Ph—≡— Co$_2$(CO)$_6$	 norbornene	CyNH$_2$	DME, 70 °C, 13.5 h	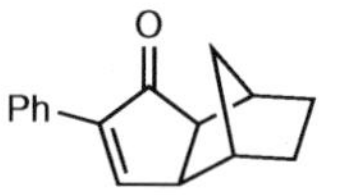 Ph-tricyclic enone	97	[23]
3	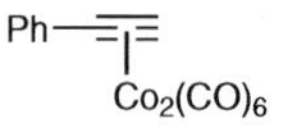 Ph—≡— Co$_2$(CO)$_6$	 cycloheptene	BuSMe	1,2-dichloroethane, 83 °C, 1.5 h	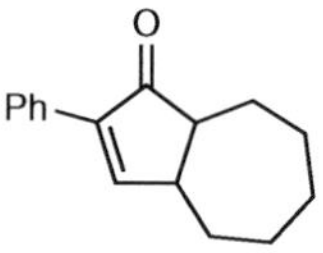 Ph-bicyclic enone 11	85	[24]
4	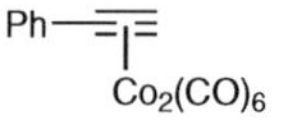 Ph—≡— Co$_2$(CO)$_6$	 cycloheptene	Me(CH$_2$)$_{11}$SMe	1,2-dichloroethane, 83 °C, 1.5 h	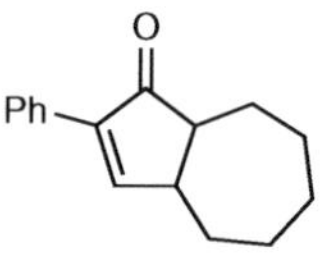 Ph-bicyclic enone	66	[25]
5	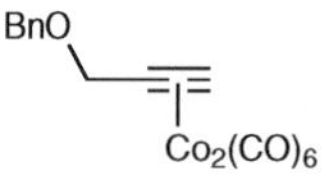 BnO—≡— Co$_2$(CO)$_6$	 norbornene	Me(CH$_2$)$_{11}$SMe	1,2-dichloroethane, 83 °C, 0.5 h	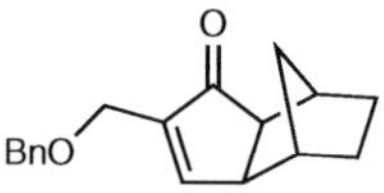 BnO-tricyclic enone	86	[25]
6	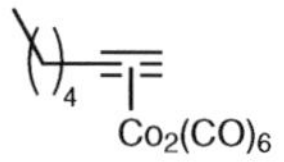 C$_4$H$_9$—≡— Co$_2$(CO)$_6$	 norbornene	Me(CH$_2$)$_{11}$SMe	1,2-dichloroethane, 83 °C, 2.0 h	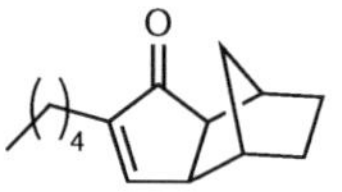 tricyclic enone	89	[25]

The Pauson–Khand reaction can be performed under microwave irradiation; these conditions shorten reaction times remarkably, use only 20 mol% octacarbonyldicobalt(0), and additional carbon monoxide is not required, but the yields are only moderate.[26] Interestingly, the use of the stable radical 2,2,6,6-tetramethylpiperidin-1-oxyl (TEMPO) as an additive is highly beneficial in the Pauson–Khand reaction. Theoretical calculations indicate a possible one-electron promotion of the reaction with the formation of paramagnetic complexes with the cobalt that are able to undergo smooth decarbonylation.[27]

The Pauson–Khand reaction can be performed using a flow microreactor. The reaction is performed at ambient temperature for a short time and uses strained alkenes, but the yields are satisfactory in only a few cases. However, this shows promise for the future development of a scalable catalytic protocol.[28]

The molybdenum-mediated Pauson–Khand reaction is an alternative to the cobalt-mediated version. Hexacarbonylmolybdenum(0) is used to mediate the Pauson–Khand re-

2.3.1 General Overview of the Pauson–Khand Reaction

action using dimethyl sulfoxide as an activating agent (Table 3, entry 1);[29] alkene substrates include allene derivatives.[30,31] Alternatively, the Pauson–Khand reaction promoted by tricarbonyltris(dimethylformamide)molybdenum(0) takes place under very mild conditions in the absence of a promoter (Table 3, entries 2–6).[29] High yields of Pauson–Khand adducts are obtained for the cyclization of a wide variety of functionalized 1,6- and 1,7-enynes. Enynes bearing electron-withdrawing groups at the alkene terminus are particularly good substrates.

Table 3 Intramolecular Pauson–Khand Reactions Mediated by Molybdenum Complexes[29,31]

Entry	Starting Materials		Conditions	Product	Yield (%)	Ref
	Enyne	Metal Complex				
1	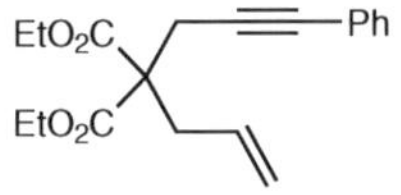 EtO$_2$C, EtO$_2$C, Ph	Mo(CO)$_6$	DMSO, toluene, 100 °C, 16 h	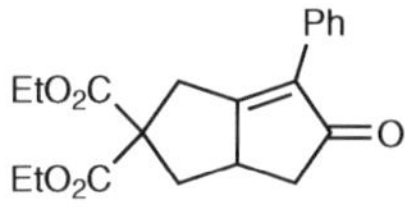 EtO$_2$C, EtO$_2$C, Ph, O	53	[29]
2	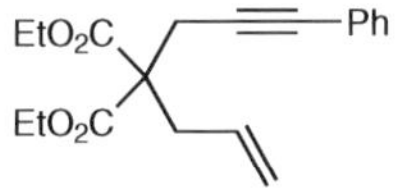 EtO$_2$C, EtO$_2$C, Ph	Mo(CO)$_3$(DMF)$_3$	CH$_2$Cl$_2$, 25 °C, 3 h	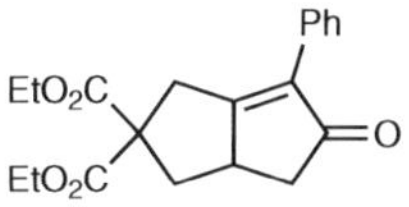 EtO$_2$C, EtO$_2$C, Ph, O **12**	88	[29]
3	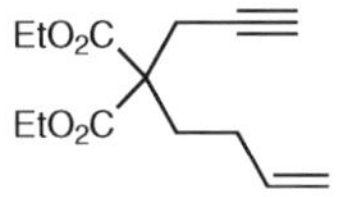 EtO$_2$C, EtO$_2$C	Mo(CO)$_3$(DMF)$_3$	CH$_2$Cl$_2$, 25 °C, 4 h	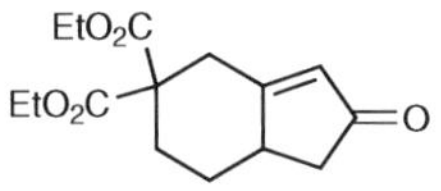 EtO$_2$C, EtO$_2$C, O	61	[29]
4	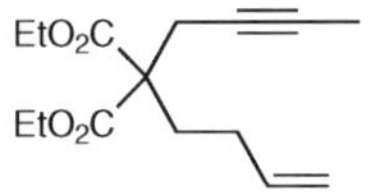 EtO$_2$C, EtO$_2$C	Mo(CO)$_3$(DMF)$_3$	CH$_2$Cl$_2$, 25 °C, 5 h	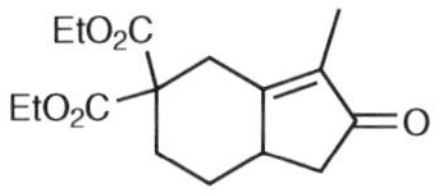 EtO$_2$C, EtO$_2$C, O	60	[29]
5	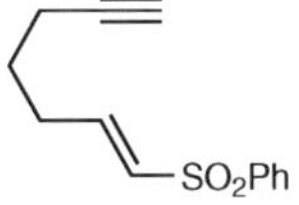 SO$_2$Ph	Mo(CO)$_3$(DMF)$_3$	CH$_2$Cl$_2$, 25 °C, 3 h	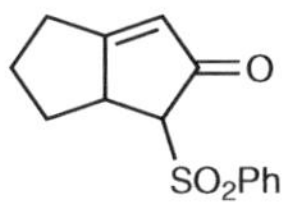 SO$_2$Ph, O	80	[29]
6	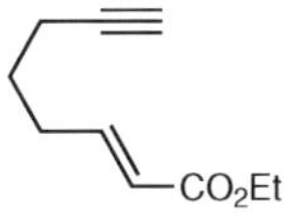 CO$_2$Et	Mo(CO)$_3$(DMF)$_3$	CH$_2$Cl$_2$, 25 °C, 3 h	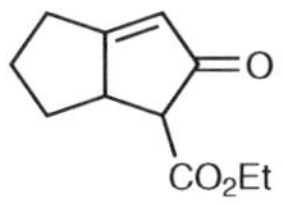 CO$_2$Et, O	72	[29]

Cyclopentenones, e.g. 2-Phenyl-4,5,6,6a-tetrahydropentalen-1(3aH)-one (10); General Procedure by N-Oxide/Ultrasound Promotion:[17]

To a soln of cobalt complex (10–12 mg/mL) and TMANO (6 equiv) in toluene (3 mL) was added a soln of norbornene (1 equiv) in toluene (1 mL). A titanium ultrasonic horn (Vibracell VC 50) was immersed in the resultant red soln and ultrasound was applied (50 W/20 kHz output) until the starting complex had been consumed (TLC analysis). The resulting purple mixture was then filtered through a small pad of kieselguhr and the solvent was removed under reduced pressure. Purification was achieved by flash column chromatography (silica gel, petroleum ether or Et_2O) to give the corresponding cyclopentenone. When less reactive alkenes (2,5-dihydrofuran and cyclopentene) were employed, alkene (1 mL) and toluene (3 mL) were added instead of the soln of norbornene in solvent. With 4-fluorostyrene, alkene (1 mL) was employed with toluene (4 mL) and with cycloheptene, alkene (10 equiv) was used with solvent (5 mL).

2-Phenyl-4,5,6,7,8,8a-hexahydroazulen-1(3aH)-one (11); Typical Procedure:[24]
To a soln of hexacarbonyl(phenylacetylene)dicobalt(0) (0.12 g, 0.3 mmol) and butyl methyl sulfide (0.15 mL, 1.2 mmol) in 1,2-dichloroethane (2 mL) was added a soln of cycloheptene (0.07 mL, 0.6 mmol) in 1,2-dichloroethane (1 mL) under an argon atmosphere. The mixture was then heated to reflux (83 °C) for 90 min. After removal of the solvent under reduced pressure, the residue was purified by flash chromatography (silica gel, Et_2O) to give the product as a colorless oil; yield: 58 mg (85%).

Diethyl 5-Oxo-6-phenyl-3,3a,4,5-tetrahydropentalene-2,2(1H)-dicarboxylate (12); Typical Procedure:[29]
An oven-dried Schlenk flask was charged under argon with $Mo(CO)_3(DMF)_3$ (230 mg, 0.57 mmol), capped with a rubber septum, and twice evacuated and backfilled with argon. A soln of diethyl 2-allyl-2-(3-phenylprop-2-ynyl)malonate (181 mg, 0.57 mmol) in CH_2Cl_2 (5 mL) was added via syringe and the mixture was stirred at rt for 3 h. The crude mixture was filtered through a plug of Celite (CH_2Cl_2) and purified by flash chromatography (hexane/EtOAc 4:1) to afford the product as a colorless oil; yield: 173 mg (88%).

2.3.1.4 Scope and Limitations

The intermolecular Pauson–Khand reaction has wide tolerance with respect to the alkyne component with few limitations. With regard to the alkene partner, terminal alkenes, including enamines and ynamines, are generally good substrates.[32] However, 1,1-disubstituted alkenes in general give lower yields, with some exceptions such as exocyclic alkenes or strained cycloalkenes. Thus, methylenecyclohexanes, methylenecyclopropanes, methylenepyrans, and norbornene derivatives are good alkenes for both the inter- and intramolecular versions. A theoretical DFT study compared the Pauson–Khand reaction of cyclohexene, cyclopentene, and norbornene with hexacarbonyl(μ-propyne)dicobalt. The calculations give the highest barrier for the reaction of cyclohexene (ca. 15 kcal·mol^{-1}), followed by cyclopentene (11 kcal·mol^{-1}), and then norbornene (6 kcal·mol^{-1}).[33] In connection with this, medium-sized *E*-cycloalkenes, due to the ring strain imparted by the *E* stereochemistry, are unusually reactive in the intermolecular Pauson–Khand reaction compared to typical monocyclic alkenes, an example is the reaction of (*E*)-1-methylcycloct-1-ene **13** to give the *trans*-fused products **14** (Scheme 5).[34]

Scheme 5 Intermolecular Pauson–Khand Reaction with (*E*)-1-Methylcycloct-1-ene[34]

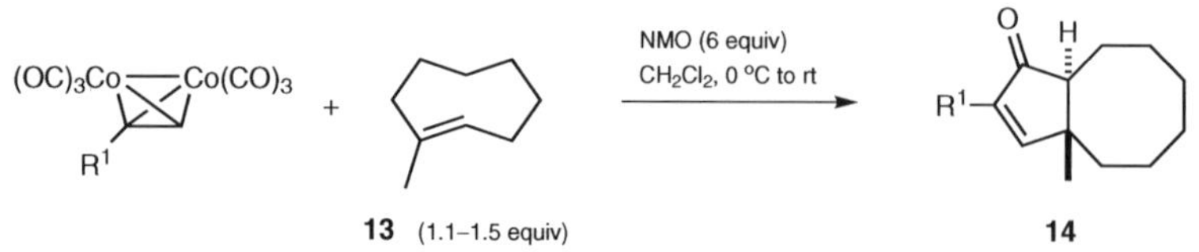

2.3.1 General Overview of the Pauson–Khand Reaction

R^1	Yield (%)	Ref
Ph	22	[34]
TMS	62	[34]

Norbornene systems (and derivatives) are highly reactive alkene partners in the Pauson–Khand reaction, efficiently giving polycyclic adducts that can be further transformed into useful products. A versatile way to take advantage of the reactivity of strained alkenes is to perform a retro-Diels–Alder reaction after a Pauson–Khand reaction with norbornadiene. The steric hindrance imposed by the norbornene ring enforces diastereoselective addition to the *exo* face of the cyclopentenone; retro-Diels–Alder reaction leads to cyclopentenone rings with various substituents and defined stereochemistry.[35,36] As an example, strained alkene 1,2,3,4-tetramethylnorbornadiene undergoes the Pauson–Khand reaction with alkynes to give the adducts **15**. Conjugate addition to the Pauson–Khand adducts followed by retro-Diels–Alder reaction under mild conditions gives functionalized cyclopentenones **16** and **17** in improved yields (47–76% yield for the retro-Diels–Alder step) (Scheme 6).[37]

Scheme 6 Synthesis of Functionalized Cyclopentenones by the Cobalt-Catalyzed Intermolecular Pauson–Khand Reaction/Conjugate Addition/Retro-Diels–Alder Reaction[37]

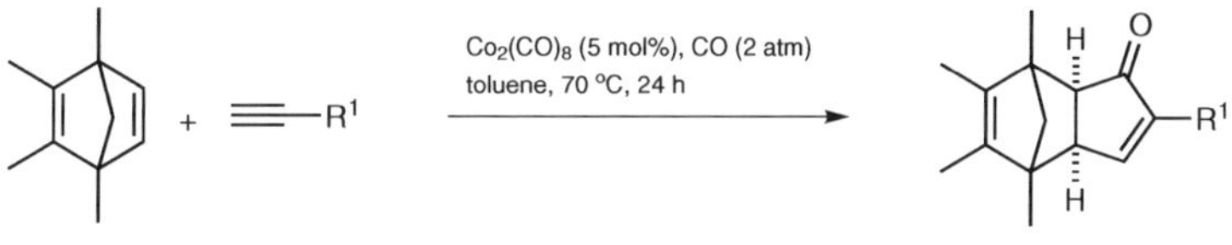

R^1	Yield (%)	Ref
Bu	85	[37]
Ph	85	[37]
CH$_2$OTBDPS	79	[37]
TMS	89	[37]

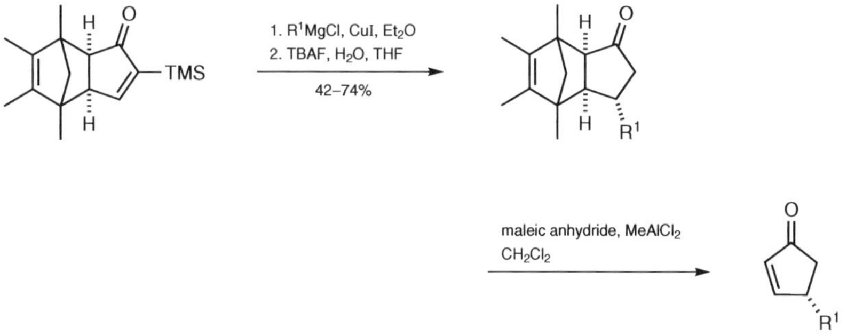

R^1 = iPr, Ph, 4-MeOC$_6$H$_4$

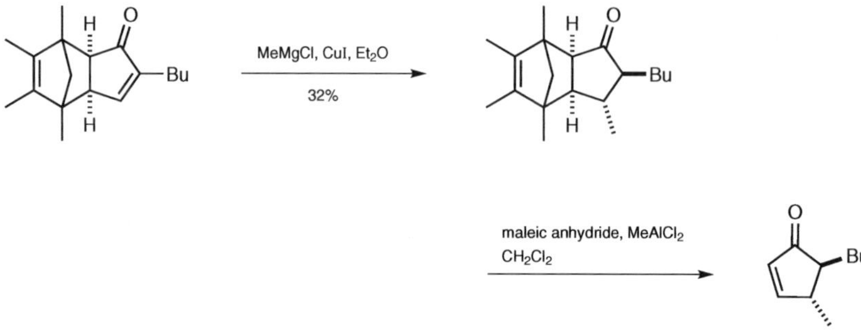

In view of the advantage of using strained alkenes, cyclopropenes are also used as the alkene partners to give interesting adducts that can easily be transformed upon further reaction of the cyclopropane ring. In the Pauson–Khand reaction of ethyl (R)-2-methyl-3-(trimethylsilyl)cycloprop-2-ene-1-carboxylate with alkyne complex **18**, formation of cyclopentenone **19** is accompanied by red cobalt complex **20**. The conditions can be tuned to favor the formation of **19** or **20** (Scheme 7). The complex **20** results from fragmentation of the cyclopropane ring after alkene insertion and it has been characterized by X-ray crystallography. This structure is a useful proof of the accepted mechanism for the Pauson–Khand reaction.[38]

Scheme 7 Intermolecular Pauson–Khand Reaction with a Cyclopropene Derivative[38]

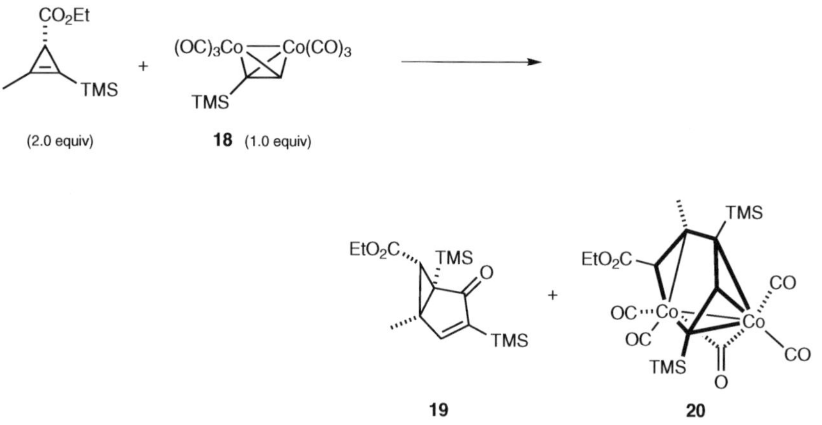

Conditions	Yield (%) of 19	Yield (%) of 20	Ref
NMO (5 equiv), CH$_2$Cl$_2$, 0 °C to rt, 45 min, then rt, 45 min	19	13	[38]
BuSMe (30 equiv), 1,4-dioxane, 100 °C, 2 h	64	–	[38]

Related methodology is used with compound **21**, which acts as a cyclobutadiene equivalent in a catalytic Pauson–Khand reaction to give **22**; retro-Diels–Alder reaction of **22** gives bicyclo[3.2.0]hepta-3,6-dien-2-ones **23** (Scheme 8).[39] In a different approach, tricarbonyl-(cyclobutadiene)iron substrates **24** containing a tethered alkynyl unit give Pauson–Khand adducts **25** by treatment with ammonium cerium(IV) nitrate and carbon monoxide. In some cases, competing (4+2) cycloaddition leading to dihydrobenzo[c]furans **26** also takes place (Scheme 8).[40]

2.3.1 General Overview of the Pauson–Khand Reaction

Scheme 8 Pauson–Khand Reactions with Cyclobutadiene Surrogates[39,40]

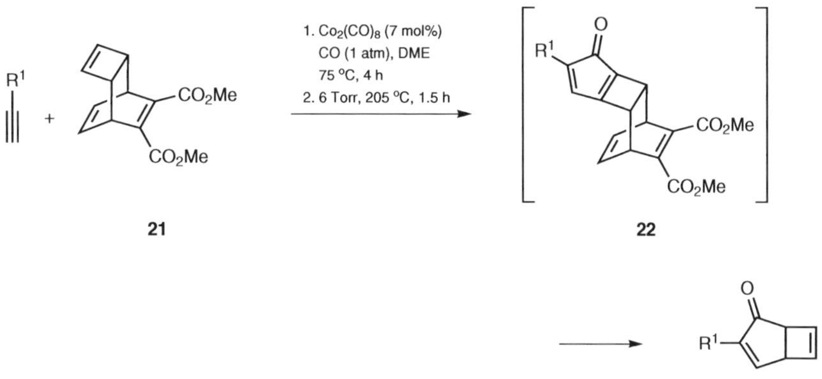

R^1	Yield (%)	Ref
$(CH_2)_5Me$	98	[39]
Ph	75	[39]
$CH_2OTBDMS$	70	[39]
CH_2NHBoc	68	[39]
TMS	73	[39]

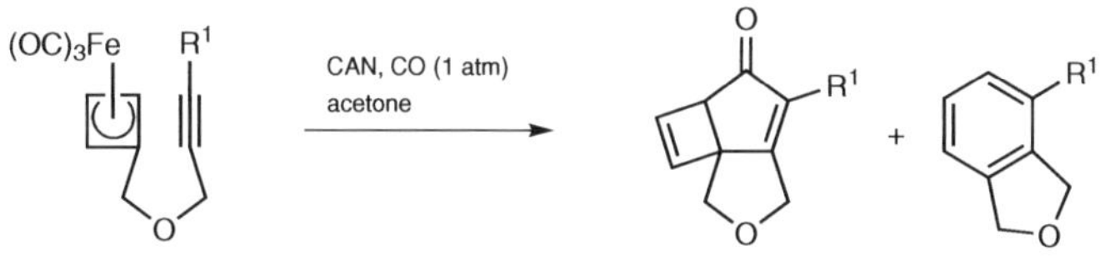

R^1	Yield (%) of 25	Yield (%) of 26	Ref
H	60	–	[40]
Me	91	–	[40]
t-Bu	78	–	[40]
TMS	77	–	[40]
CO_2Me	–	67	[40]
Ph	69	24	[40]

Ethene is a possible partner in the Pauson–Khand reaction, although initial reports worked under high pressures and at high temperatures (Table 4, entries 1 and 2).[41,42] A catalytic protocol uses supercritical ethene, which is introduced into a vessel containing the reagents under a carbon monoxide pressure of 110 atm (Table 4, entry 3).[43] Good conversions can be achieved under lower pressures of ethene by using promotion by 4-methylmorpholine N-oxide, which allows the use of lower temperatures and pressures with a great improvement in yield (Table 4, cf. entries 4 and 5).[44] Improved results are achieved upon combination of 4-methylmorpholine N-oxide with zeolites (Table 4, entry 6).[44] Alternatively, the use of ethene surrogates allows the use of mild conditions and gives the products in good conversions in a reaction where the benzoyloxy group is cleaved in the same operational step (Table 4, entry 7).[45]

for references see p 162

Table 4 Ethene and Surrogates in the Intermolecular Pauson–Khand Reaction[41–45]

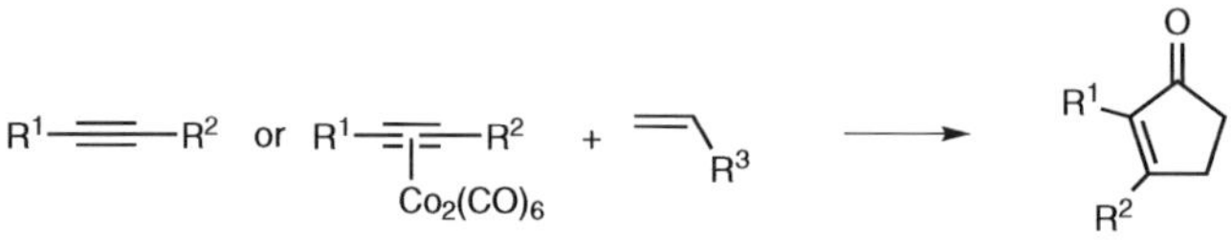

Entry	Starting Materials		Conditions	Product	Yield (%)	Ref
	Alkyne	Alkene				
1	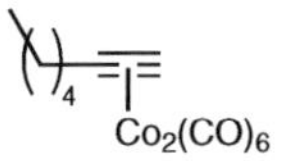 (CH₂)₄ alkyne, Co₂(CO)₆	H₂C=CH₂	toluene, 120 atm, 85 °C, 35 h	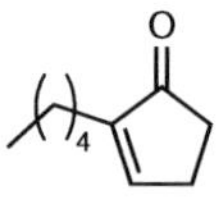 cyclopentenone with (CH₂)₄ ring	55	[41]
2	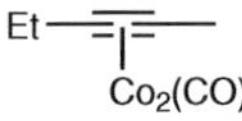 Et—≡—, Co₂(CO)₆	H₂C=CH₂	toluene, 35 atm, 110 °C, 36 h	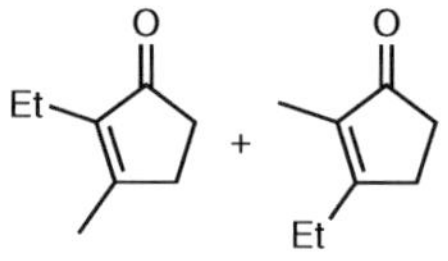 Et/Me cyclopentenone isomers 24:3	27	[42]
3	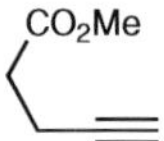 CO₂Me-alkyne	H₂C=CH₂	Co₄(CO)₁₂ (5 mol%), CO (110 atm), THF, 85 °C, 12 h	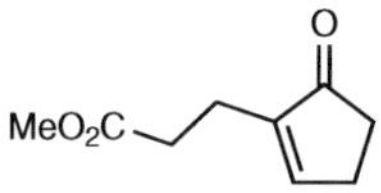 MeO₂C-cyclopentenone	87	[43]
4	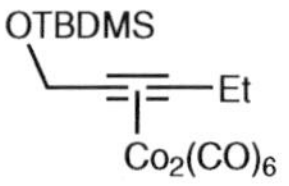 OTBDMS–≡–Et, Co₂(CO)₆	H₂C=CH₂	toluene, 90 °C	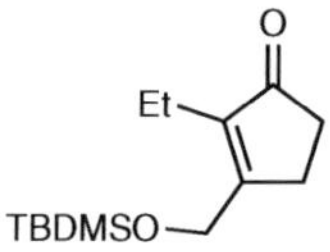 Et, TBDMSO cyclopentenone	25	[44]
5	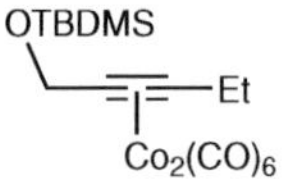 OTBDMS–≡–Et, Co₂(CO)₆	H₂C=CH₂	NMO, CH₂Cl₂, 25 °C	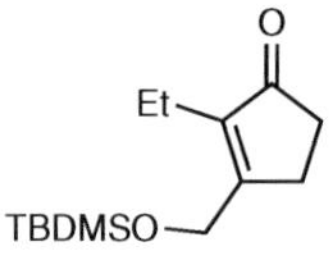 Et, TBDMSO cyclopentenone	54	[44]
6	OTBDMS, CO₂Me alkyne, Co₂(CO)₆	H₂C=CH₂	NMO, 4-Å molecular sieves, CH₂Cl₂, 25 °C	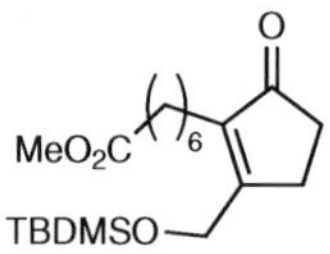 MeO₂C, TBDMSO cyclopentenone 27	70	[44]
7	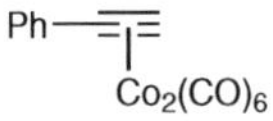 Ph—≡—, Co₂(CO)₆	H₂C=CH(OBz)	CH₂Cl₂, 25 °C, 16 h	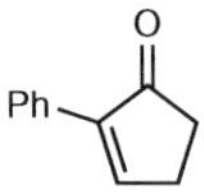 Ph-cyclopentenone	80ᵃ	[45]

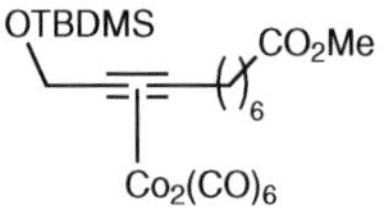

ᵃ 1 equiv of alkene was used.

From these examples the great functional-group compatibility of this reaction is clear; it tolerates ethers, alcohols, tertiary amines, acetals, esters, amides, and heterocycles. In the intramolecular version, most substrates are derived from hept-1-en-6-yne or allyl propargyl ethers or amines. Other interesting substrates are enynes connected through aromatic rings.[46] In addition, allenes can act as the alkene in the Pauson–Khand reaction.[47] Allenynes are attractive substrates as either double bond of the allene can react and the products contain an extra C=C bond that can be functionalized. The allenes can have axial chirality, and this paves the way for asymmetric reactions. Early examples by Cazes[48] and Livinghouse[49] used cobalt mediation and showed generally low yields and poor selectivity between the allene double bonds. Improvements in the reaction used hexacarbonylmolybdenum(0) to establish the substitution patterns of the reaction with either the external or the internal bond of the allene to give products with varying ring sizes, e.g. **28** and **29** (Scheme 9).[50] This route is utilized in the synthesis of hydroxymethylacylfulvene.[51]

Scheme 9 Allenynes in the Intramolecular Pauson–Khand Reaction: Selectivity as a Function of the Substitution[50]

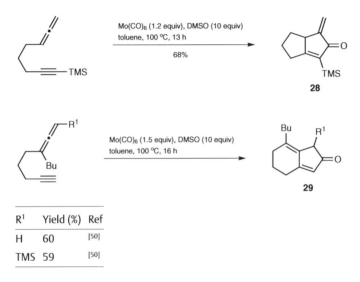

R¹	Yield (%)	Ref
H	60	[50]
TMS	59	[50]

The synthesis of medium-sized rings fused to cyclopentenones by the intramolecular Pauson–Khand reaction is limited to templated substrates in which the population of the reactive conformation is increased by bringing the alkene and alkyne moieties close to each other, e.g. certain enynes templated by aromatic nuclei.[52] The intramolecular allenic Pauson–Khand reaction allows the efficient synthesis of seven membered rings.[53]

Methyl 7-{2-[(*tert*-Butyldimethylsiloxy)methyl]-5-oxocyclopent-1-enyl}heptanoate (27); Typical Procedure:[44]

A flask purged with N_2 was charged with $Co_2(CO)_8$ (1.29 g, 3.76 mmol) and a soln of methyl 10-(*tert*-butyldimethylsiloxy)dec-8-ynoate (1.07 g, 3.42 mmol) in hexane (5 mL) was transferred via cannula to this flask. The resulting mixture was stirred for 1 h until total complexation of the alkyne. The mixture was then filtered through silica gel (hexanes/EtOAc 9:1) to give the (alkyne)hexacarbonyldicobalt in quantitative yield. A pressure reactor provided with magnetic stirring and a N_2 atmosphere was charged with a soln of the (alkyne)-hexacarbonyldicobalt (1.55 g, 2.52 mmol) in dry CH_2Cl_2 (125 mL) and powdered 4-Å molecular sieves (oven dried for 4 h at 180 °C; 8 times the mass of the starting alkyne). The reactor was immediately sealed, purged with ethene, and charged to the desired ethene pressure (6 atm). The mixture was stirred vigorously at rt and a soln of NMO (2.95 g, 25.20 mmol) in dry CH_2Cl_2 (10 mL) was added in aliquots of 1 mL every 20 min. The mixture was then stirred overnight. After filtration through alumina, the solvent was evaporated and the crude residue was purified by flash chromatography [Et_3N pretreated (2.5% v/v) silica gel, hexanes/EtOAc 95:5] to give a colorless oil; yield: 669 mg (70%).

2.3.2 The Reaction Pathway

The Pauson–Khand reaction is a complex process in which three entities are combined. The main problem with the study of its mechanism is that beyond the initial hexacarbonyldicobalt–alkyne complex, it is difficult to detect further intermediates. The basis of the mechanistic studies is the pathway proposed by Magnus in 1985, which explains most experimental results and is used as starting assumption in theoretical calculations.[54]

2.3.2.1 Stoichiometric Reaction

The so-called Magnus mechanism has four main steps (Schemes 10 and 11). Initially, the hexacarbonyldicobalt complex **30** loses one carbon monoxide in a dissociative manner, giving **31**. Sulfur-ligated decarbonylated intermediates of type **31** have been isolated and this supports the dissociative mechanism.[55] This step, which is strongly endothermic, is the rate-determining step and, consequently, promoters that act at this point make one of the carbon monoxide ligands more labile. As discussed in Section 2.3.1.3, a good method to promote the Pauson–Khand reaction is by the addition of amine N-oxides that oxidize a carbon monoxide ligand into carbon dioxide, thus favoring dissociation and lowering the enthalpy associated with this step.

Next, the alkene coordinates with the cobalt to give **32**. This is a highly favorable kinetically controlled step and coordination takes place from the sterically less hindered face of the alkene. The resulting intermediate has been detected only if an intramolecularly tethered alkene is used that is unable to continue the Pauson–Khand process due to strain in the possible product.[56] The next step, insertion of the alkene into a Co–C bond to form cobaltacycle **35**, has a low activation energy, giving an intermediate in the form of **33**; these intermediates have not been detected.

The formation of the cobaltacycle **35** is an important stage in the reaction course as it determines the stereochemical outcome of the process. In this second step, the strain within the cyclic alkene **33** is released to give **34**, thus favoring the process. The importance of facilitating carbon monoxide dissociation is shown by DFT calculations, but these indicate that the energy of the second step is also important, and this is the reason why strained alkenes react well. Difficulties in predicting the regio- and stereochemical outcome of the Pauson–Khand reaction may arise from the facile rotation of the coordinated alkene in **32** prior to alkene insertion. Then, the complex **34** adds carbon monoxide or a coordinating ligand giving **35**, which is more stable, thus making this step irreversible.

Scheme 10 Energy Associated with the Two First Steps of the Pauson–Khand Reaction[57]

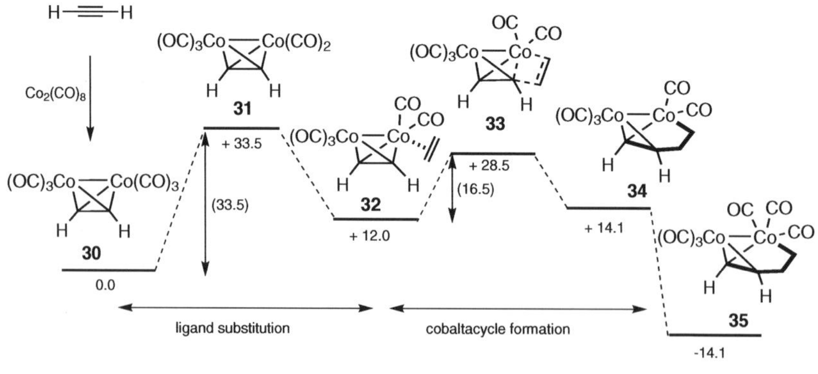

Continuing with the reaction pathway, from complex **36**, insertion of carbon monoxide follows. Theoretical studies demonstrate that the insertion occurs in the Co–C bond in

2.3.2 The Reaction Pathway

which the carbon is from the alkene to give **37**, rather than into the Co—C bond of a former alkyne carbon, which would give **38**. In addition a mass spectrometry study using an ion-molecular reaction with ^{13}C-carbon monoxide, shows that the carbon monoxide incorporated into the cyclopentenone is one that has been retained within the complex.[57] Finally, reductive elimination leads to the cyclopentenone, which is in principle coordinated with the cobalt unit in **39**, but that readily decoordinates to give the final product.

Scheme 11 Last Steps of the Course of the Pauson–Khand Reaction[58]

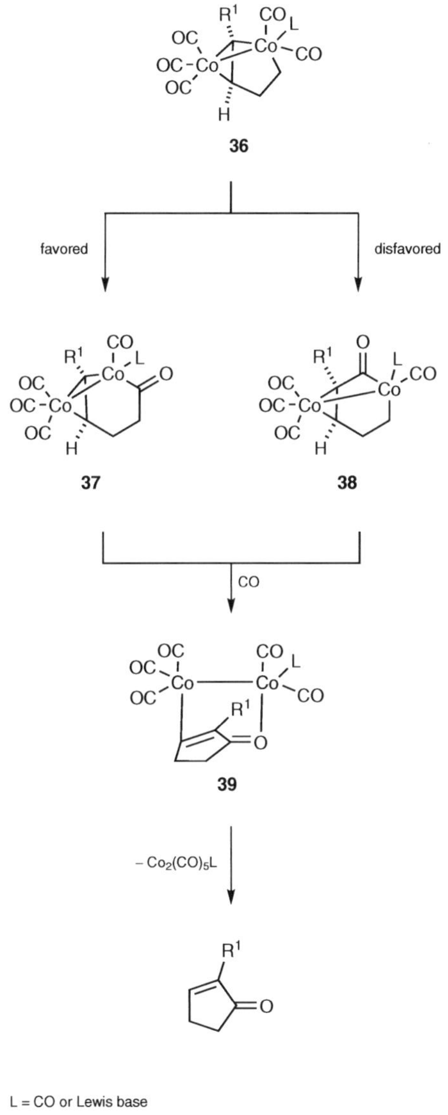

L = CO or Lewis base

Several theoretical studies support this mechanism, which explains the regio- and stereochemical results of numerous examples; there are studies available at the DFT level.[58,59] There is supporting evidence available for the main features of the Magnus mechanism that show the importance of the dissociation step for the kinetics of the reaction, and the cobaltacycle formation step as the critical stereo- and regiochemical-determining step.

The theoretical studies mentioned above indicate that the regioselectivity observed in the intermolecular Pauson–Khand reaction with unsymmetrical alkynes probably originates from steric effects. The transition state in the alkene insertion step using ethene shows that formation of **40** is more favorable than **41** because of steric hindrance. In contrast, in the case of unsymmetrical alkenes, both **42** and **43** have similar energies and this explains the lack of regioselectivity in the reaction with unsymmetrical alkenes (Scheme 12).[58] Further studies have calculated the energy of these transition states when bearing polar substituents in an attempt to explain the electronic effects that affect regioselectivity.[8,59,60]

Scheme 12 Transition States of the Regioselectivity Determining Step[58]

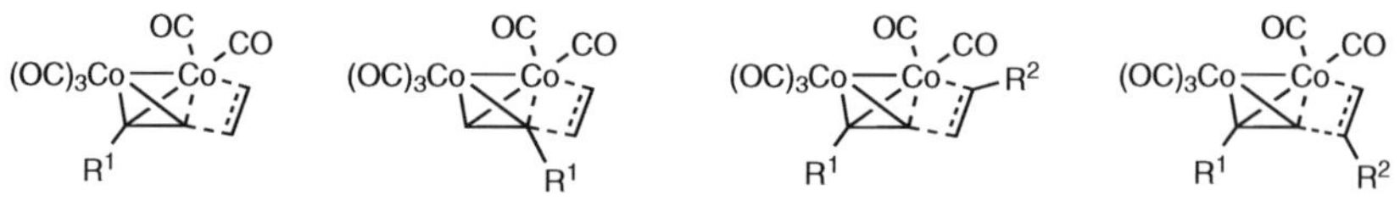

2.3.2.2 Catalytic Reaction

The cobalt-catalyzed Pauson–Khand reaction remains relatively unstudied from a theoretical point of view. It is necessary to understand both the decomplexation step and how the cobalt cluster re-enters the catalytic cycle. This would clarify the role of carbon monoxide gas in the reaction and the way in which chemical additives facilitate the process. A kinetic study of the catalytic Pauson–Khand reaction by in situ FTIR ends with a rate equation with negative dependence on carbon monoxide concentration (−1.9 order, Figure 1), indicating a beneficial effect if using low carbon monoxide pressures. The general Magnus mechanism appears to be operating in the catalytic version, although the alkene insertion step could be the rate-determining step in this case (Scheme 13).[61]

Scheme 13 Cobalt Cluster Regeneration in the Catalytic Pauson–Khand Reaction[61]

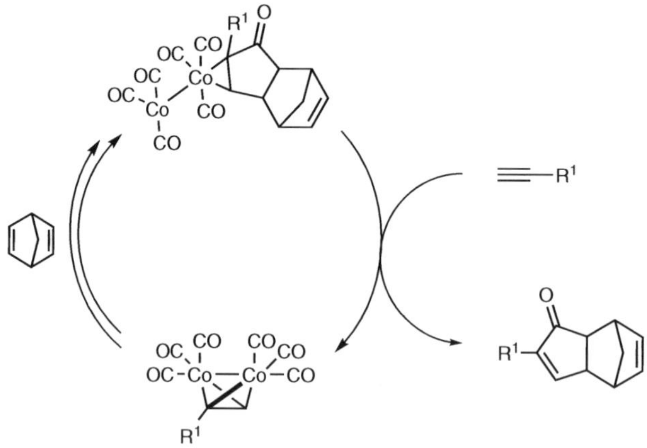

Figure 1 Rate Equation for the Catalytic Pauson–Khand Reaction[61]

$$v = k\,[Co_2(CO)_8]^{1.3}\,[\text{norbornadiene}]^{0.3-1.2}\,[CO]^{-1.9}$$

2.3.3 The Catalytic Pauson–Khand Reaction and Pauson–Khand-Type Reaction

The cobalt-catalyzed Pauson–Khand reaction was first developed in the mid-1990s.[62] Initially, the main problem in this reaction was that the coordinatively unsaturated hexacarbonyldicobalt complex, which decoordinates after the reaction, has a great tendency to polymerize and thus cannot re-enter the catalytic cycle. A large number of catalytic protocols have now been reported that use various cobalt or other metal complexes, but some of these lack general scope. There are only a few examples of intermolecular reactions performed under catalytic conditions. In terms of green chemistry, it is not possible to consider an industrial chemical reaction involving transition-metal complexes that is not catalytically efficient. In terms of synthetic applications of the Pauson–Khand reaction, great effort will be necessary to improve the available catalytic methodologies.[63]

2.3.3.1 Cobalt-Catalyzed Reactions

Initial examples of the catalytic Pauson–Khand protocol used additives as a route to stabilize the cobalt complexes that arise from the first reaction cycle. The idea was to substitute in situ one or more carbon monoxide ligands with different coordinating groups in order to form a more stable complex that could re-enter the catalytic cycle before decomposing into an inactive species. Phosphites[64] and phosphines[65] were first used as additives, albeit with limited substrate scope (Table 5, entries 1 and 2). Very good results are described using tributylphosphine sulfide as an additive with only 1 atmosphere of carbon monoxide and only 3–5 mol% octacarbonyldicobalt(0) (Table 5, entries 3 and 4).[66] Tetramethylthiourea is also an efficient additive under atmospheric pressure of carbon monoxide; the use of this additive with 3–10 mol% octacarbonyldicobalt(0) gives the corresponding products in high yields (Table 5, entries 5 and 6).[67] A cobalt–tetramethylthiourea complex, generated in situ by reduction of cobalt(II) bromide with zinc in the presence of tetramethylthiourea, can be used to catalyze Pauson–Khand reactions at a balloon pressure of carbon monoxide; this catalytic system works efficiently for both inter- and intramolecular examples.[68] 1,2-Dimethoxyethane is also used as an additive, although a relatively high carbon monoxide pressure (7 atm) is required in this case (Table 5, entries 7 and 8).[69] To avoid the use of carbon monoxide gas, a catalytic protocol utilizes carbon monoxide adsorbed in molecular sieves (zeolites) (Table 5, entry 9).[70] Modified carbonyl cobalt complexes with two carbon monoxide groups substituted by phosphines immobilized onto resins are available, these anchored cobalt complexes also catalyze the reaction with good yields (Table 5, entry 10).[71]

for references see p 162

Table 5 Intramolecular Cobalt-Catalyzed Pauson–Khand Reactions Using Additives[64–71]

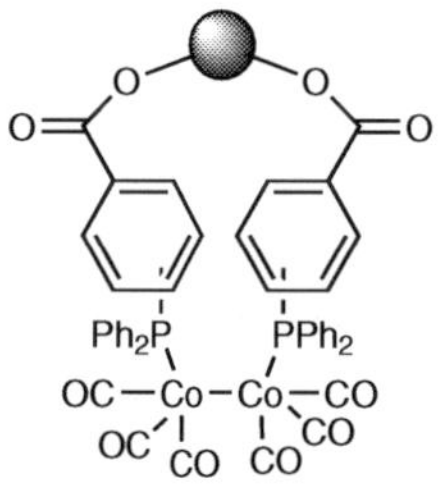

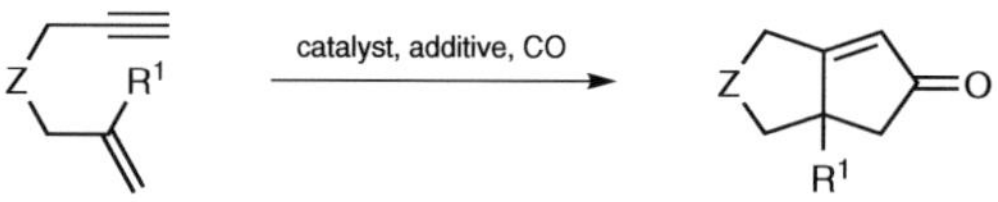

Entry	Starting Material	Catalyst (mol%)	Additive (mol%)	CO (atm)	Conditions	Product	Yield (%)	Ref
1	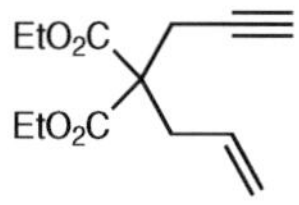 EtO₂C / EtO₂C (enyne)	$Co_2(CO)_8$ (3–5)	$P(OPh)_3$ (10–20)	3	DME, 120 °C	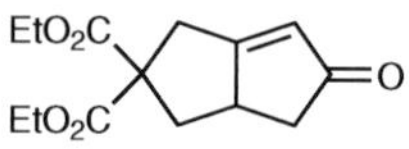 bicyclic enone	82	[64]
2	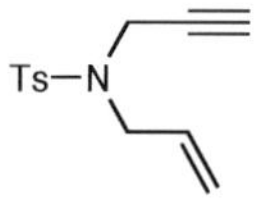 Ts–N (enyne)	$Co_2(CO)_8$ (6)	$(4\text{-MeOC}_6H_4)_3P$ (6)	1.05	THF, 50 °C	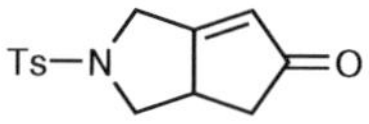 bicyclic enone	91	[65]
3	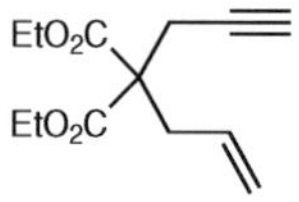 EtO₂C / EtO₂C	$Co_2(CO)_8$ (3–5)	Bu_3PS (18–30)	1	benzene, 70 °C	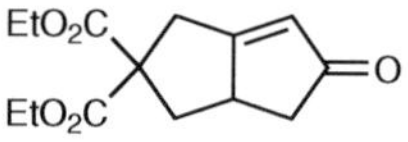 bicyclic enone	90	[66]
4	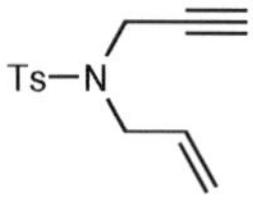 Ts–N	$Co_2(CO)_8$ (3–5)	Bu_3PS (18–30)	1	benzene, 70 °C	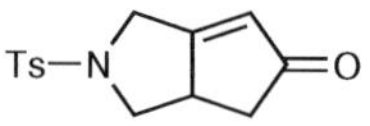 bicyclic enone	87	[66]
5	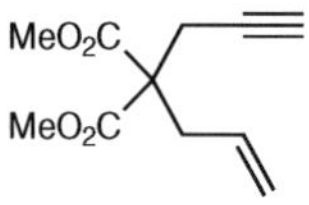 MeO₂C / MeO₂C	$Co_2(CO)_8$ (3–10)	$(Me_2N)_2C{=}S$ (18–60)	1	benzene, 70 °C	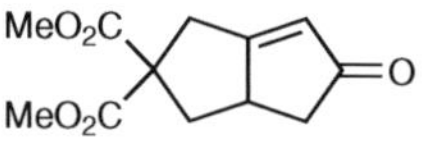 bicyclic enone **45**	92	[67]
6	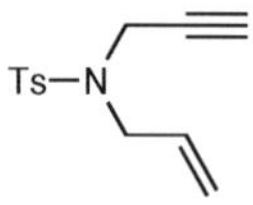 Ts–N	$Co_2(CO)_8$ (3–10)	$(Me_2N)_2C{=}S$ (18–60)	1	benzene, 70 °C	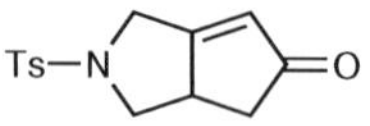 bicyclic enone	95	[67]
7	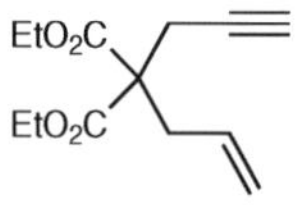 EtO₂C / EtO₂C	$Co_2(CO)_8$ (2–3)	DME (8–12)	7	toluene, 120 °C	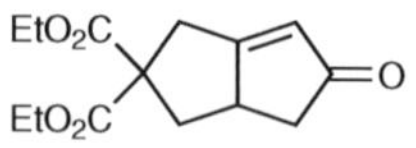 bicyclic enone	94	[69]

○ = dendronized support

Catalyst **44**: dendronized support with two benzoate linkers bearing Ph_2P groups coordinated to $Co_2(CO)_6$ cluster.

2.3.3 The Catalytic Pauson–Khand Reaction and Pauson–Khand-Type Reaction

Table 5 (cont.)

Entry	Starting Material	Catalyst (mol%)	Additive (mol%)	CO (atm)	Conditions	Product	Yield (%)	Ref
8	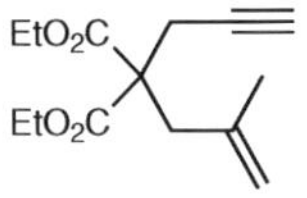 EtO₂C, EtO₂C (alkyne-alkene)	Co₂(CO)₈ (2–3)	DME (8–12)	7	toluene, 120 °C	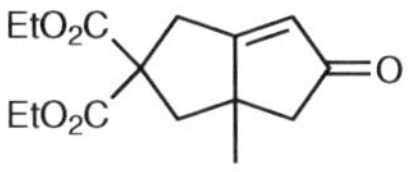 bicyclic enone (EtO₂C, EtO₂C)	86	[69]
9	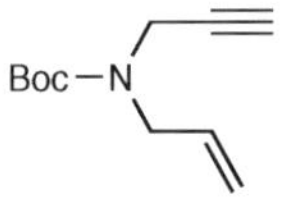 Boc–N (alkyne-alkene)	Co₂(CO)₈ (10)	4-Å molecular sieves[a]	–	toluene, 65 °C	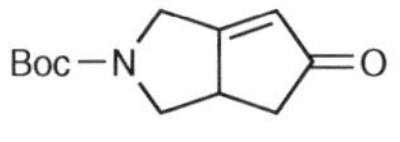 Boc–N bicyclic enone **46**	70	[70]
10	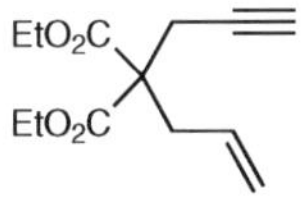 EtO₂C, EtO₂C (alkyne-alkene)	**44**	–	1	THF, 70 °C	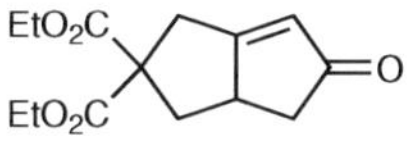 bicyclic enone (EtO₂C, EtO₂C)	68	[71]

[a] Pretreated at 200 °C for 3 h under CO atmosphere.

In an important advance in the development of a catalytic Pauson–Khand reaction, purification of the octacarbonyldicobalt(0) gives a much better catalytic performance and allows the use of only 1 atmosphere of carbon monoxide;[72] it is assumed that impurities induce inactivation of the catalyst. The readily obtained, purified, and shelf-stable alkyne–hexacarbonyldicobalt initial complex **47** is also utilized as a catalyst for the Pauson–Khand reaction; treatment of the alkyne–hexacarbonyldicobalt initial complex with triethylsilane/cyclohexylamine gives the active catalyst (Table 6, entries 1–4).[73] A similar crystalline alkyne–hexacarbonyldicobalt **48** requires either no additive or only cyclohexylamine as the additive (Table 6, entries 5–8).[74]

Table 6 Intramolecular Pauson–Khand Reactions Catalyzed by Preformed Alkyne–Hexacarbonyldicobalt Complexes[73,74]

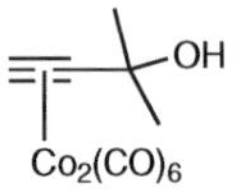

Entry	Enyne	Catalyst	Additive (mol%)	Product	Yield (%)	Ref
1	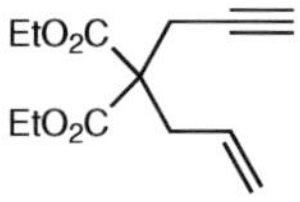 EtO$_2$C, EtO$_2$C (enyne)	47	Et$_3$SiH (5), CyNH$_2$ (15)	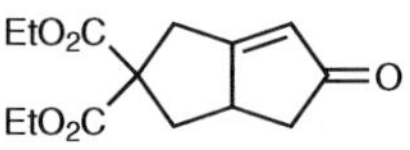 EtO$_2$C, EtO$_2$C bicyclic enone	92	[73]
2	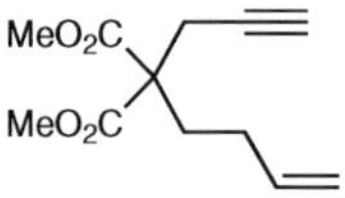 MeO$_2$C, MeO$_2$C (enyne)	47	Et$_3$SiH (5), CyNH$_2$ (15)	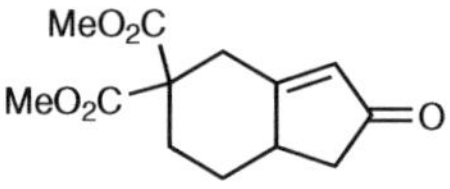 MeO$_2$C, MeO$_2$C bicyclic enone	86	[73]
3	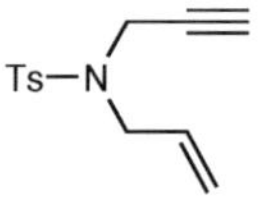 Ts–N (enyne)	47	Et$_3$SiH (5), CyNH$_2$ (15)	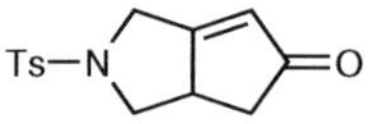 Ts–N bicyclic enone	77	[73]
4	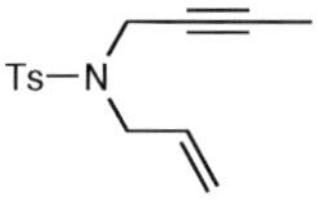 Ts–N (enyne)	47	Et$_3$SiH (5), CyNH$_2$ (15)	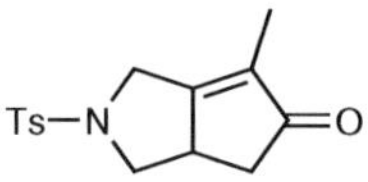 Ts–N methyl bicyclic enone	92	[73]
5	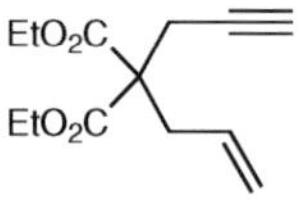 EtO$_2$C, EtO$_2$C (enyne)	48	–	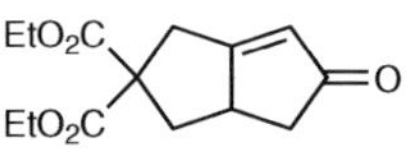 EtO$_2$C, EtO$_2$C bicyclic enone	78	[74]
6	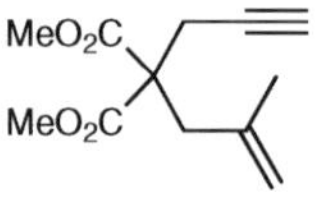 MeO$_2$C, MeO$_2$C (enyne)	48	–	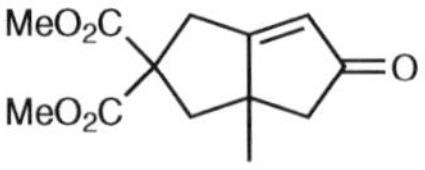 MeO$_2$C, MeO$_2$C bicyclic enone	84	[74]
7	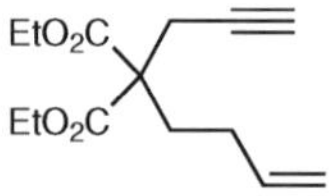 EtO$_2$C, EtO$_2$C (enyne)	48	CyNH$_2$ (20)	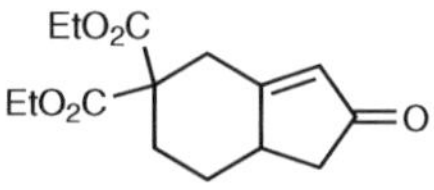 EtO$_2$C, EtO$_2$C bicyclic enone	81	[74]
8	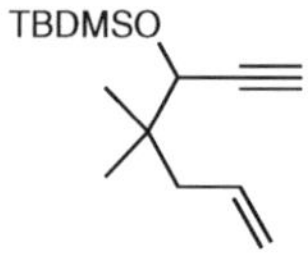 TBDMSO (enyne)	48	CyNH$_2$ (20)	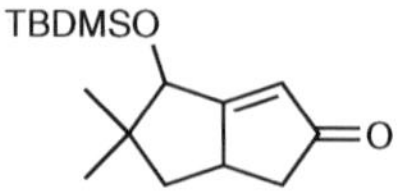 TBDMSO bicyclic enone	83	[74]

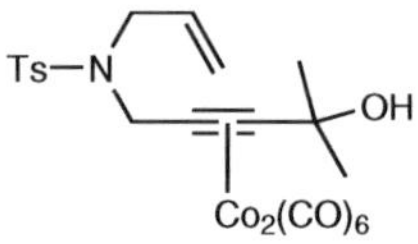

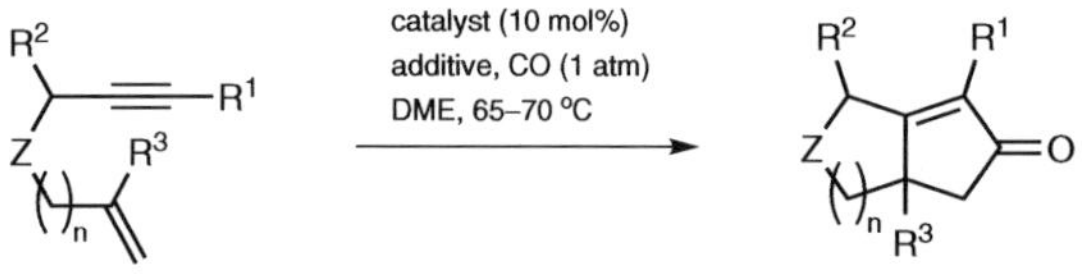

Conditions: catalyst (10 mol%), additive, CO (1 atm), DME, 65–70 °C

Compound 47: HC≡C–C(OH)– with Co$_2$(CO)$_6$
Compound 48: Ts–N(CH$_2$CH=CH$_2$)–CH$_2$–C≡C–C(OH)– with Co$_2$(CO)$_6$

General transformation: enyne with R^1, R^2, R^3, Z, n → bicyclic enone 49

Other cobalt clusters, such as nonacarbonyl(methylidyne)tricobalt(0)[75] or dodecacarbonyltetracobalt(0), exhibit high reactivity in the catalytic Pauson–Khand reaction. Dodecacarbonyltetracobalt(0) also functions as a catalyst with enynes under aqueous conditions.[76]

Dimethyl 5-Oxo-3,3a,4,5-tetrahydropentalene-2,2(1*H*)-dicarboxylate (45); Typical Procedure:[67]

> **CAUTION:** *Carbon monoxide is extremely flammable and toxic, and exposure to higher concentrations can quickly lead to a coma.*

To a soln of $Co_2(CO)_8$ (5.1 mg, 0.015 mmol) and tetramethylthiourea (11.9 mg, 0.09 mmol) in dry benzene (**CAUTION:** *carcinogen*) charged with CO (supplied from a balloon) was added dropwise a soln of dimethyl 2-allyl-2-(prop-2-ynyl)malonate (105 mg, 0.5 mmol) in benzene (10 mL) and the mixture was stirred at 70 °C for 2 h (the color of the mixture changed from orange/red to black). The solvent was removed under reduced pressure, and the residue was purified by flash chromatography (silica gel, petroleum ether/EtOAc 6:1 then 2:1) to give the product as a colorless oil; yield: 109.5 mg (92%).

tert-Butyl 5-Oxo-3,3a,4,5-tetrahydrocyclopenta[*c*]pyrrole-2(1*H*)-carboxylate (46); Typical Procedure:[70]

> **CAUTION:** *Carbon monoxide is extremely flammable and toxic, and exposure to higher concentrations can quickly lead to a coma.*

A flask containing powdered 4-Å molecular sieves was heated at 200 °C for 3 h and then cooled under a CO atmosphere. These molecular sieves (twice the mass of the enyne) were added to a flask containing a soln of *tert*-butyl allyl(prop-2-ynyl)carbamate (195 mg, 1.00 mmol) in toluene (10 mL) at rt. To this soln, $Co_2(CO)_8$ (34.2 mg, 0.10 mmol) was added and the resulting mixture was stirred for 18 h at 65 °C. After filtration through Celite, the solvent was evaporated under reduced pressure and the crude product was purified by flash chromatography (hexane/EtOAc) to give the product as a colorless oil; yield: 156 mg (70%).

Fused Cyclopentenones 49 (Table 6, Entries 5–8); General Procedure:[74]

> **CAUTION:** *Carbon monoxide is extremely flammable and toxic, and exposure to higher concentrations can quickly lead to a coma.*

Synthesis of the catalyst: To a soln of *N*-allyl-4-hydroxy-4-methyl-*N*-tosylpent-2-ynylamine (159 mg, 0.488 mmol) in petroleum ether/CH_2Cl_2 (1:1; 5 mL) was added solid $Co_2(CO)_8$ (220 mg, 0.643 mmol) at rt. The mixture was stirred for 30 min, and then it was directly flash chromatographed (silica gel, petroleum ether/Et_2O 1:1). Concentration of the eluent at rt under reduced pressure gave **48** as a red-brown solid; yield: 260 mg (90%).

Pauson–Khand reaction: A carefully base-washed 10-mL flask, equipped with a three-way stopper, a condenser, and a balloon of CO containing a mixture of the enyne (0.213 mmol) and the catalyst **48** (12.1 mg, 0.0204 mmol) was pumped briefly and purged with CO (3 ×). Freshly distilled 1,2-dimethoxyethane (1.0 mL) was added and the resulting soln was heated to 70 °C and stirred for 2 h. Upon completion of the reaction, the mixture was cooled to rt, quenched with water, and extracted with EtOAc. The organic layer was dried (Na_2SO_4) and concentrated under reduced pressure. The residue was flash chromatographed (silica gel, hexane/EtOAc 2:1) to give the final enones **49**; yield: 78–84%. If appropriate, cyclohexylamine was added prior to heating. Alternatively, the reaction could be worked up by dilution with EtOAc/hexanes (1:9) and filtered through a pad of silica gel (hexane/EtOAc). It was further observed that inadvertent introduction of air into the system from frequent checking of the reaction by TLC resulted in incomplete consumption of the starting material.

2.3.3.2 Rhodium-Catalyzed Reactions

Rhodium complexes are effective catalysts for the Pauson–Khand reaction. Dicarbonylchlororhodium(I) dimer {RhCl(CO)$_2$}$_2$[77] is used as catalyst, and several new species are also utilized, some of which need activation with silver(I) trifluoromethanesulfonate.[78] The diastereoselectivity of the reaction is strongly dependent on the carbon monoxide pressure; the best results are achieved at 1 atmosphere.[79] An explanation of the reaction pathway uses DFT computational calculations. The key transition state **50** is highly polarized, and this is why the reaction works well with nonterminal alkynes. Surprisingly, the polarization is opposite in the transition states *cis*-**50** and *trans*-**50** leading to the formation of diastereomers. It is suggested that the best way to achieve good yields and diastereoselectivities is to use chloro-substituted enynes (Scheme 14).[80]

Scheme 14 DFT Study on the Polarization and Regioselectivity of the Intramolecular Rhodium-Catalyzed Pauson–Khand-Type Reaction of an Enyne[80]

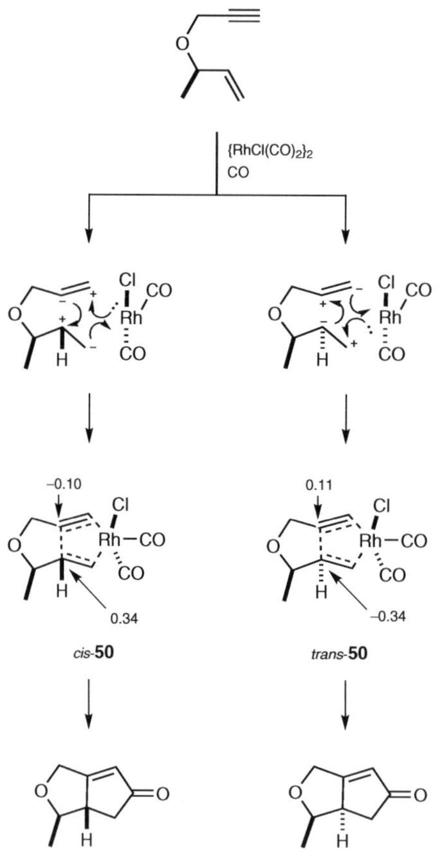

The enantioselective version of the rhodium-catalyzed Pauson–Khand reaction is performed using chiral ligands, generally binaphthyl-derived phosphines. A careful choice of conditions, including carbon monoxide pressure, activation, solvent, and ligands, is essential to obtain high enantioselectivity (Table 7). Coordinating solvents such as tetrahydrofuran are the best choice in these reactions. The use of (*S*)-BINAP with dicarbonylchlororhodium(I) dimer as the catalyst system with activation by silver(I) trifluoromethanesulfonate achieves good yields under relatively mild conditions albeit with low enantiomeric excess (Table 7, entries 1–4).[81] Changing the ligand to the more demanding phosphines

2.3.3 The Catalytic Pauson–Khand Reaction and Pauson–Khand-Type Reaction

(R)-XylBINAP (**51**) or (S)-DTMB-MeO-BIPHEP (**52**) gives a notable improvement to the enantiomeric excess (Table 7, entries 5–8).[82,83] Other ligands such as (R)-SIPHOS (**53**) and (S)-SYNPHOS (**54**) also give good enantioselectivities, although yields are significantly lower (Table 7, entries 9–11).[84,85]

Table 7 Enantioselective Intramolecular Pauson–Khand Reactions of Enynes Catalyzed by Rhodium Complexes[81–85]

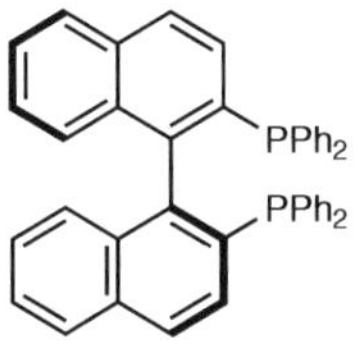

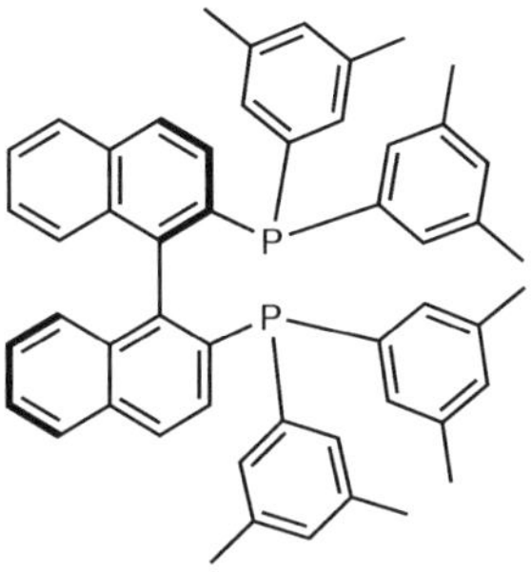

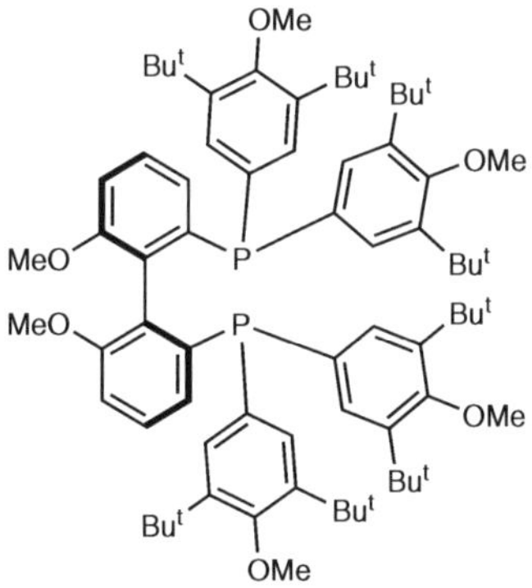

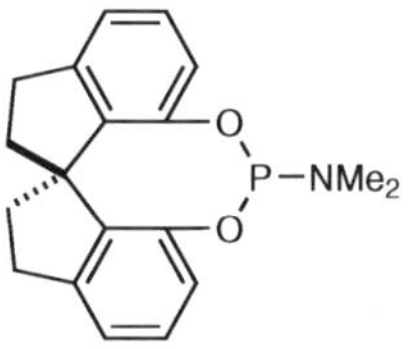

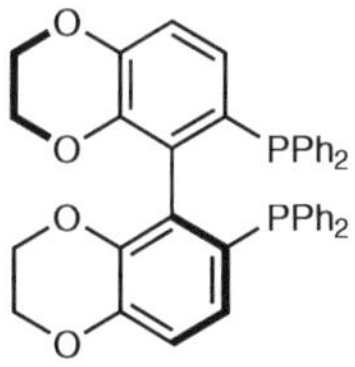

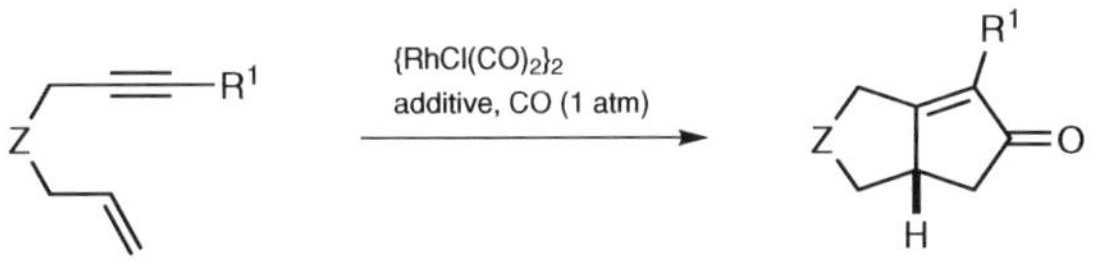

Entry	Starting Material	mol% of Catalyst	Additive (mol%)	Conditions	Product	ee (%)	Yield (%)	Ref
1	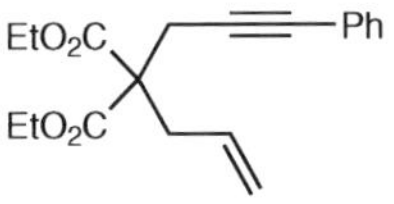 EtO₂C, EtO₂C, ≡Ph	3	(S)-BINAP (6), AgOTf (12)	THF, 90 °C	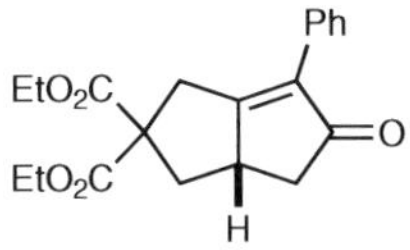	61	67	[81]
2	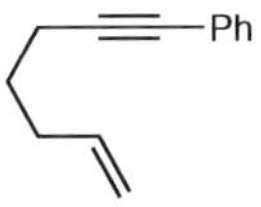 ≡Ph	3	(S)-BINAP (6), AgOTf (12)	THF, 90 °C	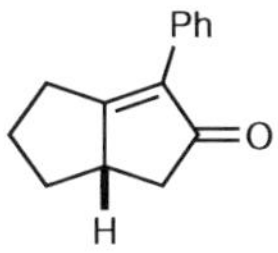	51	61	[81]

Table 7 (cont.)

Entry	Starting Material	mol% of Catalyst	Additive (mol%)	Conditions	Product	ee (%)	Yield (%)	Ref
3	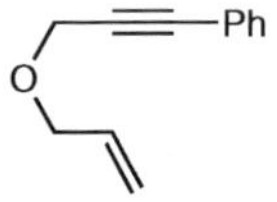 (enyne with O, Ph)	3	(S)-BINAP (6), AgOTf (12)	THF, 90 °C	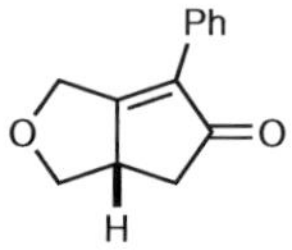 bicyclic enone with Ph, O	81	88	[81]
4	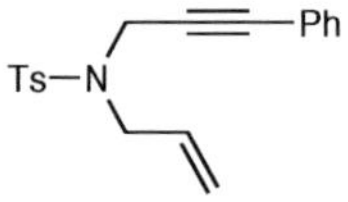 Ts–N enyne with Ph	3	(S)-BINAP (6), AgOTf (12)	THF, 90 °C	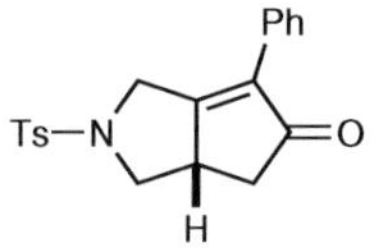 Ts–N bicyclic enone with Ph	74	93	[81]
5	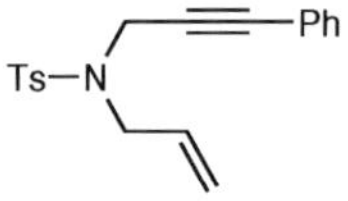 Ts–N enyne with Ph	5	52 (10), AgOTf (15)	THF, 80 °C	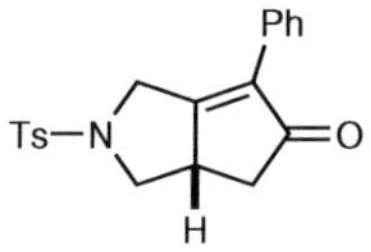 Ts–N bicyclic enone with Ph	91	96	[82]
6	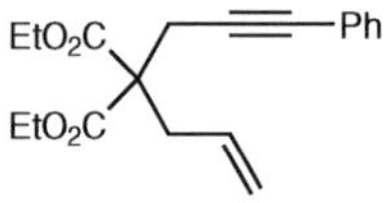 EtO₂C, EtO₂C enyne with Ph	5	52 (10), AgOTf (15)	THF, 75 °C	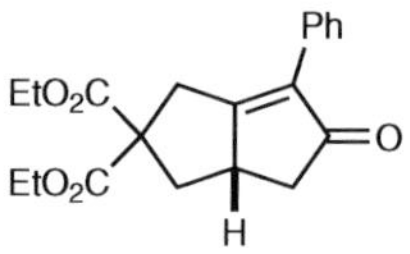 EtO₂C, EtO₂C bicyclic enone with Ph	93	98	[82]
7	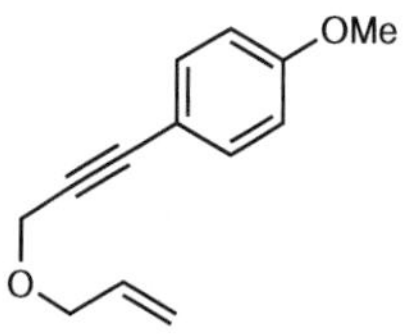 enyne with OMe-aryl and O	5	52 (10), AgOTf (15)	THF, 70 °C	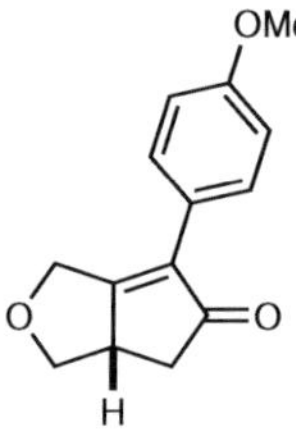 bicyclic enone with p-OMe-C₆H₄	95	92	[82]
8	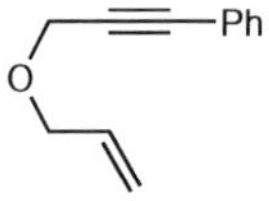 enyne with O, Ph	5	51 (10), AgOTf (12)	THF, 20 °C	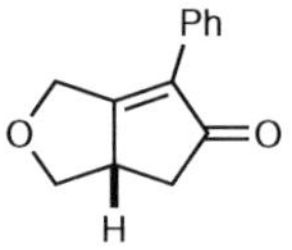 bicyclic enone with Ph, O	92	99	[83]
9	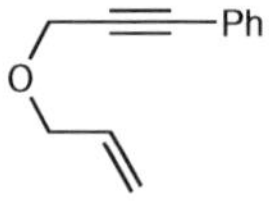 enyne with O, Ph	3	53 (13.2), AgSbF₆ (12)	1,2-dichloro-ethane, 90 °C	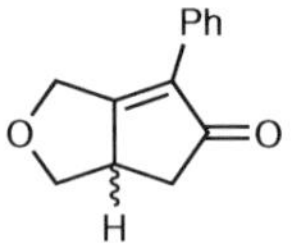 bicyclic enone with Ph, O	84	56	[84]
10	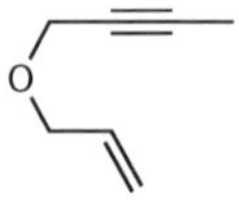 enyne with O, Me	3	54 (9), AgSbF₆ (12)	THF, 80 °C	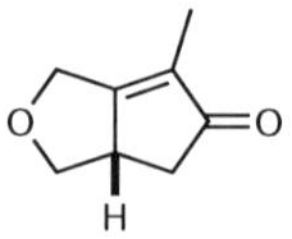 bicyclic enone with Me, O	98	42	[85]
11	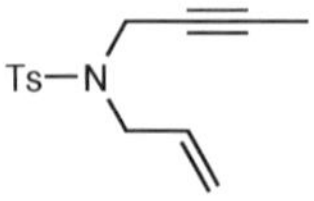 Ts–N enyne with Me	3	54 (9), AgSbF₆ (12)	THF, 80 °C	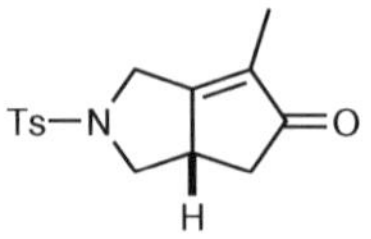 Ts–N bicyclic enone with Me	93	57	[85]

Fused Cyclopentenones 55 (Table 7, Entries 5–7); General Procedure:[82]

> CAUTION: *Carbon monoxide is extremely flammable and toxic, and exposure to higher concentrations can quickly lead to a coma.*

A soln of {RhCl(CO)$_2$}$_2$ (1.8 mg, 0.005 mmol, 5 mol%) and (S)-DTBM-MeO-BIPHEP (**52**; 10.6 mg, 0.009 mmol, 10 mol%) in THF (1 mL) was stirred for 30 min at 20 °C under atmospheric pressure of argon. A soln of AgOTf (3.5 mg, 0.014 mmol, 15 mol%) in THF (1 mL) was added, and the resulting mixture was stirred for a further 30 min at 20 °C. The argon atmosphere was replaced by CO in argon (1:10; 1 atm), and then a soln of the enyne (0.092 mmol) in THF (1 mL) was introduced. The mixture was heated at 70–80 °C. After completion of the reaction, the gases were released in the hood. The crude mixture was concentrated under reduced pressure, and then the residue was flash chromatographed (silica gel, hexane/EtOAc) to give the product. The enantiomeric excess of the product was determined by chiral HPLC analysis using Daicel columns.

(S)-6-Phenyl-3a,4-dihydro-1*H*-cyclopenta[c]furan-5(3*H*)-one (Table 7, Entry 8); Typical Procedure:[83]

> CAUTION: *Carbon monoxide is extremely flammable and toxic, and exposure to higher concentrations can quickly lead to a coma.*

A soln of {RhCl(CO)$_2$}$_2$ (3.4 mg, 0.009 mmol, 5 mol%) and ligand **51** (12.8 mg, 0.017 mmol, 10 mol%) in THF (2 mL) was stirred for 30 min at 20 °C under atmospheric pressure of argon. A soln of AgOTf (12 mol%) in THF (1 mL) was added, and the resulting mixture was stirred for a further 30 min at 20 °C. The argon atmosphere was replaced with CO in argon (1:10; 1 atm), and then a soln of [3-(allyloxy)prop-1-ynyl]benzene (30 mg, 0.174 mmol) in THF (1 mL) was introduced. The mixture was stirred at 20 °C. After completion of the reaction, the gases were released in the hood. The crude mixture was concentrated under reduced pressure, and then the residue was flash chromatographed (silica gel, hexane/EtOAc) to give the product as a colorless oil; yield: 34 mg (99%); 92% ee [HPLC (Daicel Chiralpak AD-H, hexane/iPrOH 4:1, flow 0.8 mL·min^{-1}, detection at 254 nm); t_R = 10.63 (major), 13.01 min (minor)].

2.3.3.2.1 Dienes as Alkene Partners: The Diene Effect

Extensive studies show dienes to be interesting alkene partners for both the intra- and intermolecular Pauson–Khand reaction.[86] In principle, the reaction of dienynes with rhodium catalysts gives the corresponding organometallic intermediates that can proceed to a (4+2) reaction to give **58** or may incorporate a molecule of carbon monoxide to produce either a (2+2+1) or (4+2+1) reaction to give **56** or **57**, respectively (Scheme 15). Under atmospheric pressure of carbon monoxide at 40 °C in tetrahydrofuran, (4+2) cycloaddition proceeds to give **58** as the major pathway, and (2+2+1) and (4+2+1) cycloadducts **56** and **57** are minor products. On the other hand, when the reaction is performed at room temperature in 1,2-dichloroethane, (2+2+1) cycloaddition to give **56** proceeds predominantly. This reaction works best with carbonylchlorobis(triphenylphosphine)rhodium(I) relative to other rhodium(I) catalysts.[87]

Scheme 15 Intramolecular Pauson–Khand Reactions of Dienynes under Rhodium(I) Catalysis[87]

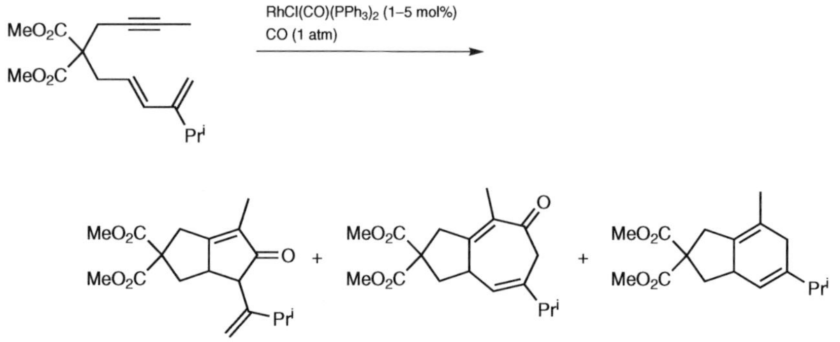

56 **57** **58**

Conditions	Yield (%) of **56**	Yield (%) of **57**	Yield (%) of **58**	Ref
THF, 40 °C	25	18	45	[87]
1,2-dichloroethane, rt	89	–	7	[87]

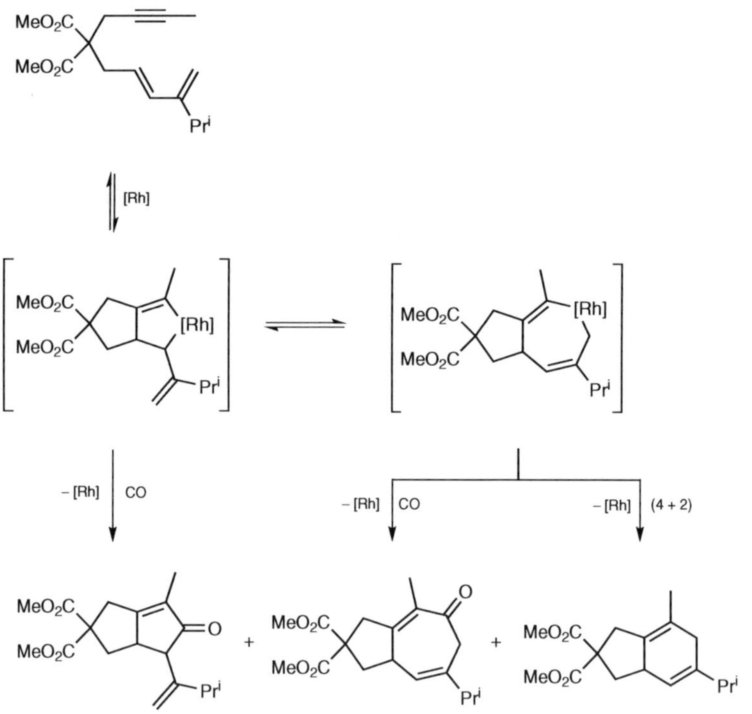

56 **57** **58**

Significantly, the intermolecular diene-Pauson–Khand reaction proceeds efficiently to give products such as **59**. The dienes provide the necessary reactivity enhancement for the intermolecular version similar to strained alkenes. The reaction can be conducted in the absence of solvent, and it can also be scaled up with similar efficiency (Scheme 16).[88]

2.3.3 The Catalytic Pauson–Khand Reaction and Pauson–Khand-Type Reaction

Scheme 16 Intermolecular Pauson–Khand Reaction of 2,3-Dimethylbuta-1,3-diene and Alkynes under Rhodium(I) Catalysis[88]

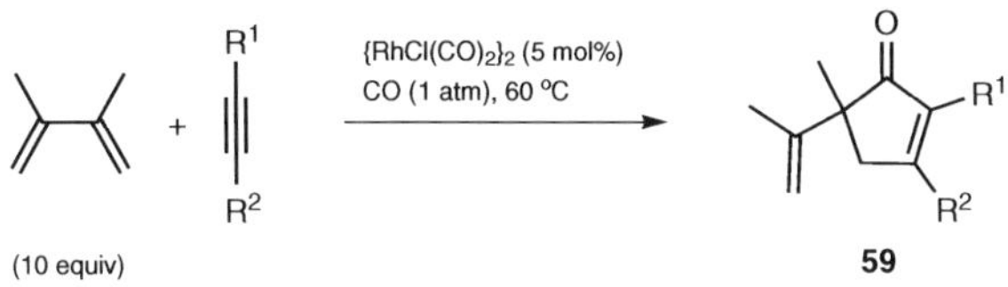

R[1]	R[2]	Conditions	Yield (%)	Ref
CH$_2$OMe	CH$_2$OMe	1,2-dichloroethane, 6 h	98	[88]
CH$_2$OMe	CH$_2$OMe	1,1,2,2-tetrachloroethane, 7 h	96	[88]
CH$_2$OMe	CH$_2$OMe	neat, 9 h	86	[88]
TMS	CO$_2$Et	1,2-dichloroethane/1,1,2,2-tetrachloroethane (1:1), 24 h	95	[88]

Mechanistic studies suggest that the unique reactivity observed with dienes arises from their participation in the putative rate-determining reductive elimination step by providing an additional energy-lowering coordination site for the transition-metal catalyst.[89]

2.3.3.2.2 Allenes as Alkene Partners

The use of allenynes has expanded the utility of rhodium catalysts. In particular, the synthesis of the bicyclo[5.3.0] ring system is possible with these substrates, while typical enynes only give fused seven-membered rings if specific structural motifs are present. A high-yielding procedure for the preparation of bicyclo[5.3.0] compounds is shown in Table 8. Treatment of 3,3-disubstituted allenynes with 2.5 or 5 mol% of dicarbonylchlororhodium(I) dimer in refluxing toluene under 1 atmosphere of carbon monoxide affords the desired cycloaddition products **60** in moderate to high yields (Table 8, entries 1–5).[90] The parent [4.3.0] systems are also formed under similar reaction conditions (Table 8, entries 6–9).[91]

Table 8 Rhodium-Catalyzed Intramolecular Pauson–Khand Reactions of Allenynes[90,91]

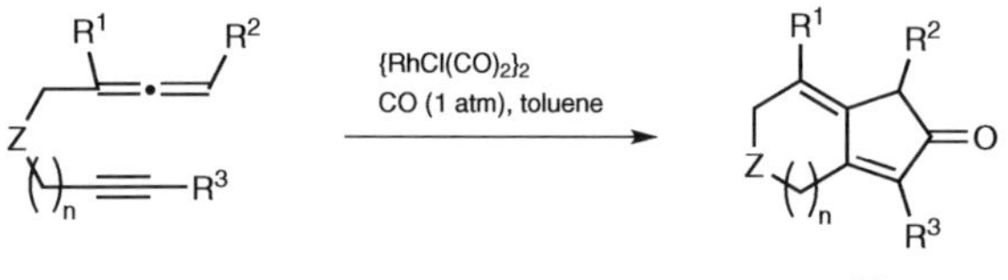

Entry	Starting Material	Catalyst (mol%)	Product	Yield (%)	Ref
1	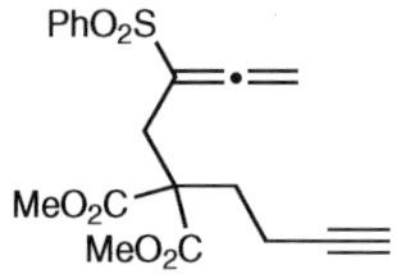	2.5	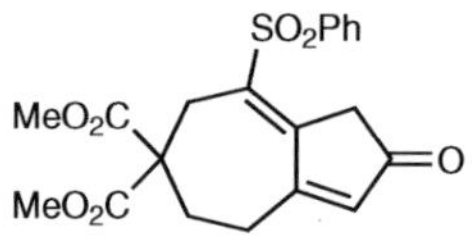	84	[90]
2	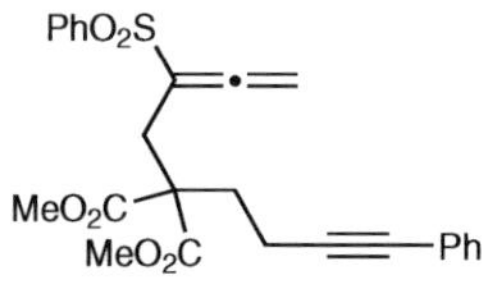	2.5	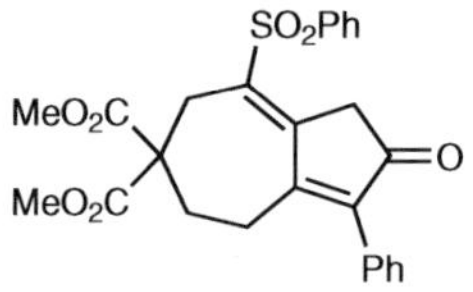	70	[90]
3	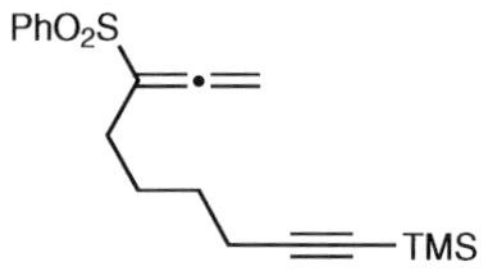	5.0	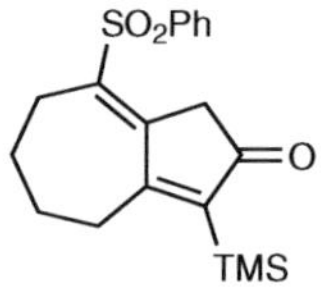	45	[90]
4	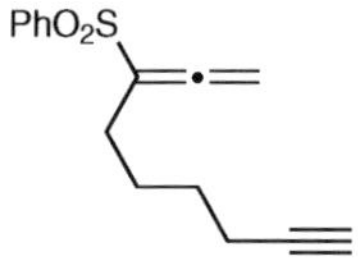	5.0	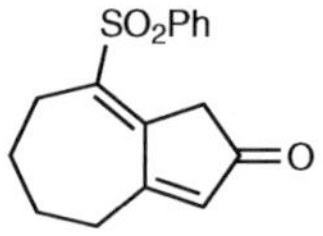	58	[90]
5	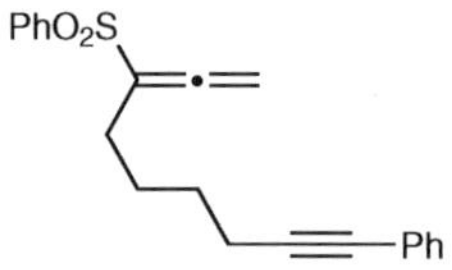	2.5	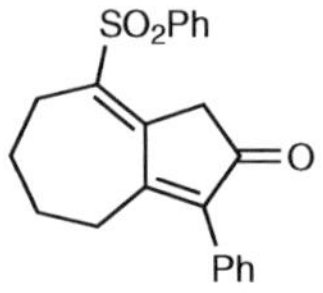	51	[90]
6	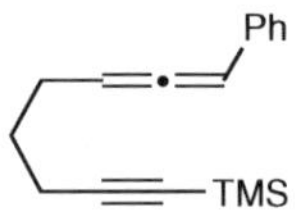	5.0	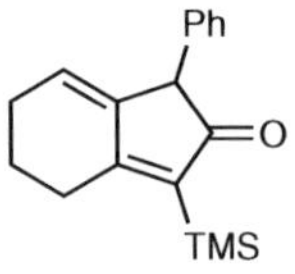	64	[91]

Table 8 (cont.)

Entry	Starting Material	Catalyst (mol%)	Product	Yield (%)	Ref
7	(allene-ene with TMS alkyne, 6)	5.0	bicyclic enone with TMS, 6	88	[91]
8	(allene-ene with Ph alkyne, 6)	5.0	bicyclic enone with Ph, 6	60	[91]
9	(TMS-allene with CO₂Me alkyne, TMS)	5.0	bicyclic enone with TMS, CO₂Me, TMS	75	[91]

2.3.3.2.3 Reaction in the Absence of Carbon Monoxide

See also *Science of Synthesis*, Vol. 40 (Section 40.1.1.5.5.3.1.4). The major drawback to the development of efficient catalytic Pauson–Khand protocols is the use of carbon monoxide. Manipulation of such a highly toxic gas discourages many groups and it is a great problem in scaling up the reaction. The introduction of rhodium complexes in the catalytic Pauson–Khand reactions has resulted in the use of the decarbonylation of aldehydes to produce carbon monoxide. The same catalyst is able to mediate both the decarbonylation and the transfer of the carbon monoxide to the enyne. Although rhodium complexes are preferred for this concurrent tandem process, ruthenium- and iridium-catalyzed reactions are also used.[92]

Using pentafluorobenzaldehyde with chloro(cycloocta-1,5-diene)rhodium(I) dimer (5 mol%) and 1,3-bis(diphenylphosphino)propane (11 mol%) as the catalytic system gives a high-yielding reaction (Table 9, entry 1). Chloro(cycloocta-1,5-diene)iridium(I) dimer and dodecacarbonyltriruthenium(0) are also able to transfer carbon monoxide from the aldehyde although with lower efficiency.[92] The use of a solvent-free reaction enables cinnamaldehyde or benzaldehyde to be used as the aldehyde component (Table 9, entries 2–4).[93,94] When the reaction is performed under an atmosphere of ^{13}C-carbon monoxide, very little incorporation of ^{13}C into the final product is observed. This shows that there is no free carbon monoxide in the reaction medium available to be incorporated into the Pauson–Khand product.

The use of formaldehyde as the carbon monoxide source in aqueous media (Table 9, entries 5 and 6) and the introduced chiral ligands in the rhodium complex (Table 9, entry 7) gives high yields and high enantioselectivities. A hydrophilic phosphine and a surfactant can also be added to enable the reaction.[95,96] An efficient reaction in 2-methylbutan-2-ol uses (S)-SYNPHOS (**54**) as ligand (Table 9, entry 8).[97] Primary alcohols can also be used as a source of carbon monoxide under rhodium catalysis.[98] In this process, initial dehydrogenation of the alcohol forms the aldehyde, which proceeds as in previous methods. The yields are high in some cases (Table 9, entry 9), although side products of reductive cyclization are sometimes isolated.

Table 9 Rhodium-Catalyzed Intramolecular Pauson–Khand Reactions of Enynes Using Aldehydes or Alcohols as the Carbon Monoxide Source[92–98]

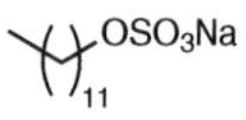

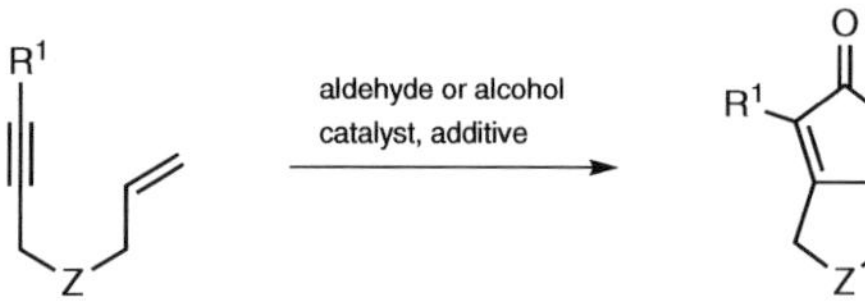

Entry	Starting Material	CO Source (Equiv)	Catalyst (mol%)	Conditions	Product	Yield (%)	Ref
1	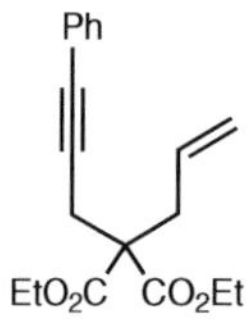 Ph-alkyne-allyl with C(CO₂Et)₂	C₆F₅CHO (2)	[RhCl(cod)]₂ (5), dppp (11)	xylene, 130 °C	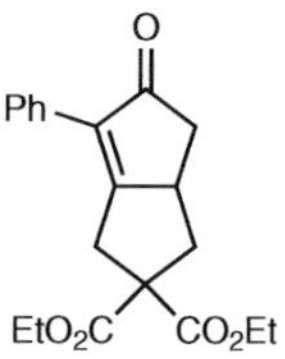 bicyclic enone with Ph, C(CO₂Et)₂	97	[92]
2	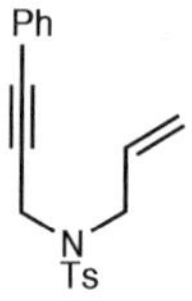 Ph-alkyne-allyl with NTs	PhCH=CHCHO (20)	RhCl(dppp)₂ (5)	120 °C	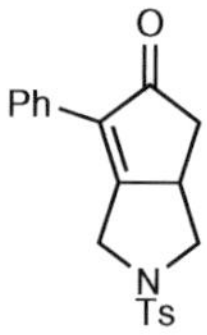 bicyclic enone with Ph, NTs	98	[93]
3	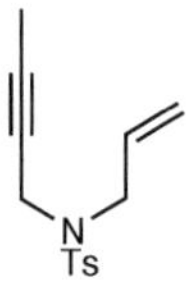 Me-alkyne-allyl with NTs	PhCH=CHCHO (20)	RhCl(dppp)₂ (5)	120 °C	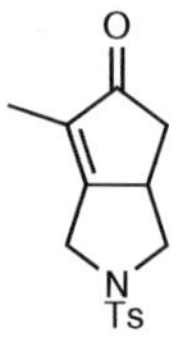 bicyclic enone with Me, NTs	90	[93]
4	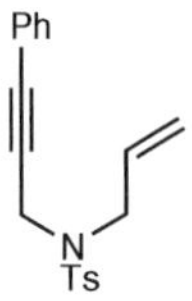 Ph-alkyne-allyl with NTs	PhCHO (20)	[RhCl(cod)]₂ (5), (S)-TolBINAP (10)	120 °C	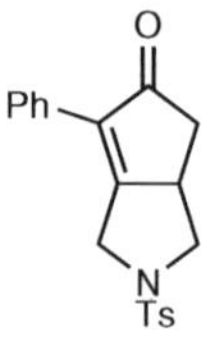 bicyclic enone with Ph, NTs	82[a]	[94]
5	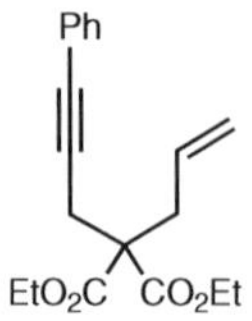 Ph-alkyne-allyl with C(CO₂Et)₂	HCHO (2.5)	[RhCl(cod)]₂ (5), dppp (10)	TPPTS (10 mol%), SDS (2 equiv), H₂O, 100 °C	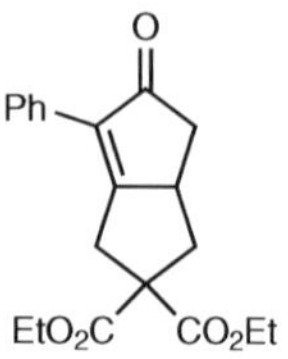 bicyclic enone with Ph, C(CO₂Et)₂	96	[95]

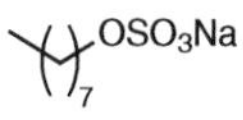

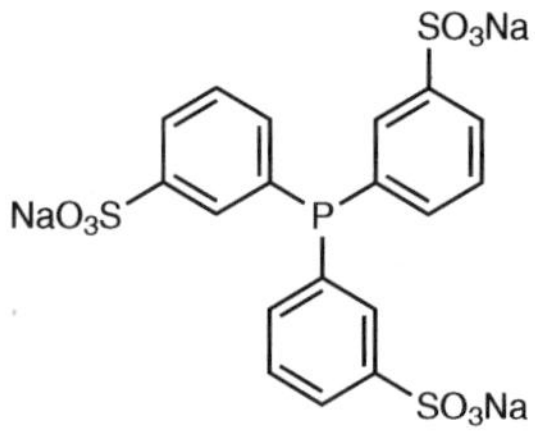

Surfactants and ligand: SDS (C₁₁H₂₃OSO₃Na), SOS (C₇H₁₅OSO₃Na), TPPTS (tris(3-sulfonatophenyl)phosphine trisodium salt).

General scheme: enyne with R¹ on alkyne and tether Z (C(CO₂Et)₂ or NTs) + aldehyde or alcohol, catalyst, additive → bicyclopentenone product.

2.3.3 The Catalytic Pauson–Khand Reaction and Pauson–Khand-Type Reaction

Table 9 (cont.)

Entry	Starting Material	CO Source (Equiv)	Catalyst (mol%)	Conditions	Product	Yield (%)	Ref
6	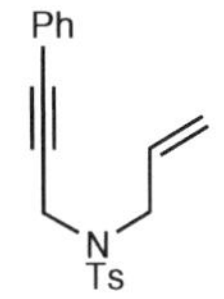 Ph, N-Ts, enyne	HCHO (10)	[RhCl(cod)]$_2$ (5), dppp (10)	TPPTS (10 mol%), SDS (2 equiv), H$_2$O, 100 °C	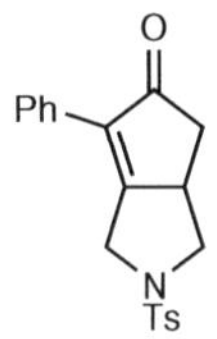 Ph-cyclopentenone-N-Ts	96	[95]
7	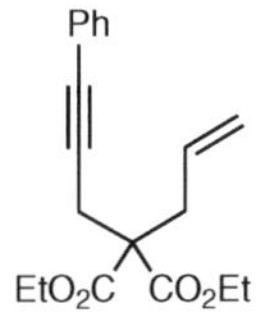 Ph, EtO$_2$C CO$_2$Et enyne	HCHO (5)	[RhCl(cod)]$_2$ (5), (S)-4-TolBINAP (10)	TPPTS (10 mol%), SOS (0.5 equiv), H$_2$O, 100 °C	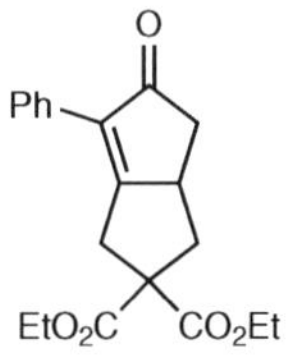 Ph-cyclopentenone, EtO$_2$C CO$_2$Et	83[b]	[96]
8	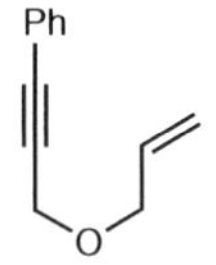 Ph, O-tethered enyne	PhCH=CHCHO (1.5)	[RhCl(cod)]$_2$ (5), (S)-SYNPHOS (10)	EtCMe$_2$OH, 100 °C	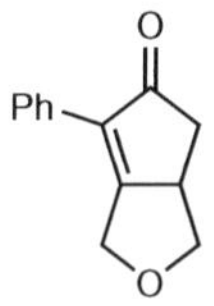 Ph-cyclopentenone, O	81[c]	[97]
9	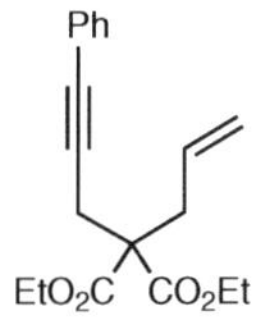 Ph, EtO$_2$C CO$_2$Et enyne	PhCH=CHCH$_2$OH (4)	[RhCl(CO)(dppp)]$_2$ (4)	130 °C	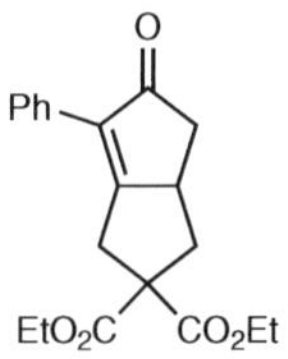 Ph-cyclopentenone, EtO$_2$C CO$_2$Et	93	[98]

[a] 69% ee.
[b] 81% ee.
[c] 85% ee.

2.3.3.3 Iridium-Catalyzed Reactions

Iridium complexes combined with phosphines catalyze the Pauson–Khand reaction [see also *Science of Synthesis: Stereoselective Synthesis*, Vol. 3 (Section 3.3.1.3)]. An initial example of the asymmetric Pauson–Khand reaction uses chiral phosphines, such as (S)-2,2'-bis(di-4-tolylphosphino)-1,1'-binaphthyl [(S)-TolBINAP], and gives high chemical yields and enantioselectivities for intramolecular reactions to give products **63**; however, the reaction is not totally regioselective for intermolecular reactions. For example, in the reaction of 1-phenylpropyne with norbornene a mixture of **61** and **62** is obtained (>10:1 ratio) with 93% ee for the major isomer (Scheme 17).[99] The chiral iridium complex is an efficient catalyst for the decarbonylation of aldehydes and subsequent Pauson–Khand cyclizations used aldehydes in a similar manner to the rhodium-catalyzed reactions discussed in Section 2.3.3.2.3.[100]

Scheme 17 Asymmetric Inter- and Intramolecular Pauson–Khand Reactions Catalyzed by Iridium[99]

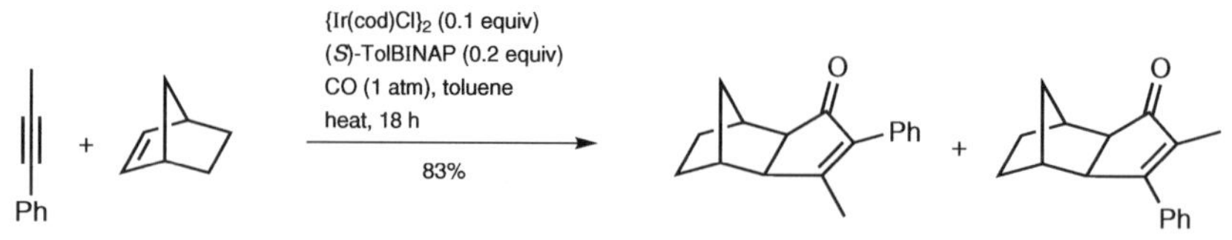

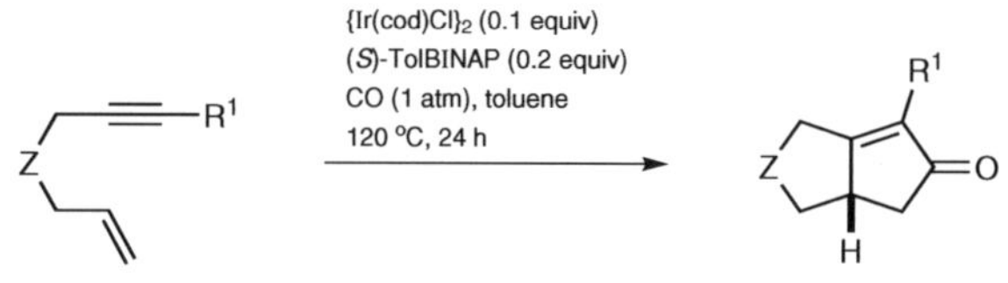

Z	R^1	Time (h)	ee (%)	Yield (%)	Ref
O	4-MeOC$_6$H$_4$	20	96	80	[100]
O	Me	20	98	60	[100]
NTs	Ph	12	95	85	[100]
C(CO$_2$Et)$_2$	Ph	36	88	51	[100]
C(CO$_2$Et)$_2$	CH$_2$OBn	72	84	15	[100]

The use of iridium complexes derived from chiral 2-(2-phosphinophenyl)-4,5-dihydrooxazoles (PHOX-type ligands) in the intramolecular Pauson–Khand reaction provides the products in high yields and with enantioselectivities of >90% ee with 2 mol% of catalyst under the optimized conditions.[101] The chiral iridium catalyst is also used for the desymmetrization of *meso*-dienynes.[102]

2.3.3.4 Other Metal-Catalyzed Pauson–Khand-Type Reactions

Titanium species are also efficient catalysts in the Pauson–Khand reaction and in Pauson–Khand-like reactions with cyanides. A practical procedure uses commercial dicarbonyltitanocene (Table 10, entries 1–3).[103] Dicarbonyltitanocene is able to catalyze the reaction of various 1,6- and 1,7-enynes **66** to give products **67** with excellent functional-group tolerance and under low carbon monoxide pressure, but fails to react with sterically hindered alkenes and alkynes. An enantioselective version of this methodology uses chiral titanocene **64** in the Pauson–Khand reaction and gives good chemical yields with moderate enantiomeric excesses (87–96%) (Table 10, entries 4–6).[104] A dichloro(cyclopentadienyl)(1,1′:3′,1″-terphenyl-2′-yloxy)titanium complex **65** promotes cyclizations with some sterically hindered enynes (Table 10, entries 7–9).[105]

2.3.3 The Catalytic Pauson–Khand Reaction and Pauson–Khand-Type Reaction

Table 10 Titanium-Catalyzed Intramolecular Pauson–Khand Reactions of Enynes[103–105]

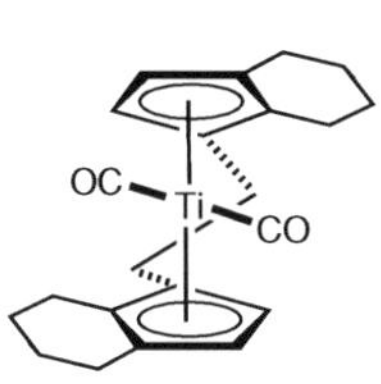

Entry	Starting Material	Catalyst	CO (atm)	Product	ee (%)	Yield (%)	Ref
1	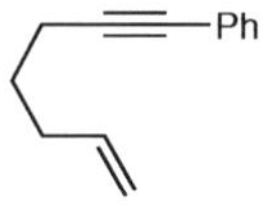 (enyne with Ph)	TiCp$_2$(CO)$_2$	1.22	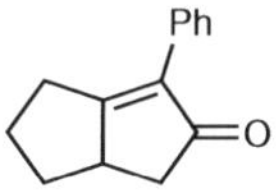 bicyclic enone with Ph	–	87	[103]
2	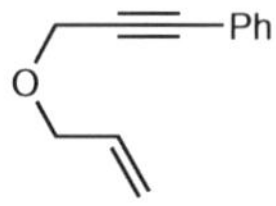 (O-tethered enyne with Ph)	TiCp$_2$(CO)$_2$	1.22	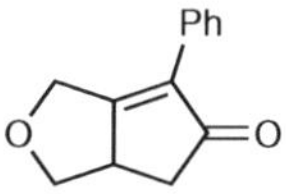 bicyclic enone with Ph, O	–	92	[103]
3	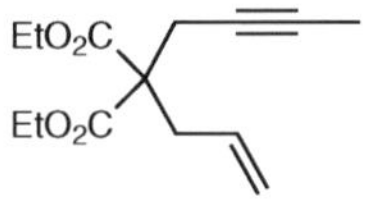 (EtO$_2$C)$_2$C-tethered enyne	TiCp$_2$(CO)$_2$	1.22	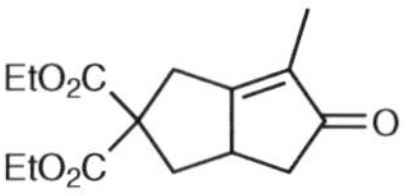 bicyclic enone with (EtO$_2$C)$_2$	–	91	[103]
4	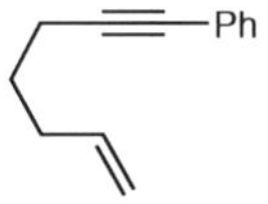 (enyne with Ph)	**64** (20 mol%)	0.95	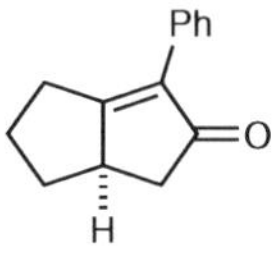 bicyclic enone with Ph, H	87	70	[104]
5	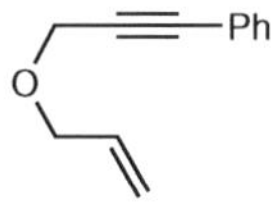 (O-tethered enyne with Ph)	**64** (20 mol%)	0.95	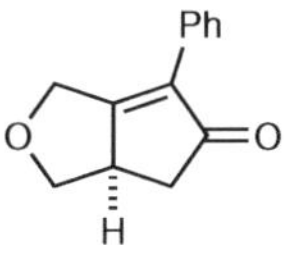 bicyclic enone with Ph, O, H	96	85	[104]
6	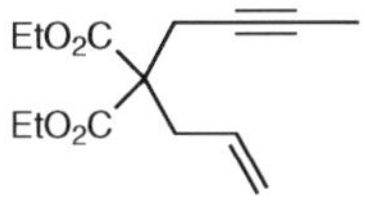 (EtO$_2$C)$_2$C-tethered enyne	**64** (5 mol%)	0.95	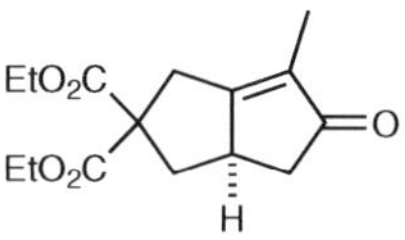 bicyclic enone with (EtO$_2$C)$_2$, H	87	90	[104]

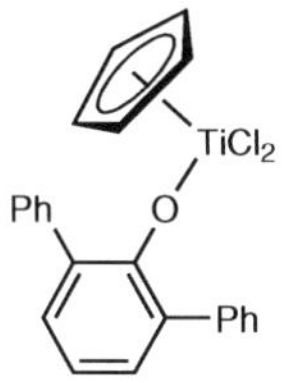

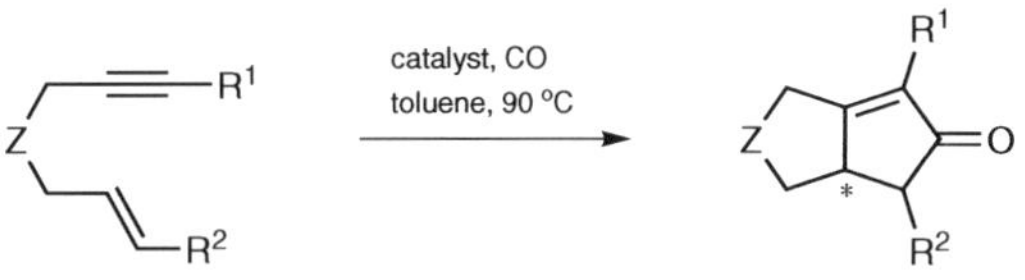

for references see p 162

Table 10 (cont.)

Entry	Starting Material	Catalyst	CO (atm)	Product	ee (%)	Yield (%)	Ref
7	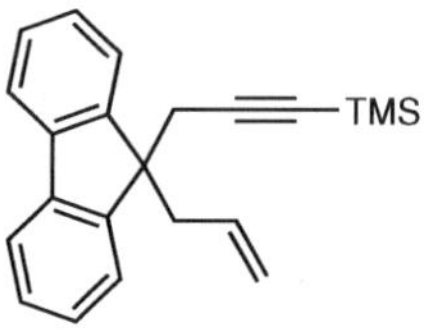 (fluorenyl with TMS-alkyne and allyl)	**65** (0.3 equiv)	2.1	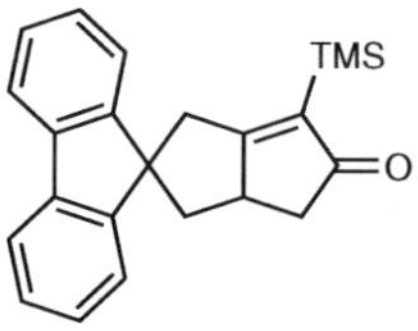 (fluorene-spiro cyclopentenone with TMS)	–	48	[105]
8	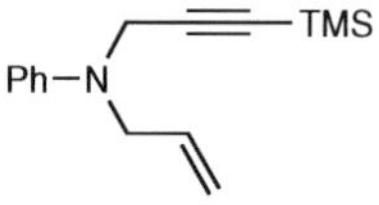 Ph–N(propargyl-TMS)(allyl)	**65** (0.3 equiv)	2.1	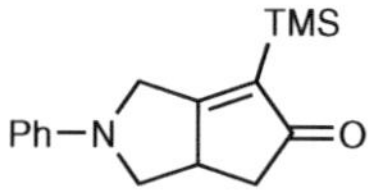 Ph–N bicyclic cyclopentenone with TMS	–	38	[105]
9	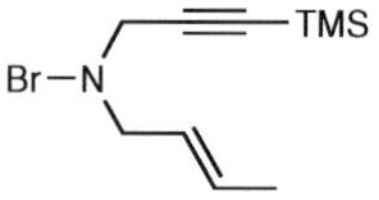 Br–N(propargyl-TMS)(allyl)	**65** (0.3 equiv)	2.1	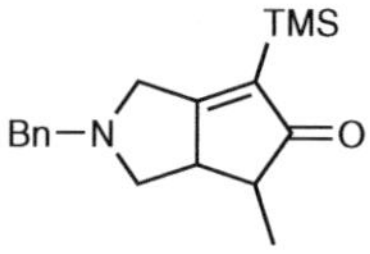 Bn–N bicyclic cyclopentenone with TMS	–	81[a]	[105]

[a] Ratio (*trans/cis*) 3:1.

Dodecacarbonyltriruthenium(0) [Ru$_3$(CO)$_{12}$] is used as a catalyst with enynes bearing disubstituted alkynes.[106,107] However, harsh reaction conditions with high carbon monoxide pressures (10–15 atm) and temperatures (140–160 °C) are required. The use of 2-pyridylsilyl-substituted alkenes **68** enables the ruthenium-catalyzed Pauson–Khand reaction under mild conditions. The pyridyl group directs the Pauson–Khand reaction by possible coordination of the nitrogen with the metal, which accelerates the process and gives complete regioselectivity; the directing group is eliminated in the reaction. This avoids the use of ethene or strained alkenes in the intermolecular Pauson–Khand reaction and solves the problem of the regioselectivity with unsymmetrical alkenes (Scheme 18).[108]

2.3.3 The Catalytic Pauson–Khand Reaction and Pauson–Khand-Type Reaction

Scheme 18 Ruthenium-Catalyzed Intermolecular Pauson–Khand Reaction of 2-Pyridylsilyl-Substituted Alkenes[108]

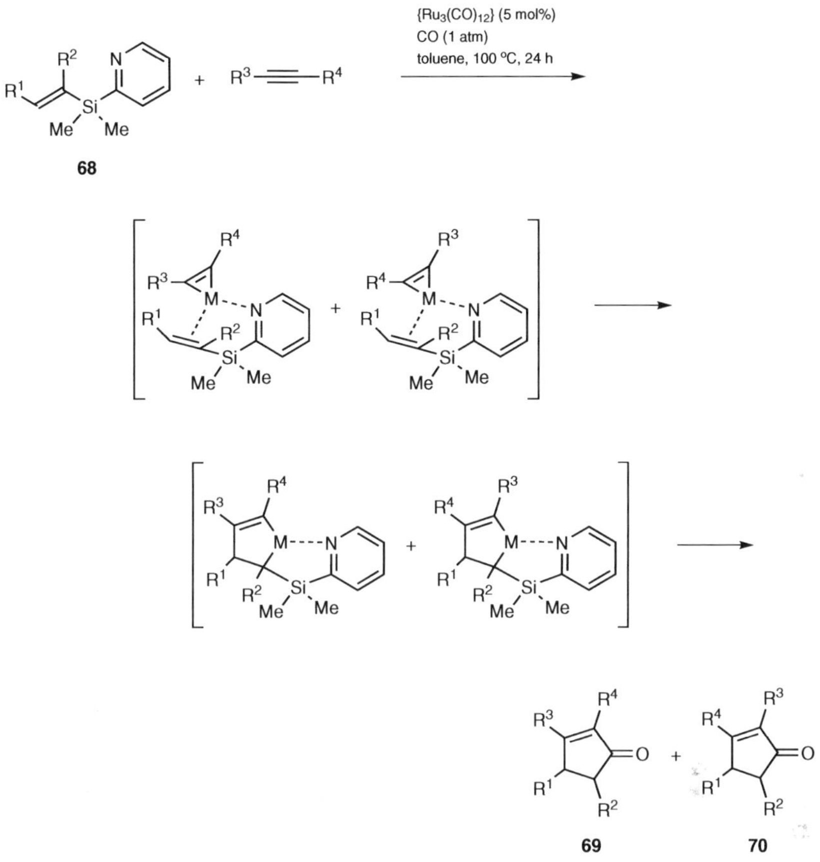

R^1	R^2	R^3	R^4	Ratio (**69/70**)	Yield (%) of **69** + **70**	Ref
H	H	Ph	Ph	–	88	[108]
H	H	H	Ph	100:0	55	[108]
H	H	H	(CH$_2$)$_5$Me	59:41	91	[108]
Bu	H	H	Ph	100:0	41	[108]
H	H	Me	Ph	100:0	75	[108]
H	Me	H	(CH$_2$)$_5$Me	62:38	40	[108]

Tri- and tetranuclear heterobimetallic ruthenium clusters, such as **71**, are used as precatalysts for an intramolecular Pauson–Khand-type reaction. The best results are obtained in the presence of 2 mol% precatalyst under 8 atmospheres of carbon monoxide at 70 °C giving cyclopentenones, such as **72**, in up to 93% yield (Scheme 19).[109]

Scheme 19 A Heterobimetallic Cluster as Precatalyst for a Pauson–Khand-Type Reaction[109]

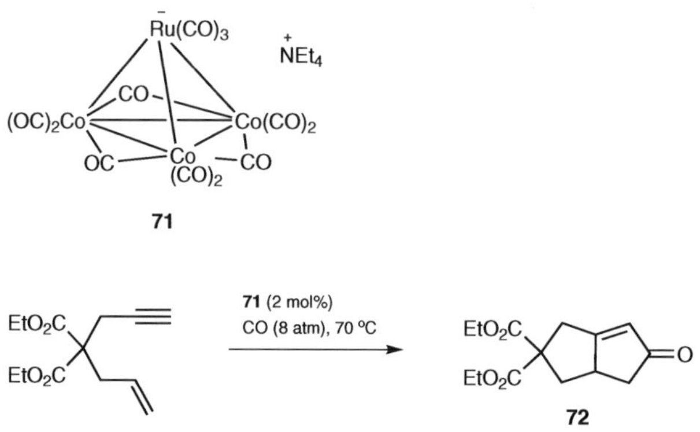

71

72

A successful palladium-catalyzed intramolecular Pauson–Khand-type cyclization utilizes palladium complexes obtained from tetramethylthiourea and palladium(II) chloride and gives the products in up to 96% yield. However, only N-tethered enynes give the corresponding adducts while C- and O-tethered 1,6-enynes are unreactive. The results are influenced by the presence of lithium chloride.[110] The regioselectivity is reversed in the intermolecular reaction of unsymmetrical alkynoates **73** with norbornene using tetramethylthiourea and palladium(II) chloride in the presence or absence of lithium chloride; in the absence of lithium chloride the major regioisomer is **75**, while in the presence of lithium chloride it is **74** (Scheme 20).[111]

Scheme 20 Palladium-Catalyzed Intermolecular Pauson–Khand-Type Cyclization[111]

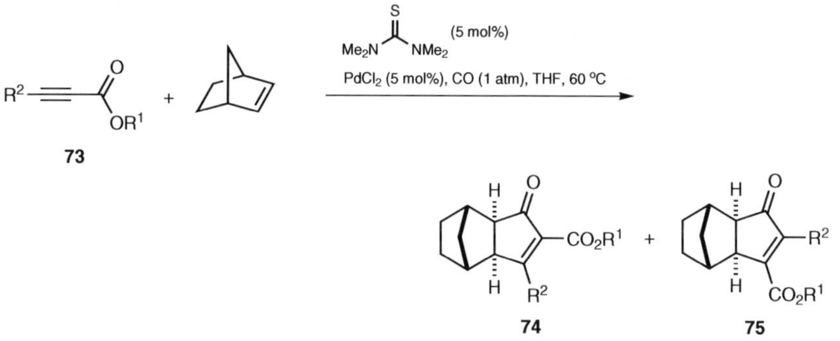

73

74

75

R^1	R^2	Additive	Ratio (**74**/**75**)	Combined Yield (%)	Ref
Et	Me	–	0:100	71	[111]
Et	Me	LiCl	95:5	54	[111]
Et	Bu	–	0:100	32	[111]
Et	Bu	LiCl	96:4	93	[111]
Et	iPr	–	0:100	89	[111]
Et	iPr	LiCl	100:0	80	[111]
Et	Cy	–	10:90	24	[111]
Et	Cy	LiCl	100:0	71	[111]
Bu	Me	–	0:100	18	[111]

R¹	R²	Additive	Ratio (**74**/**75**)	Combined Yield (%)	Ref
Bu	Me	LiCl	100:0	76	[111]
Bn	Me	–	10:90	53	[111]
Bn	Me	LiCl	95:5	70	[111]

The mechanism proposed for the palladium-catalyzed Pauson–Khand reaction is based on both DFT calculations and experimental studies; it involves *cis*-halometalation of the alkyne moiety followed by sequential alkene and carbonyl insertions. The rate-determining step is an intramolecular C–Cl oxidative addition.[112]

2.3.3.5 Heterogeneous Catalysts

Efficient methodologies result from using homogeneous catalysts anchored to solids, as the resulting catalyst is reusable and easily removed from the reaction medium.[6] Different procedures are used to immobilize metals on the surface or within the interior structure of solid supports, such as polymers, silica gel, and zeolites. Table 11 summarizes the most efficient heterogeneous catalytic systems, which are tested with typical Pauson–Khand substrates; the results are generally excellent in terms of yield. However, harsh conditions are still required with some systems. Mesoporous silicas are applied as supports for cobalt metal with high pressure of carbon monoxide and the results are satisfactory with intramolecular examples (Table 11, entry 1).[113] The use of cobalt supported on charcoal gives excellent results, but a high carbon monoxide pressure is still required (Table 11, entry 2).[114] A different immobilization method, an entrapment of catalysts by the sol-gel process, is used with rhodium in order to obtain a catalyst that can be used under milder conditions. Silica sol-gel entrapped chloro(cycloocta-1,5-diene)rhodium(I) dimer [{RhCl(cod)}$_2$] gives the Pauson–Khand product in 90% yield at 100 °C and at a carbon monoxide pressure of 5 atmospheres (Table 11, entry 3).[115] Cobalt nanoparticles, heterogeneous catalytic systems with high surface-to-volume ratio, also give good results under 5 atmosphere of carbon monoxide (Table 11, entry 4).[114] Combining the merits of conventional heterogeneous catalysts with the high catalytic activity of cobalt nanoparticles, cobalt nanoparticles on charcoal (CNC) give similar results to colloidal cobalt (Table 11, entry 5).[116] Polyethylene glycol stabilized cobalt nanoparticles are efficient in the Pauson–Khand reaction and they can be used in tetrahydrofuran or water (Table 11, entries 6 and 7).[117] Dodecacarbonyltriruthenium(0) is combined with cobalt nanoparticles on charcoal (CNC) to catalyze a Pauson–Khand-type reaction that uses 2-pyridylmethyl formate as the carbon monoxide source (Table 11, entry 8).[118] The use of immobilized heterobimetallic nanoparticle catalysts also catalyzes the Pauson–Khand reaction using either carbon monoxide (Table 11, entry 9)[119] or but-2-enal as a carbon monoxide source (Table 11, entry 10).[120] Raney cobalt also shows activity as a Pauson–Khand catalyst, albeit under high carbon monoxide pressure (Table 11, entry 11).[121]

for references see p 162

Table 11 Pauson–Khand Reactions with Heterogeneous Catalysts[113–121]

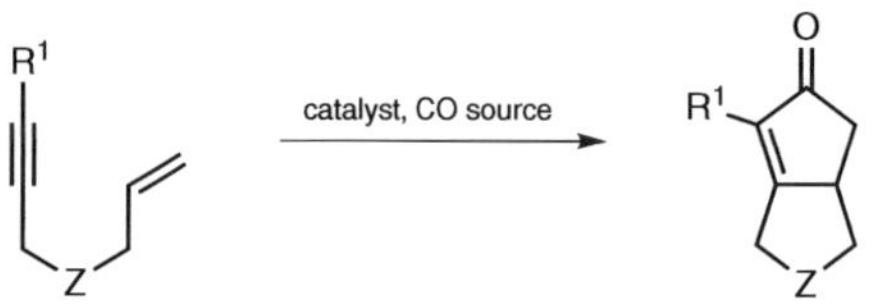

Entry	Starting Material	Catalytic System	CO Source	Conditions	Product	Yield (%)	Ref
1	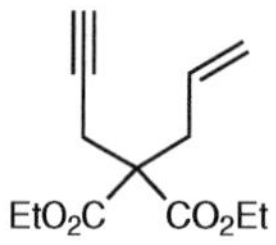 (R¹=H, Z=C(CO₂Et)₂)	Co/silica, (9–10 wt%)	CO (20 atm)	THF, 130 °C, 6.5 h	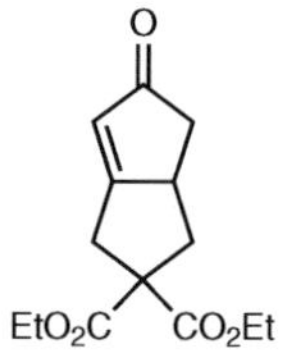 bicyclic enone	95	[113]
2	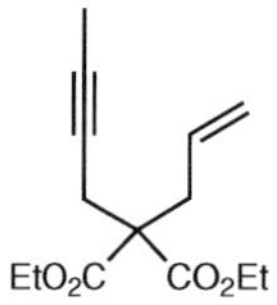 (R¹=Me, Z=C(CO₂Et)₂)	Co/charcoal (12 wt%)	CO (20 atm)	THF, 130 °C, 7–18 h	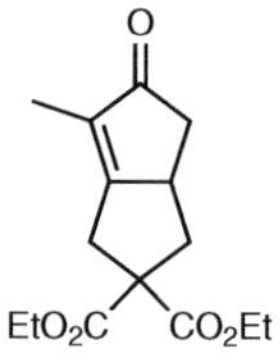 bicyclic enone	98	[114]
3	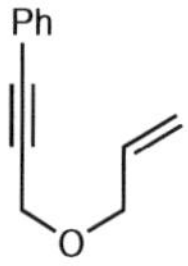 (R¹=Ph, Z=O)	entrapped Rh (0.1 equiv)	CO (5 atm)	THF, 100 °C, 12 h	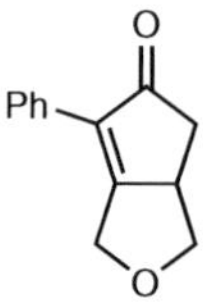 bicyclic enone	90	[115]
4	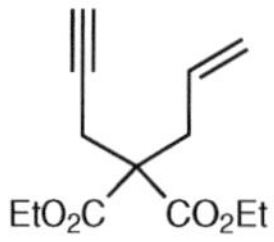 (R¹=H, Z=C(CO₂Et)₂)	colloidal Co (45 wt%)	CO (5 atm)	THF, 130 °C, 12 h	 bicyclic enone	97	[114]
5	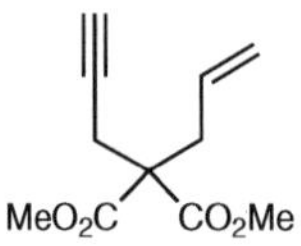 (R¹=H, Z=C(CO₂Me)₂)	Co nanoparticles on charcoal (12 wt%)	CO (5 atm)	THF, 130 °C, 18 h	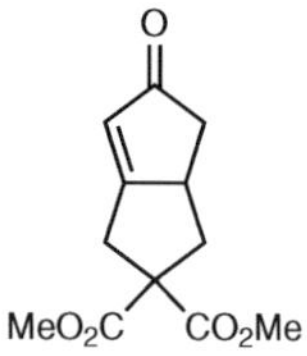 bicyclic enone	98	[116]
6	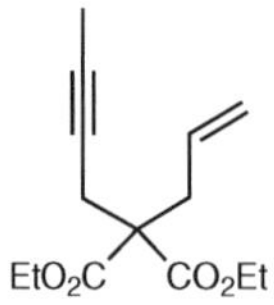 (R¹=Me, Z=C(CO₂Et)₂)	PEG₅₀₀₀-stabilized Co nanoparticles (3 mol%)	CO (23 atm)	THF, 130 °C, 16 h	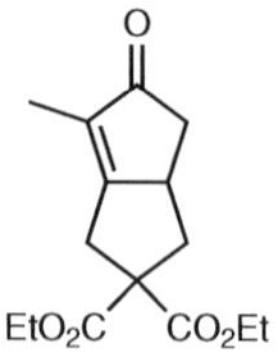 bicyclic enone	90	[117]

2.3.3 The Catalytic Pauson–Khand Reaction and Pauson–Khand-Type Reaction

Table 11 (cont.)

Entry	Starting Material	Catalytic System	CO Source	Conditions	Product	Yield (%)	Ref
7	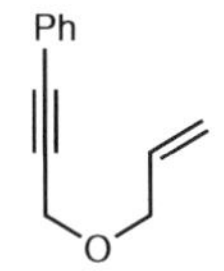 Ph-alkyne allyl ether	PEG$_{1000}$-stabilized Co nanoparticles (35 mol%)	CO (15–25 atm)	H$_2$O, 130°C, 16 h	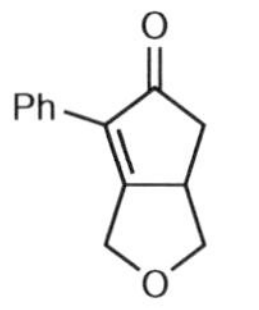 bicyclic enone	92	[117]
8	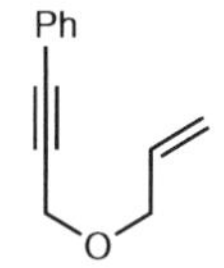 Ph-alkyne allyl ether	Ru$_3$(CO)$_{12}$, Co nanoparticles on charcoal	2-pyridylmethyl formate (1.5 equiv)	THF, 130°C, 12 h	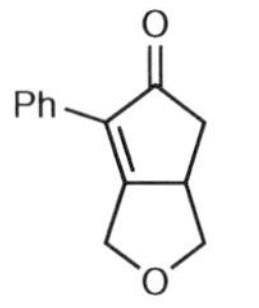 bicyclic enone	97	[118]
9	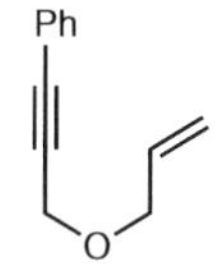 Ph-alkyne allyl ether	Co$_2$Rh$_2$	CO (1 atm)	THF, 130°C, 18 h	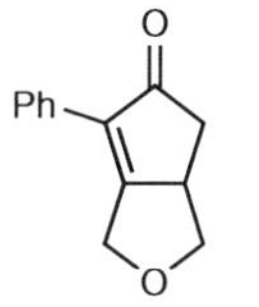 bicyclic enone	87	[119]
10	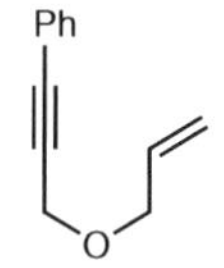 Ph-alkyne allyl ether	Co$_2$Rh$_2$	MeCH=CHCHO (2.5 equiv)	THF, 130°C, 18 h	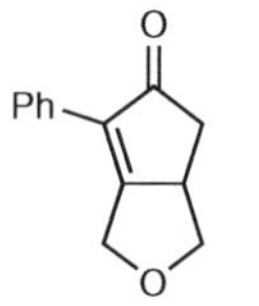 bicyclic enone	93	[120]
11	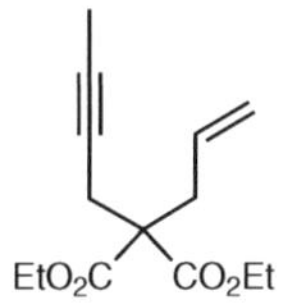 diethyl propargylallylmalonate	Raney Co (6 mol%)	CO (23 atm)	THF, 130°C, 16 h	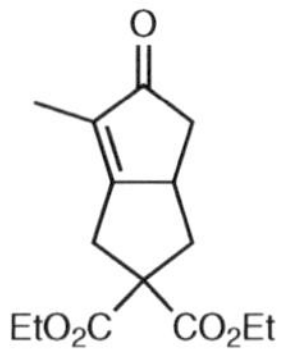 bicyclic enone with EtO$_2$C, CO$_2$Et	94	[121]

A similar approach to a heterogeneous and carbon monoxide free Pauson–Khand reaction uses a mesoporous graphitic carbon nitride (mpg-C$_3$N$_4$) as a metal-free catalyst that enables the direct activation of carbon dioxide and the oxidation of benzene to phenol. The formed carbon monoxide is used in situ for the subsequent Pauson–Khand reaction.[122]

Dimethyl 5-Oxo-3,3a,4,5-tetrahydropentalene-2,2(1H)-dicarboxylate (Table 11, Entry 5); Typical Procedure:[116]

CAUTION: *Carbon monoxide is extremely flammable and toxic, and exposure to higher concentrations can quickly lead to a coma.*

Immobilization of cobalt nanoparticles on charcoal (CNC): To 1,2-dichlorobenzene (24 mL) were added oleic acid (0.20 mL) and trioctyl phosphate (0.40 g). The soln was heated to 189°C. To the hot soln was added a soln of Co$_2$(CO)$_8$ in 1,2-dichlorobenzene (5 mL). The resulting soln was heated to 189°C for 2 h and then concentrated to a volume of 5 mL. The concentrated soln was cooled to rt. To the cooled soln was added MeOH (25 mL). The soln was well stirred for 10 min, and then charcoal (2.40 g) was added. The resulting mixture was refluxed for 12 h, and the precipitate was collected by filtration, washed with MeOH (20 mL), CH_2Cl_2 (20 mL), and Et_2O (20 mL), and dried under vacuum to give CNC as a black solid; yield: 2.50 g.

Pauson–Khand reaction: Dimethyl 2-allyl-2-(prop-2-ynyl)malonate (0.10 g, 0.48 mmol), CNC (0.10 g), and THF (5 mL) were put in a stainless-steel bomb. The bomb was pressurized under 5 atm of CO and heated at 130 °C for 18 h. After cooling and releasing the pressure, the soln was filtered and the filtrate was concentrated to dryness. The residue was purified by flash chromatography (silica gel, hexane/Et_2O 5:1); yield: 0.112 mg (98%).

2.3.4 Inducing Asymmetry in the Pauson–Khand Reaction

Several possibilities to induce asymmetry in the Pauson–Khand reaction are summarized in Scheme 21. These approaches have been reviewed[123] and only selected examples are discussed here to illustrate each approach.

Scheme 21 Possible Ways To Induce Asymmetry in the Pauson–Khand Reaction

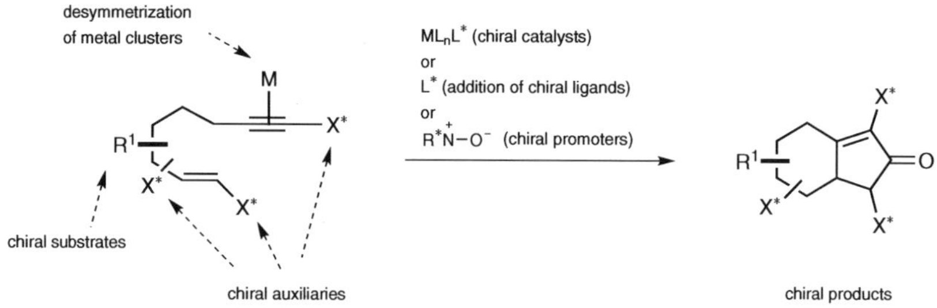

2.3.4.1 Transferring Chirality from the Substrate

This approach involves using chiral precursors that transfer their chirality to the final cyclopentenone. The synthesis of the chiral substrates is generally from classic chiral-pool materials, such as amino acids[124] and carbohydrates.[125] In a representative example, functionalized pyrrolidines **76** are used as starting materials to obtain functionalized indolizidine scaffolds **77** (Scheme 22). The substrates are prepared from homoallylamines through a 2-aza-Cope-(3+2) dipolar cycloaddition.[126]

Scheme 22 Chiral Substrates in the Cobalt-Catalyzed Intramolecular Pauson–Khand Reaction[126]

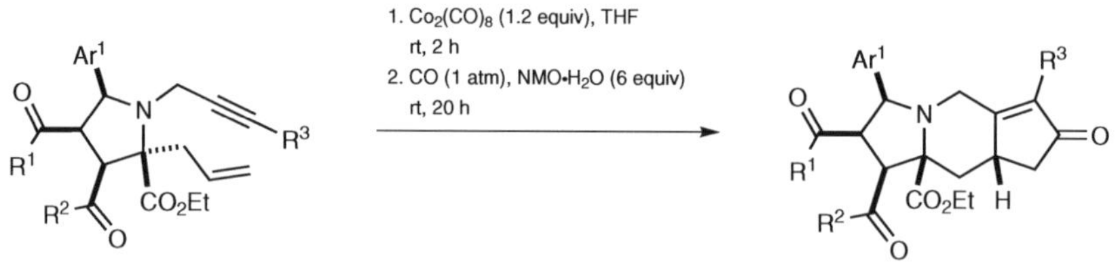

R^1	R^2	R^3	Ar^1	Yield (%)	Ref
	NPh	Me	Ph	96	[126]
	NPh	Me	4-MeOC$_6$H$_4$	95	[126]
	NPh	Me	4-F$_3$CC$_6$H$_4$	81	[126]

2.3.4 Inducing Asymmetry in the Pauson–Khand Reaction

R^1	R^2	R^3	Ar^1	Yield (%)	Ref
OMe	OMe	H	4-MeOC$_6$H$_4$	82	[126]
OMe	OMe	H	4-F$_3$CC$_6$H$_4$	73	[126]
OMe	OMe	Me	Ph	80	[126]
OMe	OMe	Me	4-MeOC$_6$H$_4$	86	[126]

In an asymmetric version of the allenyne reaction, chirality is efficiently transferred from a nonracemic, axially chiral allene to an α-alkyl- or an α-silyl-substituted cyclopentenone in a molybdenum-mediated reaction.[127] This methodology is applied to access enantioselectively tetrahydroazulenones, tetrahydrocyclopenta[c]azepinones, and dihydrocyclopenta[c]oxepinones **79**. Complete transfer of chirality is obtained for all trisubstituted allenes **78** using dicarbonylchlororhodium(I) dimer as the catalyst, but poor enantiomeric excesses are observed with disubstituted allenes (Scheme 23).[128]

Scheme 23 Axially Chiral Allenes in the Rhodium-Catalyzed Intramolecular Pauson–Khand Reaction[128]

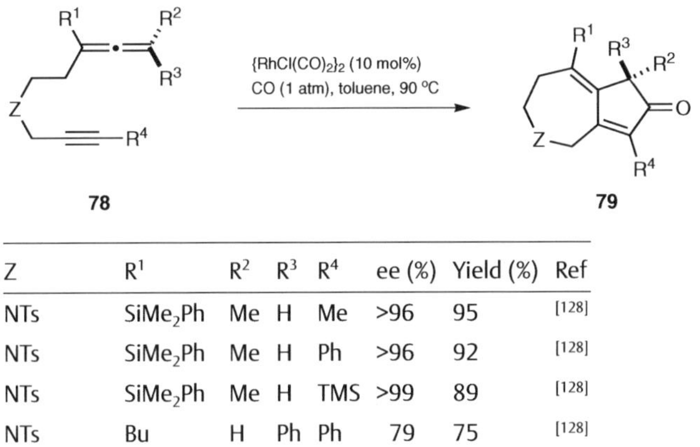

Z	R^1	R^2	R^3	R^4	ee (%)	Yield (%)	Ref
NTs	SiMe$_2$Ph	Me	H	Me	>96	95	[128]
NTs	SiMe$_2$Ph	Me	H	Ph	>96	92	[128]
NTs	SiMe$_2$Ph	Me	H	TMS	>99	89	[128]
NTs	Bu	H	Ph	Ph	79	75	[128]
O	Bu	H	Ph	TMS	77	91	[128]
C(CO$_2$Et)$_2$	Bu	H	Ph	Ph	76	92	[128]

2.3.4.2 Using Chiral Auxiliaries

The chiral-auxiliary approach has been of great importance in the development of the Pauson–Khand reaction for the total synthesis of complex molecules. The chiral auxiliary may be attached to the alkyne or the alkene moieties both in inter- and intramolecular versions. Some of the most effective chiral auxiliaries are shown in Scheme 24. Both enantiomers of *trans*-phenylcyclohexanol can be used as a chirality-inducing group, which is readily removed by reductive cleavage using samarium(II) iodide after the Pauson–Khand reaction (Table 12, entries 1–4).[129–132] The introduction of camphor-derived alcohols in an attempt to enhance stereocontrol in these processes results in enhanced diastereoselectivities (Table 12, entries 5 and 6),[132,133] but better results are achieved with 10-(methylsulfanyl)isoborneol (Table 12, entries 7 and 8)[134,135] or chiral oxazolidinones (Table 12, entries 9 and 10),[135,136] which give excellent results in terms of stereocontrol and yield. One of the best chiral auxiliaries is a derivative of Oppolzer's camphorsultam; its use results in near total stereocontrol (Table 12, entry 11).[137] Chiral sulfoxides attached to the alkene are also efficient auxiliaries due to the proximity of the chiral sulfur to the reaction center (Table 12, entry 12). Reductive cleavage of the sulfoxide is carried out by treatment with activated zinc.[138] Related chiral sulfoxides are also used in the intermolecular version (Table 12, entry 13).[139]

Scheme 24 Chiral Auxiliaries for the Pauson–Khand Reaction[129–139]

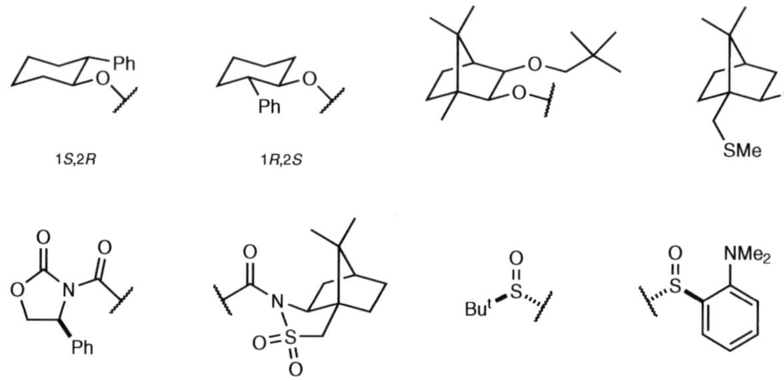

Table 12 The Chiral Auxiliary Approach in Pauson–Khand Reactions[129–139]

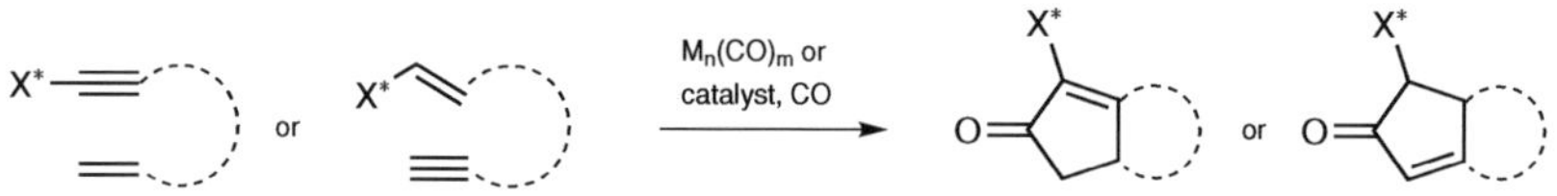

X* = chiral auxiliary

Entry	Starting Material(s)		Conditions	Product (Major Diastereomer)	dr	Yield (%)	Ref
	Alkyne	Alkene					
1	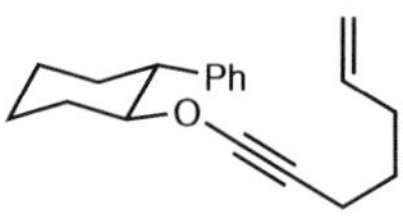		Co$_2$(CO)$_8$, isooctane, reflux, 16 h	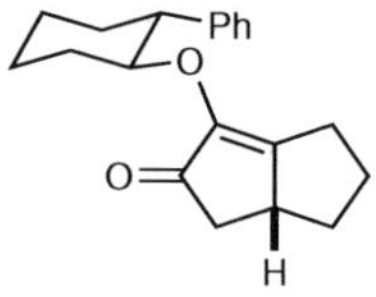	3.2:1	38	[129]
2	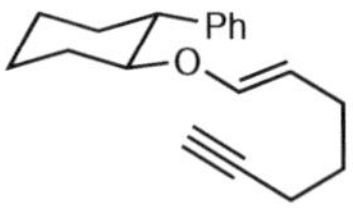		Co$_2$(CO)$_8$, isooctane, 20°C, 1.5 h, then reflux, 1.5 h	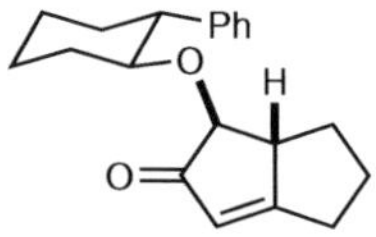	7:1	55	[130]
3	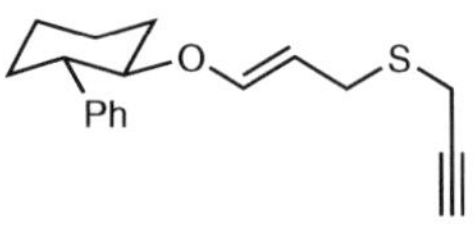		Co$_2$(CO)$_8$, isooctane, 20°C, 0.5 h, then reflux, 2.5 h	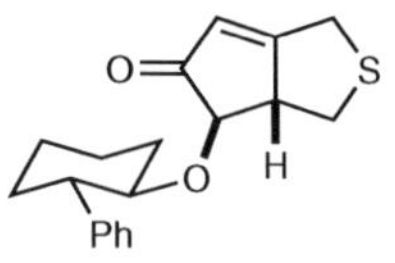	12:1	38	[131]
4	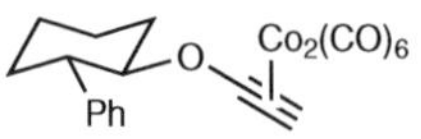	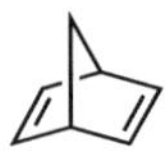	isooctane, 50–60°C, 18 h	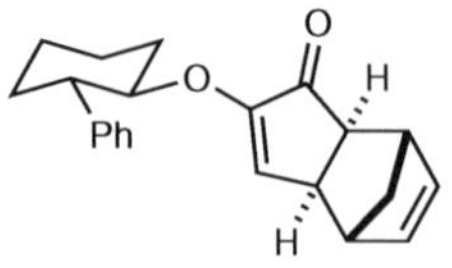	3:1	62	[132]

2.3.4 Inducing Asymmetry in the Pauson–Khand Reaction

Table 12 (cont.)

Entry	Starting Material(s)		Conditions	Product (Major Diastereomer)	dr	Yield (%)	Ref
	Alkyne	Alkene					
5	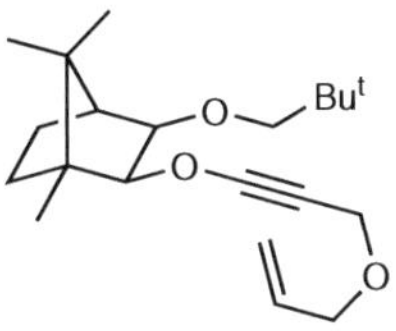		Co$_2$(CO)$_8$, hexane, 18 °C, 2 h, then 25 °C, 2 h	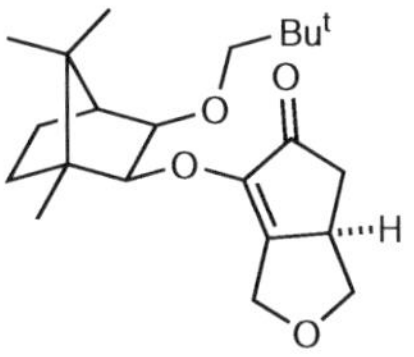	15.7:1	54	[133]
6	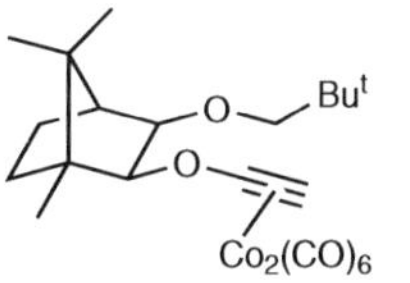		isooctane, 60–65 °C, 18 h	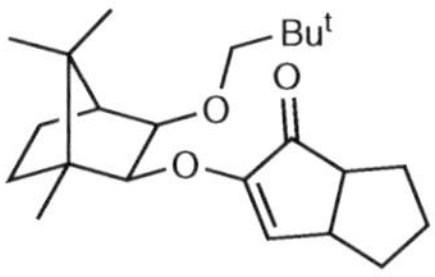	3:1	30	[132]
7	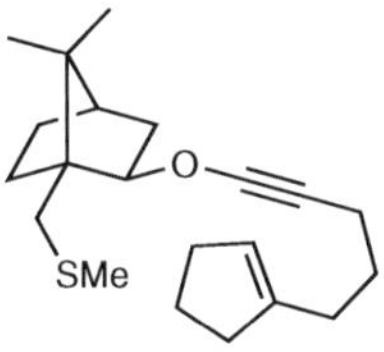		Co$_2$(CO)$_8$, isooctane, 80 °C, 10 h	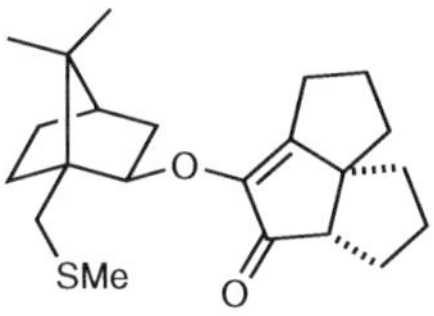	12:1	34	[134]
8	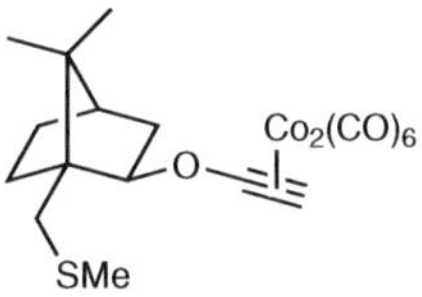		NMO, CH$_2$Cl$_2$, 25 °C, 10 min, then norbornadiene, −20 °C, overnight	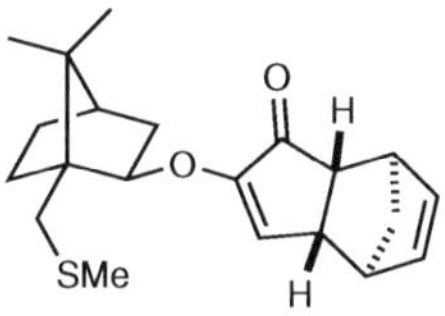	24:1	82	[135]
9	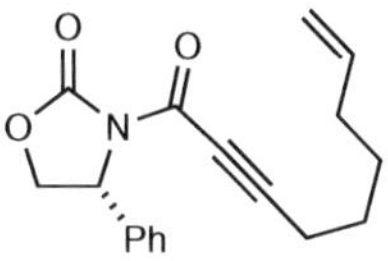		Co$_2$(CO)$_8$, toluene, 25 °C, 0.5 h, then DMSO, 60 °C, 2.5 h	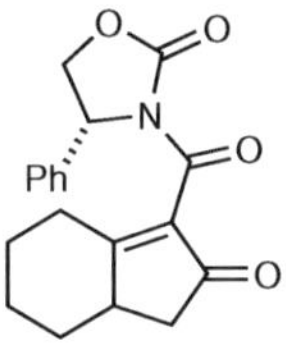	1.2:1	80	[136]
10	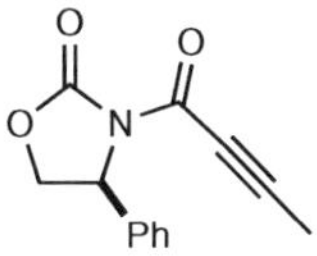		Co$_2$(CO)$_8$, toluene, 25 °C, 1 h, then norbornadiene, 25 °C, 72 h	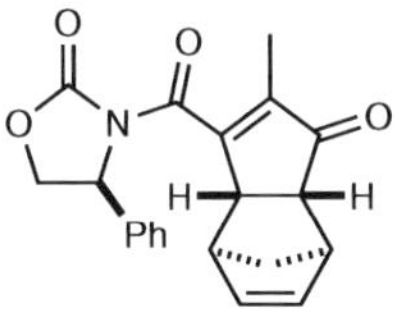	7.6:1	53	[136]

Table 12 (cont.)

Entry	Starting Material(s)		Conditions	Product (Major Diastereomer)	dr	Yield (%)	Ref
	Alkyne	Alkene					
11	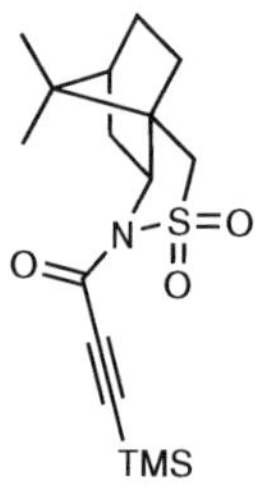 (TMS-alkyne with N-sulfinyl camphor-derived auxiliary)	 norbornadiene	$Co_2(CO)_8$, CH_2Cl_2, 25 °C, 1 h, then norbornadiene, NMO, 0 °C to 25 °C, 18 h	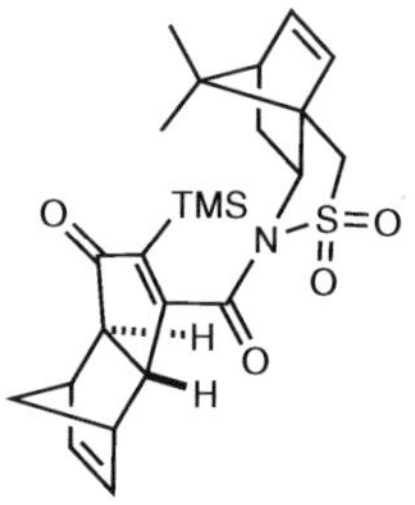 (tricyclic PK adduct)	800:1	78	[137]
12	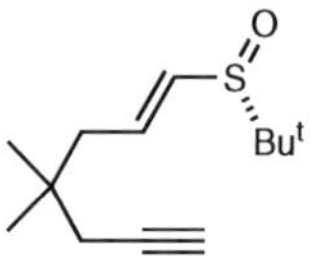 (enyne with Bu^t-sulfinyl group)		$Co_2(CO)_8$, CH_2Cl_2, 25 °C, 10 min, then MeCN, 60 °C, 15 min	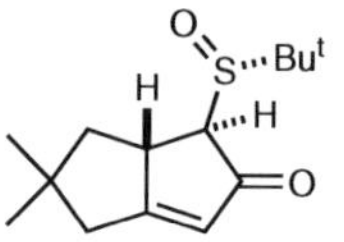 (bicyclopentenone with Bu^t-S(O))	49:1	65	[138]
13	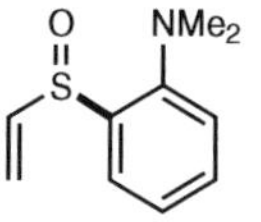 (vinyl aryl sulfoxide with ortho-NMe$_2$ carbamoyl)	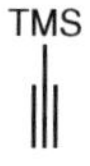 TMS-alkyne	$Co_2(CO)_8$, MeCN, 25 °C, 1 h, then NMO, 25 °C, 16 h	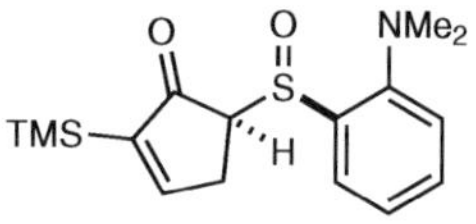 (cyclopentenone product)	49:1	59	[139]

In addition, the chiral auxiliaries can be tethered to an enyne; auxiliaries such as the Schöllkopf bislactim or chiral acetals have been used. In an example, aromatic chiral sulfoxide **80** is used to generate 4-aryl-4-cyano-1,6-enynes **81** or 4-aryl-4-carbamoyl-1,6-enynes with an additional quaternary stereocenter. This new stereogenic carbon is formed with high diastereoselectivity. Asymmetric intramolecular Pauson–Khand reaction reactions of **81** give enantiomerically enriched bicyclo[3.3.0]octenones **82**. The quaternary stereocenter and also the sulfur function located at the *ortho* position of the aryl group are responsible for the stereocontrol; it is assumed that the lone electron pair at sulfur associates with the cobalt–alkyne complexes (Scheme 25).[140]

2.3.4 Inducing Asymmetry in the Pauson–Khand Reaction

Scheme 25 Chiral Sulfoxides as Auxiliaries in the Cobalt-Catalyzed Asymmetric Intramolecular Pauson–Khand Reaction[140]

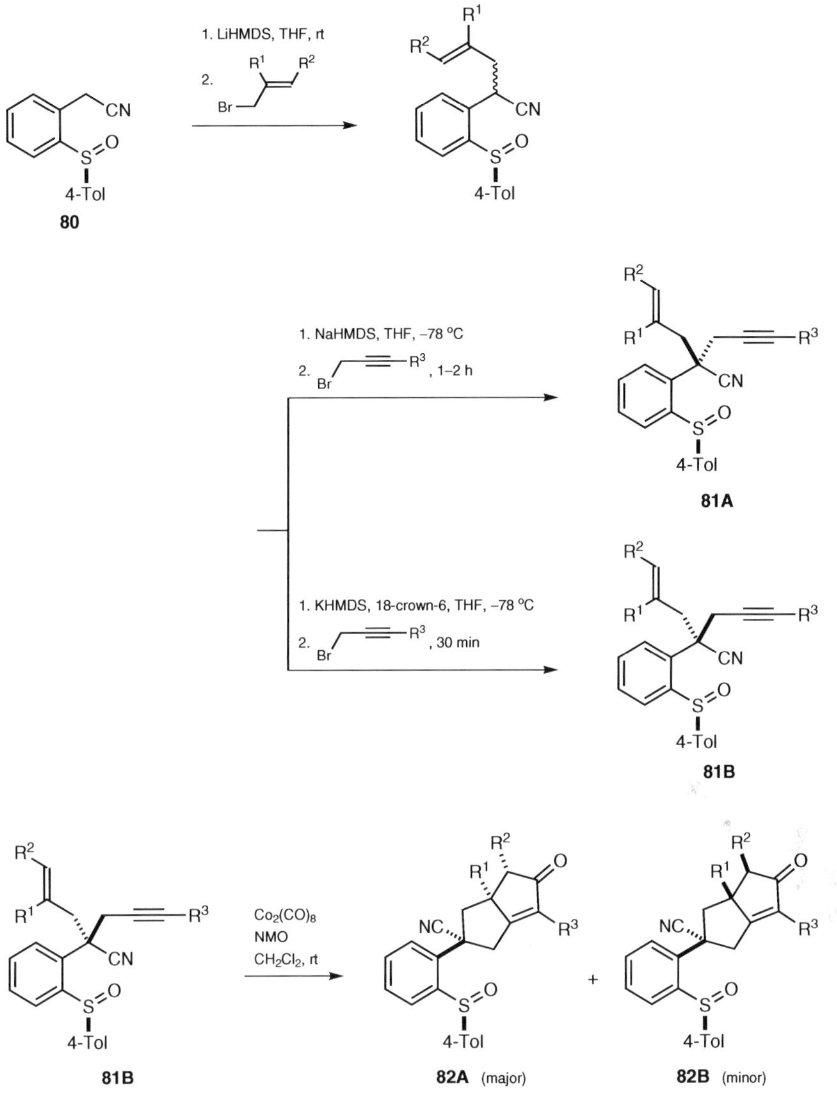

R¹	R²	R³	Time	dr (**82A/82B**)	Combined Yield (%)	Ref
H	H	H	5 min	72:28	79	[140]
H	H	Me	5 min	70:30	70	[140]
Me	H	H	10 min	81:19	65	[140]
Me	H	Me	10 min	78:22	76	[140]
H	Me	H	1.5 h	80:20	74	[140]
H	Me	Me	1.5 h	82:18	80	[140]
Ph	H	H	12 h	>98:2	56	[140]
Ph	H	Me	12 h	>98:2	58	[140]
H	Ph	H	3.5 h	>98:2	74	[140]

2.3.4.3 Using Chiral Metal Complexes

The stoichiometric reaction of ligands with metals to form chiral metal clusters is a reliable procedure, providing ready access to useful synthetic intermediates. Various bidentate chiral ligands, upon coordination with cobalt, produce isolable complexes such as **84/85** and **89** that are capable of mediating a stereocontrolled reaction. As an example, very effective results are achieved using the ligands PuPHOS (**83**) and CamPHOS (**88**), which are readily obtained from (+)-pulegone and camphorsulfonic acid, respectively. The reaction of these ligands with octacarbonyldicobalt(0) gives chiral complexes that react with norbornadiene in high yields and with high enantiomeric excess (Scheme 26).[141] The main problem is the low diastereoselectivity of these ligands in the formation of the chiral complexes, e.g. to give mixtures of **84** and **85**. This drawback is circumvented by the use of terminal propynamides as alkyne partners. With these substrates, a non-classical hydrogen bonding interaction between the amide carbonyl group and the highly polarized methine group in the ligand is established in only one of the diastereomers. This weak stabilizing interaction leads, in some cases, to total diastereoselection. The resulting complexes react with norbornadiene to yield the enantioenriched cyclopentenones (e.g., **86** and **87**).[142] A catalytic version uses 5 mol% of the chiral cobalt complex **89** in an intermolecular Pauson–Khand reaction in which the adduct **90** is obtained in high yield albeit with low enantiomeric excess (Scheme 26).[143]

Scheme 26 Asymmetric Intermolecular Pauson–Khand Reactions Using Chiral-Metal Cobalt Complexes[141–143]

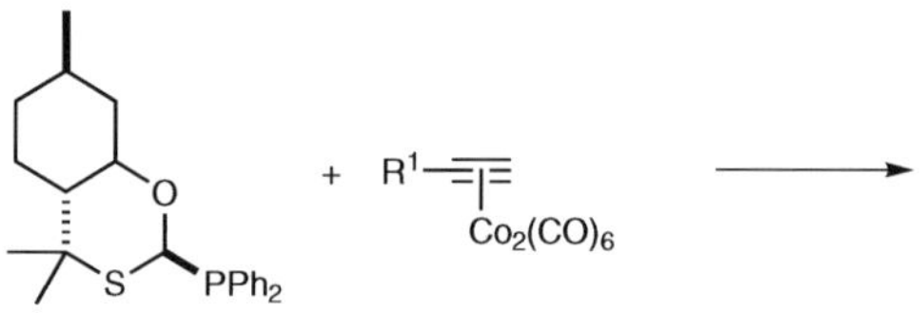

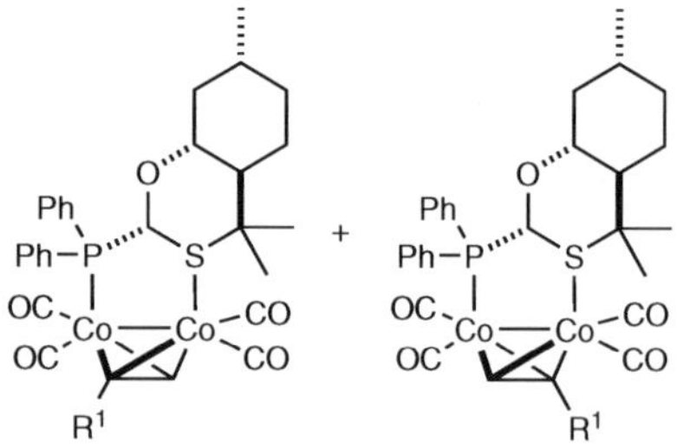

R^1	Ratio (**84/85**)	Yield (%)	Ref
Ph	1:1	91	[141]
TMS	3:1	91	[141]
$CONEt_2$	1:33	68	[142]

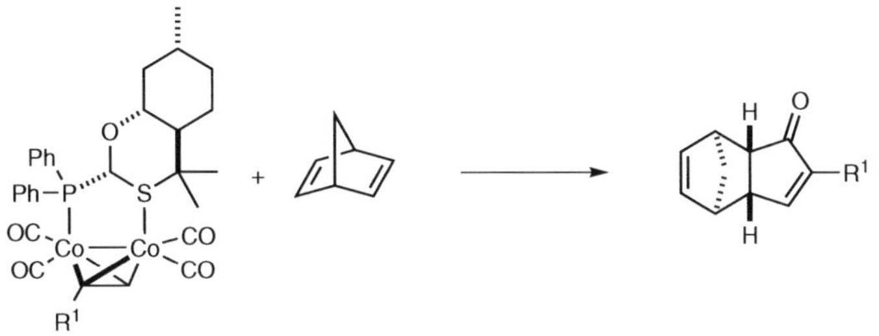

2.3.4 Inducing Asymmetry in the Pauson–Khand Reaction

R¹	Conditions	ee (%)	Yield (%)	Ref
Ph	toluene, 50 °C	99	99	[141]
TMS	NMO, CH_2Cl_2, rt	97	93	[141]

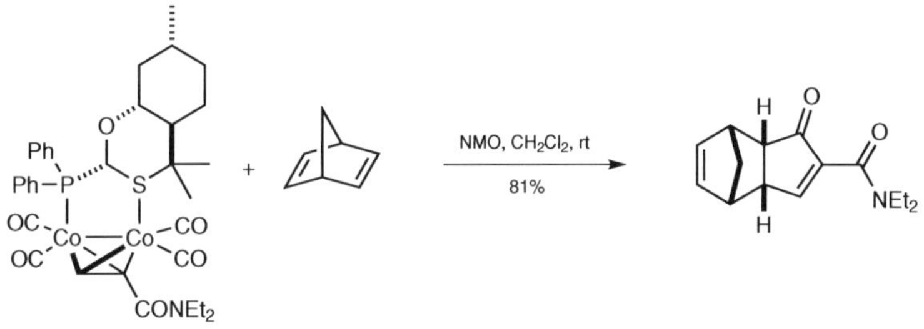

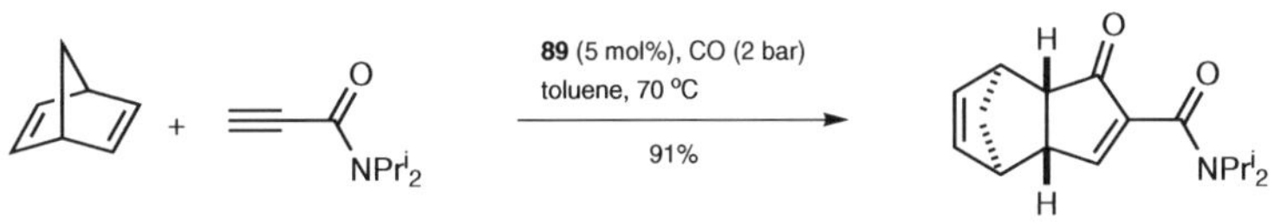

87 73% ee

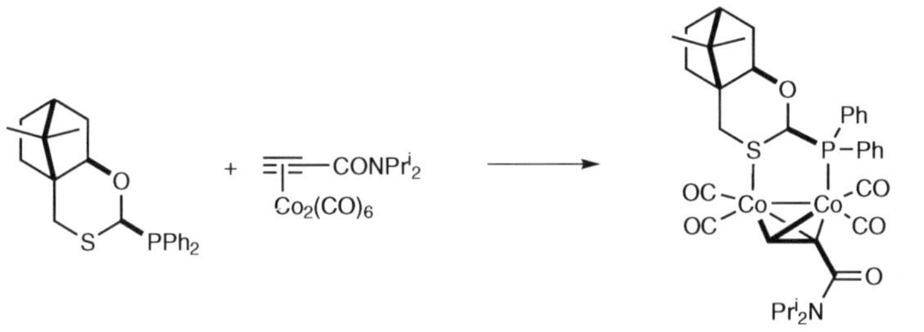

90 23% ee

2.3.4.4 Using Chiral Ligands

As discussed in Sections 2.3.3.2 and 2.3.3.3, the combination of chiral phosphines with rhodium or iridium catalysts gives excellent results in the enantioselective Pauson–Khand reaction. These catalytic systems are used in the kinetic resolution of 1-arylallyl propargyl ethers by enantioselective Pauson–Khand reaction. While cationic rhodium(I) with a BINAP-based ligand is the choice for the slow reacting substrates, neutral iridium(I) with a BINAP-based ligand provides excellent results for more reactive substrates.[144] The BINAP–cobalt combination has become a workbench system for the study of catalytic Pauson–Khand reactions and is generally restricted to the intramolecular version. One excellent example is the use of the BINOL-derived bis(phosphite) **91** and catalytic amounts (ca. 6 mol%) of octacarbonyldicobalt(0). The intramolecular Pauson–Khand reaction of diverse enynes reaches good enantiomeric excess values only with certain substrates **92** (Scheme 27).[145]

Scheme 27 A Cobalt–Chiral Ligand System for the Enantioselective Intramolecular Pauson–Khand Reaction of Enynes[145]

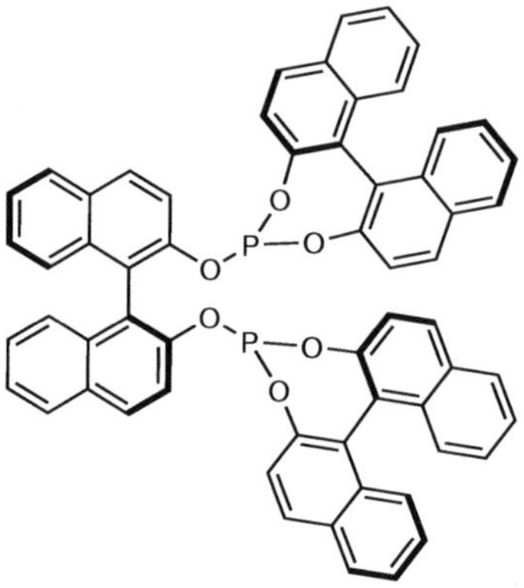

R^1	ee (%)	Yield (%)	Ref
H	22	87	[145]
Me	7	97	[145]
CH_2OMe	14	80	[145]
Ph	75	75	[145]
4-MeOC_6H_4	64	82	[145]

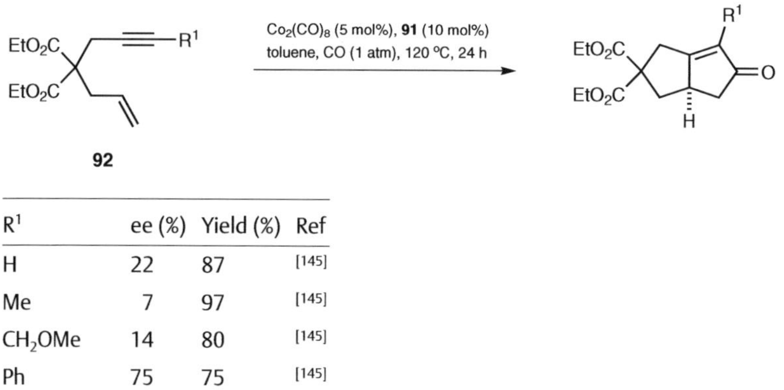

A study on the origin of the asymmetric induction concludes that coordination of BINAP to cobalt gives the catalyst–ligand combination responsible for the catalytic activity in the cyclization of enynes. The structure of this entity is not the bridged complex **93**, but rather the chelated structure **94** (Scheme 28).[146] A theoretical study has rationalized the outcome of asymmetric intramolecular Pauson–Khand reactions using the cobalt–BINAP system.[147]

2.3.5 Cascade Reactions Involving Pauson–Khand Reactions

Scheme 28 Coordination Modes of Chiral Phosphines with Cobalt Complexes[146]

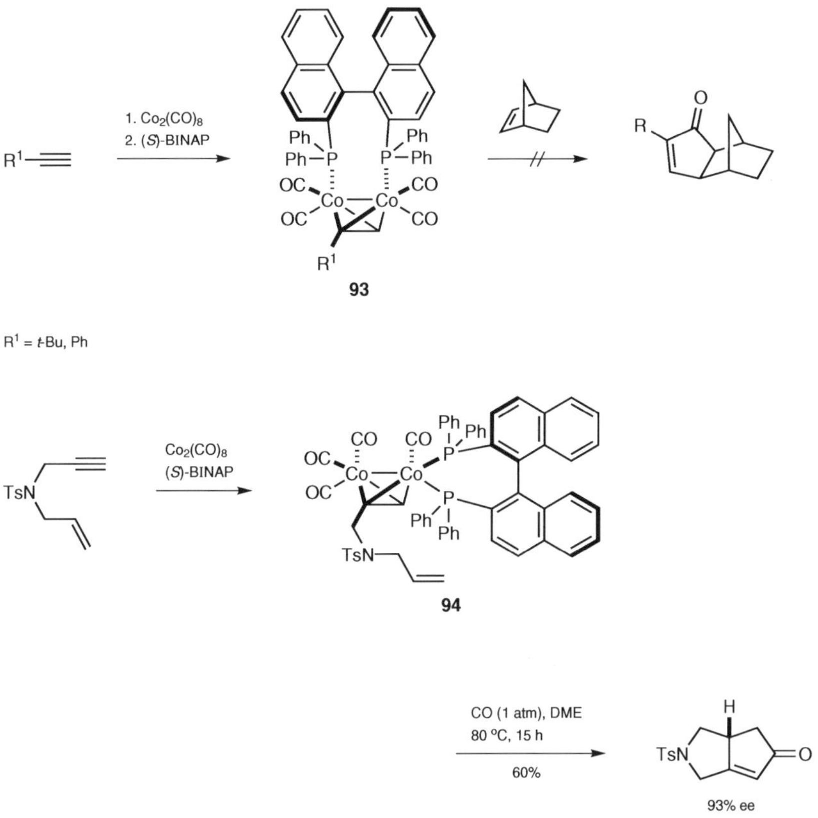

2.3.4.5 Using Chiral Promoters

Chiral promoters, generally natural alkaloid N-oxides, are potentially able to selectively decarbonylate one carbonyl of the cobalt cluster leading to a chiral activated complex, but they give poor results in terms of enantiomeric excess. A computational study on the role of brucine N-oxide in the enantioselective Pauson–Khand reaction of norbornene with 2-methylbut-3-yn-2-ol predicts enantiomeric excess values in different solvents of 68–76% (R), which compares well with the experimental values (58–78%). The relatively low enantiomeric excesses are due to competition between the racemization process of the activated chiral cobalt complex and complexation with the alkene reagent that leads to the products.[148]

2.3.5 Cascade Reactions Involving Pauson–Khand Reactions

The synthetic power of the Pauson–Khand reaction has been enhanced by combining the (2+2+1) cycloaddition with other processes. For instance, the synthesis of the starting enyne can be combined with the Pauson–Khand reaction in one operational step, preferably using the same catalyst. In other cases, various cycloaddition reactions are performed consecutively, giving complex polycyclic molecules in one synthetic step. Some tandem processes arise from unexpected results in which successive events occur after the cycloaddition. After careful optimization of the reaction conditions these side reactions have become useful operations. Scheme 29 is a summary of these transformations, which have been reviewed.[149]

Scheme 29 Summary of the Combination of the Pauson–Khand Reaction with Other Reactions

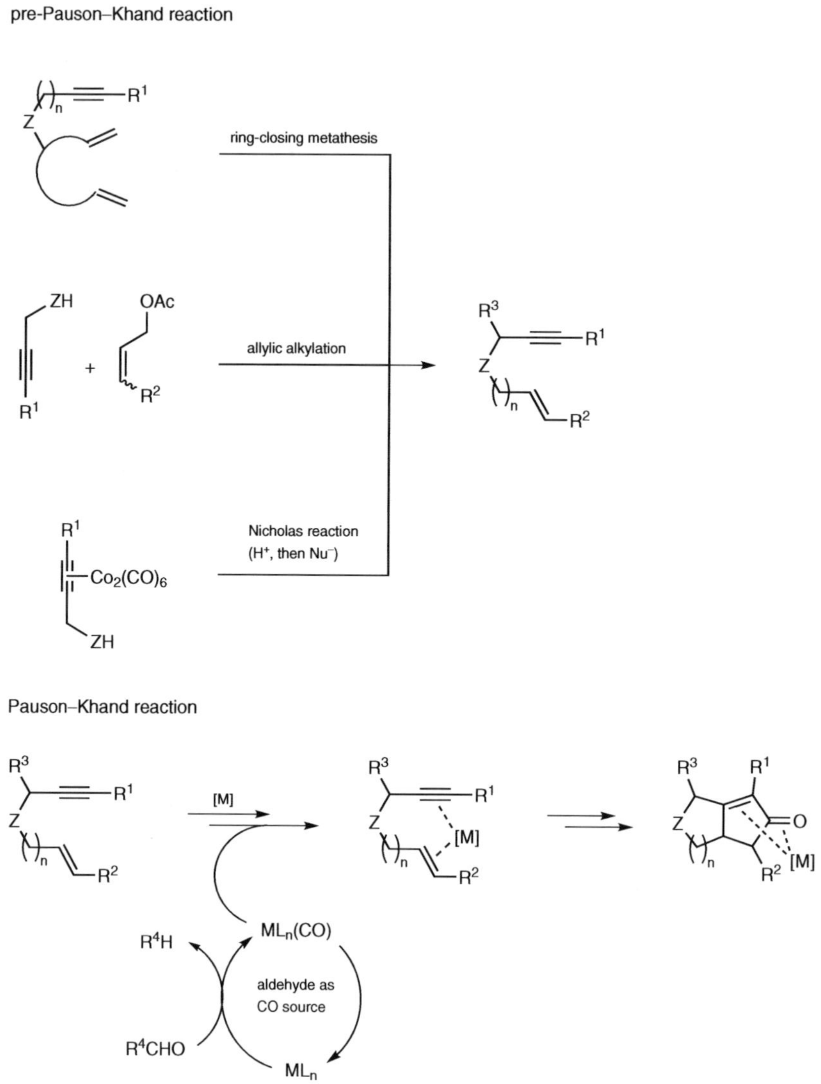

pre-Pauson–Khand reaction

Pauson–Khand reaction

2.3.5 Cascade Reactions Involving Pauson–Khand Reactions

post-Pauson–Khand reaction

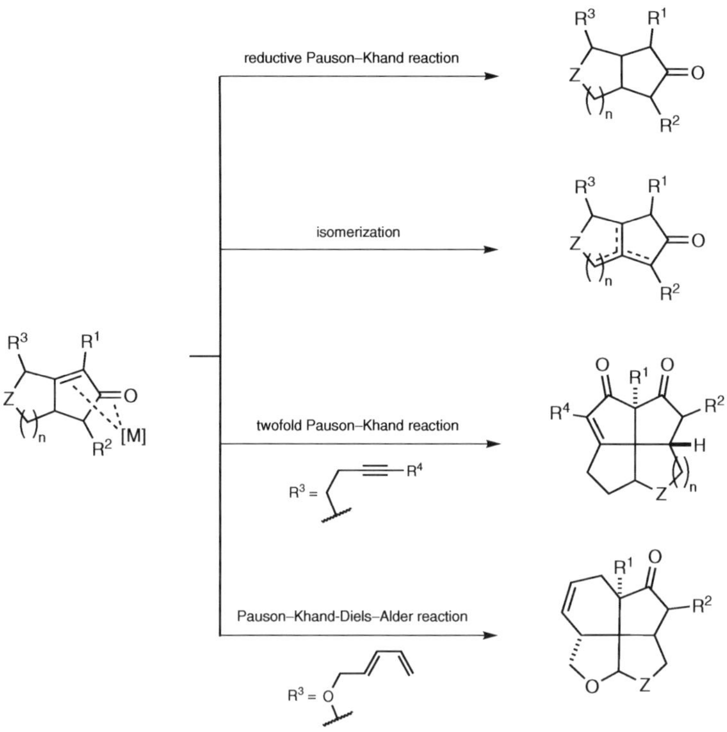

2.3.5.1 Cascade Synthesis of Enynes/Pauson–Khand Reaction

The synthesis of enynes is occasionally mediated by transition metals and this step can be combined with the Pauson–Khand reaction in a one-pot manner or a cascade reaction. Thus, the second-generation Grubbs ruthenium–carbene is a catalyst for the combination of a hetero-Pauson–Khand cycloaddition and a ring-closing metathesis to give complex polycyclic products in one step from acyclic precursors. An example is the hetero-Pauson–Khand reaction and ring-closing metathesis of **95** to give the fused tricyclic system **96** (Scheme 30).[150]

Scheme 30 Combination of the Rhodium-Catalyzed Hetero-Pauson–Khand Reaction with Ring-Closing Metathesis[150]

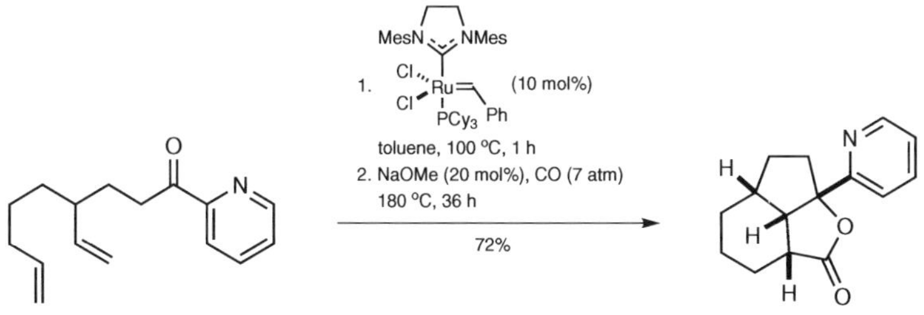

The Nicholas reaction[151] is used to generate enynes that undergo, in a domino fashion, a Pauson–Khand reaction. Following an initial exploration of this process,[152] the tandem Nicholas/Pauson–Khand reaction process is used as a key step in the synthesis of diterpene (+)-epoxydictimene, starting from natural (R)-pulegone.[153] Tricyclic [5.6.5]- to [5.9.5]-systems **98** are prepared by a tandem intramolecular Nicholas/Pauson–Khand strategy (Scheme 31).[154]

Scheme 31 Synthesis of [5.n.5]-Systems via a Tandem Nicholas/Pauson–Khand Reaction[154]

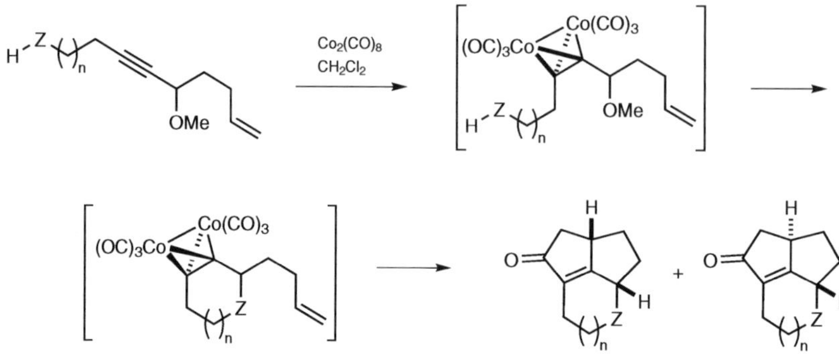

n	Z	Conditions (Step 2)	Yield (%) of 97	Conditions (Step 3)	Ratio (98A/98B)	Yield (%) of 98	Ref
1	O	BF$_3$•OEt$_2$, CH$_2$Cl$_2$, 0°C, 3.5 h	9	MeCN, air, 100°C, 1 h	66:34	17	[154]
2	O	BF$_3$•OEt$_2$, CH$_2$Cl$_2$, 0°C, 3.5 h	93	CyNH$_2$, DME, 60°C, 3 h	58:42	91	[154]
3	O	BF$_3$•OEt$_2$, CH$_2$Cl$_2$, 0°C, 3.5 h	93	MeCN, air, 100°C, 1 h	45:55	86	[154]
4	O	BF$_3$•OEt$_2$, CH$_2$Cl$_2$, 0°C, 0.5 h	43	MeCN, air, 100°C, 1 h	9:91	19	[154]
2	NTs	BF$_3$•OEt$_2$, CH$_2$Cl$_2$, 0°C, 1.5 h	73	CyNH$_2$, DME, 60°C, 3 h	100:0	98	[154]
3	NTs	HBF$_4$, CH$_2$Cl$_2$, 0°C, 1 h	27	MeCN, air, 100°C, 15 min	91:9	69	[154]

An excellent example of concurrent tandem catalysis involving Pauson–Khand-type reactions is that in which a single catalyst performs successively the synthesis of an enyne by means of an allylic alkylation followed by the Pauson–Khand reaction. One example of an allylation-Pauson–Khand reaction sequence only requires changing the reaction temperature for each step, thus, the rhodium catalyst mediates first the allylation to produce the enyne and the subsequent Pauson–Khand reaction is favored by elevating the temperature to 80°C, giving the adduct **99** in high yield (Scheme 32).[155]

2.3.5 Cascade Reactions Involving Pauson–Khand Reactions

Scheme 32 Combined Allylation/Pauson–Khand Reaction Sequence[155]

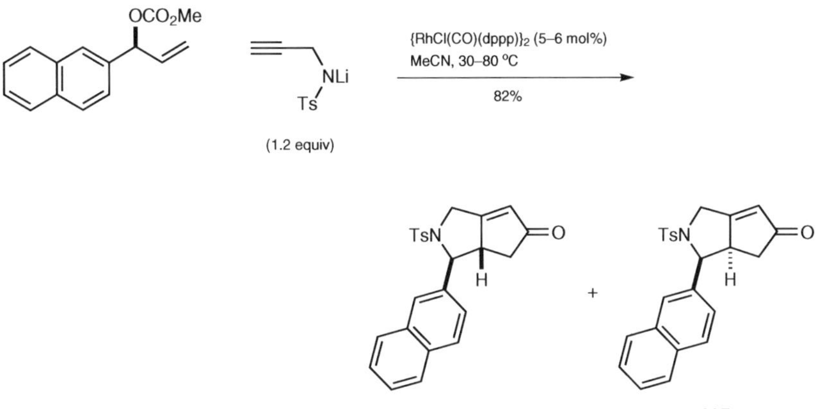

Another application allows the synthesis of [5.5.5.5] systems **101**, called fenestranes, by a three-step sequential action of cobalt nanoparticles and a palladium catalyst. The cascade reaction starts with an intramolecular Pauson–Khand reaction using the cobalt catalyst, followed by the formation of η³-allyl–palladium, which reacts with a nucleophile derived from dimethyl malonate, to give a new enyne **100**; the final step is a second Pauson–Khand reaction (Scheme 33).[156]

Scheme 33 Synthesis of a Fenestrane by the Pauson–Khand Reaction/Allylic Alkylation/Pauson–Khand Reaction[156]

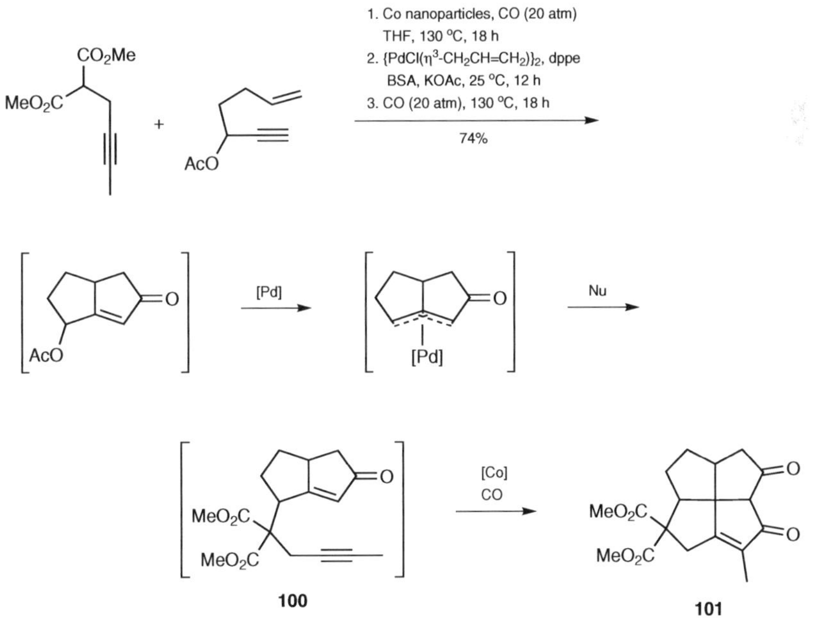

2.3.5.2 Tandem Carbocyclizations Involving (2+2+1) Reactions

The Pauson–Khand reaction has been combined with (2+2), (4+2), or (5+1) cycloadditions. The combination of the Pauson–Khand reaction with the Diels–Alder reaction applies this cascade process to the synthesis of fenestranes and triquinanes; it is postulated that the Diels–Alder reaction occurs after the Pauson–Khand reaction.[157] Enantioenriched propargylic alcohols, prepared from asymmetric enyne addition to aldehydes, are used to prepare trienynes **102/104**. These optically active substrates give a domino Pauson–Khand/Diels–Alder cycloaddition, using 10 mol% of dicarbonylchlororhodium(I) dimer and 1 atmosphere of carbon monoxide, to generate multicyclic products **103** or **105/106**, respectively (Scheme 34).[158]

Scheme 34 Synthesis of Fenestranes by a Pauson–Khand/Diels–Alder Reaction[158]

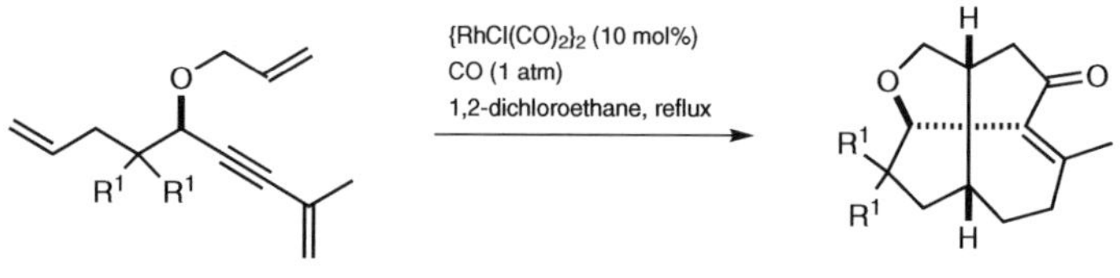

R^1	Time (h)	ee (%)	Yield (%)	Ref
H	26	90	75	[158]
Me	72	83	56	[158]

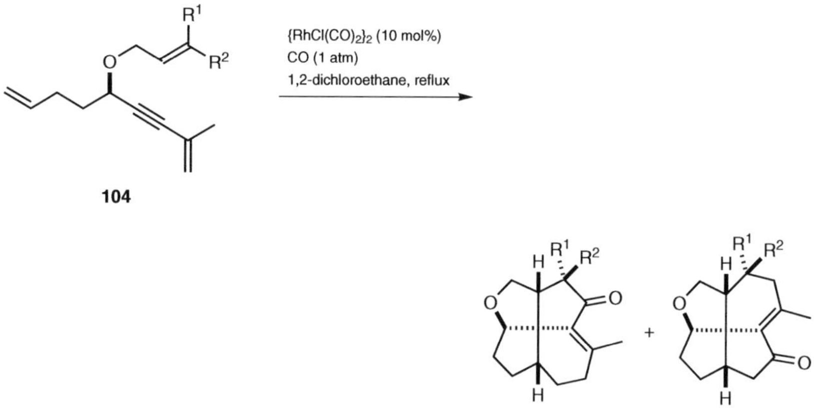

R^1	R^2	Time (h)	ee (%) of **106**	Yield (%) of **105**	Yield (%) of **106**	Ref
H	Me	24	88	34[a]	26	[158]
Me	Me	24	90	–	67	[158]
Me	$(CH_2)_2CH=CMe_2$	48	93	–	45	[158]
H	Ph	36	89	–	33	[158]

[a] 88% ee.

An octacarbonyldicobalt(0)-mediated tandem (5+1)/(2+2+1)-cycloaddition reaction gives tricyclic δ-lactones **110** starting from *cis*-epoxy enynes **107**. This process possibly involves initial opening of the epoxide in the hexacarbonyldicobalt complex to from the complexed allene **108**. Further coordination of the tethered alkene and oxidative cyclization gives cobaltacycle **109**, which inserts carbon monoxide leading to the final compound **110** (Scheme 35).[159]

2.3.5 Cascade Reactions Involving Pauson–Khand Reactions

Scheme 35 Cobalt-Mediated Tandem (5+1)/(2+2+1) Cycloaddition[159]

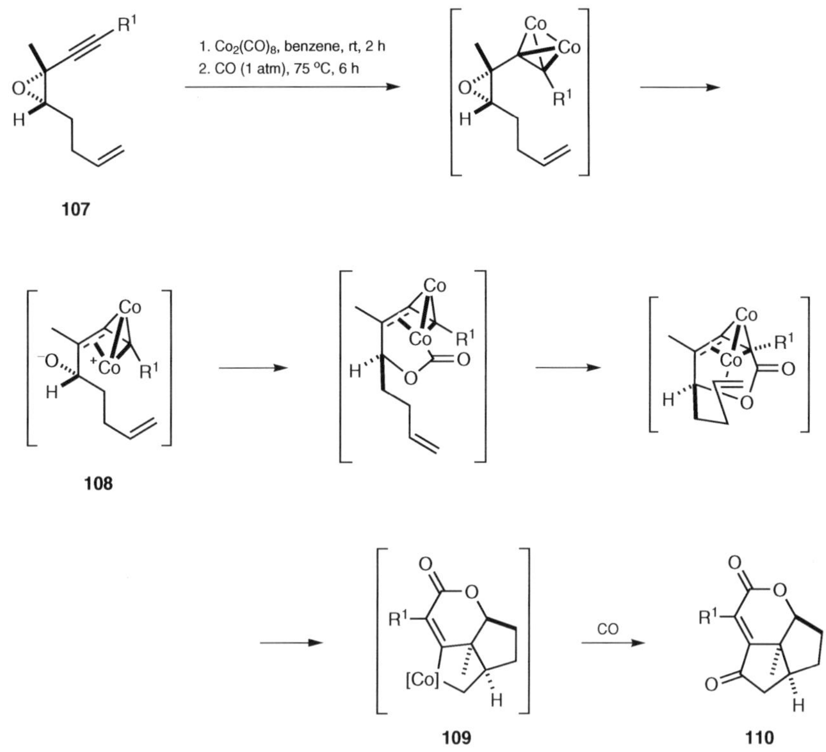

R[1]	Yield (%)	Ref
Bu	77	[159]
Ph	74	[159]
TMS	65	[159]

2.3.5.3 Tandem Pauson–Khand Reactions

The combination of two or more Pauson–Khand reactions or Pauson–Khand-type reactions in the same reaction step has been developed. The multiplication of the synthetic power of this transformation has found immediate application for the synthesis of natural [5.5.5.5] fenestranes[160] and pentalenes, that is, linear [5.5.5.5] systems.[161] Starting materials are dienediynes or diallenediynes that give two (2+2+1) cycloadditions.[31] Scheme 36 shows the different catalytic conditions used for the Pauson–Khand reaction of dienediyne **111** mediated by rhodium, iridium, or cobalt complexes. Further transformations of the pentacycle **112** allow the synthesis of the tetracyclic cyclooctene **113**, a structure that is present in many sesquiterpenes.

Scheme 36 Double Pauson–Khand Reaction for the Synthesis of a Linear [5.5.5.5] System[161]

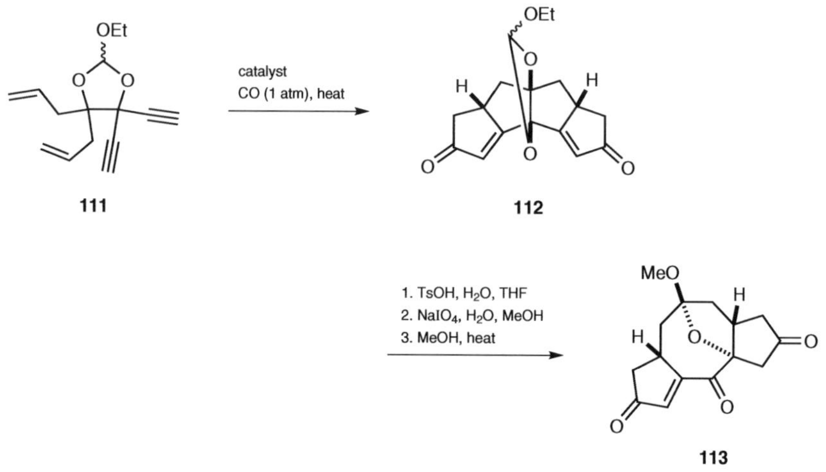

Catalyst (mol%)	Solvent	Yield[a] (%) of 112	Ref
[Ir(cod)Cl]$_2$ (20), Ph$_3$P (40)	toluene	76	[161]
[RhCl(CO)(dppp)]$_2$ (20)	MeCN	68	[161]
Co$_2$(CO)$_8$ (25)	DME	65	[161]

[a] Mixtures of isomers.

A tandem Pauson–Khand-type reaction of 1,4-enyne **114** tethered by a cyclopropyl group catalyzed by rhodium gives hydroxyindenone **115** in one step. The process involves the incorporation of two molecules of carbon monoxide and cleavage of the cyclopropane ring (Scheme 37).[162]

Scheme 37 Rhodium-Catalyzed Tandem Pauson–Khand-Type Reaction of a Cyclopropyl 1,4-Enyne[162]

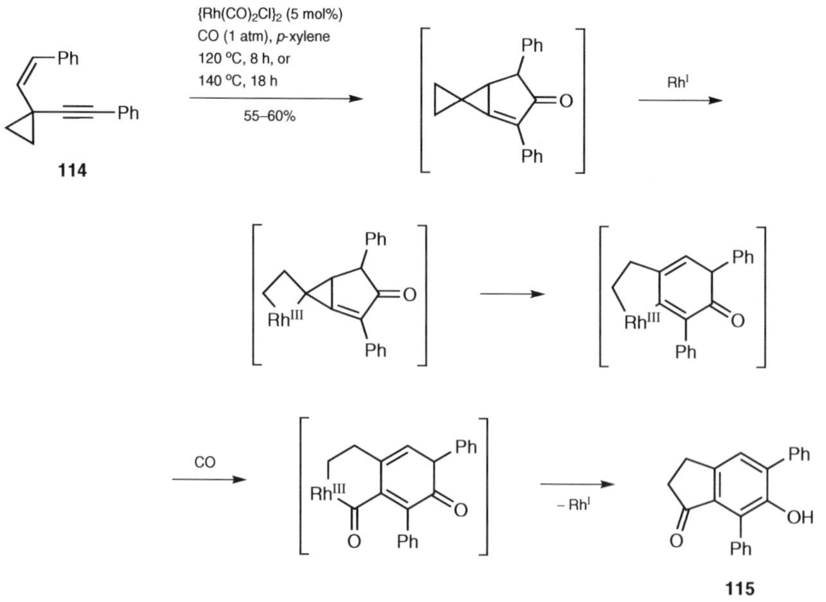

2.3.5.4 Reactions Occurring After the Pauson–Khand Process

Several reactions can occur after the Pauson–Khand reaction. Some have synthetic utility, either because they were planned or because they have been optimized after a first unexpected finding.[163] Thus, the formation of cyclopentanones from enynes occurs under Pauson–Khand reaction conditions generally as a side reaction; this process involves the tandem reduction of the cyclopentenone formed in the Pauson–Khand reaction. The reaction conditions for the reaction of enynes to give cyclopentanones **116** have been modified and the best results are obtained with dodecacarbonyltetracobalt(0) in alcohols under a hydrogen or nitrogen atmosphere. The reduction of the cyclopentenone is thought to be mediated by a cobalt hydride generated from residual cobalt species and takes place only with terminal alkynes. The role of the alcohol is to improve the generation of such hydrides. Nonterminal alkynes give cyclopentenones **117** (and other minor products) under these conditions (Scheme 38).[164]

Scheme 38 The Cobalt-Catalyzed Intramolecular Reductive Pauson–Khand Reaction[164]

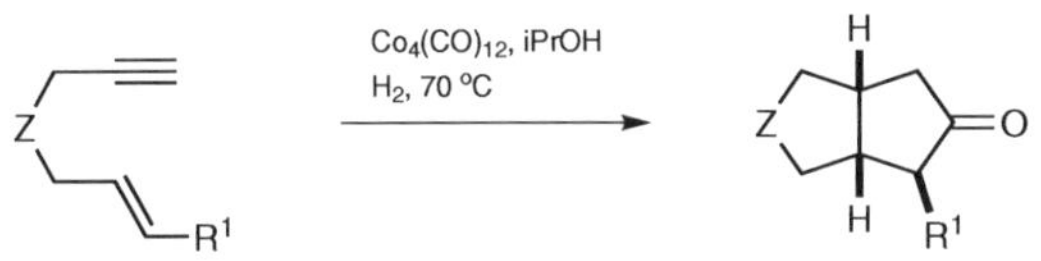

Z	R^1	Yield (%)	Ref
C(CO$_2$Et)$_2$	H	67	[164]
C(CO$_2$Et)$_2$	Me	61	[164]
C(CO$_2$Et)$_2$	Ph	45	[164]
NTs	H	62	[164]
NTs	Me	53	[164]

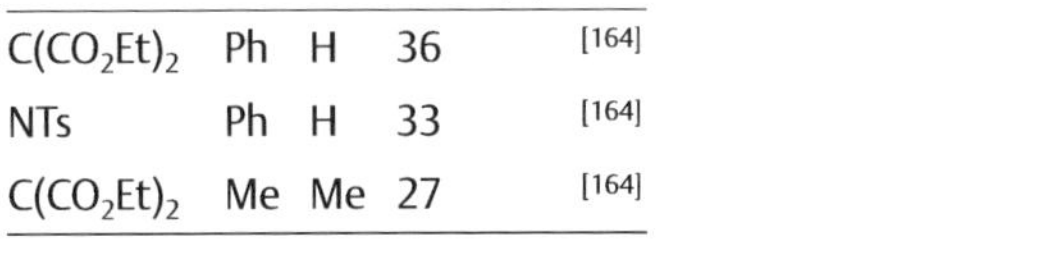

Z	R^1	R^2	Yield (%)	Ref
C(CO$_2$Et)$_2$	Ph	H	36	[164]
NTs	Ph	H	33	[164]
C(CO$_2$Et)$_2$	Me	Me	27	[164]

The migration or isomerization of double bonds and epimerization are frequently observed under Pauson–Khand reactions. Studies of these processes have led to the proposal that they take place during the Pauson–Khand reaction or with participation of the metal species present in the reaction.[165]

The cleavage of C—O, C—N, and other carbon–heteroatom bonds is a frequent side reaction observed in many Pauson–Khand reactions, especially those performed under harsh conditions or dry state adsorption conditions. The cleavage of ethers probably occurs, in most cases, after the Pauson–Khand reaction, because if the cleavage occurs prior to cyclization, the Pauson–Khand reaction then becomes an intermolecular reaction and, in most cases, it does not take place.

2.3.6 Hetero-Pauson–Khand Reaction

Heteroatom-containing multiple bond substrates can also be used in the Pauson–Khand reaction. Carbonyl groups, which give γ-butyrolactones, and various types of hetero π-systems, such as imines, nitriles, and heterocumulenes, can be utilized.[47]

The hetero-(2+2+1) cycloaddition of substituted δ-alkynyl aldehydes **118** (X = O) with carbon monoxide catalyzed by dodecacarbonyltriruthenium(0) gives γ-butyrolactones **119** (X = O);[166] molybdenum catalysis can also be used.[167] Yne–imines react with carbon monoxide in the presence of dodecacarbonyltriruthenium(0) to give lactams. For example, **118** (X = 4-MeOC$_6$H$_4$N) gives bicyclic lactam **119** (X = 4-MeOC$_6$H$_4$N) (Scheme 39).[168]

Scheme 39 Hetero-Pauson–Khand Reaction with C=X as the Alkene Partner[166–168]

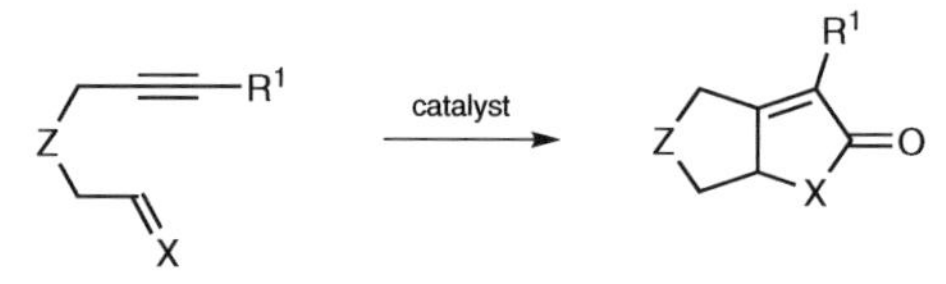

Z	R^1	X	Conditions	Yield (%)	Ref
C(CO$_2$Et)$_2$	TMS	O	Ru$_3$(CO)$_{12}$ (2 mol%), CO (10 atm), toluene, 160 °C	82	[166]
C(CO$_2$Et)$_2$	Ph	O	Mo(CO)$_3$(DMF)$_3$ (1 equiv), THF, rt	71	[167]
C(CO$_2$Et)$_2$	Me	O	Mo(CO)$_3$(DMF)$_3$ (1 equiv), THF, rt	63	[167]
NBoc	Ph	O	Mo(CO)$_3$(DMF)$_3$ (1 equiv), THF, rt	70	[167]
CH$_2$	Ph	O	Mo(CO)$_3$(DMF)$_3$ (1 equiv), THF, rt	68	[167]
C(CO$_2$Et)$_2$	TMS	4-MeOC$_6$H$_4$N	Ru$_3$(CO)$_{12}$ (5 mol%), CO (5 atm), toluene, 160 °C	66	[168]

The intramolecular hetero (2+2+1) cycloadditions of yne–carbodiimide derivatives, such as **120**, are performed using several catalytic systems (Scheme 40). Thus, molybdenum catalysis[169,170] is used to produce pyrrolo[2,3-b]indol-2-one ring systems **121** (Z = N). Octacarbonyldicobalt(0) activated by tetramethylthiourea is used to produce the heterocyclic systems **121** (Z = N) in moderate to good yields.[171] The hetero-Pauson–Khand reaction also uses the catalyst prepared in situ from chloro(cycloocta-1,5-diene)rhodium(I) dimer and 1,3-bis(diphenylphosphino)propane.[169] The hetero (2+2+1) cycloaddition of yne-isothiocyanates, using a stoichiometric amount of octacarbonyldicobalt(0) or hexacarbonylmolybdenum(0) gives **121** (Z = S).[172] Triarylketenimines are used for intermolecular hetero (2+2+1) cycloadditions with alkyne–hexacarbonyldicobalt complexes affording γ-methylene-γ-lactam derivatives.[173]

2.3.6 Hetero-Pauson–Khand Reaction

Scheme 40 Hetero-Pauson–Khand Reaction with Carbodiimides and Related Derivatives[169–172]

Z	R^1	Conditions	Yield (%)	Ref
NPr	Ph	Mo(CO)$_6$ (1 equiv), DMSO (5 equiv), toluene, 120 °C	55	[169,170]
4-MeC$_6$H$_4$N	TMS	Mo(CO)$_6$ (1.1 equiv), DMSO (5 equiv), toluene, 120 °C	70	[170]
4-MeOC$_6$H$_4$N	Pr	Co$_2$(CO)$_8$ (30 mol%), (Me$_2$N)$_2$C=S (30 mol%), CO (1 atm), toluene, 70 °C	65	[171]
4-MeOC$_6$H$_4$N	(CH$_2$)$_2$CH=CMe$_2$	Co$_2$(CO)$_8$ (30 mol%), (Me$_2$N)$_2$C=S (30 mol%), CO (1 atm), toluene, 70 °C	49	[171]
4-MeOC$_6$H$_4$N	(CH$_2$)$_2$OTBDMS	Co$_2$(CO)$_8$ (30 mol%), (Me$_2$N)$_2$C=S (30 mol%), CO (1 atm), toluene, 70 °C	64	[171]
4-PhOC$_6$H$_4$N	TMS	Co$_2$(CO)$_8$ (30 mol%), (Me$_2$N)$_2$C=S (30 mol%), CO (1 atm), toluene, 70 °C	77	[171]
NPh	Ph	[RhCl(cod)]$_2$ (5 mol%), dppp (11 mol%), CO (1 atm), toluene, 120 °C	50	[169]
NPr	Ph	[RhCl(cod)]$_2$ (5 mol%), dppp (11 mol%), CO (1 atm), toluene, 120 °C	77	[169]
4-TolN	TMS	[RhCl(cod)]$_2$ (5 mol%), dppp (11 mol%), CO (1 atm), toluene, 120 °C	77	[169]
4-ClC$_6$H$_4$N	TMS	[RhCl(cod)]$_2$ (5 mol%), dppp (11 mol%), CO (1 atm), toluene, 120 °C	77	[169]
S	t-Bu	Mo(CO)$_6$ (2 equiv), DMSO (10 equiv), toluene, 115 °C	75	[172]
S	TBDMS	Mo(CO)$_6$ (2 equiv), DMSO (10 equiv), toluene, 115 °C	64	[172]

The reaction of isocyanates and alkynes catalyzed by dodecacarbonyltriruthenium(0) gives trisubstituted maleimide derivatives **122** in high yields (Scheme 41).[174]

Scheme 41 Synthesis of Maleimides via the Ruthenium-Catalyzed Hetero-Pauson–Khand Reaction of Isocyanates[174]

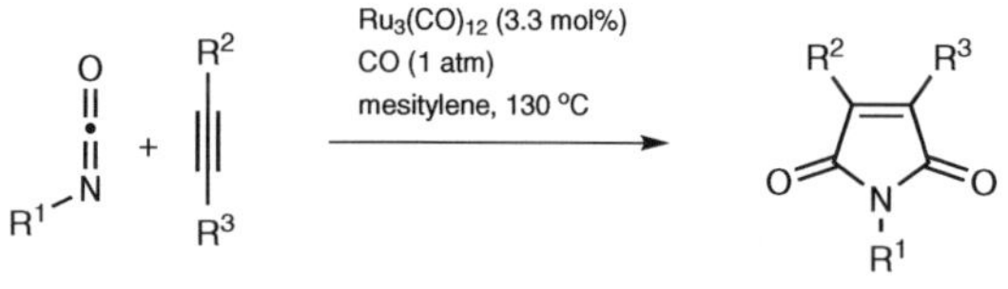

R^1	R^2	R^3	Yield (%)	Ref
Ph	Pr	Pr	82	[174]
Ph	Ph	Ph	98	[174]
Ph	Ph	TMS	96	[174]
4-MeOC$_6$H$_4$	Ph	Me	95	[174]
4-F$_3$CC$_6$H$_4$	Ph	Me	98	[174]
1-adamantyl	Ph	Me	96	[174]
t-Bu	Ph	Me	97	[174]

The alkyne moiety can be substituted by a nitrile group in the Pauson–Khand-type reaction. An intramolecular reaction of allene–nitrile derivatives **123** using carbonyl(chloro)-[1,3-bis(diphenylphosphino)propane]rhodium(I) dimer [{RhCl(CO)dppp}$_2$] gives 1,3-dihydro-2H-benzo[f]indol-2-ones **125** (Scheme 42).[175] It is proposed that the reaction is initiated by isomerization of the nitrile group to give the phenylketenimine **124** followed by a typical (2+2+1)-cycloaddition process to give **125**.

Scheme 42 Rhodium-Catalyzed Pauson–Khand-Type Reaction of Allene–Nitrile Derivatives[175]

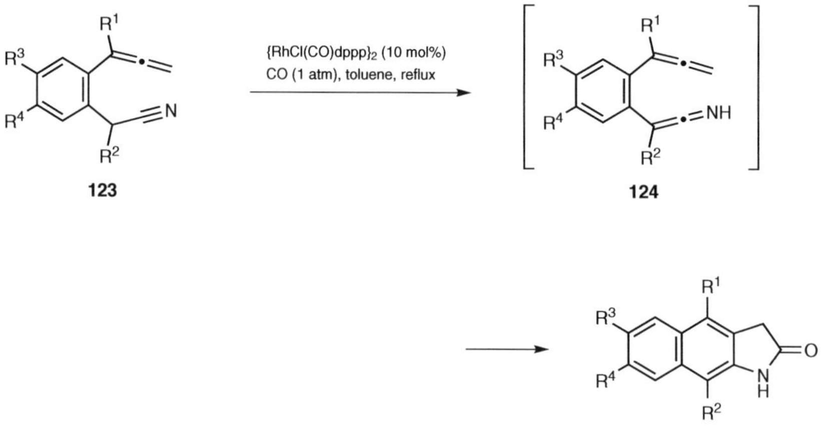

R^1	R^2	R^3	R^4	Yield (%)	Ref
Me	H	H	H	88	[175]
Me	Me	H	H	68	[175]
Me	H	NO$_2$	H	72	[175]
Me	H	H	OMe	66	[175]

R¹	R²	R³	R⁴	Yield (%)	Ref
Me	H	H	NO_2	80	[175]
SO_2Ph	H	H	H	60	[175]
SO_2Ph	H	OMe	H	79	[175]
SO_2Ph	H	Cl	H	54	[175]

2.3.7 Total Synthesis Using the Pauson–Khand Reaction as the Key Step

The cyclopentane ring is common in nature. Pauson–Khand adducts are easily functionalized and hence there has been a continued increase in the use of the Pauson–Khand reaction for the synthesis of natural products. A review on the use of the Pauson–Khand in natural product synthesis is available;[176] the most outstanding contributions are summarized in Scheme 43 and Table 13. Most examples use stoichiometric octacarbonyldicobalt(0) to mediate the Pauson–Khand reaction; this shows that the catalytic methodologies developed so far have not generally been adopted. However, some catalytic Pauson–Khand reactions are reported, such as a cobalt-catalyzed reaction in the synthesis of pentalenene (Table 13, entry 1) or the syntheses of achalensolide and ingenol, which use rhodium-based catalysts (Table 13, entries 4 and 6).

Scheme 43 Natural Products Synthesized by the Pauson–Khand-Type Reaction[177–186]

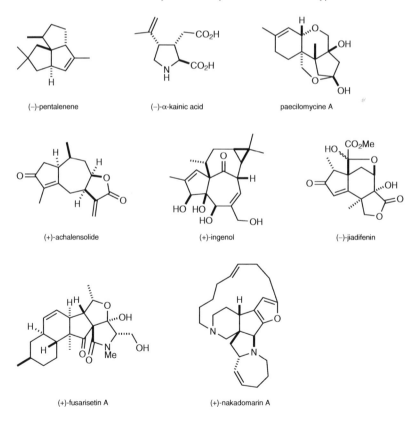

(−)-pentalenene   (−)-α-kainic acid   paecilomycine A

(+)-achalensolide   (+)-ingenol   (−)-jiadifenin

(+)-fusarisetin A   (+)-nakadomarin A

Table 13 Pauson–Khand Reactions in the Total Syntheses of Natural Products[177–186]

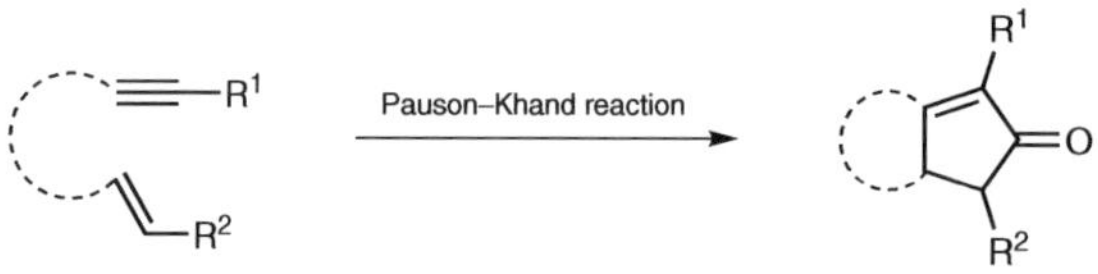

Entry	Substrate	Conditions	Product	Yield (%)	Ref
1	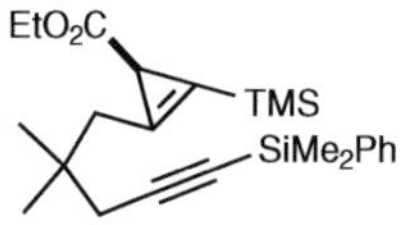 (structure with EtO₂C, TMS, SiMe₂Ph)	1. Co₂(CO)₈ (0.6 equiv), toluene 2. (Me₂N)₂C=S, CO (1 atm), 45 °C	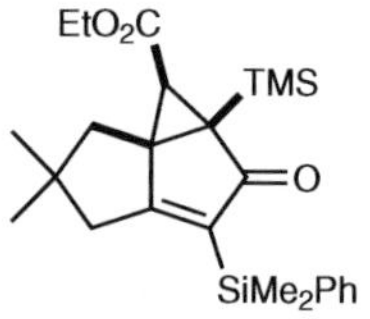 (product)	64	[177]
2	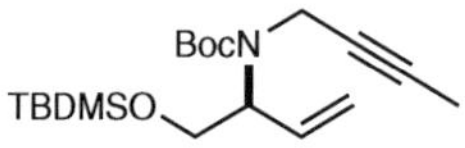 (structure with BocN, TBDMSO)	1. Co₂(CO)₈ (1.1 equiv), CH₂Cl₂, rt 2. TMANO, 4-Å molecular sieves, rt	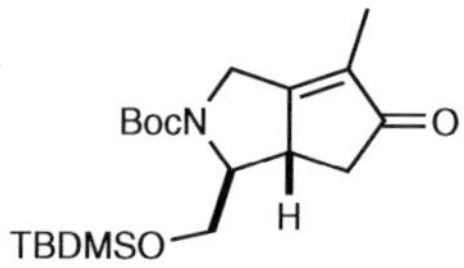 (product)	65[a]	[178]
3	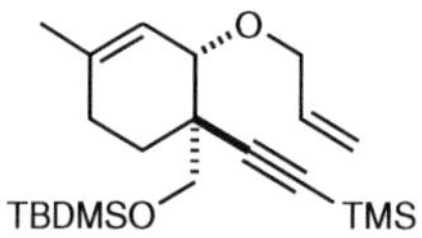 (structure with TBDMSO, TMS)	Co₂(CO)₈ (1.2 equiv), 4-Å molecular sieves, toluene, rt then 100 °C	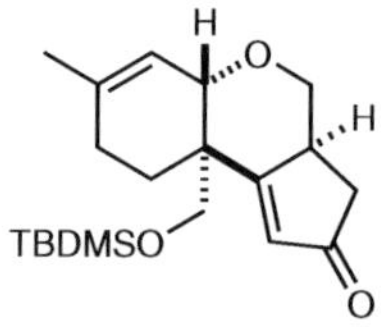 (product)	37	[179]
4	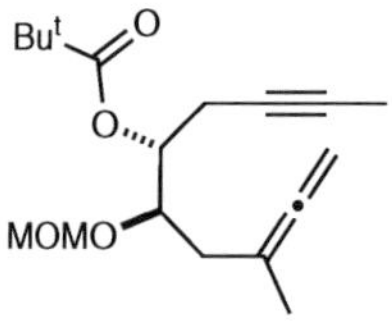 (structure with Bu^t, MOMO)	[RhCl(cod)]₂ (10 mol%), dppp (50 mol%), CO (1 atm), toluene, reflux	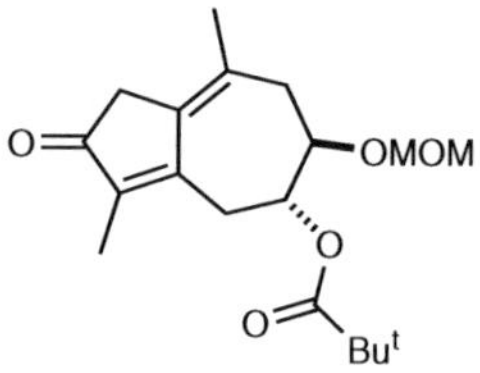 (product with OMOM, Bu^t)	96	[180]
5	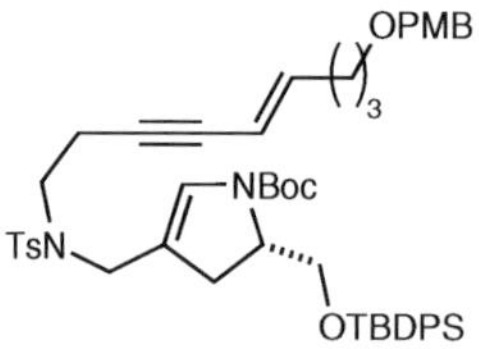 (structure with OPMB, NBoc, TsN, OTBDPS)	1. Co₂(CO)₈ (1.3 equiv), Et₂O 2. BuSMe, toluene, 110 °C	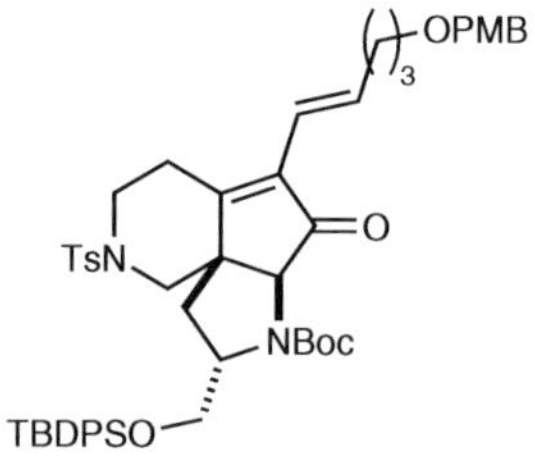 (product)	60	[181]
6	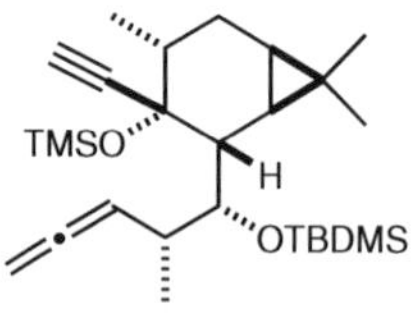 (structure with TMSO, OTBDMS)	[RhCl(CO)₂]₂ (0.1 equiv), CO (1 atm), xylenes, 140 °C	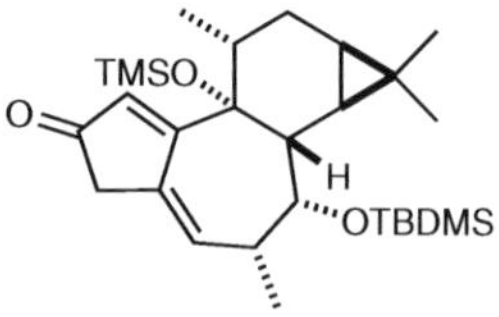 (product)	72	[182]

2.3.8 Conclusions

Table 13 (cont.)

Entry	Substrate	Conditions	Product	Yield (%)	Ref
7		$Co_2(CO)_8$ (1.1 equiv), toluene, 120 °C		82	[183]
8		1. $Co_2(CO)_8$ (1.2 equiv), toluene, rt, 2 h 2. Bu_3PS, 75 °C, 16 h		67	[184]
9		$Co_2(CO)_8$ (1.1 equiv), CH_2Cl_2, rt, then evaporation, MeCN, 60 °C		97	[185]
10		1. Me_2SiCl_2, Et_3N, DMAP, 1,2-dichloroethane, 0 °C, 2 h 2. $Co_2(CO)_8$ (1.5 equiv), CO, toluene, rt to 100 °C, 17 h 3. HCl, MeOH, 0 °C, 1 h		92	[186]

[a] Ratio (trans/cis) 88:12.

2.3.8 Conclusions

The Pauson–Khand reaction produces great molecular complexity in one step. Since its beginnings, when limited scope and harsh conditions precluded its development as a synthetically useful procedure, many methodological advances have meant that it can now be used with preparative efficiency. This section discusses many experimental procedures that give very good levels of efficiency and selectivity in both an intra- and intermolecular sense. In particular, in depth understanding of the reaction pathway allows improvements to reaction kinetics and the prediction of the regioselectivity. Cobalt-catalyzed reactions are now reliable processes and include several reusable heterogeneous systems. In addition, the introduction of other metals as catalysts has triggered a number of mild catalytic procedures that allow enantioselective reactions with the use of chiral complexes or the addition of chiral ligands. Synthetic processes have been developed in which the Pauson–Khand reaction is combined with other transformations in a single step, with promising synthetic applications. Some aspects, such as the limitation in scope of the intermolecular version, or the lack of generality of some catalytic protocols, are still not completely solved. This reaction is one of the best examples of how organometallic chemistry is useful in modern organic synthesis, and can serve as a key tool for the synthesis of natural products, in this case those possessing cyclopentane units. The number of papers dedicated to total synthesis using the Pauson–Khand reaction as a key step shows that it is gaining in importance, and exciting applications are to come in the future.

References

[1] *The Pauson–Khand Reaction: Scope, Variations and Applications*, Torres, R. R., Ed.; Wiley: Hoboken, NJ, (2012).
[2] Omae, I., *Coord. Chem. Rev.*, (2011) **255**, 139.
[3] Strübing, D.; Beller, M., *Top. Organomet. Chem.*, (2006) **18**, 165.
[4] Blanco-Urgoiti, J.; Añorbe, L.; Pérez-Serrano, L.; Domínguez, G.; Pérez-Castells, J., *Chem. Soc. Rev.*, (2004) **33**, 32.
[5] Khand, I. U.; Knox, G. R.; Pauson, P. L.; Watts, W. E., *J. Chem. Soc. D*, (1971), 36.
[6] Park, J. H.; Chang, K.-M.; Chung, Y. K., *Coord. Chem. Rev.*, (2009) **253**, 2461.
[7] Schore, N. E.; Croudace, M. C., *J. Org. Chem.*, (1981) **46**, 5436.
[8] Robert, F.; Milet, A.; Gimbert, Y.; Konya, D.; Greene, A. E., *J. Am. Chem. Soc.*, (2001) **123**, 5396.
[9] Fager-Jokela, E.; Muuronen, M.; Patzschke, M.; Helaja, J., *J. Org. Chem.*, (2012) **77**, 9134.
[10] Moulton, B. E.; Whitwood, A. C.; Duhme-Klair, A. K.; Lynam, J. M.; Fairlamb, I. J. S., *J. Org. Chem.*, (2011) **76**, 5320.
[11] Aiguabella, N.; del Pozo, C.; Verdaguer, X.; Fustero, S.; Riera, A., *Angew. Chem.*, (2013) **125**, 5463; *Angew. Chem. Int. Ed.*, (2013) **52**, 5355.
[12] Krafft, M. E.; Scott, I. L.; Romero, R. H.; Feibelmann, S.; Van Pelt, C. E., *J. Am. Chem. Soc.*, (1993) **115**, 7199.
[13] Kędzia, J. L.; Kerr, W. J.; McPherson, A. R., *Synlett*, (2010), 649.
[14] Smit, W. A.; Kireev, S. L.; Nefedov, O. M.; Tarasov, V. A., *Tetrahedron Lett.*, (1989) **30**, 4021.
[15] Shambayati, S.; Crowe, W. E.; Schreiber, S. L., *Tetrahedron Lett.*, (1990) **31**, 5289.
[16] Jeong, N.; Chung, Y. K.; Lee, B. Y.; Lee, S. H.; Yoo, S.-E., *Synlett*, (1991), 204.
[17] Ford, J. G.; Kerr, W. J.; Kirk, G. G.; Lindsay, D. M.; Middlemiss, D., *Synlett*, (2000), 1415.
[18] Pérez-Serrano, L.; Casarrubios, L.; Domínguez, G.; Pérez-Castells, J., *Org. Lett.*, (1999) **1**, 1187.
[19] Marchueta, I.; Verdaguer, X.; Moyano, A.; Pericàs, M. A.; Riera, A., *Org. Lett.*, (2001) **3**, 3193.
[20] Krafft, M. E.; Wright, J. A.; Boñaga, L. V. R., *Tetrahedron Lett.*, (2003) **44**, 3417.
[21] Perez del Valle, C.; Milet, A.; Gimbert, Y.; Greene, A. E., *Angew. Chem.*, (2005) **117**, 5863; *Angew. Chem. Int. Ed.*, (2005) **44**, 5717.
[22] Billington, D. C.; Helps, I. M.; Pauson, P. L.; Thomson, W.; Willison, D., *J. Organomet. Chem.*, (1988) **354**, 233.
[23] Krafft, M. E.; Boñaga, L. V. R.; Hirosawa, C., *J. Org. Chem.*, (2001) **66**, 3004.
[24] Sugihara, T.; Yamada, M.; Yamaguchi, M.; Nishizawa, M., *Synlett*, (1999), 771.
[25] Brown, J. A.; Irvine, S.; Kerr, W. J.; Pearson, C. M., *Org. Biomol. Chem.*, (2005) **3**, 2396.
[26] Fischer, S.; Groth, U.; Jung, M.; Schneider, A., *Synlett*, (2002), 2023.
[27] Lagunas, A.; Mairata i Payeras, A.; Jimeno, C.; Pericàs, M. A., *Org. Lett.*, (2005) **7**, 3033.
[28] Asano, K.; Uesugi, Y.; Yoshida, J.-i., *Org. Lett.*, (2013) **15**, 2398.
[29] Adrio, J.; Rodríguez Rivero, M.; Carretero, J. C., *Org. Lett.*, (2005) **7**, 431.
[30] Brummond, K. M.; Mitasev, B., *Org. Lett.*, (2004) **6**, 2245.
[31] Cao, H.; Van Ornum, S. G.; Deschamps, J.; Flippen-Anderson, J.; Laib, F.; Cook, J. M., *J. Am. Chem. Soc.*, (2005) **127**, 933.
[32] de Meijere, A.; Becker, H.; Stolle, A.; Kozhushkov, S. I.; Bes, M. T.; Salaün, J.; Noltemeyer, M., *Chem.–Eur. J.*, (2005) **11**, 2471.
[33] de Bruin, T. J. M.; Milet, A.; Greene, A. E.; Gimbert, Y., *J. Org. Chem.*, (2004) **69**, 1075.
[34] Lledó, A.; Fuster, A.; Revés, M.; Verdaguer, X.; Riera, A., *Chem. Commun. (Cambridge)*, (2013) **49**, 3055.
[35] Kizirian, J.-C.; Aiguabella, N.; Pesquer, A.; Fustero, S.; Bello, P.; Verdaguer, X.; Riera, A., *Org. Lett.*, (2010) **12**, 5620.
[36] Aiguabella, N.; Pesquer, A.; Verdaguer, X.; Riera, A., *Org. Lett.*, (2013) **15**, 2696.
[37] Revés, M.; Lledó, A.; Ji, Y.; Blasi, E.; Riera, A.; Verdaguer, X., *Org. Lett.*, (2012) **14**, 3534.
[38] Pallerla, M. K.; Yap, G. P. A.; Fox, J. M., *J. Org. Chem.*, (2008) **73**, 6137.
[39] Gibson, S. E.; Mainolfi, N.; Kalindjian, S. B.; Wright, P. T., *Angew. Chem.*, (2004) **116**, 5798; *Angew. Chem. Int. Ed.*, (2004) **43**, 5680.
[40] Seigal, B. A.; An, M. H.; Snapper, M. L., *Angew. Chem.*, (2005) **117**, 5009; *Angew. Chem. Int. Ed.*, (2005) **44**, 4929.
[41] Schore, N. E., *Org. React. (N. Y.)*, (1991) **40**, 1.
[42] Billington, D. C.; Bladon, P.; Helps, I. M.; Pauson, P. L.; Thomson, W.; Willison, D., *J. Chem. Res., Miniprint*, (1988), 2601.

[43] Jeong, N.; Hwang, S. H., *Angew. Chem.*, (2000) **112**, 650; *Angew. Chem. Int. Ed.*, (2000) **39**, 636.
[44] Vázquez-Romero, A.; Cárdenas, L.; Blasi, E.; Verdaguer, X.; Riera, A., *Org. Lett.*, (2009) **11**, 3104.
[45] Kerr, W. J.; McLaughlin, M.; Pauson, P. L.; Robertson, S. M., *Chem. Commun. (Cambridge)*, (1999), 2171.
[46] Pérez-Serrano, L.; Blanco-Urgoiti, J.; Casarrubios, L.; Domínguez, G.; Pérez-Castells, J., *J. Org. Chem.*, (2000) **65**, 3513.
[47] Kitagaki, S.; Inagaki, F.; Mukai, C., *Chem. Soc. Rev.*, (2014) **43**, 2956.
[48] Ahmar, M.; Locatelli, C.; Colombier, D.; Cazes, B., *Tetrahedron Lett.*, (1997) **38**, 5281.
[49] Pagenkopf, B. L.; Belanger, D. B.; O'Mahony, D. J. R.; Livinghouse, T., *Synthesis*, (2000), 1009.
[50] Brummond, K. M.; Wan, H.; Kent, J. L., *J. Org. Chem.*, (1998) **63**, 6535.
[51] Brummond, K. M.; Lu, J.; Petersen, J., *J. Am. Chem. Soc.*, (2000) **122**, 4915.
[52] Pérez-Serrano, L.; Casarrubios, L.; Domínguez, G.; Pérez-Castells, J., *Chem. Commun. (Cambridge)*, (2001), 2602.
[53] Mukai, C.; Nomura, I.; Kitagaki, S., *J. Org. Chem.*, (2003) **68**, 1376.
[54] Magnus, P.; Exon, C.; Albaugh-Robertson, P., *Tetrahedron Lett.*, (1985) **26**, 5861.
[55] Pericàs, M. A.; Balsells, J.; Castro, J.; Marchueta, I.; Moyano, A.; Riera, A.; Vázquez, J.; Verdaguer, X., *Pure Appl. Chem.*, (2002) **74**, 167.
[56] Banide, E. V.; Müller-Bunz, H.; Manning, A. R.; Evans, P.; McGlinchey, M. J., *Angew. Chem.*, (2007) **119**, 2965; *Angew. Chem. Int. Ed.*, (2007) **46**, 2907.
[57] Lesage, D.; Milet, A.; Memboeuf, A.; Blu, J.; Greene, A. E.; Tabet, J.-C.; Gimbert, Y., *Angew. Chem.*, (2014) **126**, 1970; *Angew. Chem. Int. Ed.*, (2014) **53**, 1939.
[58] Yamanaka, M.; Nakamura, E., *J. Am. Chem. Soc.*, (2001) **123**, 1703.
[59] de Bruin, T. J. M.; Milet, A.; Robert, F.; Gimbert, Y.; Greene, A. E., *J. Am. Chem. Soc.*, (2001) **123**, 7184.
[60] Schulte, J. H.; Gleiter, R.; Rominger, F., *Org. Lett.*, (2002) **4**, 3301.
[61] Cabot, R.; Lledó, A.; Revés, M.; Riera, A.; Verdaguer, X., *Organometallics*, (2007) **26**, 1134.
[62] Rautenstrauch, V.; Mégard, P.; Conesa, J.; Küster, W., *Angew. Chem.*, (1990) **102**, 1441; *Angew. Chem. Int. Ed. Engl.*, (1990) **29**, 1413.
[63] Shibata, T., *Adv. Synth. Catal.*, (2006) **348**, 2328.
[64] Jeong, N.; Hwang, S. H.; Lee, Y.; Chung, Y. K., *J. Am. Chem. Soc.*, (1994) **116**, 3159.
[65] Comely, A. C.; Gibson, S. E.; Stevenazzi, A.; Hales, N. J., *Tetrahedron Lett.*, (2001) **42**, 1183.
[66] Hayashi, M.; Hashimoto, Y.; Yamamoto, Y.; Usuki, J.; Saigo, K., *Angew. Chem.*, (2000) **112**, 645; *Angew. Chem. Int. Ed.*, (2000) **39**, 631.
[67] Tang, Y.; Deng, L.; Zhang, Y.; Dong, G.; Chen, J.; Yang, Z., *Org. Lett.*, (2005) **7**, 593.
[68] Wang, Y.; Xu, L.; Yu, R.; Chen, J.; Yang, Z., *Chem. Commun. (Cambridge)*, (2012) **48**, 8183.
[69] Sugihara, T.; Yamaguchi, M., *Synlett*, (1998), 1384.
[70] Blanco-Urgoiti, J.; Casarrubios, L.; Domínguez, G.; Pérez-Castells, J., *Tetrahedron Lett.*, (2002) **43**, 5763.
[71] Dahan, A.; Portnoy, M., *Chem. Commun. (Cambridge)*, (2002), 2700.
[72] Pagenkopf, B. L.; Livinghouse, T., *J. Am. Chem. Soc.*, (1996) **118**, 2285.
[73] Belanger, D. B.; Livinghouse, T., *Tetrahedron Lett.*, (1998) **39**, 7641.
[74] Krafft, M. E.; Boñaga, L. V. R.; Hirosawa, C., *Tetrahedron Lett.*, (1999) **40**, 9177.
[75] Sugihara, T.; Yamaguchi, M., *J. Am. Chem. Soc.*, (1998) **120**, 10782.
[76] Boñaga, L. V. R.; Wright, J. A.; Krafft, M. E., *Chem. Commun. (Cambridge)*, (2004), 1746.
[77] Kobayashi, T.; Koga, Y.; Narasaka, K., *J. Organomet. Chem.*, (2001) **624**, 73.
[78] Jeong, N.; Sung, B. K.; Kim, J. S.; Park, S. B.; Seo, S. D.; Shin, J. Y.; In, K. Y.; Choi, Y. K., *Pure Appl. Chem.*, (2002) **74**, 85.
[79] Wang, H.; Sawyer, J. R.; Evans, P. A.; Baik, M.-H., *Angew. Chem.*, (2008) **120**, 348; *Angew. Chem. Int. Ed.*, (2008) **47**, 342.
[80] Baik, M.-H.; Mazumder, S.; Ricci, P.; Sawyer, J. R.; Song, Y.-G.; Wang, H.; Evans, P. A., *J. Am. Chem. Soc.*, (2011) **133**, 7621.
[81] Jeong, N.; Sung, B. K.; Choi, Y. K., *J. Am. Chem. Soc.*, (2000) **122**, 6771.
[82] Kim, D. E.; Ratovelomanana-Vidal, V.; Jeong, N., *Adv. Synth. Catal.*, (2010) **352**, 2032.
[83] Kim, D. E.; Kim, I. S.; Ratovelomanana-Vidal, V.; Genêt, J.-P.; Jeong, N., *J. Org. Chem.*, (2008) **73**, 7985.
[84] Fan, B.-M.; Xie, J.-H.; Li, S.; Tu, Y.-Q.; Zhou, Q.-L., *Adv. Synth. Catal.*, (2005) **347**, 759.
[85] Kim, D. E.; Choi, C.; Kim, I. S.; Jeulin, S.; Ratovelomanana-Vidal, V.; Genêt, J.-P.; Jeong, N., *Adv. Synth. Catal.*, (2007) **349**, 1999.

[86] Croatt, M. P.; Wender, P. A., *Eur. J. Org. Chem.*, (2010), 19.
[87] Wender, P. A.; Deschamps, N. M.; Gamber, G. G., *Angew. Chem.*, (2003) **115**, 1897; *Angew. Chem. Int. Ed.*, (2003) **42**, 1853.
[88] Wender, P. A.; Deschamps, N. M.; Williams, T. J., *Angew. Chem.*, (2004) **116**, 3138; *Angew. Chem. Int. Ed.*, (2004) **43**, 3076.
[89] Pitcock, W. H., Jr.; Lord, R. L.; Baik, M.-H., *J. Am. Chem. Soc.*, (2008) **130**, 5821.
[90] Mukai, C.; Nomura, I.; Yamanishi, K.; Hanaoka, M., *Org. Lett.*, (2002) **4**, 1755.
[91] Brummond, K. M.; Chen, H.; Fisher, K. D.; Kerekes, A. D.; Rickards, B.; Sill, P. C.; Geib, S. J., *Org. Lett.*, (2002) **4**, 1931.
[92] Morimoto, T.; Fuji, K.; Tsutsumi, K.; Kakiuchi, K., *J. Am. Chem. Soc.*, (2002) **124**, 3806.
[93] Shibata, T.; Toshida, N.; Takagi, K., *Org. Lett.*, (2002) **4**, 1619.
[94] Shibata, T.; Toshida, N.; Takagi, K., *J. Org. Chem.*, (2002) **67**, 7446.
[95] Fuji, K.; Morimoto, T.; Tsutsumi, K.; Kakiuchi, K., *Angew. Chem.*, (2003) **115**, 2511; *Angew. Chem. Int. Ed.*, (2003) **42**, 2409.
[96] Fuji, K.; Morimoto, T.; Tsutsumi, K.; Kakiuchi, K., *Tetrahedron Lett.*, (2004) **45**, 9163.
[97] Kwong, F. Y.; Lee, H. W.; Qiu, L.; Lam, W. H.; Li, Y.-M.; Kwong, H. L.; Chan, A. S. C., *Adv. Synth. Catal.*, (2005) **347**, 1750.
[98] Park, J. H.; Cho, Y.; Chung, Y. K., *Angew. Chem.*, (2010) **122**, 5264; *Angew. Chem. Int. Ed.*, (2010) **49**, 5138.
[99] Shibata, T.; Takagi, K., *J. Am. Chem. Soc.*, (2000) **122**, 9852.
[100] Shibata, T.; Toshida, N.; Yamasaki, M.; Maekawa, S.; Takagi, K., *Tetrahedron*, (2005) **61**, 9974.
[101] Lu, Z.-L.; Neumann, E.; Pfaltz, A., *Eur. J. Org. Chem.*, (2007), 4189.
[102] Jeong, N.; Kim, D. H.; Choi, J. H., *Chem. Commun. (Cambridge)*, (2004), 1134.
[103] Hicks, F. A.; Kablaoui, N. M.; Buchwald, S. L., *J. Am. Chem. Soc.*, (1999) **121**, 5881.
[104] Hicks, F. A.; Buchwald, S. L., *J. Am. Chem. Soc.*, (1999) **121**, 7026.
[105] Sturla, S. J.; Buchwald, S. L., *Organometallics*, (2002) **21**, 739.
[106] Morimoto, T.; Chatani, N.; Fukumoto, Y.; Murai, S., *J. Org. Chem.*, (1997) **62**, 3762.
[107] Kondo, T.; Suzuki, N.; Okada, T.; Mitsudo, T.-a., *J. Am. Chem. Soc.*, (1997) **119**, 6187.
[108] Itami, K.; Mitsudo, K.; Yoshida, J.-i., *Angew. Chem.*, (2002) **114**, 3631; *Angew. Chem. Int. Ed.*, (2002) **41**, 3481.
[109] Paolillo, R.; Gallo, V.; Mastrorilli, P.; Nobile, C. F.; Rosé, J.; Braunstein, P., *Organometallics*, (2008) **27**, 741.
[110] Tang, Y.; Deng, L.; Zhang, Y.; Dong, G.; Chen, J.; Yang, Z., *Org. Lett.*, (2005) **7**, 1657.
[111] Wu, N.; Deng, L.; Liu, L.; Liu, Q.; Li, C.; Yang, Z., *Chem.–Asian J.*, (2013) **8**, 65.
[112] Lan, Y.; Deng, L.; Liu, J.; Wang, C.; Wiest, O.; Yang, Z.; Wu, Y.-D., *J. Org. Chem.*, (2009) **74**, 5049.
[113] Kim, S.-W.; Son, S. U.; Lee, S. I.; Hyeon, T.; Chung, Y. K., *J. Am. Chem. Soc.*, (2000) **122**, 1550.
[114] Kim, S.-W.; Son, S. U.; Lee, S. S.; Hyeon, T.; Chung, Y. K., *Chem. Commun. (Cambridge)*, (2001), 2212.
[115] Park, K. H.; Son, S. U.; Chung, Y. K., *Tetrahedron Lett.*, (2003) **44**, 2827.
[116] Son, S. U.; Park, K. H.; Chung, Y. K., *Org. Lett.*, (2002) **4**, 3983.
[117] Muller, J.-L.; Klankermayer, J.; Leitner, W., *Chem. Commun. (Cambridge)*, (2007), 1939.
[118] Park, K. H.; Son, S. U.; Chung, Y. K., *Chem. Commun. (Cambridge)*, (2003), 1898.
[119] Park, J. H.; Chung, Y. K., *Dalton Trans.*, (2008), 2369.
[120] Park, K. H.; Chung, Y. K., *Adv. Synth. Catal.*, (2005) **347**, 854.
[121] Muller, J.-L.; Rickers, A.; Leitner, W., *Adv. Synth. Catal.*, (2007) **349**, 287.
[122] Goettmann, F.; Thomas, A.; Antonietti, M., *Angew. Chem.*, (2007) **119**, 2773; *Angew. Chem. Int. Ed.*, (2007) **46**, 2717.
[123] *The Pauson–Khand Reaction: Scope, Variations and Applications*, Torres, R. R., Ed.; Wiley: Hoboken, NJ, (2012); pp 69–167.
[124] Brummond, K. M.; Curran, D. P.; Mitasev, B.; Fischer, S., *J. Org. Chem.*, (2005) **70**, 1745.
[125] Ghosh, S. K.; Hsung, R. P.; Liu, J., *J. Am. Chem. Soc.*, (2005) **127**, 8260.
[126] McCormack, M. P.; Waters, S. P., *J. Org. Chem.*, (2013) **78**, 1176.
[127] Brummond, K. M.; Curran, D. P.; Mitasev, B.; Fischer, S., *J. Org. Chem.*, (2005) **70**, 1745.
[128] Grillet, F.; Brummond, K. M., *J. Org. Chem.*, (2013) **78**, 3737.
[129] Poch, M.; Valentí, E.; Moyano, A.; Pericàs, M. A.; Castro, J.; DeNicola, A.; Greene, A. E., *Tetrahedron Lett.*, (1990) **31**, 7505.
[130] Castro, J.; Sörensen, H.; Riera, A.; Morin, C.; Moyano, A.; Pericàs, M. A.; Greene, A. E., *J. Am. Chem. Soc.*, (1990) **112**, 9388.

[131] Castro, J.; Moyano, A.; Pericàs, M. A.; Riera, A.; Greene, A. E.; Alvarez-Larena, A.; Piniella, J. F., *J. Org. Chem.*, (1996) **61**, 9016.
[132] Bernardes, V.; Verdaguer, X.; Kardos, N.; Riera, A.; Moyano, A.; Pericàs, M. A.; Greene, A. E., *Tetrahedron Lett.*, (1994) **35**, 575.
[133] Verdaguer, X.; Moyano, A.; Pericàs, M. A.; Riera, A.; Greene, A. E.; Piniella, J. F.; Alvarez-Larena, A., *J. Organomet. Chem.*, (1992) **433**, 305.
[134] Tormo, J.; Verdaguer, X.; Moyano, A.; Pericàs, M. A.; Riera, A., *Tetrahedron*, (1996) **52**, 14021.
[135] Verdaguer, X.; Vázquez, J.; Fuster, G.; Bernardes-Génisson, V.; Greene, A. E.; Moyano, A.; Pericàs, M. A.; Riera, A., *J. Org. Chem.*, (1998) **63**, 7037.
[136] Fonquerna, S.; Rios, R.; Moyano, A.; Pericàs, M. A.; Riera, A., *Eur. J. Org. Chem.*, (1999), 3459.
[137] Fonquerna, S.; Moyano, A.; Pericàs, M. A.; Riera, A., *J. Am. Chem. Soc.*, (1997) **119**, 10225.
[138] Adrio, J.; Carretero, J. C., *J. Am. Chem. Soc.*, (1999) **121**, 7411.
[139] Rodríguez Rivero, M.; de la Rosa, J. C.; Carretero, J. C., *J. Am. Chem. Soc.*, (2003) **125**, 14992.
[140] Garcia Ruano, J. L.; Torrente, E.; Parra, A.; Alemán, J.; Martín-Castro, A. M., *J. Org. Chem.*, (2012) **77**, 6583.
[141] Verdaguer, X.; Lledó, A.; López-Mosquera, C.; Maestro, M. A.; Pericàs, M. A.; Riera, A., *J. Org. Chem.*, (2004) **69**, 8053.
[142] Solà, J.; Riera, A.; Verdaguer, X.; Maestro, M. A., *J. Am. Chem. Soc.*, (2005) **127**, 13629.
[143] Lledó, A.; Solà, J.; Verdaguer, X.; Riera, A.; Maestro, M. A., *Adv. Synth. Catal.*, (2007) **349**, 2121.
[144] Kim, D. E.; Kwak, J.; Kim, I. S.; Jeong, N., *Adv. Synth. Catal.*, (2009) **351**, 97.
[145] Sturla, S. J.; Buchwald, S. L., *J. Org. Chem.*, (2002) **67**, 3398.
[146] Gibson, S. E.; Hardick, D. J.; Haycock, P. R.; Kaufmann, K. A. C.; Miyazaki, A.; Tozer, M. J.; White, A. J. P., *Chem.–Eur. J.*, (2007) **13**, 7099.
[147] Fjermestad, T.; Pericàs, M. A.; Maseras, F., *J. Mol. Catal. A: Chem.*, (2010) **324**, 127.
[148] Fjermestad, T.; Pericàs, M. A.; Maseras, F., *Chem.–Eur. J.*, (2011) **17**, 10050.
[149] Pérez-Castells, J., *Top. Organomet. Chem.*, (2006) **19**, 207.
[150] Finnegan, D. F.; Snapper, M. L., *J. Org. Chem.*, (2011) **76**, 3644.
[151] Díaz, D.; Betancort, J. M.; Martín, V. S., *Synlett*, (2007), 343.
[152] Smit, W. A.; Gybin, A. S.; Shashkov, A. S.; Struchkov, Y. T.; Kyz'mina, L. G.; Mikaelian, G. S.; Caple, R.; Swanson, E. D., *Tetrahedron Lett.*, (1986) **27**, 1241.
[153] Jamison, T. F.; Shambayati, S.; Crowe, W. E.; Schreiber, S. L., *J. Am. Chem. Soc.*, (1997) **119**, 4353.
[154] Closser, K. D.; Quintal, M. M.; Shea, K. M., *J. Org. Chem.*, (2009) **74**, 3680.
[155] Evans, P. A.; Robinson, J. E., *J. Am. Chem. Soc.*, (2001) **123**, 4609.
[156] Son, S. U.; Park, K. H.; Chung, Y. K., *J. Am. Chem. Soc.*, (2002) **124**, 6838.
[157] Kim, D. H.; Chung, Y. K., *Chem. Commun. (Cambridge)*, (2005), 1634.
[158] Chen, W.; Tay, J.-H.; Ying, J.; Yu, X.-Q.; Pu, L., *J. Org. Chem.*, (2013) **78**, 2256.
[159] Odedra, A.; Wu, C.-J.; Madhushaw, R. J.; Wang, S.-L.; Liu, R.-S., *J. Am. Chem. Soc.*, (2003) **125**, 9610.
[160] van der Waals, A.; Keese, R., *J. Chem. Soc., Chem. Commun.*, (1992), 570.
[161] Cao, H.; Mundla, S. R.; Cook, J. M., *Tetrahedron Lett.*, (2003) **44**, 6165.
[162] Chen, G.-Q.; Shi, M., *Chem. Commun. (Cambridge)*, (2013) **49**, 698.
[163] Boñaga, L. V. R.; Krafft, M. E., *Tetrahedron*, (2004) **60**, 9795.
[164] Krafft, M. E.; Boñaga, L. V. R.; Wright, J. A.; Hirosawa, C., *J. Org. Chem.*, (2002) **67**, 1233.
[165] Pérez-Serrano, L.; Domínguez, G.; Pérez-Castells, J., *J. Org. Chem.*, (2004) **69**, 5413.
[166] Chatani, N.; Morimoto, T.; Fukumoto, Y.; Murai, S., *J. Am. Chem. Soc.*, (1998) **120**, 5335.
[167] Adrio, J.; Carretero, J. C., *J. Am. Chem. Soc.*, (2007) **129**, 778.
[168] Chatani, N.; Morimoto, T.; Kamitani, A.; Fukumoto, Y.; Murai, S., *J. Organomet. Chem.*, (1999) **579**, 177.
[169] Saito, T.; Sugizaki, K.; Otani, T.; Suyama, T., *Org. Lett.*, (2007) **9**, 1239.
[170] Saito, T.; Shiotani, M.; Otani, T.; Hasaba, S., *Heterocycles*, (2003) **60**, 1045.
[171] Aburano, D.; Yoshida, T.; Miyakoshi, N.; Mukai, C., *J. Org. Chem.*, (2007) **72**, 6878.
[172] Saito, T.; Nihei, H.; Otani, T.; Suyama, T.; Furukawa, N.; Saito, M., *Chem. Commun. (Cambridge)*, (2008), 172.
[173] Saito, T.; Sugizaki, K.; Osada, H.; Kutsumura, N.; Otani, T., *Heterocycles*, (2010) **80**, 207.
[174] Kondo, T.; Nomura, M.; Ura, Y.; Wada, K.; Mitsudo, T.-a., *J. Am. Chem. Soc.*, (2006) **128**, 14816.
[175] Iwata, T.; Inagaki, F.; Mukai, C., *Angew. Chem.*, (2013) **125**, 11344; *Angew. Chem. Int. Ed.*, (2013) **52**, 11138.
[176] Van Ornum, S. G.; Hoerner, S.; Cook, J. M., *The Pauson–Khand Reaction: Scope, Variations and Applications*, Torres, R. R., Ed.; Wiley: Hoboken, NJ, (2012); pp 211–231.

[177] Pallerla, M. K.; Fox, J. M., *Org. Lett.*, (2007) **9**, 5625.
[178] Farwick, A.; Helmchen, G., *Org. Lett.*, (2010) **12**, 1108.
[179] Min, S.-J.; Danishefsky, S. J., *Angew. Chem.*, (2007) **119**, 2249; *Angew. Chem. Int. Ed.*, (2007) **46**, 2199.
[180] Hirose, T.; Miyakoshi, N.; Mukai, C., *J. Org. Chem.*, (2008) **73**, 1061.
[181] Inagaki, F.; Kinebuchi, M.; Miyakoshi, N.; Mukai, C., *Org. Lett.*, (2010) **12**, 1800.
[182] McKerrall, S. J.; Jørgensen, L.; Kuttruff, C. A.; Ungeheuer, F.; Baran, P. S., *J. Am. Chem. Soc.*, (2014) **136**, 5799.
[183] Huang, J.; Fang, L.; Long, R.; Shi, L.-L.; Shen, H.-J.; Li, C.-c.; Yang, Z., *Org. Lett.*, (2013) **15**, 4018.
[184] Yang, Y.; Fu, X.; Chen, J.; Zhai, H., *Angew. Chem.*, (2012) **124**, 9963; *Angew. Chem. Int. Ed.*, (2012) **51**, 9825.
[185] Fujioka, K.; Yokoe, H.; Yoshida, M.; Shishido, K., *Org. Lett.*, (2012) **14**, 244.
[186] Nakayama, A.; Kogure, N.; Kitajima, M.; Takayama, H., *Angew. Chem.*, (2011) **123**, 8175; *Angew. Chem. Int. Ed.*, (2011) **50**, 8025.

2.4 1,3-Dipolar Cycloadditions Involving Carbonyl or Azomethine Ylides

X. Xu and W. Hu

General Introduction

Carbonyl ylides, which behave as active 1,3-dipole species, have numerous applications in organic synthesis, and, in particular, their use in the formation of heterocycles is well documented.[1] Among the versatile transformations of carbonyl ylides, 1,3-dipolar cycloadditions with π-bonds, including (3+2)-cycloaddition reactions with carbon–carbon π-bonds, aldehydes, and imines, are ubiquitous and important reactions.[2,3] This section focuses on advances in these catalytic (3+2)-cycloaddition reactions, with an emphasis on chemo-, diastereo-, and enantiocontrol. A few selected examples of cycloaddition reactions with azomethine ylides, which behave with similar reactivity to carbonyl ylides, to generate nitrogen heterocycles are also discussed. In addition, some selected applications of these (3+2)-cycloaddition reactions in natural product syntheses are highlighted.

The earliest report related to carbonyl ylides was in 1885;[4] Huisgen was the first to examine these intermediates in detail by trapping them with various dipolarophiles.[5] There are many strategies to generate carbonyl ylide intermediates (Scheme 1), including the catalytic ring opening of epoxides,[6] nitrogen extrusion from dihydro-1,3,4-oxadiazoles,[7] carbon dioxide extrusion from 1,3-dioxolan-4-ones,[8] the catalytic cleavage of acetals,[9,10] and others.[11] Of these methods, the reaction of metal carbenes with carbonyl groups is a very efficient method for the generation of carbonyl ylides,[1–3] and the reactions discussed in this section focus on carbonyl ylides generated by this method.

Scheme 1 General Approaches to Carbonyl Ylide Generation

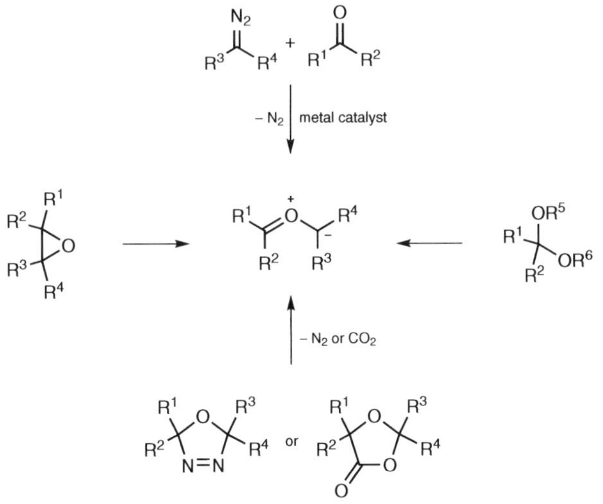

Carbonyl ylides, generated from diazo and carbonyl compounds, are usually nonisolable reactive intermediates that can provide epoxides or 1,3-dioxolanes; the former are formed if the rate of electrocyclic ring closure of the carbonyl ylide is faster than (3+2) cycloaddition with a further equivalent of the carbonyl. It is possible to control the decomposition of diazomalonates to generate either epoxides or dioxolanes by adjusting the stability of the corresponding carbonyl ylide dipole, although competition between these two transformations exists (Scheme 2).[12] Only epoxide formation is observed when 4-methoxybenzaldehyde is used, and a 1,3-dioxolane is formed exclusively when 4-nitrobenzaldehyde is involved. For further discussion see *Science of Synthesis*, Vol. 37 (Section 37.2.3.8.1). A diastereoselectivity-controlled version of this reaction uses rhodium(II) pivalate dimer [$Rh_2(OCOt\text{-}Bu)_4$ which is found abbreviated as both $Rh_2(Piv)_4$ or $Rh_2(OPiv)_4$ in the literature] as the catalyst (Section 2.4.2.1),[13] and an enantioselectivity-controlled (3+2)-cycloaddition reaction uses chiral dirhodium carboxylate catalysts (Section 2.4.2.2).[14] Carbonyl ylide transformations have been used for intramolecular and intermolecular cycloaddition reactions in the synthesis of natural products (Section 2.4.5).[15] Selectivity-controlled intramolecular cycloaddition reactions are also well studied with dirhodium catalysts (Section 2.4.1.1);[16,17] ruthenium catalysts are also utilized for these transformations.[18]

Scheme 2 Competition between Epoxidation and Cycloaddition of Carbonyl Ylides[12]

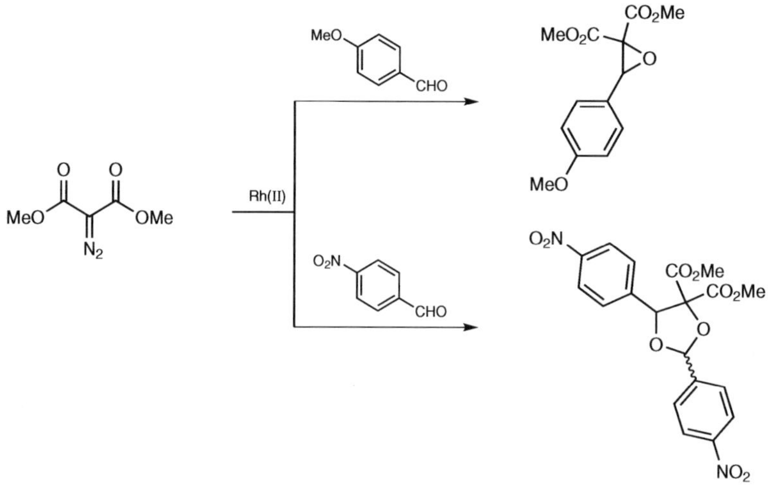

In this chapter, most typical selectivity-controlled reactions use catalysts or catalytic strategies including dirhodium catalysts **1–11** (Scheme 3), Lewis acid catalysts, or cooperative catalysis which uses both a dirhodium catalyst and a Lewis acid catalyst (Scheme 4). The content of the chapter is selective and the reaction examples are representative of these (3+2)-cycloaddition reactions; contemporary reviews give a comprehensive discussion of the development of carbonyl ylide transformations and their potential applications.[1–3]

Scheme 3 Representative Dirhodium Catalysts for Selective Metal Carbene Transformations

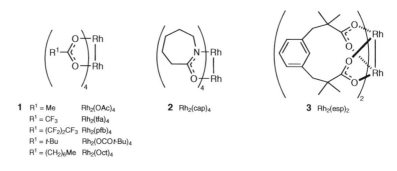

1 R^1 = Me $Rh_2(OAc)_4$
   R^1 = CF_3 $Rh_2(tfa)_4$
   R^1 = $(CF_2)_2CF_3$ $Rh_2(pfb)_4$
   R^1 = t-Bu $Rh_2(OCOt\text{-}Bu)_4$
   R^1 = $(CH_2)_6Me$ $Rh_2(Oct)_4$

2 $Rh_2(cap)_4$

3 $Rh_2(esp)_2$

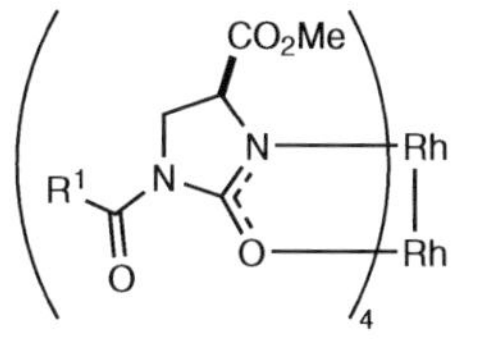

General Introduction

4 R^1 = Me   Rh$_2${(4S)-MACIM}$_4$
    R^1 = CH$_2$Cy   Rh$_2${(4S)-MCHIM}$_4$
    R^1 = (CH$_2$)$_2$Ph Rh$_2${(4S)-MPPIM}$_4$
    R^1 = Ph   Rh$_2${(4S)-MBOIM}$_4$

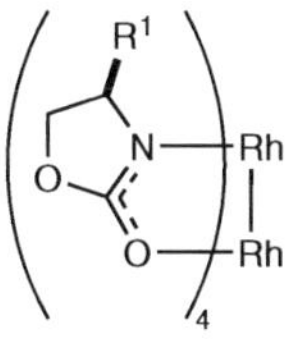

5 R^1 = Bn   Rh$_2${(4S)-BNOX}$_4$
    R^1 = Ph   Rh$_2${(4S)-PHOX}$_4$
    R^1 = CO$_2$Me Rh$_2${(4S)-MEOX}$_4$

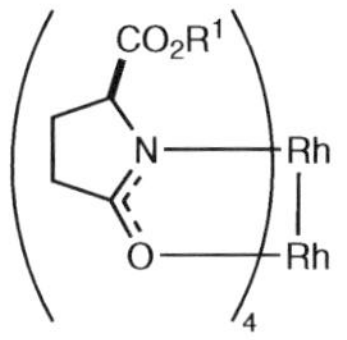

6 R^1 = Me   Rh$_2${(5S)-MEPY}$_4$
    R^1 = CH$_2$$t$-Bu   Rh$_2${(5$S$)-NEPY}$_4$
    R^1 = (CH$_2$)$_{17}$Me Rh$_2${(5S)-ODPY}$_4$

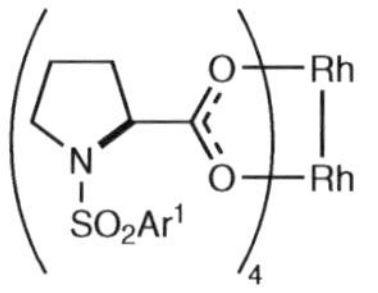

7 Ar^1 = 4-[Me(CH$_2$)$_{11}$]C$_6$H$_4$ Rh$_2${(S)-DOSP}$_4$

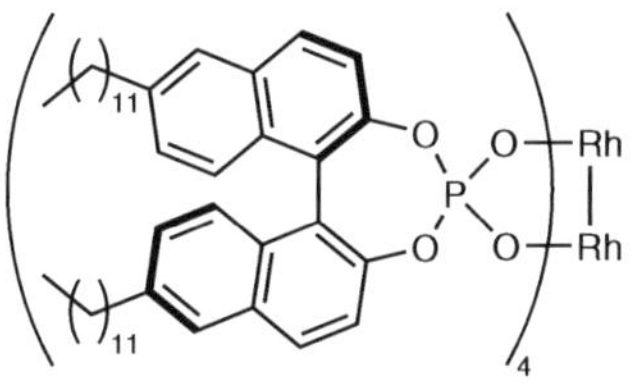

8 Rh$_2${(R)-DDBNP}$_4$

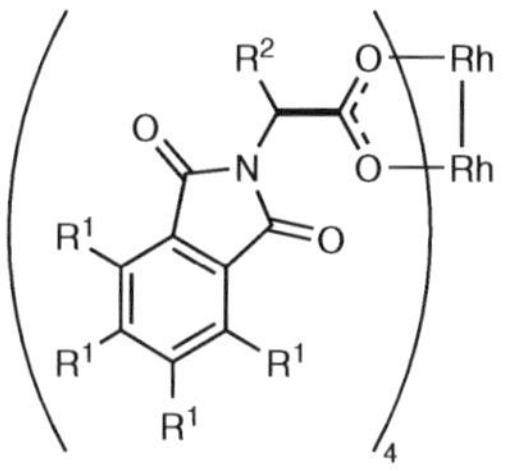

9 R^1 = H; R^2 = Me   Rh$_2${(S)-PTA}$_4$
    R^1 = H; R^2 = Bn   Rh$_2${(S)-PTPA}$_4$
    R^1 = H; R^2 = iPr   Rh$_2${(S)-PTV}$_4$
    R^1 = H; R^2 = t-Bu   Rh$_2${(S)-PTTL}$_4$
    R^1 = F; R^2 = t-Bu   Rh$_2${(S)-TFPTTL}$_4$
    R^1 = Cl; R^2 = t-Bu   Rh$_2${(S)-TCPTTL}$_4$
    R^1 = Br; R^2 = t-Bu   Rh$_2${(S)-TBPTTL}$_4$
    R^1 = H; R^2 = 1-adamantyl Rh$_2${(S)-PTAD}$_4$

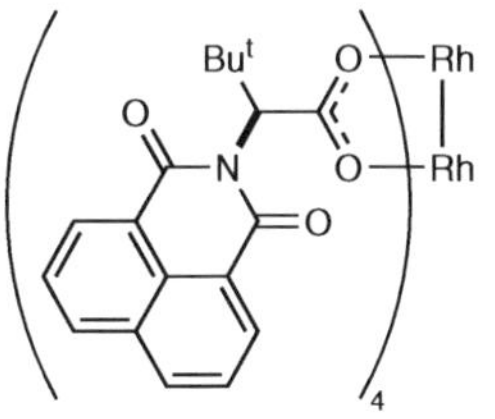

10 Rh$_2${(S)-NTTL}$_4$

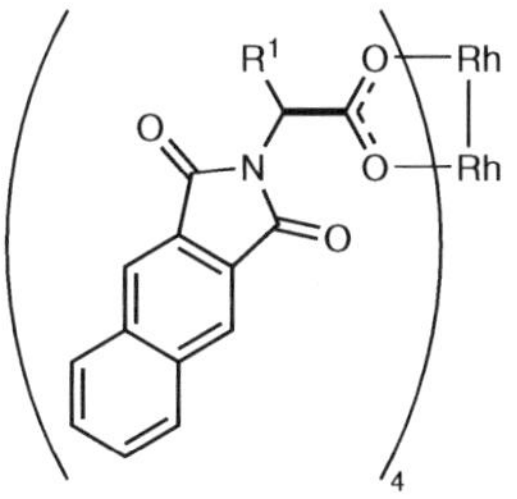

11 R^1 = iPr   Rh$_2${(S)-BPTV}$_4$
     R^1 = t-Bu Rh$_2${(S)-BPTL}$_4$

for references see p 203

Scheme 4 Catalytic Models for Selective (3+2)-Cycloaddition Reactions of Carbonyl Ylides

rhodium catalyst controlled

Lewis acid catalyst controlled

cooperative catalysis

LA = Lewis acid

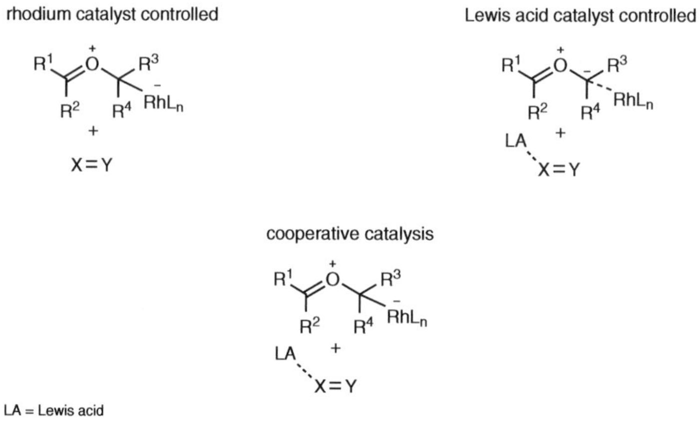

SAFETY: All reactions discussed in this chapter employ diazo substrates and release nitrogen; thus, reactions should not be performed in a sealed system.

2.4.1 (3+2)-Cycloaddition Reactions of Carbonyl Ylides with Carbon–Carbon π-Bonds

Due to the lack of lone-pair electrons in alkenes and alkynes, the control of selectivity in (3+2)-cycloaddition reactions with carbonyl ylides mainly depends on the metal catalyst that is used to decompose the diazo compound and associate with the carbonyl ylide intermediate. Stereoselective control in the (3+2)-cycloaddition reactions of carbonyl ylides with carbon–carbon π-bonds is achieved using chiral dirhodium carboxylate catalysts for control of diastereoselectivity or enantioselectivity in intramolecular or intermolecular reactions. Cooperative catalysis using Lewis acids with dirhodium catalysts is also utilized to control selectivity for reactions where the corresponding alkenes have a built-in coordination site.

2.4.1.1 Diastereoselective (3+2)-Cycloaddition Reactions of Carbonyl Ylides with Carbon–Carbon π-Bonds

The intramolecular cycloaddition reaction of carbonyl ylides with alkenes, alkynes, and allenes is an effective method for the construction of polycyclic and endocyclic structures.[16–19]

Hexane is the optimal solvent for the selective cycloaddition reactions of α-diazo-β,ε-dioxo esters; however, the phenyl sulfone substrate **12** is insoluble in hexane, and toluene is used to give improved solubility. Reaction of phenyl sulfone **12** catalyzed by rhodium(II) acetate dimer [Rh$_2$(OAc)$_4$; **1**, R^1 = Me] gives cycloadduct **13** in 12% yield, whereas the more electron-deficient rhodium(II) trifluoroacetate dimer [Rh$_2$(tfa)$_4$; **1**, R^1 = CF$_3$] gives **13** in an improved 39% yield. In contrast, Rh$_2$(cap)$_4$ (**2**) fails to decompose phenyl sulfone substrate **12**, even after 20 hours. The rhodium(II) octanoate dimer [Rh$_2$(Oct)$_4$; **1**, R^1 = (CH$_2$)$_6$Me], which is electronically very similar to rhodium(II) acetate dimer but has higher solubility in nonpolar solvents, gives cycloadduct **13** in 72% yield (Scheme 5). Asymmetric catalysis using various chiral dirhodium catalysts gives only moderate enantioselectivity.[16]

2.4.1 (3+2) Cycloadditions of Carbonyl Ylides with Carbon–Carbon π-Bonds

Scheme 5 An Octahydro-6H-3a,7-epoxyazulen-6-one by Rhodium-Catalyzed Diastereoselective Intramolecular (3+2)-Cycloaddition Reaction of a Carbonyl Ylide with an Alkene[16]

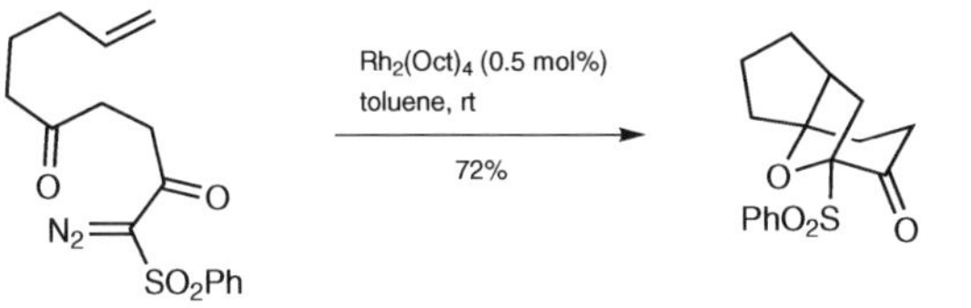

6-Diazoheptane-2,5-dione (**14**) undergoes rhodium(II) acetate dimer catalyzed tandem carbonyl ylide formation–cycloaddition with propargyl bromide to give cycloadduct **15** in 84% yield (Scheme 6). This cycloaddition shows high chemo- and regioselectivity; the regioselectivity was determined by NOE studies. Product **15** is a useful, pivotal intermediate in the synthesis of *cis*-nemorensic acids.[17]

1-Diazohexane-2,5-dione (**16**) undergoes a similar reaction with the electronic-deficient alkyne dimethyl acetylenedicarboxylate. The tandem carbonyl ylide formation–cycloaddition reaction is catalyzed by a recyclable ruthenium(II) porphyrin catalyst (1 mol%) over 30 hours at room temperature to afford **17** in 83% yield with a TON of 850.[18]

Scheme 6 8-Oxabicyclo[3.2.1]oct-6-en-2-ones by Diastereoselective Intramolecular (3+2)-Cycloaddition Reaction of Carbonyl Ylides with Alkynes[17,18]

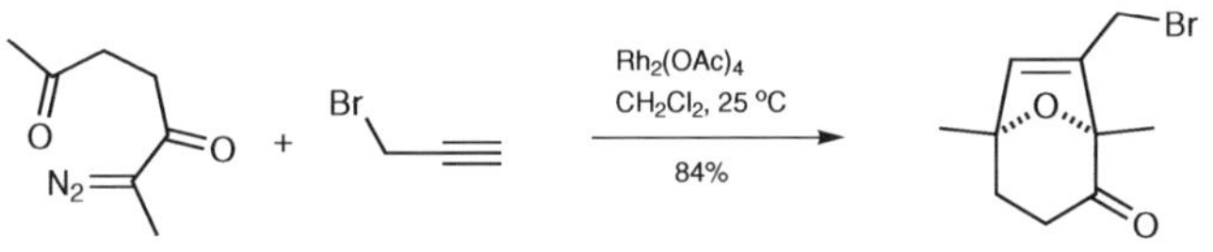

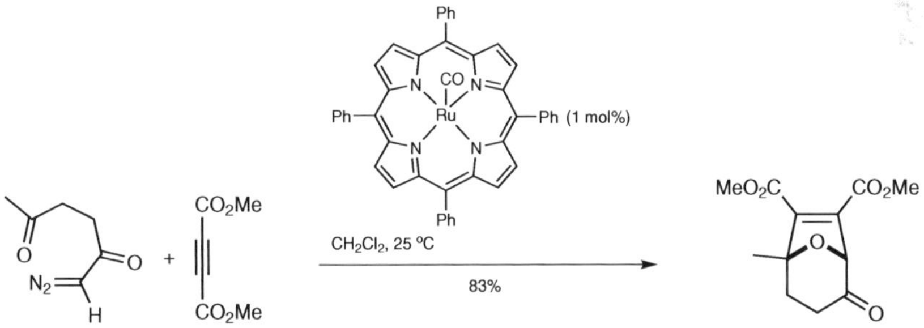

Both terminal and substituted allenes are compatible in the cycloaddition of carbonyl ylides with allenes. Only two of the possible diastereomers are formed (Scheme 7), but the observed levels of diastereoselectivity are moderate (2.7:1 to 4.1:1). In general, the size of the R^2 group in **19** appears to have a relatively minor effect on the diastereoselectivity.[19]

Scheme 7 7-Methylene-8-oxabicyclo[3.2.1]octan-2-ones by Rhodium-Catalyzed Intramolecular (3+2)-Cycloaddition Reaction of Carbonyl Ylides with Buta-2,3-dienoates[19]

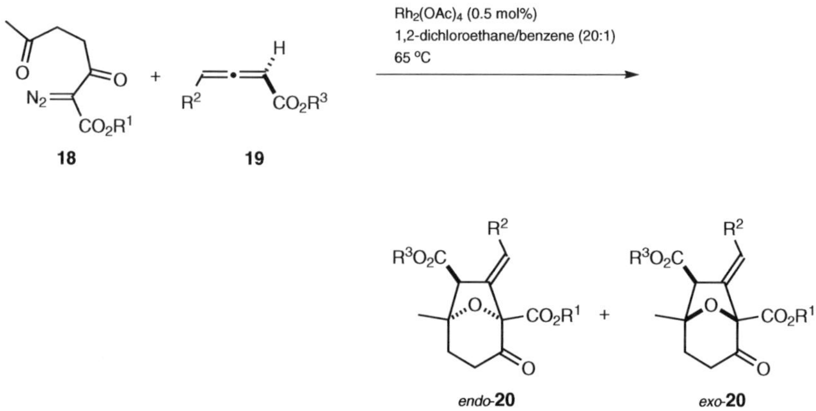

R¹	R²	R³	Ratio[a] (endo/exo)	Yield[b] (%) of 20	Ref
Et	H	Bn	73:27	50[c]	[19]
Et	Me	Bn	75:25	93	[19]
Et	Pr	Bn	>75:25	55[c]	[19]
Et	H	Et	74:26	64[c]	[19]
Et	Pr	Et	>75:25	78[c]	[19]
t-Bu	H	Et	74:26	82	[19]
t-Bu	Me	Et	74:26	82	[19]
t-Bu	Pr	Et	>75:25	72[c]	[19]
t-Bu	Me	Bn	80:20	90	[19]
t-Bu	Bn	Bn	80:20	78	[19]
t-Bu	Pr	Bn	>75:25	60[c]	[19]
t-Bu	iPr	Bn	74:26	50[c]	[19]

[a] Determined by ¹H NMR spectroscopy of the crude reaction mixture.
[b] Combined isolated yield of products.
[c] Isolated yield of major product endo-20.

The intermolecular (3+2) cycloaddition of a carbonyl ylide to an alkene is a three-component reaction utilizing diazo compounds, aldehydes (or ketones), and alkenes to give multisubstituted tetrahydrofuran derivatives with structure diversity (Scheme 8).[20,21] Although other metals, for example a ruthenium catalyst,[18] can be used, dirhodium carboxylate catalysts are the catalyst of choice; the diastereoselectivity control mainly depends on the steric and electronic properties of the substrates.[12,20] Cooperative catalysis by a dirhodium/Lewis acid system is highlighted by one example (Scheme 9); very high diastereoselectivity is obtained for the corresponding (3+2)-cycloaddition reaction with enantiomerically pure alkenes that have a built-in coordination site.[21]

In the intermolecular (3+2) cycloaddition of a carbonyl ylide to an alkene, compared to the intramolecular version, competitive reactions, including epoxidation and cycloaddition with a further equivalent of aldehyde, are the main issues; the chemoselectivity mainly depends on the steric and electronic properties of the substrates.[12] Using 2-diazo-1H-indene-1,3(2H)-dione (**21**) as the metal–carbene precursor in this three-component reaction with aldehydes and N-arylmaleimides **22** gives good yields of spirocyclic

2.4.1 (3+2) Cycloadditions of Carbonyl Ylides with Carbon–Carbon π-Bonds

products **23** (Scheme 8).[20] The electronic effects of substituents both on the N-arylmaleimide and the aromatic aldehyde have little influence on the diastereoselectivity, and satisfactory yields are obtained in most cases. The method was extended by using cinnamaldehyde as the aldehyde component and this also results in ylide-derived tetrahydrofuran products **23** in moderate yields.[20]

Scheme 8 Spiro-Fused Tetracycles by Rhodium-Catalyzed Diastereoselective Intermolecular (3+2)-Cycloaddition Reaction of an Indane-Based Carbonyl Ylide with Alkenes[20]

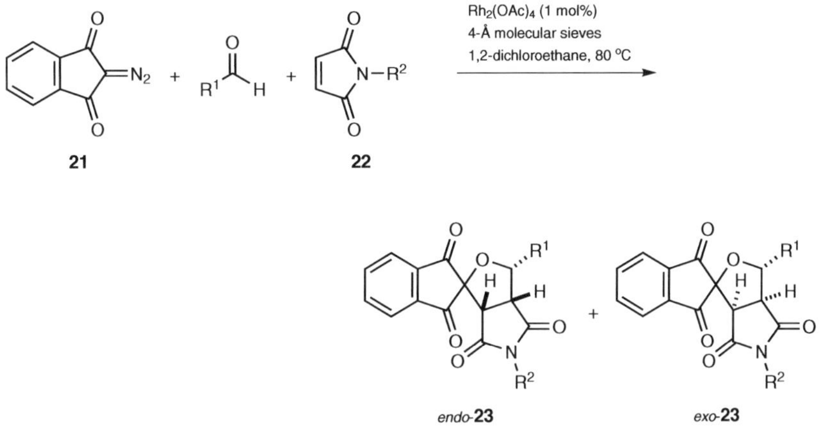

R¹	R²	Ratio[a] (endo/exo)	Yield[b] (%)	Ref
Ph	Et	75:25	47	[20]
Ph	Ph	67:33	65	[20]
Ph	4-O$_2$NC$_6$H$_4$	70:30	40	[20]
Ph	4-MeOC$_6$H$_4$	66:34	51	[20]
Ph	3-ClC$_6$H$_4$	69:31	58	[20]
Ph	3,4-Cl$_2$C$_6$H$_3$	69:31	78	[20]
4-MeOC$_6$H$_4$	3,4-Cl$_2$C$_6$H$_3$	80:20	61	[20]
4-O$_2$NC$_6$H$_4$	3,4-Cl$_2$C$_6$H$_3$	74:26	60	[20]
(benzodioxole-substituted aryl)	3,4-Cl$_2$C$_6$H$_3$	74:26	63	[20]
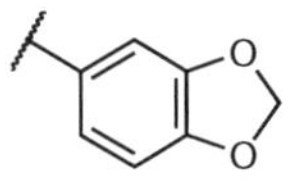 (benzodioxole-substituted aryl)	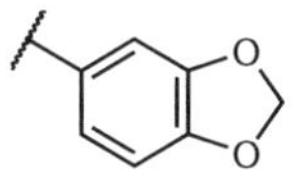 (benzodioxole-substituted aryl)	67:33	60	[20]
(E)-CH=CHPh	3,4-Cl$_2$C$_6$H$_3$	72:28	43	[20]
(E)-CH=CHPh	4-AcC$_6$H$_4$	74:26	53	[20]

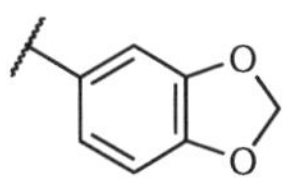

[a] Determined by ¹H NMR spectroscopy of the crude reaction mixture.
[b] Combined isolated yield of products.

An effective strategy to overcome the inherent side reactions in these three-component cycloaddition reactions is via catalysis. Cooperative catalysis, using a second catalyst to activate the dipolarophile **25**, can directly promote the cycloaddition reaction and give the desired products **26** in high yields (53–82%) with excellent selectivities (dr up to 99:1) via the favored transition state shown in Scheme 9.[21]

Scheme 9 Cooperative Rhodium/Nickel-Catalyzed Diastereoselective Intermolecular (3 + 2)-Cycloaddition Reaction of Carbonyl Ylides with Enantiomerically Pure Alkenes[21]

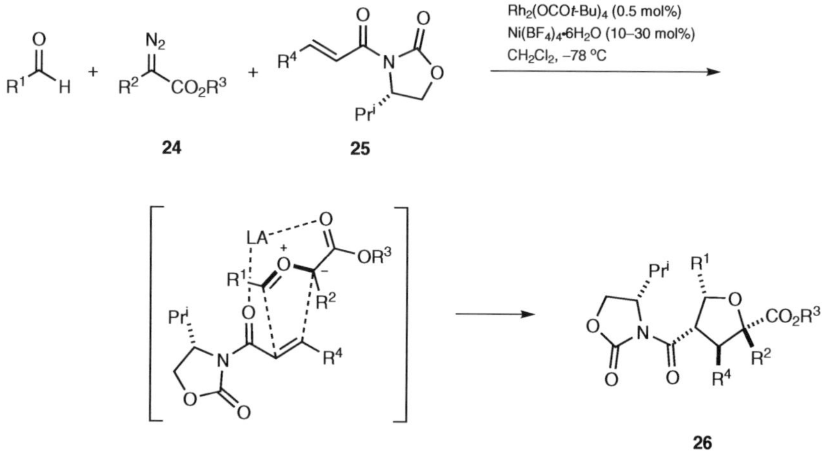

LA = Lewis acid

R^1	R^2	R^3	R^4	dr[a]	Yield[b] (%)	Ref
Ph	Ph	Et	Me	99:1	82	[21]
Ph	Ph	Et	Et	99:1	80	[21]
Ph	Ph	Et	Pr	99:1	78	[21]
Ph	Ph	Et	Ph	99:1	62	[21]
4-Tol	Ph	Et	Me	99:1	84	[21]
4-ClC$_6$H$_4$	Ph	Et	Me	99:1	67	[21]
4-O$_2$NC$_6$H$_4$	Ph	Et	Me	97:3	65	[21]
Ph	Bn	Et	Me	99:1	80	[21]
Ph	Et	Et	Me	99:1	75	[21]
Ph	Bn	Me	Me	99:1	78	[21]
Ph	Bn	t-Bu	Me	99:1	73	[21]

[a] Diastereomeric ratio was determined by ^{1}H NMR spectroscopy.
[b] Combined isolated yield of products.

7-(Phenylsulfonyl)-11-oxatricyclo[5.3.1.0^{1,5}]undecan-8-one (13); Typical Procedure:[16]

A flame-dried flask containing a magnetic stirrer bar was charged with 1-diazo-1-(phenylsulfonyl)dec-9-ene-2,5-dione (**12**; 67 mg, 0.2 mmol) in toluene (3.0 mL). Rh$_2$(Oct)$_4$ (0.5 mol%) was added at rt and the mixture was stirred until the substrate had been consumed [typically within 2 h, TLC monitoring (petroleum ether/Et$_2$O 1:2; R_f 0.1)]. The mixture was then concentrated under reduced pressure and the residue was purified by column chromatography (petroleum ether/Et$_2$O 1:1) to give pure product as white crystals; yield: 44 mg (72%); mp 196–198 °C.

Dimethyl 1-Methyl-4-oxo-8-oxabicyclo[3.2.1]oct-6-ene-6,7-dicarboxylate (17); Typical Procedure:[18]

A flame-dried flask containing a magnetic stirrer bar was charged with DMAD (51 mg, 0.36 mmol) and carbonyl(5,10,15,20-tetraphenylporphyrinato)ruthenium(II) (2.4 mg, 0.003 mmol) in CH$_2$Cl$_2$ (2.0 mL). A soln of 1-diazohexane-2,5-dione (**16**; 42 mg, 0.3 mmol) was added dropwise at rt over 2 h using a syringe pump. After the substrate had been consumed [TLC monitoring, (hexanes/EtOAc 4:1; R_f 0.4)], the mixture was then concentrated under reduced pressure and the residue was purified by column chromatography (hexanes/EtOAc 4:1) to give pure product as colorless crystals; yield: 63 mg (83%); mp 62–64 °C.

Dialkyl 7-Alkylidene-5-methyl-2-oxo-8-oxabicyclo[3.2.1]octane-1,6-dicarboxylates 20; General Procedure:[19]

A flame-dried flask containing a magnetic stirrer bar was charged with buta-2,3-dienoate **19** (1.1 mmol) and 1,2-dichloroethane (0.5 mL). To this soln was added a soln of $Rh_2(OAc)_4$ (2.2 mg, 0.005 mmol, 0.5 mol%) in 1,2-dichloroethane (0.5 mL). The mixture was heated to 65 °C and **18** (1.1 mmol) was added as a soln in 1,2-dichloroethane (0.5 mL), followed by benzene (100 µL) (**CAUTION:** *carcinogen*). The mixture was stirred at 65 °C for 12 h (TLC monitoring). When the reaction was complete, the mixture was concentrated under reduced pressure and the residue was purified by column chromatography (hexanes/EtOAc 83:17) to give the pure product.

3,5-Diaryl-3a,6a-dihydrospiro[furo[3,4-c]pyrrole-1,2′-indene]-1′,3′,4,6(3H,5H)-tetrones 23; General Procedure:[20]

A flame-dried flask containing a magnetic stirrer bar was charged with 2-diazo-1H-indene-1,3(2H)-dione (**21**; 1.0 mmol), aldehyde (4.0 mmol), N-arylmaleimide **22** (4.0 mmol), $Rh_2(OAc)_4$ (4.4 mg, 0.01 mmol), and 4-Å molecular sieves (1.0 g) in 1,2-dichloroethane (15 mL), and the mixture was stirred at 80 °C for 2–3 h under an argon atmosphere. Then, the mixture was filtered through Celite, the filtrate was concentrated under reduced pressure, and the residue was subjected to column chromatography (silica gel, petroleum ether/EtOAc 4:1 to 2:1) to give the pure product.

Ethyl 2-Benzyl-4-[(S)-4-isopropyl-2-oxooxazolidine-3-carbonyl]-3-methyl-5-phenyltetrahydrofuran-2-carboxylate (26, R^1 = Ph; R^2 = Bn; R^3 = Et; R^4 = Me); Typical Procedure:[21]

A flame-dried flask containing a magnetic stirrer bar was charged with (S)-3-[(E)-but-2-enoyl]-4-isopropyloxazolidin-2-one (**25**, R^4 = Me; 86.8 mg, 0.44 mmol), $Rh_2(OCOt\text{-}Bu)_4$ (0.002 mmol), $Ni(BF_4)_2 \cdot 6H_2O$ (0.040 mmol), and 4-Å molecular sieves (500 mg) in CH_2Cl_2 (4.0 mL) at −78 °C (dry ice/acetone bath). A soln of ethyl 2-diazo-3-phenylpropanoate (**24**, R^2 = Bn; R^3 = Et; 0.52 mmol) and benzaldehyde (0.40 mmol) in CH_2Cl_2 (5.0 mL) was added to the mixture over a period of 1 h using a syringe pump. The syringe was washed with CH_2Cl_2 (1.0 mL), and then stirring was continued for 10 min at −78 °C. The mixture was warmed to rt, and then it was filtered through Celite (1 cm) and a plug of silica gel (3 cm), eluting with hexanes/EtOAc (1:1; 80 mL). The solvent was removed under reduced pressure, and the residue was purified by column chromatography (hexanes/EtOAc 98:2 to 95:5) to provide pure product as a colorless solid; yield: 153.5 mg (80%); mp 125–127 °C.

2.4.1.2 Enantioselective (3+2)-Cycloaddition Reactions of Carbonyl Ylides with Carbon–Carbon π-Bonds

Enantioselective (3+2)-cycloaddition reactions of carbonyl ylides with carbon–carbon π-bonds show broad substrate scope, including reactions with alkenes, alkynes, allenes, and indoles (Schemes 10–16).[22–27] Control of the enantioselectivity mainly relies on chiral dirhodium catalysts, including polymer-supported chiral catalysts (Scheme 14).[25] The cooperative catalysis strategy using a chiral Lewis acid and achiral dirhodium carboxylate catalyst can be successfully applied in these reactions using alkenes that have a built-in coordination site; high enantioselectivity is achieved in these cases (Schemes 17 and 18).[28,29]

There has been sporadic investigation of catalytic enantioselective cycloadditions involving carbonyl ylides generated in situ from diazo compounds and carbonyl substrates and these investigations met with comparatively little success at an early stage. In part, this reflects the challenging role required of the chiral catalyst in which, for good enantioselectivity, it must smoothly transform the diazo substrate into a catalyst-associated ylide and then efficiently mediate facial selectivity in the cycloaddition event; cycloaddition via the catalyst-free carbonyl ylide would lead to a racemic cycloadduct. Pioneering work by the Hodgson,[30] Davies,[31] and Hashimoto groups[24–27] identified various chiral dirhodium catalysts as being very effective for (3+2)-cycloaddition reactions. Among these, $Rh_2\{(R)\text{-DDBNP}\}_4$ (**8**) is capable of generating intramolecular cycloadducts **28** in up to 90% ee from 2-diazo-3,6-dioxoalk-10-enoates **27** (Scheme 10).[22,23] $Rh_2\{(R)\text{-DDBNP}\}_4$ (**8**) also performs well with tethered acetylenic dipolarophile **29**, and the reaction gives the desired product **30** with moderate yield and high selectivity (Scheme 11).[23]

Scheme 10 Octahydro-6H-3a,7-epoxyazulen-6-ones by Rhodium-Catalyzed Enantioselective Intramolecular (3+2)-Cycloaddition Reaction of Carbonyl Ylides with Alkenes[22,23]

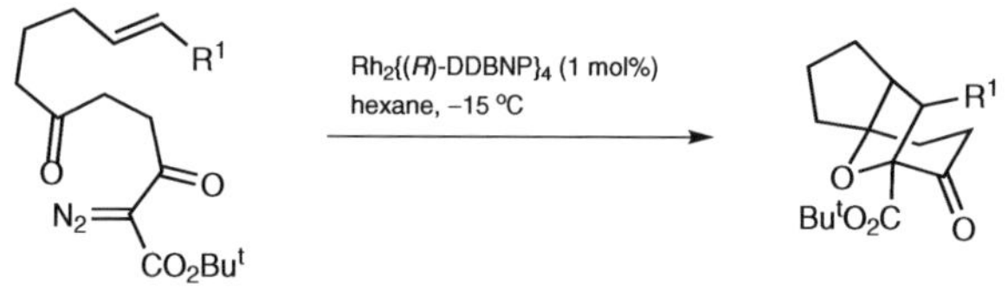

R^1	ee (%)	Yield (%)	Ref
H	90	66	[22,23]
Me	80	53	[22,23]

Scheme 11 Rhodium-Catalyzed Enantioselective Intramolecular (3+2)-Cycloaddition Reaction of a Carbonyl Ylide with an Alkyne[23]

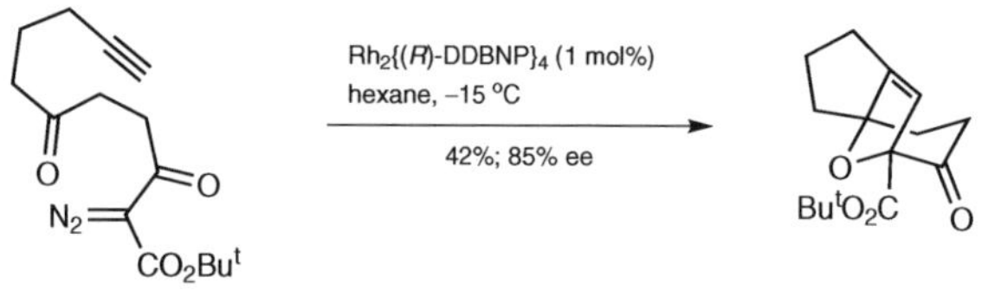

Chiral dirhodium carboxylate catalysts show robust catalytic activity and selectivity control in (3+2)-cycloaddition reactions, including *exo*/*endo* diastereocontrol and enantiocontrol. For example, $Rh_2\{(S)\text{-TCPTTL}\}_4$ (**9**, R^1 = Cl; R^2 = *t*-Bu) is a catalyst for the intermolecular (3+2)-cycloaddition reaction of substituted 2-diazo-3,6-dioxoalkanoates **31** with a styrene derivative (3 equiv) to produce cycloadducts **32** in high yields with perfect *exo* diastereoselectivity; up to 99% ee is observed (Scheme 12).[24]

2.4.1 (3+2) Cycloadditions of Carbonyl Ylides with Carbon–Carbon π-Bonds

Scheme 12 8-Oxabicyclo[3.2.1]octan-2-ones by Rhodium-Catalyzed Enantioselective Intermolecular (3+2)-Cycloaddition Reaction of Carbonyl Ylides with Alkenes[24]

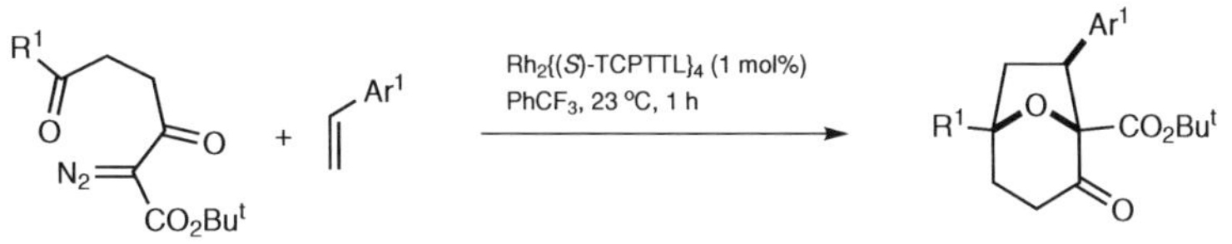

R^1	Ar^1	ee[a] (%)	Yield[b] (%)	Ref
Me	Ph	97	89	[24]
Ph	Ph	99	85	[24]
Ph	2-MeOC$_6$H$_4$	90	75	[24]
Ph	3-MeOC$_6$H$_4$	98	88	[24]
Ph	4-MeOC$_6$H$_4$	99	94	[24]
Ph	4-Tol	99	90	[24]
Ph	4-BrC$_6$H$_4$	99	80	[24]
Ph	4-F$_3$CC$_6$H$_4$	99	73	[24]

[a] Determined by chiral HPLC analysis.
[b] Isolated yield.

The chiral catalyst Rh$_2${(S)-TCPTTL}$_4$ (**9**, R^1 = Cl; R^2 = t-Bu) is also utilized for the intermolecular (3+2) cycloaddition of 2-diazo-3,6-dioxoalkanoates **33** and arylacetylene dipolarophiles, giving the products **34** in moderate to high yields and with high enantioselectivity (Scheme 13).[24] High enantioselectivity (92% ee) is maintained with (3-methoxyphenyl)- or (2-methoxyphenyl)acetylenes, though a slight decrease in product yield is observed. A polymer-supported catalyst **35** promotes the intermolecular (3+2) cycloaddition of tert-butyl 2-diazo-3,6-dioxoheptanoate (**36**) with phenylacetylene to give bicyclic system **37** in 78% yield with up to 97% ee (Scheme 14).[25]

Scheme 13 Rhodium-Catalyzed Enantioselective Intermolecular (3+2)-Cycloaddition Reaction of Carbonyl Ylides with Alkynes[24]

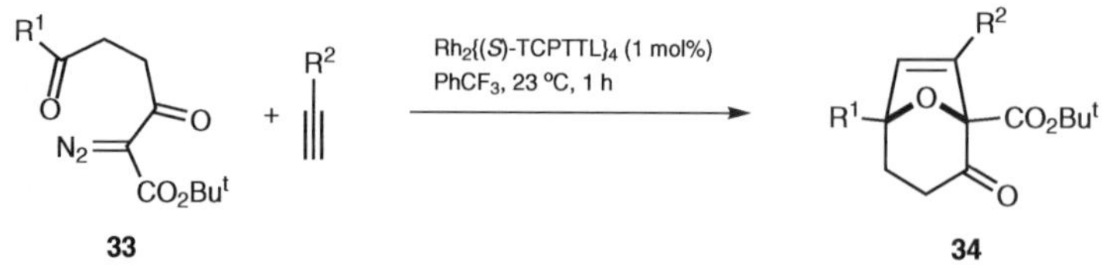

R^1	R^2	ee^a (%)	Yield^b (%)	Ref
Me	4-Tol	97	77	[24]
Me	2-MeOC$_6$H$_4$	92	53	[24]
Me	3-MeOC$_6$H$_4$	92	64	[24]
Me	4-MeOC$_6$H$_4$	91	80	[24]
Me	4-BrC$_6$H$_4$	88	70	[24]
Ph	Ph	98	64	[24]
Me	OEt	85	93	[24]
Ph	OEt	89	85	[24]
4-Tol	OEt	92	75	[24]
4-MeOC$_6$H$_4$	OEt	94	60	[24]
4-ClC$_6$H$_4$	OEt	88	84	[24]

^{a} Determined by chiral HPLC analysis.
^{b} Isolated yield.

Scheme 14 Enantioselective Intermolecular (3+2)-Cycloaddition Reaction of a Carbonyl Ylide with an Alkyne Catalyzed by a Polymer-Supported Dirhodium Catalyst[25]

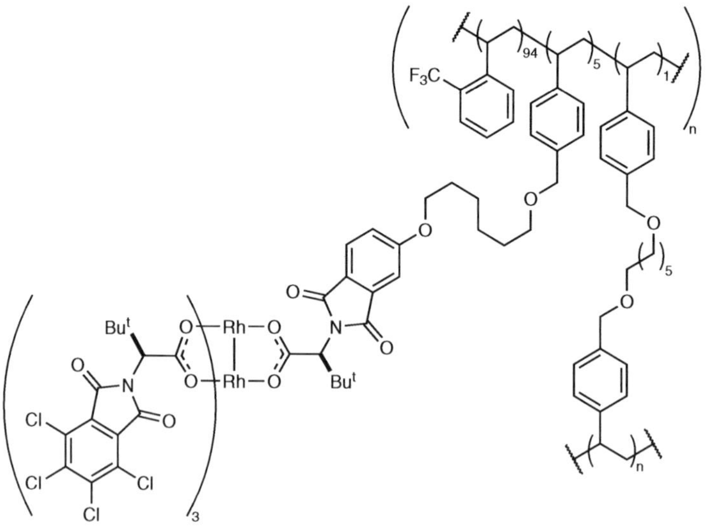

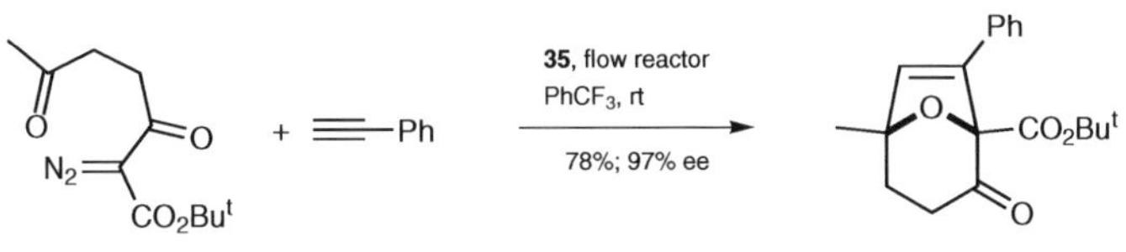

2.4.1 (3+2) Cycloadditions of Carbonyl Ylides with Carbon–Carbon π-Bonds

Allenes also work well in intermolecular (3+2)-cycloaddition reactions using the catalyst $Rh_2\{(S)\text{-TCPTTL}\}_4$ (**9**, R^1 = Cl; R^2 = t-Bu) (Scheme 15).[26] The reaction of tert-butyl 2-diazo-3,6-dioxoheptanoate (**38**, R^1 = Me) with aryl-substituted allenes gives the corresponding products **39** with complete chemo- and regioselectivity and also perfect *exo* diastereoselectivity and high enantioselectivity (92–97% ee) for 4-methyl-, 4-methoxy-, 4-bromo-, and 4-(trifluoromethyl)-substituted phenylallenes. These high levels of asymmetric induction are maintained with 3- or 2-substituted phenylallenes; a decrease in enantioselectivity of ca. 10% is observed with (2-methoxyphenyl)allene. The cycloaddition of other diazo compounds **38** (R^1 = H, Ph) with arylallenes also proceeds in a completely chemo-, and regioselective manner to give the corresponding cycloadducts **39** in good to high yields with high to excellent enantioselectivities.

Scheme 15 6-Methylene-8-oxabicyclo[3.2.1]octan-2-ones by Rhodium-Catalyzed Enantioselective Intermolecular (3+2)-Cycloaddition Reaction of Carbonyl Ylides with Allenes[26]

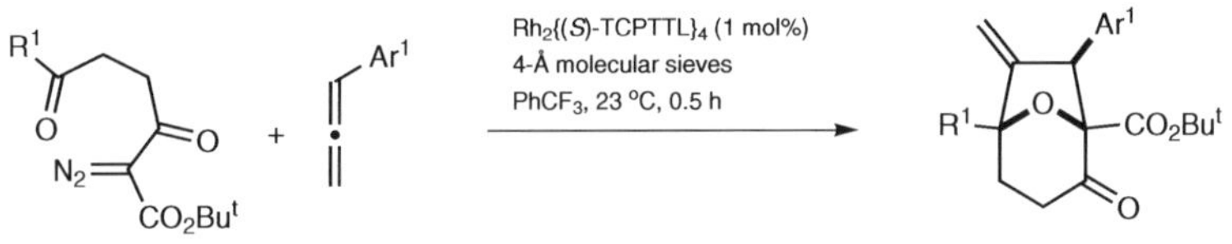

R^1	Ar^1	ee[a] (%)	Yield[b] (%)	Ref
H	Ph	96	80	[26]
H	2-Tol	99	76	[26]
H	4-F$_3$CC$_6$H$_4$	98	68	[26]
H	4-MeOC$_6$H$_4$	96	75	[26]
H	4-BrC$_6$H$_4$	92	73	[26]
Me	Ph	99	82	[26]
Me	2-Tol	99	82	[26]
Me	3-Tol	92	65	[26]
Me	4-Tol	92	77	[26]
Me	4-F$_3$CC$_6$H$_4$	97	72	[26]
Me	2-MeOC$_6$H$_4$	83	63	[26]
Me	3-MeOC$_6$H$_4$	98	65	[26]
Me	4-MeOC$_6$H$_4$	97	89	[26]
Me	4-BrC$_6$H$_4$	96	71	[26]
Ph	Ph	97	96	[26]

R¹	Ar¹	ee^a (%)	Yield^b (%)	Ref
Ph	2-Tol	99	98	[26]
Ph	4-F₃CC₆H₄	99	88	[26]
Ph	4-MeOC₆H₄	94	94	[26]

^a Determined by chiral HPLC analysis.
^b Isolated yield.

The dirhodium catalyst $Rh_2\{(S)\text{-TCPTTL}\}_4$ (**9**, R^1 = Cl; R^2 = *t*-Bu) is also effective in the highly enantio- and diastereoselective intermolecular (3+2) cycloaddition of 2-diazo-3,6-dioxoalkanoates **40** with 1-methylindoles **41**, giving functionalized dihydroindole derivatives **42** in high yields with excellent enantioselectivity (Scheme 16); optically active hexahydrocarbazole derivatives are also available by this route.[27]

Scheme 16 Indole-Fused Tetrahydropyranones by Rhodium-Catalyzed Enantioselective Intermolecular (3+2)-Cycloaddition Reaction of Carbonyl Ylides with Indoles[27]

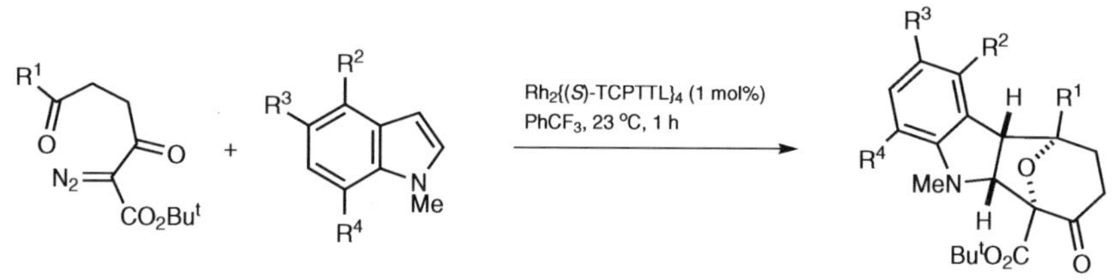

R¹	R²	R³	R⁴	ee^a (%)	Yield^b (%)	Ref
Ph	H	H	H	97	83	[27]
Me	Me	H	H	95	85	[27]
Me	H	Me	H	97	80	[27]
Me	H	OMe	H	97	86	[27]
Me	H	Br	H	99	76	[27]
Me	H	H	Me	94	84	[27]

^a Determined by chiral HPLC analysis.
^b Isolated yield.

High levels of asymmetric induction are observed in dipole-LUMO/dipolarophile-HOMO controlled cycloaddition reactions catalyzed by rhodium(II) acetate dimer between the carbonyl ylides derived from diazo compound **44** and vinyl ether derivatives activated by chiral Lewis acid **43**–nickel(II) perchlorate (Scheme 17).[28] In this catalytic system, higher enantioselectivity is observed at high temperatures, e.g. at reflux in dichloromethane. The unusual dependence of the selectivity on the reaction temperature probably arises from the coordination of the chiral Lewis acid to the carbonyl ylide via dissociation of the rhodium-associated species to a free carbonyl ylide, which may occur at higher temperatures. The cyclic ethers **45** are obtained with high enantioselectivity from vinyl ethers under these conditions.[28]

2.4.1 (3 + 2) Cycloadditions of Carbonyl Ylides with Carbon–Carbon π-Bonds

Scheme 17 Cooperative Rhodium/Lewis Acid Catalyzed Enantioselective Intermolecular (3 + 2)-Cycloaddition Reaction of Carbonyl Ylides with an Electron-Rich Alkene[28]

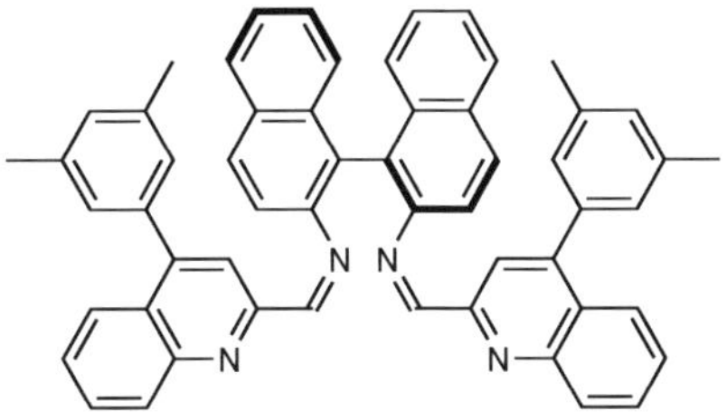

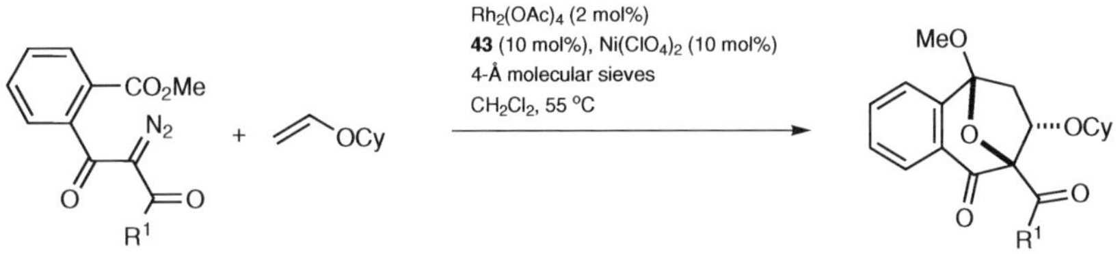

R^1	ee[a] (%)	Yield[b] (%)	Ref
Pr	93	96	[28]
iPr	97	86	[28]
Bu	93	87	[28]
iBu	88	82	[28]
$(CH_2)_4Me$	84	66	[28]
Cy	96	78	[28]
Bn	92	85	[28]
$(CH_2)_2Ph$	77	87	[28]

[a] Determined by chiral HPLC analysis.
[b] Isolated yield.

A similar cooperative-catalysis strategy is utilized for the highly enantioselective 1,3-dipolar cycloaddition reactions of the carbonyl ylides from **47** with 3-(alk-2-enoyl)oxazolidin-2-ones **48**. N-(Diazoacetyl)lactams **47** that possess five-, six-, and seven-membered rings are transformed into the corresponding epoxy-bridged indolizidines, quinolizidines, and 1-azabicyclo[5.4.0]undecanes **49** with good to high enantioselectivities (Scheme 18).[29] The obtained cycloadducts undergo regio- and stereoselective ring-opening of the epoxy-bridged structure to give the corresponding alcohol as a single diastereomer, these compounds are applied in the syntheses of several chiral indolizidine derivatives, including (+)-tashiromine.

Scheme 18 Cooperative Rhodium/Lewis Acid Catalyzed Enantioselective Intermolecular (3+2)-Cycloaddition Reaction of Carbonyl Ylides with Electron-Deficient Alkenes To Give Epoxy-Bridged Nitrogen Heterocycles[29]

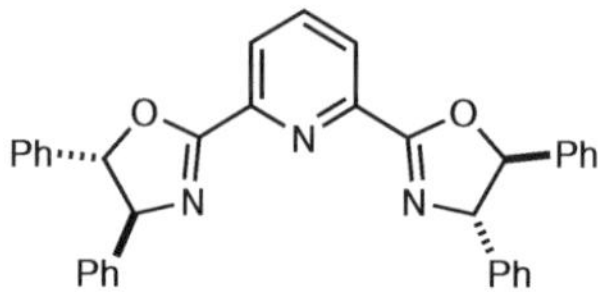

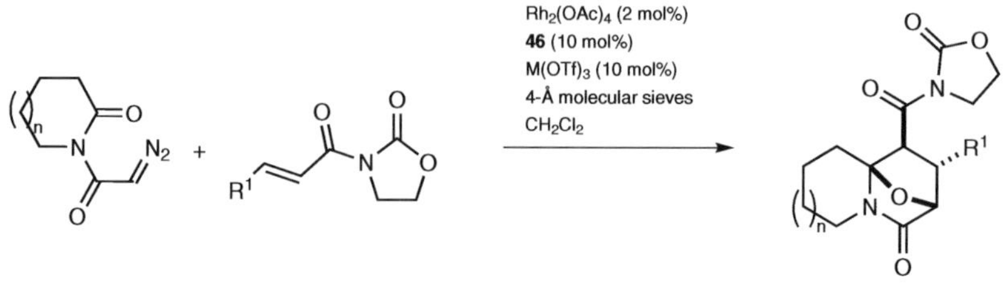

n	R^1	M	Ratio^a (exo/endo)	ee^b (%) of exo-**49**	Yield^c (%)	Ref
1	H	Lu	79:21	87	55	[29]
1	Me	La	85:15	87	78	[29]
1	Et	La	87:13	86	68	[29]
1	Pr	La	87:13	84	62	[29]
1	OAc	La	87:13	78	54	[29]
2	H	Tm	93:7	85	53	[29]
2	Me	La	93:7	88	76	[29]
2	Et	La	91:9	87	75	[29]
2	Pr	La	83:17	84	64	[29]

^{a} Determined by ^{1}H NMR spectroscopy.
^{b} Determined by chiral HPLC analysis.
^{c} Isolated yield of *exo*- and *endo*-isomers.

tert-Butyl endo-6-Methyl-8-oxo-11-oxatricyclo[5.3.1.0^{1,5}]undecane-7-carboxylate (28, R^1 = Me); Typical Procedure:[22]
A flame-dried flask containing a magnetic stirrer bar was charged with degassed soln of *tert*-butyl (*E*)-2-diazo-3,6-dioxododec-10-enoate (**27**, R^1 = Me; 100 mg, 0.32 mmol) in dry hexane (5.0 mL). Rh$_2${(*R*)-DDBNP}$_4$ (**8**; 9.5 mg, 0.003 mmol) was added at −15 °C. When the reaction was complete [ca. 50 min; TLC monitoring (petroleum ether/Et$_2$O 4:1; R_f 0.17)], the mixture was concentrated under reduced pressure. The residue was purified by flash chromatography (petroleum ether/Et$_2$O 4:1) to afford the product as a colorless oil; yield: 60.2 mg (53%); 80% ee.

tert-Butyl (1R,5R,7S)-2-Oxo-5,7-diphenyl-8-oxabicyclo[3.2.1]octane-1-carboxylate (32, R^1 = Ar^1 = Ph); Typical Procedure:[24]
A flame-dried flask containing a magnetic stirrer bar was charged with Rh$_2${(*S*)-TCPTTL}$_4$ (3.95 mg, 0.0020 mmol, 1 mol%) in PhCF$_3$ (1.0 mL). A soln of *tert*-butyl 2-diazo-3,6-dioxo-6-phenylhexanoate (**31**, R^1 = Ph; 60.5 mg, 0.20 mmol) and styrene (62.5 mg, 0.60 mmol) in PhCF$_3$ (1.0 mL) was added using a syringe pump over 1 h. When the reaction was complete [TLC monitoring (hexanes/EtOAc 4:1; R_f 0.33)], the mixture was concentrated under re-

duced pressure, and the residue was purified by column chromatography (hexanes/EtOAc 6:1) to provide the product as a white solid; yield: 64.5 mg (85%); mp 190.5–192.0 °C; 99% ee [HPLC (Chiralpak AD column: hexanes/iPrOH 25:1; flow: 1.0 mL/min; detection: 220 nm); t_R = 18.3 (major), 23.4 min (minor)].

tert-Butyl (1R,5R)-2-Oxo-5,7-diphenyl-8-oxabicyclo[3.2.1]oct-6-ene-1-carboxylate (34, R¹ = R² = Ph); Typical Procedure:[24]

A flame-dried flask containing a magnetic stirrer bar was charged with $Rh_2\{(S)\text{-TCPTTL}\}_4$ (3.95 mg, 0.0020 mmol, 1 mol%) in $PhCF_3$ (1.0 mL). A soln of tert-butyl 2-diazo-3,6-dioxo-6-phenylhexanoate (33, R¹ = Ph; 60.5 mg, 0.20 mmol) and phenylacetylene (102.1 mg, 1.00 mmol) in $PhCF_3$ (1.0 mL) was added via syringe pump over 1 h. When the reaction was complete [TLC monitoring (hexanes/EtOAc 4:1; R_f 0.42)], the mixture was concentrated under reduced pressure, and the residue was purified by column chromatography (hexanes/EtOAc 6:1) to provide the product as a pale yellow oil; yield: 48.5 mg (64%); 98% ee [HPLC (Chiralpak OD column; hexanes/iPrOH 40:1; flow: 1.0 mL/min; detection: 254 nm); t_R = 16.5 (minor), 21.0 min (major)].

tert-Butyl (1R,5R)-5-Methyl-2-oxo-7-phenyl-8-oxabicyclo[3.2.1]oct-6-ene-1-carboxylate (37); Typical Procedure:[25]

A soln of tert-butyl 2-diazo-3,6-dioxoheptanoate (36; 21.6 g, 90 mmol) and phenylacetylene (27.6 g, 270 mmol, 3 equiv) in $PhCF_3$ (450 mL) was passed through the flow reactor packed with polymer-supported dirhodium(II) complex 35 (32 mg, 0.006 mmol) and sea sand (4 g) at 30 mL/h for 15 h. When the reaction was complete, the flow reactor was washed with additional $PhCF_3$ (2.0 mL). After evaporation of the solvent under reduced pressure, the residue was purified by column chromatography (silica gel, hexanes/EtOAc 6:1) to provide the product as a white solid; yield: 22.0 g (78%); mp 133.5–135.0 °C; 97% ee [HPLC (Chiralpak OD column; hexanes/iPrOH 9:1; flow: 1.0 mL/min; detection: 254 nm); t_R = 7.9 (minor), 9.8 min (major)].

tert-Butyl (1R,5R,7R)-5-Methyl-6-methylene-2-oxo-7-phenyl-8-oxabicyclo[3.2.1]octane-1-carboxylate (39, R¹ = Me; Ar¹ = Ph); Typical Procedure:[26]

A flame-dried flask containing a magnetic stirrer bar was charged with powdered 4-Å molecular sieves (50 mg), tert-butyl 2-diazo-3,6-dioxoheptanoate (38, R¹ = Me; 48.0 mg, 0.20 mmol), and phenylallene (46.4 mg, 0.40 mmol) in $PhCF_3$ (2.0 mL). $Rh_2\{(S)\text{-TCPTTL}\}_4$ (3.95 mg, 0.0020 mmol) was added at 23 °C. The mixture was stirred for 30 min, and then it was filtered through Celite. The filtrate was concentrated under reduced pressure and the residue was purified by column chromatography (hexanes/EtOAc 4:1) to provide the product as a white solid; yield: 54 mg (82%); mp 133.5–135.0 °C; 99% ee [HPLC (Chiralpak IC column; hexanes/iPrOH 9:1; flow: 1.0 mL/min; detection: 226 nm); t_R = 18.9 (minor), 20.4 min (major)].

tert-Butyl (5aR,6R,10R,10aS)-5-Methyl-7-oxo-10-phenyl-5a,7,8,9,10,10a-hexahydro-6,10-epoxycyclohepta[b]indole-6(5H)-carboxylate (42, R¹ = Ph; R² = R³ = R⁴ = H); Typical Procedure:[27]

A flame-dried flask containing a magnetic stirrer bar was charged with $Rh_2\{(S)\text{-TCPTTL}\}_4$·2EtOAc (3.95 mg, 0.0020 mmol, 1 mol%) in $PhCF_3$ (1.0 mL). A soln of tert-butyl 2-diazo-3,6-dioxo-6-phenylhexanoate (40, R¹ = Ph; 60.5 mg, 0.20 mmol) and 1-methyl-1H-indole (41, R² = R³ = R⁴ = H; 52.5 mg, 0.40 mmol) in $PhCF_3$ (1.0 mL) was added over 1 h, the mixture was concentrated under reduced pressure, and the residue was purified by column chromatography (hexanes/EtOAc 6:1) to provide the product as a yellow solid (hexanes/EtOAc 2:1; R_f 0.44); yield: 67.5 mg (83%); mp 138.5–140.0 °C; 97% ee [HPLC (Chiralpak OD-H column; hexanes/iPrOH 9:1; flow: 1.0 mL/min; t_R = 13.7 (major), 19.6 min (minor)].

(5R,7R,8R)-8-Butanoyl-7-(cyclohexyloxy)-5-methoxy-5,6,7,8-tetrahydro-9H-5,8-epoxybenzo[7]annulen-9-one (45, R^1 = Pr); Typical Procedure:[28]
A flame-dried, two-necked, round-bottomed flask containing a magnetic stirrer bar was charged with powdered 4-Å molecular sieves (500 mg) and Ni(ClO$_4$)$_2$·6H$_2$O (18.3 mg, 0.05 mmol), and equipped with a reflux condenser. A soln of **43** (29.5 mg, 0.05 mmol) in CH$_2$Cl$_2$ (2.5 mL) was added and the resulting mixture was stirred at rt for 6 h. Cyclohexyl vinyl ether (126 mg, 1.00 mmol), Rh$_2$(OAc)$_4$ (4.4 mg, 0.01 mmol), and CH$_2$Cl$_2$ (1.5 mL) were added successively, followed by the addition of a soln of methyl 2-(2-diazo-3-oxohexanoyl)-benzoate (**44**, R^1 = Pr; 137 mg, 0.50 mmol) in CH$_2$Cl$_2$ (5.0 mL) over a period of 1 h using a syringe pump under reflux (bath temp 55 °C); the syringe was washed with CH$_2$Cl$_2$ (1.0 mL). After removal of 4-Å molecular sieves through Celite, the mixture was filtered through a plug of silica gel [3 cm, hexanes/EtOAc 1:1 (80 mL)]. The solvent was removed under reduced pressure, and the residue was purified by column chromatography (hexanes/EtOAc 99:1) to provide pure product as a colorless viscous oil; yield: 172 mg (96%); 93% ee [HPLC (Chiralpak AD-H column; hexanes/iPrOH 99:1; flow: 0.5 mL/min, 35 °C); t_R = 13.8 (minor), 17.8 min (major)].

(1S,3R,9aS)-1-(2-Oxooxazolidin-3-ylcarbonyl)hexahydro-3,9a-epoxyquinolizin-4(1H)-one (49, n = 1; R^1 = H); Typical Procedure:[29]
A flame-dried flask containing a magnetic stirrer bar was charged with Lu(OTf)$_3$ (0.05 mmol) and iPrOH (3.8 µL, 0.05 mmol) in THF (1.0 mL). A soln of (4S,5S)-pybox-Ph$_2$ **46** (26.1 mg, 0.05 mmol) in THF (1.5 mL) was added. After stirring for 2 h, the solvent was removed under reduced pressure, and resulting solid was dried under vacuum at rt for 5 h. 3-Acryloyloxazolidin-2-one (**48**, R^1 = H; 141.1 mg, 1.0 mmol), 4-Å molecular sieves (0.5 g), Rh$_2$(OAc)$_4$ (4.4 mg, 0.01 mmol), and CH$_2$Cl$_2$ (4.0 mL) were successively added to the prepared Lu(III)–pybox complex. After cooling the mixture to 10 °C, a soln of diazo compound **47** (n = 1; 0.50 mmol) in CH$_2$Cl$_2$ (5.0 mL) was added over a period of 6 h using a syringe pump. The syringe was washed with CH$_2$Cl$_2$ (1.0 mL). After removal of 4-Å molecular sieves through Celite, the mixture was filtered through a plug of silica gel [3.0 cm, EtOAc (80 mL)]. The solvent was removed under reduced pressure, and the residue was purified by column chromatography (silica gel, CHCl$_3$) to provide the product; yield: 55%; ratio (exo/endo) 79:21 (^{1}H NMR analysis); exo-**49**: colorless prisms (R_f 0.37, EtOAc); mp 121–123 °C; endo-**49**: colorless prisms (R_f 0.29, EtOAc); mp 151–153 °C.

2.4.2 (3+2)-Cycloaddition Reactions of Carbonyl Ylides with Aldehydes

The transition-metal-catalyzed reaction of diazo compounds with aldehydes is a well-established route to substituted 1,3-dioxolanes. Intramolecular epoxidation is favored for carbonyl ylides[12] and intermolecular (3+2)-cycloaddition reactions require the use of excess aldehyde (Scheme 20, 4 equivalents)[13] or the presence of two aldehydes with inverse electronic properties (Scheme 19 and Scheme 21);[32,33] the products are multisubstituted 1,3-dioxolane derivatives. Similar transformations using epoxides as the carbonyl ylide precursor catalyzed by a Lewis acid are not discussed here further.[34] Enantioselective versions of these transformations remain rare; a selected example of the reaction of carbonyl ylides with aldehydes catalyzed by chiral dirhodium catalysts gives moderate to high enantioselectivities (Section 2.4.2.2).[14]

2.4.2.1 Diastereoselective (3+2)-Cycloaddition Reactions of Carbonyl Ylides with Aldehydes

The selectivity of the rhodium-catalyzed preparation of dioxolanes from one molecule of a diazo compound and two different aldehyde molecules is highly dependent on the structure of the diazo compound and the electronic properties of the aldehydes. Generally, an

2.4.2 (3+2) Cycloadditions of Carbonyl Ylides with Aldehydes

electron-rich aldehyde reacts with a metal–carbene to form the carbonyl ylide that undergoes cycloaddition with an electron-deficient aldehyde to give the corresponding dioxolane in high yield and with high diastereoselectivity. An example is the reaction of methyl diazo(phenyl)acetate (1.5 equiv) with electron-rich 4-methoxybenzaldehyde (2 equiv) and electron-deficient 2,4-dinitrobenzaldehyde (1 equiv) to give dioxolane **50** (Scheme 19).[32] Performing the cycloaddition reaction using excess aldehyde (4 equiv) that not only reacts with the diazo compound to form the carbonyl ylide but also functions as the dipolarophile gives dioxolanes **51** in moderate to high yields and with high diastereoselectivity (up to >95:5 dr) (Scheme 20).[13]

Scheme 19 Rhodium(II) Acetate Dimer Catalyzed Diastereoselective Intermolecular (3+2)-Cycloaddition Reaction of a Carbonyl Ylide with an Electron-Deficient Aldehyde[32]

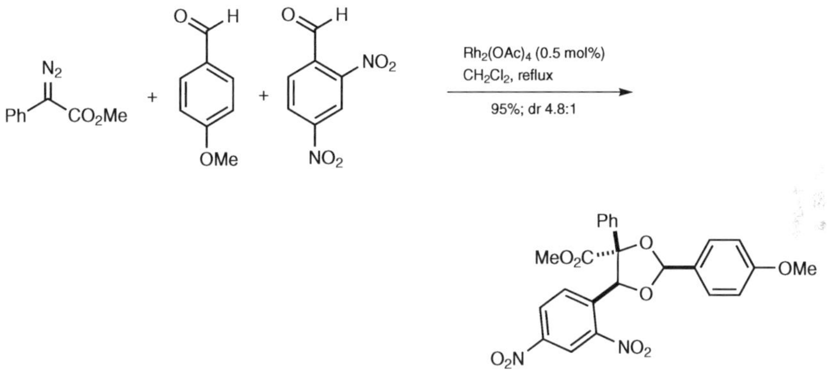

Scheme 20 Rhodium-Catalyzed Diastereoselective Intermolecular (3+2)-Cycloaddition Reaction of Carbonyl Ylides with an Excess of Aldehyde[13]

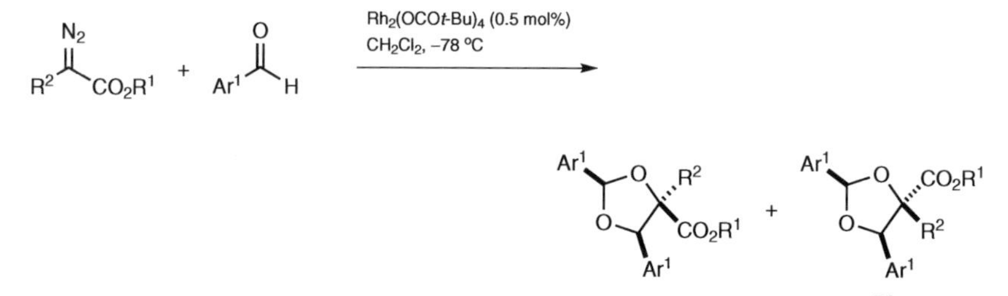

cis-**51** major   trans-**51** minor

R^1	R^2	Ar^1	dr[a] (cis/trans)	Yield[b] (%)	Ref
Et	Me	Ph	>95:5	65	[13]
Et	Me	4-$O_2NC_6H_4$	94:6	64	[13]
Et	Me	2-thienyl	94:6	58	[13]
Et	Et	Ph	95:5	53	[13]
Et	Et	4-FC_6H_4	>95:5	66	[13]
Et	Bn	3-ClC_6H_4	>95:5	54	[13]
t-Bu	Bn	2-Tol	94:6	60	[13]
Et	Bn	Ph	>92:8	60	[13]

[a] Diastereomeric ratio was determined by ^{1}H NMR spectroscopy.
[b] Isolated yields; average of two runs.

A highly diastereoselective intermolecular (3+2)-cycloaddition reaction of carbonyl ylides derived from 3-diazoindol-2-ones **52** with aldehydes provides the spirodioxolanes **53** in high yields; only one diastereomer is obtained (Scheme 21).[33]

Scheme 21 Diastereoselective Intermolecular (3+2)-Cycloaddition Reaction of Carbonyl Ylides with Aldehydes for the Synthesis of Spirodioxolanes[33]

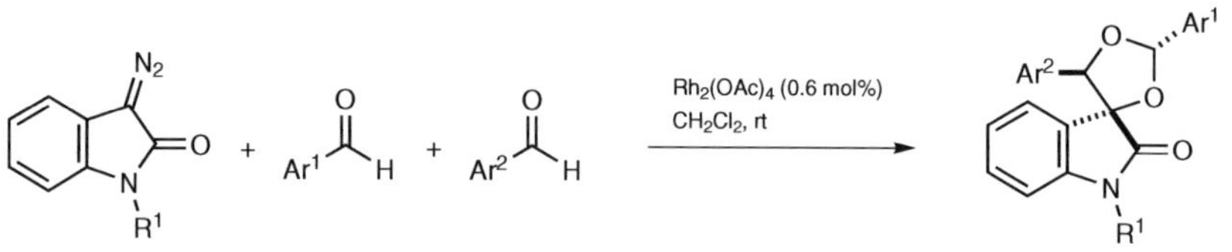

52   **53**

R^1	Ar^1	Ar^2	Yield[a] (%)	Ref
Me	4-$MeOC_6H_4$	2-$O_2NC_6H_4$	82	[33]
Me	4-$MeOC_6H_4$	3-$O_2NC_6H_4$	78	[33]
Me	4-$MeOC_6H_4$	4-$O_2NC_6H_4$	80	[33]
Me	4-$MeOC_6H_4$	2-ClC_6H_4	55	[33]
Me	4-$MeOC_6H_4$	3-FC_6H_4	60	[33]
Me	4-$MeOC_6H_4$	3-BrC_6H_4	60	[33]
Me	3,4-$(MeO)_2C_6H_3$	2-$O_2NC_6H_4$	80	[33]
Me	3,4-$(MeO)_2C_6H_3$	4-$O_2NC_6H_4$	75	[33]
Me	1-naphthyl	4-$O_2NC_6H_4$	80	[33]

R¹	Ar¹	Ar²	Yield[a] (%)	Ref
Me	1-naphthyl	3-O$_2$NC$_6$H$_4$	70	[33]
Bn	Ph	2-O$_2$NC$_6$H$_4$	70	[33]
Bn	4-MeOC$_6$H$_4$	2-O$_2$NC$_6$H$_4$	85	[33]
Bn	4-MeOC$_6$H$_4$	4-O$_2$NC$_6$H$_4$	85	[33]

[a] Isolated yield.

Methyl (2S*,4R*,5S*)-5-(2,4-Dinitrophenyl)-2-(4-methoxyphenyl)-4-phenyl-1,3-dioxolane-4-carboxylate (50); Typical Procedure:[32]

A flame-dried, two-necked, round-bottomed flask containing a magnetic stirrer bar was charged with Rh$_2$(OAc)$_4$ (4.4 mg, 0.01 mmol), 4-methoxybenzaldehyde (544 mg, 4.0 mmol), and 2,4-dinitrobenzaldehyde (392 mg, 2.0 mmol) in CH$_2$Cl$_2$ (8.0 mL). A soln of methyl diazo(phenyl)acetate (528 mg, 3.0 mmol) in CH$_2$Cl$_2$ (4 mL) was added using a syringe pump over 1 h under reflux. When the addition was complete, the mixture was cooled to rt. Solvent was removed, and a portion of the crude product was subjected to ¹H NMR analysis to determine the isomeric ratio. The crude product was purified by flash chromatography (Et$_2$O/EtOAc 10:1) to give the product as white crystals; yield: 912 mg (95%); mp 182.1–183.2 °C.

Ethyl (2R*,4R*,5R*)-4-Ethyl-2,5-bis(4-fluorophenyl)-1,3-dioxolane-4-carboxylate (51, R¹ = R² = Et; Ar¹ = 4-FC$_6$H$_4$); Typical Procedure:[13]

A flame-dried, two-necked, round-bottomed flask containing a magnetic stirrer bar was charged with Rh$_2$(OCOt-Bu)$_4$ (2.0 mg, 0.003 mmol) and the flask was evacuated and filled with N$_2$. Anhyd CH$_2$Cl$_2$ (5.0 mL) and 4-fluorobenzaldehyde (332 mg, 2.68 mmol) were added, and the flask was cooled in a dry ice/acetone bath. Ethyl 2-diazobutanoate (95 mg, 0.67 mmol) was dissolved in anhyd CH$_2$Cl$_2$ (3.0 mL) and added to the mixture using a syringe pump, at a rate of 1 mL/h. Following the addition, the mixture was allowed to warm to rt and mesitylene (80 mg, 0.67 mmol) was added. ¹H NMR analysis was then performed and the diastereomeric ratio was determined to be >95:5. The solvent was then removed and the residue was purified by column chromatography (hexanes/EtOAc 99:1 to 95:5) to give pure product as a clear oil; yield: 168 mg (70%).

1-Benzyl-2′-(4-methoxyphenyl)-5′-(4-nitrophenyl)spiro[indoline-3,4′-[1,3]dioxolan]-2-one (53, R¹ = Bn; Ar¹ = 4-MeOC$_6$H$_4$; Ar² = 4-O$_2$NC$_6$H$_4$); Typical Procedure:[33]

A flame-dried, two-necked, round-bottomed flask containing a magnetic stirrer bar was charged with 1-benzyl-3-diazo-1,3-dihydro-2H-indole-2-one (52, R¹ = Bn; 225 mg, 0.92 mmol), 4-methoxybenzaldehyde (150 mg, 1.10 mmol), and 4-nitrobenzaldehyde (181 mg, 1.20 mmol) in CH$_2$Cl$_2$ (20 mL). Rh$_2$(OAc)$_4$ (2.5 mg, 0.62 mol%) was added to the mixture at rt and the mixture was stirred for 3 h. Chromatographic purification (hexanes/EtOAc 3:1) gave the product as a colorless solid; yield: 390 mg (85%); mp 138–140 °C.

2.4.2.2 Enantioselective (3+2)-Cycloaddition Reactions of Carbonyl Ylides with Aldehydes

There are no reported enantioselective three-component cycloaddition reactions of carbonyl ylides with aldehydes; however, examples of enantioselective intermolecular 1,3-dipolar cycloadditions of carbonyl ylides associated with a chiral dirhodium(II) catalyst with an aromatic aldehyde dipolarophile are available (Scheme 22).[14] The tandem carbonyl ylide formation/cycloaddition reaction of 5-diazo-1-phenylpentane-1,4-dione with aromatic aldehydes using Rh$_2${(S)-BPTV}$_4$•2THF (11•2THF, R¹ = iPr) as the catalyst provides exclusively *exo*-cycloadducts **54** with up to 92% ee.[14] The enantioselectivity is highly sensitive to the substitution pattern at the ylide carbonyl and the electronic nature of the dipolarophiles, as well as the length of the tether.

Scheme 22 6,8-Dioxabicyclo[3.2.1]octan-2-ones by Enantioselective Intramolecular (3+2)-Cycloaddition Reaction of a Carbonyl Ylide with Aromatic Aldehydes[14]

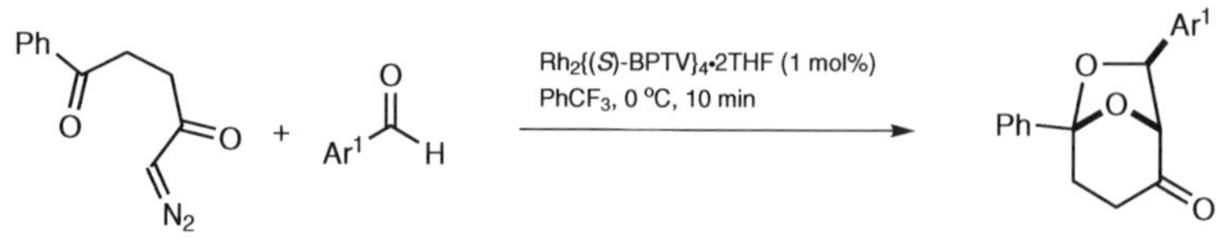

Ar^1	ee (%)	Yield (%)	Ref
4-O$_2$NC$_6$H$_4$	92	71	[14]
4-F$_3$CC$_6$H$_4$	91	69	[14]
4-ClC$_6$H$_4$	88	68	[14]
4-MeOC$_6$H$_4$	76	26	[14]

(1R,5S,7S)-7-(4-Nitrophenyl)-5-phenyl-6,8-dioxabicyclo[3.2.1]octan-2-one (54, Ar^1 = 4-O$_2$NC$_6$H$_4$); Typical Procedure:[14]
A flame-dried, two-necked, round-bottomed flask containing a magnetic stirrer bar was charged with 5-diazo-1-phenylpentane-1,4-dione (60.6 mg, 0.30 mmol) and 4-nitrobenzaldehyde (90.7 mg, 0.60 mmol) in PhCF$_3$ (3.0 mL). Rh$_2${(S)-BPTV}$_4$•2THF (4.6 mg, 0.003 mmol, 1.0 mol%) was added to the mixture at 0°C. The mixture was stirred for 10 min, and then it was concentrated and the residue was purified by column chromatography [hexanes/benzene 1:2 to 0:100 (**CAUTION: carcinogen**)] to give the product as a white solid; yield: 69 mg (71%); mp 160.0–161.0°C; R_f 0.31 (benzene/Et$_2$O 5:1); 92% ee [HPLC (Chiralpak OD-H column; hexanes/iPrOH 5:1; flow 1.0 mL/min); t_R = 18.1 (minor), 21.6 min (major)].

2.4.3 Diastereoselective (3+2)-Cycloaddition Reactions of Carbonyl Ylides with Imines

In contrast to reactions with other dipolarophiles, (3+2)-cycloaddition reactions of carbonyl ylides with imines are rare. Activation of the imine in the transition state by coordination of the lone electron pair on the nitrogen to a Lewis acid catalyst means that the control of selectivity in these reactions is effectively promoted by the second, Lewis acid, catalyst (Scheme 23 and Scheme 24).[35–37] There are no asymmetric catalytic versions of these transformations, but diastereoselective versions using chiral diazo compounds are reported (Scheme 25 and Scheme 26).[38,39]

1,3-Dipolar cycloaddition reactions between carbonyl ylides and imines are effectively promoted by Lewis acids; in some cases, no 1,3-dipolar cycloaddition product is observed when the reaction is catalyzed by rhodium(II) acetate dimer alone. In the presence of Lewis acids such as ytterbium(III) trifluoromethanesulfonate, 1,3-dipolar cycloaddition reactions proceed smoothly with several imines and give the *exo*-product *exo*-**55** in most cases, without formation of the dimeric product (Scheme 23). Using this Lewis acid as a cocatalyst, a fundamental catalytic effect is also observed in the cycloaddition reactions of imines with other carbonyl ylides. This efficient catalytic effect can be satisfactorily explained by calculation studies in terms of the energetics of the cycloaddition in the absence and the presence of Lewis acid.[35]

2.4.3 Diastereoselective (3+2) Cycloadditions of Carbonyl Ylides with Imines

Scheme 23 Dihydro-1H-1,4-epoxybenzo[c]azepin-5(2H)-ones by Cooperative Rhodium/Lewis Acid Catalyzed (3+2)-Cycloaddition Reaction of a Carbonyl Ylide with Imines[35]

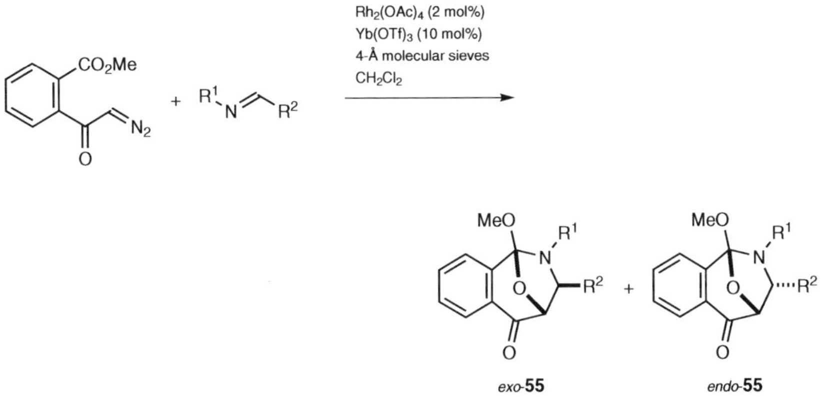

R^1	R^2	Ratio[a] (exo/endo)	Yield[b] (%)	Ref
Cy	Cy	12:88	31	[35]
Cy	Ph	81:19	21	[35]
$CHPh_2$	CO_2Et	58:42	45	[35]
Ph	Cy	84:16	47	[35]
Ph	Ph	62:38	84	[35]
Ph	2-$MeOC_6H_4$	85:15	94	[35]
Ph	4-$MeOC_6H_4$	77:23	75	[35]
Ph	2-$BnOC_6H_4$	92:8	92	[35]
Ph	4-ClC_6H_4	73:27	62	[35]
2-$MeOC_6H_4$	Ph	84:16	72	[35]
4-$MeOC_6H_4$	Ph	88:12	76	[35]
4-ClC_6H_4	Ph	76:24	42	[35]

[a] Determined by ^{1}H NMR spectroscopy.
[b] Isolated yield.

The β-amino alcohol and α-hydroxy-β-amino acid moieties are found in a large variety of bioactive compounds and natural products, including the C13 side chain of taxol. One straightforward access to these motifs is by the hydrolysis of oxazolidines, which can be directly synthesized from a three-component reaction of an imine, benzaldehyde, and a diazo compound. An efficient protocol for the synthesis of *syn*-β-amino alcohols and *syn*-α-hydroxy-β-amino acid derivatives is based on a highly diastereoselective (3+2) cycloaddition of a carbonyl ylide with imines catalyzed by rhodium(II) acetate dimer, and this is followed by acid-promoted hydrolysis. The reaction produces the corresponding β-amino acid derivatives **56** in high yields with high diastereoselectivity (Scheme 24).[36] The use of a Lewis acid additive promotes this transformation and the corresponding products are obtained in high yields and with good selectivity; the dipolarophile, the imine, is potentially activated by coordination with the Lewis acid catalyst.[37]

Scheme 24 One-Pot Process for the Synthesis of α-Hydroxy-β-amino Esters from Carbonyl Ylides[36,37]

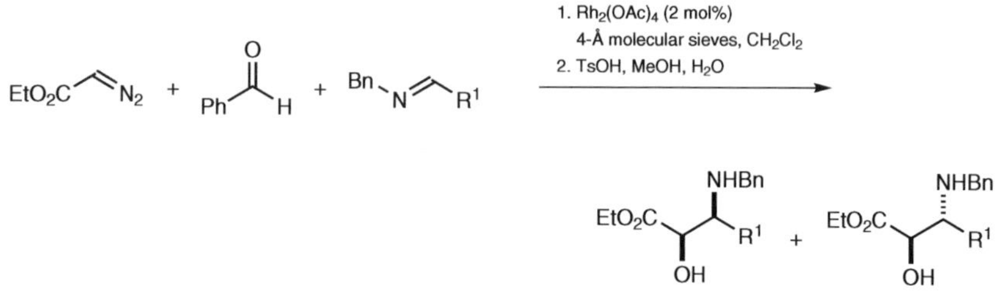

R¹	Ratio[a] (syn/anti)	Yield[b] (%)	Ref
Ph	93:7	82	[36]
4-Tol	97:3	87	[36]
3-MeOC$_6$H$_4$	98:2	77	[36]
4-MeOC$_6$H$_4$	94:6	78	[36]
4-O$_2$NC$_6$H$_4$	91:9	61	[36]
4-FC$_6$H$_4$	97:3	78	[36]
4-ClC$_6$H$_4$	98:2	75	[36]
2-naphthyl	98:2	83	[36]
2-furyl	92:8	75	[36]
CO$_2$Et	83:17	64	[36]

[a] Determined by ^{1}H NMR spectroscopy.
[b] Isolated yield.

There are a few examples of the diastereoselective cycloaddition of carbonyl ylides with imines, but no reported asymmetric catalytic version. However, enantiomerically pure cycloadducts can be accessed by diastereoselective cycloaddition starting from enantiopure materials. An example is the rhodium(II) acetate dimer catalyzed reaction of enantiomerically pure diazo ketone **57** with imines **58** to give the cycloadducts **59** in moderate yields for a limited number of examples (Scheme 25).[38] The three-component reaction of an imine, 4-bromobenzaldehyde, and chiral diazo ketone **60** catalyzed by the combined action of rhodium(II) acetate dimer and silver(I) hexafluoroantimonate generates the cycloadducts **61** in high yields; only one diastereomer is detected (Scheme 26).[39]

Scheme 25 Tetrahydro-1H-3,9a-epoxypyrimido[2,1-c][1,4]oxazin-4(9H)-ones by Diastereoselective (3+2)-Cycloaddition Reaction of a Carbonyl Ylide with Imines[38]

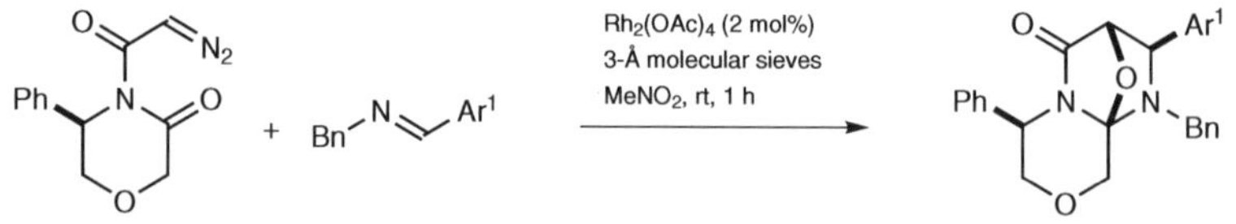

2.4.3 Diastereoselective (3+2) Cycloadditions of Carbonyl Ylides with Imines

Ar^1	Yield^a (%)	Ref
Ph	70	[38]
4-MeOC$_6$H$_4$	43	[38]
4-Me$_2$NC$_6$H$_4$	21	[38]
4-O$_2$NC$_6$H$_4$	47	[38]
4-ClC$_6$H$_4$	30	[38]

^{a} Isolated yield.

Scheme 26 Three-Component Diastereoselective Intermolecular (3+2)-Cycloaddition Reaction of a Carbonyl Ylide with Imines and an Aldehyde To Give Oxazolidines[39]

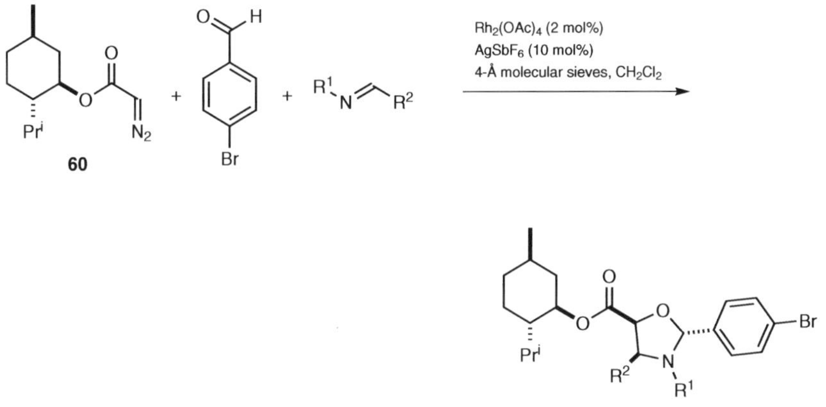

R^1	R^2	Yield (%)	Ref
Ph	Ph	62	[39]
4-MeOC$_6$H$_4$	Ph	66	[39]
4-MeOC$_6$H$_4$	4-BrC$_6$H$_4$	73	[39]

3-[2-(Benzyloxy)phenyl]-1-methoxy-2-phenyl-3,4-dihydro-1H-1,4-epoxybenzo[c]azepin-5(2H)-one (55, R^1 = Ph; R^2 = 2-BnOC$_6$H$_4$); Typical Procedure:[35]

A flame-dried, two-necked, round-bottomed flask containing a magnetic stirrer bar was charged with N-[2-(benzyloxy)benzylidene]aniline (287.4 mg, 1.0 mmol), Rh$_2$(OAc)$_4$ (4.4 mg, 0.01 mmol), Yb(OTf)$_3$ (31.1 mg, 0.05 mmol), and powdered 4-Å molecular sieves (0.50 g) in CH$_2$Cl$_2$ (5.0 mL). A soln of methyl 2-(diazoacetyl)benzoate (102.1 mg, 0.5 mmol) in CH$_2$Cl$_2$ (5.0 mL) was added over a period of 1 h. After removal of the 4-Å molecular sieves through Celite, the mixture was filtered through a plug of silica gel [3 cm, hexanes/EtOAc 1:1 (80 mL)]. The solvent was removed under reduced pressure. The resulting mixture was purified by medium-pressure liquid chromatography (MPLC) (hexanes/EtOAc 99:1) to give *exo*-**55** (R^1 = Ph; R^2 = 2-BnOC$_6$H$_4$) as colorless prisms; yield: 199.2 mg (85%); mp 173.5–174.5 °C; and *endo*-**55** (R^1 = Ph; R^2 = 2-BnOC$_6$H$_4$) as colorless prisms; yield: 16.1 mg (7%); mp 168.0–168.5 °C.

Ethyl *syn*-3-(Benzylamino)-2-hydroxypropanoates 56; General Procedure:[36]

A flame-dried, two-necked, round-bottomed flask containing a magnetic stirrer bar was charged with imine (0.256 mmol), benzaldehyde (39.0 µL, 0.384 mmol), Rh$_2$(OAc)$_4$ (2.3 mg, 0.005 mmol), and powdered 4-Å molecular sieves (250 mg) in CH$_2$Cl$_2$ (3.0 mL). A soln of ethyl diazoacetate (44 mg, 0.384 mmol) in CH$_2$Cl$_2$ (1.0 mL) was added over 1 or

10 h. The mixture was stirred for an additional hour, filtered through a plug (pipet) of basic alumina (activity grade I), eluted with CH_2Cl_2 (ca. 4–5 mL), and concentrated. The residue was dissolved in $MeOH/H_2O$ (95:5; 4.0 mL) and 1.5 M TsOH in MeOH (0.340 mL, 0.512 mmol) was added. The resultant mixture was stirred at rt for 2 h. The solvents were removed under reduced pressure and the residue was dissolved in CH_2Cl_2 and washed with sat. $NaHCO_3$ soln. The aqueous phase was extracted with CH_2Cl_2 (2×) and the combined organic extracts were dried ($MgSO_4$) and concentrated under reduced pressure. The residue was purified by column chromatography to give the pure product.

(2R,3S,6R,9aR)-2-Aryl-1-benzyl-6-phenyltetrahydro-1H-3,9a-epoxypyrimido[2,1-c]-[1,4]oxazin-4(9H)-ones 59; General Procedure:[38]

CAUTION: *Nitromethane is flammable, a shock- and heat-sensitive explosive, and an eye, skin, and respiratory tract irritant.*

A flame-dried, two-necked, round-bottomed flask containing a magnetic stirrer bar was charged with $Rh_2(OAc)_4$ (2 mol%) and 3-Å molecular sieves in $MeNO_2$ (1.0 mL) under an inert atmosphere. A soln of imine **58** (0.81 mmol) in $MeNO_2$ (1.0 mL) was added followed by the addition of a soln of diazo ketone **57** (0.27 mmol) in $MeNO_2$ (1.0 mL) dropwise. The reaction was stirred until TLC analysis showed no evidence of the diazo compound. The molecular sieves were removed by filtration and washed with CH_2Cl_2 (3–5 mL), the solvent was removed under reduced pressure, and the residue was purified by gradient column chromatography (Et_2O/petroleum ether 30:70 to 40:60 with 2% Et_3N) to give the pure product as a single diastereomer.

(1R,2S,5R)-2-Isopropyl-5-methylcyclohexyl (2R,4S,5S)-2-(4-Bromophenyl)-3,4-diphenyloxazolidine-5-carboxylate (61, $R^1 = R^2 = Ph$); Typical Procedure:[39]

A flame-dried, two-necked, round-bottomed flask containing a magnetic stirrer bar was charged with 4-bromobenzaldehyde (41 mg, 0.22 mmol), N-benzylideneaniline (36 mg, 0.20 mmol), 4-Å molecular sieves (0.1 g), $Rh_2(OAc)_4$ (1.8 mg, 2 mol%), and $AgSbF_6$ (6.7 mg, 10 mol%) in CH_2Cl_2 (1.5 mL) under an argon atmosphere. The flask was cooled to 0 °C, and diazo compound **60** (49 mg, 0.22 mmol) in CH_2Cl_2 (0.5 mL) was added to the mixture over a 1 h period using a syringe pump. When the addition was complete, the mixture was stirred for a further 30 min. The crude products were subjected to ^1H NMR analysis to determine the diastereoselectivity; only one diastereomer was observed. The mixture was purified by flash chromatography (silica gel, petroleum ether/EtOAc 50:1 to 20:1) to give the pure product as a yellow oil; yield: 70 mg (62%).

2.4.4 1,3-Dipolar (3+2)-Cycloaddition Reactions of Azomethine Ylides

The chemistry of azomethine ylides is well known and cycloaddition reactions related to these species have been extensively studied;[40–43] see also *Science of Synthesis*, Vol. 27 (Section 27.11). However, there are few examples of the reactions of azomethine ylides generated from metal–carbenes and imines;[44] this is mainly due to the side reactions, in particular aziridination, that follow azomethine ylide formation. The range of dipolarophiles that can be utilized is limited to alkenes, alkynes, and diazenes. Other dipolarophiles, for example aldehydes, can compete to form the corresponding carbonyl ylide in the first step, thus they are not reliable dipolarophiles for azomethine ylides. In this section, only reactions of azomethine ylides generated in situ from metal–carbene intermediates and imines are discussed; selected examples for the formation of N-heterocyclic compounds are given.

An example of interception of an azomethine ylide is the intramolecular cyclization reaction of methyl 2-diazo-4-phenylbut-3-enoate and an imine catalyzed by rhodium(II)

2.4.4 1,3-Dipolar (3 + 2) Cycloadditions of Azomethine Ylides

acetate dimer to generate dihydropyrrole **62** in high yield and with high stereocontrol (Scheme 27).[45]

Scheme 27 Intramolecular Cyclization of an Azomethine Ylide To Give a 4,5-Dihydropyrrole[45]

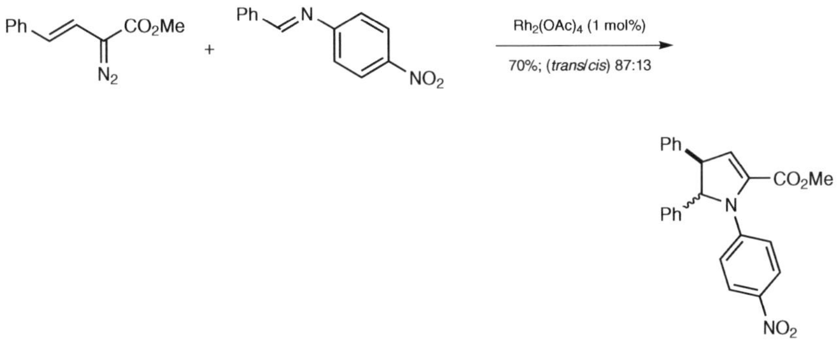

Although dirhodium(II) and copper catalysts are remarkably effective for metal–carbene reactions of diazo compounds, copper(II) trifluoromethanesulfonate is a much stronger Lewis acid than rhodium(II) acetate dimer and, consequently, its preferred association is with the more basic imine. The rhodium(II) acetate dimer may also associate with the imine, but its preferred reaction is at the diazo carbon of the 2-diazobut-3-enoate, which becomes an irreversible reaction following the loss of nitrogen. For example, reactions of methyl 2-diazobut-3-enoates **63** with imines **64** catalyzed by rhodium(II) acetate dimer occur through metal–carbene formation and subsequent electrophilic reaction by the metal–carbene at the basic imine nitrogen, i.e. intermediate **65** (Scheme 28), to give 4,5-dihydropyrroles **66**. In contrast, reactions catalyzed by copper(II) trifluoromethanesulfonate involve initial iminium ion formation followed by electrophilic addition by the iminium ion at the diazo carbon, i.e. via intermediate **67**, to give regioisomeric 2,5-dihydropyrroles **68** in moderate to high yields (Scheme 28).[46]

It is worthy of note that when two equivalents of diazoacetate are used in the reaction catalyzed by rhodium(II) acetate dimer, intermediate **65** could be intercepted by another molecule of the metal–carbene intermediate to generate bicyclic pyrrolidines in moderate to high yields. Although the mechanism is not clear, the outcome of the reaction strongly suggests that the "free" ylide intermediate derived from **65** exists long enough to undergo a vinylogous reaction with another metal–carbene intermediate.[47]

Scheme 28 Divergent Pathways to Isomeric Dihydropyrroles Catalyzed by Rhodium or Copper Catalysts[46]

metal–carbene pathway

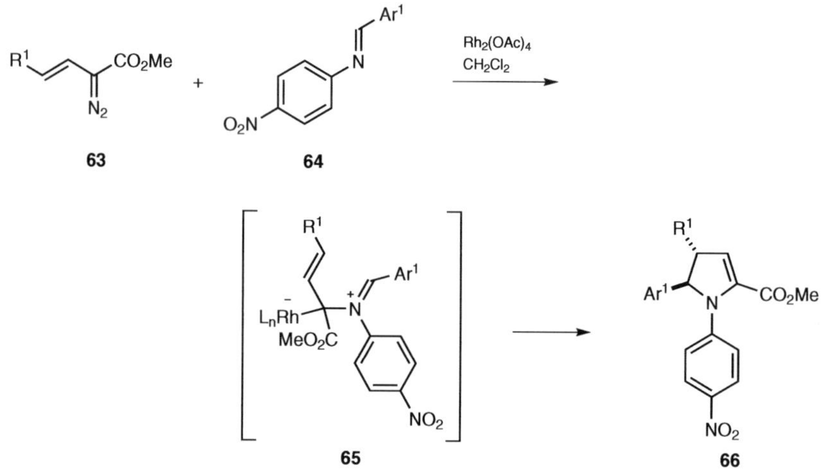

R^1	Ar^1	Yield[a] (%)	Ref
Me	Ph	44	[46]
Ph	Ph	62	[46]

[a] Isolated yields.

Lewis acid pathway

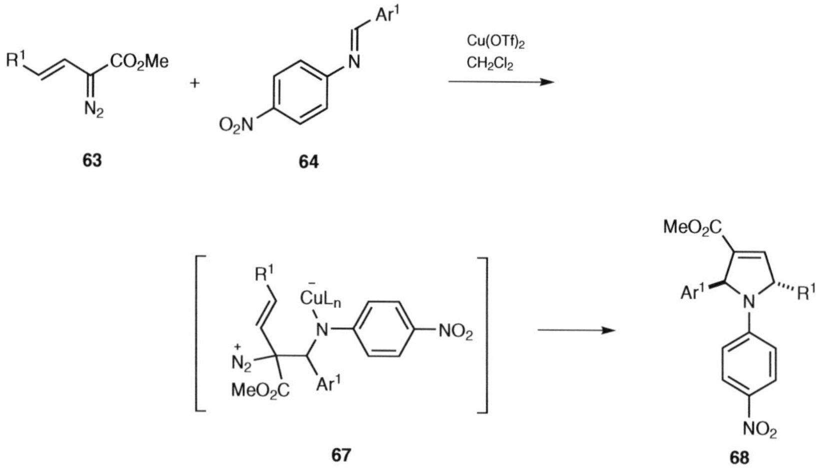

R^1	Ar^1	Yield[a] (%)	Ref
Ph	Ph	67	[46]
Ph	4-Tol	63	[46]
Ph	4-MeOC$_6$H$_4$	74	[46]

[a] Isolated yields.

Ruthenium porphyrins also work well in the three-component cycloaddition reaction of azomethine ylides with dipolarophiles, including alkenes and alkynes.[44] For example, cy-

2.4.4 1,3-Dipolar (3 + 2) Cycloadditions of Azomethine Ylides

cloaddition of azomethine ylides with methyl acrylate catalyzed by Ru(TMP)(CO) (TMP = 5,10,15,20-tetramesitylporphyrin) forms functionalized pyrrolidines **69** with excellent diastereoselectivity in moderate to high yields (Scheme 29).[48]

The scope of the substrates in the cycloaddition of azomethine ylides extends to dimethyl acetylenedicarboxylate and the optimized catalyst for this reaction is Ru(TDCPP)-(CO) [TDCPP = 5,10,15,20-tetrakis(2,6-dichlorophenyl)porphyrin]. All imines that were employed in this reaction gave the desired cycloadducts **70** in good to excellent yields. N-(4-Nitrobenzylidene)aniline, an imine containing an electron-withdrawing NO$_2$ group, gives the lowest yield (59%); however, almost quantitative yields (98%) are obtained for the reaction using N-(4-methylbenzylidene)aniline, an imine containing an electron-donating substituent (Scheme 30).[48]

Scheme 29 Pyrrolidines by Ruthenium-Catalyzed Three-Component Cycloaddition Reaction of Azomethine Ylides with an Alkene[48]

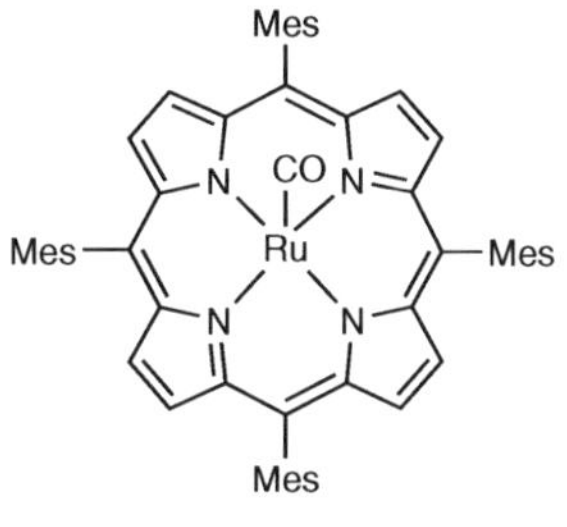

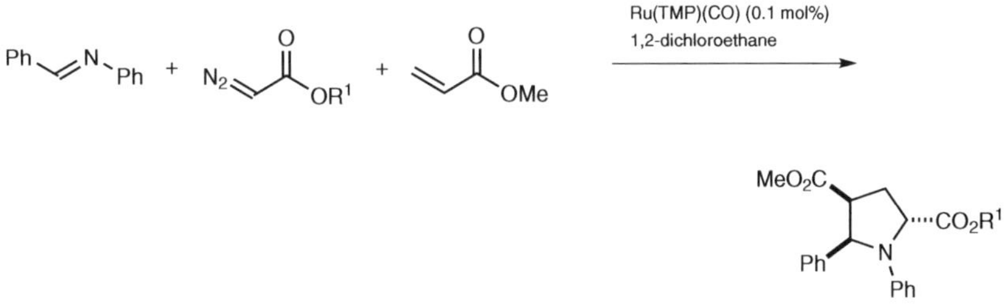

R^1	Yield (%)	Ref
Me	76	[48]
Et	56	[48]
Bn	74	[48]

Scheme 30 2,5-Dihydropyrroles by Ruthenium-Catalyzed Three-Component Cycloaddition Reaction of Azomethine Ylides with an Alkyne[48]

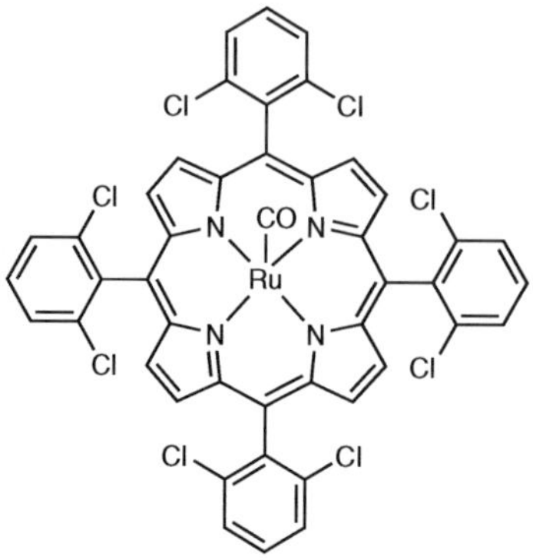

Ru(TDCPP)(CO)

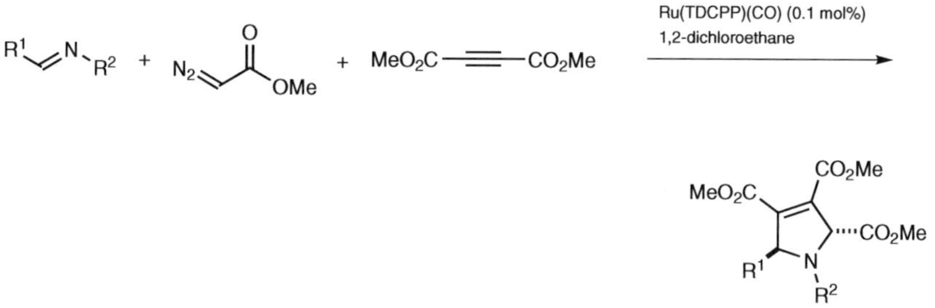

R^1-N=R^2 + N$_2$=CHC(O)OMe + MeO$_2$C≡CO$_2$Me → [Ru(TDCPP)(CO) (0.1 mol%), 1,2-dichloroethane] → **70**

R^1	R^2	Yield[a] (%)	Ref
Ph	Ph	90	[48]
Ph	4-MeOC$_6$H$_4$	84	[48]
Ph	3-ClC$_6$H$_4$	69	[48]
Ph	4-ClC$_6$H$_4$	85	[48]
4-Tol	Ph	98	[48]
3-MeOC$_6$H$_4$	Ph	90	[48]
4-MeOC$_6$H$_4$	Ph	87	[48]
4-O$_2$NC$_6$H$_4$	Ph	59	[48]
4-ClC$_6$H$_4$	Ph	92	[48]

[a] Isolated yield.

An asymmetric 1,3-dipolar cycloaddition reaction of chiral azomethine ylides, formed from an imine and diazo ester **71**, with an alkyne exhibits remarkable solvent effects on the diastereoselectivity; significant improvements in diastereoselectivity (60–85% de) are achieved by using aromatic solvents, for example toluene. In this one-pot, three-component reaction, chiral multifunctionalized 2,5-dihydropyrroles **72** are obtained in moderate to good yields (Scheme 31).[49]

2.4.4 1,3-Dipolar (3+2) Cycloadditions of Azomethine Ylides

Scheme 31 2,5-Dihydropyrroles by Ruthenium-Catalyzed Three-Component Cycloaddition Reaction of Chiral Azomethine Ylides with an Alkyne[49]

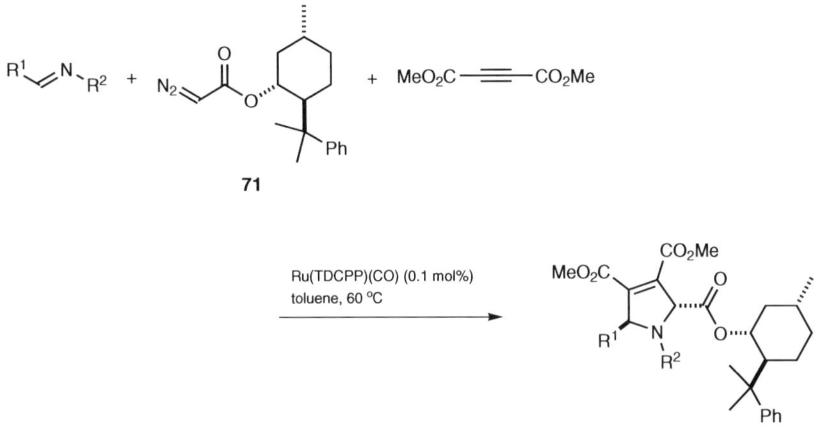

R¹	R²	de[a] (%)	Yield[b] (%)	Ref
Ph	Ph	79	70	[49]
Ph	4-MeOC$_6$H$_4$	74	61	[49]
Ph	4-BrC$_6$H$_4$	60	30	[49]
Ph	3-ClC$_6$H$_4$	72	32	[49]
Ph	4-ClC$_6$H$_4$	70	47	[49]
4-Tol	Ph	74	67	[49]
3-MeOC$_6$H$_4$	Ph	70	58	[49]
4-MeOC$_6$H$_4$	Ph	77	68	[49]
4-O$_2$NC$_6$H$_4$	Ph	85	39	[49]
4-ClC$_6$H$_4$	Ph	73	45	[49]

[a] Determined by ¹H NMR analysis of the crude products.
[b] Isolated yield.

As illustrated in the examples given in Schemes 30 and 31, (3+2) cycloaddition of azomethine ylides with alkynes affords 2,5-dihydro-1H-pyrrole derivatives. However, in the case of azomethine ylides generated in situ from diazoacetonitrile and an imine, the corresponding cycloadducts react directly with alkynes to produce aromatic pyrroles **73** via an autoelimination step in moderate to high yields (Scheme 32).[50] Notably, this process works efficiently at low catalyst loadings, and substitution on the starting imine is well tolerated.

Scheme 32 Synthesis of Pyrroles by Rhodium-Catalyzed Intermolecular Cyclization of In Situ Generated Azomethine Ylides[50]

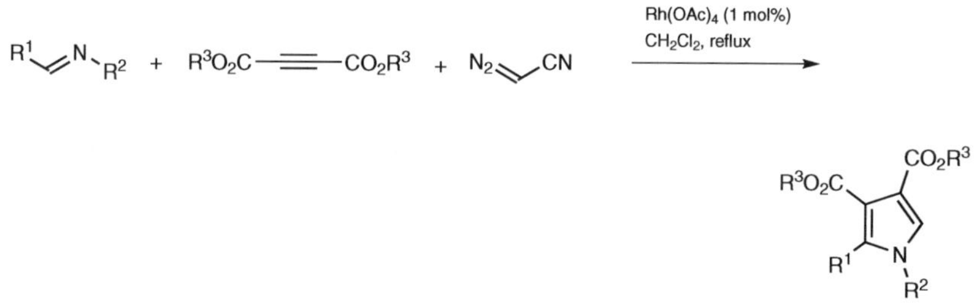

R^1	R^2	R^3	Yield[a] (%)	Ref
Ph	Ph	Me	71	[50]
Ph	3-Tol	Me	63	[50]
Ph	4-IC$_6$H$_4$	Me	61	[50]
4-Tol	Ph	Me	66	[50]
4-BrC$_6$H$_4$	Ph	Me	53	[50]
4-ClC$_6$H$_4$	Ph	Me	59	[50]
4-ClC$_6$H$_4$	Ph	Et	55	[50]
3,4-Cl$_2$C$_6$H$_3$	Ph	Me	47	[50]

[a] Isolated yield.

Multifunctionalized 1,2,4-triazolidines **74** are synthesized by ruthenium porphyrin [Ru(TDCPP)(CO)] catalyzed three-component cycloaddition of azomethine ylides with dialkyl azodicarboxylates (Scheme 33).[51] Imines containing either electron-donating or electron-withdrawing substituents show comparable reactivity and give the corresponding 1,2,4-triazolidines **74** in good yields (up to 86%). 5-Furyl-substituted 1,2,4-triazolidine **74** (R^1 = 2-furyl; R^2 = 3-MeOC$_6$H$_4$; R^3 = Et) is obtained in 72% yield and reactions with diisopropyl azodicarboxylate and di-*tert*-butyl azodicarboxylate afford cycloadducts **74** (R^3 = iPr, *t*-Bu) in 82 and 76% yields, respectively. Notably, some of the 1,2,4-triazolidines produced by this one-pot reaction exhibit good cytotoxicity against human nasopharyngeal carcinoma (SUNE1) (IC$_{50}$ = 10.4 μM) and human cervical carcinoma (Hela) (IC$_{50}$ = 10.7 μM) cell lines.

Scheme 33 Ruthenium-Catalyzed Intermolecular Cyclization of Azomethine Ylides To Give 1,2,4-Triazolidines[51]

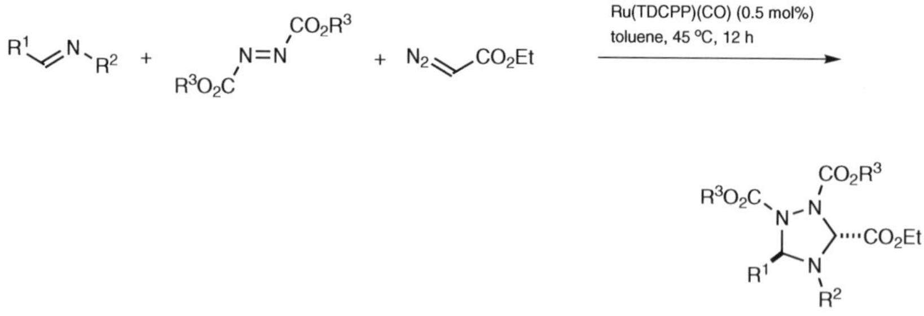

2.4.4 1,3-Dipolar (3+2) Cycloadditions of Azomethine Ylides

R¹	R²	R³	Yield[a] (%)	Ref
Ph	Ph	Et	82	[51]
4-Tol	Ph	Et	85	[51]
3-MeOC₆H₄	Ph	Et	78	[51]
4-MeOC₆H₄	Ph	Et	85	[51]
4-O₂NC₆H₄	Ph	Et	62	[51]
4-ClC₆H₄	Ph	Et	80	[51]
Ph	4-ClC₆H₄	Et	86	[51]
2-furyl	3-MeOC₆H₄	Et	72	[51]
4-MeOC₆H₄	Ph	iPr	82	[51]
4-MeOC₆H₄	Ph	t-Bu	76	[51]

[a] Isolated yield.

Methyl trans-1-(4-Nitrophenyl)-4,5-diphenyl-4,5-dihydro-1H-pyrrole-2-carboxylate (62); Typical Procedure:[45]

A flame-dried, round-bottomed flask containing a magnetic stirrer bar was charged with Rh₂(OAc)₄ (5.5 mg, 0.0125 mmol) and N-benzylidene-4-nitroaniline (283 mg, 1.25 mmol) in CH₂Cl₂ (10 mL). A soln of methyl 2-diazo-4-phenylbut-3-enoate (253 mg, 1.25 mmol) in CH₂Cl₂ (5 mL) was added using a syringe pump (5 mL/h) over 1 h under reflux. When the addition was complete, the mixture was cooled to rt, and then passed through a short silica plug, which was subsequently washed with CH₂Cl₂ (20 mL). The solvent was removed and a portion of the crude product was subjected to ¹H NMR analysis to determine the product ratio (trans/cis 87:13). The crude product was purified by column chromatography (silica gel, hexanes/EtOAc 10:1) to give the product as a mixture of cis- and trans-isomers; yield: 350 mg (70%).

2-Benzyl 4-Methyl (2R*,4S*,5R*)-1,5-Diphenylpyrrolidine-2,4-dicarboxylate (69, R¹ = Bn); Typical Procedure:[48]

A flame-dried, round-bottomed flask containing a magnetic stirrer bar was charged with N-benzylideneaniline (1.0 mmol), methyl acrylate (4.0 mmol), and Ru(TMP)(CO) catalyst (0.001 mmol) in 1,2-dichloroethane (8 mL). A soln of benzyl diazoacetate (1.0 mmol) in 1,2-dichloroethane (10 mL) was added over 10 h using a syringe pump at rt. After the addition, the resulting soln was stirred for a further 0.5 h. The mixture was concentrated and purified by flash chromatography (EtOAc/petroleum ether) to give the product; yield: 307 mg (74%); mp 122–123 °C.

Trimethyl (2R*,5S*)-1,5-Diphenyl-2,5-dihydro-1H-pyrrole-2,3,4-tricarboxylate (70, R¹ = R² = Ph); Typical Procedure:[48]

A flame-dried, round-bottomed flask containing a magnetic stirrer bar was charged with N-benzylideneaniline (1.0 mmol), DMAD (1.3 mmol), and Ru(TDCPP)(CO) catalyst (0.001 mmol) in 1,2-dichloroethane (8 mL). A soln of ethyl diazoacetate (1.0 mmol) in 1,2-dichloroethane (10 mL) was added over 10 h using a syringe pump at rt. After the addition, the resulting soln was stirred for a further 0.5 h. The mixture was concentrated and purified by flash chromatography (EtOAc/petroleum ether) to give the product; yield: 90%; mp 146–147 °C.

3,4-Dimethyl 2-[(1R,2S,5R)-5-Methyl-2-(2-phenylpropan-2-yl)cyclohexyl] (2R,5S)-1,5-Diphenyl-2,5-dihydro-1H-pyrrole-2,3,4-tricarboxylate (72, R¹ = R² = Ph); Typical Procedure:[49]

A flame-dried, round-bottomed flask containing a magnetic stirrer bar was charged with N-benzylideneaniline (1.2 mmol), DMAD (2.0 mmol), and Ru(TDCPP)(CO) catalyst

for references see p 203

(0.001 mmol) in toluene (8 mL). A soln of diazoacetate **71** (1.0 mmol) in toluene (10 mL) was added over 10 h using a syringe pump at 60 °C. After the addition, the resulting soln was stirred for a further 1 h. The mixture was concentrated and purified by flash chromatography (EtOAc/petroleum ether) to give the product; yield: 70%; de 79%; mp 62–64 °C.

Dimethyl 1,2-Diphenyl-1*H*-pyrrole-3,4-dicarboxylate (73, R¹ = R² = Ph; R³ = Me); Typical Procedure:[50]
A flame-dried, two-necked, round-bottomed flask containing a magnetic stirrer bar was charged with $Rh_2(OAc)_4$ (5.3 mg, 1 mol%), *N*-benzylideneaniline (213 mg, 1.2 mmol), and DMAD (49 µL, 0.4 mmol) in CH_2Cl_2 (3 mL). A soln of diazoacetonitrile (1.8 mmol) in CH_2Cl_2 (4 mL) was added to the mixture under reflux, using a syringe pump (1.5 mL/h). When the addition was complete, the mixture was heated for a further 1–2 h and then allowed to cool to 25 °C. The unpurified mixture was filtered through a silica plug [CH_2Cl_2 (50 mL)]. The solvent was evaporated, and the residue was purified by flash chromatography (silica gel, EtOAc/toluene 6:94 to 12:88 gradient) to afford pure product as a viscous white foam; yield: 94 mg (71%); R_f 0.24 (EtOAc/hexanes 3:1).

Triethyl (3*R,5*R**)-4,5-Diphenyl-1,2,4-triazolidine-1,2,3-tricarboxylate (74, R¹ = R² = Ph; R³ = Et); Typical Procedure:**[51]
A flame-dried, round-bottomed flask containing a magnetic stirrer bar was charged with *N*-benzylideneaniline (1.0 mmol), DEAD (1.2 mmol), and Ru(TDCPP)(CO) catalyst (0.005 mmol) in toluene (3 mL). A soln of ethyl diazoacetate (1.2 mmol) in toluene (3 mL) was added over 10 h using a syringe pump at 45 °C. After the addition, the resulting soln was stirred for a further 2 h. The mixture was concentrated and purified by flash chromatography (EtOAc/petroleum ether 10:1 to 15:1) to give the product; yield: 362 mg (82%).

2.4.5 (3+2)-Cycloaddition Reactions of Carbonyl Ylides in Natural Product Synthesis

Natural products with polycyclic structures are ubiquitous in nature,[52] and intramolecular (3+2)-cycloaddition reactions of carbonyl ylides with carbon–carbon π-bonds are a direct and effective access to these motifs.[15,53] Two examples of the application of this transformation as a key step in the construction of the core structures in bioactive compounds are highlighted (Scheme 34 and Scheme 35).[54,55]

The aspidosperma alkaloids occupy an important place in natural product chemistry because of their wide range of complex structural variations and diverse biological activity;[56,57] this family of indole alkaloids have a common pentacyclic ABCDE framework. Within this framework, the C ring is of critical importance because all of the stereogenic centers and the majority of functional groups are located in this ring. A direct method to access these motifs uses the diazopropanoate derivative **75** for an intramolecular (3+2)-cycloaddition of a carbonyl ylide, with the C=C bond in the indole used as the dipolarophile, to give the fully functionalized polycyclic structure of the ABCDE framework **76** in 97% yield (Scheme 34).[54]

Scheme 34 Synthesis of Aspidophytine[54]

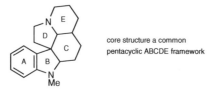

core structure a common pentacyclic ABCDE framework

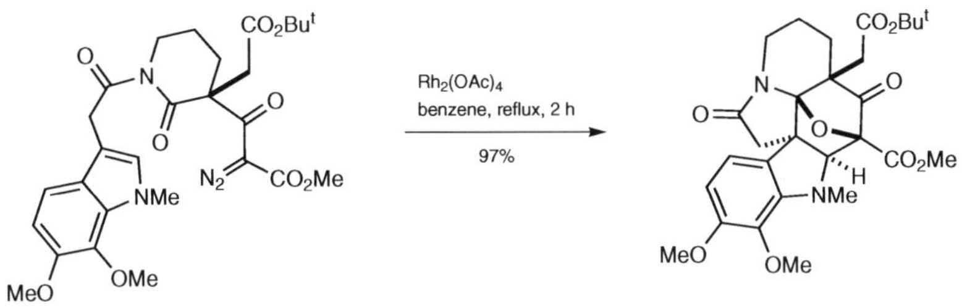

2.4.5 (3+2) Cycloadditions of Carbonyl Ylides in Natural Product Synthesis

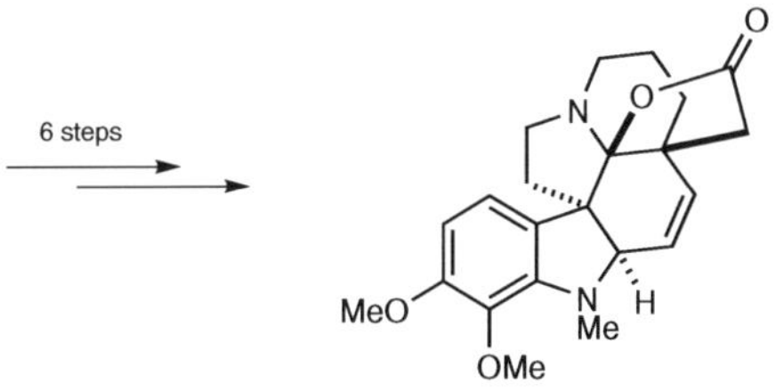

An asymmetric catalytic (3+2)-cycloaddition reaction of a carbonyl ylide is used in the synthesis of an enantiomerically pure natural product from *Ligusticum chuanxiong*. Cycloaddition of a carbonyl ylide from *tert*-butyl 2-diazo-3,6-dioxohexanoate with (4-hydroxy-3-methoxyphenyl)acetylene, catalyzed by chiral dirhodium carboxylate catalyst Rh$_2${(S)-TCPTTL}$_4$ (**9**, R^1 = Cl; R^2 = *t*-Bu), gives the cycloadduct **77** in 73% yield with 95% ee; a further 8 steps gives the natural product in 18% overall yield (Scheme 35).[55]

Scheme 35 Asymmetric Catalytic Synthesis of a Natural Product from *Ligusticum chuanxiong*[55]

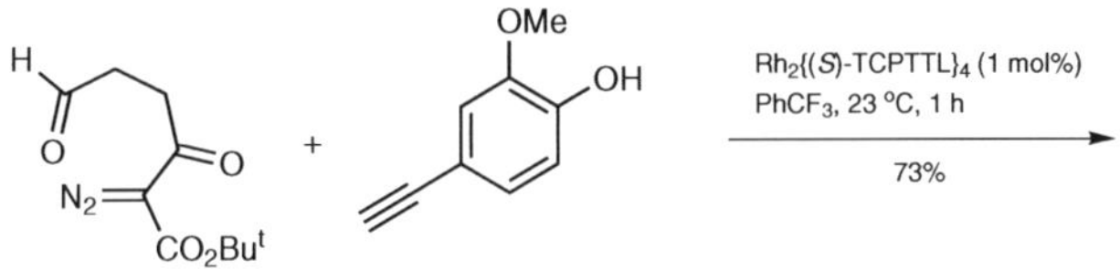

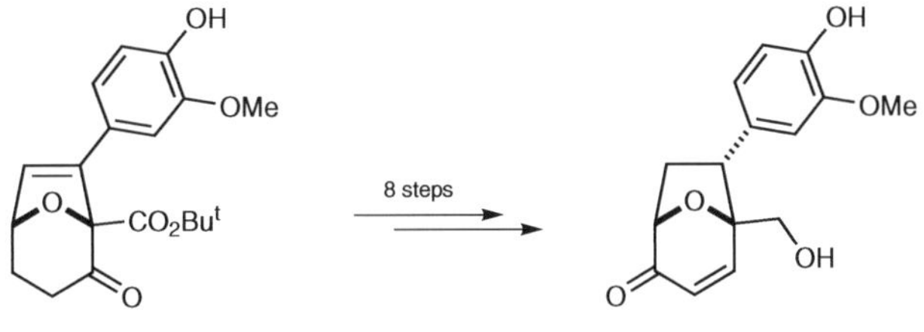

Methyl (3aR,3a^{1}R,5S,5aS,10bR)-3a-[(*tert*-Butoxycarbonyl)methyl]-7,8-dimethoxy-6-methyl-4,12-dioxo-2,3,3a,4,5a,6,11,12-octahydro-3a^1,5-epoxyindolizino[8,1-cd]carbazole-5(1H)-carboxylate (76):[54]

A flame-dried, two-necked, round-bottomed flask containing a magnetic stirrer bar was charged with diazo compound **75** (1.20 g, 2.2 mmol), Rh$_2$(OAc)$_4$ (7.5 mg, 0.8 mol%), and 4-Å molecular sieves (8.0 g) in benzene (100 mL) (**CAUTION:** *carcinogen*). The mixture was heated under reflux with stirring for 2 h. Then, the mixture was allowed to cool to rt and filtered through a pad of Celite. The solvent was removed under reduced pressure to give the product as a colorless oil, which was used directly in the next step without further purification; yield: 1.16 g (97%).

tert-Butyl (1R,5R)-7-(4-Hydroxy-3-methoxyphenyl)-2-oxo-8-oxabicyclo[3.2.1]oct-6-ene-1-carboxylate (77):[55]
A flame-dried, two-necked, round-bottomed flask containing a magnetic stirrer bar was charged with $Rh_2\{(S)\text{-TCPTTL}\}_4$ (39.5 mg, 0.020 mmol, 1 mol%) in $PhCF_3$ (10 mL), a soln of tert-butyl 2-diazo-3,6-dioxohexanoate (452 mg, 2.0 mmol) and (4-hydroxy-3-methoxyphenyl)acetylene (889 mg, 6.0 mmol) in $PhCF_3$ (10 mL) was added using a syringe pump, over 1 h at 23 °C. When the addition was complete, the mixture was concentrated under reduced pressure, and the residue was purified by column chromatography (hexanes/EtOAc 3:1) to give pure product as a pale yellow amorphous liquid; yield: 505.0 mg (73%); R_f 0.45 (hexanes/EtOAc 1:1); 95% ee [HPLC (Chiralpak IA column; hexanes/iPrOH 9:1; 1.0 mL/min); t_R = 17.8 (major), 22.9 min (minor)].

2.4.6 Conclusions

In summary, (3+2)-cycloaddition reactions of carbonyl ylides with dipolarophiles have been well studied, and their applications in natural product synthesis are also well demonstrated. These transformations offer a direct and effective access to five-membered heterocyclic compounds, including tetrahydrofurans, furans, 1,3-dioxolanes, 4,5-dihydrooxazoles (oxazolidines), pyrrolidines, pyrroles, 1,2,4-triazolidines, etc. However, the reactions of azomethine ylides, as analogues of carbonyl ylides, are less well-explored. The main focus of research is now the control of selectivity in these cycloaddition reactions, and this includes the development of new catalysts and strategies for cooperative catalysis. A core challenge in these cycloaddition reactions remains the development of catalytic versions that give access to enantiomerically pure products. For example, the asymmetric catalysis of three-component cycloaddition reactions, especially for (3+2)-cycloaddition reactions of carbonyl ylides with imines that generate α-hydroxy-β-amino acid moieties, useful building blocks that are present in many bioactive molecules, is of particular importance.

References

[1] Muthusamy, S.; Krishnamurthi, J., *Top. Heterocycl. Chem.*, (2008) **12**, 147.
[2] Padwa, A., *Helv. Chim. Acta*, (2005) **88**, 1357.
[3] Wang, J., In *Comprehensive Organometallic Chemistry III*, Crabtree, R. H.; Mingos, D. M. P., Eds.; Elsevier: Oxford, (2007); Vol 11, pp 159–163.
[4] Buchner, E.; Curtius, T., *Ber. Dtsch. Chem. Ges.*, (1885) **18**, 2371.
[5] de March, P.; Huisgen, R., *J. Am. Chem. Soc.*, (1982) **104**, 4952.
[6] Bentabed-Ababsa, G.; Hamza-Reguig, S.; Derdour, A.; Domingo, L. R.; Sáez, J. A.; Roisnel, T.; Dorcet, V.; Nassar, E.; Mongin, F., *Org. Biomol. Chem.*, (2012) **10**, 8434.
[7] Elliott, G. I.; Velcicky, J.; Ishikawa, H.; Li, Y.; Boger, D. L., *Angew. Chem.*, (2006) **118**, 636; *Angew. Chem. Int. Ed.*, (2006) **45**, 620.
[8] Bèkhazi, M.; Warkentin, J., *J. Am. Chem. Soc.*, (1983) **105**, 1289.
[9] Sammakia, T.; Smith, R. S., *J. Am. Chem. Soc.*, (1992) **114**, 10998.
[10] Woodall, E. L.; Simanis, J. A.; Hamaker, C. G.; Goodell, J. R.; Mitchell, T. A., *Org. Lett.*, (2013) **15**, 3270.
[11] Kusama, H.; Funami, H.; Shido, M.; Hara, Y.; Takaya, J.; Iwasawa, N., *J. Am. Chem. Soc.*, (2005) **127**, 2709.
[12] Doyle, M. P.; Hu, W.; Timmons, D. J., *Org. Lett.*, (2001) **3**, 933.
[13] DeAngelis, A.; Panne, P.; Yap, G. P. A.; Fox, J. M., *J. Org. Chem.*, (2008) **73**, 1435.
[14] Tsutsui, H.; Shimada, N.; Abe, T.; Anada, M.; Nakajima, M.; Nakamura, S.; Nambu, H.; Hashimoto, S., *Adv. Synth. Catal.*, (2007) **349**, 521.
[15] Padwa, A.; Cheng, B.; Zou, Y., *Aust. J. Chem.*, (2014) **67**, 343.
[16] Hodgson, D. M.; Glen, R.; Redgrave, A. J., *Tetrahedron: Asymmetry*, (2009) **20**, 754.
[17] Hodgson, D. M.; Avery, T. D.; Donohue, A. C., *Org. Lett.*, (2002) **4**, 1809.
[18] Zhou, C.-Y.; Yu, W.-Y.; Che, C.-M., *Org. Lett.*, (2002) **4**, 3235.
[19] Rout, L.; Harned, A. M., *Chem.–Eur. J.*, (2009) **15**, 12926.
[20] Lu, C.-D.; Chen, Z.-Y.; Liu, H.; Hu, W.-H.; Mi, A.-Q.; Doyle, M. P., *J. Org. Chem.*, (2004) **69**, 4856.
[21] Hashimoto, Y.; Itoh, K.; Kakehi, A.; Shiro, M.; Suga, H., *J. Org. Chem.*, (2013) **78**, 6182.
[22] Hodgson, D. M.; Labande, A. H.; Glen, R.; Redgrave, A. J., *Tetrahedron: Asymmetry*, (2003) **14**, 921.
[23] Hodgson, D. M.; Labande, A. H.; Pierard, F. Y. T. M.; Expósito Castro, M. Á., *J. Org. Chem.*, (2003) **68**, 6153.
[24] Shimada, N.; Anada, M.; Nakamura, S.; Nambu, H.; Tsutsui, H.; Hashimoto, S., *Org. Lett.*, (2008) **10**, 3603.
[25] Takeda, K.; Oohara, T.; Shimada, N.; Nambu, H.; Hashimoto, S., *Chem.–Eur. J.*, (2011) **17**, 13992.
[26] Krishnamurthi, J.; Nambu, H.; Takeda, K.; Anada, M.; Yamano, A.; Hashimoto, S., *Org. Biomol. Chem.*, (2013) **11**, 5374.
[27] Shimada, N.; Oohara, T.; Krishnamurthi, J.; Nambu, H.; Hashimoto, S., *Org. Lett.*, (2011) **13**, 6284.
[28] Suga, H.; Ishimoto, D.; Higuchi, S.; Ohtsuka, M.; Arikawa, T.; Tsuchida, T.; Kakehi, A.; Baba, T., *Org. Lett.*, (2007) **9**, 4359.
[29] Suga, H.; Hashimoto, Y.; Yasumura, S.; Takezawa, R.; Itoh, K.; Kakehi, A., *J. Org. Chem.*, (2013) **78**, 10840.
[30] Hodgson, D. M.; Stupple, P. A.; Johnstone, C., *Tetrahedron Lett.*, (1997) **38**, 6471.
[31] Davies, H. M. L.; Bruzinski, P. R.; Lake, D. H.; Kong, N.; Fall, M. J., *J. Am. Chem. Soc.*, (1996) **118**, 6897.
[32] Lu, C.-D.; Chen, Z.-Y.; Liu, H.; Hu, W.-H.; Mi, A.-Q., *Org. Lett.*, (2004) **6**, 3071.
[33] Muthusamy, S.; Ramkumar, R.; Mishra, A. K., *Tetrahedron Lett.*, (2011) **52**, 148.
[34] Chen, Z.; Wei, L.; Zhang, J., *Org. Lett.*, (2011) **13**, 1170.
[35] Suga, H.; Ebiura, Y.; Fukushima, K.; Kakehi, A.; Baba, T., *J. Org. Chem.*, (2005) **70**, 10782.
[36] Torssell, S.; Kienle, M.; Somfai, P., *Angew. Chem.*, (2005) **117**, 3156; *Angew. Chem. Int. Ed.*, (2005) **44**, 3096.
[37] Torssell, S.; Somfai, P., *Adv. Synth. Catal.*, (2006) **348**, 2421.
[38] Gan, Y.; Harwood, L. M.; Richards, S. C.; Smith, I. E. D.; Vinader, V., *Tetrahedron: Asymmetry*, (2009) **20**, 723.
[39] Xu, X.; Guo, X.; Han, X.; Yang, L.; Hu, W., *Org. Chem. Front.*, (2014) **1**, 181.
[40] Narayan, R.; Potowski, M.; Jia, Z.-J.; Antonchick, A. P.; Waldmann, H., *Acc. Chem. Res.*, (2014) **47**, 1296.
[41] Yu, J.; Shi, F.; Gong, L.-Z., *Acc. Chem. Res.*, (2011) **44**, 1156.

[42] Randjelovic, J.; Simic, M.; Tasic, G.; Husinec, S.; Savic, V., *Curr. Org. Chem.*, (2014) **18**, 1073.
[43] Tao, H.-Y.; Wang, C.-J., *Synlett*, (2014) **25**, 461.
[44] Zhou, C.-Y.; Huang, J.-S.; Che, C.-M., *Synlett*, (2010), 2681.
[45] Doyle, M. P.; Hu, W.; Timmons, D. J., *Org. Lett.*, (2001) **3**, 3741.
[46] Doyle, M. P.; Yan, M.; Hu, W.; Gronenberg, L. S., *J. Am. Chem. Soc.*, (2003) **125**, 4692.
[47] Yan, M.; Jacobsen, N.; Hu, W.; Gronenberg, L. S.; Doyle, M. P.; Colyer, J. T.; Bykowski, D., *Angew. Chem.*, (2004) **116**, 6881; *Angew. Chem. Int. Ed.*, (2004) **43**, 6713.
[48] Li, G.-Y.; Chen, J.; Yu, W.-Y.; Hong, W.; Che, C.-M., *Org. Lett.*, (2003) **5**, 2153.
[49] Xu, H.-W.; Li, G.-Y.; Wong, M.-K.; Che, C.-M., *Org. Lett.*, (2005) **7**, 5349.
[50] Galliford, C. V.; Scheidt, K. A., *J. Org. Chem.*, (2007) **72**, 1811.
[51] Wang, M.-Z.; Xu, H.-W.; Liu, Y.; Wong, M.-K.; Che, C.-M., *Adv. Synth. Catal.*, (2006) **348**, 2391.
[52] Geldenhuys, W. J.; Malan, S. F.; Bloomquist, J. R.; Marchand, A. P.; Van der Schyf, C. J., *Med. Res. Rev.*, (2005) **25**, 21.
[53] Padwa, A., *Tetrahedron*, (2011) **67**, 8057.
[54] Mejía-Oneto, J. M.; Padwa, A., *Helv. Chim. Acta*, (2008) **91**, 285.
[55] Shimada, N.; Hanari, T.; Kurosaki, Y.; Takeda, K.; Anada, M.; Nambu, H.; Shiro, M.; Hashimoto, S., *J. Org. Chem.*, (2010) **75**, 6039.
[56] Herbert, R. B., In *The Chemistry of Heterocyclic Compounds*, Saxton, J. E., Ed.; Wiley: Chichester, UK, (1994); Vol. 25, Part 4, Chapter 1.
[57] Toyota, M.; Ihara, M., *Nat. Prod. Rep.*, (1998), 327.

2.5 Metal-Catalyzed Asymmetric Diels–Alder and Hetero-Diels–Alder Reactions

H. Du

General Introduction

Since the seminal work published by Otto P. H. Diels and Kurt Alder in 1928, the Diels–Alder (DA) reaction has become one of the most useful atom-economic reactions in synthetic chemistry (Scheme 1).[1,2] The stereoselective construction of two σ-bonds and up to four contiguous stereocenters in a single step makes the Diels–Alder reaction a powerful and reliable method for the synthesis of six-membered cyclic compounds. As a result of tremendous efforts on the study of substrate scope and mechanism, the development of an asymmetric transformation, and application in total synthesis, great progress has been achieved over the past over 80 years. Numerous versions of the Diels–Alder reaction have been successfully developed, including intramolecular Diels–Alder reactions, hetero-Diels–Alder (HDA) reactions, and dehydro-Diels–Alder reactions. Furthermore, enzymatic catalysis of Diels–Alder reactions, and Lewis acid catalysis or organocatalysis of Diels–Alder and hetero-Diels–Alder reactions have been implemented.

Scheme 1 The Diels–Alder Reaction[1,2]

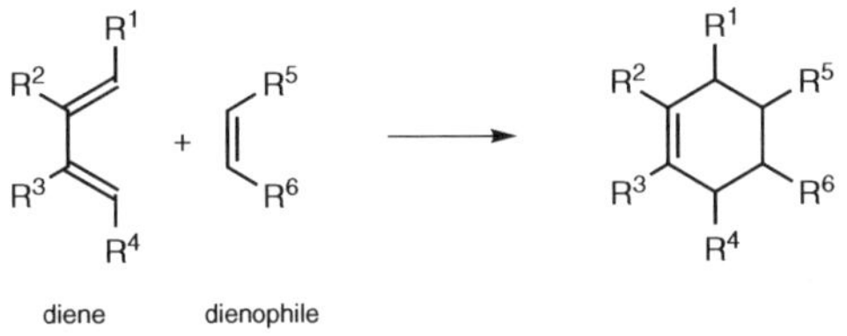

The asymmetric Diels–Alder reaction provides a powerful approach for the synthesis of optically active compounds. The first asymmetric Diels–Alder reaction was reported by Korolev and Mur in 1948,[3] and, after the discovery by Walborsky and co-workers in 1963 that Lewis acids can efficiently catalyze the Diels–Alder reaction,[4] the chiral Lewis acid catalyzed asymmetric Diels–Alder reaction has been extensively investigated.[5] Moreover, since the first organocatalytic asymmetric Diels–Alder reaction was reported by Kagan and co-workers in 1989,[5] the enantioselective organocatalysis of Diels–Alder processes has also become an important method by which to access optically active compounds. Chiral catalysts undoubtedly play a key role in determining both reactivity and enantioselectivity, and the following chapter will focus on the asymmetric Diels–Alder reaction, the oxa-Diels–Alder reaction, and the aza-Diels–Alder reaction based on the category of chiral Lewis acid catalyst used.

2.5.1 Diels–Alder Reactions

2.5.1.1 Reaction Using Chiral Boron Complexes

2.5.1.1.1 (Acyloxy)borane Catalysts

Chiral (acyloxy)borane (CAB) complex **1**, derived from monoacylated tartaric acid and borane–tetrahydrofuran complex, is an efficient catalyst for the Diels–Alder reaction of α,β-unsaturated aldehydes and cyclopentadiene to give the corresponding *exo-* and *endo-*adducts **2A** and **2B** in moderate to good yield and with high enantiomeric excess (Scheme 2).[6,7] An α-substituent on the dienophile increases the enantioselectivity (in favor of the *exo*-product **2A**), while β-substitution dramatically decreases the selectivity. In the case of a substrate having substituents at both the α- and β-positions, a high enantiomeric excess is obtained, indicating that the α-substituent effect appears to overcome the β-substituent effect. A similar chiral oxazaborolidinone complex derived from (S)-tryptophan is also an efficient catalyst for this reaction.[8,9]

Scheme 2 Asymmetric Diels–Alder Reactions Catalyzed by a Chiral (Acyloxy)borane[7]

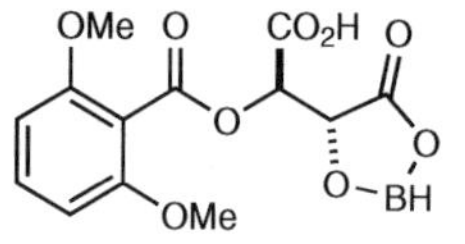

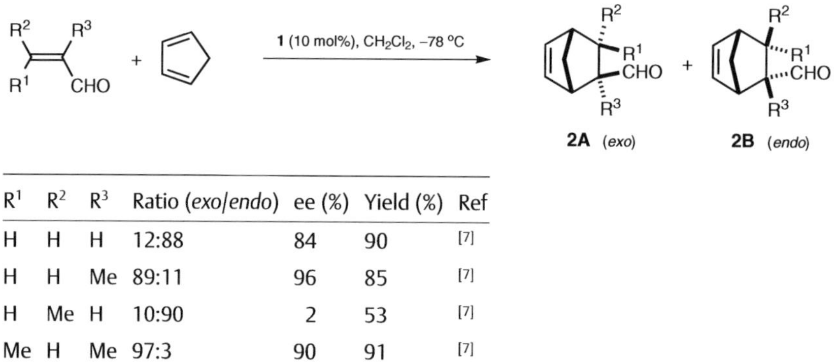

R^1	R^2	R^3	Ratio (exo/endo)	ee (%)	Yield (%)	Ref
H	H	H	12:88	84	90	[7]
H	H	Me	89:11	96	85	[7]
H	Me	H	10:90	2	53	[7]
Me	H	Me	97:3	90	91	[7]

2-Methylbicyclo[2.2.1]hept-5-ene-2-carbaldehydes 2A and 2B ($R^1 = R^2 = H$; $R^3 = Me$); Typical Procedure:[7]

A 1 M soln of BH$_3$•THF in THF (0.20 mL, 0.2 mmol) was added at 0 °C to a stirred suspension of the monoacylated tartaric acid (63 mg, 0.2 mmol) in CH$_2$Cl$_2$ (2.0 mL). After 15 min, during which time the mixture became a clear soln with evolution of H$_2$ gas, 2-methylpropenal (140 mg, 2.0 mmol) and cyclopentadiene (396 mg, 6.0 mmol) were introduced successively to the catalyst soln at −78 °C, and the resulting mixture was stirred at the same temperature for 3 h. The desired product was isolated after concentration and purification by chromatography (silica gel); yield: 85%; ratio (*exo/endo*) 89:11; 96% ee.

2.5.1.1.2 Superacidic Lewis Acid Catalysts

The application of superacidic Lewis acid catalysts **3** to the Diels–Alder reaction of cyclopentadiene with a variety of α,β-unsaturated aldehydes gives the cycloadduct **4** (n = 1) as the major product in high yield and with high enantiomeric excess (Scheme 3).[10] In general, for highly reactive cyclopentadiene, there is no difference between catalyst **3** with a bromide (X = Br) or tetrakis[3,5-bis(trifluoromethyl)phenyl]borate (X = BARF) counterion. However, for unreactive dienes (e.g., n = 2), catalyst **3** with a bromide counterion is not an active catalyst, while that with the borate counterion gives an excellent result.

Scheme 3 Asymmetric Diels–Alder Reactions Catalyzed by Superacidic Chiral Borane Lewis Acids[10]

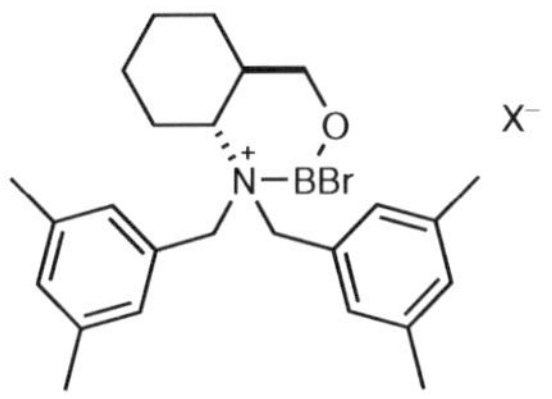

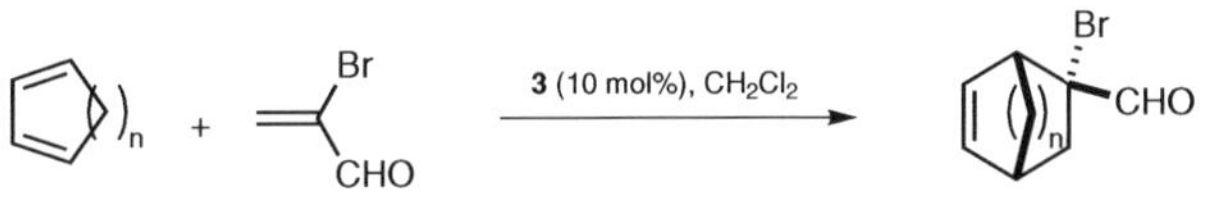

n	X[a]	Ratio (exo/endo)	ee (%)	Yield[b] (%)	Ref
1	Br	94:6	95	99	[10]
1	BARF	91:9	98	99	[10]
2	Br	–	–	n.r.	[10]
2	BARF	4:96	93	99	[10]

[a] BARF = tetrakis[3,5-bis(trifluoromethyl)phenyl]borate.
[b] n.r. = not reported.

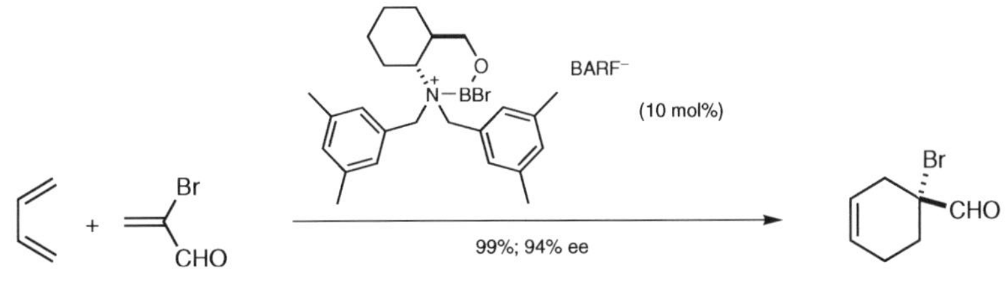

In a hypothetical transition state **5**, one of the benzyl groups on nitrogen serves to block attack on the lower face of the *s-trans*-coordinated dienophile while the other screens off another region in space and limits the rotational position of both the dienophile and the *N*-benzyl moieties (Scheme 4).

Scheme 4 Hypothetical Transition State for Superacidic Borane Catalyzed Asymmetric Diels–Alder Reactions[10]

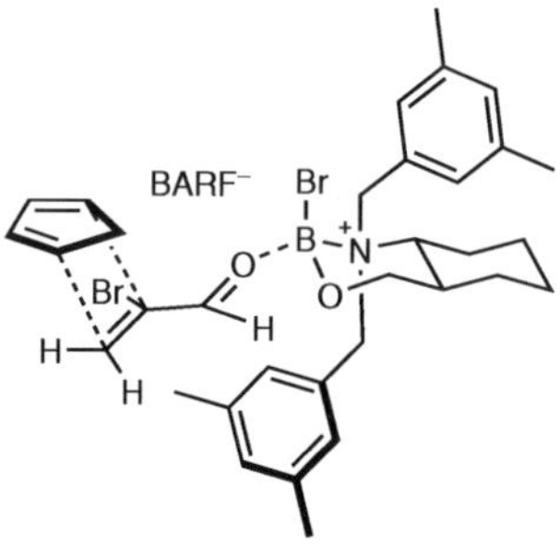

5

(1R,2R,4R)-2-Bromobicyclo[2.2.1]hept-5-ene-2-carbaldehyde (4, n = 1);
Typical Procedure:[10]

A soln of catalyst **3** (X = Br; 0.115 mmol) in CH$_2$Cl$_2$ (1.25 mL) was cooled to −94 °C and 2-bromopropenal (0.093 mL, 1.15 mmol) and a soln of cyclopentadiene (0.355 mL) in CH$_2$Cl$_2$ (0.5 mL) were added successively and slowly. The resulting mixture was stirred for 1 h at −94 °C and then the reaction was quenched by addition of Et$_3$N (0.15 mL, 1 mmol). The solvent was removed under reduced pressure, and the residue was purified by chromatography (silica gel); yield: 229 mg (99%).

2.5.1.1.3 Brønsted Acid Activated Borane Catalysts

Chiral boron complex **6** is prepared from the corresponding 1,1′-binaphthalene-2,2′-diol derivative and trimethyl borate, and is a highly effective catalyst for the reaction of α,β-unsaturated aldehydes with dienes to give cycloaddition products such as **7** with both high enantioselectivity and high *exo* selectivity (Scheme 5).[11]

Scheme 5 Brønsted Acid Assisted Asymmetric Diels–Alder Reactions[11]

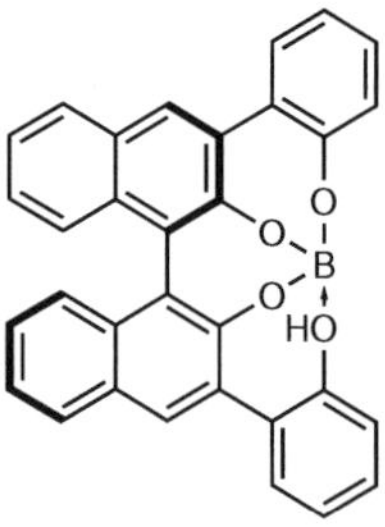

6

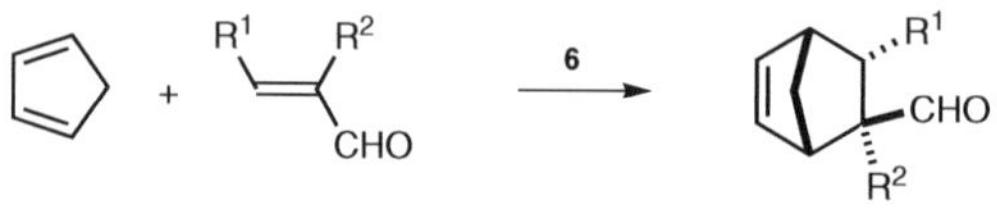

7

2.5.1 Diels–Alder Reactions

R¹	R²	Ratio (exo/endo)	ee (%)	Yield (%)	Ref
H	Br	>99:1	99	>99	[11]
H	Me	>99:1	99	>99	[11]
H	Et	97:3	92	>99	[11]
Me	Me	>99:1	98	>99	[11]

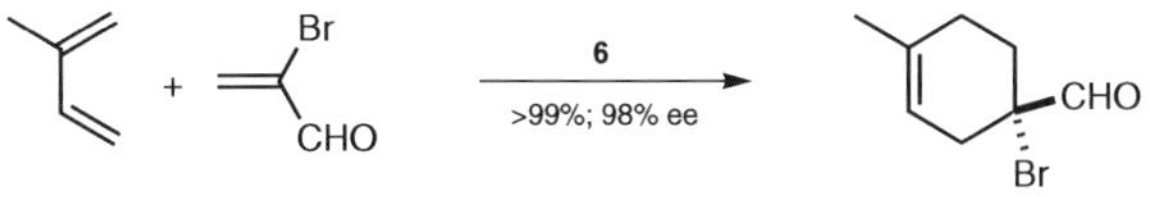

The absolute stereochemical preference in this reaction can be easily understood in terms of the transition state **8**, in which an attractive donor–acceptor interaction favors coordination of the dienophile at the face of boron which is *cis* to the 2-hydroxyphenyl substituent (Scheme 6).[11] It is supposed that the coordination of a proton on the 2-hydroxyphenyl group with an oxygen of the adjacent B—O bond causes the Lewis acidity of boron and the π-basicity of the phenoxy moiety to increase, which stabilizes the transition state. In this conformation, the hydroxyphenyl group blocks the *si*-face of the dienophile, leaving the *re*-face open to approach by the diene.

Scheme 6 Transition State Model for Brønsted Acid Assisted Asymmetric Diels–Alder Reactions[11]

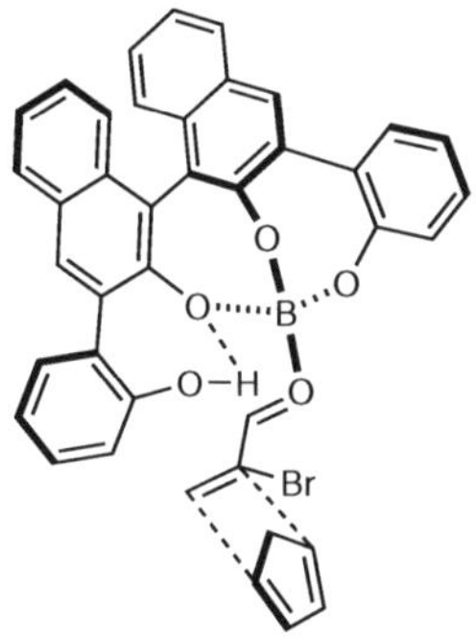

Chiral superacid **9** is generated by the condensation of an amino alcohol and an arylboroxin with subsequent activation by trifluoromethanesulfonic acid. An unusually wide range of dienophiles, such as α,β-unsaturated aldehydes, ketones, and esters, undergo reaction with cyclopentadiene catalyzed by this species, to give the cycloaddition products in excellent yield and with excellent enantiomeric excess (e.g., Table 1, entries 1 and 2).[12] To avoid issues around the decomposition of catalyst **9**, a more stable catalyst, catalyst **10**, has been generated using the strong acid 1,1,1-trifluoro-*N*-[(trifluoromethyl)sulfonyl]-methanesulfonamide (triflimide) instead of trifluoromethanesulfonic acid. This species is effective for the Diels–Alder reaction of a wide range of dienophiles, including diethyl fumarates, acrylates, enones, and quinones, with both cyclic and acyclic dienes to give the corresponding cycloadducts (e.g., Table 1, entries 3 and 4).[13,14]

for references see p 262

Table 1 Brønsted Acid Activated Chiral Oxazaborolidines for Asymmetric Diels–Alder Reactions[12–14]

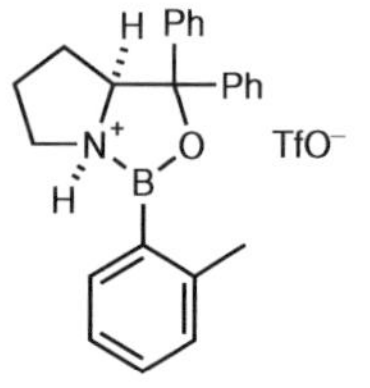

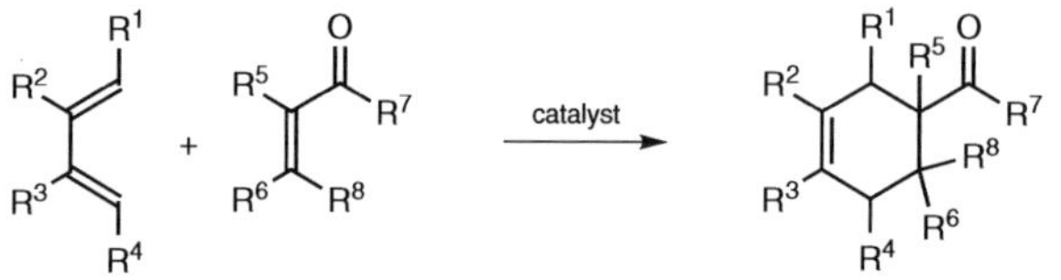

Entry	Starting Material		Catalyst	Product	Ratio (exo/endo)	ee (%)	Yield (%)	Ref
	Diene	Dienophile						
1	 cyclopentadiene	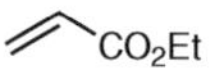 CH$_2$=CHCO$_2$Et	9	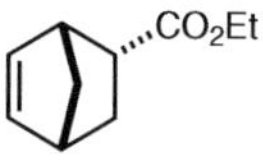	3:97	98	94	[12]
2	 cyclopentadiene	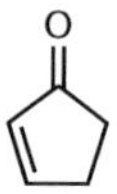 cyclopentenone	9	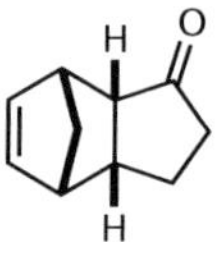	5:95	92	99	[12]
3	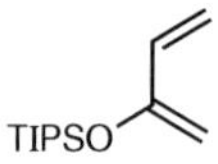 TIPSO-diene	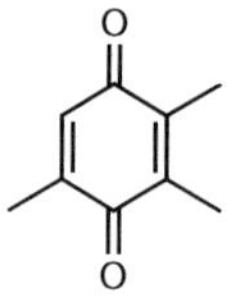 dimethylbenzoquinone	10	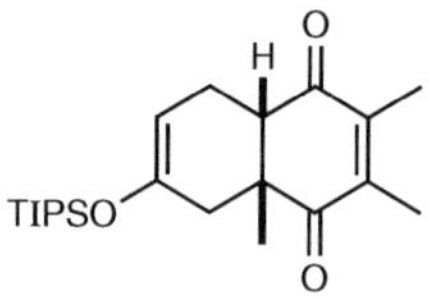	–	99	98	[13]
4	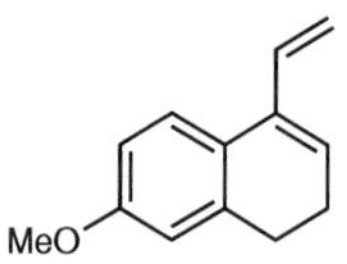 MeO-vinyl dihydronaphthalene	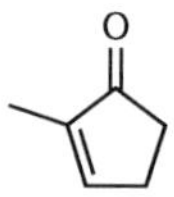 methylcyclopentenone	10	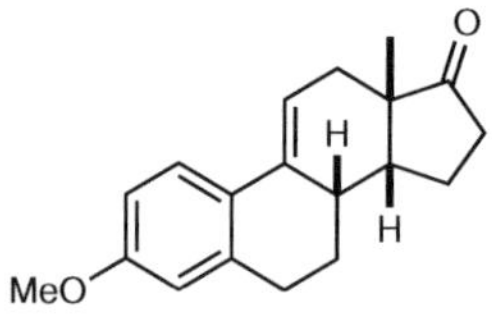	–	82	98	[14]

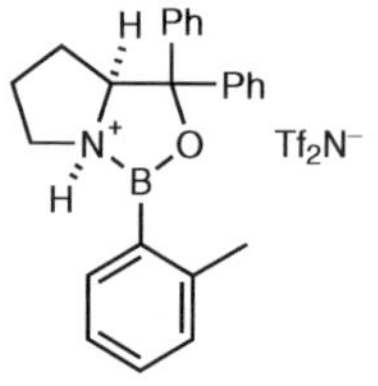

(1S,2S,4S)-2-Bromobicyclo[2.2.1]hept-5-ene-2-carbaldehyde (7, R^1 = H; R^2 = Br);
Typical Procedure:[11]
Catalyst **6** was prepared instantly by reaction of (R)-3,3′-bis(2-hydroxyphenyl)-1,1′-binaphthalene-2,2′-diol (23.5 mg, 0.05 mmol) with a 0.1 M soln of B(OMe)$_3$ in CH$_2$Cl$_2$ (0.5 mL, 0.05 mmol) in THF (0.05 mL) at −78 °C. 2-Bromopropenal (0.0808 mL, 1.0 mmol) and cyclopentadiene (0.332 mL, 4 mmol) were added dropwise to this catalyst soln and, after 4 h, H$_2$O (0.05 mL) was added and the mixture was warmed to 25 °C, dried (MgSO$_4$), filtered, and purified by chromatography (silica gel, hexanes/EtOAc 10:1); yield: 201 mg (>99%); ratio (exo/endo) >99:1; 99% ee.

2.5.1 Diels–Alder Reactions

Ethyl (1R,2R,4R)-Bicyclo[2.2.1]hept-5-ene-2-carboxylate (Table 1, Entry 1); Typical Procedure:[12]

A freshly prepared 0.20 M soln of TfOH in CH_2Cl_2 (0.667 mL, 0.133 mmol) was added dropwise to a soln of the oxazaborolidine precursor (0.160 mmol) in CH_2Cl_2 (2 mL) at −78 °C. After 10–15 min at −78 °C, a colorless homogeneous soln was obtained. Ethyl acrylate (0.667 mmol) was then added, followed by a soln of cyclopentadiene (0.55 mL, 3.33 mmol) in CH_2Cl_2 (0.55 mL). The resulting mixture was stirred for 15 h at −20 °C and then the reaction was quenched by addition of Et_3N (0.2 mL). This mixture was allowed to warm to rt, the solvent was removed on a rotary evaporator, and the residue obtained was purified by chromatography (silica gel, Et_2O/pentane 1:99 to 3:97); yield: 94%; ratio (exo/endo) 3:97; 98% ee.

2.5.1.1.4 Lewis Acid Activated Borane Catalysts

The combination of a chiral oxazaborolidine and tin(IV) chloride leads to an extremely reactive chiral, Lewis acid activated catalyst **11** for the enantioselective Diels–Alder reaction of a variety of dienophiles, such as α,β-unsaturated aldehydes, esters, enones, and quinones, with cyclopentadiene to give the cycloaddition products **12–14** (Scheme 7).[15] The reactivity and asymmetric induction can be maintained even in the presence of a small amount of water as well as other Lewis bases by adding a slightly larger amount of tin(IV) chloride.

Scheme 7 A Lewis Acid Activated Chiral Oxazaborolidine for Asymmetric Diels–Alder Reactions[15]

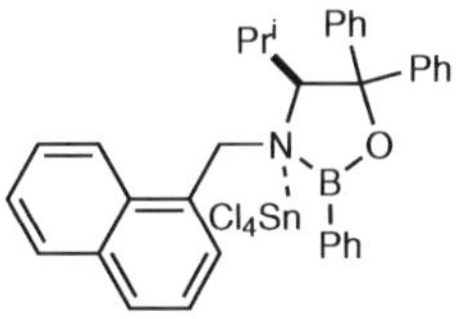

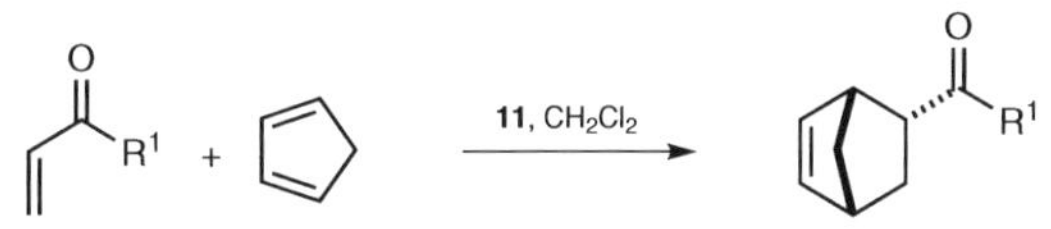

R^1	Ratio (exo/endo)	ee (%)	Yield (%)	Ref
Et	1:99	96	90	[15]
OEt	1:99	95	96	[15]

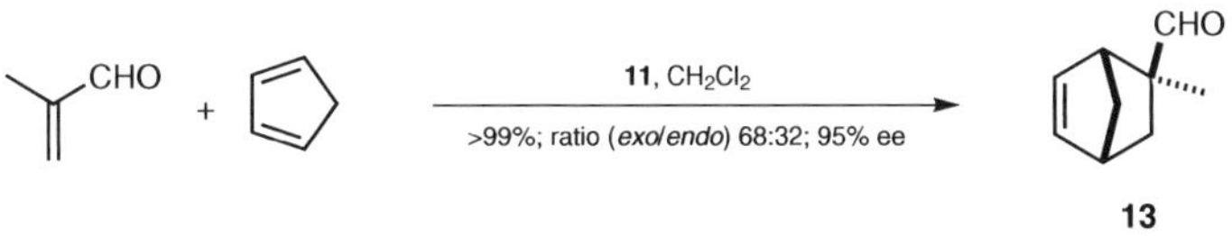

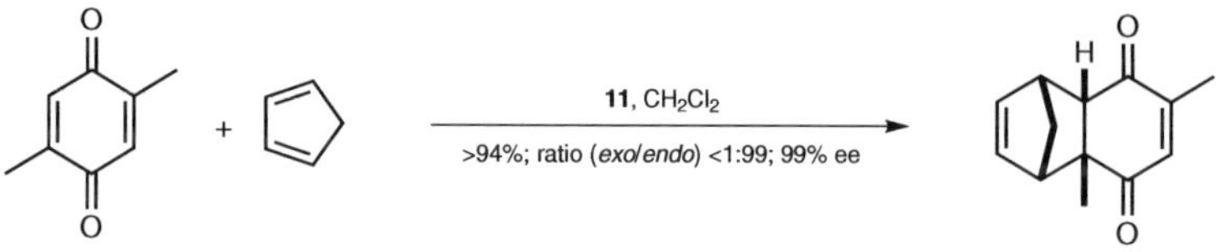

(1R,4S,4aR,8aS)-4a,7-Dimethyl-1,4,4a,8a-tetrahydro-1,4-methanonaphthalene-5,8-dione (14); Typical Procedure:[15]

A 1.0 M soln of the oxazaborolidine in CH_2Cl_2 (1.0 mL, 1.0 mmol) and CH_2Cl_2 (35 mL) were added to a dried, three-necked flask under argon. The resulting mixture was cooled to −78 °C, a 1.0 M soln of $SnCl_4$ in CH_2Cl_2 (1.0 mL, 1.0 mmol) was added dropwise, and the resulting pale yellow soln was stirred for 15 min at the same temperature. A soln of 2,5-dimethylbenzo-1,4-quinone (1.4 g, 10 mmol) in CH_2Cl_2 (5 mL) was then added dropwise and, after 1–2 min, a soln of cyclopentadiene (4.1 mL, 25 mmol) in CH_2Cl_2 (4.1 mL), precooled to −78 °C, was added dropwise over 10 min from a syringe. The resulting mixture was stirred for 90 min at −78 °C and the reaction was then quenched by addition of Et_3N (0.7 mL). This mixture was then allowed to warm slowly to rt, the solvent was removed under reduced pressure, and the crude residue obtained was purified by chromatography (silica gel, hexane/EtOAc 10:1); yield: 94%; 99% ee.

2.5.1.2 Reaction Using Chiral Aluminum Complexes

Chiral aluminum complex **15** (10 mol%), derived from a diamide ligand, can be used as a catalyst in the enantioselective Diels–Alder reaction of cyclopentadienes and 3-acryloyloxazolidin-2-ones to give the corresponding adducts **16** in high yield and with enantiomeric excesses in the range of 91–95% (Scheme 8).[16]

Scheme 8 Asymmetric Diels–Alder Reactions Catalyzed by a Diamide–Aluminum Complex[16]

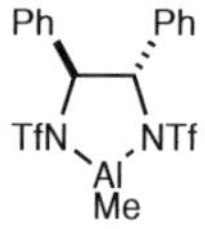

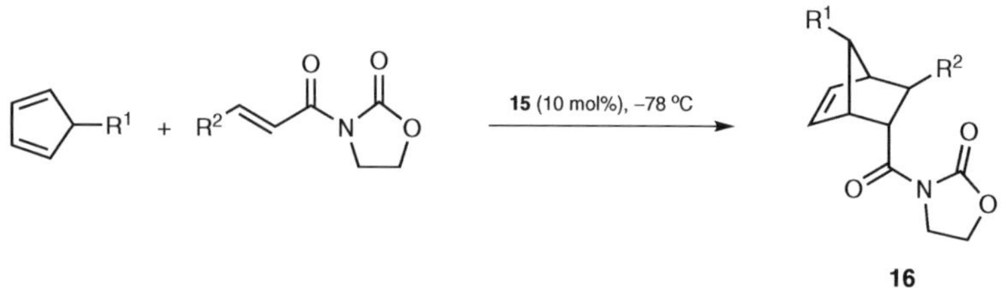

R^1	R^2	Ratio (endo/exo)	ee (%)	Yield (%)	Ref
H	H	>50:1	91	92	[16]
H	Me	96:4	94	88	[16]
CH_2OBn	H	>99:1	95	94	[16]

2.5.1 Diels–Alder Reactions

The absolute stereochemical preference in this Diels–Alder reaction is believed to be the result of catalyst **15** binding to the acryloyl carbonyl of the dienophile via the lone pair *anti* to nitrogen, fixing the acryloyl group in the *s-trans* conformation (see transition state **17** where $R^1 = CH_2OBn$; $R^2 = H$) (Scheme 9). Moreover, a 2,2′-diphenyl-3,3′-biphenanthrene-4,4′-diol (VAPOL) derived chiral aluminum complex is also efficient for the Diels–Alder reaction of cyclopentadiene and 2-methylpropenal.[17]

Scheme 9 Transition State for Diamide–Aluminum Complex Catalyzed Asymmetric Diels–Alder Reactions[16]

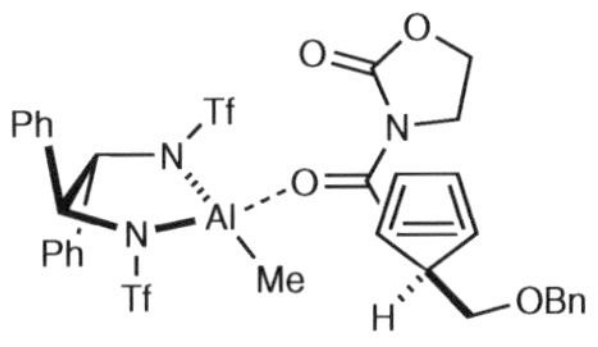

3-{(1R,2S,4S,7R)-7-[(Benzyloxy)methyl]bicyclo[2.2.1]hept-5-ene-2-carbonyl}oxazolidin-2-one (16, $R^1 = CH_2OBn$; $R^2 = H$); Typical Procedure:[16]

Complex **15** (0.5 mmol) was dissolved in freshly dried CH_2Cl_2 (5 mL) and the resulting soln was cooled to −78 °C. 3-(Acryloyl)oxazolidin-2-one (705.7 mg, 5 mmol) in CH_2Cl_2 (5 mL) was then added and the resulting homogeneous soln was stirred for 5 min before 5-[(benzyloxy)methyl]cyclopenta-1,3-diene (10 mmol) in CH_2Cl_2 (5 mL) was introduced slowly. The resulting mixture was stirred at −78 °C for 10 h and monitored by TLC. On completion of the reaction, 1 M HCl (10 mL) was added and the resulting mixture was allowed to warm to rt, the phases were separated, and the aqueous phase was extracted with CH_2Cl_2. The combined organic phases were washed with aq Na_2CO_3 and brine, dried ($MgSO_4$), and filtered. The solvents were then removed and the residue obtained was purified by chromatography (silica gel, hexanes/EtOAc 2:1) to afford a colorless oil; yield: 1.538 g (94%).

2.5.1.3 Reaction Using Chiral Indium Complexes

A chiral indium catalyst, generated by activation of an (S)-1,1′-binaphthalene-2,2′-diol [(S)-BINOL]/indium(III) chloride precatalyst with allyltributylstannane, is effective for the asymmetric Diels–Alder reaction of a variety of cyclic and acyclic dienes with 2-methylpropenal and 2-bromopropenal to afford cycloadducts such as **18** and **19** in good yield and with excellent enantioselectivity (Scheme 10).[18] This reaction can also be carried out in aqueous media to give the products with a slightly lower yield or enantiomeric excess.

Scheme 10 Asymmetric Diels–Alder Reactions Catalyzed by a Chiral Indium Complex[18]

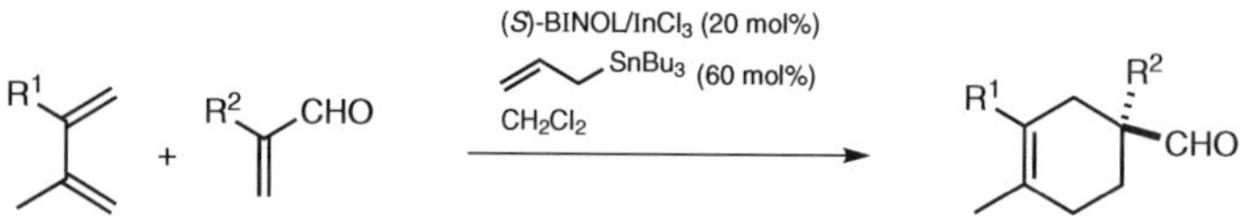

R^1	R^2	Temp	ee (%)	Yield (%)	Ref
H	Br	–20 °C	96	70	[18]
Me	Me	rt	98	63	[18]

R^1	ee (%)	Yield (%)	Ref
Br	98	71	[18]
Me	98	72	[18]

64%; ratio (exo/endo) 98:2; 94% ee

As shown in the transition state **20**, the aromatic rings of (S)-1,1′-binaphthalene-2,2′-diol effectively screen the re-face of the complexed s-trans-α,β-unsaturated aldehyde from attack by the diene component, which facilitates addition of the diene to the si-face of the double bond (Scheme 11).[18]

Scheme 11 Transition State for an Indium-Catalyzed Asymmetric Diels–Alder Reaction[18]

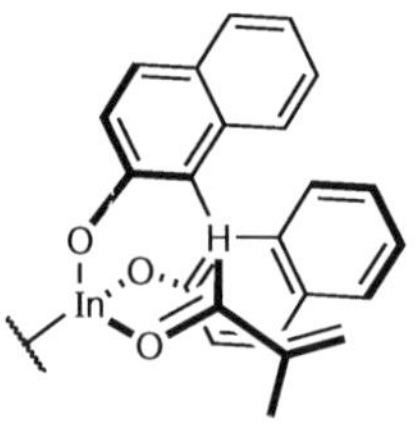

(R)-1-Bromo-4-methylcyclohex-3-ene-1-carbaldehyde (18, R^1 = H; R^2 = Br);
Typical Procedure:[18]
$InCl_3$ (22 mg, 0.1 mmol) was added to an oven-dried, round-bottomed flask and the solid was azeotropically dried twice with anhyd THF (2 × 2 mL) prior to the addition of CH_2Cl_2 (1.5 mL). (S)-BINOL (31 mg, 0.11 mmol) was added and the resulting mixture was stirred

under N$_2$ at rt for 2 h. Allyltributylstannane (0.093 mL, 0.3 mmol) was added and the resulting mixture was stirred for 10 min, followed by addition of H$_2$O (10.5 µL) to afford a white suspension. The preformed catalyst was then cooled to −20 °C for 15 min, 2-bromopropenal (67.5 mg, 0.5 mmol) and isoprene (1.5 mmol) were added successively, and the resulting mixture was stirred at −20 °C for 20 h. The reaction was then quenched by addition of sat. aq NaHCO$_3$ (5 mL) and the mixture was extracted with Et$_2$O (3 × 10 mL). The combined organic extracts were washed with brine, dried (MgSO$_4$), filtered, and concentrated under reduced pressure. The residue obtained was purified by chromatography (silica gel) to afford a colorless solid; yield: 70%; 96% ee.

2.5.1.4 Reaction Using Chiral Copper Complexes

2.5.1.4.1 Bis(oxazoline)–Copper Catalysts

The bis(oxazoline)–copper complex **21** can act as a Lewis acid to promote the enantioselective Diels–Alder reaction of 3-propenoyloxazolidin-2-one and a range of substituted dienes to give the corresponding cycloadducts **22–24** with high enantioselectivity (Scheme 12).[19] It has been found that the structure of the catalyst counterion and the use of a chelating dienophile are critically important for this reaction. Moreover, a variety of dienophiles including acrylimides, α′-hydroxy enones, 2-alkenoylpyridine 1-oxides, pyrazolidinones, and α′-arylsulfonyl enones are suitable achiral templates for asymmetric Diels–Alder reactions.[20–23]

Scheme 12 Asymmetric Diels–Alder Reactions Catalyzed by Chiral Copper Complexes[19]

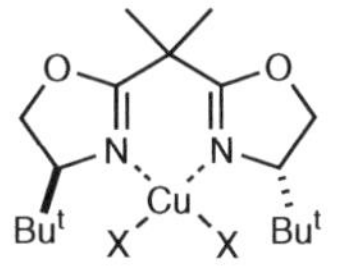

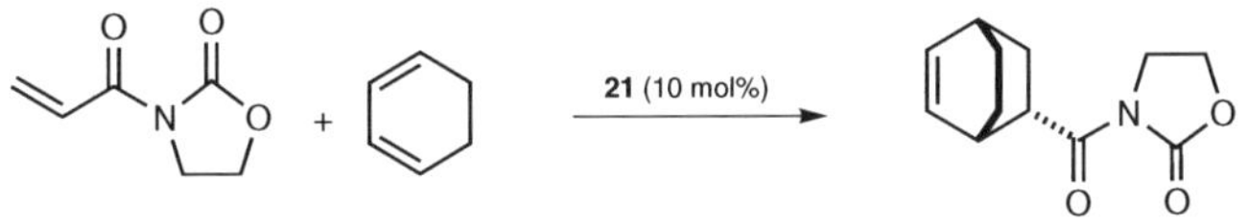

X	Ratio (endo/exo)	ee (%)	Yield (%)	Ref
OTf	98:2	82	90	[19]
SbF$_6$	95:5	93	90	[19]

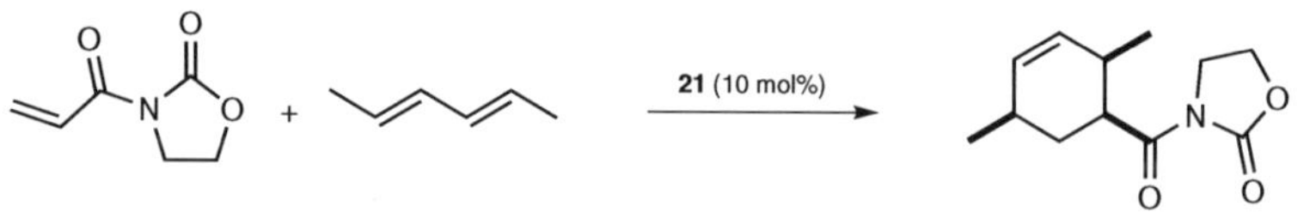

X	Ratio (cis/trans)	ee (%)	Yield (%)	Ref
OTf	78:22	84	66	[19]
SbF$_6$	77:23	93	59	[19]

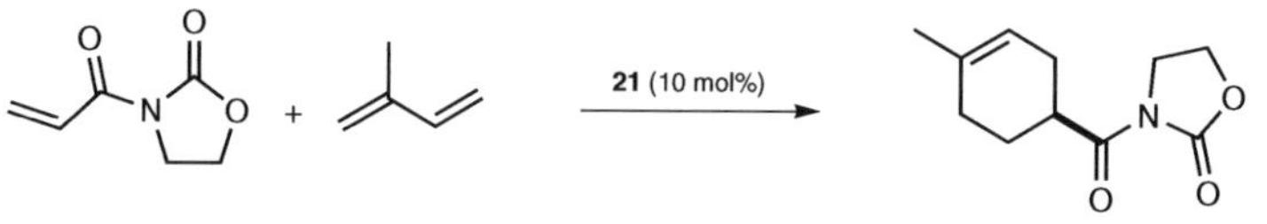

X	Ratio (1,4/1,3)^a	ee (%)	Yield (%)	Ref
OTf	97:3	60	95	[19]
SbF$_6$	96:4	59	81	[19]

^{a} Ratio of 1,4-/1,3-substitution on the cyclohexene ring.

3-{(1S,2S,4S)-Bicyclo[2.2.2]oct-5-ene-2-carbonyl}oxazolidin-2-one (22); Typical Procedure:[19]

Cyclohexa-1,3-diene (1.0 mL, 5.0 mmol) was added to a soln of 3-propenoyloxazolidin-2-one (71 mg, 0.5 mmol) in CH$_2$Cl$_2$ and the resulting soln was immediately cooled to −78 °C. A 0.05 M soln of catalyst **21** (X = SbF$_6$) in CH$_2$Cl$_2$ (1 mL, 0.05 mmol) was added dropwise via syringe and the reaction flask was then placed in a 25 °C oil bath for 5 h. EtOAc (5 mL) was then added and the resulting soln was loaded onto a plug of silica gel (60 mL), eluted with EtOAc (250 mL), and concentrated under reduced pressure; yield: 199 mg (90%); ratio (endo/exo) 95:5; 93% ee.

2.5.1.4.2 Bis(sulfinyl)imidoamidine–Copper Catalysts

A readily available bis(sulfinyl)imidoamidine ligand **25** is effective for the copper-catalyzed asymmetric Diels–Alder reaction of 3-propenoyloxazolidin-2-ones with cyclopentadiene to provide cycloadducts **26** with high levels of asymmetric induction (Scheme 13).[24] Notably, the copper complex exists as a unique M$_2$L$_4$ helicate in the solid state, while in solution ligand **25** predominately coordinates with copper through the sulfinyl oxygen atom. The use of monodentate propenal as the dienophile shows markedly decreased selectivity.

Scheme 13 Bis(sulfinyl)imidoamidine–Copper Catalyzed Asymmetric Diels–Alder Reactions[24]

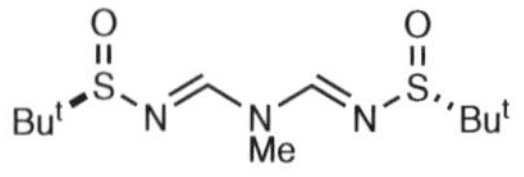

2.5.1 Diels–Alder Reactions

R¹	Time (h)	Temp (°C)	Ratio (endo/exo)	ee (%)	Yield (%)	Ref
H	0.1	−78	>99:1	>98	96	[24]
Me	8	−40	98:2	97	76	[24]
Ph	16	0	95:5	94	58	[24]
CO_2Et	2	−78	97:3	96	85	[24]

3-[(1S,2S,4S)-Bicyclo[2.2.1]hept-5-ene-2-carbonyl]oxazolidin-2-one (26, R¹ = H); Typical Procedure:[24]

Ligand **25** (0.55 mmol) was added to a 10-mL round-bottomed flask (protected from light) charged with $CuCl_2$ (0.05 mmol) and $AgSbF_6$ (0.05 mmol) in CH_2Cl_2 (2.5 mL). This was followed by addition of activated 3-Å molecular sieves (50 mg) and the resulting green mixture was stirred for 1 h. The 3-propenoyloxazolidin-2-one (0.5 mmol) was then added and the mixture was immediately cooled to −78 °C and freshly distilled cyclopentadiene (0.3 mL, ~5 mmol) was added. Upon completion of the reaction, the mixture was filtered directly through a plug of silica gel (hexanes/EtOAc 1:1). The solvent was removed and the residue obtained was purified by chromatography (silica gel, CH_2Cl_2); yield: 96%; dr >99:1; >98% ee.

2.5.1.5 Reaction Using Chiral Titanium Complexes

The molecular-sieve-free, chiral 1,1′-binaphthalene-2,2′-diol-derived titanium complex **27** is employed for the asymmetric Diels–Alder reaction of 5-hydroxynaphthalene-1,4-dione or 2-methylpropenal with 1,3-dienol ethers or esters to provide the corresponding cycloadducts with high levels of *endo* selectivity and enantioselectivity (Table 2, entries 1–3).[25] However, if molecular sieves are involved, a significantly lower enantioselectivity is obtained. A surprisingly interesting phenomenon in this reaction is that the molecular-sieve-free complex **27** exhibits not only a linear relationship between the enantiomeric excess of the catalyst and the product but also an asymmetric amplification depending simply on the manner of mixing of (*R*)-**27** with (*S*)-**27** or *rac*-**27**. Moreover, the chiral, helical titanium complex **28** has also been utilized as an efficient chiral template for the conformational fixation of α,β-unsaturated aldehydes, thereby allowing efficient enantiofacial recognition of the substrates for achievement of uniformly high asymmetric induction in the asymmetric Diels–Alder reaction with dienes, regardless of temperature (Table 2, entries 4 and 5).[26]

Table 2 Chiral Titanium Complex Catalyzed Asymmetric Diels–Alder Reactions[25,26]

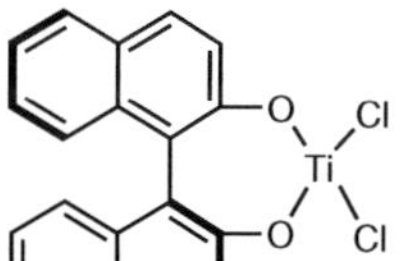

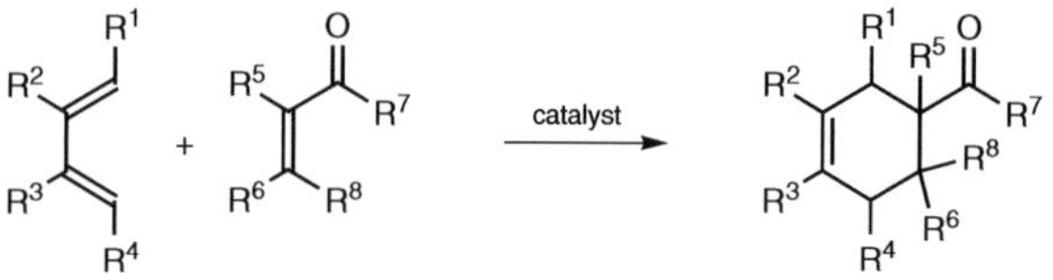

Entry	Starting Material		Catalyst	Product	Ratio (endo/exo)	ee (%)	Yield (%)	Ref
	Diene	Dienophile						
1	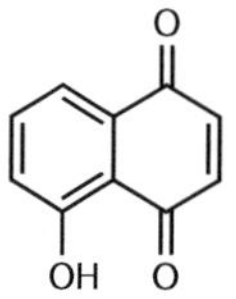 5-hydroxynaphthalene-1,4-dione	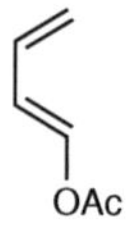 OAc-butadiene	27	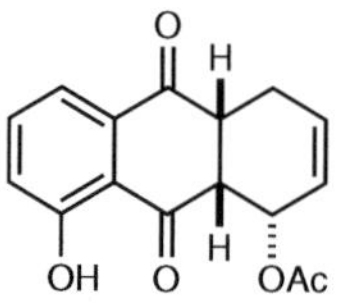 hexahydroanthracenyl acetate	–	96	86	[25]
2	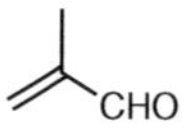 methacrolein	 OMe-diene	27	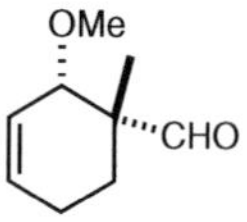 OMe-CHO cyclohexene	93:7	85	40	[25]
3	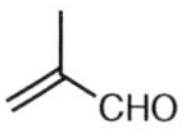 methacrolein	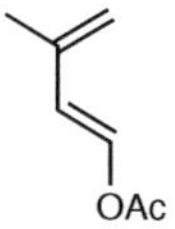 OAc-diene	27	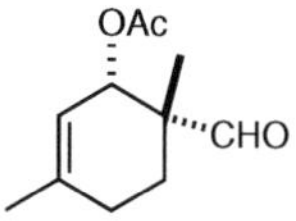 OAc-CHO cyclohexene	89:11	80	98	[25]
4	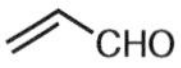 acrolein	 cyclopentadiene	28	 norbornene-CHO	85:15	96	70	[26]
5	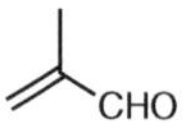 methacrolein	 cyclopentadiene	28	 norbornene-CHO	1:99	94	75	[26]

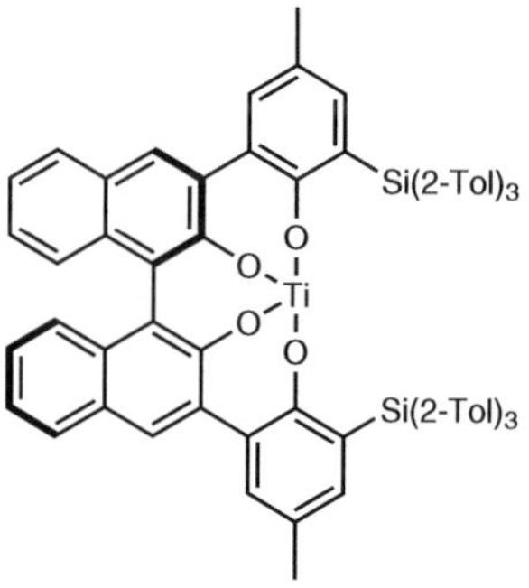

(1S,4aR,9aR)-8-Hydroxy-9,10-dioxo-1,4,4a,9,9a,10-hexahydroanthracen-1-yl Acetate (Table 2, Entry 1); Typical Procedure:[25]

The molecular-sieve-free complex **27** (43 mg, 0.1 mmol) was dissolved in toluene (3 mL) and freshly distilled 5-hydroxynaphthalene-1,4-dione (174 mg, 1 mmol) and a freshly dis-

2.5.1 Diels–Alder Reactions

tilled soln of 1-acetoxybuta-1,3-diene (356 mg, 3.18 mmol) in CH_2Cl_2 (1 mL) were added at rt. The resulting mixture was stirred for 18 h at rt and then diluted with Et_2O (5 mL). The reaction was then quenched by the addition of sat. aq $NaHCO_3$ (10 mL) and this soln was filtered through a pad of Celite and Florisil. The filtrate was extracted with Et_2O (3×) and the combined organic layers were washed with brine. The organic extracts were then dried ($MgSO_4$), filtered, and concentrated under reduced pressure. The residue obtained was purified by chromatography (silica gel, hexane/EtOAc); yield: 86%; 96% ee.

(1S,2S,4S)-Bicyclo[2.2.1]hept-5-ene-2-carbaldehyde (Table 2, Entry 4); Typical Procedure:[26]
$Ti(OiPr)_4$ (0.029.8 mL, 0.1 mmol) was added to a soln of the silylated BINOL derivative (110 mg, 0.1 mmol) in dry CH_2Cl_2 (20 mL) at rt under argon. The resulting yellow soln was heated and CH_2Cl_2 (16 mL) was removed azeotropically with iPrOH to leave a soln of **28** in CH_2Cl_2. This soln was cooled to −78 °C and freshly distilled propenal (0.067 mL, 1 mmol) was added dropwise, followed by cyclopentadiene (0.163 mL, 2 mmol). The resulting mixture was stirred at −78 °C for 3.5 h, poured into 1 M HCl, and extracted with CH_2Cl_2. The CH_2Cl_2 extracts were dried (Na_2SO_4), filtered, and concentrated. The residue obtained was purified by chromatography (silica gel, Et_2O/pentane 1:30); yield: 85.5 mg (70%); ratio (endo/exo) 85:15; 96% ee.

2.5.1.6 Reaction Using Chiral Chromium Complexes

A highly enantioselective Diels–Alder reaction of a substituted 1-aminobuta-1,3-diene with various α,β-unsaturated aldehydes has been realized, using chiral salen–chromium complex **29**, to produce the corresponding functionalized cyclohexene derivatives **30** bearing a chiral quaternary center in excellent yield and with excellent enantiomeric excess (Scheme 14).[27]

Scheme 14 Chiral Chromium Complex Catalyzed Asymmetric Diels–Alder Reactions[27]

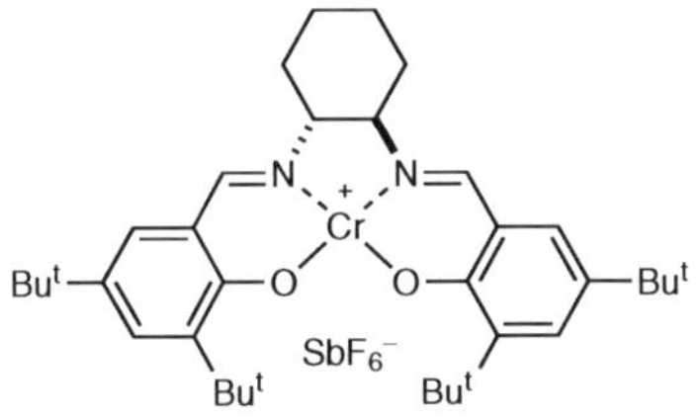

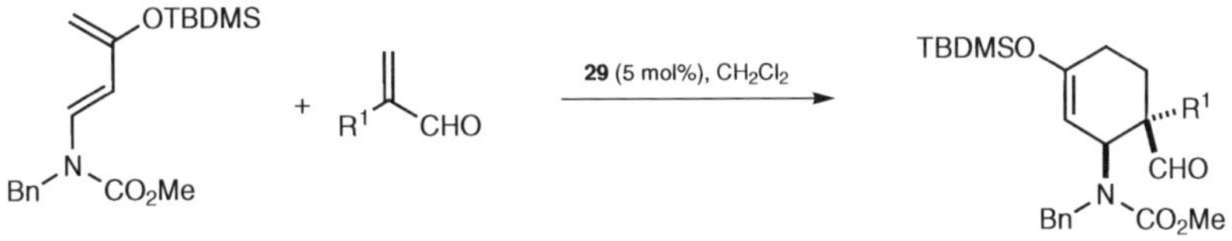

R^1	Time (d)	ee (%)	Yield (%)	Ref
Me	2	97	93	[27]
Et	2	97	91	[27]
iPr	5	>97	92	[27]
$(CH_2)_2OTBDMS$	2	95	93	[27]
OTBDMS	2	>97	86	[27]

As shown in the transition state **31**, the coordination of chromium with the carbonyl lone pair *anti* to the R[1] group is expected to activate the dienophile. The diene can then approach the dienophile from the more open surface on the scaffold, which is believed to be over the imine linkages and away from the bulky *tert*-butyl groups (Scheme 15).[27]

Scheme 15 Transition State in a Chromium-Catalyzed Asymmetric Diels–Alder Reaction[27]

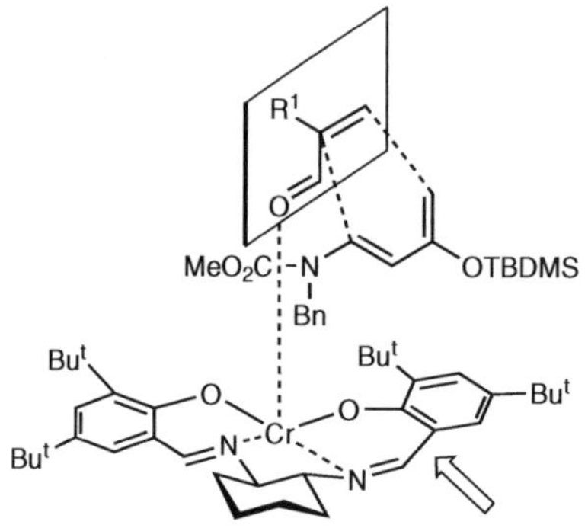

Methyl Benzyl[(1S,6R)-3-(*tert*-butyldimethylsiloxy)-6-formyl-6-methylcyclohex-2-en-1-yl]carbamate (30, R¹ = Me); Typical Procedure:[27]

2-Methylpropenal (2 mmol) was added to a stirred mixture of catalyst **29** (0.05 mmol) and oven-dried, powdered 4-Å molecular sieves (0.8 g) in CH_2Cl_2 (1 mL) cooled to −40 °C. The carbamate (1 mmol) was added in one portion, and the resulting mixture was stirred at −40 °C until completion of the reaction (2 d). Any solid material was removed by filtration through Celite, which was then washed with CH_2Cl_2. The filtrate was concentrated and the residue obtained was purified by chromatography (silica gel); yield: 93%; 97% ee.

2.5.1.7 Reaction Using Chiral Cobalt Complexes

The cobalt complex **32** is effective for the asymmetric Diels–Alder reaction of a 1-aminobuta-1,3-diene with various 2-substituted propenals. Simple concentration of the reaction mixture followed by trituration to remove the remaining reactants affords the pure cycloadducts **33** in high yield and with high enantiomeric excess (Scheme 16).[28] Importantly, the reaction is conveniently carried out at room temperature under an air atmosphere with a very low catalyst loading and minimal solvent, which are desirable conditions for industrial application.

Scheme 16 Chiral Cobalt Complex Catalyzed Asymmetric Diels–Alder Reactions[28]

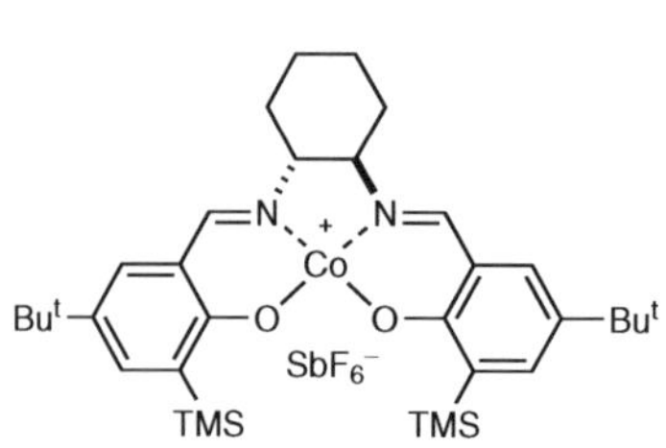

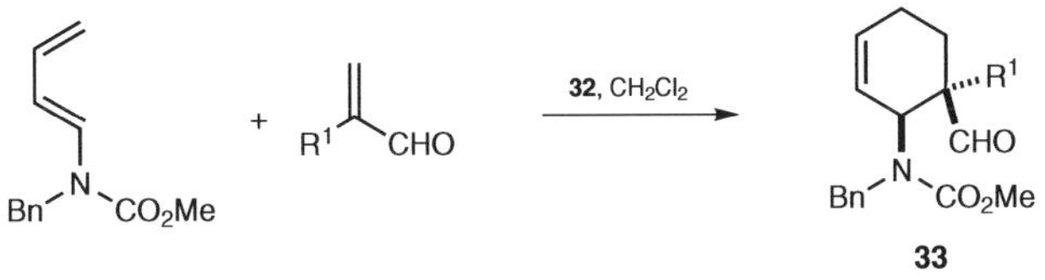

2.5.1 Diels–Alder Reactions

R¹	mol% of 32	ee (%)	Yield (%)	Ref
Me	0.1	98	98	[28]
Et	0.1	>97	93	[28]
(CH$_2$)$_2$OTBDMS	0.5	>97	100	[28]
H	0.1	85	100	[28]

Methyl Benzyl[(1S,6R)-6-formyl-6-methylcyclohex-2-enyl]carbamate (33, R¹ = Me);
Typical Procedure:[28]
2-Methylpropenal (10 mmol) was added to a stirred soln of catalyst **32** (0.005 mmol) in CH$_2$Cl$_2$ (2.5 mL) at rt. The carbamate (5.0 mmol) was added in one portion and the mixture was stirred at rt for 16 h. No purification beyond concentration was required and this gave the pure cycloadduct; yield: 98%; 98% ee.

2.5.1.8 Reaction Using Chiral Ruthenium Complexes

The chiral ruthenium complex **34** is used as a Lewis acid catalyst for the asymmetric Diels–Alder reaction between dienes (such as cyclopentadiene, methylcyclopentadiene, isoprene, and 2,3-dimethylbuta-1,3-diene) and a wide range of α,β-unsaturated ketones (such as methyl vinyl ketone, ethyl vinyl ketone, divinyl ketone, 1-bromovinyl methyl ketone, and 1-chlorovinyl methyl ketone) to afford the corresponding cycloadducts such as **35** in 50–90% yield and with up to 96% enantiomeric excess (Scheme 17).[29] This is the first example of a one-point-binding, transition metal Lewis acid catalyst capable of coordinating and activating an α,β-unsaturated ketone for an enantioselective Diels–Alder reaction.

Scheme 17 Chiral Ruthenium Complex Catalyzed Asymmetric Diels–Alder Reactions[29]

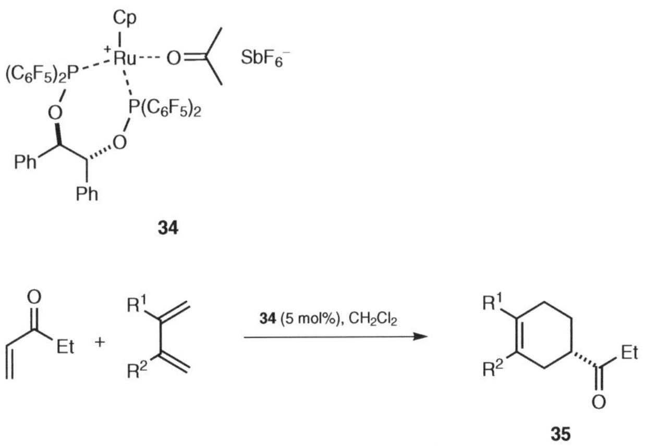

R¹	R²	1,4-Isomer (%)	ee (%)	Yield (%)	Ref
Me	Me	–	91	76	[29]
Me	H	91	92	61	[29]

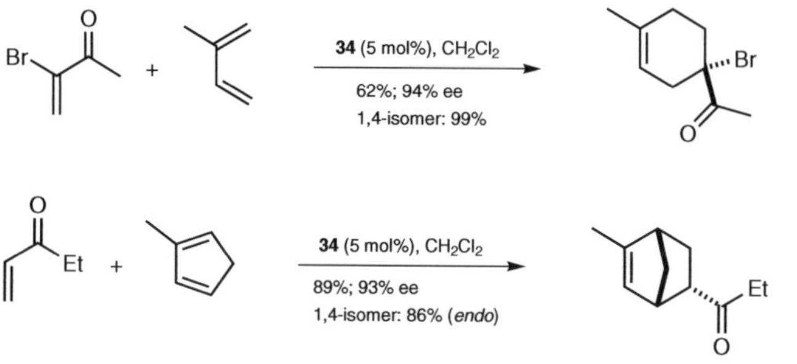

(S)-1-(3,4-Dimethylcyclohex-3-enyl)propan-1-one (35, $R^1 = R^2 = Me$); Typical Procedure:[29]
CH_2Cl_2 (0.30 mL), 2,6-lutidine (4 µL, 0.035 mmol), decane (33 µL, 0.17 mmol), and ethyl vinyl ketone (0.66 mmol) were added to a Schlenk tube charged with catalyst **34** (46 mg, 0.033 mmol). The resulting mixture was stirred at −20 °C and 2,3-dimethylbuta-1,3-diene (0.97 mmol) in CH_2Cl_2 (0.35 mL) was added dropwise over 10 min. The reaction was monitored by GC and, on completion, hexane (8 mL) was added. This mixture was filtered through a Celite plug and the solvent was removed under reduced pressure. The residue obtained was dissolved in CH_2Cl_2 and passed through a silica gel plug. The solvent was then removed under reduced pressure; yield: 76%; 91% ee.

2.5.1.9 Reaction Using Chiral Rare Earth Complexes

Chiral ytterbium catalyst **36**, prepared from (R)-1,1′-binaphthalene-2,2′-diol, ytterbium(III) trifluoromethanesulfonate, and a tertiary amine (cis-1,2,6-trimethylpiperidine), is effective for the enantioselective Diels–Alder reaction of 3-(propenoyl)oxazolidin-2-ones with cyclopentadiene to yield predominantly the endo-cycloadducts, e.g. **37**, with up to 93% enantiomeric excess (Scheme 18). Tertiary amines play a very important role for both diastereo- and enantioselectivity. It has been noted that chiral catalysts with reverse selectivity can be prepared using the same chiral source and a choice of achiral ligands.[30] The same catalyst is also effective for an asymmetric inverse-electron-demand Diels–Alder reaction of pyran-2-one derivatives with electron-rich dienophiles.[31]

Scheme 18 Ytterbium-Catalyzed Asymmetric Diels–Alder Reactions of Cyclopentadiene[30]

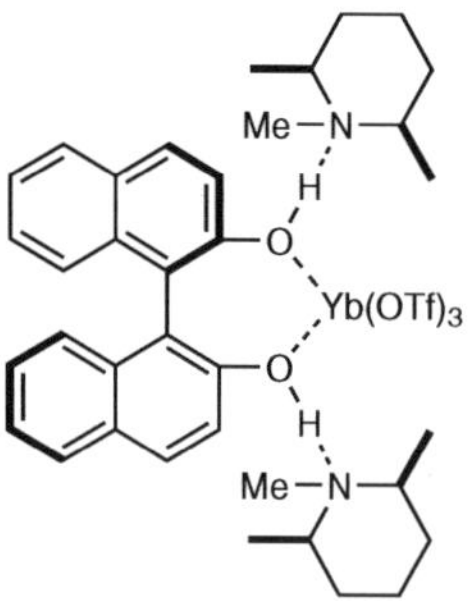

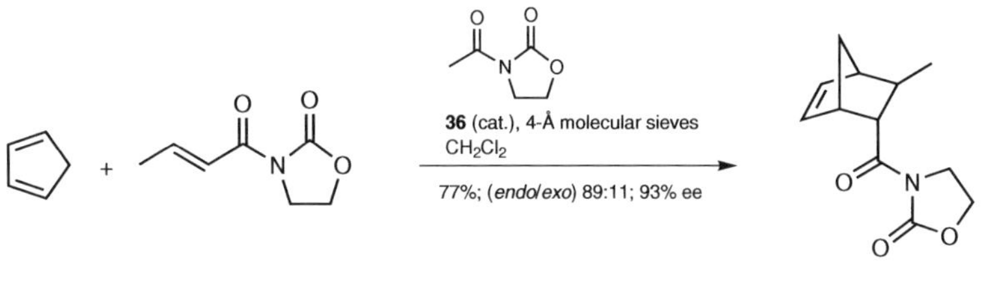

2.5.1 Diels–Alder Reactions

37: 77%; (endo/exo) 89:11; 93% ee
(with **36** (cat.), 4-Å molecular sieves, CH$_2$Cl$_2$)

69%; (endo/exo) 88:12; 69% ee
(with **36** (cat.), 4-Å molecular sieves, CH$_2$Cl$_2$)

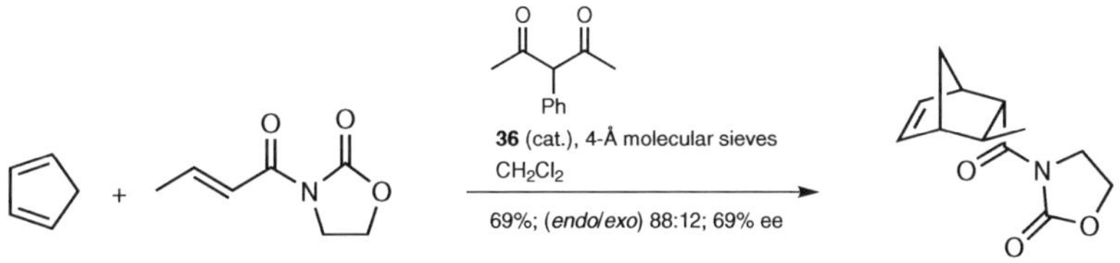

A highly enantioselective Diels–Alder reaction of a Danishefsky-type diene with electron-deficient alkenes such as 3-(propenoyl)oxazolidin-2-ones has also been realized to give cycloadducts **39** using a catalyst obtained from ligand **38** and ytterbium(III) trifluoromethanesulfonate with 1,8-diazabicyclo[5.4.0]undec-7-ene as an additive (Scheme 19).[32] Apparent positive nonlinear effects in this reaction may suggest the possible formation of a reservoir of nonreactive aggregates.

Scheme 19 Ytterbium-Catalyzed Asymmetric Diels–Alder Reactions of a Danishefsky-Type Diene[32]

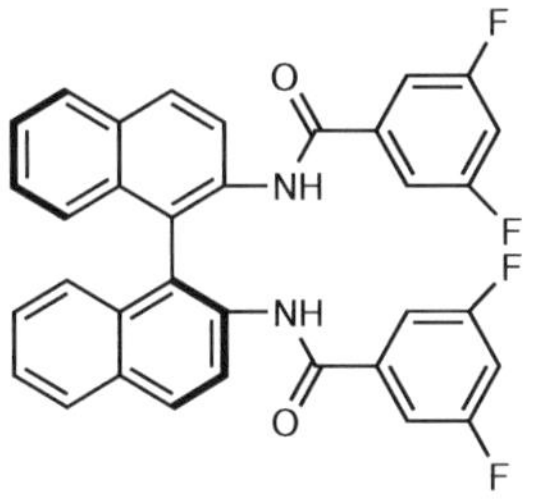

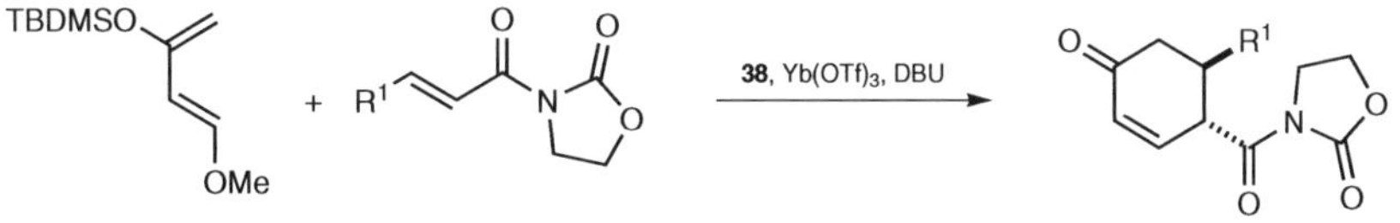

R^1	ee (%)	Yield (%)	Ref
Me	94	94	[32]
CO$_2$Me	92	93	[32]

3-[(1S,2S,3R,4R)-3-Methylbicyclo[2.2.1]hept-5-ene-2-carbonyl]oxazolidin-2-one (37);
Typical Procedure:[30]
cis-1,2,6-Trimethylpiperidine (0.24 mmol) in CH$_2$Cl$_2$ (0.75 mL) was added to a mixture of Yb(OTf)$_3$ (0.10 mmol), (R)-(+)-BINOL (0.12 mmol), and 4-Å molecular sieves (125 mg) at 0 °C. The resulting mixture was stirred for 30 min at this temperature, and then 3-acetyl-

oxazolidin-2-one (0.1 mmol) in CH_2Cl_2 (0.25 mL) was added. (E)-3-(But-2-enoyl)oxazolidin-2-one (0.5 mmol) in CH_2Cl_2 (0.25 mL) and cyclopentadiene (1.5 mmol) in CH_2Cl_2 (0.25 mL) were added successively to the resulting soln of catalyst **36** at 0 °C. The resulting mixture was stirred for 20 h and then H_2O was added to quench the reaction. Any insoluble materials were then removed by filtration and, after the usual workup, the crude product was purified by chromatography (silica gel); yield: 77%; ratio (*endo/exo*) 89:11; 93% ee.

3-[(1S,6R)-6-Methyl-4-oxocyclohex-2-ene-1-carbonyl]oxazolidin-2-one (39, R^1 = Me); Typical Procedure:[32]

A mixture of $Yb(OTf)_3$ (18.6 mg, 0.030 mmol) and ligand **38** (20.3 mg, 0.036 mmol) was dried under reduced pressure (90 °C/<0.1 Torr) for 0.5 h with stirring and then allowed to cool to rt. CH_2Cl_2 (2.0 mL) and DBU (0.011 mL, 0.074 mmol) were added successively and the resulting mixture was stirred for 2 h. (E)-3-(But-2-enoyl)oxazolidin-2-one (93 mg, 0.60 mmol) in CH_2Cl_2 (1.0 mL) and the (E)-1-methoxy-3-(*tert*-butyldimethylsiloxy)buta-1,3-diene (0.30 mL, 1.20 mmol) were added successively at 0 °C, and this mixture was stirred for 5 h at the same temperature. H_2O (1 mL) was then added to quench the reaction, and any insoluble materials were removed by filtration through a pad of Celite. The usual workup gave a residue, which was dissolved in a mixture of 1,2-dichloroethane (6 mL) and TFA (0.12 mL). The resulting soln was stirred at 60 °C for 0.5 h and the reaction was then quenched with sat. aq $NaHCO_3$ (2 mL). This mixture was extracted with CH_2Cl_2 and the usual workup gave a residue, which was purified by chromatography (silica gel, hexane/EtOAc 1:1) to give a colorless solid; yield: 125.6 mg (94%); 94% ee.

2.5.2 Oxa-Diels–Alder Reactions

Generally, two different mechanistic pathways are taken into account for the Lewis acid catalyzed oxa-Diels–Alder reaction: one is a traditional concerted Diels–Alder cycloaddition, and the other is a stepwise Mukaiyama aldol reaction pathway (Scheme 20). For example, the reaction of benzaldehyde with 1-methoxy-2-methyl-3-(trimethylsiloxy)penta-1,3-diene using boron trifluoride as a Lewis acid catalyst proceeds via the stepwise Mukaiyama pathway, while the reaction proceeds via the concerted Diels–Alder route using zinc(II) chloride or lanthanides.[33] In the stepwise Mukaiyama pathway the major product has a 2,3-*trans* configuration, while in the concerted Diels–Alder pathway the reaction affords the 2,3-*cis*-product exclusively. In general, the reaction pathway in hetero-Diels–Alder reactions using chiral Lewis acid catalysts is proposed on the basis of the intermediate formed and the stereochemistry observed.

Scheme 20 Reaction Pathways for an Oxa-Diels–Alder Reaction

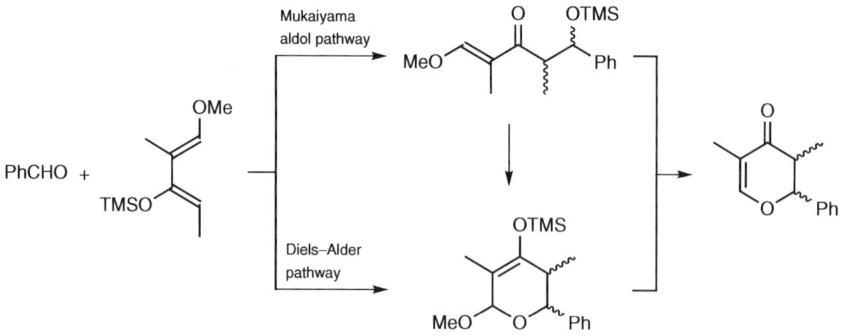

2.5.2 Oxa-Diels–Alder Reactions

2.5.2.1 Reaction Using Chiral Aluminum Complexes

The enantioselective oxa-Diels–Alder reaction between various Danishefsky-type dienes and aldehydes, catalyzed by chiral organoaluminum complex **40**, which is derived from a 3,3′-disubstituted 1,1′-binaphthalene-2,2′-diol and trimethylaluminum, gives the desired 2,3-dihydro-4H-pyran-4-one derivatives **41** in high yield and with up to 95% enantiomeric excess (Scheme 21).[34]

Scheme 21 Chiral Aluminum Complex Catalyzed Asymmetric Oxa-Diels–Alder Reactions[34]

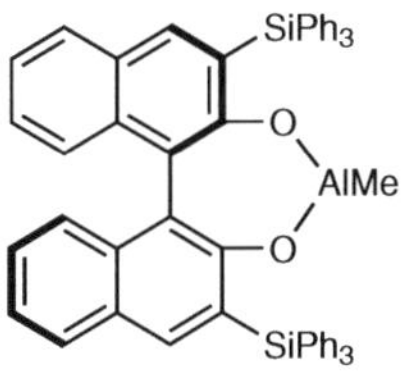

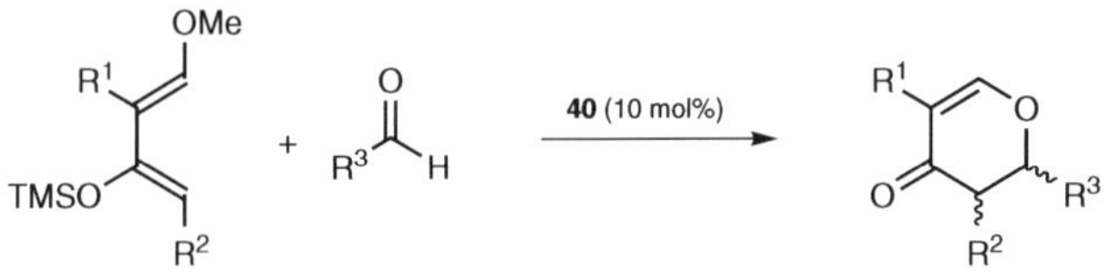

R^1	R^2	R^3	Ratio (cis/trans)	ee (%)	Yield[a] (%)	Ref
Me	Me	Ph	11:1	95	84	[34]
H	Me	Ph	19:1	95	96	[34]
OAc	H	Ph	–	86	83	[34]
Me	Me	Cy	100:0	91	65	[34]
Me	Me	Bu	3.5:1	86	80	[34]

[a] Combined yield of cis- and trans-products.

As shown in transition state **42**, the organoaluminum complex **40** forms a stable 1:1 complex with the aldehyde (e.g., benzaldehyde) (Scheme 22).[34] The diene approaches the aldehyde with an *endo*-alignment between the phenyl substituent and the diene in order to minimize steric repulsion between the diene and the front triphenylsilyl moiety, thereby yielding predominately the *cis*-adduct.

Scheme 22 Transition State for Aluminum-Catalyzed Asymmetric Oxa-Diels–Alder Reactions[34]

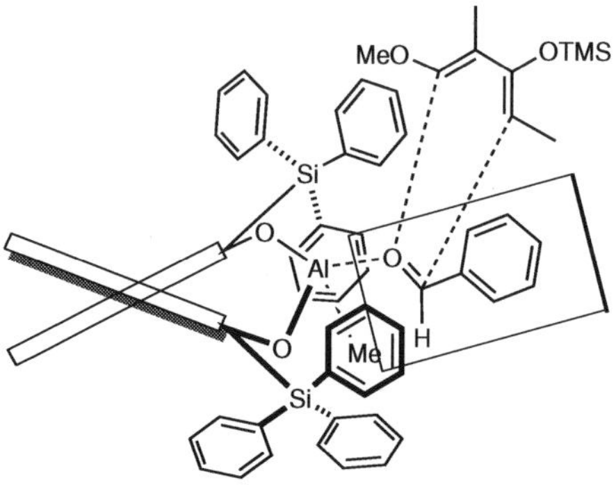

42

(2R,3R)-3,5-Dimethyl-2-phenyl-2,3-dihydro-4H-pyran-4-one (41, R¹ = R² = Me; R³ = Ph); Typical Procedure:[34]

CAUTION: *Neat trimethylaluminum is highly pyrophoric.*

A 0.5 M soln of Me₃Al in hexane (0.2 mL, 0.1 mmol) was added to a degassed soln of (R)-(+)-3,3′-bis(triphenylsilyl)-1,1′-binaphthalene-2,2′-diol (88 mg, 0.10 mmol) in dry toluene (5.0 mL), and the resulting mixture was stirred at rt for 1 h and then cooled to −20 °C. PhCHO (0.102 mL, 1.0 mmol) and (1E,3Z)-1-methoxy-2-methyl-3-(trimethylsiloxy)penta-1,3-diene (220 mg, 1.1 mmol) were added and the mixture was stirred for 2 h at −20 °C and then poured into 10% aq HCl and extracted with Et₂O. The combined extracts were concentrated under reduced pressure to give a crude residue, which was redissolved in CH₂Cl₂ and treated with TFA (0.092 mL, 1.2 mmol) at 0 °C for 1 h. This mixture was then poured into sat. aq NaHCO₃, extracted with CH₂Cl₂, and dried (Na₂SO₄). Removal of the solvent followed by chromatography of the residue (silica gel, Et₂O/hexane 1:3) gave the *cis* product; yield: 156 mg (77%); 95% ee and the *trans*-product; yield: 14 mg (7%); ratio (*cis/trans*) 11:1.

2.5.2.2 Reaction Using Chiral Indium Complexes

The chiral complex obtained from N,N′-dioxide ligand **43** and indium(III) trifluoromethanesulfonate is a highly efficient catalyst for the asymmetric oxa-Diels–Alder reaction. A very broad range of Danishefsky-type dienes and aldehydes are well tolerated by this catalytic system to give optically active 5-methyl-2,3-dihydro-4H-pyran-4-ones **44** in good to excellent yield with up to 99% enantiomeric excess (Scheme 23).[35] The activity of the catalyst is found to remain unchanged even after more than half a year, which facilitates its practical usage. This reaction has been further applied successfully on a sub-gram scale to the synthesis of triketide **45**.

2.5.2 Oxa-Diels–Alder Reactions

Scheme 23 Chiral Indium Complex Catalyzed Asymmetric Oxa-Diels–Alder Reactions[35]

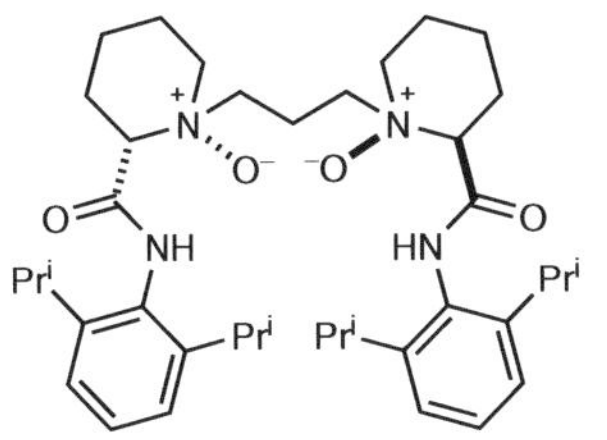

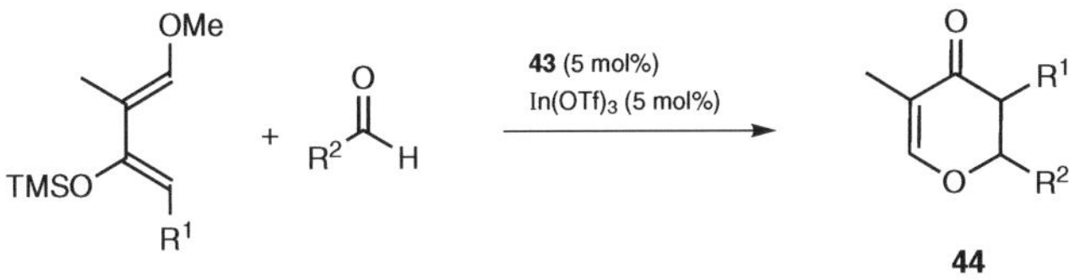

R^1	R^2	Ratio (cis/trans)	ee (%)	Yield (%)	Ref
Me	Ph	>20:1	98	96	[35]
Me	2-furyl	9:1	99	89	[35]
Me	$(CH_2)_4Me$	11:1	97	52	[35]
Me	(E)-CH=CHPh	17:1	98	92	[35]
H	Ph	–	98	99	[35]

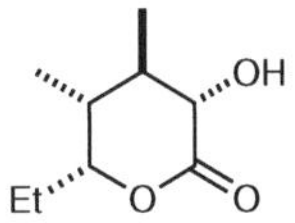

3,5-Dimethyl-2-phenyl-2,3-dihydro-4H-pyran-4-one (44, R^1 = Me; R^2 = Ph); Typical Procedure:[35]

An N_2-protected mixture of N,N′-dioxide ligand **43** (8.2 mg, 0.0125 mmol), In(OTf)$_3$ (7.0 mg, 0.0125 mmol), and 4-Å molecular sieves (25 mg) in THF (0.1 mL) was stirred at rt for 20 min. The resulting mixture was concentrated under reduced pressure, the reaction vessel was refilled with N_2 (× 3), and anisole (2 mL) was added. This mixture was cooled to 0 °C for 10 min, PhCHO (0.026 mL, 0.25 mmol) and (1E,3Z)-1-methoxy-2-methyl-3-(trimethylsilyloxy)penta-1,3-diene (0.100 mL, 0.375 mmol) were added, and the resulting mixture was stirred at 0 °C for 48 h. TFA (0.1 mL) was then added and the mixture was stirred for 10 min before sat. aq NaHCO$_3$ (10 mL) was added. The resulting soln was stirred for 5 min, diluted with Et$_2$O (10 mL), and filtered through a plug of Celite. The layers were separated and the aqueous layer was extracted with Et$_2$O (3 × 10 mL). The combined organic layers were washed with brine, dried (MgSO$_4$), and concentrated to give a residual oil, which was purified by chromatography (silica gel, petroleum ether/Et$_2$O 10:1) to afford predominantly the cis-isomer of the product as a colorless, viscous liquid; yield: 48.5 mg (96%); ratio (cis/trans) >20:1; 98% ee.

2.5.2.3 Reaction Using Chiral Titanium Complexes

2.5.2.3.1 1,1′-Binaphthalene-2,2′-diol/Titanium (1:1) Catalysts

The 1:1 complexes formed between (R)-1,1′-binaphthalene-2,2′-diol [46; (R)-BINOL] or its derivatives 47 and 48 and titanium(IV) isopropoxide are one type of efficient catalyst for the asymmetric oxa-Diels–Alder reaction between Danishefsky-type dienes and aldehydes to furnish 2,3-dihydro-4H-pyran-4-ones 49 with excellent enantioselectivity (Scheme 24).[36–39] This reaction is thought to proceed via a stepwise Mukaiyama aldol pathway and, in most cases, 20 mol% of catalyst is required to ensure satisfactory yields and enantiomeric excesses.

Scheme 24 1,1′-Binaphthalene-2,2′-diol/Titanium(IV) Isopropoxide (1:1) Catalyzed Oxa-Diels–Alder Reaction of Danishefsky-Type Dienes[37,39]

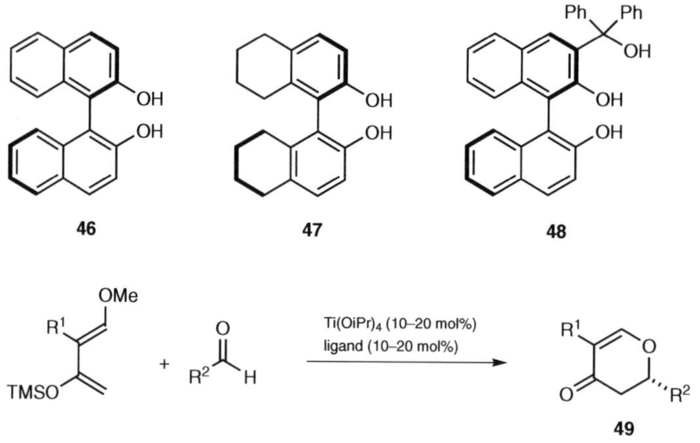

R^1	R^2	Ligand	ee (%)	Yield (%)	Ref
H	Ph	47	97	92	[37]
H	2-pyridyl	47	92	55	[37]
H	(E)-CH=CHPh	47	98	80	[37]
Me	Ph	48	94	98	[39]
Me	(CH$_2$)$_6$Me	48	97	92	[39]

The complex of (R)-1,1′-binaphthalene-2,2′-diol [46; (R)-BINOL] and titanium(IV) isopropoxide together with 4-(chloromethyl)pyridine hydrochloride is effective for the asymmetric oxa-Diels–Alder reaction between a Brassard-type diene and aliphatic aldehydes. The corresponding α,β-unsaturated δ-lactone derivatives 50 are obtained in 46–79% yield with up to 88% enantiomeric excess (Scheme 25).[40] (R)-(+)-Kavain (70% ee) and (S)-(+)-dihydrokavain (84% ee) are also prepared in a single step using this methodology. A strong positive nonlinear effect is observed, which implies that the reaction occurs in the presence of a polymeric titanium active species.

2.5.2 Oxa-Diels–Alder Reactions

Scheme 25 1,1'-Binaphthalene-2,2'-diol/Titanium (1:1) Catalyzed Oxa-Diels–Alder Reaction of a Brassard-Type Diene[40]

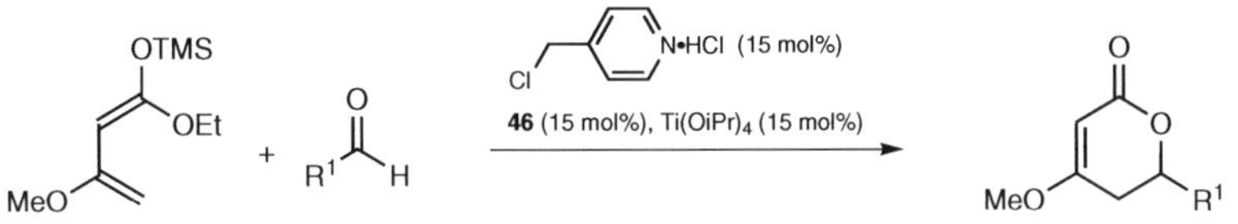

R^1	ee (%)	Yield (%)	Ref
(CH$_2$)$_4$Me	83	74	[40]
Et	88	72	[40]
Cy	81	65	[40]
Ph	87	56	[40]
(E)-CH=CHPh	70	56	[40]
(CH$_2$)$_2$Ph	84	57	[40]

(S)-2-Phenyl-2,3-dihydro-4H-pyran-4-one (49, R^1 = H; R^2 = Ph); Typical Procedure:[37]

A mixture of ligand **47** (16.2 mg, 0.05 mmol), a 1.0 M soln of Ti(OiPr)$_4$ in CH$_2$Cl$_2$ (0.05 mL, 0.05 mmol), and oven-dried, powdered 4-Å molecular sieves (120 mg) in toluene (1.0 mL) was heated at 35 °C for 1 h. The resulting yellow mixture was allowed to cool to rt and PhCHO (26 µL, 0.25 mmol) was added. This mixture was stirred for 10 min and then cooled to 0 °C. 1-Methoxy-3-(trimethylsiloxy)buta-1,3-diene (60 µL, 0.30 mmol) was added and the resulting mixture was stirred at 0 °C for 24 h before being treated with TFA (5 drops). This mixture was stirred for 15 min at 0 °C, sat. aq NaHCO$_3$ (1.5 mL) was added, and the whole was then stirred for 10 min and filtered through a plug of Celite. The organic layer was separated and the aqueous layer was extracted with Et$_2$O (5 × 3 mL). The combined organic layers were dried (Na$_2$SO$_4$) and concentrated and the crude residue was purified by flash chromatography (petroleum ether/EtOAc 4:1) to afford a clear oil; yield: 40 mg (92%); 97% ee.

4-Methoxy-6-pentyl-5,6-dihydro-2H-pyran-2-one [50, R^1 = (CH$_2$)$_4$Me]; Typical Procedure:[40]

(R)-BINOL (**46**; 10.8 mg, 0.0375 mmol) and 4-(chloromethyl)pyridine hydrochloride (6.1 mg, 0.0375 mmol) in toluene (1.0 mL) were stirred at rt for 2 h. A 1.0 M soln of Ti(OiPr)$_4$ in toluene (0.0375 mL, 0.0375 mmol) was added and the resulting mixture was stirred at 35 °C for 1 h under a N$_2$ atmosphere. The mixture was then cooled to 28 °C and hexanal (0.031 mL, 0.25 mmol) and (E)-1-ethoxy-3-methoxy-1-(trimethylsiloxy)buta-1,3-diene (0.085 mL, 0.375 mmol) were added. This mixture was stirred at 28 °C for 115 h before the reaction was quenched with 5 drops of TFA and the whole was stirred for 15 min. The resulting mixture was neutralized with sat. aq NaHCO$_3$ and then extracted with CH$_2$Cl$_2$ (× 2). The combined organic layers were washed with sat. brine, dried (Na$_2$SO$_4$), and concentrated under reduced pressure. The residue obtained was purified by chromatography (silica gel, petroleum ether/EtOAc 5:1); yield: 74%; 83% ee.

2.5.2.3.2 1,1'-Binaphthalene-2,2'-diol/Titanium (2:1) Catalysts

An exceptionally efficient enantioselective oxa-Diels–Alder reaction of Danishefsky's diene with aldehydes has been realized using titanium complexes generated in situ from chiral diols **47** and/or **51** with titanium(IV) isopropoxide. Catalyst systems titanium(IV) isopropoxide/**47**/**47** and titanium(IV) isopropoxide/**47**/**51** can promote the solvent-free oxa-Diels–Alder reaction with only 0.5–0.05 mol% catalyst loading to give 2,3-dihydro-4*H*-pyran-4-ones **52** in up to quantitative yield and >99% enantiomeric excess (Scheme 26).[41] In particular, only 0.005 mol% of catalyst system titanium(IV) isopropoxide/**47**/**51** is effective for the reaction between furan-2-carbaldehyde and Danishefsky's diene to give the desired product in 63% yield and with 96.2% enantiomeric excess.

Scheme 26 1,1′-Binaphthalene-2,2′-diol/Titanium (2:1) Catalyzed Oxa-Diels–Alder Reaction of Danishefsky's Diene[41]

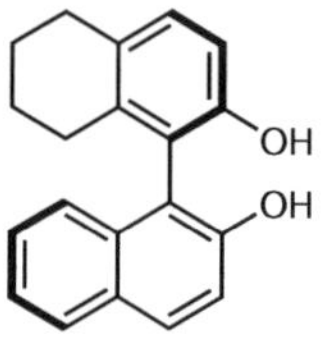

R¹	Catalyst	Catalyst Loading (mol%)	ee (%)	Yield (%)	Ref
Ph	Ti(OiPr)₄/**47**/**47**	0.05	99.3	>99	[41]
Ph	Ti(OiPr)₄/**47**/**51**	0.05	99.4	82	[41]
(CH₂)₂Ph	Ti(OiPr)₄/**47**/**47**	0.05	97.9	>99	[41]
(*E*)-CH=CHPh	Ti(OiPr)₄/**47**/**51**	0.05	96.6	56.6	[41]
2-furyl	Ti(OiPr)₄/**47**/**51**	0.005	96.2	63	[41]

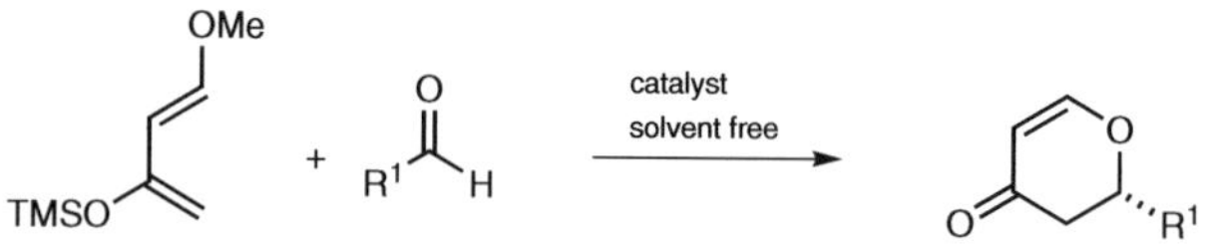

(*S*)-2-Phenyl-2,3-dihydro-4*H*-pyran-4-one (52, R¹ = Ph); Typical Procedure:[41]
A dried Schlenk tube was charged with a 0.1 M soln of catalyst Ti(OiPr)₄/**47**/**47** (0.08 mL, 0.008 mmol) in toluene [generated in situ by mixing ligand **47** with Ti(OiPr)₄ in a 2:1 molar ratio]. Freshly distilled PhCHO (1.70 g, 16 mmol) and Danishefsky's diene (4.13 g, 24 mmol) were added sequentially, and the resulting mixture was stirred at 20 °C for 48 h, diluted with Et₂O (10 mL), and then treated with TFA (3 mL). This mixture was then stirred for 0.5 h, sat. aq NaHCO₃ (20 mL) was added, and the resulting mixture was stirred for 10 min. The layers were then separated and the aqueous layer was extracted with Et₂O (3 × 50 mL). The combined organic layers were dried (Na₂SO₄) and concentrated and the crude residue was purified by flash chromatography (silica gel, hexanes/EtOAc 4:1) to afford a colorless liquid; yield: 2.77 g (>99%); >99% ee.

2.5.2.3.3 Chiral Schiff Base/Titanium Catalysts

Titanium complexes modified by chiral tridentate Schiff base **53** are efficient catalysts for the oxa-Diels–Alder reaction of Danishefsky's diene with aldehydes via a concerted (4+2)-cycloaddition process. A variety of 2-substituted 2,3-dihydro-4*H*-pyran-4-ones **55** are obtained in >99% yield and in up to 97% enantiomeric excess using the Ti(OiPr)₄/**53** system as a catalyst precursor in the presence of Naproxen (**54**) as an activator (Scheme 27).[42,43]

2.5.2 Oxa-Diels–Alder Reactions

Scheme 27 Titanium–Schiff Base Complex Catalyzed Asymmetric Oxa-Diels–Alder Reactions[42]

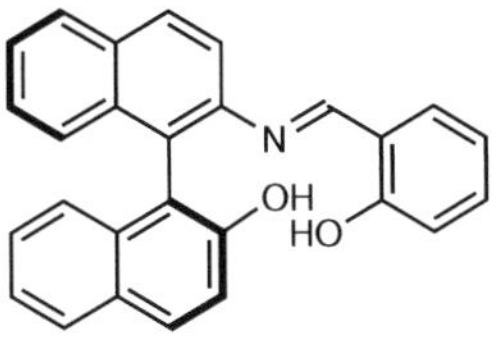

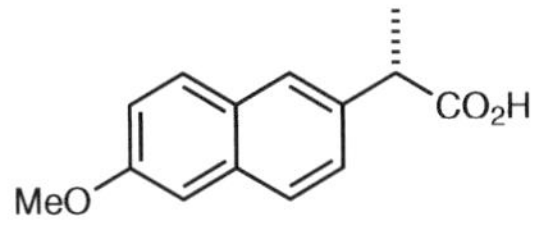

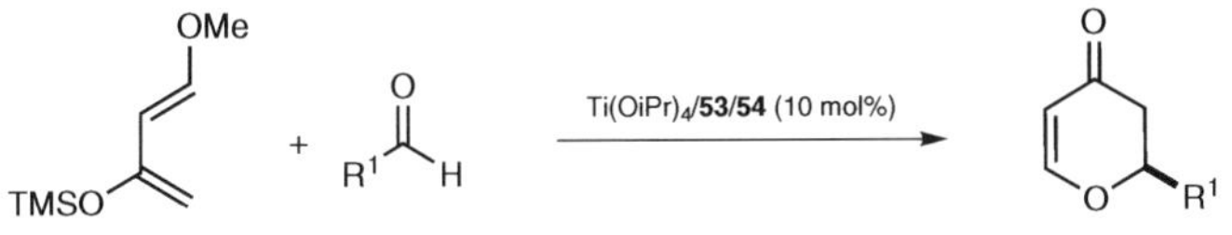

R¹	ee (%)	Yield (%)	Ref
Ph	97	>99	[42]
4-BrC$_6$H$_4$	95.2	98	[42]
2-furyl	74.9	90	[42]
(E)-CH=CHPh	83.4	96	[42]
(CH$_2$)$_2$Ph	75	83	[42]

A possible transition-state model **56** illustrates the asymmetric induction pathway. The aldehyde binds in the remaining equatorial coordination site of the titanium(IV) atom and the R¹ group adopts a down arrangement to prevent repulsive interaction with the naphthalene ring. This allows the attack of the diene from the *si*-face to give the product with an *S* configuration (Scheme 28).[42] Some dendritic catalysts have also proved to be effective for the same reaction, and these catalysts can be recycled and reused at least three times without significant loss of activity or enantioselectivity.[44]

Scheme 28 Transition State for Titanium–Schiff Base Catalyzed Oxa-Diels–Alder Reactions[42]

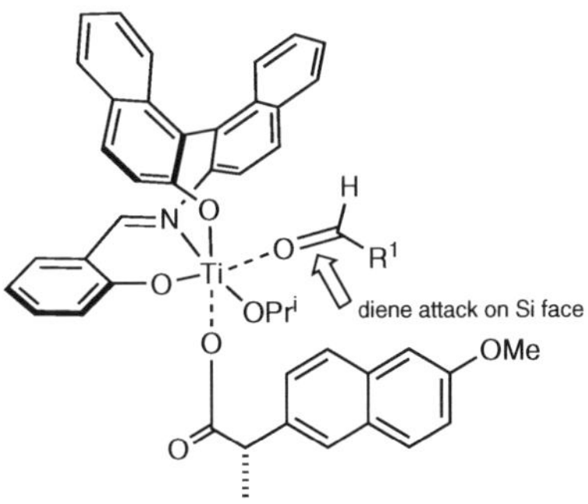

(S)-2-Phenyl-2,3-dihydro-4H-pyran-4-one (55, R¹ = Ph); Typical Procedure:[42]
A mixture of Schiff base **53** (0.05 mmol), a 0.5 M soln of Ti(OiPr)$_4$ in CH$_2$Cl$_2$ (0.05 mL, 0.025 mmol), and activated powdered 4-Å molecular sieves (30 mg) in toluene (1.0 mL) was stirred for 2 h at 50 °C. The resulting red soln was allowed to cool to rt and acid **54**

(0.025 mmol), benzaldehyde (0.25 mmol), and Danishefsky's diene (0.060 mL, 0.3 mmol) were added sequentially. The resulting mixture was stirred for 20 h and then the reaction was quenched with TFA (10 drops). After stirring for an additional 5 min, the mixture was neutralized with sat. aq NaHCO$_3$ (3 mL) and, after filtration through a plug of Celite, the organic layer was separated and the aqueous layer was extracted with EtOAc (3 × 5 mL). The combined organic layers were dried (Na$_2$SO$_4$) and concentrated, and the residue obtained was purified by flash chromatography (silica gel); yield: >99%; 97% ee.

2.5.2.4 Reaction Using Chiral Zirconium Complexes

The chiral zirconium complex generated in situ from 3,3′-diiodo-1,1′-binaphthalene-2,2′-diol derivatives **57**, zirconium(IV) *tert*-butoxide, a primary alcohol, and a small amount of water is highly active for the asymmetric oxa-Diels–Alder reaction of Danishefsky-type dienes with aldehydes via a stepwise cycloaddition pathway. The corresponding 2,3-dihydro-4*H*-pyran-4-ones **58** are obtained in high yield, with high *cis/trans* selectivity, and with high enantioselectivity (Scheme 29).[45,46] The reaction has also been applied to a concise synthesis of (+)-prelactone C (**59**).

Scheme 29 Chiral Zirconium Complex Catalyzed Asymmetric Oxa-Diels–Alder Reactions[46]

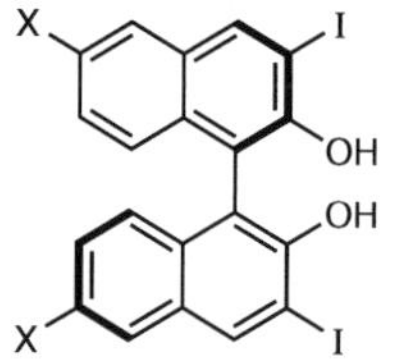

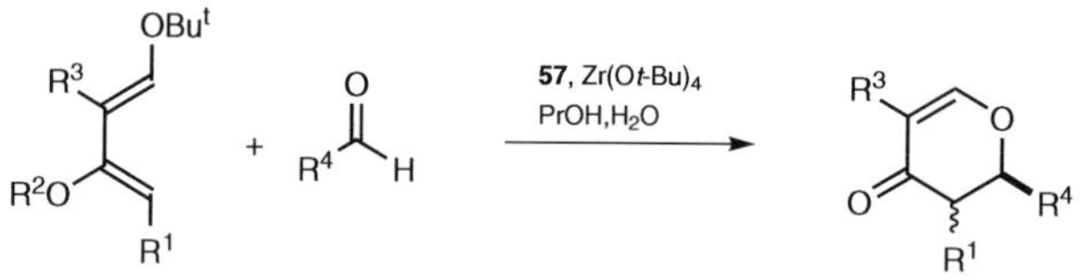

R^1	R^2	R^3	R^4	X	Ratio (cis/trans)	ee (%)	Yield (%)	Ref
H	SiEtMe$_2$	H	Ph	H	–	97	>99	[46]
Me	TMS	Me	(E)-CH=CHPh	CF$_2$CF$_3$	1:9	90	96	[46]
Bn	TMS	H	Ph	I	>30:1	97	95	[46]
Bn	TMS	H	4-ClC$_6$H$_4$	I	>30:1	97	>99	[46]

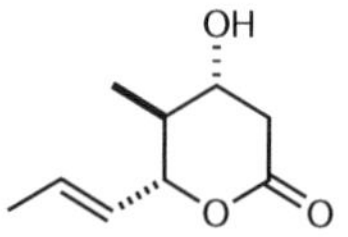

(R)-2-Phenyl-2,3-dihydro-4H-pyran-4-one (58, R^1 = R^3 = H; R^4 = Ph); Typical Procedure:[46]
Zr(O*t*-Bu)$_4$ (0.040 mmol) in toluene (0.5 mL) was added to a suspension of (R)-3,3′-diiodo-1,1′-binaphthalene-2,2′-diol (**57**, X = H; 0.048 mmol) in toluene (0.5 mL) at rt and the result-

ing soln was stirred for 30 min. A mixture of PrOH (0.32 mmol) and H$_2$O (0.080 mmol) in toluene (0.3 mL) was then added, and this mixture was stirred at rt for 3 h and then cooled to −78 °C. PhCHO (0.40 mmol) in t-BuOMe (0.35 mL) and the diene (R^1 = R^3 = H; R^2 = SiEtMe$_2$; 0.48 mmol) in t-BuOMe (0.35 mL) were added successively and the mixture was warmed to −20 °C and stirred for 18 h. The reaction was then quenched by addition of sat. aq NaHCO$_3$ (10 mL) followed by CH$_2$Cl$_2$ (10 mL). The organic layer was separated and the aqueous layer was extracted with CH$_2$Cl$_2$ (2 × 10 mL). The organic layers were combined, dried (Na$_2$SO$_4$), filtered, and concentrated under reduced pressure. The residue obtained was treated with TFA (0.5 mL) in CH$_2$Cl$_2$ (8 mL) at 0 °C for 1 h. The soln was basified with sat. aq NaHCO$_3$ (20 mL) and the organic layer was separated. The aqueous layer was extracted with CH$_2$Cl$_2$ (2 × 10 mL) and the organic layers were combined, dried (Na$_2$SO$_4$), filtered, and concentrated under reduced pressure. The residue was purified by preparative TLC (benzene/EtOAc 20:1); yield: >99%; 97% ee.

2.5.2.5 Reaction Using Chiral Chromium Complexes

2.5.2.5.1 Chromium–Salen Catalysts

Chiral chromium–salen complexes **60** (2 mol%) catalyze the enantioselective oxa-Diels–Alder reaction of Danishefsky's diene with a variety of aldehydes in the presence of 4-Å molecular sieves to give the corresponding 2,3-dihydro-4H-pyran-4-ones **61** in 65–98% yield and 62–93% enantiomeric excess (Scheme 30).[47] The ^{1}H NMR spectrum of the crude reaction prior to workup reveals the exclusive presence of cycloadduct intermediates, while a synthesized Mukaiyama aldol condensation product does not give the cycloadducts under the reaction conditions, which indicates that this reaction proceeds via a concerted (4+2)-cycloaddition pathway.

Scheme 30 Chiral Chromium Complex Catalyzed Asymmetric Oxa-Diels–Alder Reactions[47]

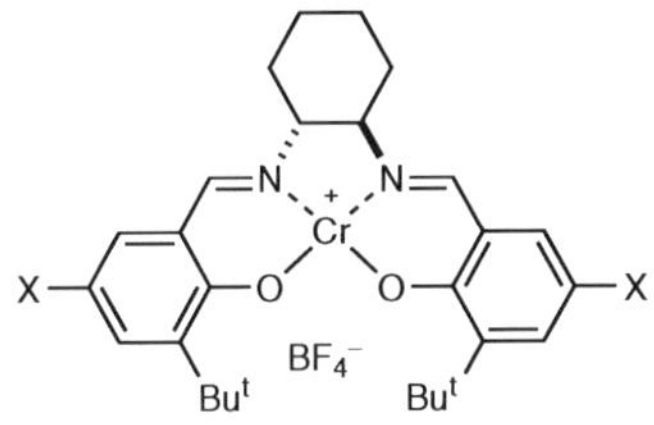

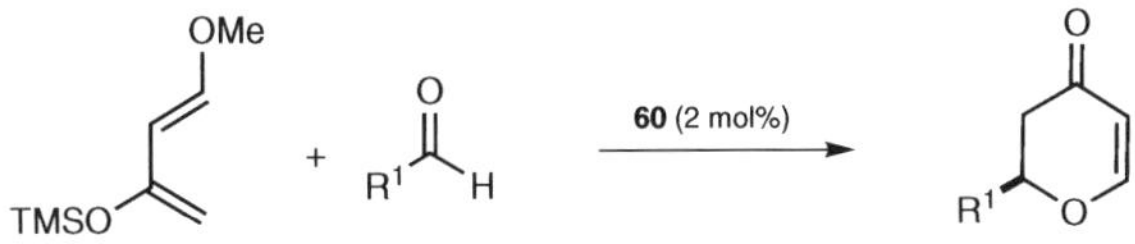

R^1	X	ee (%)	Yield (%)	Ref
Ph	t-Bu	87	85	[47]
2-furyl	t-Bu	76	89	[47]
(E)-CH=CHPh	t-Bu	70	65	[47]
Ph	OMe	65	98	[47]
2-furyl	OMe	68	80	[47]
(E)-CH=CHPh	OMe	73	96	[47]

(R)-2-Phenyl-2,3-dihydro-4H-pyran-4-one (61, R¹ = Ph); Typical Procedure:[47]

An oven-dried 10-mL flask equipped with a stirrer bar was charged with catalyst **60** (X = t-Bu; 13 mg, 0.02 mmol) and oven-dried powdered 4-Å molecular sieves (0.3 g). The flask was sealed with a rubber septum and purged with N_2. The catalyst was dissolved in t-BuOMe (0.2 mL) and PhCHO (0.1 mL, 1.0 mmol) was added via a syringe at rt. The resulting mixture was then cooled to −30 °C and Danishefsky's diene (0.195 mL, 1.0 mmol) was added. This mixture was stirred at −30 °C for 24 h, at which time it was removed from the bath, diluted with CH_2Cl_2 (2 mL), treated with a drop of TFA, and stirred for 10 min at rt. The resulting mixture was concentrated under reduced pressure and the crude residue was purified by flash chromatography (silica gel, hexanes/EtOAc 7:3) to give a clear oil; yield: 151 mg (85%, as reported); 87% ee.

2.5.2.5.2 Tridentate Chromium Catalysts

Chromium complexes **62** bearing a chiral tridentate Schiff base can act as catalysts for the asymmetric oxa-Diels–Alder reaction of 3-siloxy-substituted hexa-2,4-dienes with aldehydes to give cis-substituted 3,6-dimethyltetrahydro-4H-pyran-4-ones **63** in 28–97% yield and with 90–>99% enantiomeric excess (Scheme 31).[48] The catalyst loading can be reduced to 0.5 mol% for the reaction of 1-methoxy-substituted 1,3-dienes with (tert-butyldimethylsiloxy)acetaldehyde without a decrease in selectivity.

Scheme 31 Asymmetric Oxa-Diels–Alder Reactions of 3-Siloxy-Substituted Hexa-2,4-dienes with Aldehydes[48]

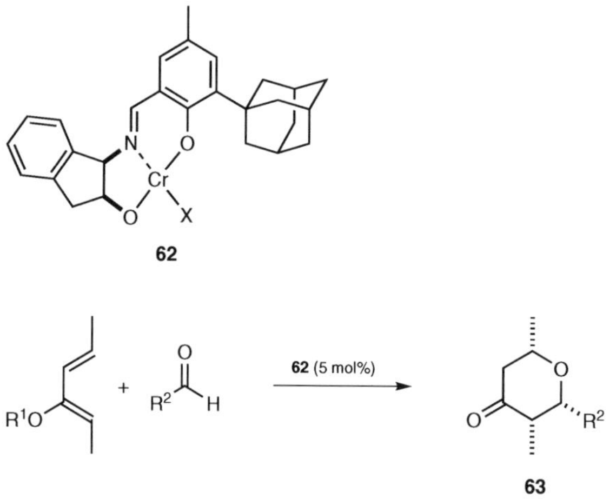

R¹	R²	X	ee (%)	Yield (%)	Ref
TES	$CH_2OTBDMS$	Cl	98	88	[48]
TMS	$(CH_2)_4Me$	SbF_6	98	81	[48]
TBDMS	$(CH_2)_4Me$	SbF_6	96	93	[48]
TES	$(CH_2)_4Me$	SbF_6	94	77	[48]

Complex **64** is also a highly efficient catalyst for the inverse-electron-demand oxa-Diels–Alder reaction of α,β-unsaturated aldehydes with ethyl vinyl ether to afford cis-substituted 3,4-dihydro-2H-pyrans **65** in good yields with a high level of enantioselectivity (Scheme 32).[49] In the solid state, catalyst **64** exists as a dimeric structure, bridged through a single water molecule and bearing one terminal water ligand on each chromium center, and this dimeric structure is maintained in the catalytic cycle. Dissociation of a terminal water

2.5.2 Oxa-Diels–Alder Reactions

molecule to open a coordination site for the aldehyde is energetically difficult, which may explain the crucial role of molecular sieves in these reactions.

Scheme 32 Asymmetric Oxa-Diels–Alder Reactions of α,β-Unsaturated Aldehydes with Ethyl Vinyl Ether[49]

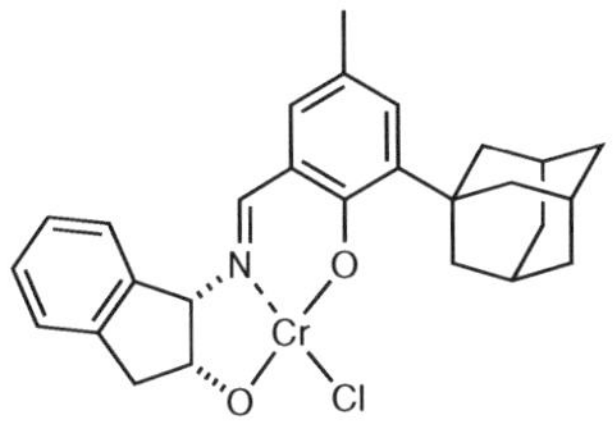

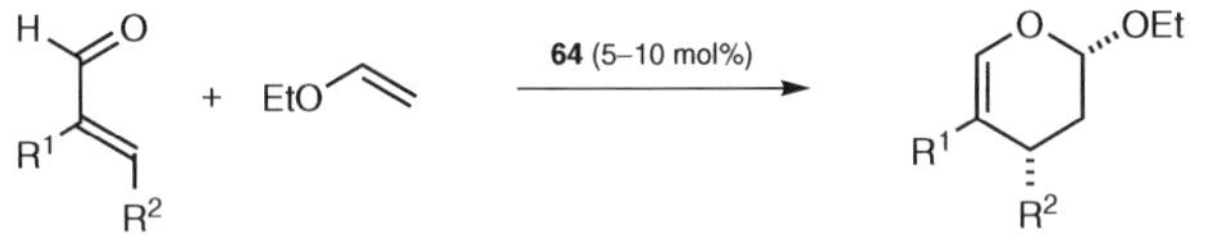

R^1	R^2	ee (%)	Yield (%)	Ref
H	Me	94	75	[49]
H	Pr	94	73	[49]
H	Ph	98	75	[49]
H	CO$_2$Et	95	90	[49]
Br	Ph	98	75	[49]

2-[(*tert*-Butyldimethylsiloxy)methyl]-3,6-dimethyltetrahydro-4*H*-pyran-4-one (63, R^1 = TES; R^2 = CH$_2$OTBDMS); Typical Procedure:[48]
3-(Triethylsiloxy)hexa-2,4-diene (72% pure; 0.342 mL, 1.38 mmol) was added to a stirred mixture of (*tert*-butyldimethylsiloxy)acetaldehyde (0.2 mL, 1.0 mmol), chromium complex **62** (X = Cl; 15.0 mg, 0.03 mmol), and 4-Å molecular sieves (200 mg) under N$_2$ at ambient temperature. The resulting mixture was stirred for 16 h, diluted with THF (4 mL), and cooled to 0 °C. AcOH (0.114 mL, 2.0 mmol) and a 1.0 M soln of TBAF in THF (1.5 mL, 1.5 mmol) were added, and the mixture was allowed to stir for 30 min, diluted with hexanes/Et$_2$O (2:1; 60 mL), and washed with H$_2$O (2 × 30 mL), sat. aq NaHCO$_3$ (30 mL), and brine (30 mL). The pale yellow organic soln was dried (MgSO$_4$), filtered, concentrated under reduced pressure, and purified by flash chromatography (silica gel, EtOAc/hexanes 5:95) to afford a colorless oil; yield: 273 mg (88%, as reported); 98% ee.

(2*S*,4*S*)-2-Ethoxy-4-methyl-3,4-dihydro-2*H*-pyran (65, R^1 = H; R^2 = Me); Typical Procedure:[49]
Freshly distilled ethyl vinyl ether (0.96 mL, 10.0 mmol) and (*E*)-but-2-enal (70.0 mg, 0.083 mL, 1.0 mmol) were added to an oven-dried, 10-mL round-bottomed flask containing a stirrer bar. Catalyst **64** (X = Cl; 24 mg, 0.050 mmol) and freshly oven-dried powdered 4-Å molecular sieves (150 mg) were added to this soln and the flask was sealed with a septum or stopper and allowed to stir for 2 d. The resulting mixture was then diluted with pentane and filtered through Celite. The pentane was removed and the product was isolated by vacuum transfer to a flask cooled to −78 °C at 0.5 Torr to give a clear oil; yield: 106 mg (75%); 94% ee.

2.5.2.6 Reaction Using Chiral Copper Complexes

2.5.2.6.1 Normal-Electron-Demand Oxa-Diels–Alder Reactions

A highly enantioselective oxa-Diels–Alder reaction of Danishefsky-type dienes with various ketones (including α-oxo esters and 1,2-diketones) has been realized using copper complexes **66** as catalysts to give the corresponding cycloadducts **67** in moderate to good yield and with up to 99% enantiomeric excess (Scheme 33).[50]

Scheme 33 Asymmetric Oxa-Diels–Alder Reaction of Danishefsky-Type Dienes with α-Oxo Esters or 1,2-Diones[50]

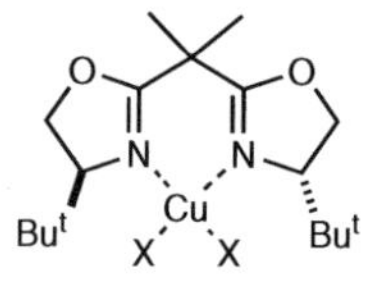

X	R^1	R^2	R^3	ee (%)	Yield (%)	Ref
OTf	H	Me	OEt	99	78	[50]
SbF$_6$	H	Me	OEt	89	37	[50]
OTf	H	Me	OMe	99	96	[50]
OTf	H	Me	Ph	94	95	[50]
OTf	Me	Ph	OEt	99	57	[50]
OTf	H	Me	Et	98	77	[50]

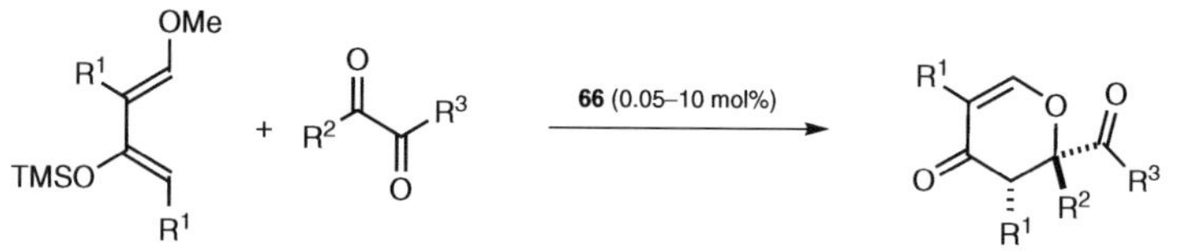

Further improvement has led to a more general catalytic process for the oxa-Diels–Alder reaction of 1,2-diketones which exhibits broad substrate scope, low catalyst loading, and excellent reactivity, diastereoselectivity, and enantioselectivity. For example, only 0.05 mol% of a catalyst of type **66** can catalyze the reaction of pentane-2,3-dione with Danishefsky's diene efficiently. The reaction occurs at the methyl ketone fragment in 76% yield and with 97.8% enantiomeric excess. It is assumed, as shown in transition state **68**, that both the chiral ligand and the α-oxo ester are bidentate in their coordination to the copper(II) center during the reaction, and the carbonyl *re*-face is effectively shielded by the ligand *tert*-butyl groups (Scheme 34).[50]

2.5.2 Oxa-Diels–Alder Reactions

Scheme 34 Transition State of an Oxa-Diels–Alder Reaction between a Danishefsky-Type Diene and an α-Oxo Ester[50]

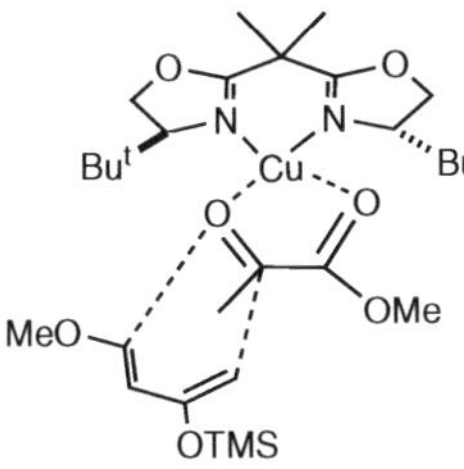

68

Methyl (S)-2-Methyl-4-oxo-3,4-dihydro-2H-pyran-2-carboxylate (67, R^1 = H; R^2 = Me; R^3 = OMe); Typical Procedure:[50]
Cu(OTf)$_2$ (36 mg, 0.1 mmol) and (4S,4′S)-2,2′-(propane-2,2-diyl)bis(4-tert-butyl-4,5-dihydrooxazole) (31.5 mg, 0.1 mmol) were added to a flame-dried Schlenk tube under N$_2$. The resulting mixture was dried under vacuum for 1–2 h, anhyd THF (1.5–2.0 mL) was added, and the resulting suspension was stirred vigorously for 1–5 h. The catalyst soln was then cooled to −78 °C and methyl 2-oxopropanoate (0.090 mL, 1.0 mmol) was added, followed by addition of 1-methoxy-3-(trimethylsiloxy)buta-1,3-diene (0.240 mL, 1.2 mmol). This mixture was kept stirring at −78 °C for 30 h, a soln of TFA (0.1 mL) in CH$_2$Cl$_2$ (20 mL) was added, and the resulting mixture was stirred at 0 °C for 1 h. The soln was then neutralized with sat. aq NaHCO$_3$ and filtered through a plug of cotton. The organic phase was separated and the aqueous phase was extracted with CH$_2$Cl$_2$ (2 ×). The combined organic phases were dried, filtered, and concentrated to give a crude residue, which was purified by flash chromatography (silica gel, EtOAc/pentane 1:4) to afford a light yellow oil; yield: 164 mg (96%); 99% ee.

2.5.2.6.2 Inverse-Electron-Demand Oxa-Diels–Alder Reactions

An inverse-electron-demand oxa-Diels–Alder reaction of β,γ-unsaturated α-oxo esters bearing alkyl, aryl, and alkoxy substituents at the γ-position with vinyl ethers has been realized using chiral complexes **66** as catalysts to give the corresponding cycloadducts in 51–96% yield and with 90.4–99.5% enantiomeric excess. For example, γ-ethoxy-substituted dienes react with functionalized vinyl ethers to give cycloadducts **69** in the presence of catalyst **66** (X = OTf) (Scheme 35).[51,52] Mechanistic studies suggest that the alkene approaches the si-face of the reacting α,β-unsaturated carbonyl functionality when coordinated to the catalyst.

Scheme 35 Asymmetric Oxa-Diels–Alder Reaction of β,γ-Unsaturated α-Oxo Esters with Vinyl Ethers[51,52]

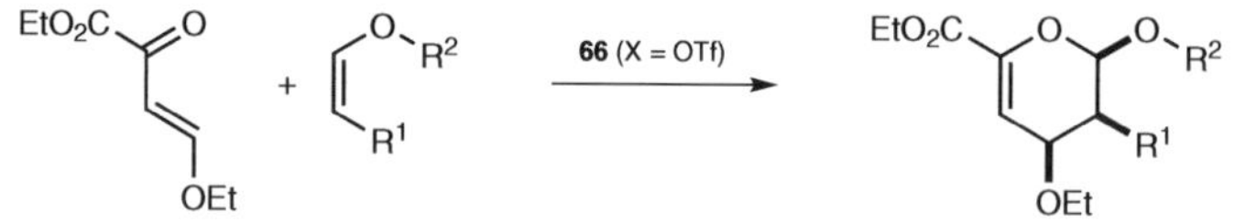

R^1	R^2	mol% of Catalyst	ee (%)	Yield (%)	Ref
H	Et	10	>99.5	93	[51]
OAc	Et	20	99.5	70	[52]
OCMe$_2$		20	99	63	[52]
CH$_2$OCH$_2$		20	96	65	[52]
(CH$_2$)$_2$		10	97.5	84	[51]

Ethyl (3a*S*,4*S*,7a*R*)-4-Ethoxy-2,3,3a,7a-tetrahydro-4*H*-furo[2,3-*b*]pyran-6-carboxylate [69, R¹,R² = (CH₂)₂]; Typical Procedure:[51]

Catalyst **66** (X = OTf) was prepared by the addition of Cu(OTf)₂ (36 mg, 0.1 mmol) to (4*S*,4′*S*)-2,2′-(propane-2,2-diyl)bis(4-*tert*-butyl-4,5-dihydrooxazole) (31 mg, 0.1 mmol) dissolved in dry THF (2 mL) under N₂, followed by stirring for 1 h. Ethyl (*E*)-4-ethoxy-2-oxobut-3-enoate (172 mg, 1.0 mmol) was added to this catalyst soln, which was then cooled to −78 °C and treated with freshly distilled 2,3-dihydrofuran (0.15 mL, 2 mmol). The resulting soln was warmed to −45 °C and stirred for 50 h, and then pentane (2 mL) was added. The resulting mixture was filtered through a plug of silica gel (pentane/Et₂O 1:1) and, after removal of the solvent, the crude residue was purified by flash chromatography (silica gel, pentane/Et₂O 3:1) to give a highly crystalline colorless solid; yield: 84%; 97.5% ee.

Ethyl (2*R*,3*S*,4*S*)-3-Acetoxy-2,4-diethoxy-3,4-dihydro-2*H*-pyran-6-carboxylate (69, R¹ = OAc; R² = Et); Typical Procedure:[52]

Cu(OTf)₂ (36.1 mg, 0.1 mmol) and (4*S*,4′*S*)-2,2′-(propane-2,2-diyl)bis(4-*tert*-butyl-4,5-dihydrooxazole) (30.9 mg, 0.1 mmol) were added to a flame-dried Schlenk tube under N₂. The resulting mixture was dried under vacuum for 1–2 h, anhyd Et₂O (2.0 mL) was added, and the resulting suspension was stirred vigorously for 1–5 h. The catalyst soln was cooled to 0 °C and ethyl (*E*)-4-ethoxy-2-oxobut-3-enoate (86 mg, 0.5 mmol) and (*Z*)-1-acetoxy-2-ethoxyethene (100 mg, 0.75 mmol, 1.5 equiv) were added. The resulting mixture was stirred at rt for 16 h and purification by flash chromatography (silica gel, pentane/EtOAc 7:1) gave a pale yellow oil; yield: 70%; 99.5% ee.

2.5.2.7 Reaction Using Chiral Zinc Complexes

The 3,3′-dibromo-1,1′-binaphthalene-2,2′-diol–zinc complex **70** can promote the oxa-Diels–Alder reaction between Danishefsky's diene and aldehydes to give 2-substituted 2,3-dihydro-4*H*-pyran-4-ones **71** in up to quantitative yield and with up to 98% enantiomeric excess (Scheme 36).[53]

2.5.2 Oxa-Diels–Alder Reactions

Scheme 36 Substituted 1,1′-Binaphthalene-2,2′-diol–Zinc Complex Catalyzed Asymmetric Oxa-Diels–Alder Reactions[53]

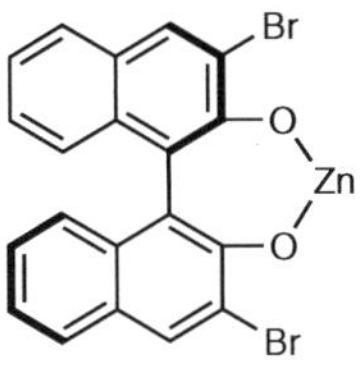

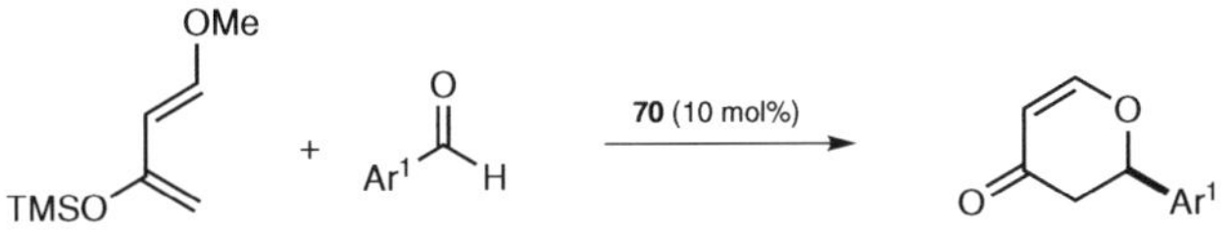

Ar¹	ee (%)	Yield (%)	Ref
Ph	97	>99	[53]
2-furyl	96	>99	[53]
4-NCC$_6$H$_4$	96	>99	[53]
2,6-Cl$_2$C$_6$H$_3$	89	82	[53]

Further studies have shown that a series of chiral diimine activated 1,1′-binaphthalene-2,2′-diol–zinc complexes are also effective at promoting the oxa-Diels–Alder reaction with excellent enantioselectivity.[54,55] Interestingly, two distinct asymmetric reactions, the oxa-Diels–Alder reaction of Danishefsky's diene with aldehydes, and diethylzinc addition to aldehydes, have been successfully integrated into a one-pot procedure in the presence of a single catalyst **72** to give the corresponding cycloaddition derivatives **73-75** in up to 97.4% enantiomeric excess and up to 95% diastereomeric excess (Scheme 37).[54,55] Moreover, chiral 1,1′-binaphthalene-2,2′-diol–magnesium complexes are also highly efficient catalysts for the same oxa-Diels–Alder reaction to afford the products in up to 99% enantiomeric excess.[56]

Scheme 37 Chiral Diimine Activated Zinc Complex Catalyzed Tandem Reactions[54,55]

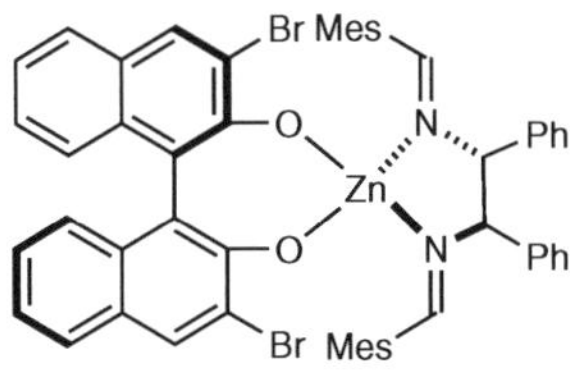

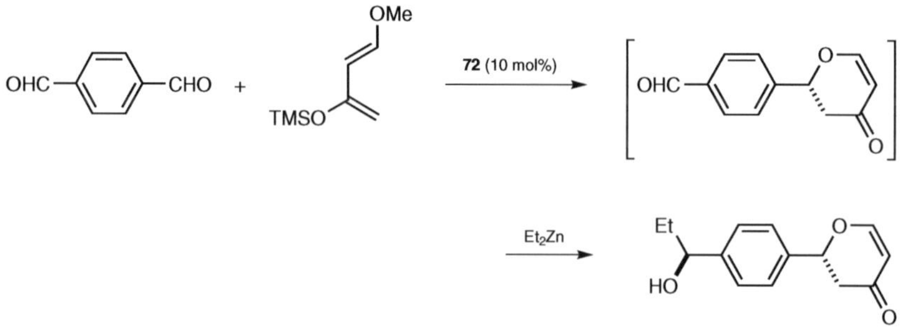

73 92%; 95% de; 97.4% ee

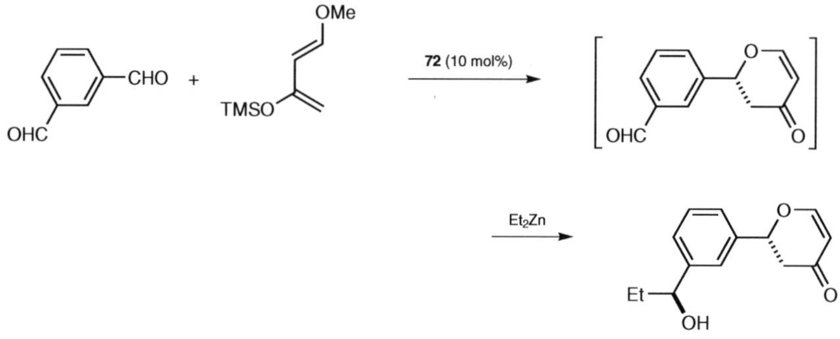

74 82%; 94.9% de; 95.9% ee

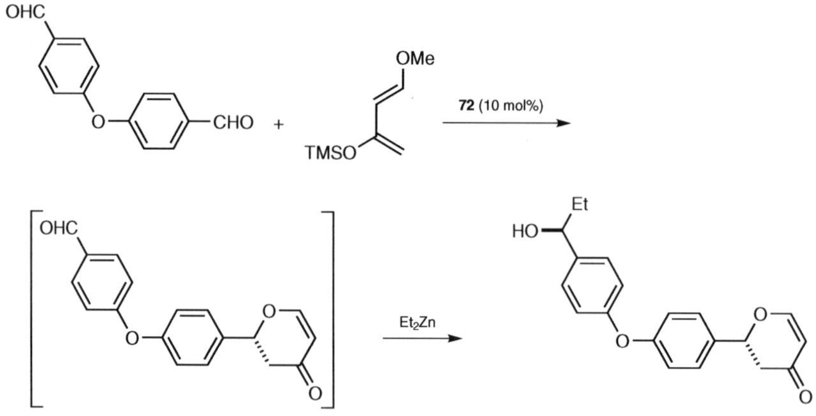

75 74%; 92% de; 94.1% ee

(R)-2-Phenyl-2,3-dihydro-4H-pyran-4-one (71, Ar¹ = Ph); Typical Procedure:[53]
A 0.025 M soln of 3,3′-dibromo-1,1′-binaphthalene-2,2′-diol (0.8 mL, 0.02 mmol) in toluene and a 1.0 M soln of Et$_2$Zn in hexane (0.024 mL, 0.024 mmol) were added to a 1.5-mL poly-(propylene) microtube, and the resulting mixture was kept at rt for 0.5 h. Freshly distilled PhCHO (21.7 mg, 0.20 mmol) was then added, the mixture was kept at −25 °C for 30 min, and then Danishefsky's diene (34.4 mg, 0.20 mmol) was added. The reaction was finally quenched after 24 h by introduction of TFA (10 drops), and then sat. aq NaHCO$_3$ (0.8 mL) was added. The aqueous layer was extracted with Et$_2$O (3 × 15 mL) and the combined organic layers were dried (Na$_2$SO$_4$), filtered, and concentrated. The crude residue was purified by flash chromatography (silica gel, hexanes/EtOAc 4:1) to give a colorless liquid; yield: 34.8 mg (>99%); 97% ee.

2.5.2 Oxa-Diels–Alder Reactions

(R)-2-{4-[(S)-1-Hydroxypropyl]phenyl}-2,3-dihydro-4H-pyran-4-one (73);
Typical Procedure:[54]

(R)-3,3'-Dibromo-1,1'-binaphthalene-2,2'-diol (8.9 mg, 0.02 mmol) and (1R,2R)-1,2-diphenyl-N,N'-bis(2,4,6-trimethylbenzylidene)ethane-1,2-diamine (9.4 mg, 0.02 mmol) were introduced into a 1.5-mL poly(propylene) microtube. Toluene (0.8 mL), a 1 M soln of Et_2Zn in hexane (0.28 mL, 0.28 mmol), and terephthalaldehyde (26.8 mg, 0.20 mmol) were added to the microtube in a glovebox using a microsyringe. The microtube was then set up in a cooling bath to maintain the temperature at −20°C for 30 min before Danishefsky's diene (41.2 mg, 0.24 mmol) was added quickly. After agitation at −20°C for 30 h, a 3 M soln of Et_2Zn in toluene (0.2 mL, 0.6 mmol) was introduced to the reaction system, which was then agitated for an additional 24 h. The microtube was opened and TFA was added to quench the reaction. The aqueous layer was neutralized with aq $NaHCO_3$ and extracted with EtOAc (3 ×). The combined organic phases were washed with brine and dried (Na_2SO_4), and the solvent was removed. The residue obtained was purified by flash column chromatography (silica gel, EtOAc/hexanes 1:2) to give a colorless oil; yield: 42.8 mg (92%); 95% de; 97.4% ee.

2.5.2.8 Reaction Using Chiral Rhodium Complexes

A variety of dirhodium complexes **76–78** are highly effective catalysts for the enantioselective oxa-Diels–Alder reaction of aldehydes with Danishefsky-type dienes to give the corresponding dihydropyranones **80** in excellent yield and with good enantiomeric excess values.[57,58] When dirhodium(II) carboxamidate complex **79** is used as a catalyst for this reaction, high levels of both *endo* selectivity and enantioselectivity are obtained (Scheme 38). In particular, in the reaction of a phenylethynyl-substituted aldehyde with a Danishefsky-type diene, the catalyst loading can be reduced to as low as 0.002 mol%, and the product is obtained in 74% yield and 91% enantiomeric excess. The turnover number in this case reaches 48 000.[59]

Scheme 38 Chiral Rhodium Complex Catalyzed Asymmetric Oxa-Diels–Alder Reactions[59]

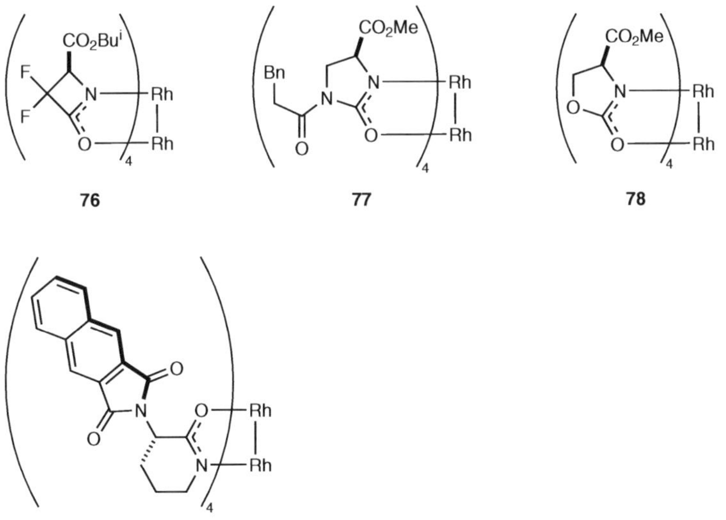

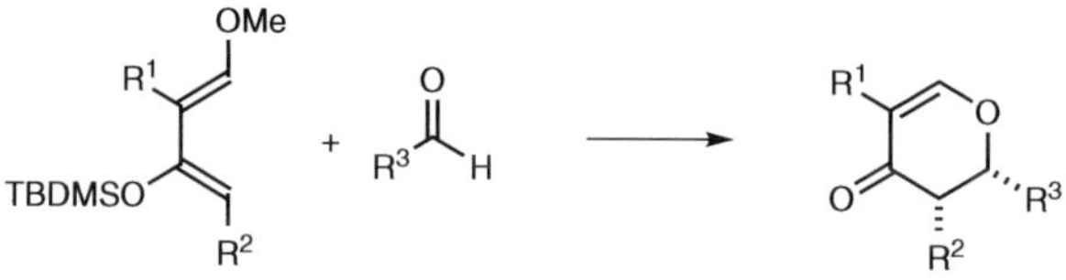

R¹	R²	R³	Catalyst	Catalyst Loading (mol%)	Temp (°C)	ee (%)	Yield (%)	Ref
H	H	4-Tol	79	1.0	23	96	97	[59]
H	H	4-O₂NC₆H₄	79	0.0075	0	94	96	[59]
H	H	(CH₂)₂Ph	79	1	23	94	89	[59]
H	H	C≡CPh	79	0.002	0	91	74	[59]
H	Me	4-O₂NC₆H₄	79	1.0	23	96[a]	97	[59]
Me	Me	4-O₂NC₆H₄	79	1.0	23	97[a]	92	[59]

[a] Only the *cis*-product is obtained.

(S)-2-(Phenylethynyl)-2,3-dihydro-4H-pyran-4-one (80, R¹ = R² = H; R³ = C≡CPh); Typical Procedure:[59]

A soln of 1-methoxy-3-(*tert*-butyldimethylsiloxy)buta-1,3-diene (2.57 g, 12.0 mmol) in CH₂Cl₂ (2.6 mL) was added to a mixture of an ice-cooled 0.5 mM soln of catalyst 79 in CH₂Cl₂ (0.40 mL, 0.0002 mmol) and 3-phenylpropynal (1.30 g, 10 mmol) in CH₂Cl₂ (17 mL). The resulting mixture was stirred at this temperature (0 °C) for 64 h before the reaction was quenched with a 10% soln of TFA in CH₂Cl₂ (ca. 0.5 mL) and stirred at 23 °C for an additional 0.5 h. This mixture was partitioned between EtOAc (100 mL) and sat. aq NaHCO₃ (10 mL), and the separated organic layer was successively washed with H₂O and brine, dried (Na₂SO₄), filtered, and concentrated under reduced pressure. The residue obtained was purified by column chromatography [silica gel (60 g), hexanes/EtOAc 2:1] to give a pale yellow oil; yield: 1.46 g (74%); 91% ee.

2.5.2.9 Reaction Using Chiral Palladium Complexes

The enantioselective oxa-Diels–Alder reaction of arylglyoxals with cyclic or acyclic substituted buta-1,3-dienes has been realized using chiral 2,2′-bis(diphenylphosphino)-1,1′-binaphthalene–palladium complex 81 (2 mol%) as the catalyst to give the corresponding cycloadducts, e.g. 82, in 20–88 yield and with 38–99% enantiomeric excess (Scheme 39).[60] The addition of 3-Å molecular sieves is found to be critically important for the enantioselectivity of the reaction. When glyoxylate esters are employed as a substrate, the ene adducts are also obtained besides the oxa-Diels–Alder adducts.

Scheme 39 Chiral Palladium Complex Catalyzed Asymmetric Oxa-Diels–Alder Reactions[60]

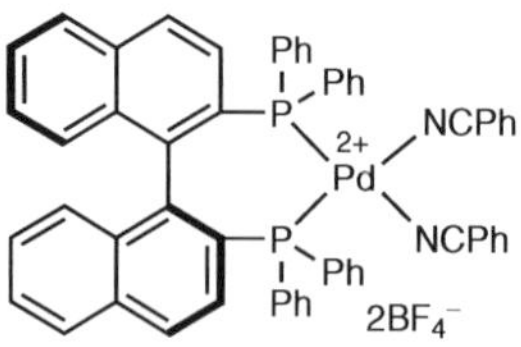

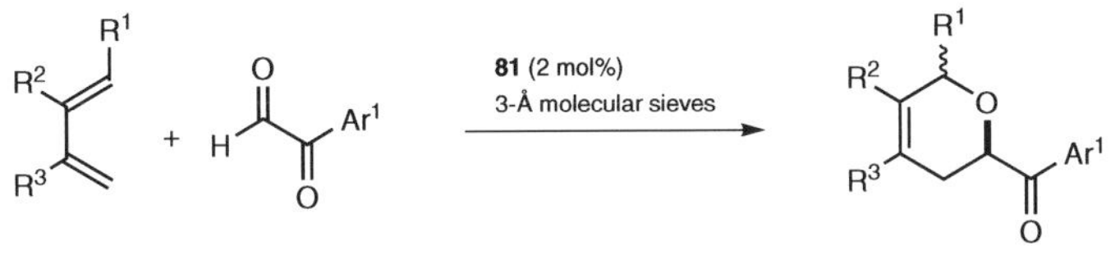

2.5.2 Oxa-Diels–Alder Reactions

R¹	R²	R³	Ar¹	ee (%)	Yield (%)	Ref
H	Me	Me	Ph	99	67	[60]
Me	H	H	Ph	38[a]	46[a]	[60]
H	Me	Me	4-ClC$_6$H$_4$	97	57	[60]
H	Me	Me	4-MeOC$_6$H$_4$	98	64	[60]

[a] The yield and ee are quoted for the *cis*-product; however, the *trans*-product is also obtained (20%; 80% ee).

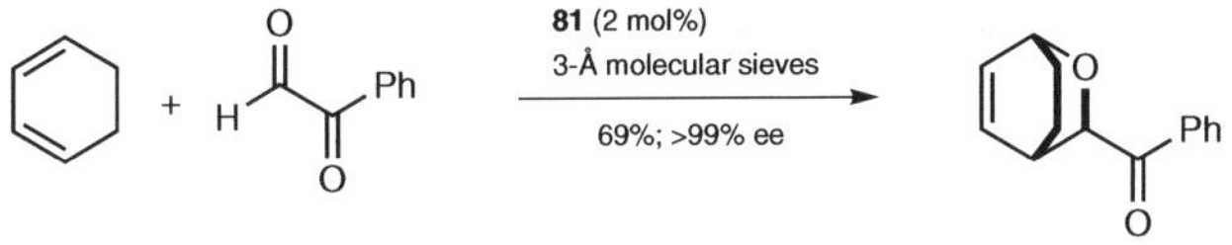

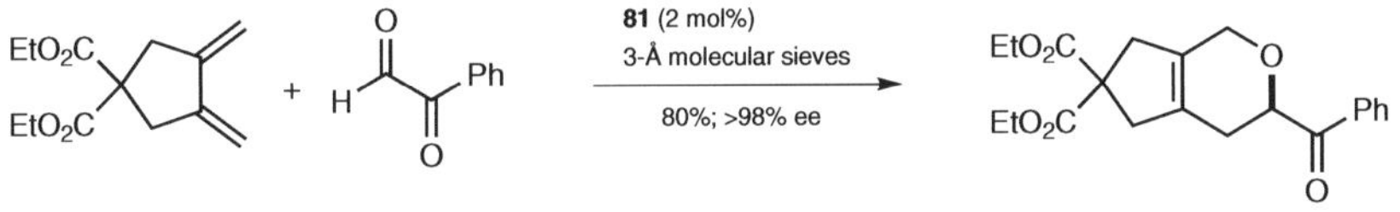

As shown in a proposed asymmetric induction model **83**, the arylglyoxal replaces the benzonitrile ligands by coordinating to the palladium center via the two carbonyl oxygen atoms (Scheme 40). Attack on the *si*-face of the arylgloxyal is obstructed by the phenyl substituents of the diphenylphosphino groups, and thus the attack of the diene on the *re*-face is favored to afford the cycloadduct in an *R* configuration.

Scheme 40 Asymmetric Induction Model for Palladium-Catalyzed Oxa-Diels–Alder Reactions[60]

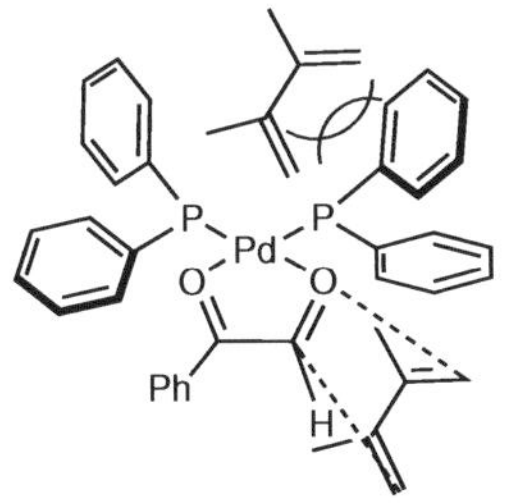

(*R*)-2-Benzoyl-4,5-dimethyl-3,6-dihydro-2*H*-pyran (82, R¹ = H; R² = R³ = Me; Ar¹ = Ph); Typical Procedure:[60]

Phenylglyoxal (268 mg, 2.0 mmol) and 2,3-dimethylbuta-1,3-diene (246 mg, 3.0 mmol) were added to a mixture of catalyst **81** (44.3 mg, 0.040 mmol), powdered 3-Å molecular sieves (100 mg), and CHCl$_3$ (4.0 mL). The resulting mixture was stirred at 0 °C for 24 h under a N$_2$ atmosphere and then Et$_2$O (25 mL) was added. This soln was filtered through a short silica gel column, eluting with Et$_2$O, and the solvent was then removed. The residue obtained was purified by column chromatography (silica gel, hexanes/EtOAc 10:1) to give a colorless oil; yield: 290 mg (67%); 99% ee.

2.5.2.10 Reaction Using Chiral Rare Earth Metal Complexes

A chiral erbium complex generated in situ from norephedrine derivative **84** (10–20 mol%) and erbium(III) trifluoromethanesulfonate (1.5 equiv) in the presence of N,N-diisopropylethylamine promotes the asymmetric oxa-Diels–Alder reaction of α,β-unsaturated acid chlorides with a variety of aromatic aldehydes to afford 5,6-dihydro-2H-pyran-2-ones **85** efficiently with up to 98% enantiomeric excess (Scheme 41).[61,62] This reaction proceeds via an aldol reaction pathway followed by an intramolecular acylation. The chiral erbium complex acts as a bifunctional catalyst which binds both substrates at the same metal center.

Scheme 41 Erbium-Catalyzed Asymmetric Oxa-Diels–Alder Reactions[61]

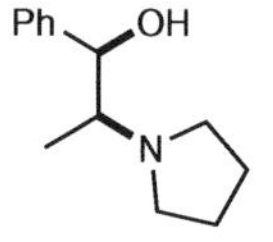

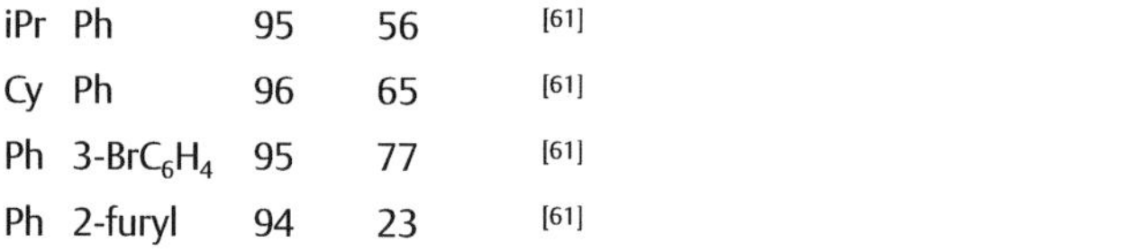

R^1	Ar^1	ee (%)	Yield (%)	Ref
iPr	Ph	95	56	[61]
Cy	Ph	96	65	[61]
Ph	3-BrC$_6$H$_4$	95	77	[61]
Ph	2-furyl	94	23	[61]
Ph	2-thienyl	95	40	[61]

(R)-4-Isopropyl-6-phenyl-5,6-dihydro-2H-pyran-2-one (85, R^1 = iPr; Ar^1 = Ph); Typical Procedure:[61]

THF (0.9 mL), toluene (0.8 mL), and a soln of amine **84** (0.068 mmol) in THF (0.3 mL) were added to a reaction flask charged with activated Er(OTf)$_3$ (0.51 mmol). The resulting mixture was stirred at rt for 15 min and, after cooling to −10 °C, iPr$_2$NEt (0.85 mmol) and PhCHO (0.34 mmol) were added successively. A soln of 3,4-dimethylpent-2-enoyl chloride (0.34 mmol) in toluene (0.5 mL) was then added over 30 min using a syringe pump and, after being stirred for an additional 2 h at −10 °C, the mixture was filtered through a short (2 cm) plug of silica gel, eluting with hexanes/EtOAc (1:1). The solvent was removed under reduced pressure and the residue obtained was purified by flash chromatography (silica gel, hexanes/EtOAc 4:1) to give a white solid; yield: 41.2 mg (56%); 95% ee.

2.5.3 Aza-Diels–Alder Reactions

There are two possible mechanistic pathways for Lewis acid catalyzed aza-Diels–Alder reactions: a concerted (4+2)-cycloaddition (Diels–Alder) pathway and a stepwise Mannich-type pathway (Scheme 42). The influence of Lewis acids on the reaction pathway is dependent on the nature of the Lewis acid and the reactant structure.

Scheme 42 Two Mechanistic Pathways for Aza-Diels–Alder Reactions

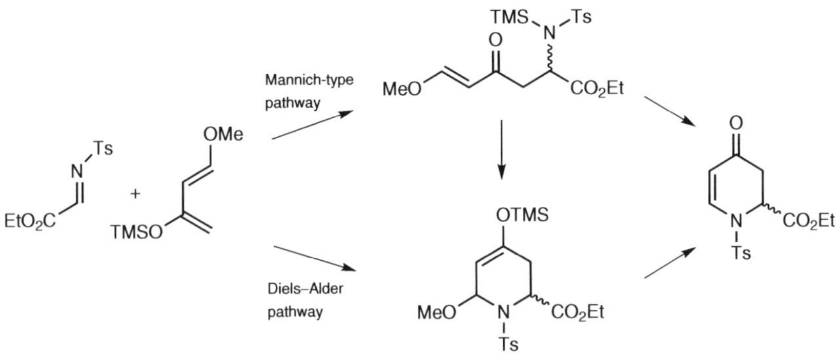

2.5.3.1 Reaction Using Chiral Copper Complexes

2.5.3.1.1 Normal-Electron-Demand Aza-Diels–Alder Reactions

Chiral copper catalyst **86**, based on a planar chiral 1-phosphino-2-sulfanylferrocene, exhibits excellent activity and high enantioselectivity for the asymmetric aza-Diels–Alder reaction between N-sulfonyl-substituted aldimines and Danishefsky-type dienes, resulting in the formation of 1-sulfonyl-substituted 2,3-dihydropyridin-4(1H)-ones **87** (Scheme 43).[63] A wide range of imines are well tolerated in this reaction and, importantly, the products are relatively stable, easy-to-handle crystalline solids which give rise to enantiopure samples after a single recrystallization.

Scheme 43 Copper-Catalyzed Aza-Diels–Alder Reactions[63]

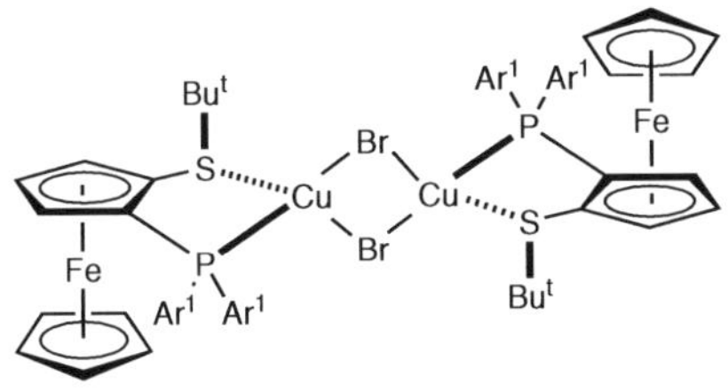

86 Ar¹ = 1-naphthyl

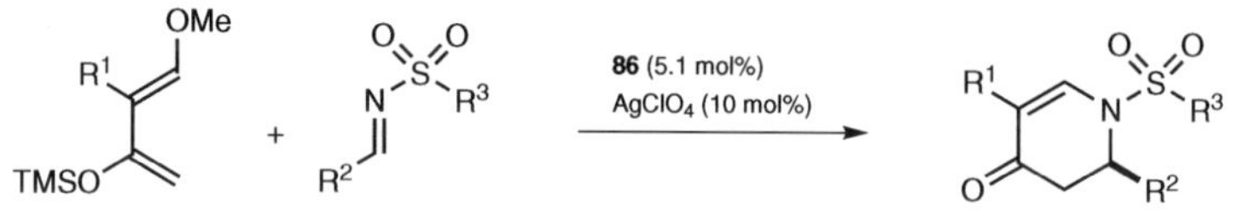

R^1	R^2	R^3	ee (%)	Yield (%)	Ref
H	Ph	4-Tol	93	90	[63]
H	4-Me$_2$NC$_6$H$_4$	4-Tol	93	39	[63]
H	(E)-CH=CHPh	4-Tol	83	66	[63]
H	Ph	4-O$_2$NC$_6$H$_4$	90	58	[63]
Me	Ph	4-Tol	87	64	[63]
Me	(E)-CH=CHPh	4-Tol	88	57	[63]

(R)-2-Phenyl-1-tosyl-2,3-dihydropyridin-4(1H)-one (87, R^1 = H; R^2 = Ph; R^3 = 4-Tol); Typical Procedure:[63]

A soln of catalyst **86** (0.0056 mmol) and AgClO$_4$ (0.011 mmol) in CH$_2$Cl$_2$ (0.6 mL) was stirred in the dark under argon at rt for 2 h and then treated with a soln of N-benzylidene-4-toluenesulfonamide (0.11 mmol) in CH$_2$Cl$_2$ (1 mL). The resulting mixture was stirred at rt for 5 min before Danishefsky's diene (0.15 mmol) was added. Once the starting material was consumed, TFA (0.1–0.2 mL) was added and this mixture was stirred at rt for 30–60 min, neutralized with sat. aq NaHCO$_3$, and extracted with CH$_2$Cl$_2$. The organic extracts were dried (MgSO$_4$) and filtered, and the solvent was removed. The residue obtained was purified by flash chromatography (silica gel); yield: 90%; 93% ee.

2.5.3.1.2 Inverse-Electron-Demand Aza-Diels–Alder Reactions

An inverse-electron-demand aza-Diels–Alder reaction of indoles with in situ formed azoalkenes, catalyzed by a chiral complex derived from ligand **88** and a copper(I) salt, provides a variety of 2,3-fused indole-containing heterocycles **89** in good yield and with high levels of regioselectivity, diastereoselectivity (dr >20:1), and enantioselectivity (up to 99% ee) (Scheme 44).[64]

Scheme 44 Copper-Catalyzed Inverse-Electron-Demand Aza-Diels–Alder Reactions[64]

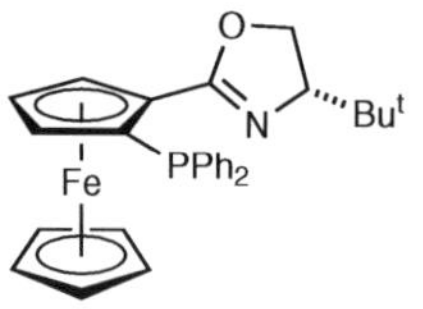

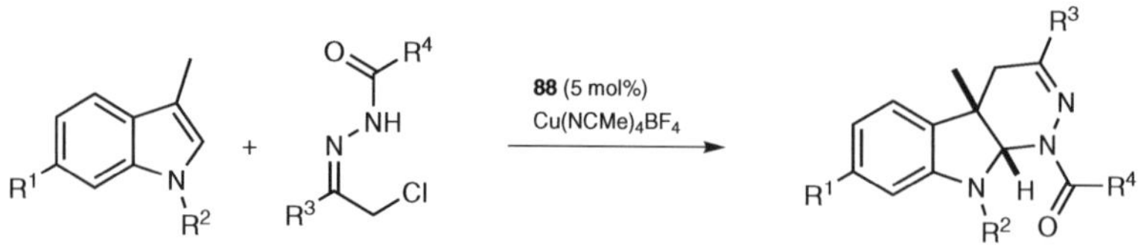

2.5.3 Aza-Diels–Alder Reactions

R¹	R²	R³	R⁴	ee (%)	Yield (%)	Ref
H	Me	Ph	Ph	97	95	[64]
Cl	Me	Ph	Ph	98	90	[64]
H	CH$_2$CH=CH$_2$	Ph	Ph	97	89	[64]
H	Me	(E)-CH=CHPh	Ph	93	88	[64]
H	Me	Ph	O*t*-Bu	94	84	[64]

(4a*S*,9a*S*)-1-(Benzoyl)-4a,9-dimethyl-3-phenyl-4,4a,9,9a-tetrahydro-1*H*-pyridazino[3,4-*b*]indole (89, R¹ = H; R² = Me; R³ = R⁴ = Ph); Typical Procedure:[64]

Ligand **88** (5.2 mg, 0.0165 mmol) and Cu(NCMe)$_4$BF$_4$ (4.8 mg, 0.015 mmol) were dissolved in CH$_2$Cl$_2$ (4.0 mL) under an argon atmosphere, and the resulting mixture was stirred at rt for about 30 min. The reaction temperature was then lowered to −20 °C and the benzohydrazide (R³ = R⁴ = Ph; 0.3 mmol) and Na$_2$CO$_3$ (0.4 mmol) were added sequentially, followed by 1,3-dimethyl-1*H*-indole (0.5 mmol). Once the starting material had been consumed, the organic solvent was removed and the residue was purified by column chromatography (silica gel); yield: 95%; 97% ee.

2.5.3.2 Reaction Using Chiral Zinc Complexes

A chiral zinc complex derived from (*R*)-1,1′-binaphthalene-2,2′-diol [**46**; (*R*)-BINOL] and diethylzinc has been used to promote the enantioselective inverse-electron-demand aza-Diels–Alder reaction of electron-deficient 1-benzopyran-4-one-derived dienes and various cyclic imines to give ring-fused quinolizines **90** in moderate to good yield and with up to 95% enantiomeric excess (Scheme 45).[65]

Scheme 45 Zinc-Catalyzed Inverse-Electron-Demand Aza-Diels–Alder Reactions[65]

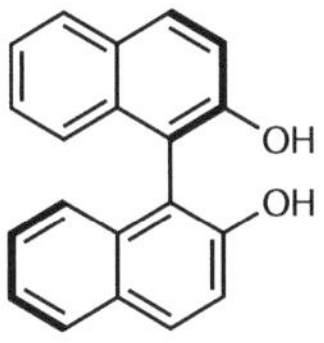

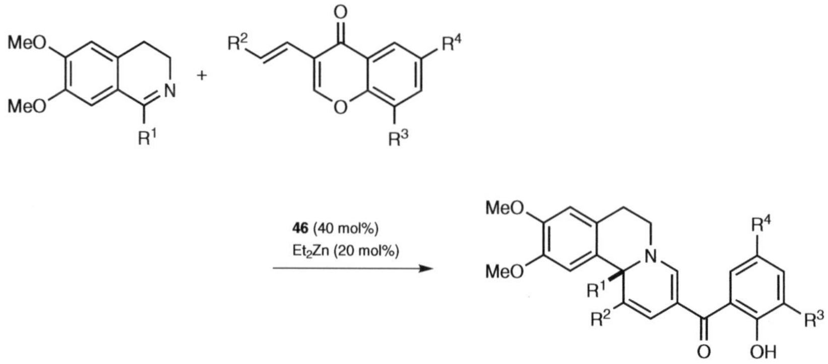

R^1	R^2	R^3	R^4	ee (%)	Yield (%)	Ref
Cy	CN	H	H	84	41	[65]
Me	CO$_2$Me	Cl	Cl	88	90	[65]
Me	CO$_2$Me	H	Cl	87	86	[65]
Me	CN	H	Cl	83	90	[65]
Et	CO$_2$Me	H	Me	82	67	[65]

In the ensuing transition state **91**, the zinc complex coordinates to the imine nitrogen and the vinylogous ester oxygen atom of the diene. One naphthalene ring is oriented orthogonal to the plane of the hetero-dienophile and thus shields the *si*-face of the imine, which allows attack of the diene from the *re*-face of the imine (Scheme 46).[65]

Scheme 46 Transition State for Zinc-Catalyzed Inverse-Electron-Demand Aza-Diels–Alder Reactions[65]

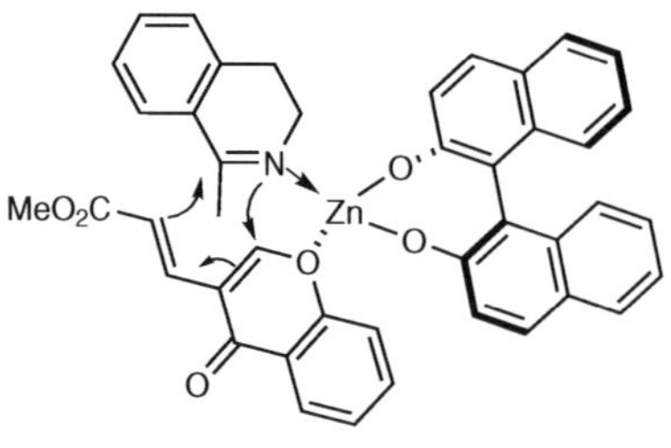

11b-Cyclohexyl-3-(2-hydroxybenzoyl)-9,10-dimethoxy-7,11b-dihydro-6H-pyrido[2,1-a]isoquinoline-1-carbonitrile (90, R^1 = Cy; R^2 = CN; R^3 = R^4 = H); Typical Procedure:[65]
Et$_2$Zn (0.01 mmol) and ligand **46** (0.02 mmol) were dissolved in dry toluene (5 mL) under argon and stirred for 15 min. 1-Cyclohexyl-6,7-dimethoxy-3,4-dihydroisoquinoline (0.05 mmol) was added and the resulting mixture was cooled to −78 °C. (*E*)-3-(4-Oxo-4*H*-1-benzopyran-3-yl)acrylonitrile (0.10 mmol) was then added and this mixture was allowed to react for 12 h at −78 °C in a sealed tube under argon. On completion of the reaction, the solvent was removed and the residue was purified by flash chromatography (silica gel); yield: 41%; 84% ee.

2.5.3.3 Reaction Using Chiral Silver Complexes

The chiral P,N-ligand **92**, derived from an inexpensive amino acid, is effective for the silver-catalyzed asymmetric cycloaddition of aryl imines with Danishefsky's diene. With the use of 0.1–1 mol% of the catalyst, the cycloadducts **93** are obtained in >78% yield and with up to 95% enantiomeric excess (Scheme 47).[66] The presence of the 2-methoxy group in the imine is required for high enantioselectivity but not for reactivity. In the absence of propan-2-ol significantly lower conversions and enantioselectivities are obtained.

Scheme 47 Silver-Catalyzed Aza-Diels–Alder Reactions[66]

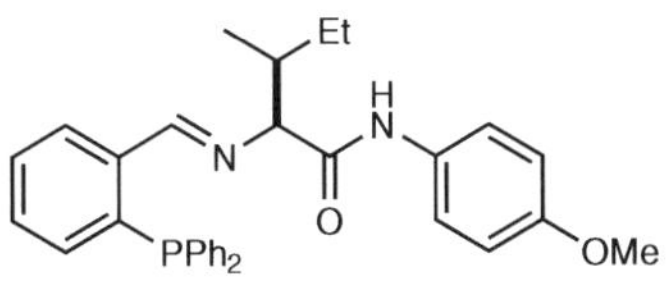

Ar¹	Catalyst Loading (mol%)	ee (%)	Yield (%)	Ref
Ph	1	93	94	[66]
Ph	0.1	88	78	[66]
1-naphthyl	1	90	94	[66]
2-naphthyl	0.5	95	>98	[66]
4-ClC$_6$H$_4$	1	90	98	[66]
2-furyl	1	92	89	[66]

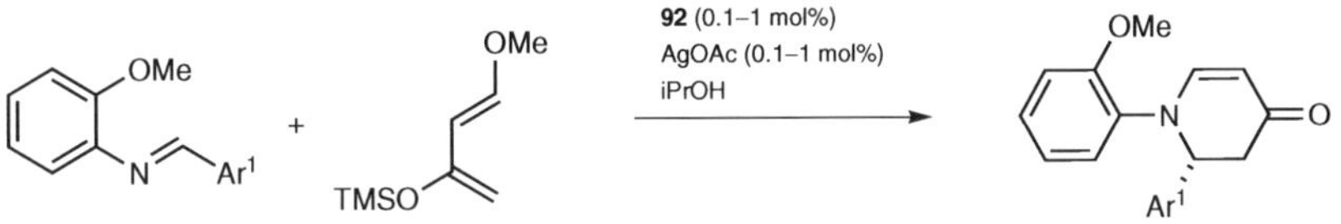

(R)-1-(2-Methoxyphenyl)-2-phenyl-2,3-dihydropyridin-4(1H)-one (93, Ar¹ = Ph); Typical Procedure:[66]

Ligand **92** (6.0 mg, 0.012 mmol) and AgOAc (2.0 mg, 0.012 mmol) were weighed into a test tube (16 × 150 mm) under a N$_2$ atmosphere in a glovebox. The contents were dissolved in THF (0.5 mL) and stirred for 15 min at 22 °C. N-Benzylidene-2-methoxyaniline (1.20 mmol) was added, followed immediately by THF (1.5 mL) and iPrOH (0.1 mL, 1.3 mmol). The test tube was capped with a septum, sealed with Parafilm, removed from the glovebox, and allowed to stir at 4 °C for 15 min. Danishefsky's diene (0.36 mL, 1.9 mmol) was added by syringe and the mixture was stirred at 4 °C for 12 h. The reaction was quenched by addition of 10% aq HCl (2 mL) followed by vigorous stirring for 5 min. CH$_2$Cl$_2$ (5 mL) was added to the test tube and the mixture was stirred vigorously for 1 min. The layers were allowed to separate and the bottom layer was removed using a pipet and placed into a round-bottomed flask. The aqueous layer was washed with CH$_2$Cl$_2$ (2 × 5 mL) and the combined organic layers were concentrated under reduced pressure and purified by chromatography (silica gel, EtOAc/hexanes 7:3); yield: 94%; 93% ee.

2.5.3.4 Reaction Using Chiral Zirconium Complexes

Chiral zirconium complex **94**, prepared from zirconium(IV) *tert*-butoxide, (R)-6,6′-dibromo-1,1′-binaphthalene-2,2′-diol, and 1-methyl-1*H*-imidazole, is an efficient catalyst for the enantioselective aza-Diels–Alder reaction of imine dienophiles and Danishefsky-type dienes to give optically active piperidine derivatives **95** with high yield and with high levels of enantioselectivity (Scheme 48).[67–69] The achiral ligand and the solvent have a strong influence on both the reactivity and enantioselectivity. To prevent the isomerization of the *Z*- and *E*-forms of the cyclohexyl-substituted aldimine (R^2 = Cy), additional methyl-substitution (R^1 = Me) is required and results in a satisfactory enantiomeric excess.

Scheme 48 Zirconium-Catalyzed Aza-Diels–Alder Reactions[67]

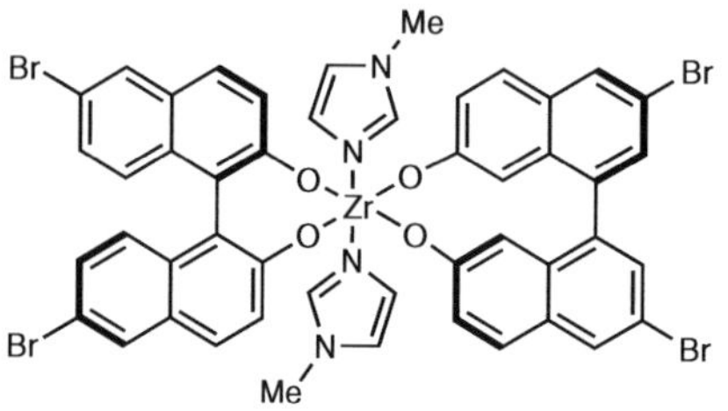

R^1	R^2	R^3	Catalyst Loading (mol%)	ee (%)	Yield (%)	Ref
H	2-Tol	H	10	76	81	[67]
H	2-Tol	Me	20	77	97	[67]
H	1-naphthyl	H	5	67	72	[67]
H	2-thienyl	H	10	64	86	[67]
Me	Cy	H	20	86	51	[67]

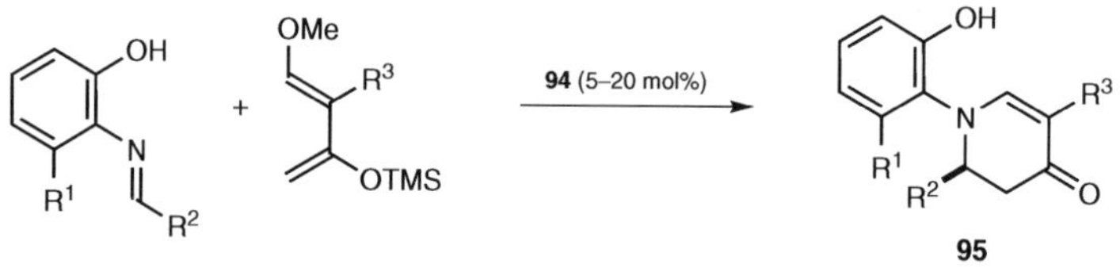

(S)-1-(2-Hydroxyphenyl)-2-(2-tolyl)-2,3-dihydropyridin-4(1H)-one (95, R^1 = R^3 = H; R^2 = 2-Tol); Typical Procedure:[67]

(R)-6,6′-Dibromo-1,1′-binaphthalene-2,2′-diol (0.088 mmol) in toluene (0.5 mL) and 1-methyl-1*H*-imidazole (0.12 mmol) in toluene (0.25 mL) were added to Zr(O*t*-Bu)$_4$ (0.04 mmol) in toluene (0.25 mL) at rt. The mixture was stirred for 1 h at the same temperature and then cooled to −45 °C. The imine (R^1 = H; R^2 = 2-Tol; 0.4 mmol) in toluene (0.75 mL) and Danishefsky's diene (0.6 mmol) in toluene (0.75 mL) were added successively, and the mixture was stirred for 35 h at the same temperature. Sat. aq NaHCO$_3$ was added to quench the reaction and the aqueous layer was extracted with CH$_2$Cl$_2$. The crude adduct was treated with THF/1 M HCl (20:1) at 0 °C for 30 min and, after the usual workup, the crude residue was purified by chromatography (silica gel); yield: 81%; 76% ee.

2.5.3.5 Reaction Using Chiral Niobium Complexes

The chiral complex derived from niobium(V) methoxide and ligand **96** is an effective catalyst for the enantioselective aza-Diels–Alder reaction of aromatic and aliphatic imines with a Danishefsky-type diene to afford the adducts **97** in high yield and with high enantioselectivity (Scheme 49).[70] The phenolic hydroxy group *ortho* to the nitrogen atom is necessary for high yields and selectivities. When this group is substituted by a methoxy group, very low yields and enantiomeric excesses are obtained.

Scheme 49 Niobium-Catalyzed Aza-Diels–Alder Reactions[70]

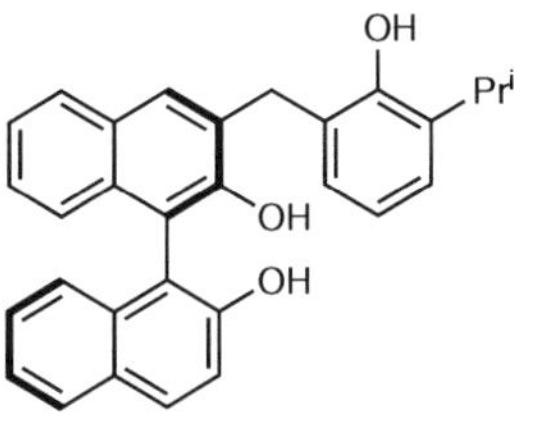

R^1	R^2	ee (%)	Yield (%)	Ref
H	Ph	96	81	[70]
H	4-Tol	94	90	[70]
H	1-naphthyl	92	89	[70]
H	2-F$_3$CC$_6$H$_4$	99	94	[70]
H	3-pyridyl	92	74	[70]
Me	Cy	90	67	[70]
Me	iPr	92	63	[70]

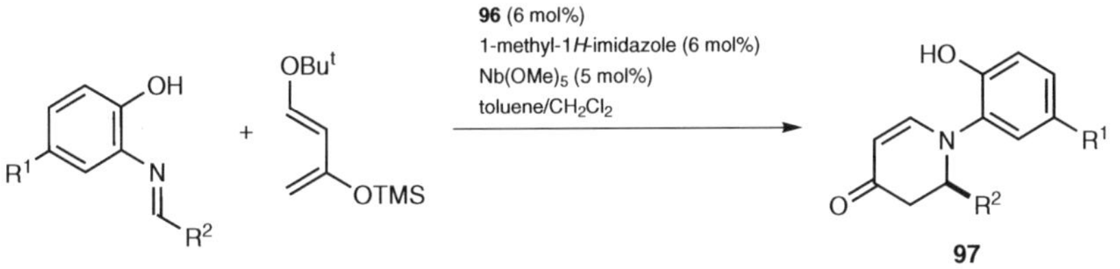

(R)-1-(2-Hydroxyphenyl)-2-phenyl-2,3-dihydropyridin-4(1H)-one (97, R^1 = H; R^2 = Ph); Typical Procedure:[70]

1-Methyl-1H-imidazole (1.5 µL, 0.018 mmol) was added to a soln of ligand **96** (7.8 mg, 0.018 mmol) in toluene (1.0 mL) and, after 5 min of stirring, Nb(OMe)$_5$ (3.7 mg, 0.015 mmol) was added as a solid under a gentle stream of argon. A trace of Nb(OMe)$_5$ which remained on the vessel wall was washed with an additional portion of toluene (0.5 mL) and the resulting mixture was heated to 60 °C for 3 h and then allowed to cool to rt. At this point, 3-Å molecular sieves (25 mg) were added and the catalyst soln was cooled to −20 °C. The imine (R^1 = H; R^2 = Ph; 0.3 mmol) in CH$_2$Cl$_2$ (1.5 mL) was then added, followed by the diene (0.1 mL, 0.4 mmol), and, after 48 h, the reaction was quenched by addition of sat. aq NaHCO$_3$ (3 mL). The resulting mixture was extracted with EtOAc (4 × 5 mL) and the combined organic fractions were dried (Na$_2$SO$_4$). The solvent was removed under reduced pressure and the crude material was cooled to 0 °C, treated with 0.1 M HCl in THF (10 mL), and, after 15 min, basified by addition of sat. aq NaHCO$_3$. This mixture was extracted with EtOAc (4 × 5 mL) and the combined organic fractions were dried (Na$_2$SO$_4$) and filtered, and the solvent was removed under reduced pressure. The crude residue obtained was purified by chromatography (silica gel, hexanes/EtOAc 1:1); yield: 81%; 96% ee.

2.5.3.6 Reaction Using Chiral Nickel Complexes

An inverse-electron-demand aza-Diels–Alder reaction of 1-azadienes, bearing an (8-quinolyl)sulfonyl moiety at the imine nitrogen, and vinyl ethers has been realized using the chiral complex derived from ligand **98** and nickel(II) perchlorate hexahydrate as the catalyst. This process gives the highly functionalized piperidine derivatives **99** in good yield with excellent *endo* selectivity, and with enantioselectivities typically in the range of 77–92% (Scheme 50).[71] In contrast to the good tolerance of alkene substituents (R^1), substituent compatibility at the imine carbon (Ar^1) has proved to be more limited.

Scheme 50   Nickel-Catalyzed Aza-Diels–Alder Reactions[71]

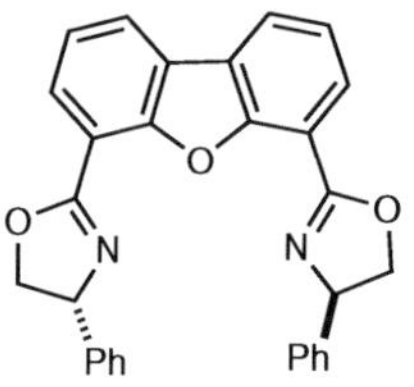

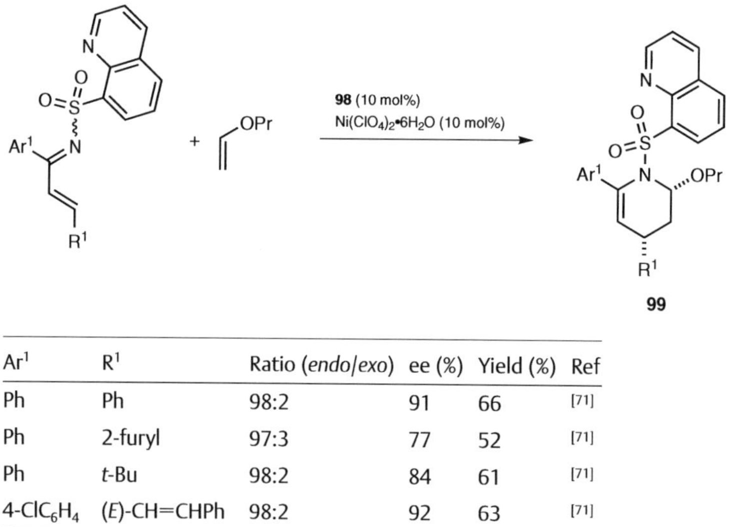

Ar^1	R^1	Ratio (*endo/exo*)	ee (%)	Yield (%)	Ref
Ph	Ph	98:2	91	66	[71]
Ph	2-furyl	97:3	77	52	[71]
Ph	*t*-Bu	98:2	84	61	[71]
4-ClC$_6$H$_4$	(*E*)-CH=CHPh	98:2	92	63	[71]

(2*R*,4*S*)-4,6-Diphenyl-2-propoxy-1-(quinolin-8-ylsulfonyl)-1,2,3,4-tetrahydropyridine (99, Ar^1 = R^1 = Ph); Typical Procedure:[71]

A soln of Ni(ClO$_4$)$_2$·6H$_2$O (7.2 mg, 0.02 mmol) and ligand **98** (9.2 mg, 0.02 mmol) in CH$_2$Cl$_2$ (1 mL) was stirred at rt for 4 h in a Schlenk flask. A soln of the (*E*)-1-azadiene (Ar^1 = R^1 = Ph; 0.2 mmol) in CH$_2$Cl$_2$ (1 mL) was then added, followed by propyl vinyl ether (5 equiv). The resulting mixture was stirred at rt until complete consumption of the starting azadiene was observed. The reaction was then quenched with sat. aq NH$_4$Cl and extracted several times with CH$_2$Cl$_2$. The combined organic phases were dried (Na$_2$SO$_4$), filtered, and concentrated, and the crude residue was purified by flash chromatography (silica gel, CH$_2$Cl$_2$) to give the product; yield: 66%; ratio (*endo/exo*) 98:2; 91% ee.

2.5.3.7 Reaction Using Chiral Rhodium Complexes

The enantioselective cycloaddition reaction of 3-(silyloxy)-substituted 2-azabuta-1,3-dienes with a variety of aldehydes proceeds in an *endo*-selective mode, using chiral rhodium complex **79** (1 mol%) as the catalyst, to give all-*cis*-substituted 1,3-oxazinan-4-ones **100** in 64–99% yield and with 90–98% enantiomeric excess (Scheme 51).[72] A hydrogen bond between the formyl hydrogen atom and the carboamide oxygen is proposed to exist in the rhodium catalyst–aldehyde complex.

Scheme 51 Rhodium-Catalyzed Aza-Diels–Alder Reactions[72]

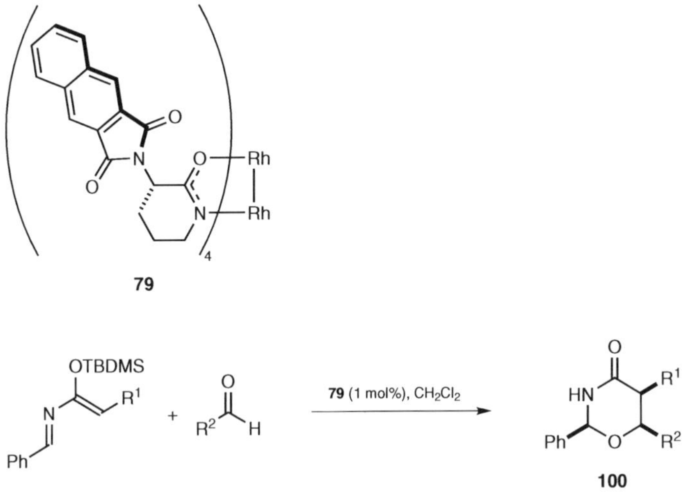

R¹	R²	ee (%)	Yield (%)	Ref
H	Ph	98	95	[72]
H	2-furyl	95	93	[72]
H	(E)-CH=CHPh	96	88	[72]
H	C≡CPh	95	99	[72]
H	Pr	93	92	[72]
Me	C≡CPh	91	95	[72]

(2S,6S)-2,6-Diphenyl-1,3-oxazinan-4-one (100, R¹ = H; R² = Ph); Typical Procedure:[72]
Catalyst **79** (4.3 mg, 0.003 mmol) was added to a soln of PhCHO (0.45 mmol) in CH₂Cl₂ (0.4 mL) and the color of the soln changed from colorless to brown. The resulting mixture was stirred for 5 min, a soln of the 2-azadiene (R¹ = H; 78.4 mg, 0.30 mmol) in CH₂Cl₂ (0.2 mL) was added at rt, and this mixture was stirred at rt for 24 h. MeOH (ca. 0.1 mL) was then added and the mixture was stirred for an additional 0.5 h, concentrated under reduced pressure, and purified by column chromatography (silica gel, CHCl₃/EtOAc 9:1); yield: 95%; 98% ee.

2.5.3.8 Reaction Using Chiral Rare Earth Metal Complexes

2.5.3.8.1 Normal-Electron-Demand Aza-Diels–Alder Reactions

The enantioselective aza-Diels–Alder reaction between a buta-1,3-diene and various aldimines has been realized using the chiral complex derived from scandium(III) trifluoromethanesulfonate and N,N′-dioxide ligand **101** to afford 2-substituted 1-(2-hydroxyphenyl)-5-methyl-2,3-dihydropyridin-4(1H)-ones **102** in moderate to good yield and with high enantioselectivity (Scheme 52).[73] A wide range of imines containing aromatic, heteroaromatic, conjugated, and aliphatic substituents are effective for this reaction.

Scheme 52 Scandium-Catalyzed Aza-Diels–Alder Reactions[73]

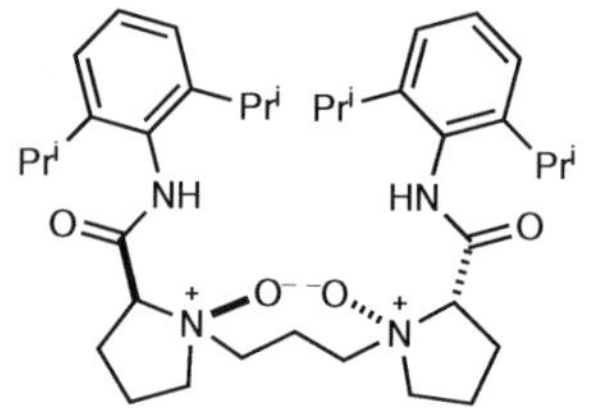

R^1	ee (%)	Yield (%)	Ref
Ph	81	74	[73]
4-NCC$_6$H$_4$	90	84	[73]
2-ClC$_6$H$_4$	83	89	[73]
(E)-CH=CHPh	84	92	[73]
2-thienyl	82	46	[73]
iPr	71	67	[73]

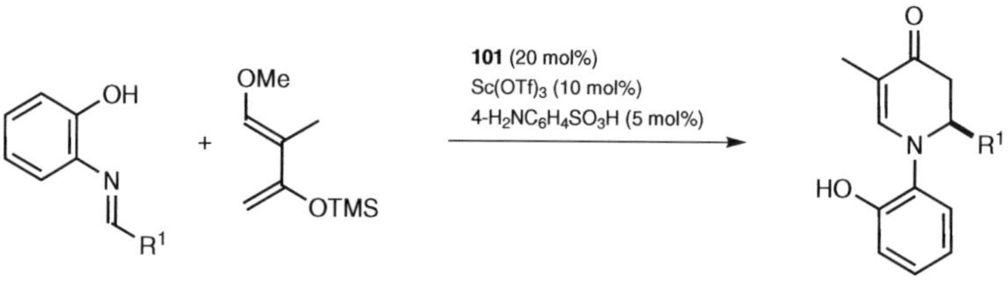

(S)-1-(2-Hydroxyphenyl)-5-methyl-2-phenyl-2,3-dihydropyridin-4(1H)-one (102, R^1 = Ph); Typical Procedure:[73]

Ligand **101** (24.8 mg, 0.04 mmol), Sc(OTf)$_3$ (10.0 mg, 0.02 mmol), 4-H$_2$NC$_6$H$_4$SO$_3$H (sulfanilic acid; 1.7 mg, 0.01 mmol), and the imine (R^1 = Ph; 39.4 mg, 0.2 mmol) were dissolved in THF (3 mL), and the resulting mixture was stirred in a test tube under an argon atmosphere for 15 min. The diene (0.1 mL, 0.4 mmol) was then added and the resulting mixture was stirred at rt for 24 h, cooled to 0 °C, treated with 1 M HCl (1 mL), and stirred for 1 h. Sat. aq NaHCO$_3$ (5 mL) was added and the soln was stirred for an additional 5 min and then diluted with CH$_2$Cl$_2$ (5 mL). The aqueous layer was extracted with CH$_2$Cl$_2$ (3 × 5 mL) and the combined organic layers were washed with brine, dried (Na$_2$SO$_4$), filtered, and concentrated under reduced pressure. The crude residue obtained was purified by flash chromatography (silica gel, hexanes/EtOAc 3:2); yield: 74%; 81% ee.

2.5.3.8.2 Inverse-Electron-Demand Aza-Diels–Alder Reactions

A highly efficient asymmetric three-component inverse-electron-demand aza-Diels–Alder reaction of aldehydes, 2-aminophenols, and cyclopentadiene has been realized, using 0.5–5 mol% of the complex derived from N,N′-dioxide ligand **103** and scandium(III) trifluoromethanesulfonate as the catalyst, to give ring-fused 3a,4,5,9b-tetrahydro-3H-cyclopenta-[c]quinolin-6-ols **104** containing three contiguous stereocenters in good to high yield with up to >99:1 diastereoselectivity and with up to >99% enantiomeric excess (Scheme 53).[74] Other dienophiles including cyclohexadiene, ethyl vinyl ether, and 2,3-dihydrofuran are not suitable substrates for this reaction.

Scheme 53 Scandium-Catalyzed Inverse-Electron-Demand Aza-Diels–Alder Reactions[74]

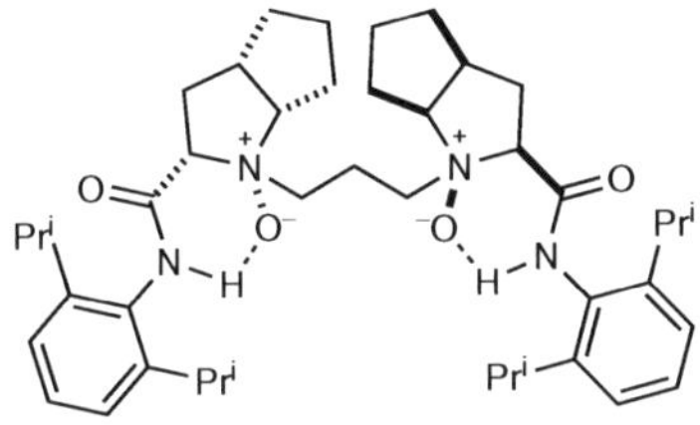

R^1	R^2	R^3	Ratio (cis/trans)	ee (%)	Yield (%)	Ref
Ph	H	H	96:4	97	90	[74]
Cy	H	H	>99:1	99	99	[74]
Ph	H	Me	96:4	98	85	[74]
Ph	Me	H	92:8	96	70	[74]
Ph	Cl	H	92:8	90	96	[74]
3-furyl	H	H	98:2	95	82	[74]
$CHEt_2$	H	H	>95:5	94	62	[74]

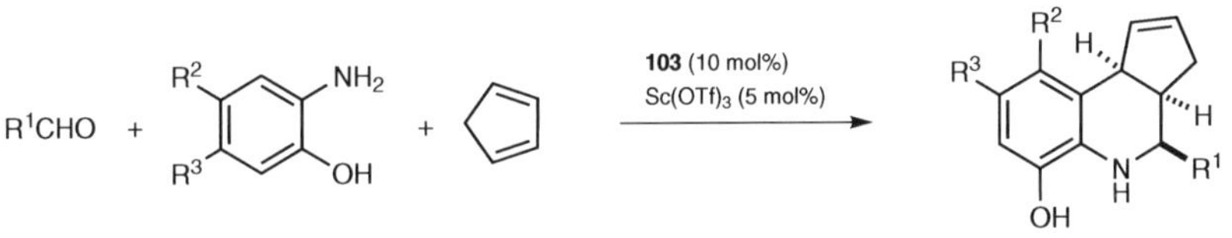

The combination of the Lewis acid yttrium(III) chloride with simple chiral silver phosphate **105** efficiently promotes the three-component inverse-electron-demand aza-Diels–Alder reaction of cyclic ketones with enones and arylamines to give the corresponding cycloadducts such as alkyl 1,4-diaryl-1,4,5,6,7,8-hexahydroquinoline-2-carboxylates **106** with a high level of reactivity and enantioselectivity (Scheme 54).[75] Replacing the arylamine with an aliphatic amine or the chiral Lewis acid with a Brønsted acid leads to a very messy reaction, suggesting that the use of both an arylamine and a metal Lewis acid is necessary in this reaction.

Scheme 54 Yttrium-Catalyzed Inverse-Electron-Demand Aza-Diels–Alder Reactions[75]

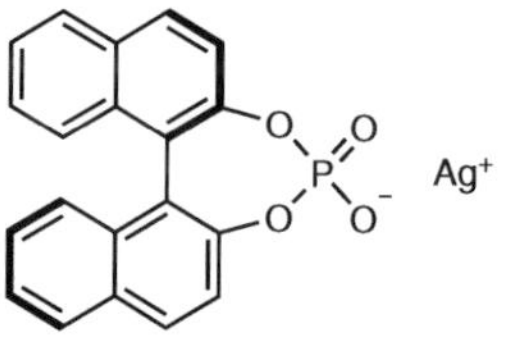

R^1	R^2	R^3	Z	ee (%)	Yield (%)	Ref
H	Bn	OMe	CH_2	89	72	[75]
Cl	Bn	OMe	CH_2	93	90	[75]
Br	Me	OMe	CH_2	95	85	[75]
NO_2	Et	OMe	CH_2	92	68	[75]
Cl	Me	Cl	CH_2	96	68	[75]
Cl	Me	OMe	S	93	73	[75]

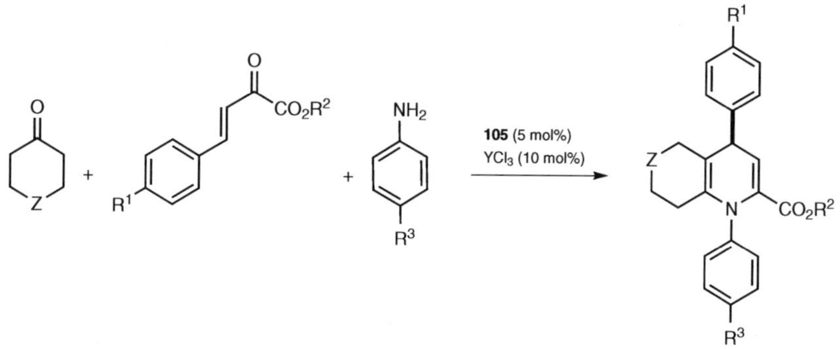

(3a*R*,4*R*,9b*S*)-4-Phenyl-3a,4,5,9b-tetrahydro-3*H*-cyclopenta[*c*]quinolin-6-ol (104, R^1 = Ph; R^2 = R^3 = H); Typical Procedure:[74]
Sc(OTf)$_3$ (2.5 mg, 0.005 mmol), ligand **103** (7.1 mg, 0.01 mmol), PhCHO (0.1 mmol), 2-aminophenol (10.9 mg, 0.1 mmol), and 4-Å molecular sieves (10.0 mg) were stirred in CH$_2$Cl$_2$ (0.3 mL) under N$_2$ at 25 °C for 0.5 h. Cyclopentadiene (0.15 mL) was added at 0 °C and the mixture was stirred until the 2-aminophenol was consumed. The residue was purified by flash chromatography (silica gel); yield: 90%; ratio (*cis/trans*) 96:4; 97% ee.

Benzyl (*R*)-1-(4-Methoxyphenyl)-4-phenyl-1,4,5,6,7,8-hexahydroquinoline-2-carboxylate (106, R^1 = H; R^2 = Bn; R^3 = OMe; Z = CH_2); Typical Procedure:[75]
A mixture of YCl$_3$ (3.9 mg, 0.02 mmol), phosphate **105** (4.6 mg, 0.01 mmol), and cyclohexanone (0.1 mL) in toluene (1 mL) was stirred at rt for 3 h. The enone (R^1 = H; R^2 = Bn; 0.2 mmol) and 4-methoxyaniline (0.2 mmol) were then added, and the resulting suspension was stirred at rt until the reaction was complete. The mixture was then filtered through a silica gel plug and the filtrate was concentrated. The residue obtained was purified by flash chromatography (silica gel); yield: 72%; 89% ee.

2.5.4 Applications in the Syntheses of Natural Products and Drug Molecules

Chiral Lewis acid catalyzed enantioselective Diels–Alder and hetero-Diels–Alder reactions have been successfully applied to the synthesis of various natural products and drug molecules. Some representative examples are briefly discussed in this section to illustrate these applications. For example, the enantioselective Diels–Alder reaction of benzo-1,4-quinone with a 1,3-diene under the catalysis of chiral oxazaborolidine complex **10** (20 mol%) in toluene at −78 °C affords the cycloadduct **107** in 95% yield with 90% enantiomeric excess and 100% diastereomeric excess. From this product, a further three-step transformation gives a key intermediate in the synthesis of cortisone (Scheme 55).[76]

Scheme 55 Synthesis of an Optically Active Key Intermediate toward Cortisone[76]

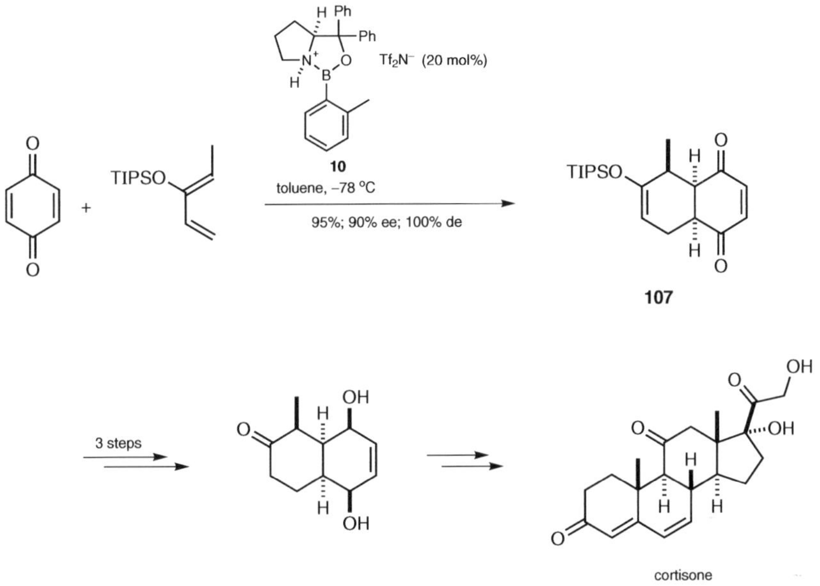

Tamiflu (oseltamivir phosphate) is a very important anti-influenza drug and several synthetic routes have been successfully developed using (−)-shikimic acid as a starting material. A facile approach to the total synthesis of Tamiflu employing the catalytic asymmetric Diels–Alder reaction as a key step has been reported. This process avoids the use of the relatively expensive and sparingly available (−)-shikimic acid, as well as avoiding potentially hazardous and explosive azide-containing intermediates. Thus, the enantioselective Diels–Alder reaction of buta-1,3-diene with 2,2,2-trifluoroethyl acrylate in the presence of chiral boron complex **10** (10 mol%) gives the cycloadduct **108** in 97% yield with >97% enantiomeric excess. Tamiflu is then efficiently synthesized in 11 steps from this key intermediate (Scheme 56).[77]

Scheme 56 Synthesis of Tamiflu Using an Enantioselective Diels–Alder Reaction[77]

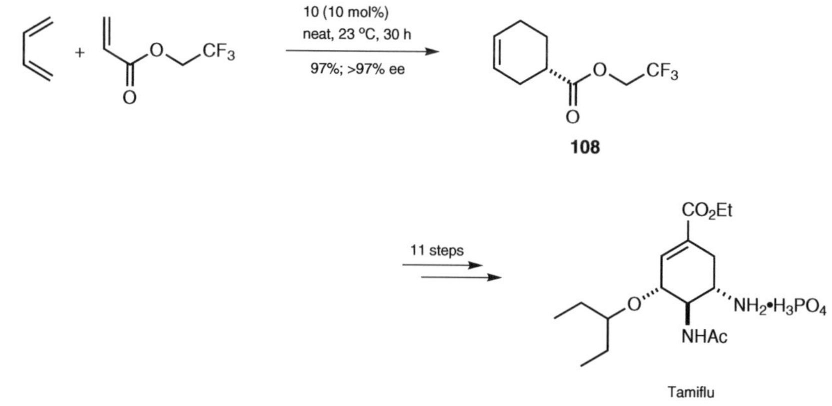

The total synthesis of fluostatin C, a member of a class of biologically important compounds with antibiotic and antitumor activities, has been developed utilizing an enantioselective Diels–Alder reaction as a key step. The cycloaddition of a 3-vinyl-1*H*-indene with 7-methyl-1,4-dioxaspiro[4.5]deca-6,9-dien-8-one under the catalysis of 1,1′-binaphthalene-2,2′-diol–titanium(IV) complex **27** affords the key intermediate **109** for the synthesis of fluostatin C in 93% yield with 65% enantiomeric excess (Scheme 57). Fluostatin C can be further converted into fluostatin E on treatment with hydrogen chloride.[78]

Scheme 57 Total Synthesis of Fluostatin C and Fluostatin E Using an Enantioselective Diels–Alder Reaction[78]

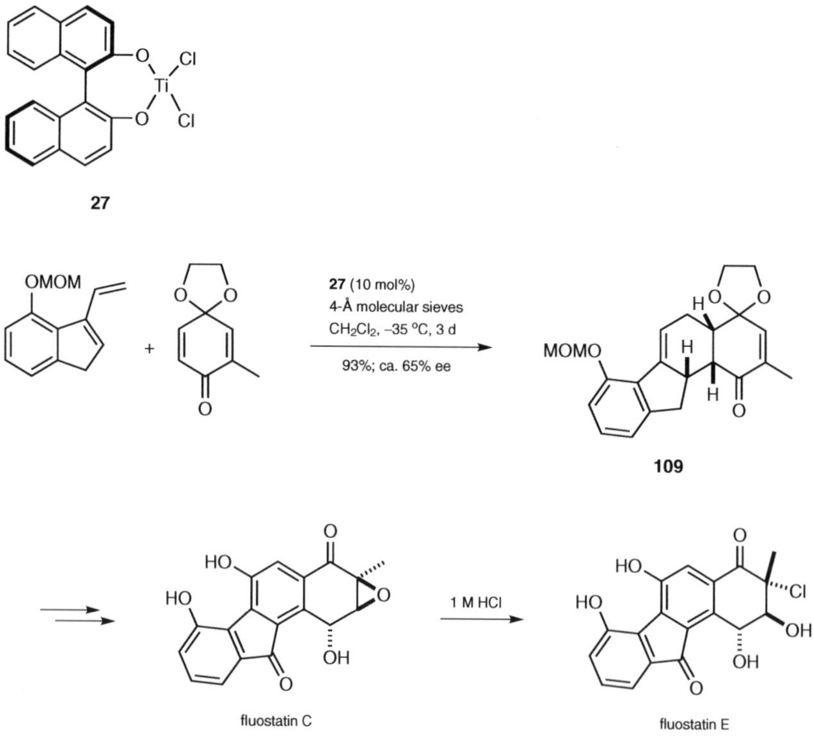

2.5.4 Applications in the Syntheses of Natural Products and Drug Molecules

A key intermediate for the total synthesis of the immunogen marine toxin (−)-gymnodimine has been obtained using a catalytic asymmetric Diels–Alder reaction as a key step in the presence of copper complex **110** (20 mol%). The highly functionalized cycloadduct **111** is constructed in 85% yield with 95% enantiomeric excess and a diastereomeric ratio of >19:1 (Scheme 58).[79]

Scheme 58 Total Synthesis of (−)-Gymnodimine Using a Catalytic Asymmetric Diels–Alder Reaction[79]

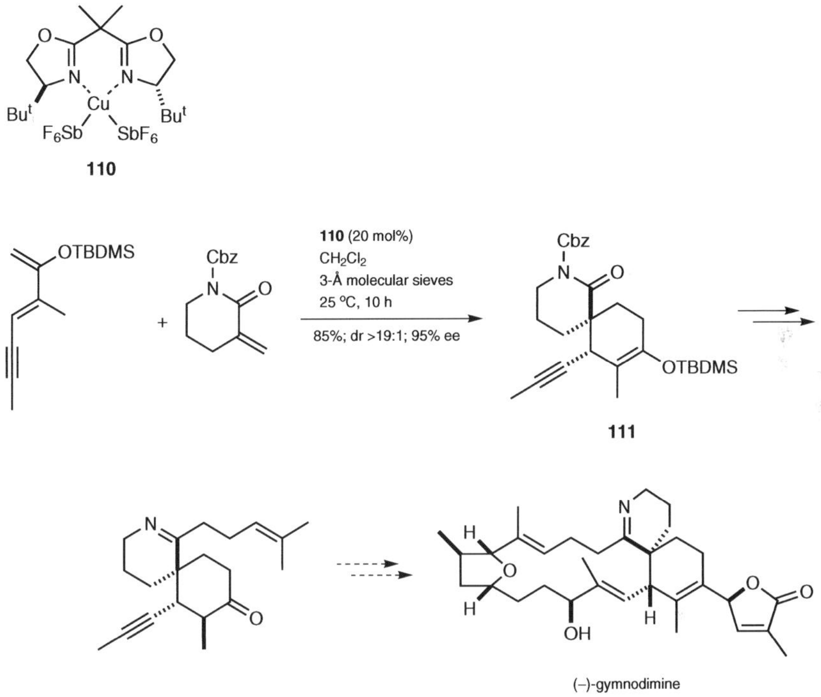

Enantioselective hetero-Diels–Alder reactions promoted by chiral chromium(III) complexes are often used as the key step in the total synthesis of natural products [e.g., FR901464,[80] (−)-colombiasin A,[81] elisapterosin B,[81] (−)-lasonolide A,[82] (+)-neopeltolide,[83] and platencin[84]]. In the total synthesis of antitumor polyketide anguinomycin C, chiral chromium(III) complex **112** catalyzes the enantioselective oxa-Diels–Alder reaction of 1-methoxybuta-1,3-diene with a triethylsilyl-protected propargylic aldehyde to construct the key dihydropyran fragment **113** in 86% yield and with 96% enantiomeric excess. A further seven-step reaction sequence is required to give fragment **114**, which undergoes cross coupling with fragment **115** followed by deprotection to accomplish the total synthesis of anguinomycin C (Scheme 59).[85]

Scheme 59 Total Synthesis of Anguinomycin C Using an Enantioselective Oxa-Diels–Alder Reaction[85]

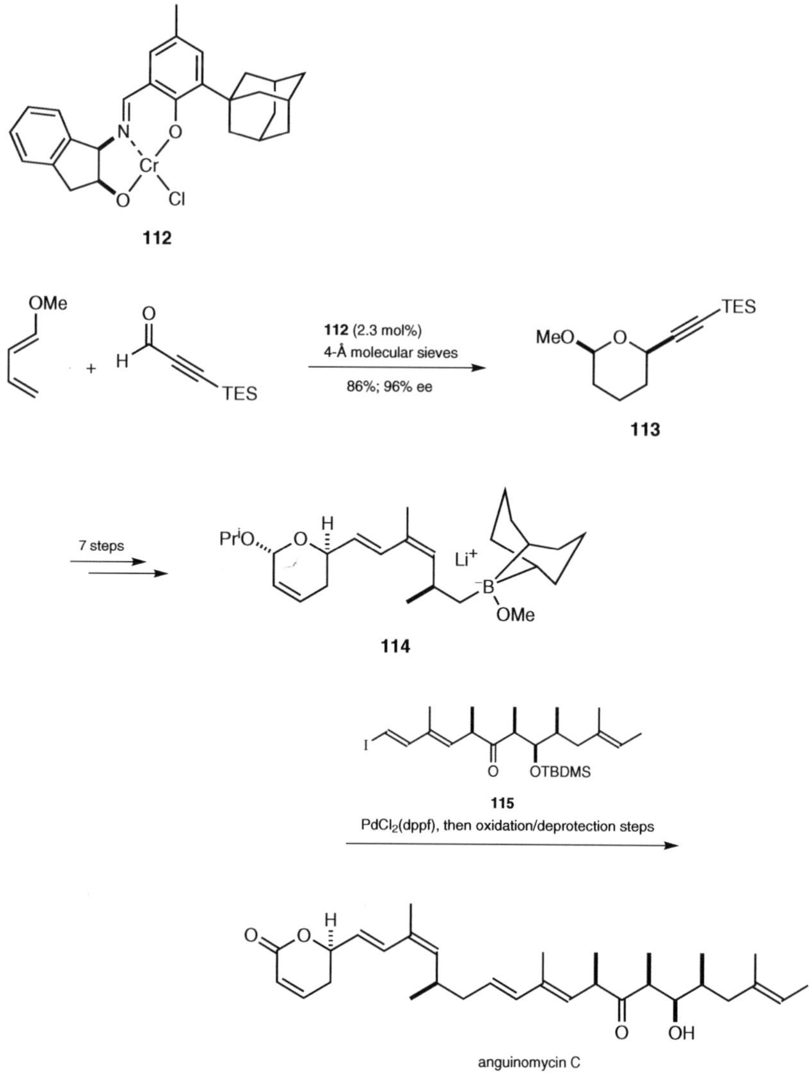

The chiral complex derived from niobium(IV) methoxide and ligand **96** catalyzes the enantioselective aza-Diels–Alder reaction of an imine derived from 3-pyridylcarbaldehyde (nicotinaldehyde) with a Danishefsky-type diene to afford the adduct **116** in 74% yield and with 92% enantiomeric excess (Scheme 60). The synthesis of the minor nicotine alkaloid (+)-anabasine from this compound is achieved in five steps including etherification, reduction, deoxygenation, and removal of a protecting group.[70]

Scheme 60 Synthesis of (+)-Anabasine Using an Enantioselective Aza-Diels–Alder Reaction[70]

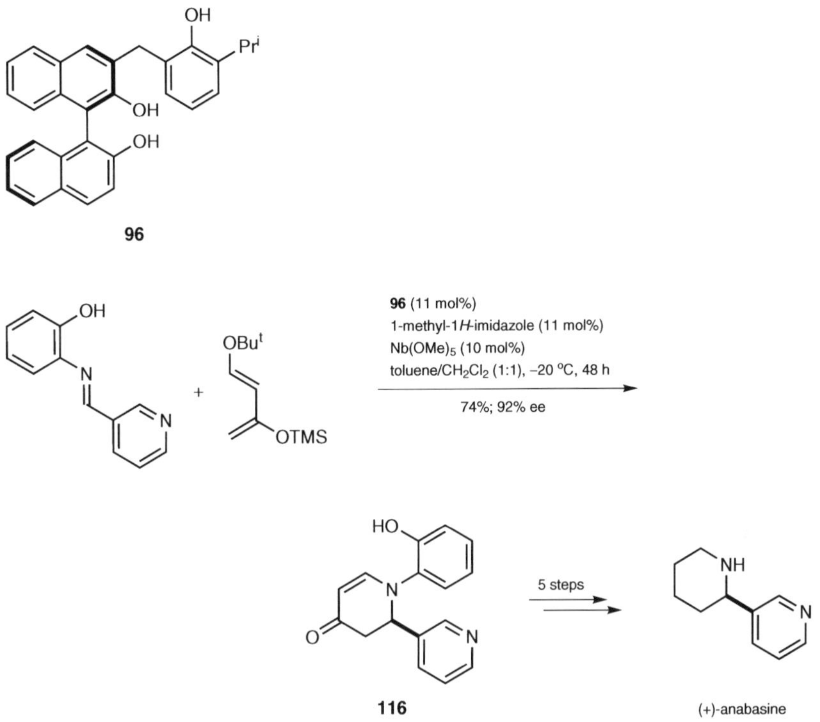

2.5.5 Conclusions and Future Perspectives

Numerous chiral Lewis acid catalysts have been successfully developed for the enantioselective Diels–Alder and hetero-Diels–Alder reactions. These reactions provide diverse optically active six-membered carbocyclic, oxacyclic, and azacyclic compounds in a convenient fashion. Significantly, in some cases, the catalyst loading can reach an extremely low level, which makes these catalytic systems more practical. Notably, some chiral Lewis acid catalysts have been applied in the total synthesis of various natural products and biologically important compounds. Despite these advances, the exploration of novel catalyst systems with high chemo-, regio-, and enantioselectivity at low catalyst loadings, and the practical application of Diels–Alder or hetero-Diels–Alder reactions in total synthesis and industrial processes are still very important subject areas in this field.

References

[1] Diels, O.; Alder, K., *Justus Liebigs Ann. Chem.*, (1928) **460**, 98.
[2] Berson, J. A., *Tetrahedron*, (1992) **48**, 3.
[3] Korolev, A.; Mur, V., *Dokl. Akad. Nauk SSSR*, (1948) **51**, 251.
[4] Walborsky, H. M.; Barash, L.; Davis, T. C., *Tetrahedron*, (1963) **19**, 2333.
[5] Riant, O.; Kagan, H. B., *Tetrahedron Lett.*, (1989) **30**, 7403.
[6] Furuta, K.; Miwa, Y.; Iwanaga, K.; Yamamoto, H., *J. Am. Chem. Soc.*, (1988) **110**, 6254.
[7] Furuta, K.; Shimizu, S.; Miwa, Y.; Yamamoto, H., *J. Org. Chem.*, (1989) **54**, 1481.
[8] Corey, E. J.; Loh, T.-P., *J. Am. Chem. Soc.*, (1991) **113**, 8966.
[9] Corey, E. J.; Loh, T.-P.; Roper, T. D.; Azimioara, M. D.; Noe, M. C., *J. Am. Chem. Soc.*, (1992) **114**, 8290.
[10] Hayashi, Y.; Rohde, J. J.; Corey, E. J., *J. Am. Chem. Soc.*, (1996) **118**, 5502.
[11] Ishihara, K.; Yamamoto, H., *J. Am. Chem. Soc.*, (1994) **116**, 1561.
[12] Ryu, D. H.; Lee, T. W.; Corey, E. J., *J. Am. Chem. Soc.*, (2002) **124**, 9992.
[13] Ryu, D. H.; Corey, E. J., *J. Am. Chem. Soc.*, (2003) **125**, 6388.
[14] Canales, E.; Corey, E. J., *Org. Lett.*, (2008) **10**, 3271.
[15] Futatsugi, K.; Yamamoto, H., *Angew. Chem. Int. Ed.*, (2005) **44**, 1484.
[16] Corey, E. J.; Imwinkelried, R.; Pikul, S.; Xiang, Y. B., *J. Am. Chem. Soc.*, (1989) **111**, 5493.
[17] Bao, J.; Wulff, W. D.; Rheingold, A. L., *J. Am. Chem. Soc.*, (1993) **115**, 3814.
[18] Teo, Y.-C.; Loh, T.-P., *Org. Lett.*, (2005) **7**, 2539.
[19] Evans, D. A.; Barnes, D. M.; Johnson, J. S.; Lectka, T.; von Matt, P.; Miller, S. J.; Murry, J. A.; Norcross, R. D.; Shaughnessy, E. A.; Campos, K. R., *J. Am. Chem. Soc.*, (1999) **121**, 7582.
[20] Palomo, C.; Oiarbide, M.; García, J. M.; González, A.; Arceo, E., *J. Am. Chem. Soc.*, (2003) **125**, 13942.
[21] Barroso, S.; Blay, G.; Pedro, J. R., *Org. Lett.*, (2007) **9**, 1983.
[22] Sibi, M. P.; Stanley, L. M.; Nie, X.; Venkatraman, L.; Liu, M.; Jasperse, C. P., *J. Am. Chem. Soc.*, (2007) **129**, 395.
[23] Barroso, S.; Blay, G.; Al-Midfa, L.; Muñoz, M. C.; Pedro, J. R., *J. Org. Chem.*, (2008) **73**, 6389.
[24] Owens, T. D.; Hollander, F. J.; Oliver, A. G.; Ellman, J. A., *J. Am. Chem. Soc.*, (2001) **123**, 1539.
[25] Mikami, K.; Motoyama, Y.; Terada, M., *J. Am. Chem. Soc.*, (1994) **116**, 2812.
[26] Maruoka, K.; Murase, N.; Yamamoto, H., *J. Org. Chem.*, (1993) **58**, 2938.
[27] Huang, Y.; Iwama, T.; Rawal, V. H., *J. Am. Chem. Soc.*, (2000) **122**, 7843.
[28] Huang, Y.; Iwama, T.; Rawal, V. H., *J. Am. Chem. Soc.*, (2002) **124**, 5950.
[29] Rickerby, J.; Vallet, M.; Bernardinelli, G.; Viton, F.; Kündig, E. P., *Chem.–Eur. J.*, (2007) **13**, 3354.
[30] Kobayashi, S.; Ishitani, H.; Hachiya, I.; Araki, M., *Tetrahedron*, (1994) **50**, 11623.
[31] Markó, I. E.; Evans, G. R.; Seres, P.; Chellé, I.; Janousek, Z., *Pure Appl. Chem.*, (1996) **68**, 113.
[32] Sudo, Y.; Shirasaki, D.; Harada, S.; Nishida, A., *J. Am. Chem. Soc.*, (2008) **130**, 12588.
[33] Danishefsky, S. J.; Larson, E.; Askin, D.; Kato, N., *J. Am. Chem. Soc.*, (1985) **107**, 1246.
[34] Maruoka, K.; Itoh, T.; Shirasaka, T.; Yamamoto, H., *J. Am. Chem. Soc.*, (1988) **110**, 310.
[35] Yu, Z.; Liu, X.; Dong, Z.; Xie, M.; Feng, X., *Angew. Chem. Int. Ed.*, (2008) **47**, 1308.
[36] Keck, G. E.; Li, X.-Y.; Krishnamurthy, D., *J. Org. Chem.*, (1995) **60**, 5998.
[37] Wang, B.; Feng, X.; Cui, X.; Liu, H.; Jiang, Y., *Chem. Commun. (Cambridge)*, (2000), 1605.
[38] Kii, S.; Hashimoto, T.; Maruoka, K., *Synlett*, (2002), 931.
[39] Yang, X.-B.; Feng, J.; Zhang, J.; Wang, N.; Wang, L.; Liu, J.-L.; Yu, X.-Q., *Org. Lett.*, (2008) **10**, 1299.
[40] Lin, L.; Chen, Z.; Yang, X.; Liu, X.; Feng, X., *Org. Lett.*, (2008) **10**, 1311.
[41] Long, J.; Hu, J.; Shen, X.; Ji, B.; Ding, K., *J. Am. Chem. Soc.*, (2002) **124**, 10.
[42] Yuan, Y.; Long, J.; Sun, J.; Ding, K., *Chem.–Eur. J.*, (2002) **8**, 5033.
[43] Yuan, Y.; Li, X.; Sun, J.; Ding, K., *J. Am. Chem. Soc.*, (2002) **124**, 14866.
[44] Ji, B.; Yuan, Y.; Ding, K.; Meng, J., *Chem.–Eur. J.*, (2003) **9**, 5989.
[45] Yamashita, Y.; Saito, S.; Ishitani, H.; Kobayashi, S., *Org. Lett.*, (2002) **4**, 1221.
[46] Yamashita, Y.; Saito, S.; Ishitani, H.; Kobayashi, S., *J. Am. Chem. Soc.*, (2003) **125**, 3793.
[47] Schaus, S. E.; Brånalt, J.; Jacobsen, E. N., *J. Org. Chem.*, (1998) **63**, 403.
[48] Dossetter, A. G.; Jamison, T. F.; Jacobsen, E. N., *Angew. Chem. Int. Ed.*, (1999) **38**, 2398.
[49] Gademann, K.; Chavez, D. E.; Jacobsen, E. N., *Angew. Chem. Int. Ed.*, (2002) **41**, 3059.
[50] Yao, S.; Johannsen, M.; Audrain, H.; Hazell, R. G.; Jørgensen, K. A., *J. Am. Chem. Soc.*, (1998) **120**, 8599.
[51] Thorhauge, J.; Johannsen, M.; Jørgensen, K. A., *Angew. Chem. Int. Ed.*, (1998) **37**, 2404.

[52] Audrain, H.; Thorhauge, J.; Hazell, R. G.; Jørgensen, K. A., *J. Org. Chem.*, (2000) **65**, 4487.
[53] Du, H.; Long, J.; Hu, J.; Li, X.; Ding, K., *Org. Lett.*, (2002) **4**, 4349.
[54] Du, H.; Ding, K., *Org. Lett.*, (2003) **5**, 1091.
[55] Du, H.; Zhang, X.; Wang, Z.; Ding, K., *Tetrahedron*, (2005) **61**, 9465.
[56] Du, H.; Zhang, X.; Wang, Z.; Bao, H.; You, T.; Ding, K., *Eur. J. Org. Chem.*, (2008), 2248.
[57] Doyle, M. P.; Phillips, I. M.; Hu, W., *J. Am. Chem. Soc.*, (2001) **123**, 5366.
[58] Valenzuela, M.; Doyle, M. P.; Hedberg, C.; Hu, W.; Holmstrom, A., *Synlett*, (2004), 2425.
[59] Anada, M.; Washio, T.; Shimada, N.; Kitagaki, S.; Nakajima, M.; Shiro, M.; Hashimoto, S., *Angew. Chem. Int. Ed.*, (2004) **43**, 2665.
[60] Oi, S.; Terada, E.; Ohuchi, K.; Kato, T.; Tachibana, Y.; Inoue, Y., *J. Org. Chem.*, (1999) **64**, 8660.
[61] Tiseni, P. S.; Peters, R., *Org. Lett.*, (2008) **10**, 2019.
[62] Tiseni, P. S.; Peters, R., *Chem.–Eur. J.*, (2010) **16**, 2503.
[63] Mancheño, O. G.; Arrayás, R. G.; Carretero, J. C., *J. Am. Chem. Soc.*, (2004) **126**, 456.
[64] Tong, M.-C.; Chen, X.; Li, J.; Huang, R.; Tao, H.; Wang, C.-J., *Angew. Chem. Int. Ed.*, (2014) **53**, 4680.
[65] Eschenbrenner-Lux, V.; Küchler, P.; Ziegler, S.; Kumar, K.; Waldmann, H., *Angew. Chem. Int. Ed.*, (2014) **53**, 2134.
[66] Josephsohn, N. S.; Snapper, M. L.; Hoveyda, A. H., *J. Am. Chem. Soc.*, (2003) **125**, 4018.
[67] Kobayashi, S.; Komiyama, S.; Ishitani, H., *Angew. Chem. Int. Ed.*, (1998) **37**, 979.
[68] Kobayashi, S.; Kusakabe, K.; Komiyama, S.; Ishitani, H., *J. Org. Chem.*, (1999) **64**, 4220.
[69] Kobayashi, S.; Kusakabe, K.; Ishitani, H., *Org. Lett.*, (2000) **2**, 1225.
[70] Jurčík, V.; Arai, K.; Salter, M. M.; Yamashita, Y.; Kobayashi, S., *Adv. Synth. Catal.*, (2008) **350**, 647.
[71] Esquivias, J.; Arrayás, R. G.; Carretero, J. C., *J. Am. Chem. Soc.*, (2007) **129**, 1480.
[72] Watanabe, Y.; Washio, T.; Krishnamurthi, J.; Anada, M.; Hashimoto, S., *Chem. Commun. (Cambridge)*, (2012) **48**, 6969.
[73] Shang, D.; Xin, J.; Liu, Y.; Zhou, X.; Liu, X.; Feng, X., *J. Org. Chem.*, (2008) **73**, 630.
[74] Xie, M.; Chen, X.; Zhu, Y.; Gao, B.; Lin, L.; Liu, X.; Feng, X., *Angew. Chem. Int. Ed.*, (2010) **49**, 3799.
[75] Deng, Y.; Liu, L.; Sarkisian, R. G.; Wheeler, K.; Wang, H.; Xu, Z., *Angew. Chem. Int. Ed.*, (2013) **52**, 3663.
[76] Hu, Q.-Y.; Zhou, G.; Corey, E. J., *J. Am. Chem. Soc.*, (2004) **126**, 13708.
[77] Yeung, Y.-Y.; Hong, S.; Corey, E. J., *J. Am. Chem. Soc.*, (2006) **128**, 6310.
[78] Yu, M.; Danishefsky, S. J., *J. Am. Chem. Soc.*, (2008) **130**, 2783.
[79] Kong, K.; Moussa, Z.; Lee, C.; Romo, D., *J. Am. Chem. Soc.*, (2011) **133**, 19844.
[80] Thompson, C. F.; Jamison, T. F.; Jacobsen, E. N., *J. Am. Chem. Soc.*, (2001) **123**, 9974.
[81] Boezio, A. A.; Jarvo, E. R.; Lawrence, B. M.; Jacobsen, E. N., *Angew. Chem. Int. Ed.*, (2005) **44**, 6046.
[82] Ghosh, A. K.; Gong, G., *Org. Lett.*, (2007) **9**, 1437.
[83] Paterson, I.; Miller, N. A., *Chem. Commun. (Cambridge)*, (2008), 4708.
[84] Nicolaou, K. C.; Tria, G. S.; Edmonds, D. J., *Angew. Chem. Int. Ed.*, (2008) **47**, 1780.
[85] Bonazzi, S.; Güttinger, S.; Zemp, I.; Kutay, U.; Gademann, K., *Angew. Chem. Int. Ed.*, (2007) **46**, 8707.

2.6 Metal-Catalyzed (2+2+2) Cycloadditions

K. Tanaka and Y. Shibata

General Introduction

The transition-metal-catalyzed (2+2+2) cycloaddition is a useful and atom-economical method for the synthesis of substituted six-membered-ring molecules. The reaction has been extensively investigated and this research topic has been thoroughly reviewed.[1–13] Compared with the well-established transition-metal-catalyzed cross-coupling approach, the transition-metal-catalyzed (2+2+2)-cycloaddition approach is advantageous for the synthesis of densely substituted six-membered carbocycles and heterocycles. In the (2+2+2) cycloaddition, many different substituents can be introduced through the formation of the six-membered ring. This chapter discusses the metal-catalyzed (2+2+2) cycloaddition of not only alkynes but also heterocumulenes, alkenes, and carbonyl compounds for the synthesis of six-membered carbocycles and heterocycles. Asymmetric variants of these reactions will also be discussed. Although this chapter focuses on catalytic reactions, (2+2+2) cycloadditions mediated by stoichiometric amounts of metal complexes have also been reported.[1–13]

Reppe reported the first transition-metal-catalyzed (2+2+2) cycloaddition of alkynes using a nickel complex.[14] Vollhardt smartly utilized the cobalt-mediated (2+2+2) cycloaddition in organic synthesis[15,16] and Yamazaki examined the general reactivity of cobaltacycles in detail.[17] After these pioneering works, many transition-metal complexes were utilized as catalysts for the (2+2+2) cycloaddition.[1–13] Among them, the most common catalysts are cobalt [CoCpL$_2$, CoCp(alkene)(L), and CoX$_2$/M/L], nickel [Ni(0)/phosphine], ruthenium [Ru(Cp*)Cl(cod)], rhodium [RhCl(PPh$_3$)$_3$ and Rh(I)$^+$/diphosphine], and iridium [Ir(I)/diphosphine] complexes.[1–13]

General mechanisms of the transition-metal-catalyzed (2+2+2) cycloaddition of alkynes are shown in Schemes 1 and 2. Two alkynes react with the transition-metal complex to generate metallacyclopentadiene **1**. The subsequent (4+2) cycloaddition of **1** with the alkyne affords metallabicyclo[2.2.0]heptadiene **2**. Reductive elimination affords the desired benzene (Scheme 1). Alternatively, insertion of the alkyne into **1** leading to metallacycloheptatriene **3** followed by reductive elimination also affords the desired benzene (Scheme 1).[18] On the other hand, a mechanism through metallacyclopentatriene **4** is proposed in the (2+2+2) cycloaddition of alkynes catalyzed by Ru(Cp*)Cl(cod).[19] The (2+2) cycloaddition of metallacyclopentatriene **4** with the alkyne affords metallabicyclo-[3.2.0]heptatriene **5**. Skeletal rearrangement leading to metallacycloheptatetraene **6** followed by reductive elimination affords the desired benzene (Scheme 2).[19]

Scheme 1 Mechanism of Transition-Metal-Catalyzed (2+2+2) Cycloaddition through a Metallacyclopentadiene Intermediate

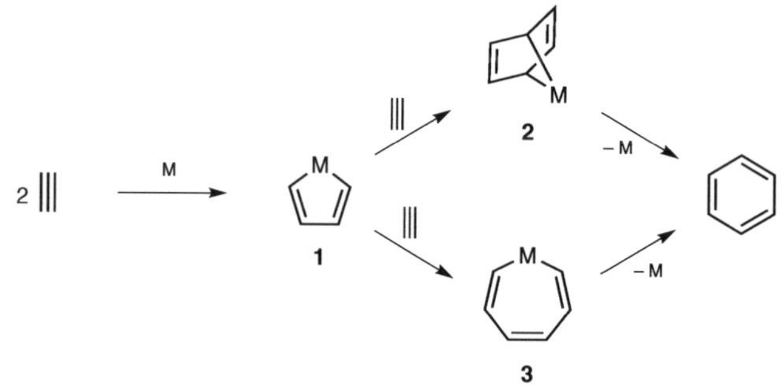

Scheme 2 Mechanism of Transition-Metal-Catalyzed (2+2+2) Cycloaddition through a Metallacyclopentatriene Intermediate

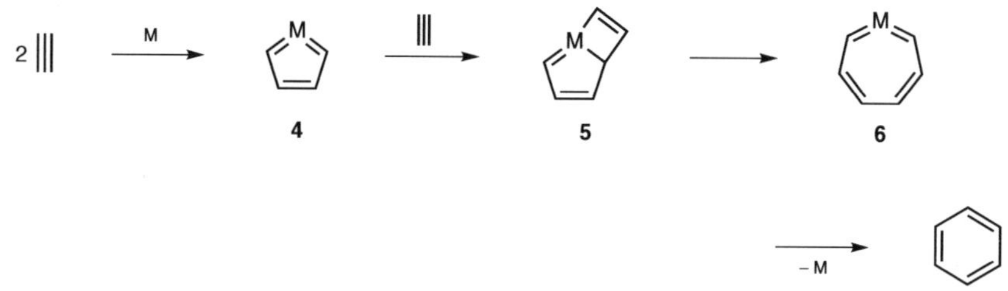

2.6.1 (2+2+2) Cycloadditions of Alkynes

The transition-metal-catalyzed (2+2+2) cycloaddition of alkynes enables the synthesis of densely substituted benzenes and has been most actively investigated. A wide variety of transition-metal catalysts and alkynes have been employed in this transformation. When the reactions allow the use of two or three different alkynes, a diverse array of substituted benzenes can be synthesized in a single step by simply changing the combination of substituted alkynes. Furthermore, various non-centrochiralities (axial, planar, and helical chiralities) can be constructed through the ring formation by using chiral transition-metal complexes as catalysts.

2.6.1.1 Intermolecular Reactions

In the intermolecular (2+2+2) cycloaddition of unsymmetrical monoynes, it is difficult to control chemo- and regioselectivity. A cationic rhodium(I)/H$_8$-BINAP complex is a highly active and selective catalyst for the chemo- and regioselective cross-(2+2+2) cycloaddition of dialkyl acetylenedicarboxylates with two molecules of an electron-rich terminal alkyne (Scheme 3).[20]

2.6.1 (2+2+2) Cycloadditions of Alkynes

Scheme 3 Phthalates by Rhodium-Catalyzed (2+2+2) Cycloaddition of Terminal Alkynes with Acetylenedicarboxylates[20]

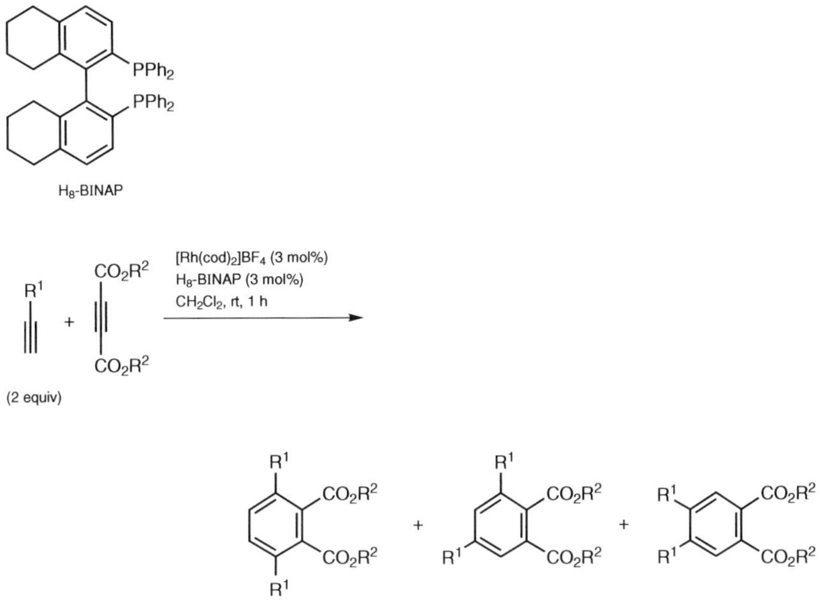

R[1]	R[2]	Ratio (7/8/9)	Yield (%)	Ref
$(CH_2)_9Me$	Me	91:8:1	78	[20]
$(CH_2)_9Me$	Et	92:6:2	88	[20]
$(CH_2)_9Me$	t-Bu	94:4:2	94	[20]
$(CH_2)_3Cl$	Et	91:8:1	92	[20]
cyclohex-1-enyl	Et	91:4:5	90	[20]
Ph	Et	89:9:2	90	[20]
2-Tol	Et	89:9:2	77	[20]
CH_2OAc	Et	87:10:3	63	[20]
TMS	Et	99:1:0	57	[20]

When diynes with a long tether are used, the cyclophane skeletons **10** and **11** can be constructed by alkyne cyclotrimerization (Scheme 4).[20] The length of the carbon tether controls the *meta/para* selectivity.

Scheme 4 Cyclophanes by Rhodium-Catalyzed (2+2+2) Cycloaddition of Terminal Diynes with Acetylenedicarboxylates[20]

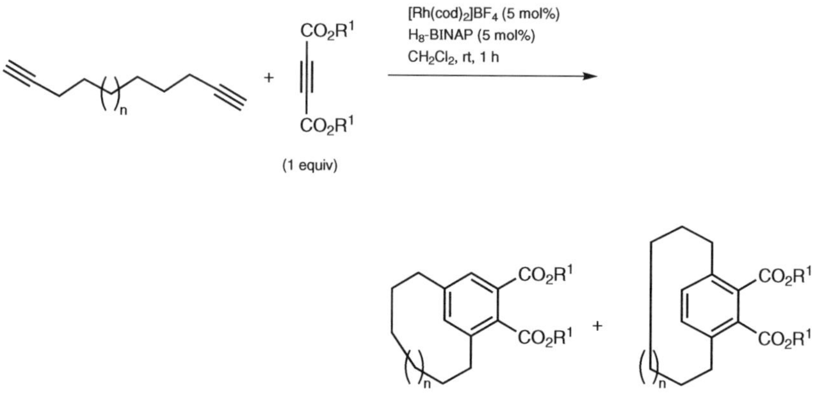

n	R[1]	Ratio (10/11)	Yield (%)	Ref
1	Et	1:0	50	[20]
2	Me	0.8:1	23	[20]
2	Et	0.8:1	20	[20]
2	t-Bu	0.8:1	17	[20]
3	Et	0:1	46	[20]
4	Et	0:1	53	[20]
5	Et	0:1	36	[20]
6	Et	0:1	45	[20]
7	Et	0:1	35	[20]

The chemo- and regioselectivity problems can be solved by the (2+2+2) cycloaddition of diynes with monoynes, although the accessible products are limited to bicyclic compounds. The use of heteroatom-linked diynes enables the efficient synthesis of substituted heterofluorenes **12** and **13**. For example, substituted carbazoles **12** and **13** (Z = NTs) are synthesized by the chlorotris(triphenylphosphine)rhodium-catalyzed (2+2+2) cycloaddition of tosylamide-linked 1,6-diynes with terminal alkynes (Scheme 5).[21] Substituted dibenzofurans **12/13** (Z = O) are also synthesized by the cationic rhodium(I)/H$_8$-BINAP catalyzed (2+2+2) cycloaddition of phenol-linked 1,6-diynes with alkynes.[22] The diiridium–triphenylphosphine catalyzed (2+2+2) cycloaddition of silicon-linked diynes with monoynes affords 9-silafluorenes (siloles) **12** (Z = SiMe$_2$).[23]

2.6.1 (2+2+2) Cycloadditions of Alkynes

Scheme 5 Heterofluorenes by (2+2+2) Cycloadditions of Heteroatom-Linked Diynes with Monoynes[21–23]

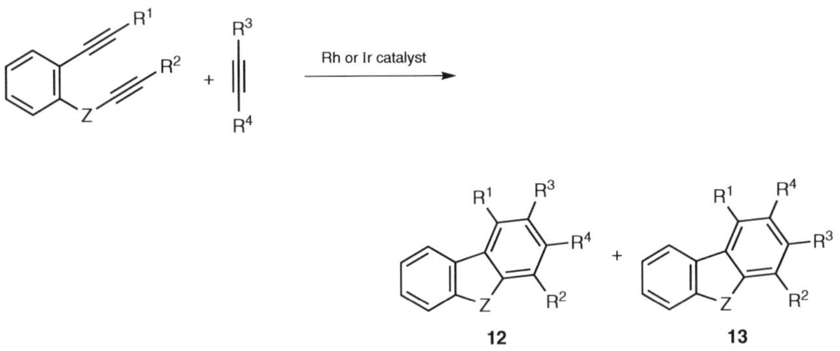

Z	R^1	R^2	R^3	R^4	Conditions	Ratio (12/13)	Yield (%)	Ref
NTs	H	H	H	H	RhCl(PPh$_3$)$_3$ (3–5 mol%), toluene, rt, 1–12 h	–	90	[21]
NTs	H	Ph	CH$_2$OH	H	RhCl(PPh$_3$)$_3$ (3–5 mol%), toluene, rt, 1–12 h	5:1	95	[21]
NTs	Me	H	CH$_2$OH	H	RhCl(PPh$_3$)$_3$ (3–5 mol%), toluene, rt, 1–12 h	3.5:1	89	[21]
NTs	H	H	CO$_2$Me	H	RhCl(PPh$_3$)$_3$ (3–5 mol%), toluene, rt, 1–12 h	4:1	72	[21]
NTs	H	Ph	CO$_2$Et	Me	RhCl(PPh$_3$)$_3$ (3–5 mol%), toluene, rt, 1–12 h	6:1	91	[21]
O	Bu	H	Ph	CO$_2$Et	[Rh(cod)$_2$]BF$_4$/H$_8$-BINAP (5 mol%), CH$_2$Cl$_2$, rt, 1 h	1:2	49	[22]
O	Ph	H	Ph	CO$_2$Et	[Rh(cod)$_2$]BF$_4$/H$_8$-BINAP (5 mol%), CH$_2$Cl$_2$, rt, 1 h	2:1	78	[22]
O	Ph	H	Me	CO$_2$Et	[Rh(cod)$_2$]BF$_4$/H$_8$-BINAP (5 mol%), CH$_2$Cl$_2$, rt, 1 h	3:1	82	[22]
O	TMS	H	Ph	CO$_2$Et	[Rh(cod)$_2$]BF$_4$/H$_8$-BINAP (5 mol%), CH$_2$Cl$_2$, rt, 1 h	1:1	72	[22]
O	TMS	H	CH$_2$OH	CH$_2$OH	[Rh(cod)$_2$]BF$_4$/H$_8$-BINAP (5 mol%), THF, rt, 1 h	–	69	[22]
O	Bu	H	CH$_2$OH	CH$_2$OH	[Rh(cod)$_2$]BF$_4$/H$_8$-BINAP (5 mol%), THF, rt, 1 h	–	48	[22]
SiMe$_2$	Ph	Ph	CH$_2$OMe	CH$_2$OMe	[IrCl(cod)]$_2$ (2.5 mol%), Ph$_3$P (10 mol%), Bu$_2$O, 110 °C, 24 h	–	86	[23]
SiMe$_2$	Ph	Ph	Pr	Pr	[IrCl(cod)]$_2$ (5 mol%), Ph$_3$P (20 mol%), Bu$_2$O, 110 °C, 24 h	–	75	[23]
SiMe$_2$	Ph	(CH$_2$)$_4$Me	CH$_2$OMe	CH$_2$OMe	[IrCl(cod)]$_2$ (2.5 mol%), Ph$_3$P (10 mol%), Bu$_2$O, 110 °C, 24 h	–	81	[23]
SiMe$_2$	(CH$_2$)$_4$Me	(CH$_2$)$_4$Me	CH$_2$OMe	CH$_2$OMe	[IrCl(cod)]$_2$ (2.5 mol%), Ph$_3$P (10 mol%), Bu$_2$O, 145 °C, 24 h	–	69	[23]

Alkynylboronates **14** participate in a cobalt-catalyzed (2+2+2) cycloaddition with terminal alkynes to produce arylboronates **16**, which are valuable intermediates for palladium-catalyzed transformations such as Suzuki–Miyaura cross coupling (Scheme 6).[24]

Scheme 6 Cobalt-Catalyzed (2+2+2) Cycloaddition of Diboryldiynes with Monoynes and Their Subsequent Use in Suzuki–Miyaura Cross Couplings[24]

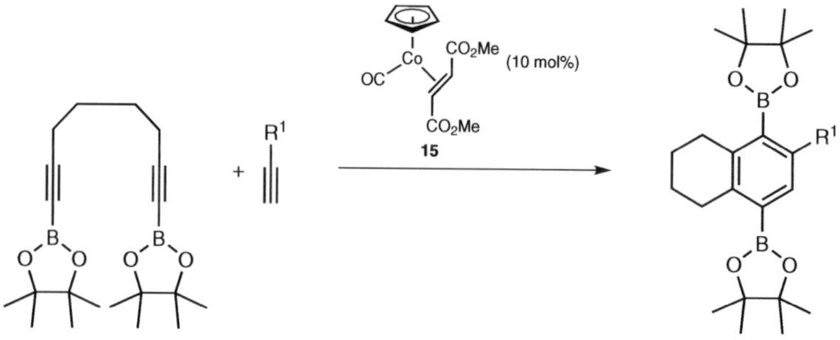

14 (5 equiv) 16

R^1	Conditions	Yield (%)	Ref
CH_2OMe	microwave, DMF, 200 °C, 10 min	48	[24]
Ph	o-xylene, reflux, 1 h	43	[24]
CO_2Me	o-xylene, reflux, 1 h	41	[24]
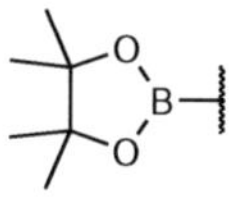	o-xylene, reflux, 1 h	11	[24]

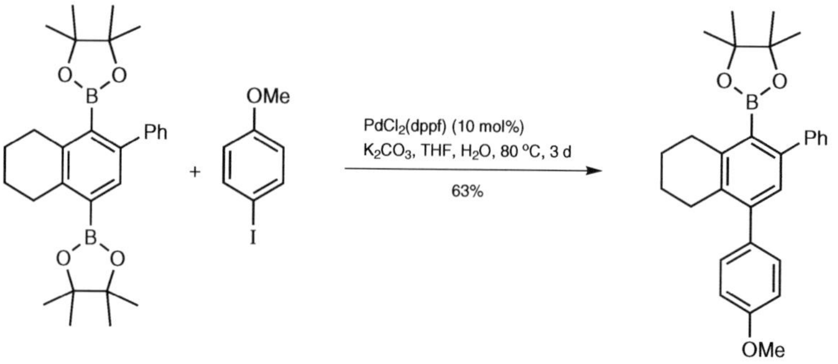

(1 equiv) 17

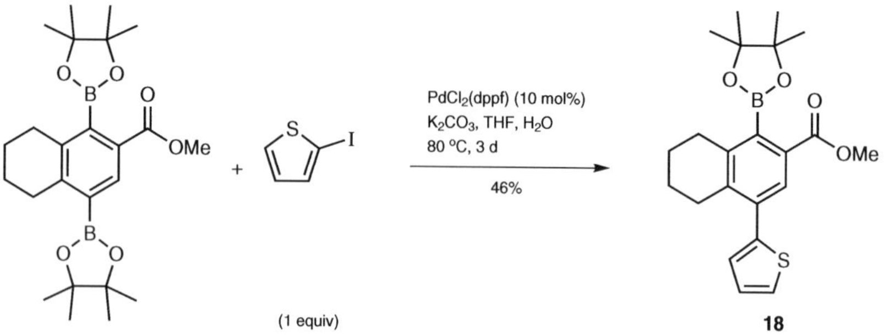

(1 equiv) 18

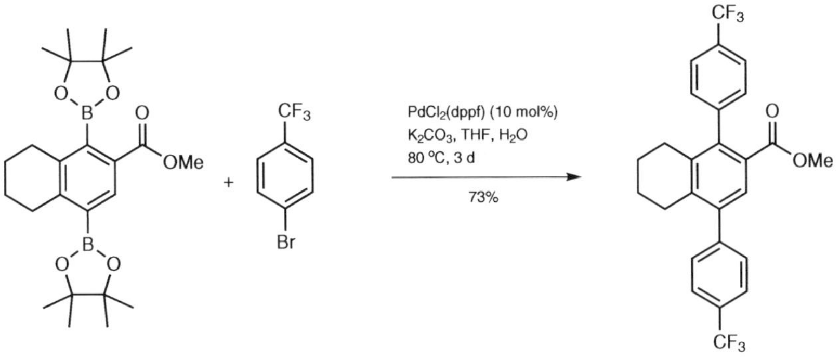

2.6.1 (2+2+2) Cycloadditions of Alkynes

Diiododiynes can be employed in ruthenium catalysis to produce iodoarenes **20**, which are also valuable intermediates for palladium-catalyzed transformations (Scheme 7).[25]

Scheme 7 Iodoarenes by Ruthenium-Catalyzed (2+2+2) Cycloaddition of Diiododiynes with Acetylene[25]

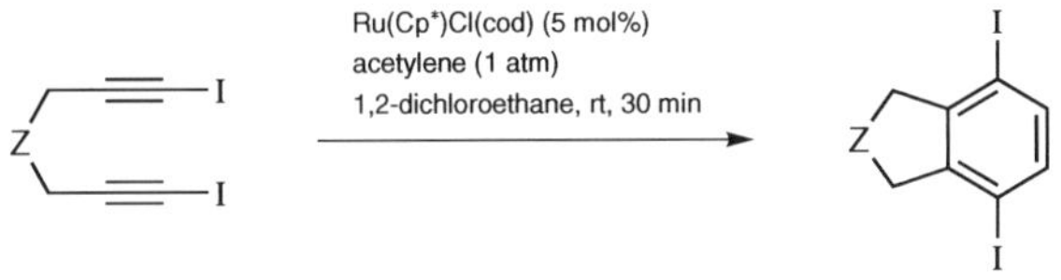

Z	Yield (%)	Ref
$C(CO_2Me)_2$	83	[25]
$C(CN)_2$	70	[25]
O	83	[25]
NTs	80	[25]
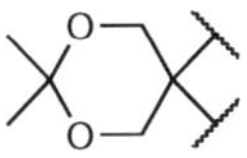 (dioxolane group)	87	[25]
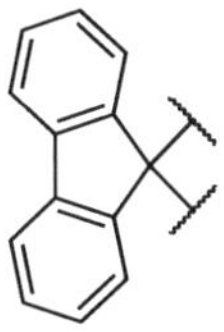 (fluorene group)	93	[25]

The ruthenium-catalyzed double (2+2+2) cycloaddition of diiodotetrayne **21** with acetylene affords 4,4′-diiodobiaryl **22**.[25] Biaryl **22** was subjected to palladium-catalyzed double Suzuki–Miyaura cross coupling with boronate **24**, prepared from diynylboronate **23** and acetylene,[26] to give the corresponding hexaphenylene **25** (Scheme 8).[25]

Scheme 8 Synthesis of a Hexaphenylene via Ruthenium-Catalyzed (2+2+2) Cycloadditions[25,26]

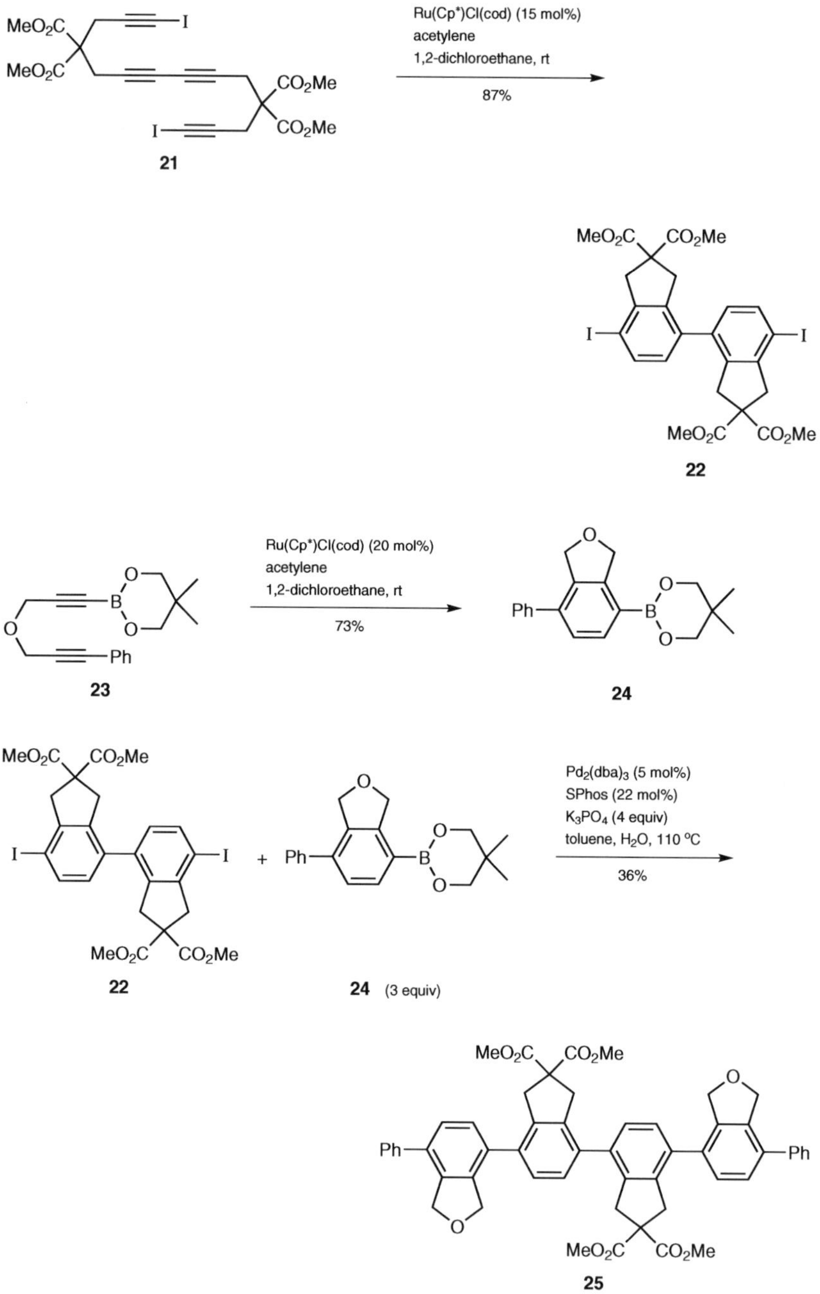

SPhos = 2-(dicyclohexylphosphino)-2',6'-dimethoxybiphenyl

The highly enantioselective synthesis of axially chiral biaryl phosphine oxides and esters **28** has been achieved by the cationic rhodium(I) complex catalyzed (2+2+2) cycloadditions of diynes **26** with alkynyl phosphonates, phosphine oxides, and esters **27** (Scheme 9).[27,28] In the reactions of alkynyl phosphine oxides and phosphonates, the use of the ligand (R)-H$_8$-BINAP affords the corresponding biaryl phosphine oxides and phosphonates

2.6.1 (2+2+2) Cycloadditions of Alkynes

with up to 98% ee.[27] On the other hand, (S)-BINAP is a suitable ligand in the reactions of alkynyl esters and the corresponding biaryl esters are obtained with up to 96% ee.[28]

Scheme 9 Rhodium-Catalyzed Enantioselective Synthesis of Axially Chiral Biaryls[27,28]

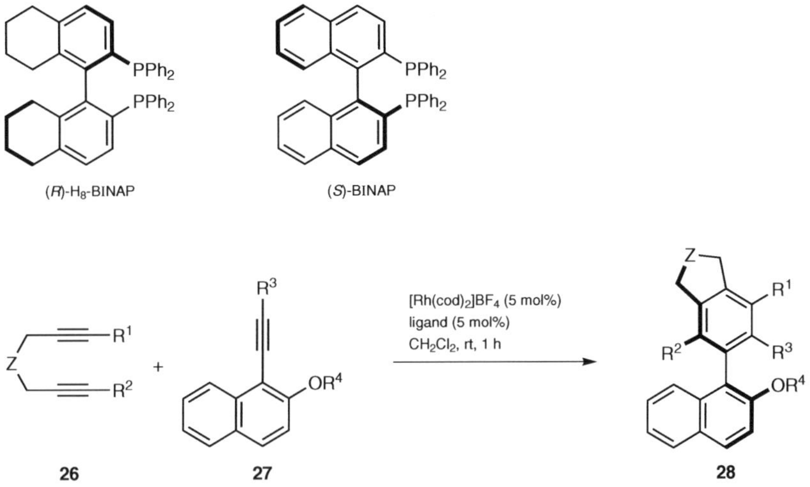

Z	R^1	R^2	R^3	R^4	Equiv of Diyne	Ligand	ee (%)	Yield (%)	Ref
O	Me	Me	P(O)(OEt)$_2$	Me	1.5	(R)-H$_8$-BINAP	97	>99	[27]
CH$_2$	Me	Me	P(O)(OEt)$_2$	Me	1.5	(R)-H$_8$-BINAP	97	>99	[27]
NTs	Me	Me	P(O)(OEt)$_2$	Me	1	(R)-H$_8$-BINAP	95	96	[27]
O	Me	Me	P(O)(OEt)$_2$	CH$_2$OMe	1.5	(R)-H$_8$-BINAP	98	99	[27]
O	Me	Me	P(O)Ph$_2$	Me	1.5	(R)-H$_8$-BINAP	91	92	[27]
O	Me	Me	P(O)Cy$_2$	Me	3	(R)-H$_8$-BINAP	95	>99	[27]
O	Ph	H	P(O)(OEt)$_2$	Me	3	(R)-H$_8$-BINAP	96	73	[27]
O	Me	Me	CO$_2$iPr	Me	1.2	(S)-BINAP	96	>99	[28]
NTs	Me	Me	CO$_2$iPr	Me	1.2	(S)-BINAP	90	>99	[28]
(CH$_2$)$_2$	Me	Me	CO$_2$iPr	Me	1.2	(S)-BINAP	93	71	[28]
O	Me	Me	CO$_2$iPr	Bn	1.2	(S)-BINAP	89	91	[28]
O	Ph	Me	CO$_2$Et	Me	1.2	(S)-H$_8$-BINAP	74	96^a	[28]

^{a} As a 94:6 ratio with the product regioisomer where R^1 = Me and R^2 = Ph.

C_2-Symmetric axially chiral biaryl diphosphonates and diesters **31** are obtained with excellent enantioselectivity by using a phosphonate- or ester-substituted buta-1,3-diyne **30** and internal 1,6-diynes **29** (Scheme 10).[29] As shown in Schemes 9 and 10, the cationic rhodium(I)/biaryl bisphosphine complexes are highly effective catalysts for the enantioselective biaryl synthesis via (2+2+2) cycloaddition.

Scheme 10 Rhodium-Catalyzed Enantioselective Synthesis of Axially Chiral Biaryls[29]

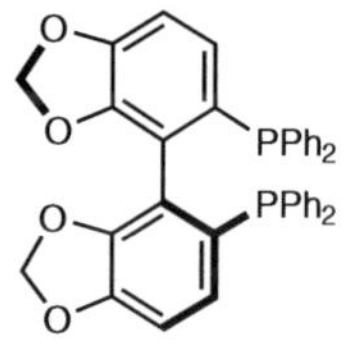

(*S*)-SEGPHOS

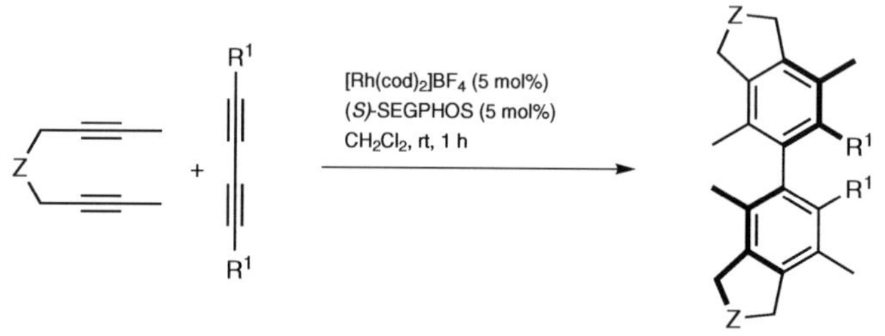

29 (3 equiv) **30** **31**

Z	R^1	ee (%)	Yield (%)	Ref
O	$P(O)(OEt)_2$	>99	65	[29]
NTs	$P(O)(OEt)_2$	>99	74	[29]
NTs	CO_2Et	98	54	[29]

On the other hand, a neutral iridium(I)/bisphosphine complex is a highly effective catalyst for diastereo- and enantioselective teraryl synthesis via (2+2+2) cycloaddition. As shown in Scheme 11, 1,4-teraryls **34** with two atropoisomeric chiralities are synthesized by a neutral iridium(I)/(*S*,*S*)-Me-DuPhos complex catalyzed (2+2+2) cycloaddition of but-2-ynes **33** and diynes **32** with excellent enantio- and diastereoselectivity.[30]

Scheme 11 Iridium-Catalyzed Enantioselective Synthesis of Axially Chiral 1,4-Teraryls[30]

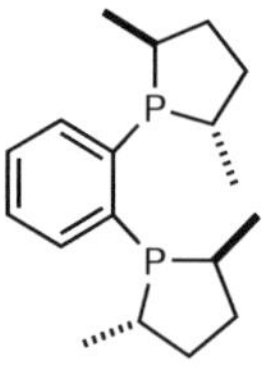

(*S*,*S*)-Me-DuPhos

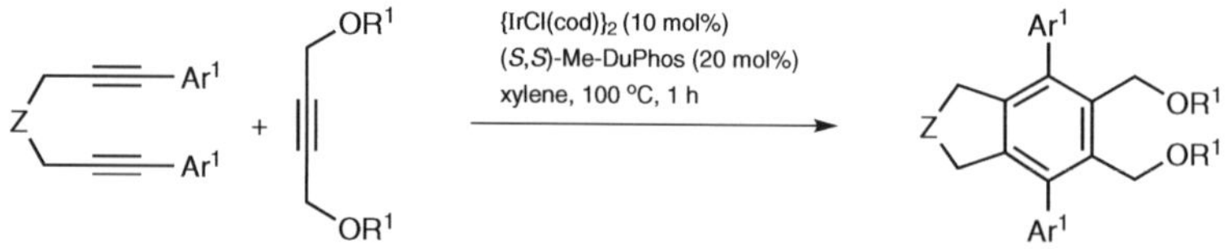

32 **33** (3 equiv) **34**

2.6.1 (2+2+2) Cycloadditions of Alkynes

Z	Ar[1]	R[1]	ee (%)	Yield (%)	Ref
O	1-naphthyl	Me	99.6	83	[30]
O	1-naphthyl	THP	99.5	76[a]	[30]
O	1-naphthyl	TBDMS	99.5	74	[30]
O	1-naphthyl	CH_2OMe	98.5	76	[30]
O	2-Tol	Me	99.6	85	[30]
O	2-ClC_6H_4	THP	97.7	85[a]	[30]
NTs	1-naphthyl	Me	99.4	92	[30]
CH_2	1-naphthyl	Me	>99.8	96	[30]

[a] Yields of the corresponding diol after deprotection.

Substituted tetraphenylenes are known as interesting biaryl-based chiral cyclic scaffolds. The cationic rhodium(I)/Cy-BINAP or rhodium(I)/QuinoxP* catalyzed enantioselective double homo-(2+2+2) cycloaddition of triynes **35** affords chiral tetraphenylenes **36** with high enantioselectivity (Scheme 12).[31]

Scheme 12 Rhodium-Catalyzed Enantioselective Synthesis of Chiral 1,4-Tetraphenylenes by Double Homo-(2+2+2) Cycloaddition of Triynes[31]

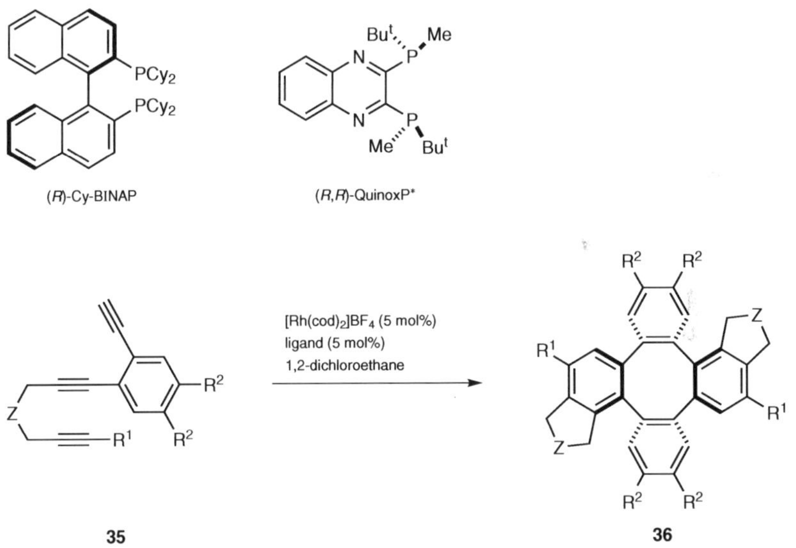

Z	R[1]	R[2]	Ligand	Conditions	ee (%)	Yield (%)	Ref
NTs	Ph	H	(R)-Cy-BINAP	reflux, 24 h	95	62	[31]
$C(CO_2Me)_2$	Me	H	(R)-Cy-BINAP	60°C, 24 h	97	86	[31]
$C(CO_2Me)_2$	Ph	H	(R)-Cy-BINAP	60°C, 24 h	96	45	[31]
$C(CO_2Me)_2$	Me	OMe	(R)-Cy-BINAP	60°C, 24 h	82	64	[31]
O	Me	H	(R)-Cy-BINAP	rt to 60°C, 5 h	75	80	[31]
O	Me	H	(R,R)-QuinoxP*	rt to 60°C, 5 h	83	59	[31]

Anilides and benzamides bearing a sterically demanding *ortho*-substituent are known to exist as atropisomers. The cationic rhodium(I)/axially chiral biaryl bisphosphine catalyzed enantioselective (2+2+2) cycloaddition of 1,6-diynes with (trimethylsilyl)ynamides affords axially chiral anilides with good to excellent enantioselectivity.[32] The use of *N*,*N*-dialkylalkynamides **38** in place of (trimethylsilyl)ynamides affords axially chiral benzamides **39** with perfect enantioselectivity (Scheme 13).[33]

Scheme 13 Rhodium-Catalyzed Enantioselective Synthesis of Axially Chiral Benzamides from 1,6-Diynes and Ynamides[33]

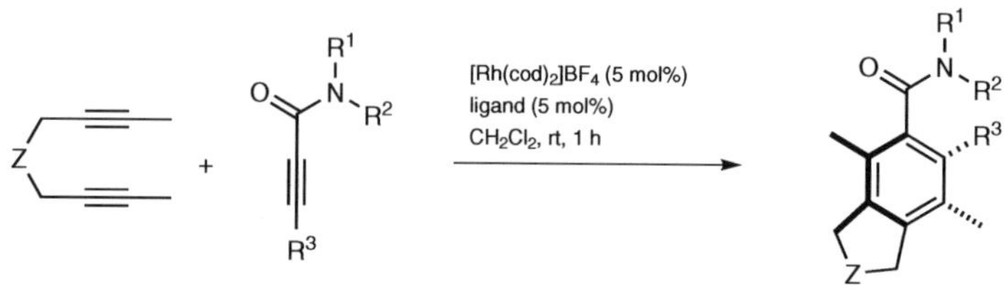

37 **38** (1.1 equiv) **39**

Z	R^1	R^2	R^3	Ligand	ee (%)	Yield (%)	Ref
$C(CO_2Bn)_2$	iPr	iPr	$C(OMe)Me_2$	(*S*)-SEGPHOS	>99	92	[33]
$C(CO_2Bn)_2$	Me	Me	$C(OMe)Me_2$	(*S*)-SEGPHOS	>99	90	[33]
$C(CO_2Bn)_2$	$(CH_2)_5$		$C(OMe)Me_2$	(*S*)-SEGPHOS	>99	98	[33]
$C(CO_2Bn)_2$	iPr	iPr	*t*-Bu	(*S*)-SEGPHOS	>99	90[a]	[33]
$C(CO_2Bn)_2$	iPr	iPr	iPr	(*S*)-BINAP	>99	>99	[33]
$C(CO_2Bn)_2$	iPr	iPr	Bu	(*S*)-BINAP	>99	96	[33]
NTs	iPr	iPr	$C(OMe)Me_2$	(*S*)-SEGPHOS	>99	85	[33]
O	iPr	iPr	$C(OMe)Me_2$	(*S*)-SEGPHOS	>99	81[a]	[33]

[a] 2 equiv of ynamide **38** was used.

The high-yielding and highly enantioselective synthesis of carbaparacyclophanes **42** has been achieved by the cationic rhodium(I)/(2*S*,4*S*)-2,4-bis(diphenylphosphino)pentane [(*S*,*S*)-BDPP] catalyzed (2+2+2) cycloaddition of cyclic diynes **40** with terminal monoynes **41** (Scheme 14).[34]

Scheme 14 Rhodium-Catalyzed Enantioselective Synthesis of Planar Chiral Cyclophanes by Cycloaddition of Cyclic Diynes and Terminal Monoynes[34]

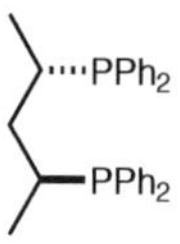

(*S*,*S*)-BDPP

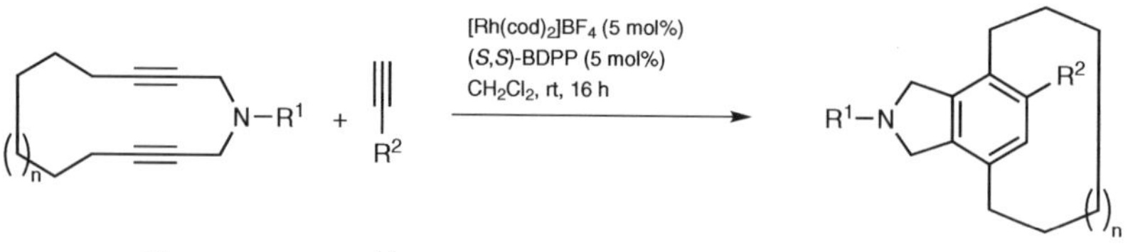

40 **41** **42**

2.6.1 (2+2+2) Cycloadditions of Alkynes

n	R^1	R^2	Equiv of 41	ee (%)	Yield (%)	Ref
5	Ts	CO$_2$Me	1.2	92	87	[34]
5	Ts	CO$_2$t-Bu	1.2	85	88	[34]
5	Ts	CO(CH$_2$)$_2$Ph	1.2	84	54	[34]
5	Ts	CH$_2$CH$_2$-phthalimide	1.2	81	85	[34]
5	Ts	CH$_2$OH	1.2	74	91	[34]
5	Ts	(CH$_2$)$_3$OH	1.2	71	46	[34]
5	Ts	Bu	5	64	52	[34]
5	2-O$_2$NC$_6$H$_4$SO$_2$	CO$_2$Me	2	93	90	[34]
6	Ts	CO$_2$Me	5	74	91	[34]

The cationic rhodium(I)/axially chiral biaryl bisphosphine complex catalyzed (2+2+2) cycloaddition of biaryl-linked tetraynes **43** with dialkynyl ketones **44** (Z = CO) or dialkynylphosphinates **44** [Z = P(O)OMe] affords helically chiral 1,1′-bitriphenylenes **45**, containing a densely substituted fluorenone or phosphafluorene core (Scheme 15).[35]

Scheme 15 Rhodium-Catalyzed Enantioselective Synthesis of [7]Helicenes by Cycloaddition of Biaryl-Linked Tetraynes and Diynes[35]

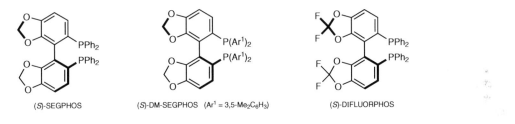

R^1	R^2	Z	mol% of Catalyst	Ligand (mol%)	ee (%)	Yield (%)	Ref
H	Me	CO	20	(S)-DM-SEGPHOS (20)	91	63[a]	[35]
H	Me	CO	10	(S)-DM-SEGPHOS (10)	91	59	[35]
H	$(CH_2)_6Me$	CO	20	(S)-DM-SEGPHOS (20)	92	62	[35]
H	Ph	CO	20	(S)-DM-SEGPHOS (20)	91	60	[35]
H	Cl	CO	20	(S)-DM-SEGPHOS (20)	93	59	[35]
CO_2Bu	Me	CO	10	(S)-DIFLUORPHOS (10)	65	73[a]	[35]
CO_2Bu	Cl	CO	20	(S)-DIFLUORPHOS (20)	53	73[a]	[35]
H	Me	P(O)OMe	20	(S)-SEGPHOS (20)	68	46[a]	[35]
H	Ph	P(O)OMe	20	(S)-SEGPHOS (20)	75	43[a]	[35]

[a] 1.2 equiv of diyne **44** was used.

Diethyl 3,6-Didecylphthalate [7, R^1 = $(CH_2)_9Me$; R^2 = Et]; Typical Procedure:[20]
Under an argon atmosphere, H_8-BINAP (5.7 mg, 0.009 mmol) and [Rh(cod)$_2$]BF$_4$ (3.7 mg, 0.009 mmol) were dissolved in CH$_2$Cl$_2$ (1.0 mL) and the mixture was stirred for 5 min. H$_2$ was then introduced into the resulting soln in a Schlenk tube. After stirring at rt for 0.5 h, the resulting soln was concentrated to dryness and the residue was redissolved in CH$_2$Cl$_2$ (2.0 mL). A soln of dodec-1-yne (99.8 mg, 0.60 mmol) and diethyl acetylenedicarboxylate (51.0 mg, 0.30 mmol) in CH$_2$Cl$_2$ (0.5 mL) was then added dropwise to this soln over 1 min, and any substrates remaining in the syringe were rinsed into the mixture with further CH$_2$Cl$_2$ (0.5 mL). The mixture was stirred at rt (20–25 °C) for 1 h. The resulting soln was then concentrated and the residue was purified by preparative TLC (hexane/EtOAc 10:1), which furnished a mixture of diethyl 3,6-didecylphthalate [7, R^1 = $(CH_2)_9Me$; R^2 = Et], diethyl 3,5-didecylphthalate [8, R^1 = $(CH_2)_9Me$; R^2 = Et], and diethyl 4,5-didecylphthalate [9, R^1 = $(CH_2)_9Me$; R^2 = Et]; combined yield: 133 mg (88%); ratio (**7**/**8**/**9**) 92:6:2. This mixture could be purified by preparative TLC (hexane/EtOAc 10:1), which furnished pure ester **7** [R^1 = $(CH_2)_9Me$; R^2 = Et]; yield: 120 mg (79%).

N-Tosylcarbazoles 12 (Z = NTs); General Procedure:[21]
To a soln of the diyne (1 equiv) in dry toluene saturated with ethene (1 atm) or purged by bubbling argon through it and containing the alkyne (5–6 equiv) was added RhCl(PPh$_3$)$_3$ (3–5 mol%). The mixture was stirred at rt until completion of the reaction. The solvent was evaporated and the resulting crude product was dissolved in CH$_2$Cl$_2$ and filtered through a pad of alumina to remove the catalyst residue. Purification by column chromatography (silica gel, petroleum ether/Et$_2$O) afforded pure **12**.

2.6.1 (2+2+2) Cycloadditions of Alkynes

Ethyl 1-Butyl-3-phenyldibenzofuran-2-carboxylate (13, Z = O; R^1 = Bu; R^2 = H; R^3 = Ph; R^4 = CO$_2$Et); Typical Procedure:[22]

H$_8$-BINAP (6.3 mg, 0.010 mmol) and [Rh(cod)$_2$]BF$_4$ (4.1 mg, 0.010 mmol) were dissolved in CH$_2$Cl$_2$ (1.0 mL) and the mixture was stirred for 5 min. H$_2$ was introduced to the resulting soln in a Schlenk tube. After stirring at rt for 0.5 h, the resulting mixture was concentrated to dryness. To a CH$_2$Cl$_2$ (0.5 mL) soln of the residue and the alkyne (69.7 mg, 0.40 mmol) was added a CH$_2$Cl$_2$ (1.5 mL) soln of the diyne (39.7 mg, 0.20 mmol). The mixture was stirred at rt for 1 h. The resulting mixture was concentrated and purified by preparative TLC (hexane/EtOAc 4:1), which furnished compound **12** (Z = O; R^1 = Bu; R^2 = H; R^3 = Ph; R^4 = CO$_2$Et) as a yellow solid; yield: 12.5 mg (17%); and the title compound as a pale yellow oil; yield: 23.9 mg (32%).

2,3-Bis(methoxymethyl)-9,9-dimethyl-1,4-diphenyl-9-silafluorene (12, Z = SiMe$_2$; R^1 = R^2 = Ph; R^3 = R^4 = CH$_2$OMe); Typical Procedure:[23]

To a soln of {IrCl(cod)}$_2$ (5.0 mg, 7.5 µmol) and Ph$_3$P (7.9 mg, 0.030 mmol) in dry Bu$_2$O (0.6 mL) was added a soln of the diyne (101.0 mg, 0.30 mmol) and the alkyne (68.5 mg, 0.60 mmol) in dry Bu$_2$O (3.0 mL). After being stirred under an argon atmosphere for 24 h at 110 °C, the volatile material was removed under reduced pressure. The residue was purified by preparative TLC (hexane/EtOAc 10:1) to give the title compound; yield: 116.8 mg (86%).

5,8-Bis(4,4,5,5-tetramethyl-1,3,2-dioxaborolan-2-yl)-1,2,3,4-tetrahydronaphthalenes 16; General Procedure Using Microwave Irradiation:[24]

In a microwave vial, a soln of diyne **14** (0.68 mmol), an alkyne (3.4 mmol), and Co catalyst **15** (0.06 mmol) in DMF (4 mL) was heated at 200 °C for 10 min. After being cooled to rt, the mixture was extracted with Et$_2$O and washed several times with H$_2$O. The combined organic extracts were dried (MgSO$_4$), and the solvent was removed under reduced pressure after filtration. The crude product was purified by flash column chromatography (silica gel, pentane/Et$_2$O gradient) to give the corresponding product **16**.

5,8-Bis(4,4,5,5-tetramethyl-1,3,2-dioxaborolan-2-yl)-1,2,3,4-tetrahydronaphthalenes 16; General Procedure by Refluxing in o-Xylene:[24]

A soln of diyne **14** (0.68 mmol), an alkyne (3.4 mmol), and Co catalyst **15** (0.06 mmol) in o-xylene (10 mL) was heated at reflux for 1 h. After being cooled to rt, the crude product was purified by flash column chromatography (silica gel, pentane/Et$_2$O gradient) to give the corresponding product **16**.

Monoaryl- or Diaryltetrahydronaphthalenes 17–19; General Procedure:[24]

PdCl$_2$(dppf) (0.013 mmol) and the aryl halide (0.13 or 0.27 mmol) were added to a soln of compound **16** (0.13 mmol) in dry THF (10 mL). K$_2$CO$_3$ (0.81 mmol) and H$_2$O were added, and the mixture was heated at reflux for 72 h. After cooling to rt, the soln was treated with H$_2$O and CH$_2$Cl$_2$, and the aqueous layer was extracted with CH$_2$Cl$_2$. The combined organic extracts, washed with H$_2$O, dried (MgSO$_4$), and concentrated. The residue was purified by flash column chromatography (silica gel, pentane/Et$_2$O gradient) to give the corresponding monoaryl- or diaryltetrahydronaphthalenes **17–19**.

Dimethyl 4,7-Diiodo-1,3-dihydro-2H-indene-2,2-dicarboxylate [20, Z = C(CO$_2$Me)$_2$]; Typical Procedure:[25]

To a soln of Ru(Cp*)Cl(cod) (5.7 mg, 0.015 mmol) in dry degassed 1,2-dichloroethane (1.5 mL) was added a soln of the diiododiyne (138.2 mg, 0.30 mmol) in dry degassed 1,2-dichloroethane (2.0 mL) over 15 min at rt under an acetylene atmosphere and the soln was stirred for 30 min. The solvent was removed under reduced pressure, and the residue was purified by flash column chromatography (silica gel, hexane/EtOAc 20:1) to give the title compound as a colorless solid; yield: 120.0 mg (83%).

(−)-Diethyl [6-(2-Methoxynaphthalen-1-yl)-4,7-dimethyl-1,3-dihydrobenzo[c]furan-5-yl]phosphonate [28, Z = O; $R^1 = R^2 = R^4$ = Me; R^3 = P(O)(OEt)$_2$]; Typical Procedure:[27]

Under an argon atmosphere, (R)-H$_8$-BINAP (6.3 mg, 0.010 mmol) and [Rh(cod)$_2$]BF$_4$ (4.1 mg, 0.010 mmol) were dissolved in CH$_2$Cl$_2$ (1.0 mL) in a Schlenk tube, and the soln was stirred at rt for 5 min. H$_2$ (1 atm) was introduced into the resulting soln, which was then stirred at rt for 1 h. The mixture was concentrated to dryness, the residue was redissolved in CH$_2$Cl$_2$ (0.4 mL), and a soln of phosphonate **27** (318.3 mg, 1.00 mmol) in CH$_2$Cl$_2$ (1.6 mL) was added. A soln of the diyne **26** (183.2 mg, 1.50 mmol) in CH$_2$Cl$_2$ (3.0 mL) was then added dropwise over 20 min at rt, and the resulting mixture was stirred at rt for 1 h, and then concentrated and purified by column chromatography (silica gel, hexane/EtOAc/Et$_3$N 3:1:1) to furnish the title compound; yield: 439.6 mg (>99%); 97% ee.

(+)-Isopropyl 6-(2-Methoxynaphthalen-1-yl)-4,7-dimethyl-1,3-dihydrobenzo[c]furan-5-carboxylate (28, Z = O; $R^1 = R^2 = R^4$ = Me; R^3 = CO$_2$iPr); Typical Procedure:[28]

(S)-BINAP (6.2 mg, 0.010 mmol) and [Rh(cod)$_2$]BF$_4$ (4.1 mg, 0.010 mmol) were dissolved in CH$_2$Cl$_2$ (1.0 mL), and the mixture was stirred at rt for 5 min. H$_2$ (1 atm) was introduced to the resulting soln in a Schlenk tube. After stirring at rt for 1 h, the resulting soln was concentrated and dissolved in CH$_2$Cl$_2$ (0.4 mL). To this soln was added a CH$_2$Cl$_2$ (0.4 mL) soln of ester **27** (53.7 mg, 0.200 mmol), and then a CH$_2$Cl$_2$ (1.2 mL) soln of the diyne **26** (29.3 mg, 0.240 mmol) was added dropwise over 20 min at rt. After stirring at rt for 1 h, the resulting soln was concentrated and purified by preparative TLC (toluene/EtOAc 10:1), which furnished the title compound; yield: 77.9 mg (>99%); 96% ee.

5,6-Bis(methoxymethyl)-4,7-di(naphthalen-1-yl)-1,3-dihydrobenzo[c]furan (34, Z = O; Ar^1 = 1-Naphthyl; R^1 = Me); Typical Procedure:[30]

(S,S)-Me-DuPhos (6.4 mg, 0.021 mmol) and {IrCl(cod)}$_2$ (7.1 mg, 0.0105 mmol) were stirred in degassed xylene (1.0 mL) at rt to give a reddish yellow soln. After the addition of a xylene soln (1.5 mL) of 1,4-dimethoxybut-2-yne (**33**, R^1 = Me; 36.0 mg, 0.315 mmol) and a xylene soln (1.5 mL) of diyne **32** (36.5 mg, 0.105 mmol), the resulting mixture was further stirred under reflux for 1 h. The solvent was removed under reduced pressure and purification of the crude products by TLC (toluene/EtOAc 15:1) gave pure **34** (Z = O; Ar^1 = 1-naphthyl; R^1 = Me); yield: 40.3 mg (83%).

Tetraphenylenes 36; General Procedure:[31]

[Rh(cod)$_2$]BF$_4$ (2.0 mg, 0.005 mmol) and Cy-BINAP or QuinoxP* (0.005 mmol) were placed in a Schlenk tube, which was then evacuated and backfilled with argon (3 ×). CH$_2$Cl$_2$ (1.0 mL) was added to the reaction vessel, which was then filled with H$_2$. The mixture was then stirred at rt for 30 min under H$_2$. After removal of the solvent and H$_2$ under reduced pressure, the reaction vessel was filled with argon. 1,2-Dichloroethane (0.1 mL) was added to the flask and the mixture was stirred, giving a red soln. Then, a 1,2-dichloroethane soln (0.4 mL) of triyne **35** (0.05 mmol) was added to the red soln and the mixture was stirred at the appropriate temperature. After completion of the reaction, the volatiles were removed under reduced pressure, and the crude mixture was purified by preparative TLC.

(−)-Dibenzyl 5-(Diisopropylcarbamoyl)-6-(2-methoxypropan-2-yl)-4,7-dimethyl-1,3-dihydro-2H-indene-2,2-dicarboxylate [39, Z = C(CO$_2$Bn)$_2$; $R^1 = R^2$ = iPr; R^3 = C(OMe)Me$_2$]; Typical Procedure:[33]

A soln of (S)-SEGPHOS (4.6 mg, 0.0075 mmol) in CH$_2$Cl$_2$ (0.5 mL) was added to a soln of [Rh(cod)$_2$]BF$_4$ (3.0 mg, 0.0075 mmol) in CH$_2$Cl$_2$ (0.5 mL) at rt, and the mixture was stirred

2.6.1 (2+2+2) Cycloadditions of Alkynes

for 5 min. The resulting soln was stirred under H_2 (1 atm) at rt for 1 h, concentrated to dryness, and dissolved in CH_2Cl_2 (0.5 mL). To this soln was added a soln of ynamide **38** (37.2 mg, 0.165 mmol) in CH_2Cl_2 (0.5 mL). Then a soln of 1,6-diyne **37** (58.3 mg, 0.15 mmol) in CH_2Cl_2 (1.0 mL) was added dropwise over 5 min at rt. The soln was stirred at rt for 1 h. The resulting soln was concentrated and purified by chromatography (silica gel, hexane/EtOAc 5:1 to 2:1), which furnished the title compound; yield: 85.1 mg (92%); >99% ee.

Helically Chiral 1,1′-Bitriphenylene (45, R^1 = H; R^2 = Me; Z = CO); Typical Procedure:[35]
(S)-DM-SEGPHOS (7.2 mg, 0.010 mmol) and [Rh(cod)$_2$]BF$_4$ (4.1 mg, 0.010 mmol) were dissolved in CH_2Cl_2 (2.0 mL) and the mixture was stirred at rt for 30 min. H_2 was introduced to the resulting soln in a Schlenk tube. After stirring at rt for 1 h, the resulting mixture was concentrated to dryness. To a 1,2-dichloroethane (1.0 mL) soln of the residue and dialkynyl ketone **44** (11.4 mg, 0.060 mmol) was added a 1,2-dichloroethane (1.0 mL) soln of tetrayne **43** (20.1 mg, 0.050 mmol). The mixture was stirred at rt for 16 h. The resulting soln was concentrated and purified by preparative TLC (hexane/toluene/CHCl$_3$ 2:1:1) to give the title compound; yield: 18.7 mg (63%); 91% ee.

2.6.1.2 Intramolecular Reactions

The intramolecular (2+2+2) cycloaddition of triynes can afford tricyclic compounds that are not readily accessible by other methods. In particular, the intramolecular approach is effective for the synthesis of sterically demanding polycyclic compounds, such as helicenes and helicene-like molecules. For example, the novel direct synthesis of fully aromatic [n]helicenes **47** was achieved in the nickel-catalyzed (2+2+2) cycloaddition of the cis,cis-dienetriynes **46** (Scheme 16).[36]

Scheme 16 Nickel-Catalyzed Synthesis of [n]Helicenes from cis,cis-Dienetriynes[36]

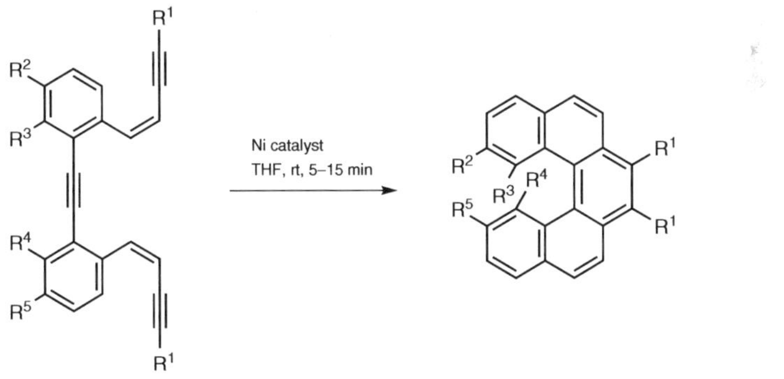

R^1	R^2	R^3	R^4	R^5	Conditions	Yield (%)	Ref
H	H	H	H	H	Ni(cod)$_2$ (20 mol%), Ph$_3$P (40 mol%)	64	[36]
H	H	H	H	H	Ni(cod)$_2$ (100 mol%)	83	[36]
Bu	H	H	H	H	Ni(cod)$_2$ (20 mol%), Ph$_3$P (40 mol%)	76	[36]
H	(CH=CH)$_2$	H		H	Ni(cod)$_2$ (100 mol%)	86	[36]
H	(CH=CH)$_2$	H		H	Ni(cod)$_2$ (10 mol%), Ph$_3$P (20 mol%)	54	[36]
H	(CH=CH)$_2$	(CH=CH)$_2$			Ni(cod)$_2$ (100 mol%)	60	[36]
H	(CH=CH)$_2$	(CH=CH)$_2$			Ni(cod)$_2$ (10 mol%), Ph$_3$P (20 mol%)	51	[36]

The highly enantioselective synthesis of [7]helicene-like molecules **49** has been achieved with the cationic rhodium(I)/(R,R)Me-DuPhos catalyzed (2+2+2) cycloaddition of triynes **48** (Scheme 17).[37]

Scheme 17 Rhodium-Catalyzed Enantioselective Synthesis of [7]Helicene-Like Molecules from Triynes[37]

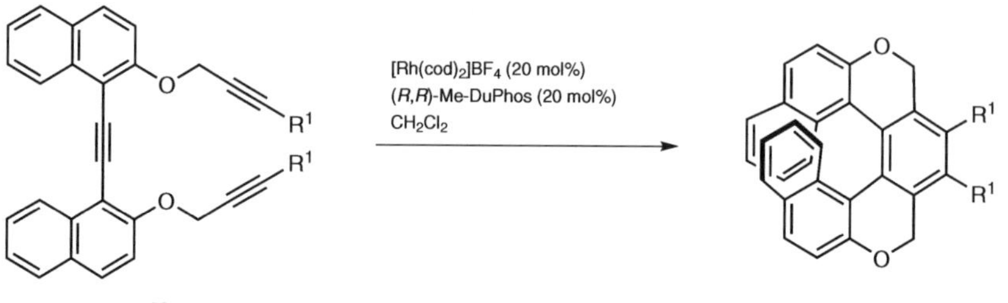

R^1	Conditions	ee (%)	Yield (%)	Ref
CO_2Me	rt, 15 h	71	80	[37]
CO_2Bu	rt, 15 h	77	71	[37]
Bu	40 °C, 140 h	85	71	[37]

[n]Helicenes 47; General Procedure:[36]
Stoichiometric method: A Schlenk flask was charged with *cis,cis*-dienetriyne **46** (0.100 mmol) and flushed with argon. The substrate was dissolved in THF (2 mL), a stock soln of Ni(cod)$_2$ in THF (0.06 M, 1.70 mL) was added, and the mixture was stirred at rt for 15 min. The solvent was removed under reduced pressure and the residue was chromatographed (silica gel) to obtain **47**.

Catalytic method: In a Schlenk flask, Ph$_3$P (10.5 mg, 0.040 mmol) was dissolved in THF (1 mL) under argon, a stock soln of Ni(cod)$_2$ in THF (0.06 M, 330 µL) was added, and the mixture was stirred at rt for 5 min. *cis,cis*-Dienetriyne **46** (0.200 mmol) in THF (2 mL) was added and the mixture was stirred at rt for 5–15 min. The solvent was evaporated in vacuo and the residue was chromatographed (silica gel) to obtain **47**.

(M)-(–)-Dimethyl 8H,11H-Naphtho[2,1-b]naphtho[1″,2″:5′,6′]pyrano[3′,4′:5,6]-benzo[d]pyran-9,10-dicarboxylate (49, R^1 = CO$_2$Me); Typical Procedure:[37]
Under an argon atmosphere, (R,R)-Me-DuPhos (6.1 mg, 0.02 mmol) and [Rh(cod)$_2$]BF$_4$ (8.1 mg, 0.02 mmol) were dissolved in CH$_2$Cl$_2$ (1.0 mL) and the mixture was stirred for 5 min. H$_2$ was introduced to the resulting soln in a Schlenk tube. After stirring at rt for 1 h, the resulting mixture was concentrated to dryness. To a CH$_2$Cl$_2$ (1.0 mL) soln of the residue was added a CH$_2$Cl$_2$ (3.0 mL) soln of triyne **48** (50.3 mg, 0.10 mmol) and the remaining substrate was transferred by rinsing with CH$_2$Cl$_2$ (1.0 mL). The mixture was stirred at rt for 15 h. The resulting mixture was concentrated and purified by preparative TLC (hexane/EtOAc/CH$_2$Cl$_2$ 10:1:1), which furnished the title compound; yield: 40.3 mg (80%); 71% ee.

2.6.2 (2+2+2) Cycloadditions of Alkynes with Nitriles

The transition-metal-catalyzed (2+2+2) cycloaddition of alkynes with nitriles has been actively investigated for the synthesis of substituted pyridines. Because different substituents can be introduced through the formation of the pyridine ring in the transition-

2.6.2 (2+2+2) Cycloadditions of Alkynes with Nitriles

metal-catalyzed (2+2+2) cycloaddition, this method is occasionally more advantageous than conventional substitution or cross-coupling methods for the synthesis of densely substituted pyridines. Since the pioneering work using cobalt catalysts by Yamazaki and Wakatsuki,[38] Vollhardt,[39] and Bönnemann,[40] a number of transition-metal catalysts and nitriles have been employed in this transformation.

2.6.2.1 Intermolecular Reactions

Various (η^5-cyclopentadienyl)cobalt(I) complexes have been employed for the (2+2+2) cycloaddition of alkynes with nitriles, but they are unstable and/or require harsh reaction conditions (high temperature or visible light irradiation). Recently, the new catalytic systems shown in Scheme 18 have been developed for pyridine synthesis. In the case of the cobalt complex **15**, although the yields are significantly improved under visible light irradiation (Scheme 18), the formation of bicyclic pyridines **52** could also be achieved in toluene that was neither degassed nor distilled.[41]

The highly reactive cobalt–trimethyl(vinyl)silane complex **50** can be used to catalyze the formation of pyridines under mild conditions.[42] The co-cyclization of diynes and nitriles **51** is carried out at room temperature without irradiation in good yields.

A low-valent cobalt catalyst generated in situ, prepared from cobalt(II) chloride, zinc, and bis(diphenylphosphino)ethane, was found to be a highly efficient catalyst for the synthesis of pyridines **52** from diynes and nitriles **51** (Scheme 18).[43] The reactions are carried out in N-methylpyrrolidin-2-one (NMP) at room temperature to give the expected bicyclic pyridines **52** in good yields. Importantly, the incorporation of the nitrile proceeds in a regioselective fashion when unsymmetrical diynes are used.

Scheme 18 Fused Pyridines by Cobalt-Catalyzed (2+2+2) Cycloaddition of Diynes with Nitriles[41–43]

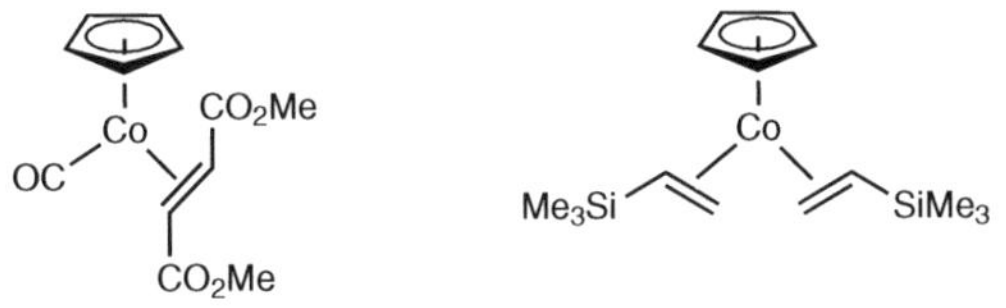

Z	R¹	R²	R³	Equiv of **51**	Conditions	Yield (%)	Ref
(CH$_2$)$_2$	H	H	Et	1.3	**15** (5 mol%), toluene, reflux, $h\nu$, 3 h	65	[41]
(CH$_2$)$_2$	H	H	Ph	1.3	**15** (5 mol%), toluene, reflux, $h\nu$, 3 h	70	[41]
NTs	H	H	Et	1.3	**15** (5 mol%), toluene, reflux, $h\nu$, 3 h	63	[41]
NTs	H	H	Ph	1.3	**15** (5 mol%), toluene, reflux, $h\nu$, 3 h	66	[41]
CH$_2$	H	H	Me	2	**50** (5 mol%), THF, rt, 2–10 min	57	[41]
CH$_2$	H	H	iPr	2	**50** (5 mol%), THF, rt, 2–10 min	86	[42]
CH$_2$	H	H	Ph	2	**50** (5 mol%), THF, rt, 2–10 min	92	[42]
CH$_2$	t-Bu	t-Bu	Ph	2	**50** (5 mol%), THF, rt, 2–10 min	53	[42]
(CH$_2$)$_2$	H	H	Ph	2	**50** (5 mol%), THF, rt, 2–10 min	46	[42]
C(CO$_2$Me)$_2$	H	H	Me	20	CoCl$_2$·6H$_2$O (5 mol%), dppe (6 mol%), Zn (10 mol%), NMP, rt, 1 h	84	[43]
C(CO$_2$Me)$_2$	H	H	iPr	20	CoCl$_2$·6H$_2$O (5 mol%), dppe (6 mol%), Zn (10 mol%), NMP, rt, 1 h	90	[43]
C(CO$_2$Me)$_2$	H	H	(CH$_2$)$_2$OH	20	CoCl$_2$·6H$_2$O (5 mol%), dppe (6 mol%), Zn (10 mol%), NMP, rt, 2 h	80	[43]
C(CO$_2$Me)$_2$	H	H	Ph	20	CoCl$_2$·6H$_2$O (5 mol%), dppe (6 mol%), Zn (10 mol%), NMP, rt, 1 h	97	[43]
NBn	H	H	Ph	20	CoCl$_2$·6H$_2$O (5 mol%), dppe (6 mol%), Zn (10 mol%), NMP, rt, 6 h	83	[43]
C(CO$_2$Me)$_2$	H	Bu	Me	20	CoCl$_2$·6H$_2$O (5 mol%), dppe (6 mol%), Zn (10 mol%), NMP, 50 °C, 24 h	78[a]	[43]
C(CO$_2$Me)$_2$	TMS	Ph	Me	20	CoCl$_2$·6H$_2$O (5 mol%), dppe (6 mol%), Zn (10 mol%), NMP, 50 °C, 24 h	91	[43]
O	Bu	Bu	Me	80	CoCl$_2$·6H$_2$O (5 mol%), dppe (6 mol%), Zn (10 mol%), MeCN, 50 °C, 24 h	83	[43]

[a] Regioselectivity 95:5.

The nickel(0)/Xantphos catalyst system is highly efficient for the (2+2+2) cycloaddition of diynes **53** with nitriles **54**.[44] The reactions are carried out in toluene at room temperature to give the expected bicyclic pyridines **55** in high yields (Scheme 19). Importantly, the incorporation of the nitrile proves regioselective when unsymmetrical diynes are used.[44]

2.6.2 (2+2+2) Cycloadditions of Alkynes with Nitriles

Scheme 19 Fused Pyridines by Nickel-Catalyzed (2+2+2) Cycloaddition of Diynes with Nitriles[44]

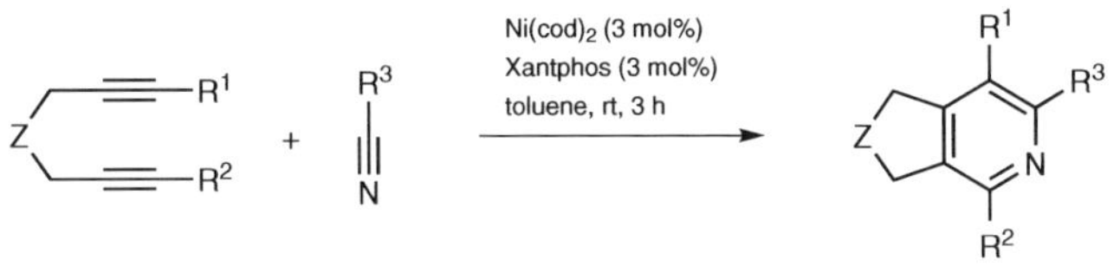

Xantphos = 4,5-bis(diphenylphosphino)-9,9-dimethylxanthene

Z	R^1	R^2	R^3	Yield (%)	Ref
$C(CO_2Me)_2$	Me	Me	Ph	92	[44]
$C(CO_2Me)_2$	Me	Me	$CH=CH_2$	>99	[44]
$C(CO_2Me)_2$	Me	Me	iPr	90	[44]
$C(CO_2Me)_2$	Me	H	Ph	79	[44]
O	Me	Me	Ph	80	[44]
$[C(CO_2Et)_2]_2$	Me	Me	Ph	98	[44]
$[C(CO_2Et)_2]_2$	Me	Me	$N(CH_2CH=CH_2)_2$	76	[44]
$[C(CO_2Et)_2]_2$	Me	Me	morpholino	75	[44]

The catalyst chloro(cyclooctadiene)(η^5-pentamethylcyclopentadienyl)ruthenium(II) [Ru(Cp*)Cl(cod)] is highly efficient for the (2+2+2) cycloaddition of diynes with electron-deficient nitriles.[45] Various activated nitriles can participate in this reaction to give the expected bicyclic pyridines **56** in high yields (Scheme 20). Moreover, when unsymmetrical 1,6-diynes are employed, pyridines **56** are formed as the predominant products along with regioisomers **57** (Scheme 20).[46]

Scheme 20 Fused Pyridines by Ruthenium-Catalyzed (2+2+2) Cycloaddition of Diynes with Nitriles[45,46]

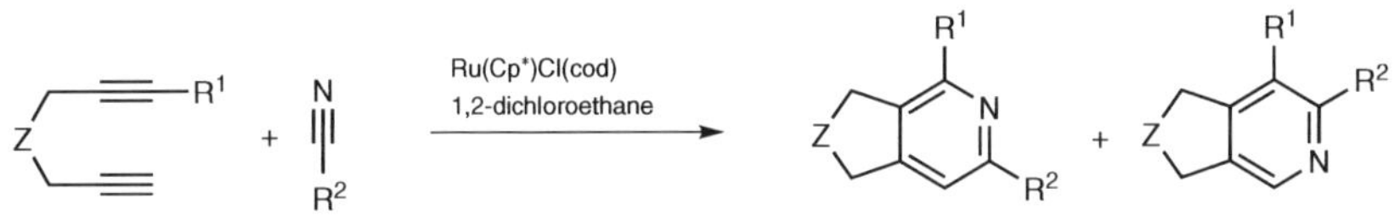

Z	R^1	R^2	mol% of Catalyst	Conditions	Ratio (56/57)	Yield (%)	Ref
$C(CO_2Me)_2$	H	CO_2Et	2	60°C, 0.5 h	–	83	[45]
$C(CN)_2$	H	CO_2Et	2	60°C, 0.5 h	–	80	[45]
CH_2	H	CO_2Et	2	60°C, 0.5 h	–	89	[45]
NTs	H	CO_2Et	2	60°C, 0.5 h	–	75	[45]
S	H	CO_2Et	2	60°C, 17 h	–	64	[45]
$C(CO_2Me)_2$	Me	CO_2Et	5	60°C, 2 h	88:12	87	[45]
$C(CO_2Me)_2$	Ph	CO_2Et	20	60°C, 6 h	100:0	64	[45]
$C(CO_2Me)_2$	H	Bz	2	60°C, 0.5 h	–	84	[45]
$C(CO_2Me)_2$	H	C_6F_5	5	60°C, 1 h	–	67	[45]
$C(CO_2Me)_2$	H	C_6F_5	5	60°C, 1 h	–	80ª	[45]

Z	R[1]	R[2]	mol% of Catalyst	Conditions	Ratio (56/57)	Yield (%)	Ref
C(CO$_2$Me)$_2$	H	CH$_2$CN	2	rt, 22 h	–	92	[46]
C(CO$_2$Me)$_2$	Ph	CH$_2$CN	10	60 °C, 24 h	100:0	78	[46]
C(CO$_2$Me)$_2$	Me	(CH$_2$)$_2$CN	2	rt, 10 h	100:0	83	[46]
C(CO$_2$Me)$_2$	H	2-NCC$_6$H$_4$	2	rt, 4 h	–	61	[46]
C(CO$_2$Me)$_2$	H	(E)-CH=CHCN	2	rt, 22 h	–	88	[46]
C(CO$_2$Me)$_2$	H	CH$_2$Cl	2	rt, 2 h	–	93	[46]
C(CO$_2$Me)$_2$	H	CH$_2$Br	2	rt, 3 h	–	42	[46]
C(CO$_2$Me)$_2$	H	CH(Cl)CH$_2$NHPh	2	rt, 6 h	–	71	[46]
C(CO$_2$Me)$_2$	Me	CH$_2$OMe	2	rt, 10 h	100:0	79	[46]
C(CO$_2$Me)$_2$	Me	CH$_2$C≡CSiMe$_3$	5	rt, 7 h	100:0	75	[46]

[a] 3 equiv of nitrile was used.

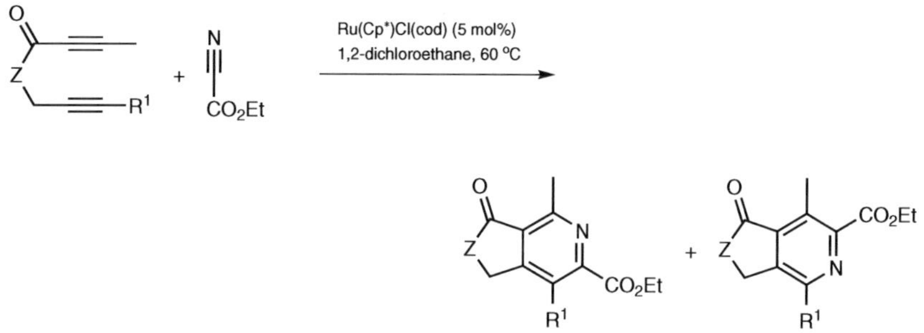

Z	R[1]	Time (h)	Ratio (58/59)	Yield (%)	Ref
O	H	0.5	98:2	84	[46]
NBn	Me	6	80:20	82	[46]

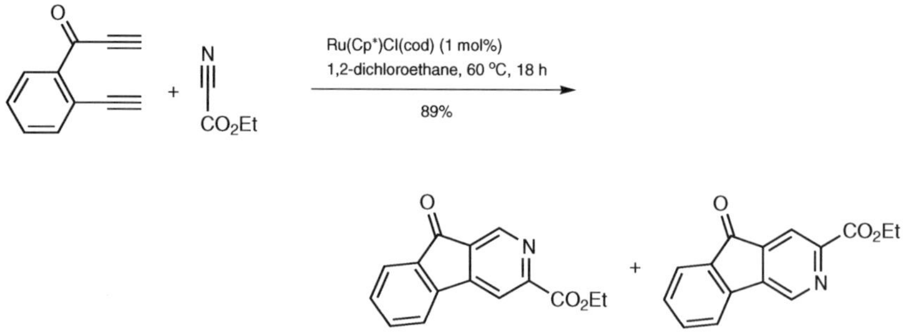

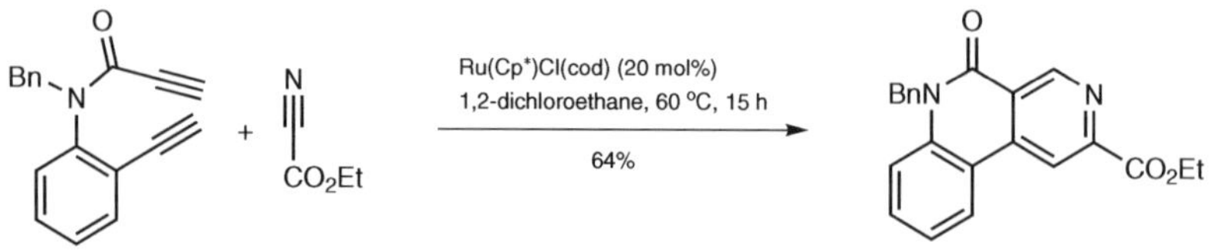

2.6.2 (2+2+2) Cycloadditions of Alkynes with Nitriles

Cationic rhodium(I)/biaryl bisphosphine complexes are highly active and selective catalysts for the (2+2+2) cycloaddition of alkynes with nitriles.[47] Both activated and unactivated nitriles can participate in this reaction to give the expected bicyclic pyridines in good to high yields (Scheme 21). Moreover, a commercially available electron-deficient perfluoroalkanenitrile **61** [$R^2 = (CF_2)_6CF_3$] was found to be a suitable cycloaddition partner with 1,6-diynes **60** to give the corresponding perfluoroalkylated pyridines **62** at room temperature in good yields (Scheme 21).[48]

Scheme 21 Fused Pyridines by Rhodium-Catalyzed (2+2+2) Cycloaddition of Diynes with Nitriles[47,48]

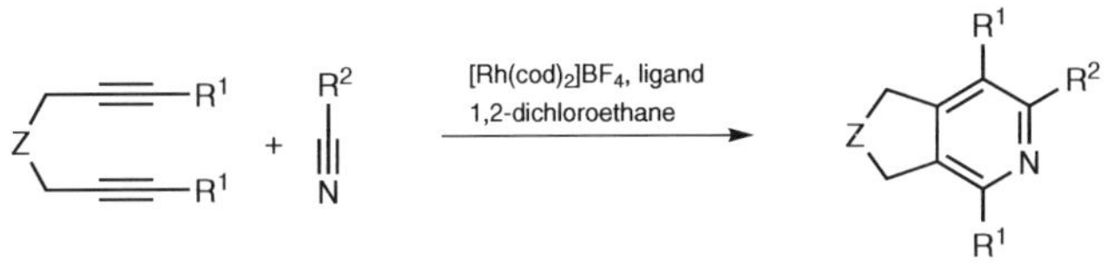

Z	R^1	R^2	Equiv of **61**	Ligand (mol%)	mol% of Catalyst	Conditions	Yield (%)	Ref
C(CO$_2$Me)$_2$	Me	CO$_2$Et	1.1	BINAP (3)	3	rt, 3 h	>99[a]	[47]
NTs	Me	CO$_2$Et	1.1	BINAP (3)	3	rt, 1 h	>99[a]	[47]
O	Et	CO$_2$Et	1.1	BINAP (3)	3	rt, 1 h	>99[a]	[47]
C(CO$_2$Me)$_2$	H	CO$_2$Et	2	BINAP (3)	3	60 °C, 6 h	69	[47]
C(CO$_2$Me)$_2$	H	Bz	1.1	BINAP (3)	3	60 °C, 16 h	>99	[47]
C(CO$_2$Me)$_2$	H	Ph	5	SEGPHOS (3)	3	80 °C, 1 h	87	[47]
C(CO$_2$Me)$_2$	H	morpholino	2	H$_8$-BINAP (3)	3	40 °C, 5 h	47[a]	[47]
C(CO$_2$Me)$_2$	Me	CH$_2$CN	1.1	H$_8$-BINAP (5)	5	60 °C, 18 h	84	[47]
C(CO$_2$Me)$_2$	Me	(CF$_2$)$_6$CF$_3$	1.1	4-Tol-BINAP (5)	5	rt, 3 h	85	[48]
C(CO$_2$Me)$_2$	H	(CF$_2$)$_6$CF$_3$	1.1	4-Tol-BINAP (5)	5	rt, 1 h	62	[48]
NTs	Me	(CF$_2$)$_6$CF$_3$	1.1	4-Tol-BINAP (5)	5	rt, 1 h	92	[48]

[a] The solvent was CH$_2$Cl$_2$.

Iridium(I)/bisphosphine complexes are highly effective catalysts for the (2+2+2) cycloaddition of diynes **63** with nitriles **64**.[49] Both aromatic and aliphatic nitriles can be used for this reaction (Scheme 22). Functional groups such as nitro, aldehyde, and ketone moieties are compatible with the reaction.

Scheme 22 Fused Pyridines by Iridium-Catalyzed (2+2+2) Cycloaddition of Diynes with Nitriles[49]

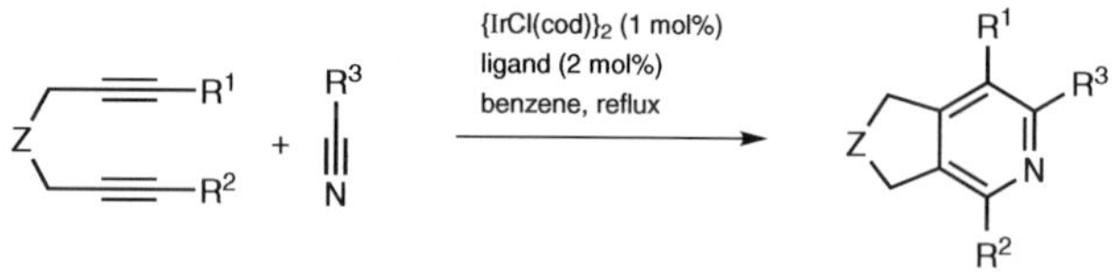

Z	R^1	R^2	R^3	Equiv of **64**	Ligand	Time (h)	Yield (%)	Ref
C(CO$_2$Me)$_2$	Me	Me	Ph	3	dppf	3	91	[49]
C(CO$_2$Me)$_2$	Me	Me	4-OHCC$_6$H$_4$	3	dppf	24	73	[49]
C(CO$_2$Me)$_2$	Me	Me	Me	10	dppf	2	75	[49]
C(CO$_2$Me)$_2$	Me	Me	Cy	10	dppf	1	54	[49]
C(CO$_2$Me)$_2$	Me	Me	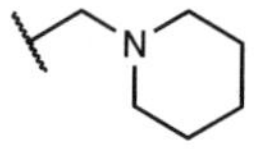 (piperidinylmethyl)	3	dppf	1	86	[49]
C(CO$_2$Me)$_2$	Me	Me	2-pyridyl	3	dppf	2	>99	[49]
C(CO$_2$Me)$_2$	Et	Et	Ph	3	dppf	2	71	[49]
C(CH$_2$OMe)$_2$	Me	Me	Ph	3	dppf	3	83	[49]
[C(CO$_2$Et)$_2$]$_2$	Me	Me	Ph	3	dppf	1	95	[49]
C(CO$_2$Me)$_2$	Me	Ph	Me	10	BINAP	1	94[a]	[49]
C(CO$_2$Me)$_2$	TMS	Me	Bn	10	dppf	4	59[a]	[49]

[a] Regioselectivity >99:1.

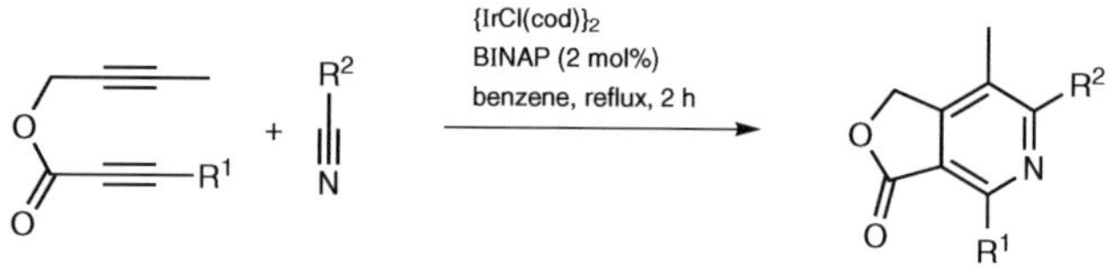

R^1	R^2	mol% of Catalyst	Yield[a] (%)	Ref
Me	Ph	1	73	[49]
Ph	Bn	2	83	[49]

[a] Regioselectivity >99:1.

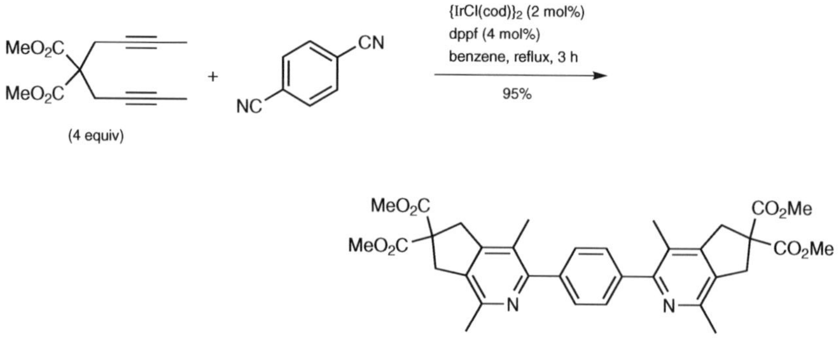

2.6.2 (2+2+2) Cycloadditions of Alkynes with Nitriles

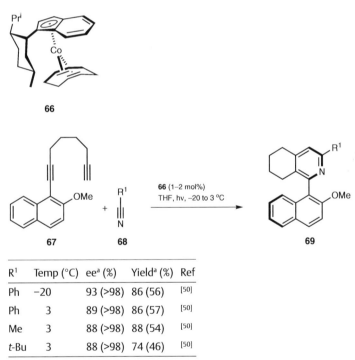

The enantioselective synthesis of axially chiral arylpyridines **69** has been achieved by the chiral (η[5]-cyclopentadienyl)cobalt **66** catalyzed (2+2+2) cycloaddition of diynes **67** and nitriles **68** (Scheme 23).[50] Lowering the reaction temperature (−20 °C) affords the arylpyridine with greater than 90% ee.[50]

Scheme 23 Cobalt-Catalyzed Enantioselective Synthesis of Axially Chiral Arylpyridines[50]

R^1	Temp (°C)	ee[a] (%)	Yield[a] (%)	Ref
Ph	−20	93 (>98)	86 (56)	[50]
Ph	3	89 (>98)	86 (57)	[50]
Me	3	88 (>98)	88 (54)	[50]
t-Bu	3	88 (>98)	74 (46)	[50]

[a] Figures in parentheses are values after recrystallization.

Bicyclic Pyridines 52; General Procedure Using Cobalt Catalyst 15:[41]
A soln of diyne (1 mmol), nitrile **51** (1.3 mmol), and Co catalyst **15** (0.05 mmol) in toluene (10 mL) was refluxed under visible light irradiation for 3 h. After being cooled, the solvent was removed under reduced pressure and the crude product was purified by flash column chromatography (pentane/EtOAc gradient) to give the corresponding product **52**.

Bicyclic Pyridines 52; General Procedure Using Cobalt Catalyst 50:[42]
The diyne (1.0 mmol) and nitrile **51** (2.0 mmol) were dissolved in THF (4 mL) and stirred at rt. Co catalyst **50** (0.016 g, 0.05 mmol) dissolved in Et_2O (0.25 mL) was added dropwise, and the red soln was stirred for 2–10 min, after which the diyne starting material had disappeared. The product was separated via column chromatography (hexane/EtOAc 6:1), yielding the pyridines **52**.

Bicyclic Pyridines 52; General Procedure Using an In Situ Prepared Low-Valent Cobalt Catalyst:[43]
To a mixture of the diyne (1.0 mmol), nitrile **51** (1.5–20 mmol), and Zn powder (6.5 mg, 0.10 mmol) in NMP (0.6 mL) was added a soln of $CoCl_2 \cdot 6H_2O$ (11.9 mg, 0.05 mmol) and dppe (23.9 mg, 0.06 mmol) in NMP (0.4 mL). [Note: A 0.125 M soln that was prepared from $CoCl_2 \cdot 6H_2O$ (476 mg, 2.0 mmol), dppe (956 mg, 2.4 mmol), and NMP (16.0 mL) could be stored under an argon atmosphere for a month.] The resulting mixture was stirred at rt. After completion of the reaction (checked by TLC analysis), Et_2O (5 mL) was added and the mixture was filtered through a pad of Celite with Et_2O (5 mL). The filtrate was concentrated under reduced pressure and the residue was chromatographed (silica gel) to give the corresponding substituted pyridine **52**.

Bicyclic Pyridines 55; General Procedure:[44]
In a N_2 filled glovebox, diyne **53** (1 equiv, 0.1 M) and nitrile **54** (1.5 equiv) were added to an oven-dried screw-cap vial equipped with a magnetic stirrer bar. In a separate vial, $Ni(cod)_2$ and Xantphos were weighed (in a 1:1 molar ratio) and dissolved in toluene. A soln of the catalyst (3 mol%) was added to the mixture. The vial was sealed and brought out of the glovebox. The reaction was stirred at rt for 3 h. The resulting mixture was concentrated and purified by flash column chromatography to give **55**.

3-Ethyl 6,6-Dimethyl 5,7-Dihydro-6H-cyclopenta[c]pyridine-3,6,6-tricarboxylate [56, Z = C(CO$_2$Me)$_2$; R^1 = H; R^2 = CO$_2$Et]; Typical Procedure:[45]
To a soln of ethyl cyanoformate (74 mg, 0.75 mmol) and Ru(Cp*)Cl(cod) (3.8 mg, 0.01 mmol) in dry degassed 1,2-dichloroethane (2 mL) was added a soln of the 1,6-diyne (104 mg, 0.5 mmol) in dry, degassed 1,2-dichloroethane (3 mL) over 15 min under an argon atmosphere at rt. The mixture was stirred at 60 °C for 0.5 h. The solvent was evaporated and the crude product was purified by flash column chromatography (silica gel, hexane/EtOAc 1:2) to give the pyridine **56**; yield: 128 mg (83%).

3-Ethyl 6,6-Dimethyl 1,4-Dimethyl-5,7-dihydro-6H-cyclopenta[c]pyridine-3,6,6-tricarboxylate [62, Z = C(CO$_2$Me)$_2$; R^1 = Me; R^2 = CO$_2$Et]; Typical Procedure:[47]
Under argon, BINAP (7.5 mg, 0.012 mmol) and [Rh(cod)$_2$]BF$_4$ (4.9 mg, 0.012 mmol) were dissolved in CH_2Cl_2 (2.0 mL), and the mixture was stirred at rt for 5 min. H_2 was introduced to the resulting soln in a Schlenk tube. After stirring at rt for 0.5 h, the resulting soln was concentrated to dryness and the residue was dissolved in CH_2Cl_2 (2.0 mL). To this soln was added dropwise over 1 min a soln of dimethyl 2,2-dibut-2-ynylmalonate [**60**, Z = C(CO$_2$Me)$_2$; R^1 = Me; 94.5 mg, 0.400 mmol] and ethyl cyanoformate (**61**, R^2 = CO$_2$Et; 43.6 mg, 0.440 mmol) in CH_2Cl_2 (1.0 mL) at rt. Undissolved substrate was dissolved by the addition of CH_2Cl_2 (1.0 mL). The mixture was stirred at rt for 3 h. The resulting soln was concentrated and purified by column chromatography (silica gel, Et_2O), which furnished the title compound; yield: 134.1 mg (99%).

Dimethyl 1,4-Dimethyl-3-phenyl-5,7-dihydro-6H-cyclopenta[c]pyridine-6,6-dicarboxylate [65, Z = C(CO$_2$Me)$_2$; R^1 = R^2 = Me; R^3 = Ph]; Typical Procedure:[49]
A flask was charged with {IrCl(cod)}$_2$ (7.2 mg, 0.01 mmol) and dppf (11.3 mg, 0.02 mmol). The flask was evacuated and filled with argon. To the flask were added benzene (5 mL)

(**CAUTION:** *carcinogen*) and benzonitrile (**64**, R^3 = Ph; 325 mg, 3.2 mmol). Diyne **63** (235 mg, 1.0 mmol) was added to the mixture. The mixture was stirred under reflux for 3 h. The progress of the reaction was monitored by GLC. After the reaction was complete, the solvent was removed under reduced pressure. Column chromatography (hexane/EtOAc 7:3) of the residue gave the title compound; yield: 305 mg (91%).

1-(2-Methoxynaphthalen-1-yl)-3-phenyl-5,6,7,8-tetrahydroisoquinoline (69, R^1 = Ph); Typical Procedure:[50]

A thermostated (3 °C) reaction vessel was loaded with diyne **67** (524 mg, 2 mmol), Co catalyst **66** (8.4 mg, 0.02 mmol), THF (20 mL), and benzonitrile (**68**, R^1 = Ph; 412 µL, 4 mmol) under an argon atmosphere. The mixture was stirred and irradiated by two 460-W lamps (λ 420 nm) for 24 h. The reaction was quenched by switching off the lamps and simultaneously letting in air. The extent of the reaction, i.e. conversion of the starting diyne **67**, was determined by GC. The mixture was filtered through a thin pad of silica gel, eluting with THF. The solvent was removed under reduced pressure to give an oily residue, which was further dissolved in Et_2O (10 mL). Colorless crystals of **69** were filtered off and washed with Et_2O; yield: 413 mg (57%). The optical purity was determined to be >98% ee by HPLC.

2.6.2.2 Intramolecular Reactions

Heteroatom-containing C_2-symmetric axially chiral spiranes are valuable compounds as efficient chiral ligands. The enantioselective synthesis of C_2-symmetric spirobipyridines **71** has been achieved by the intramolecular double (2+2+2) cycloaddition of bis(diyneni-triles) **70** using cationic rhodium(I)/axially chiral biaryl bisphosphine catalysts (Scheme 24).[51]

Scheme 24 Rhodium-Catalyzed Enantioselective Synthesis of Spirobipyridines[51]

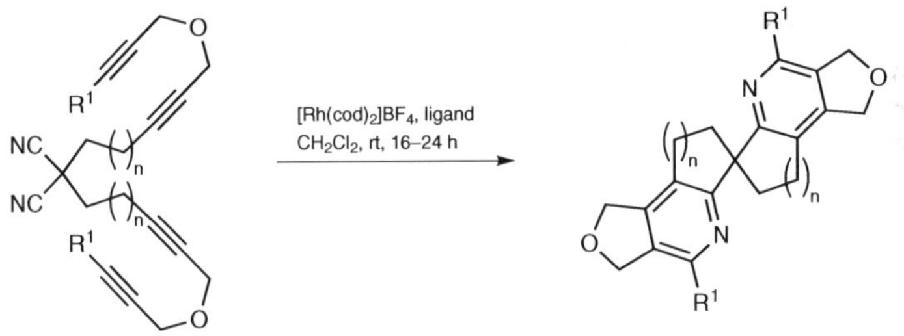

n	R^1	Ligand (mol%)	mol% of Catalyst	ee (%)	Yield (%)	Ref
1	Ph	(*R*)-SEGPHOS (10)	10	64	99	[51]
1	4-ClC$_6$H$_4$	(*R*)-SEGPHOS (5)	5	71	99	[51]
1	Me	(*R*)-H$_8$-BINAP (10)	10	49	98	[51]
1	H	(*R*)-H$_8$-BINAP (10)	10	47	70	[51]
2	Ph	(*R*)-SEGPHOS (10)	10	45	90	[51]
2	Me	(*R*)-SEGPHOS (10)	10	40	98	[51]

4,4′-Diphenyl-1,1′,3,3′,7,7′,8,8′-octahydro-6,6′-spirobi[cyclopenta[b]furo[3,4-d]pyridine] (71, n = 1; R^1 = Ph); Typical Procedure:[51]

(*R*)-SEGPHOS (6.1 mg, 0.01 mmol) and [Rh(cod)$_2$]BF$_4$ (4.1 mg, 0.01 mmol) were dissolved in CH$_2$Cl$_2$ (1.0 mL) and the mixture was stirred at rt for 5 min. H$_2$ was introduced to the resulting soln in a Schlenk tube. After stirring at rt for 0.5 h, the resulting soln was concentrated and the residue was dissolved in CH_2Cl_2 (0.5 mL). To this soln was added a soln of bis(diyne-nitrile) **70** (45.8 mg, 0.10 mmol) in CH_2Cl_2 (0.25 mL) at rt and the remaining substrate was transferred by rinsing with CH_2Cl_2 (0.25 mL). The mixture was stirred at rt for 16 h. The mixture was concentrated and purified by column chromatography (silica gel, CH_2Cl_2/EtOAc 100:3), which furnished bipyridine (−)-**71**; yield: 45.4 mg (99%); 64% ee.

2.6.3 (2+2+2) Cycloadditions Involving Heterocumulenes

The transition-metal-catalyzed (2+2+2) cycloaddition of alkynes with isocyanates has also been actively investigated for the synthesis of substituted pyridin-2-ones. The catalytic formation of pyridin-2-ones was first reported by Yamazaki using cobalt catalysts[52] and by Hoberg using nickel catalysts.[53] Subsequently, Vollhardt reported the cobalt-catalyzed partially intramolecular (2+2+2) cycloaddition of 5-isocyanatoalkynes.[54] Since these pioneering works, a number of transition-metal catalysts and heterocumulenes have been developed for this transformation.

2.6.3.1 Isocyanates

A nickel(0)/1,3-bis(2,6-diisopropylphenyl)-4,5-dihydroimidazol-2-ylidene (SIPr) complex is a highly efficient catalyst for the (2+2+2) cycloaddition of diynes **72** with isocyanates **73**.[55] The reaction proceeds at room temperature to give the expected bicyclic pyridin-2-ones **74** in good yields (Scheme 25). [6,6]-Fused pyridinone **74** [Z = $(CH_2)_2$] is obtained in excellent yield when a diyne lacking a Thorpe–Ingold effect is reacted with cyclohexyl isocyanate.

Scheme 25 Bicyclic Pyridin-2-ones by Nickel-Catalyzed (2+2+2) Cycloaddition of Diynes with Isocyanates[55]

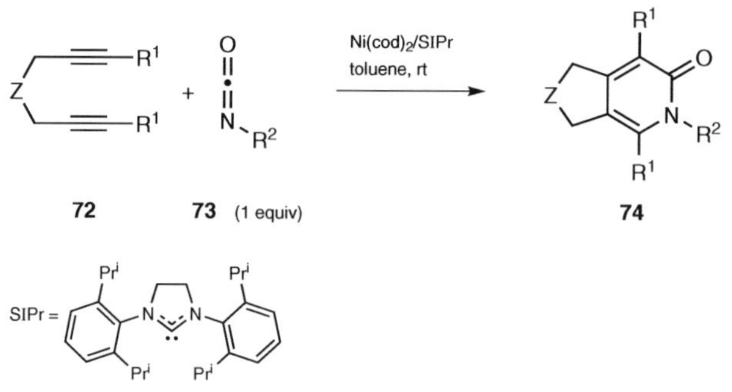

Z	R^1	R^2	mol% of Catalyst	Conc (M)	Yield (%)	Ref
$C(CO_2Me)_2$	Me	Ph	3	0.1	86	[55]
$C(CO_2Me)_2$	Me	4-MeOC$_6$H$_4$	3	0.1	84	[55]
$C(CO_2Me)_2$	Me	4-F$_3$CC$_6$H$_4$	5	0.1	82[a]	[55]
$C(CO_2Me)_2$	Me	2,6-Me$_2$C$_6$H$_3$	3	0.1	82	[55]
$C(CO_2Me)_2$	Me	Cy	3	0.1	91	[55]
$C(CO_2Me)_2$	Me	Bu	3	0.1	92	[55]
$C(CO_2Me)_2$	iPr	Bu	5	0.1	85	[55]
$C(CO_2Me)_2$	H	Ph	5	0.05	71[b]	[55]
$C(CO_2Me)_2$	H	Bu	5	0.05	77[b]	[55]
$C(CO_2Me)_2$	iPr	Bu	5	0.1	85	[55]

2.6.3 (2+2+2) Cycloadditions Involving Heterocumulenes

Z	R^1	R^2	mol% of Catalyst	Conc (M)	Yield (%)	Ref
(CH$_2$)$_2$	Et	Cy	3	0.1	99	[55]
NTs	Me	Ph	3	0.1	78	[55]
O	H	Ph	3	0.1	31	[55]

^a At 80 °C.
^b 4 equiv of isocyanate was used.

a At 80 °C.
b 4 equiv of isocyanate was used.

The Ru(Cp*)Cl(cod) catalyst was found to be highly efficient for the (2+2+2) cycloaddition of diynes **75** with isocyanates **76**.[45] Various terminal diynes and isocyanates can participate in this reaction to give the expected bicyclic pyridin-2-ones **77** in high yields (Scheme 26). When unsymmetrical 1,6-diynes are employed, a single regioisomeric product can be formed.

Scheme 26 Bicyclic Pyridin-2-ones by Ruthenium-Catalyzed (2+2+2) Cycloaddition of Diynes with Isocyanates[45]

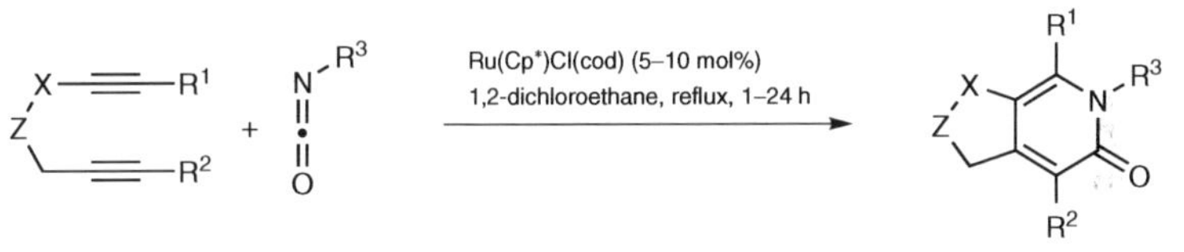

X	Z	R^1	R^2	R^3	mol% of Catalyst	Time (h)	Yield (%)	Ref
CH$_2$	C(CO$_2$Me)$_2$	H	H	Ph	5	1	87	[45]
CH$_2$	C(CO$_2$Me)$_2$	H	H	2-furyl	5	1	87	[45]
CH$_2$	C(CO$_2$Me)$_2$	H	H	Pr	5	1	89	[45]
CH$_2$	C(CO$_2$Me)$_2$	H	H	Cy	5	1	85	[45]
CH$_2$	CAc$_2$	H	H	Ph	10	1	75	[45]
CH$_2$	CH$_2$	H	H	Ph	5	2	62	[45]
CH$_2$	S	H	H	Ph	5	24	60	[45]
CH$_2$	C(CO$_2$Me)$_2$	Me	H	Pr	5	3	85^a	[45]
CH$_2$	NTs	Me	H	Pr	5	6	80^a	[45]
CH$_2$	O	Me	H	Pr	5	15	88^a	[45]
CH$_2$	C(CO$_2$Me)$_2$	Ph	H	Pr	10	20	80^a	[45]
CH$_2$	NTs	Me	H	Ph	5	3	85^a	[45]
CO	CMe$_2$	H	H	Pr	5	18	75^b	[45]
CO	NBn	Me	H	Pr	5	6	82^a	[45]
CO	NBn	Me	Me	Pr	5	6	77^c	[45]

^a Regioselectivity >99:1.
^b At 60 °C; regioselectivity 97:3.
^c Regioselectivity 83:17.

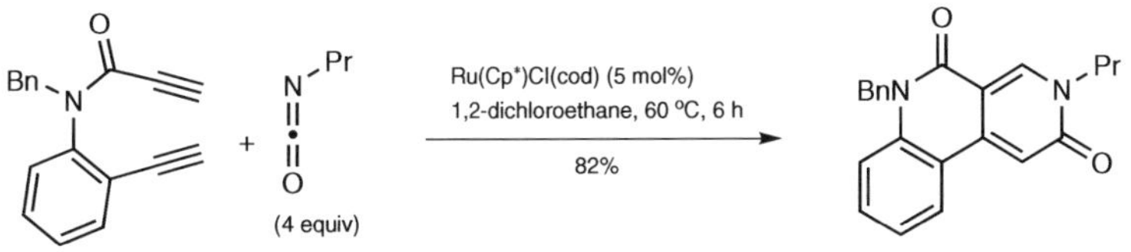

a Regioselectivity >99:1.
b At 60 °C; regioselectivity 97:3.
c Regioselectivity 83:17.

A cationic rhodium(I)/H$_8$-BINAP complex is a highly effective catalyst for the (2+2+2) cycloaddition of both internal and terminal 1,6-diynes **78** with isocyanates **79** (Scheme 27).[56] Furthermore, 1,7- and 1,8-diynes without the Thorpe–Ingold effect can also be employed for this reaction.

Scheme 27 Bicyclic Pyridin-2-ones by Rhodium-Catalyzed (2+2+2) Cycloaddition of Diynes with Isocyanates[56]

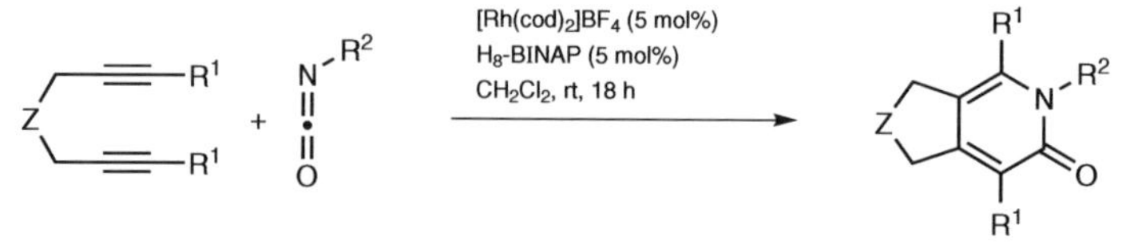

Z	R^1	R^2	Equiv of **79**	Yield (%)	Ref
C(CO$_2$Me)$_2$	Me	Bn	1.1	99	[56]
C(CO$_2$Me)$_2$	Me	Ph	1.1	87	[56]
C(CO$_2$Me)$_2$	H	Bn	2	84	[56]
C(CO$_2$Me)$_2$	H	Cy	2	81	[56]
NTs	Me	Bn	1.1	93	[56]
CH$_2$	H	Bn	2	64	[56]
(CH$_2$)$_2$	Et	Bn	2	98	[56]
(CH$_2$)$_3$	H	Bn	2	48	[56]

An iridium(I)/BINAP complex has been reported to be an efficient catalyst for the reaction of α,ω-diynes **81** with isocyanates **82**.[57] Both aliphatic and aromatic isocyanates react smoothly with 1,6-diynes to give the expected pyridin-2-ones **83** in good to high yields, while electron-rich isocyanates give the products in higher yields than electron-deficient isocyanates (Scheme 28).

Scheme 28 Bicyclic Pyridin-2-ones by Iridium-Catalyzed (2+2+2) Cycloaddition of Diynes with Isocyanates[57]

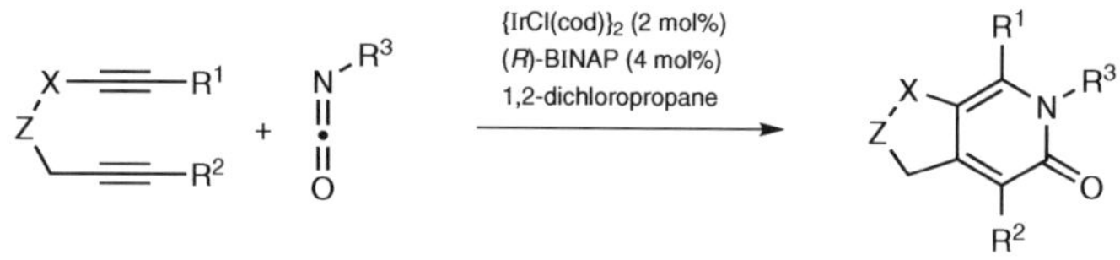

X	Z	R^1	R^2	R^3	Conditions	Yield (%)	Ref
CH$_2$	C(CO$_2$Me)$_2$	Me	Me	Ph	reflux, 1 h	97	[57]
CH$_2$	C(CO$_2$Me)$_2$	Me	Me	4-BrC$_6$H$_4$	reflux, 4 h	87	[57]
CH$_2$	C(CO$_2$Me)$_2$	Me	Me	4-AcC$_6$H$_4$	reflux, 24 h	71	[57]
CH$_2$	C(CO$_2$Me)$_2$	Me	Me	Bu	rt, 20 min	98	[57]
CH$_2$	C(CO$_2$Me)$_2$	Me	Me	Cy	rt, 20 min	96	[57]
CH$_2$	C(CO$_2$Me)$_2$	Me	Me	CH$_2$CH=CH$_2$	reflux, 24 h	80	[57]
CH$_2$	CAc$_2$	Me	Me	Bu	rt, 30 min	98	[57]

2.6.3 (2+2+2) Cycloadditions Involving Heterocumulenes

X	Z	R¹	R²	R³	Conditions	Yield (%)	Ref
CH$_2$	[C(CO$_2$Et)$_2$]$_2$	Me	Me	Bu	reflux, 24 h	94	[57]
CH$_2$	NTs	Me	Me	Bu	rt, 30 min	96	[57]
CH$_2$	C(CO$_2$Me)$_2$	H	H	Bu	reflux, 30 min	56	[57]
CH$_2$	C(CO$_2$Me)$_2$	Me	Ph	Bu	reflux, 1 h	80[a]	[57]
CH$_2$	C(CO$_2$Me)$_2$	Me	TMS	Bu	reflux, 24 h	86[a]	[57]
CO	O		Me	Me	Bu	reflux, 30 min	91[a]	[57]

[a] Single regioisomers (>99:1).

The use of a cationic rhodium(I)/(R)-DTBM-SEGPHOS (**84**) catalyst enables an atropselective pyridin-2-one synthesis (Scheme 29).[56] The reactions of 2-chlorophenyl- or 2-bromophenyl-substituted unsymmetrical 1,6-diynes **85** with alkyl isocyanates **86** proceed to give axially chiral pyridin-2-ones **87** with good yields and enantiomeric excess values.

Scheme 29 Rhodium-Catalyzed Enantioselective Synthesis of Axially Chiral Arylpyridin-2-ones[56]

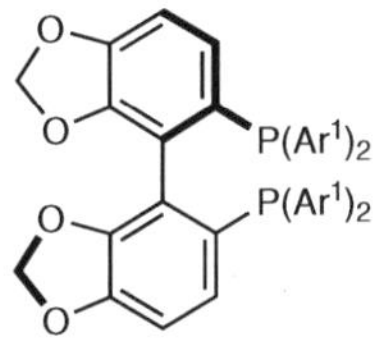

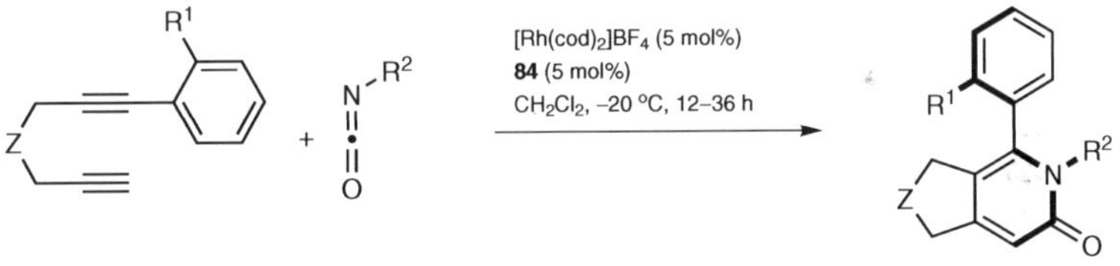

Z	R¹	R²	Time (h)	ee (%)	Yield (%)	Ref
CH$_2$	Cl	Bn	12	87	81	[56]
CH$_2$	Cl	Bu	12	88	79	[56]
CH$_2$	Cl	(CH$_2$)$_7$Me	12	90	75	[56]
CH$_2$	Br	Bn	12	85	83	[56]
O	Cl	Bn	15	91	58	[56]
C(CO$_2$Me)$_2$	Cl	Bn	36	92	89	[56]

Quinolizine and indolizine units are found in various biologically active natural compounds. The regio- and enantioselective rhodium(I)/chiral phosphoramidite complex catalyzed (2+2+2) cycloaddition of alkenyl isocyanates **91** with alkynes **90** affords quinolizine derivatives **92** and CO migration product **93** with high yields and enantiomeric excess values (Scheme 30).[58]

Scheme 30 Enantioselective Synthesis of Indolizines and Quinolizines by Rhodium-Catalyzed (2+2+2) Cycloaddition of Alkenyl Isocyanates with Alkynes[58]

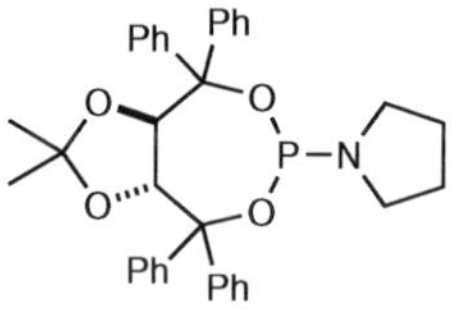

R¹	n	Ligand	Ratio (92/93)	ee (%) of 92	ee (%) of 93	Yield (%)	Ref
3,4-(MeO)$_2$C$_6$H$_3$	1	88	<1:>20	–	94	72	[58]
2-MeOC$_6$H$_4$	1	88	<1:>20	–	94	64	[58]
3-thienyl	1	88	1:9	–	86	64	[58]
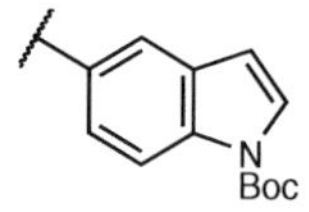 N-Boc-indolyl	1	88	<1:>20	–	91	85	[58]
4-BrC$_6$H$_4$	1	88	1:3.2	90	89	72	[58]
3-FC$_6$H$_4$	1	88	1:1.8	94	94	68	[58]
cyclohex-1-enyl	1	88	<1:>20	–	92	96	[58]
3,4-(MeO)$_2$C$_6$H$_3$	2	88	<1:>20	–	98	62	[58]
(CH$_2$)$_5$Me	1	89	5:1	80	–	78	[58]
Bn	1	89	>20:1	84	–	50	[58]
(CH$_2$)$_2$OTBDMS	1	89	>20:1	87	–	65	[58]
Cy	1	89	1.2:1	77	95	82	[58]

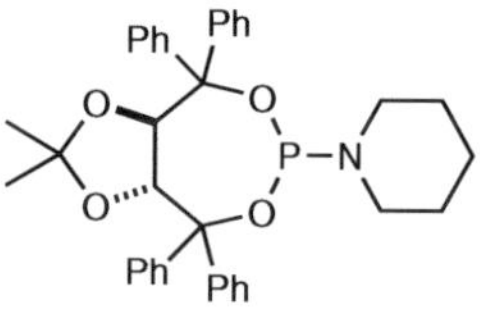

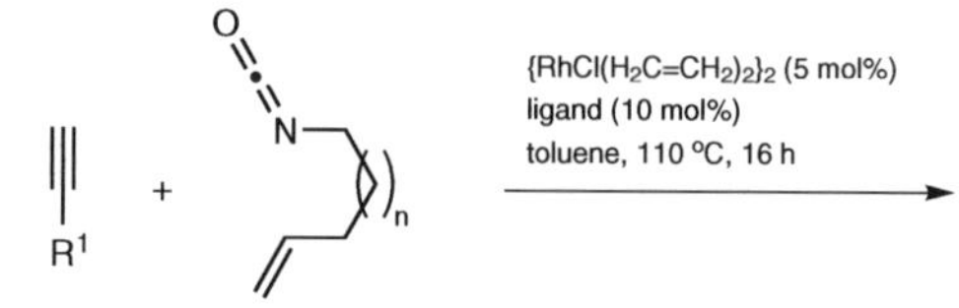

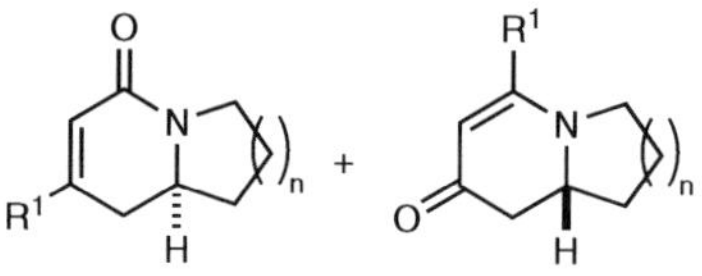

Enantioenriched piperidines **98** can also be synthesized, from indolizinones and quinolizinones **96** (Z = O), by the use of a temporary tether arising from the alkenyl isocyanate moiety **95** (Scheme 31).[59]

2.6.3 (2+2+2) Cycloadditions Involving Heterocumulenes

Scheme 31 Enantioselective Synthesis of Piperidine Derivatives by Rhodium-Catalyzed (2+2+2) Cycloaddition of Alkenyl Isocyanates with Alkynes[59]

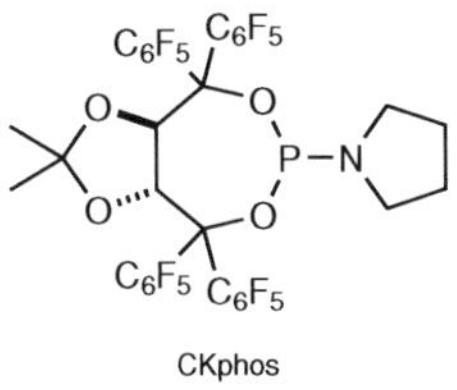

CKphos

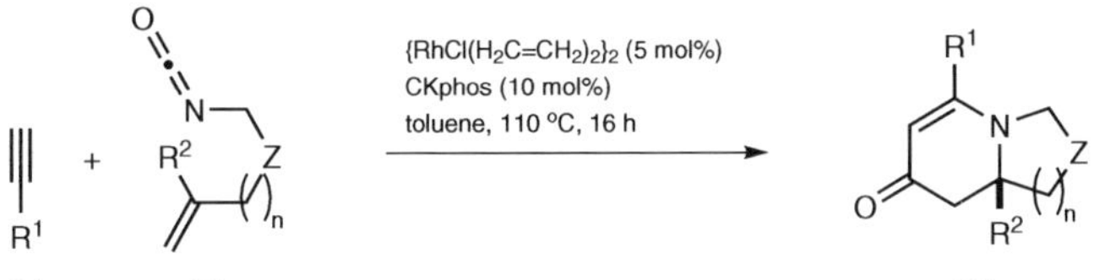

R^1	R^2	Z	n	ee (%)	Yield (%)	Ref
4-MeOC$_6$H$_4$	H	O	1	94	77	[59]
Ph	H	O	1	97	74	[59]
4-BrC$_6$H$_4$	H	O	1	96	68	[59]
4-F$_3$CC$_6$H$_4$	H	O	1	96	46	[59]
3-FC$_6$H$_4$	H	O	1	96	64	[59]
2-ClC$_6$H$_4$	H	O	1	97	52	[59]
3-thienyl	H	O	1	97	50	[59]
cyclohex-1-enyl	H	O	1	97	68	[59]
(CH$_2$)$_4$Me	H	O	1	97	68	[59]
Ph	H	NCbz	1	96	80	[59]
(CH$_2$)$_4$Me	H	NBoc	1	94	66	[59]
Ph	Me	O	1	84	61	[59]
Ph	H	O	2	95	57	[59]

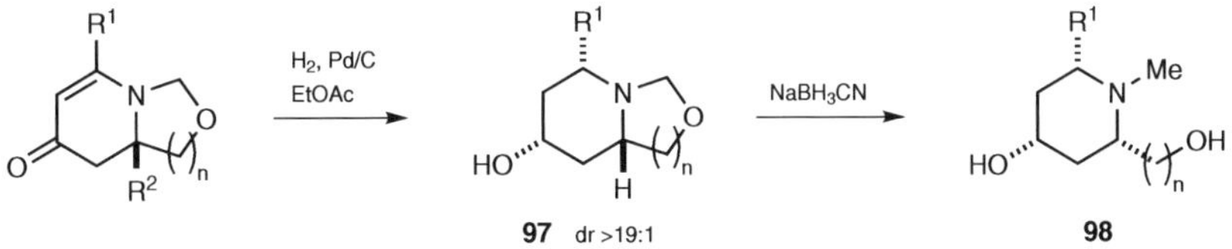

97 dr >19:1

R^1	n	Yield (%) of **97**	Yield (%) of **98**	Ref
4-MeOC$_6$H$_4$	1	70	80	[59]
Ph	1	71	80	[59]
4-F$_3$CC$_6$H$_4$	1	50	72	[59]
cyclohex-1-enyl	1	52	52	[59]
Ph	2	50	80	[59]

Pyridin-2-ones 74; General Procedure:[55]
A toluene soln of Ni(cod)$_2$ and SIPr was prepared and allowed to equilibrate for at least 6 h. In a glovebox, a soln of diyne **72** and isocyanate **73** in toluene was added to an oven-dried vial equipped with a stirrer bar. To this stirring soln, the soln of Ni(cod)$_2$ and SIPr was added and the reaction was stirred at rt for 30 min (or until complete consumption of starting material was observed as judged by GC). The mixture was then concentrated and purified by column chromatography (silica gel) to give **74**.

Dimethyl 3-Oxo-2-phenyl-2,3,5,7-tetrahydro-6H-cyclopenta[c]pyridine-6,6-dicarboxylate [77, X = CH$_2$; Z = C(CO$_2$Me)$_2$; R^1 = R^2 = H; R^3 = Ph]; Typical Procedure:[45]
To a soln of Ru(Cp*)Cl(cod) (5.7 mg, 0.015 mmol) and phenyl isocyanate (**76**, R^3 = Ph; 71 mg, 0.60 mmol) in dry degassed 1,2-dichloroethane (1 mL) was added dropwise a soln of diyne **75** (62 mg, 0.30 mmol) and isocyanate (**76**, R^3 = Ph; 71 mg, 0.60 mmol) in dry degassed 1,2-dichloroethane (4 mL) at 90 °C under an argon atmosphere. The soln was stirred for 1 h at this temperature. The solvent was removed and the residue was purified by flash column chromatography (silica gel, hexane/EtOAc 1:1) to give the pyridin-2-one **77**; yield: 87%.

Dimethyl 2-Benzyl-1,4-dimethyl-3-oxo-2,3,5,7-tetrahydro-6H-cyclopenta[c]pyridine-6,6-dicarboxylate [80, Z = C(CO$_2$Me)$_2$; R^1 = Me; R^2 = Bn]; Typical Procedure:[56]
Under an argon atmosphere, H$_8$-BINAP (15.8 mg, 0.0250 mmol) and [Rh(cod)$_2$]BF$_4$ (10.2 mg, 0.0250 mmol) were dissolved in CH$_2$Cl$_2$ (2.0 mL) and the mixture was stirred at rt for 5 min. H$_2$ was introduced to the resulting soln in a Schlenk tube. After stirring at rt for 0.5 h, the resulting mixture was concentrated to dryness. To a CH$_2$Cl$_2$ (3.5 mL) soln of the residue was added a CH$_2$Cl$_2$ (0.5 mL) soln of dimethyl 2,2-dibut-2-ynylmalonate (118.1 mg, 0.500 mmol) and benzyl isocyanate (73.2 mg, 0.550 mmol) at rt, and the remaining substrates were transferred by rinsing with CH$_2$Cl$_2$ (1.0 mL). The mixture was stirred at rt for 18 h. The resulting mixture was concentrated and purified by preparative TLC (hexane/EtOAc 1:1), which furnished the title compound as a colorless oil; yield: 183.6 mg (99%).

Dimethyl 1,4-Dimethyl-3-oxo-2-phenyl-2,3,5,7-tetrahydro-6H-cyclopenta[c]pyridine-6,6-dicarboxylate [83, X = CH$_2$; Z = C(CO$_2$Me)$_2$; R^1 = R^2 = Me; R^3 = Ph]; Typical Procedure:[57]
A flask was charged with {IrCl(cod)}$_2$ (14.0 mg, 0.02 mmol) and (R)-BINAP (24.9 mg, 0.04 mmol). The flask was evacuated and filled with argon. To the flask were added 1,2-dichloropropane (5 mL) and phenyl isocyanate (**82**, R^3 = Ph; 164 mg, 1.4 mmol). Diyne **81** (236 mg, 1.0 mmol) was added to the mixture. The mixture was stirred under reflux for 1 h. The progress of the reaction was monitored by GLC. After the reaction was complete, the solvent was removed under reduced pressure. Column chromatography of the residue (hexane/EtOAc 3:7) gave the title compound; yield: 344 mg (97%).

(+)-2-Benzyl-1-(2-chlorophenyl)-2,5,6,7-tetrahydro-3H-cyclopenta[c]pyridin-3-one (87, Z = CH$_2$; R^1 = Cl; R^2 = Bn); Typical Procedure:[56]
Under an argon atmosphere, (R)-DTBM-SEGPHOS (**84**; 29.5 mg, 0.025 mmol) and [Rh(cod)$_2$]BF$_4$ (10.2 mg, 0.025 mmol) were dissolved in CH$_2$Cl$_2$ (3.0 mL) and the mixture was stirred at rt for 5 min. H$_2$ was introduced to the resulting soln in a Schlenk tube. After stirring at rt for 0.5 h, the resulting mixture was concentrated to dryness. To a CH$_2$Cl$_2$ (3.5 mL) soln of the residue was added a CH$_2$Cl$_2$ (0.5 mL) soln of 1-chloro-2-(hepta-1,6-diynyl)benzene (**85**, Z = CH$_2$; R^1 = Cl; 101.3 mg, 0.500 mmol) and benzyl isocyanate (**86**, R^2 = Ph; 133.2 mg, 1.000 mmol) below −20 °C. The remaining substrates were washed into the vessel using CH$_2$Cl$_2$ (1.0 mL). The soln was kept at −20 °C for 12 h. The resulting soln was concentrated and purified by preparative TLC (hexane/EtOAc 2:1) to furnish the product; yield: 135.5 mg (81%); 87% ee.

Tetrahydroindolizinones 92 and 93; General Procedure:[58]

A flame-dried, round-bottomed flask was charged with {RhCl(H_2C=CH_2)$_2$}$_2$ (0.05 equiv) and the phosphoramidite ligand 88 or 89 (0.1 equiv), and was fitted with a flame-dried reflux condenser in an inert atmosphere (N_2) glovebox. Upon removal from the glovebox, toluene (1.0 mL) was added via syringe and the resulting yellow soln was stirred at ambient temperature under an argon flow for 15 min. To this soln was added a soln of alkyne 90 (2.0 equiv) and isocyanate 91 (0.270 mmol) in toluene (2 mL) via syringe or cannula. After adding more toluene (1 mL) to wash in the remaining residue, the resulting soln was heated to 110 °C in an oil bath, and maintained at reflux for ca. 16 h. The mixture was cooled to ambient temperature, concentrated under reduced pressure, and purified by flash column chromatography (gradient elution typically 100% EtOAc). Evaporation of solvent afforded the analytically pure products 92 and 93.

Dihydropyridin-4-ones 96; General Procedure:[59]

In a glovebox, a round-bottomed flask was charged with {RhCl(H_2C=CH_2)$_2$}$_2$ (0.005 mmol) and CKphos (0.01 mmol). The flask was equipped with a reflux condenser and septum. Outside the glovebox, toluene (1 mL) was added, and the mixture was stirred for 15 min after which time alkenyl isocyanate 95 (0.10 mmol) and alkyne 94 (0.16 mmol) in toluene (1 mL) were added dropwise. The mixture was heated to reflux and stirred for 16 h. Upon completion of the reaction, the flask was cooled to 23 °C, solvent was removed via rotary evaporation, and the crude material was subjected to column chromatography (EtOAc to EtOAc/MeOH 20:1) to give 96.

2.6.3.2 Carbodiimides and Carbon Dioxide

The (2+2+2) cycloadditions of diynes with carbodiimides can be catalyzed by cationic rhodium(I)/biaryl bisphosphine catalysts at room temperature (Scheme 32).[60] Both alkyl and aryl carbodiimides 100, and 1,6- and 1,7-diynes 99 can be employed for this reaction. When using unsymmetrical diynes, good to high regioselectivities are observed.

Scheme 32 Bicyclic Pyridin-2-imines by Rhodium-Catalyzed (2+2+2) Cycloaddition of Diynes with Carbodiimides[60]

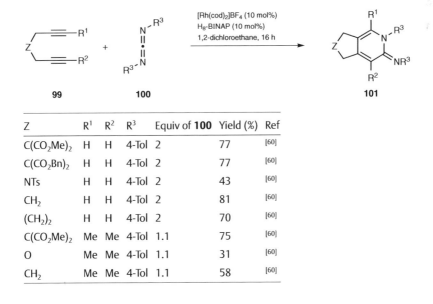

Z	R^1	R^2	R^3	Equiv of 100	Yield (%)	Ref
C(CO$_2$Me)$_2$	H	H	4-Tol	2	77	[60]
C(CO$_2$Bn)$_2$	H	H	4-Tol	2	77	[60]
NTs	H	H	4-Tol	2	43	[60]
CH$_2$	H	H	4-Tol	2	81	[60]
(CH$_2$)$_2$	H	H	4-Tol	2	70	[60]
C(CO$_2$Me)$_2$	Me	Me	4-Tol	1.1	75	[60]
O	Me	Me	4-Tol	1.1	31	[60]
CH$_2$	Me	Me	4-Tol	1.1	58	[60]

Z	R^1	R^2	R^3	Equiv of **100**	Yield (%)	Ref
C(CO$_2$Me)$_2$	Me	Me	Cy	1.1	80	[60]
C(CO$_2$Me)$_2$	Me	H	4-Tol	1.1	85^a	[60]
C(CO$_2$Me)$_2$	Ph	Me	4-Tol	1.1	80^b	[60]

^{a} Regioselectivity 91:9.
^{b} Regioselectivity 72:28.

The (2+2+2) cycloadditions of internal diynes **102** with carbon dioxide at atmospheric pressure can be catalyzed by cationic rhodium(I)/biaryl bisphosphine[60] and nickel(0)/1,3-bis(2,6-diisopropylphenyl)imidazol-2-ylidene (IPr) catalysts (Scheme 33).[61] The rhodium-catalyzed reactions proceed at room temperature, where slow addition of the diyne is necessary. Slow addition of the diyne is not necessary in the nickel-catalyzed reactions, but an elevated temperature (60 °C) is required. When using unsymmetrical diynes, good to high regioselectivities are observed.

Scheme 33 Fused Pyran-2-ones by Rhodium and Nickel-Catalyzed (2+2+2) Cycloaddition of Diynes with Carbon Dioxide[60,61]

Z	R^1	R^2	Conditions	Ratio (**103**/**104**)	Yield (%)	Ref
C(CO$_2$Me)$_2$	Me	Me	[Rh(cod)$_2$]BF$_4$/H$_8$-BINAP (5 mol%), 1,2-dichloroethane, rt, 1 h	–	90	[60]
C(CO$_2$Bn)$_2$	Me	Me	[Rh(cod)$_2$]BF$_4$/H$_8$-BINAP (5 mol%), 1,2-dichloroethane, rt, 1 h	–	85	[60]
CBz$_2$	Me	Me	[Rh(cod)$_2$]BF$_4$/H$_8$-BINAP (5 mol%), 1,2-dichloroethane, rt, 1 h	–	70	[60]
C(CH$_2$OMe)$_2$	Me	Me	[Rh(cod)$_2$]BF$_4$/H$_8$-BINAP (5 mol%), 1,2-dichloroethane, rt, 1 h	–	>99	[60]
O	Me	Me	[Rh(cod)$_2$]BF$_4$/H$_8$-BINAP (5 mol%), 1,2-dichloroethane, rt, 1 h	–	27	[60]
C(CO$_2$Me)$_2$	Et	Et	[Rh(cod)$_2$]BF$_4$/H$_8$-BINAP (5 mol%), 1,2-dichloroethane, rt, 1 h	–	78	[60]
C(CO$_2$Me)$_2$	iPr	Me	[Rh(cod)$_2$]BF$_4$/H$_8$-BINAP (5 mol%), 1,2-dichloroethane, rt, 1 h	96:4	81	[60]
C(CO$_2$Me)$_2$	iPr	iPr	Ni(cod)$_2$ (5 mol%), IPr (10 mol%), toluene, 60 °C, 2 h	–	86	[61]
C(CH$_2$OBn)$_2$	Me	Me	Ni(cod)$_2$ (5 mol%), IPr (10 mol%), toluene, 60 °C, 2 h	–	93	[61]
CHCO$_2$Me	Me	Me	Ni(cod)$_2$ (5 mol%), IPr (10 mol%), toluene, 60 °C, 2 h	–	82	[61]
(CH$_2$)$_2$	Me	Me	Ni(cod)$_2$ (5 mol%), IPr (10 mol%), toluene, 60 °C, 2 h	–	75	[61]
[C(CO$_2$Et)$_2$]$_2$	Me	Me	Ni(cod)$_2$ (5 mol%), IPr (10 mol%), toluene, 60 °C, 2 h	–	97	[61]
C(CO$_2$Me)$_2$	Me	TMS	Ni(cod)$_2$ (5 mol%), IPr (10 mol%), toluene, 60 °C, 2 h	100:0	83	[61]

Dimethyl 2-(4-Tolyl)-3-(4-tolylimino)-2,3,5,7-tetrahydro-6H-cyclopenta[c]pyridine-6,6-dicarboxylate [101, Z = C(CO$_2$Me)$_2$; R^1 = R^2 = H; R^3 = 4-Tol]; Typical Procedure:[60]

H$_8$-BINAP (6.3 mg, 0.010 mmol) and [Rh(cod)$_2$]BF$_4$ (4.1 mg, 0.010 mmol) were dissolved in CH$_2$Cl$_2$ (2.0 mL) and the mixture was stirred at rt for 10 min. H$_2$ was introduced to the resulting soln in a Schlenk tube. After stirring at rt for 1 h, the resulting soln was concentrated to dryness and a 1,2-dichloroethane (1.0 mL) soln of carbodiimide **100** (44.5 mg, 0.200 mmol) was added. To this soln was added dropwise over 5 min a 1,2-dichloroethane (1.0 mL) soln of diyne **99** (20.8 mg, 0.100 mmol) at rt. The mixture was stirred at rt for 16 h. The resulting mixture was concentrated and purified by preparative TLC (hexane/EtOAc/Et$_3$N 15:15:1), which furnished the title compound; yield: 33.1 mg (77%).

Dimethyl 1,4-Dimethyl-3-oxo-3,5-dihydro-7H-cyclopenta[c]pyran-6,6-dicarboxylate [103, Z = C(CO$_2$Me)$_2$; R^1 = R^2 = Me]; Typical Procedure Using a Rhodium Catalyst:[60]

H$_8$-BINAP (6.3 mg, 0.010 mmol) and [Rh(cod)$_2$]BF$_4$ (4.1 mg, 0.010 mmol) were dissolved in CH$_2$Cl$_2$ (2.0 mL) and the mixture was stirred at rt for 10 min. H$_2$ was introduced to the resulting soln in a Schlenk tube. After stirring at rt for 1 h, the resulting soln was concentrated to dryness. The residue was dissolved in 1,2-dichloroethane (2.5 mL) and transferred to a two-necked flask equipped with a balloon. The flask was filled with atmospheric pressure CO$_2$ and the soln was stirred at rt for 5 min. To this soln was added dropwise over 120 min a 1,2-dichloroethane (1.0 mL) soln of diyne **102** (47.3 mg, 0.200 mmol) at rt. The mixture was stirred at rt for 1 h. The resulting mixture was concentrated and purified by preparative TLC (hexane/EtOAc 2:1), which furnished the title compound; yield: 50.3 mg (90%).

6,7-Dihydrocyclopenta[c]pyran-3(5H)-ones 103; General Procedure Using a Nickel Catalyst:[61]

An oven-dried, two-necked, round-bottomed flask equipped with a magnetic stirrer bar, septum, gas adapter, and balloon was evacuated and filled with CO$_2$. A soln of diyne **102** was added and the flask was submerged into a 60 °C oil bath. To the stirring soln, a soln of Ni(cod)$_2$ and IPr was added. The dark greenish-black mixture was then heated for 2 h (or until complete consumption of starting material was observed as judged by GC), cooled to rt, concentrated, and purified by column chromatography (silica gel) to give **103**.

2.6.4 (2+2+2) Cycloadditions Involving C(sp^2) Multiple Bonds

Transition-metal-catalyzed (2+2+2)-cycloaddition reactions involving C(sp^2) multiple bonds enable the facile synthesis of non-aromatic six-membered compounds and have been actively investigated. Importantly, asymmetric variants of these reactions are capable of constructing one or more central chiralities.

2.6.4.1 Alkenes

A cationic rhodium(I)/Xyl-BINAP complex catalyzes the enantioselective (2+2+2) cycloadditions of 1,6-diynes **105** with *exo*-methylene lactones and cyclic ketones, and α,β-unsaturated esters **106** to give spirocyclic derivatives **107** with high yields and enantiomeric excesses (Scheme 34).[62]

Scheme 34 Rhodium-Catalyzed Enantioselective (2+2+2) Cycloaddition of 1,6-Diynes with Alkenes[62]

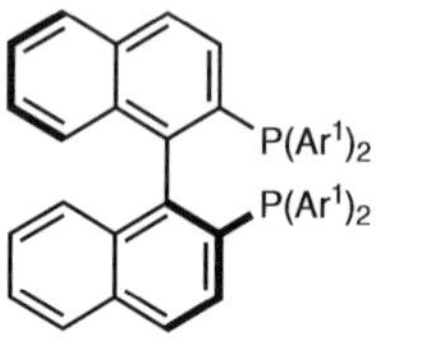

(S)-Xyl-BINAP (Ar^1 = 3,5-Me$_2$C$_6$H$_3$)

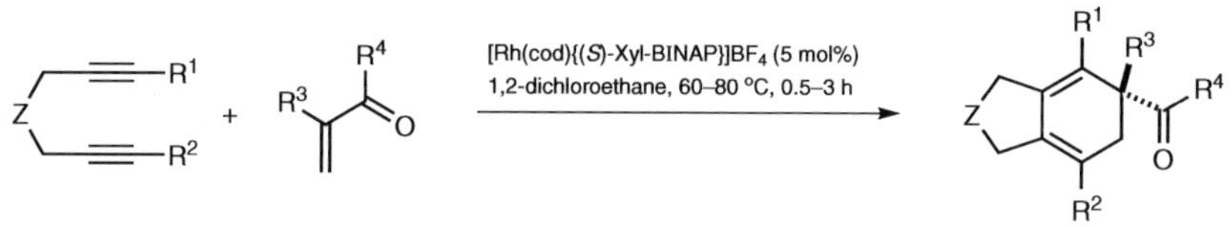

Z	R^1	R^2	R^3	R^4	Equiv of 106	Temp (°C)	Time (h)	ee (%)	Yield (%)	Ref
C(CO$_2$Bn)$_2$	Me	Me		(CH$_2$)$_2$O	3	80	0.5	99	94	[62]
C(CO$_2$Bn)$_2$	Me	Me		(CH$_2$)$_3$O	3	80	0.5	98	93	[62]
C(CO$_2$Bn)$_2$	Me	Me		(CH$_2$)$_4$O	3	80	0.5	97	88	[62]
C(CO$_2$Bn)$_2$	Me	Me		(CH$_2$)$_3$	3	60	3	96	85	[62]
C(CO$_2$Bn)$_2$	Me	Me	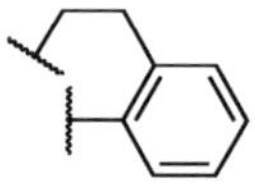	benzo-fused	3	60	3	96	85	[62]
C(CO$_2$Bn)$_2$	Me	Me	Me	OMe	3	60	0.5	>99	92	[62]
C(CO$_2$Bn)$_2$	Me	Me	Ph	OMe	10	80	0.5	93	54	[62]
C(CO$_2$Bn)$_2$	Me	Me	H	OMe	3	60	0.5	91	87	[62]
C(CO$_2$Bn)$_2$	H	H		(CH$_2$)$_2$O	3	40	0.5	95	81	[62]
C(CO$_2$Bn)$_2$	Me	Ph		(CH$_2$)$_2$O	10	80	0.5	>99	80^a	[62]
NTs	Me	Me		(CH$_2$)$_2$O	10	80	0.5	97	92	[62]
O	Et	Et		(CH$_2$)$_2$O	20	60	0.5	92	50	[62]

^{a} Regioselectivity >20:1.

Cationic rhodium(I)/axially chiral biaryl bisphosphine complexes catalyze the enantioselective (2+2+2) cycloadditions of 1,6-enynes **108** with alkynes **109** to give bicyclic cyclohexadiene derivatives **110** and **111** with high yields and enantiomeric excesses (Scheme 35).[63,64]

2.6.4 (2+2+2) Cycloadditions Involving C(sp²) Multiple Bonds

Scheme 35 Bicyclic Cyclohexadienes by Rhodium-Catalyzed Enantioselective (2+2+2) Cycloaddition of 1,6-Enynes with Alkynes[63,64]

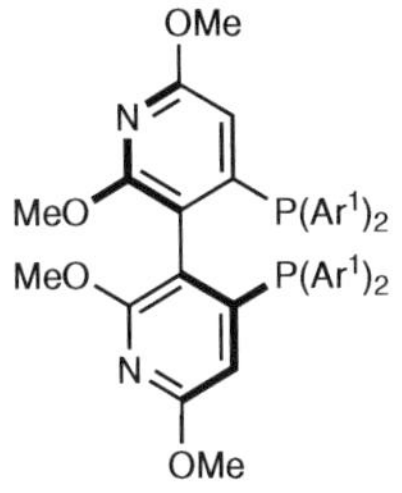

(S)-Xyl-P-PHOS (Ar¹ = 3,5-Me₂C₆H₃)

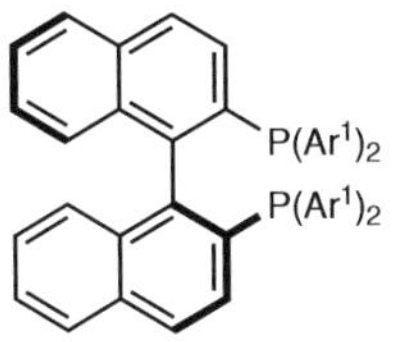

(S)-Tol-BINAP (Ar¹ = 4-Tol)

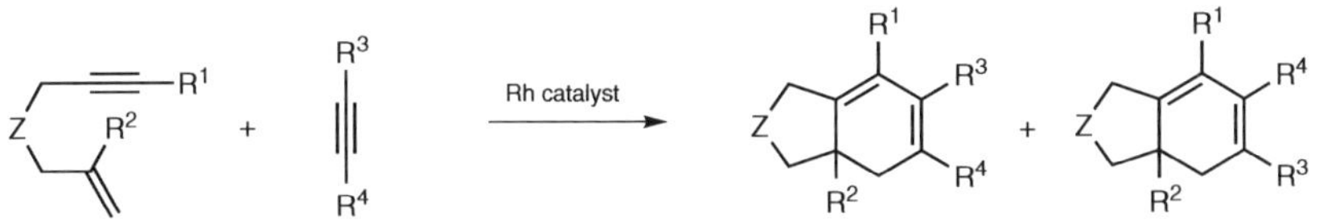

Z	R¹	R²	R³	R⁴	Equiv of 109	Conditions	Ratio (110/111)	ee (%)	Yield (%)	Ref
NTs	H	H	CO₂Me	Ph	3	[RhCl(cod)]₂ (5 mol%), (S)-Xyl-P-PHOS (12 mmol), AgBF₄ (20 mol%), THF, 60 °C, 2.5 h	10:1	97	98	[63]
NTs	H	H	CO₂Me	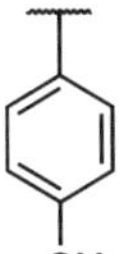 4-MeOC₆H₄	3	[RhCl(cod)]₂ (5 mol%), (S)-Xyl-P-PHOS (12 mmol), AgBF₄ (20 mol%), THF, 60 °C, 2.5 h	14:1	97	84	[63]
NTs	H	H	CO₂Me	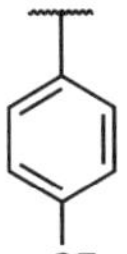 4-CF₃C₆H₄	3	[RhCl(cod)]₂ (5 mol%), (S)-Xyl-P-PHOS (12 mmol), AgBF₄ (20 mol%), THF, 60 °C, 2.5 h	10:1	98	86	[63]
NTs	H	H	Ac	Ph	3	[RhCl(cod)]₂ (5 mol%), (S)-Xyl-P-PHOS (12 mmol), AgBF₄ (20 mol%), THF, 60 °C, 2.5 h	>19:1	95	86	[63]
NTs	H	Me	Ac	Ph	3	[RhCl(cod)]₂ (5 mol%), (S)-Xyl-P-PHOS (12 mmol), AgBF₄ (20 mol%), THF, 60 °C, 2.5 h	10:1	>99	84	[63]
C(CO₂Me)₂	H	H	CO₂Me	Ph	3	[RhCl(cod)]₂ (5 mol%), (S)-Xyl-P-PHOS (12 mmol), AgBF₄ (20 mol%), THF, 60 °C, 2.5 h	9:1	>99	88	[63]
O	H	H	CO₂Me	Ph	3	[RhCl(cod)]₂ (5 mol%), (S)-Xyl-P-PHOS (12 mmol), AgBF₄ (20 mol%), THF, 60 °C, 2.5 h	>19:1	>99	86	[63]
NTs		Me	Me	CH₂OMe	CH₂OMe	2	[Rh(cod)₂]BF₄/(S)-Tol-BINAP (10 mol%), 1,2-dichloroethane, 60 °C, 12 h	–	97	81	[64]

Z	R¹	R²	R³	R⁴	Equiv of 109	Conditions	Ratio (110/111)	ee (%)	Yield (%)	Ref
NTs	Ph	Me	CH$_2$OMe	CH$_2$OMe	3	[Rh(cod)$_2$]BF$_4$/(S)-Tol-BINAP (10 mol%), 1,2-dichloroethane, reflux, 4 h	–	88	96	[64]
NTs	Me	Ph	CH$_2$OMe	CH$_2$OMe	3	[Rh(cod)$_2$]BF$_4$/(S)-Tol-BINAP (10 mol%), 1,2-dichloroethane, reflux, 24 h	–	89	61	[64]
NTs	H	Me	CH$_2$OMe	CH$_2$OMe	10	[Rh(cod)$_2$]BF$_4$/(S)-Tol-BINAP (10 mol%), 1,2-dichloroethane, 40 °C, 15 h	–	98	72	[64]
C(CO$_2$Me)$_2$	H	Me	CH$_2$OMe	CH$_2$OMe	10	[Rh(cod)$_2$]BF$_4$/(S)-Tol-BINAP (10 mol%), 1,2-dichloroethane, 40 °C, 12 h	–	92	60	[64]
O	H	Me	CH$_2$OMe	CH$_2$OMe	10	[Rh(cod)$_2$]BF$_4$/(S)-Tol-BINAP (10 mol%), 1,2-dichloroethane, 80 °C, 1 h	–	97	65	[64]
NTs	Me	Me	CH$_2$OH	CH$_2$OH	10	[Rh(cod)$_2$]BF$_4$/(S)-Tol-BINAP (10 mol%), 1,2-dichloroethane, 60 °C, 3 h	–	98	50	[64]
NTs	Me	Me	CH$_2$OH	H	10	[Rh(cod)$_2$]BF$_4$/(S)-Tol-BINAP (10 mol%), 1,2-dichloroethane, 60 °C	7:1	98ᵃ	63	[64]

ᵃ ee of **110**; ee of **111** was 97%.

Although the intermolecular (2+2+2) cycloaddition of two equivalents of an alkyne with one alkene has been disclosed, that of two different alkynes with one alkene has rarely been reported. The cationic rhodium(I)/(R)-Tol-BINAP complex is capable of catalyzing the (2+2+2) cycloaddition of silylacetylenes **112**, acetylenecarboxylates **113**, and acrylamides **114** with excellent chemo-, regio-, and enantioselectivities (Scheme 36).[65]

Scheme 36 Rhodium-Catalyzed Chemo-, Regio-, and Enantioselective (2+2+2) Cycloaddition of Silylacetylenes, Acetylenecarboxylates, and Acrylamides[65]

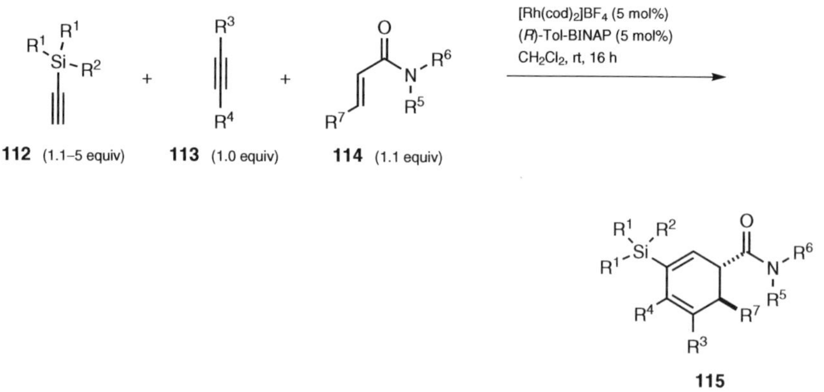

2.6.4 (2+2+2) Cycloadditions Involving C(sp²) Multiple Bonds

R^1	R^2	R^3	R^4	R^5	R^6	R^7	Equiv of 112	mol% of Catalyst	ee (%)	Yield (%)	Ref
Me	Me	CO₂t-Bu	CO₂t-Bu	Ph	Me	H	1.1	5	99	92	[65]
Me	Me	CO₂t-Bu	CO₂t-Bu	Me	Me	H	1.1	5	99	92	[65]
Me	Me	CO₂t-Bu	CO₂t-Bu	OMe	Me	H	1.1	5	98	90	[65]
Me	Me	CH₂OMe	CO₂Me	Ph	Me	H	5	10	>99	53[a]	[65]
Me	Me	Me	CO₂Et	Ph	Me	H	5	20	>99	44	[65]
Me	Me	CF₃	CO₂Et	Ph	Me	H	1.1	20	>99	32[a,b]	[65]
Me	Bn	CO₂t-Bu	CO₂t-Bu	Ph	Me	H	1.1	5	99	95	[65]
Me	Me	CO₂t-Bu	CO₂t-Bu	Me	Me	Me	1.1	5	>99	76	[65]
Me	Me	CO₂t-Bu	CO₂t-Bu	(CH₂)₄		Me	5	1.1	67	>98	[63]
Me	Me	CO₂t-Bu	CO₂t-Bu	Me	Me	Pr	1.1	5	>99	78	[65]
Me	Me	CO₂t-Bu	CO₂t-Bu	Me	Me	iPr	1.1	5	95	75	[65]
Me	Me	CO₂t-Bu	CO₂t-Bu	Me	Me	Ph	1.1	20	99	21[a]	[65]
Me	Me	Me	CO₂Et	Me	Me	Me	5	10	>99	63[a]	[65]
Et	Et	CO₂t-Bu	CO₂t-Bu	Me	Me	Me	1.1	5	>99	55	[65]

[a] 72 h.
[b] Ratio (112/113/114) 1.1:1.5:1.0.

The enantioselective (2+2+2) cycloaddition of 1,6-enynes 116 with alkenes leading to bicyclic cyclohexenes 118 is achieved by using acrylamides 117 as alkenes and a cationic rhodium(I)/(R)-H₈-BINAP complex as catalyst (Scheme 37).[66]

Scheme 37 Bicyclic Cyclohexenes by Rhodium-Catalyzed Enantioselective (2+2+2) Cycloaddition of 1,6-Enynes with Acrylamides[66]

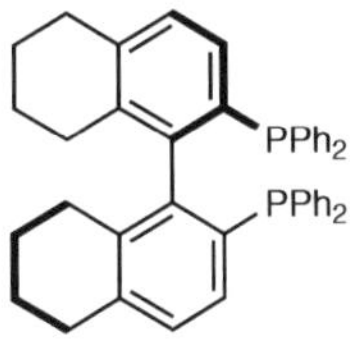

(R)-H₈-BINAP

2.6 (2+2+2) Cycloadditions

Z	R¹	R²	R³	R⁴	Equiv of 117	mol% of Catalyst	Temp	ee (%)	Yield (%)	Ref
NTs	Me	Me	Me	Me	1.1	3	rt	>99	85	[66]
NTs	Me	Me	(CH$_2$)$_4$		1.1	5	rt	>99	>99	[66]
NTs	Me	Me	Ph	Ph	1.1	3	rt	>99	97	[66]
NTs	Me	Me	OMe	Me	1.1	10	rt	>99	92	[66]
C(CO$_2$Me)$_2$	Me	Me	Ph	Me	1.1	10	rt	>99	72	[66]
O	(CH$_2$)$_4$Me	Me	Ph	Me	1.1	10	40 °C	93	59	[66]
NTs	Ph	Me	Ph	Ph	1.1	10	40 °C	95	90	[66]
NTs	H	Me	(CH$_2$)$_4$		3	5	rt	79	>99	[66]
NTs	Me	H	Ph	Ph	1.1	5	rt	91	62	[66]
NTs	Ph	H	Ph	Ph	1.1	5	rt	87	88	[66]
NTs	H	H	Ph	Ph	3	5	rt	96	57	[66]

The novel enantioselective intramolecular (2+2+2) cycloaddition of dienynes **119** or **121** to produce strained polycyclic compounds **120** bearing two quaternary centers or bicyclic compounds **122** is achieved using a cationic rhodium(I)/(S)-Tol-BINAP complex as catalyst (Scheme 38).[67]

Scheme 38 Rhodium-Catalyzed Enantioselective (2+2+2) Cycloaddition of Dienynes[67]

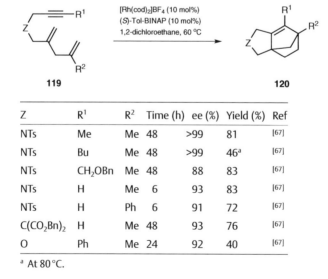

Z	R¹	R²	Time (h)	ee (%)	Yield (%)	Ref
NTs	Me	Me	48	>99	81	[67]
NTs	Bu	Me	48	>99	46ᵃ	[67]
NTs	CH$_2$OBn	Me	48	88	83	[67]
NTs	H	Me	6	93	83	[67]
NTs	H	Ph	6	91	72	[67]
C(CO$_2$Bn)$_2$	H	Me	48	93	76	[67]
O	Ph	Me	24	92	40	[67]

ᵃ At 80 °C.

2.6.4 (2+2+2) Cycloadditions Involving C(sp²) Multiple Bonds

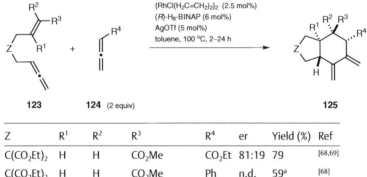

Z	R¹	Time (h)	ee (%)	Yield (%)	Ref
NTs	Me	12	99	91	[67]
NTs	H	12	>99	79a	[67]
C(CO₂Bn)₂	H	6	90	80	[67]
O	Ph	48	92	55	[67]
O	(CH₂)₂Ph	6	94	64	[67]

a At 40 °C.

Functionalized cyclohexanes **125** bearing four stereogenic centers can be synthesized via the novel enantioselective (2+2+2) cycloaddition of ene–allenes **123** with allenoates **124** using a cationic rhodium(I)/(R)-H$_8$-BINAP catalyst generated in situ (Scheme 39).[68,69]

Scheme 39 Functionalized Cyclohexanes by Rhodium-Catalyzed Enantioselective (2+2+2) Cycloaddition of Ene–Allenes with Allenoates[68,69]

Z	R¹	R²	R³	R⁴	er	Yield (%)	Ref
C(CO₂Et)₂	H	H	CO₂Me	CO₂Et	81:19	79	[68,69]
C(CO₂Et)₂	H	H	CO₂Me	Ph	n.d.	59a	[68]
C(CO₂Et)₂	H	H	CO₂t-Bu	CO₂Et	83:17	80	[69]
C(CO₂Et)₂	H	H	Bz	CO₂Et	n.d.	61	[68]
C(CO₂Et)₂	H	H	4-MeO₂CC₆H₄	CO₂Et	84:16	63	[68,69]
C(CO₂Et)₂	H	H	4-F₃CC₆H₄	CO₂Et	n.d.	58	[68]
C(CO₂Et)₂	H	CO₂Me	H	CO₂Et	n.d.	76b	[68]
C(CO₂Et)₂	H	Me	CO₂Et	CO₂Et	78:22	72	[68,69]
C(CO₂Et)₂	CO₂Et	H	H	CO₂Et	86:14	62	[68,69]
C(CO₂Et)₂	Me	H	H	CO₂Et	n.d.	27	[68]
NTs	H	H	CO₂Me	CO₂Et	81:19	46	[68,69]

a AgBF₄ was used instead of AgOTf.
b dr 1.2:1.

Dibenzyl 4′,7′-Dimethyl-2-oxo-1′,4,5,6′-tetrahydro-2H-spiro[furan-3,5′-indene]-2′,2′(3′H)-dicarboxylate [107, Z = C(CO$_2$Bn)$_2$; R^1 = R^2 = Me; R^3,R^4 = (CH$_2$)$_2$O];
Typical Procedure:[62]
[Rh(cod){(S)-Xyl-BINAP}]BF$_4$ (5.4 mg, 0.005 mmol) and 3-methylenedihydrofuran-2(3H)-one [**106**, R^3,R^4 = (CH$_2$)$_2$O; 29.8 mg, 0.30 mmol] were stirred in 1,2-dichloroethane (0.3 mL) at 80 °C. To the yellow-orange soln, dibenzyl 2,2-dibut-2-ynylmalonate [**105**, Z = C(CO$_2$Bn)$_2$; R^1 = R^2 = Me; 38.4 mg, 0.10 mmol] in 1,2-dichloroethane (0.7 mL) was added dropwise over 30 min at 80 °C and the mixture was stirred. After completion of the reaction, the solvent was removed under reduced pressure, and the crude products were purified by TLC [benzene (**CAUTION:** *carcinogen*)/EtOAc 10:1] to give the analytically pure title compound; yield: 45.1 mg (94%); 99% ee.

Methyl (7aS)-6-Phenyl-2-tosyl-2,3,7,7a-tetrahydro-1H-isoindole-5-carboxylate (110, Z = NTs; R^1 = R^2 = H; R^3 = CO$_2$Me; R^4 = Ph); Typical Procedure:[63]
{RhCl(cod)}$_2$ (6.2 mg) and AgBF$_4$ (9.7 mg) were suspended in anhyd THF (1.0 mL) and stirred at rt under an atmosphere of argon for ca. 10 min. (S)-Xyl-P-PHOS (22.7 mg) in anhyd THF (3.0 mL) was then added to the yellow suspension, and the mixture was stirred at rt for an additional ca. 30 min. Methyl 3-phenylpropynoate (**109**, R^3 = CO$_2$Me; R^4 = Ph; 120.1 mg, 0.75 mmol) was added in one portion, followed by the addition of 1,6-enyne **108** (62.3 mg, 0.25 mmol) in anhyd THF (2.0 mL) using a syringe pump over ca. 2 h at 60 °C, and the mixture was then stirred for an additional ca. 30 min (TLC control). The mixture was allowed to cool to rt, and the resultant mixture was filtered through a short pad of silica gel (EtOAc/hexanes 1:1) and concentrated under reduced pressure to afford the crude product. Purification by flash chromatography (silica gel, EtOAc/hexanes 1:9 to 3:7 gradient) afforded the bicyclohexadienes **110/111**; yield: 94.1 mg (98%); ratio (**110/111**) 10:1; 97% ee.

Fused Cyclohexadienes 110 (R^3 = R^4 = CH$_2$OMe); General Procedure:[64]
Under an atmosphere of argon, (S)-Tol-BINAP (6.8 mg, 0.010 mmol) and [Rh(cod)$_2$]BF$_4$ (4.1 mg, 0.010 mmol) were stirred in 1,2-dichloroethane (0.25 mL) at rt to give a yellow soln. Then, 1,4-dimethoxybut-2-yne (**109**, R^3 = R^4 = CH$_2$OMe; 22.8 mg, 0.20 mmol or 114.1 mg, 1.00 mmol) and enyne **108** (0.10 mmol) in 1,2-dichloroethane (0.75 mL) were added to the soln and the mixture was stirred at the appropriate temperature. After completion of the reaction, the solvent was removed under reduced pressure and the crude products were purified by TLC to give chiral cycloadduct **110**.

(+)-Di-*tert*-butyl 5-[Methyl(phenyl)carbamoyl]-3-(trimethylsilyl)cyclohexa-1,3-diene-1,2-dicarboxylate (115, R^1 = R^2 = R^6 = Me; R^3 = R^4 = CO$_2$t-Bu; R^5 = Ph; R^7 = H); Typical Procedure:[65]
A CH$_2$Cl$_2$ (0.5 mL) soln of (R)-Tol-BINAP (6.8 mg, 0.010 mmol) was added to a CH$_2$Cl$_2$ (0.5 mL) soln of [Rh(cod)$_2$]BF$_4$ (4.1 mg, 0.010 mmol) at rt, and the mixture was stirred at rt for 10 min. The resulting soln was stirred under H$_2$ (1 atm) at rt for 1 h, concentrated to dryness, and dissolved in CH$_2$Cl$_2$ (0.5 mL). To this soln was added a CH$_2$Cl$_2$ (1.5 mL) soln of silylacetylene **112** (21.6 mg, 0.220 mmol), acetylenecarboxylate **113** (45.3 mg, 0.200 mmol), and acrylamide **114** (35.5 mg, 0.220 mmol) at rt. After stirring at rt for 16 h, the resulting soln was concentrated and purified by preparative TLC (hexane/EtOAc 2:1), which furnished the title compound; yield: 88.9 mg (92%); 99% ee.

(5S,7aR)-(−)-N,N,4,7a-Tetramethyl-2-tosyl-2,3,5,6,7,7a-hexahydro-1H-isoindole-5-carboxamide (118, Z = NTs; R^1 = R^2 = R^3 = R^4 = Me); Typical Procedure:[66]
(R)-H$_8$-BINAP (3.8 mg, 0.0060 mmol) and [Rh(cod)$_2$]BF$_4$ (2.4 mg, 0.0060 mmol) were dissolved in CH$_2$Cl$_2$ (1.5 mL) and the mixture was stirred at rt for 30 min. H$_2$ was introduced to the resulting soln in a Schlenk tube. After stirring at rt for 1 h, the resulting mixture was concentrated to dryness. To the residue was added a soln of 1,6-enyne **116** (55.5 mg,

2.6.4 (2+2+2) Cycloadditions Involving C(sp²) Multiple Bonds

0.200 mmol) and acrylamide **117** (21.8 mg, 0.220 mmol) in CH_2Cl_2 (1.5 mL) at rt. The mixture was stirred at rt for 16 h. The resulting soln was concentrated and purified by preparative TLC (hexane/EtOAc 1:1), which furnished the title compound; yield: 64.2 mg (85%); >99% ee.

Cycloadducts 120 or 122; General Procedure:[67]
Under an atmosphere of argon, Tol-BINAP (6.8 mg, 0.010 mmol) and [Rh(cod)$_2$]BF$_4$ (4.1 mg, 0.010 mmol) were stirred in 1,2-dichloroethane (0.25 mL) at rt to give a yellow soln. Then, dienyne **119** or **121** (0.10 mmol) in 1,2-dichloroethane (0.75 mL) was added to the soln and the mixture was stirred at the appropriate temperature. After completion of the reaction, the solvent was removed under reduced pressure and the crude products were purified by TLC.

Fused Dimethylenecyclohexanes 125; General Procedure:[69]
To a vial in a glovebox was added {RhCl(H$_2$C=CH$_2$)$_2$}$_2$ (0.9 mg, 0.002 mmol), (R)-H$_8$-BINAP (3.3 mg, 0.0053 mmol), and toluene (0.7 mL). This soln was allowed to stir at rt for 30 min, by which time a clear red soln was obtained. To this, a soln of AgOTf (1.1 mg, 0.0044 mmol) in toluene (0.7 mL) was added. This caused the soln to immediately turn orange and gradually become cloudy. Finally, a soln of the substrate **123** (0.088 mmol) and allene **124** (0.18 mmol) in toluene (0.7 mL) was added. The soln gradually turned clear and yellow, and within 10 min, the soln faded to pale yellow. At this point the reactions were heated in an aluminum reaction block preheated to 100 °C and were monitored by TLC. Upon completion of the reaction, the soln was cooled to rt, diluted with EtOAc, and filtered through Celite. The crude reaction product was observed by NMR spectroscopy, and the mixture was purified via flash chromatography (EtOAc/hexanes 1:9 to 3:7) to give the title compound.

2.6.4.2 Carbonyl Compounds

Although a number of transition-metal complexes can catalyze the (2+2+2) cycloadditions of diynes with compounds containing C(sp)–heteroatom multiple bonds, such as nitriles and heterocumulenes, analogous (2+2+2) cycloadditions involving C(sp²)–heteroatom multiple bonds such as aldehydes and ketones have been reported in a limited number of examples. Among the transition-metal catalysts developed to date, a cationic rhodium(I)/H$_8$-BINAP complex shows the highest activity, selectivity, and generality. For example, this complex catalyzes the (2+2+2) cycloadditions of a variety of 1,6- and 1,7-diynes, e.g. **126**, with both electron-deficient and electron-rich carbonyl compounds, e.g. **127**, leading to dienones **128** (after electrocyclic ring opening) or oxacyclohexadienes, in high yields under mild reaction conditions (Scheme 40).[70]

Scheme 40 Rhodium-Catalyzed (2+2+2) Cycloadditions of Diynes with Carbonyl Compounds [70]

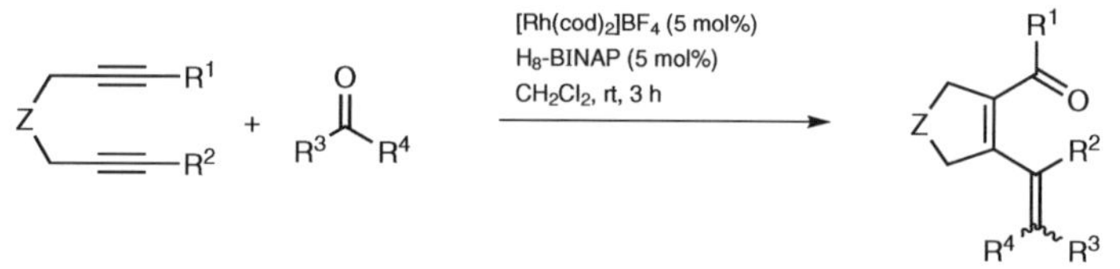

Z	R^1	R^2	R^3	R^4	Ratio (E/Z)	Yield (%)	Ref
$C(CO_2Me)_2$	Me	Me	Me	CO_2Et	14:86	99	[70]
$C(CO_2Me)_2$	Me	Me	Ph	CO_2Et	29:71	98	[70]
$C(CO_2Me)_2$	Me	Me	Me	Ac	<5:>95	90[a]	[70]
$C(CO_2Me)_2$	Me	Me	H	Ph	90:10	97	[70]
$C(CO_2Me)_2$	Me	Me	H	t-Bu	60:40	82	[70]
NTs	Me	Me	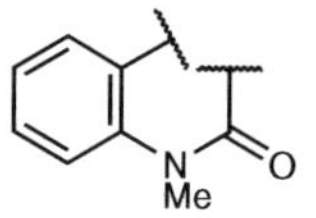		16:84	>99	[70]
$C(CO_2Me)_2$	Me	Me	CH_2CO_2Me	Me	27:73	67	[70]
$C(CO_2Me)_2$	Me	Me	C≡CPh	CO_2Et	<5:>95	96	[70]
NTs	Me	Me	Ph	CO_2Et	23:77	98	[70]
O	Et	Et	CO_2Et	CO_2Et	–	84	[70]
$(CH_2)_2$	Me	Me	Ph	CO_2Et	13:87	84	[70]
$(CH_2)_2$	H	H	Ph	CO_2Et	36:64	55	[70]
$C(CO_2Me)_2$	CO_2Et	CO_2Et	Ph	CO_2Et	29:71	98	[70]
NTs	Me	Me	Me	$P(O)(OEt)_2$	12:88	84	[70]
NTs	Me	H	Me	$P(O)(OEt)_2$	40:60	78[b]	[70]
$C(CO_2Me)_2$	Me	Me	Ph	$P(O)(OEt)_2$	51:49	96[c]	[70]

[a] 2 equiv of ketone was used.
[b] Regioselectivity >99:1.
[c] Additive: $(CO_2Et)_2$ (0.2 equiv).

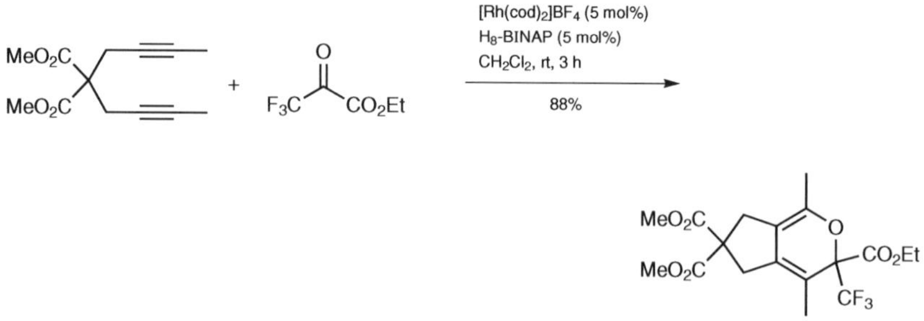

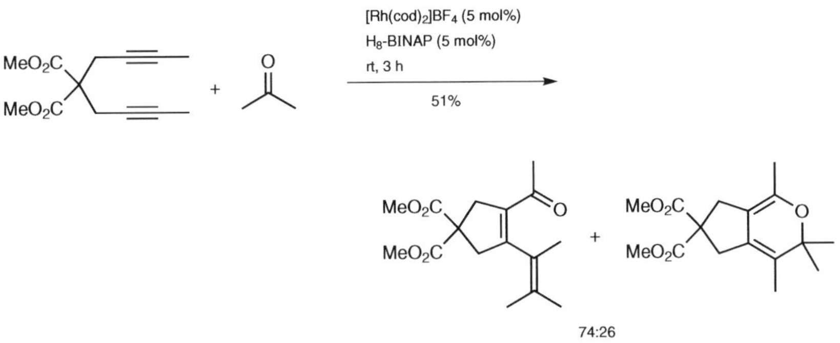

2.6.4 (2+2+2) Cycloadditions Involving C(sp²) Multiple Bonds

The cationic rhodium(I)/H₈-BINAP complex also catalyzes the enantioselective (2+2+2) cycloadditions of 1,6-enynes **129** with dicarbonyl compounds **130**, leading to fused dihydropyrans **131**, with excellent regio-, diastereo-, and enantioselectivity (Scheme 41).[71]

Scheme 41 Fused Dihydropyrans by Rhodium-Catalyzed Enantioselective (2+2+2) Cycloadditions of 1,6-Enynes with Dicarbonyl Compounds[71]

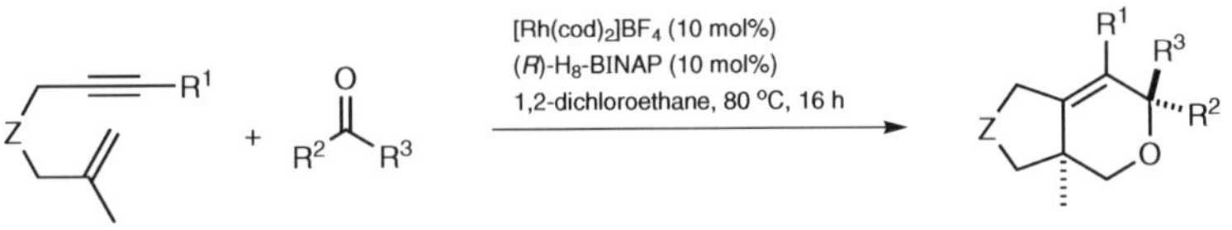

Z	R^1	R^2	R^3	ee (%)	Yield (%)	Ref
NTs	Me	Me	CO_2Et	95	>99	[71]
NTs	Me	Me	Ac	98	73[a]	[71]
NTs	Me	Ph	CO_2Et	97	24	[71]
NTs	Me	CO_2Et	CO_2Et	97	89	[71]
$C(CO_2Me)_2$	4-BrC₆H₄	Me	CO_2Et	>99	82[b]	[71]
O	Ph	CO_2Et	CO_2Et	96	64	[71]
O	Ph	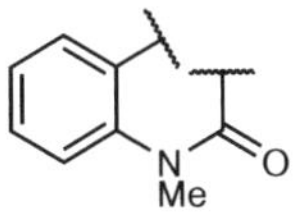 *(N-Me benzamide)*		92	85[a]	[71]
O	CO_2Me	CO_2Et	CO_2Et	93	61	[71]

[a] 2 equiv of ketone was used.
[b] 20 mol% catalyst was used at rt in CH_2Cl_2.

Dimethyl 3-Acetyl-4-(4-ethoxy-4-oxo-3-phenylbut-2-en-2-yl)cyclopent-3-ene-1,1-dicarboxylate [128, Z = C(CO₂Me)₂; R¹ = R² = Me; R³ = Ph; R⁴ = CO₂Et]; Typical Procedure:[70]

H₈-BINAP (9.5 mg, 0.015 mmol) and [Rh(cod)₂]BF₄ (6.1 mg, 0.015 mmol) were dissolved in CH₂Cl₂ (2.0 mL), and the mixture was stirred at rt for 5 min. H₂ was introduced into the resulting soln in a Schlenk tube. After stirring at rt for 1 h, the soln was concentrated to dryness and the residue was dissolved in CH₂Cl₂ (0.5 mL). To this soln was added dropwise a soln of diyne **126** (70.9 mg, 0.300 mmol) and ketone **127** (58.8 mg, 0.330 mmol) in CH₂Cl₂

(1.5 mL) at rt. The mixture was stirred at rt for 3 h. The resulting soln was concentrated and purified by preparative TLC (hexane/EtOAc 3:1), which furnished the title compound; yield: 121.4 mg (98%); ratio (E/Z) 29:71.

(3aR,6R)-(+)-1′,3a-Dimethyl-7-phenyl-3a,4-dihydro-1H,3H-spiro[furo[3,4-c]pyran-6,3′-indolin]-2′-one (131, Z = O; R¹ = Ph; R²,R³ = 2-C$_6$H$_4$NMeCO); Typical Procedure:[71]
(R)-H$_8$-BINAP (18.9 mg, 0.030 mmol) and [Rh(cod)$_2$]BF$_4$ (12.2 mg, 0.030 mmol) were dissolved in CH$_2$Cl$_2$ (2.0 mL) in a Schlenk tube under an argon atmosphere, and the mixture was stirred at rt for 5 min. H$_2$ was introduced into the resulting soln, which was then stirred at rt for 0.5 h, concentrated to dryness, and dissolved in 1,2-dichloroethane (0.5 mL). A soln of **129** (55.9 mg, 0.300 mmol) and **130** (96.7 mg, 0.600 mmol) in 1,2-dichloroethane (1.5 mL) was added at rt, and the resulting mixture was stirred at 80 °C for 16 h. The mixture was then concentrated, and the product was purified by preparative TLC (hexane/EtOAc 2:1) to furnish the title compound; yield: 88.8 mg (85%); 92% ee (single diastereomer).

2.6.5 Applications in the Syntheses of Natural Products and Drug Molecules

Transition-metal-catalyzed (2+2+2)-cycloaddition reactions have been applied in the synthesis of biologically active polycyclic molecules. For example, the rhodium-catalyzed (2+2+2) cycloaddition of a nitrogen-linked diyne **132** with monoyne **133** has been applied to the total synthesis of antiostatin A1 (Scheme 42).[72]

Scheme 42 Total Synthesis of Antiostatin A1 via Rhodium-Catalyzed (2+2+2) Cycloaddition of a Nitrogen-Linked Diyne with a Monoyne[72]

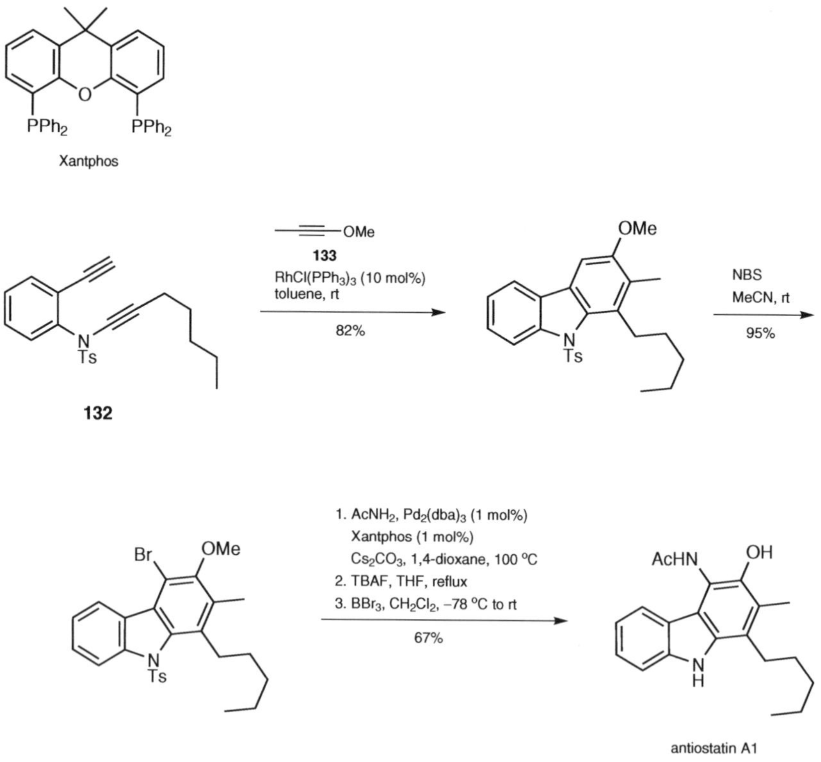

2.6.5 Applications in the Syntheses of Natural Products and Drug Molecules

The ruthenium-catalyzed (2+2+2) cycloaddition of functionalized chlorodiyne **134** with functionalized propargylic alcohol **135** was successfully applied to the total synthesis of sporolide B (Scheme 43).[73]

Scheme 43 Total Synthesis of Sporolide B via Ruthenium-Catalyzed (2+2+2) Cycloaddition[73]

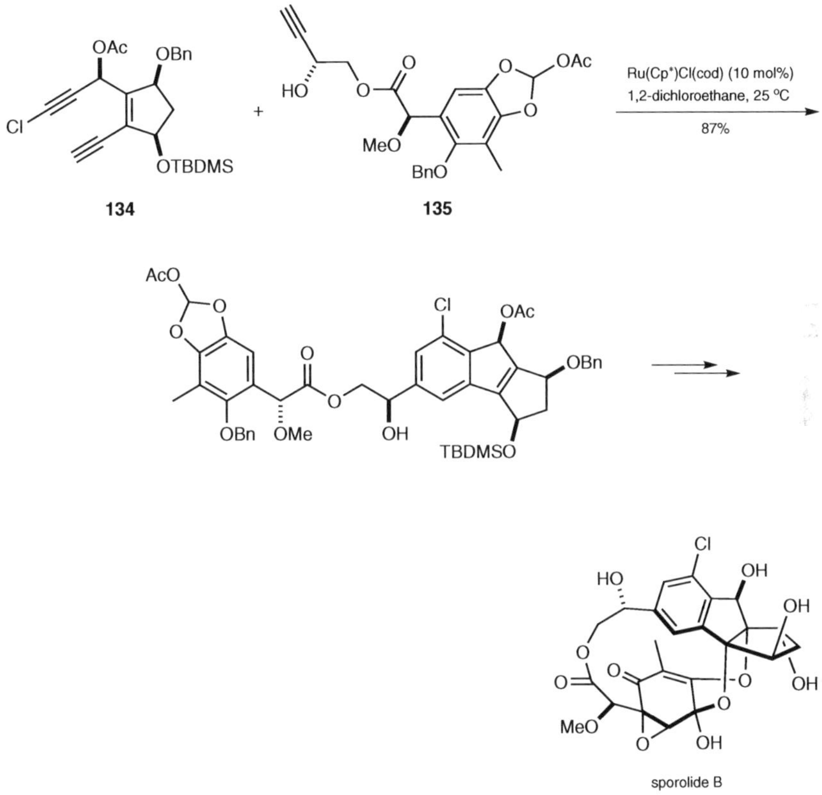

The synthesis of arylnaphthalenes **138** has been achieved by the palladium-catalyzed (2+2+2) cycloaddition of diynes **136** with arynes from **137** (Scheme 44).[74] Arylnaphthalenes **138** thus obtained could be transformed into natural arylnaphthalene lignans, taiwanin E or taiwanin C.

Scheme 44 Palladium-Catalyzed (2+2+2) Cycloaddition of Diynes with Arynes[74]

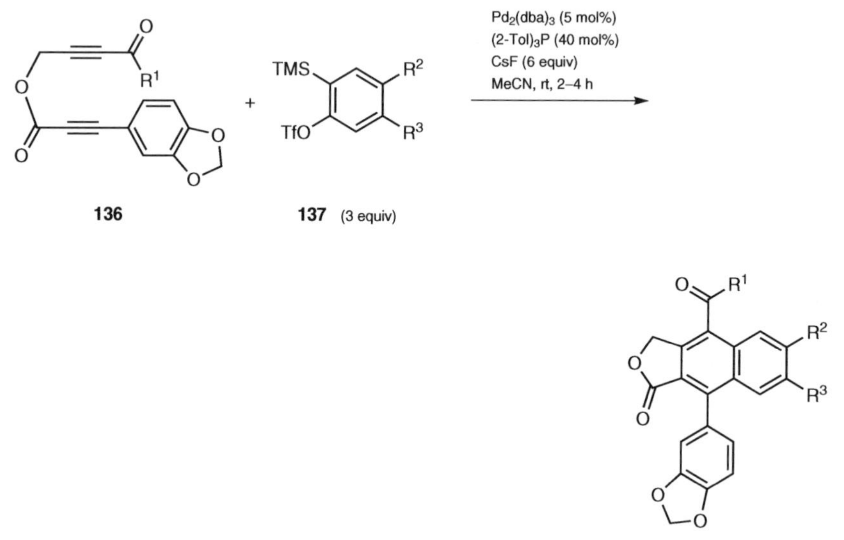

R¹	R²	R³	Time (h)	Yield (%)	Ref
OMe	H	H	2	57	[74]
N(OMe)Me	H	H	2	78	[74]
N(OMe)Me	OCH$_2$O		4	61	[74]

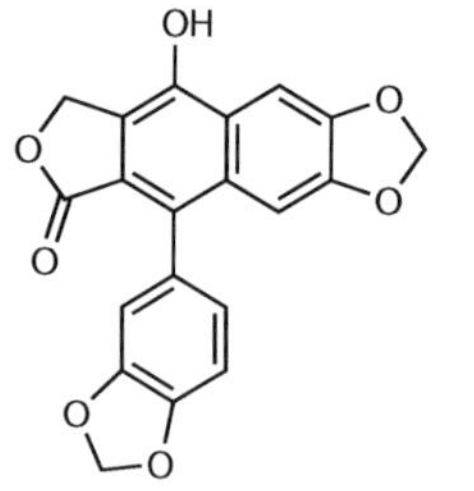

taiwanin E

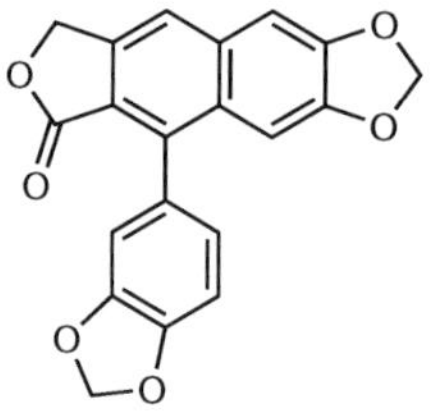

taiwanin C

10-(Benzo[*d*][1,3]dioxol-5-yl)-*N*-methoxy-*N*-methyl-1-oxo-1,3,6,8-tetrahydro-2,6,8-trioxa-indeno[5,6-*f*]inden-4-carboxamide [138, R¹ = N(OMe)Me; R²,R³ = OCH$_2$O]:
Typical Procedure:[74]
Pd$_2$(dba)$_3$·CHCl$_3$ (26 mg, 0.025 mmol) and (2-Tol)$_3$P (61 mg, 0.20 mmol) were dissolved in MeCN (1.2 mL), and the mixture was stirred at rt for 15 min. The catalyst soln was added through a cannula to a soln of diyne **136** (160 mg, 0.51 mmol), arene **137** (530 mg, 1.6 mmol), and CsF (472 mg, 3.1 mmol) in MeCN (1.8 mL) at 0 °C. More MeCN (1.0 mL) was used to wash the catalyst through. The mixture was stirred at rt for 4 h and then quenched with sat. aq NH$_4$Cl. The mixture was extracted with EtOAc, and the organic layer was washed with brine and dried (Na$_2$SO$_4$). After removal of the solvent, the residue was purified by column chromatography (silica gel, hexane/EtOAc 3:2) to give the title compound; yield: 134 mg (61%). Unconverted arene **137** was recovered; yield: 257 mg (48%).

2.6.6 Conclusions and Future Perspectives

In this chapter, typical examples of metal-catalyzed (2+2+2) cycloadditions are described. Their application in the syntheses of functional and biologically active molecules has also been described. A number of transition-metal catalysts have been developed for (2+2+2)-cycloaddition reactions. In particular, highly active transition-metal catalysts have enabled the syntheses of polycyclic and sterically hindered compounds, and chiral transition-metal catalysts have enabled the enantioselective construction of not only centrochiralities but also non-centrochiralities, such as axial, planar, and helical chiralities. We believe that metal-catalyzed (2+2+2)-cycloaddition reactions will be widely employed for the synthesis of complex organic molecules from now on. Further development of efficient and powerful transition-metal-catalyzed (2+2+2)-cycloaddition reactions is awaited for more facile and diverse syntheses of complex organic molecules.

References

[1] *Transition-Metal-Mediated Aromatic Ring Construction*, Tanaka, K., Ed.; Wiley: Hoboken, NJ, (2013).
[2] Okamoto, S.; Sugiyama, Y., *Synlett*, (2013) **24**, 1044.
[3] Broere, D. L. J.; Ruijter, E., *Synthesis*, (2012) **44**, 2639.
[4] Shibata, Y.; Tanaka, K., *Synthesis*, (2012) **44**, 323.
[5] Weding, N.; Hapke, M., *Chem. Soc. Rev.*, (2011) **40**, 4525.
[6] Domínguez, G.; Pérez-Castells, J., *Chem. Soc. Rev.*, (2011) **40**, 3430.
[7] Tanaka, K., *Chem.–Asian J.*, (2009) **4**, 508.
[8] Leboeuf, D.; Gandon, V.; Malacria, M., In *Handbook of Cyclization Reactions*, Ma, S., Ed.; Wiley-VCH: Weinheim, Germany, (2010); p 367.
[9] Hess, W.; Treutwein, J.; Hilt, G., *Synthesis*, (2008), 3537.
[10] Shibata, T.; Tsuchikama, K., *Org. Biomol. Chem.*, (2008) **6**, 1317.
[11] Agenet, N.; Buisine, O.; Slowinski, F.; Gandon, V.; Aubert, C.; Malacria, M., *Org. React. (Hoboken, NJ, U. S.)*, (2007) **68**, 1.
[12] Chopade, P. R.; Louie, J., *Adv. Synth. Catal.*, (2006) **348**, 2307.
[13] Kotha, S.; Brahmachary, E.; Lahiri, K., *Eur. J. Org. Chem.*, (2005), 4741.
[14] Reppe, W.; Schlichting, O.; Klager, K.; Toepel, T., *Justus Liebigs Ann. Chem.*, (1948) **560**, 1.
[15] Vollhardt, K. P. C., *Acc. Chem. Res.*, (1977) **10**, 1.
[16] Vollhardt, K. P. C., *Angew. Chem. Int. Ed. Engl.*, (1984) **23**, 539.
[17] Yamazaki, H., *J. Synth. Org. Chem., Jpn.*, (1987) **45**, 244.
[18] Agenet, N.; Gandon, V.; Vollhardt, K. P. C.; Malacria, M.; Aubert, C., *J. Am. Chem. Soc.*, (2007) 129, 8860.
[19] Yamamoto, Y.; Arakawa, T.; Ogawa, R.; Itoh, K., *J. Am. Chem. Soc.*, (2003) **125**, 12143.
[20] Tanaka, K.; Toyoda, K.; Wada, A.; Shirasaka, K.; Hirano, M., *Chem.–Eur. J.*, (2005) **11**, 1145.
[21] Witulski, B.; Alayrac, C., *Angew. Chem. Int. Ed.*, (2002) **41**, 3281.
[22] Komine, Y.; Kamisawa, A.; Tanaka, K., *Org. Lett.*, (2009) **11**, 2361.
[23] Matsuda, T.; Kadowaki, S.; Goya, T.; Murakami, M., *Org. Lett.*, (2007) **9**, 133.
[24] Iannazzo, L.; Vollhardt, K. P. C.; Malacria, M.; Aubert, C.; Gandon, V., *Eur. J. Org. Chem.*, (2011), 3283.
[25] Yamamoto, Y.; Hattori, K.; Nishiyama, H., *J. Am. Chem. Soc.*, (2006) **128**, 8336.
[26] Yamamoto, Y.; Hattori, K.; Ishii, J.; Nishiyama, H., *Tetrahedron*, (2006) **62**, 4294.
[27] Nishida, G.; Noguchi, K.; Hirano, M.; Tanaka, K., *Angew. Chem. Int. Ed.*, (2007) **46**, 3951.
[28] Ogaki, S.; Shibata, Y.; Noguchi, K.; Tanaka, K., *J. Org. Chem.*, (2011) **76**, 1926.
[29] Nishida, G.; Ogaki, S.; Yusa, Y.; Yokozawa, T.; Noguchi, K.; Tanaka, K., *Org. Lett.*, (2008) **10**, 2849.
[30] Shibata, T.; Fujimoto, T.; Yokota, K.; Takagi, K., *J. Am. Chem. Soc.*, (2004) **126**, 8382.
[31] Shibata, T.; Chiba, T.; Hirashima, H.; Ueno, Y.; Endo, K., *Angew. Chem. Int. Ed.*, (2009) **48**, 8066.
[32] Tanaka, K.; Takeishi, K.; Noguchi, K., *J. Am. Chem. Soc.*, (2006) **128**, 4586.
[33] Suda, T.; Noguchi, K.; Hirano, M.; Tanaka, K., *Chem.–Eur. J.*, (2008) **14**, 6593.
[34] Araki, T.; Noguchi, K.; Tanaka, K., *Angew. Chem. Int. Ed.*, (2013) **52**, 5617.
[35] Sawada, Y.; Furumi, S.; Takai, A.; Takeuchi, M.; Noguchi, K.; Tanaka, K., *J. Am. Chem. Soc.*, (2012) **134**, 4080.
[36] Teplý, F.; Stará, I. G.; Starý, I.; Kollárovič, A.; Šaman, D.; Rulíšek, L.; Fiedler, P., *J. Am. Chem. Soc.*, (2002) **124**, 9175.
[37] Tanaka, K.; Kamisawa, A.; Suda, T.; Noguchi, K.; Hirano, M., *J. Am. Chem. Soc.*, (2007) **129**, 12078.
[38] Wakatsuki, Y.; Yamazaki, H., *J. Chem. Soc., Chem. Commun.*, (1973), 280.
[39] Vollhardt, K. P. C.; Bergman, R. G., *J. Am. Chem. Soc.*, (1974) **96**, 4996.
[40] Bönnemann, H.; Brinkmann, R.; Schenkluhn, H., *Synthesis*, (1974), 575.
[41] Geny, A.; Agenet, N.; Iannazzo, L.; Malacria, M.; Aubert, C.; Gandon, V., *Angew. Chem. Int. Ed.*, (2009) **48**, 1810.
[42] Hapke, M.; Weding, N.; Spannenberg, A., *Organometallics*, (2010) **29**, 4298.
[43] Kase, K.; Goswami, A.; Ohtaki, K.; Tanabe, E.; Saino, N.; Okamoto, S., *Org. Lett.*, (2007) **9**, 931.
[44] Kumar, P.; Prescher, S.; Louie, J., *Angew. Chem. Int. Ed.*, (2011) **50**, 10694.
[45] Yamamoto, Y.; Kinpara, K.; Saigoku, T.; Takagishi, H.; Okuda, S.; Nishiyama, H.; Itoh, K., *J. Am. Chem. Soc.*, (2005) **127**, 605.
[46] Yamamoto, Y.; Kinpara, K.; Ogawa, R.; Nishiyama, H.; Itoh, K., *Chem.–Eur. J.*, (2006) **12**, 5618.
[47] Tanaka, K.; Suzuki, N.; Nishida, G., *Eur. J. Org. Chem.*, (2006), 3917.
[48] Tanaka, K.; Hara, H.; Nishida, G.; Hirano, M., *Org. Lett.*, (2007) **9**, 1907.

References

[49] Onodera, G.; Shimizu, Y.; Kimura, J.; Kobayashi, J.; Ebihara, Y.; Kondo, K.; Sakata, K.; Takeuchi, R., *J. Am. Chem. Soc.*, (2012) **134**, 10515.

[50] Gutnov, A.; Heller, B.; Fischer, C.; Drexler, H.-J.; Spannenberg, A.; Sundermann, B.; Sundermann, C., *Angew. Chem. Int. Ed.*, (2004) **43**, 3795.

[51] Wada, A.; Noguchi, K.; Hirano, M.; Tanaka, K., *Org. Lett.*, (2007) **9**, 1295.

[52] Hong, P.; Yamazaki, H., *Synthesis*, (1977), 50.

[53] Hoberg, H.; Oster, B. W., *Synthesis*, (1982), 324.

[54] Earl, R. A.; Vollhardt, K. P. C., *J. Org. Chem.*, (1984) **49**, 4786.

[55] Duong, H. A.; Cross, M. J.; Louie, J., *J. Am. Chem. Soc.*, (2004) **126**, 11438.

[56] Tanaka, K.; Wada, A.; Noguchi, K., *Org. Lett.*, (2005) **7**, 4737.

[57] Onodera, G.; Suto, M.; Takeuchi, R., *J. Org. Chem.*, (2012) **77**, 908.

[58] Yu, R. T.; Rovis, T., *J. Am. Chem. Soc.*, (2006) **128**, 12370.

[59] Martin, T. J.; Rovis, T., *Angew. Chem. Int. Ed.*, (2013) **52**, 5368.

[60] Ishii, M.; Mori, F.; Tanaka, K., *Chem.–Eur. J.*, (2014) **20**, 2169.

[61] Louie, J.; Gibby, J. E.; Farnworth, M. V.; Tekavec, T. N., *J. Am. Chem. Soc.*, (2002) **124**, 15188.

[62] Tsuchikama, K.; Kuwata, Y.; Shibata, T., *J. Am. Chem. Soc.*, (2006) **128**, 13686.

[63] Evans, P. A.; Lai, K. W.; Sawyer, J. R., *J. Am. Chem. Soc.*, (2005) **127**, 12466.

[64] Shibata, T.; Arai, Y.; Tahara, Y., *Org. Lett.*, (2005) **7**, 4955.

[65] Hara, J.; Ishida, M.; Kobayashi, M.; Noguchi, K.; Tanaka, K., *Angew. Chem. Int. Ed.*, (2014) **53**, 2956.

[66] Masutomi, K.; Sakiyama, N.; Noguchi, K.; Tanaka, K., *Angew. Chem. Int. Ed.*, (2012) **51**, 13031.

[67] Shibata, T.; Tahara, Y., *J. Am. Chem. Soc.*, (2006) **128**, 11766.

[68] Brusoe, A. T.; Alexanian, E. J., *Angew. Chem. Int. Ed.*, (2011) **50**, 6596.

[69] Brusoe, A. T.; Edwankar, R. V.; Alexanian, E. J., *Org. Lett.*, (2012) **14**, 6096.

[70] Otake, Y.; Tanaka, R.; Tanaka, K., *Eur. J. Org. Chem.*, (2009), 2737.

[71] Tanaka, K.; Otake, Y.; Sagae, H.; Noguchi, K.; Hirano, M., *Angew. Chem. Int. Ed.*, (2008) **47**, 1312.

[72] Alayrac, C.; Schollmeyer, D.; Witulski, B., *Chem. Commun. (Cambridge)*, (2009), 1464.

[73] Nicolaou, K. C.; Tang, Y.; Wang, J., *Angew. Chem. Int. Ed.*, (2009) **48**, 3449.

[74] Sato, Y.; Tamura, T.; Mori, M., *Angew. Chem. Int. Ed.*, (2004) **43**, 2436.

2.7 Metal-Catalyzed (4 + 3) Cycloadditions Involving Allylic Cations

D. E. Jones and M. Harmata

General Introduction

Cycloaddition reactions exhibit high atom economy and are often stereoselective, high-yielding processes leading to complex and highly functionalized ring systems. Many of these ring systems are common motifs found in a multitude of natural products, as well as pharmaceutical and agricultural agents. This makes the efficient preparation of these ring systems an important area of study. The (4 + 3)-cycloaddition reaction is one such reaction that is used to prepare medium-sized ring systems. This is a reaction of a four-carbon-atom diene and a three-carbon-atom allylic cation, resulting in the formation of a cycloheptenyl cation (Scheme 1).[1–3] The name of the reaction is derived from the carbon-atom count of the reactants and not the number of electrons involved in the bond-forming process. However, it is electronically analogous to the Diels–Alder cycloaddition, a [4π + 2π]-electron process. (4 + 3) Cycloadditions can be performed with equal facility either intermolecularly or in an intramolecular fashion, with the diene attached to the allylic cation by a tether.[4,5] It should also be noted that the products formed can serve as a platform for the preparation of other ring systems, either larger or smaller. There are many excellent reviews available on the (4 + 3)-cycloaddition reaction, including several recent ones.[1–3] There are many reported methods for the generation of allylic cations for (4 + 3)-cycloaddition reactions including solvolytic approaches and non-metal Lewis acid catalysis. Within this chapter, only metal-catalyzed or metal-mediated (4 + 3)-cycloaddition reactions between an allylic cationic species and a diene are described.

Scheme 1 General (4 + 3) Cycloaddition of a Diene and an Allylic Cation[1–3]

The (4 + 3)-cycloaddition reaction was serendipitously discovered by Arthur Fort in 1962 while studying mechanistic aspects of the Favorskii rearrangement (Scheme 2).[6,7] He found that treating α-chloro ketone **1** with a base in the presence of a large excess of furan resulted in the formation of adduct **3**. The reaction proceeds through the oxyallyl intermediate **2**, which is formed by an enolization/ionization process with no metal required. While the yield of this reaction is meager at best, it initiated a flurry of research over the following fifty years that transformed the (4 + 3) cycloaddition into the powerful tool for chemical synthesis that it is today.

Scheme 2 The First (4 + 3)-Cycloaddition Reaction[6,7]

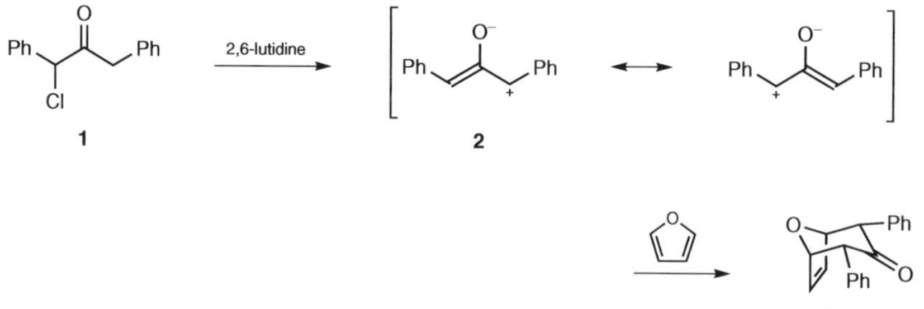

The mechanism of the (4 + 3)-cycloaddition reaction is highly dependent on many factors (solvent, structure of the allylic cation and diene, etc.). The nature of these various factors greatly influences the mechanism for a particular reaction. The mechanistic subtleties of the specific reactions presented will often require additional experimentation for true elucidation and are beyond the scope of this article. With that statement, the general mechanisms of the reaction can be viewed on a continuum of two fundamental extremes: concerted and stepwise. As a simplified example, nucleophilic attack of furan on the resonance-stabilized oxyallylic cation **4**, formed by enolization/ionization of an appropriate precursor, affords zwitterion **5** (Scheme 3).[2,3] Cycloadduct **6A** is then formed by intramolecular attack of the enolate on the cation. Alternatively, the oxyallyl cation **4** can undergo cycloaddition with furan in a concerted manner. It is important to note the relative orientation of the reactants: Oxyallyl species **4** and furan can react through an extended or "*exo*" transition state **7A** to afford the cycloadduct **6B** after a ring flip. Conversely, the reaction can progress through compact or "*endo*" transition state **7B** to form the diastereomeric cycloadduct **6C**. This is the pathway that is energetically preferred for the concerted process.

Scheme 3 General Mechanisms of (4 + 3) Cycloaddition[2,3]

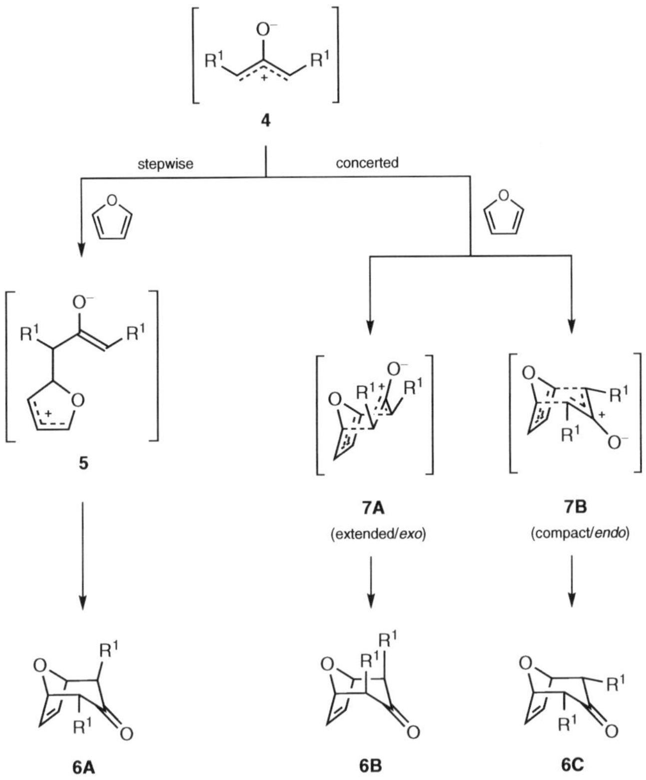

2.7.1 Carbon-Substituted Allylic Cations in Cycloaddition Reactions

The generation of simple allylic cations (with only carbon substituents at positions 1 and 3 of the allylic unit) for use in (4 + 3)-cycloaddition reactions is an important synthetic method. The nature of substitution of the cation and diene allows for diversity of molecular targets.[8] A variety of allylic cation precursors have been developed that upon activation readily undergo both inter- and intramolecular cycloaddition with a variety of dienes.

2.7.1.1 Reduction of α,α′-Dihalo Ketones

There are a variety of approaches for the generation of oxyallylic cations by the reduction of α,α′-dihalo ketones. Many were developed and refined by the earlier pioneers of (4 + 3)-cycloaddition reactions.[9] α,α′-Dihalo ketones are well suited as precursors to allylic cations. The halogen substitution facilitates the reduction leading to enolate formation. Secondly, halogens are good leaving groups and, under the appropriate conditions, will readily dissociate, resulting in oxyallylic cation formation.

2.7.1.1.1 Reductive Cycloaddition with Copper Bronze

Treatment of 2,4-dibromopentan-3-one (**8**) with excess furan in the presence of sodium iodide and copper bronze results in the formation of cycloadduct **9A** as a single diastereomer on a multigram scale (Scheme 4).[10,11] Cycloadduct **9A** presumably arises through a compact/*endo* transition state. This method has been applied to a variety of α,α′-dihalo ketones, including cyclic ketones. Using furan and cyclopentadiene as the allylic cation trap, relatively complex tricyclic ring systems have been prepared.[12] It is important to note that the ring size of the cyclic α,α′-dihalo ketone has a considerable influence on the stereochemical outcome of the reaction.[13,14]

Scheme 4 Cycloaddition of 2,4-Dibromopentan-3-one with Furan Mediated by Copper Bronze[10,11]

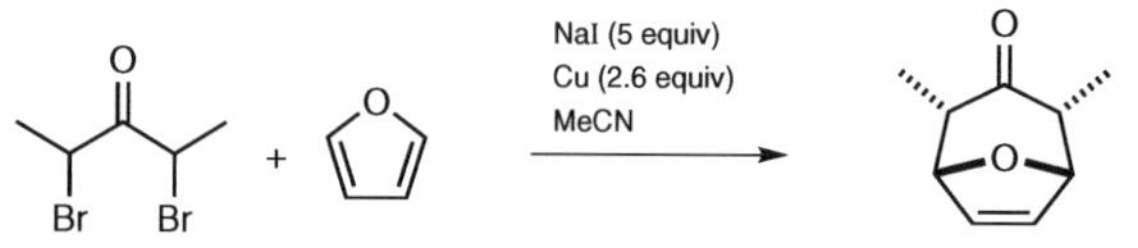

endo,endo-2,4-Dimethyl-8-oxabicyclo[3.2.1]oct-6-en-3-one (9A); Typical Procedure:[11]

A 1-L, two-necked, round-bottomed flask was fitted with a 100-mL dropping funnel and dry MeCN (200 mL) was added. Dry NaI (90 g, 0.6 mol) was added with vigorous stirring under a slow stream of N_2. Then, powdered Cu (20 g, 0.314 mmol) was added, followed by furan (30 mL, 0.4 mol). A soln of 2,4-dibromopentan-3-one (**8**; 29.04 g, 0.12 mol) in dry MeCN (50 mL) was added by means of the dropping funnel over 50 min at 0 °C. The mixture was allowed to warm to rt and was stirred for an additional 12 h. After that time, the flask was cooled to 0 °C and CH_2Cl_2 (150 mL) was added. The resultant mixture was then poured into a conical flask containing H_2O (500 mL) and crushed ice (500 mL) and thoroughly stirred to allow the precipitation of Cu salts. After filtration through a Celite pad, the mother liquor was washed with 45% aq NH_3 (3 × 20 mL) and brine (20 mL), dried ($MgSO_4$), and concentrated under reduced pressure to leave a pale yellow oil, which was purified by flash chromatography (petroleum ether/Et_2O 2:1); yield: 12.2 g (67%). A checked procedure, with a yield of 40–48%, is available for this reaction.[10]

2.7.1.1.2 Reductive Cycloaddition with Zinc/Copper Couple

It is well known that dibromo and dichloro ketones are notorious lachrymators, a property that can make their use problematic, particularly for reactions performed on a larger scale. Special precautions are made to mitigate this effect. Montaña and Grima reported a reductive procedure using zinc/copper couple with α,α′-diiodo ketones, which are non-lachrymatory.[15,16] A mixture of α,α′-diiodo ketone **10** and furan in the presence of zinc/copper couple is sonicated for 15 minutes to afford *endo*-cycloadduct **9A** and *exo*-cycloadduct **9B** in a ratio of 91:9 and 90% overall yield (Scheme 5).[16] The reaction can also be performed under thermal conditions with a slight decrease in yield and *endo/exo* selectivity. The generality of this reaction has not been demonstrated.

Scheme 5 Cycloaddition of 2,4-Diiodopentan-3-one and Furan Mediated by Zinc/Copper Couple[16]

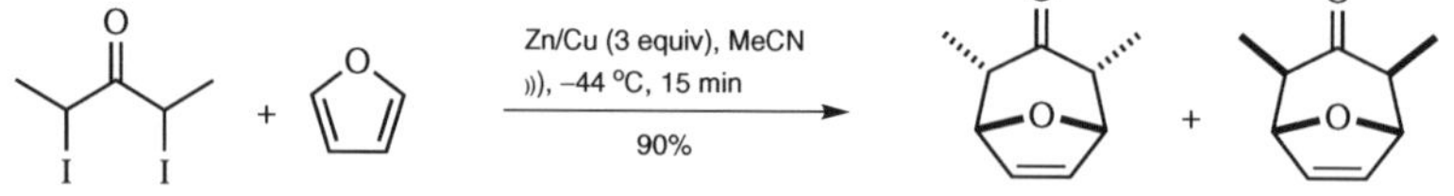

endo,endo-2,4-Dimethyl-8-oxabicyclo[3.2.1]oct-6-en-3-one (9A):[16]

In a three-necked sonication reactor,[17] fitted with septa and a 1/8" sonication probe (Branson EDP ref. no. 101–148–062) was placed Zn/Cu (freshly prepared by deposition of Cu^0, from $CuSO_4$, on Zn powder).[18] Dry MeCN (5 mL) was added, and the resulting black suspension was partially immersed in a cooling bath until a working temperature of −44 °C [the inner temperature was monitored by a flexible thermocouple (PT-100)]. Then, furan (4.24 mmol) was added, at once, followed by a soln of 2,4-diiodopentan-3-one (**10**; 360 mg, 1.06 mmol) in MeCN (6 mL), which was added dropwise by cannula. The mixture was then sonicated at 40% of maximum power for 15 min. The cold mixture was quenched with H$_2$O (5 mL) and the solids were removed by filtration. The filtered liquid was extracted with CH$_2$Cl$_2$ (5 × 15 mL). The organic phases were combined and washed with cold 25% w/w aq NH$_3$ soln (2 × 50 mL), followed by cold distilled H$_2$O (2 × 50 mL), until no blue color of Cu(NH$_3$)$_4^{2+}$ was observed. The resulting organic soln was dried (MgSO$_4$), filtered through neutral alumina, and concentrated to dryness under vacuum without heating. The crude product was purified by flash chromatography (hexanes/EtOAc mixtures of increasing polarity) to afford pure diastereomeric products **9A** and **9B** in a ratio of 91:9; yield: 90%.

2.7.1.1.3 Reductive Cycloaddition with Iron Carbonyls

Noyori and co-workers discovered that iron carbonyls efficiently mediate the cycloaddition of α,α′-dihalo ketones with a large variety of cyclic and acyclic dienes to afford cycloheptenones (Scheme 6).[19] Treatment of dibromo ketone **11** (R^1 = Me) and a large excess of 1,3-diene **12** (R^2 = R^3 = Me) with nonacarbonyldiiron(0) in benzene results in the formation of cycloadduct **13** (R^1 = R^2 = R^3 = Me) in 71% (GLC analysis). This reaction works with a variety of simple acyclic 1,3-dienes. However, removal of the methyl groups from either dibromo ketone **11** or diene **12** results in a dramatic loss in yield of the cycloadduct. The reaction of the simple cyclic dienes furan and cyclopentadiene with ketone **11** (R^1 = Me) provides the cycloadducts in high yield, 96 and 82%, respectively. It is also interesting to note that the reaction between the preformed iron carbonyl complex of 1,3-diene **12** (R^2 = R^3 = H) and dibromo ketone **11** (R^1 = Me) increases the yield of **13** (R^1 = Me; R^2 = R^3 = H) from 33 to 90%.

Scheme 6 Cycloaddition of 2,4-Dibromopentan-3-ones with Buta-1,3-dienes Mediated by Nonacarbonyldiiron(0)[19]

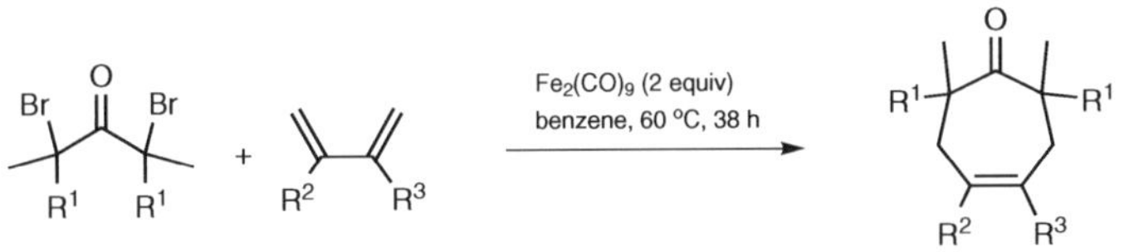

R^1	R^2	R^3	Yield (%)	Ref
Me	Me	Me	71	[19]
Me	Me	H	47	[19]
Me	H	H	33	[19]
H	Me	Me	46	[19]
H	H	Me	36	[19]
H	H	H	30	[19]

2,2,4,5,7,7-Hexamethylcyclohept-4-en-1-one (13, R^1 = R^2 = R^3 = Me); Typical Procedure:[19]

A mixture of Fe$_2$(CO)$_9$ (1.60 g, 4.40 mmol), the dibromo ketone **11** (R^1 = Me; 540 mg, 2.00 mmol), 2,3-dimethylbuta-1,3-diene (**12**, R^2 = R^3 = Me; 7.28 g, 88.5 mmol), dry benzene (5.0 mL) (**CAUTION:** *carcinogen*), and a small amount of hydroquinone was heated at 60 °C for 38 h with occasional shaking. The mixture was cooled to rt. The resulting precipitates were removed by filtration through a pad of Celite 545 (ca. 2 g) and the filtrate was concentrated under reduced pressure to give a yellow oil; yield: 290 mg. GLC analysis (100 °C, tetradecane as an internal standard) indicated the production of 2,2,4,5,7,7-hexamethyl-cyclohept-4-en-1-one in 71% yield.

2.7.1.1.4 Reduction with Diethylzinc

Lautens and co-workers have developed a highly diastereoselective (4 + 3) cycloaddition with the use of chiral furfuryl alcohols as dienes.[20] The treatment of chiral (but racemic) alcohol 14 and 3 equivalents of dibromo ketone 8 with 2 equivalents of diethylzinc provides cycloadduct 16 in a diastereomeric ratio of 95:5 (16/all other isomers) and a yield of 80% (Scheme 7). The high facial and *exo* selectivity is explained by a chelation-controlled process proceeding through the extended (*exo*) transition state 15. The use of 1 equivalent of diethylzinc results in lower yields without compromising the diastereoselectivity (54%, dr 98:2). Importantly, this could provide access to diastereomers which are enantiomerically pure and were previously inaccessible. However, only a select few chiral furfuryl alcohols have been submitted to this reaction and, as such, the generality of this process has not been demonstrated.

Scheme 7 Cycloaddition of 2,4-Dibromopentan-3-one with a Chiral Furfuryl Alcohol Mediated by Diethylzinc[20]

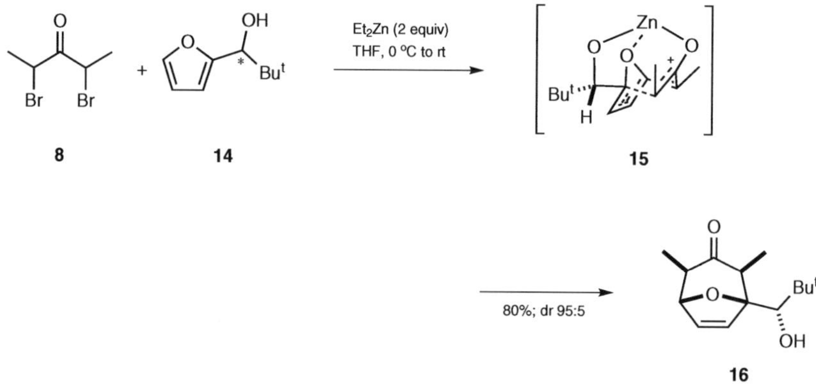

(1S*,2S*,4R*,5R*)-1-[(S*)-1-Hydroxy-2,2-dimethylpropyl]-2,4-dimethyl-8-oxabicyclo-[3.2.1]oct-6-en-3-one (16):[20]

Prior to use, dibromo ketone 8 was freshly distilled and then filtered through basic alumina, eluting with pentane, and the solvent was removed under reduced pressure to give a clear and colorless oil, free of any residual acid. To a soln of furan 14 (86 mg, 0.56 mmol) in THF (1.5 mL, 0.3 M final concentration) was added neat Et$_2$Zn (0.11 mL, 1.12 mmol) under N$_2$ at 0 °C. The resulting mixture was stirred for 10 min at 0 °C prior to the dropwise addition of dibromo ketone 8 (410 mg, 1.68 mmol). **WARNING:** *the reaction is exothermic, therefore care should be taken not to add the dibromo ketone rapidly, especially on a large scale.* The mixture was then stirred at 0 °C (1 d) then rt (1 d), followed by quenching with EtOAc and sat. aq Na$_2$(edta) (1:1). The products were extracted into the organic layer (2 × EtOAc), and the organic layer was dried (MgSO$_4$) and filtered and the solvent was removed under reduced pressure to give an oil. The crude oil was purified by flash chromatography (Et$_2$O/toluene 1:9) to give a 95:5 mixture of 16 to all other isomers; yield: 107 mg (80%, combined isolated yield).

2.7.1.2 Ionization of α-Halo Enol Ethers and Related Species

The methods discussed in Section 2.7.1.1 all require an enolization/ionization process to generate the allylic cation for cycloaddition. In the chemistry described in the following sections, the allylic precursor only requires the dissociation of a leaving group, most commonly a halogen ion, to provide an allylic cation for (4+3) cycloaddition.

2.7.1.2.1 Silver-Mediated Ionization of Enol Ethers

One of the earliest reports using α-halo enol ethers as a precursor to allylic cations was by Shimizu and co-workers.[21,22] They demonstrated that silver ions effectively facilitate the dissociation of chloride ion from chlorinated silyl enol ethers to form allylic cations, which are readily trapped by simple 1,3-dienes in good to excellent yields. The addition of a solution of silyl enol ether **17** to a mixture of silver(I) perchlorate, calcium carbonate, and furan (**18**, X=O) in nitromethane results in a 92% yield of cycloadduct **19** (X=O) as a single isomer (Scheme 8). The reaction works equally well using cyclopentadiene (**18**, X=CH_2) as the diene. When the reaction is run in a mixture of tetrahydrofuran and diethyl ether (1:3), the yield of the cycloadduct decreases slightly. However, the solvent effect can be very dramatic. Using nitromethane as the solvent and 2-methylfuran (**20**) as the diene, cycloadducts **21** and **22** are formed in 92% yield and a ratio of 1.9:1, respectively (Scheme 9). Upon changing the solvent to a mixture of tetrahydrofuran and diethyl ether (1:3), the Friedel–Crafts alkylation products **23** and **24** (4:1) are the major products in 43% yield. Cycloadducts **21** and **22** are formed in only 28% yield and in a ratio of 19:1. The decrease in solvent polarity presumably results in a change in reaction mechanism. Zinc(II) chloride has been used to catalyze the (4+3)-cycloaddition reaction of α-halo enol ethers with 1,3-dienes with comparable results.[23]

Scheme 8 Cycloaddition of a Trimethylsiloxy-Substituted Allylic Chloride with Furan and Cyclopentadiene Mediated by Silver(I) Perchlorate[22]

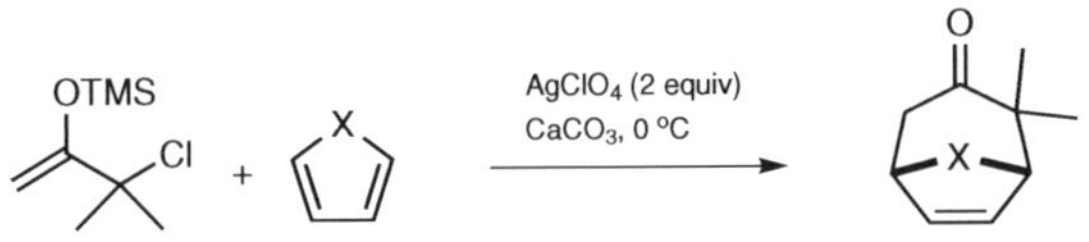

X	Solvent	Yield (%)	Ref
O	$MeNO_2$	92	[22]
O	THF/Et_2O	85	[22]
CH_2	$MeNO_2$	91	[22]
CH_2	THF/Et_2O	71	[22]

Scheme 9 Cycloaddition of a Trimethylsiloxy-Substituted Allylic Chloride with 2-Methylfuran Mediated by Silver(I) Perchlorate[22]

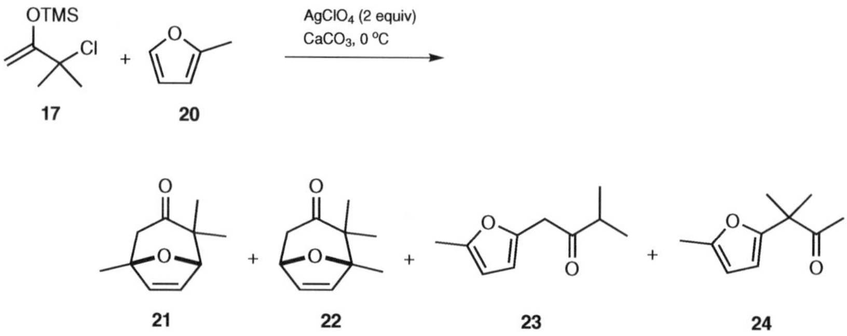

Solvent	Ratio (21/22)	Yield (%) of 21 + 22	Ratio (23/24)	Yield (%) of 23 + 24	Ref
MeNO$_2$	1.9:1	92	–	0	[22]
THF/Et$_2$O	19:1	28	4:1	43	[22]

(1S*,5S*)-2,2-Dimethyl-8-oxabicyclo[3.2.1]oct-6-en-3-one (19, X = O):[22]

CAUTION: *Nitromethane is flammable, a shock- and heat-sensitive explosive, and an eye, skin, and respiratory tract irritant.*

To a well-stirred mixture of AgClO$_4$ (10 mmol), CaCO$_3$ (2 g), and furan (25 mmol) in MeNO$_2$ (20 mL) was added a soln of allyl chloride **17** (1 g, 5.2 mmol) in MeNO$_2$ (5 mL) over a period of 15 min at 0 °C. The resulting mixture was stirred for 15 min at 0 °C and the mixture was diluted with Et$_2$O (50 mL). Then, a NaCl soln was added until the inorganic materials aggregated. The organic layer was separated by decantation, washed with aq NaHCO$_3$, and dried. Evaporation of the solvent and purification afforded pure product as colorless crystals; yield: 727 mg (92%); mp 46–47 °C.

2.7.1.2.2 Silver-Mediated Ionization of α-Chloroenamines

Allylic cations are efficiently generated from α-chloroenamines. These cations are also effective dienophiles in (4 + 3) cycloadditions. Treatment of a mixture of silver(I) tetrafluoroborate and furan with crude allylic chloride **25** affords iminium salt **26**. Cycloadduct **27** is obtained in excellent yield after hydrolysis (Scheme 10).[24,25] The research group of Kende began to explore the use of chiral α-chloroimines, as precursors to chiral allylic cations by an enolization/ionization process, for (4 + 3)-cycloaddition reactions.[26] The yields of cycloaddition with simple, cyclic dienes are low to moderate with equally low to moderate enantiomeric excesses. More important is the demonstration of the principle that asymmetric induction using chiral allylic cations is indeed possible, but further research is required to identify the subtleties of the reaction.

2.7.1 Carbon-Substituted Allylic Cations in Cycloaddition Reactions

Scheme 10 Cycloaddition of 6-Chloro-1-pyrrolidinocyclohex-1-ene and Furan Mediated by Silver(I) Tetrafluoroborate[24,25]

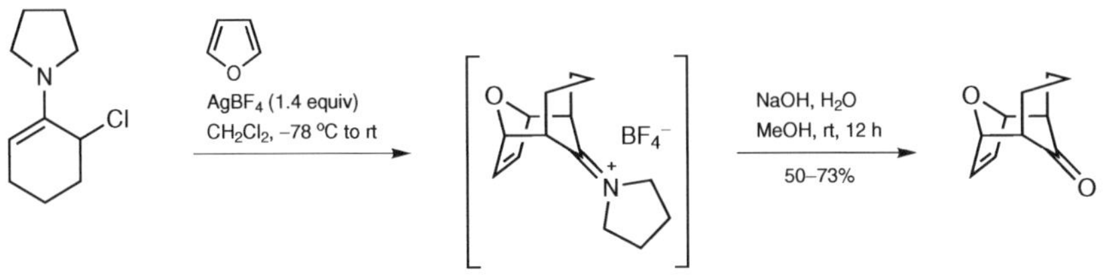

25   26   27

11-Oxatricyclo[4.3.1.1^{2,5}]undec-3-en-10-one (27); Typical Procedure:[25]

A 1-L, three-necked, round-bottomed flask, equipped with a magnetic stirrer bar, was charged quickly with AgBF$_4$ (45 g, 0.23 mol) and wrapped in aluminum foil. After the flask had been dried under vacuum and purged with N$_2$, dry CH$_2$Cl$_2$ (210 mL) was added, followed by freshly distilled furan (31 mL, 0.42 mol). The mixture was then cooled to −78 °C and a soln of freshly prepared enamine **25**[24] (29.61 g, 0.16 mmol) in dry CH$_2$Cl$_2$ (150 mL) was added dropwise via a syringe pump over a period of 2 h. The mixture was then slowly warmed to rt, and stirred for an additional 12 h. The inorganic salt was removed by filtration through Celite, which was rinsed thoroughly with CH$_2$Cl$_2$ (3 × 70 mL). The combined filtrates were concentrated under reduced pressure to afford the iminium salt **26** as a dark brown oil. The iminium salt **26** was dissolved in distilled H$_2$O (400 mL) and MeOH (200 mL), and NaOH (25 g) was added. The mixture was stirred at rt for 12 h, and then extracted with Et$_2$O (6 × 250 mL). The combined organic layers were washed with brine (1 × 600 mL), dried (MgSO$_4$), and concentrated under reduced pressure to afford the crude product as a dark orange oil. Purification by column chromatography (silica gel, EtOAc/hexane 1:3) gave pure ketone **27** as a white solid; yield: 18.31 g (70%). A checked procedure is also available.[24]

2.7.1.3 Other Approaches to Carbon-Substituted Allylic Cations

A variety of innovative and efficient precursors to simple allylic cations have been developed that each fall into their own unique category. The following are several excellent examples, but these should not be considered to be all-inclusive.

2.7.1.3.1 Furfuryl Alcohols as Allylic Cation Precursors

The activation of furfuryl alcohols **28** in dichloromethane solution with a slight excess of titanium(IV) chloride in the presence of cyclohexa-1,3-diene results in (4 + 3) cycloaddition, forming preferentially the *exo*-diastereomer **29** as shown (Scheme 11).[27] The reaction is largely dependent on the structure of the furfuryl alcohol; the furfuryl alcohol needs to be minimally a secondary alcohol and the furan ring requires substitution at the 5-position for good to excellent yields and good diastereoselectivity. As well as simple furfuryl alcohols, 2-(benzofuran-2-yl)propan-2-ol also cyclizes with a variety of 1,3-dienes.

for references see p 347

Scheme 11 Cycloaddition of Furfuryl Alcohols with Cyclohexa-1,3-diene Mediated by Titanium(IV) Chloride[27]

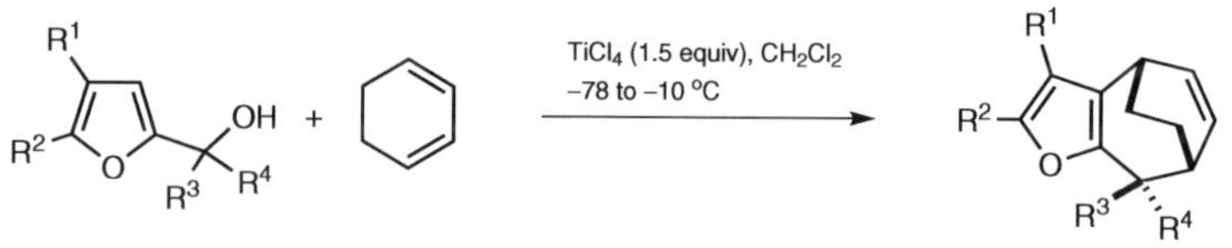

R^1	R^2	R^3	R^4	dr	Yield (%)	Ref
H	H	H	H	–	–[a]	[27]
H	Me	H	H	–	17	[27]
H	Me	Me	H	4:1	82	[27]
H	Me	Et	H	4:1	89	[27]
H	Me	Ph	H	10:1	72	[27]
H	Me	Me	Me	–	96	[27]
(CH=CH)$_2$	Me	Me	–	97	[27]

[a] Polymer obtained.

(4S*,7R*)-2,8,8-Trimethyl-7,8-dihydro-4H-4,7-ethanocyclohepta[b]furan (29, R^1 = H; R^2 = R^3 = R^4 = Me):[27]

A soln of 2-(5-methylfuran-2-yl)propan-2-ol (**28**, R^1 = H; R^2 = R^3 = R^4 = Me; 70 mg, 0.5 mmol) and cyclohexa-1,3-diene (56 μL, 0.6 mmol) in dry CH$_2$Cl$_2$ (2.0 mL) was cooled to −78 °C. At this temperature, a freshly prepared soln of TiCl$_4$ (83 μL, 0.75 mmol) in CH$_2$Cl$_2$ (300 μL) was added dropwise over 1 min. The resulting mixture was stirred and allowed to warm gradually to −10 °C over a 2 h period. Then, sat. aq NaHCO$_3$ (6 mL) was added all at once. After addition of t-BuOMe (10 mL), the layers were separated and the aqueous phase was extracted with t-BuOMe (2 × 5 mL). The combined organic layer was washed with brine (5 mL), dried (Na$_2$SO$_4$), and concentrated under reduced pressure. The residue was dissolved in CH$_2$Cl$_2$ and Florisil (0.5 g) was added. After removal of CH$_2$Cl$_2$ on a rotary evaporator, the resulting powder was placed on a plug of Florisil (1.0 g), and washed with a mixture of petroleum ether (bp 40–60 °C) and t-BuOMe. The filtrate was concentrated under reduced pressure to give pure product as a clear colorless oil; yield: 96%.

2.7.1.3.2 Dienones as Allylic Cation Precursors

The Lewis acid activation of substituted 1,4-dien-3-ones results in the electrocyclization to a cyclic allylic cation, the Nazarov intermediate.[28] West and co-workers have developed novel methodology to capture this cyclic allylic cation with an internal 1,3-diene to afford complex (4 + 3) cycloadducts in good yield and diastereoselectivity.[29] Treatment of a cooled solution of tetraenones **30** in dichloromethane with 1 equivalent of iron(III) chloride results in an intramolecular (4 + 3) cycloaddition to afford the cycloadducts **31A** and **31B** in good to excellent yield (Scheme 12). The diastereoselectivity of the reaction is highly dependent on the tether length between the dienone and the internal 1,3-diene. In the cycloaddition of substrates **30** (R^1 = Me; n = 1) and **30** (R^1 = Ph; n = 1), each with a three-carbon tether, the *endo*-cycloadducts **31A** are slightly preferred over the *exo*-cycloadducts **31B**. When the tether length is extended to four carbon atoms, **30** (R^1 = Me; n = 2) and **30** (R^1 = Ph; n = 2), **31B** is formed as a single diastereomer in good yield. It is interesting to note that substoichiometric amounts of iron(III) chloride (0.2 equiv) could be used in the cycloaddition of **30** (R^1 = Ph; n = 1) with no detrimental effects on yield or diastereoselec-

2.7.1 Carbon-Substituted Allylic Cations in Cycloaddition Reactions

tivity. The generality of this reaction has not been demonstrated. However, an intermolecular version has been reported that is catalyzed by non-metal Lewis acids.[30]

Scheme 12 Intramolecular Cycloaddition of Dienones Mediated by Iron(III) Chloride[29]

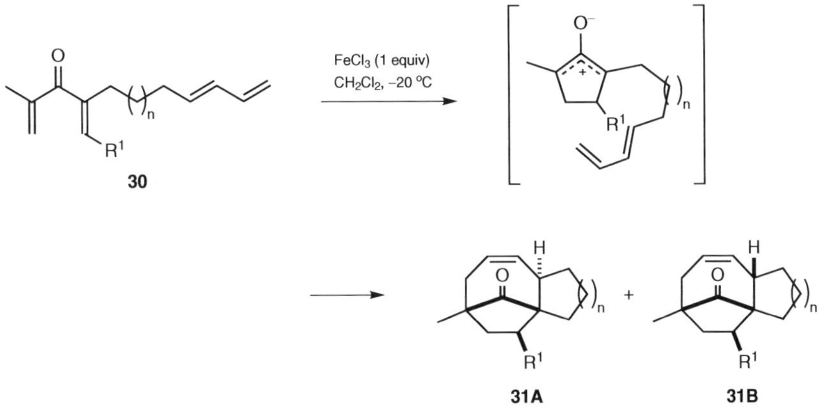

R^1	n	Ratio (31A/31B)	Yield (%)	Ref
Me	1	1.3:1	65	[29]
Ph	1	1.3:1	72	[29]
Me	2	0:1	67	[29]
Ph	2	0:1	75	[29]

(4aR*,5R*,7S*,10aS*)-7-Methyl-5-phenyl-1,3,4,5,6,7,8,10a-octahydro-2H-4a,7-methanobenzo[8]annulen-11-one (31B, R^1 = Ph; n = 2):[29]

To a soln of tetraenone 30 (R^1 = Ph; n = 2; 54 mg, 0.19 mmol) in CH_2Cl_2 (10 mL) at −20 °C was added anhyd $FeCl_3$ (31 mg, 0.19 mmol). After the mixture had been stirred at −20 °C for 30 min, sat. aq $NaHCO_3$ (10 mL) was added and the mixture was allowed to warm to rt. The aqueous phase was extracted with CH_2Cl_2 (2 × 10 mL) and the combined organic phase was washed with sat. aq $NaHCO_3$ (10 mL) and brine (10 mL), dried ($MgSO_4$), and concentrated. The crude product was purified by flash chromatography [silica gel (30 g, 20 × 2 cm column), Et_2O/hexanes 1:40] to give a white solid; yield: 43 mg (79%).

2.7.1.3.3 Alkoxy Allylic Sulfones as Allylic Cation Precursors

It has been established that sulfones, upon Lewis acid activation, behave as good leaving groups.[31] Harmata and co-workers have developed an intramolecular (4 + 3) cycloaddition utilizing this unique characteristic of sulfones. Stirring a solution of allylic sulfone 32 in the presence of titanium(IV) chloride results in the formation of a single diastereomer 33 in high yield (Scheme 13).[32] The reaction requires highly substituted allylic sulfones for good yields of cycloadduct and consequently lacks generality. Interestingly, the initial geometry of the allylic cation is inconsequential; both (E)-34 and (Z)-34 lead to the single diastereomer 33 in 58% yield.[33]

Scheme 13 Intramolecular Cycloaddition of Alkoxy Allylic Sulfones Mediated by Titanium(IV) Chloride[32,33]

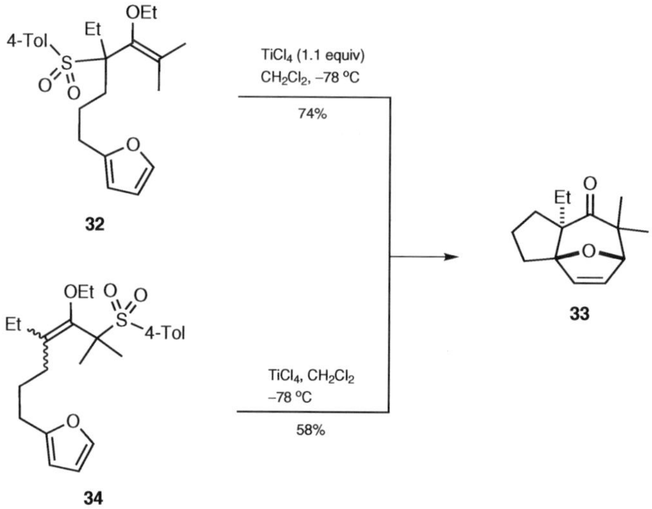

(3aR^*,6S^*,8aR^*)-8a-Ethyl-7,7-dimethyl-1,2,3,6,7,8a-hexahydro-8H-3a,6-epoxyazulen-8-one (33):[32]

To a soln of furan **32** (163 mg, 0.40 mmol) in CH_2Cl_2 (4 mL, 0.1 M) cooled to –78 °C, $TiCl_4$ (49 μL, 0.44 mmol, 1.1 equiv) was added. The reaction was complete immediately. The mixture was quenched by adding H_2O, and the products were extracted into Et_2O (2 × 20 mL). The combined Et_2O layers were washed with H_2O and brine and dried ($MgSO_4$). Solvents were removed under reduced pressure and the residue was purified by flash chromatography (EtOAc/hexanes 1:19) to afford a colorless oil; yield: 74%.

2.7.1.3.4 Allenes as Allylic Cation Precursors

The gold(I)- and platinum(II)-catalyzed (4 + 3) cycloaddition of allenedienes is a relatively recent innovation.[1] The groups of Mascareñas,[34–37] Toste,[38] Fürstner,[39] and Gung[40] have studied the scope of this reaction (Scheme 14). Exposure of a solution of an allenediene **38** to a gold(I) or platinum(II) catalyst results in intramolecular cycloaddition to afford (4 + 2) product **40** along with (4 + 3) products **41** and **42**. Theoretical calculations and experimental results support a mechanism that involves metal activation of the allene forming a metal–allyl cation species, which undergoes (4 + 3) cycloaddition to form metal-carbene intermediate **39** via an extended (*exo*) type transition state.[41,42] Intermediate **39** can undergo a 1,2-alkyl shift resulting in a ring contraction to afford **40**. Alternatively **39** can undergo a 1,2-hydrogen shift or alkyl shift to afford **41** and/or **42**, respectively. The reaction is tolerant of alkyl substitution on both the allene and diene. The nature of the ligand has a dramatic effect on the course of the gold(I)-catalyzed reactions. Donating σ-ligands (as in gold catalyst **36**) are selective for (4 + 3) cycloadducts **41/42** and π-accepting ligands (as in catalyst **35**) favor the formation of (4 + 2) cycloadducts **40**. The reactions proceed with high diastereoselectivity in good to high yield. Under chiral catalysis conditions (e.g., using catalyst **37**), excellent enantioselectivities are achieved. It is also important to note that a transannular (4 + 3) cycloaddition has been accomplished using gold(I) catalysis.[40,43]

2.7.1 Carbon-Substituted Allylic Cations in Cycloaddition Reactions

Scheme 14 Platinum(II)- and Gold(I)-Catalyzed Intramolecular Cycloaddition of Allenedienes[34–37]

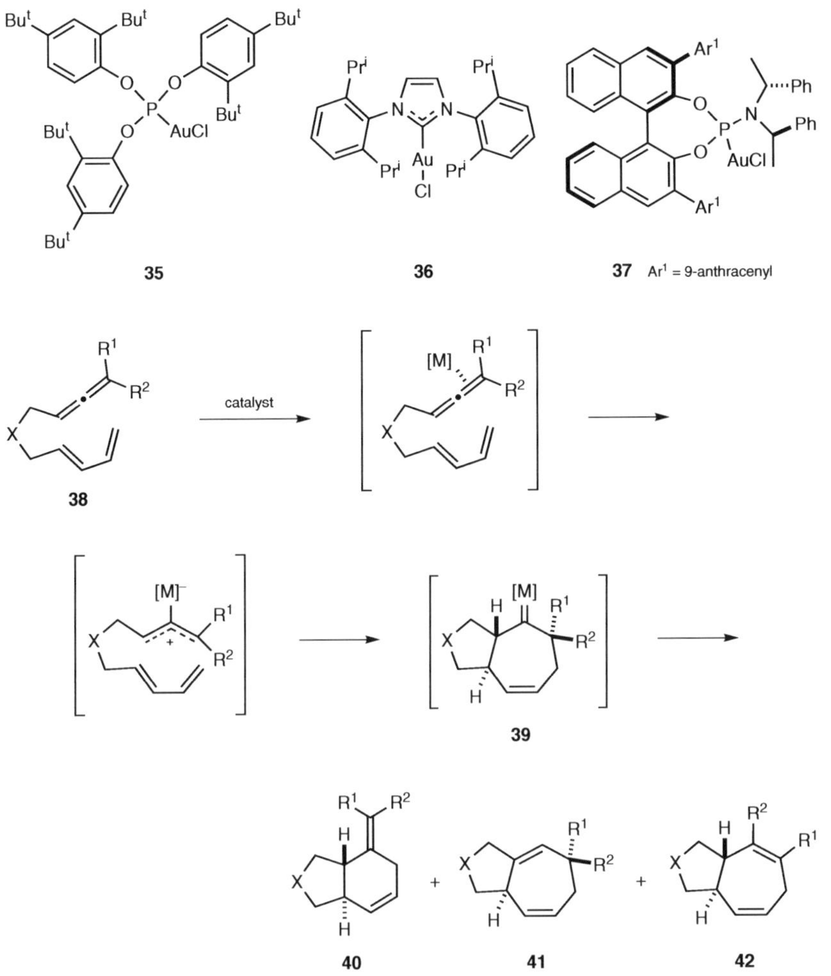

37 Ar¹ = 9-anthracenyl

R¹	R²	X	Catalyst (mol%)	Solvent	Temp (°C)	Ratio (40/41/42)	ee (%)	Yield (%)	Ref
Me	H	C(CO$_2$Et)$_2$	PtCl$_2$ (10)	toluene	110	0:0:1	–	62	[35]
Ph	H	C(CO$_2$Et)$_2$	PtCl$_2$ (10)	toluene	110	0:4:6	–	95	[35]
Ph	H	C(CO$_2$Et)$_2$	PtCl$_2$ (10)	toluene	65	0:3:7	–	87	[35]
Me	Me	C(CO$_2$Me)$_2$	35/AgSbF$_6$ (10)	CH$_2$Cl$_2$	0	13:0:1	–	79	[34]
Me	Me	C(CO$_2$Me)$_2$	36/AgSbF$_6$ (10)	CH$_2$Cl$_2$	rt	0:1:0	–	91	[37]
Ph	H	C(CO$_2$Me)$_2$	37/AgSbF$_6$ (5)	CH$_2$Cl$_2$	−15 to rt	0:1:9	87	64	[36]
Ph	H	NTs	37/AgSbF$_6$ (5)	CH$_2$Cl$_2$	−15 to rt	0:0:1	95	74	[36]

(3aR,8aR)-5-Phenyl-2-tosyl-1,2,3,3a,6,8a-hexahydrocyclohepta[c]pyrrole (42, R¹ = Ph; R² = H; X = NTs); Typical Procedure:[36]

A soln of allenediene **38** (R¹ = Ph; R² = H; X = NTs; 0.32 mmol, 1.00 equiv) in CH$_2$Cl$_2$ (0.8 mL) was added to a suspension of gold complex **37** (17.7 mg, 0.016 mmol, 0.05 equiv) and AgSbF$_6$ (5.4 mg, 0.016 mmol, 0.05 equiv) in CH$_2$Cl$_2$ (1.2 mL) in a dried Schlenk tube under argon at −15 °C. The mixture was allowed to slowly reach rt and, after completion of the reaction (progress was easily monitored by TLC), it was filtered through a short pad of Florisil, eluting with Et$_2$O. The filtrate was concentrated and purified by flash chromatography (Et$_2$O/hexanes 1:19 to 1:9) to afford pure cycloadduct as a white solid; yield: 74%; 95% ee [by HPLC analysis on Chiralpak IA (hexane/iPrOH 97:3, 0.5 mL/min].

2.7.1.3.5 Decomposition of Vinyl Diazoacetates

Other metal–carbenoid species have also been used for (4 + 3) cycloadditions. Rhodium-catalyzed decomposition of vinyl diazoacetates in the presence of 1,3-dienes affords seven-membered ring systems. This is a formal (4 + 3) cycloaddition that occurs through a tandem cyclopropanation/Cope rearrangement.[44] While this reaction falls outside the boundaries of a "true" (4 + 3)-cycloaddition reaction, it occurs with high stereocontrol and yield. This has resulted in a general asymmetric synthesis.[45,46] As such, it warrants special consideration.

Treatment of a solution of vinyl diazoacetate **46** and a moderate excess of 1,3-diene **47** with a catalytic amount of rhodium carboxylate **43** [Rh$_2${(S)-DOSP}$_4$] results in the enantioselective formation of cycloheptadienes **49A** or **49B** (Scheme 15). Other rhodium-based chiral catalysts such as **44** [Rh$_2${(S)-PTAD}$_4$] are also equally effective. The reaction mechanism involves initial cyclopropanation to give divinylcyclopropane intermediate **48**, which after Cope rearrangement affords the product. The reaction occurs with high diastereoselectivity and enantioselectivity and with high regioselectivity, which is dependent on both reactant and catalyst structure. Substitution on both reactants is well tolerated. Cyclic dienes such as furan,[47,48] cyclopentadiene,[45] and N-substituted pyrroles[49] readily undergo reaction as well. Using the chiral catalyst **45** [Rh$_2${(S)-BTPCP}$_4$] results in high diastereoselectivity and enantioselectivity but complete reversal of the commonly observed regiochemistry, giving product **50**. This advancement greatly broadens the scope of the reaction.

Scheme 15 Rhodium-Catalyzed Formal (4 + 3) Cycloaddition of Vinyl Diazoacetates with 1,3-Dienes[45,46]

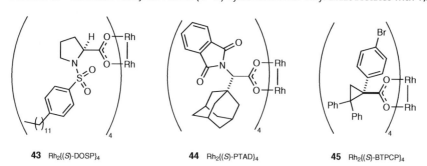

2.7.2 Heteroatom-Substituted Allylic Cations in Cycloaddition Reactions

R^1	R^2	R^3	R^4	R^5	Catalyst	Solvent	Temp (°C)	Ratio (49A/49B/50)	ee (%)	Yield (%)	Ref
H	Me	H	Ph	H	43	hexanes	−78	1:0:0	98	87	[45]
H	Ph	H	Ph	H	43	hexanes	−78	1:0:0	98	83	[45]
H	CH=CHPh	Me	H	H	43	hexanes	−78	0:1:0	98	62	[45]
H	CH=CHPh	H	Me	H	43	hexanes	−78	1:0:0	95	82	[45]
H	Ph	H	OTBDMS	H	43	hexanes	−78	1:0:0	93	63	[45]
OTBDMS	H	H	Me	OTIPS	45	pentane	rt	5:0:95	96	59	[46]

Methyl Cyclohepta-1,5-dienecarboxylates 49; General Procedure:[45]
A soln of vinyldiazoacetate **46** (0.5–1.5 mmol) in dry hexanes (20 mL) was added dropwise over 0.5–2 h to a 100-mL, oven-dried Kjeldahl flask containing a stirred soln of Rh$_2${(S)-DOSP}$_4$ (**43**; 0.01 equiv) and the diene **47** (5–16 equiv) in dry hexane (30 mL) cooled to −78 °C. After the addition was complete, the mixture was maintained at −78 °C for an additional 36–48 h and then slowly warmed to rt. The solvent was removed under reduced pressure, and the residue was purified by flash chromatography (silica gel).

2.7.2 Heteroatom-Substituted Allylic Cations in Cycloaddition Reactions

The use of a heteroatom, particularly nitrogen, oxygen, or sulfur, to stabilize an allylic cation for use in (4 + 3) cycloadditions affords several advantages. The heteroatom facilitates the ionization process, resulting in milder conditions to produce the allylic cation. The heteroatom stabilizes the resultant allylic cation via resonance, providing access to allylic cations that are not readily accessible or are inaccessible with the unsubstituted version of the precursor. Lastly, the heteroatom increases the complexity of the cycloadduct products, providing additional functionality, which can be used for further elaboration. It is important to mention that the generation of halogen-substituted allylic cations is commonly performed under solvolytic conditions (rather than metal mediation/catalysis) due to the reactive nature of the precursors. Consequently, it is not discussed here other than to mention that halogen-substituted allylic cations, particularly oxyallylic cations, are efficient dienophiles in (4 + 3) cycloaddition.[2]

2.7.2.1 Nitrogen-Stabilized Allylic Cations

Nitrogen-substituted allylic cations, or vinyliminium ions, have become efficient dienophiles in diastereoselective (4 + 3)-cycloaddition reactions since the first report over twenty years ago.[50] Since that time, several solvolytic, non-metal-catalyzed, and metal-catalyzed methods for the generation of nitrogen-substituted allylic cations for (4 + 3)-cycloaddition reactions have been reported.[51] A selection of synthetically useful metal-catalyzed processes are discussed below.

2.7.2.1.1 Allenamides as Vinyliminium Ion Precursors

For nearly fifteen years, Hsung and co-workers have been developing methodology employing allenamides as nitrogen-substituted allylic cation precursors for both inter- and intramolecular (4 + 3)-cycloaddition reactions.[52] These reactions in general afford cycloaddition products in moderate to good yields and with good diastereoselectivity. From various experimental and theoretical studies, a predictive model for the regioselectivity and stereoselectivity in (4 + 3)-cycloaddition reactions has been developed.[53–55] The regioselectivity and stereoselectivity of the reaction is influenced by the stereoelectronic effects of both the allylic cation and the diene. As an example, treatment of a solution of chiral allenamide **51** and zinc(II) chloride with excess dimethyldioxirane (DMDO) generates a chiral allene oxide **52**, which opens to the nitrogen-stabilized allylic cation and is trapped, in this case by 2-methylfuran, to provide cycloadducts **54/56** in 47% yield in a ratio of 80:20, respectively (Scheme 16).[54] The major product is formed from a compact (*endo*) approach of 2-methylfuran on the most sterically hindered face of the allylic cation, the face of the phenyl group on the oxazolidinone, which is stabilized via a CH–π interaction. This results in the nitrogen of the allylic cation and the methyl substituent of the furan being in a *syn* relationship, hence a *syn/endo* transition state **53**. The minor diastereomer arises from the attack of the opposite face of 2-methylfuran in an *endo* approach on the allylic cation, resulting in an *anti* relationship between the diene substituent and the nitrogen atom (*anti/endo* transition state **55**). As shown in Table 1, cycloadditions of substituted furans with small electron-donating substituents (e.g., Me) at position 2 prefer to proceed through transition states of type **53** (entry 1). Reactions with small electron-withdrawing substituents (e.g., CO_2Me) at position 2 of the furan prefer to proceed through transition states of type **57** (*syn/endo*) leading to cycloadducts of type **58** preferentially (entry 2). When the furan is substituted at position 3, reactions prefer to proceed via transition states of type **55** (*anti/endo*) in order to maintain CH–π interaction, regardless of the electronic nature of the substituent (entries 3 and 4). Furans that are 2,3- or 2,5-disubstituted (entries 5–7) proceed exclusively via transition states of type **57** (*syn/endo*), regardless of the electronic nature of the substituents. It is interesting that no reaction affords the cycloadduct **60** that is derived from *anti/endo* transition state **59**. Walters and co-workers have also used chiral nitrogen-substituted allylic cations generated from oxazolidinone-substituted α-bromo ketones for (4 + 3)-cycloaddition reactions with comparable selectivities and yields.[56]

2.7.2 Heteroatom-Substituted Allylic Cations in Cycloaddition Reactions

Scheme 16 Zinc-Catalyzed Cycloaddition of an Allenamide with Substituted Furans and Transition States Potentially Involved in This Reaction[54,55]

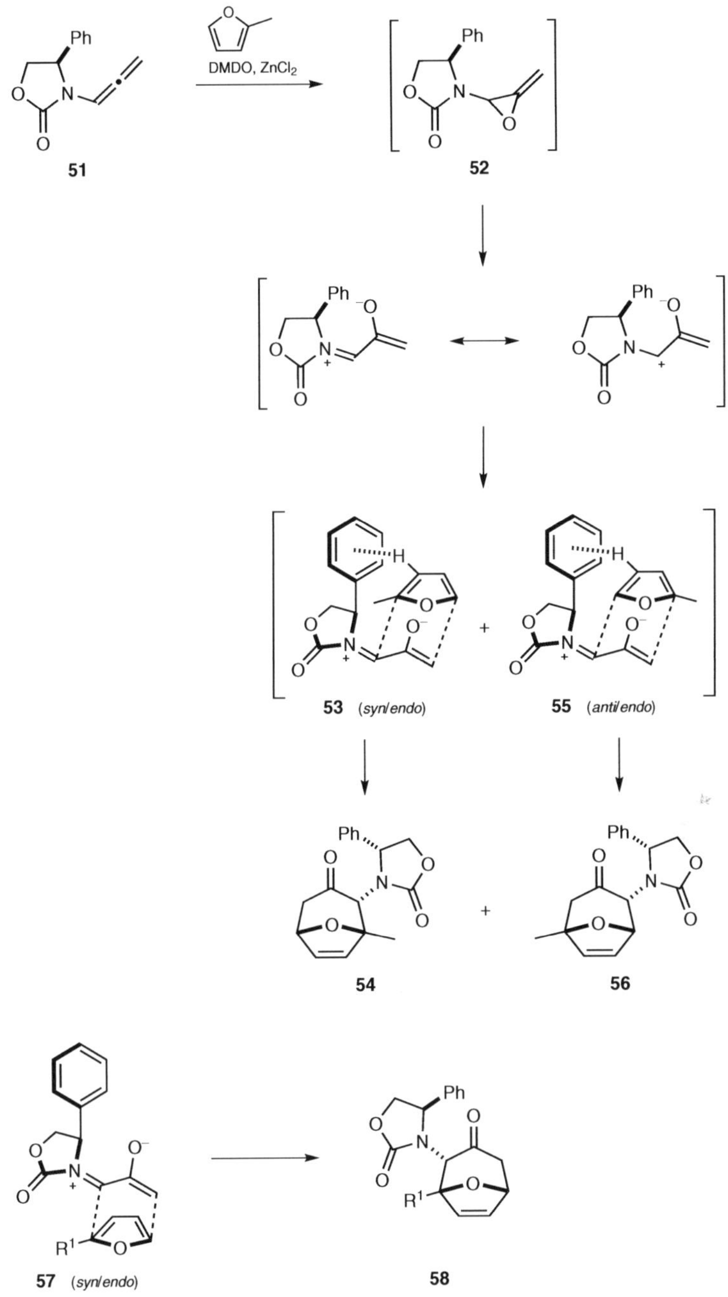

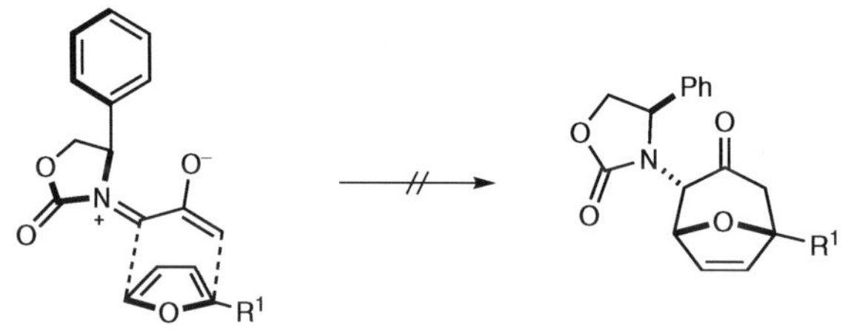

59 (anti/endo) → **60**

Table 1 Zinc-Catalyzed Cycloaddition of Allenamides with Substituted Furans[54,55]

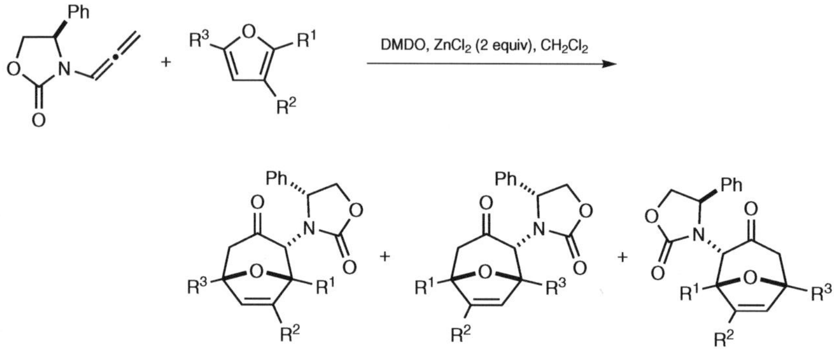

Entry	Starting Materials		Product(s)	Yield (%)	Ref
	Allenamide	Furan			
1	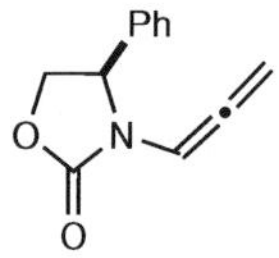 Ph-oxazolidinone allenamide	 2-methylfuran	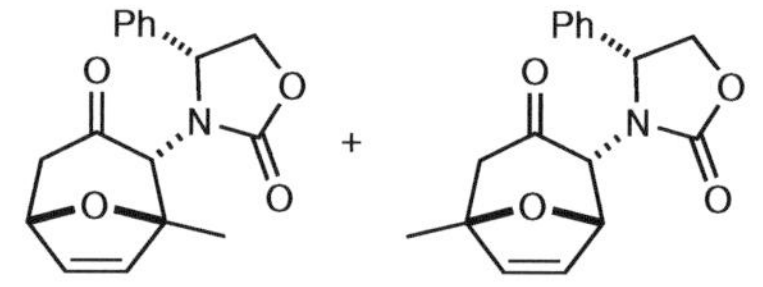 80:20	47	[54]
2	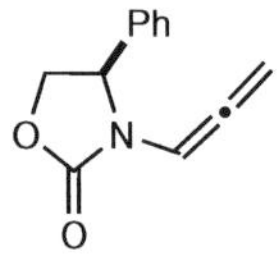 Ph-oxazolidinone allenamide	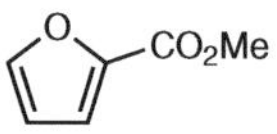 methyl furan-2-carboxylate	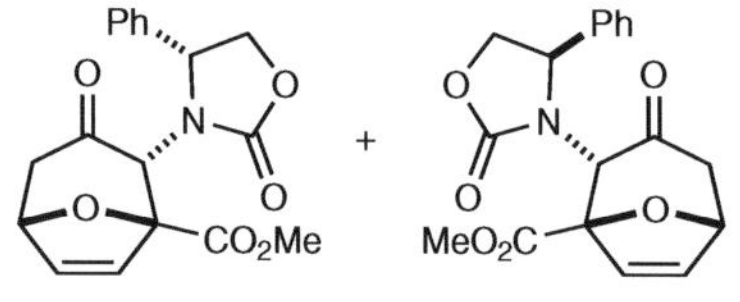 30:70	51	[54]
3	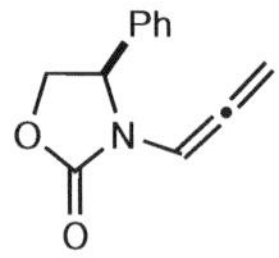 Ph-oxazolidinone allenamide	 3-methylfuran	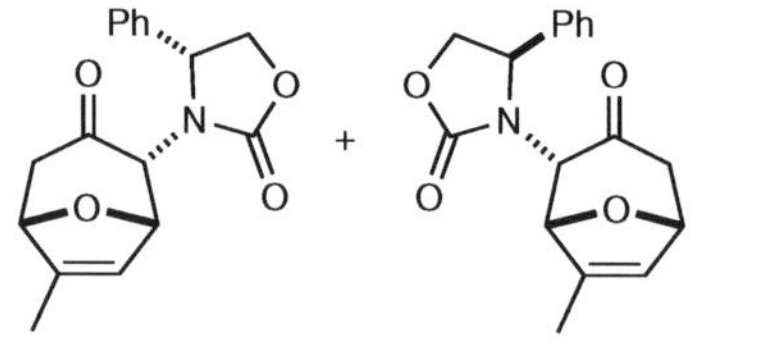 88:12	62	[54]

Conditions: DMDO, ZnCl$_2$ (2 equiv), CH$_2$Cl$_2$

2.7.2 Heteroatom-Substituted Allylic Cations in Cycloaddition Reactions 337

Table 1 (cont.)

Entry	Starting Materials		Product(s)	Yield (%)	Ref
	Allenamide	Furan			
4	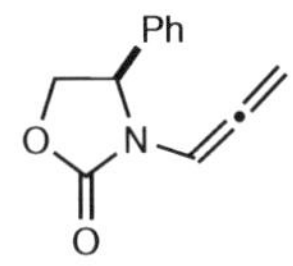 (Ph-oxazolidinone allenamide)	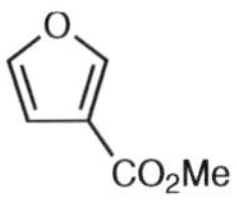 (furan-CO₂Me)	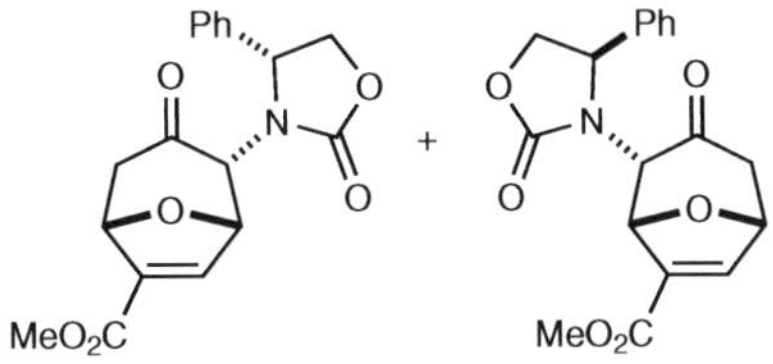 (two bicyclic adducts, 90:10)	60	[54]
5	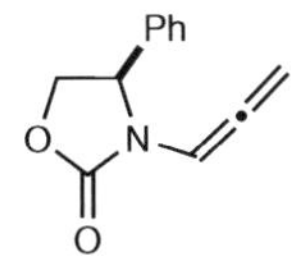 (Ph-oxazolidinone allenamide)	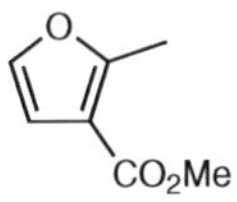 (methylfuran-CO₂Me)	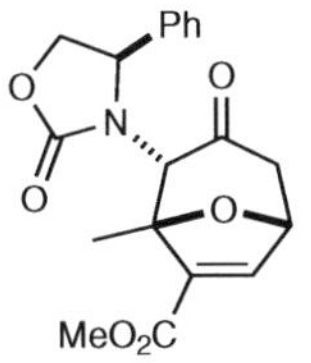 (bicyclic adduct)	60	[55]
6	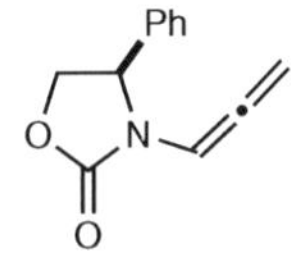 (Ph-oxazolidinone allenamide)	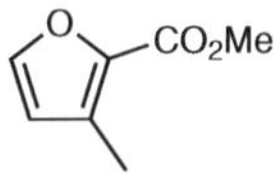 (methylfuran-CO₂Me isomer)	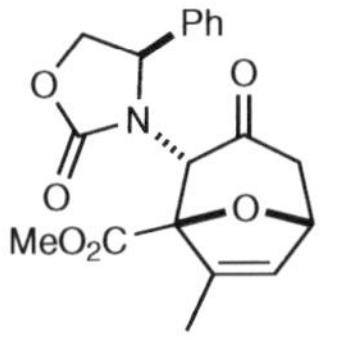 (bicyclic adduct)	67	[55]
7	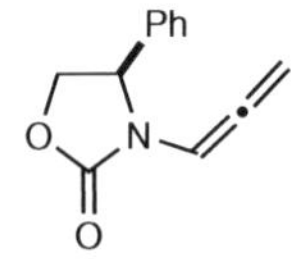 (Ph-oxazolidinone allenamide)	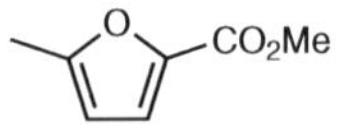 (methylfuran-CO₂Me isomer)	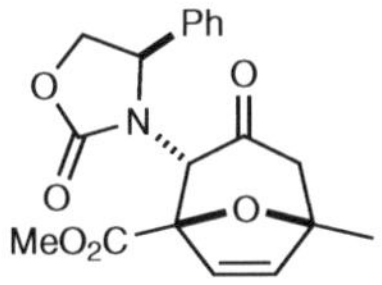 (bicyclic adduct)	40	[55]

Methyl (1R,2S,5R)-7-Methyl-3-oxo-2-[(R)-2-oxo-4-phenyloxazolidin-3-yl]-8-oxabicyclo-[3.2.1]oct-6-ene-1-carboxylate (Table 1, Entry 6):[55]

To a soln of (R)-4-phenyl-3-(propa-1,2-dien-1-yl)oxazolidin-2-one (0.5 mmol) in CH_2Cl_2 (0.1 M) were added methyl 3-methylfuran-2-carboxylate (3–9 equiv) and powdered 4-Å molecular sieves (0.5 g). The mixture was cooled to −78 °C, and 1.0 M $ZnCl_2$ in Et_2O (2.0 equiv) was added. A chilled soln (−78 °C) of dimethyldioxirane (DMDO; 4.0–6.0 equiv) in acetone was then added via syringe pump over 3–4 h. The syringe pump was cooled using dry ice over the entire addition time. After the addition, the mixture was stirred for another 14 h. The reaction was then quenched with sat. aq $NaHCO_3$ and the mixture was filtered through Celite, concentrated under reduced pressure, and extracted with CH_2Cl_2 (4 × 20 mL). The combined organic layers were dried (Na_2SO_4) and concentrated under reduced pressure. The crude residue was purified by column chromatography (silica gel, EtOAc/hexane 1:9 to 3:1 gradient) to afford the cycloadduct as a single diastereomer; yield: 67%.

2.7.2.1.2 Indole-3-methanols as Allylic Cation Precursors

In a strategy similar to that developed using furfuryl alcohols as cation precursors (see Section 2.7.1.3.1),[27] the groups of Wu and Li have recently reported the generation of nitrogen-stabilized allylic cations from indole-3-methanols and their use in (4 + 3)-cycloaddition reactions under metal-catalyzed conditions.[57,58] A creative three-component process has been developed where the indole-3-methanol **62** is prepared in situ and then undergoes ionization to the nitrogen-substituted allylic cation **63**. The nitrogen-substituted allylic cation is ultimately trapped by a 1,3-diene to afford cycloadducts **64A/64B** (Scheme 17). The reaction is performed at room temperature and proceeds with good diastereoselectivity and yield. The reaction also appears to be quite broad in scope. Indoles and aldehydes containing both electron-donating and electron-withdrawing groups are effective substrates. A variety of cyclic and acyclic dienes are also suitable for the reaction. It is also noteworthy that cyclic ketones and symmetrical ketals are effective carbonyl components in the reaction.

Scheme 17 Gallium-Catalyzed Three-Component (4 + 3)-Cycloaddition Reactions[57]

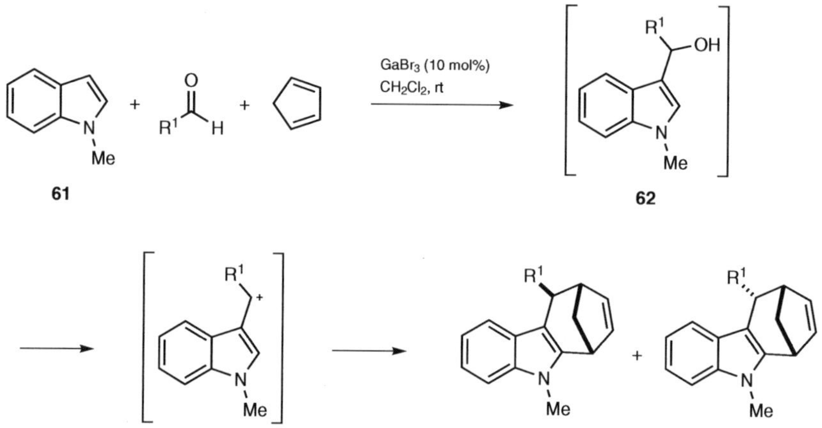

R^1	Yield (%)	Ratio (**64A/64B**)	Ref
Ph	94	4:1	[57]
4-MeOC$_6$H$_4$	75	10:1	[57]
4-F$_3$CC$_6$H$_4$	85	2:1	[57]
2-furyl	83	>10:1	[57]
2-thienyl	83	10:1	[57]
Et	82	1:1	[57]

(6R*,9S*,10R*)-5-Methyl-10-phenyl-5,6,9,10-tetrahydro-6,9-methanocyclohepta[b]indole (64A, R^1 = Ph):[57]

To a round-bottomed flask charged with indole **61** (19.7 mg, 0.150 mmol), PhCHO (31.6 mg, 0.298 mmol), and cyclopentadiene (49.6 mg, 0.750 mmol) was added CH$_2$Cl$_2$ (1.0 mL), followed by the addition of GaBr$_3$ (4.6 mg, 0.015 mmol). The mixture was stirred at rt until the reaction was judged complete, as determined by TLC analysis. The volatiles were removed under reduced pressure and the residue was purified by chromatography (silica gel, EtOAc/hexanes) to afford a colorless oil; yield: 40.4 mg (94%). The NMR spectrum of the crude reaction mixture showed a 4:1 mixture of the two diastereomers.

2.7.2.2 Oxygen-Stabilized Allylic Cations

The generation and use of oxygen-stabilized allylic cations, or vinyloxocarbenium ions, in (4+3) cycloadditions have earnestly studied since the early 1990s, although the first reports were nearly a decade earlier.[59,60] Of particular interest is the development of chiral oxygen-stabilized cations for use in (4+3) cycloadditions.[61]

2.7.2.2.1 2-Siloxyacroleins as Vinyloxocarbenium Ion Precursors

Independently, the groups of Harmata[62] and Funk[63] have reported the (4+3)-cycloaddition reaction of 2-siloxyacroleins with simple furans and 1,3-dienes under metal catalysis (Table 2). For example, treatment of a solution of 2-(triisopropylsiloxy)acrolein and excess furan in dichloromethane at room temperature with a catalytic amount of scandium(III) trifluoromethanesulfonate results in the formation of the cycloadduct as a single stereoisomer in 90% yield (entry 1). The cycloadduct presumably arises from a compact (*endo*) transition state if the reaction is concerted in nature. The reaction proceeds in moderate to good yield with simple cyclic dienes with moderate diastereoselection (entries 1–3). The reactions of acyclic dienes under these conditions give low yield or no yield due to competing (4+2)-cycloaddition reaction (entry 4). With the conditions developed by Funk and co-workers (use of aluminum-based Lewis acids and low temperature), a variety of acyclic dienes as well as cyclic dienes undergo (4+3) cycloaddition in moderate to good yield and diastereoselection (entries 5 and 6).

Table 2 Metal-Catalyzed (4 + 3)-Cycloaddition Reactions of 2-Siloxyacroleins with Dienes[62,63]

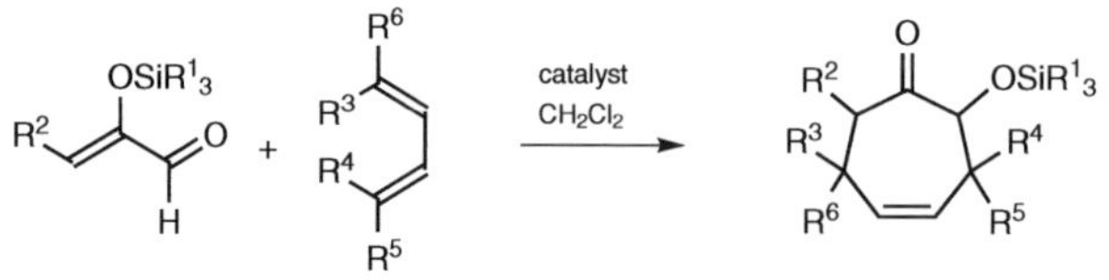

Entry	Starting Materials		Catalyst (mol%)	Product(s)	Yield (%)	Ref
	Acrolein	Diene				
1	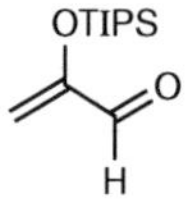 OTIPS acrolein	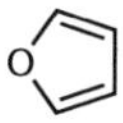 furan	Sc(OTf)$_3$ (10)	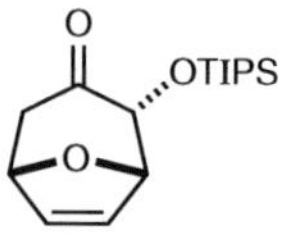 8-oxabicyclic ketone with OTIPS	90[a]	[62]
2	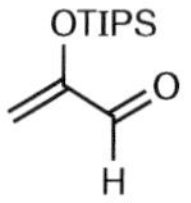 OTIPS acrolein	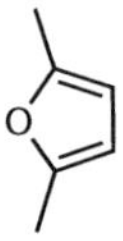 2-methylfuran	Sc(OTf)$_3$ (10)	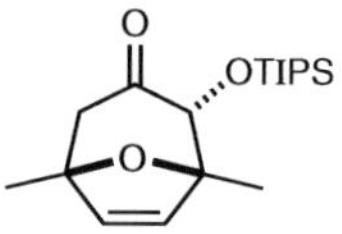 methylated oxabicyclic ketone with OTIPS	55[a]	[62]
3	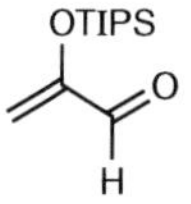 OTIPS acrolein	 cyclopentadiene	Sc(OTf)$_3$ (10)	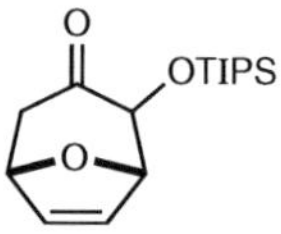 bicyclic ketone with OTIPS	72[a,b]	[62]
4	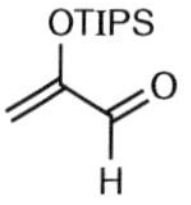 OTIPS acrolein	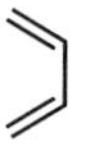 2,3-dimethylbutadiene	Sc(OTf)$_3$ (10)	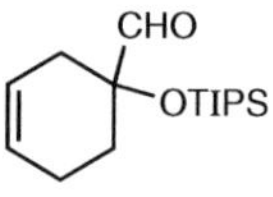 cyclohexene-CHO with OTIPS	67[a]	[62]
5	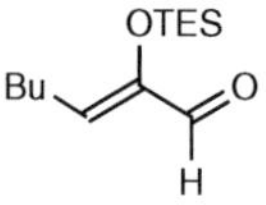 Bu-OTES acrolein	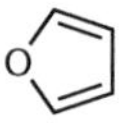 furan	Me$_2$AlCl (100)	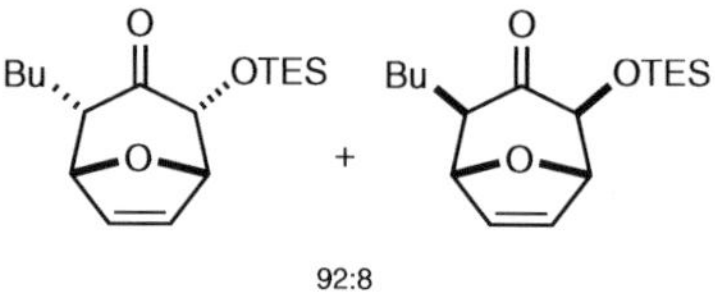 two diastereomers Bu/OTES oxabicyclic ketones (92:8)	78[c]	[63]
6	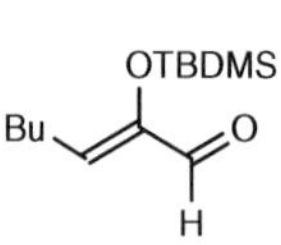 Bu-OTBDMS acrolein	 (E)-pentadiene	EtAlCl$_2$ (100)	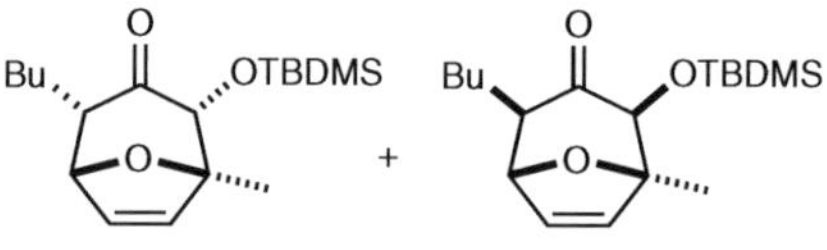 two diastereomers Bu/OTBDMS oxabicyclic ketones (3:97)	50[c]	[63]

[a] Reaction at rt.
[b] dr 3:1.
[c] Reaction at −78 °C.

(1S*,2R*,5S*)-2-(Triisopropylsiloxy)-8-oxabicyclo[3.2.1]oct-6-en-3-one (Table 2, Entry 1):[62]

A 10-mL, oven-dried, round-bottomed flask equipped with a stirrer bar, rubber septum, and N$_2$ inlet was charged with 2-(triisopropylsiloxy)acrolein (0.075 g, 0.328 mmol) and dry CH$_2$Cl$_2$ (1.5 mL). The flask was placed in an ice bath, furan (0.056 g, 0.82 mmol) was added, and the soln was stirred for 5 min. Solid Sc(OTf)$_3$ (Aldrich; 0.016 mg, 0.033 mmol) was added in one portion. The mixture was stirred for 2 h as it slowly warmed to rt. TLC showed complete consumption of the starting material. The mixture was transferred to a separatory funnel with the aid of Et$_2$O (20 mL) and washed with H$_2$O (2 × 15 mL) and brine (15 mL). Drying (Na$_2$SO$_4$) and removal of the solvent under reduced pressure afforded a crude product, which was purified by flash chromatography (silica gel, hexane/EtOAc 20:1) to afford the product as a single stereoisomer; yield: 0.087 g (90%).

2.7.2.2.2 Allylic Acetals as Vinyloxocarbenium Ion Precursors

It has been demonstrated that achiral allylic acetals readily undergo (4+3) cycloaddition with simple dienes.[64] Harmata has expanded on this report by using allylic dioxanes and allylic dioxolanes to access chiral oxocarbenium ion precursors for (4+3)-cycloaddition reactions.[65–67] Treatment of the chiral 1,3-dioxane **65** with titanium(IV) chloride in the presence of excess furan results in the formation of cycloadduct **66** (X=O) in 71% yield with a dr of 17.3:1 (Scheme 18). Several Lewis acids were surveyed, both metal and non-metal, as were several solvents. The high level of diastereoselection has yet to be explained. Interestingly, the cycloaddition of **65** with cyclopentadiene affords a good yield of the cycloadduct **66** (X=CH$_2$), but with nearly complete loss of diastereoselectivity. Allylic dioxolanes also effectively participate in (4+3) cycloadditions in moderate yield and diastereoselectivity.[67] Lastly, acyclic chiral acetals have been employed in (4+3) cycloadditions using non-metal Lewis acids.[68]

Scheme 18 Titanium-Mediated (4+3)-Cycloaddition Reactions of a Chiral Allylic Acetal with Furan and Cyclopentadiene[65,66]

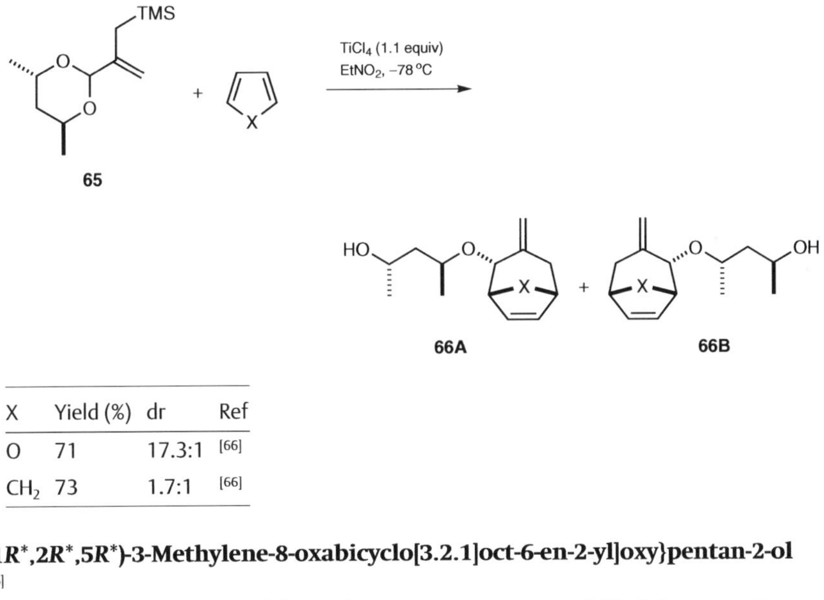

X	Yield (%)	dr	Ref
O	71	17.3:1	[66]
CH$_2$	73	1.7:1	[66]

(2S*,4S*)-4-{[(1R*,2R*,5R*)-3-Methylene-8-oxabicyclo[3.2.1]oct-6-en-2-yl]oxy}pentan-2-ol (66A, X=O):[66]

A round-bottomed flask equipped with a stirrer bar, septum, and N$_2$ inlet was charged with dioxane **65** (1 equiv) followed by EtNO$_2$ to give a 0.2 M soln. The soln was cooled to −78 °C and furan (10 equiv) was added. After stirring for 5 min, the soln was treated with TiCl$_4$ (1.1 equiv). The reaction was complete within 5 min and was quenched with 1 M NaOH and MeOH. The mixture was then diluted with Et$_2$O/H$_2$O (1:1). The aqueous layer was extracted with Et$_2$O (3×). The combined organic phase was washed with 1 M HCl (1×), H$_2$O (3×), and brine (1×), and then dried (Na$_2$SO$_4$) and filtered and the solvent was removed under reduced pressure without heating. The residue was purified by flash chromatography (EtOAc/hexanes 1:1); yield: 71%.

2.7.2.3 Sulfur-Stabilized Allylic Cations

Regarding the use of heteroatom-substituted allylic cations in (4 + 3) cycloadditions, sulfur is easily the least common of the heteroatoms employed. In that context, the most common methods for the generation of sulfur-stabilized allylic cations, or vinylthionium ions, involve non-metal Lewis acid catalysis. However, there are a few reports using metal-mediated processes to generate vinylthionium ions for (4 + 3) cycloadditions.

2.7.2.3.1 Sulfur-Substituted Allylic Acetals as Vinylthionium Ion Precursors

Harmata and co-workers have explored the use of sulfur-substituted allylic acetals in (4 + 3) cycloadditions as an extension of their work on allylic acetals. They reported a limited optimization study of solvent, temperature, and titanium-based Lewis acid on the cycloaddition of the phenylsulfanyl-substituted allylic acetal **67** and furan.[69] Treating a solution of **67** and excess furan with 4 equivalents of isopropoxytitanium(IV) chloride affords cycloadduct **68** in 70% yield as a single diastereomer (Scheme 19). Attenuation of Lewis acid strength appears to be critical to appreciable reaction yield as stronger Lewis acids result in decomposition of the starting materials. Presumably, the resultant allylic cation is stabilized by both sulfur and oxygen.

Scheme 19 Titanium-Catalyzed (4 + 3) Cycloaddition of a Sulfur-Substituted Allylic Acetal with Furan: A Sulfur- and Oxygen-Stabilized Allylic Cation[69]

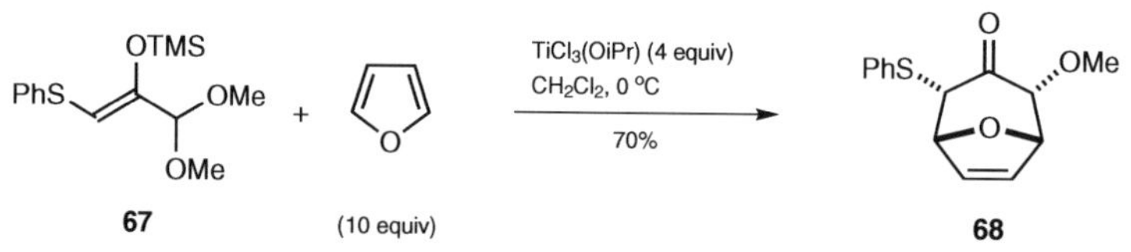

(1S*,2R*,4S*,5R*)-2-Methoxy-4-(phenylsulfanyl)-8-oxabicyclo[3.2.1]oct-6-en-3-one (68):[69]
To a flame-dried, round-bottomed flask equipped with a stirrer bar, septum, and N$_2$ balloon were added CH$_2$Cl$_2$ (0.20 M), silyl enol ether **67** (1.0 equiv), and furan (10.0 equiv). To this cooled (0 °C), colorless soln was added freshly distilled TiCl$_3$(OiPr) (4 equiv). The resulting dark soln was stirred at 0 °C for 10–15 min, and then quenched by the addition of MeOH/HCl (1:1). The mixture was poured into H$_2$O and extracted several times with CH$_2$Cl$_2$. The extracts were washed with brine (1 ×), dried (Na$_2$SO$_4$), and filtered. The filtrate was concentrated under reduced pressure to give a crude oil. The oil was purified by flash chromatography (EtOAc/hexanes) to afford **68** as a single isomer; yield: 70%.

2.7.2.3.2 Sulfur-Substituted Allylic Sulfones as Vinylthionium Ion Precursors

As discussed in Section 2.7.1.3.3, the intramolecular cycloaddition of alkoxy allylic sulfones proceeds in good yield and with good diastereoselectivity.[32] However, the reaction is limited in scope due to the requirement of highly substituted allylic sulfones for good yields of cycloadduct. In order to broaden the scope of this reaction, Harmata studied the use of phenylsulfanyl-substituted alkoxy allylic sulfones.[70,71] Treatment of a solution of (*E*)-**69** in dichloromethane with titanium(IV) chloride results in formation of cycloadduct **70** in 67% yield as a 1:1 mixture of diastereomers which is epimeric at the sulfur-bearing carbon (Scheme 20). Interestingly, treatment of (*Z*)-**69** under the same conditions results in a meager 12% yield of **70**. The authors suggest that the alkoxy and the phenylsulfanyl group behave as a bidentate ligand for the titanium Lewis acid, preventing generation of the vinylthionium ion.

2.7.3 Applications to the Synthesis of Natural Products

Scheme 20 Titanium-Catalyzed (4 + 3)-Cycloaddition Reactions of Sulfur-Substituted Allylic Sulfones[70,71]

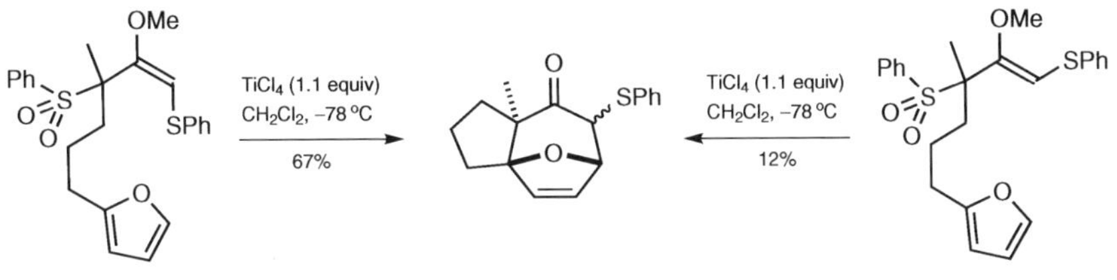

(3aR*,6S*,8aR*)-8a-Methyl-7-(phenylsulfanyl)-1,2,3,6,7,8a-hexahydro-8H-3a,6-epoxyazulen-8-one (70):[70]

A 50-mL, round-bottomed flask equipped with a stirrer bar, septum, and N_2 balloon was charged with dry CH_2Cl_2 (20 mL) and $TiCl_4$ (63 µL, 109 mg, 0.57 mmol). The flask was placed in a dry ice/iPrOH bath and a soln of (E)-**69** (220 mg, 0.5 mmol) in CH_2Cl_2 (2 mL) was added over 7–8 min. This was followed by a CH_2Cl_2 (0.75 mL) rinse. The reaction was monitored by TLC and quenched after 20 min with a mixture of MeOH and 1 M HCl (1:1; 10 mL). The mixture was stirred until clear and EtOAc (100 mL) was used in transferring the mixture to a separatory funnel. The layers were separated and the organic phase was washed with 1 M HCl (2 × 10 mL) and brine (2 × 10 mL). Drying (Na_2SO_4) and removal of solvent under reduced pressure gave a residue, which was purified by HPLC (hexanes/EtOAc 3.5:1) to give a 1:1 mixture of isomers as determined by NMR; yield: 96.2 mg (67%).

2.7.3 Applications of Metal-Catalyzed (4 + 3) Cycloadditions to the Synthesis of Natural Products

A common characteristic among cycloaddition reactions is that they rapidly produce molecular structures of markedly increased complexity from simple starting materials and often with high stereoselectivity. This is a powerful tool that is often brought to bear on tough synthetic problems. That is certainly the case for the (4 + 3)-cycloaddition reaction.[72] The following sections discuss select examples using metal-catalyzed (4 + 3)-cycloaddition reactions as the critical step in the synthesis of natural products.

2.7.3.1 (±)-Frondosin B

The low micromolar interleukin-8 receptor (IL-8) antagonist (+)-frondosin B (**71**) was isolated from a marine sponge in 1997.[73] Its anti-inflammatory activity generated considerable interest in a laboratory preparation of the natural product. Winne and co-workers recently developed a (4 + 3)-cycloaddition route utilizing the furfuryl alcohol methodology developed in their laboratories (see Section 2.7.1.3.1).[74] Treatment of a solution of benzofuranyl alcohol **72** and 2 equivalents of diene **73** in dichloromethane with titanium(IV) chloride results in an inseparable mixture of cycloadducts **74** and **75** in a ratio of 2.5:1 in 26% yield (Scheme 21). Careful analysis of the minor cycloadduct **75** revealed that it was mixture of minor positional alkene isomers with a dr of 2:1. This mixture of **74** and **75** had been previously converted into frondosin B under demethylation reaction conditions.[75] The Davies group has also reported an asymmetric formal synthesis of (+)-frondosin B using a formal (4 + 3) cycloaddition between a vinyl diazoacetate and a diene.[76]

Scheme 21 Titanium-Mediated (4 + 3)-Cycloaddition Route to (±)-Frondosin B[74]

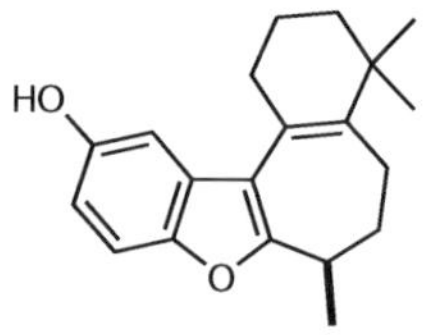

71 (+)-frondosin B

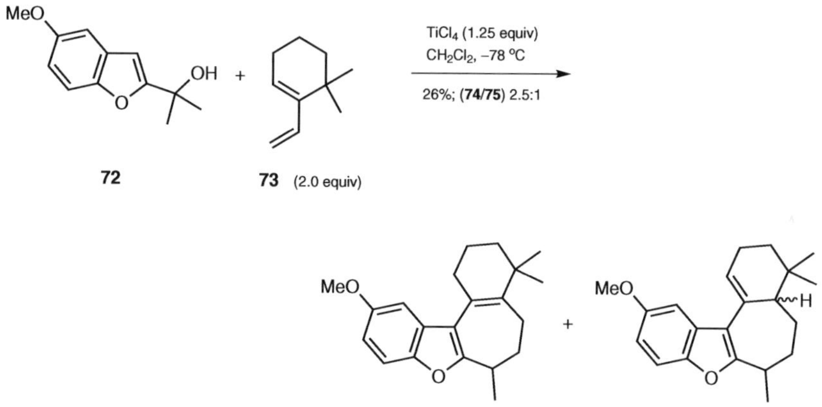

72

73 (2.0 equiv)

TiCl$_4$ (1.25 equiv)
CH$_2$Cl$_2$, −78 °C

26%; (**74/75**) 2.5:1

74

75 dr 2:1

2.7.3.2 (−)-5-*epi*-Vibsanin E

The highly functionalized cycloheptane-containing natural product (−)-5-*epi*-vibsanin E (**76**) was isolated from the flowering shrub *Viburnum awabuki* in 2002.[77] The interesting architecture of this molecule has drawn the interest of many, including the Davies and Williams laboratories. They employed a formal (4 + 3)-cycloaddition reaction between a vinyl diazoacetate and a diene to prepare (−)-5-*epi*-vibsanin E (**76**). Upon treatment with a catalytic amount of rhodium catalyst **44** [Rh$_2${(S)-PTAD}$_4$], a solution of vinyl diazoacetate **77** and a moderate excess of 1,3-diene **78** affords cycloadduct **79** in 65% yield and 90% ee (Scheme 22).[78] Formation of the quaternary center is essential and this allows the controlled formation of the remaining stereocenters. The cycloaddition reaction is performed on a 10-gram scale, providing appreciable material for the several remaining steps to complete the total asymmetric synthesis of (−)-5-*epi*-vibsanin E (**76**).

Scheme 22 Asymmetric, Rhodium-Catalyzed (4 + 3)-Cycloaddition Route to (−)-5-*epi*-Vibsanin E[78]

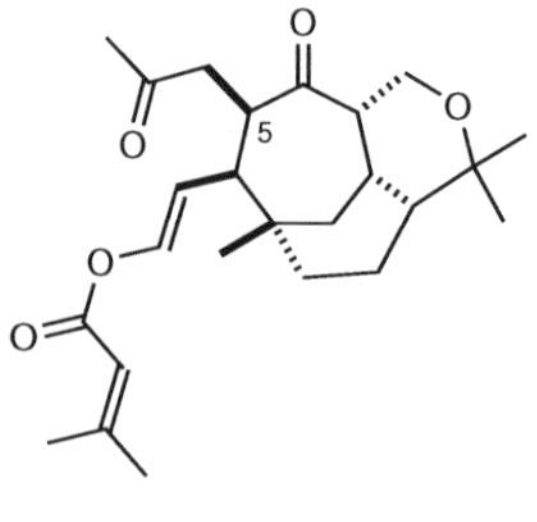

76 (−)-5-*epi*-vibsanin E

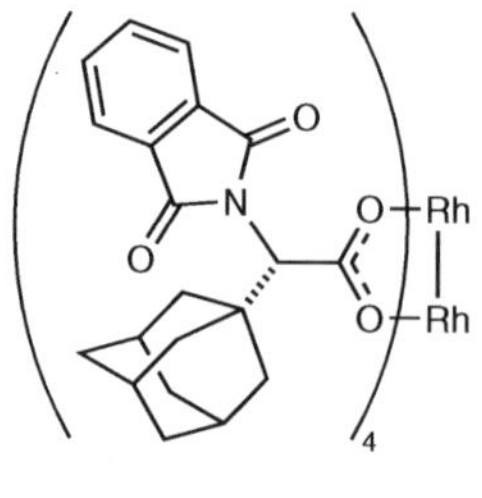

44 Rh$_2${(S)-PTAD}$_4$

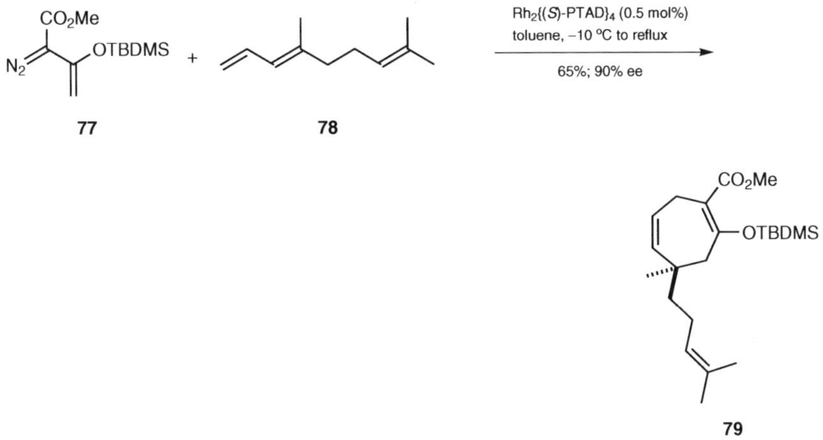

2.7.3 Applications to the Synthesis of Natural Products

2.7.3.3 (±)-Widdrol

The natural product widdrol (**80**), containing a 6,7-fused carbon skeleton, was isolated over fifty years ago.[79] It would have likely been forgotten, but recently widdrol was discovered to exhibit a variety of biological activities, including antifungal and anticancer activity.[80,81] The Harmata group prepared widdrol (**80**) using vinylthionium ions generated from alkoxy-substituted allylic sulfones. The inverse addition of the alkoxy-bearing allylic sulfone **81** to a cold solution of titanium(IV) chloride in dichloromethane generates a sulfur-stabilized allylic cation, which is trapped by the furan to afford cycloadducts **82A** and **82B** in 86% yield (Scheme 23).[82] Practically no diastereoselection is observed for this reaction, which is not unexpected; allylic cations and dienes linked by a four-carbon tether commonly exhibit poor diastereoselection.[83] The cycloadducts were chromatographically separated and cycloadduct **82B**, which has the angular methyl group in the proper orientation, was ultimately transformed into **80**.

Scheme 23 Titanium-Mediated Vinylthionium Intramolecular (4 + 3)-Cycloaddition Route to (±)-Widdrol[82]

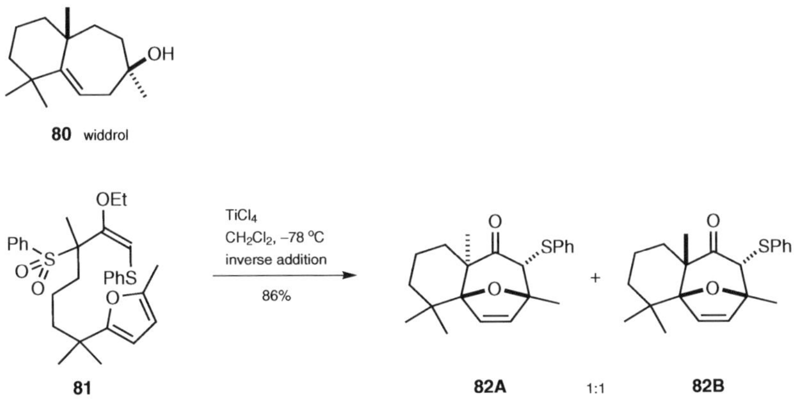

2.7.3.4 (±)-Urechitol A

The trinorsesquiterpenoid urechitol A (**83**) was isolated from the root extract of *Pentalinon andrieuxii*.[84] It is interesting that the people of Yucatan use this root extract to treat parasitic infections; however, **83** appears to be devoid of biological activity.[85] Urechitol A presents several interesting structural features including a highly functionalized and oxidized cycloheptane ring bearing two ether bridges.[84] These synthetic challenges were addressed and overcome by Watanabe and co-workers using an oxygen-stabilized allylic cation that was generated from an acyclic allylic acetal (Scheme 24).[86] The treatment of siloxy allylic acetal **84** and 3-furylmethanol **85** in nitroethane with titanium(IV) chloride results in the formation of cycloadduct **86** in 46% yield. Although the yield is moderate at best, the cycloadduct is formed as a single regioisomer and a single diastereomer resulting from an *endo* transition state. Cycloadduct **86** was then converted into (±)-urechitol A (**83**) using standard chemistry.

Scheme 24 Synthesis of (±)-Urechitol A Using a Vinyloxocarbenium Ion[86]

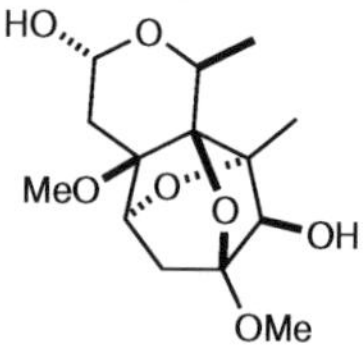

2.7.4 Conclusions and Future Perspectives

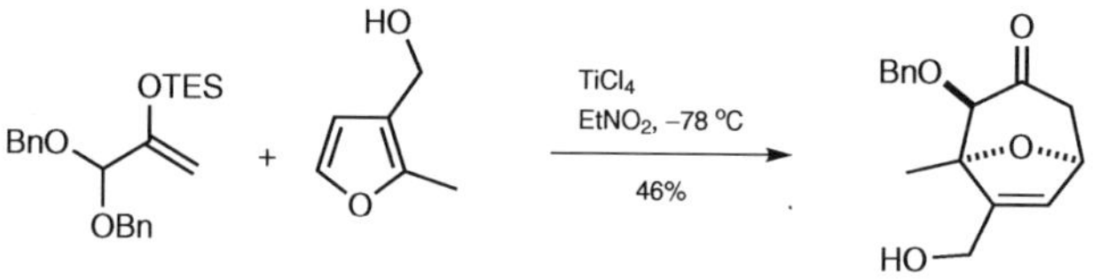

The metal-catalyzed or metal-mediated generation of allylic cations remains a challenge. More emphasis on catalysis, particularly asymmetric catalysis, is necessary. However, in the context of this work, it is not merely the generation of allylic cations that is important, it is the generation of cations that will productively undergo a real or formal (4 + 3) cycloaddition that is vital. This generally means formation of an allylic cation with an electron donor at the 2-position, an entity that is certainly accessible, but whose generation is not trivial. Further complications arise because of mechanistic ambiguities in the "cycloaddition" reactions of such species with dienes. It is no doubt the case that many of these processes, though not all, are certainly stepwise cyclizations, not cycloadditions in a mechanistically rigorous sense. This does not necessarily detract from their synthetic utility, but it does mean that caution is necessary in the design of any process based on this chemistry, particularly when stereocontrol is desired. Notwithstanding these caveats, this type of reaction has been quite successful both from basic and applied perspectives. Catalytic, asymmetric processes need to be developed; chemistry already exists that serves as the basis for investigations in this direction and both inter- and intramolecular processes can be used to investigate such chemistry. Given Nature's use of allylic cations, it may be possible to apply biocatalysis to productively effect asymmetric cycloadditions catalytically using intact microbes or biomolecules derived therefrom.

References

[1] Fernández, I.; Mascareñas, J. L., *Org. Biomol. Chem.*, (2012) **10**, 699.
[2] Harmata, M., *Chem. Commun. (Cambridge)*, (2010) **46**, 8904.
[3] Harmata, M., *Chem. Commun. (Cambridge)*, (2010) **46**, 8886.
[4] Harmata, M., *Acc. Chem. Res.*, (2001) **34**, 595.
[5] Harmata, M., *Tetrahedron*, (1997) **53**, 6235.
[6] Fort, A. W., *J. Am. Chem. Soc.*, (1962) **84**, 2620.
[7] Fort, A. W., *J. Am. Chem. Soc.*, (1962) **84**, 4979.
[8] Battiste, M. A.; Pelphrey, P. M.; Wright, D. L., *Chem.–Eur. J.*, (2006) **12**, 3438.
[9] Mann, J., *Tetrahedron*, (1986) **42**, 4611.
[10] Ashcroft, M. R.; Hoffmann, H. M. R., *Org. Synth., Coll. Vol. VI*, (1978), 512.
[11] Costa, A. V.; Barbosa, L. C. de A.; Demuner, A. J.; Silva, A. A., *J. Agric. Food Chem.*, (1999) **47**, 4807.
[12] Hoffmann, H. M. R.; Wagner, D.; Wartchow, R., *Chem. Ber.*, (1990) **123**, 2131.
[13] Goodman, J. M.; Vinter, J. G.; Hoffmann, H. M. R., *Tetrahedron Lett.*, (1995) **36**, 7757.
[14] Vinter, J. G.; Hoffmann, H. M. R., *J. Am. Chem. Soc.*, (1974) **96**, 5466.
[15] Montaña, A. M.; Grima, P. M., *Tetrahedron Lett.*, (2001) **42**, 7809.
[16] Montaña, A. M.; Grima, P. M., *Synth. Commun.*, (2003) **33**, 265.
[17] Montaña, A. M.; Grima, P. M., *J. Chem. Educ.*, (2000) **77**, 754.
[18] Smith, R. D.; Simmons, H. E., *Org. Synth.*, (1961) **41**, 72.
[19] Takaya, H.; Makino, S.; Hayakawa, Y.; Noyori, R., *J. Am. Chem. Soc.*, (1978) **100**, 1765.
[20] Lautens, M.; Aspiotis, R.; Colucci, J., *J. Am. Chem. Soc.*, (1996) **118**, 10930.
[21] Shimizu, N.; Tsuno, Y., *Chem. Lett.*, (1979) **8**, 103.
[22] Shimizu, N.; Tanaka, M.; Tsuno, Y., *J. Am. Chem. Soc.*, (1982) **104**, 1330.
[23] Sakurai, H.; Shirahata, A.; Hosomi, A., *Angew. Chem. Int. Ed. Engl.*, (1979) **18**, 163.
[24] Oh, J.; Ziani-Cherif, C.; Choi, J.-R.; Cha, J. K., *Org. Synth., Coll. Vol. X*, (2004), 584.
[25] Kim, H.; Ziani-Cherif, C.; Oh, J.; Cha, J. K., *J. Org. Chem.*, (1995) **60**, 792.
[26] Kende, A. S.; Huang, H., *Tetrahedron Lett.*, (1997) **38**, 3353.
[27] Winne, J. M.; Catak, S.; Waroquier, M.; Van Speybroeck, V., *Angew. Chem. Int. Ed.*, (2011) **50**, 11990.
[28] Grant, T. N.; Rieder, C. J.; West, F. G., *Chem. Commun. (Cambridge)*, (2009), 5676.
[29] Wang, Y.; Arif, A. M.; West, F. G., *J. Am. Chem. Soc.*, (1999) **121**, 876.
[30] Wang, Y.; Schill, B. D.; Arif, A. M.; West, F. G., *Org. Lett.*, (2003) **5**, 2747.
[31] Trost, B. M.; Ghadiri, M. R., *J. Am. Chem. Soc.*, (1984) **106**, 7260.
[32] Harmata, M.; Gamlath, C. B., *J. Org. Chem.*, (1988) **53**, 6154.
[33] Harmata, M.; Gamlath, C. B.; Barnes, C. L., *Tetrahedron Lett.*, (1990) **31**, 5981.
[34] Alonso, I.; Trillo, B.; López, F.; Montserrat, S.; Ujaque, G.; Castedo, L.; Lledós, A.; Mascareñas, J. L., *J. Am. Chem. Soc.*, (2009) **131**, 13020.
[35] Trillo, B.; López, F.; Gulías, M.; Castedo, L.; Mascareñas, J. L., *Angew. Chem. Int. Ed.*, (2008) **47**, 951.
[36] Alonso, I.; Faustino, H.; López, F.; Mascareñas, J. L., *Angew. Chem. Int. Ed.*, (2011) **50**, 11496.
[37] Trillo, B.; López, F.; Montserrat, S.; Ujaque, G.; Castedo, L.; Lledós, A.; Mascareñas, J. L., *Chem–Eur. J.*, (2009) **15**, 3336.
[38] Mauleón, P.; Zeldin, R. M.; González, A. Z.; Toste, F. D., *J. Am. Chem. Soc.*, (2009) **131**, 6348.
[39] Teller, H.; Flügge, S.; Goddard, R.; Fürstner, A., *Angew. Chem. Int. Ed.*, (2010) **49**, 1949.
[40] Gung, B. W.; Craft, D. T., *Tetrahedron Lett.*, (2009) **50**, 2685.
[41] Fernández, I.; Cossío, F. P.; de Cózar, A.; Lledós, A.; Mascareñas, J. L., *Chem.–Eur. J.*, (2010) **16**, 12147.
[42] Montserrat, S.; Alonso, I.; López, F.; Mascareñas, J. L.; Lledós, A.; Ujaque, G., *Dalton Trans.*, (2011) **40**, 11095.
[43] Gung, B. W.; Craft, D. T.; Bailey, L. N.; Kirschbaum, K., *Chem.–Eur. J.*, (2010) **16**, 639.
[44] Davies, H. M. L., *Tetrahedron*, (1993) **49**, 5203.
[45] Davies, H. M. L.; Stafford, D. G.; Doan, B. D.; Houser, J. H., *J. Am. Chem. Soc.*, (1998) **120**, 3326.
[46] Guzmán, P. E.; Lian, Y.; Davies, H. M. L., *Angew. Chem. Int. Ed.*, (2014) **53**, 13083.
[47] Krainz, T.; Chow, S.; Korica, N.; Bernhardt, P. V.; Boyle, G. M.; Parsons, P. G.; Davies, H. M. L.; Williams, C. M., *Eur. J. Org. Chem.*, (2016), 41.
[48] Davies, H. M. L.; Ahmed, G.; Churchill, M. R., *J. Am. Chem. Soc.*, (1996) **118**, 10774.
[49] Reddy, R. P.; Davies, H. M. L., *J. Am. Chem. Soc.*, (2007) **129**, 10312.
[50] Walters, M. A.; Arcand, H. R.; Lawrie, D. J., *Tetrahedron Lett.*, (1995) **36**, 23.

[51] Lohse, A. G.; Hsung, R. P., *Chem.–Eur. J.*, (2011) **17**, 3812.
[52] Xiong, H.; Hsung, R. P.; Berry, C. R.; Rameshkumar, C., *J. Am. Chem. Soc.*, (2001) **123**, 7174.
[53] Lohse, A. G.; Krenske, E. H.; Antoline, J. E.; Houk, K. N.; Hsung, R. P., *Org. Lett.*, (2010) **12**, 5506.
[54] Antoline, J. E.; Krenske, E. H.; Lohse, A. G.; Houk, K. N.; Hsung, R. P., *J. Am. Chem. Soc.*, (2011) **133**, 14 443.
[55] Du, Y.; Krenske, E. H.; Antoline, J. E.; Lohse, A. G.; Houk, K. N.; Hsung, R. P., *J. Org. Chem.*, (2013) **78**, 1753.
[56] MaGee, D. I.; Godineau, E.; Thornton, P. D.; Walters, M. A.; Sponholtz, D. J., *Eur. J. Org. Chem.*, (2006), 3667.
[57] Han, X.; Li, H.; Hughes, R. P.; Wu, J., *Angew. Chem. Int. Ed.*, (2012) **51**, 10 390.
[58] Gong, W.; Liu, Y.; Zhang, J.; Jiao, Y.; Xue, J.; Li, Y., *Chem.–Asian J.*, (2013) **8**, 546.
[59] Sasaki, T.; Ishibashi, Y.; Ohno, M., *Tetrahedron Lett.*, (1982) **23**, 1693.
[60] Blackburn, C.; Childs, R. F.; Kennedy, R. A., *Can. J. Chem.*, (1983) **61**, 1981.
[61] Harmata, M.; Rashatasakhon, P., *Tetrahedron*, (2003) **59**, 2371.
[62] Harmata, M.; Sharma, U., *Org. Lett.*, (2000) **2**, 2703.
[63] Aungst, R. A., Jr.; Funk, R. L., *Org. Lett.*, (2001) **3**, 3553.
[64] Murray, D. H.; Albizati, K. F., *Tetrahedron Lett.*, (1990) **31**, 4109.
[65] Harmata, M.; Jones, D. E., *J. Org. Chem.*, (1997) **62**, 1578.
[66] Harmata, M.; Jones, D. E.; Kahraman, M.; Sharma, U.; Barnes, C. L., *Tetrahedron Lett.*, (1999) **40**, 1831.
[67] Harmata, M.; Brackley, J. A., III; Barnes, C. L., *Tetrahedron Lett.*, (2006) **47**, 8151.
[68] Stark, C. B. W.; Eggert, U.; Hoffmann, H. M. R., *Angew. Chem. Int. Ed.*, (1998) **37**, 1266.
[69] Harmata, M.; Carter, K. W., *ARKIVOC*, (2002), viii, 62.
[70] Harmata, M.; Fletcher, V. R.; Claassen, R. J., II, *J. Am. Chem. Soc.*, (1991) **113**, 9861.
[71] Harmata, M.; Kahraman, M., *Tetrahedron Lett.*, (1998) **39**, 3421.
[72] Jones, D. E.; Harmata, M., In *Methods and Applications of Cycloaddition Reactions in Organic Syntheses*, Nishiwaki, N., Ed.; Wiley: Hoboken, NJ, (2013); p 599.
[73] Patil, A. D.; Freyer, A. J.; Killmer, L.; Offen, P.; Carte, B.; Jurewicz, A. J.; Johnson, R. K., *Tetrahedron*, (1997) **53**, 5047.
[74] Laplace, D. R.; Verbraeken, B.; Van Hecke, K.; Winne, J. M., *Chem.–Eur. J.*, (2014) **20**, 253.
[75] Inoue, M.; Carson, M. W.; Frontier, A. J.; Danishefsky, S. J., *J. Am. Chem. Soc.*, (2001) **123**, 1878.
[76] Olson, J. P.; Davies, H. M. L., *Org. Lett.*, (2008) **10**, 573.
[77] Fukuyama, Y.; Minami, H.; Matsuo, A.; Kitamura, K.; Akizuki, M.; Kubo, M.; Kodama, M., *Chem. Pharm. Bull.*, (2002) **50**, 368.
[78] Schwartz, B. D.; Denton, J. R.; Lian, Y.; Davies, H. M. L.; Williams, C. M., *J. Am. Chem. Soc.*, (2009) **131**, 8329.
[79] Enzell, C., *Acta Chem. Scand.*, (1962) **16**, 1553.
[80] Nuñez, Y. O.; Salabarria, I. S.; Collado, I. G.; Hernández-Galán, R., *J. Agric. Food Chem.*, (2006) **54**, 7517.
[81] Kwon, H. J.; Hong, Y. K.; Park, C.; Choi, Y. H.; Yun, H. J.; Lee, E. W.; Kim, B. W., *Cancer Lett.*, (2010) **290**, 96.
[82] Harmata, M.; Kahraman, M.; Adenu, G.; Barnes, C. L., *Heterocycles*, (2004) **62**, 583.
[83] Harmata, M.; Jones, D. E., *Tetrahedron Lett.*, (1996) **37**, 783.
[84] Yam-Puc, A.; Escalante-Erosa, F.; Pech-López, M.; Chan-Bacab, M. J.; Arunachalampillai, A.; Wendt, O. F.; Sterner, O.; Peña-Rodríguez, L. M., *J. Nat. Prod.*, (2009) **72**, 745.
[85] Lezama-Dávila, C. M.; Pan, L.; Isaac-Márquez, A. P.; Terrazas, C.; Oghumu, S.; Isaac-Márquez, R.; Pech-Dzib, M. Y.; Barbi, J.; Calomeni, E.; Parinandi, P.; Kinghorn, A. D.; Satoskar, A. R., *Phytother. Res*, (2014) **28**, 909.
[86] Sumiya, T.; Ishigami, K.; Watanabe, H., *Angew. Chem. Int. Ed.*, (2010) **49**, 5527.

2.8 Metal-Catalyzed (5 + 1), (5 + 2), and (5 + 2 + 1) Cycloadditions

X. Li and W. Tang

General Introduction

Transition-metal-catalyzed cycloaddition has emerged as a versatile strategy to access highly functionalized carbocycles and heterocycles that are present in diverse natural products and pharmaceutical reagents.[1–11] These methods are particularly attractive owing to their nature of constructing multiple carbon–carbon and carbon–heteroatom bonds in a single step with a high degree of regioselectivity and stereoselectivity. Many cycloadditions that can be catalyzed by a metal are difficult or are not feasible under thermal and photolytic conditions. The complexation of the metal with an alkyne, alkene, allene, or diene significantly modifies the reactivity of the π-systems. A plethora of synthetically useful synthons have been discovered for highly selective cycloaddition reactions with the aid of transition-metal catalysts or promoters. To date, extensive efforts have been made to prepare three- to six-membered rings by (1 + 2)-, (2 + 2)-, (3 + 2)-, and (4 + 2)-cycloaddition reactions involving one- to four-carbon synthons. In contrast, five-carbon synthons have received much less attention. Significant progress has been recently made to demonstrate that (5 + n) cycloaddition is a powerful strategy for the synthesis of six-, seven-, and eight-membered carbocycles. For example, transition-metal-catalyzed vinylcyclopropane (5 + 2) cycloadditions have emerged as a rich and highly active area for the construction of seven-membered rings.[12–14] These cycloadditions usually involve oxidative cyclization to form a metallacycle; insertion of carbon monoxide, an alkyne, an alkene, an allene, or a combination of these, into C—M bonds; and reductive elimination. The focus of this chapter is on recent advances in transition-metal-catalyzed (5 + 1)-, (5 + 2)-, and (5 + 2 + 1)-cycloaddition reactions.

2.8.1 Metal-Catalyzed (5 + 1) Cycloadditions

2.8.1.1 Iron-Catalyzed (5 + 1) Cycloaddition

The first published example of metal-mediated (5 + 1) cycloaddition involving vinylcyclopropanes was reported by Sarel in 1969.[15] Initial studies indicated that the desired cyclohexenone was obtained as a mixture of two isomers in low yield. Later investigations demonstrated that the formation of (5 + 1) cycloadduct **2** from vinylcyclopropanes **1** and carbon monoxide could be promoted by pentacarbonyliron(0) or nonacarbonyldiiron(0) under photoirradiation (Scheme 1).[16] This transformation has limited substrate scope and uses stoichiometric amounts of an iron complex. Aumann has reported the isolation and characterization of some reaction intermediates, and his work provides insight into the mechanism of this process.[17]

Scheme 1 Iron-Mediated (5 + 1) Cycloaddition of Vinylcyclopropanes To Give Cyclohexenones[16]

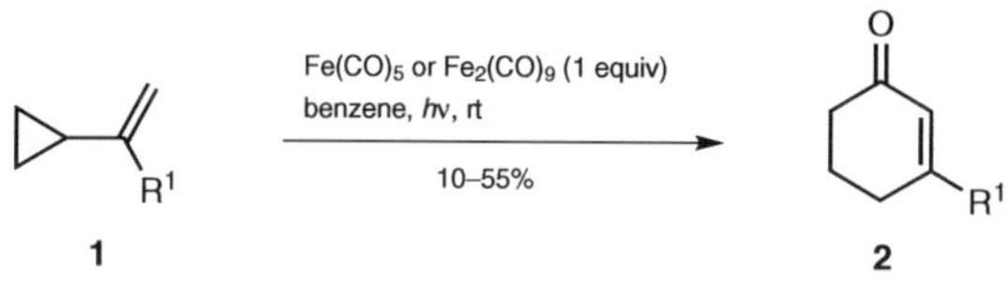

R^1 = Me, aryl, cyclopropyl

It was not until Taber reported the iron-catalyzed carbonylation of vinylcyclopropanes with photoirradiation under a carbon monoxide atmosphere that the (5 + 1)-cycloaddition reaction was developed into a synthetically useful method (Scheme 2).[18] The substrate scope has since been expanded, and good regioselectivity is realized, along with efficient transfer of chirality (>99% ee). Cleavage of bond b in cyclopropanes **3** is generally preferred owing to potential steric interactions with the R^3 substituent. This process offers an exciting possibility to produce enantioenriched 5-alkylcyclohexenones **4**, which are important building blocks for natural product synthesis; however, they are obtained along with small amounts of byproducts **5** that result from cleavage of bond a.

Scheme 2 Iron-Catalyzed (5 + 1) Cycloaddition of Vinylcyclopropanes and Carbon Monoxide To Give Cyclohexenones[18]

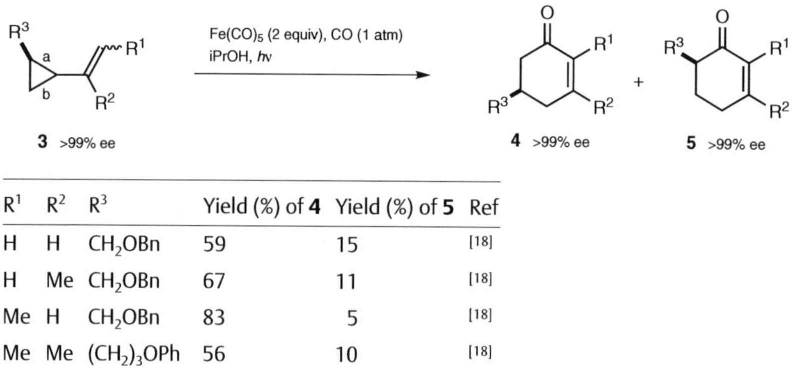

R^1	R^2	R^3	Yield (%) of **4**	Yield (%) of **5**	Ref
H	H	CH$_2$OBn	59	15	[18]
H	Me	CH$_2$OBn	67	11	[18]
Me	H	CH$_2$OBn	83	5	[18]
Me	Me	(CH$_2$)$_3$OPh	56	10	[18]
Me	H	(CH$_2$)$_3$OPh	83	5	[18]

Enantioenriched 5-Alkylcyclohexenones 4; General Procedure:[18]

CAUTION: *Carbon monoxide is extremely flammable and toxic, and exposure to higher concentrations can quickly lead to a coma.*

A mixture of 0.05 M cyclopropane in iPrOH and Fe(CO)$_5$ (2 equiv) under an atmosphere of CO (1 atm) was irradiated in a Pyrex tube in a Rayonet apparatus.

2.8.1.2 Chromium- or Molybdenum-Catalyzed (5 + 1) Cycloaddition

In 1987, Semmelhack and co-workers reported the hexacarbonylchromium(0)- and hexacarbonylmolybdenum(0)-promoted rearrangement of arylcyclopropenes **6** into naphthols **7** (Scheme 3).[19] It is proposed that a vinylketene is formed from reaction of the cyclopropene with the metal carbonyl, and this is followed by electrocyclization to produce the naphthol derivatives, which are obtained as a mixture of regioisomers **7** and **8**. It was

2.8.1 Metal-Catalyzed (5 + 1) Cycloadditions

also shown that the reaction can be accomplished by using a catalytic amount of hexacarbonylmolybdenum(0), although the products are obtained in much lower yields.

Scheme 3 Chromium/Molybdenum-Mediated (5 + 1) Cycloaddition of Arylcyclopropenes To Give Naphthols[19]

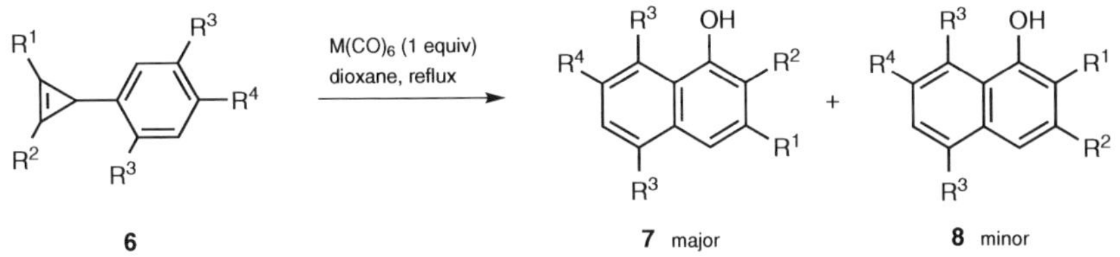

R^1	R^2	R^3	R^4	M	Ratio (7/8)	Yield[a] (%)	Ref
Et	Et	H	OMe	Cr	–	44	[19]
t-Bu	Me	H	H	Mo	100:0	40	[19]
iPr	Me	H	H	Mo	72:28	35	[19]
iPr	Me	H	H	Cr	75:25	40	[19]
Ph	Me	H	H	Mo	95:5	44	[19]
Me	H	OMe	H	Mo	100:0	63	[19]
Et	Me	OMe	H	Mo	60:40	76	[19]

[a] Combined yield of isolated products **7** and **8**.

2.8.1.3 Cobalt-Catalyzed (5 + 1) Cycloaddition

Iwasawa and co-workers have reported the preparation of 2-substituted or 2,3-disubstituted hydroquinone derivatives **13** from allenylcyclopropanols **9** using a cobalt complex (Scheme 4).[20] Upon treatment of **9** with octacarbonyldicobalt(0), carbonyl insertion complex **10** is formed. The cyclopropane ring of this complex undergoes ring expansion to generate intermediate **11**, and this is followed by reductive elimination to give **12**. Only activated allenylcyclopropanes are tolerated. de Meijere's group have compared the reactivity of octacarbonyldicobalt(0) with that of dicarbonyl(chloro)rhodium(I) dimer in the conversion of vinylcyclopropanes **14** into cyclohexenones **15** (Scheme 5).[21] The cobalt catalyst works well for most substrates, whereas the rhodium catalyst is only effective for several special vinylcyclopropanes under a carbon monoxide atmosphere.

Scheme 4 Cobalt-Mediated (5 + 1) Cycloaddition of 1-Allenylcyclopropan-1-ols To Give Hydroquinone Derivatives[20]

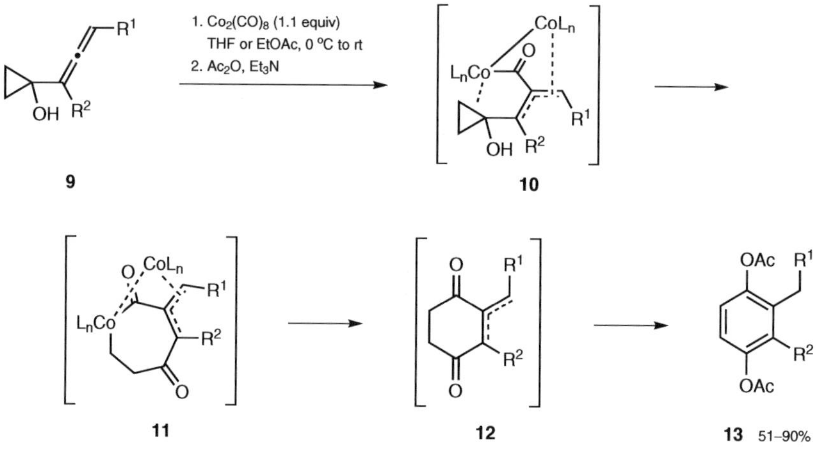

R^1 = H, $(CH_2)_5Me$, Ph; R^2 = H, $(CH_2)_5Me$, TMS, TBDMS, Ph

Scheme 5 Cobalt-Catalyzed (5 + 1) Cycloaddition of Vinylcyclopropanes To Give Cyclohexenones[21]

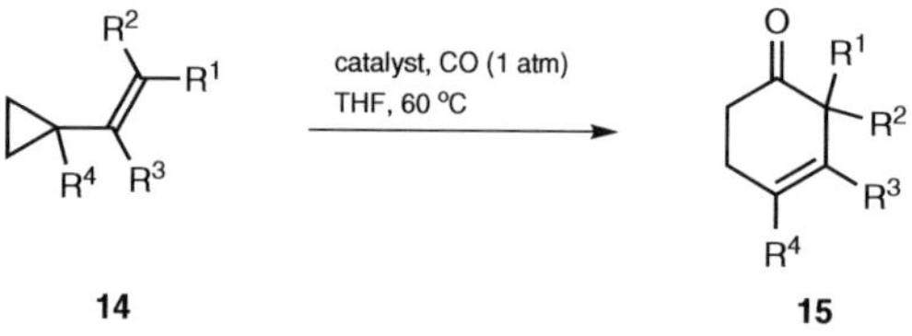

R^1	R^2	R^3	R^4	Catalyst (mol%)	Time (h)	Yield (%)	Ref
$(CH_2)_2$		Me	H	$Co_2(CO)_8$ (5)	12	68	[21]
$(CH_2)_2$		Me	H	$Rh_2Cl_2(CO)_4$ (2.5)	12	87	[21]
$(CH_2)_2$		H	H	$Co_2(CO)_8$ (5)	12	89	[21]
$(CH_2)_2$		H	H	$Rh_2Cl_2(CO)_4$ (2.5)	12	93	[21]
$(CH_2)_2$		Ph	H	$Co_2(CO)_8$ (5)	48	25[a]	[21]
$(CH_2)_2$		Ph	H	$Rh_2Cl_2(CO)_4$ (2.5)	48	33[a]	[21]
$(CH_2)_2$		cyclopropyl	H	$Co_2(CO)_8$ (5)	24	75	[21]
$(CH_2)_2$		cyclopropyl	H	$Rh_2Cl_2(CO)_4$ (2.5)	24	78	[21]
$(CH_2)_2$		H	Me	$Co_2(CO)_8$ (5)	12	72	[21]
$(CH_2)_2$		H	Me	$Rh_2Cl_2(CO)_4$ (2.5)	12	95	[21]
H	H	H	$O(CH_2)_2OMe$	$Co_2(CO)_8$ (5)	24	54	[21]
H	H	H	$O(CH_2)_2OMe$	$Rh_2Cl_2(CO)_4$ (2.5)	24	<1	[21]
H	H	H	$CH=CH_2$	$Co_2(CO)_8$ (5)	24	16	[21]
H	H	H	$CH=CH_2$	$Rh_2Cl_2(CO)_4$ (2.5)	24	<1	[21]
H	H	H	H	$Co_2(CO)_8$ (5)	24	62	[21]
H	H	H	H	$Rh_2Cl_2(CO)_4$ (2.5)	24	<1	[21]

[a] Conditions: dioxane, 100 °C.

2.8.1 Metal-Catalyzed (5 + 1) Cycloadditions

8-Methylspiro[2.5]oct-7-en-4-one [15, R^1,R^2 = (CH$_2$)$_2$; R^3 = Me; R^4 = H]; Typical Procedure:[21]

> **CAUTION:** *Carbon monoxide is extremely flammable and toxic, and exposure to higher concentrations can quickly lead to a coma.*

(1-Cyclopropylethylidene)cyclopropane (108 mg, 1 mmol) was added to a soln of Co$_2$(CO)$_8$ (17 mg, 0.05 mmol) in THF (10 mL), and the mixture was heated at 60 °C for 12 h under an atmosphere of CO (balloon). The cooled mixture was diluted with Et$_2$O (20 mL) and stirred under air for 1 h. Filtration through a pad of Celite and purification by Kugelrohr distillation gave a pale yellow oil; yield: 92 mg (68%).

2.8.1.4 Iridium-Catalyzed (5 + 1) Cycloaddition

Murakami and Ito have demonstrated that allenylcyclopropanes **16** can serve as five-carbon synthons in iridium-catalyzed (5 + 1)-cycloaddition reactions to give alkene-appended cyclohexenones **17** (Scheme 6).[22] No reaction occurs with vinylcyclopropanes under analogous conditions, which implies the importance of the allene functionality.

Scheme 6 Iridium-Catalyzed (5 + 1) Cycloaddition of Allenylcyclopropanes with Carbon Monoxide To Give Methylenecyclohexenones[22]

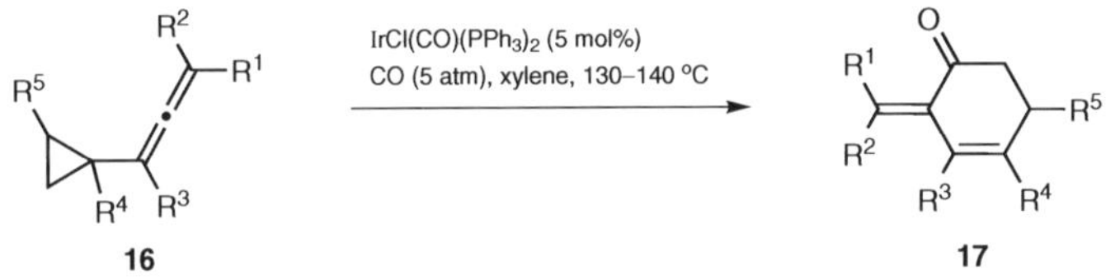

R^1	R^2	R^3	R^4	R^5	Yield (%)	Ref
Et	Et	H	H	H	81	[22]
Me	Me	H	OEt	H	74	[22]
Ph	Me	H	H	H	83	[22]
Me	Me	Ph	H	H	81	[22]
Me	Me	H	H	Ph	63	[22]
Ph	H	H	H	H	28	[22]

2-(Pentan-3-ylidene)cyclohex-3-en-1-one (17, R^1 = R^2 = Et; R^3 = R^4 = R^5 = H); Typical Procedure:[22]

> **CAUTION:** *Carbon monoxide is extremely flammable and toxic, and exposure to higher concentrations can quickly lead to a coma.*

A mixture of 1-cyclopropyl-3-ethylpenta-1,2-diene (80.0 mg, 590 µmol) and IrCl(CO)(PPh$_3$)$_2$ (23 mg, 29 µmol) in xylene (2 mL) under an atmosphere of CO (5 atm) in an autoclave was stirred in an oil bath at 130 °C for 35 h. After the mixture was cooled down to rt, the solvent was removed under reduced pressure. The residue was passed through a short pad of Florisil, concentrated, and subjected to preparative TLC (silica gel, Et$_2$O/hexane 1:5) to afford the title compound; yield: 78 mg (81%).

2.8.1.5 Ruthenium-Catalyzed (5 + 1) Cycloaddition

Murai and co-workers have reported that dodecacarbonyltriruthenium(0) is an active catalyst for the carbonylative (5 + 1) cycloaddition of cyclopropylimines **18** to give functionalized six-membered lactams **19** (Scheme 7).[23] The reaction scope is limited to alkyl-substituted imines, and poor regioselectivity is observed for disubstituted cyclopropane substrates, with a total yield of 30%. The mechanism involves the oxidative cyclization of the cyclopropylimine to form a ruthenium metallacycle, and this process is facilitated by coordination of the nitrogen atom to the ruthenium center; this is followed by carbon monoxide insertion and reductive elimination to produce the lactam.

Scheme 7 Ruthenium-Catalyzed (5 + 1) Cycloaddition of Cyclopropylimines with Carbon Monoxide To Give Lactams[23]

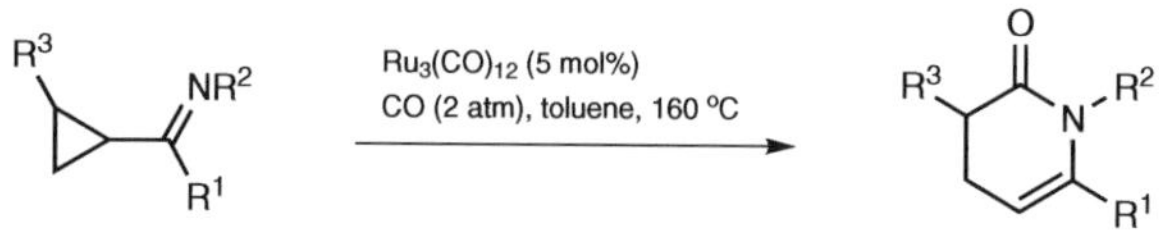

R^1	R^2	R^3	Yield (%)	Ref
Me	t-Bu	H	34	[23]
Me	Cy	H	71	[23]
H	Cy	H	64	[23]
2-thienyl	t-Bu	H	27	[23]
4-MeOC$_4$H$_4$	t-Bu	H	64	[23]
H	Cy	Ph	–[a]	[23]
Ph	t-Bu	H	76	[23]

[a] More than three products were isolated.

1-tert-Butyl-6-phenyl-3,4-dihydropyridin-2(1H)-one (19, R^1 = Ph; R^2 = t-Bu; R^3 = H); Typical Procedure:[23]

> **CAUTION:** *Carbon monoxide is extremely flammable and toxic, and exposure to higher concentrations can quickly lead to a coma.*

A 50-mL stainless-steel autoclave was charged with N-tert-butyl-1-cyclopropyl-1-phenylmethanimine (201 mg, 1 mmol), toluene (3 mL), and Ru$_3$(CO)$_{12}$ (13 mg, 0.02 mmol). The system was flushed with CO (10 atm, 3×), after which it was pressurized to 2 atm and immersed in an oil bath at 160 °C. After 60 h, the autoclave was removed from the oil bath and cooled for 1 h, and this was followed by release of CO. The contents were transferred to a round-bottomed flask using Et$_2$O, and the volatiles were removed under reduced pressure. The residue was subjected to column chromatography (silica gel, hexane/Et$_2$O 3:1) to give a white solid; yield: 175 mg (76%).

2.8.1.6 Rhodium-Catalyzed (5 + 1) Cycloaddition

Cyclopropenyl esters and ketones **20** can also be employed in rhodium-catalyzed (5 + 1) carbonylation; substituted α-pyrones **21** are delivered with good regioselectivity (Scheme 8).[24] It is proposed that electrophilic attack of the dicarbonyl(chloro)rhodium(I) dimer on the π-bond of the cyclopropene results in a cyclopropyl carbocation that is stabilized by the adjacent carbonyl group; this is followed by cyclopropane opening, which leads to the α-pyrone. If vinylcyclopropenes **22** are used as substrates instead, substituted phenols **23** are synthesized under the same conditions.

2.8.1 Metal-Catalyzed (5 + 1) Cycloadditions

Scheme 8 Rhodium-Catalyzed (5 + 1) Carbonylation of Cyclopropenyl Esters or Ketones, and Vinylcyclopropenes with Carbon Monoxide To Give α-Pyrones or Phenols[24]

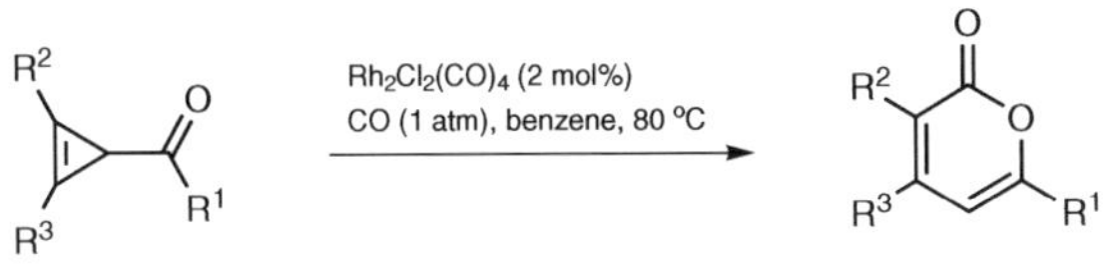

R^1	R^2	R^3	Yield (%)	Ref
OEt	Et	Et	78	[24]
OEt	H	t-Bu	89	[24]
OEt	Ph	Me	60	[24]
OEt	Me	TMS	77	[24]
Ph	Et	Et	40	[24]
Me	Et	Et	47	[24]

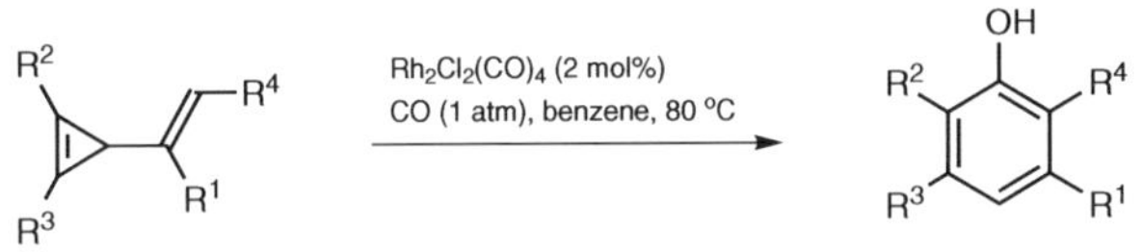

R^1	R^2	R^3	R^4	Yield (%)	Ref
Ph	Et	Et	H	31	[24]
Ph	Et	Et	Ph	51	[24]
Ph	Et	Et	Pr	82	[24]
iPr	Et	Et	H	46	[24]

Besides common vinylcyclopropanes, Fukuyama, Ryu, Fensterbank, and Malacria have demonstrated that 3-acyloxy-1,4-enynes **24** can act as novel five-carbon synthons for the synthesis of resorcinols **25** through a rhodium-catalyzed (5 + 1) carbonylative strategy (Scheme 9).[25,26] Two pathways are proposed for this transformation: One pathway involves rhodium-catalyzed 1,2-acyloxy migration of the 3-acyloxy-1,4-enyne to give a rhodacyclohexadiene, followed by carbon monoxide insertion and reductive elimination to release the product. The other pathway involves the formation of a rhodium–carbene after 1,2-acyloxy migration of the 3-acyloxy-1,4-enyne, followed by carbon monoxide insertion to generate a ketene and electrocyclization to afford the product. Both pathways involve rhodium-catalyzed 1,2-acyloxy migration of propargylic esters, which was first discovered by Rautenstrauch in 1984 by using a palladium(II) catalyst.[27–31]

Scheme 9 Rhodium-Catalyzed (5 + 1) Cycloaddition of 3-Acyloxy-1,4-enynes with Carbon Monoxide To Give Resorcinols[25]

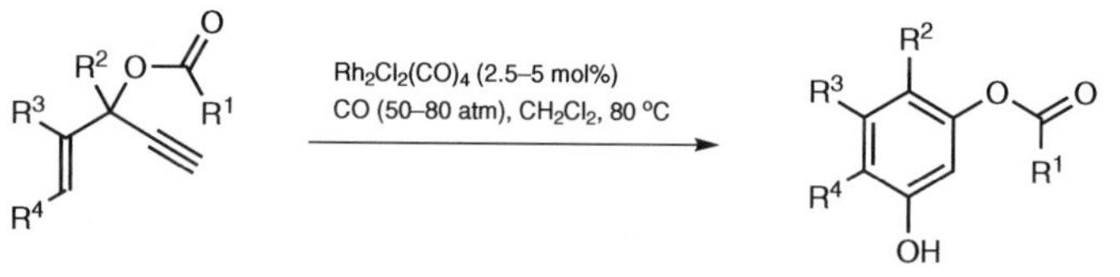

R¹	R²	R³	R⁴	Yield (%)	Ref
t-Bu	H	H	Ph	76	[25]
Me	H	H	Ph	67	[25]
t-Bu	H	H	4-F$_3$CC$_6$H$_4$	56	[25]
t-Bu	H	H	4-MeOC$_6$H$_4$	56	[25]
t-Bu	H	Me	Ph	53	[25]
t-Bu	Me	H	Ph	68	[25]
t-Bu	H	H	Me	58	[25]
t-Bu	H	H	iPr	74	[25]

Recently, Tang and co-workers developed a rhodium-catalyzed tandem 1,3-acyloxy migration/(5 + 1) cycloaddition to produce highly functionalized cyclohexenones **27** and **28** from cyclopropyl-substituted propargylic esters **26** (Scheme 10).[32,33] The reaction proceeds with good regioselectivity and the yields range from 74 to 95%. The chirality of the cyclopropane ring is fully transferred to the product, and thus, this is an attractive method to prepare optically pure cyclohexenones. The mechanism involves rhodium-catalyzed Saucy–Marbet 1,3-acyloxy migration of the propargylic ester to form an allenylcyclopropane and (5 + 1) cycloaddition with carbon monoxide. Shortly after this finding was published, the same group reported a complementary version of this (5 + 1)-cycloaddition reaction with substrates **29**.[33]

Scheme 10 Rhodium-Catalyzed (5 + 1) Cycloaddition To Give Methylenecyclohexenones[32,33]

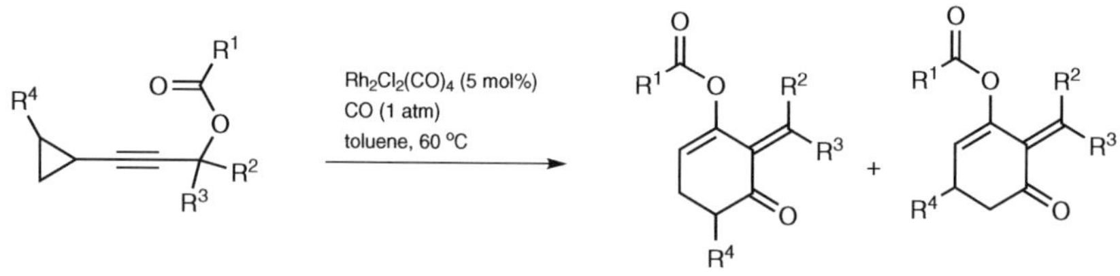

R¹	R²	R³	R⁴	Config of Cyclopropane 26	Ratio (27/28)	Yield (%)	Ref
t-Bu	Me	Me	H	–	–	88	[32]
Me	Me	Me	H	–	–	85	[32]
t-Bu	Ph	H	H	–	–	95	[32]
t-Bu	iPr	H	H	–	–	87	[32]
t-Bu	Me	Me	Ph	trans	2.5:1	90	[32]
t-Bu	Me	Me	Ph	cis	1:3.5	95	[32]

2.8.1 Metal-Catalyzed (5 + 1) Cycloadditions

R^1	R^2	R^3	R^4	Config of Cyclopropane **26**	Ratio (**27/28**)	Yield (%)	Ref
t-Bu	Me	Me	CH$_2$OTIPS	trans	1:10	91	[32]
t-Bu	Me	Me	CH$_2$OTIPS	cis	1:20	93	[32]
t-Bu	Me	Me	CH$_2$OH	cis	1:12	79	[32]
t-Bu	Me	Me	CO$_2$Me	cis	1:20	76	[32]

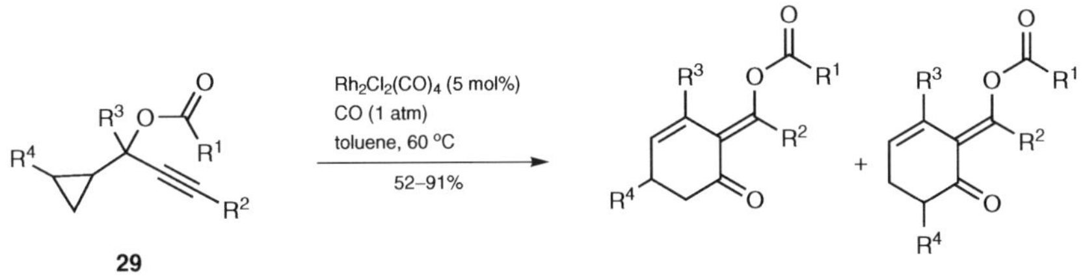

R^1 = Me, t-Bu; R^2 = Pr, t-Bu, (CH$_2$)$_3$Ph; R^3 = R^4 = H, Me

After de Meijere's group demonstrated that the dicarbonyl(chloro)rhodium(I) dimer is an active catalyst for (5 + 1) cycloaddition,[21] Yu reported a more general method that employs cationic rhodium complexes (Scheme 11).[34] A mixture of cyclohex-3-enones **31** and cyclohex-2-enones **32** in a 5:1 ratio is produced from vinylcyclopropanes **30**. Cyclohex-3-enones **31** can be converted into cyclohex-2-enones **32** by the addition of 1,8-diazabicyclo[5.4.0]undec-7-ene. However, only terminal (1,1-disubstituted) vinylcyclopropanes work in this reaction.

Scheme 11 Rhodium-Catalyzed (5 + 1) Cycloaddition of Vinylcyclopropanes with Carbon Monoxide[34]

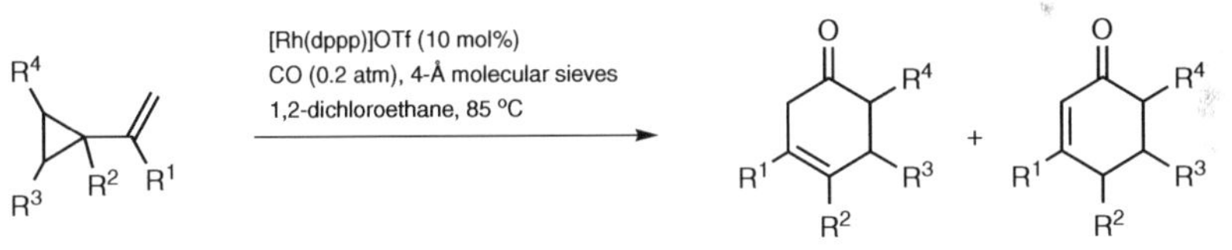

Reaction scheme: Vinylcyclopropane **30** → Cyclohex-2-en-1-one **32**

Conditions:
1. [Rh(dppp)]OTf (10 mol%), CO (0.2 atm), 4-Å molecular sieves, 1,2-dichloroethane, 85 °C
2. DBU, rt

R¹	R²	R³	R⁴	Yield (%) of 32	Ref
Ph	H	H	H	73	[34]
4-MeOC$_6$H$_4$	H	H	H	76	[34]
4-FC$_6$H$_4$	H	H	H	64	[34]
(CH$_2$)$_2$Ph	H	H	H	85	[34]
H	Ph	H	H	80	[34]
H	CH$_2$OBn	H	H	38	[34]
H	H	H	CH$_2$OH	35[a]	[34]

[a] The starting cyclopropane had a cis configuration.

Cyclohex-2-en-1-ones 32; General Procedure:[34]

CAUTION: *Carbon monoxide is extremely flammable and toxic, and exposure to higher concentrations can quickly lead to a coma.*

Under an atmosphere of argon, a soln of the Rh(I)⁺ catalyst in 1,2-dichloroethane [10 µmol·mL⁻¹, 1.5 mL, 15.0 µmol; with either SbF$_6^-$ or OTf⁻ as the counterion, prepared from Rh$_2$Cl$_2$(CO)$_4$ and AgOTf or AgSbF$_6$] was added to a flame-dried reaction tube containing dppp [7.4 mg, 18.0 µmol, 1.2 equiv to Rh(I)⁺] and freshly activated 4-Å molecular sieves (100 mg). The resulting orange suspension was stirred at rt for 10 min, and then a soln of the vinylcyclopropane substrate **30** (0.15 mmol) in 1,2-dichloroethane (1.5 mL) was added. Then, the mixture was bubbled with a mixture of CO/N$_2$ (balloon; 20% CO and 80% N$_2$) for 5 min. The reaction tube was immersed in an oil bath (75 or 85 °C) and was treated under atmospheric pressure with CO/N$_2$ (1:4 v/v). Upon disappearance of the starting material, as indicated by TLC, the mixture was cooled to rt and DBU (22.0 µL, 0.15 mmol) was added. The system was stirred at 85 °C. When TLC indicated complete transformation of the cyclohex-3-enone into the cyclohex-2-enone, the mixture was filtered through a thin pad of silica gel. The filter cake was washed (petroleum ether/EtOAc), and the combined filtrate was concentrated. The crude product was purified by flash column chromatography (silica gel).

2.8.1.7 Applications of Metal-Catalyzed (5 + 1) Cycloaddition in Natural Product Synthesis

Encouraged by the photochemical iron-catalyzed carbonylation of vinylcyclopropanes (see Section 2.8.1.1), Taber's group applied it in a total synthesis of the natural product (−)-delobanone (Scheme 12).[35] Alkenyl cyclopropane **33** undergoes (5 + 1) cycloaddition to give (−)-delobanone using pentacarbonyliron(0) as the catalyst, with photoirradiation under an atmosphere of carbon monoxide.

2.8.2 Metal-Catalyzed (5 + 2) Cycloadditions

Scheme 12 (5 + 1) Cycloaddition in the Total Synthesis of (−)-Delobanone[35]

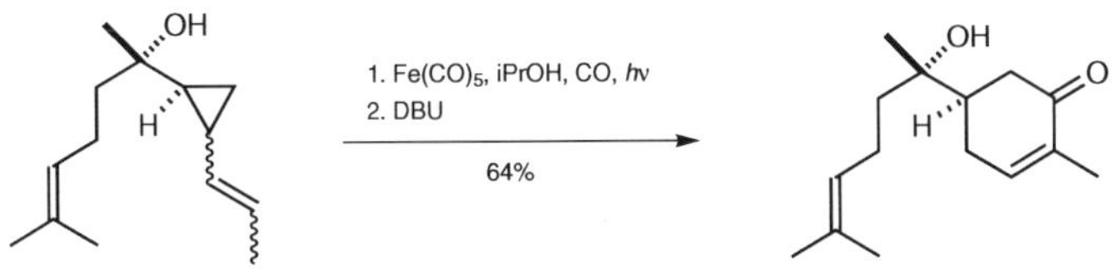

33 → (−)-delobanone

Reagents: 1. Fe(CO)$_5$, iPrOH, CO, hν; 2. DBU; 64%

2.8.2 Metal-Catalyzed (5 + 2) Cycloadditions

2.8.2.1 Rhodium-Catalyzed (5 + 2) Cycloaddition

Diels–Alder cycloaddition is the most powerful method to prepare six-membered rings with diverse functionality, high regioselectivity, and high stereoselectivity. Extensive efforts[36–38] have been made to develop a (5 + 2) homo-Diels–Alder cycloaddition that can match the scope and impact of the Diels–Alder reaction, because seven-membered rings are ubiquitous in bioactive natural products and pharmaceutical agents.[39,40] In addition, the discovery of new five-carbon synthons would also lead to the development of a series of novel cycloaddition reactions.

Metal-catalyzed (5 + 2) cycloadditions generally involve the formation of a six-membered metallacycle that can undergo an insertion reaction with an alkyne, alkene, or allene to form an eight-membered metallacycle. Finally, reductive elimination affords the seven-membered-ring product.

2.8.2.1.1 Rhodium-Catalyzed Cycloaddition with Vinylcyclopropanes

In 1995, Wender and co-workers first reported the rhodium-catalyzed intramolecular (5 + 2) cycloaddition of vinylcyclopropanes **34** possessing a tethered alkyne (Scheme 13).[41] Initial work examined the nature of the tether, alkene substitutions, and steric and electronic effects of the alkyne. The reaction proceeds with excellent regio- and stereoselectivity for substituted cyclopropanes and provides 5,7-fused products **35**.[42]

Scheme 13 Rhodium-Catalyzed Intramolecular (5 + 2) Cycloaddition of Vinylcyclopropanes and Alkynes To Give Fused Cycloheptadienes[41]

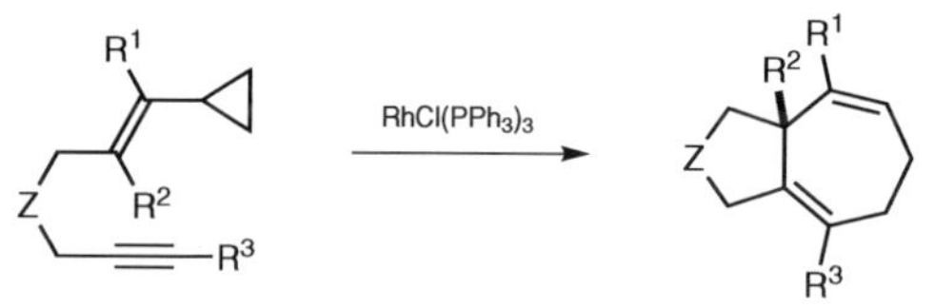

R^1	R^2	R^3	Z	RhCl(PPh$_3$)$_3$ (mol%)	Conditions	Yield (%)	Ref
H	H	H	O	10	THF, 100 °C	50	[41]
H	H	Me	O	10	toluene, 110 °C	88	[41]
H	H	TMS	O	10	toluene, 110 °C	83	[41]
H	H	CO$_2$Me	O	10	toluene, 110 °C	74	[41]
H	H	Ph	O	10	toluene, 110 °C	80	[41]
H	H	Me	C(CO$_2$Me)$_2$	0.5	AgOTf (0.5 mol%), toluene, 110 °C	83	[41]
Me	H	H	C(CO$_2$Me)$_2$	10	toluene, 110 °C	82	[41]
Me	H	CO$_2$Me	C(CO$_2$Me)$_2$	10	toluene, 110 °C	81	[41]
Me	H	TMS	C(CO$_2$Me)$_2$	10	toluene, 110 °C	71	[41]
H	Me	Me	C(CO$_2$Me)$_2$	10	AgOTf (10 mol%), toluene, 110 °C	82	[41]

This (5 + 2) cycloaddition is not limited to tethered alkynes. The intramolecular (5 + 2) cycloaddition of vinylcyclopropanes **36** possessing an alkene functionality (Scheme 14) was achieved shortly after the original report, despite the fact that alkenes are much less reactive than alkynes.[43] The cycloaddition with alkenes is also completely diastereoselective, with *cis*-fused 5,7-bicyclic products **37** delivered as the only isomer. Interestingly, *trans*-isomer **38** is observed for a six-membered tether. Furthermore, allenes **39** can also participate in this intramolecular (5 + 2) cycloaddition (Scheme 15).[44] The allene system displays complete regioselectivity for the internal alkene of the allene moiety. Complete transfer of chirality from the allene to the bicyclic product is also observed and provides a potential substrate-controlled route for the asymmetric synthesis of 5,7-fused bicyclic compounds **40**. There is no doubt that a catalyst-controlled asymmetric synthesis is more general and attractive. The enantioselective intramolecular (5 + 2) cycloaddition of vinylcyclopropanes with alkenes/alkynes can be performed using chiral BINAP[45,46] or chiral phosphoramidite ligands.[47]

Scheme 14 Rhodium-Catalyzed Intramolecular (5 + 2) Cycloaddition of Vinylcyclopropanes and Alkenes To Give Fused Cycloheptenes[43]

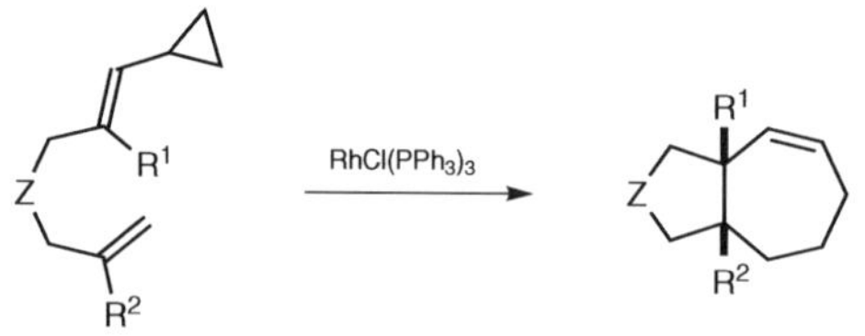

2.8.2 Metal-Catalyzed (5 + 2) Cycloadditions

R¹	R²	Z	RhCl(PPh₃)₃ (mol%)	Conditions	Yield (%)	Ref
H	H	C(CO₂Me)₂	0.1	AgOTf (0.1 mol%), toluene (1.0 M), 110 °C	86	[43]
H	H	O	5	AgOTf (5 mol%), THF (0.01 M), 65 °C	70	[43]
H	Me	C(CO₂Me)₂	10	AgOTf (10 mol%), toluene (0.01 M), 110 °C	92	[43]
Me	H	C(CO₂Me)₂	10	AgOTf (10 mol%), toluene (0.01 M), 110 °C	94	[43]
H	H	C(CO₂Me)₂	10	AgOTf (10 mol%), toluene (0.02 M), 100 °C	0	[43]

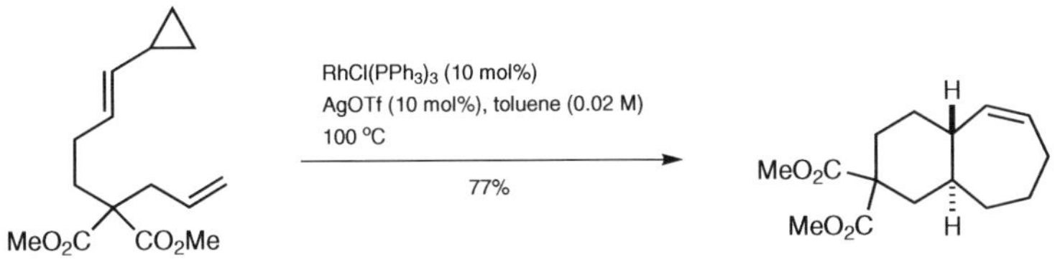

Scheme 15 Rhodium-Catalyzed Intramolecular (5 + 2) Cycloaddition of Vinylcyclopropanes and Allenes To Give Fused Methylenecycloheptenes[44]

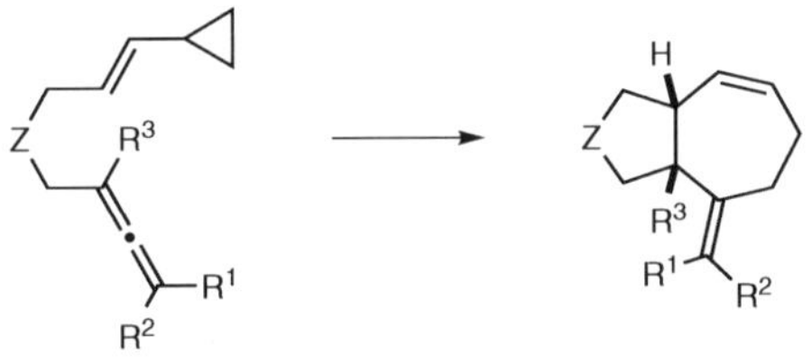

R¹	R²	R³	Z	Conditions	Yield (%)	Ref
t-Bu	H	H	C(CO₂Me)₂	RhCl(PPh₃)₃ (0.2 mol%), toluene (1.0 M), 100 °C	90[a]	[44]
H	H	H	C(CO₂Me)₂	Rh₂Cl₂(CO)₄ (5 mol%), 1,2-dichloroethane (0.003–0.01 M), 90 °C	83	[44]
Me	Me	H	C(CO₂Me)₂	RhCl(CO)₂ (10 mol%), toluene (0.01 M), 110 °C	90	[44]
Me	Me	H	CH(CO₂Me)	Rh₂Cl₂(CO)₄ (5 mol%), toluene (0.01 M), 100 °C	93	[44]
t-Bu	H	H	CH(CO₂Me)	Rh₂Cl₂(CO)₄ (5 mol%), toluene (0.01 M), 100 °C	91[a]	[44]
H	H	H	NTs	Rh₂Cl₂(CO)₄ (5 mol%), toluene (0.01 M), 100 °C	90	[44]
t-Bu	H	Me	C(CO₂Me)₂	RhCl(PPh₃)₃ (5 mol%), AgOTf (5 mol%), toluene (0.01 M), 110 °C	70[a]	[44]

[a] Z-Isomer at external methylene group.

Relative to intramolecular cycloaddition, the intermolecular version is much more versatile because of the easier access to both reactants. However, it is known that intermolecular cycloaddition is much more difficult to realize, as there is less control of the chemoselectivity, regioselectivity, and reactivity after removing the tether between the two reactants. Binger, de Meijere, and co-workers have reported one isolated example of a rhodium-catalyzed intermolecular (5 + 2) cycloaddition of vinylcyclopropane **41** with an alkyne to afford 5,7-fused bicyclic product **42** (Scheme 16).[48]

Scheme 16 Rhodium-Catalyzed Intermolecular (5 + 2) Cycloaddition of a Vinylcyclopropane and But-2-yne[48]

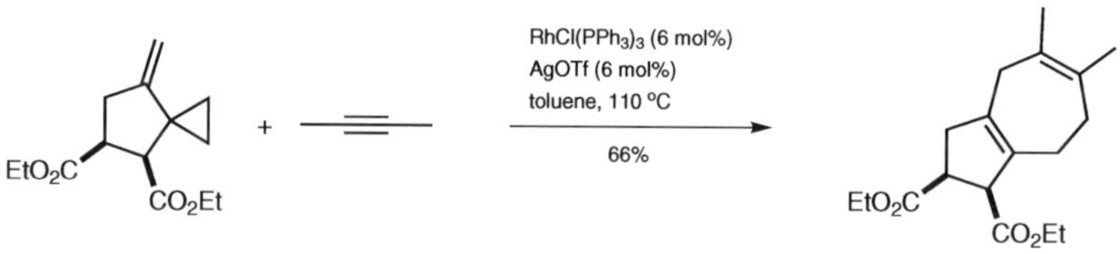

Shortly thereafter, Wender and co-workers reported the first rhodium-catalyzed intermolecular (5 + 2) cycloaddition of activated vinylcyclopropane **43** with alkynes (Scheme 17).[49] Heteroatom-activated cyclopropanes show higher reactivity. This process provides easy access to diverse siloxycycloheptadienes, which are hydrolyzed in situ to afford ketones **44**. Unactivated vinylcyclopropanes **45** also partake in this cycloaddition, and this provides an economical route for the production of synthetically useful cycloheptadienes **46**.[50]

Scheme 17 Rhodium-Catalyzed Intermolecular (5 + 2) Cycloaddition of Vinylcyclopropanes with Alkynes To Give Cycloheptenones and Cycloheptadienes[49,50]

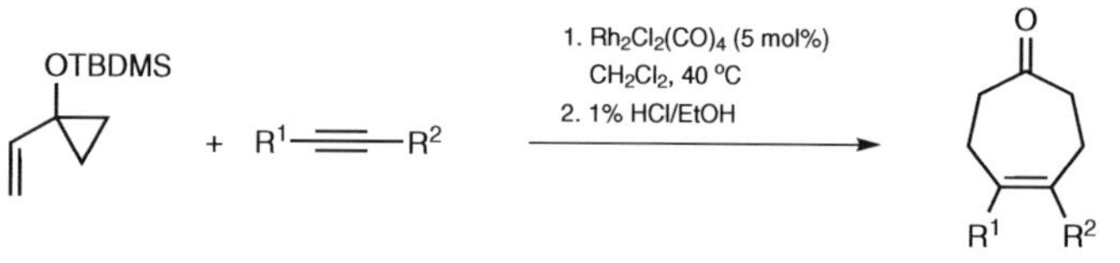

R^1	R^2	Yield (%)	Ref
CO_2Me	H	93	[49]
CO_2Me	Me	92	[49]
CH_2OMe	H	88	[49]
TMS	H	77	[49]
Ph	H	81	[49]
cyclopropyl	H	88	[49]
iPr	H	84	[49]
Et	Et	65	[49]
H	H	79	[49]

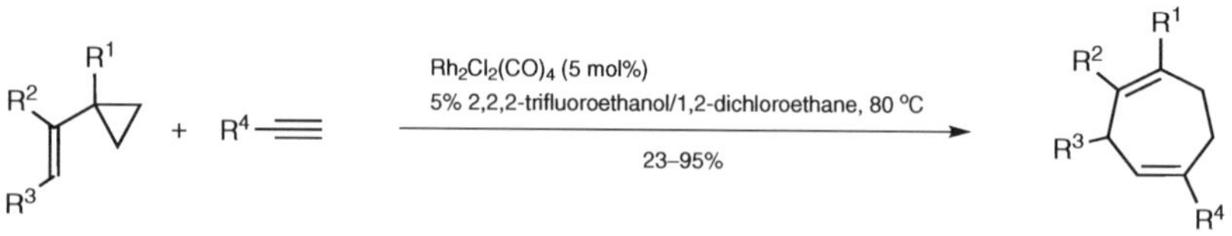

R^1 = H, Me, iPr, TMS, $CH_2OTBDMS$, CH(OH)Me, CH(OTBDMS)Me; R^2 = H, Me; R^3 = H, CH_2OH, $CH_2OTBDMS$, CH_2OMe; R^4 = Pr, Ph, TMS, CO_2Me, CH_2OMe, CH_2OH

2.8.2 Metal-Catalyzed (5 + 2) Cycloadditions

Furthermore, tandem (5 + 2)/(4 + 2) cycloaddition[51] and (5 + 2)/Nazarov cyclization[52] for the convenient preparation of complex polycyclic compounds have been reported. Wender and co-workers have very recently disclosed another elegant example involving a rhodium-catalyzed (5 + 2) cycloaddition/vinylogous Peterson elimination/(4 + 2) cycloaddition cascade to form complex carbocycles **47** or **48** with diverse functionalities in one step (Scheme 18).[53]

Scheme 18 Rhodium-Catalyzed (5 + 2) Cycloaddition/Vinylogous Peterson Elimination/(4 + 2) Cycloaddition Cascade[53]

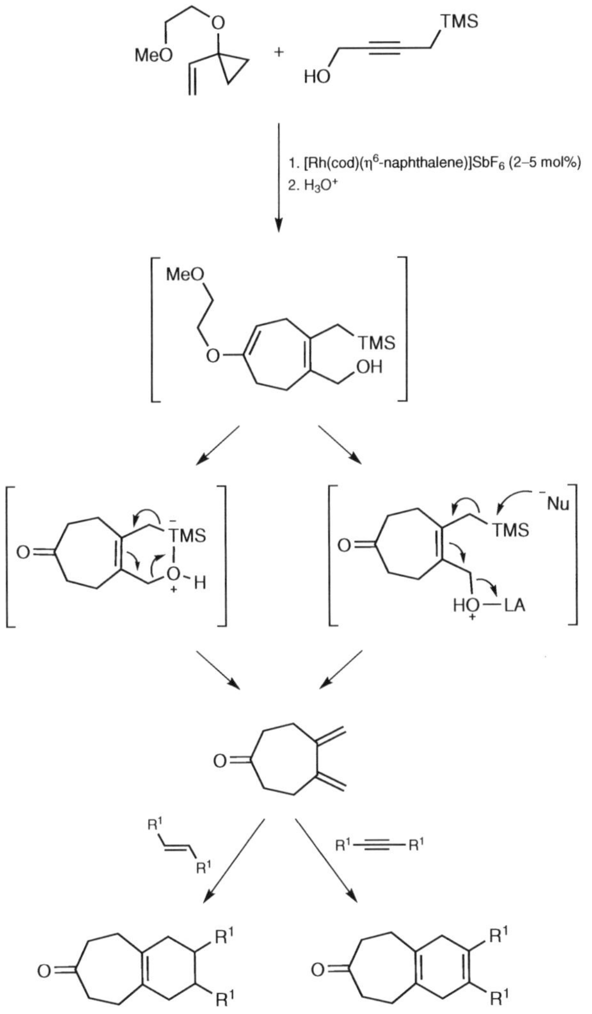

The intermolecular (5 + 2) cycloaddition of vinylcyclopropanes with allenes has only been realized for activated vinylcyclopropane **49** (Scheme 19).[54] Allenes with various coordinating groups have been examined, including alkynes, alkenes, and nitriles, with seven-membered rings **50** produced in good to excellent yields; however, the products are obtained as a mixture of E/Z-isomers with low selectivity.

Scheme 19 Rhodium-Catalyzed Intermolecular (5 + 2) Cycloaddition of Vinylcyclopropanes with Allenes To Give Methylenecycloheptanones[54]

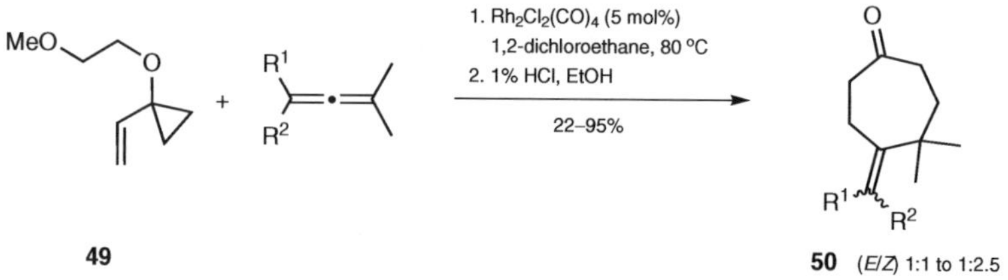

R¹	R²	Yield (%)	Ref
C≡CTMS	H	95	[54]
C≡CTMS	t-Bu	80	[54]
C≡CPh	H	83	[54]
C≡CPh	t-Bu	80	[54]
C≡CPh	CH₂CO₂Et	92	[54]
C≡C(CH₂)₂OH	t-Bu	65	[54]
C≡CCH₂OMe	H	45	[54]
C≡CCH₂NBn₂	t-Bu	22	[54]

Recently, Yu discovered that either (5 + 2) or (3 + 2) cycloadditions can be achieved depending on the stereochemistry of the cyclopropane (Scheme 20).[55,56] *cis*-Substituted substrate **51** leads to (5 + 2) product **52**, whereas *trans*-substituted **53** produces (3 + 2) product **54** (Scheme 20). DFT calculations suggest that after C–C bond activation and alkene insertion, the configuration of the reaction intermediate dictates the reaction pathway: If the C1 and C7 carbon atoms are proximal, the (5 + 2)-cycloaddition product is obtained; if the C1 and C5 carbon atoms are proximal, the (3 + 2)-cycloaddition product is obtained.

Scheme 20 Rhodium-Catalyzed (5 + 2) versus (3 + 2) Cycloaddition[55,56]

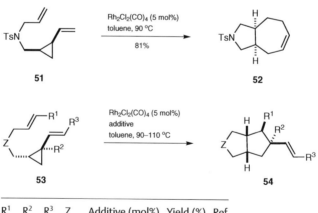

R¹	R²	R³	Z	Additive (mol%)	Yield (%)	Ref
H	H	H	NTs	–	83	[55]
H	H	Me	NTs	–	67	[55]
H	H	Ph	NTs	AgOTf (10)	71	[55]

2.8.2 Metal-Catalyzed (5 + 2) Cycloadditions

R^1	R^2	R^3	Z	Additive (mol%)	Yield (%)	Ref
H	H	Ph	O	AgOTf (10)	49	[55]
Me	H	H	NTs	–	62	[55]
H	Me	H	NTs	AgOTf (10)	80	[55]

Mukai has extended the five-carbon synthon to allenylcyclopropanes **55** instead of standard vinylcyclopropanes (Scheme 21).[57] Functionalized 6,7-fused bicyclic compounds **56**, which are not easily accessed through rhodium-catalyzed vinylcyclopropane cycloaddition, are realized in good yields.

Scheme 21 Rhodium-Catalyzed Intramolecular (5 + 2) Cycloaddition of Allenylcyclopropanes with Alkynes[57]

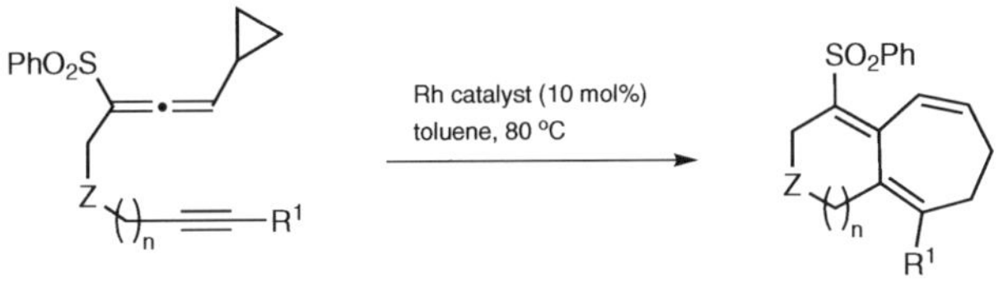

n	R^1	Z	Catalyst	Time (h)	Yield (%)	Ref
1	H	C(CO$_2$Me)$_2$	Rh$_2$Cl$_2$(CO)$_4$	0.5	86	[57]
1	H	C(SO$_2$Ph)$_2$	Rh$_2$Cl$_2$(CO)$_4$	0.2	80	[57]
1	H	C(CN)$_2$	Rh$_2$Cl$_2$(CO)$_2$(dppp)$_2$	1	53	[57]
1	H	CH$_2$	Rh$_2$Cl$_2$(CO)$_4$	0.2	65	[57]
1	H	O	Rh$_2$Cl$_2$(CO)$_2$(dppp)$_2$	1	52	[57]
1	Bu	NTs	Rh$_2$Cl$_2$(CO)$_4$	0.2	76	[57]
1	TMS	NTs	Rh$_2$Cl$_2$(CO)$_4$	0.2	59	[57]
2	H	NTs	Rh$_2$Cl$_2$(CO)$_2$(dppp)$_2$	2	74	[57]

Wender's group has also reported the intermolecular hetero-(5 + 2) cycloaddition of cyclopropylimines and alkynes (Scheme 22).[58] Versatile dihydroazepines are prepared. However, the reaction appears to be limited to electron-deficient alkynes.

Scheme 22 Rhodium-Catalyzed Intermolecular (5 + 2) Cycloaddition of Cyclopropylimines with Alkynes To Give Dihydroazepines[58]

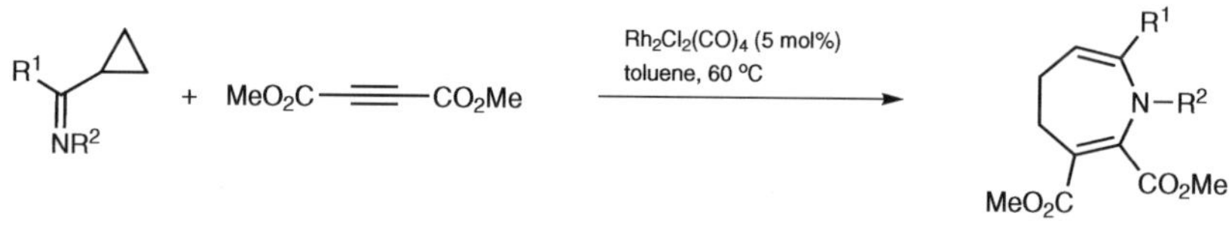

R^1	R^2	Yield (%)	Ref
H	Cy	83	[58]
H	$(CH_2)_5Me$	91	[58]
H	cyclopentyl	79	[58]
H	cyclopropyl	61	[58]
cyclopropyl	Bu	85	[58]
Ph	Bu	88	[58]

In 2011, vinyl epoxides **57** were found to be new five-carbon synthons in (5 + 2) cycloadditions realized by Feng and Zhang (Scheme 23).[59] A rhodium/N-heterocyclic carbene complex is effective for this transformation. The authors propose a tandem intramolecular hetero-(5 + 2)-cycloaddition/Claisen rearrangement process for the regioselective and stereoselective formation of [3.1.0] bicyclic compounds **58**.

Scheme 23 Rhodium-Catalyzed Intramolecular (5 + 2) Cycloaddition of Vinyl Epoxides with Alkynes[59]

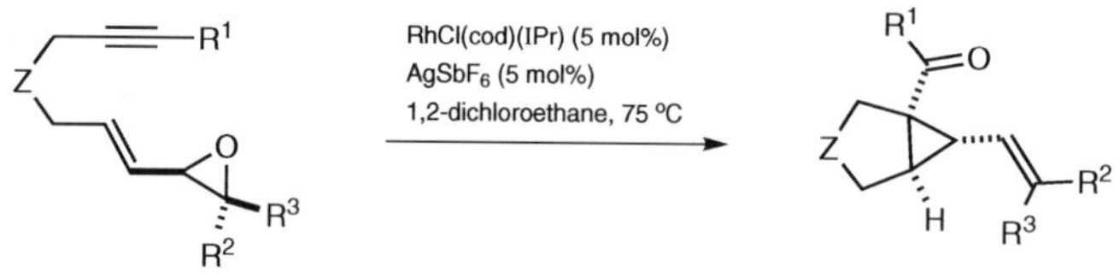

IPr = 1,3-bis(2,6-diisopropylphenyl)imidazol-2-ylidene

R^1	R^2	R^3	Z	Time (h)	Yield (%)	Ref
Ph	Ph	H	$C(CO_2Me)_2$	2	92	[59]
4-MeOC$_6$H$_4$	Ph	H	$C(CO_2Me)_2$	5	91	[59]
4-O$_2$NC$_6$H$_4$	Ph	H	$C(CO_2Me)_2$	8	55	[59]
Ph	Me	H	$C(CO_2Me)_2$	2	80	[59]
Ph	Me	Me	$C(CO_2Me)_2$	2	47[a]	[59]
Ph	Ph	H	NTs	7	86	[59]
Ph	Me	Me	NTs	7	67	[59]
Ph	Ph	H	O	2	85	[59]

[a] The reaction was run at 80 °C.

2.8.2 Metal-Catalyzed (5 + 2) Cycloadditions

Dimethyl 8-Methyl-3,3a,6,7-tetrahydroazulene-2,2(1H)-dicarboxylate [35, $R^1 = R^2 = H$; $R^3 = Me$; $Z = C(CO_2Me)_2$]; Typical Procedure:[41]

RhCl(PPh$_3$)$_3$ (0.5 mol%) and AgOTf (0.5 mol%) were added sequentially under an argon atmosphere to O$_2$-free toluene (4 mL) in an oven-dried, base-washed Schlenk flask. After 5 min at rt, a soln of vinylcyclopropane **34** (0.9 mmol) in toluene (5 mL) was added, and the resulting soln was heated at 110 °C for 20 min. After cooling, the mixture was passed through neutral alumina and concentrated. Chromatographic purification (silica gel, 5% EtOAc in hexanes) gave a colorless oil; yield: 83%.

Dimethyl 3,3a,4,5,6,8a-Hexahydroazulene-2,2(1H)-dicarboxylate [37, $R^1 = R^2 = H$; $Z = C(CO_2Me)_2$]; Typical Procedure:[43]

Under an argon atmosphere, RhCl(PPh$_3$)$_3$ (3.65 mg, 0.004 mmol, 0.1 mol%) and AgOTf (1.02 mg, 0.004 mmol, 0.1 mol%) were added sequentially, each in one batch, to a base-washed, oven-dried Schlenk flask containing freshly distilled, oxygen-free toluene (2 mL). The soln was stirred for 5 min at rt, after which ene–vinylcyclopropane **36** (1.00 g, 3.95 mmol, 1.0 equiv) in toluene (2 mL) was added over 10 s. The resultant soln was heated at 110 °C for 17 h. After cooling, the mixture was filtered through a plug of alumina and concentrated under reduced pressure. Purification by flash chromatography (silica gel, 5% EtOAc in hexane) gave a clear, colorless oil; yield: 0.857 g (86%).

Dimethyl 4-(2,2-Dimethylpropylidene)-3,3a,4,5,6,8a-hexahydroazulene-2,2(1H)-dicarboxylate [40, $R^1 = t$-Bu; $R^2 = R^3 = H$; $Z = C(CO_2Me)_2$]; Typical Procedure:[44]

Under an argon atmosphere, RhCl(PPh$_3$)$_3$ (0.2 mol%) was added in one batch to a base-washed, oven-dried Schlenk flask and was dissolved in freshly distilled, oxygen-free toluene (1.0 mL). The soln was stirred for 5 min at rt, after which ene–vinylcyclopropane **39** (0.132 mmol) in toluene (0.3 mL) was added over 10 s. The soln was heated to 100 °C for 5 h. After cooling, the mixture was filtered through a plug of alumina and concentrated. HPLC analysis of this mixture indicated that the product had formed with >99% selectivity. Flash chromatography (silica gel, 10% EtOAc in hexane) gave a colorless oil; yield: 90%.

Fused Ethenylcyclopentanes 54; General Procedure Using Rh$_2$Cl$_2$(CO)$_4$:[55]

A soln of vinylcyclopropane **53** (0.2 mmol) in toluene (4 mL) was added to a soln of Rh$_2$Cl$_2$(CO)$_4$ (3.9 mg, 0.01 mmol, 5 mol%) in toluene (4 mL) at rt under an atmosphere of argon. The resulting soln was immersed into an oil bath and was stirred at 90 °C. Once TLC indicated disappearance of the starting material, the mixture was cooled to rt and was filtered through a thin pad of silica gel. The filter cake was washed with Et$_2$O, and the combined filtrate was concentrated. The crude product was purified by flash column chromatography (silica gel).

Fused Ethenylcyclopentanes 54; General Procedure Using Rh$_2$Cl$_2$(CO)$_4$ and AgOTf:[55]

A mixture of Rh$_2$Cl$_2$(CO)$_4$ (3.9 mg, 0.01 mmol, 5 mol%) and AgOTf (5.1 mg, 0.02 mmol, 10 mol%) in toluene (4 mL) was stirred at rt under an atmosphere of argon for 5 min. A soln of vinylcyclopropane **53** (0.2 mmol) in toluene (4 mL) was added at rt, and the resulting soln was immersed into an oil bath and was stirred at 90 °C. Once TLC indicated disappearance of the starting material, the mixture was cooled to rt and was filtered through a thin pad of silica gel. The filter cake was washed with Et$_2$O, and the combined filtrate was concentrated. The crude product was purified by flash column chromatography (silica gel).

[3.1.0] Bicyclic Compounds 58; General Procedure:[59]

A mixture of RhCl(cod)(IPr) (7.0 mg, 0.01 mmol, 5 mol%) and AgSbF$_6$ (3.4 mg, 0.01 mmol, 5 mol%) in 1,2-dichloroethane (1 mL) was stirred at rt under an atmosphere of N$_2$ for 30 min. A soln of oxirane **57** (0.2 mmol) in 1,2-dichloroethane (1.5 mL) was added at rt, and the resulting mixture was stirred at 75 °C until the reaction was complete (monitored by TLC). After concentration, the residue was purified by column chromatography (silica gel, hexanes/EtOAc 10:1 to 5:1).

2.8.2.1.2 Rhodium-Catalyzed Cycloaddition with 3-Acyloxy-1,4-enynes

Recently, Tang and co-workers developed the rhodium-catalyzed intramolecular (5 + 2) cycloaddition of 3-acyloxy-1,4-enynes **59** with tethered alkynes (Scheme 24).[60] Highly functionalized 5,7-fused bicyclic products **60** are synthesized. It is proposed that rhodium catalyzes 1,2-acyloxy migration of the propargylic ester to afford a rhodium carbene or a metallacyclohexadiene. This is then followed by alkyne insertion and reductive elimination to produce the final product.

Scheme 24 Rhodium-Catalyzed Intramolecular (5 + 2) Cycloaddition of 3-Acyloxy-1,4-enynes and Alkynes To Give Fused Cycloheptatrienes[60]

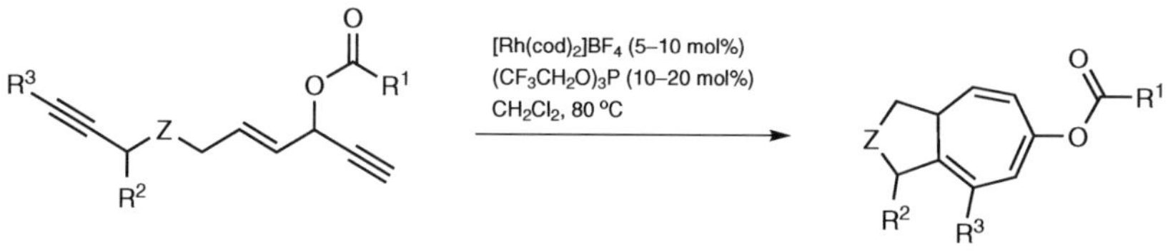

R^1	R^2	R^3	Z	Yield (%)	Ref
t-Bu	H	H	O	85	[60]
Me	H	H	O	81	[60]
t-Bu	H	H	NTs	96	[60]
t-Bu	H	H	C(CO$_2$Me)$_2$	75	[60]
t-Bu	H	Ph	NTs	88	[60]
t-Bu	H	Me	C(CO$_2$Me)$_2$	82	[60]
t-Bu	H	H	CH$_2$	76	[60]
t-Bu	Ph	H	O	90	[60]

The intermolecular (5 + 2) cycloaddition of 3-acyloxy-1,4-enynes **61** with alkynes **62** can be realized (Scheme 25).[61] This is a versatile strategy for the synthesis of substituted cycloheptatrienes **63**, for which the three double bonds can be selectively functionalized. High regioselectivity is observed for terminal alkynes **62**. The rate of this transformation can be increased with an electron-rich ester in the 3-acyloxy-1,4-enyne, and the catalyst loading can be decreased to 0.5 mol%.[62]

Scheme 25 Rhodium-Catalyzed Intermolecular (5 + 2) Cycloaddition of 3-Acyloxy-1,4-enynes and Alkynes To Give Cycloheptatrienes[61]

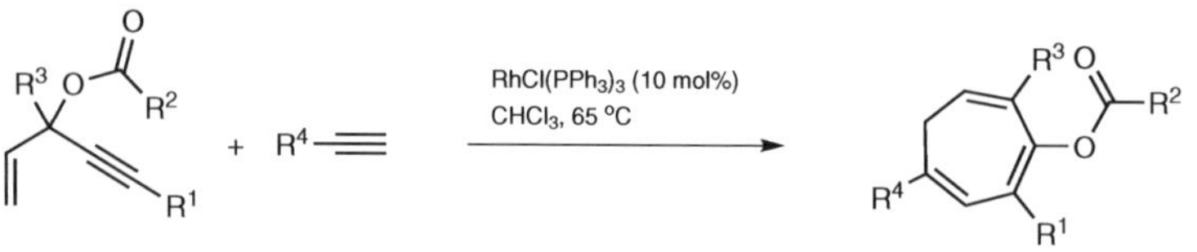

2.8.2 Metal-Catalyzed (5 + 2) Cycloadditions

R^1	R^2	R^3	R^4	Yield (%)	Ref
H	t-Bu	Me	CH$_2$OH	81	[61]
H	Me	Me	CH$_2$OH	71	[61]
Br	t-Bu	Me	CH$_2$OH	34	[61]
H	t-Bu	Ph	CH$_2$OH	81	[61]
H	t-Bu	CH$_2$OTBDMS	CH$_2$OH	92	[61]
H	t-Bu	Me	CH(OH)Ph	83	[61]
H	t-Bu	Me	CMe$_2$OH	87	[61]
H	t-Bu	Me	CH$_2$OMe	86	[61]
H	t-Bu	Me	CH$_2$NHTs	91	[61]
H	t-Bu	Me	CH$_2$CH(CO$_2$Me)$_2$	85	[61]

The rhodium-catalyzed intramolecular (5 + 2) cycloaddition of 3-acyloxy-1,4-enynes **64** with alkynes is also enantiospecific (Scheme 26).[63] For substrates bearing terminal alkynes, the stereochemistry from the propargylic esters can be relayed to the products with excellent efficiency. The absolute stereochemistry of **65** implicates approach of the rhodium catalyst to the 3-acyloxy-1,4-enyne from the face opposite to the acyloxy group.

Scheme 26 Chirality Transfer for the Rhodium-Catalyzed (5 + 2) Cycloaddition of 3-Acyloxy-1,4-enynes and Alkynes[63]

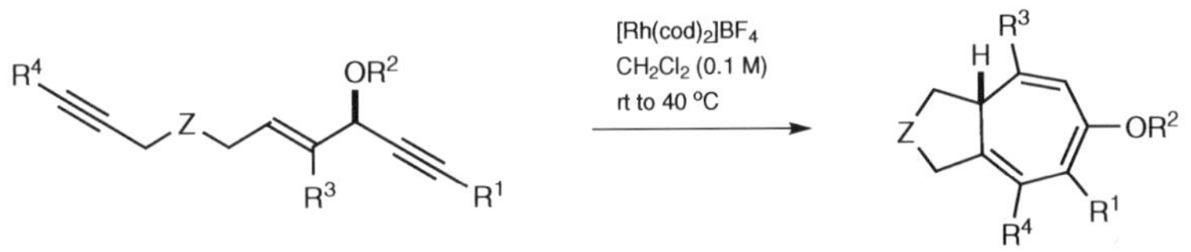

64 86–99% ee

65 59–95% ee

R^1	R^2	R^3	R^4	Z	[Rh(cod)$_2$]BF$_4$ (mol%)	ee (%)	Yield (%)	Ref
H	Ac	H	H	O	7	95	88	[63]
H	COt-Bu	H	H	4-BrC$_6$H$_4$SO$_2$N	7	92	70	[63]
H	COt-Bu	H	H	C(CO$_2$Me)$_2$	10	90	61	[63]
CO$_2$Et	COt-Bu	H	H	O	7	90	67	[63]
CO$_2$Et	COt-Bu	H	H	NTs	7	92	85	[63]
CO$_2$Et	COt-Bu	H	H	C(CO$_2$Me)$_2$	10	59	46	[63]
H	COt-Bu	Me	H	NTs	7	93	60	[63]
H	COt-Bu	H	Ph	NTs	7^a	84	80	[63]
H	COt-Bu	H	H	CH$_2$	10	91	52	[63]

^{a} (CF$_3$CH$_2$O)$_3$P (14 mol%) was also used.

Fused Cycloheptatrienes 65; General Procedure:[63]

Propargylic ester **64** (0.2 mmol) was added to a soln of [Rh(cod)$_2$]BF$_4$ (6 mg, 7 mol%) in CH$_2$Cl$_2$ (0.1 M). The soln was stirred at rt or 40 °C until the reaction was complete, as determined by TLC analysis (12–36 h). The solvent was removed under reduced pressure and the residue was purified by chromatography (silica gel).

2.8.2.1.3 Mechanisms of the Two Types of Rhodium-Catalyzed (5 + 2) Cycloaddition

Extensive DFT calculations have been conducted by the research groups of Houk, Yu, and Wender for rhodium-catalyzed (5 + 2) cycloadditions involving vinylcyclopropanes.[64–70] It is proposed that (5 + 2) cycloadditions with vinylcyclopropanes involve the formation of rhodium–allyl complex **66**, alkyne insertion to form metallacycle **67**, and reductive elimination to release product **68** (Scheme 27). Alkyne insertion into complex **66** takes place at the rhodium–allyl bond (C1) and regioselectively forms the first C—C bond with the terminal alkenyl carbon atom of the vinylcyclopropane.

The groups of Xu, Houk, and Tang have also worked together to study the mechanism of (5 + 2) cycloadditions involving five-carbon 3-acyloxy-1,4-enyne building blocks (Scheme 27).[71] It is proposed that metallacycle **69** is generated from concerted rhodium-catalyzed 1,2-acyloxy migration and oxidative cyclization, followed by insertion of the alkyne into the rhodium–C5 bond for which the R^1 group is distal to the forming C—C bond; subsequent reductive elimination releases product **70**.

Although the formation of intermediate **69** is much more complicated than that of rhodium–allyl complex **66**, the latter half of the second proposed mechanism appears to be similar to the mechanism for the (5 + 2) cycloaddition with vinylcyclopropanes. DFT investigations have revealed several fundamental differences in the nature of the rate- and regioselectivity-determining steps and in the order of formation of the two new C—C bonds upon employing 3-acyloxy-1,4-enynes in place of vinylcyclopropanes as the five-carbon synthons. The change in mechanism results in different regiochemical control as well as unique substituent effects on reactivity.

Scheme 27 (5 + 2) Cycloaddition Mechanistic Studies Based on DFT Calculations[71]

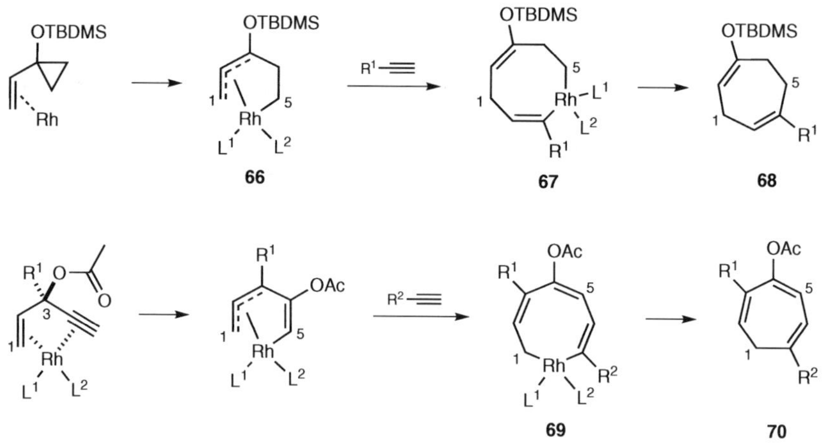

2.8.2.2 Ruthenium-Catalyzed (5 + 2) Cycloaddition with Vinylcyclopropanes

Shortly after Wender's discovery of the (5 + 2) cycloaddition, Trost reported that tris(acetonitrile)(η^5-cyclopentadienyl)ruthenium(II) hexafluorophosphate can also promote the cyclization of alkyne-tethered vinylcyclopropanes **71** to generate 5,7-fused bicyclic products **72** (Scheme 28).[72–74] The initial study indicated that the substrate scope is similar to that of the rhodium-catalyzed reaction. A mechanism is proposed that involves oxidative cyclization to form a metallacyclopentene intermediate, followed by ring opening of the cyclopropane; this is different to that proposed for the rhodium-catalyzed reaction.[64–70]

2.8.2 Metal-Catalyzed (5 + 2) Cycloadditions

Scheme 28 Ruthenium-Catalyzed Intramolecular (5 + 2) Cycloaddition of Vinylcyclopropanes with Alkynes[72]

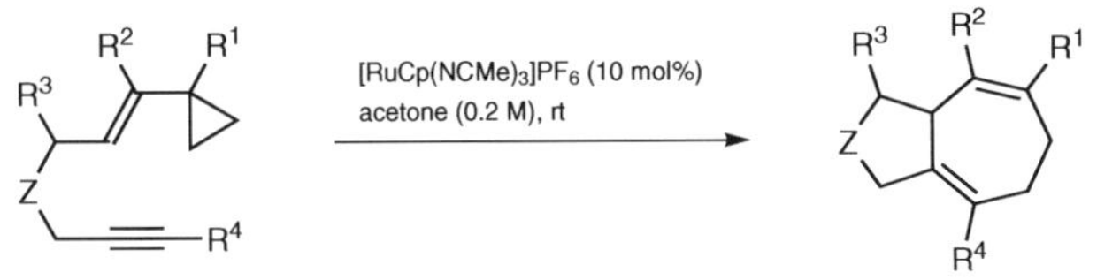

R^1	R^2	R^3	R^4	Z	Yield (%)	Ref
H	H	H	H	C(CO$_2$Me)$_2$	83	[72]
H	H	H	Ph	C(CO$_2$Me)$_2$	82	[72]
H	Me	H	Me	C(CO$_2$Me)$_2$	87	[72]
Me	H	H	Me	C(CO$_2$Me)$_2$	75[a]	[72]
H	H	H	Ph	O	77	[72]
H	H	H	TMS	NTs	84	[72]
H	H	OTBDMS	Me	CH$_2$	92[b]	[72]

[a] Reaction performed at 60 °C.
[b] dr 3.1:1.

Dimethyl 3,3a,6,7-Tetrahydroazulene-2,2(1*H*)-dicarboxylate [72, $R^1 = R^2 = R^3 = R^4 = H$; $Z = C(CO_2Me)_2$]; Typical Procedure:[72]

[RuCp(NCMe)$_3$]PF$_6$ (6 mg, 0.0138 mmol) was weighed into a dry test tube fitted with a rubber septum. The test tube was evacuated and filled with argon. A soln of vinylcyclopropane **71** (35 mg, 0.139 mmol) in dry acetone (0.7 mL) was added. The resulting dark orange soln was stirred at rt under argon for 2 h. Then, the mixture was concentrated and the resulting residue was purified by flash chromatography (silica gel, petroleum ether/Et$_2$O 3:1) to afford a colorless liquid; yield: 29 mg (83%).

2.8.2.3 Nickel-Catalyzed (5 + 2) Cycloaddition with Vinylcyclopropanes

The nickel/N-heterocyclic carbene complex catalyzed (5 + 2) cycloaddition of vinylcyclopropanes **73** has been reported by Zuo and Louie to provide tetrahydrofuran-fused products **74** (Scheme 29).[75] This process works smoothly for bulky alkyne substituents. In contrast, a mixture of five- and seven-membered rings is obtained for smaller substituents.

Scheme 29 Nickel-Catalyzed Intramolecular (5 + 2) Cycloaddition of Vinylcyclopropanes with Alkynes To Give Fused Tetrahydrofurans[75]

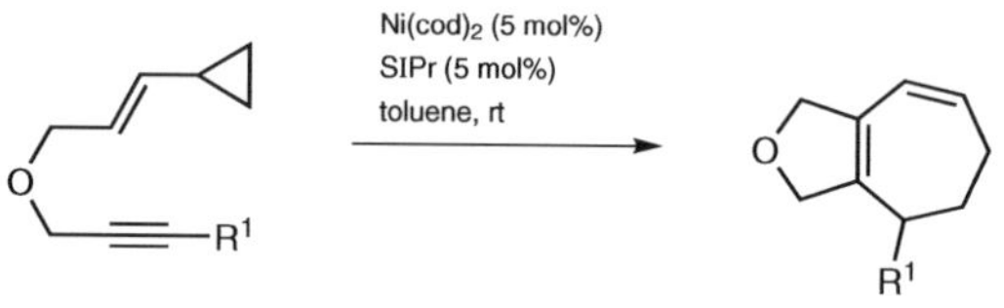

SIPr = 1,3-bis(2,6-diisopropylphenyl)-4,5-dihydroimidazol-2-ylidene

R^1	Yield (%)	Ref
t-Bu	82	[75]
TMS	88	[75]

4-*tert*-Butyl-3,4,5,6-tetrahydro-1*H*-cyclohepta[*c*]furan (74, R^1 = *t*-Bu); Typical Procedure:[75]
A mixture of **73** (100 mg, 0.5 mmol), Ni(cod)$_2$ (8 mg, 0.03 mmol), SIPr (11 mg, 0.03 mmol), and toluene (0.5 mL) was allowed to react at rt. Upon completion of the reaction, the mixture was purified by column chromatography (silica gel, 10% Et$_2$O/pentanes) to afford the title compound as a colorless oil; yield: 82 mg (82%, as reported).

2.8.2.4 Iron-Catalyzed (5 + 2) Cycloaddition with Vinylcyclopropanes

Fürstner and co-workers have demonstrated that iron can be an effective catalyst for the (5 + 2) cycloaddition of vinylcyclopropanes (Scheme 30).[76] An unusual anionic iron complex produces the (5 + 2) cycloadducts with good diastereoselectivities.

Scheme 30 Iron-Catalyzed (5 + 2) Intramolecular Cycloaddition of Vinylcyclopropanes with Alkynes[76]

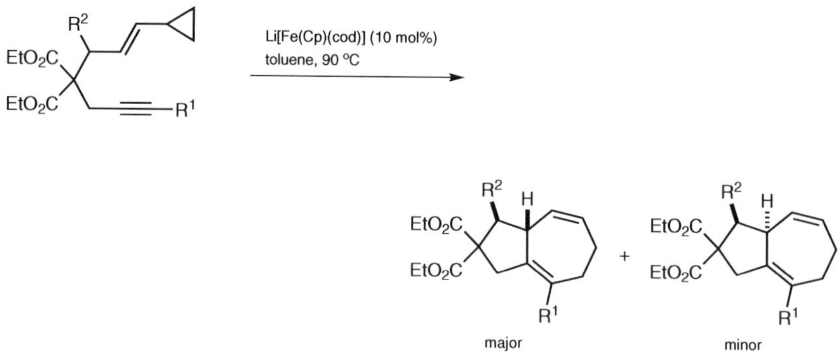

R^1	R^2	Yield (%)	Ref
H	H	66	[76]
TMS	H	66	[76]
Ph	H	75	[76]
H	Me	56	[76]
Me	Me	92	[76]
CO$_2$Et	Me	76	[76]
TMS	Me	99	[76]

2.8.2.5 Applications of Metal-Catalyzed (5 + 2) Cycloaddition in Natural Product Synthesis

The metal-catalyzed (5 + 2) cycloaddition is a powerful method that has been applied to the synthesis of several natural products.[77–86] Wender's group reported the synthesis of (+)-dictamnol (Scheme 31)[77] and (+)-aphanamol I (Scheme 32)[78] by using intramolecular rhodium-catalyzed (5 + 2) cycloadditions of vinylcyclopropanes with allenes as the key steps to afford the bicyclic cores efficiently. The same group also applied the (5 + 2) cycloaddition of a vinylcyclopropane and an alkyne in the synthesis of the tricyclic core of (+)-allocyathin B_2 (Scheme 33).[79]

Scheme 31 Synthesis of (+)-Dictamnol[77]

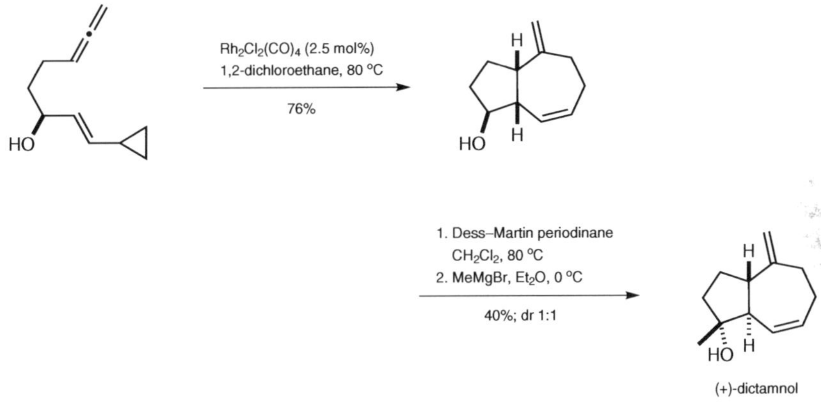

Scheme 32 Synthesis of (+)-Aphanamol I[78]

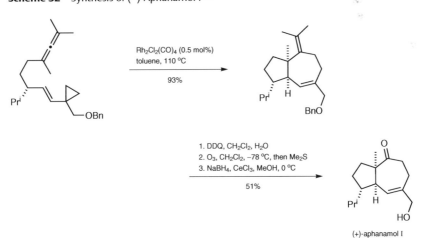

Scheme 33 Synthesis of (+)-Allocyathin B$_2$[79]

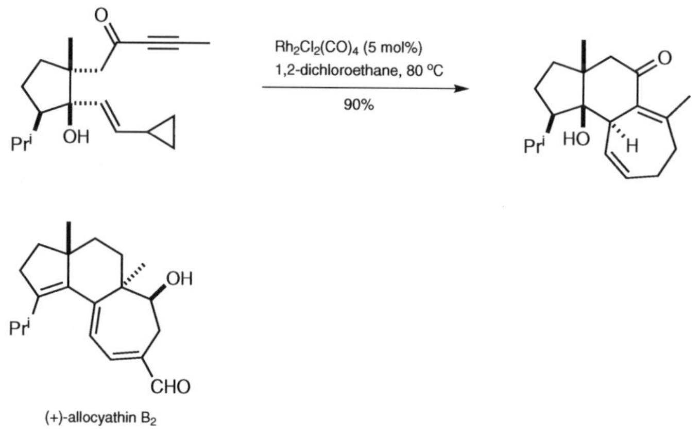

Rh$_2$Cl$_2$(CO)$_4$ (5 mol%)
1,2-dichloroethane, 80 °C
90%

(+)-allocyathin B$_2$

Ashfield and Martin have reported the synthesis of tremulenediol A and tremulenolide A by the rhodium-catalyzed (5 + 2) cycloaddition of a vinylcyclopropane and an alkyne (Scheme 34).[80,81] As shown in Scheme 35, Trost's group have applied the rhodium-catalyzed (5 + 2) cycloaddition of a vinylcyclopropane and an alkyne to the synthesis of (−)-pseudolaric acid B by using densely functionalized precursor **75**.[83,84] In this case, the ruthenium-catalyzed reaction leads to a mixture.

Scheme 34 Synthesis of Tremulenediol A and Tremulenolide A[80]

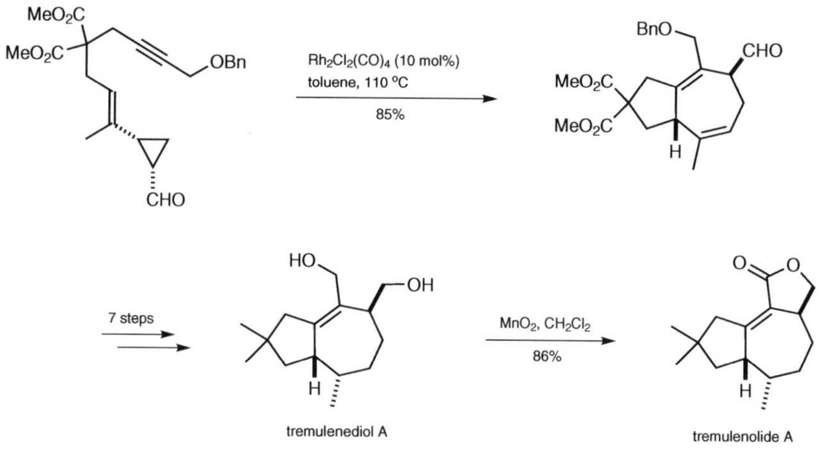

Rh$_2$Cl$_2$(CO)$_4$ (10 mol%)
toluene, 110 °C
85%

7 steps

MnO$_2$, CH$_2$Cl$_2$
86%

tremulenediol A

tremulenolide A

2.8.2 Metal-Catalyzed (5 + 2) Cycloadditions

Scheme 35 Synthesis of (−)-Pseudolaric Acid B[83]

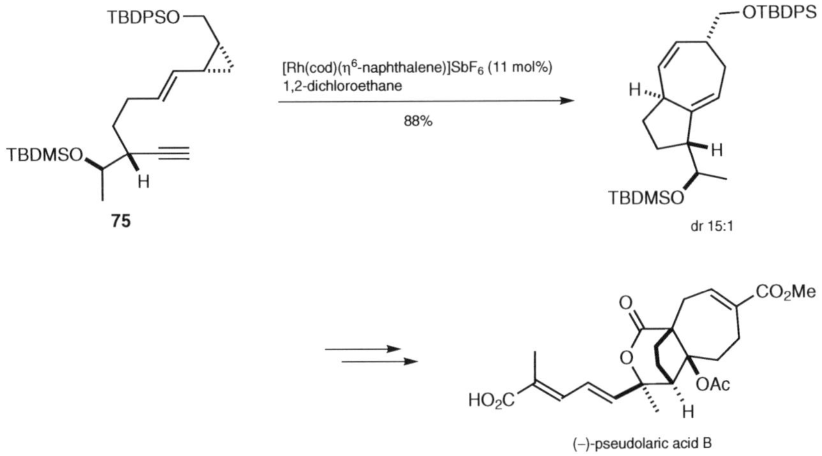

A ruthenium-catalyzed (5 + 2) cycloaddition is the key step in the synthesis of two natural products. One is the synthesis of (+)-frondosin A (Scheme 36),[82] and the other is the preparation of tricyclic core **77** toward rameswaralide from vinylcyclopropane **76** (Scheme 37).[86]

Scheme 36 Synthesis of (+)-Frondosin A[82]

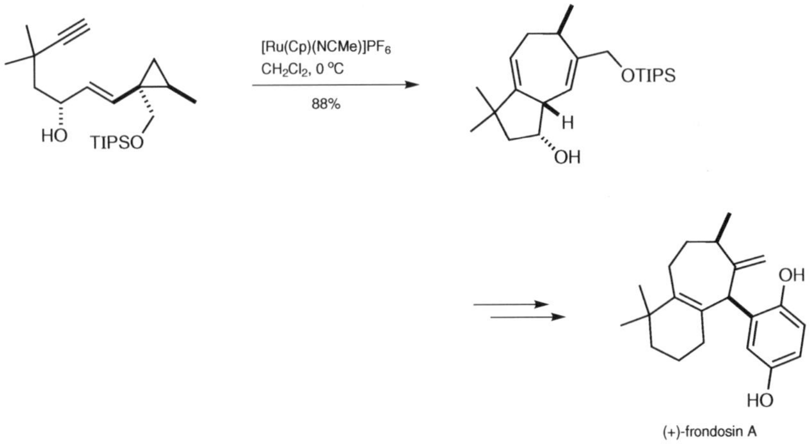

Scheme 37 Synthesis of a Rameswaralide Precursor[86]

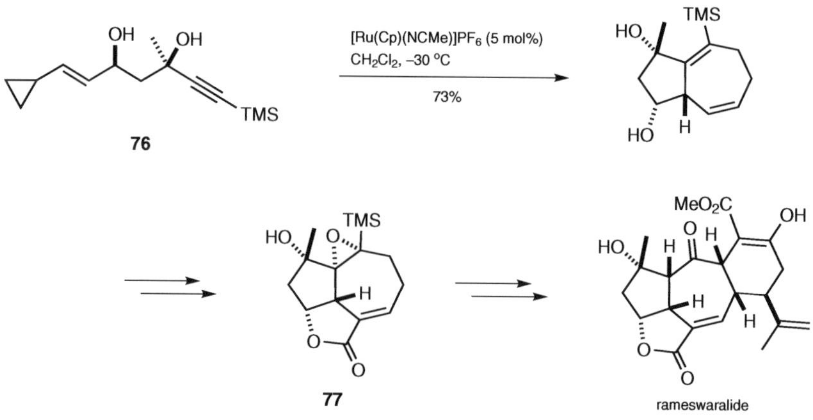

2.8.3 Metal-Catalyzed (5+2+1) Cycloadditions

Eight-membered carbocycles are present in a variety of natural products and pharmaceutical agents with a broad range of biological and medicinal activities. However, the synthesis of eight-membered carbocycles is quite challenging. Transition-metal-catalyzed (5+2+1) cycloaddition represents an elegant and efficient method for their construction. The reaction may involve either the insertion of an alkene or alkyne followed by the insertion of carbon monoxide, or insertion of carbon monoxide followed by insertion of the alkene or alkyne. This cycloaddition has also been applied in the total synthesis of natural products (Section 2.8.3.2).

2.8.3.1 Rhodium-Catalyzed (5+2+1) Cycloaddition with Vinylcyclopropanes

Wender and co-workers disclosed the first example of the rhodium-catalyzed intermolecular (5+2+1) cycloaddition of activated vinylcyclopropane **49**, alkynes, and carbon monoxide (Scheme 38).[87] The reaction proceeds through the initial generation of eight-membered-ring intermediate **78**, which forms regioselectively from the (5+2+1) cycloaddition; intermediate **78** undergoes stereoselective transannular ring closure to afford bicyclo[3.3.0]octadienes, which are hydrolyzed to bicyclo[3.3.0]octenone products **79**.

2.8.3 Metal-Catalyzed (5 + 2 + 1) Cycloadditions

Scheme 38 Rhodium-Catalyzed Intermolecular (5 + 2 + 1) Cycloaddition of a Vinylcyclopropane, Alkynes, and Carbon Monoxide To Give Bicyclooctenones[87]

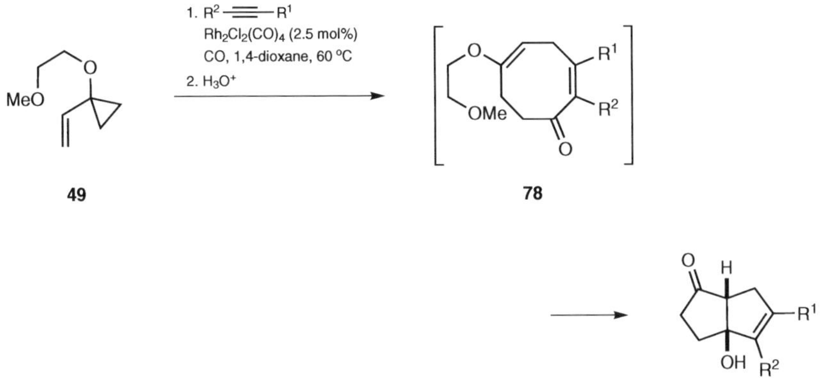

R^1	R^2	CO (atm)	Time (h)	Yield (%)	Ref
Ac	Et	2	20	97	[87]
Ac	TMS	1	42	54	[87]
Ac	Ph	1	26	88	[87]
CONH$_2$	Ph	1	40	96	[87]
CHO	Ph	2	26	69	[87]
CO$_2$Et	Ph	1	24	79	[87]
CO$_2$Et	TMS	1	26	67	[87]
CO$_2$Et	Me	1	20	85	[87]
CO$_2$Me	CO$_2$Me	1	30	48	[87]

Early studies indicated that many rhodium catalysts are not very effective for the intramolecular (5 + 2) cycloaddition of vinylcyclopropanes with alkenes. DFT calculations suggest that this might be attributed to a difficult reductive elimination step. It is much more difficult to form a C(sp^3)—C(sp^3) bond than a C(sp^2)—C(sp^3) bond via reductive elimination. On the basis of these studies, Wender, Yu, and co-workers have developed the rhodium-catalyzed (5 + 2 + 1) cycloaddition of ene–vinylcyclopropanes **80** with carbon monoxide (Scheme 39).[88] This reaction provides a general means to prepare 5,8- and 6,8-fused bicyclic cyclooctenones **81** and **82**, respectively, in good to excellent yields with a variety of tether types and substitution patterns.

Scheme 39 Rhodium-Catalyzed Intramolecular (5 + 2 + 1) Cycloaddition of Ene–Vinylcyclopropanes and Carbon Monoxide To Give Fused Cyclooctenones[88]

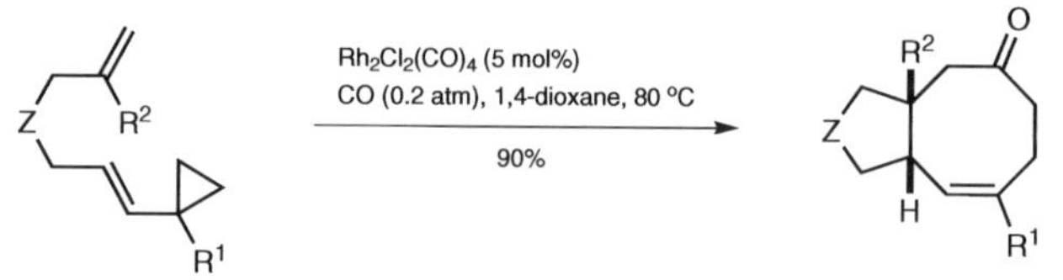

R^1	R^2	Z	Yield (%)	Ref
H	H	C(CO$_2$Me)$_2$	70	[88]
H	H	NTs	81	[88]
H	H	O	90	[88]
H	Me	C(CO$_2$Me)$_2$	29	[88]
H	Me	NTs	71	[88]
Ph	Me	O	78	[88]
Ph	H	NTs	92	[88]
Ph	H	O	73	[88]

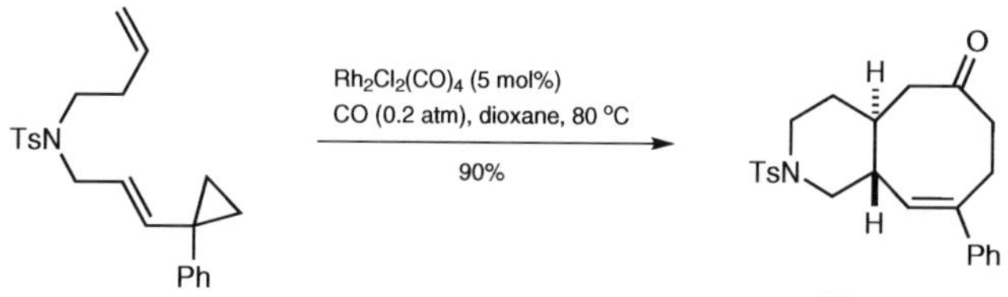

(3aR^*,6aS^*)-5-Acetyl-4-ethyl-3a-hydroxy-3,3a,6,6a-tetrahydropentalen-1(2H)-one (79, R^1 = Ac; R^2 = Et); Typical Procedure:[87]

CAUTION: *Carbon monoxide is extremely flammable and toxic, and exposure to higher concentrations can quickly lead to a coma.*

An oven-dried, septum-capped test tube was charged with Rh$_2$Cl$_2$(CO)$_4$ (0.025 mmol) and 1,4-dioxane (2.0 mL). Vinylcyclopropane **49** (1.0 mmol) was added by syringe. The mixture was purged with a stream of CO from a balloon for 15 min. Hex-3-yn-2-one (1.2 mmol) was added by syringe. The septum was pierced with a needle, and the test tube was placed in an autoclave pressurized to 2 atm and heated to 60 °C in a thermostat-controlled oil bath (for 1 atm reactions the test tube was heated directly in an oil bath under a balloon of CO). The reaction was monitored by TLC or GC, and upon completion, the mixture was cooled to rt. H$_2$O (200 µL) and 1% HCl in EtOH (25 µL) were added, and the hydrolysis was monitored by TLC (24 h). After hydrolysis, the mixture was concentrated under reduced pressure and the residue was purified by flash chromatography (silica gel, EtOAc/pentane 1:1); yield: 97%.

Bicyclic Cyclooctenones 81; General Procedure:[88]

CAUTION: *Carbon monoxide is extremely flammable and toxic, and exposure to higher concentrations can quickly lead to a coma.*

A base-washed, oven-dried Schlenk flask was charged with $Rh_2Cl_2(CO)_4$ (5 mol%) and degassed 1,4-dioxane (2 mL) under N_2, and the mixture was stirred at rt for 5 min. Then, the catalyst mixture was bubbled with CO for 10 min. A 0.05 M soln of the substrate in degassed 1,4-dioxane was added, and then CO was bubbled again through the soln for 5 min. The mixture was stirred at 80 °C for the indicated time under the CO atmosphere. After cooling to rt, the mixture was concentrated, and the residue was purified by flash column chromatography (silica gel, EtOAc/pentane).

2.8.3.2 Applications of Metal-Catalyzed (5 + 2 + 1) Cycloaddition in Natural Product Synthesis

Eight-membered carbocycles are fundamentally important building blocks, and they have garnered much attention as a result of their efficient preparation.[9] A tandem rhodium-catalyzed (5 + 2 + 1)/aldol cyclization has been reported to produce a triquinane skeleton, and this method is also effective for the preparation of hirsutene and 1-desoxyhypnophilin (Scheme 40).[85] Yu and co-workers have applied this (5 + 2 + 1) cycloaddition to the synthesis of the natural product (+)-asteriscanolide (Scheme 41).[89]

Scheme 40 Synthesis of Hirsutene and 1-Desoxyhypnophilin[85]

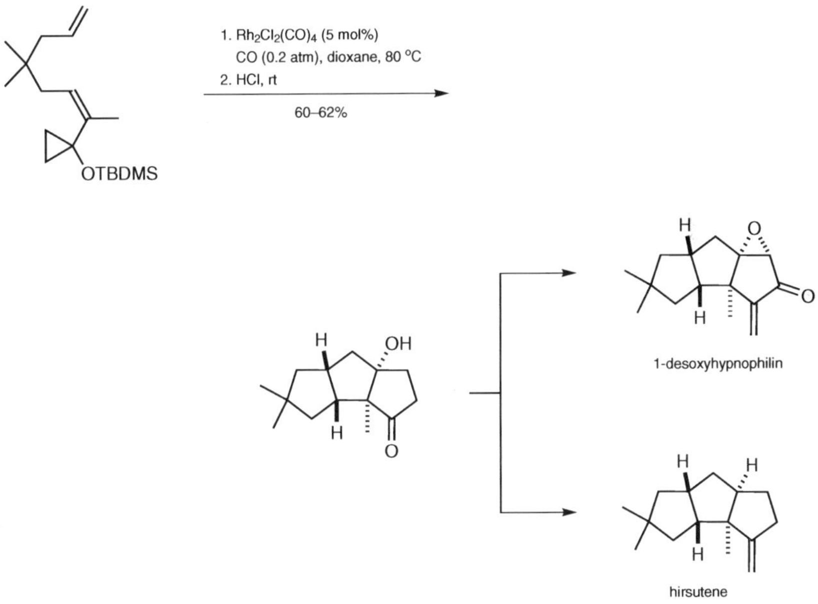

Scheme 41 Synthesis of (+)-Asteriscanolide[89]

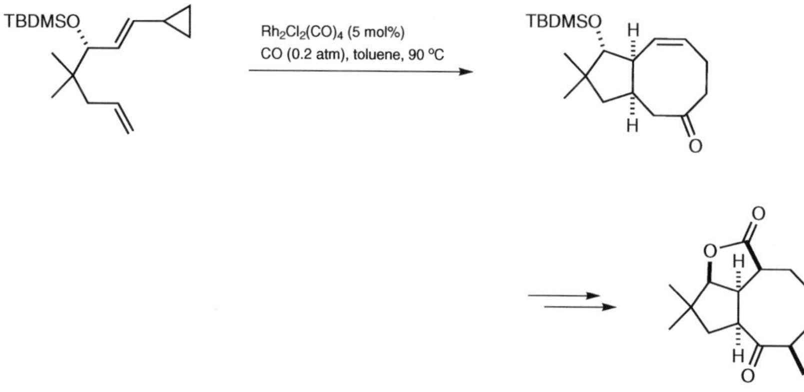

2.8.4 Conclusions

Transition-metal-catalyzed cycloaddition involving vinylcyclopropane as a five-carbon synthon has become one of the most powerful methods to access highly functionalized six-, seven-, and eight-membered carbocycles and heterocycles. Extensive research pioneered by Wender's group has demonstrated its utility in different areas and has led to the discovery of different metal catalysts and the development of new reactions. Recent reports from Tang's group have disclosed a new type of five-carbon component (3-acyloxy-1,4-enynes) that clearly offers an exciting opportunity for a new wave of metal-catalyzed (5 + n)- and (5 + m + n)-cycloaddition reactions. However, multistep substrate preparation and the high cost of the rhodium catalysts used by both five-carbon synthons limit their potential application, especially in industry. The development of less expensive and environmentally benign metal catalysts and the discovery of new five-carbon synthons would address these issues. There is no doubt that great opportunities exist in this field, and this should encourage chemists to pursue truly general and powerful metal-catalyzed cycloaddition reactions.

References

[1] Lautens, M.; Klute, W.; Tam, W., *Chem. Rev.*, (1996) **96**, 49.
[2] Ojima, I.; Tzamarioudaki, M.; Li, Z. Y.; Donovan, R. J., *Chem. Rev.*, (1996) **96**, 635.
[3] Frühauf, H.-W., *Chem. Rev.*, (1997) **97**, 523.
[4] Trost, B. M.; Krische, M. J., *Synlett*, (1998), 1.
[5] Yet, L., *Chem. Rev.*, (2000) **100**, 2963.
[6] Aubert, C.; Buisine, O.; Malacria, M., *Chem. Rev.*, (2002) **102**, 813.
[7] *Modern Rhodium-Catalyzed Organic Reactions*, Evans, P. A., Ed.; Wiley-VCH: Weinheim, Germany, (2005).
[8] Michelet, V.; Toullec, P. Y.; Genêt, J.-P., *Angew. Chem. Int. Ed.*, (2008) **47**, 4268.
[9] Yu, Z.-X.; Wang, Y.; Wang, Y., *Chem.–Asian J.*, (2010) **5**, 1072.
[10] Inglesby, P. A.; Evans, P. A., *Chem. Soc. Rev.*, (2010) **39**, 2791.
[11] Aubert, C.; Fensterbank, L.; Garcia, P.; Malacria, M.; Simonneau, A., *Chem. Rev.*, (2011) **111**, 1954.
[12] Battiste, M. A.; Pelphrey, P. M.; Wright, D. L., *Chem.–Eur. J.*, (2006) **12**, 3438.
[13] Butenschön, H., *Angew. Chem. Int. Ed.*, (2008) **47**, 5287.
[14] Nguyen, T. V.; Hartmann, J. M.; Enders, D., *Synthesis*, (2013) **45**, 845.
[15] Ben-Shoshan, R.; Sarel, S., *J. Chem. Soc. D*, (1969), 883.
[16] Sarel, S., *Acc. Chem. Res.*, (1978) **11**, 204.
[17] Aumann, R., *J. Am. Chem. Soc.*, (1974) **96**, 2631.
[18] Taber, D. F.; Kanai, K.; Jiang, Q.; Bui, G., *J. Am. Chem. Soc.*, (2000) **122**, 6807.
[19] Semmelhack, M. F.; Ho, S.; Steigerwald, M.; Lee, M. C., *J. Am. Chem. Soc.*, (1987) **109**, 4397.
[20] Iwasawa, N.; Owada, Y.; Matsuo, T., *Chem. Lett.*, (1995) **24**, 115.
[21] Kurahashi, T.; de Meijere, A., *Synlett*, (2005), 2619.
[22] Murakami, M.; Itami, K.; Ubukata, M.; Tsuji, I.; Ito, Y., *J. Org. Chem.*, (1998) **63**, 4.
[23] Kamitani, A.; Chatani, N.; Morimoto, T.; Murai, S., *J. Org. Chem.*, (2000) **65**, 9230.
[24] Cho, S. H.; Liebeskind, L. S., *J. Org. Chem.*, (1987) **52**, 2631.
[25] Brancour, C.; Fukuyama, T.; Ohta, Y.; Ryu, I.; Dhimane, A.-L.; Fensterbank, L.; Malacria, M., *Chem. Commun. (Cambridge)*, (2010) **46**, 5470.
[26] Fukuyama, T.; Ohta, Y.; Brancour, C.; Miyagawa, K.; Ryu, I.; Dhimane, A.-L.; Fensterbank, L.; Malacria, M., *Chem.–Eur. J.*, (2012) **18**, 7243.
[27] Rautenstrauch, V., *J. Org. Chem.*, (1984) **49**, 950.
[28] Marion, N.; Nolan, S. P., *Angew. Chem. Int. Ed.*, (2007) **46**, 2750.
[29] Marco-Contelles, J.; Soriano, E., *Chem.–Eur. J.*, (2007) **13**, 1350.
[30] Rudolph, M.; Hashmi, A. S. K., *Chem. Soc. Rev.*, (2012) **41**, 2448.
[31] Shu, X.-z.; Shu, D.; Schienebeck, C. M.; Tang, W., *Chem. Soc. Rev.*, (2012) **41**, 7698.
[32] Shu, D.; Li, X.; Zhang, M.; Robichaux, P. J.; Tang, W., *Angew. Chem. Int. Ed.*, (2011) **50**, 1346.
[33] Shu, D.; Li, X.; Zhang, M.; Robichaux, P. J.; Guzei, I. A.; Tang, W., *J. Org. Chem.*, (2012) **77**, 6463.
[34] Jiang, G.-J.; Fu, X.-F.; Li, Q.; Yu, Z.-X., *Org. Lett.*, (2012) **14**, 692.
[35] Taber, D. F.; Bui, G.; Chen, B., *J. Org. Chem.*, (2001) **66**, 3423.
[36] Singh, V.; Krishna, U. M.; Vikrant; Trivedi, G. K., *Tetrahedron*, (2008) **64**, 3405.
[37] Pellissier, H., *Adv. Synth. Catal.*, (2011) **353**, 189.
[38] Ylijoki, K. E. O.; Stryker, J. M., *Chem. Rev.*, (2013) **113**, 2244.
[39] Fraga, B. M., *Nat. Prod. Rep.*, (2012) **29**, 1334.
[40] Fraga, B. M., *Nat. Prod. Rep.*, (2013) **30**, 1226.
[41] Wender, P. A.; Takahashi, H.; Witulski, B., *J. Am. Chem. Soc.*, (1995) **117**, 4720.
[42] Wender, P. A.; Dyckman, A. J.; Husfeld, C. O.; Kadereit, D.; Love, J. A.; Rieck, H., *J. Am. Chem. Soc.*, (1999) **121**, 10442.
[43] Wender, P. A.; Husfeld, C. O.; Langkopf, E.; Love, J. A., *J. Am. Chem. Soc.*, (1998) **120**, 1940.
[44] Wender, P. A.; Glorius, F.; Husfeld, C. O.; Langkopf, E.; Love, J. A., *J. Am. Chem. Soc.*, (1999) **121**, 5348.
[45] Wender, P. A.; Husfeld, C. O.; Langkopf, E.; Love, J. A.; Pleuss, N., *Tetrahedron*, (1998) **54**, 7203.
[46] Wender, P. A.; Haustedt, L. O.; Lim, J.; Love, J. A.; Williams, T. J.; Yoon, J.-Y., *J. Am. Chem. Soc.*, (2006) **128**, 6302.
[47] Shintani, R.; Nakatsu, H.; Takatsu, K.; Hayashi, T., *Chem.–Eur. J.*, (2009) **15**, 8692.
[48] Binger, P.; Wedemann, P.; Kozhushkov, S. I.; de Meijere, A., *Eur. J. Org. Chem.*, (1998), 113.
[49] Wender, P. A.; Rieck, H.; Fuji, M., *J. Am. Chem. Soc.*, (1998) **120**, 10976.
[50] Wender, P. A.; Barzilay, C. M.; Dyckman, A. J., *J. Am. Chem. Soc.*, (2001) **123**, 179.

[51] Wender, P. A.; Gamber, G. G.; Scanio, M. J. C., *Angew. Chem. Int. Ed.*, (2001) **40**, 3895.
[52] Wender, P. A.; Stemmler, R. T.; Sirois, L. E., *J. Am. Chem. Soc.*, (2010) **132**, 2532.
[53] Wender, P. A.; Fournogerakis, D. N.; Jeffreys, M. S.; Quiroz, R. V.; Inagaki, F.; Pfaffenbach, M., *Nat. Chem.*, (2014) **6**, 448.
[54] Wegner, H. A.; de Meijere, A.; Wender, P. A., *J. Am. Chem. Soc.*, (2005) **127**, 6530.
[55] Jiao, L.; Ye, S.; Yu, Z.-X., *J. Am. Chem. Soc.*, (2008) **130**, 7178.
[56] Li, Q.; Jiang, G.-J.; Jiao, L.; Yu, Z.-X., *Org. Lett.*, (2010) **12**, 1332.
[57] Inagaki, F.; Sugikubo, K.; Miyashita, Y.; Mukai, C., *Angew. Chem. Int. Ed.*, (2010) **49**, 2206.
[58] Wender, P. A.; Pedersen, T. M.; Scanio, M. J. C., *J. Am. Chem. Soc.*, (2002) **124**, 15 154.
[59] Feng, J.-J.; Zhang, J., *J. Am. Chem. Soc.*, (2011) **133**, 7304.
[60] Shu, X.-z.; Huang, S.; Shu, D.; Guzei, I. A.; Tang, W., *Angew. Chem. Int. Ed.*, (2011) **50**, 8153.
[61] Shu, X.-z.; Li, X.; Shu, D.; Huang, S.; Schienebeck, C. M.; Zhou, X.; Robichaux, P. J.; Tang, W., *J. Am. Chem. Soc.*, (2012) **134**, 5211.
[62] Schienebeck, C. M.; Robichaux, P. J.; Li, X.; Chen, L.; Tang, W., *Chem. Commun. (Cambridge)*, (2013) **49**, 2616.
[63] Shu, X.-z.; Schienebeck, C. M.; Song, W.; Guzei, I. A.; Tang, W., *Angew. Chem. Int. Ed.*, (2013) **52**, 13 601.
[64] Yu, Z.-X.; Wender, P. A.; Houk, K. N., *J. Am. Chem. Soc.*, (2004) **126**, 9154.
[65] Yu, Z.-X.; Cheong, P. H.-Y.; Liu, P.; Legault, C. Y.; Wender, P. A.; Houk, K. N., *J. Am. Chem. Soc.*, (2008) **130**, 2378.
[66] Liu, P.; Cheong, P. H.-Y.; Yu, Z.-X.; Wender, P. A.; Houk, K. N., *Angew. Chem. Int. Ed.*, (2008) **47**, 3939.
[67] Liu, P.; Sirois, L. E.; Cheong, P. H.-Y.; Yu, Z.-X.; Hartung, I. V.; Rieck, H.; Wender, P. A.; Houk, K. N., *J. Am. Chem. Soc.*, (2010) **132**, 10 127.
[68] Xu, X.; Liu, P.; Lesser, A.; Sirois, L. E.; Wender, P. A.; Houk, K. N., *J. Am. Chem. Soc.*, (2012) **134**, 11 012.
[69] Hong, X.; Liu, P.; Houk, K. N., *J. Am. Chem. Soc.*, (2013) **135**, 1456.
[70] Hong, X.; Trost, B. M.; Houk, K. N., *J. Am. Chem. Soc.*, (2013) **135**, 6588.
[71] Xu, X.; Liu, P.; Shu, X.-z.; Tang, W.; Houk, K. N., *J. Am. Chem. Soc.*, (2013) **135**, 9271.
[72] Trost, B. M.; Toste, F. D.; Shen, H., *J. Am. Chem. Soc.*, (2000) **122**, 2379.
[73] Trost, B. M.; Shen, H. C., *Angew. Chem. Int. Ed.*, (2001) **40**, 2313.
[74] Trost, B. M.; Shen, H. C.; Horne, D. B.; Toste, F. D.; Steinmetz, B. G.; Koradin, C., *Chem.–Eur. J.*, (2005) **11**, 2577.
[75] Zuo, G.; Louie, J., *J. Am. Chem. Soc.*, (2005) **127**, 5798.
[76] Fürstner, A.; Majima, K.; Martín, R.; Krause, H.; Kattnig, E.; Goddard, R.; Lehmann, C. W., *J. Am. Chem. Soc.*, (2008) **130**, 1992.
[77] Wender, P. A.; Fuji, M.; Husfeld, C. O.; Love, J. A., *Org. Lett.*, (1999) **1**, 137.
[78] Wender, P. A.; Zhang, L., *Org. Lett.*, (2000) **2**, 2323.
[79] Wender, P. A.; Bi, F. C.; Brodney, M. A.; Gosselin, F., *Org. Lett.*, (2001) **3**, 2105.
[80] Ashfeld, B. L.; Martin, S. F., *Org. Lett.*, (2005) **7**, 4535.
[81] Ashfeld, B. L.; Martin, S. F., *Tetrahedron*, (2006) **62**, 10 497.
[82] Trost, B. M.; Hu, Y.; Horne, D. B., *J. Am. Chem. Soc.*, (2007) **129**, 11 781.
[83] Trost, B. M.; Waser, J.; Meyer, A., *J. Am. Chem. Soc.*, (2007) **129**, 14 556.
[84] Trost, B. M.; Waser, J.; Meyer, A., *J. Am. Chem. Soc.*, (2008) **130**, 16 424.
[85] Jiao, L.; Yuan, C.; Yu, Z.-X., *J. Am. Chem. Soc.*, (2008) **130**, 4421.
[86] Trost, B. M.; Nguyen, H. M.; Koradin, C., *Tetrahedron Lett.*, (2010) **51**, 6232.
[87] Wender, P. A.; Gamber, G. G.; Hubbard, R. D.; Zhang, L., *J. Am. Chem. Soc.*, (2002) **124**, 2876.
[88] Wang, Y.; Wang, J.; Su, J.; Huang, F.; Jiao, L.; Liang, Y.; Yang, D.; Zhang, S.; Wender, P. A.; Yu, Z.-X., *J. Am. Chem. Soc.*, (2007) **129**, 10 060.
[89] Liang, Y.; Jiang, X.; Yu, Z.-X., *Chem. Commun. (Cambridge)*, (2011) **47**, 6659.

2.9 Intramolecular Free-Radical Cyclization Reactions

M. Mondal and U. Bora

2.9.1 Introduction to Radical Cyclization

Over the past five decades, radical chemistry has developed into one of the most powerful tools in synthetic organic chemistry. Initially considered too reactive and extremely uncontrollable, radicals now play a significant role in the development of novel methodology and have found extensive use in the synthesis of complex natural products.[1–8]

The first free-radical-based polycyclizations were reported by the groups of Breslow and Julia.[9–11] Pioneering work performed by the Julia,[12] Walling,[13] and Ingold and Beckwith[14–19] groups has carved a new direction in the field of controlled radical chemistry. Advances in this area outlined by Hart,[20] Stork,[21,22] Curran,[23,24] and others have revolved around applications of radical reactions. Houk[25] and RajanBabu[26–29] have provided mechanistic insight into the control of stereochemistry in intramolecular free-radical cyclizations, and these findings have led to widespread advances in the area of monocyclization and polycyclization reactions. Conventionally, tributyltin hydride or hexabutyldistannane is chiefly employed to generate carbon-centered radicals through cleavage of C—halogen, C—S, or C—OR^1 bonds. In these reactions, the product radical centers are normally reduced by hydrogen-atom transfer, which often results in a loss of functionality on going from the reactant material to the product. To avoid this disadvantage, transition-metal-based radical reactions have been adopted, mainly as a result of the pioneering efforts of Kharasch,[30,31] Kochi,[32] and Minisci,[33] who showed that carbon-centered radicals may be generated by using organometallic reagents. The advantages of using transition-metal-based reagents are that they provide selective free-radical reactions under mild conditions and tolerate organic functionalities during the process of the reaction.[34,35] Moreover, owing to their high stereo- and regioselectivity, these processes have been extensively employed to construct different ring sizes and structures[36–39] and, furthermore, have been used in the synthesis of numerous complex organic molecules.[40–43]

Although many free-radical cyclization strategies are available for the construction of organic molecules, the use of annulation and related cascade reactions have received the most consideration. These reactions mimic the enzymatic processes that occur in the biosynthesis of organic compounds such as terpenoids, alkaloids, sterols, and so on and, thus, provide a potential synthetic tool to perform multiple ring-closing reactions in a single step or in a minimum number of steps.[24,44–46]

Radical cyclization reactions can be mediated by transition metals such as manganese,[47–53] copper,[54,55] lead,[56–58] cerium,[59–61] iron,[62–64] samarium,[65–68] titanium,[69–72] cobalt,[73] and ruthenium;[74] these elements are considered the most efficient and suitable for organic synthesis, as they provide unique single or tandem modes for the synthesis of numerous biologically active complex molecules in a single operation or in a minimum of steps. Although these metals are different in their nature and in the fundamental chemistry that they can perform, they are often found to deliver similar free-radical mechanisms with identical reactivities.

Radical reactions offer a wide range of advantages over their ionic counterparts. Reactions involving anions or cations generally proceed under conditions of high basicity or high acidity, respectively, whereas radical reactions usually proceed under mild or neutral conditions. This property allows the efficient transformation of pH-sensitive and chiral substrates without decomposition or racemization. Moreover, in reactions involving ions, the solvent often interferes, whereas radicals are generally less solvated and can react identically in a wide range of different solvents. With broad functional group scope, stability under protic conditions, and less solvation character, free-radical reactions are often employed to assemble sterically hindered centers within complex target molecules. Numerous notable methods for radical cyclization have been reported, amongst which are George's total synthesis of (+)-garcibracteatone,[75,76] Chen's total syntheses of ageliferins[77,78] and palau'amine,[79] and Yamashita's concise synthesis of the tetracyclic framework of azadiradione.[80] Various carbonyl compounds, such as β-diketones, acetoacetamides, β-keto esters, and acylacetonitriles, have been used as substrates to generate carbon-centered radicals in the presence of manganese(III) acetate; the use of these radicals in a number of cyclization reactions,[48] including tandem and cascade processes, allows the synthesis of products such as furan derivatives,[52] γ-lactones,[52] oxepin derivatives,[81] (+)-fusarisetin A,[82] (±)-mersicarpine,[83] ambroxide,[84] isosteviol,[85] wentilactone B,[86] and (−)-triptolide.[87,88] Herein, we provide a brief description of the chemistry of the commonly used metal complexes that are primarily employed in intramolecular free-radical cyclization reactions.

2.9.1.1 Manganese(III) Acetate Based Radical Reactions

Manganese(III) acetate is a one-electron oxidant. It is recognized as the most prominent reagent in the field of contemporary free-radical chemistry, as it allows extensive versatility in its application in the formation of various unique organic molecules, owing to its moderate reactivity and high selectivity under mild reaction conditions. Since the late 1960s, manganese(III) acetate mediated radical reactions have led to significant advances in the field of oxidative chemistry, as manganese(III) acetate plays a significant role as a single-electron-transfer reagent and provides chemists with an overwhelming variety of carbon-centered radicals that can be obtained from carbonyl compounds such as β-diketones, acetoacetamides, β-keto esters, and acylacetonitriles; phosphorus-centered radicals from organophosphorus compounds such as phosphites and phosphine oxides; sulfur-centered radicals from thiols, α-sulfanyl ketones, arylthioformanilides, thioglycolic acid, and ammonium thiocyanate; and nitrogen-based radicals from metal and vinyl azides. Thus, there is plethora of alternatives available for the synthesis of vital organic compounds.[53]

The chemistry of manganese(III) acetate started in the late nineteenth century, when Christensen reported its synthesis on the basis of the electrochemical oxidation of manganese(II) acetate tetrahydrate with potassium permanganate and chlorine.[89,90] Nowadays, manganese(III) acetate is commercially available in two distinguishable forms: manganese(III) acetate dihydrate (cinnamon brown powder) and anhydrous manganese(III) acetate (dark brown). Manganese(III) acetate dihydrate can be prepared in the laboratory by adding potassium permanganate to a preheated mixture of manganese(II) acetate and glacial acetic acid, followed by the addition of a suitable amount of water (Scheme 1). The addition of acetic anhydride during the process produces anhydrous manganese(III) acetate having solubility identical to that of the hydrated form.[91] The amount of water used during the synthesis may alter the yield and the nature of the manganese(III) acetate complex, in addition to the rate of the subsequent alkene–lactonization reaction.[91] The catalytic manganese(II) acetate/cobalt(II) acetate/oxygen redox system, in which manganese(II) is oxidized to manganese(III) in the presence of cobalt(II)/oxygen,[92–94] can be used to convert dialkyl phosphites and ketones into the corresponding phosphorus radicals and α-keto radicals.

2.9.1 Introduction to Radical Cyclization

Scheme 1 Synthesis of Manganese(III) Acetate[91]

$$Mn(OAc)_2 + KMnO_4 + AcOH \xrightarrow{110\ °C,\ 20\ min} Mn(OAc)_3 + MnO_2 + AcOK$$

In the 1920s, Weinland proposed that the structural formula of manganese(III) acetate dihydrate is actually a trimetallic $[Mn_3(OAc)_6(H_2O)_2](OAc)_3 \cdot 4H_2O$ complex.[95] The structure of an anhydrous crystal of manganese(III) acetate was proposed by Hessel and Romers in 1969.[96] They proposed an oxo-centered trinuclear complex having three manganese atoms bridged with acetate units within the consecutive metal units. In 1968, Heiba[97] and Bush[91] independently reported the earliest method for a manganese(III) acetate mediated oxidative free-radical cyclization reaction. It was observed that, in the presence of acetic acid, manganese(III) acetate promotes the cyclization of alkenes to γ-lactones **1** in excellent yields with only trace amounts of oxidized side products (Scheme 2).

Scheme 2 Manganese(III) Acetate Mediated Oxidative Free-Radical Cyclization of Alkenes to γ-Lactones[91,97]

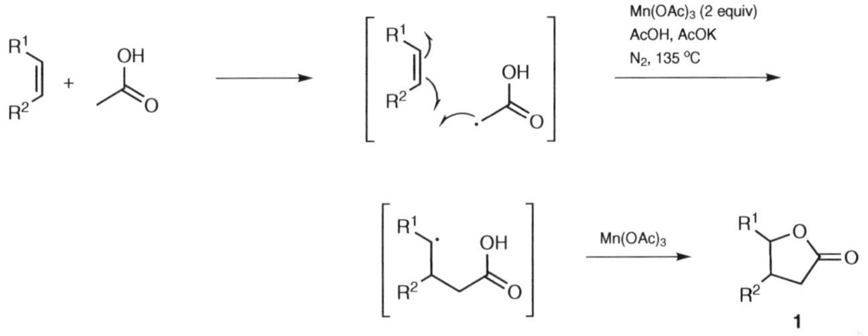

The alkylation of benzene with acetone can be performed using manganese(III) acetate. It generates a radical from acetone by one-electron oxidation, and this radical finally adds to benzene (Scheme 3).[98]

Scheme 3 Manganese(III) Acetate Mediated Alkylation of Benzene with Acetone[98]

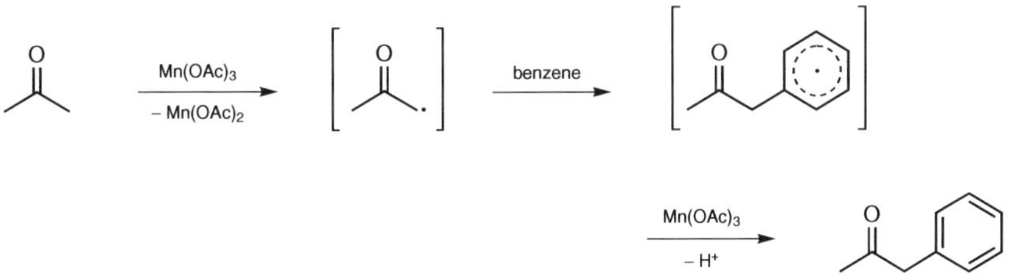

Manganese(III) acetate has also been employed in the coupling of β-dicarbonyl compounds **2** and styrene to produce the corresponding substituted dihydrofurans **3** (Scheme 4).[99]

Scheme 4 Synthesis of Dihydrofuran Derivatives by Manganese(III) Acetate Mediated Coupling of β-Dicarbonyl Compounds and Styrene[99]

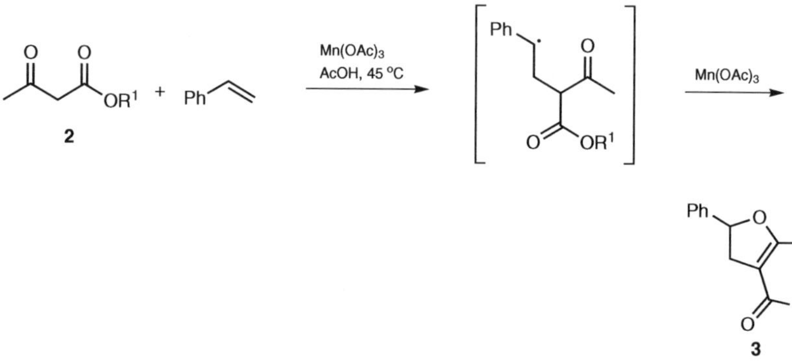

The enolizability and C—H acidity of substrates affect the rate at which a radical can be generated by manganese(III) acetate,[100–102] and thus, C—O bonds are predominantly formed highly efficiently through the generation of a manganese(III)–enolate complex followed by addition of a carbon-centered radical to the unsaturated system.[48,49,103]

Mechanistically, the reaction proceeds with ligand exchange between manganese(III) acetate and dicarbonyl **4** to generate manganese(III)–enolate complex **5** in situ. Through a one-electron oxidation process, complex **5** produces carbon-centered, electron-deficient radical **6**, which undergoes oxidative addition with the nucleophilic C=C bond of **7** to produce cyclic derivative **10** (Scheme 5).[99,104–108] Copper(II) acetate is also added as a co-oxidant in some reactions to assist the oxidation of carbon radical **8** to carbocation **9**.[49] Two equivalents of manganese(III) acetate (both the anhydrous and dihydrate forms can be used) are often required to generate the carbocation intermediate. Two equivalents are also necessary if copper(II) acetate is used, as manganese(III) acetate reoxidizes copper(I) to copper(II).

2.9.1 Introduction to Radical Cyclization

Scheme 5 Mechanism for the Formation of the Manganese(III)–Enolate Complex and the Synthesis of Dihydrofurans[99,104–108]

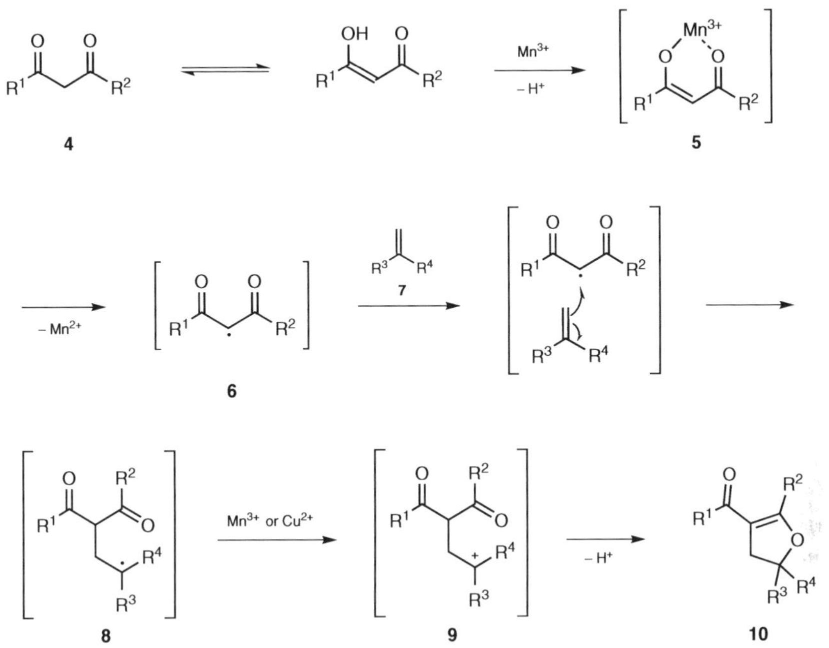

Manganese(III) acetate exhibits excellent catalytic activity in acetic acid, whereas in other solvents, such as ethanol, benzene, toluene, and dichloromethane,[109–111] its reactivity is slightly decreased. The lower oxidizing efficiency is possibly due to destruction of the oxygen-centered trinuclear manganese(III) acetate complex in these solvents. This property has been used in many reactions to control selectivity.[111–114]

2.9.1.2 Titanocene(III)-Based Radical Reactions

Titanium, the seventh most abundant metal on Earth, is found in almost all living things, rocks, water bodies, and soils. It is one of the least-expensive transition metals, and generally, titanium-based compounds are nontoxic and ecofriendly. They act as powerful radical promoters in many synthetic routes owing to their selective involvement during the initiation and termination steps of radical chains. Their low Lewis acidity and their reducing properties provide excellent coordination with electron-rich organic substrates.

In the titanium family, the most popular reagent used to generate radicals is chlorobis(η^5-cyclopentadienyl)titanium(III) [titanocene(III) chloride]. Titanocene(III) complexes are well established as powerful, soft, single-electron-transfer reagents in organic synthesis. They are capable of promoting a diverse range of transformations, such as homolytic epoxide[115–118] and oxetane ring opening,[119] Wurtz-type reactions,[120] Barbier-type reactions,[121] Reformatsky-type reactions,[122,123] reduction reactions,[124–126] and pinacol-coupling reactions.[127–132]

Titanocene(III) complexes are highly sensitive to oxygen, and similar to any other low-valent titanium species, they are not very stable. Thus, they are normally prepared in situ from commercially available dichlorobis(η^5-cyclopentadienyl)titanium(IV) by using a reductant such as aluminum,[133] zinc,[134] or manganese.[135,136] Skrydstrup and coworkers have studied the nature of the reducing agent in solution and have provided strong evidence for the existence of an equilibrium between monomers **12/13** and dimer **11** (Scheme 6).[137]

Scheme 6 Equilibrium between Monomeric and Dimeric Species of Chlorobis(η^5-cyclopentadienyl)titanium(III)[137]

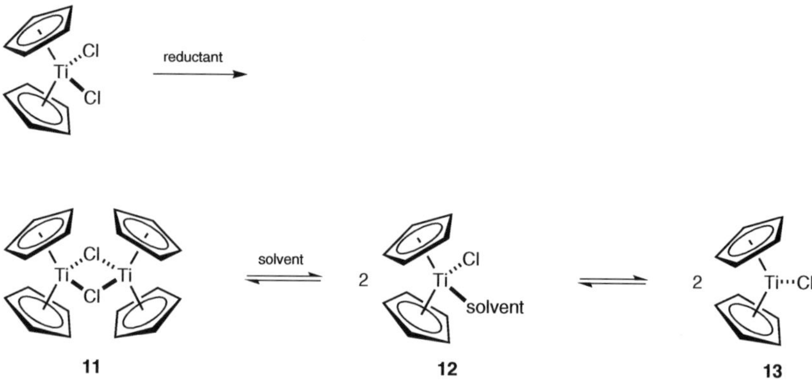

The chemistry of titanocene(III) complexes has been extensively investigated in the oxidation of epoxides, allylic and propargylic halides, and α,β-unsaturated and aromatic aldehydes and ketones. Epoxides are commonly used as radical precursors owing to their numerous attractive features relative to other conventional precursors. They are easily synthesized from carbonyls or alkenes in enantiomerically pure form and retain adjacent functionality, even after the generation of the radical.

Traditionally, epoxides are known to undergo heterolytic ring opening to facilitate C—C bond formation through carbocation- or carbanion-type chemistry upon treatment with acidic reagents[138] or carbon nucleophiles,[139] respectively. In 1988, RajanBabu and Nugent established that homolytic ring opening of epoxides **14** can be achieved in the presence of titanocene(III) complexes.[140–143] Initially, the reaction of the titanocene(III) species with epoxides leads to homolytic ring opening, which yields carbon-centered β-titanoxyl radical **15** (Scheme 7).

2.9.1 Introduction to Radical Cyclization

Scheme 7 Titanocene(III)-Mediated Homolytic Ring Opening of Epoxides[140–143]

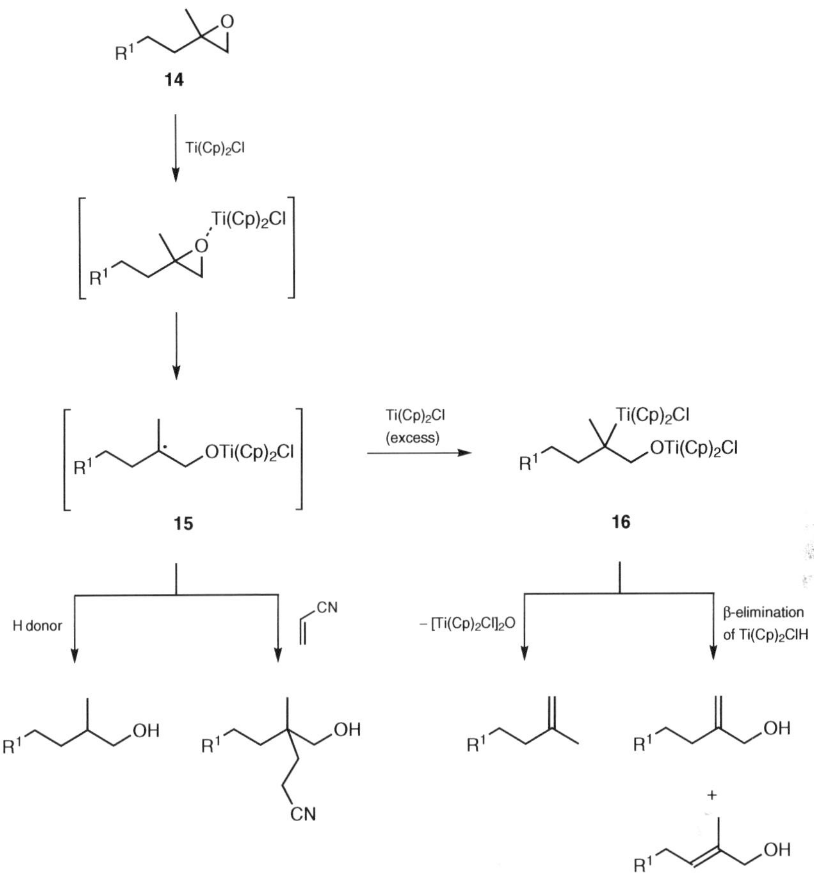

Radical **15** can be reduced with a good hydrogen-atom donor, such as cyclohexa-1,4-diene, to the corresponding alcohol with anti-Markovnikov regioselectivity.[142,143] This reduction has been utilized for the synthesis of both 1,2- and 1,3-diols from Sharpless epoxides[143–148] and for the synthesis of β-hydroxy ketones from α,β-epoxy ketones.[149] In the presence of an excess amount of the titanocene(III) species, radical **15** reacts with a second molecule of chlorobis(η^5-cyclopentadienyl)titanium(III) to give alkyltitanium complex **16**. This species has been used to achieve a number of significant organic transformations. Alkyltitanium complex **16** has been used to prepare alkenes by elimination of oxo-bridged dichlorotetrakis(η^5-cyclopentadienyl)dititanium(III)[142] and to generate allylic alcohols by β-elimination of chlorobis(η^5-cyclopentadienyl)hydridotitanium(III).[150] Another remarkable feature of β-titanoxyl radical **15** is its reactivity toward alkenes activated with unsaturated groups such as nitriles and esters to form new C—C bonds with either equimolar[141,143] or catalytic quantities[135,151,152] of titanium(III). Similarly, epoxide-derived radicals are also known to undergo addition to α,β-unsaturated carbene complexes.[153–155] Titanoxyl radicals formed by the reaction of an epoxide bearing suitable unsaturated groups, for example, alkenes, alkynes, carbonyl groups, and nitrile groups, undergo intramolecular cyclization in the presence of chlorobis(η^5-cyclopentadienyl)titanium(III).

Gansäuer and Rinker compared the reactivities of several commonly used low-valent electron-transfer reagents toward the reductive ring opening of epoxides and showed that titanocene(III) reagents were optimal for this purpose.[156] Upon performing experiments with chlorobis(η^5-cyclopentadienyl)titanium(III),[137] samarium(II) iodide, chromium(II) chloride, and the vanadium(II) complex [V$_2$Cl$_3$(thf)$_6$]$_2$[Zn$_2$Cl$_6$], obtained by the reduction of trichlorotris(tetrahydrofuran)vanadium(III)[157] with zinc dust in tetrahydrofuran, as the actual reducing agent, it was revealed that the optimum reagent shows a good balance of Lewis acidity versus reducing tendency. These results suggest that the reason for the high efficacy of the chlorobis(η^5-cyclopentadienyl)titanium(III) reagent is due to its unique combination of low Lewis acidity, which prevents epoxide opening by nucleophilic substitution (S$_N$1 or S$_N$2), and low reducing power toward the β-metal oxy radical (e.g., **15**). However, samarium(II) iodide is an unsuitable reagent for the reductive opening of epoxides. This is because the high Lewis acidity of samarium metal, along with the high nucleophilicity of the iodide ions, produces iodohydrins, which are further reduced by samarium. In the case of the vanadium(II) reagent, which is less Lewis acidic, epoxide deoxygenation is a particularly prevailing side reaction. Although the vanadium(II) complex is not a powerful reductant, this species can react through ligand exchange to give a metallaoxetane that eliminates a vanadium oxo species to yield the allylic ether. This anomalous behavior has been explained by assuming a dimeric nature of the vanadium complex in solution as the cause of the deoxygenation. Chromium(II) chloride is the only other reagent that shows the necessary balance of properties to allow C—C bond formation; alas, it is severely limited by low reactivity.

The majority of reactions promoted by chlorobis(η^5-cyclopentadienyl)titanium(III) require the use of stoichiometric quantities of titanium (a frequent environmental concern), which results in a decrease in the yield because of reductive trapping of the intermediate radical. The catalytic cycle necessary for titanocene(III) chemistry can be obtained by adding a reagent to regenerate a titanocene(IV) species along with an electron source, such as zinc dust or manganese. The most commonly used regenerating agents are a combination of chlorotrimethylsilane and 2,4,6-trimethylpyridine for reactions performed under aprotic conditions and 2,4,6-trimethylpyridine hydrochloride for reactions performed under aqueous conditions. The earliest employed catalytic combination was dichlorobis(η^5-cyclopentadienyl)titanium(IV)/manganese/2,4,6-trimethylpyridine hydrochloride.[135] In this system, 2,4,6-trimethylpyridine hydrochloride protonates the Ti—O and Ti—C bonds formed during the reaction to give the alcohol (O—H) and the alkane (C—H). In a subsequently developed system, under aprotic conditions, the combination of dichlorobis(η^5-cyclopentadienyl)titanium(IV)/manganese/2,4,6-trimethylpyridine/chlorotrimethylsilane prevents reduction of the putative alkyltitanium(IV) species to the simple alkane and allows β-elimination to furnish the alkene.[158] An alternative nonreductive catalytic system based on dichlorobis(η^5-cyclopentadienyl)titanium(IV)/manganese/2,6-dimethylpyridine hydrochloride/triethylborane has also been reported. Triethylborane is believed to reduce the chlorobis(η^5-cyclopentadienyl)hydridotitanium(IV) species to the active titanium(III) species, which thereby facilitates alkyl radical disproportionation to yield an alkene rather than reduction to the simple alkane.[159] Terminal radicals are reduced by employing hydrogen gas as the hydrogen-atom source in combination with the catalytic dichlorobis(η^5-cyclopentadienyl)titanium(IV)/manganese/2,6-dimethylpyridine hydrochloride system.[160] Notably, strong Lewis acids (chlorosilanes) are incompatible with epoxides, whereas the weak acid 2,4,6-trimethylpyridine hydrochloride is not a suitable reagent to regenerate dichlorobis(η^5-cyclopentadienyl)titanium(IV) from acetoxytitanium derivatives.

A titanocene(III)-mediated epoxide ring-opening/radical cyclization strategy can be employed for the synthesis of most ring sizes.[161] Five-membered rings such as **19** are the simplest rings to create (Scheme 8).[162] The reaction furnishes either bicyclic furofurans **18** through a cascade radical cyclization/ionic cyclization mechanism or monocyclic alcohols **19** by protonation of the intermediate Ti—C bond of **17**. The cascade radical/ionic cyclization reaction is achieved by quenching the intermediate organotitanium complex with iodine followed by intramolecular S$_N$2-like cyclization. This methodology has been utilized in the synthesis of numerous natural products.[163]

2.9.1 Introduction to Radical Cyclization

Scheme 8 Chlorobis(η^5-cyclopentadienyl)titanium(III)-Mediated Ring Formation of Furans and Furofurans[162]

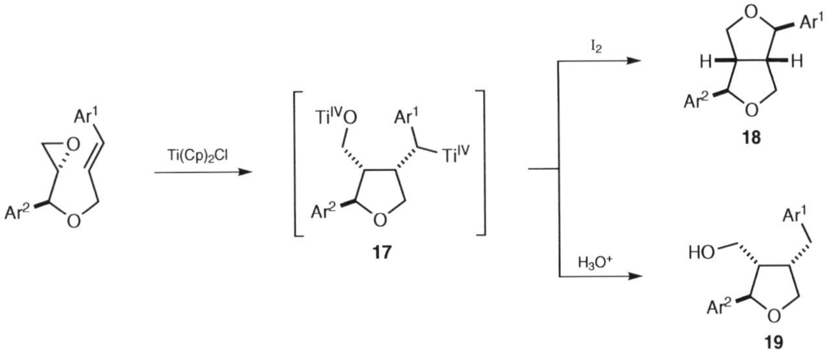

Traditionally, radical transformations are performed using toxic hydrogen donors, such as cyclohexa-1,4-diene, tributyltin hydride, or thiols. This limitation has been somewhat removed by the use of water. Water is most abundant, nontoxic, and noncorrosive; additionally, it is a nonflammable solvent. Moreover, water shows excellent chemical reactivity and in several cases increases the solubility of salts and polar compounds, which thereby enhances the rate of the reaction.[164] In radical reactions, pure water is often regarded as an inert medium because of the strength of the H—OH bond, which prevents hydrogen-atom transfer. However, the hydrogen atom can be generated from water by adding a Lewis acid (e.g., trimethylborane–water complex).[165] Reductive cyclization of **20** to **21** appears to be the first example of the use of water as a hydrogen donor in a radical transformation (Scheme 9).[166] The reaction is believed to proceed through the formation of aqua complex **22**, and the hydrogen atom in the 4-position of **21** comes from water (Scheme 9). Subsequently, treatment of a wide range of epoxides with chlorobis(η^5-cyclopentadienyl)-titanium(III) and an excess of water has been shown to allow efficient reductive ring opening to yield alcohols.[124]

Scheme 9 Chlorobis(η^5-cyclopentadienyl)titanium(III)-Mediated Ring Formation Using Water as the Hydrogen Donor[166]

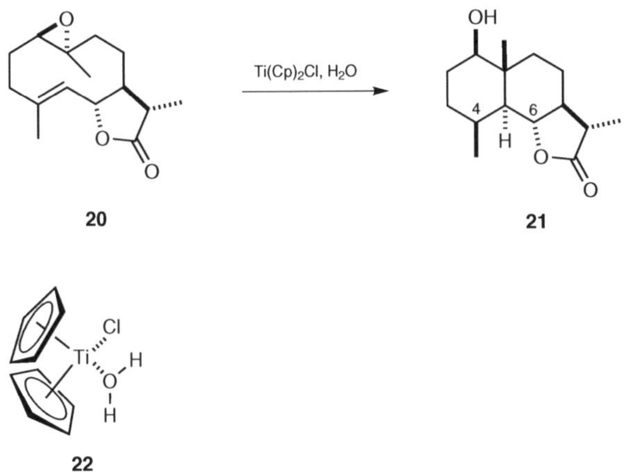

Apart from the use of epoxides, alkyl bromides,[167] unactivated alkynes and alkenes,[168] and unactivated iodides have been utilized as radical precursors in chlorobis(η^5-cyclopentadienyl)titanium(III)-mediated reactions. Thus, reactions mediated by titanocene(III) species are invaluable in the synthetic chemist's arsenal. Titanocene(III) reagents are easily

for references see p 533

accessible, easily removed after completion of the reaction, and are not toxic. Many reactions mediated by chlorobis(η⁵-cyclopentadienyl)titanium(III) proceed far more efficiently than reactions performed with the traditional samarium(II) iodide single-electron-transfer reagent. Reactions based on functional group interconversions are often employed in the synthesis of complex natural products such as sesterstatin,[143] barekoxide, and sicannin, owing to their outstanding chemoselective behavior. Within this context, the aim of this chapter is to present the best intramolecular radical cyclization reactions promoted by chlorobis(η⁵-cyclopentadienyl)titanium(III), together with the use of this species as a versatile tool to construct complex and biologically relevant natural products.

2.9.1.3 Samarium(II) Iodide Based Radical Reactions

Coverage within *Science of Synthesis* is limited to the best methods available for practical use in both academia and industry. However, in the quest to find the ideal method for organic synthesis, high yields do not always signify efficiency and are rather dependent on other factors. For instance, organic reactions that proceed without functional group protection, that generate single isomeric products, and that form multiple bonds and stereocenters in a single operation are the most efficient or ideal ones. Thus, sequential or tandem one-pot reaction processes that proceed without isolation of any intermediates represent powerful approaches to meet these criteria. In the late 1970s, a new paradigm for free-radical chemistry emerged that has considerably enhanced the ability of synthetic chemists to assemble complex molecular frameworks to provide new opportunities, particularly in total synthesis, under operationally efficient steps.[169–171] Prominent among these transition-metal-based reagents is the exceptional prowess of samarium(II) iodide for sequential reactions (Scheme 10).[65,172–177]

Scheme 10 Samarium(II) Iodide Mediated Synthesis of a Polycyclic System by Sequential Reactions[177]

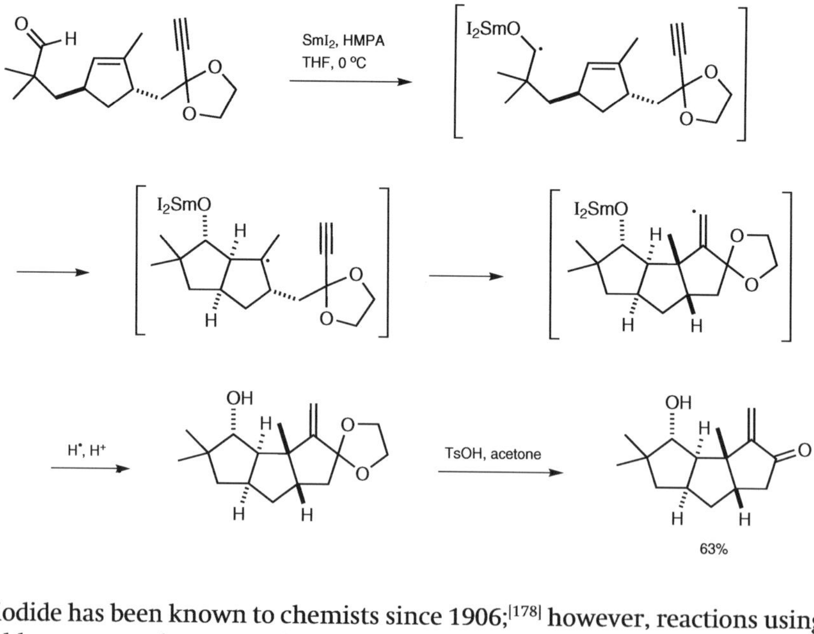

Samarium(II) iodide has been known to chemists since 1906;[178] however, reactions using this ether-soluble reagent only appeared on the scene in the late 1970s.[169–171] Since then, it has played a growing role in organic synthesis owing to its versatility in one- and two-electron-transfer reactions (Scheme 11). The attractiveness of samarium(II) iodide is also due to its unique ability to selectively access both radical[179–183] and ionic reactions during

2.9.1 Introduction to Radical Cyclization

the course of the same reaction.[184,185] This fact makes samarium(II) iodide a widely employed and recognized single-electron reducing agent in synthetic chemistry with tunable reactivity and selectivity. Numerous synthetic transformations ranging from functional group interconversions to C—C bond forming reactions have been performed using samarium(II) iodide. Samarium(II) iodide mediated reactions exhibit high stereoselectivity owing to the excellent Lewis acid character of this reagent, which generates highly oxophilic intermediate species during the reactions. With its excellent tendency to perform sequential transformations, samarium(II) iodide based reactions require considerably less time, materials, and effort than traditional multistep procedures.[65,186]

Scheme 11 Reactions Proceeding by One-Electron- and Two-Electron-Transfer Mechanisms[169–186]

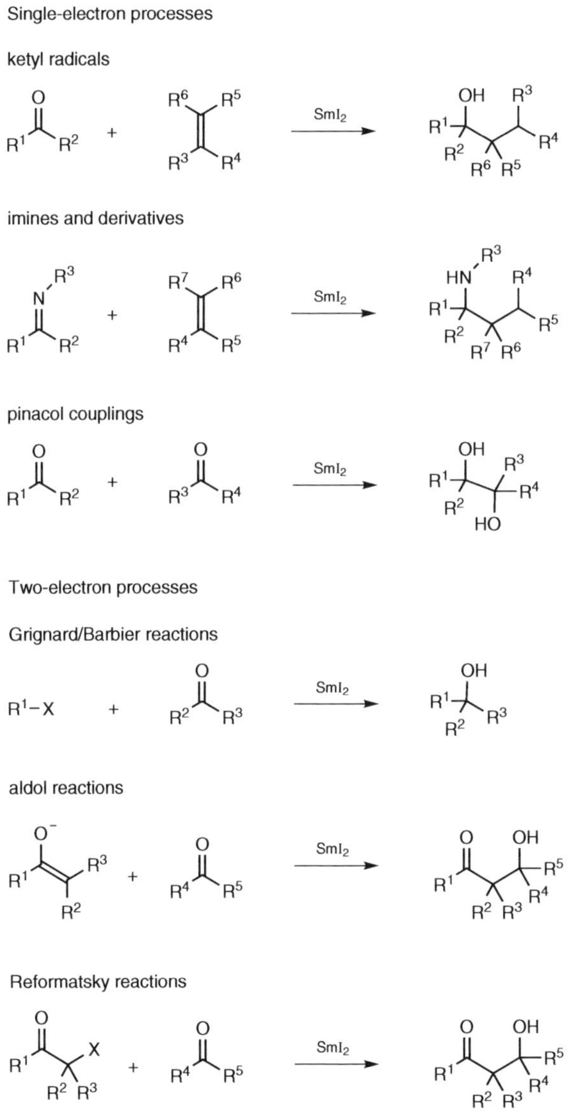

Single-electron processes

ketyl radicals

imines and derivatives

pinacol couplings

Two-electron processes

Grignard/Barbier reactions

aldol reactions

Reformatsky reactions

Owing to intense scientific interest, samarium(II) iodide is now commercially available as a 0.1 M solution in tetrahydrofuran. In the laboratory, it can be prepared conveniently by treatment of samarium metal with diiodomethane,[187] 1,2-diiodoethane,[170] or iodine[188]

for references see p 533

at low concentrations (0.1 M) of tetrahydrofuran. In tetrahydrofuran (0.1 M), samarium(II) iodide is often obtained quantitatively as a deep-blue solution. If an excess of iodine is present, samarium(III) iodide is formed and the solution does not take on the intense dark blue/almost black color of samarium(II) iodide. Sonication of samarium metal with various iodine sources, including iodoform, over a short reaction time is an alternative method that is effective.[189] Samarium(II) iodide can be prepared in acetonitrile by reaction of samarium metal with chlorotrimethylsilane and sodium iodide.[190] Samarium(II) iodide can also be readily produced in pivalonitrile (2,2-dimethylpropanenitrile) and 1,2-dimethoxyethane.[191–194]

Given that samarium(II) iodide is sensitive to air, it should be handled using standard syringe techniques. Samarium(II) iodide based reactions are usually performed in tetrahydrofuran, although other solvents have also been investigated.[195] The reactivity, stereoselectivity, and chemoselectivity of samarium(II) iodide mediated reactions can be widely modulated by the addition of different cosolvents and additives.[196] Owing to chelating effects involving samarium(3+) and samarium(2+) ions, numerous C—C bond forming reactions, particularly those involved in the preparation of polyfunctional molecules, have been achieved with high levels of diastereoselectivity.[197]

Reactions performed with a stoichiometric amount of samarium(II) iodide can be transformed into catalytic reactions by the addition of a stoichiometric amount of a co-reducing agent (e.g., zinc,[197] magnesium,[198] or mischmetal[199]). The addition of simple additives (e.g., water, alcohols, amines, etc.) also modulates the reactivity of samarium(II) iodide; these additives can induce specific reactions in polyfunctional systems.[174,195] Within the family of alcohols, those most commonly used in samarium(II) iodide mediated reductions include methanol and *tert*-butyl alcohol.[170,200] In the samarium(II) iodide mediated reduction of ketones, Flowers has shown that the presence of water and methanol significantly affect the reduction rate of ketones by coordinating with samarium(II).[201] The effects of Lewis bases such as hexamethylphosphoric triamide have also been extensively studied in recent years.[202–206] Other inorganic salts such as lithium chloride and lithium bromide have been used as additives in samarium(II) iodide mediated reactions, as they give rise to notable increases in reactivity.[207] Lithium halide additives also influence samarium(II) iodide mediated chemoselective reactions.[208] Other transition-metal salts,[209] such as nickel(II) iodide and iron(III) complexes, have also been used in samarium(II) iodide mediated reactions to increase reaction rates. Visible-light irradiation of samarium(II) iodide based reactions has also been used to increase the rates of reactions, including sequential radical anionic processes and carbonylation reactions.[210]

The historic, mechanistic, theoretical, and practical aspects of these processes have been sufficiently discussed,[211] and our focus in this chapter is on selected intramolecular cyclizations of the most efficient samarium(II) iodide catalyzed reactions and their corresponding use in total synthesis processes. Cyclization substrates include halides and pseudohalides; carboxylic acid derivatives; aldehydes, ketones, and imines; arenes, alkenes, and alkynes; N—O, N—N, and N—S bonds; and C—O, C—N, and C—C bonds.

2.9.2 Intramolecular Free-Radical Cyclization Routes to N-Heterocycles

2.9.2.1 Synthesis of Pyrrole-Containing Moieties

Nitrogen heterocycles are ubiquitous structural motifs that are found across various biologically active organic molecules. Recently developed methods for the synthesis of nitrogen heterocycles by metal-catalyzed intramolecular free-radical cyclization strategies have expanded the scope of the organic chemist's toolbox.

2.9.2.1.1 Synthesis of 2-Arylpyrrole Derivatives

The 2-arylpyrrole motif is widespread in drugs and scaffolds, including in antipsychotic drugs, chromogenic and fluorescent anion sensors, and ligands for transition-metal complexes.[212–215] Numerous methods have been developed for the synthesis of pyrroles;[216–218] however, there are only a few reports describing a general and convenient approach for the synthesis of 2-arylpyrroles.[219–222]

Copper halides and copper acetate are often considered efficient electron-transfer reagents that are able to promote radical-based nucleophilic substitution reactions in organic systems. Guan and co-workers report an efficient copper(I) bromide catalyzed anti-Baldwin 5-*endo-trig* cyclization method for the synthesis of 2-arylpyrrole derivatives **23** in high yields in toluene under an argon atmosphere (Scheme 12).[223] The use of ketoxime carboxylates is advantageous because of easy separation of the product and the lower probability of the formation of byproducts. Various 2-arylpyrrole derivatives can be synthesized by 5-*endo-trig* cyclization of a wide range of ketoxime acetates. This transformation displays excellent functional group tolerance and is nearly insensitive to the electronic effects of the substrates; thus, ketoxime acetates with electronically diverse substituents on the aryl ring, including methyl, methoxy, halo, and cyano groups, give the corresponding 2-arylpyrroles in good to excellent yields. Reactions of *ortho*-substituted ketoxime acetates also proceed efficiently.

Scheme 12 Copper-Catalyzed Cyclization of Ketoxime Acetates To Give 2-Arylpyrroles[223]

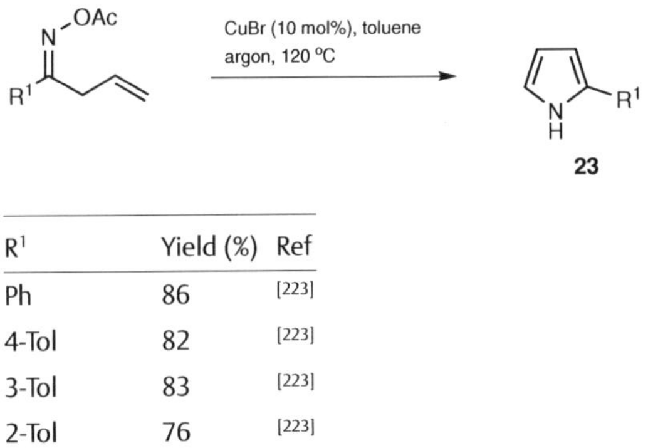

R^1	Yield (%)	Ref
Ph	86	[223]
4-Tol	82	[223]
3-Tol	83	[223]
2-Tol	76	[223]
4-MeOC$_6$H$_4$	52	[223]
2-MeOC$_6$H$_4$	96	[223]
4-FC$_6$H$_4$	90	[223]
3-FC$_6$H$_4$	89	[223]
4-ClC$_6$H$_4$	89	[223]
4-BrC$_6$H$_4$	82	[223]
4-NCC$_6$H$_4$	82	[223]
2-thienyl	55	[223]
t-Bu	51	[223]

Ketoxime carboxylates bearing different substituents on the alkene moiety also produce pyrroles in high yields under the same reaction conditions (Table 1).

Table 1 Copper-Catalyzed Cyclization of Ketoxime Carboxylates To Give Pyrroles[223]

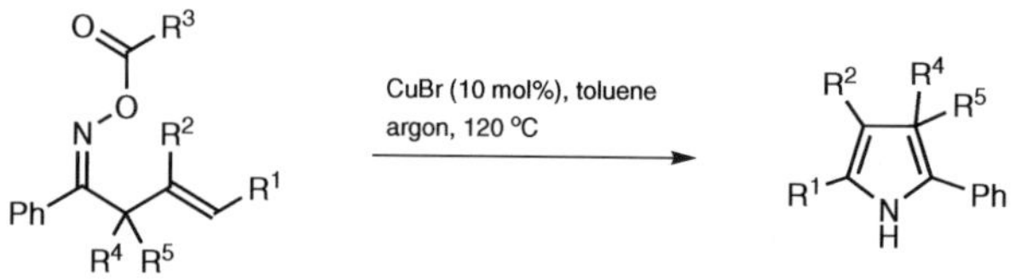

Entry	Substrate	Product	Yield (%)	Ref
1	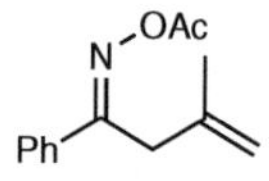	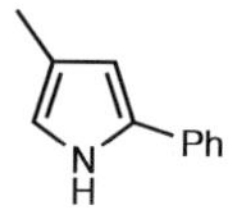	64	[223]
2	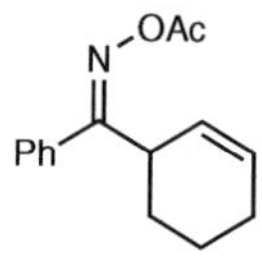	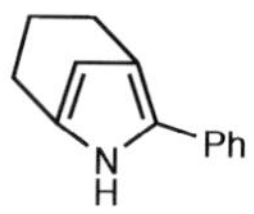	0	[223]
3	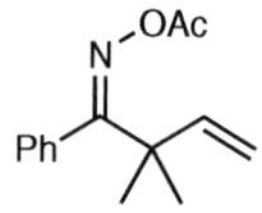	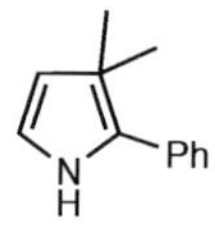	0	[223]
4	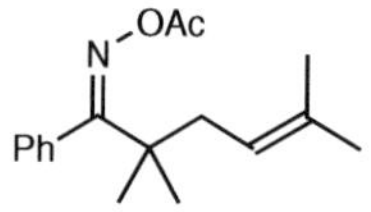	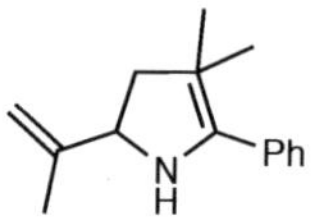	80	[223]
5	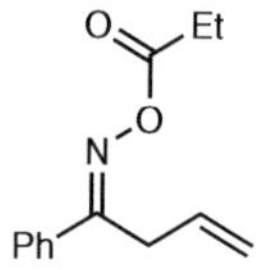	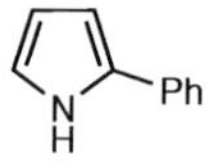	84	[223]
6	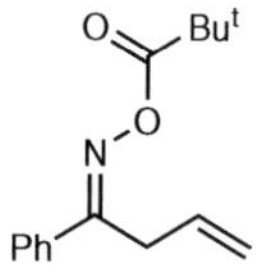	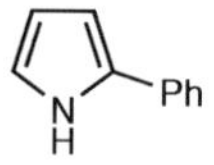	82	[223]
7	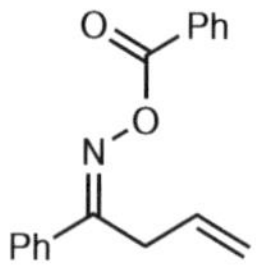	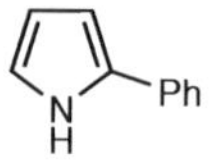	80	[223]
8	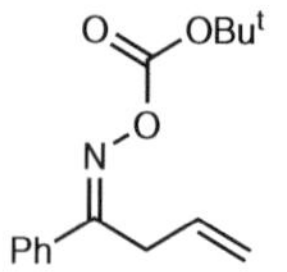	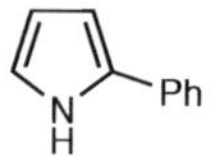	0	[223]

2-Arylpyrroles 23; General Procedure:[223]

A 25-mL round-bottomed flask was charged with CuBr (4.3 mg, 10 mol%), the ketoxime carboxylate (0.3 mmol), and toluene (3 mL). Then, the flask was evacuated and backfilled with argon (3×, balloon), and the mixture was stirred at 120 °C. Upon completion of the reaction (as detected by TLC), the mixture was cooled to rt. The resulting mixture was quenched with H_2O (10 mL) and extracted with EtOAc (3 × 10 mL). The organic phase was dried (Na_2SO_4) and concentrated under reduced pressure. The residue was purified by column chromatography (silica gel, hexane/EtOAc).

2.9.2.1.2 Synthesis of Fused Pyrrole Derivatives

The photoredox catalyst tris(2,2′-bipyridyl)dichlororuthenium(II) catalyzes the intramolecular functionalization of substituted pyrroles upon irradiation with a household lightbulb.[224] Under irradiation by a 14-W compact fluorescent lamp, N-(bromoalkyl)pyrroles undergo cyclization in the presence of tris(2,2′-bipyridyl)dichlororuthenium(II) (1.0 mol%) and triethylamine (2 equiv) in dimethylformamide to provide functionalized pyrroles **24** (Scheme 13). The photoredox system displays good functional group tolerance and allows substitution at both the C2- and C3-positions; however, a 3-phenyl-substituted pyrrole provides an equimolar mixture of two regioisomeric cyclization products. The cyclization reactions proceed more efficiently for more electron-rich pyrroles, which results in the formation of fewer reductive dehalogenation products.

Scheme 13 Photoredox Intramolecular Cyclization of Pyrrole Substrates[224]

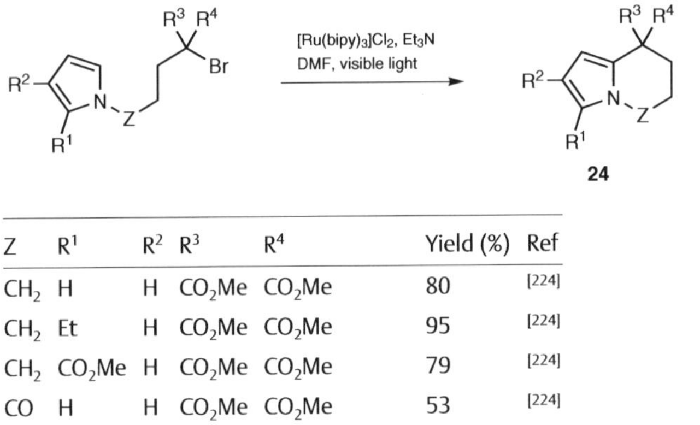

Z	R^1	R^2	R^3	R^4	Yield (%)	Ref
CH_2	H	H	CO_2Me	CO_2Me	80	[224]
CH_2	Et	H	CO_2Me	CO_2Me	95	[224]
CH_2	CO_2Me	H	CO_2Me	CO_2Me	79	[224]
CO	H	H	CO_2Me	CO_2Me	53	[224]
CH_2	H	H	H	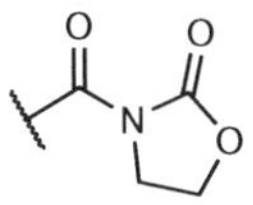 (oxazolidinone)	89	[224]

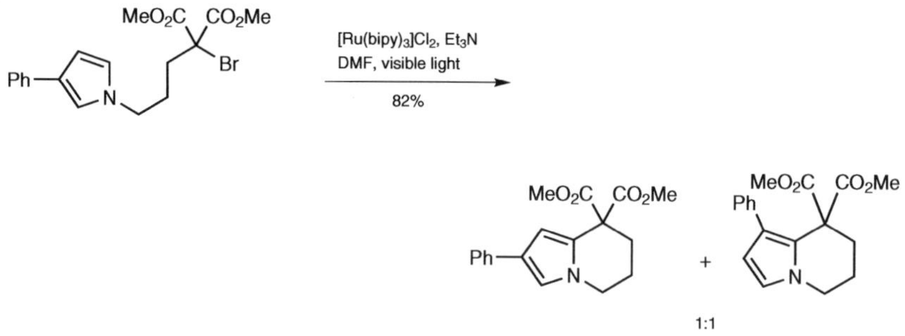

A plausible mechanism for the ruthenium-catalyzed photoredox cyclization is outlined in Scheme 14. Visible-light irradiation generates the metal-to-ligand charge-transfer excited tris(2,2′-bipyridyl)ruthenium(2+)* species, which is then reductively quenched by triethylamine to generate the electron-rich ruthenium(I) complex. This ruthenium(I) species then acts as a single-electron-transfer agent and selectively reduces the activated C—Br bond, which generates electron-deficient alkyl radical **25** and regenerates tris(2,2′-bipyridyl)ruthenium(2+). Thereafter, intramolecular cyclization of the electron-rich aromatic species with the electrophilic radical affords radical **26**. The resultant bicyclic product is then oxidized, probably by the excited-state ruthenium(II)* species or by the bromomalonate starting material to give intermediate **27** or **28**, respectively. Elimination affords final product **24** along with triethylammonium hydrobromide.

2.9.2 Intramolecular Free-Radical Cyclization Routes to N-Heterocycles

Scheme 14 Probable Mechanism for Photoredox Intramolecular Cyclization[224]

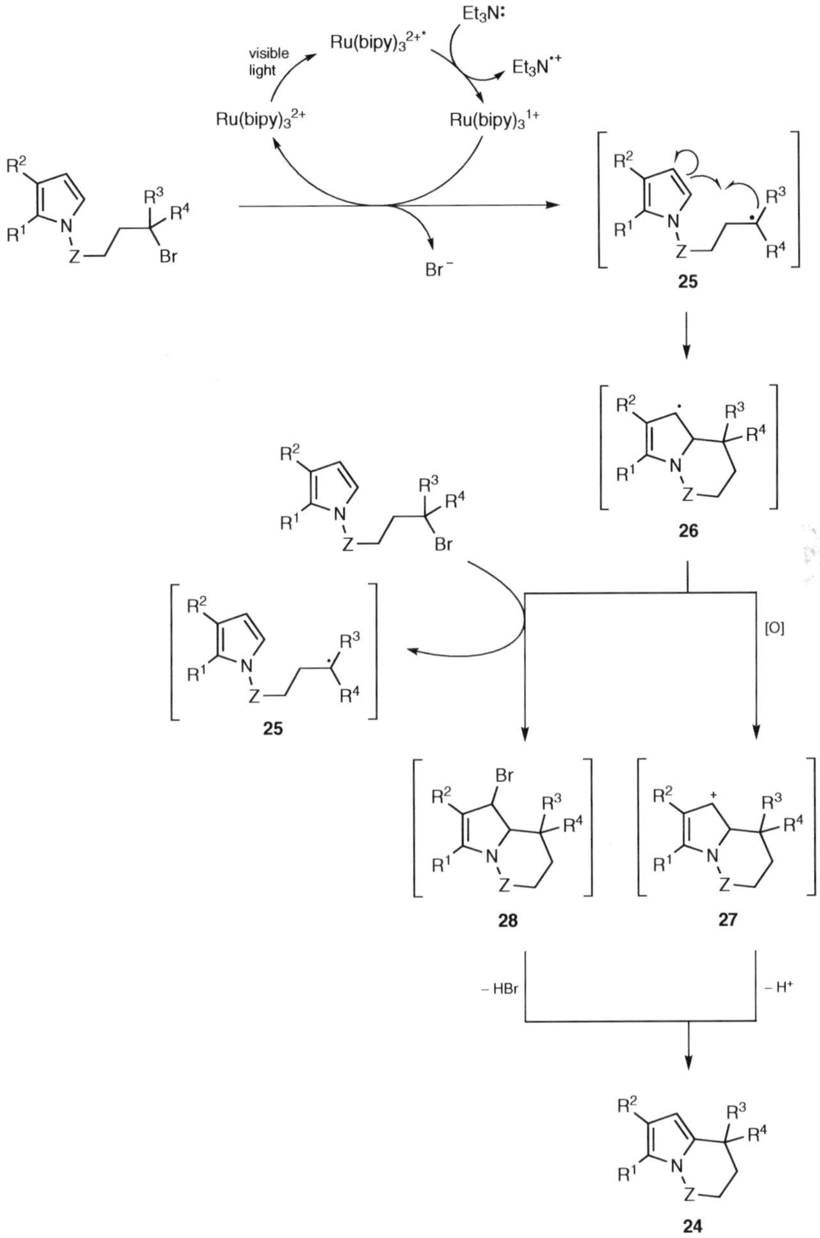

Fused-Pyrrole Derivatives 24; General Procedure:[224]

A flame-dried, 10-mL round-bottomed flask was charged with [Ru(bipy)$_3$]Cl$_2$·6H$_2$O (1.0 µmol, 0.01 equiv), the bromide (0.10 mmol, 1.0 equiv), Et$_3$N (0.20 mmol, 2.0 equiv), and DMF (2.0 mL). The mixture, after degassing by a freeze–pump–thaw cycle, was placed at a distance of ~10 cm from a 15-W compact fluorescent lamp. Upon completion of the reaction (as monitored by TLC), the mixture was poured into a separatory funnel containing H$_2$O (25 mL) and Et$_2$O (25 mL). The organic and aqueous layers were separated, and the aqueous layer was extracted with Et$_2$O (2 × 50 mL). The organic extract was dried (Na$_2$SO$_4$) and concentrated under reduced pressure. The residue was purified by column chromatography (silica gel).

2.9.2.1.3 Synthesis of Pyrrolidine Derivatives

Pyrrolidine derivatives are significant heterocycles that are present in several natural products;[225–231] they are used in the field of organocatalysis and display potential pharmacological activities.[232–234] Numerous methods have been developed to synthesize pyrrolidines.[235–240] In particular, the intramolecular addition of nitrogen-centered radicals onto unsaturated bonds provides an attractive technique to produce five-membered azaheterocycles.[241–243] However, halogenated amines, which are commonly employed as N-radical precursors, are highly unstable in the synthesis of pyrrolidines.[244–246] Thus, hydroxylamine derivatives, such as amidyl, iminyl, and aminyl radicals, have been employed as alternative N-radical precursors.[247–250] The radical cyclization of diester internal unsaturated acyclic systems allows the efficient synthesis of fused azaheterocycles. For instance, Burton and co-workers have devised an efficient method for oxidative radical cyclization starting from α-amidomalonate **29** to give intermediate **30** in the total synthesis of (−)-salinosporamide A (Scheme 15).[251]

Scheme 15 Synthesis of (−)-Salinosporamide A[251]

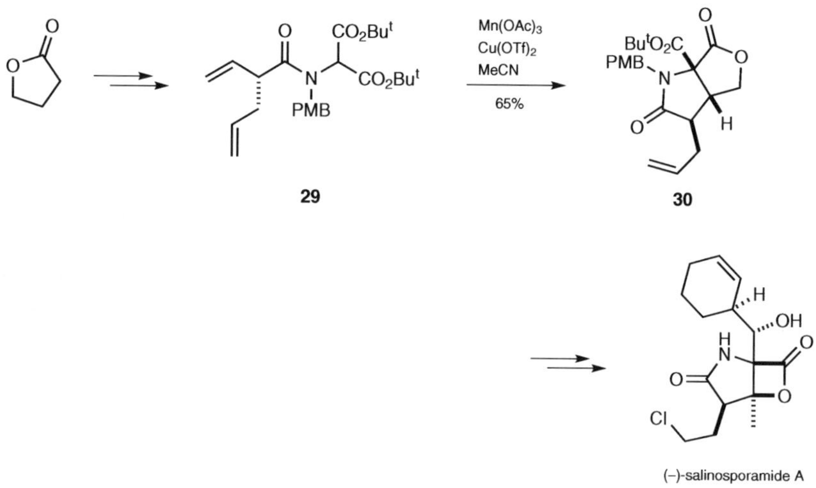

***tert*-Butyl (3R,3aR,6aS)-1-(4-Methoxybenzyl)-2,6-dioxo-3-(prop-2-en-1-yl)tetrahydro-1*H*-furo[3,4-*b*]pyrrole-6a(6*H*)-carboxylate (30); Typical Procedure:**[251]
N$_2$-purged MeCN (0.40 M) was rapidly added to a degassed mixture of terminal alkene **29** (69.0 mg, 0.150 mmol), Mn(OAc)$_3$•2H$_2$O (80.4 mg, 0.300 mmol), and Cu(OTf)$_2$ (108 mg, 0.300 mmol). The mixture was stirred at 40 °C for 1 h, cooled to rt, and then quenched with H$_2$O (10 mL). The organic layer was extracted with CHCl$_3$ (3 × 10 mL). The combined organic extract was washed with brine (15 mL), dried (Na$_2$SO$_4$), and concentrated under reduced pressure. The crude product was purified by flash column chromatography (petroleum ether/EtOAc) to give a colorless oil; yield: 39.0 mg (65%); R_f 0.32 [petroleum ether (bp 40–60 °C)/EtOAc 2:1].

2.9.2.1.3.1 Manganese(III) Acetate Mediated Cyclization of N-Substituted Internal Alkyne Esters

Burton has studied the selectivity involved in the manganese(III) acetate based synthesis of *exo*-alkylidene heterocycles.[252] Free-radical cyclization of N-substituted internal alkyne ester **31** in ethanol at room temperature gives product **32** with 2.8:1 (Z/E) selectivity (Scheme 16).

Scheme 16 Manganese(III) Acetate Mediated Synthesis of a 3-Alkylidene-1-(ethoxycarbonyl)pyrrolidine[252]

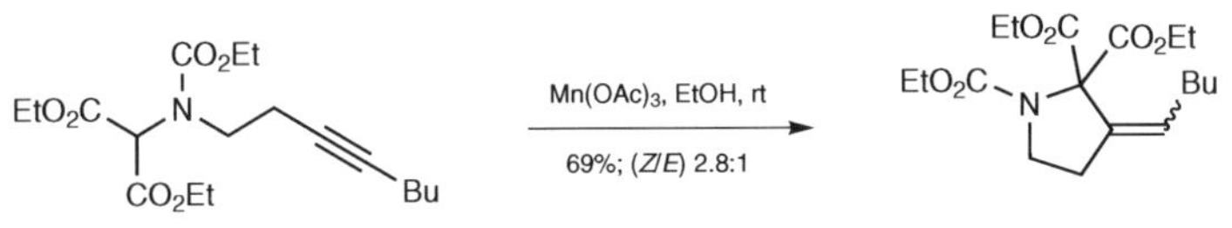

Triethyl 3-Pentylidenepyrrolidine-1,2,2-tricarboxylate (32); Typical Procedure:[252]
Mn(OAc)$_3$ (2.0 equiv) and alkyne **31** (1.0 equiv) were dissolved in degassed EtOH (20 mL/mmol malonate). The mixture was stirred at rt for 2 h and then filtered through a pad of silica gel (Et$_2$O). The solvent was removed under reduced pressure to yield the crude product, which was purified by flash column chromatography (petroleum ether/Et$_2$O 1:1) to afford a colorless oil; yield: 103 mg (69%); ratio (Z/E) 2.8:1; R_f 0.30 (petroleum ether/Et$_2$O 1:1).

2.9.2.1.3.2 Copper(I) Trifluoromethanesulfonate–Benzene Complex Mediated Synthesis of 2,5-Disubstituted Pyrrolidines

The commercially available copper(I) trifluoromethanesulfonate–benzene complex can be used to synthesize 2,5-disubstituted pyrrolidines **34** by cyclization of unsaturated N-(benzoyloxy)sulfonamides **33** in refluxing 1,2-dichloroethane under a nitrogen atmosphere (Scheme 17).[253] The cyclization proceeds with a low catalyst loading (4–20 mol%) through the intramolecular addition of sulfonamidoyl radicals to unsaturated bonds or through allylic hydrogen abstraction with radical intermediates. Although the reaction proceeds rapidly within 1–5 hours, the formation of *cis/trans*-isomers often hampers purification of the desired products. Moreover, the probable generation of trifluoromethanesulfonic acid by the hydrolysis of the copper(I) trifluoromethanesulfonate–benzene complex does not cause any nucleophilic substitution at the nitrogen atom.

Scheme 17 Pyrrolidines by Cyclization of N-(Alk-4-enyl)-N-(benzoyloxy)sulfonamides[253]

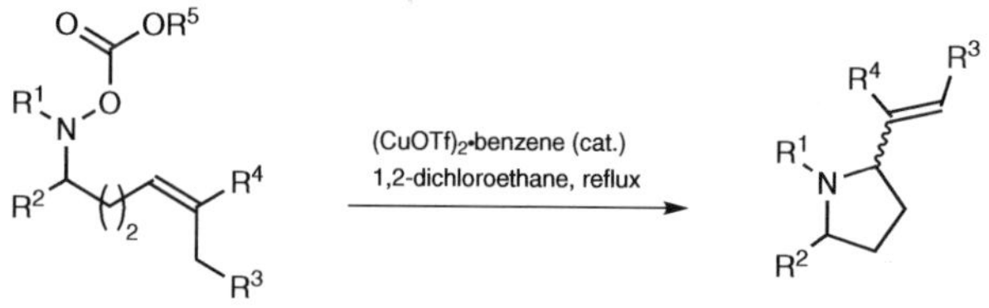

33 → **34**

(CuOTf)$_2$·benzene (cat.)
1,2-dichloroethane, reflux

R^1	R^2	R^3	R^4	R^5	Mol% of Catalyst	Time (h)	Ratio (cis/trans)	Yield (%)	Ref
Ms	(CH$_2$)$_2$Ph	H	Me	Ph	10	1	1:1	95	[253]
Ms	(CH$_2$)$_2$Ph	H	Me	Me	10	4	6:7	91	[253]
Ms	Bn	H	Me	Ph	10	1.5	2:7	64	[253]
Ms	(CH$_2$)$_2$Ph	H	H	Ph	10	1	1:2	56	[253]
Ms	Me	H	Me	Ph	4	1	1:1	84	[253]
SO$_2$Ph	Me	H	Me	Ph	10	4	4:5	60	[253]
4-O$_2$NC$_6$H$_4$SO$_2$	Me	H	Me	Ph	10	48	5:4	30	[253]
Ts	Me	H	Me	Ph	15	3	4:5	55	[253]

The copper(I) trifluoromethanesulfonate–benzene complex also catalyzes the cyclization of other N-alkenyl-N-(benzoyloxy)sulfonamides and N-alkynyl-N-(benzoyloxy)sulfonamides.[253] If N-(alk-5-enyl)-N-(benzoyloxy)sulfonamides **35** are subjected to cyclization, the *cis*-isomers of pyrrolidine derivatives **36** are usually preferentially obtained (Scheme 18).[253]

Scheme 18 Cyclization of N-(Alk-5-enyl)-N-(benzoyloxy)sulfonamides To Give 2-Alkenylpyrrolidines[253]

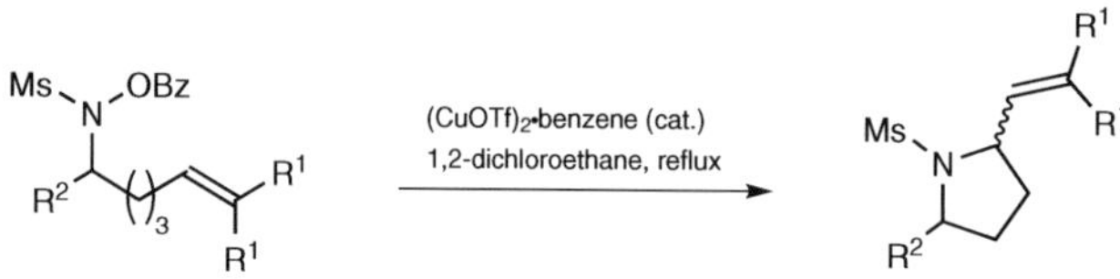

35 → **36**

(CuOTf)$_2$·benzene (cat.)
1,2-dichloroethane, reflux

R^1	R^2	Mol% of Catalyst	Time (h)	Ratio (cis/trans)	Yield (%)	Ref
Me	(CH$_2$)$_2$Ph	10	2	10:1	66	[253]
Me	Me	10	7	6:1	47	[253]
H	(CH$_2$)$_2$Ph	20	52	3:4	61	[253]
H	Me	20	1	1:1	82	[253]

Cyclization of N-alkynyl-N-(benzoyloxy)sulfonamide **37** [R^1 = (CH$_2$)$_2$Ph] under standard conditions gives pyrrolidine **38** and 2H-pyrrole **39** in low yields of 25 and 20%, respectively (Scheme 19).[253] If cyclohexa-1,4-diene is added, the pyrrolidine **38** is obtained in 56% yield, and the formation of 2H-pyrrole **39** is not observed. Substrates **40** containing a longer tether give 2-alkynylpyrrolidines **41** instead of 2-alkylidenepyrrolidines.

2.9.2 Intramolecular Free-Radical Cyclization Routes to N-Heterocycles

Scheme 19 Cyclization of a N-Alkynyl-N-(benzoyloxy)sulfonamides To Give 2-Alkylidene- and 2-Alkynylpyrrolidines[253]

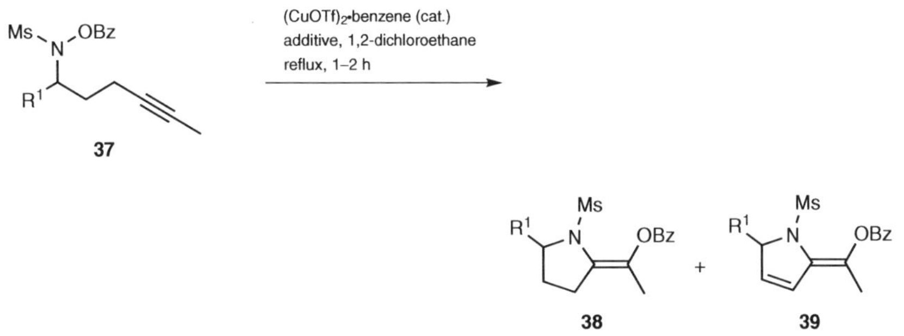

R^1	Additive (equiv)	Yield (%) of 38	Yield (%) of 39	Ref
$(CH_2)_2Ph$	–	25	20	[253]
$(CH_2)_2Ph$	cyclohexa-1,4-diene (10)	56	0	[253]
Bn	cyclohexa-1,4-diene (10)	33	0	[253]

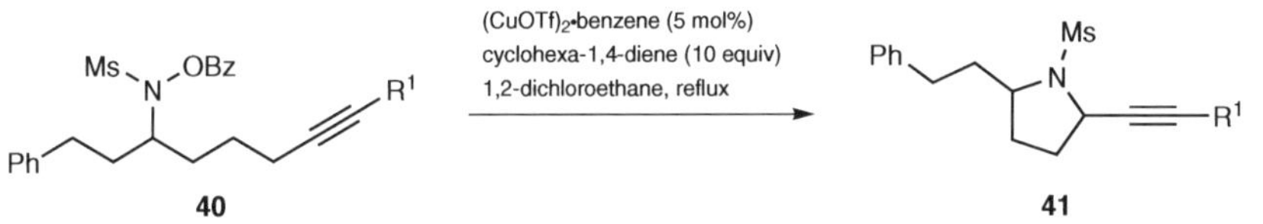

R^1	Time (h)	Ratio (cis/trans)	Yield (%)	Ref
Me	4	6:1	38	[253]
Ph	5	6:7	74	[253]

1-(Methanesulfonyl)-2-(2-phenylethyl)-5-(prop-1-en-2-yl)pyrrolidine [34, R^1 = Ms; R^2 = $(CH_2)_2Ph$; R^3 = H; R^4 = Me]; Typical Procedure:[253]

(CuOTf)$_2$•benzene (10 mg, 0.02 mmol) was added to a soln of N-(benzoyloxy)-N-(7-methyl-1-phenyloct-6-en-3-yl)methanesulfonamide (0.4 mmol) in 1,2-dichloroethane (10 mL) at rt under a N_2 atmosphere. The mixture was heated at reflux for 1 h, and then the reaction was quenched by the addition of sat. aq NaHCO$_3$. The mixture was extracted with EtOAc (3×), and the combined organic layer was dried (Na$_2$SO$_4$). The solvent was removed under reduced pressure, and the crude residue was purified by preparative TLC (hexane/EtOAc 3:1); yield: 0.112 g (95%); ratio (cis/trans) 1:1.

2.9.2.1.3.3 Bis(η^5-cyclopentadienyl)dimethyltitanium(IV) (Petasis Reagent) Mediated Intramolecular Hydroamination/Cyclization of Alkynamines

In the presence of 5 mol% of the Petasis reagent [TiMe$_2$(Cp)$_2$; bis(η^5-cyclopentadienyl)dimethyltitanium(IV)} the intramolecular hydroamination/cyclization of alkynamines proceeds smoothly to give five-membered cyclic imines within 4–6 hours in toluene at 100–110°C. Subsequent reduction of the cyclic imine with zinc-modified sodium cyanoborohydride in tetrahydrofuran at room temperature provides efficient access to pyrrolidines **42** (Scheme 20).[254]

Scheme 20 Synthesis of Pyrrolidine Derivatives from Alkynamines Using the Petasis Reagent[254]

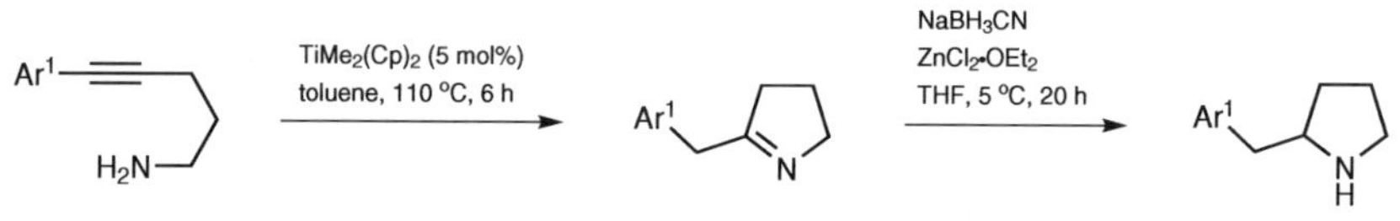

Ar[1]	Yield (%)	Ref
Ph	90	[254]
4-MeOC$_6$H$_4$	94	[254]
2-BrC$_6$H$_4$	88	[254]
2-F$_3$CC$_6$H$_4$	82	[254]
3,5-(F$_3$C)$_2$C$_6$H$_3$	62	[254]

2-Benzylpyrrolidines 42; General Procedure:[254]

Preparation of Petasis reagent:[255,256] A suspension of Ti(Cp)$_2$Cl$_2$ (770 mg, 3.09 mmol, 2.2 equiv) in toluene (5 mL) was prepared at −5 °C. Then, a soln of titrated 1.58 M MeLi in Et$_2$O (4.4 mL, 6.95 mmol, 4.8 equiv) was slowly added. The resulting mixture was stirred at −5 °C for 1 h and then at rt for 1 h. The mixture was cooled to 0 °C and quenched carefully by the addition of an ice-cold soln of 6% aq NH$_4$Cl (10 mL). The resulting aqueous and organic phases were separated, and the organic phase was washed with H$_2$O (10 mL) and brine (10 mL), dried (Na$_2$SO$_4$), filtered, and concentrated under reduced pressure.

Reaction using Petasis reagent: A flame-dried Schlenk tube equipped with a Teflon stopcock was charged with the alkynamine (1.0 mmol), a 0.25 M soln of TiMe$_2$(Cp)$_2$ in toluene (0.20 mL, 0.05 mmol, 5.0 mol%) and toluene (0.3 mL) under an argon atmosphere. The resulting suspension was stirred at 110 °C for 6 h. The mixture was then cooled to rt, and the solvent was removed under reduced pressure. Subsequently, a suspension of NaBH$_3$CN (126 mg, 2.0 mmol) and 1.0 M ZnCl$_2$·OEt$_2$ in Et$_2$O (1.0 mL, 1.0 mmol) in THF (5.0 mL) was added. The resulting mixture was then stirred at rt for another 20 h. The mixture was diluted with CH$_2$Cl$_2$ (10.0 mL) and 2.0 M aq HCl (10.0 mL). Stirring at rt was continued for another 2 h. The resulting layers were separated and 2.0 M aq KOH was added to the aqueous layer to achieve pH 7. The aqueous mixture was extracted with CH$_2$Cl$_2$ (3 × 50 mL). The combined organic layer was washed with brine, dried (Na$_2$SO$_4$), and concentrated under reduced pressure. The resultant residue was purified by flash chromatography (silica gel, EtOAc/MeOH 8:1 to 1:1).

2.9.2.1.4 Synthesis of Pyrrolidinone Derivatives

2.9.2.1.4.1 Manganese(III)-Mediated Synthesis of *exo*-Alkylidene Heterocycles

Burton and co-workers have investigated the selectivity involved in the manganese(III)-mediated synthesis of *exo*-alkylidene heterocycles.[252] Conditions essential for the selective synthesis of 2-alkylidenepyrrolidinones by the manganese(III) acetate oxidation of 2-(alkynamido)malonates have also been studied. Consequently, in the presence of manganese(III) acetate in ethanol, the *exo* cyclization of 2-(alkynamido)malonates predominates and provides the Z-isomers as the major products at room or high temperature. The radical cyclization of internal alkyne substrates **43** possessing a strong tendency to form stereoisomeric products yields a high quantity of Z-selective products **44** (Scheme 21).[252] Ethyl-substituted 4-alkylidenepyrrolidinone **44** (R[1] = Et) delivers maximum Z selectivity, even at room temperature. Upon increasing the bulk of the esters **43**, cyclization produces (Z)-4-alkylidenepyrrolidinones **44** more selectively (Scheme 21).[252]

Scheme 21 Manganese(III) Acetate Mediated Cyclization of 2-(Alkynamido)-malonates To Give 4-Alkylidenepyrrolidinones[252]

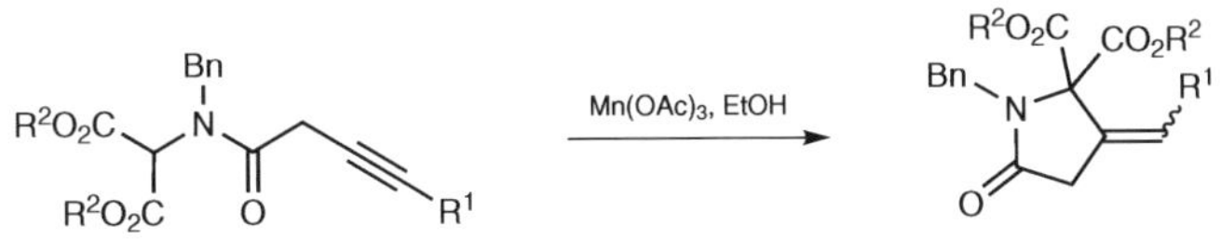

R^1	R^2	Conditions	Ratio (E/Z)	Yield (%)	Ref
Me	Et	80°C, 15 h	1:3.8	78	[252]
Me	Et	rt, 15 min	1:3.9	75	[252]
Et	Et	rt, 2 h	1:4.7	86	[252]
Pr	Et	rt, 2 h	1:2.9	72	[252]
Ph	Et	rt, 2 h	1:3.0	65	[252]
Bu	Et	80°C, 15 h	1:3.0	73	[252]
Bu	Bn	rt, 2 h	1:3.5	79	[252]
Bu	t-Bu	80°C, 2 h	1:6.6	82	[252]

***exo*-Alkylidene Heterocycles 44; General Procedure:**[252]
Synthesis of amidomalonates **43**: DMF (0.02 mL/mmol carboxylic acid) was added to an ice-cooled soln of the desired carboxylic acid (1.2 equiv) in CH_2Cl_2 (1 mL/mol malonate). $(COCl)_2$ (1.1 equiv) was added, and the mixture was stirred at rt for 1 h. The mixture was concentrated at 40°C at atmospheric pressure to afford the corresponding acid chloride. Sat. $NaHCO_3$ (1 mL/mmol malonate) and a soln of the dialkyl N-benzylaminomalonate (1.0 equiv) in CH_2Cl_2 (1.5 mL/mmol malonate) were added dropwise to the soln of the crude acid chloride. After stirring for 30 min at rt, the mixture was filtered through a pad of silica gel (Et_2O). The residue was washed with 5 M HCl, sat. aq $NaHCO_3$, and brine, then dried, concentrated under reduced pressure, and purified by chromatography (silica gel).[257]

Cyclization at 80°C: A soln of the amidomalonate **43** (1.0 equiv) and $Mn(OAc)_3$ (2.0 equiv) in degassed alcohol solvent (20 mL/mmol malonate) was heated to 80°C and stirred for 15 h. The mixture was then cooled to rt, and the soln was concentrated under reduced pressure. The resulting residue was filtered through a pad of silica gel (Et_2O). The solvent was removed under reduced pressure, and the residue was purified by flash chromatography (hexane/Et_2O).

Cyclization at room temperature: A soln of the amidomalonate **43** (1.0 equiv) and $Mn(OAc)_3$ (2.0 equiv) in degassed alcohol solvent (20 mL/mmol malonate) was stirred at rt for 2 h. The resultant residue was filtered through a pad of silica gel (Et_2O). The solvent was removed under reduced pressure, and the residue was purified by flash chromatography (hexane/Et_2O).

2.9.2.1.4.2 Tandem Visible-Light-Mediated Radical Cyclization/Rearrangement to Tricyclic Pyrrolidinones

A novel route for the tandem radical cyclization/sigmatropic rearrangement can be performed by visible-light-promoted single-electron reduction/cyclization. Thus, photoredox of bromocyclopropane derivatives generates tricyclic pyrrolidinones having considerable molecular complexity from simple and readily available starting materials. Visible-light photocatalysis is emerging as a powerful tool in organic synthesis.[258–260] In particular, metal-mediated single-electron transfer has received widespread application in a variety of redox reactions.[261–268] The (4,4-di-*tert*-butyl-2,2′-bipyridyl)bis(2-phenylpyridyl)iridium(III) hexafluorophosphate {[Ir(ppy)$_2$(dtbbipy)]PF$_6$; ppy = 2-(2-pyridyl)phenyl (commonly called 2-phenylpyridyl), dtbbipy = 4,4′-di-*tert*-butyl-2,2′-bipyridyl} based photoredox cyclization of functionalized cyclopropane derivatives **45** under visible-light irradiation affords highly complex fused tricyclic pyrrolidinones **46**. The method proceeds efficiently in dimethylformamide at 40 °C and in the presence of triethylamine as the base.[269] The transformation displays a wide substrate scope (Table 2) and is insensitive to substitutions on the aryl ring. Electronically diverse aromatic rings undergo this tandem reaction efficiently. When 2,2-bis(3-methoxyphenyl)-substituted cyclopropanecarboxamide **47** or 2-(4-fluorophenyl)-2-(4-methoxyphenyl)-substituted cyclopropanecarboxamide **48** are employed as the substrates a 1:1 mixture of the corresponding regioisomers is obtained (Scheme 22).[269]

Table 2 Light-Mediated Radical Cyclization of Cyclopropane Derivatives[269]

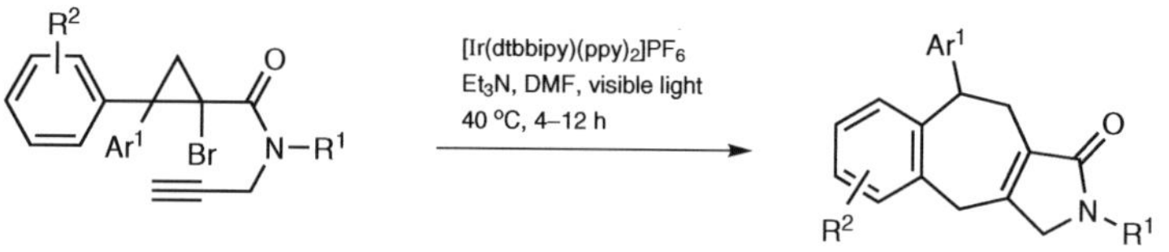

Entry	Substrate	Product	Yield (%)	Ref
1	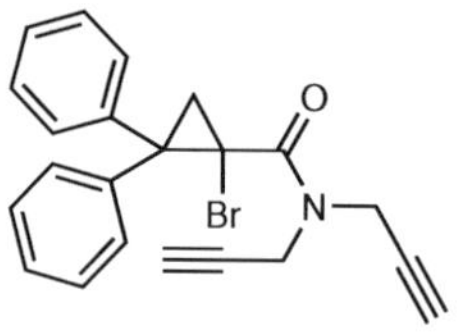	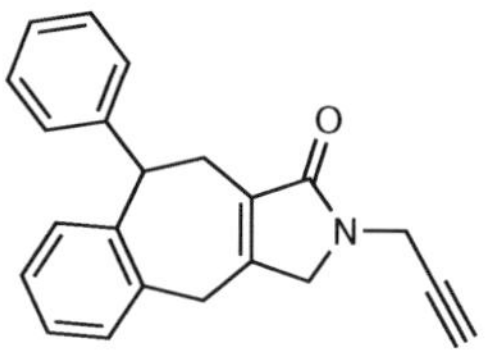	69	[269]
2	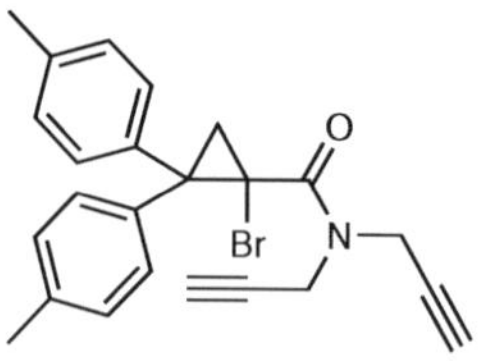	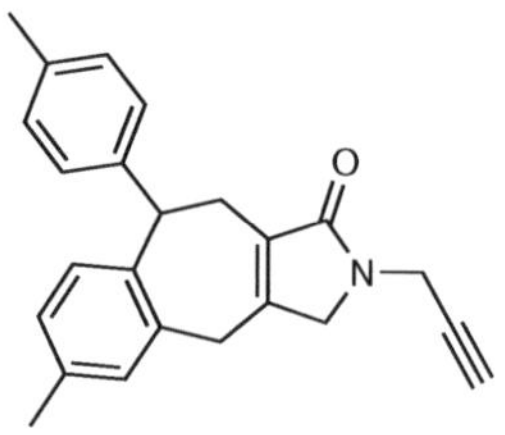	91	[269]

Table 2 (cont.)

Entry	Substrate	Product	Yield (%)	Ref
3	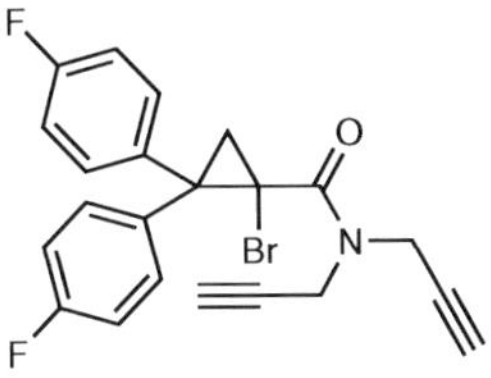	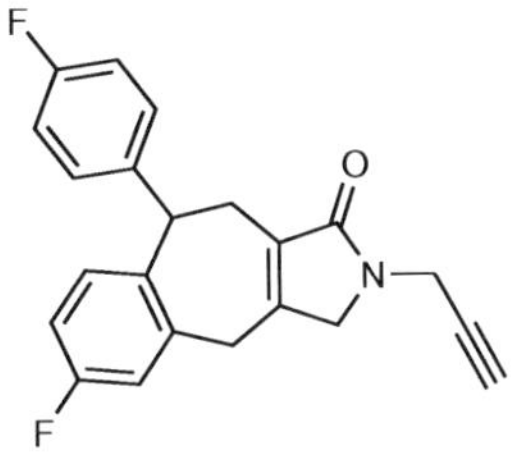	89	[269]
4	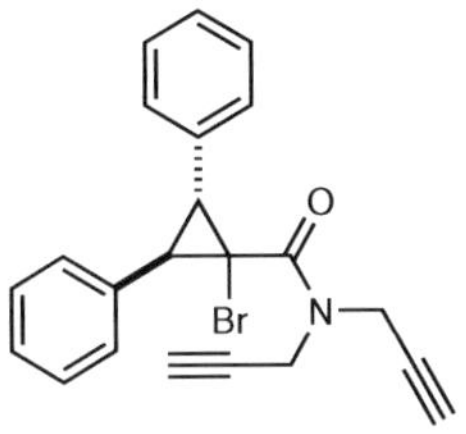	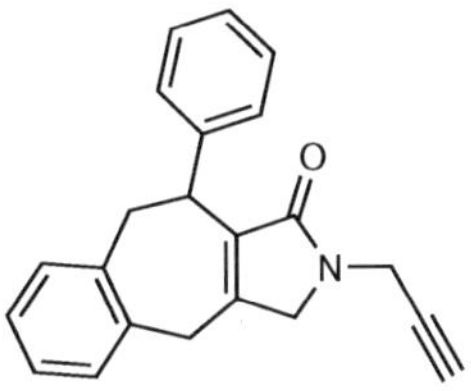	85	[269]
5	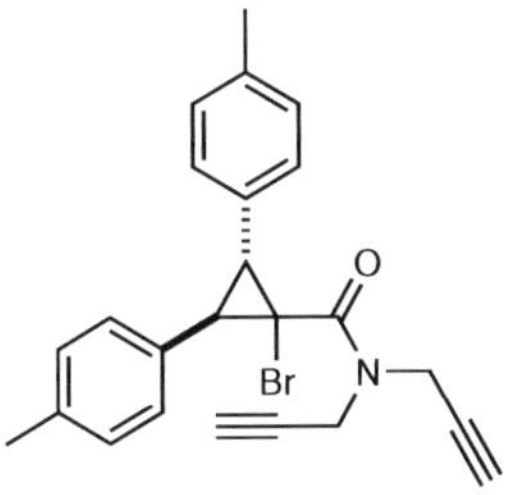	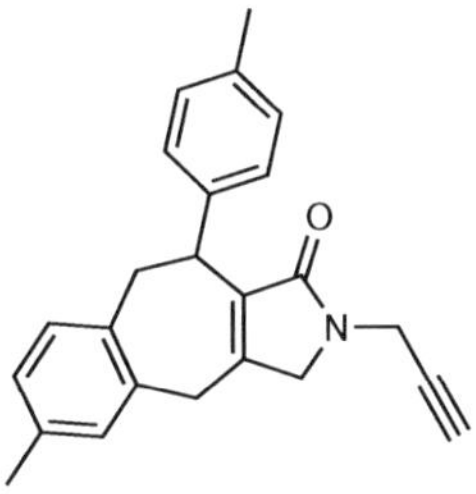	82	[269]
6	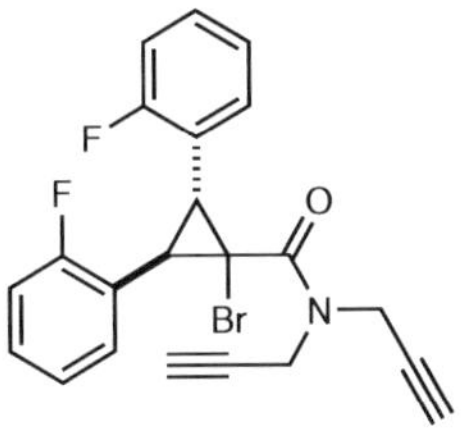	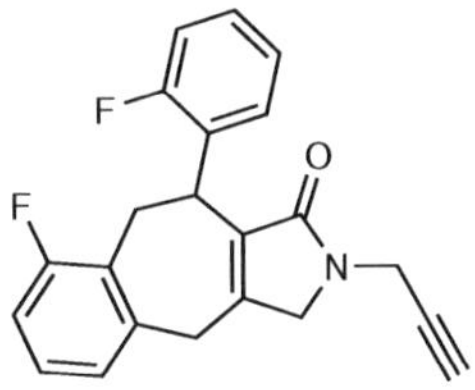	88	[269]
7	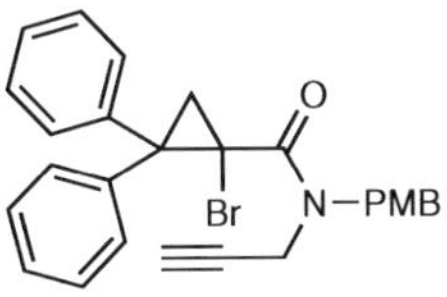	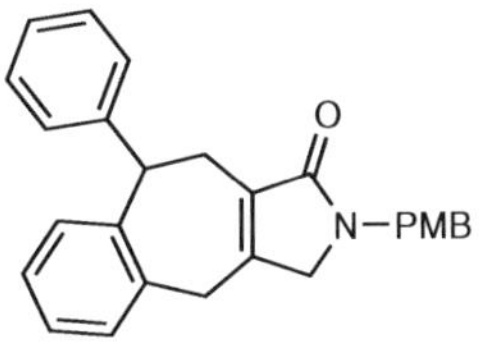	32[a]	[269]
8	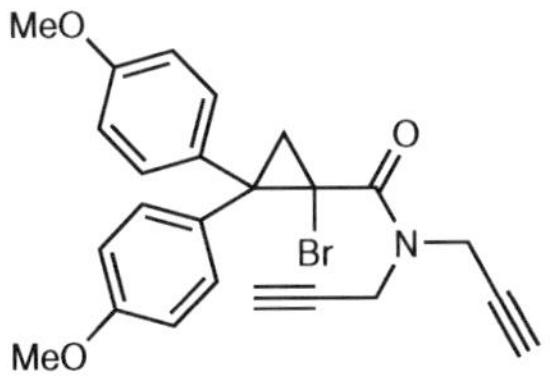	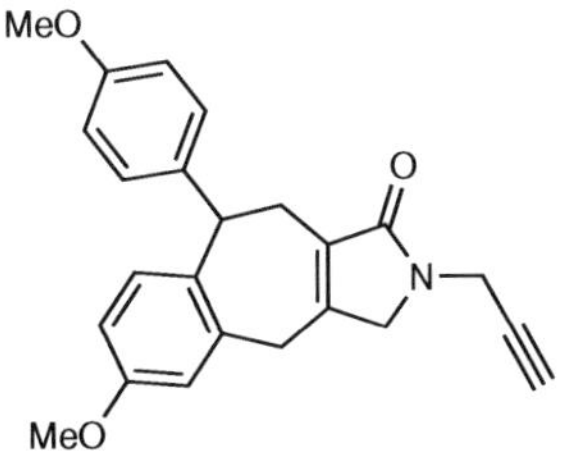	79	[269]

[a] The product of electrocyclic ring opening was isolated in 41% yield.

Scheme 22 Iridium-Catalyzed Light-Mediated Cyclization of 1-Bromocyclopropanecarboxamides[269]

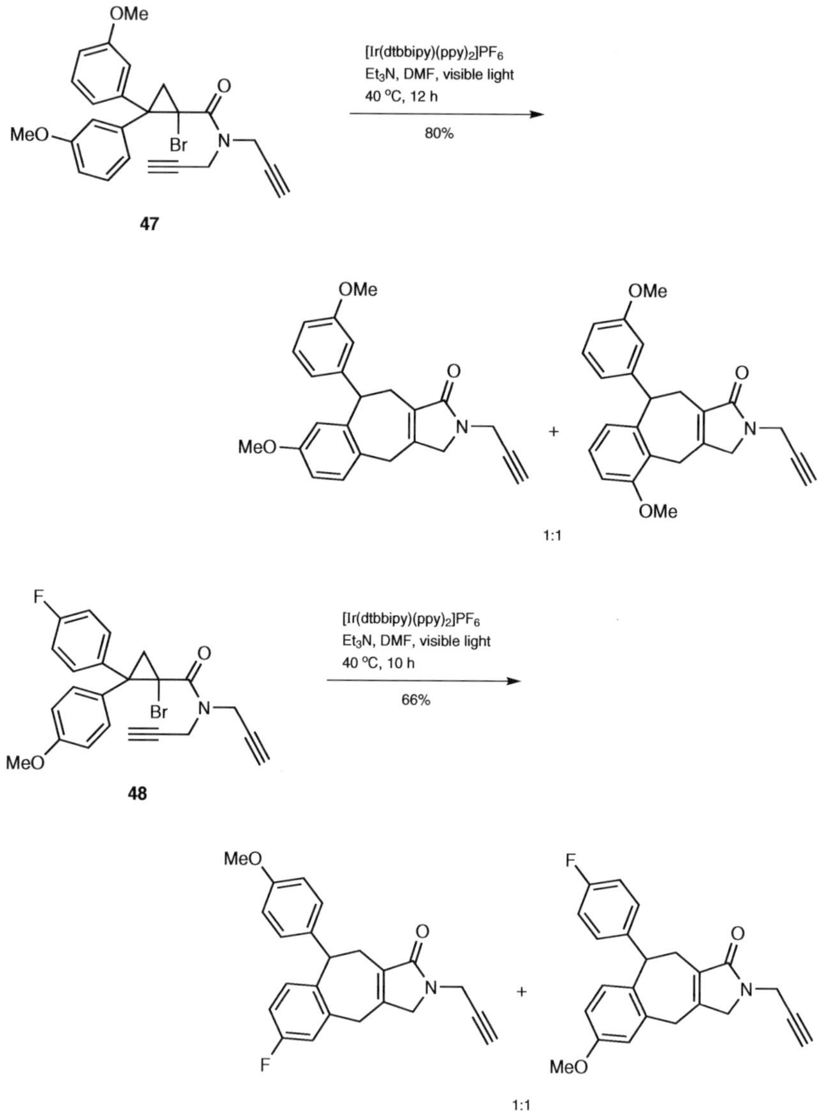

A possible mechanism for the radical cyclization of cyclopropane derivatives is outlined in Scheme 23. The reaction starts with a 5-*exo-dig* radical cyclization followed by quenching of the vinyl radical by abstraction of a hydrogen atom. The seven-membered cyclic product may then form by a Cope rearrangement (Scheme 23, path a).[270,271] Alternatively, the same product may form by single-electron reduction of the vinyl cyclopropane intermediate, followed by a β-scission process and subsequent cyclization and oxidation steps (Scheme 23, path b).[272] Moreover, there is a possibility that rearrangement occurs before quenching of the vinyl radical. In this case, the (high-energy) intermediate may initially cyclize onto the phenyl ring, which initiates a β-scission/isomerization process to form the seven-membered ring (Scheme 23, path c).[273–275]

2.9.2 Intramolecular Free-Radical Cyclization Routes to N-Heterocycles

Scheme 23 Possible Mechanisms for the Photolysis of Cyclopropane Derivatives[269–275]

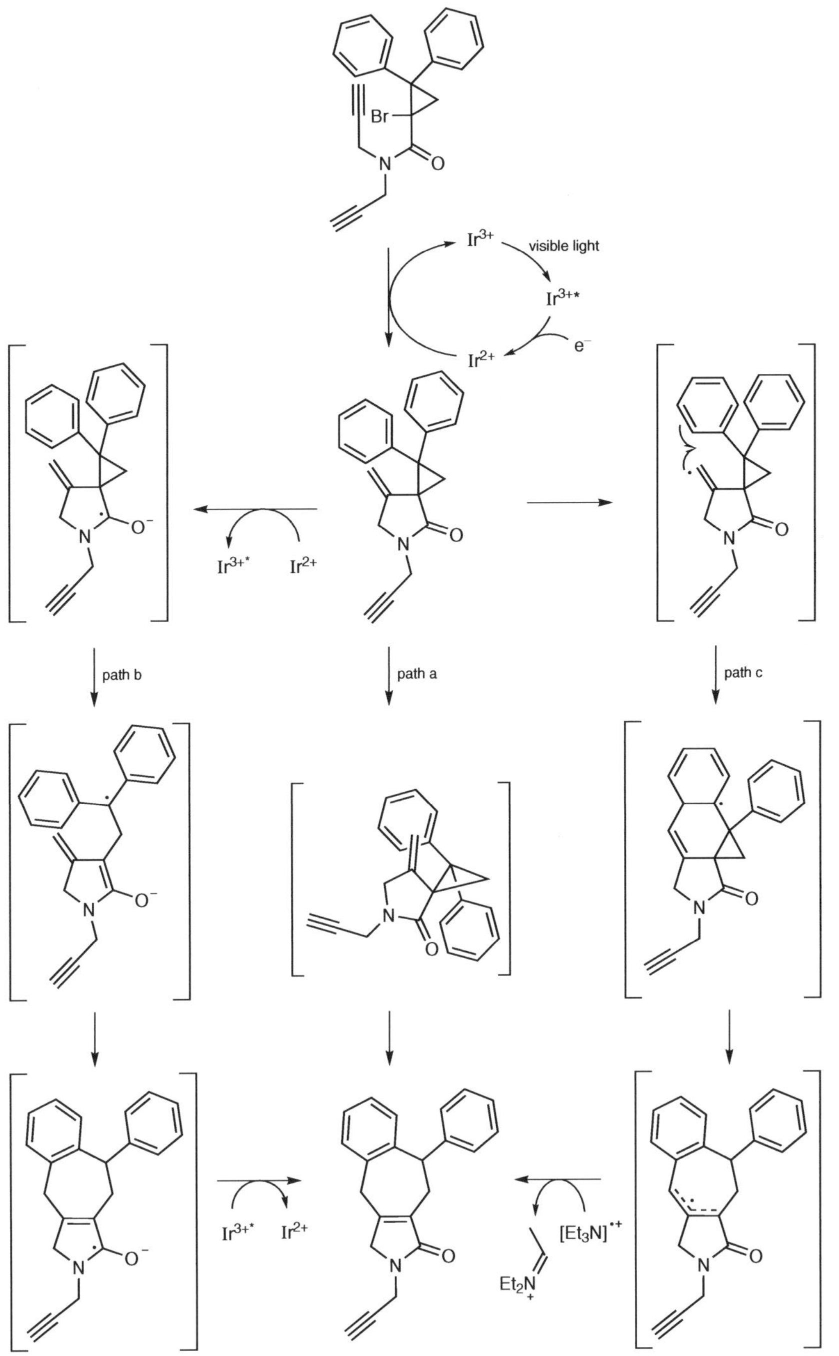

Tricyclic Pyrrolidinones 46; General Procedure:[269]

A 25-mL round-bottomed flask was charged with bromocyclopropane **45** (1.0 mmol, 1.0 equiv), Et₃N (2.0 mmol, 2.0 equiv), DMF (15 mL), and [Ir(ppy)₂(dtbbipy)]PF₆ (0.010 mmol, 0.010 equiv). The mixture was shielded from light and was degassed by purging with argon for 10 min. The mixture and the light source were surrounded by a layer of aluminum foil and then the flask was irradiated using a commercially available 14-W compact fluorescent lamp (to heat the reaction medium) under an argon atmosphere for 6–24 h. Upon completion of the reaction (monitored by TLC), the mixture was diluted with EtOAc (25 mL) and H_2O (50 mL). The combined extracts were separated and the aqueous layer was extracted with EtOAc (2 × 50 mL). The combined organic layer was dried (Na_2SO_4) and concentrated under reduced pressure. The residue was purified by column chromatography (silica gel).

2.9.2.2 Synthesis of the Indole Moiety

The most commonly employed method for the synthesis of indoles remains the Fischer indole synthesis.[276,277] However, the major drawback of the Fischer indole synthesis is the reversible tautomerism of the enehydrazine in acidic media. The enehydrazine is derived from tautomerization of the fully enolizable ketone; this leads to the formation of regio-isomeric mixtures at the enehydrazine stage and, thus, results in regioisomeric indoles. Despite extensive applications, significant efforts are still centered on discovering synthetic routes that can be performed under mild reaction conditions and that have good regioselectivity.[278–281]

2.9.2.2.1 Synthesis of Indole Derivatives

2.9.2.2.1.1 Synthesis of N-Methylindole Derivatives

Indoles can be synthesized by alkylidenation of acyl phenylhydrazides using phosphoranes and the Petasis reagent, followed by in situ thermal rearrangement of the enehydrazine products. Murphy and co-workers have found that by using acyl phenylhydrazides as the starting substrates, the Petasis reagent provides a neutral equivalent of the acid-catalyzed Fischer indole synthesis.[282] They have synthesized indoles **50** by intramolecular cyclization of acyl phenylhydrazides **49** with the Petasis reagent in toluene (Scheme 24). In the presence of a titanium reagent, there is no trace of the oxindole product, which would, however, arise from the Brunner reaction.

Scheme 24 Synthesis of Indoles from Acyl Hydrazides Using the Petasis Reagent[282]

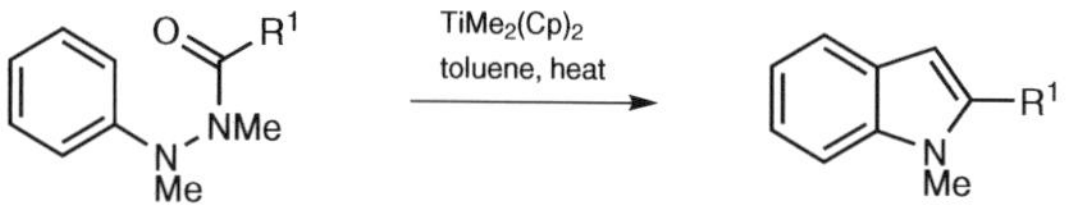

R^1	Yield (%)	Ref
Me	53	[282]
Et	47	[282]
Ph	48	[282]
2-furyl	8	[282]
3-pyridyl	38	[282]

Indoles 50; General Procedure:[282]
A soln of $TiMe_2(Cp)_2$ (2.2 equiv) in toluene (5 mL) was transferred by cannula to a soln of **49** (1.0 equiv) in toluene (~5 mL). The mixture was then heated to reflux for 72 h (48 h for R^1 = Me). The solvent was removed under reduced pressure and Et_2O (25 mL) was added. The mixture was filtered through Celite, which was washed with Et_2O (25 mL). The organic

2.9.2.2.1.2 Photoredox Cyclization of Indole Substrates

The photoredox catalyst tris(2,2′-bipyridyl)dichlororuthenium(II) catalyzes the intramolecular functionalization of substituted indoles upon irradiation with a household lightbulb.[224] Under irradiation by a 14-W compact fluorescent lamp, *N*-(bromoalkyl)indoles containing a malonate group on the tether undergo cyclization in the presence of tris(2,2′-bipyridyl)dichlororuthenium(II) and triethylamine in dimethylformamide to provide functionalized indoles **51** (Scheme 25). The syntheses of both five- and six-membered fused rings is possible, although five-membered rings often require a substituent at the C3-position to stabilize the tertiary radical resulting from cyclization. The photoredox system is chemoselective for the reduction of activated C—Br bonds in the presence of aryl C—Br bonds [**51**, Z = (CH$_2$)$_2$; R^1 = R^3 = H; R^2 = Br]. The reaction tolerates both electron-rich and electron-deficient substrates, including ester, amide, and cyano groups. The reaction can be employed for the synthesis of linear tricyclic systems, which are formed by *endo* cyclization of the malonyl radical.

Scheme 25 Ruthenium(II)-Mediated Photoredox Cyclization of Indoles[224]

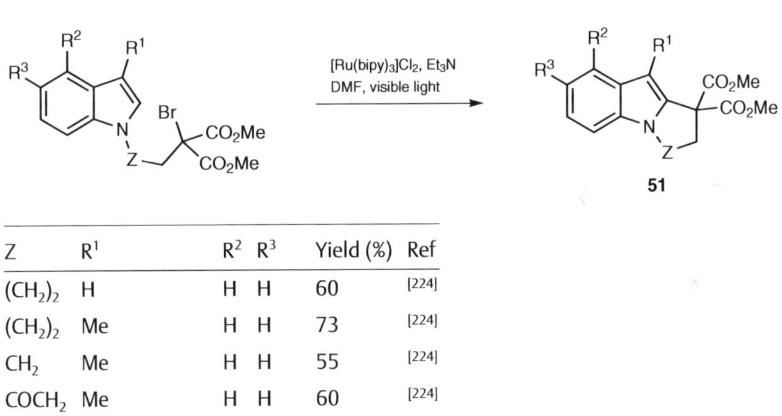

Z	R^1	R^2	R^3	Yield (%)	Ref
(CH$_2$)$_2$	H	H	H	60	[224]
(CH$_2$)$_2$	Me	H	H	73	[224]
CH$_2$	Me	H	H	55	[224]
COCH$_2$	Me	H	H	60	[224]
(CH$_2$)$_2$	H	Br	H	55	[224]
(CH$_2$)$_2$	(CH$_2$)$_2$CO$_2$Me	H	H	61	[224]
(CH$_2$)$_2$	H	H	OMe	59	[224]
(CH$_2$)$_2$	CH$_2$CN	H	H	60	[224]
(CH$_2$)$_2$	CN	H	H	40	[224]

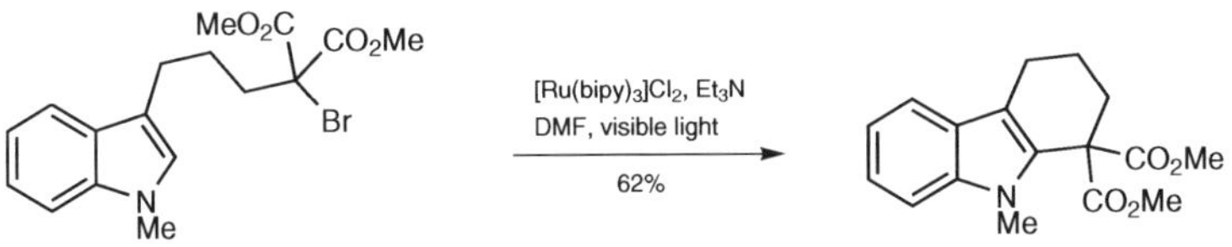

Fused [1,2-a]Indoles 51; General Procedure:[224]

A flame-dried, 10-mL round-bottomed flask was charged with [Ru(bipy)$_3$]Cl$_2$·6H$_2$O (1.0 µmol, 0.01 equiv), the bromide (0.10 mmol, 1.0 equiv), Et$_3$N (0.20 mmol, 2.0 equiv), and DMF (2.0 mL). The mixture, after degassing by a freeze–pump–thaw procedure, was placed at a distance of ~10 cm from a 15-W compact fluorescent lamp. Upon completion of the reaction (as monitored by TLC), the mixture was poured into a separatory funnel containing H_2O (25 mL) and Et_2O (25 mL). The organic and aqueous layers were separated, and the aqueous layer was extracted with Et_2O (2 × 50 mL). The organic extract was dried (Na_2SO_4) and concentrated under reduced pressure. The residue was purified by column chromatography (silica gel).

2.9.2.2.1.3 Synthesis of Indole-3-carbaldehyde Derivatives

Irradiation of tertiary amines **52** with a blue-light-emitting diode in the presence of a catalytic amount of an iridium catalyst {[Ir(ppy)$_2$(dtbbipy)]PF$_6$; ppy = 2-(2-pyridyl)phenyl (commonly called 2-phenylpyridyl), dtbbipy = 4,4′-di-tert-butyl-2,2′-bipyridyl} in chloroform in air results in the formation of indole-3-carbaldehyde derivatives **53** at room temperature within 20–36 hours (Scheme 26).[283] These types of compounds are important precursors for many biologically and medicinally active compounds.

Scheme 26 Iridium-Mediated Photoredox-Catalyzed Cascade Reaction To Give Indole-3-carbaldehydes[283]

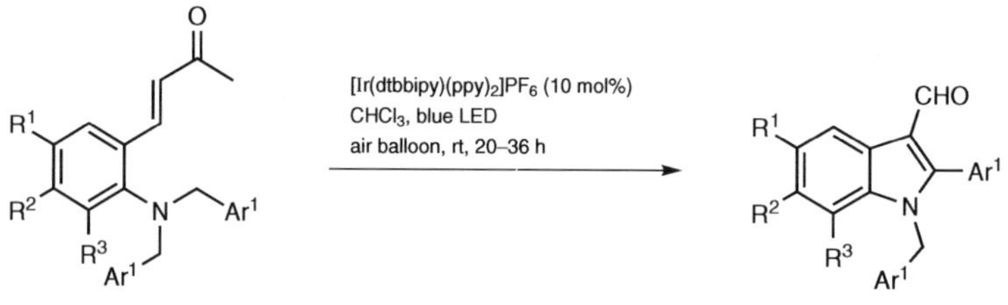

Ar¹	R¹	R²	R³	Yield (%)	Ref
Ph	H	H	H	63	[283]
4-Tol	H	H	H	59	[283]
Ph	Cl	H	Cl	41	[283]
Ph	F	H	H	42	[283]
Ph	Cl	H	H	51	[283]
3-ClC$_6$H$_4$	H	H	H	47	[283]
4-Tol	F	H	H	51	[283]
Ph	H	F	H	41	[283]
4-ClC$_6$H$_4$	Cl	H	H	56	[283]
4-FC$_6$H$_4$	H	H	H	67	[283]
4-FC$_6$H$_4$	Cl	H	H	51	[283]
4-NCC$_6$H$_4$	H	H	H	43	[283]

2.9.2 Intramolecular Free-Radical Cyclization Routes to N-Heterocycles

A possible mechanism for the radical cyclization of these tertiary amine derivatives is outlined in Scheme 27. Visible-light-mediated single-electron transfer of **52** forms radical cation **54**. This is followed by facile deprotonation of radical cation **54** to give α-amino alkyl radical **55**. Intramolecular radical transfer and addition give intermediate radical **56**. Single-electron reduction of radical **56** followed by protonation results in intermediate **57** and the regeneration of iridium(III) to continue the catalytic cycle. Intermediate **57** undergoes oxidative aromatization[264] to form **58**. Single-electron oxidation of **58** produces **59**, which gives captodative α-radical carbonyl intermediate **60**. Subsequent electron-transfer and carbon–carbon bond cleavage reactions result in product **53**.

Scheme 27 Proposed Reaction Mechanism for the Photoredox Cascade Reaction To Give Indole-3-carbaldehydes[264,283]

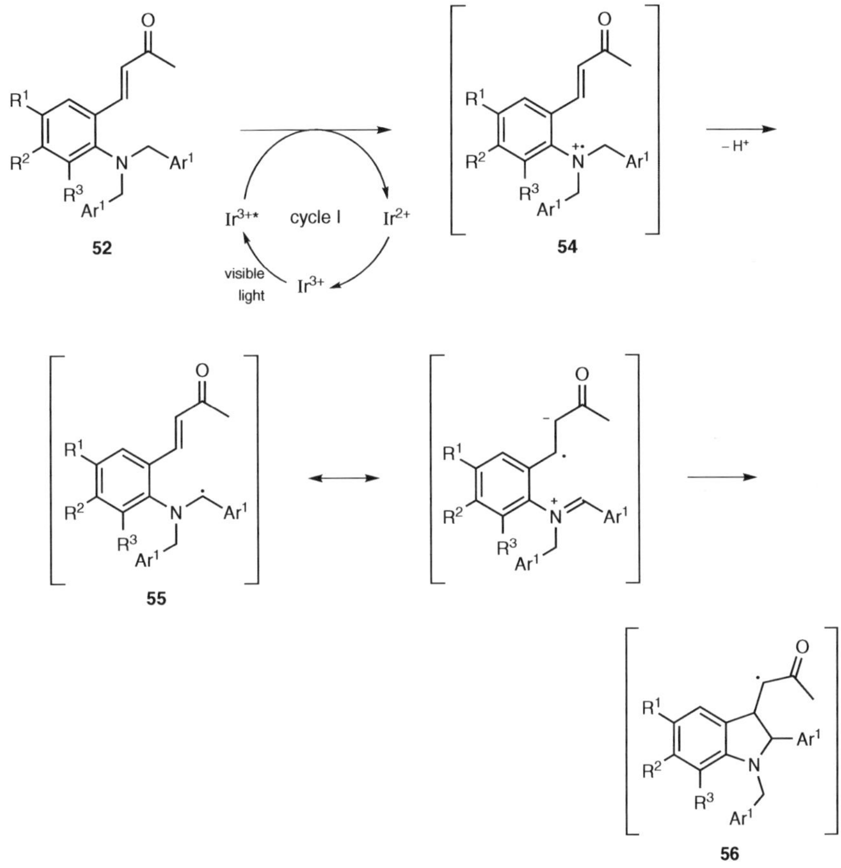

Indole-3-carboxaldehydes 53; General Procedure:[283]
A flame-dried, sealed Schlenk tube was charged with [Ir(ppy)$_2$(dtbbipy)]PF$_6$ (10 mol%), and the tube was fitted with an air balloon. A soln of starting compound **52** (0.2 mmol) in anhyd CHCl$_3$ (2 mL) was added by syringe. After stirring for the required time at rt, the mixture was filtered through a short plug of silica gel, and the plug was washed with copious amounts of CH$_2$Cl$_2$. The solvent was evaporated under reduced pressure and the product was purified by column chromatography.

2.9.2.2.2 Synthesis of Indolines and Azaindolines

Epoxides **61** undergo reductive intramolecular cyclization in the presence of 3 mol% of dichlorobis(η^5-cyclopentadienyl)titanium(IV) and a stoichiometric amount of manganese metal by radical annulation to produce 3,3-disubstituted indoline and azaindoline derivatives **62** (Table 3). In the case of benzyloxycarbonyl-protected epoxide substrates, deprotection is acheived using palladium on carbon under an atmosphere of hydrogen.[284] However, palladium on carbon can also dechlorinate the substrate (Table 3, entry 9).

Table 3 Preparation of Indoline Derivatives by Intramolecular Cyclization of Epoxides[284]

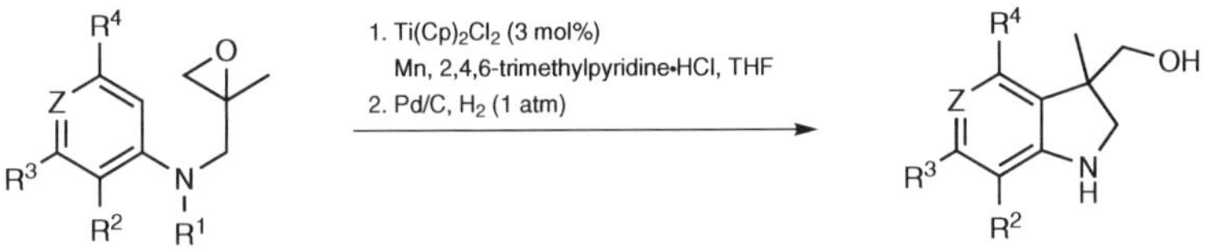

Entry	Substrate	Product	Yield (%)	Ref
1	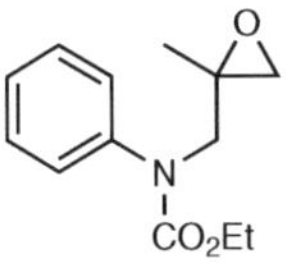	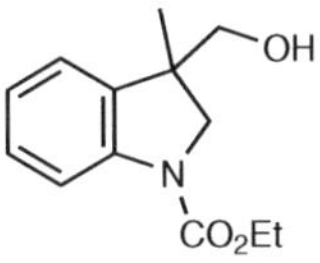	65	[284]
2	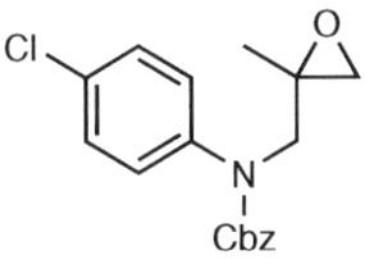	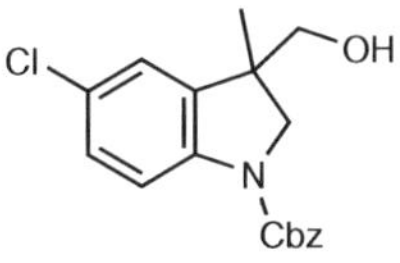	41	[284]
3	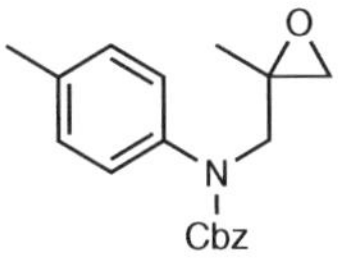	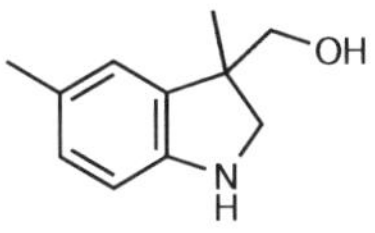	62	[284]
4	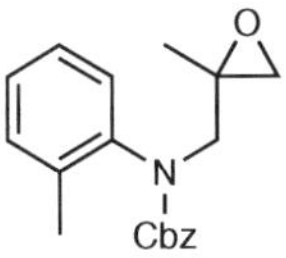	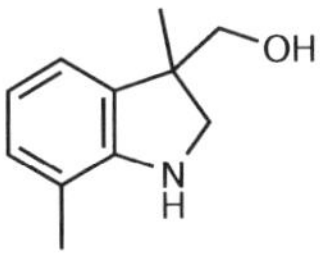	35	[284]

Table 3 (cont.)

Entry	Substrate	Product	Yield (%)	Ref
5	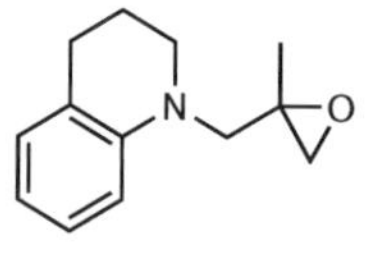	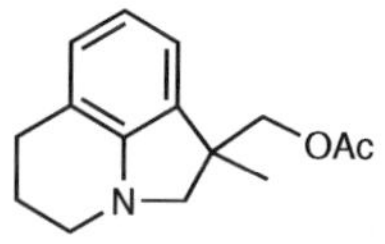	69	[284]
6	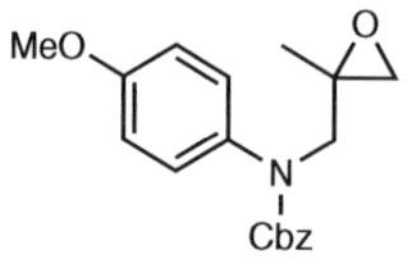	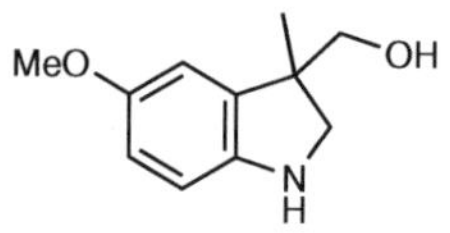	21	[284]
7	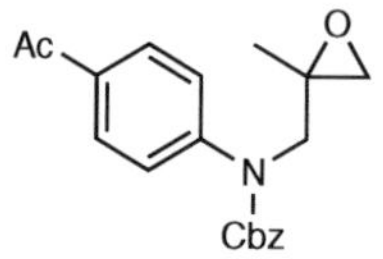	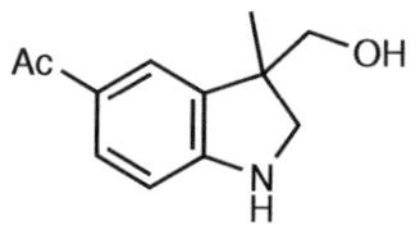	56	[284]
8	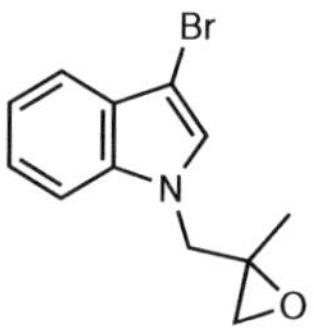	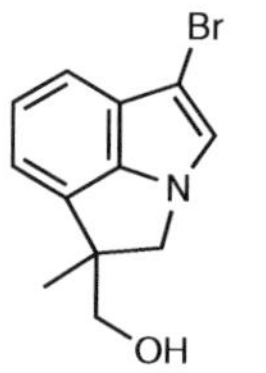	–[a]	[284]
9	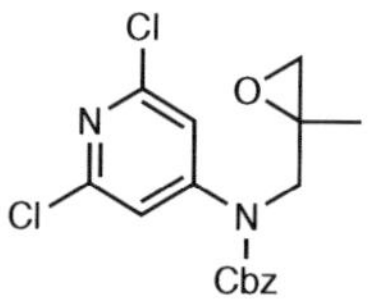	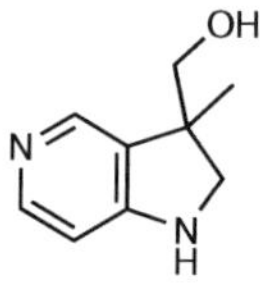	52	[284]

[a] The product was not formed.

The mechanism of epoxide cyclization is presented in Scheme 28. The reaction proceeds with the formation of β-titanoxyl radical **63**,[135,143,285,286] which is followed by reversible cyclization onto the aromatic ring to form cyclohexadienyl radical intermediate **64**.[287–289] Oxidation of the dienyl radical by a trace amount of oxygen followed by loss of a proton affords indoline **65**.[290–292] Finally, protodemetalation of **65** by 2,4,6-trimethylpyridine hydrochloride leads to product **62** and regeneration of the active dichlorobis-(η^5-cyclopentadienyl)titanium(IV) precatalyst.

2.9.2 Intramolecular Free-Radical Cyclization Routes to N-Heterocycles

Scheme 28 Mechanism for the Epoxide-Opening Annulation To Give Indolines[135,143,285–292]

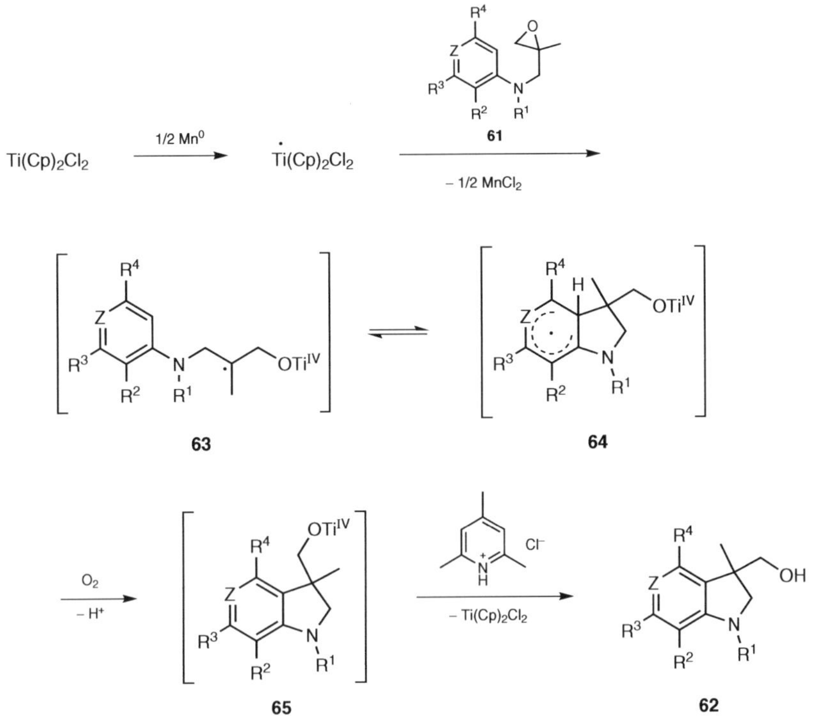

Ethyl 3-(Hydroxymethyl)-3-methyl-2,3-dihydro-1H-indole-1-carboxylate (Table 3, Entry 1); Typical Procedure:[284]

The epoxide (125 mg, 0.53 mmol), Ti(Cp)$_2$Cl$_2$ (3.9 mg, 0.01 mmol), 2,4,6-trimethylpyridine•HCl (125 mg, 0.79 mmol), and Mn(0) (43 mg, 0.79 mmol) were added to a three-necked flask. The vessel was fitted with a reflux condenser and was purged with argon (3×). THF (12.1 mL, 0.1 M) was added, and the mixture was heated at reflux for 45 min. The color gradually changed from light orange to deep violet/blue, which indicated completion of the reaction. The mixture was cooled to rt and then quenched with sat. aq NH$_4$Cl and extracted with Et$_2$O (3 × 10 mL). The combined extracts were dried (MgSO$_4$) and concentrated under reduced pressure. The residue was purified by column chromatography (neutral alumina, hexane/EtOAc 3:1 to 1:3); yield: 81 mg (65%).

(3,5-Dimethyl-2,3-dihydro-1H-indol-3-yl)methanol (Table 3, Entry 3); Typical Procedure:[284]

The epoxide (163 mg, 0.52 mmol), Ti(Cp)$_2$Cl$_2$ (3.9 mg, 0.01 mmol), 2,4,6-trimethylpyridine•HCl (123 mg, 0.78 mmol), and Mn(0) (43 mg, 0.78 mmol) were added to a two-necked flask. The vessel was fitted with a reflux condenser and was purged with argon (3 ×). THF (4.3 mL, 0.1 M) was added, and the mixture was heated at reflux for 3 h. The color gradually changed from light pink to dark violet. The mixture was cooled to rt and then quenched with sat. aq NH$_4$Cl and extracted with Et$_2$O (3 × 10 mL). The combined extracts were washed with brine, dried (MgSO$_4$), and concentrated under reduced pressure. The residue was dissolved in MeOH (5 mL) and treated with 25% w/w Pd/C (30 mg, 0.01 mmol). The mixture was stirred at rt under an atmosphere of hydrogen (1 atm), and the disappearance of the starting material was monitored by TLC (hexane/EtOAc 1:1). The reaction was quenched with Celite, and the mixture was filtered and purified by column chromatography (silica gel, hexane/EtOAc) to afford an oil; yield: 57 mg (62%, 2 steps).

2.9.2.2.3 Synthesis of Oxindole Derivatives

2.9.2.2.3.1 Oxindoles and Indole-2,3-diones

The intramolecular cyclization of 2-(methylsulfanyl)-N-phenylacetamides **66** to prepare oxindoles and indole-2,3-diones (isatins) can be performed using manganese(III) acetate and/or copper(II) acetate at 90 °C.[293] The reaction proceeds through five-membered ring cyclization of 2-(methylsulfanyl)-N-phenylacetamides **66** and is mostly dependent on the reaction conditions (Scheme 29).[293] Treatment of acetamides **66** with both manganese(III) acetate and copper(II) acetate in different solvents leads to the product oxindoles **67** and **68** or acetamide **69**, depending on the solvent used, whereas treatment of 2-(methylsulfanyl)-N-phenylacetamides with only copper(II) acetate in formic acid leads to oxindole **70** as the major product (Scheme 29).[293]

Scheme 29 Synthesis of Oxindoles by Cyclization of 2-(Methylsulfanyl)-N-phenylacetamides[293]

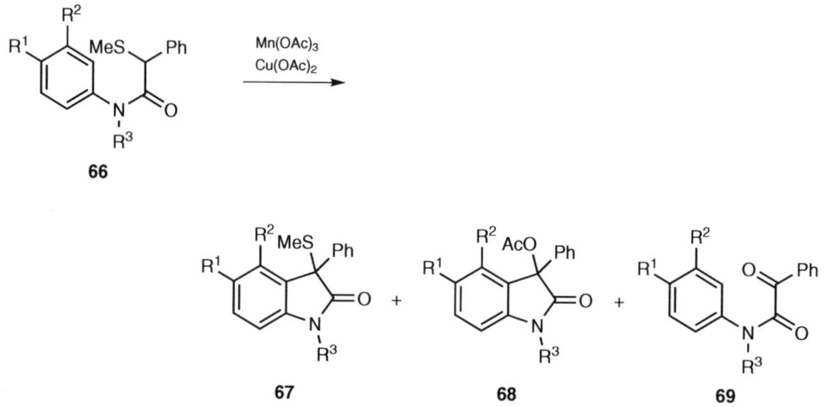

R^1	R^2	R^3	Solvent	Yield (%) 67	Yield (%) 68	Yield (%) 69	Ref
H	H	Et	2,2,2-trifluoroethanol	87	–	5	[293]
H	H	Et	MeCN	74	–	11	[293]
H	H	Et	AcOH	–	74	12	[293]
Me	H	Et	2,2,2-trifluoroethanol	84	–	5	[293]
Me	H	Et	MeCN	68	–	10	[293]
Me	H	Et	AcOH	–	65	13	[293]
OMe	H	Et	2,2,2-trifluoroethanol	69	–	3	[293]
OMe	H	Et	AcOH	–	66	11	[293]
H	Me	Et	2,2,2-trifluoroethanol	87	–	5	[293]
H	Me	Et	MeCN	77	–	8	[293]
H	Me	Et	AcOH	–	60	12	[293]
Br	H	Et	2,2,2-trifluoroethanol	79	–	9	[293]
Br	H	Et	MeCN	71	–	16	[293]
Br	H	Et	AcOH	–	30	12	[293]
Cl	H	Et	2,2,2-trifluoroethanol	72	–	17	[293]
Cl	H	Et	MeCN	60	–	24	[293]

2.9.2 Intramolecular Free-Radical Cyclization Routes to N-Heterocycles

R^1	R^2	R^3	Solvent	Yield (%) 67	Yield (%) 68	Yield (%) 69	Ref
Cl	H	Et	AcOH	–	33	10	[293]
CO_2Et	H	Et	2,2,2-trifluoroethanol	77	–	12	[293]
CO_2Et	H	Et	AcOH	–	0	97	[293]
H	H	Ph	AcOH	–	80	10	[293]
OMe	H	4-MeOC$_6$H$_4$	AcOH	–	83	11	[293]

66 → 70 (Cu(OAc)$_2$, HCO$_2$H)

R^1	R^2	Yield (%)	Ref
H	H	53	[293]
Me	H	65	[293]
OMe	H	34	[293]
H	Me	74	[293]
Br	H	38	[293]
Cl	H	25	[293]
CO_2Et	H	0	[293]

1-Ethyl-3-(methylsulfanyl)-3-phenyl-1,3-dihydro-2H-indol-2-one (67, $R^1 = R^2 = H$; $R^3 = Et$) and N-Ethyl-2-oxo-N,2-diphenylacetamide (69, $R^1 = R^2 = H$; $R^3 = Et$); Typical Procedure Using Manganese(III) Acetate and Copper(II) Acetate in 2,2,2-Trifluoroethanol:[293]
A round-bottomed flask was charged with acetamide **66** (151 mg, 0.53 mmol), Mn(OAc)$_3$ (282 mg, 1.05 mmol), Cu(OAc)$_2$ (210 mg, 1.05 mmol), and 2,2,2-trifluoroethanol (10 mL), and the mixture was heated at 90 °C for 16 h [until disappearance of the dark brown color of Mn(OAc)$_3$]. Extra aliquots of Mn(OAc)$_3$ (285 mg, 1.06 mmol) and Cu(OAc)$_2$ (213 mg, 1.06 mmol) were then added. The mixture was stirred for 11 h and then EtOAc (100 mL) was added. The mixture was washed with sat. aq NaHSO$_4$ (50 mL) and H$_2$O (3 × 50 mL), dried (Na$_2$SO$_4$), and concentrated under reduced pressure. The residue was purified by column chromatography (silica gel, EtOAc/hexane 1:20 then 1:10) to afford **67**; yield: 130 mg (87%); and **69**; yield: 7 mg (5%).

5-Bromo-1-ethyl-3-(methylsulfanyl)-3-phenyl-1,3-dihydro-2H-indol-2-one (67, $R^1 = Br$; $R^2 = H$; $R^3 = Et$) and N-(4-Bromophenyl)-N-ethyl-2-oxo-2-phenylacetamide (69, $R^1 = Br$; $R^2 = H$; $R^3 = Et$); Typical Procedure Using Manganese(III) Acetate and Copper(II) Acetate in Acetonitrile:[293]
A round-bottomed flask was charged with acetamide **66** (149 mg, 0.41 mmol), Mn(OAc)$_3$ (221 mg, 0.82 mmol), Cu(OAc)$_2$ (165 mg, 0.82 mmol), and MeCN (10 mL), and the mixture was heated at 90 °C for 46 h. The mixture was washed with sat. aq NaHSO$_4$ (50 mL) and H$_2$O (3 × 50 mL), dried (Na$_2$SO$_4$), and concentrated under reduced pressure. The residue was purified by chromatography (silica gel, EtOAc/hexane 1:20 then 1:10) to afford **67**; yield: 105 mg (71%); and **69**; yield: 22 mg (16%).

3-Acetyl-1-ethyl-3-phenyl-1,3-dihydro-2H-indol-2-one (68, $R^1 = R^2 = H$; $R^3 = Et$) and N-Ethyl-2-oxo-N,2-diphenylacetamide (69, $R^1 = R^2 = H$; $R^3 = Et$); Typical Procedure Using Manganese(III) Acetate and Copper(II) Acetate in Acetic Acid:[293]

A round-bottomed flask was charged with acetamide **66** (150 mg, 0.53 mmol), Mn(OAc)$_3$ (706 mg, 2.63 mmol), Cu(OAc)$_2$ (210 mg, 1.05 mmol), and AcOH (10 mL), and the mixture was heated at 90 °C for 42 h. The mixture was washed with sat. aq NaHSO$_4$ (50 mL) and H$_2$O (3 × 50 mL), dried (Na$_2$SO$_4$), and concentrated under reduced pressure. The residue was purified by chromatography (silica gel, EtOAc/hexane 1:20) to give **68**; yield: 115 mg (74%); and **69**; yield: 16 mg (12%).

1-Ethyl-3-hydroxy-3-phenyl-1,3-dihydro-2H-indol-2-one (70, $R^1 = R^2 = H$); Typical Procedure Using Copper(II) Acetate in Formic Acid:[293]

A round-bottomed flask was charged with the acetamide (150 mg, 0.53 mmol) and 60% aq HCO$_2$H (10 mL), and the mixture was heated to 90 °C. Cu(OAc)$_2$ (630 mg, 3.15 mmol) was added in three portions at 48-h intervals. The mixture was heated at 90 °C for another 60 h and then diluted with EtOAc (100 mL), washed with H$_2$O (3 × 50 mL), dried (Na$_2$SO$_4$), and concentrated under reduced pressure. The residue was purified by chromatography (silica gel, EtOAc/hexane 1:10 then 1:3) to give N-ethyl-2-oxo-N,2-diphenylacetamide; yield: 16 mg (16%, as reported); and **70**; yield: 70 mg (53%).

2.9.2.2.3.2 Synthesis of 3,3-Disubstituted Oxindoles

Oxindole derivatives are a significant class of heterocycles that exist in numerous biologically active molecules, and they have utility as important synthetic building blocks.[294–296] The primary method to synthesize substituted oxindole rings involves intramolecular homolytic aromatic substitution of hydrogen on a phenyl ring by amidoalkyl radicals.[297–299] 2-Bromo-N-phenylacetamides **71** possessing two electron-withdrawing groups on the phenyl ring can be converted into 3,3-disubstituted oxindoles **72** upon treatment with visible light in the presence of an iridium(III) photoredox catalyst [fac-Ir(ppy)$_3$; ppy = 2-(2-pyridyl)phenyl (commonly called 2-phenylpyridyl)] in dimethylformamide at room temperature (Scheme 30).[300] This method is advantageous for the synthesis of oxindoles with chlorine and bromine atoms attached to the phenyl ring and, thus, has much synthetic utility. However, if compounds **73** are used as substrates, sterically more hindered compounds **75** are obtained as the major products (Scheme 31).

Scheme 30 Synthesis of 3,3-Disubstituted Oxindoles from 2-Bromo-N-phenylacetamides under Iridium Catalysis[300]

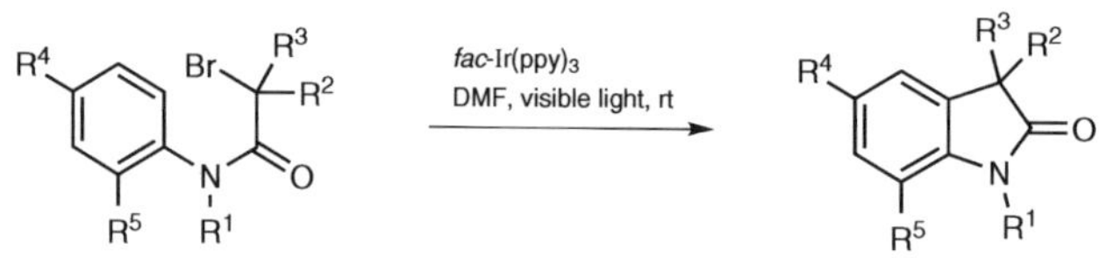

R^1	R^2	R^3	R^4	R^5	Yield (%)	Ref
Me	CO$_2$Et	Me	H	H	95	[300]
Bn	CO$_2$Et	Me	H	H	88	[300]
Me	CO$_2$Et	Me	Me	H	95	[300]
Me	CO$_2$Et	Me	OMe	H	94	[300]
Me	CO$_2$Et	Me	Cl	H	93	[300]
Me	CO$_2$Et	Me	H	Cl	80	[300]

2.9.2 Intramolecular Free-Radical Cyclization Routes to N-Heterocycles

R^1	R^2	R^3	R^4	R^5	Yield (%)	Ref
Me	CO$_2$Et	CH$_2$CH=CH$_2$	H	H	98	[300]
Bn	CO$_2$Et	CH$_2$CH=CH$_2$	OMe	H	85	[300]
Me	Ac	Me	H	H	96	[300]
Bn	Ac	Me	H	H	98	[300]
Bn	Ac	Me	Me	H	95	[300]
Bn	Ac	Me	OMe	H	98	[300]
Bn	Ac	Me	F	H	96	[300]
Bn	Ac	Me	Cl	H	92	[300]
Me	Ac	Me	Br	H	81	[300]
Me	Ac	Me	H	Me	80	[300]
Me	Ac	Me	H	OMe	93	[300]
Bn	Ac	Me	H	OMe	81	[300]
Me	Ac	Me	H	Br	mixture	[300]
Me	Ac	Et	H	H	86	[300]
Me	Ac	Bn	H	H	87	[300]
Me	CN	Me	H	H	98	[300]
Bn	CN	Me	H	H	89	[300]
Me	CN	CH$_2$CH=CH$_2$	H	H	92	[300]

Scheme 31 Cyclization of Sterically Hindered 2-Bromo-N-phenylacetamides[300]

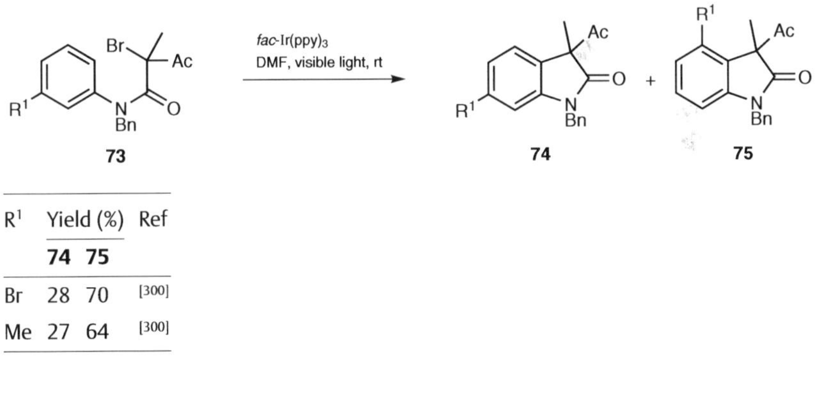

R^1	Yield (%) 74	Yield (%) 75	Ref
Br	28	70	[300]
Me	27	64	[300]

Oxindoles 72; General Procedure:[300]
A flame-dried, 10-mL, round-bottomed flask equipped with a rubber septum and a magnetic stirrer bar was charged with *fac*-Ir(ppy)$_3$ (1.5 mg, 0.02 equiv) and 2-bromo-N-phenylacetamide **71** (0.12 mmol, 1.0 equiv). The mixture was degassed and backfilled with argon (4 × 1 min). DMF (2.4 mL, previously degassed with argon for 30 min) was then added, and the soln was stirred and irradiated with a 40-W household fluorescent lamp (distance ~4 cm) under an argon atmosphere at rt for 12–16 h. Upon completion of the reaction (monitored by TLC), the mixture was diluted with EtOAc (10 mL) and poured into sat. aq NH$_4$Cl (15 mL). The combined aqueous layers were extracted with EtOAc (3 × 10 mL). The combined organic extracts were washed with brine (4 × 15 mL) and dried (Na$_2$SO$_4$). The mixture was concentrated under reduced pressure, and the residue was purified by chromatography (silica gel).

2.9.2.3 Synthesis of Lactam Derivatives

2.9.2.3.1 Manganese(III)-Mediated Diastereoselective 4-exo-trig Cyclization of Enamides

The synthesis of β-lactams by 4-*exo-trig* radical cyclization of suitable precursors is a widely utilized technique. Manganese(III)-mediated 4-*exo-trig* cyclization of enamides provides good diastereoselection if suitable chiral substituents are placed on the nitrogen atoms of the enamides.[301] Treatment of enamides **76** with manganese(III) acetate dihydrate in glacial acetic acid at 70 °C to prevent the formation of side products affords *trans*-azetidinones **77A** and **77B** in good to excellent yields (Scheme 32). Enamides with esters of α-amino acids show a significant level of diastereoselection.

Scheme 32 Manganese(III)-Mediated Cyclization of Chiral Enamides To Give *trans*-Azetidinones[301]

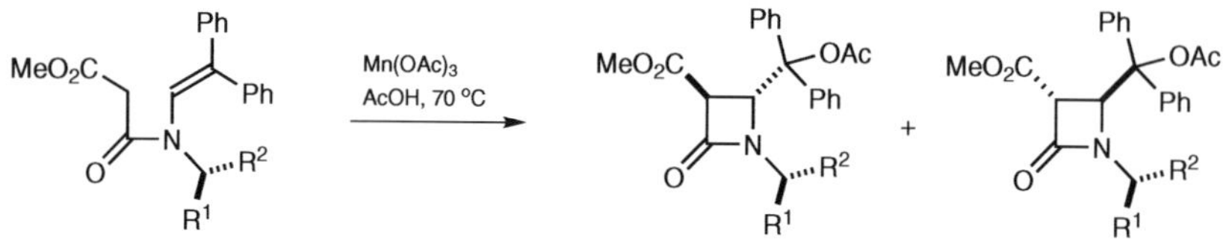

R¹	R²	dr^a	Yield^b (%)	Ref
Me	Ph	54:46	78	[301]
Ph	Me	55:45	68	[301]
Cy	Me	61:39	79	[301]
Me	1-naphthyl	62:38	72	[301]
CO₂Me	Me	57:43	55	[301]
CO₂Me	Bn	50:50	72	[301]
CO₂Me	Ph	80:20	57	[301]
CO₂Et	iPr	80:20	63	[301]
CO₂Me	t-Bu	80:20	55	[301]

^a Diastereomeric ratios were calculated by NMR spectroscopy.
^b Chromatographically isolated products.

***trans*-Azetidinones 77A and 77B; General Procedure:**[301]

Mn(OAc)₃·2H₂O (536 mg, 2 mmol) was added to a soln of enamide **76** (1 mmol) and glacial AcOH (5 mL) under an argon atmosphere. The mixture was stirred at 70 °C until the brown color of the mixture disappeared. The mixture was then cooled to rt and added to H₂O (50 mL). The resulting mixture was extracted with CH₂Cl₂, and the organic phase was washed with sat. aq NaHCO₃ and H₂O and finally dried (Na₂SO₄). The solvent was removed under reduced pressure, and the residue was purified by column chromatography (silica gel, hexane/Et₂O).

2.9.2.3.2 Synthesis of γ-Lactams

2.9.2.3.2.1 Manganese(III)-Mediated Spirolactam Synthesis

Heating a solution of diethyl 2-[2-(2,4-dimethoxyphenyl)amino-2-oxoethyl]malonate with manganese(III) acetate and acetic acid at reflux selectively produces the corresponding spirolactam **78** by oxidative 5-*exo-trig* cyclization (Scheme 33).[302]

Scheme 33 Manganese(III) Acetate Mediated 5-*exo* Cyclization of a 2-(2-Amino-2-oxoethyl)malonate for the Synthesis of a Spirolactam[302]

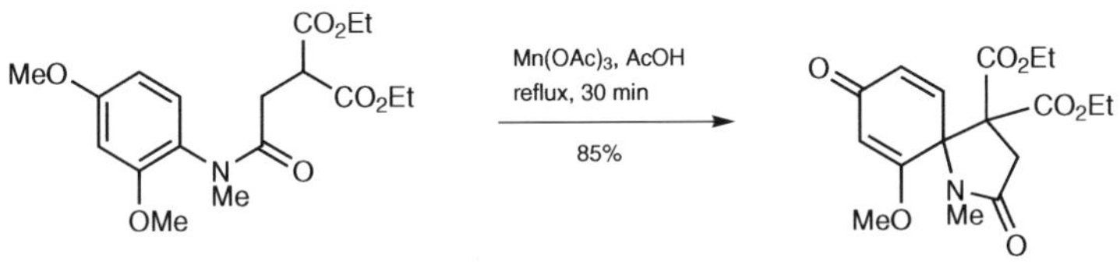

Diethyl 6-Methoxy-1-methyl-2,8-dioxo-1-azaspiro[4.5]deca-6,9-diene-4,4-dicarboxylate (78); Typical Procedure:[302]
Mn(OAc)$_3$·2H$_2$O (1.5 mmol) was added to a mixture of the malonate (0.5 mmol) and glacial AcOH (30 mL). The mixture was stirred at reflux for 30 min in air. Upon completion of the reaction, the solvent was removed under reduced pressure, and the residue was triturated with H$_2$O followed by extraction with CH$_2$Cl$_2$ (3 × 10 mL). The combined extract was dried (MgSO$_4$) and then concentrated under reduced pressure. The residue was purified by flash column chromatography or TLC (silica gel) and recrystallized (CHCl$_3$/hexane) to afford colorless prisms; yield: 85%.

2.9.2.3.2.2 γ-Lactams by Reverse Atom-Transfer Radical Cyclization of α-Polychloro-N-allylamides

The cyclization of N-substituted α-polychloro-N-allylamides **79** to γ-lactams **80** can be efficiently performed using a catalytic system based on a combination of 0.5 mol% copper(II) chloride/[nitrogen ligand] (as redox catalyst), ascorbic acid (as catalyst regenerator), and sodium carbonate (as acidity quencher).[303] Both 1,1,4,7,7-pentamethyldiethylenetriamine and tris(2-pyridylmethyl)amine as nitrogen ligands provide the γ-lactam products in excellent yields; however, the former is preferred because it is less expensive. The redox catalyst delivers excellent activity (up to 99%) in a cosolvent system (EtOAc/EtOH 3:1) at 35°C (Table 4).[303]

Table 4 Reverse Atom-Transfer Radical Cyclization of N-Substituted α-Polychloro-N-allyl-amides To Give γ-Lactams[303]

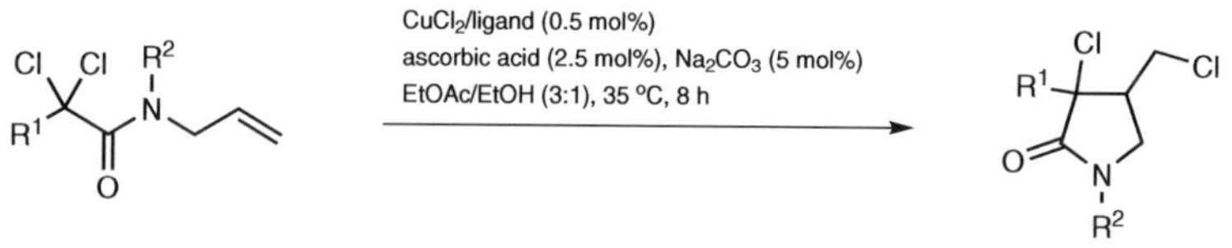

Entry	Substrate	Product	Ratio (cis/trans)	Yield[a] (%)	Ref
1	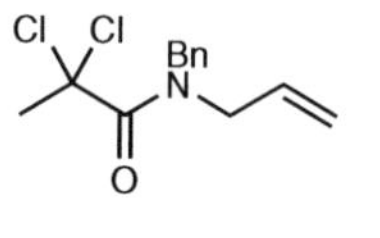 R¹=Me, R²=Bn	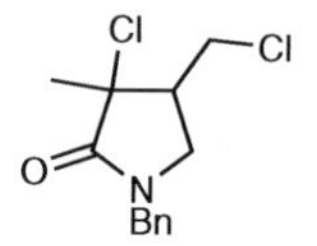	84:16	99	[303]
2	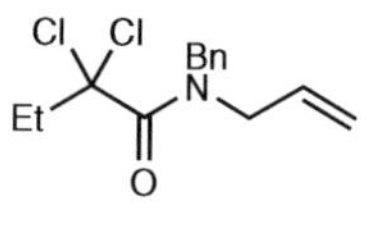 R¹=Et, R²=Bn	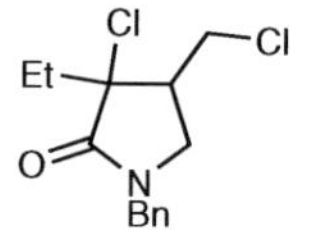	87:13	94	[303]
3	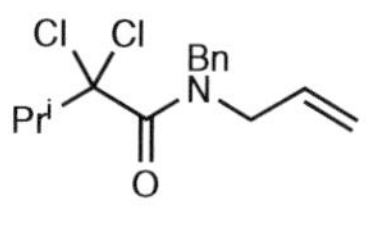 R¹=Pr^i, R²=Bn	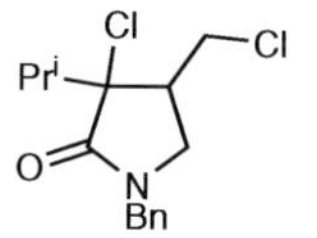	99:1	93	[303]
4	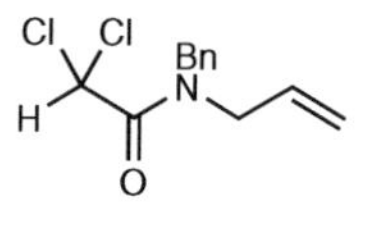 R¹=H, R²=Bn	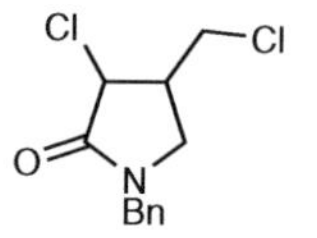	26:74	95[b]	[303]
5	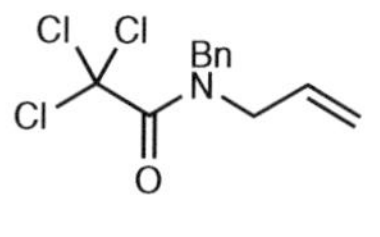 R¹=Cl, R²=Bn	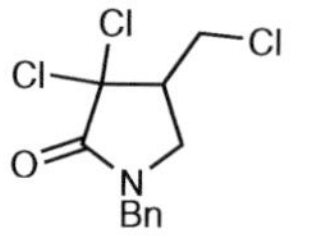	–	93	[303]
6	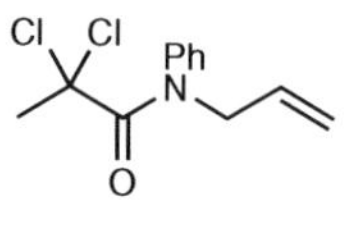 R¹=Me, R²=Ph	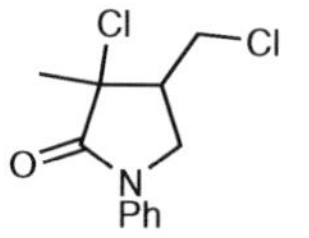	87:13	96	[303]
7	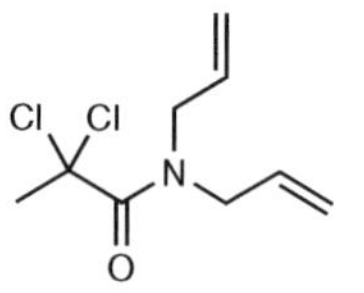 R¹=Me, R²=allyl	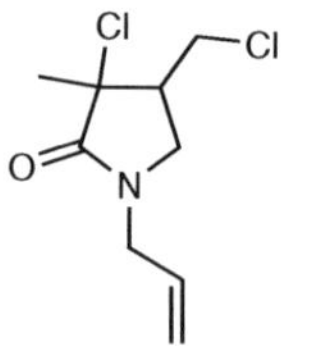	84:16	94[c]	[303]
8	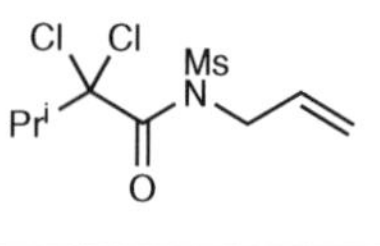 R¹=Pr^i, R²=Ms	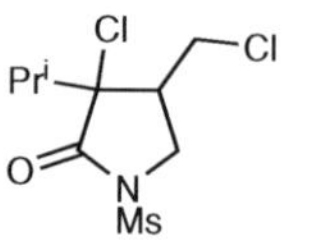	96:4	96[c]	[303]

2.9.2 Intramolecular Free-Radical Cyclization Routes to N-Heterocycles

Table 4 (cont.)

Entry	Substrate	Product	Ratio (cis/trans)	Yield[a] (%)	Ref
9	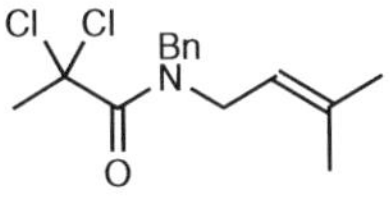 (structure)	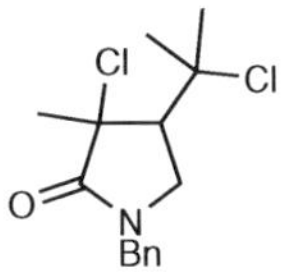 (structure)	58:42	95	[303]
10	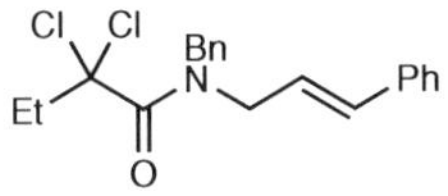 (structure)	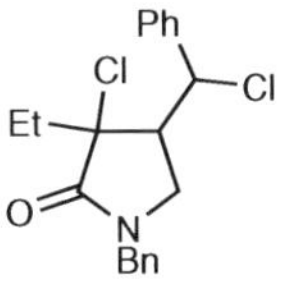 (structure)	97:3[d]	96[b]	[303]
11	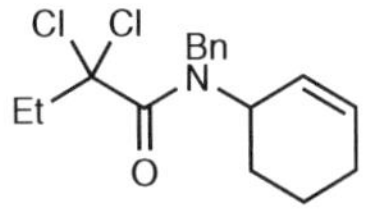 (structure)	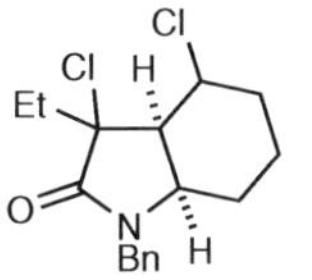 (structure)	82:18[e]	96[b]	[303]

[a] Reactions were performed with 1,1,4,7,7-pentamethyldiethylenetriamine except where noted otherwise.
[b] Reaction was performed with tris(2-pyridylmethyl)amine.
[c] Using ascorbic acid (1.5 mol%) and NaHCO$_3$ (3 mol%).
[d] dr 84:16 (cis) and 100:0 (trans).
[e] dr 59:41 (cis) and 76:24 (trans).

1-Benzyl-3-chloro-4-(chloromethyl)-3-ethylpyrrolidin-2-one (Table 4, Entry 2); Typical Procedure:[303]

An oven-dried Schlenk tube (previously washed sequentially with a soln of NH$_3$ and then H$_2$O) was charged with ascorbic acid (35.2 mg, 0.2 mmol), Na$_2$CO$_3$ (42.4 mg, 0.4 mmol), and amide **79** (2.290 g, 8 mmol). After three vacuum/argon cycles (10 min, ~8.50 min for vacuum and 1.50 min for argon), EtOAc (3 mL) was added. Upon complete solubilization of the substrates, a 0.04 M soln of CuCl$_2$/1,1,4,7,7-pentamethyldiethylenetriamine in abs EtOH (1 mL) was added under argon. The mixture was then stirred at 35 °C in a water bath. Upon completion of the reaction (8 h), the mixture was diluted with H$_2$O (8 mL), acidified with 10% w/v HCl, and extracted with CH$_2$Cl$_2$ (3 × 6 mL). The combined organic layer was concentrated under reduced pressure, and the residue was purified by flash chromatography (silica gel, hexane/Et$_2$O gradient from 100:0 to 0:100) to afford a colorless oil; yield: 2.156 g (94%); ratio (cis/trans) 87:13.

2.9.2.4 Synthesis of Piperidines

2.9.2.4.1 Synthesis of 3-Chloropiperidine Derivatives

The diastereoselective radical cyclization of N-chloro-N-pentenylamines **81** to produce 3-chloropiperidines **82** can be efficiently achieved using copper(I) chloride in tetrahydrofuran at 60 °C (Table 5).[244]

Table 5 Copper(I)-Catalyzed Radical Cyclization of N-Chloroamines To Give 3-Chloropiperidines[244]

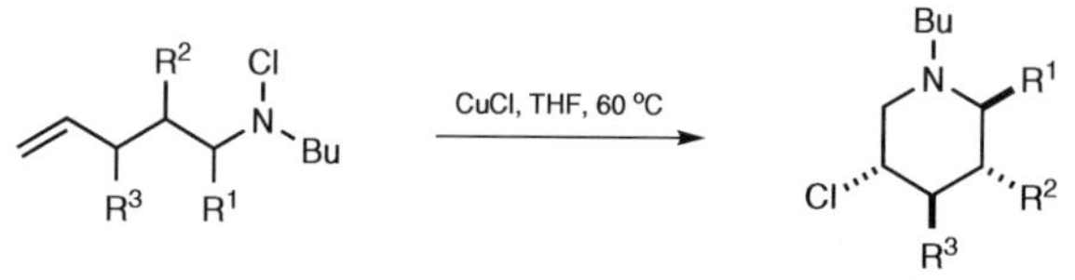

Entry	dr	Yield (%)	Ref
1	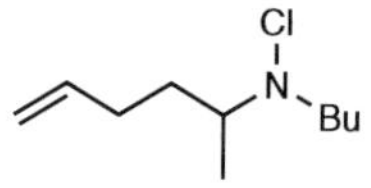 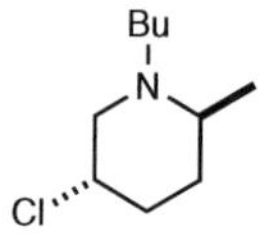 6:1	77	[244]
2	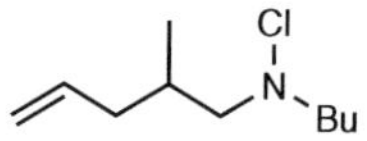 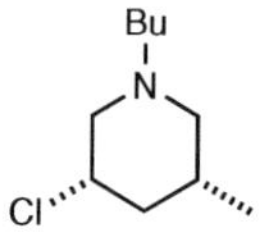 4:1	70	[244]
3	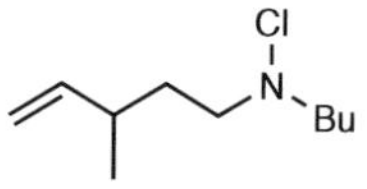 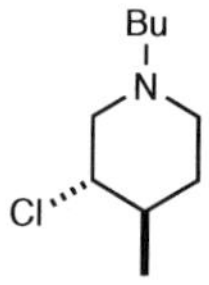 5:1	58	[244]
4	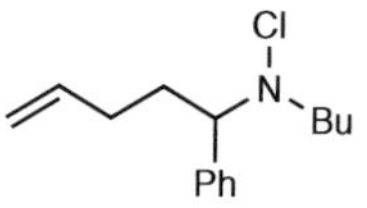 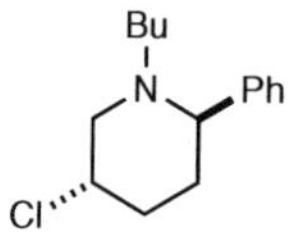 6:1	55	[244]
5	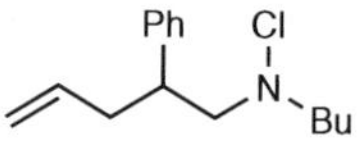 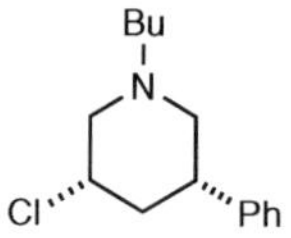 3:1	73	[244]
6	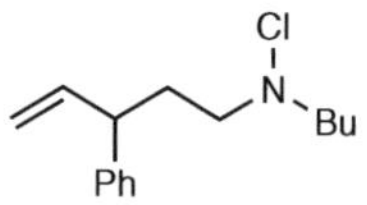 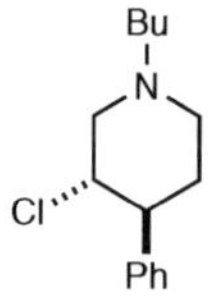 –[b]	55	[244]

[a] Structure of the major product is shown.
[b] Only one isomer was formed.

1-Butyl-3-chloropiperidines 82; General Procedure:[244]

N-Chloroamine **81** (3 mmol) was added to a suspension of CuCl (30 mg, 10 mol%) in THF (5 mL) at 60 °C. The resulting mixture immediately afforded a blue-green soln, which was then kept at 60 °C for 12 h. The mixture was cooled to rt, silica gel (~3 g) was added, and the solvent was removed under reduced pressure. The residue bound to the silica gel was purified by flash chromatography (pentane/t-BuOMe 10:1 then 3:1 and 1:1).

2.9.2.4.2 Synthesis of *exo*-Alkylidene Piperidinones

The manganese(III) acetate mediated cyclization of 2-(alkynamido)malonates **83** in ethanol gives the corresponding piperidinones **84** (Scheme 34).[252]

Scheme 34 Synthesis of Piperidinones from 2-(Alkynamido)malonates[252]

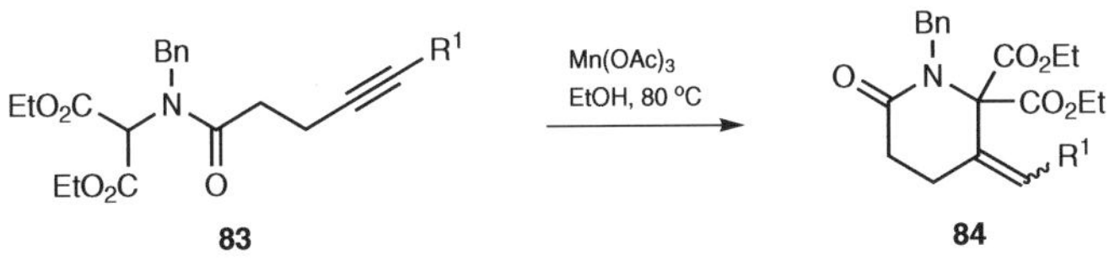

R^1	Time (h)	Ratio[a] (E/Z)	Yield (%)	Ref
H	15	–	66	[252]
Me	5	1:5	71	[252]

[a] Ratio was determined by ^{1}H NMR spectroscopy; the stereochemistry of the product was determined by NOE experiments.

Diethyl 1-Benzyl-3-methylene-6-oxopiperidine-2,2-dicarboxylate (84, R^1 = H); Typical Procedure:[252]

2-(Alkynamido)malonate **83** (R^1 = H; 1.0 equiv) and Mn(OAc)$_3$ (2.0 equiv) were dissolved in degassed EtOH (20 mL/mmol malonate). The soln was heated to 80 °C and stirred for 15 h. Upon completion of the reaction, the mixture was cooled to rt and the solvent was removed under reduced pressure. The resulting residue was dissolved in Et$_2$O and filtered through a pad of silica gel (Et$_2$O). The resulting filtrate was concentrated under reduced pressure to yield the crude product, which was purified by flash chromatography (petroleum ether/Et$_2$O 1:1) to afford a colorless oil; yield: 128 mg (66%); R_f 0.20 (petroleum ether/Et$_2$O 1:1).

Diethyl 1-Benzyl-3-ethylidene-6-oxopiperidine-2,2-dicarboxylate (84, R^1 = Me); Typical Procedure:[252]

2-(Alkynamido)malonate **83** (R^1 = Me; 1.0 equiv) and Mn(OAc)$_3$ (2.0 equiv) were dissolved in degassed EtOH (20 mL/mmol malonate). The soln was heated to 80 °C and stirred for 5 h. Upon completion of the reaction, the mixture was cooled to rt and the solvent was removed under reduced pressure. The resulting residue was dissolved in Et$_2$O and filtered through a pad of silica gel (Et$_2$O). The resulting filtrate was concentrated under reduced pressure to yield the crude product, which was purified by flash chromatography (petroleum ether/Et$_2$O 1:2) to afford a colorless oil; yield: 14.4 mg (71%); ratio (E/Z) 1:5.

2.9.2.5 Synthesis of Quinoline Derivatives

2.9.2.5.1 Manganese(III)-Mediated Oxidative 6-*endo*-trig Cyclization

Heating a mixture of 2-(2-anilino-2-oxoethyl)malonates **85**, manganese(III) acetate, and acetic acid at reflux results in an oxidative 6-*endo*-trig cyclization reaction to produce diethyl 2-oxo-2,3-dihydroquinoline-4,4(1H)-dicarboxylate derivatives **86** selectively.[302] However, if an excess amount of the oxidant is used, the carboxymethyl radical is introduced into either the dihydroquinolinone skeleton or into the N-phenyl group to produce **87** (up to 32%) or **88**, respectively (Scheme 35). 4-Substituted N-phenyl 2-(2-amino-2-oxoethyl)malonates **89** give the corresponding products **90**, but in the case of malonate **89** with R¹=Ac; R²=Me, substitution product **91** is also obtained. 3-Substituted *N*-phenyl 2-(2-amino-2-oxoethyl)malonates **92** give a mixture of cyclization products **93** and **94**.

Scheme 35 Synthesis of Dihydro-2(1*H*)-quinolinones by Manganese(III) Acetate Mediated 6-*endo-trig* Cyclization of Substituted Malonates[302]

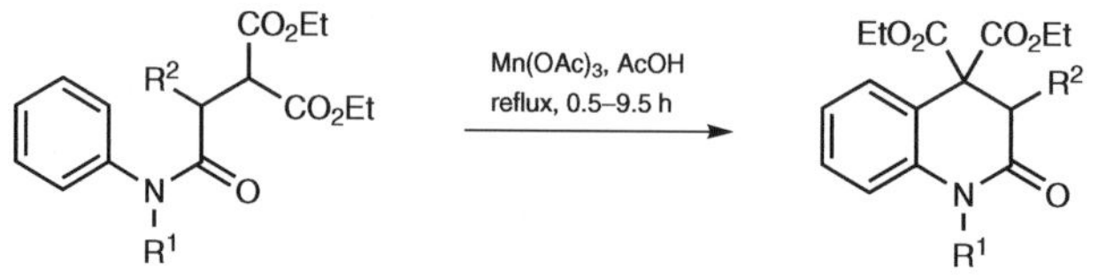

R¹	R²	Ratio [85/Mn(III)]	Time (h)	Recovered Starting Material (%)	Yield (%)	Ref
Me	H	1:3	0.5	0	97	[302]
Et	H	1:3	0.5	0	98	[302]
iPr	H	1:3	0.5	0	93	[302]
Bu	H	1:3	0.5	0	92	[302]
Ph	H	1:3	0.5	0	100	[302]
Bn	H	1:3	0.5	0	93	[302]
Me	Me	1:6	5	8	65[a]	[302]
Ph	Me	1:6	4	10	65[b]	[302]
Me	Ph	1:10	5	16	42[c]	[302]
iPr	Ph	1:10	5	16	44[c]	[302]

[a] Compound **87** was also obtained in trace amounts.
[b] Compounds **87** and **88** were also obtained in yields of 11 and 4%, respectively.
[c] Compound **87** was also obtained in 18% yield.

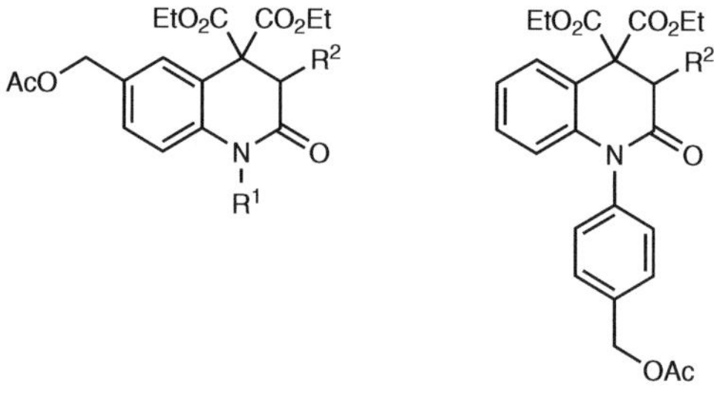

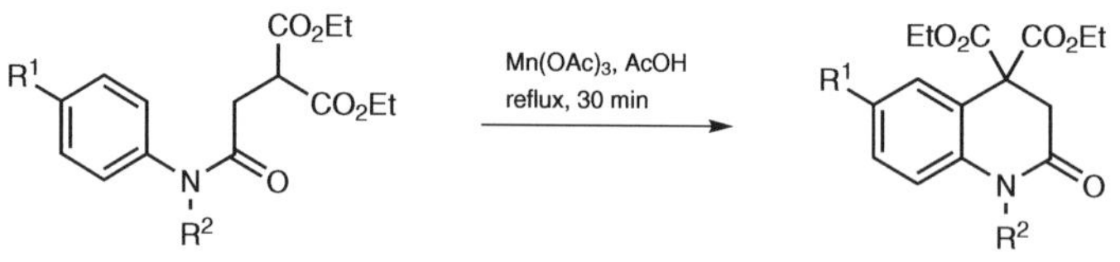

2.9.2 Intramolecular Free-Radical Cyclization Routes to N-Heterocycles

R^1	R^2	Ratio [**89**/Mn(III)]	Yield (%)	Ref
OMe	Me	1:3	97	[302]
Me	Me	1:4	97	[302]
Cl	Me	1:4	88	[302]
Cl	PMB	1:3	85	[302]
F	Me	1:3	95	[302]
NO_2	Me	1:3	92	[302]
Ac	Me	1:4	61[a]	[302]

[a] Compound **91** was also obtained in 28% yield.

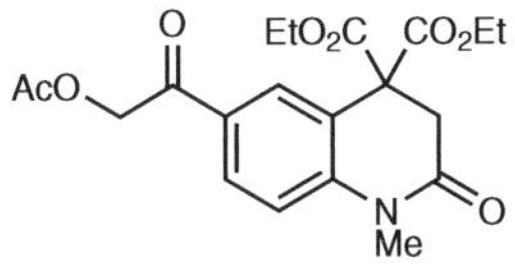

91

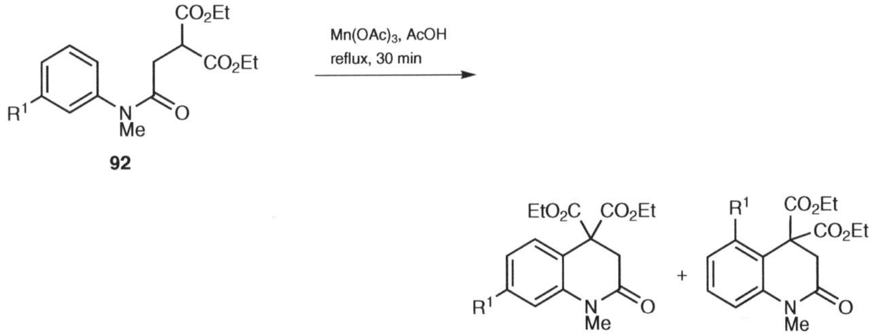

92

93 **94**

R^1	Ratio [**92**/Mn(III)]	Yield (%)		Ref
		93	94	
OMe	1:3	46	51	[302]
NO_2	1:3.5	76	15	[302]

Bis(ethoxycarbonyl)dihydroquinolin-2-ones 86; General Procedure:[302]

Mn(OAc)$_3$ (1.5–12.5 mmol) was added to a mixture of malonate **85** (0.5 mmol) and glacial AcOH (30 mL). The mixture was heated at reflux for 0.5–9.5 h in air. Upon completion of the reaction, the solvent was evaporated under reduced pressure, and the residue was triturated with H$_2$O followed by extraction with CH$_2$Cl$_2$ (3 × 10 mL). The combined extract was dried (MgSO$_4$) and then concentrated. The obtained residue was purified by silica-gel TLC or flash column chromatography. In the case of a solid sample, it was further purified by recrystallization.

2.9.2.5.2 Manganese(III)-Mediated Oxidative 6-*exo-trig* Cyclization

Highly functionalized quinolines can be prepared by free-radical cyclization of N-[2-(alk-1-enyl)aryl]-substituted enamines in the presence of a manganese-based catalytic system.[304] Imine radicals formed by the manganese(II)/cobalt(II)/oxygen (redox system) oxidation of N-[2-(alk-1-enyl)aryl]enamines undergo efficient 6-*exo-trig* cyclization onto C=C bonds.

After ring closure to give a dihydroquinoline bearing a benzylic radical in the 4-position, this radical is trapped with oxygen. Hydrogen transfer gives a hydroperoxide, which undergoes C–C bond cleavage with loss of benzaldehyde to give the products. The reaction tolerates various functional groups, including morpholinocarbonyl, benzoyl, and cyano groups. Treatment of N-[2-(alk-1-enyl)aryl]enamine carboxamides **95** with manganese(II) acetate tetrahydrate (0.3 equiv) and cobalt(II) acetate tetrahydrate (1.5 equiv) in acetic acid at 80 °C under an oxygen atmosphere results in the target cyclization products **96** in excellent yields (Scheme 36). The catalytic manganese(II) acetate/cobalt(II) acetate/oxygen redox system in which manganese(II) is reoxidized to manganese(III) by cobalt(III)/oxygen, generated in situ from cobalt(II) and oxygen,[305,306] can be used to generate carbon radicals that efficiently undergo addition to C=C bonds.

Scheme 36 Oxidative Radical Cyclization of N-[2-(Alk-1-enyl)aryl]enamines[304]

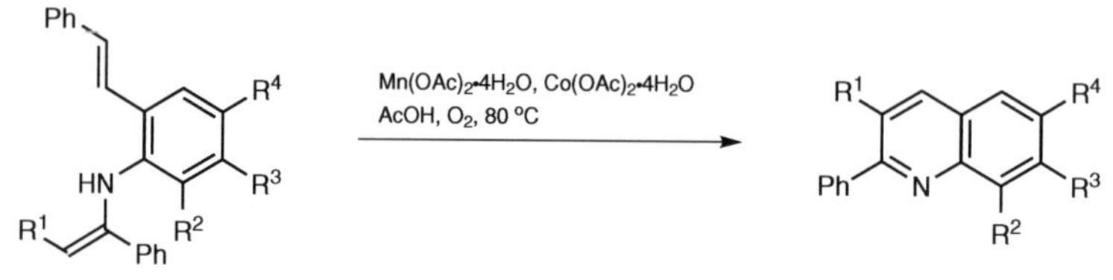

R^1	R^2	R^3	R^4	Yield (%)	Ref
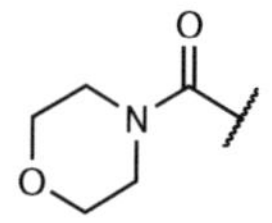 morpholinocarbonyl	H	H	H	84	[304]
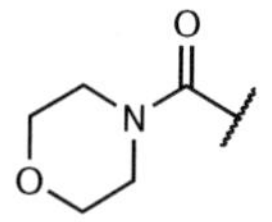 morpholinocarbonyl	H	H	Me	90	[304]
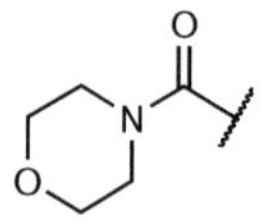 morpholinocarbonyl	Me	H	Me	89	[304]
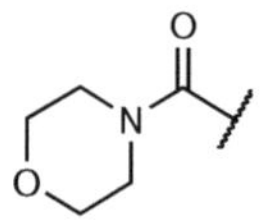 morpholinocarbonyl	H	Me	Me	86	[304]
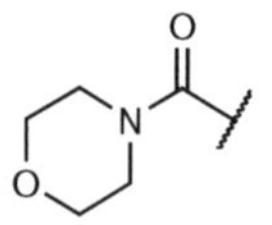 morpholinocarbonyl	H	H	Cl	86	[304]
Bz	H	H	H	87	[304]
Bz	H	H	Me	86	[304]
Bz	H	H	Cl	85	[304]
CN	H	H	H	87	[304]
CN	H	H	Me	85	[304]
CN	H	H	Cl	92	[304]

3-(Morpholinocarbonyl)-2-phenylquinoline (96, R^1 = Morpholinocarbonyl; R^2 = R^3 = R^4 = H); Typical Procedure:[304]

A mixture of enamine **95** (103 mg, 0.25 mmol), Mn(OAc)$_2$•4H$_2$O (18 mg, 0.07 mmol), and Co(OAc)$_2$•4H$_2$O (93 mg, 0.37 mmol) in AcOH (5 mL) was heated at 80 °C for 5 min under an O$_2$ atmosphere (1 atm). Upon completion of the reaction, the mixture was diluted with EtOAc (100 mL), washed with sat. aq NaHSO$_3$ (3 × 50 mL), sat. aq NaHCO$_3$ (3 × 50 mL), and H$_2$O (3 × 50 mL), dried (Na$_2$SO$_4$), and concentrated under reduced pressure. The residue was purified by column chromatography (silica gel, EtOAc/hexane 1:3) followed by crystallization (CHCl$_3$/hexane) to give colorless crystals; yield: 67 mg (84%); mp 149–150 °C.

2.9.3 Intramolecular Free-Radical Cyclization Routes to O-Heterocycles

2.9.3.1 Synthesis of Furan Derivatives

The room-temperature oxidation of alkenyl β-keto ester **97** in the presence of manganese(III) acetate dihydrate (3 equiv), copper(II) acetate monohydrate (1 equiv), and ethanol produces β,γ-unsaturated δ-hydroxy α-keto ester **98** (19%) and furan **99** (58%) within 6 hours (Scheme 37).[307] Oxidation of **97** at first produces α-keto ester **98** bearing a δ-hydroxy group, and this subsequently undergoes heterocyclization to produce furan derivative **99**.

Scheme 37 Manganese(III)-Mediated Oxidation/Heterocyclization of an Alkenyl β-Keto Ester[307]

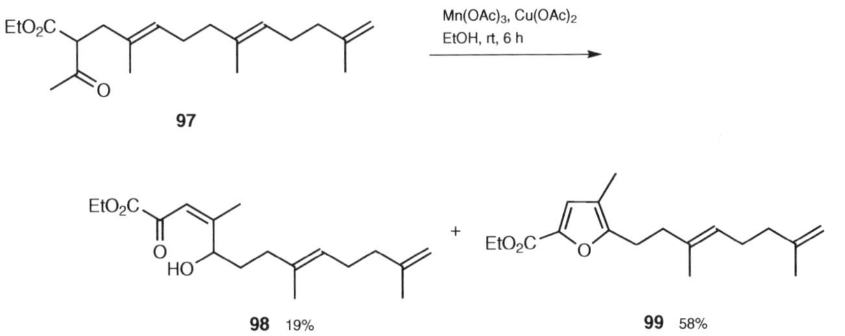

Initially, the one-electron oxidation of the manganese(III) enolate of β-keto ester **97** produces α-carbon radical **100** (Scheme 38).[308] α-Peroxy radical **101** is then formed from **100** by trapping dissolved oxygen.[309] The first oxidation is completed through formation and cleavage of 1,2-dioxetane **102** to produce α-keto ester **103**. This is followed by a second oxidation of **103** to **98** in a similar way. Finally, β,γ-unsaturated δ-hydroxy α-keto ester **98** cyclizes and aromatizes to give furan **99**.

Scheme 38 Mechanism of the Manganese(III)-Initiated Reaction of a β-Keto Ester[308,309]

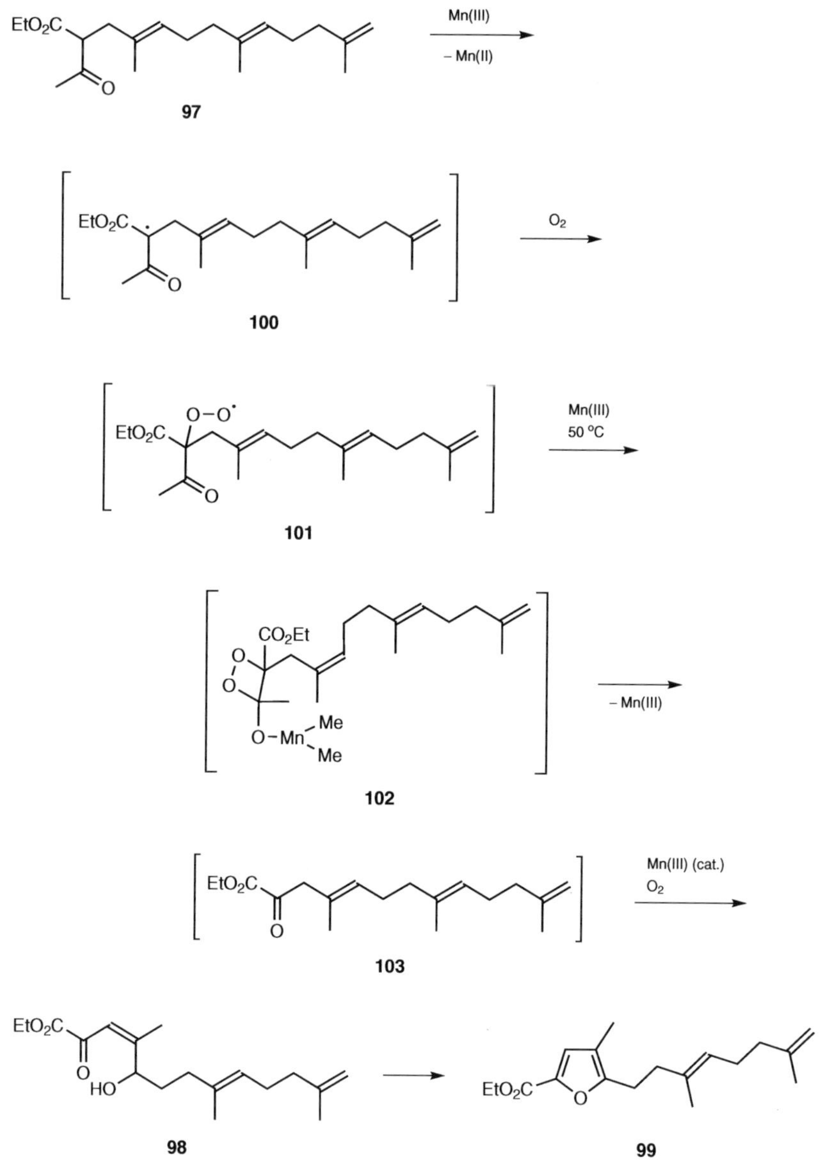

The formation of furans without the addition of copper(II) salts is possible; however, high reaction temperatures are required. Treatment of substituted β-keto esters **104** (R^1 = OEt) or analogous amides or ketones, bearing an α-allylic substituent, with manganese(III) acetate (3 equiv) in a mixture of ethanol/water (94:6) at 50 °C over 5 hours results in furan derivatives **105** in excellent yields without any trace amount of the α-keto esters (Scheme 39). This occurs because of the presence of dissolved oxygen in the solvent system, which transforms the keto ester into a δ-carbon radical; subsequent reaction of this species results in the formation of furan derivatives **105** through heterocyclization.

2.9.3 Intramolecular Free-Radical Cyclization Routes to O-Heterocycles

Scheme 39 Synthesis of Furans by Manganese(III)-Mediated Heterocyclization of β-Keto Carbonyl Compounds[307]

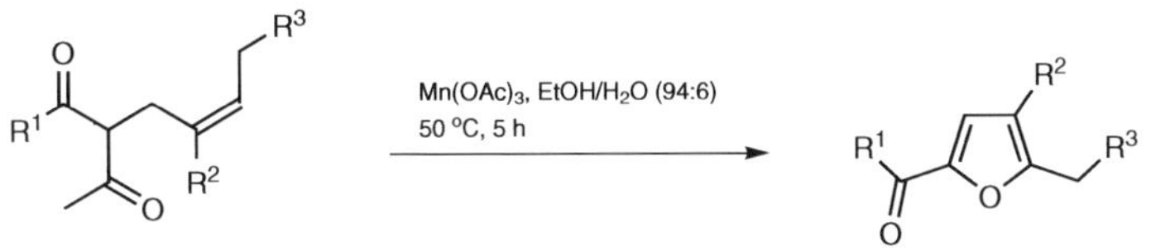

R¹	R²	R³	Yield (%)	Ref
OEt	Me	OAc	57	[307]
OEt	Ph	OAc	74	[307]
OEt	4-ClC$_6$H$_4$	OAc	65	[307]
OEt	4-MeOC$_6$H$_4$	OAc	52	[307]
OEt	2-naphthyl	OAc	64	[307]
OEt	cyclopropyl	OAc	63	[307]
OEt	Cy	OAc	55	[307]
OEt	Me	OMe	60	[307]
Me	Ph	OAc	47	[307]
NEt$_2$	Ph	OAc	41	[307]

Ethyl 5-[(Acetoxy)methyl]-4-phenylfuran-2-carboxylate (105, R¹ = OEt; R² = Ph; R³ = OAc); Typical Procedure:[307]

A mixture of **104** (0.11 g, 0.35 mmol), 94% EtOH (10 mL), and Mn(OAc)$_3$·2H$_2$O (0.29 g, 1.04 mmol) was heated at 50 °C for 5 h. Then, the mixture was cooled to rt. The mixture was diluted with CH$_2$Cl$_2$, washed with 1 M HCl, dried (Na$_2$SO$_4$), filtered, and concentrated under reduced pressure. The crude product was purified by flash column chromatography (silica gel) to afford a light-yellow oil; yield: 0.076 g (74%).

2.9.3.2 Synthesis of Tetrahydrofuran Derivatives

2.9.3.2.1 Titanocene(III)-Mediated Synthesis of Trisubstituted Tetrahydrofurans

Titanocene(III) generates radicals selectively and efficiently from α-bromocarbonyl compounds. Radical cyclization of α-bromocarbonyl alkynes **106** using chlorobis(η⁵-cyclopentadienyl)titanium(III) in tetrahydrofuran under an argon atmosphere affords trisubstituted tetrahydrofuran derivatives **107** in good to excellent yields within 1 hour (Scheme 40).[168] Starting bromides **106** can be prepared from the corresponding cinnamyl esters or acids by reaction with the appropriate alcohol in the presence of N-bromosuccinimide following a standard procedure.[310–312] The radical initiator chlorobis(η⁵-cyclopentadienyl)titanium(III) is prepared in situ by stirring commercially available dichlorobis(η⁵-cyclopentadienyl)titanium(IV) in dry tetrahydrofuran with zinc dust for 1 hour under argon. However, alkene **108** produces an inseparable mixture of isomers **109** in a 1:1 ratio with respect to the C4-methyl group.

Scheme 40 Synthesis of Furans by Chlorobis(η^5-cyclopentadienyl)titanium(III)-Mediated Cyclization of α-Bromocarbonyl Alkynes[168]

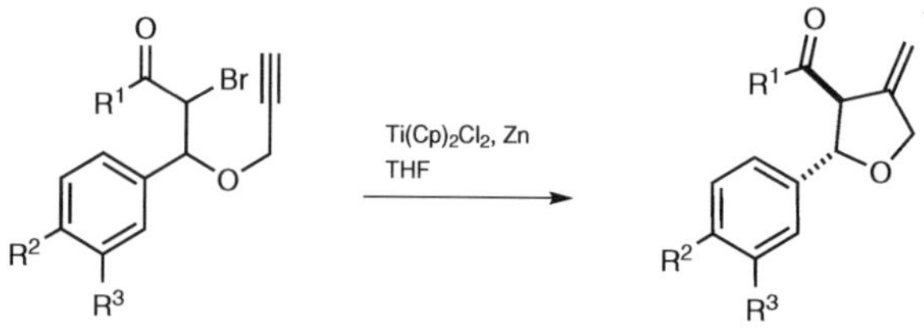

R^1	R^2	R^3	Yield (%)	Ref
OEt	OCH$_2$O		80	[168]
OH	OCH$_2$O		66	[168]
OEt	OBn	OMe	82	[168]
OEt	OMe	OMe	81	[168]
OMe	OMe	H	85	[168]
Me	OMe	H	70	[168]
Me	OMe	OMe	75	[168]

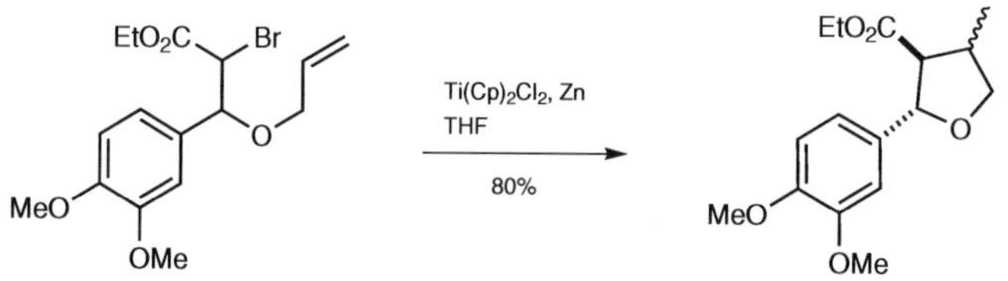

Ethyl 2-(3,4-Dimethoxyphenyl)-4-methylenetetrahydrofuran-3-carboxylate (107, R^1 = OEt; R^2 = R^3 = OMe); Typical Procedure:[168]

A soln of Ti(Cp)$_2$Cl$_2$ (250 mg, 2 mmol) in dry THF (15 mL) was stirred with activated Zn dust (7 mmol) for 1 h under an argon atmosphere. The resulting green suspension was then added dropwise to a stirred soln of bromoalkyne **106** (370 mg, 1 mmol) in dry THF (5 mL) at rt over 30 min. The mixture was stirred for an additional 30 min and was then quenched with sat. aq NaH$_2$PO$_4$ (5 mL). The mixture was concentrated under reduced pressure, and the residue was extracted with Et$_2$O (4 × 25 mL). The combined organic extracts were washed successively with H$_2$O (2 × 10 mL) and brine (2 × 10 mL) and then finally dried (Na$_2$SO$_4$). The crude residue was purified by column chromatography (silica gel, EtOAc/hexane 1:9); yield: 236 mg (81%).

2.9.3.2.2 Titanocene(III)-Mediated Synthesis of Multifunctional Tetrahydrofurans from Alkenyl Iodo Ethers

A catalytic quantity of dichlorobis(η^5-cyclopentadienyl)titanium(IV) promotes the intramolecular reductive cyclization of alkenyl iodo ethers **110** in the presence of manganese metal and chlorotrimethylsilane in tetrahydrofuran at 65 °C. This synthesis method can selectively form multisubstituted tetrahydrofurans **111** in high yields (Scheme 41).[313]

2.9.3 Intramolecular Free-Radical Cyclization Routes to O-Heterocycles

Scheme 41 Synthesis of Multifunctional Tetrahydrofurans from Alkenyl Iodo Ethers Using Dichlorobis(η^5-cyclopentadienyl)titanium(IV)/Chlorotrimethylsilane/Manganese[313]

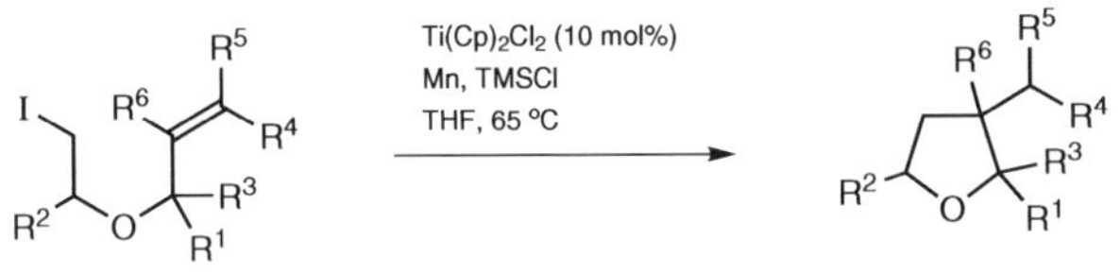

R^1	R^2	R^3	R^4	R^5	R^6	Time (h)	Ratio[a] (trans/cis)	Yield (%)	Ref
H	4-MeOC$_6$H$_4$	H	H	H	H	24	80:20	83	[313]
Me	4-MeOC$_6$H$_4$	Me	H	H	H	24	85:15	64	[313]
H	4-MeOC$_6$H$_4$	H	H	Pr	H	24	80:20	63	[313]
H	4-MeOC$_6$H$_4$	H	Pr	H	H	24	71:39	61	[313]
Me	4-MeOC$_6$H$_4$	H	H	H	H	28	–[b]	59	[313]
H	4-MeOC$_6$H$_4$	H	H	H	Me	48	–	56	[313]
H	4-MeOC$_6$H$_4$	(CH$_2$)$_3$	H	H		36	–[c]	59	[313]
Me	OBu	Me	H	H	H	30	69:31	73	[313]
H	OBu	H	H	H	Me	36	–	45	[313]
H	OBu	H	H	H	H	30	60:40	70	[313]

[a] The ratio was determined by ^{1}H NMR spectroscopy.
[b] Obtained as an isomeric mixture.
[c] Obtained as a mixture of exo/endo-isomers (60:40).

2-(4-Methoxyphenyl)-4-methyloxolane (111, $R^1 = R^3 = R^4 = R^5 = R^6 = H$; $R^2 = 4$-MeOC$_6$H$_4$); Typical Procedure:[313]

TMSCl (0.13 mL, 1.0 mmol) was added to a mixture of Mn powder (220 mg, 4.0 mmol) and Ti(Cp)$_2$Cl$_2$ (12.5 mg, 0.050 mmol) in THF (6 mL) at rt under an argon atmosphere. The mixture was stirred for 30 min, and a soln of iodo ether **110** (0.5 mmol) in THF (3 mL) was then added. The mixture was magnetically stirred for 24 h at 65 °C. Then, 1 M HCl (3 mL) and Et$_2$O (50 mL) were added, and the mixture was extracted. The combined organic layer was washed with sat. aq Na$_2$S$_2$O$_3$ (10 mL), H$_2$O (10 mL), and brine (10 mL), dried (MgSO$_4$), and concentrated under reduced pressure. The crude residue was purified by flash chromatography (silica gel, hexane/EtOAc 50:0, 48:2, 46:4, 44:6, 42:8, 40:10, 38:12, 35:15, 32:18, 28:22, 25:25, 50 mL each) to afford a colorless oil; yield: 83%; ratio (trans/cis) 80:20.

2.9.3.2.3 (2+2) Cycloadditions by Oxidative Visible Light Tris(2,2'-bipyridyl)-ruthenium(II) Bis(hexafluorophosphate) Mediated Photocatalysis

Ruthenium photocatalysts are powerful reagents that can be used for the (2+2) cycloaddition of both electron-rich and electron-deficient alkenes. Substituted bis(allylic) compounds **112** undergo efficient intramolecular cycloaddition upon irradiation with visible light in the presence of tris(2,2'-bipyridyl)ruthenium(II) bis(hexafluorophosphate) (1.0–5.0 mol%), methyl viologen bis(hexafluorophosphate) (15 mol%), magnesium sulfate (2 wt equiv), and nitromethane under a nitrogen atmosphere to afford cyclobutane–tetrahydrofuran derivatives **113** in high yields with excellent diastereoselectivity (Table 6).[314] The optimized reaction conditions are only efficient for substrates in which at least one styrene unit bears an electron-donating substituent in the 4- or 2-position (Table 6, entries 1 and 2). Given that 3-substituted and unsubstituted styrenes are negligibly electron rich and are not able to undergo one-electron oxidation, they are not able to afford the key radical cation intermediate (Table 6, entries 3 and 4). Electron-donating substituents in the 4-position effectively activate the styrene (Table 6, entries 5 and 6), whereas the presence of an electron-withdrawing 3-substituent does not seem to influence the efficiency of the cycloaddition (Table 6, entry 7). Overall, tris(2,2′-bipyridyl)ruthenium(II) bis(hexafluorophosphate) is a powerful photocatalyst for the efficient (2+2) cycloaddition of both electron-rich and electron-deficient alkenes.

Table 6 Ruthenium-Mediated (2+2) Cycloaddition of Bis(allylic) Ethers and Amines To Give Cyclobutane–Tetrahydrofuran Derivatives[314]

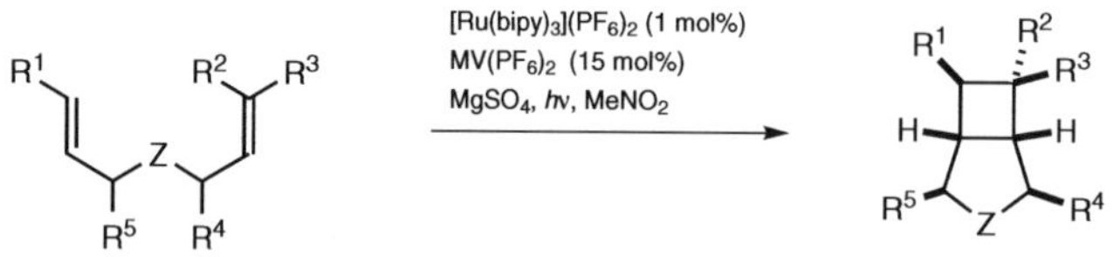

MV(PF$_6$)$_2$ = methyl viologen bis(hexafluorophosphate)

Entry	Substrate	Time (h)	Product	Yield[a] (%)	Ref
1	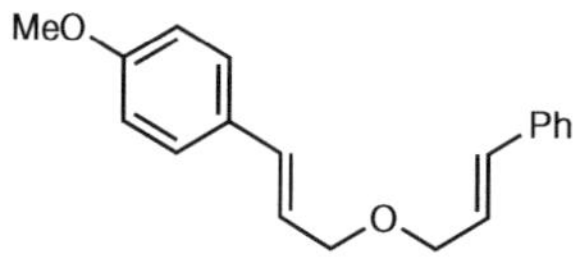	3.5	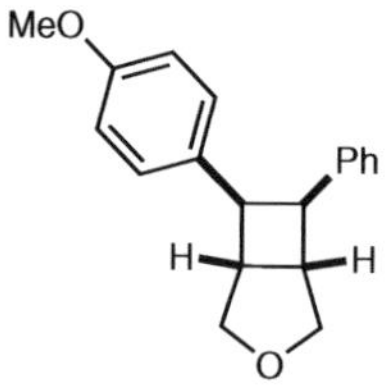	88[b]	[314]
2	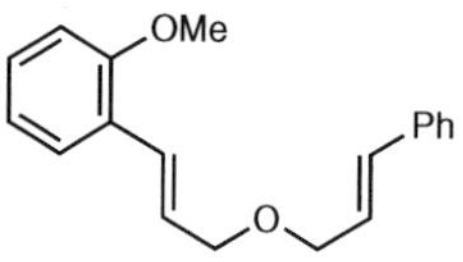	11	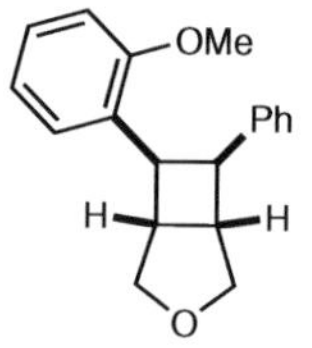	73	[314]
3	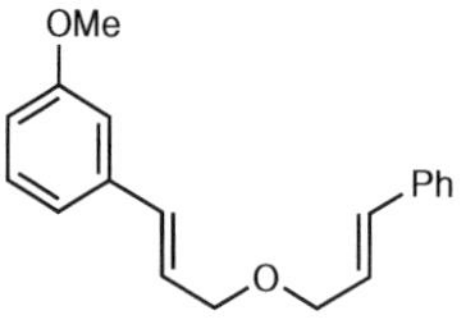	22	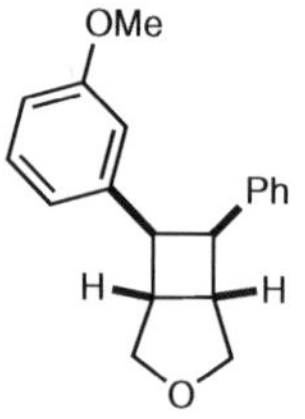	0	[314]
4	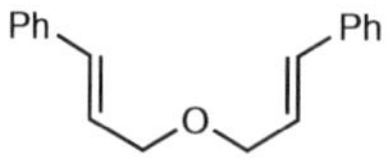	22	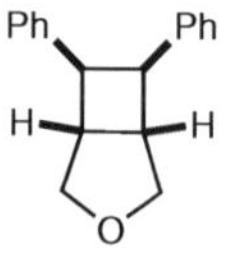	0	[314]

Table 6 (cont.)

Entry	Substrate	Time (h)	Product	Yield[a] (%)	Ref
5	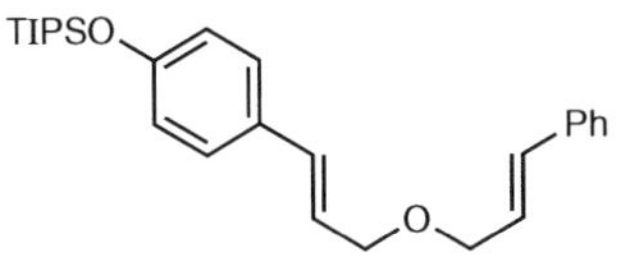	2.5	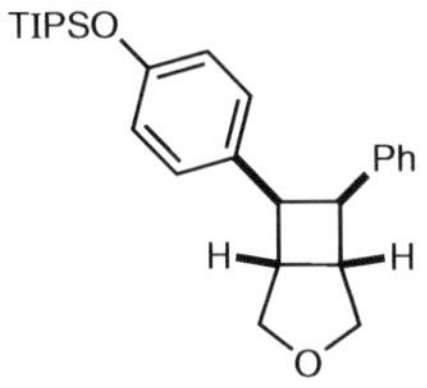	69	[314]
6	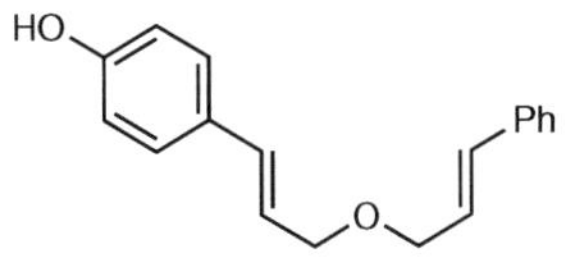	13	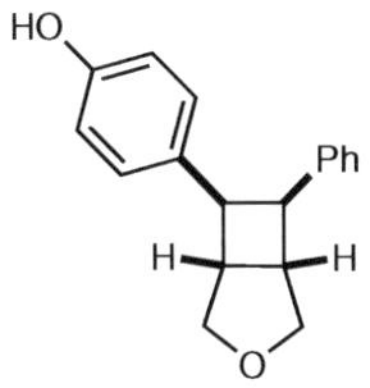	64[b]	[314]
7	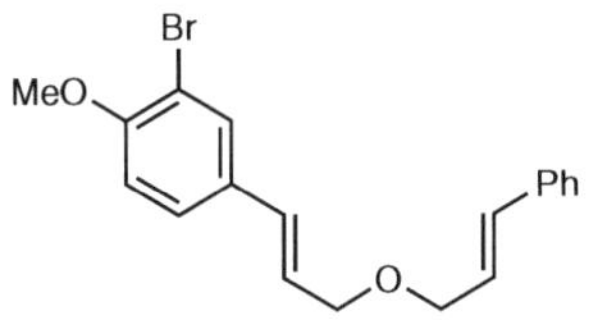	8	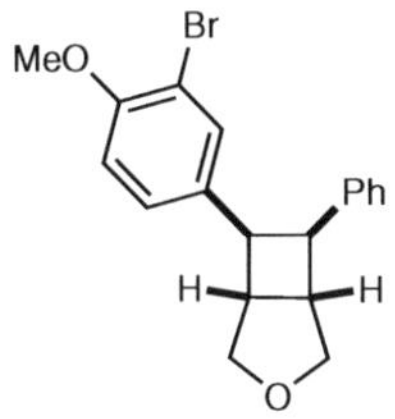	71	[314]
8	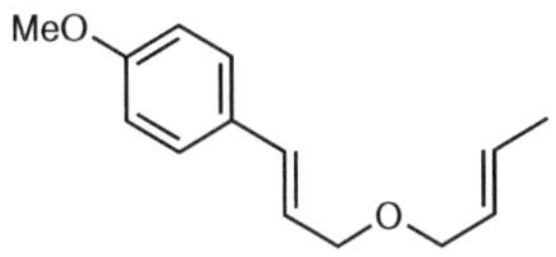	22	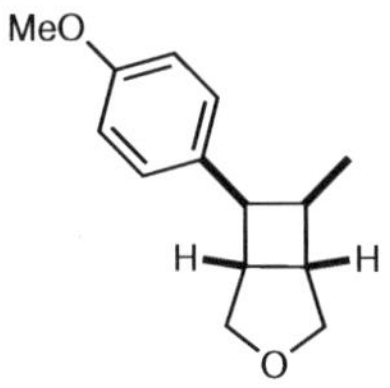	0	[314]
9	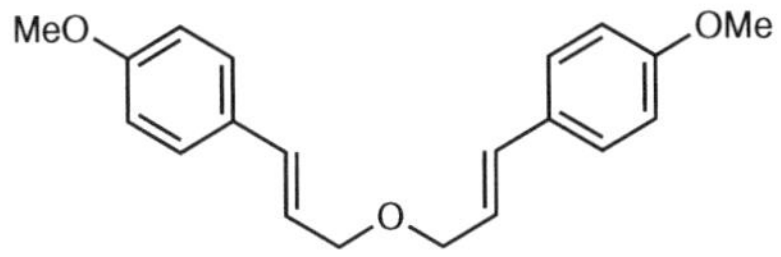	2	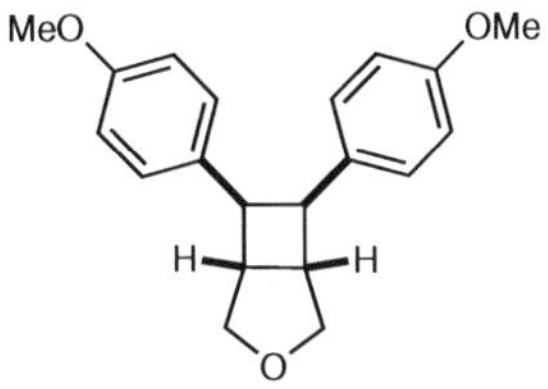	67	[314]
10	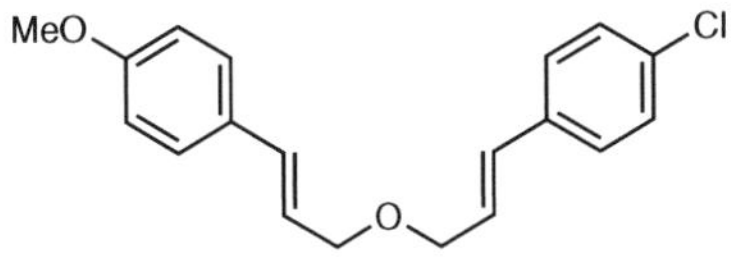	5	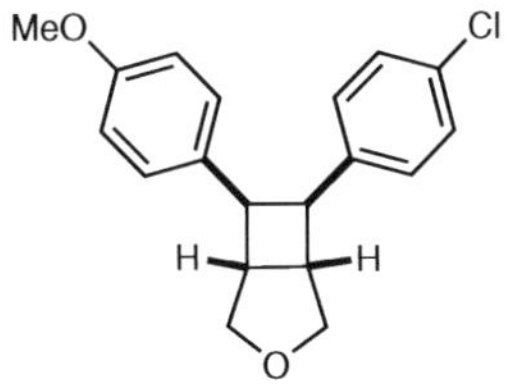	92	[314]

for references see p 533

Table 6 (cont.)

Entry	Substrate	Time (h)	Product	Yield[a] (%)	Ref
11	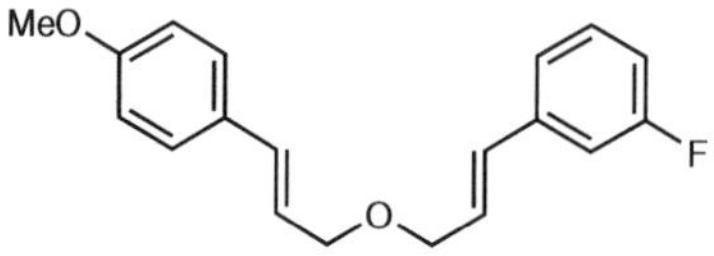	5	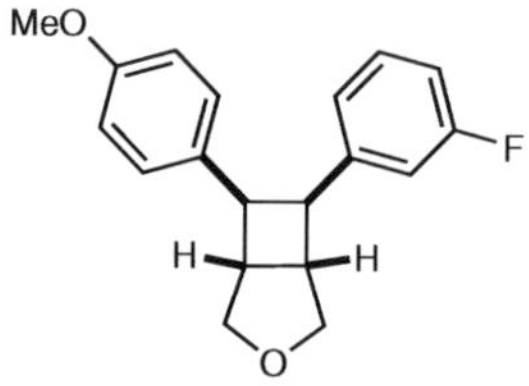	78	[314]
12	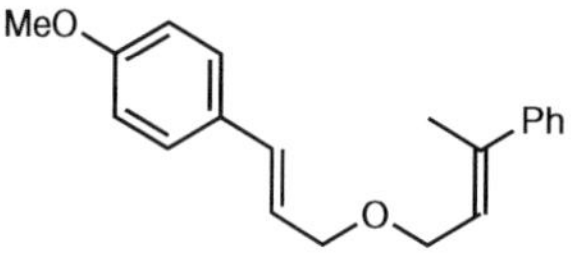	6	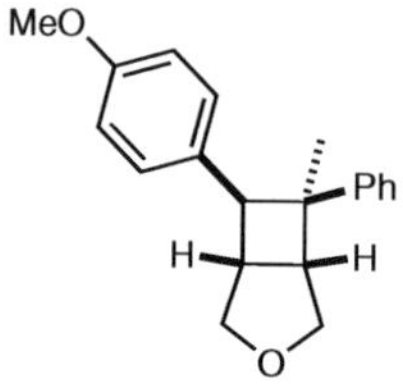	54[b]	[314]
13	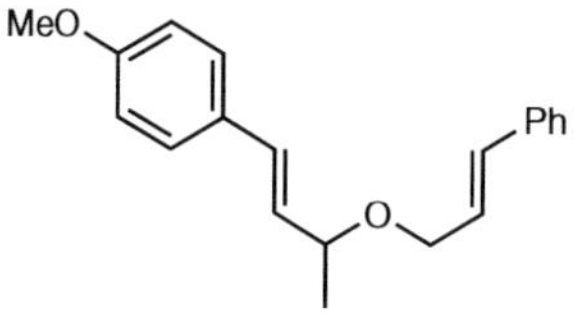	7	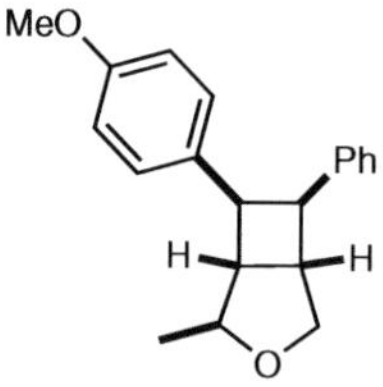	69[c]	[314]
14	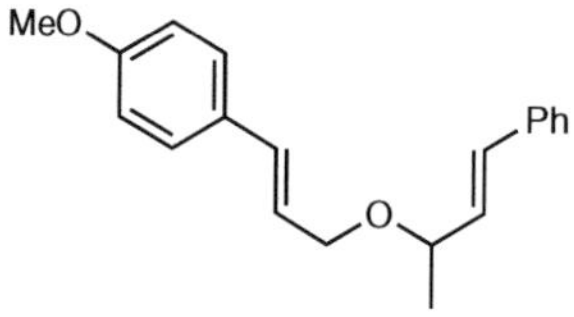	6	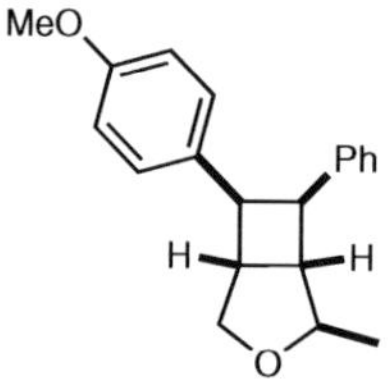	69[d]	[314]
15	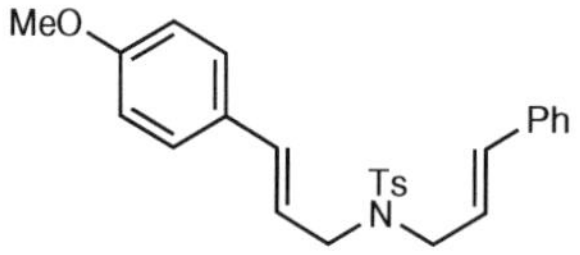	6	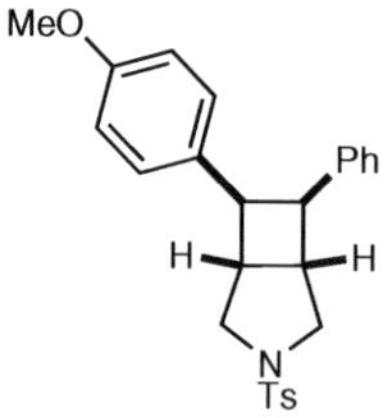	67[b]	[314]

[a] Yield is the average of two reproducible experiments.
[b] Reaction was performed with 5 mol% [Ru(bipy)$_3$](PF$_6$)$_2$.
[c] dr 5:1.
[d] dr 7:1.

3-Oxabicyclo[3.2.0]heptanes 113 (Z = O); General Procedure:[314]

CAUTION: *Nitromethane is flammable, a shock- and heat-sensitive explosive, and an eye, skin, and respiratory tract irritant.*

A 25-mL Schlenk flask was charged with MgSO$_4$ (2 wt equiv) and was then flame dried under vacuum and cooled to rt under a dry N$_2$ atmosphere. Diallyl ether **112** (1 equiv), methyl viologen bis(hexafluorophosphate) (15 mol%), and [Ru(bipy)$_3$](PF$_6$)$_2$ (1 or 5 mol%) were added. The flask was charged with MeNO$_2$, and the soln was degassed by freeze–

pump–thaw cycles (3 ×) under a N₂ atmosphere. The mixture was placed in a H₂O bath and stirred with irradiation by a light bulb. Upon completion of the reaction, the mixture was diluted with Et₂O and passed through a short pad of silica gel (EtOAc or Et₂O). The solvent was removed under reduced pressure, and the residue was purified by flash chromatography.

2.9.3.3 Synthesis of Lactone Derivatives

2.9.3.3.1 Synthesis of γ-Substituted Phthalides by Benzyl Radical Cyclization in Water

The direct synthesis of γ-substituted phthalides **115** can be achieved by the oxidation of 2-substituted benzoic acids **114** in the presence of a peroxydisulfate/copper(II) chloride system in aqueous solution at 85–90 °C (Scheme 42).[315] The reaction is highly regioselective and provides γ-butyrolactone derivatives exclusively via a very stable benzylic radical intermediate.

Scheme 42 Copper(II)-Mediated Synthesis of γ-Substituted Phthalides from 2-Arylbenzoic Acids[315]

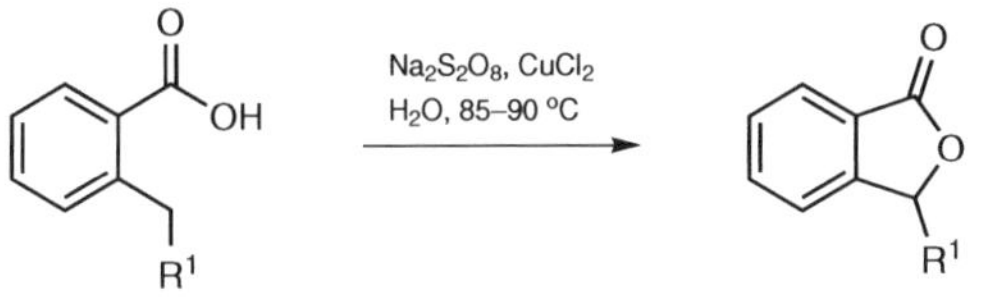

R¹	Yield (%)	Ref
H	55	[315]
Ph	56	[315]
4-Tol	62	[315]
4-MeOC₆H₄	85	[315]
1-naphthyl	48	[315]

Benzo[c]furan-1(3H)-one (115, R¹ = H); Typical Procedure:[315]
A 250-mL two-necked, round-bottomed flask was charged with benzoic acid **114** (4.08 g, 30 mmol), distilled H₂O (27 mL), and CuCl₂·2H₂O (5.1 g, 30 mmol). The flask was fitted with a reflux condenser and an addition funnel. The addition funnel was charged with a soln of Na₂S₂O₈ (8.5 g, 30 mmol) in H₂O (15 mL). The mixture in the flask was stirred vigorously keeping the temperature of the soln at 85–90 °C. The soln from the addition funnel was added dropwise over 40 min to the flask containing the substrate, and the resulting mixture was then heated at reflux for another 3 h. After 3 h, the reaction was stopped. The flask was cooled, and the contents were extracted with Et₂O (3 × 10 mL) and the extracts were dried (MgSO₄). The solvent was removed under pressure, and the residue was purified by preparative TLC (silica gel, EtOAc/petroleum ether 5:95 to 10:90); yield: 2.25 g (55%); mp 73 °C.

2.9.3.3.2 Chemoselective Synthesis of δ-Lactones through Benzyl Radical Cyclization Using Potassium Persulfate/Copper(II) Chloride

The strategy presented in Section 2.9.3.3.1 can be utilized for the direct chemoselective oxidation of substituted 8-benzyl-1-naphthoic acids **116** in the presence of potassium persulfate/copper(II) chloride to produce δ-lactones **117** in moderate to good yields (Scheme 43).[316] This method is advantageous in that water is used as the solvent and the starting materials are readily available.

Scheme 43 Synthesis of δ-Lactones Using Potassium Persulfate/Copper(II) Chloride[316]

R^1	R^2	R^3	Time (h)	Temp (°C)	Yield (%)	mp (°C)	Ref
H	H	H	3.5	85–90	67	82–85	[316]
H	Me	H	4	85–90	71	200–201	[316]
H	Cl	H	4	85–90	61	92–95	[316]
Me	H	Me	4.5	90–95	63	231–232	[316]
Me	Me	H	4.5	90–95	70	223–224	[316]
H	Me	Me	4.5	90–95	61	214–215	[316]

3-Phenyl-1*H*,3*H*-naphtho[1,8-*cd*]pyran-1-one (117, $R^1 = R^2 = R^3 = H$); Typical Procedure:[316]
A 250-mL three-necked, round-bottomed flask fitted with a reflux condenser and an addition funnel was charged with 1-naphthoic acid **116** (1.31 g, 5 mmol), H_2O (15 mL), and $CuCl_2·2H_2O$ (0.85 g, 5 mmol). A soln of $K_2S_2O_8$ (1.35 g, 5 mmol) in H_2O (10 mL) was added to the funnel. The mixture in the flask was stirred vigorously keeping the temperature of the soln at 85–90 °C. The soln from the addition funnel was added dropwise over 40 min to the flask containing the substrate, and the resulting mixture was heated at reflux for 4 h. After 4 h, the reaction was stopped. The flask was then cooled, and the contents were extracted with Et_2O (3 × 30 mL) and the extracts were dried ($MgSO_4$). The solvent was removed under reduced pressure, and the residue was purified by preparative TLC (silica gel, EtOAc/petroleum ether 5:95 to 10:90); yield: 0.87 g (67%); mp 82–85 °C.

2.9.3.3.3 Atom-Transfer Radical Cyclization Reactions of Various Trichloroacetates to Macrolactones

The atom-transfer radical cyclization of unsaturated and polyoxaalkenyl trichloroacetates **121** in the presence of complexes formed between nitrogen-based ligands **118**, **119**, and **120** and copper(I) and iron(II) species in 1,2-dichloroethane at 80 °C efficiently produces 5- to 18-membered macrocyclic lactones **122**.[317] Ligand **119** associated with copper(I) gives the products in very poor yields, but surprisingly, if it is used with iron(II) chloride very efficient conversions can be achieved with trichloroacetates. Ligands **118** and **120** work well with copper(I) chloride (Table 7).

2.9.3 Intramolecular Free-Radical Cyclization Routes to O-Heterocycles

Table 7 Cyclizations of Unsaturated and Polyoxaalkenyl Trichloroacetates[317]

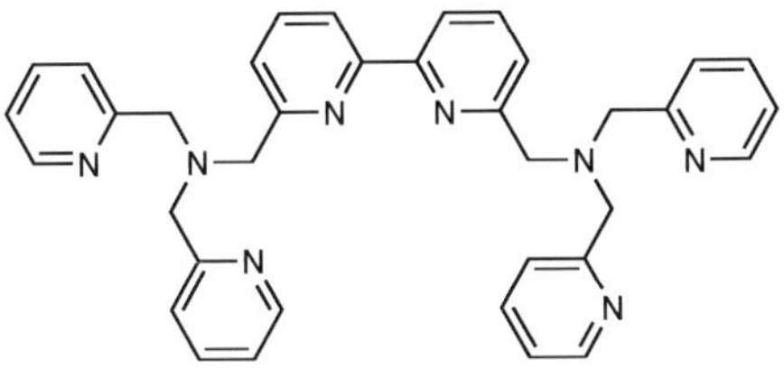

Entry	Substrate	Conditions	Product	Yield (%)	Ref
1	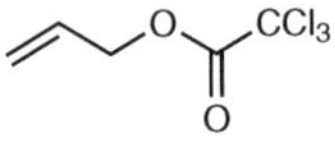 allyl trichloroacetate	FeCl$_2$/**119**	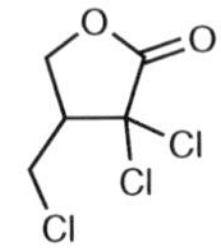 chlorinated γ-butyrolactone	55	[317]
2	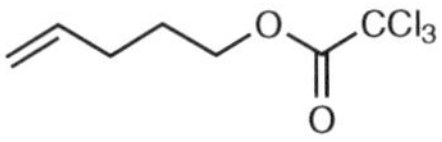 but-3-enyl trichloroacetate	CuCl/**118**	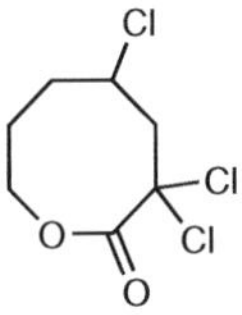 8-membered chlorinated lactone	99	[317]
3	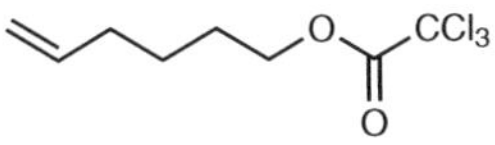 pent-4-enyl trichloroacetate	CuCl/**118**	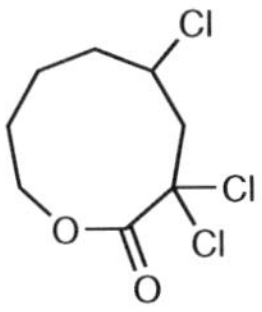 9-membered chlorinated lactone	53	[317]
4	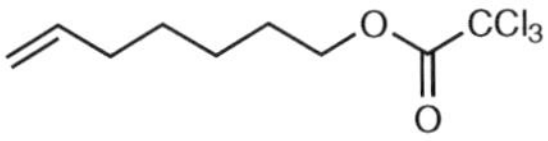 hex-5-enyl trichloroacetate	CuCl/**120**	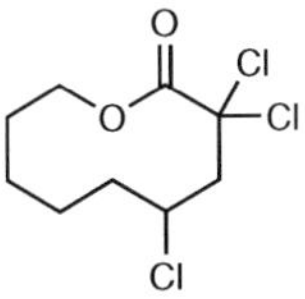 10-membered chlorinated lactone	51	[317]

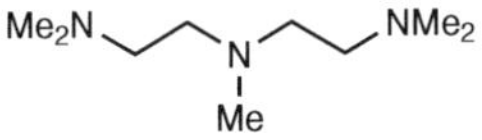

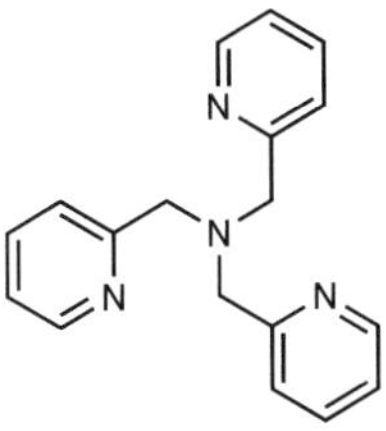

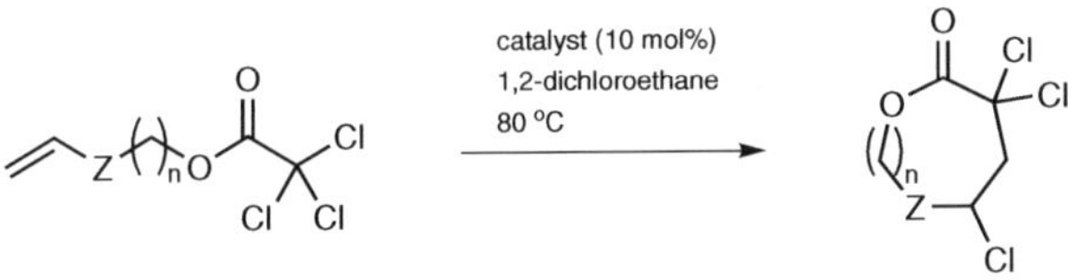

Table 7 (cont.)

Entry	Substrate	Conditions	Product	Yield (%)	Ref
5	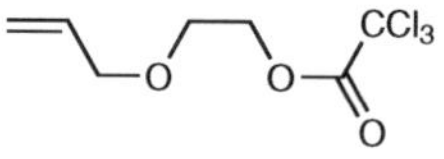 (allyl-O-CH2CH2-O-C(O)-CCl3)	FeCl2/**119**	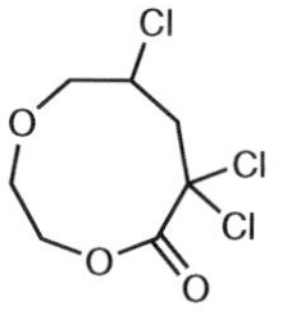 9-membered lactone with 3 Cl	56	[317]
6	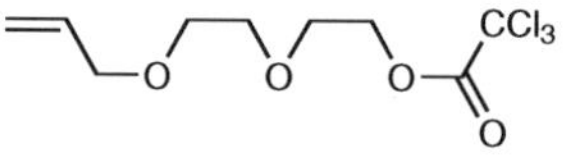 (allyl-O-(CH2CH2O)2-C(O)-CCl3)	FeCl2/**119**	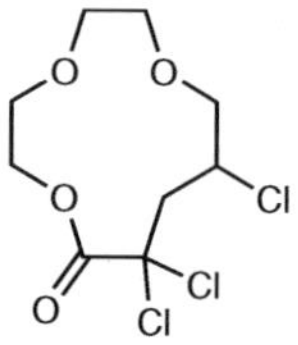 12-membered lactone with 3 Cl	60	[317]
7	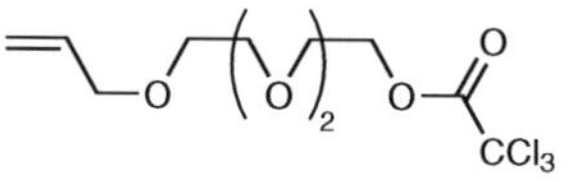 (allyl-O-(CH2CH2O)2-C(O)-CCl3)	CuCl/**120**	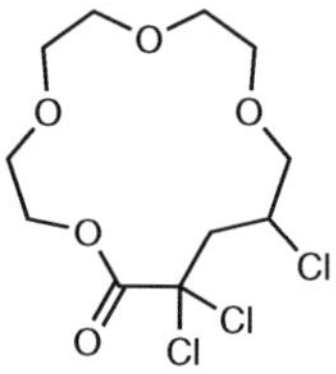 macrocyclic lactone with 3 Cl	70	[317]
8	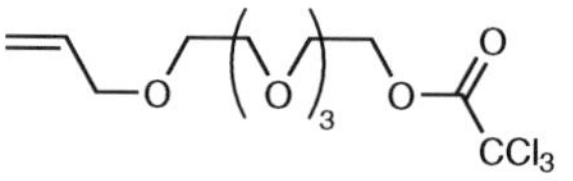 (allyl-O-(CH2CH2O)3-C(O)-CCl3)	CuCl/**120**	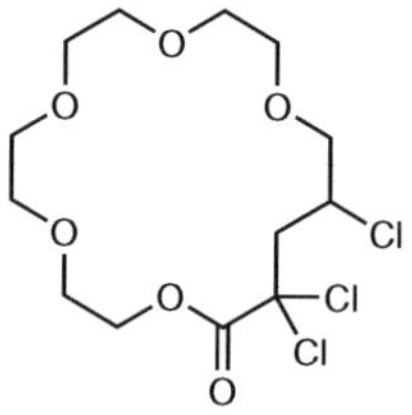 macrocyclic lactone with 3 Cl	70	[317]

Macrocyclic Lactones 122; General Procedure:[317]
The reaction was performed under argon in 1,2-dichloroethane (0.2 M). Separate vessels containing ligand **119** (10 mol%) and a mixture of substrate **121** (1 mmol) and the metal salt (10 mol%) were degassed by freeze–pump–thaw cycles (3×). The ligand was then added to generate the active catalyst, and the mixture was stirred at 80 °C. Upon completion of the reaction, the mixture was concentrated and the residue was purified by column chromatography (silica gel, petroleum ether/EtOAc).

2.9.3.3.4 Manganese(III)-Mediated Synthesis of Densely Functionalized and Sterically Crowded Lactones

2.9.3.3.4.1 Synthesis of Fused Tricyclic γ-Lactones

The manganese(III)-mediated intramolecular cyclization of cyclic alkenes **123** bearing a carboxylic acid and a malonate group in the presence of a copper(II) salt selectively provides corresponding tricyclic γ-lactones **124** in good to excellent yields.[318] The cyclic alkenes are treated with manganese(III) acetate and either copper(II) trifluoromethanesulfonate or copper(II) tetrafluoroborate in acetonitrile under reflux conditions (Table 8). The reaction proceeds with the formation of a malonyl radical, which undergoes 5-*exo-trig* cyclization onto the alkene to afford a secondary adduct radical. Finally, copper(II)-mediated oxidative C—O bond formation delivers the tricyclic γ-lactone.

2.9.3 Intramolecular Free-Radical Cyclization Routes to O-Heterocycles

Table 8 Synthesis of Fused Tricyclic γ-Lactones[318]

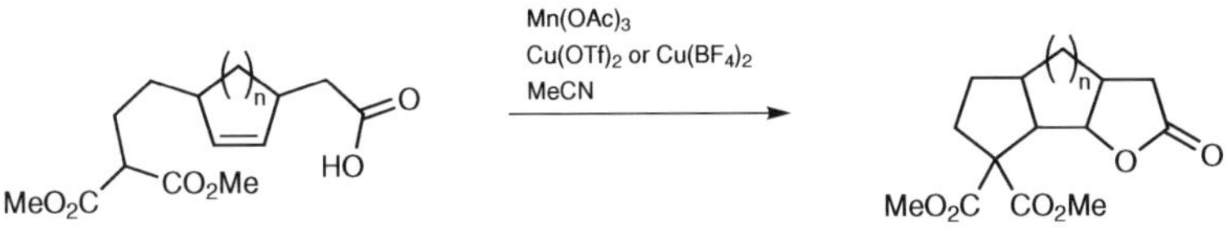

Entry	Substrate	Product(s)	Yield (%)	Ref
1	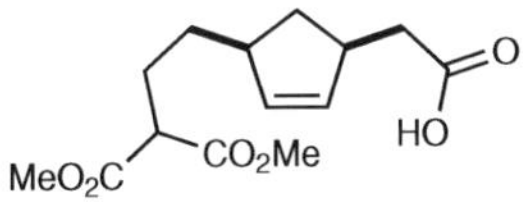	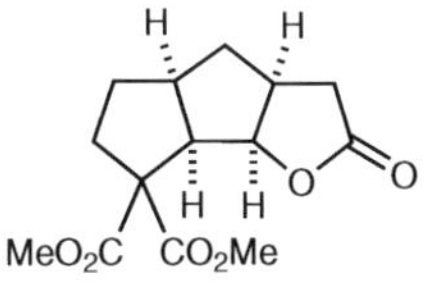	72	[318]
2	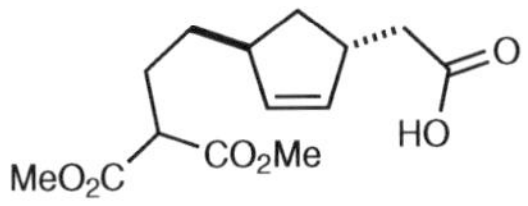	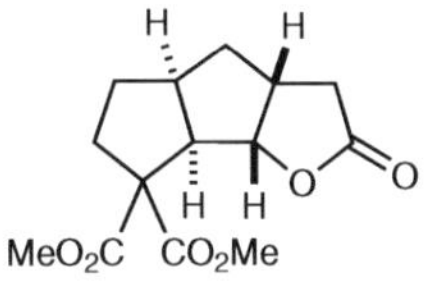	71	[318]
3	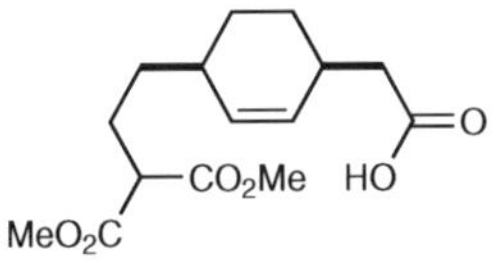	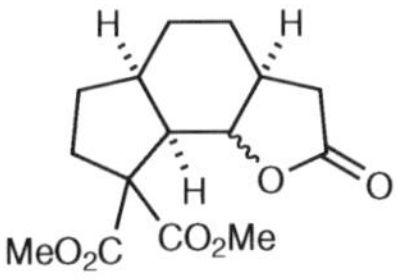	82[a]	[318]
4	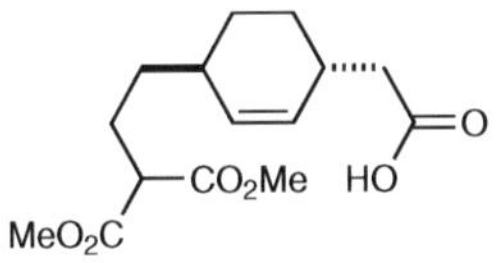	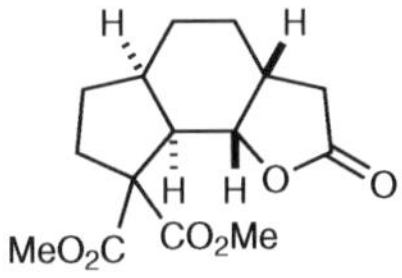	94	[318]
5	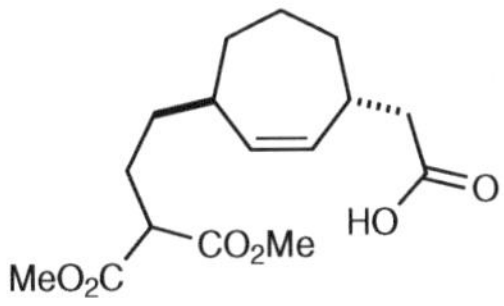	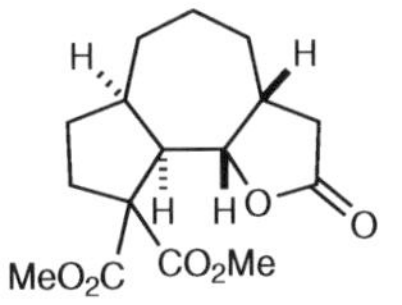	86	[318]
6	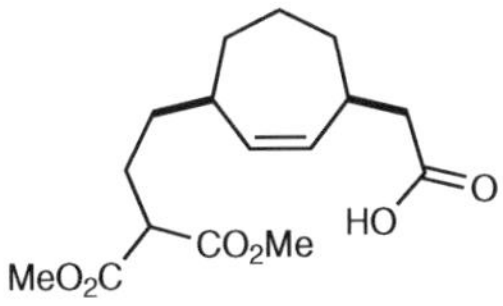	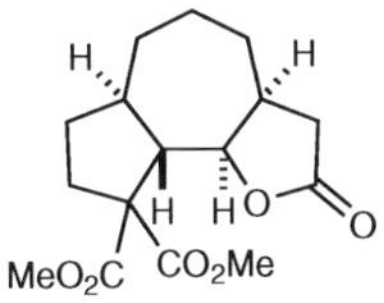 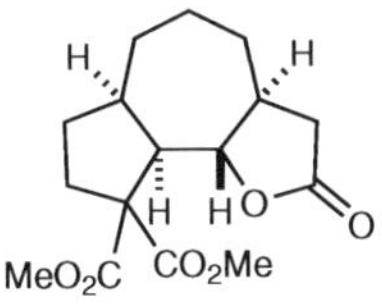 (52:23)	75	[318]

[a] Formed as a 1:1 mixture of epimers.

Fused Tricyclic γ-Lactones 124; General Procedure:[318]

Mn(OAc)$_3$·2H$_2$O (39.7 mg, 0.148 mmol) and Cu(OTf)$_2$ (26.8 mg, 1.0 mmol) were added to a round-bottomed flask containing substrate **123** (0.074 mmol) and MeCN (2 mL). The mixture was degassed and then heated at reflux overnight under N$_2$. Upon completion of the reaction, the mixture was cooled to rt and H$_2$O (4 mL) and Et$_2$O (4 mL) were added. The aqueous/organic layer was extracted and washed with Et$_2$O (5 × 4 mL). The combined organic layer was dried (MgSO$_4$), filtered, and concentrated under reduced pressure. The resulting crude material was purified by flash chromatography (petroleum ether/EtOAc 1:1).

2.9.3.3.4.2 Cyclization of 2-Alkenylmalonates for the Synthesis of Bicyclo[3.3.0] γ-Lactones

Bicyclo[3.3.0] γ-lactones are considered attractive intermediates in organic synthesis owing to the presence of adjacent quaternary and tertiary stereocenters and different oxygen functionalities. Exposure of malonates **125** to manganese(III) acetate (2 equiv) and copper(II) trifluoromethanesulfonate (1 equiv) in acetonitrile at reflux delivers bicyclo[3.3.0] γ-lactones **126** in excellent yields (Table 9).[319] The use of deoxygenated acetonitrile is essential to attain high yields of the desired γ-lactones. A 3-substituted 2-(pent-4-enyl)malonate gives the corresponding bicyclo[3.3.0] γ-lactone with good diastereocontrol (dr >13:1) in 88% yield (Table 9, entry 2). Similarly, reactions of the 1- and 2-substituted 2-(pent-4-enyl)malonates give the corresponding bicyclo[3.3.0] γ-lactones in excellent yields (Table 9, entries 3 and 4). However, in ethanol [without any copper(II) additive], the unsubstituted malonate (the starting material in Table 9, entry 1) gives the reductive methylcyclopentane product exclusively.

2.9.3 Intramolecular Free-Radical Cyclization Routes to O-Heterocycles

Table 9 Manganese(III)-Catalyzed Synthesis of Bicyclo[3.3.0] γ-Lactones from 2-Alkenylmalonates[319]

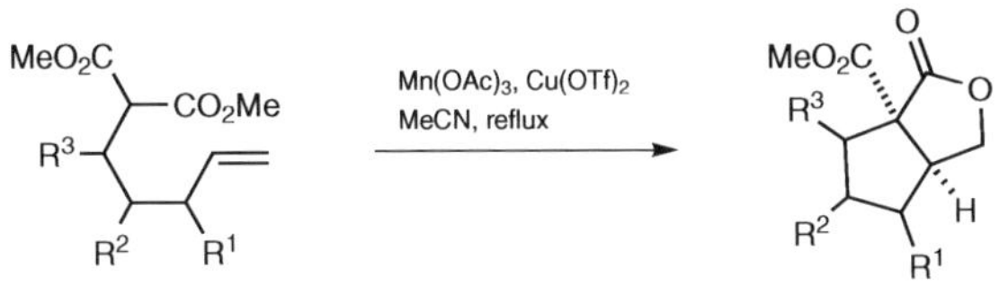

Entry	Substrate	Product	Yield (%)	Ref
1	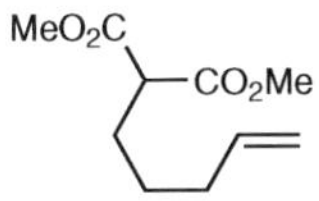	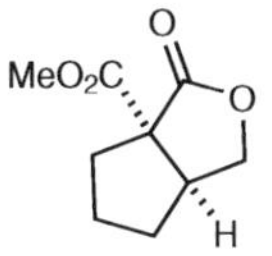	91	[319]
2	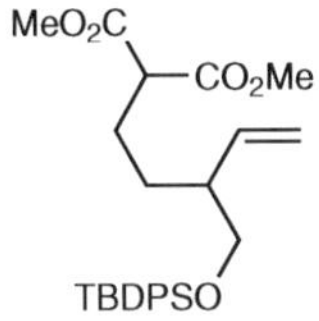	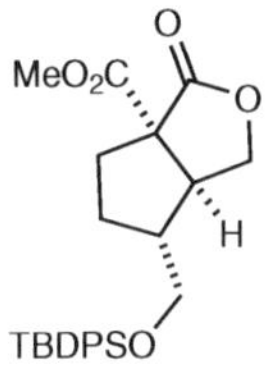	88	[319]
3	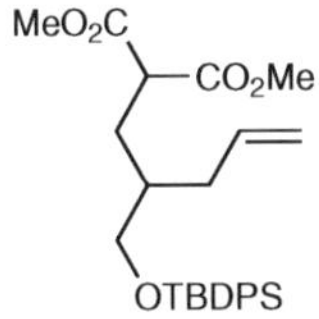	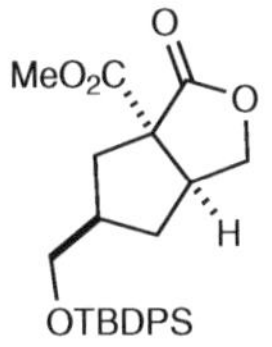	92[a]	[319]
4	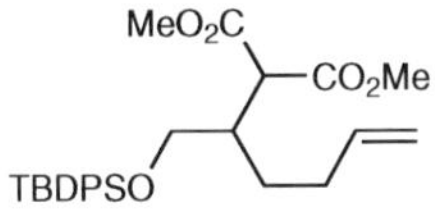	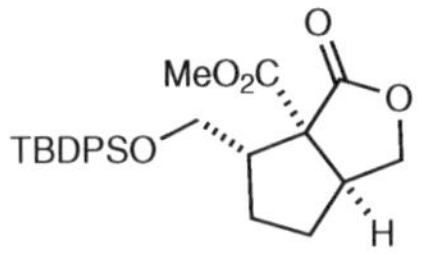	90[a]	[319]

[a] Combined yield for a 5:1 mixture of diastereomers; the major diastereomer is shown.

The reaction of linear malonates **127** and **129** with manganese(III) acetate and copper(II) trifluoromethanesulfonate in acetonitrile at reflux gives the corresponding γ-lactones **128** and **130** (Scheme 44). Cyclization–lactonization is also feasible with cyclic alkene **131** and trisubstituted alkene **133** to give γ-lactones **132** and **134**, respectively. However, to achieve effective cyclization–lactonization with 1,3-diene substrates the substituted malonic acid instead of the corresponding dimethyl malonate must be used. Thus, under the above-mentioned reaction conditions, malonic acid **135** gives bicyclo[3.3.0] γ-lactone **136** in excellent yield.

Scheme 44 Synthesis of Bi- and Tricyclic γ-Lactones from Malonates Using Manganese(III) Acetate/Copper(II) Trifluoromethanesulfonate[319]

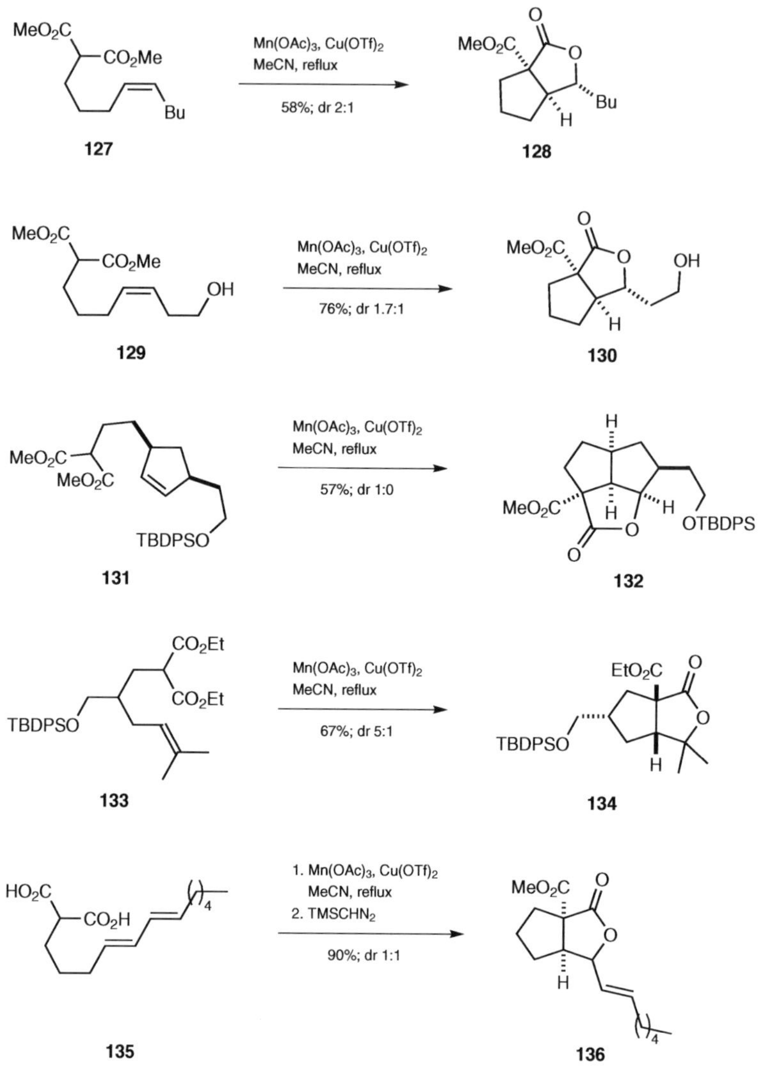

Methyl (3aR*,6aS*)-3-Oxotetrahydro-1H-cyclopenta[c]furan-3a(3H)-carboxylate (Table 9, Entry 1); Typical Procedure:[319]

Mn(OAc)$_3$·2H$_2$O (536 mg, 2.0 mmol) and Cu(OTf)$_2$ (362 mg, 1.0 mmol) were added to a soln of malonate **125** (200 mg, 1.0 mmol) in dry degassed MeCN (5 mL, N$_2$ bubbled through the solvent for 2 h prior to use). The resulting mixture was then heated to 80 °C for 24 h. Upon completion of the reaction, H$_2$O (20 mL) and Et$_2$O (20 mL) were added to the mixture, and then the mixture was extracted with Et$_2$O (3 × 100 mL). The combined organic layer was dried (MgSO$_4$), filtered, and concentrated under reduced pressure. The crude residue was then purified by flash column chromatography (Et$_2$O/petroleum ether 5:95 to 20:80) to afford a colorless oil; yield: 167.4 mg (91%).

2.9.3.3.4.3 Cyclization of γ-Lactones to Tricyclo[5.2.1.0^{1,5}] Bis(lactones)

The methodology outlined in Section 2.9.3.3.4.2 has been utilized for the synthesis of tricyclo[5.2.1.0^{1,5}] bis(lactone) **138** containing adjacent quaternary and tertiary stereocenters and different oxygen-based functionalities in a single step starting from γ-lactone **137** (Scheme 45).[319] Thus, reaction of γ-lactone **137** under the standard reaction conditions of manganese(III) acetate and copper(II) trifluoromethanesulfonate in acetonitrile at reflux delivers tricyclo[5.2.1.0^{1,5}] bis(lactone) **138** with good diastereocontrol (dr >10:1); exposure of **138** to methanol at room temperature gives bicyclo[3.3.0] γ-lactone **139** in good yield.

Scheme 45 Synthesis of a Tricyclo[5.2.1.0^{1,5}] Bis(lactone) and a Bicyclo[3.3.0] γ-Lactone[319]

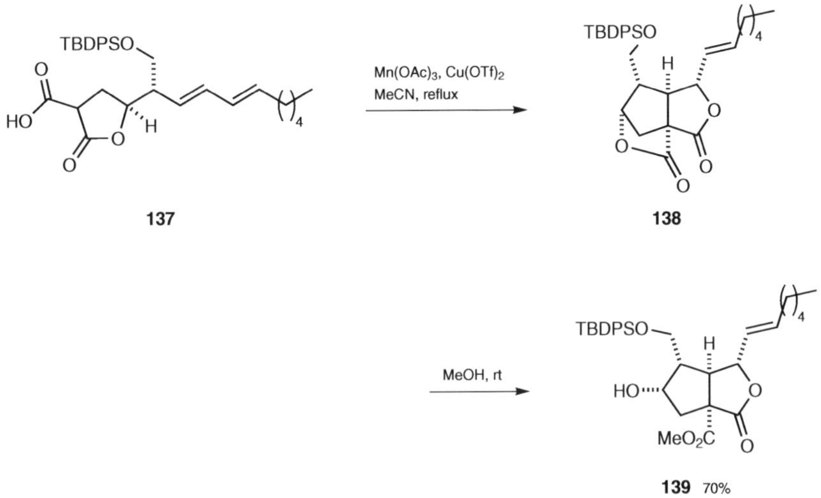

(1S,3aS,6S,7S,7aR)-7-[(tert-Butyldiphenylsiloxy)methyl]-1-[(1E)-hept-1-en-1-yl]tetrahydro-3H,4H-3a,6-methanofuro[3,4-c]pyran-3,4-dione (138); Typical Procedure:[319]

Mn(OAc)$_3$ (2.0 mmol) and Cu(OTf)$_2$ (1.0 mmol) were added to a soln of γ-lactone **137** (1.0 mmol) in dry degassed MeCN (5 mL, N$_2$ bubbled through the solvent for 2 h prior to use). The resulting mixture was then heated to 80°C for 24 h. H$_2$O (20 mL) and Et$_2$O (20 mL) were added, and the mixture was filtered and extracted with Et$_2$O (3 × 100 mL). The combined organic extract was dried (MgSO$_4$), filtered, and concentrated under reduced pressure. The resulting crude material was then purified by flash column chromatography (Et$_2$O/petroleum ether 5:95 to 20:80).

2.9.3.3.4.4 Synthesis of Densely Functionalized and Sterically Crowded Cyclopentane-Fused Lactones

Manganese(III) acetate mediated oxidative radical cyclizations have been utilized for the synthesis of densely functionalized and sterically crowded bicyclo[3.3.0] γ-lactones.[320] Treatment of 2-(alkenyl)malonates **140**, bearing a 1,2-disubstituted alkene, with manganese(III) acetate (2 equiv) and copper(II) trifluoromethanesulfonate (1 equiv) in nitrogen-purged acetonitrile (0.4 M) at 80°C produces the corresponding bicyclo[3.3.0] γ-lactones **141** in yields ranging from 54 to 84% (Scheme 46). Lactones **141** are isolated as diastereomeric mixtures, and electron-rich substrates give reduced diastereocontrol of the γ-lactone products (R^1 = 4-Tol, 2-Tol, or 2-naphthyl); however, the most electron-rich substrate (R^1 = 4-MeOC$_6$H$_4$) produces an intractable mixture of products.

2-(Alkenyl)malonates **142**, **144**, and **147**, which contain 1,2,2-trisubstituted, sterically hindered 1,2,2-trisubstituted, and fully substituted alkene units, respectively, undergo efficient and diastereoselective cyclization to give the corresponding bicyclo[3.3.0] γ-lactones **143**, **145** (in addition to cyclization product **146**), and **148** in excellent yields (Scheme 46). Internal tetrasubstituted alkene **149** yields tricyclic γ-lactone **150**, as a single diastereomeric product in excellent yield.

Scheme 46 Manganese(III) Acetate Mediated Oxidative Radical Cyclization of 2-(Alk-4-enyl)malonates To Give Cyclopentane-Fused Lactones[320]

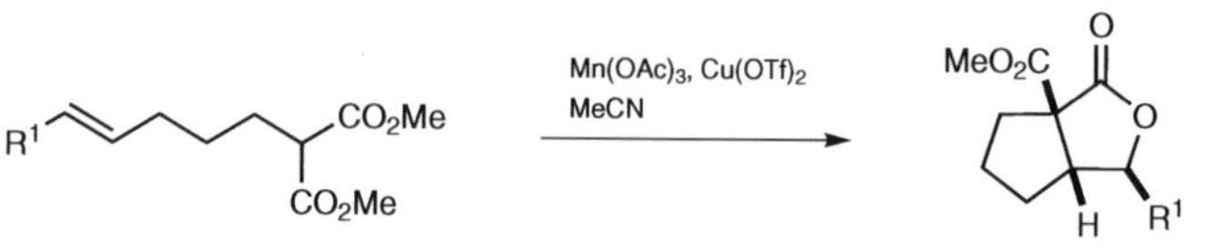

R^1	dr^a	Yield (%)	Ref
Ph	4.1:1	84	[320]
4-FC$_6$H$_4$	4.1:1	72	[320]
2-FC$_6$H$_4$	4.8:1	54	[320]
4-BrC$_6$H$_4$	4.2:1	74	[320]
4-Tol	1.5:1	58	[320]
2-Tol	1.3:1	72	[320]
3-MeOC$_6$H$_4$	6.0:1	73	[320]
4-MeOC$_6$H$_4$	–^{b}	–^{b}	[320]
3-O$_2$NC$_6$H$_4$	3.4:1	66	[320]
2-naphthyl	1.9:1	83	[320]

^{a} The dr was established by analysis of the crude mixture by ^{1}H NMR spectroscopy; the major diastereomer is shown.
^{b} Not determined.

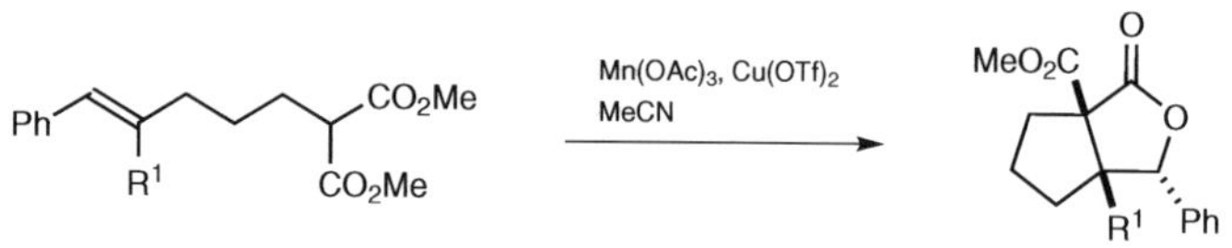

R^1	dr^a	Yield (%)	Ref
Me	8.0:1	83	[320]
Bu	10.7:1	91	[320]
CH$_2$C=CH	11.9:1	74	[320]
(CH$_2$)$_2$OTBDPS	10.2:1	96	[320]

^{a} The dr was established by analysis of the crude product by ^{1}H NMR spectroscopy.

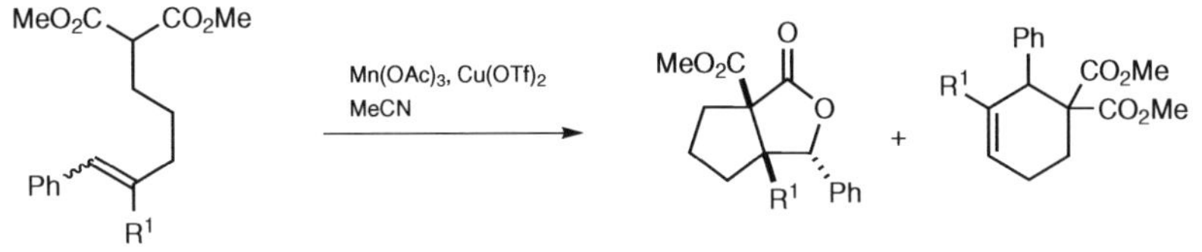

2.9.3 Intramolecular Free-Radical Cyclization Routes to O-Heterocycles

R¹	Ratio (Z/E) of 144	Yield (%) 145	Yield (%) 146	Ref
iPr	ca. 5:1	39	48	[320]
t-Bu	0:1	0	23	[320]

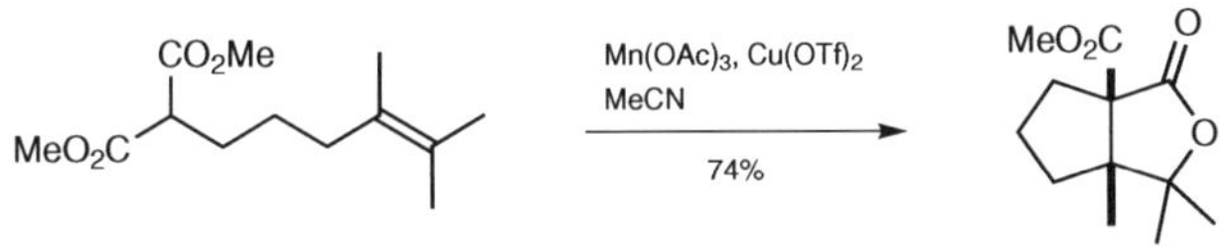

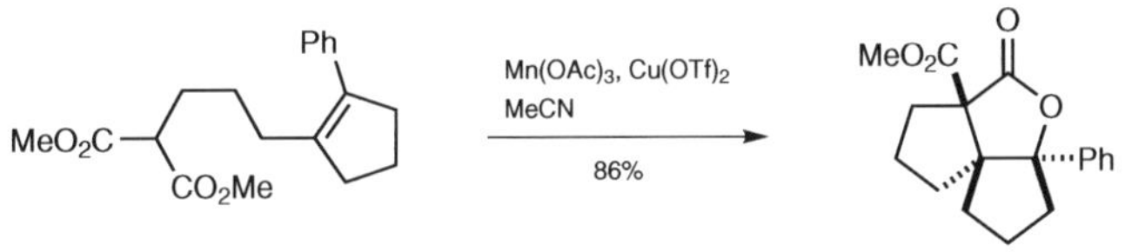

Cyclopentane-Fused Lactones 141 and 143; General Procedure:[320]

A mixture of Mn(OAc)$_3$·2H$_2$O (2 equiv) and Cu(OTf)$_2$ (1 equiv) was placed under vacuum and heated to 80 °C for 10 min and then quenched with N$_2$. A 0.4 M soln of 2-(5-arylpent-4-enyl)malonate (1 equiv) in MeCN was then added rapidly under N$_2$, and the soln was stirred overnight at 80 °C. Upon completion of the reaction, H$_2$O and EtOAc (1:1; 10 mL/mmol malonate) were added, and the layers were separated. The aqueous layer was extracted with EtOAc (3 × 5 mL/mmol), and the combined organic extract was washed with brine, dried (Na$_2$SO$_4$), filtered, and concentrated under reduced pressure. The lactones were isolated as a mixture of diastereomers, epimeric at the C1-position adjacent to the aryl group.

2.9.3.3.5 Synthesis of Tricyclic γ-Lactones

Cascade reactions involving reductive cyclization, Dieckmann condensation, and lactonization of acyclic keto diesters produce *cis*- and *trans*-bicyclic ring systems **151** containing a γ-lactone ring.[321] Upon treating a keto diester bearing an (*E*)-alkene with samarium(II) iodide in tetrahydrofuran at room temperature, a tricyclic compound comprising a *cis*-bicyclo[4.4.0]decane (decalin) ring system and a γ-lactone is obtained as the major product **153** in 63% yield, along with a spiro γ-lactone **154** (Scheme 47). Treatment of the same keto diester, but bearing a (*Z*)-alkene, with samarium(II) iodide in tetrahydrofuran produces a product possessing a *trans*-decalin ring system with a γ-lactone as the major product (Table 10, entries 2 and 3). The *E*-isomer of an analogous keto ester bearing a shorter tether between the ketone group and the two ester groups does not undergo lactonization owing to excessive strain of the corresponding lactone; thus, it stereoselectively gives bicyclic keto ester **152** possessing a *cis*-perhydroindane ring system in high yield (Table 10, entries

5 and 6). Reaction of the (Z)-alkene bearing the shorter tether with samarium(II) iodide in tetrahydrofuran provides the corresponding tricyclic compound containing a *trans*-ring junction and a γ-lactone as the major product (Scheme 47). The stereochemistry of the products was determined by X-ray analysis and/or NOE experiments.

Table 10 Reductive Cyclization/Dieckmann Condensation/Lactonization of Acyclic Keto Diesters[321]

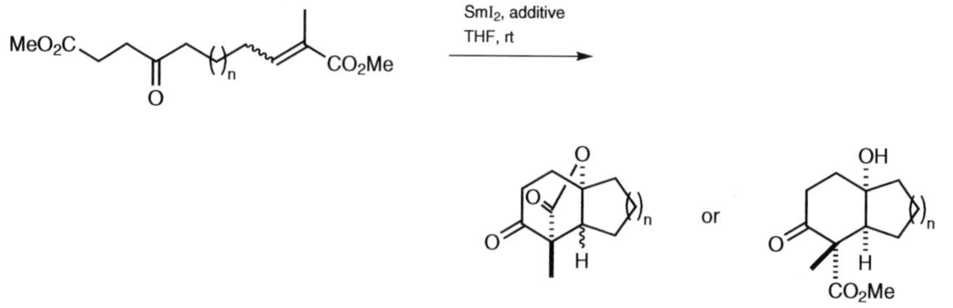

Entry	Substrate	Additive	Time (min)	Product	Yield (%)	Ref
1	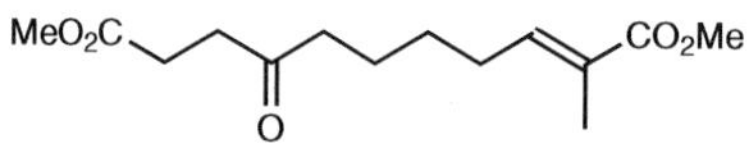 MeO₂C...CO₂Me	HMPA (5 equiv)	30	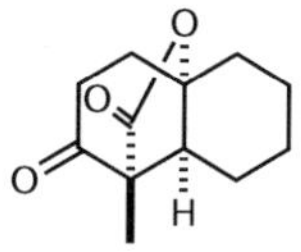 151-type	61	[321]
2	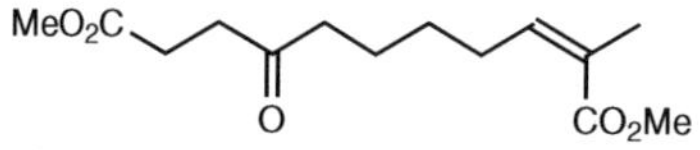 MeO₂C...CO₂Me	none	120	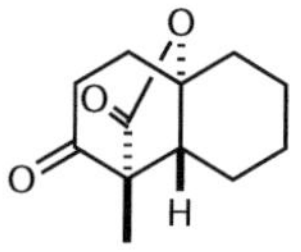 151-type	59	[321]
3	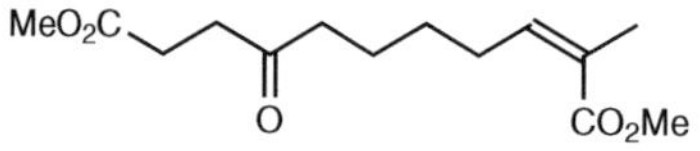 MeO₂C...CO₂Me	none	900	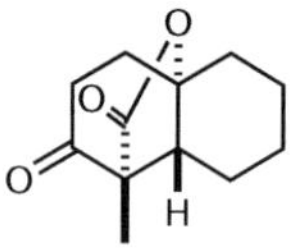 151-type	75[a]	[321]
4	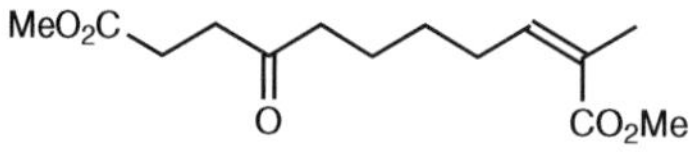 MeO₂C...CO₂Me	HMPA (5 equiv)	30	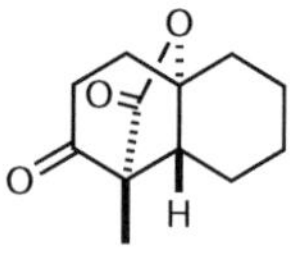 151-type	37	[321]
5	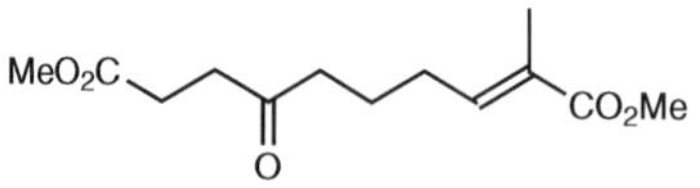 MeO₂C...CO₂Me	none	30	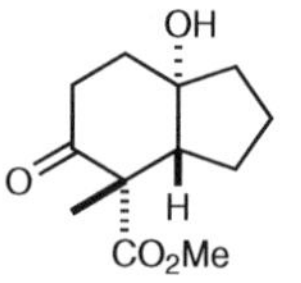 152-type	88	[321]
6	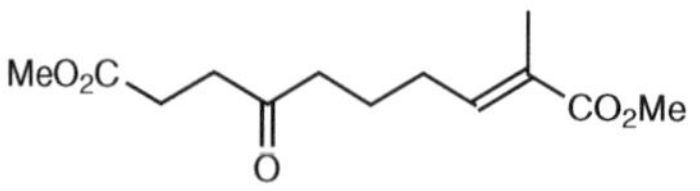 MeO₂C...CO₂Me	HMPA (5 equiv)	10	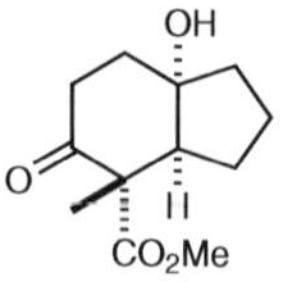 152-type	52	[321]

[a] Reaction was performed at 0 °C.

2.9.3 Intramolecular Free-Radical Cyclization Routes to O-Heterocycles

Scheme 47 Reductive Cyclization/Dieckmann Condensation/Lactonization of Acyclic Keto Diesters[321]

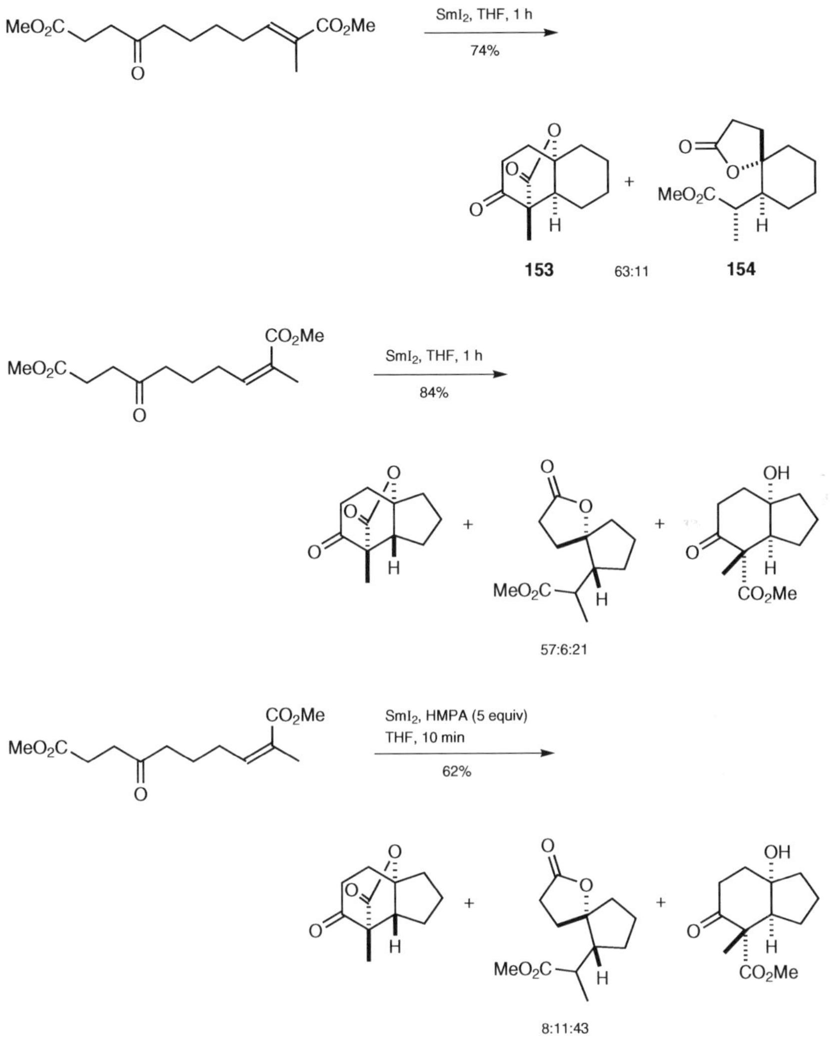

(1S*,4aS*,8aS*)-1-Methyloctahydro-2H-4a,1-(epoxymethano)naphthalene-2,10-dione (153) and Methyl (2S*)-2-[(5S*,6S*)-2-Oxo-1-oxaspiro[4.5]decan-6-yl]propanoate (154); Typical Procedure:[321]

A mixture of Sm(0) (292 mg, 1.94 mmol) and 1,2-diiodoethane (469 mg, 1.67 mmol) in THF (17 mL) was sonicated at rt for 2 h under argon and was then cooled to −78 °C. A soln of (E)-dimethyl 2-methyl-8-oxoundec-2-enedioate (150 mg, 0.556 mmol) in THF (1.4 mL) was added dropwise over 1 min, and the resulting mixture was warmed to rt. After 1 h, the reaction was quenched with bubbling air. The resulting mixture was diluted with Et$_2$O, washed with 1 M HCl, H$_2$O, sat. Na$_2$S$_2$O$_3$, sat. NaHCO$_3$, and brine, dried (MgSO$_4$), and concentrated under reduced pressure. The crude product was purified by column chromatography (silica gel, EtOAc/hexane 1:5) to afford the dione **153** as colorless crystals; yield: 73.1 mg (63%); and the spiro γ-lactone **154** as a colorless oil; yield: 15.0 mg (11%).

2.9.3.4 Synthesis of Pyrans and Derivatives

2.9.3.4.1 Synthesis of Polycyclic Dihydropyrans

The manganese(III)-mediated oxidative radical cyclization of electron-rich terpenoids bearing two alkyl substituents at the terminal sp^2-carbon atom involves an intramolecular hetero-Diels–Alder reaction with the terpenoid chain.[322] Thus, the reaction of β-keto ester terpenoid **155** with manganese(III) acetate and copper(II) acetate in 94 % ethanol at 55 °C in air leads to the formation of heterocyclic compound **156** within 3 hours. The complete pathway for the reaction is provided in Scheme 48. For an efficient intramolecular Diels–Alder reaction, substitution of the terminal sp^2-carbon atom with two electron-rich alkyl groups is essential, as more electron-poor terpenoids with only a single alkyl group at the terminal sp^2-carbon atom, or those with no substituents, afford linear β,γ-unsaturated δ-hydroxy α-keto esters in lieu of the cyclized products in yields of 50–61%. Notably, no further cyclized product is obtained even at elevated temperatures.

Scheme 48 Synthesis of Bicyclic Dihydropyrans by Manganese(III)-Mediated Oxidative Radical Cyclization of Terpenoids[322]

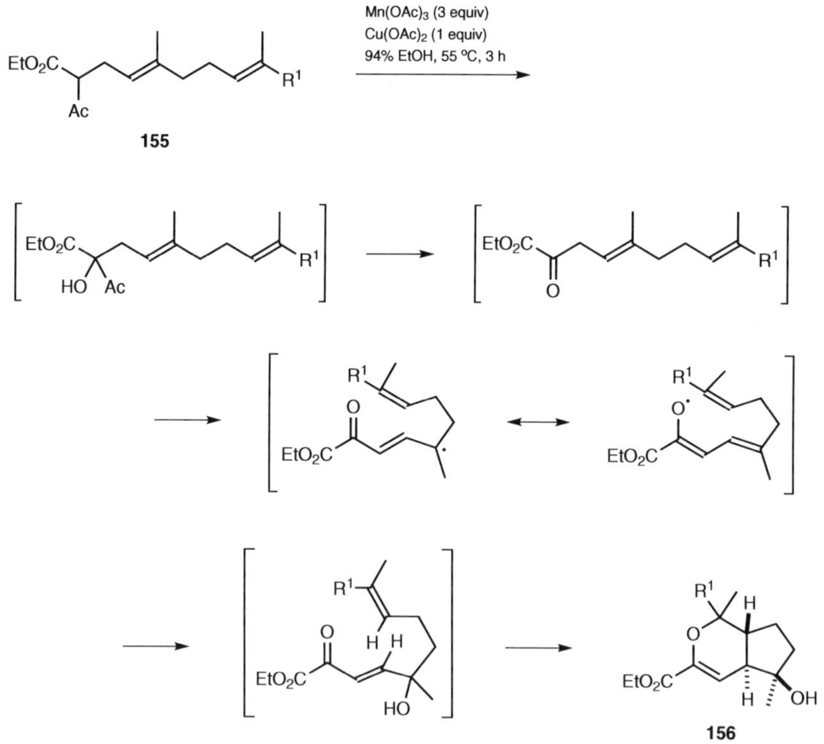

R^1	Yield (%)	Ref
Me	75	[322]
CH$_2$CH=CMe$_2$	59	[322]
![structure]	59	[322]

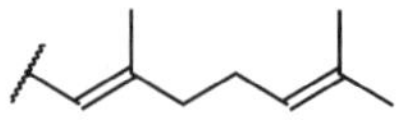

2.9.3 Intramolecular Free-Radical Cyclization Routes to O-Heterocycles

The manganese(III)-promoted cyclization of highly branched β-keto ester terpenoids **157** bearing two alkyl substituents at the terminal sp^2-carbon atom provides an efficient means to obtain polycyclic dihydropyran derivatives **158** (Table 11).

Table 11 Synthesis of Polycyclic Dihydropyrans by Cyclization of Highly Branched Terpenoids[322]

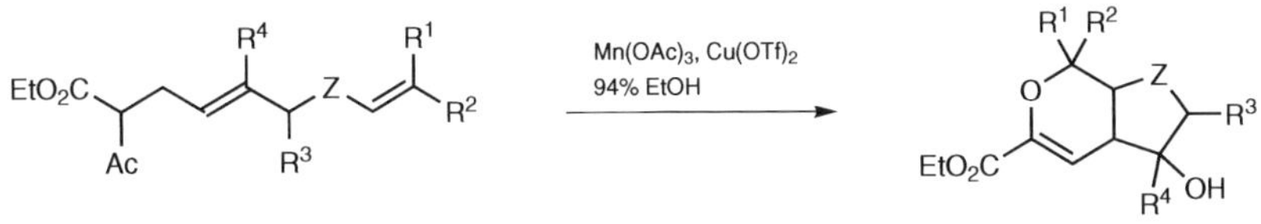

Entry	Substrate	Product	Ratio[a] (trans/cis)	Ratio[b] (β/α)	Yield (%)	Ref
1	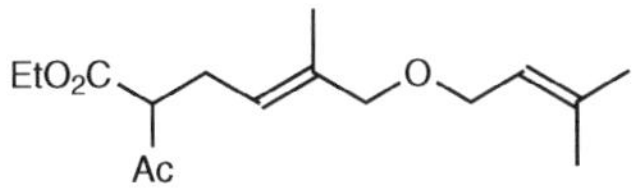	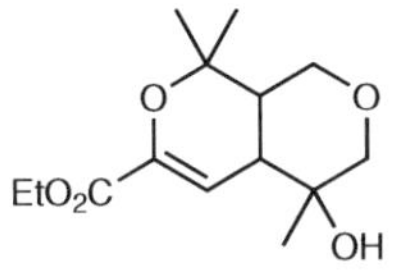	1:1.2	8:1	65	[322]
2	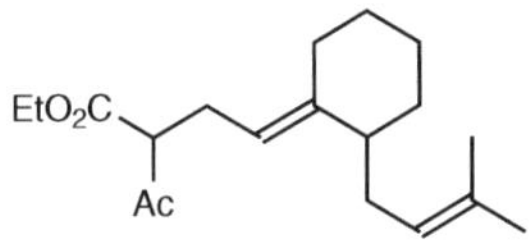	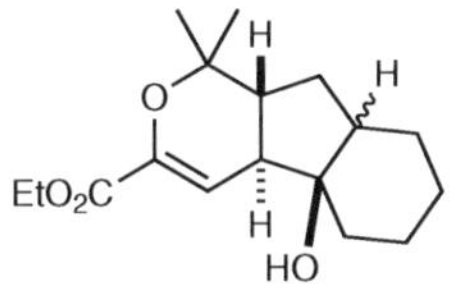	1:3	6:1	81	[322]
3	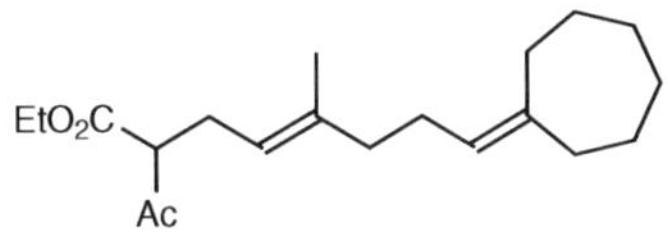	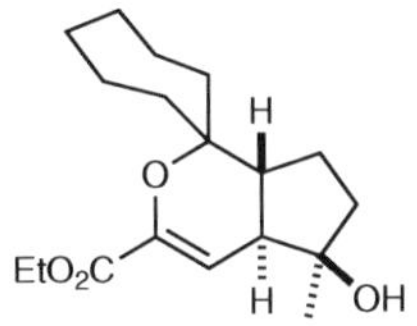	22:1	11:1	74	[322]
4	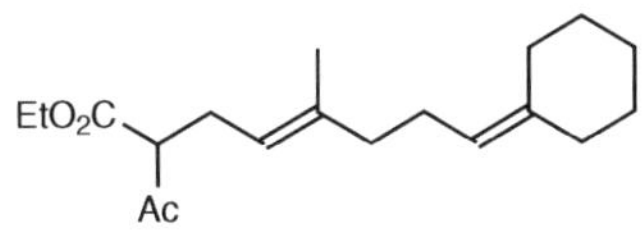	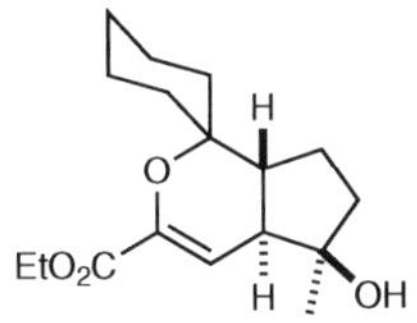	15:1	6:1	73	[322]

[a] Ratio at the ring junction.
[b] Ratio for the OH group of the *trans*-isomer.

Ethyl (4aS^*,5R^*,7aR^*)-5-Hydroxy-1,1,5-trimethyl-1,4a,5,6,7,7a-hexahydrocyclopenta[c]pyran-3-carboxylate (156, R^1 = Me); Typical Procedure:[322]

Mn(OAc)$_3$•2H$_2$O (0.78 g, 2.82 mmol) and Cu(OAc)$_2$•H$_2$O (0.19 g, 0.94 mmol) were added to a stirred soln of (E)-ethyl 2-acetyl-5,9-dimethyldeca-4,8-dienoate (0.25 g, 0.94 mmol) in 94% EtOH (10 mL). The flask was left open to air and then stirred at 55 °C for 3 h. Upon completion of the reaction, the resulting mixture was treated with 1 M HCl and extracted with CH$_2$Cl$_2$, and the extracts were dried (Na$_2$SO$_4$), filtered, and concentrated under reduced pressure. The crude residue was purified by flash chromatography (silica gel); yield: 0.18 g (75%, as reported).

2.9.3.4.2 Synthesis of Polysubstituted Tetrahydropyrans

Chlorobis(η^5-cyclopentadienyl)titanium(III) is a well-accepted reagent that can be used to generate carbon-centered radicals from epoxides.[143] Roy and Banerjee have developed an effective method for the synthesis of polysubstituted tetrahydropyrans in good yields with high diastereoselectivity.[161] In the presence of chlorobis(η^5-cyclopentadienyl)titanium(III), radical cyclization of epoxyalkyl propargyl ethers **159** affords a mixture of isomeric products **160A** and **160B** (Table 12, entries 1–6). Under the reaction conditions, the radical cyclization of epoxyalkyl propargyl ethers can also furnish spirocyclic tetrahydropyrans as the main products; in this case, it is possible that only one isomer or a mixture of two diastereomers is formed (Table 12, entries 7 and 8). The reaction proceeds efficiently in dry tetrahydrofuran at room temperature under an argon atmosphere. The titanium(III) species is prepared in situ by treatment of commercially available dichlorobis(η^5-cyclopentadienyl)titanium(IV) with activated zinc dust in dry tetrahydrofuran under argon for 1 hour. The ratio of the isomers in the products can be determined from two distinguishable multiplets in the ^{1}H NMR spectra for the 4H proton; for example, for the tetrahydropyran shown in Table 12, entry 1, two multiplets centered at δ 2.73 and 2.61 can be attributed to the minor and major isomers, respectively. Isolation of the major isomer can be achieved by preparative TLC (EtOAc/petroleum ether 15:85), but it is difficult to isolate the minor isomer in pure form and it is always contaminated by the major isomer.

2.9.3 Intramolecular Free-Radical Cyclization Routes to O-Heterocycles

Table 12 Titanium(III)-Mediated Radical Cyclization of Epoxyalkyl Propargyl Ethers To Give Tetrahydropyrans[161]

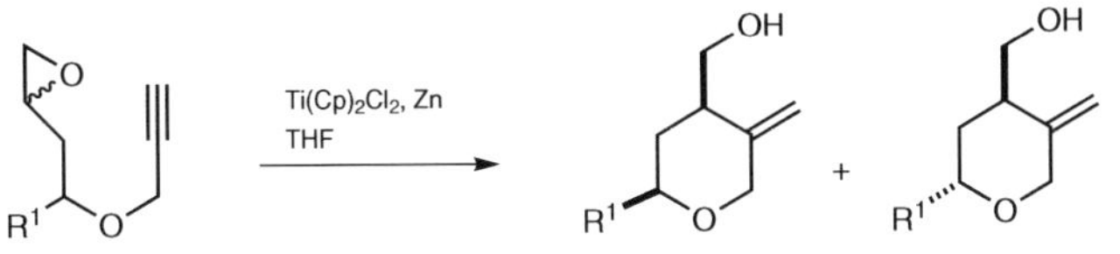

Entry	Substrate	Major Product	Ratio (160A/160B)	Combined Yield (%)	Yield (%) of Isolated Major Isomer	Ref
1	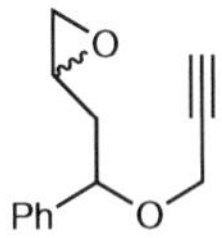 (R¹ = Ph)	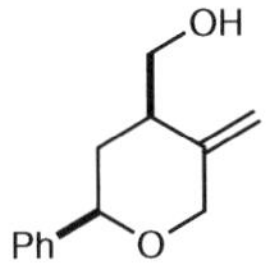	4.6:1	78	55	[161]
2	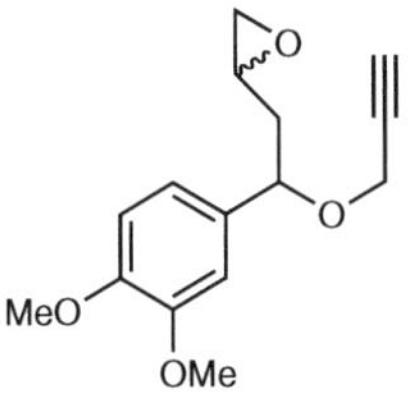 (R¹ = 3,4-dimethoxyphenyl)	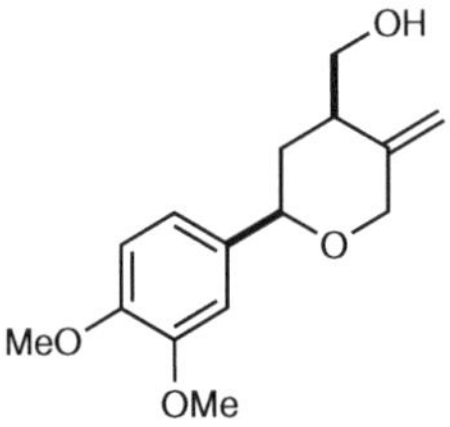	4.7:1	73	51	[161]
3	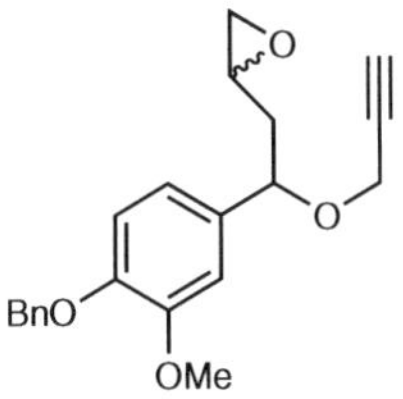 (R¹ = 4-BnO-3-MeO-phenyl)	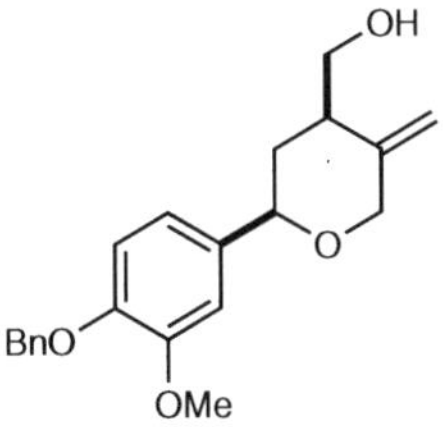	5:1	82	57	[161]
4	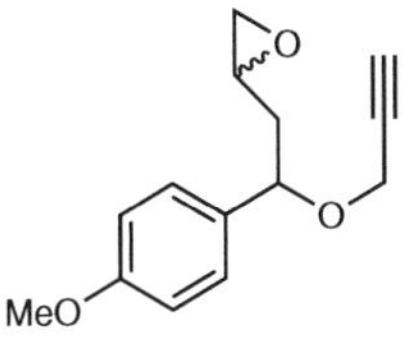 (R¹ = 4-MeO-phenyl)	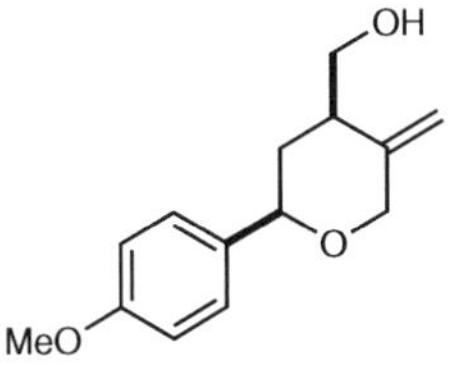	4:1	72	50	[161]
5	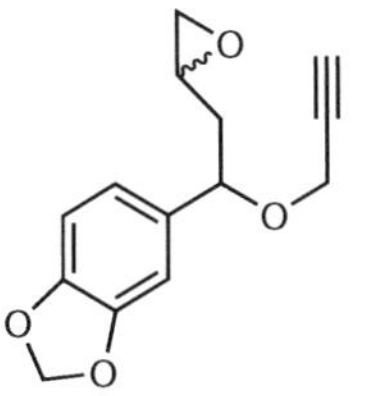 (R¹ = 3,4-methylenedioxyphenyl)	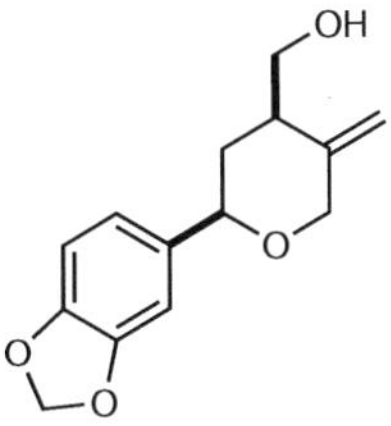	5:1	74	53	[161]

Table 12 (cont.)

Entry	Substrate	Major Product	Ratio (160A/160B)	Combined Yield (%)	Yield (%) of Isolated Major Isomer	Ref
6	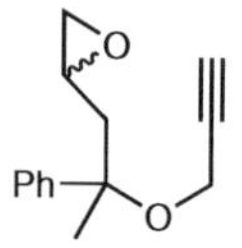	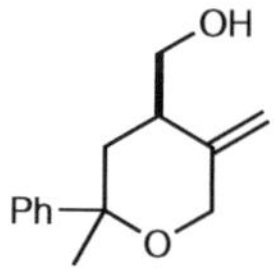	2:1	85	–[a]	[161]
7	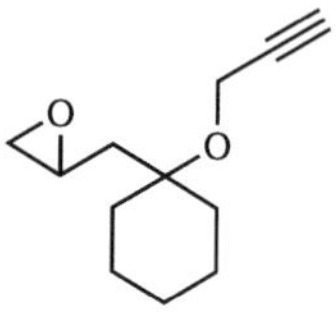	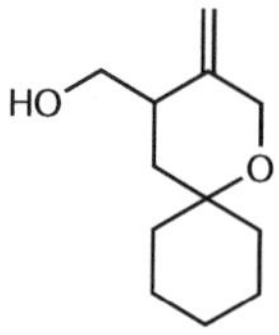	–	56	–	[161]
8	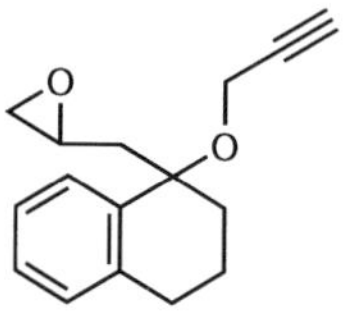	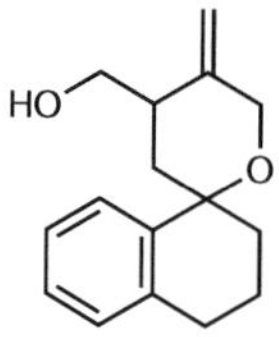	–[b]	70	–	[161]

[a] Isomers were inseparable.
[b] Ratio of isomers could not be determined; it is possible that only one isomer or a mixture of two diastereomers was formed.

On the other hand, similar radical cyclization reactions of allyl epoxyalkyl ethers **161** produce a mixture of isomeric products **162A**, **162B**, and **162C** in each case (Table 13). The ratio of the three isomers in the products can be confirmed from distinguishable doublets of the methyl protons in the ^{1}H NMR spectra; for example, for the allyl epoxyalkyl ether shown in Table 13, entry 1, the two doublets at δ 0.90 and 1.07 can be attributed to the two major isomers and a doublet at δ 0.93 can be assigned to the third minor isomer. Isolation of the major two isomers can be achieved by preparative TLC (EtOAc/petroleum ether 1:4), but the formation of a very minute quantity of the minor isomer limits its proper isolation.

2.9.3 Intramolecular Free-Radical Cyclization Routes to O-Heterocycles

Table 13 Titanium(III)-Mediated Radical Cyclization of Allyl Epoxyalkyl Propargyl Ethers[161]

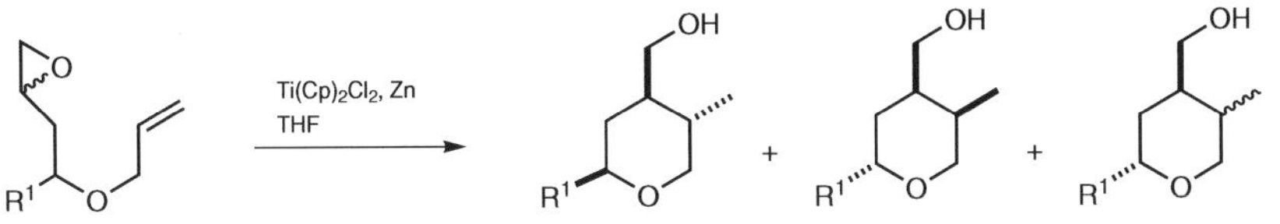

Entry	Substrate	Product	Ratio (162A/162B/162C)	Yield (%)	Isolated Yield (%) 162A	Isolated Yield (%) 162B	Ref
1	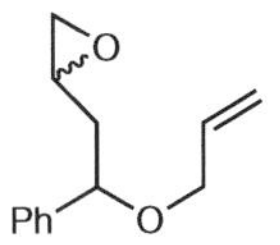	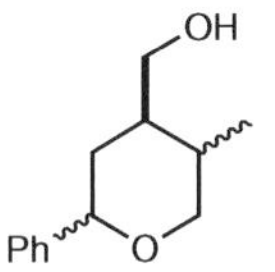	2.6:1:0.2	83	49	19	[161]
2	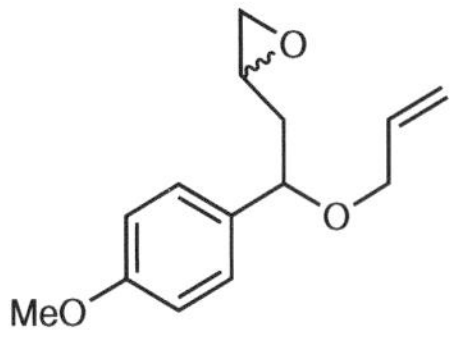	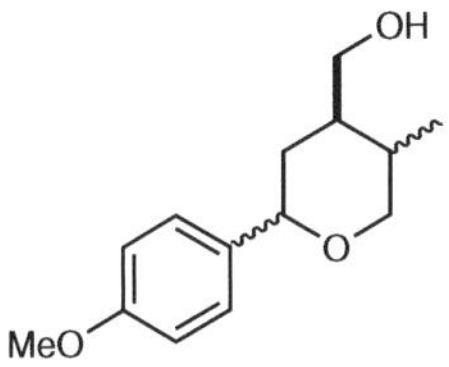	2.4:1:0.22	87	52	21	[161]
3	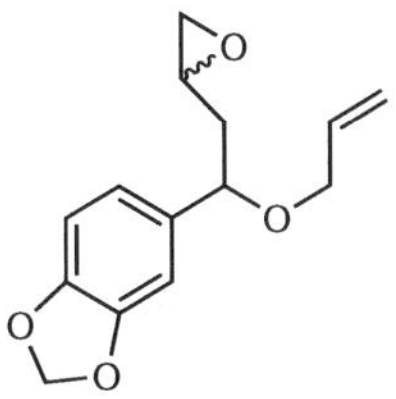	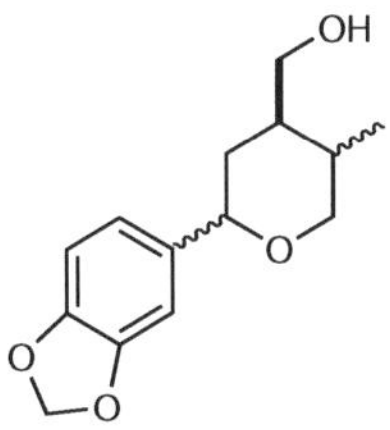	2.5:1:0.2	81	49	20	[161]

[(2R*,4R*)-5-Methylene-2-phenyltetrahydro-2H-pyran-4-yl]methanol (Table 12, Entry 1); Typical Procedure:[161]

A soln of Ti(Cp)$_2$Cl$_2$ (0.65 g, 2.6 mmol) in dry THF (35 mL) was stirred with activated Zn dust (0.482 g, 7.2 mmol) under an argon atmosphere for 1 h [activated Zn dust was prepared by washing Zn with 4 M HCl (60 mL), H$_2$O, and finally dry acetone followed by drying under reduced pressure]. The resulting green suspension was then transferred by cannula to an addition funnel and was added dropwise to a magnetically stirred soln of 2-{2-phenyl-2-[(prop-2-yn-1-yl)oxy]ethyl}oxirane (250 mg, 1.2 mmol) in dry THF (30 mL) at rt over 40 min under argon. The mixture was stirred for another 1.5 h and then quenched with 10% H$_2$SO$_4$ (15 mL). The mixture was concentrated under reduced pressure, and the residue was extracted with Et$_2$O (4 × 30 mL). The combined organic layer was washed with sat. NaHCO$_3$ (2 × 20 mL) and brine (20 mL), dried (Na$_2$SO$_4$), and concentrated under reduced pressure. The crude residue was chromatographed (silica gel, EtOAc/petroleum ether 1:4) to afford a mixture of isomers (4.6:1); yield: 196 mg (78%). The isomers were separated by preparative TLC (EtOAc/petroleum ether 15:85) to afford the major isomer as a viscous oil; yield: 139 mg (55%).

**[(2R*,4R*,5S*)-5-Methyl-2-phenyltetrahydro-2H-pyran-4-yl]methanol and [(2R*,4R*,5R*)-5-Methyl-2-phenyltetrahydro-2H-pyran-4-yl]methanol (Table 13, Entry 1);
Typical Procedure:**[161]

A soln of Ti(Cp)$_2$Cl$_2$ (0.65 g, 2.6 mmol) in dry THF (35 mL) was stirred with activated Zn dust (0.482 g, 7.2 mmol) under an argon atmosphere for 1 h [activated Zn dust was prepared by washing Zn with 4 M HCl (60 mL), H$_2$O, and finally dry acetone followed by drying under reduced pressure]. The resulting green suspension was then transferred by cannula to an addition funnel and was added dropwise to a magnetically stirred soln of 2-{2-phenyl-2-[(prop-2-en-1-yl)oxy]ethyl}oxirane (210 mg, 1.03 mmol) in dry THF (25 mL) at rt over 40 min under argon. The mixture was stirred for another 1.5 h and then quenched with 10% H$_2$SO$_4$ (15 mL). The mixture was concentrated under reduced pressure, and the residue was extracted with Et$_2$O (4 × 30 mL). The combined organic layer was washed with sat. NaHCO$_3$ (2 × 20 mL) and brine (20 mL), dried (Na$_2$SO$_4$), and concentrated under reduced pressure. The crude residue was chromatographed (silica gel, EtOAc/petroleum ether 3:7) to yield a mixture of three isomers (2.6:1:0.2); yield: 176 mg (83%). The mixture was subjected to preparative TLC (EtOAc/petroleum ether 1:4) to afford the major (2R*,4R*,5S*)-isomer as a viscous oil; yield: 105 mg (49%); and the minor (2R*,4R*,5R*)-isomer as a crystalline solid; yield: 40 mg (19%); mp 138–140 °C.

2.9.3.4.3 Synthesis of Benzopyran Derivatives

The intramolecular radical-promoted cyclization of aromatic carbonyl compounds bearing an internal radical acceptor, such as an alkene or an alkyne, using chlorobis(η^5-cyclopentadienyl)titanium(III) in tetrahydrofuran under argon furnishes the corresponding benzopyrans in good yields (Table 14).[323] The reaction strictly depends on the way in which the catalyst is added. Normal addition of compound **163** to chlorobis(η^5-cyclopentadienyl)titanium(III) provides the intermolecular coupling product, whereas slow addition of the carbonyl compound to the chlorobis(η^5-cyclopentadienyl)titanium(III) reagent yields cyclized product **164** in good yield, with no coupling product.[323] The reaction proceeds faster for aldehydes (Table 14, entries 1, 2, 4, and 5) than for ketones (Table 14, entry 6). A 7-*endo*-cyclization reaction is unsuccessful and only the self-coupling product is obtained in good yield (Table 14, entry 3). Aliphatic aldehydes are not compatible with the reaction conditions.

2.9.3 Intramolecular Free-Radical Cyclization Routes to O-Heterocycles

Table 14 Titanium(III)-Mediated Cyclization of Carbonyl Compounds To Give Dihydrobenzopyranols[323]

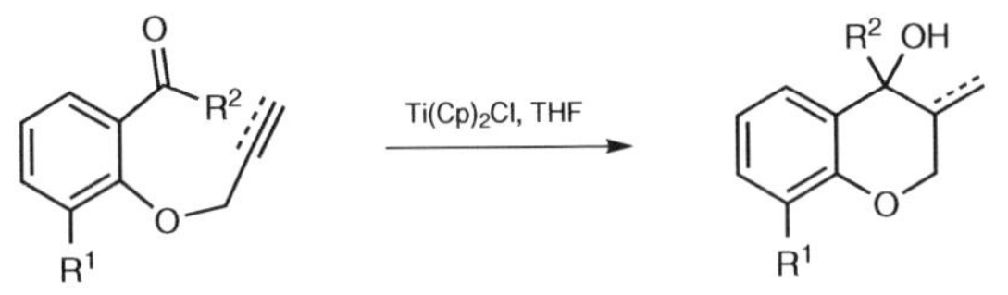

Entry	Substrate	Time (h)	Product	Yield (%)	Ref
1	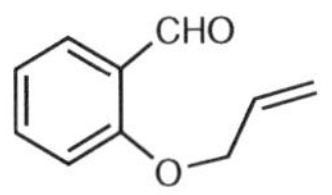	3	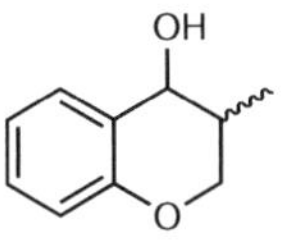	72[a]	[323]
2	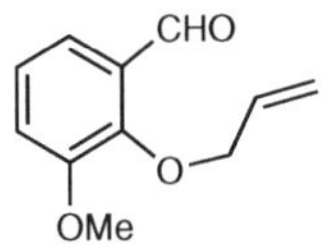	3	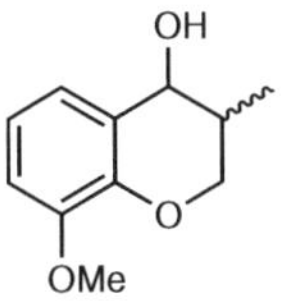	68[a]	[323]
3	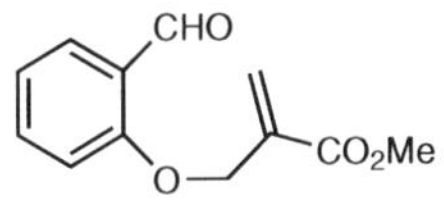	–	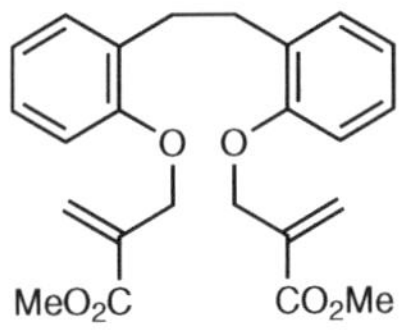	80[b]	[323]
4	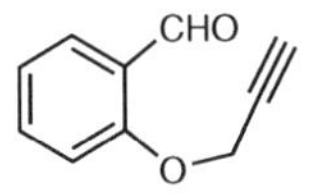	2.5	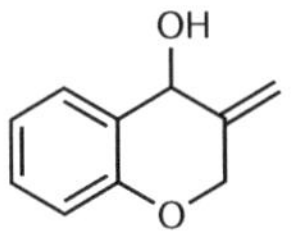	75	[323]
5	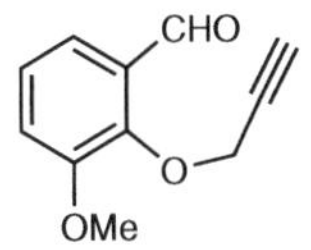	2.5	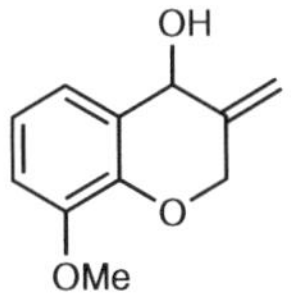	70	[323]
6	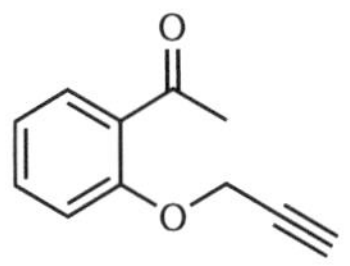	24	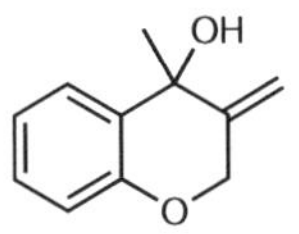	55	[323]

[a] Mixture of two isomers in a 2:1 ratio was obtained (determined by analysis of the crude product by ^{1}H NMR spectroscopy).
[b] Structure of likely product is shown.

3-Methylene-3,4-dihydro-2H-1-benzopyran-4-ol (Table 14, Entry 4); Typical Procedure:[323]
A red soln of Ti(Cp)$_2$Cl$_2$ (249 mg, 1 mmol) in dry THF (12.5 mL) was treated with activated Zn dust (130 mg, 2 mmol) under argon until the soln turned green. Then, a soln of 2-[(prop-2-yn-1-yl)oxy]benzaldehyde (160 mg, 1 mmol) in deoxygenated THF (12 mL) was added dropwise over 2.5 h at rt. The mixture was stirred for an additional 5 min and then

quenched with sat. aq NaH$_2$PO$_4$ (10 mL). THF was removed under reduced pressure, and the resulting residue was extracted with Et$_2$O (3 × 25 mL). The organic layer was washed with H$_2$O (2 × 5 mL) and brine (2 × 5 mL) and then dried (Na$_2$SO$_4$). The solvent was removed under reduced pressure, and the crude residue obtained was purified by column chromatography (silica gel, EtOAc/petroleum ether 1:9) to afford a colorless oil; yield: 120 mg (75%).

2.9.3.4.4 Synthesis of Hexahydro-2H-1-benzopyran Derivatives

Radical cyclizations with vinyl radicals for the synthesis of five- and six-membered rings have many applications in organic chemistry.[324] Vinyl radicals can be generated by reaction of samarium(II) iodide with vinyl halides;[170] electron transfer between samarium(II) and the organic fragment leads to the formation of a free radical by loss of the halide ion. Zhan and Lang have employed this technique for the efficient radical cyclization of vinyl iodides **165** (the triple bond of the vinyl iodide should be substituted with a vinyl group or an aromatic ring) in a 6-(π-exo)-exo-dig mode to produce the corresponding exocyclic dienes fused to a six-membered ring, that is, hexahydro-2H-1-benzopyran derivatives **166** (Scheme 49).[325] If the triple bond is substituted with an ethyl or trimethylsilyl group or a hydrogen atom, the yield of the corresponding exocyclic diene is markedly reduced. The reaction proceeds smoothly in the presence of isopropenyl and aromatic substituents.

Scheme 49 6-(π-exo)-exo-dig Radical Cyclization of Vinyl Iodides To Give Hexahydro-2H-1-benzopyrans[325]

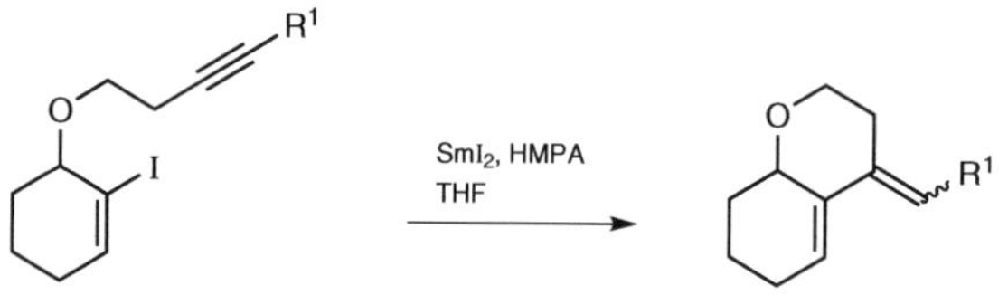

R^1	Ratio^a (E/Z)	Yield (%)	Ref
Ph	1.7:1	71	[325]
4-MeOC$_6$H$_4$	1.9:1	70	[325]
2-Tol	1.9:1	68	[325]
2-thienyl	1:1.5	58	[325]
3-furyl	1:1.1	63	[325]
Et	1.3:1	29	[325]
TMS	1.1:1	30	[325]
H	–	14	[325]
CMe=CH$_2$	1:1.2	70	[325]

^{a} The structures of the products were confirmed by NOE experiments, and the (E/Z) ratios were determined by ^{1}H NMR spectroscopy.

4-[(4-Methoxyphenyl)methylene]-3,4,6,7,8,8a-hexahydro-2H-1-benzopyran (166, R^1 = 4-MeOC$_6$H$_4$); Typical Procedure:[325]

> **CAUTION:** *Hexamethylphosphoric triamide is a possible human carcinogen and an eye and skin irritant.*

A soln of 1-{4-[(2-iodocyclohex-2-en-1-yl)oxy]but-1-yn-1-yl}-4-methoxybenzene (382 mg, 1 mmol) in THF (1 mL) was added to a 0.1 M soln of SmI$_2$ in THF (30 mL, 3 mmol); then, HMPA (1.5 mL) was added. The resulting mixture was stirred for 45 min and was then quenched with sat. aq NH$_4$Cl. Subsequently, the crude mixture was extracted with Et$_2$O and the extracts were washed with H$_2$O, 5% Na$_2$S$_2$O$_3$, and brine, and then dried (Na$_2$SO$_4$). The crude residue was purified by flash column chromatography (silica gel, hexane/CH$_2$Cl$_2$ 1:1) to afford a mixture of E/Z-isomers; yield: 179 mg (70%). The two isomers were separated by preparative TLC (silica gel, hexane/CH$_2$Cl$_2$ 1:1). The E-isomer was isolated as colorless crystals after recrystallization (hexane); mp 105–106 °C.

2.9.3.5 Synthesis of Oxepin Derivatives by an Intramolecular Ring Expansion Approach

Nishino and co-workers have investigated the efficient and selective manganese(III)-mediated synthesis of dibenz[b,f]oxepin derivatives **168** by ring expansion of substituted 9H-xanthenes **167**.[81] The oxidative ring-expansion reaction of substituted 9H-xanthenes **167** with manganese(III) acetate is performed at a molar ratio of 1:4 in boiling acetic acid/water (9:1 v/v) and gives dibenz[b,f]oxepin derivatives **168** in good yields (Scheme 50). The reaction is highly regioselective, except for system **169**, for which two regioisomers are obtained.

Scheme 50 Synthesis of Benz[b,f]oxepins by Manganese(III)-Mediated Oxidation of 2-(9-Xanthenyl)malonates[81]

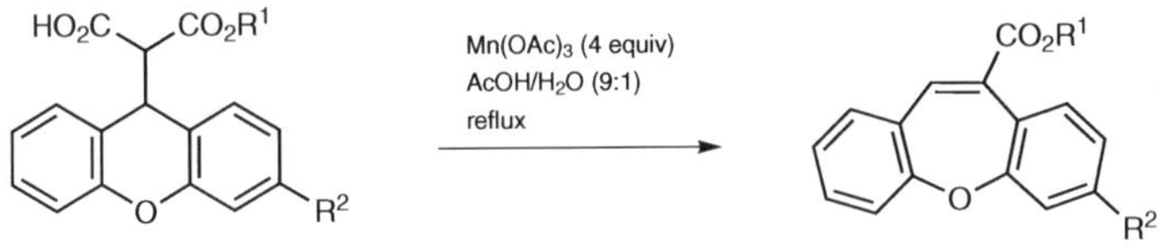

R^1	R^2	Yield (%)	Ref
Me	H	72	[81]
Et	H	63	[81]
iPr	H	68	[81]
Me	OMe	81	[81]

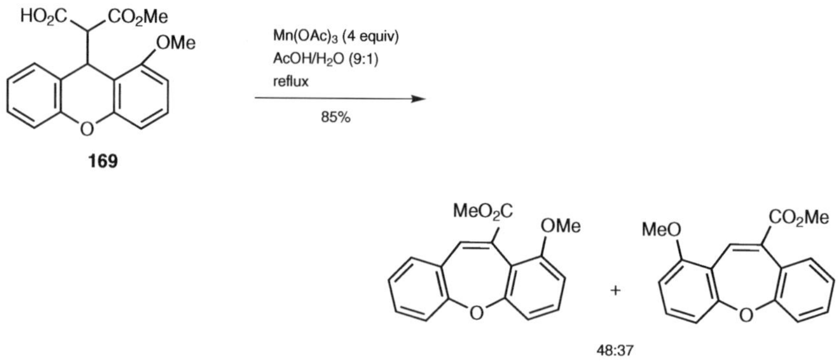

The mechanism of the ring expansion for the formation of carboxylates **168** starts with the reaction of manganese(III) acetate with malonate **167** (Scheme 51). Then, one-electron oxidation of **170** gives α-carbonyl carbon radical **171**, and 1,2-aryl radical rearrangement of this species gives more-stable benzyl-type radical **172**. The corresponding oxepin **168** is finally produced by oxidative decarboxylation.

Scheme 51 Mechanism Involved in the Synthesis of Oxepin Derivatives from 2-(9-Xanthenyl)malonates[81]

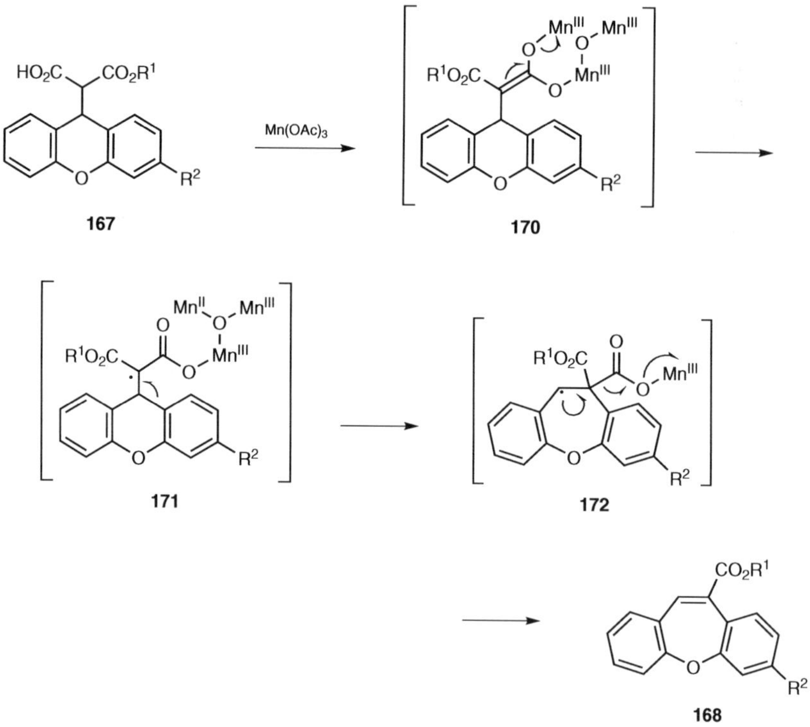

Dibenz[b,f]oxepin Derivatives 168; General Procedure:[81]
Mn(OAc)$_3$ (2 mmol) was added, just before reflux was reached, to a heated soln of **167** (0.5 mmol) in glacial AcOH (9 mL) in the presence of H$_2$O (1 mL). The progress of the reaction was monitored by the change in the color of this soln from dark brown to clear red. The resulting mixture was then cooled to rt, and the solvent was removed under reduced pressure. The residue was triturated with 2 M HCl (15 mL), which was followed by extrac-

tion with CHCl$_3$ (3 × 10 mL). The combined extract was washed with sat. aq NaHCO$_3$ (2 × 15 mL) and H$_2$O (2 × 10 mL). The combined organic layer was dried (MgSO$_4$) and concentrated to dryness under reduced pressure. The crude residue was purified by TLC (Wako B-10 silica gel, CHCl$_3$). In some cases, the xanthen-9-one and xanthene were also isolated. For the reaction of malonate **167** (R^1 = Me; R^2 = H), the aqueous layer was acidified with concd H$_2$SO$_4$ and then extracted with CHCl$_3$ (3 × 10 mL). The extract was washed with H$_2$O (2 × 20 mL), dried (MgSO$_4$), and concentrated to dryness.

2.9.4 Intramolecular Free-Radical Cyclization Routes to Carbocycles

2.9.4.1 Synthesis of Cycloalkane Derivatives

2.9.4.1.1 Synthesis of Monosubstituted and 1,1-Disubstituted Cyclopropane Derivatives

1,1-Disubstituted and monosubstituted cyclopropanes are directly obtained from their corresponding 2,2-disubstituted and 2-monosubstituted 1,3-dihalopropanes using samarium(II) iodide in tetrahydrofuran.[326] The cyclization of halopropanes proceeds through a samarium(II) iodide mediated radical 3-*exo*-*tet* cyclization pathway. Treatment of 1,3-diiodopropanes **173** bearing benzyl, 4-methylbenzyl, 4-methoxybenzyl, and 4-chlorobenzyl groups with samarium(II) iodide (2.5 equiv) at room temperature provides, after fading of the deep-blue colored solution over 2 hours, the corresponding cyclopropanes **174** in good yields (Table 15, entries 1–4). This samarium(II)-based reaction system can also be employed for the efficient cyclization of a 2-monobenzyl-1,3-diiodopropane (Table 15, entry 7). However, for 2,2-dialkyl-substituted 1,3-diiodopropanes, an efficient transformation can only be achieved under reflux conditions with a higher samarium loading (5.5 equiv for 3 h) (Table 15, entries 5 and 6).

Table 15 Samarium(II) Iodide Meditated Synthesis of Cyclopropanes from 1,3-Diiodopropanes[326]

Entry	Substrate	Product	Yield^a (%)	Ref
1	Ph, Ph (1,3-diiodo)	Ph, Ph cyclopropane	99	[326]
2	bis(4-methylbenzyl) 1,3-diiodopropane	bis(4-methylbenzyl)cyclopropane	93	[326]
3	bis(4-methoxybenzyl) 1,3-diiodopropane	bis(4-methoxybenzyl)cyclopropane	98	[326]
4	bis(4-chlorobenzyl) 1,3-diiodopropane	bis(4-chlorobenzyl)cyclopropane	98	[326]

Table 15 (cont.)

Entry	Substrate	Product	Yield[a] (%)	Ref
5	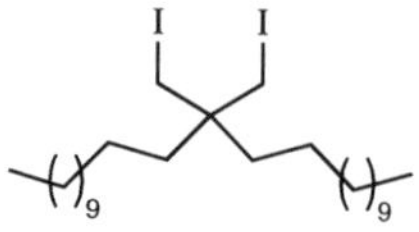 (diiodide with two C9 chains)	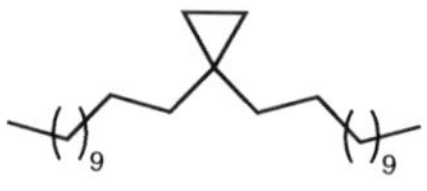 (cyclopropane with two C9 chains)	57 (23)	[326]
6	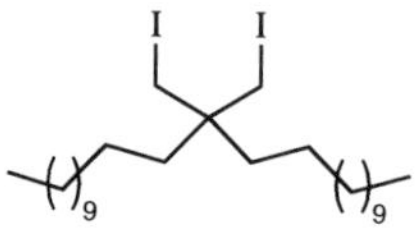 (diiodide with two C9 chains)	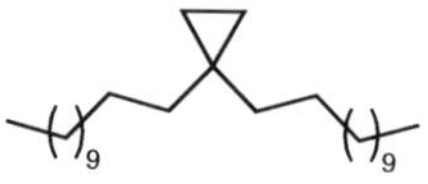 (cyclopropane with two C9 chains)	84[b]	[326]
7	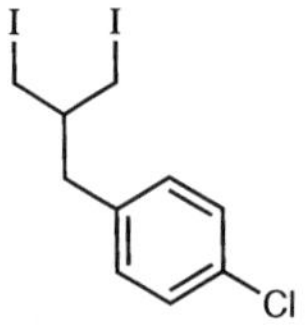 (diiodide with p-Cl-benzyl group)	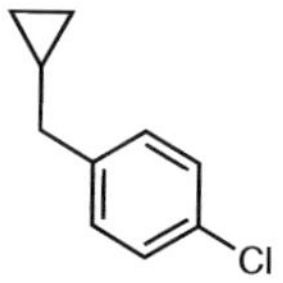 (cyclopropane with p-Cl-benzyl group)	70 (25)	[326]

[a] Yield of recovered starting material is given in parentheses.
[b] Reaction was performed at 70 °C for 3 h with SmI$_2$ (5.5 equiv).

The cyclization of 2,2-disubstituted and 2-monosubstituted 1,3-dibromopropanes **175** can be achieved by treating the substrate with an excess amount of samarium(II) iodide (5.5 equiv) under reflux conditions in tetrahydrofuran for 3 hours to obtain corresponding 1,1-disubstituted and monosubstituted cyclopropanes **176** in good yields (Table 16).

Table 16 Samarium(II) Iodide Mediated Synthesis of Cyclopropanes from 1,3-Dibromopropanes[326]

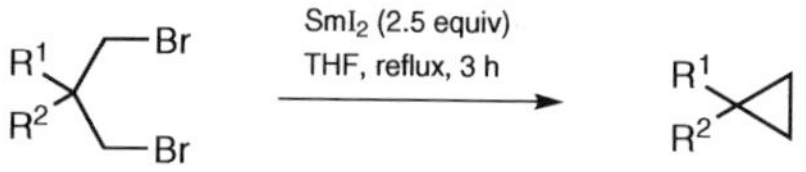

$$R^1R^2C(CH_2Br)_2 \xrightarrow{\text{SmI}_2 \text{ (2.5 equiv)}, \text{THF, reflux, 3 h}} R^1R^2C(\text{cyclopropane})$$

175 → **176**

Entry	Substrate	Product	Yield[a] (%)	Ref
1	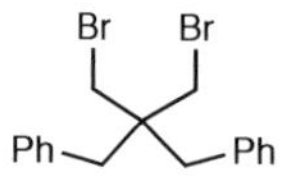 Ph$_2$C(CH$_2$Br)$_2$	 Ph$_2$C(cyclopropane)	5[b] (90)	[326]
2	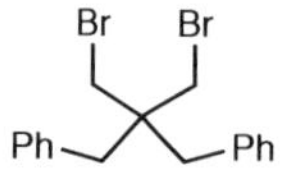 Ph$_2$C(CH$_2$Br)$_2$	 Ph$_2$C(cyclopropane)	87	[326]
3	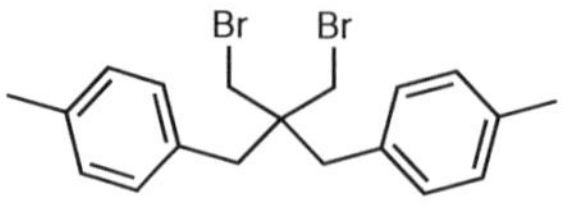 (bis(4-methylbenzyl) dibromide)	 (bis(4-methylbenzyl) cyclopropane)	80	[326]
4	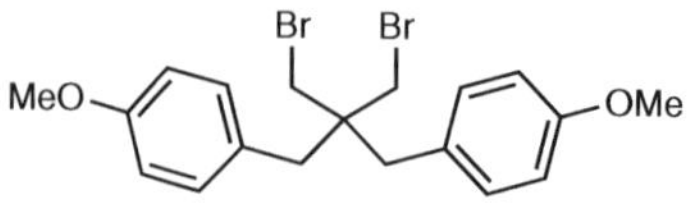 (bis(4-methoxybenzyl) dibromide)	 (bis(4-methoxybenzyl) cyclopropane)	82	[326]

2.9.4 Intramolecular Free-Radical Cyclization Routes to Carbocycles

Table 16 (cont.)

Entry	Substrate	Product	Yield[a] (%)	Ref
5	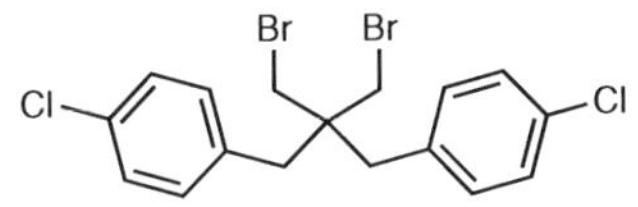 (4-ClC6H4CH2)2C(CH2Br)2 derivative	 cyclopropane with two 4-chlorobenzyl groups	80 (4)	[326]
6	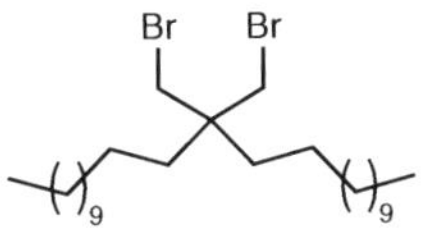 C9H19–C(CH2Br)2–C9H19 derivative	 dialkyl cyclopropane	82	[326]
7	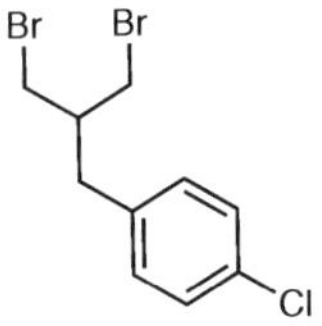 4-ClC6H4CH2CH(CH2Br)2	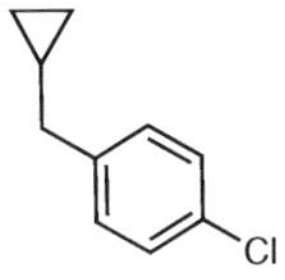 (4-chlorobenzyl)cyclopropane	52 (4)	[326]
8	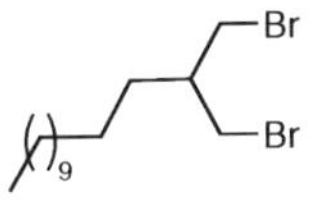 C9H19CH(CH2Br)2 derivative	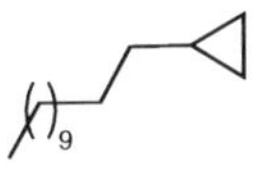 alkylcyclopropane	90	[326]

[a] Yield of recovered starting material is given in parentheses.
[b] Reaction was performed at rt.

1,1′-[Cyclopropane-1,1-diylbis(methylene)]dibenzene (Table 15, Entry 1); Typical Procedure:[326]

1,1′-[2,2-Bis(iodomethyl)propane-1,3-diyl]dibenzene (1 mmol) was added to a soln of SmI$_2$ (2.5 mmol) in THF (5 mL) under an argon atmosphere. The resulting soln was stirred at rt for 3 h. Upon completion of the reaction, the mixture was quenched by the addition of H$_2$O (0.1 mL) and was then extracted with Et$_2$O (3 × 20 mL). The organic extract was dried (Na$_2$SO$_4$), filtered, and concentrated under reduced pressure. The crude residue was purified by preparative TLC or column chromatography (silica gel); yield: 99%.

2.9.4.1.2 Photoredox-Mediated Radical Cyclization to Five- and Six-Membered Rings

Photoredox-catalyzed cyclizations onto alkenyl and alkynyl bromides **177** with unactivated π-systems proceed efficiently to afford five- and six-membered ring systems **178** (Table 17).[327] Usually, tris(2,2′-bipyridyl)dichlororuthenium(II) as the catalyst provides efficient yields (Table 17, entries 1–4); however, an iridium catalyst can be utilized for substrates that are unreactive with the ruthenium catalyst (Table 17, entries 6 and 7).

Table 17 Photoredox-Catalyzed Radical Cyclization of Unactivated π-Systems To Give Five- and Six-Membered Rings[327]

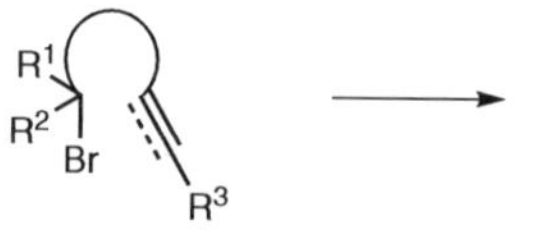

Entry	Substrate	Conditions[a]	Product	Yield (%)	Ref
1	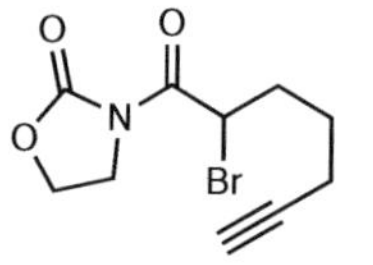 (oxazolidinone-N-CO-CH(Br)-CH$_2$CH$_2$-C≡CH)	[Ru(bipy)$_3$]Cl$_2$ (1.0 mol%), Et$_3$N (2.0 equiv), DMF, blue LED irradiation	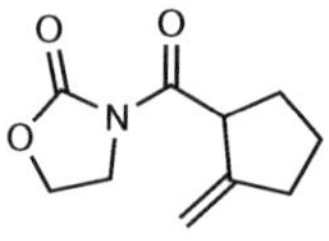 (oxazolidinone-N-CO-cyclopentyl with exo-methylene)	85	[327]
2	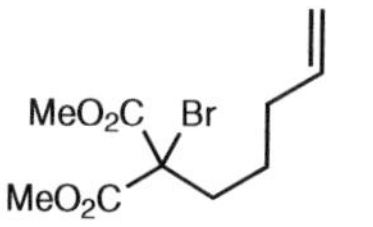 MeO$_2$C, Br, MeO$_2$C substrate with terminal alkene	[Ru(bipy)$_3$]Cl$_2$ (1.0 mol%), Et$_3$N (2.0 equiv), DMF, blue LED irradiation	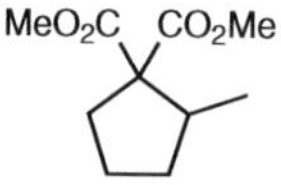 cyclopentane with gem-diester and Me	77	[327]
3	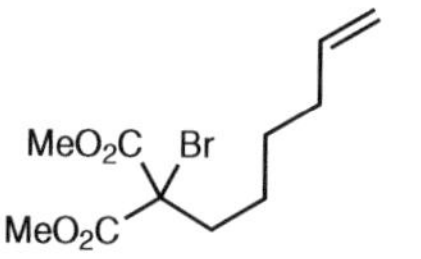 MeO$_2$C, Br, MeO$_2$C substrate with terminal alkene (longer chain)	[Ru(bipy)$_3$]Cl$_2$ (1.0 mol%), Et$_3$N (2.0 equiv), DMF, blue LED irradiation	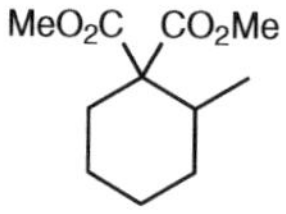 cyclohexane with gem-diester and Me	69	[327]
4	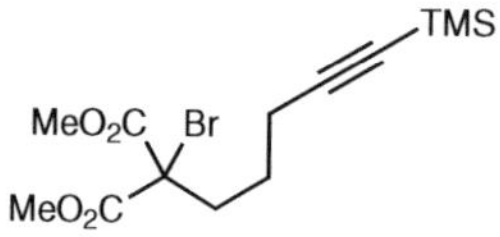 MeO$_2$C, Br, MeO$_2$C with TMS-alkyne	[Ru(bipy)$_3$]Cl$_2$ (1.0 mol%), Et$_3$N (2.0 equiv), DMF, blue LED irradiation	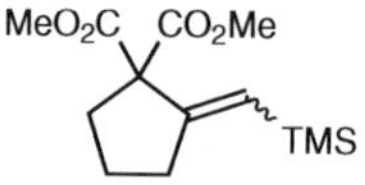 cyclopentane with gem-diester and =CHTMS	100[b]	[327]
5	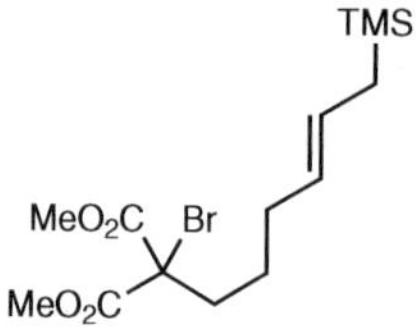 MeO$_2$C, Br, MeO$_2$C with TMS-alkene	[Ru(bipy)$_3$]Cl$_2$ (1.0 mol%), Et$_3$N (2.0 equiv), DMF, blue LED irradiation	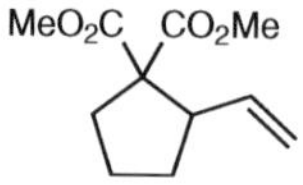 cyclopentane with gem-diester and vinyl	92	[327]
6	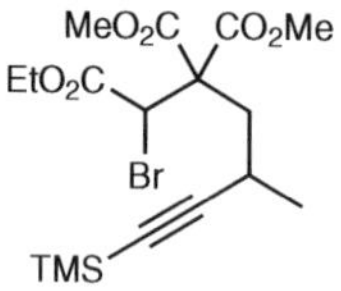 MeO$_2$C, CO$_2$Me, EtO$_2$C, Br, TMS-alkyne substrate	[Ir(ppy)$_2$(dtbbipy)]PF$_6$ (1.0 mol%), Et$_3$N (2.0 equiv), DMF, compact fluorescent lamp irradiation	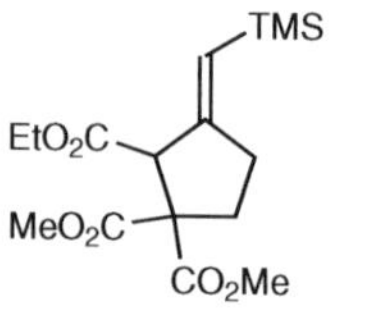 cyclopentane with EtO$_2$C, MeO$_2$C, CO$_2$Me, =CHTMS	74	[327]
7	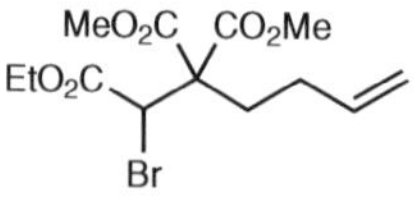 MeO$_2$C, CO$_2$Me, EtO$_2$C, Br, terminal alkene substrate	[Ir(ppy)$_2$(dtbbipy)]PF$_6$ (1.0 mol%), Et$_3$N (2.0 equiv), DMF, compact fluorescent lamp irradiation	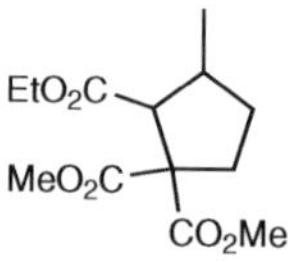 cyclopentane with EtO$_2$C, MeO$_2$C, CO$_2$Me, Me	76[b]	[327]

[a] ppy = 2-(2-pyridyl)phenyl (commonly called 2-phenylpyridyl); dtbbipy = 4,4′-di-*tert*-butyl-2,2′-bipyridyl.
[b] Yield of diastereomeric mixture.

Visible-light irradiation of the metal catalyst generates the excited ruthenium(II)* or iridium(III)* species (Scheme 52). Oxidation of triethylamine produces the active single-electron-transfer agent ruthenium(I) or iridium(II), along with the radical cation of the amine. Reduction of the activated C—Br bond by the electron-rich metal complex affords the reactive free-radical intermediate. Then, intramolecular π-addition of the radical yields the cyclized intermediate. Finally, hydrogen-atom abstraction from the triethylammonium radical cation forms the iminium ion and the cyclized product.

Scheme 52 Plausible Mechanism for Photoredox Radical Cyclization[327]

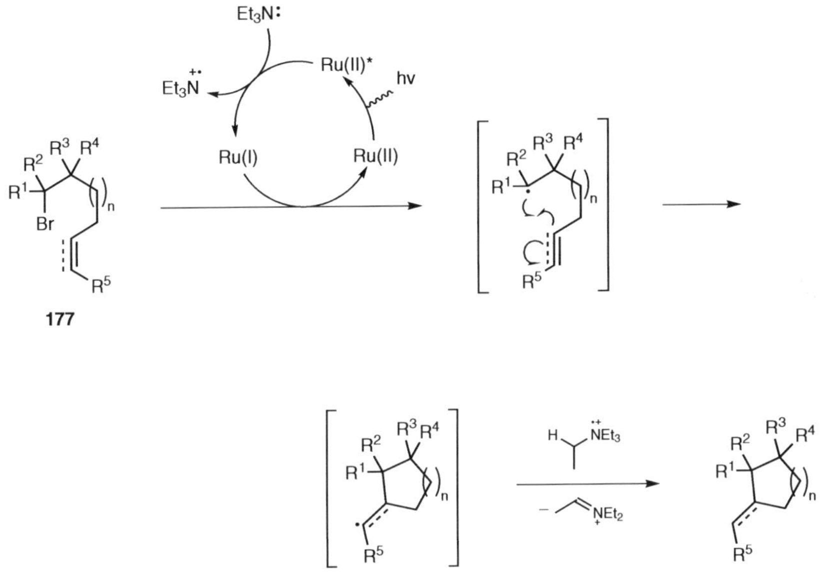

Cycloalkanes 178; General Procedure:[327]
A flame-dried, round-bottomed flask was charged with the photoredox catalyst (1.0 µmol, 0.010 equiv), the halide (0.10 mmol, 1.0 equiv), Et$_3$N (0.20 mmol, 2.0 equiv), and DMF (5.0 mL). The mixture was degassed by freeze–pump–thaw cycles and was stirred in the respective irradiation apparatus at a distance of 15 cm from the reaction vessel. A blue LED strip (λ_{max} 435 nm) was used for the [Ru(bipy)$_3$]Cl$_2$ catalyzed reactions, whereas a 14-W household compact fluorescent lamp was used for the [Ir(ppy)$_2$(dtbbipy)]PF$_6$ catalyzed reactions. Upon completion of the reaction (as monitored by TLC), the mixture was diluted with Et$_2$O (25 mL) and H$_2$O (25 mL). The layers were separated, and the aqueous layer was extracted with Et$_2$O (2 × 50 mL). The combined organic layer was dried (Na$_2$SO$_4$) and concentrated under reduced pressure. The crude residue was purified by column chromatography (silica gel).

2.9.4.1.3   5-exo Cyclization of Unsaturated Epoxides to Cyclic Derivatives

Highly substituted cyclopentane systems such as **179** can be obtained stereoselectively by the catalytic opening of epoxides through a single-electron-transfer pathway in the presence of dichlorobis(η^5-cyclopentadienyl)titanium(IV) and a metal reductant (Scheme 53).[135] The radical intermediate formed by reductive opening of the epoxide is stable toward protic conditions, and thus, the addition of 2,4,6-trimethylpyridine hydrochloride helps to protonate the consumed titanocene(III) intermediate and further to regenerate the active catalyst.

Scheme 53 Titanocene(III)-Catalyzed Reductive Cyclization of Unsaturated Epoxides To Give Cyclopentanes[135]

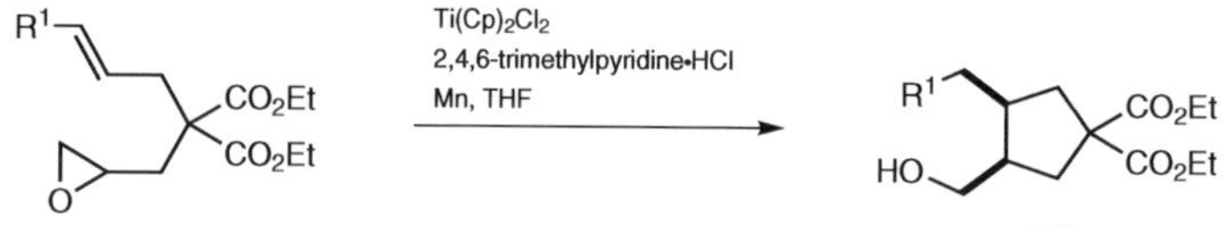

R^1	Time (h)	Ratio (cis/trans)	Yield (%)	Ref
H	30	88:12	78	[135]
Ph	30	86:14	55	[135]

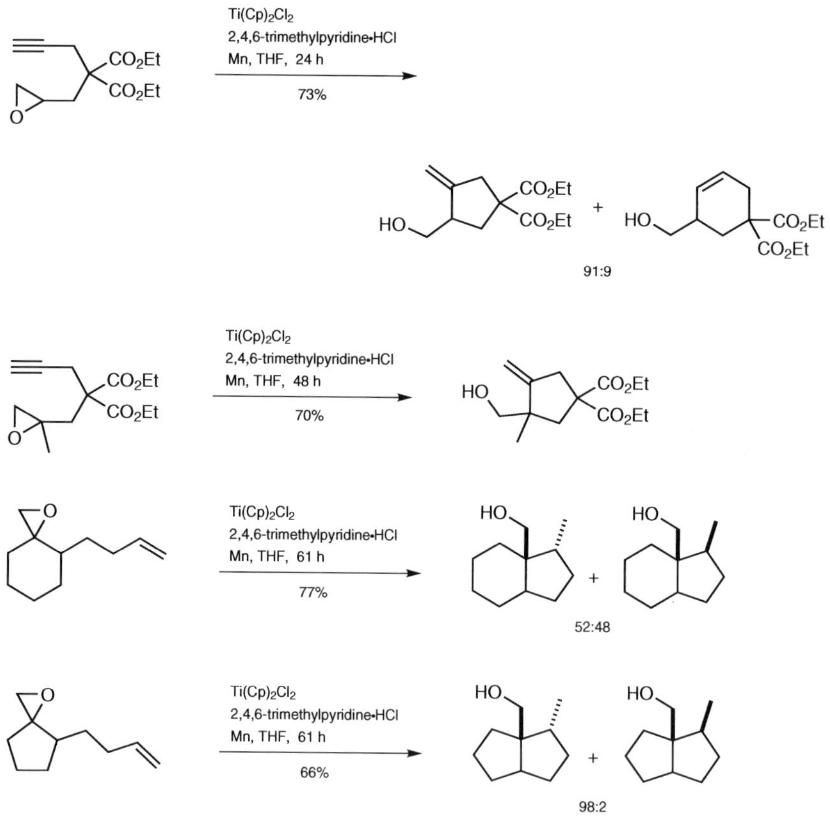

Diethyl (3R*,4S*)-3-(Hydroxymethyl)-4-methylcyclopentane-1,1-dicarboxylate (179, R^1 = H); Typical Procedure:[135]

Ti(Cp)$_2$Cl$_2$ (12.5 mg, 0.05 mmol) was added to a mixture of 2,4,6-trimethylpyridine•HCl (394 mg, 2.50 mmol), diethyl 2-allyl-2-(oxiran-2-ylmethyl)malonate (256 mg, 1.0 mmol), and Mn (82 mg, 1.50 mmol) in THF (10 mL), and the resulting mixture was stirred for 30 h. After the addition of t-BuOMe (50 mL), the mixture was washed with H$_2$O (30 mL), 2 M HCl (30 mL), H$_2$O (30 mL), sat. aq NaHCO$_3$ (30 mL), and H$_2$O (30 mL). The mixture was then dried (MgSO$_4$) and concentrated under reduced pressure. The crude residue was purified by flash chromatography (silica gel, t-BuOMe/petroleum ether 1:3 to 1:1); yield: 202 mg (78%); ratio (cis/trans) 88:12.

2.9.4.1.4 Photocatalyzed (3+2) Cycloadditions of Unsaturated Aryl Cyclopropyl Ketones To Give Cyclopentanes

Highly substituted cyclopentane ring systems can be obtained by the formal (3+2)-cycloaddition reaction of aryl cyclopropyl ketones with alkenes. Irradiation of aryl cyclopropyl ketones **180** bearing a double bond with a standard consumer 23-W compact fluorescent lamp in the presence of tris(2,2′-bipyridyl)dichlororuthenium(II), lanthanum(III) trifluoromethanesulfonate, N,N,N′,N′-tetramethylethylenediamine, and magnesium sulfate in acetonitrile at room temperature generates a ketyl radical; after electron migration, the radical undergoes intramolecular (2+3) cycloaddition to afford cyclopentanes **181** (Scheme 54).[272] Under these conditions, a variety of enone moieties, including esters, ketones, and thioesters, can participate efficiently in the (3+2) cycloaddition. Notably, the presence of a substituent increases the efficiency and stereoselectivity of the product formation; reactions of alkyl-substituted thioesters are predominantly diastereoselective. Aliphatic cyclopropyl ketones do not produce the desired products under the reaction conditions. On the other hand, a variety of substituted aryl ketones possessing three-carbon-atom, four-carbon-atom, and heteroatom linkers and aryl and aliphatic alkynes are tolerated. Electron-donating substituents decrease the rate of the reaction.

Scheme 54 Photocatalyzed Intramolecular (3+2) Cycloadditions of Unsaturated Aryl Cyclopropyl Ketones To Give Cyclopentanes[272]

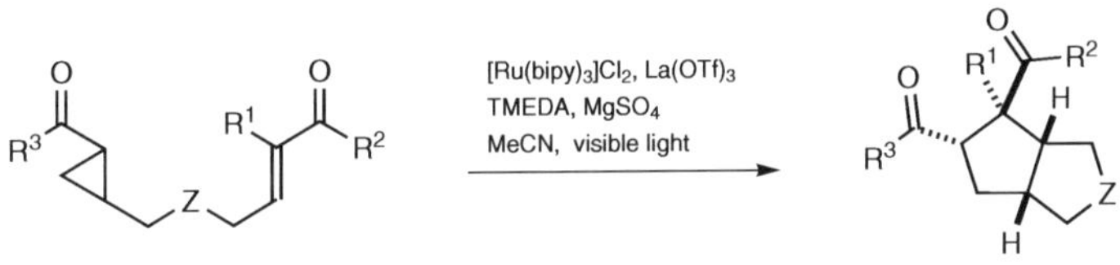

Z	R^1	R^2	R^3	Time (h)	dr^a	Yield^b (%)	Ref
CH$_2$	Me	OEt	Ph	6.5	6:1	83	[272]
CH$_2$	H	OEt	Ph	5.5	2:1	67	[272]
CH$_2$	Me	O*t*-Bu	Ph	6.5	5:1	86	[272]
CH$_2$	H	O*t*-Bu	Ph	12	2:1	84	[272]
CH$_2$	Me	*t*-Bu	Ph	5	4:1	82	[272]
CH$_2$	Me	SEt	Ph	17	>10:1	79	[272]
CH$_2$	Et	SEt	Ph	16	10:1	70	[272]
CH$_2$	Me	SEt	4-MeOC$_6$H$_4$	48	>10:1	55	[272]
CH$_2$	Me	SEt	4-ClC$_6$H$_4$	12	>10:1	73	[272]
CMe$_2$	Me	SEt	Ph	19	>10:1	82	[272]
(CH$_2$)$_2$	Me	SEt	Ph	34	1:1	76	[272]

^{a} Determined by GC.
^{b} Yield of isolated product, average of two reproducible experiments.

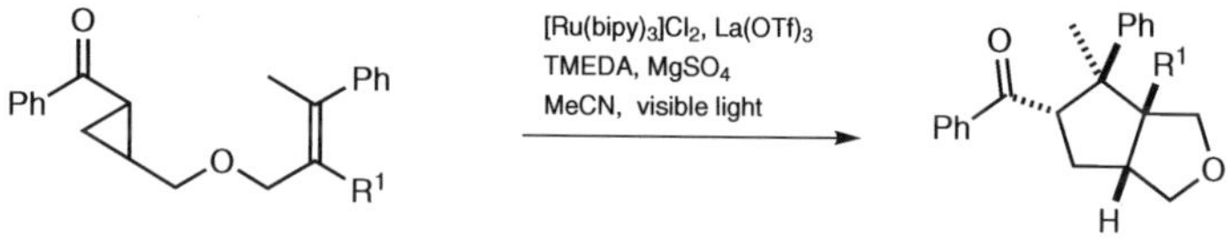

R^1	Time (h)	dr[a]	Yield[b] (%)	Ref
H	29	>10:1	63	[272]
Me	48	3:1	57	[272]

[a] Determined by GC.
[b] Yield of isolated product, average of two reproducible experiments.

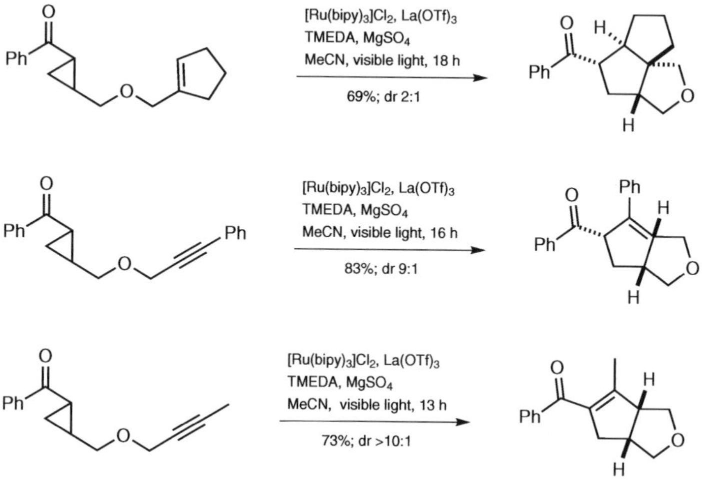

Fused Cyclopentanes 181; General Procedure:[272]
A 25-mL, flame-dried Schlenk flask was charged with $MgSO_4$ (2 w/w with respect to substrate) under dry N_2. [Ru(bipy)$_3$]Cl$_2$·6H$_2$O (2.5 mol%), La(OTf)$_3$ (1 equiv), substrate **180** (1 equiv), MeCN (0.05 M), and TMEDA (5 equiv) were then added. The suspension was degassed by freeze–pump–thaw cycles (3 ×) under N_2. The flask was placed in a water bath, and the mixture was stirred in front of a 23-W (1380 lumen) compact fluorescent lamp at a distance of 30 cm. Upon completion of the reaction, the resulting mixture was diluted with Et$_2$O and passed through a short pad of silica gel (Et$_2$O). The filtrate was concentrated under reduced pressure, and the crude residue was purified by column chromatography (silica gel).

2.9.4.2 Synthesis of Cycloalkanol Derivatives

2.9.4.2.1 Synthesis of *anti*-Cyclopropanol Derivatives

In the presence of samarium(II) iodide and *tert*-butyl alcohol as a proton source in tetrahydrofuran, γ,γ-disubstituted α,β-unsaturated δ-keto esters **182** undergo cyclization with complete diastereocontrol to give *anti*-cyclopropanol products **183** exclusively in good yields under mild conditions (Scheme 55).[328]

Scheme 55 Synthesis of *anti*-Cyclopropanol Derivatives[328]

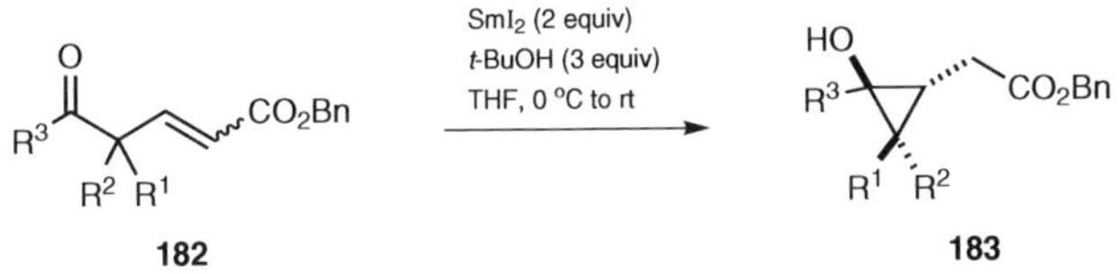

Config of 182	R^1	R^2	R^3	Yield (%)	Ref
E	Me	Me	H	87	[328]
E	Me	Me	H	90[a]	[328]
Z	Me	Me	H	60	[328]
E	(CH$_2$)$_5$		H	60	[328]
E	Me	Me	Me	81	[328]

[a] The reaction was performed in the presence of HMPA (8 equiv.)

anti-**Cyclopropanol Derivatives 183; General Procedure:**[328]
A soln of *t*-BuOH (1.3 mmol) and carbonyl compound **182** (0.43 mmol) in THF (1 mL) was added dropwise at 0 °C to a dry Schlenk tube containing a 0.1 M soln of SmI$_2$ in THF (8.6 mL, 0.86 mmol) under argon. Once the addition was complete, the suspension was stirred overnight at rt and then quenched by the addition of H$_2$O. The aqueous layer was extracted with Et$_2$O, and the combined organic layer was washed with sat. Na$_2$S$_2$O$_3$, dried (MgSO$_4$), and concentrated under reduced pressure. The crude residue was purified by column chromatography (silica gel, cyclohexane/EtOAc 4:1) to give a colorless oil.

2.9.4.2.2 4-*exo-trig* Cyclizations of Unsaturated Aldehydes to Functionalized Cyclobutanols

Highly functionalized *anti*-cyclobutanols **185** can be obtained efficiently and selectively by the samarium(II) iodide mediated 4-*exo-trig* cyclization of γ,δ-unsaturated aldehydes **184** bearing a fully substituted center in either the α- or β-position.[329] The reaction involves the addition of the aldehyde substrate to a solution of samarium(II) iodide (inverse addition) in tetrahydrofuran in the presence of an excess amount of methanol (THF/MeOH ratio 1:4) as a co-solvent. Methanol plays two roles, as it is an efficient proton source and promotes the cyclization reaction by increasing the reduction potential of samarium(II) iodide (Table 18).

Table 18 4-*exo-trig* Cyclization of Unsaturated Aldehydes[329]

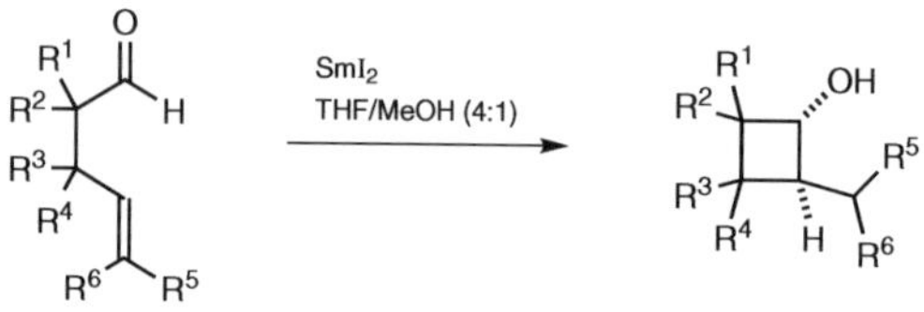

Entry	Substrate	Product	Yield (%)	Ref
1	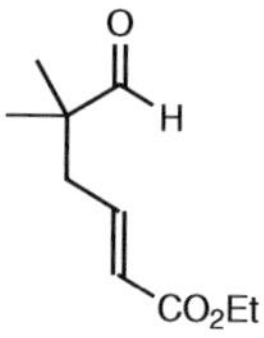	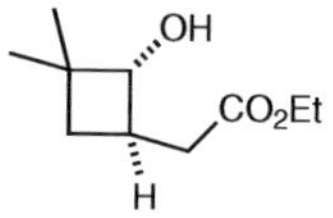	65	[329]
2	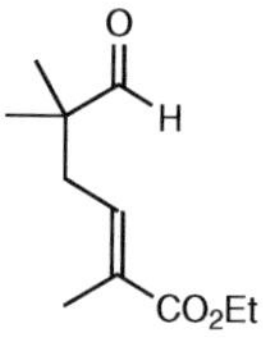	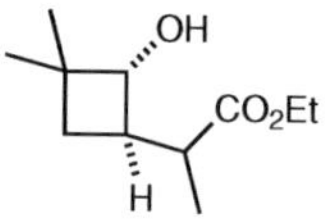	66[a]	[329]
3	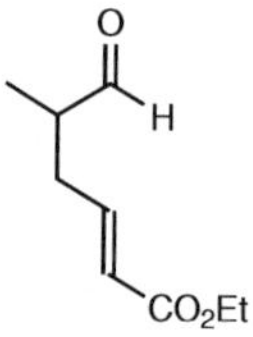	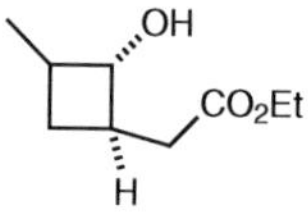	0[b]	[329]
4	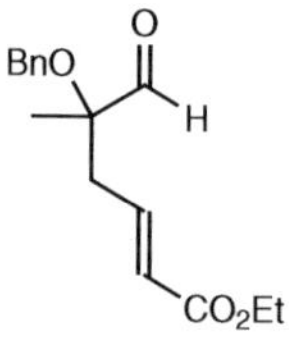	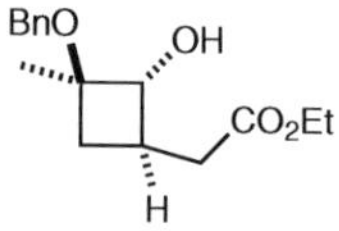	57[a,c]	[329]
5	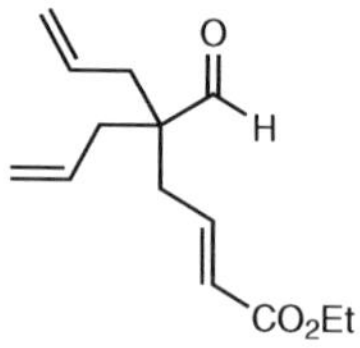	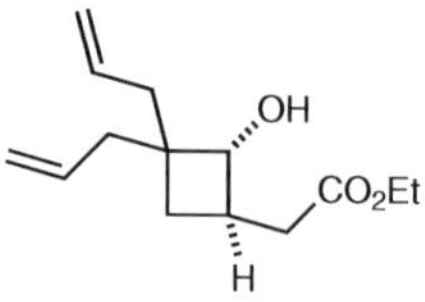	80	[329]
6	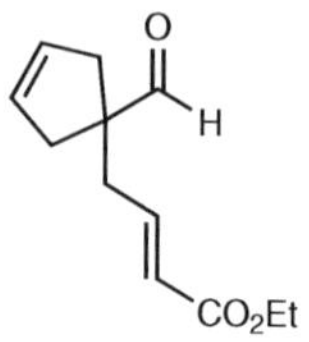	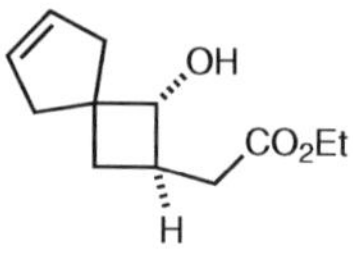	62	[329]

2.9.4 Intramolecular Free-Radical Cyclization Routes to Carbocycles

Table 18 (cont.)

Entry	Substrate	Product	Yield (%)	Ref
7	(2,2-dimethyl-5-(phenylsulfonyl)pent-4-enal)	2-hydroxy-cyclobutyl with SO₂Ph	21[d]	[329]
8	BnO-substituted aldehyde with CO₂Et alkene	BnO-cyclobutyl with OH and CO₂Et	70	[329]
9	BnO-substituted aldehyde with SO₂Ph alkene	BnO-cyclobutyl with OH and SO₂Ph	60[e]	[329]

[a] A 4:1 mixture of diastereomers was formed.
[b] Ethyl 6-hydroxy-5-methylhexanoate was the major product (31%).
[c] HMPA was added.
[d] (E)-2,2-Dimethyl-5-(phenylsulfonyl)pent-4-enol was the major product (35%).
[e] (E)-3,3-Dimethyl-5-(phenylsulfonyl)pent-4-enal (26%) was also isolated.

Ethyl [(1R*,2R*)-2-Hydroxy-3,3-dimethylcyclobutyl]acetate (Table 18, Entry 1); Typical Procedure:[329]

A soln of ethyl (2E)-5,5-dimethyl-6-oxohex-2-enoate (25.1 mg, 0.14 mmol, 1 equiv) in THF (0.5 mL) was added to a 0.1 M soln of SmI₂ in THF (2.80 mL, 0.28 mmol, 2 equiv) and MeOH (0.83 mL) at 0 °C. The mixture was stirred at 0 °C for 5 min. Then, a soln of sat. aq NaCl (1 mL) and citric acid (58.8 mg, 0.28 mmol, 2 equiv) was added, and the mixture was warmed to rt. The aqueous layer was separated and extracted with EtOAc (3 × 4 mL). The combined organic layer was dried (Na₂SO₄) and concentrated under reduced pressure. The crude residue was purified by column chromatography (EtOAc/hexane 3:7) to afford a colorless oil; yield: 17.0 mg (65%).

2.9.4.2.3 Enantioselective Reductive Cyclization of Ketonitriles to Cycloalkanol Derivatives: Synthesis of Cyclic α-Hydroxy Ketones

Numerous methods involving the use of stoichiometric amounts of reagents are reported for the successful cyclization of ketonitriles and related reactions.[330–336] However, enantioselective options for such radical cyclizations are limited and usually require stoichiometric quantities of chiral additives.[337–339] Streuff and co-workers have developed a highly efficient catalytic system involving chiral titanium catalyst **186** that facilitates the desired reductive ketonitrile cyclization to afford highly enantioenriched 2-hydroxycyclopentanones under mild reaction conditions.[340] In the presence of Brintzinger's commercially available *ansa*-bis(η⁵-cyclopentadienyl)titanium catalyst **186**, zinc dust, chlorotrimethylsilane, and triethylamine hydrochloride, ω-ketonitriles **187** undergo cyclization to yield 2-hydroxycyclopentanones **188** with high enantioselectivity (up to 94%) in tetrahydrofuran at room temperature (Table 19). The enantioselectivity of the reaction is strongly dependent on the configuration of the catalyst, as the (R,R)-catalyst produces only the (S)-configured product and vice versa. The high yield and enantioselectivity of the reaction are also controlled by the triethylamine hydrochloride additive. Moreover, the reaction can be scaled between 0.2 and 3.0 mmol with consistent yields and enantiomeric ratios. Another advantage is that the products can be easily isolated by simple acid–base extraction without chromatographic purification. However, in some cases the product (Table 19, entries 11 and 14) does require chromatographic purification. Six-membered cyclohexanone (Table 19, entry 13) and piperidinone (Table 19, entry 14) derivatives are synthesized in moderate yields, sometimes with slightly lower enantioselectivities.

Table 19 Enantioselective Synthesis of 2-Hydroxycycloalkanones from Ketonitriles[340]

Entry	Substrate	Product	Time (h)	ee (%)	Yield (%)	Ref
1			24	91	88[a,b]	[340]
2			24	86	89	[340]
3			24	88	98	[340]
4			24	93	8	[340]

2.9.4 Intramolecular Free-Radical Cyclization Routes to Carbocycles

Table 19 (cont.)

Entry	Substrate	Product	Time (h)	ee (%)	Yield (%)	Ref
5	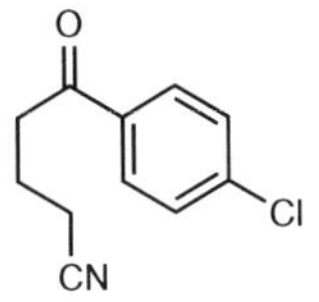	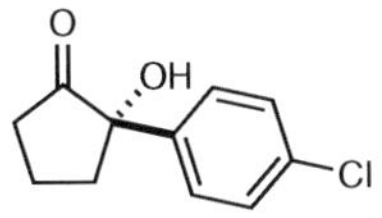	24	86	94	[340]
6	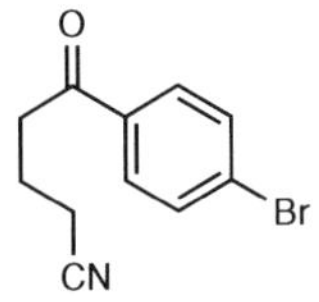	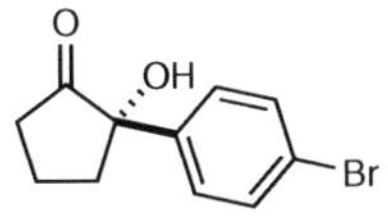	24	85	99	[340]
7	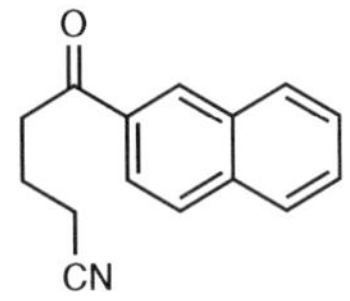	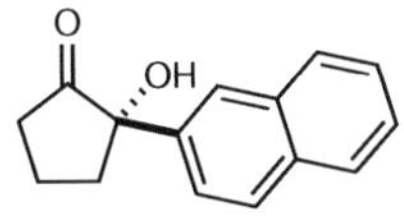	24	85	94	[340]
8	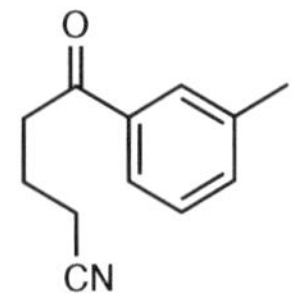	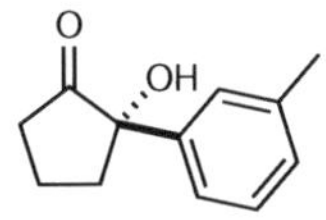	24	90	82	[340]
9	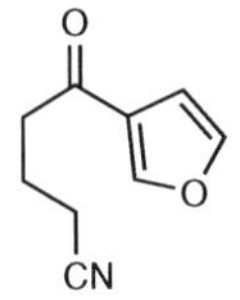	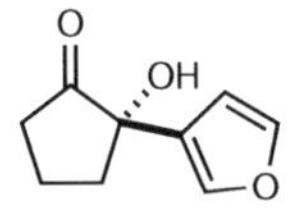	24	60	72	[340]
10	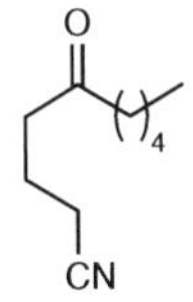	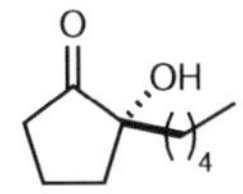	48	94	72	[340]
11	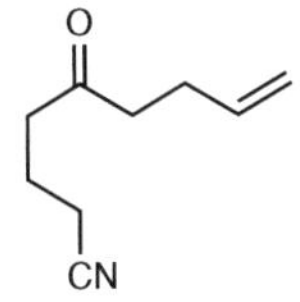	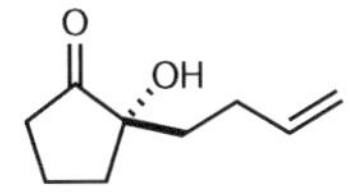	48	94	78[c]	[340]
12	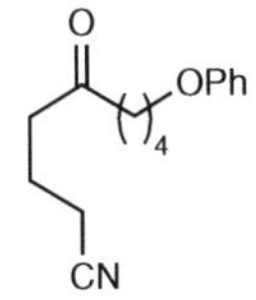	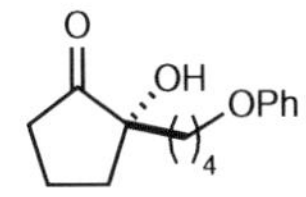	96	86	90	[340]

for references see p 533

Table 19 (cont.)

Entry	Substrate	Product	Time (h)	ee (%)	Yield (%)	Ref
13	5-oxo-5-phenylpentanenitrile (O=C(Ph)CH₂CH₂CH₂CN)	2-hydroxy-2-phenylcyclohexan-1-one	72	78	42	[340]
14	Ts-N(CH₂CN)CH₂C(O)Ph	N-Ts 3-hydroxy-3-phenylpiperidin-2-one	48	98	47[d]	[340]
15	2-(2-cyanoethyl)phenyl phenyl ketone	1-hydroxy-1-phenyl-1,2,3,4-tetrahydronaphthalen-2-one	24	82	55[e]	[340]

[a] Yield average of three experiments.
[b] Using the opposite enantiomer of catalyst **186** the opposite enantiomer of the product was obtained in identical yield and ee.
[c] Product purified by chromatography.
[d] TBAF was used in the workup instead of ice-cold 1 M aq HCl; product was purified by chromatography and crystallization (iPrOH).
[e] The corresponding imine required hydrolysis with HCl in THF/H$_2$O (55% overall yield). Zn dust (2 × 1.0 equiv) was added at the beginning of the reaction and after 7.5 h.

(2S)-2-Hydroxy-2-phenylcyclopentan-1-one (Table 19, Entry 1); Typical Procedure:[340]

A 10-mL Schlenk tube containing a magnetic needle and closed with a rubber septum was evacuated, heated with a heat gun for 1 min, and backfilled with argon. Et$_3$N•HCl (55 mg, 0.4 mmol, 2.0 equiv) was added to the tube, followed by Zn dust (26.2 mg, 0.4 mmol, 2.0 equiv) and catalyst **186** (7.7 mg, 0.02 mmol, 10 mol%). Under continuous stirring, the tube was evacuated again for 1 min and backfilled with argon. Degassed, dry THF (0.5 mL, tolerated H$_2$O content: 10–300 ppm) was added to the mixture, which was stirred until the color changed (ca. 1–2 min) from red to green (varied according to H$_2$O content). 5-Oxo-5-phenylpentanenitrile (34.6 mg, 0.2 mmol) was quickly added, followed by TMSCl (76 µL, 0.6 mmol, 3.0 equiv). The rubber septum was replaced by a greased glass stopper and a metal clip, and the mixture was stirred for 24 h at rt (23 °C). Upon completion of the reaction, CH$_2$Cl$_2$ (1 mL) was added, and the mixture was transferred to a separatory funnel filled with ice-cold 1 M aq HCl (15 mL) and ice-cold Et$_2$O (40 mL). The funnel was briefly shaken, and the separated aqueous layer was quickly collected. The aqueous extraction was repeated (6 ×). The collected aqueous layers were combined and stirred until deprotection of the TMS group was complete (about 5–10 min, monitored by TLC); then, the mixture was carefully neutralized to pH 8 (sat. aq NaHCO$_3$) and extracted with CH$_2$Cl$_2$ (3 × 10 mL). The CH$_2$Cl$_2$ layer was dried (Na$_2$SO$_4$), filtered, and concentrated; yield: 88%; 91% ee.

2.9.4.2.4 Reductive Annulations of Ketones Bearing a Distal Vinyl Epoxide Moiety: Synthesis of Allyl Alcohols

The efficient intramolecular coupling of ketones such as **189**, **191**, **193**, **195**, and **197** bearing distal epoxy alkenes can be performed in the presence of samarium(II) iodide, hexamethylphosphoric triamide, and tetrahydrofuran. The reaction proceeds through the coupling of a ketyl radical with the unsaturated epoxide. Further fragmentation of the epoxides yields carbocycles **190**, **192**, **194**, **196**, and **198** with an allyl alcohol side

2.9.4 Intramolecular Free-Radical Cyclization Routes to Carbocycles

chain in good yields with high diastereoselectivities (Scheme 56).[341] Tetramethylguanidine can also be used as an additive instead of hexamethylphosphoric triamide to obtain better yields; however, a reduction in the diastereoselectivity is observed in this case. The oxophilicity of the samarium ions allows the addition of chelating hexamethylphosphoric triamide,[342–344] which increases the steric bulk around the ketyl oxygen atom, and thus, cyclizations under these conditions provide higher diastereoselectivities.

Scheme 56 Ketyl Coupling Reactions of Vinyl Epoxides To Give Allyl Alcohols[341]

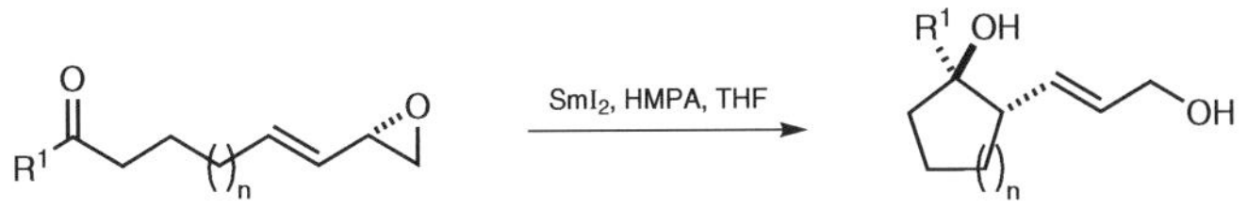

n	R^1	dr	Yield (%)	Ref
1	Me	>200:1	57	[341]
1	iBu	3:1	48	[341]

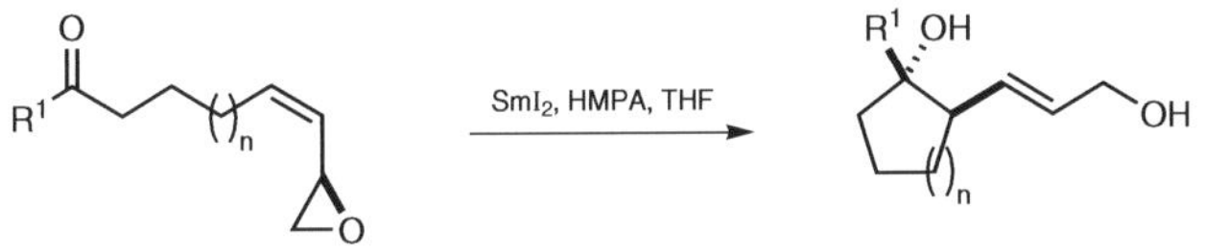

n	R^1	dr	Yield (%)	Ref
1	Me	>200:1	72	[341]
1	iBu	1:1	71[a]	[341]
2	Me	>200:1	17	[341]

[a] Tetramethylguanidine (19 equiv) was used.

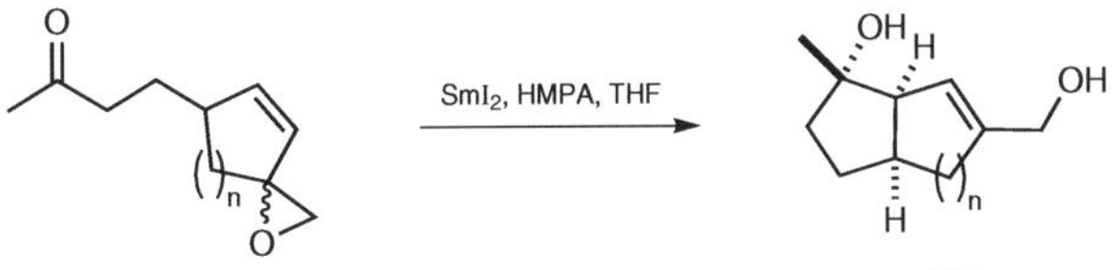

n	dr	Yield (%)	Ref
1	>200:1	72	[341]
2	>200:1	78	[341]

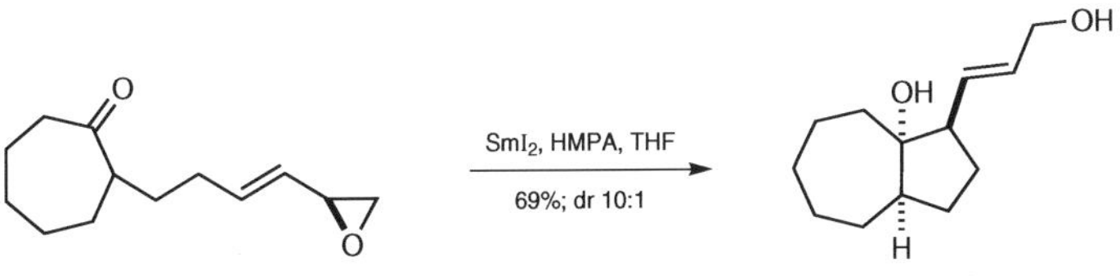

for references see p 533

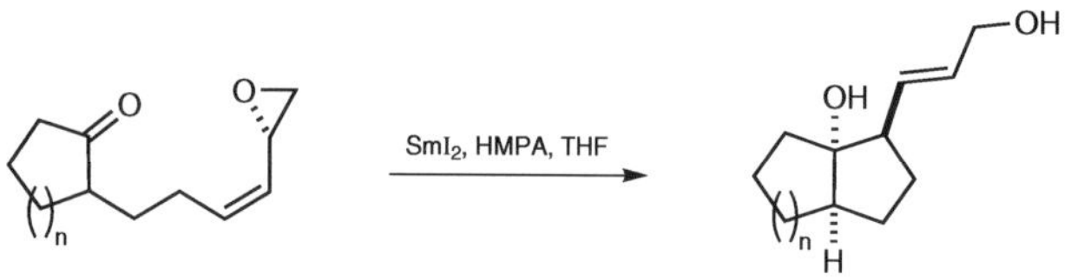

n	dr	Yield (%)	Ref
1	>200:1	78	[341]
2	>200:1	76	[341]
4	1.5:1.2:1	83[a]	[341]

[a] Tetramethylguanidine (19 equiv) was used.

(1R,2S)-2-[(1E)-3-Hydroxyprop-1-en-1-yl]-1-methylcyclopentan-1-ol (190, n = 1; R¹ = Me); Typical Procedure:[341]

> **CAUTION:** *Hexamethylphosphoric triamide is a possible human carcinogen and an eye and skin irritant.*

A soln of (6E)-7-[(2S)-oxiran-2-yl]hept-6-en-2-one (0.140 g, 0.909 mmol) in THF (40 mL) was added dropwise over 1 h using a syringe pump to a soln of freshly prepared SmI$_2$ [Sm (0.375 g, 2.50 mmol) and CH$_2$I$_2$ (0.535 g, 1.999 mmol)] in THF (20 mL) and HMPA (2.812 g, 17.26 mmol). After stirring for 15–30 min at rt, the mixture was quenched with sat. K$_2$CO$_3$ (5 mL). The resulting mixture was extracted repeatedly with EtOAc. The organic layer was dried (MgSO$_4$) and concentrated under reduced pressure. The crude residue was purified by flash column chromatography (silica gel, hexanes/EtOAc 1:8); yield: 0.081 g (57%); R_f 0.2 (hexanes/EtOAc 1:8).

2.9.4.2.5 Reductive Cyclization of Unactivated Alkenes: Synthesis of Five- and Six-Membered Cycloalkanols

The intramolecular coupling of ketones **199** possessing an unactivated alkene moiety proceeds through a reductive ketyl–alkene radical-cyclization mechanism in the presence of samarium(II) iodide and hexamethylphosphoric triamide. The reaction is quite efficient and general for the synthesis of five- and six-membered cycloalkanols **200**, which are delivered in good yields with high diastereoselectivities.[345] The reducing ability of this system is due to ligand field effects of the strong hexamethylphosphoric triamide donor ligand (Table 20).

2.9.4 Intramolecular Free-Radical Cyclization Routes to Carbocycles

Table 20 Samarium(II)-Mediated Cyclization of Alkenyl Ketones To Give Cycloalkanols[345]

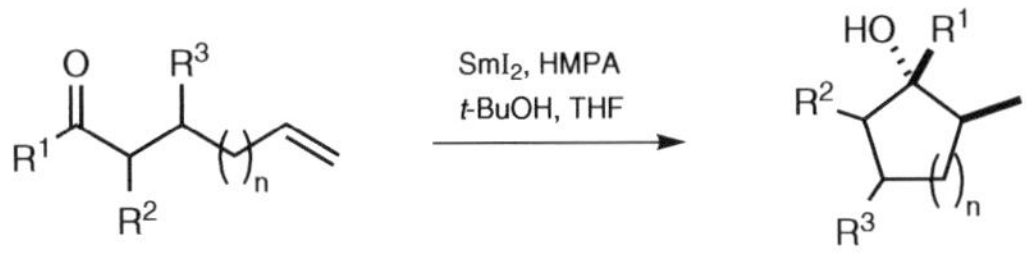

Entry	Substrate	Product	dr	Yield (%)	Ref
1	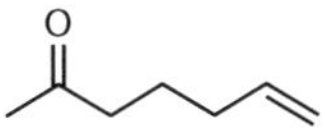	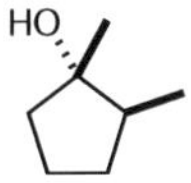	>150:1	86	[345]
2	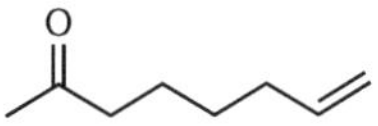	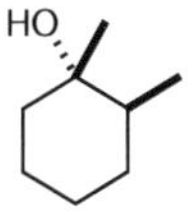	36:1	91	[345]
3	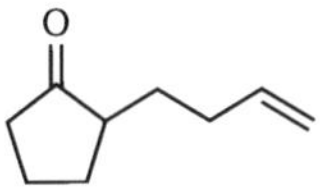	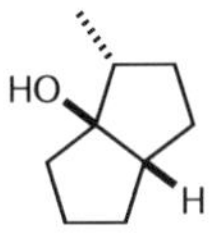	>150:1	90	[345]
4	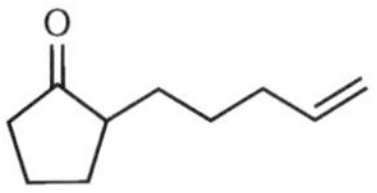	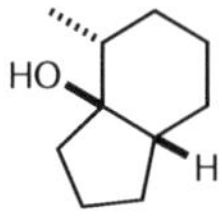	93:5.2	92	[345]
5	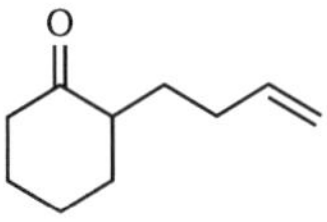	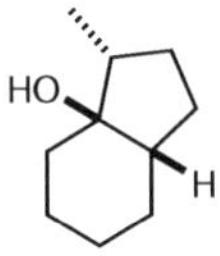	>150:1	89	[345]
6	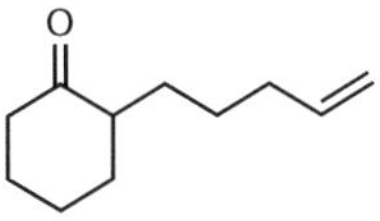	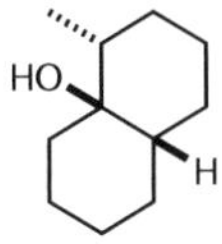	2:1:1	85	[345]
7	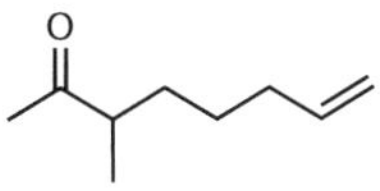	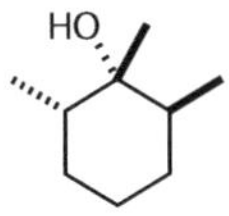	6:1	89	[345]

Table 20 (cont.)

Entry	Substrate	Product	dr	Yield (%)	Ref
8	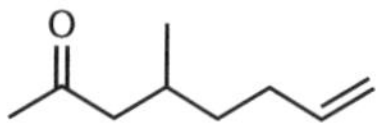	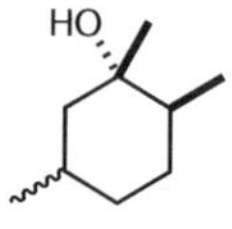	4:1	86	[345]
9	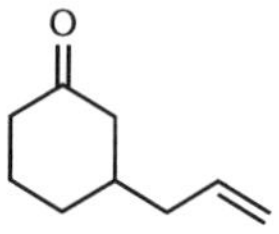	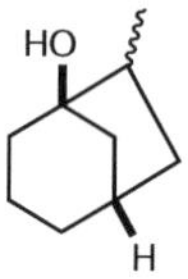	17:1[a]	88	[345]
10	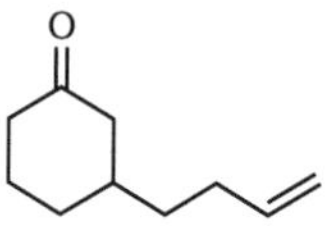	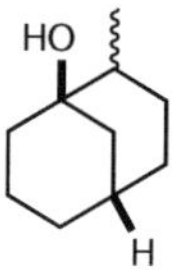	17:1[b]	66	[345]

[a] The major diastereomer corresponds to the isomer with an *endo*-methyl group.
[b] Major *endo*-isomer.

Five- and Six-Membered Cycloalkanols 200; General Procedure:[345]

CAUTION: *Hexamethylphosphoric triamide is a possible human carcinogen and an eye and skin irritant.*

A soln of alkenyl ketone **199** (0.72 mmol) and *t*-BuOH (0.160 g, 2.16 mmol) in THF (14 mL) was added to a SmI$_2$/HMPA soln [prepared from Sm (2.0 mmol) and CH$_2$I$_2$ (1.8 mmol) in THF (13 mL), followed by addition of HMPA (14.5 mmol)] over a 15-min period. After the addition was complete, the reaction was quenched with sat. aq NaHCO$_3$, and the aqueous layer was extracted with Et$_2$O. The combined organic layer was washed with H$_2$O and brine and then dried (MgSO$_4$). The extract was concentrated under reduced pressure. Residual hexamethylphosphoric triamide was removed by filtering through a short column of Florisil, and the crude material was purified by Kugelrohr distillation or flash chromatography.

2.9.4.2.6 Small-Ring 3-*exo* and 4-*exo* Cyclizations

Small-ring cyclopropane (Table 21, entries 1–8) and cyclobutane (Table 21, entries 9–10) derivatives **202** can be obtained through 3-*exo* and 4-*exo* cyclizations by nucleophilic intramolecular radical cyclization of epoxides **201**. This highly efficient method requires epoxides as radical precursors and a catalytic amount of a titanocene(III) complex as the electron-transfer catalyst.[346] Enone functional groups are highly sensitive and are rapidly reduced by other popular electron-transfer reagents such as samarium(II) iodide,[65,66,172,196,347,348] but the titanium catalyst allows mild reaction conditions for the generation of radicals from epoxides. Thus, good to excellent yields are obtained with substrates bearing an enone moiety with excellent diastereoselectivities during the cyclizations.

Table 21 Cyclopropanes and Cyclobutanes by Titanocene(III)-Mediated Intramolecular Cyclization of Epoxides[346]

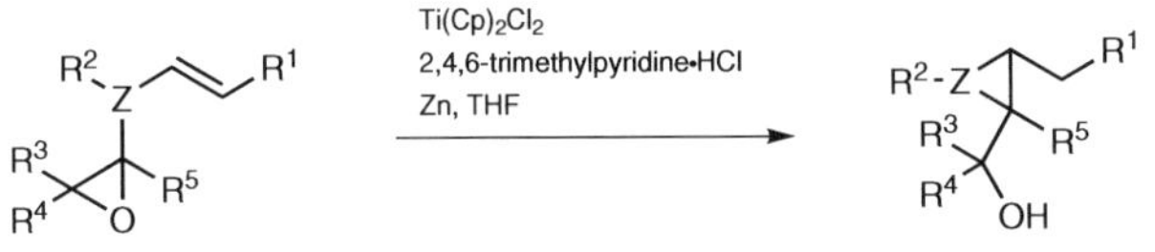

Entry	Substrate	Time (h)	Product	dr	Yield (%)	Ref
1	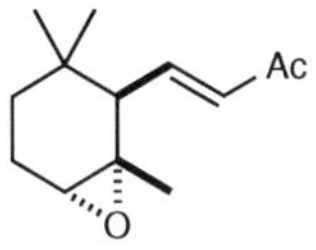	16	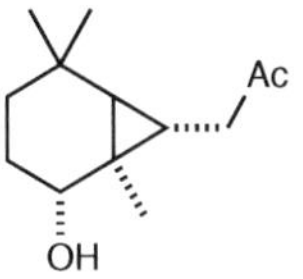	–	72	[346]
2	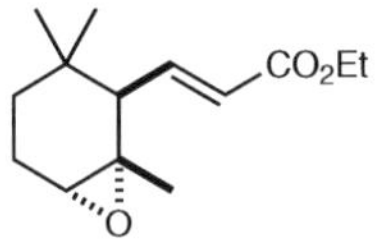	16	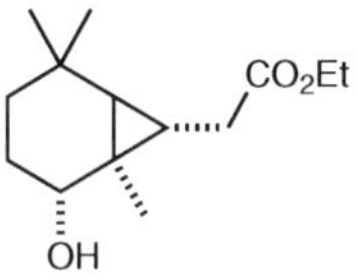	–	92	[346]
3	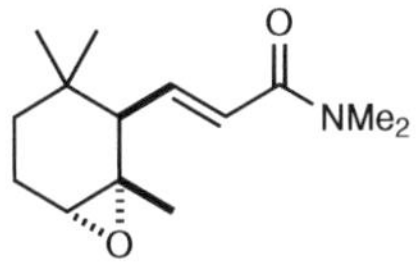	16	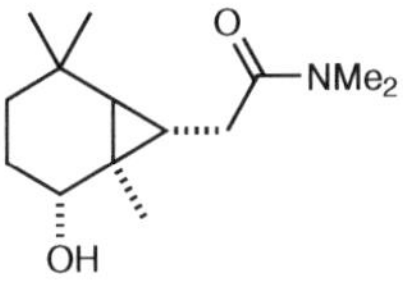	–	96	[346]
4	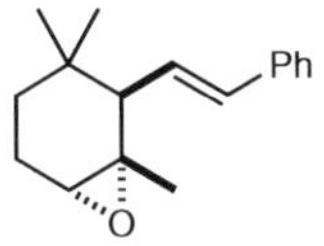	16	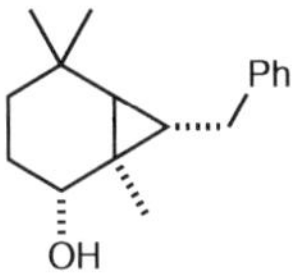	–	<10	[346]
5	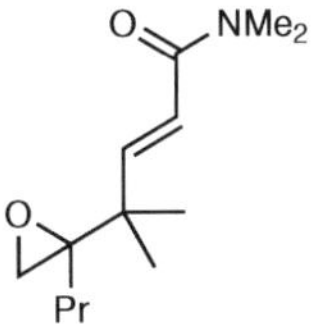	72	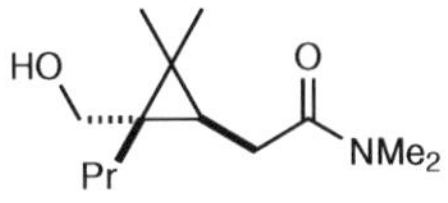	63:37	95	[346]
6	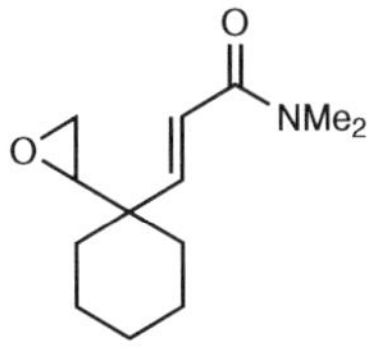	16	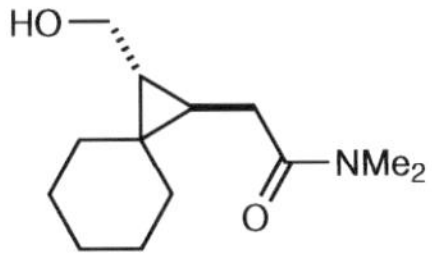	81:19	88	[346]
7	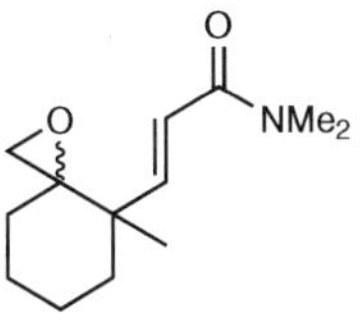	16	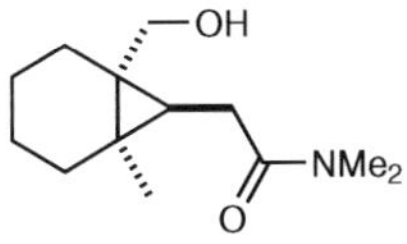	75:25	72	[346]

Table 21 (cont.)

Entry	Substrate	Time (h)	Product	dr	Yield (%)	Ref
8	(cycloheptane epoxide with CH=CH-C(O)-NMe₂ substituent)	16	(cycloheptane with OH and CH₂C(O)NMe₂ on cyclopropane)	75:25	84	[346]
9	(epoxide with CH₂CH=CH-CO₂Bu^t)	16	(cyclobutane with OH and CO₂Bu^t)	67:33	94[a]	[346]
10	(epoxide with CH₂CH=CH-C(O)NMe₂)	72	(cyclobutane with OH and CH₂C(O)NMe₂)	73:27	94[a]	[346]

[a] 4-*exo* cyclization.

Three- and Four-Membered Cycloalkanes 202; General Procedure:[346]
A suspension of Ti(Cp)₂Cl₂ (24.9 mg, 100 μmol), 2,4,6-trimethylpyridine•HCl (394 mg, 2.50 mmol), and Zn dust (131 mg, 2.00 mmol) in dry THF (10 mL) was heated under reduced pressure until 2,4,6-trimethylpyridine•HCl began to sublime slightly. The epoxide (1.00 mmol) was added to the resultant suspension, and the mixture was stirred at rt for the indicated time. Upon completion of the reaction, Et₂O (20 mL) was added to the mixture, which was washed with 2 M HCl [2 × 10 mL; for amides the mixture was washed with H₂O (2 × 10 mL)]. The combined aqueous layer was extracted with CH₂Cl₂ (10 mL), and the combined organic extract was washed with brine (5 mL), dried (MgSO₄), and concentrated under reduced pressure. The crude residue was purified by chromatography (silica gel).

2.9.4.2.7 Asymmetric Pinacol-Type Ketone *tert*-Butylsulfinyl Imine Reductive Coupling: Synthesis of *trans*-1,2-Vicinal Amino Alcohols

trans-1,2-Vicinal amino alcohols **204** can be obtained with high levels of stereoselectivity (dr >87:13) and enantioselectivity (er >99:1) by the asymmetric intramolecular reductive coupling of ketones and *tert*-butylsulfinyl imines **203** (Scheme 57).[349] In the presence of samarium(II) iodide, *tert*-butyl alcohol, and tetrahydrofuran at −78 °C, the radical reaction proceeds efficiently to produce tertiary β-amino alcohols having the hydroxy and sulfinyl-amino groups in a *trans*-relationship (diaxial substitution) to each other. Notably, the intermolecular samarium(II) iodide mediated coupling between acetophenone and *tert*-butylsulfinyl imines does not yield the desired amino alcohols.

The synthetic utility of these vicinal amino alcohols is demonstrated by the facile synthesis of piperidine derivatives **205** by removal of the *tert*-butylsulfinyl moiety followed by double N-alkylation with 1,5-dibromopentane (Scheme 58).

2.9.4 Intramolecular Free-Radical Cyclization Routes to Carbocycles

Scheme 57 Intramolecular Ketone–Imine Reductive Coupling To Give *trans*-1,2-Vicinal Amino Alcohols[349]

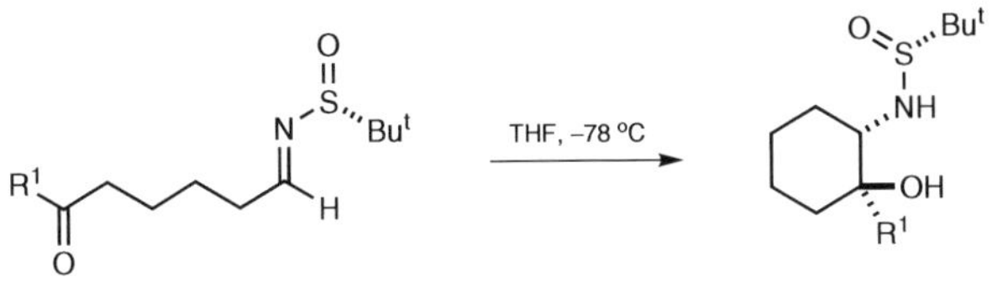

R¹	Conditions	dr[a]	Yield[b] (%)	Ref
Ph	SmI$_2$ (3 equiv), *t*-BuOH (3 equiv)	>20:1	69	[349]
iPr	SmI$_2$ (4 equiv), *t*-BuOH (4 equiv), HMPA (16 equiv)	9:1	76	[349]
4-Tol	SmI$_2$ (3 equiv), *t*-BuOH (3 equiv)	>20:1	95	[349]
4-MeOC$_6$H$_4$	SmI$_2$ (3 equiv), *t*-BuOH (3 equiv)	>20:1	96	[349]
4-BrC$_6$H$_4$	SmI$_2$ (3 equiv), *t*-BuOH (3 equiv)	>20:1	56	[349]
4-ClC$_6$H$_4$	SmI$_2$ (3 equiv), *t*-BuOH (3 equiv)	>20:1	63	[349]
3,4-Cl$_2$C$_6$H$_3$	SmI$_2$ (3 equiv), *t*-BuOH (3 equiv)	>20:1	28	[349]
2-naphthyl	SmI$_2$ (3 equiv), *t*-BuOH (3 equiv)	>20:1	35	[349]
Me	SmI$_2$ (4 equiv), *t*-BuOH (4 equiv), HMPA (16 equiv)	>20:1	70	[349]
Et	SmI$_2$ (4 equiv), *t*-BuOH (4 equiv), HMPA (16 equiv)	14:1	66	[349]
t-Bu	SmI$_2$ (4 equiv), *t*-BuOH (4 equiv), HMPA (16 equiv)	6.6:1	36	[349]
(CH$_2$)$_3$OBn	SmI$_2$ (4 equiv), *t*-BuOH (4 equiv), HMPA (16 equiv)	12:1	80	[349]

[a] The dr was determined by analysis of the crude product by ^{1}H NMR spectroscopy.
[b] Yield of isolated major diastereomer.

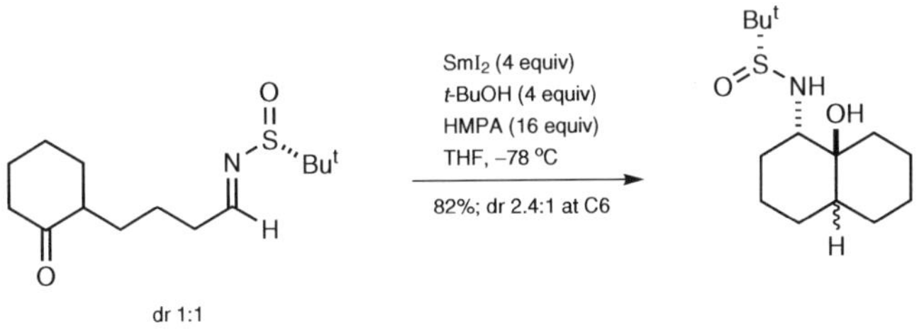

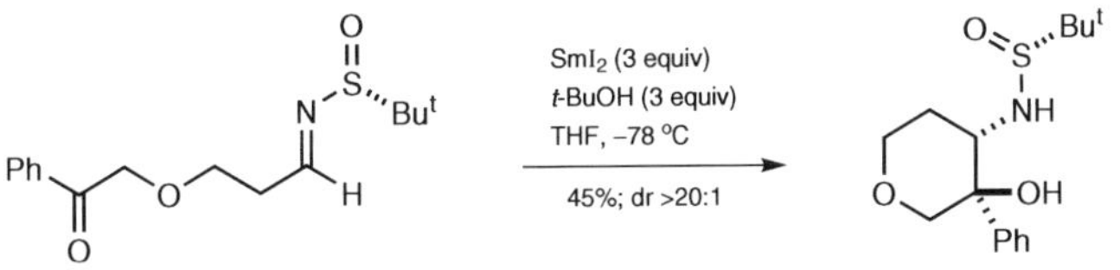

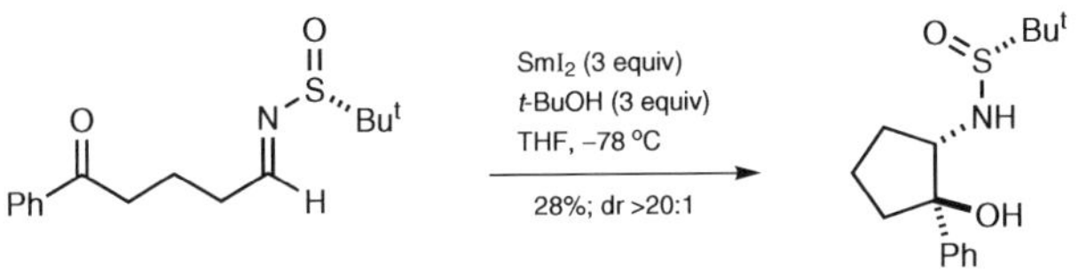

for references see p 533

Scheme 58 Synthetic Utility of Vicinal Amino Alcohols[349]

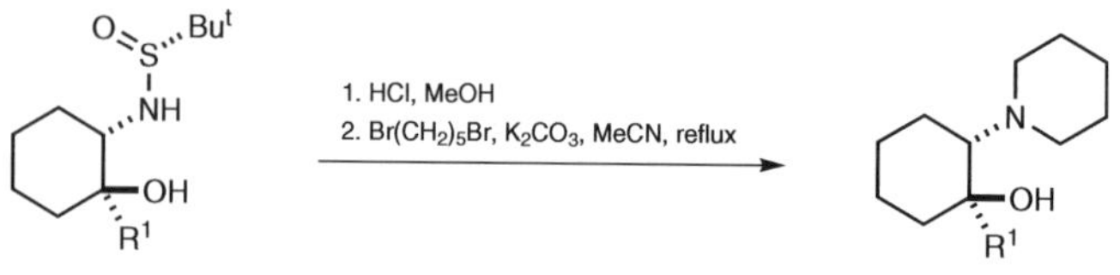

1. HCl, MeOH
2. Br(CH$_2$)$_5$Br, K$_2$CO$_3$, MeCN, reflux

205

R^1	Yield (%)	Ref
Ph	66	[349]
iPr	75	[349]

trans-1,2-Vicinal Amino Alcohols 204; General Procedure:[349]

> **CAUTION:** *Hexamethylphosphoric triamide is a possible human carcinogen and an eye and skin irritant.*

Under argon, a soln of aryl ketone **203** (0.78 mmol) and *t*-BuOH (0.23 mL, 2.4 mmol) in THF (6 mL) was added dropwise to a soln of preformed SmI$_2$ (2.4 mmol) in THF (8 mL) at −78 °C [for alkyl ketones **203**, SmI$_2$ (4.0 equiv), *t*-BuOH (4.0 equiv), and HMPA (16 equiv) were used]. The mixture was stirred at this temperature until the starting material was fully consumed. The reaction was quenched by adding 10% aq Na$_2$S$_2$O$_3$ (10 mL), and the mixture was then warmed to rt and diluted with Et$_2$O. The aqueous phase was extracted with an excess amount of Et$_2$O. The combined organic layer was dried (Na$_2$SO$_4$), filtered, and concentrated under reduced pressure. The crude residue was purified by flash chromatography (silica gel).

2.9.4.3 Synthesis of Substituted Cyclooctanols by an 8-*endo*-Radical Cyclization Process

Good yields of substituted cyclooctanols **207** can be obtained through the samarium(II) iodide promoted 8-*endo*-radical cyclization of alkenyl ketones **206** in the presence of hexamethylphosphoric triamide and *tert*-butyl alcohol in tetrahydrofuran (Table 22).[350]

2.9.4 Intramolecular Free-Radical Cyclization Routes to Carbocycles

Table 22 8-*endo*-Radical Cyclization of Substituted Alkenyl Ketones[350]

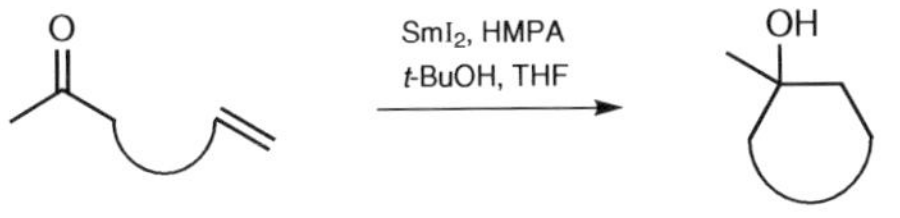

Entry	Substrate	Product	dr	Yield (%)	Ref
1	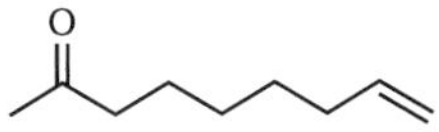	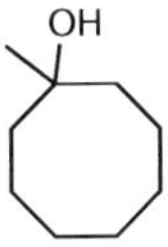	–	54	[350]
2	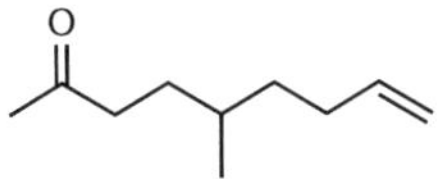	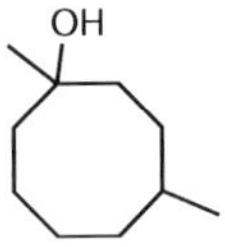	1:1	53	[350]
3	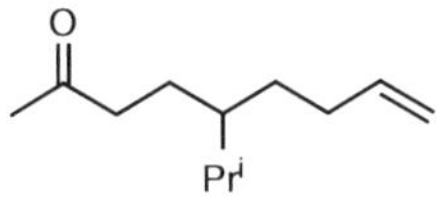	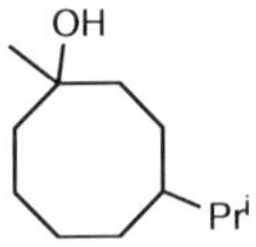	1.5:1	58	[350]
4	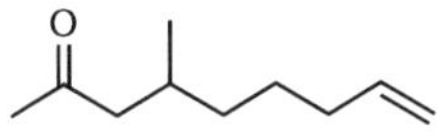	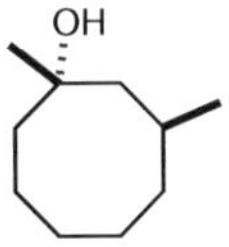	3:1	54	[350]
5	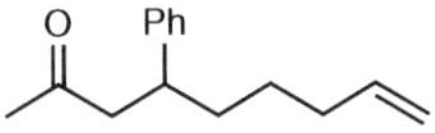	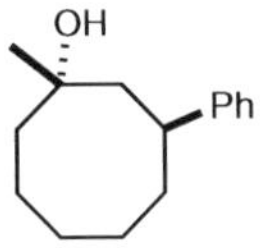	>30:1	49	[350]
6	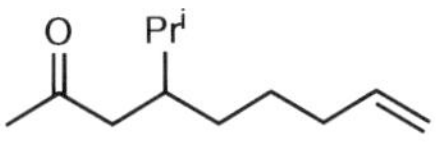	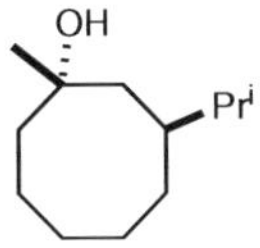	>30:1	63	[350]
7	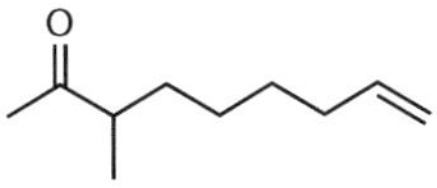	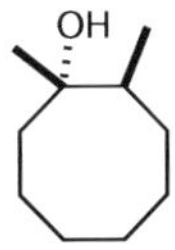	3:1[a]	24	[350]
8	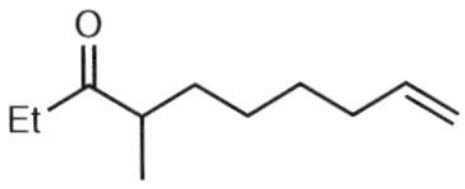	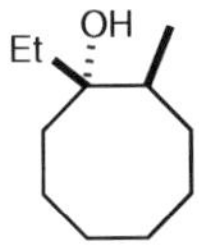	–	0	[350]
9	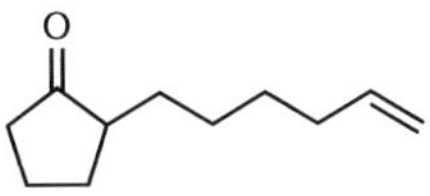	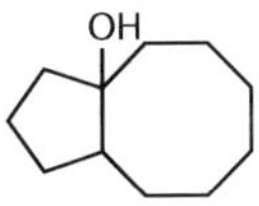	–	0	[350]

Table 22 (cont.)

Entry	Substrate	Product	dr	Yield (%)	Ref
10	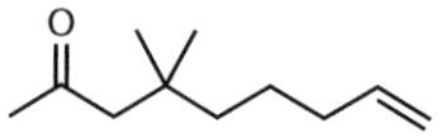	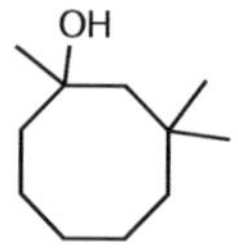	–	0	[350]
11	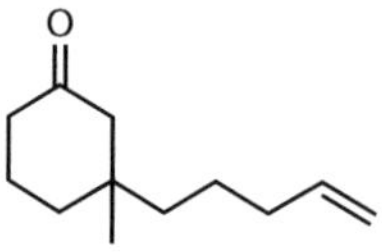	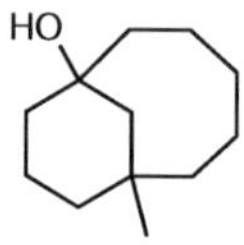	>30:1	46	[350]
12	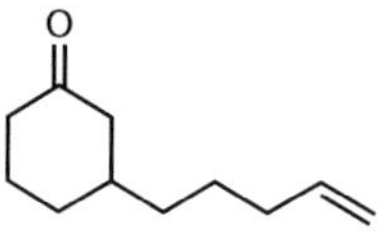	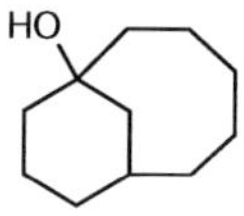	–	0	[350]
13	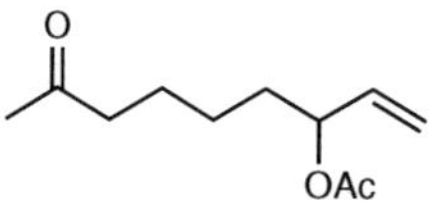	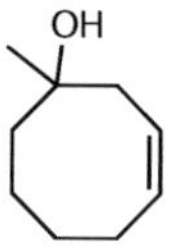	–	81	[350]
14	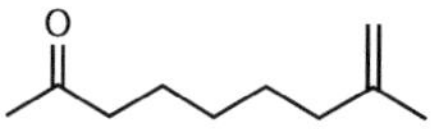	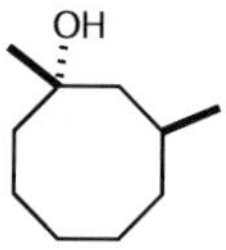	2:1	31	[350]
15	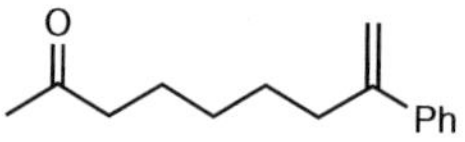	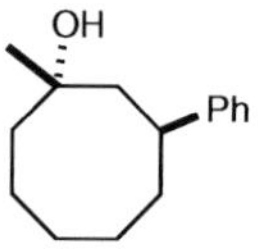	>30:1	91	[350]
16	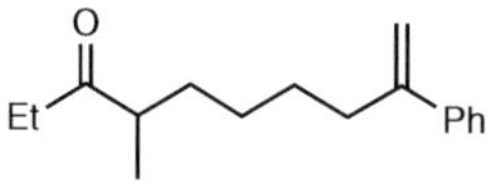	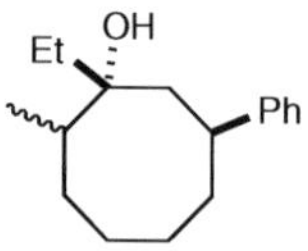	2:1	86	[350]
17	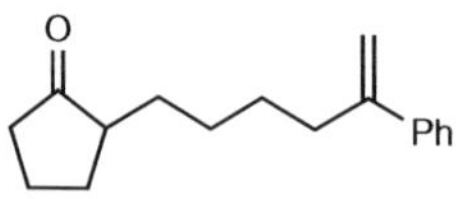	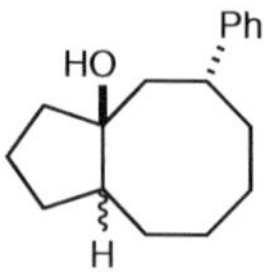	1:1	78	[350]
18	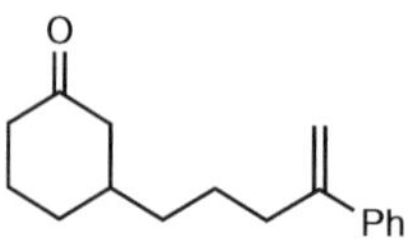	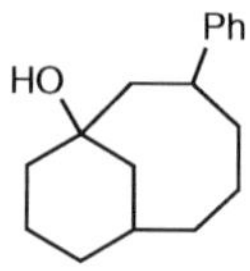	6:1[a]	78	[350]

[a] Uncertain relative stereochemistry.

Cyclooctanol Derivatives 207; General Procedure:[350]

> CAUTION: *Hexamethylphosphoric triamide is a possible human carcinogen and an eye and skin irritant.*

HMPA (2.63 g, 17.4 mmol) was added to a soln of SmI$_2$ (1.84 mmol) in THF, and argon gas was bubbled through the soln for 10 min. A soln of the alkenyl ketone (0.83 mmol) and *t*-BuOH (1.64 mmol) in THF (40 mL) was then added over 90 min. After complete consumption of the starting material, aqueous workup followed by flash chromatography and/or Kugelrohr distillation gave the product.

2.9.4.4 Synthesis of Cyclopentanone Derivatives

2.9.4.4.1 Titanium(III)-Catalyzed Synthesis of Cyclic Amino Ketones

A titanium-catalyzed reductive umpolung strategy can be utilized for the synthesis of amino ketones **209** through direct construction of a C—C bond between the nitrile carbon atom and the imine carbon atom of iminoalkanenitriles **208** (Scheme 59).[351] The corresponding iminoalkanenitriles are easily accessible from either a commercial source or literature-based standard condensation protocols. This reductive umpolung protocol can tolerate a wide range of substituents on the imine nitrogen atom.

Scheme 59 Synthesis of Cyclic Amino Ketones by Intramolecular Radical Cyclization of (Imino)alkanenitriles[351]

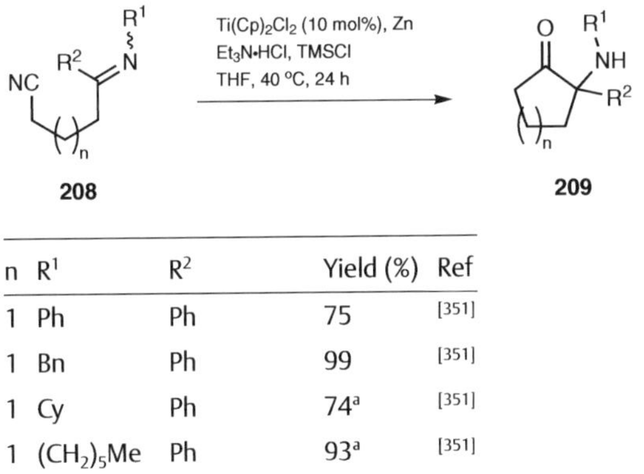

n	R^1	R^2	Yield (%)	Ref
1	Ph	Ph	75	[351]
1	Bn	Ph	99	[351]
1	Cy	Ph	74[a]	[351]
1	(CH$_2$)$_5$Me	Ph	93[a]	[351]
1	Bn	4-FC$_6$H$_4$	60[a]	[351]
1	Bn	4-ClC$_6$H$_4$	82[a]	[351]
1	Bn	4-BrC$_6$H$_4$	73	[351]
1	Bn	4-Tol	61[a,b]	[351]
1	Bn	4-MeOC$_6$H$_4$	62[a,b]	[351]
1	Bn	3-Tol	72[a,b]	[351]
1	Bn	2-thienyl	99[a]	[351]
2	Bn	Ph	79[b]	[351]

[a] No chromatography was required.
[b] Reaction time was 48 h.

Cyclic Amino Ketones 209; General Procedure:[351]

A flame-dried, argon-filled, 10-mL Schlenk tube was charged with $Et_3N \cdot HCl$ (137.7 mg, 1.0 mmol, 2 equiv), Zn dust (65.4 mg, 1.0 mmol, 2 equiv), and $Ti(Cp)_2Cl_2$ (12.4 mg, 0.05 mmol, 10 mol%). The tube was then evacuated and backfilled with argon. Deoxygenated abs THF (1.25 mL) was added, and the mixture was stirred for about 1 min to allow for a change in the color of the soln from red to lime green. Then, the imine (0.5 mmol, 1.0 equiv) was added, followed by TMSCl (190.4 µL, 1.5 mmol, 3.0 equiv), and the reaction vessel was sealed with a greased glass stopper. Depending on the substrate, the color of the mixture varied. The mixture was stirred at 40 °C for 24 h, unless otherwise noted. Upon completion of the reaction, the mixture was cooled to rt and then transferred with the help of CH_2Cl_2 (3–5 mL) to a separatory funnel containing ice-cold 1 M aq HCl (20 mL) and ice-cold Et_2O (50 mL). The Et_2O layer was extracted (by quick shaking) with ice-cold 1 M aq HCl (6 × 20 mL). Sat. aq $NaHCO_3$ was carefully added to the combined aqueous phase to pH 8. The aqueous phase was extracted with CH_2Cl_2 (3 × 30 mL), and the organic layer was dried (Na_2SO_4), filtered, and concentrated under reduced pressure. In most cases, the title compounds were obtained in analytically pure form without further purification.

2.9.4.4.2 Synthesis of Bicyclo[4.3.0]nonan-8-one and Bicyclo[3.3.0]octan-3-one Derivatives from Bis(α,β-unsaturated esters) by Samarium(II) Iodide Induced Tandem Reductive Coupling/Dieckmann Condensation Reaction

The stereoselective synthesis of bicyclic cyclopentane carboxylates **211A/211B** can be achieved by tandem cyclization of bis(α,β-unsaturated esters) **210** in the presence of samarium(II) iodide, samarium, tetrahydrofuran, and a catalytic amount of methanol (Table 23).[352]

2.9.4 Intramolecular Free-Radical Cyclization Routes to Carbocycles

Table 23 Tandem Reductive Coupling/Dieckmann Condensation of Bis(α,β-unsaturated esters)[352]

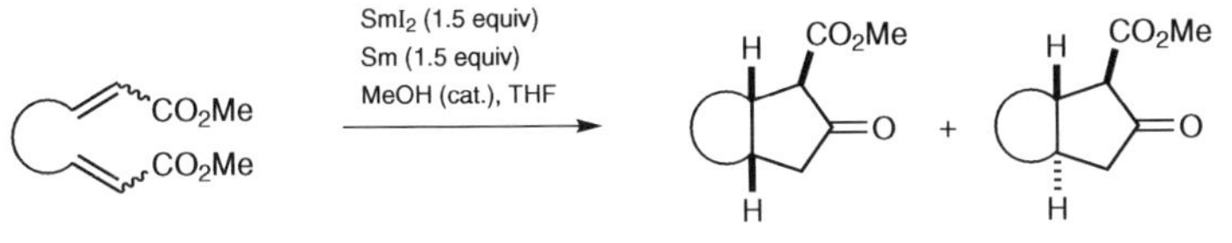

Entry	Substrate	Conditions	Product	Yield[a] (%)	Ref
1	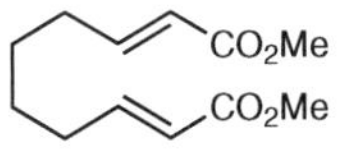	50 °C, 1.5 h	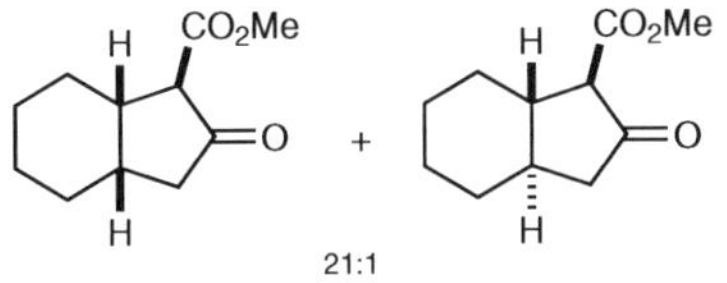 21:1	78[b]	[352]
2	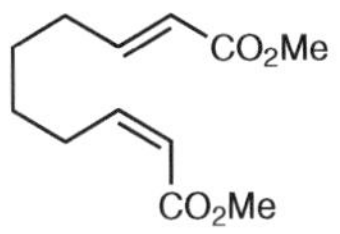	50 °C, 2 h	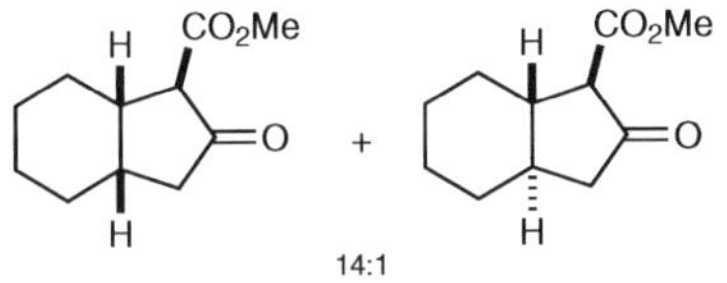 14:1	76[b]	[352]
3	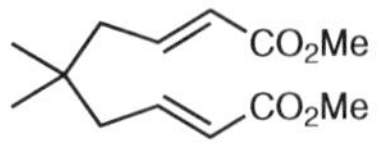	0 °C, 15 min	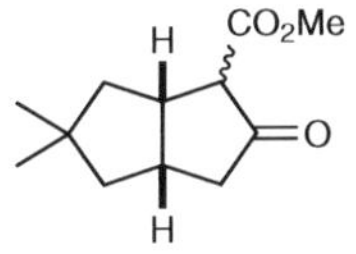	62	[352]
4	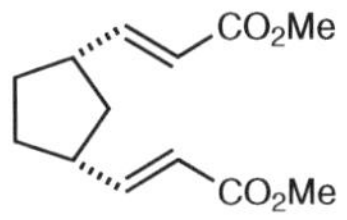	rt, 17 h	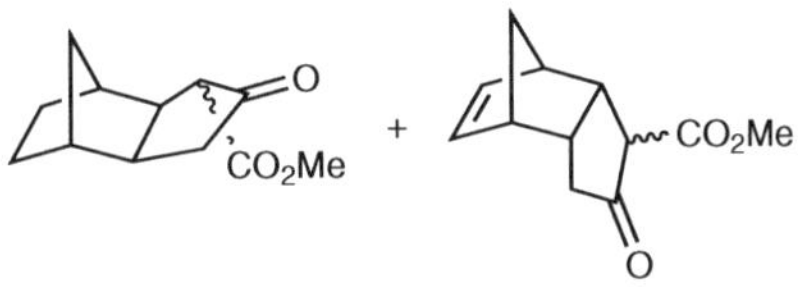 9:1	49[b]	[352]
5	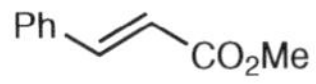	rt, 5 min	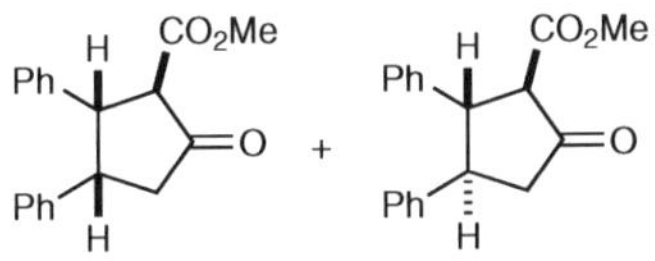 1:2	70[c]	[352]

[a] Combined yield, as determined by ^{1}H NMR spectroscopy.
[b] Products were inseparable.
[c] The reaction was performed with HMPA (3.3 equiv) and SmI_2 (1.1 equiv).

Methyl (1R*,3aS*,7a*S)-2-Oxooctahydro-1H-indene-1-carboxylate and Methyl (1R*,3aR*,7aS*)-2-Oxooctahydro-1H-indene-1-carboxylate (Table 23, Entry 1); Typical Procedure:[352]

A mixture of Sm (151 mg, 1.00 mmol) and 1,2-diiodoethane (141 mg, 1.00 mmol) in THF (3.3 mL) was sonicated for 1 h at rt under argon. A soln of dimethyl (2E,8E)-deca-2,8-dienedioate (75.3 mg, 0.333 mmol) and MeOH (40 µg) in THF (2 mL) was added to the suspension, which was followed by stirring for 1.5 h at 50 °C. Upon completion of the reaction, a few drops of 30% H_2O_2 and 1 M HCl were added. The mixture was then diluted with Et_2O, washed with sat. aq $NaHCO_3$ and sat. aq $Na_2S_2O_3$, dried, and concentrated. The crude residue was purified by column chromatography (silica gel); yield: 51.0 mg (78%); ratio (**211A/211B**) 21:1.

2.9.4.4.3 Intramolecular Cyclization of Dicarbonyls to 3-Heterobicyclo[3.1.0]hexan-2-ones

In acetic acid, manganese(III) acetate has strong oxidizing ability, whereas in alcohol solvents the oxidizing capacity of this reagent is weakened, which decreases the probability of overoxidation of certain susceptible substrates.[53] Thus, the synthesis of 3-azabicyclo-[3.1.0]hexan-2-one derivatives **213** under mild conditions can be achieved by intramolecular oxidative cyclization of substituted N-propenyl-3-oxobutanamides **212** in the presence of manganese(III) acetate in refluxing ethanol (Scheme 60).[111] Moreover, the reaction can also yield pyrrolidin-2-ones **214**, **215**, and **216** in certain cases.

Scheme 60 Synthesis of 3-Azabicyclo[3.1.0]hexan-2-ones[111]

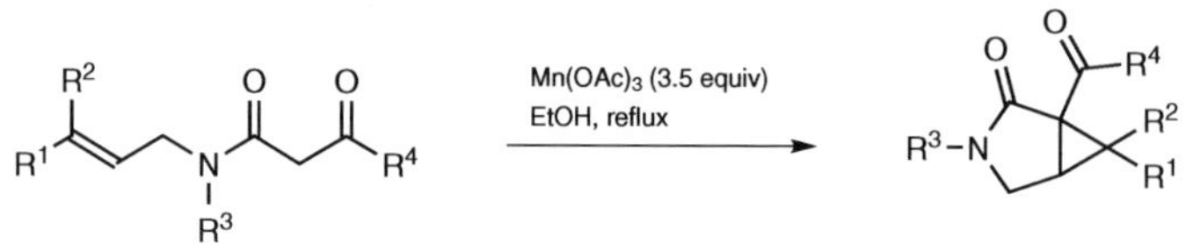

R¹	R²	R³	R⁴	Time (min)	Yield (%)	Ref
Ph	Ph	Bn	Me	1.5	88	[111]
Ph	Ph	Me	Me	1	91	[111]
Ph	Ph	Et	Me	1	98	[111]
Ph	Ph	Pr	Me	1	86	[111]
Ph	Ph	iPr	Me	1	88	[111]
Ph	Ph	Bu	Me	1	92	[111]
Ph	Ph	t-Bu	Me	2	84	[111]
Ph	Ph	PMB	Me	1.5	91	[111]
Ph	Ph	Ph	Me	1.5	46	[111]
Ph	Ph	H	Me	20	29[a]	[111]
4-Tol	4-Tol	Bn	Me	1	22[b]	[111]
4-ClC₆H₄	4-ClC₆H₄	Bn	Me	2.5	70[c]	[111]
4-FC₆H₄	4-FC₆H₄	Bn	Me	2	60[c]	[111]
4-FC₆H₄	4-FC₆H₄	Et	Me	1.5	55[c]	[111]
H	Ph	Bn	Me	2.5	23[d]	[111]
H	H	Bn	Me	2460	27	[111]
Ph	Ph	Bn	OEt	15	37[e]	[111]

[a] Mn(OAc)₃ (6 equiv) was used.
[b] **214** (16%) and **215** (34%) were also obtained.
[c] Cu(OAc)₂ (0.6 equiv) added as a co-oxidant at rt.
[d] Obtained as a 17:6 diastereomeric mixture; **216** (6%) was also obtained.
[e] **215** (36%) was also obtained.

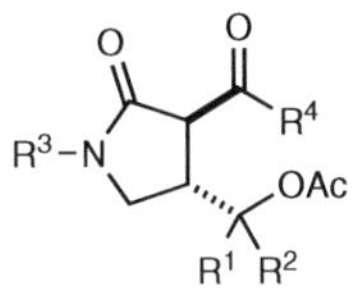

2.9.4 Intramolecular Free-Radical Cyclization Routes to Carbocycles

214 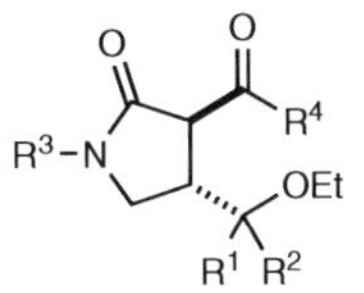 **215** 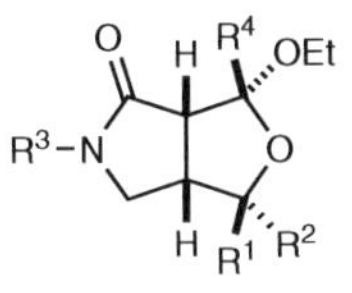 **216**

Substituted 3-oxobutanoates **217** also undergo manganese(III)-induced oxidative intramolecular cyclization to produce 3-oxabicyclo[3.1.0]hexan-2-one derivatives **218** (Scheme 61).

Scheme 61 Manganese(III)-Induced Oxidative Intramolecular Cyclization of 3-Oxobutanoates To Give 3-Oxabicyclo[3.1.0]hexan-2-ones[111]

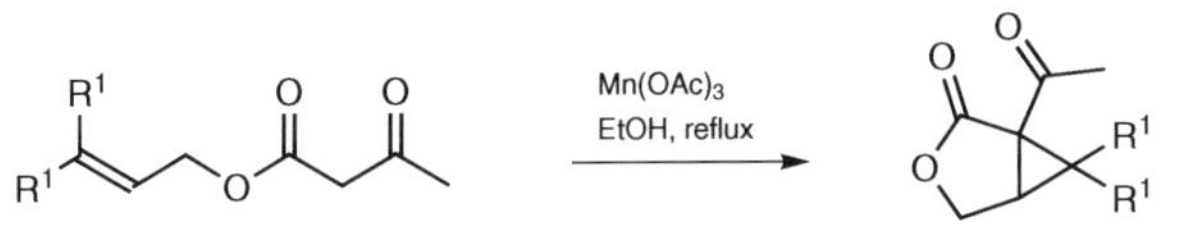

R^1	Molar Ratio [**217**/Mn(OAc)$_3$•2H$_2$O]	Time (min)	Yield (%)	Ref
Ph	1:6	30	64	[111]
4-Tol	1:8	50	51	[111]
4-ClC$_6$H$_4$	1:6	45	62	[111]
4-FC$_6$H$_4$	1:4	45	64	[111]
H	1:3	25	28[a]	[111]

[a] KOAc (2.5 mmol) and Cu(OAc)$_2$ (1.5 mmol) were added as buffers.

3-Azabicyclo[3.1.0]hexan-2-ones 213 and 3-Oxabicyclo[3.1.0]hexan-2-ones 218; General Procedure:[111]

A 50-mL flask was charged with 1,3-dicarbonyl compound **212** or **217** (0.5 mmol) and EtOH (20 mL). The mixture was degassed under reduced pressure for 30 min using an ultrasonicator, and the flask was then backfilled with argon. Mn(OAc)$_3$•2H$_2$O was quickly added, and the mixture was heated to reflux until the brown color of Mn(III) disappeared. Upon completion of the reaction, the solvent was removed under reduced pressure, and the residue was triturated with H$_2$O and then extracted with CHCl$_3$ (3 × 10 mL). The combined organic extract was dried (MgSO$_4$) and concentrated under reduced pressure. The products were isolated by TLC (Wakogel B-10 silica gel, CHCl$_3$). Recrystallization (Et$_2$O/hexane; except for **212**, $R^1 = R^2 = Ph$; $R^3 = Bn$; $R^4 = OEt$, for which CHCl$_3$/EtOAc/hexane was used) afforded a solid.

2.9.4.5 Synthesis of 6-(Trifluoromethyl)phenanthridine Derivatives

6-(Trifluoromethyl)phenanthridine derivatives **220** can be obtained by visible-light-mediated intramolecular radical cyclization of trifluoroacetimidoyl chlorides **219** under mild reaction conditions.[353] Trifluoroacetimidoyl chlorides **219** can be prepared according to a reported procedure,[354] and the cyclization reaction proceeds in the presence of tris(2,2′-bipyridyl)dichlororuthenium(II) hexahydrate as the catalyst with tributylamine as the base in acetonitrile (Table 24).

Table 24 Cyclization of *N*-Biaryltrifluoroacetimidoyl Chlorides To Give 6-(Trifluoromethyl)-phenanthridines[353]

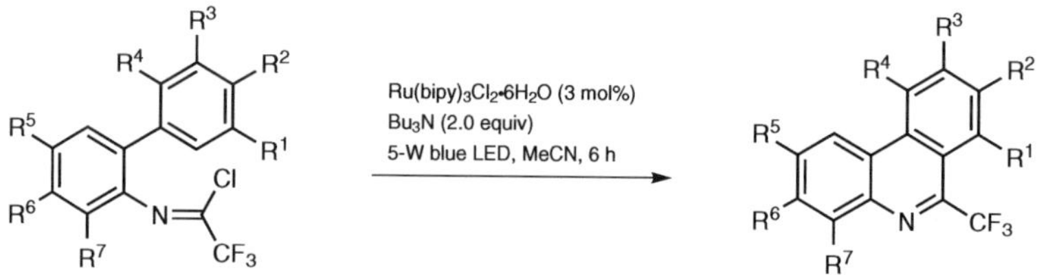

Entry	Substrate	Product(s)	Yield (%)	Ref
1	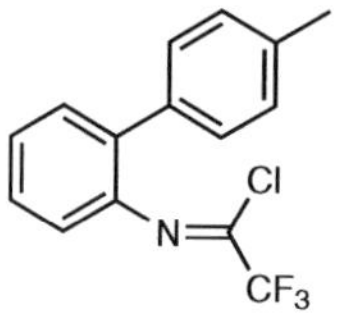	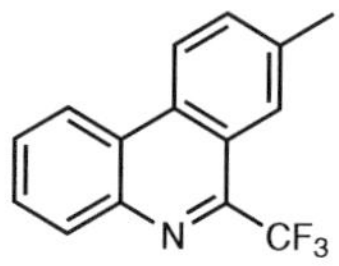	88	[353]
2	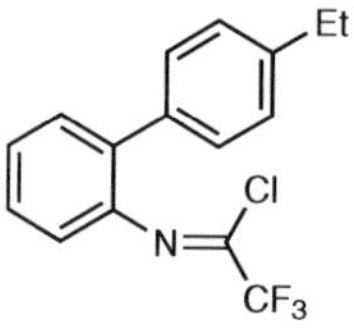	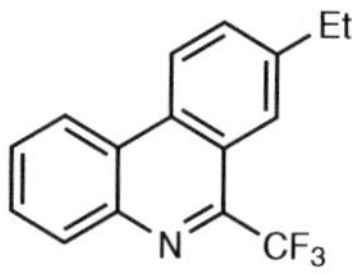	85	[353]
3	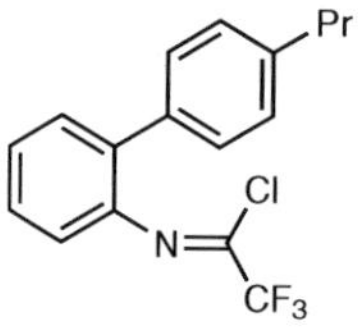	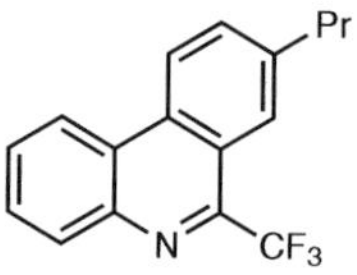	87	[353]
4	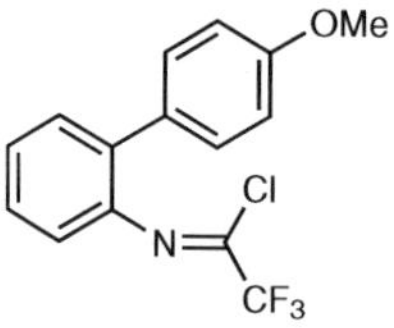	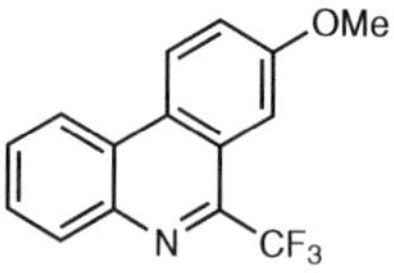	83	[353]
5	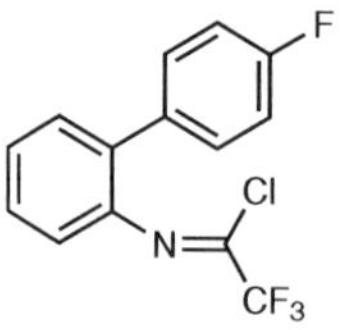	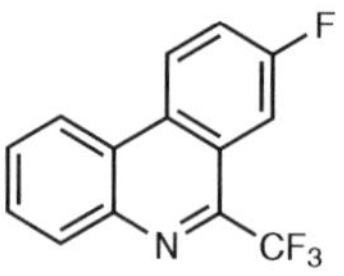	76	[353]
6	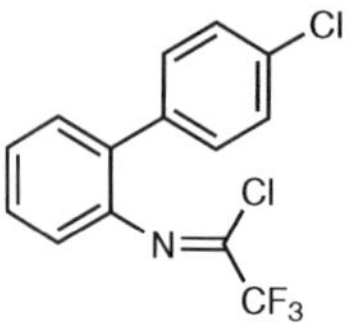	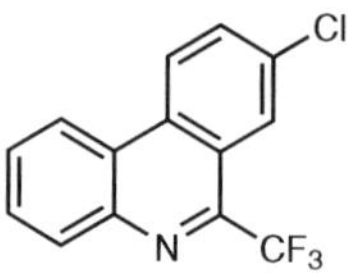	81	[353]

2.9.4 Intramolecular Free-Radical Cyclization Routes to Carbocycles

Table 24 (cont.)

Entry	Substrate	Product(s)	Yield (%)	Ref
7	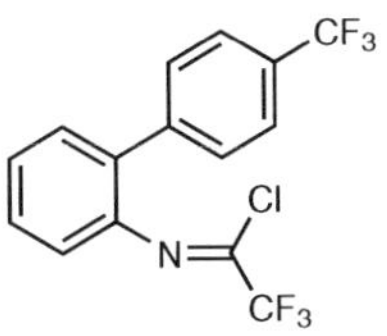 biphenyl with 4-CF₃ and N=C(Cl)CF₃ imidoyl chloride	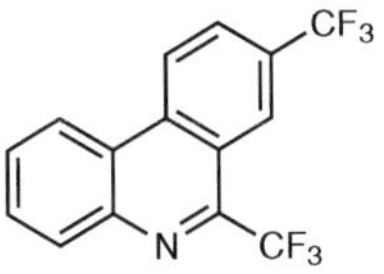 phenanthridine with 3-CF₃ and 6-CF₃	70	[353]
8	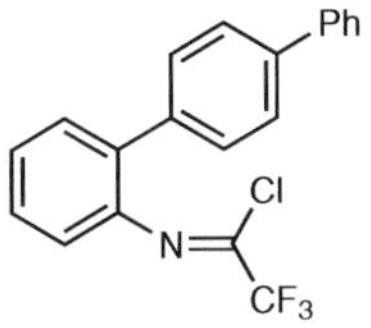 biphenyl with 4-Ph and N=C(Cl)CF₃	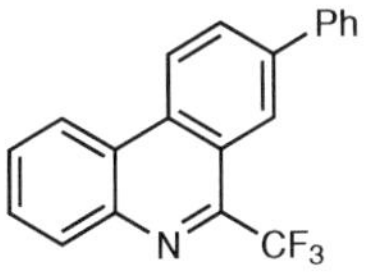 phenanthridine with 3-Ph and 6-CF₃	62	[353]
9	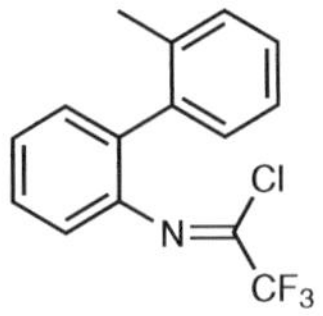 biphenyl with 2′-Me and N=C(Cl)CF₃	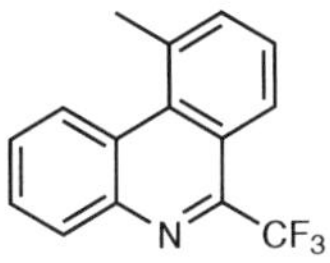 phenanthridine with Me and 6-CF₃	56	[353]
10	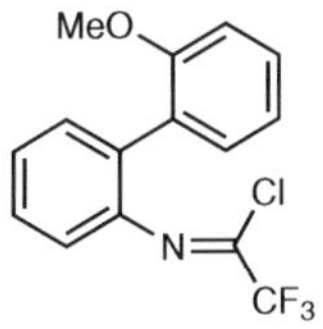 biphenyl with 2′-OMe and N=C(Cl)CF₃	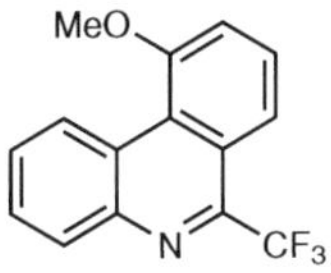 phenanthridine with OMe and 6-CF₃	53	[353]
11	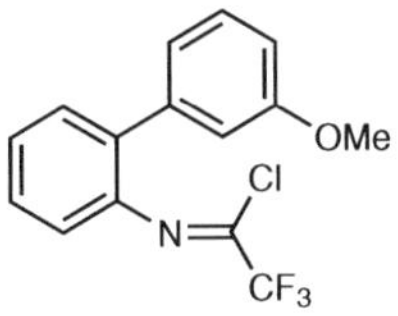 biphenyl with 3′-OMe and N=C(Cl)CF₃	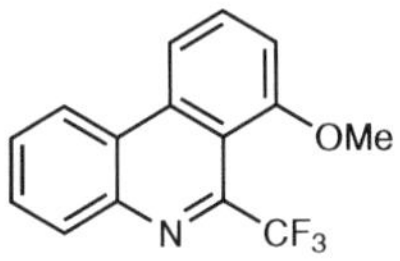 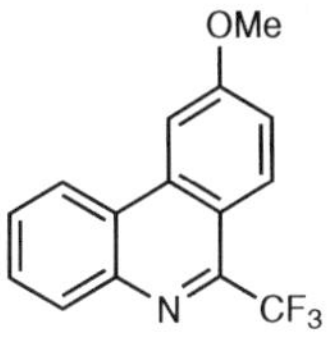 two regioisomeric OMe-phenanthridines with 6-CF₃ (3:2)	72[a]	[353]
12	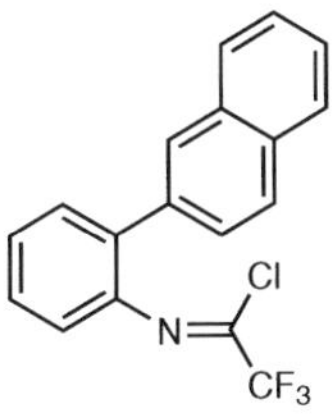 2-naphthyl aryl imidoyl chloride (CF₃)	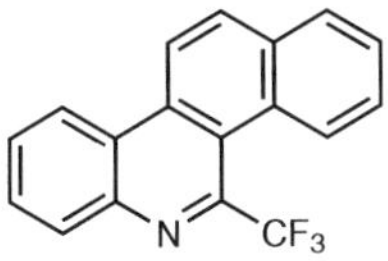 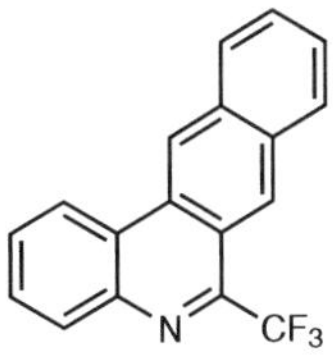 two isomeric benzo-fused phenanthridines (10:1)	67[a]	[353]
13	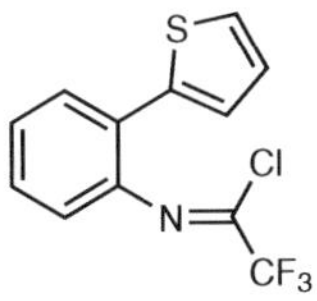 2-thienyl aryl imidoyl chloride (CF₃)	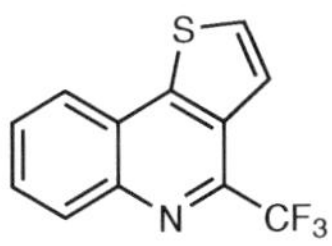 thieno-fused quinoline with 4-CF₃	42[b]	[353]

for references see p 533

Table 24 (cont.)

Entry	Substrate	Product(s)	Yield (%)	Ref
14	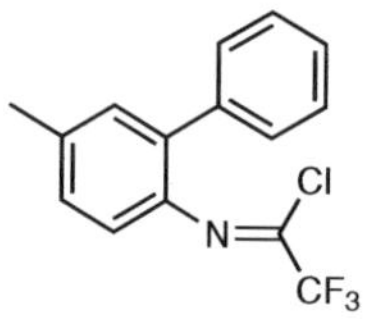	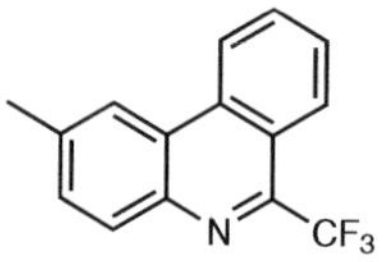	81	[353]
15	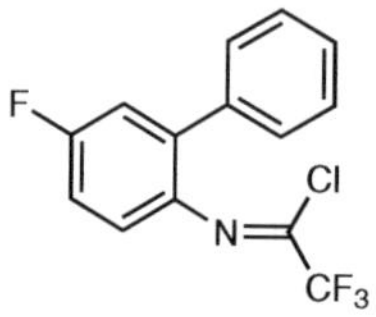	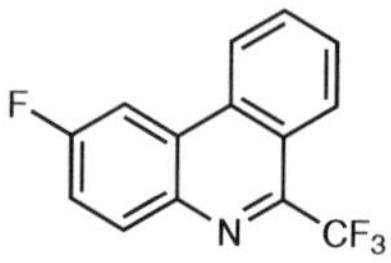	77	[353]
16	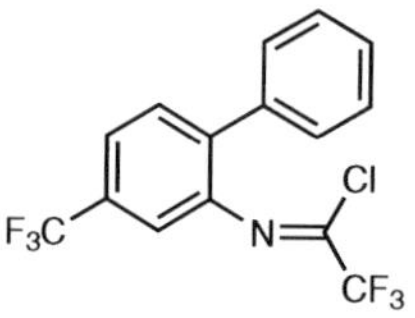	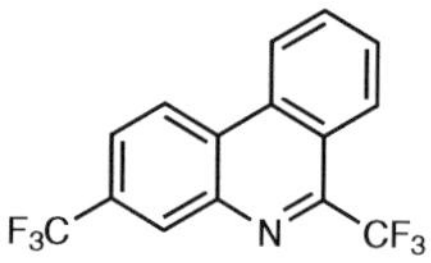	80	[353]
17	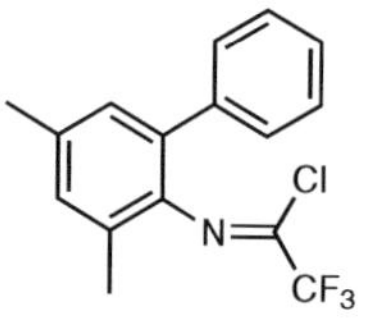	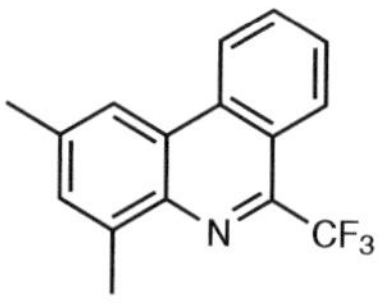	75	[353]
18	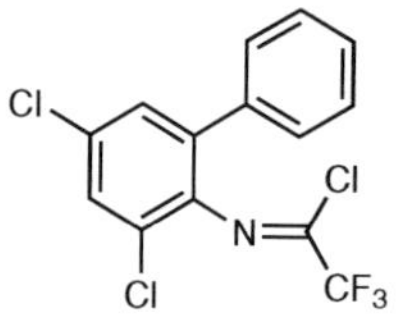	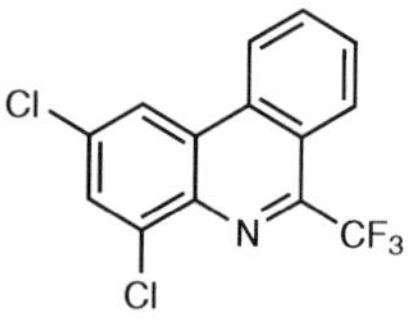	72	[353]
19	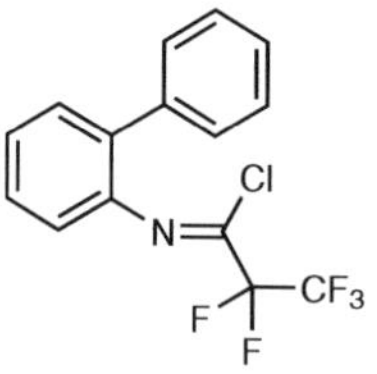	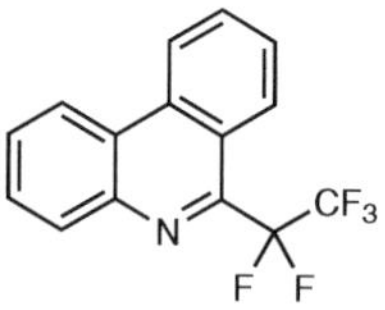	79	[353]

[a] Ratio of regioisomers determined by ^{19}F NMR spectroscopy.
[b] Treated for 12 h.

6-(Trifluoromethyl)phenanthridines 220; General Procedure:[353]

Bu$_3$N (0.6 mmol) was added to a mixture of trifluoroacetimidoyl chloride **219** (0.3 mmol) and [Ru(bipy)$_3$]Cl$_2$·6H$_2$O (3 mol%) in MeCN (3.0 mL) under a N$_2$ atmosphere. The soln was stirred for 6 h at rt under irradiation by a 5-W blue-light-emitting diode. Upon completion of the reaction, the solvent was evaporated under reduced pressure. Then, the crude product was directly purified by flash column chromatography (silica gel, petroleum ether/EtOAc 10:1).

2.9.5 Intramolecular Free-Radical Cyclization Routes to S-Heterocycles

2.9.5.1 Synthesis of Benzothiazole Derivatives

2.9.5.1.1 Synthesis of 2-Arylbenzothiazoles

2-Arylbenzothiazoles **222** can be obtained by reductive cyclization of bis[(2-benzylideneamino)phenyl]disulfides **221** mediated by the titanium(IV) chloride/samarium catalytic system (Scheme 62).[355] Substrates **221** can be easily prepared,[356] and the reaction is complete within a short timeframe (10 min), facilitating access to the arylbenzothiazole moiety.

Scheme 62 Synthesis of 2-Arylbenzothiazoles from Disulfides[355]

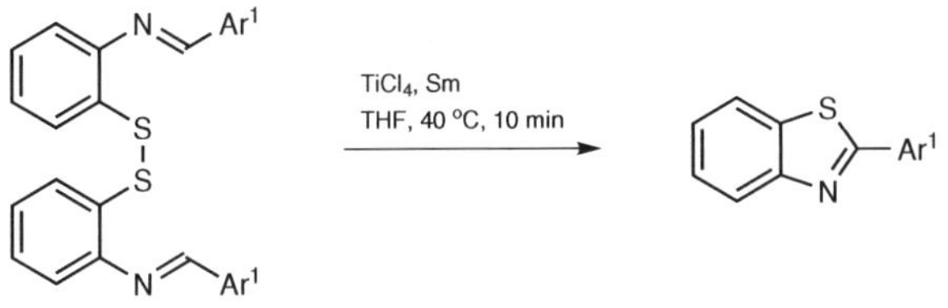

Ar^1	Yield (%)	Ref
Ph	70	[355]
4-Tol	91	[355]
4-MeOC$_6$H$_4$	81	[355]
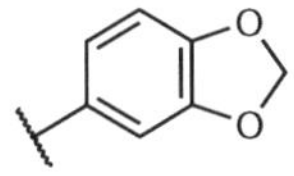 3,4-(methylenedioxy)C$_6$H$_3$	88	[355]
3,4-(MeO)$_2$C$_6$H$_3$	82	[355]
3,4,5-(MeO)$_3$C$_6$H$_2$	89	[355]
4-ClC$_6$H$_4$	79	[355]
4-BrC$_6$H$_4$	69	[355]
3,4-Cl$_2$C$_6$H$_3$	92	[355]
4-Me$_2$NC$_6$H$_4$	65	[355]
2-thienyl	52	[355]

2-Arylbenzothiazoles 222; General Procedure:[355]

TiCl$_4$ (0.4 mL, 3 mmol) was added dropwise by syringe to a stirred suspension of Sm(0) powder (0.9 g, 6 mmol) in freshly distilled anhyd THF (10 mL) at rt under N$_2$. After complete addition, the mixture was refluxed for 2 h. The suspension of the formed low-valent titanium reagent was cooled to rt, and a soln of disulfide **221** (1 mmol) in THF (10 mL) was added dropwise. The mixture was stirred at 40 °C for 10 min under a N$_2$ atmosphere. Upon completion of the reaction (detected by TLC), the reaction was quenched with 5% HCl (30 mL), and the mixture was extracted with 1,2-dichloroethane (3 × 30 mL). The combined extract was washed with H$_2$O (2 × 50 mL), dried (Na$_2$SO$_4$), filtered, and concentrated under reduced pressure. Recrystallization (acetone) of the crude material afforded the title compound.

2.9.5.1.2 Synthesis of 2-Substituted Benzothiazoles

2-Arylbenzothiazoles **224** and 2-benzoylbenzothiazoles **226** can be obtained by manganese(III)-promoted radical cyclization of the corresponding N-arylbenzothioamides **223** (Scheme 63) and benzoyl-containing N-arylthioamides **225** (Scheme 64) in acetic acid under microwave irradiation conditions.[357]

Scheme 63 Manganese(III)-Promoted Cyclization of N-Arylbenzothioamides To Give 2-Arylbenzothiazoles[357]

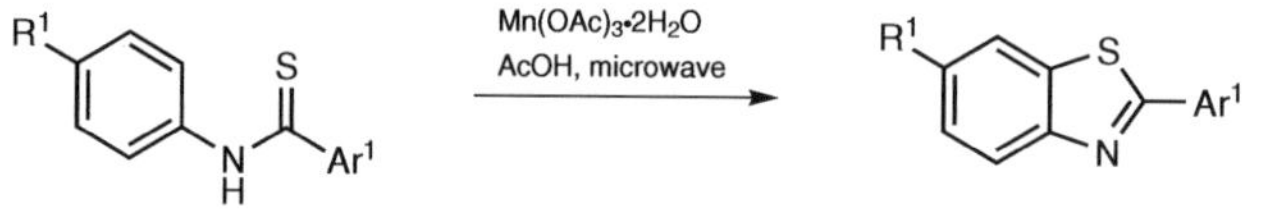

R^1	Ar^1	Yield (%)	Ref
H	Ph	80	[357]
OMe	Ph	75	[357]
H	4-FC$_6$H$_4$	62	[357]
H	2-ClC$_6$H$_4$	76	[357]
OMe	4-[Me(CH$_2$)$_7$O]C$_6$H$_4$	83	[357]
Cl	4-[Me(CH$_2$)$_7$O]C$_6$H$_4$	79	[357]
Br	4-[Me(CH$_2$)$_7$O]C$_6$H$_4$	86	[357]
Cl	4-[Me(CH$_2$)$_9$O]C$_6$H$_4$	88	[357]

Scheme 64 Manganese(III)-Promoted Cyclization of Benzoyl-Containing N-Arylthioamides To Give 2-Benzoylbenzothiazoles[357]

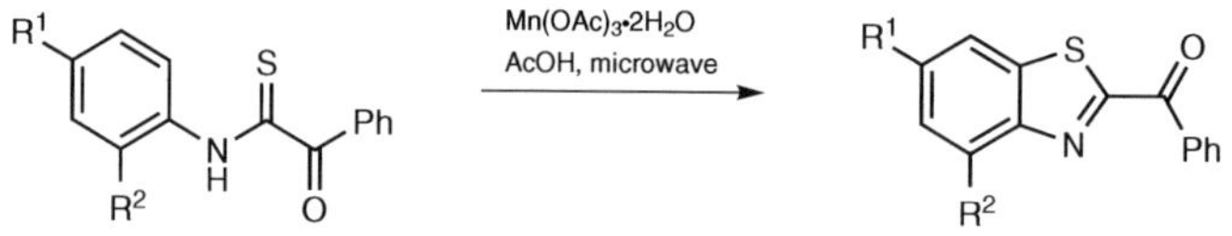

R^1	R^2	Yield (%)	Ref
H	H	60	[357]
Me	H	62	[357]
OMe	H	67	[357]
Cl	H	50	[357]
Br	H	57	[357]
H	Cl	63	[357]

2-Phenylbenzothiazole (224, R^1 = H; Ar^1 = Ph); Typical Procedure:[357]
A 100-mL, three-necked flask was charged with N-phenylbenzothioamide (0.21 g, 1 mmol) and AcOH (15 mL). Mn(OAc)$_3$•2H$_2$O (0.81 g, 3 mmol, 3 equiv) was added, and the flask was placed into a microwave vessel. The mixture was then heated to 110 °C for 6 min at 300 W in a kitchen-type microwave oven. Upon completion of the reaction, the mixture was cooled to rt, poured into H$_2$O, and extracted with CHCl$_3$ (3 × 20 mL). The organic layer

was dried (Na$_2$SO$_4$), filtered, concentrated, and purified by column chromatography (silica gel, petroleum ether/acetone 4:1); yield: 80%; mp 115–116 °C.

Benzothiazol-2-yl(phenyl)methanone (226, R^1 = R^2 = H); Typical Procedure:[357]
A 100-mL, three-necked flask was charged with 2-oxo-2,N-diphenylethanethioamide (0.241 g, 1 mmol) and AcOH (15 mL). Mn(OAc)$_3$·2H$_2$O (0.81 g, 3 mmol, 3 equiv) was added, and the flask was placed into a microwave vessel. The mixture was then heated to 110 °C for 6 min at 300 W in a kitchen-type microwave oven. Upon completion of the reaction, the mixture was cooled to rt, poured into H$_2$O, and extracted with CHCl$_3$ (3 × 20 mL). The organic layer was dried (Na$_2$SO$_4$), filtered, concentrated, and purified by column chromatography (silica gel, petroleum ether/acetone 4:1); yield: 60%; mp 98–99 °C.

2.9.5.2 Synthesis of Substituted 2,3-Dihydrothiophenes

Substituted 2,3-dihydrothiophenes **228** can be obtained in good yields by treating S-[3,3-bis(phenylsulfanyl)propyl] alkanethioates **227** with the low-valent titanium species bis(η^5-cyclopentadienyl)bis(triethyl phosphite)titanium(II) in tetrahydrofuran (Scheme 65). However, these thiophene derivatives are sensitive to photoirradiation and easily isomerize to 2-alkylidenetetrahydrothiophenes **229**.[358]

Scheme 65 Synthesis of Substituted 2,3-Dihydrothiophenes from S-[3,3-Bis(phenylsulfanyl)propyl] Alkanethioates[358]

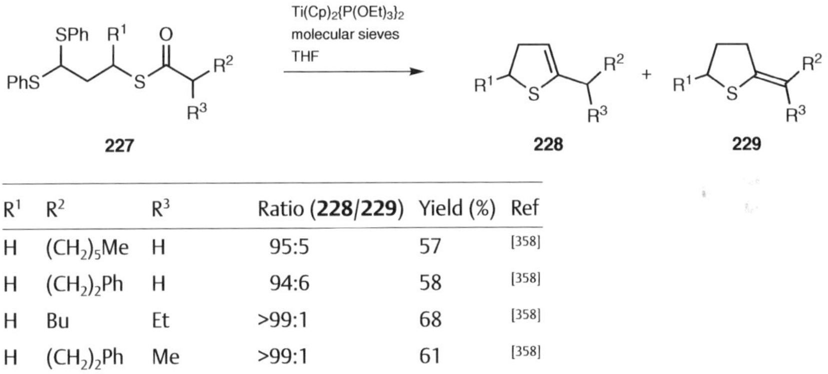

R^1	R^2	R^3	Ratio (228/229)	Yield (%)	Ref
H	(CH$_2$)$_5$Me	H	95:5	57	[358]
H	(CH$_2$)$_2$Ph	H	94:6	58	[358]
H	Bu	Et	>99:1	68	[358]
H	(CH$_2$)$_2$Ph	Me	>99:1	61	[358]
H	(CH$_2$)$_5$		97:3	72	[358]
Me	(CH$_2$)$_5$		>99:1	65	[358]
H	Bn	Bn	>99:1	55	[358]
Me	H	(CH$_2$)$_5$Me	>99:1	62	[358]

5-(Heptan-3-yl)-2,3-dihydrothiophene (228, R^1 = H; R^2 = Bu; R^3 = Et); Typical Procedure:[358]
After the preparation of Ti(Cp)$_2${P(OEt)$_3$}$_2$, all operations were performed in the dark. Mg turnings (49 mg, 2 mmol), finely powdered 4-Å molecular sieves (200 mg), and Ti(Cp)$_2$Cl$_2$ (498 mg, 2 mmol) were placed in a flask, and the contents of the flask were dried by heating with a heat gun under reduced pressure (2–3 Torr), taking care to avoid sublimation of Ti(Cp)$_2$Cl$_2$. After cooling, THF (6 mL) and P(OEt)$_3$ (0.69 mL, 4 mmol) were added successively with stirring at rt under an argon atmosphere. After 3 h, a soln of S-[3,3-bis(phenylsulfanyl)propyl] 2-ethylhexanethioate (209 mg, 0.5 mmol) in THF (10 mL) was added slowly to the mixture over 20 min, and the mixture was then stirred for another 3 h. Thereafter, the reaction was quenched with 1 M NaOH (15 mL), and the resulting insoluble matter was

filtered off through Celite. The organic material was extracted with Et_2O, and the extract was dried (Na_2SO_4), filtered, and concentrated. The crude residue was purified by preparative TLC (hexane); yield: 63 mg (68%).

2.9.6 Synthesis of the Core Frameworks of Biologically Active Molecules by Metal-Catalyzed Intramolecular Free-Radical Cyclization Reactions

2.9.6.1 Manganese(III)-Based Total Synthesis of Biologically Active Molecules

Over the past four decades, manganese(III) acetate mediated oxidative free-radical cyclization has progressed into a broadly applicable synthetic method. Owing to its broad range of properties, including a wide functional group tolerance, selectivity, spatial retention, and capability to be incorporated into complicated reaction sequences, manganese(III) acetate has attracted immense attention from chemists dealing with the synthesis and functionalization of diverse organic compounds. Specific examples of the synthesis of natural products and biologically active molecules using manganese(III)-mediated oxidative cyclization as a key step include the preparation of aloesaponol III;[359] okicenone;[359] avenaciolide;[360] epiupial;[361] fredericamycin models;[362,363] furanoditerpenes;[364,365] the CD ring system of gibberellic acid;[366] (+)- and (−)-podocarpic acid;[367] margocilin O-methyl ether;[368,369] triptoquinones B and C;[370] upial;[371] aryltetralin lignans;[372] (−)-jolkinolide E;[373] chromolaenin (laevigatin);[374] (−)-triptolide, (−)-triptonide, (+)-triptophenolide, and (+)-triptoquinonide;[87,88,375] spongidines A, B, and D;[376] (±)-annularin H;[377] araliopsine;[378] and (−)-estafiatin.[379] Certain significant total syntheses of natural and/or biologically active compounds mediated by manganese(III) acetate are discussed within this chapter.

2.9.6.1.1 Synthesis of (±)-Garcibracteatone

George and co-workers have recently utilized the manganese(III) acetate mediated oxidative radical cyclization technique for the synthesis of the polycyclic polyprenylated acylphloroglucinol natural product (±)-garcibracteatone (**232A**).[76] Garcibracteatone is a compact biologically potent polycyclic ring system with seven stereocenters, including five quaternary stereocenters.[380] The biomimetic radical cascade strategy starting from phloroglucinol (**230**) to give highly functionalized **232A** in just four steps illustrates the efficiency of manganese(III) acetate mediated oxidation in total synthesis. The oxidation of the inseparable mixture of diastereomers **231** in the presence of manganese(III) acetate/copper(II) acetate furnishes a mixture of (±)-garcibracteatone (**232A**, 14% yield) and (±)-5-*epi*-garcibracteatone (**232B**, 8% yield) (Scheme 66).

2.9.6 Synthesis of the Core Frameworks of Biologically Active Molecules

Scheme 66 Biomimetic Synthesis of (±)-Garcibracteatone[76]

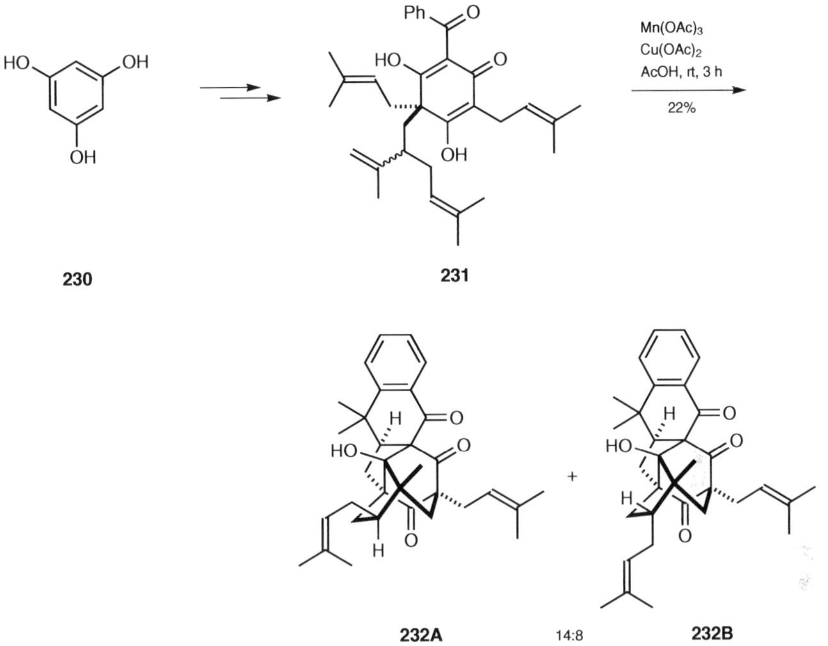

(±)-Garcibracteatone (232A):[76]

2-Benzoyl-3,5-dihydroxy-4-[5-methyl-2-(prop-1-en-2-yl)hex-4-en-1-yl]-4,6-bis(3-methylbut-2-en-1-yl)cyclohexa-2,5-dienone (**231**; 408 mg, 0.81 mmol) in degassed AcOH (12 mL) was added to a soln of Cu(OAc)$_2$•H$_2$O (154 mg, 0.81 mmol) and Mn(OAc)$_3$•2H$_2$O (457 mg, 1.70 mmol) in degassed AcOH (2 mL) at rt. The mixture was stirred at rt. After 3 h, the mixture was diluted with H$_2$O (20 mL) and extracted with EtOAc (3 × 30 mL). The combined extract was washed sequentially with H$_2$O (50 mL), sat. aq NaHCO$_3$ (50 mL), and brine (50 mL); it was then dried (MgSO$_4$), filtered, and concentrated under reduced pressure. The residue was purified by flash chromatography (silica gel, petroleum ether/EtOAc 20:1 to 15:1 gradient) to give a white solid. Crystallization (MeOH) gave crystals; yield: 56 mg (14%).

2.9.6.1.2 Synthesis of a Precursor to 7,11-Cyclobotryococca-5,12,26-triene

The manganese(III) acetate mediated oxidative lactonization of 2-(pent-4-enyl)malonate **233** to afford bicyclic intermediate **234** is a key step in the synthesis of 7,11-cyclobotryococca-5,12,26-triene (**235**), a unique botryococcene-related hydrocarbon (Scheme 67).[381]

Scheme 67 Synthesis of a Precursor to 7,11-Cyclobotryococca-5,12,26-triene[381]

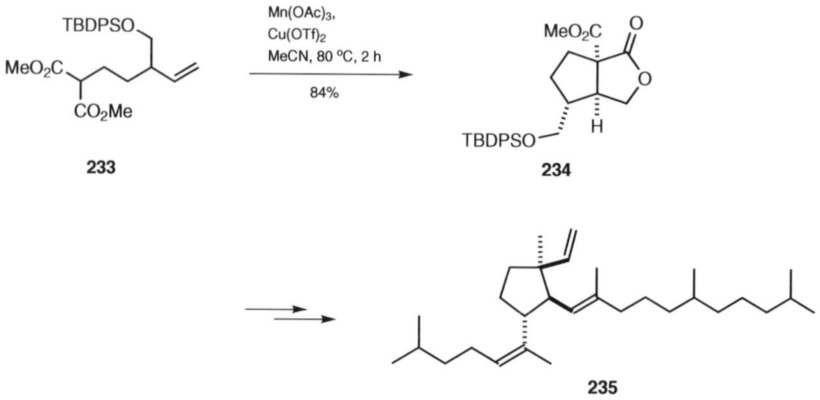

Methyl (3aR*,6R*,6aS*)-6-[(tert-Butyldiphenylsiloxy)methyl]-3-oxohexahydro-1H-cyclopenta[c]furan-3a(3H)-carboxylate (234):[381]

A soln of dimethyl 2-{3-[(tert-butyldiphenylsiloxy)methyl]pent-4-enyl}malonate (233; 2.40 g, 4.92 mmol) in MeCN (20 mL) was degassed by freeze–pump–thaw cycles (3×). Mn(OAc)$_3$·2H$_2$O (2.65 g, 9.88 mmol) and Cu(OTf)$_2$ (1.77 g, 4.89 mmol) were added in one portion at ambient temperature. The mixture was stirred vigorously at 80 °C for 2 h. The progress of the reaction was monitored by TLC, and the color of the mixture changed from dark brown to dark blue-green. Upon completion of the reaction, the mixture was cooled to ambient temperature, diluted with Et$_2$O (30 mL), quenched with H$_2$O (20 mL), opened to air, and stirred for another 12 h, which allowed the metal byproducts to oxidize, as observed by a change in the color of the soln from dark blue-green to a lighter blue with no trace of the original brown coloration. After this, the majority of the brown solid initially lying at the phase boundary eventually disappeared. At this point, the mixture was transferred to a separatory funnel, extracted with Et$_2$O (100 mL), and washed with H$_2$O (300 mL) and brine (150 mL). The organic phase was dried (MgSO$_4$), filtered, and concentrated under reduced pressure. The crude mixture was purified by flash column chromatography (hexane/Et$_2$O 1:1) to afford a colorless oil; yield: 1.87 g (84%); R_f 0.34 (hexane/Et$_2$O 1:1).

2.9.6.1.3 Total Syntheses of Bakkenolides I, J, and S

A manganese(III) acetate mediated diastereoselective cyclization approach to create a lactone moiety has been utilized by the group of Scheidt for the asymmetric total syntheses of bakkenolide I (238, R^1 = COiPr), bakkenolide J (238, R^1 = COiBu), and bakkenolide S (238, R^1 = H).[382] The bakkanes, possessing a wide array of biological activities,[383,384] are significant members of the sesquiterpene class of natural products bearing a characteristic cis-fused 6,5-bicyclic core.[385,386] Treatment of β-keto propargyl ester 236 with manganese(III) acetate results in a highly diastereoselective cyclization reaction that produces single diastereomeric spirocyclic lactone 237 in good yield (70%). Further reaction of ketone 237 leads to the formation of bakkenolides I, J, and S (Scheme 68).

2.9.6 Synthesis of the Core Frameworks of Biologically Active Molecules

Scheme 68 Syntheses of Bakkenolides I, J, and S[382]

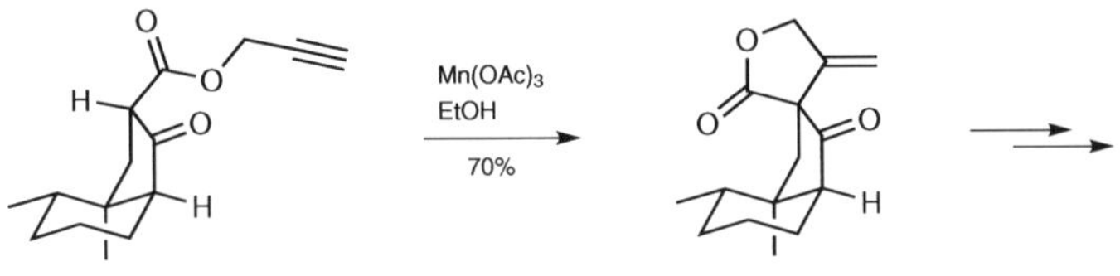

236 dr 9:1   **237** dr 20:1

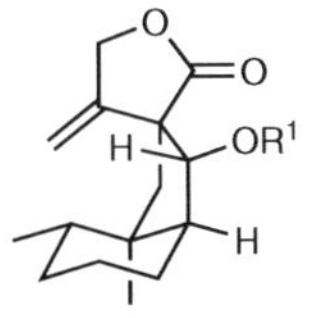

238

(2′S,3a′R,4′S,7a′S)-3a′,4′-Dimethyl-4-methyleneoctahydro-2H-spiro[furan-3,2′-indene]-1′,2(3′H)-dione (237):[382]
A flame-dried, round-bottomed flask equipped with a magnetic stirrer bar was charged with Mn(OAc)$_3$•H$_2$O (627 mg, 2.7 mmol) and degassed (freeze–pump–thaw) abs EtOH (4 mL). A soln of ester **236** (200 mg, 0.9 mmol) in degassed abs EtOH (6 mL, 0.15 M) was transferred dropwise to the reaction vessel by cannula. After stirring for 20 h, the mixture was filtered through a mixture of silica gel and Celite (Et$_2$O). The filtrate was concentrated and purified by flash column chromatography (silica gel, hexane/EtOAc 10:1 to 5:1 gradient) to afford a colorless oil; yield: 140 mg (70%, as reported).

2.9.6.1.4 Synthesis of the Core Framework of the Welwitindolinone Alkaloids

Rawal and co-workers have reported the synthesis of the core framework of the welwitindolinone family of indole alkaloids bearing a unique bicyclo[4.3.1]decane skeleton. Manganese(III) acetate promotes the oxidative cyclization of 3-substituted indoles **241** at the 4-position to give **242**[387] only if the C2-position is blocked by a weak leaving group (e.g., chloride). The authors argue that the reactivity of indoles **239** toward cyclization at the C2-position, that is, at the pyrrole part, to give **240**,[224,388] can be avoided by blocking the C2-position with a poor leaving group. By doing so, manganese(III)-mediated oxidative cyclization of **241** at the 4-position enables the construction of the distinctive framework of the welwitindolinone alkaloids [e.g., N-methylwelwitindolinone D isonitrile (**243**)] (Scheme 69).

Scheme 69 Oxidation of Indole Alkaloids[387]

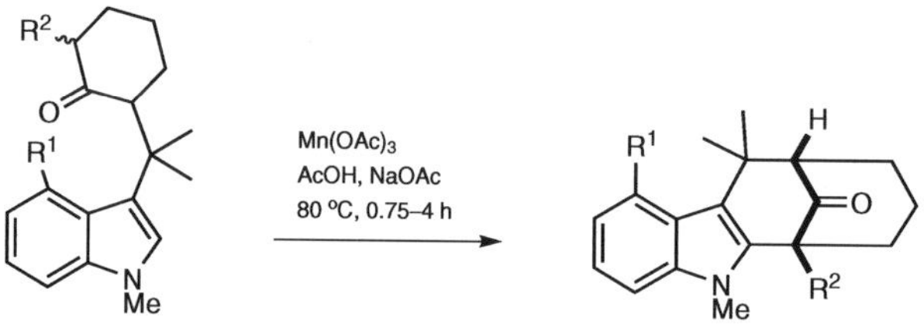

239 **240**

R1	R2	Yield (%)	Ref
H	CO_2Me	82	[387]
H	CHO	87	[387]
H	$CO_2(CH_2)_2Cl$	43	[387]
Br	CO_2Me	64	[387]

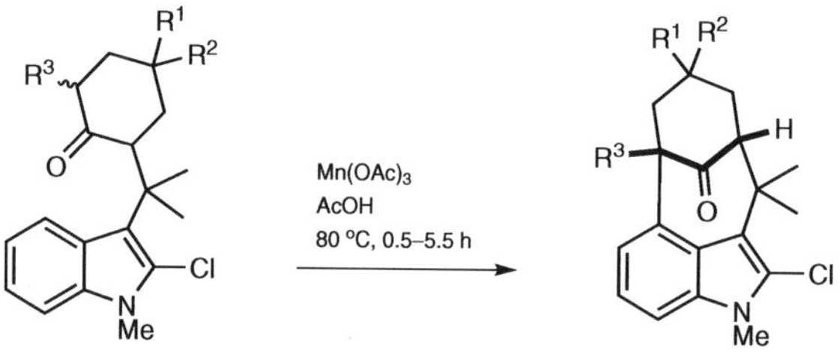

241 **242**

R1	R2	R3	Yield (%)	Ref
H	H	CHO	0	[387]
H	H	CO_2Me	66	[387]
$O(CH_2)_2O$		CO_2Me	67	[387]
$O(CH_2)_2O$		$CO_2(CH_2)_2Cl$	51	[387]
$O(CH_2)_2O$		$CO_2CH_2CCl_3$	52	[387]

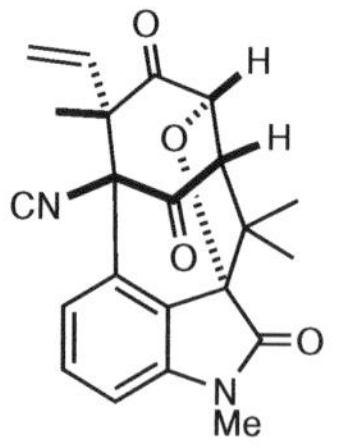

243

Indole-Fused Bicyclo[3.3.1]nonanes 240; General Procedure:[387]
$Mn(OAc)_3$•$2H_2O$ (2.0 equiv) and NaOAc (2.0 equiv) were added to a soln of 3-substituted indole **239** (1.0 equiv) in glacial AcOH. The dark brown soln was stirred at 80 °C (oil bath) until it turned into a clear orange-yellow soln. Upon completion of the reaction, the mixture was cooled to ambient temperature and carefully poured over sat. aq Na_2CO_3 and extracted with EtOAc. The crude product was purified by chromatography (silica gel).

2.9.6 Synthesis of the Core Frameworks of Biologically Active Molecules

Indole-Fused Bicyclo[4.3.1]decanes 242; General Procedure:[387]
Mn(OAc)$_3$•2H$_2$O (2.0 equiv) was added to a soln of 3-substituted indole **241** (1.0 equiv) in glacial AcOH. The dark brown soln was stirred at 80 °C (oil bath) until it turned into a clear orange-yellow soln. Upon completion of the reaction, the mixture was cooled to ambient temperature and carefully poured over sat. aq Na$_2$CO$_3$ and extracted with EtOAc. The crude product was purified by chromatography (silica gel).

2.9.6.1.5   Synthesis of Ageliferins and Palau'amine

Lactonization mediated by manganese(III) has been employed as a key step in the asymmetric synthesis of ageliferins **248**. These pyrrole–imidazole alkaloids are found in many *Agelas* and *Stylissa* sponges and bear significant biological activity.[389,390] The oxidation of β-keto ester **246** (synthesized by aldol reaction of imidazole-5-carbaldehyde **244** with aminopropanoate **245** followed by Dess–Martin periodinane oxidation) with manganese(III) acetate in acetic acid at 50–60 °C gives C9-stereoisomers **247A/247B** in a 2.5–3:1 ratio. Lactones **247A/247B**, after a series of operations, produce ageliferin (**248**, R^1 = R^2 = H) (Scheme 70).[53] This method has also been used in the synthesis of bromoageliferin (**248**, R^1 = H; R^2 = Br) and dibromoageliferin (**248**, R^1 = R^2 = Br) in their natural enantiomeric forms.[78]

Scheme 70   Synthesis of Ageliferins[53,78]

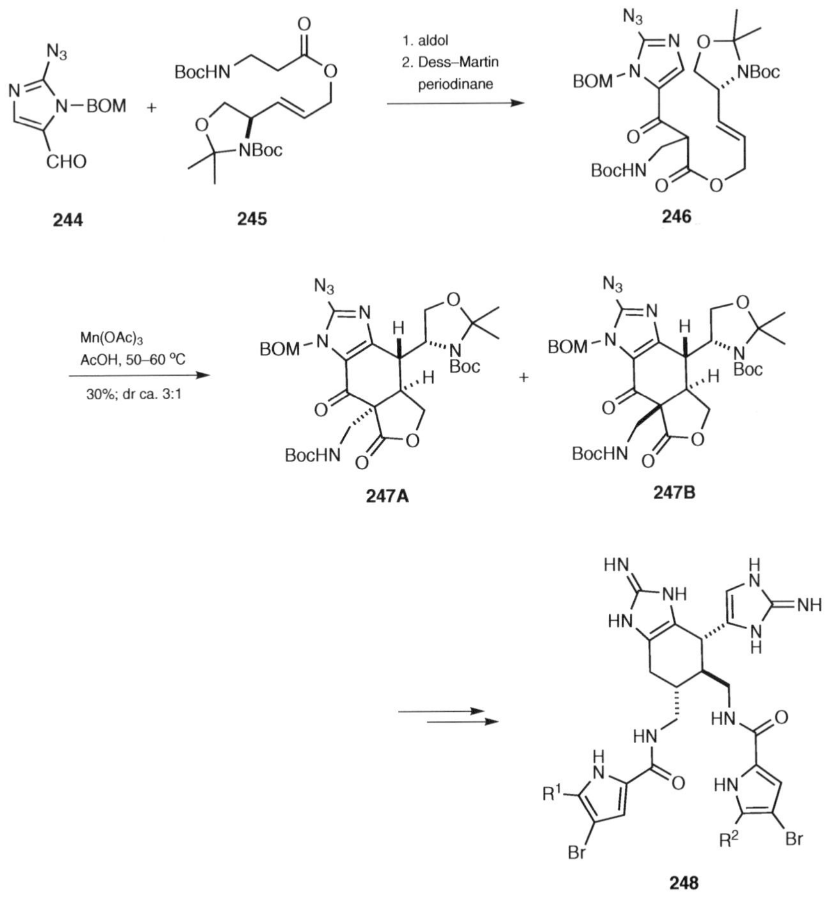

In another approach, Chen and co-workers have employed this method for the synthesis of **251** as a close analogue of the dimeric pyrrole–imidazole alkaloid palau'amine (**252**). Starting from structure **249**, a series of chemical transformations afford analogue **251** via intermediate **250** (Scheme 71).[79]

Scheme 71 Synthesis of a Palau'amine Analogue[79]

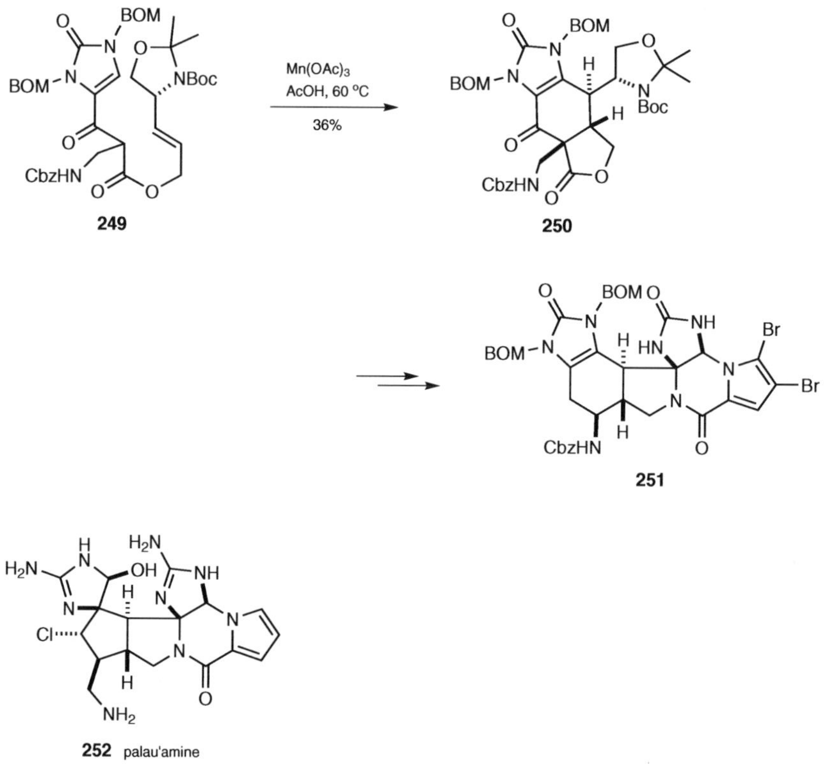

Ageliferin Precursor Lactones 247A/247B:[53,78]
A 50-mL flask was charged with Mn(OAc)$_3$·2H$_2$O (1.37 g, 5.10 mmol, 2.5 equiv) and crude **246**. AcOH (15 mL) degassed by freeze–pump–thaw cycles (3 ×) was then added. The dark brown mixture was heated to 50–60 °C under argon for 8 h. After removing AcOH, the residue was dissolved in a mixture of EtOAc (50 mL) and 10% aq NaHSO$_3$ (50 mL). The organic layer was extracted, washed with brine (2 × 30 mL), dried (Na$_2$SO$_4$), filtered, concentrated, and purified with a Biotage KP-C18-HS (39 × 157 mm, 120 g) column (MeCN/H$_2$O 50:50 to 65:35). The impure fractions were collected and again purified by preparative HPLC (Waters Atlantis dC18 OBD, 19 × 150 mm, 5 µm, solvent A: H$_2$O; solvent B: MeCN; gradient: t = 0 min: 60% B, t = 15 min: 70% B, t = 45 min: 70% B; flow rate 5.0 mL·min^{-1}), which afforded **247A** as a white solid; yield: 258 mg (18–25%, depending on amount of **246**), and **247B** as a white solid; yield: ca. 9% yield for four steps; t_R 36 min; R_f 0.58 (EtOAc/hexanes 1:1).

2.9.6.1.6 Synthesis of (±)-Ialibinones A and B

The synthesis of (±)-ialibinones A (**254A**) and B (**254B**) can be achieved by manganese(III) acetate mediated 5-*exo-trig* cyclization of acylphloroglucinol derivative **253**. Simpkins and Weller have devised this convenient route for the construction of many stereocenters and multiple bonds in the lowest number of possible steps. Treatment of **253** with manganese(III) acetate and copper(II) acetate in acetic acid allows its smooth conversion into ialibinones A and B in 35% overall yield [ratio (**254A**/**254B**) ~41:59] (Scheme 72).[391]

Scheme 72 Synthesis of (±)-Ialibinones A and B[391]

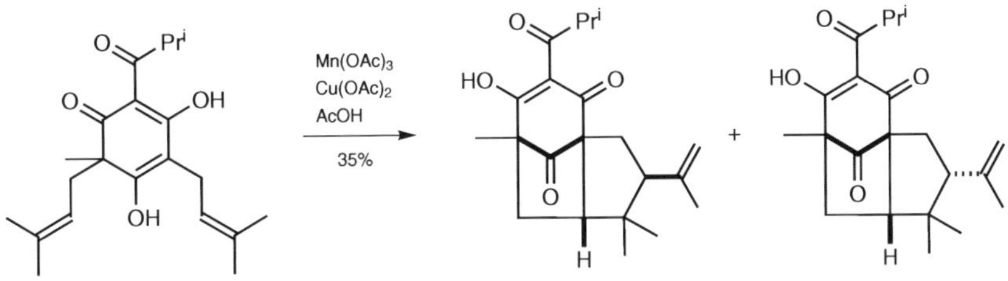

Ialibinone A (254A) and Ialibinone B (254B):[391]

Mn(OAc)$_3$•2H$_2$O (379 mg, 1.41 mmol) and Cu(OAc)$_2$•H$_2$O (141 mg, 0.71 mmol) were added to a soln of acylphloroglucinol derivative **253** (245 mg, 0.71 mmol) in glacial AcOH (15 mL) at rt. The mixture was stirred for 30 min and then diluted with petroleum ether (20 mL) and filtered through a short plug of silica gel [Et$_2$O/petroleum ether 1:4 (60 mL)]. The solvent was removed under reduced pressure, and the residue was purified by flash column chromatography (Et$_2$O/petroleum ether 1:9) to yield a 41:59 mixture of **254A**/**254B** as a yellow oil; yield: 85 mg (35%); R_f 0.50 (EtOAc/petroleum ether 1:4). The combined mixture was separated by reversed-phase HPLC using a Phenomenex Luna C18(2) semipreparative column and MeCN/H$_2$O/TFA (70:29.95:0.05) as the eluent. t_R (25 °C, flow rate: 3 mL•min^{-1}): 49.6 (for **254A**), 53.4 min (**254B**). After isolation, MeCN was removed under reduced pressure, and the aqueous layer was extracted with Et$_2$O (3 × 20 mL). The combined organic layer was washed with brine (20 mL) and dried (Na$_2$SO$_4$). Removal of the solvent under reduced pressure afforded **254A** as a yellow oil; yield: 24 mg; and **254B** as a yellow oil; yield: 36 mg.

2.9.6.1.7 Synthesis of the ABC Ring System of Zoanthenol

The stereospecific synthesis of **258** as the ABC ring core of zoanthenol (**259**), a polyfunctionalized phenanthrene carbocycle, can be achieved by utilizing the manganese(III)-mediated radical cyclization strategy. Zoanthenol is a zoanthamine alkaloid bearing a heptacyclic ring system with phenanthrene carbocycle (ABC) and amino acetal (DEFG) moieties. The manganese(III) acetate mediated oxidative radical cyclization of β-keto ester **256**, synthesized from 7-methyl-1-benzopyran-2-one (**255**), provides key tricyclic system **257**. After multiple operations, ketonic ABC ring **257** is transformed into zoanthenol ABC ring **258** (Scheme 73).[392]

Scheme 73 Synthesis of the ABC Ring of Zoanthenol[392]

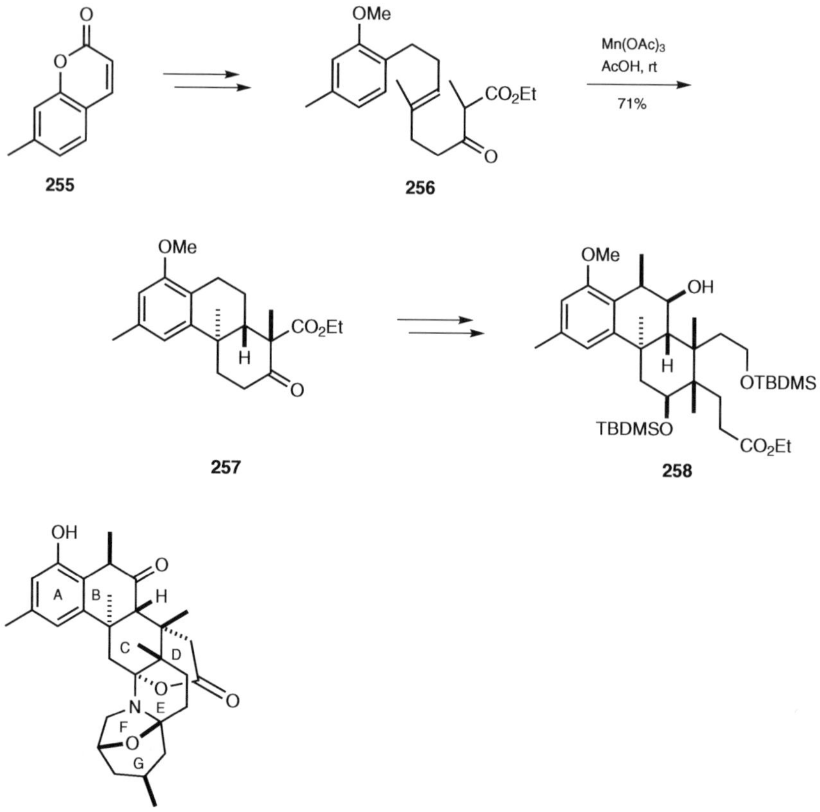

Ethyl (1S*,4aR*,10aS*)-8-Methoxy-1,4a,6-trimethyl-2-oxo-1,2,3,4,4a,9,10,10a-octahydrophenanthrene-1-carboxylate (257):[392]
Mn(OAc)$_3$·2H$_2$O (3.24 g, 12.0 mmol) was added to a soln of β-keto ester **256** (1.90 g, 5.49 mmol) in degassed glacial AcOH (22 mL) at rt. After stirring for 12 h, aq NaHSO$_4$ was added. After stirring for another 1 h, the mixture was extracted with EtOAc. The organic layer was washed with brine, dried (MgSO$_4$), and concentrated under reduced pressure. The residue was purified by flash column chromatography (hexane/EtOAc 50:1 to 30:1) to yield a white solid; yield: 1.37 g (71%); mp 131–132 °C.

2.9.6.1.8 Synthesis of the Tetracyclic Core of Tronocarpine

The manganese(III)-mediated oxidative radical cyclization of N-malonyl-functionalized indole derivatives yields 1,2-annulated products. This methodology has been utilized by Magolan and Kerr for the synthesis of **262**, the tetracyclic subunit of tronocarpine (**263**). Radical cyclization of N-acylated 2-(1H-indol-3-yl)acetonitrile **260** in the presence of manganese(III) acetate produces nitrile **261** in 72% yield. Nitrile **261** then cyclizes to amide **262** upon reduction with Raney nickel (Scheme 74).[393]

2.9.6 Synthesis of the Core Frameworks of Biologically Active Molecules

Scheme 74 Synthesis of the Tetracyclic Core of Tronocarpine[393]

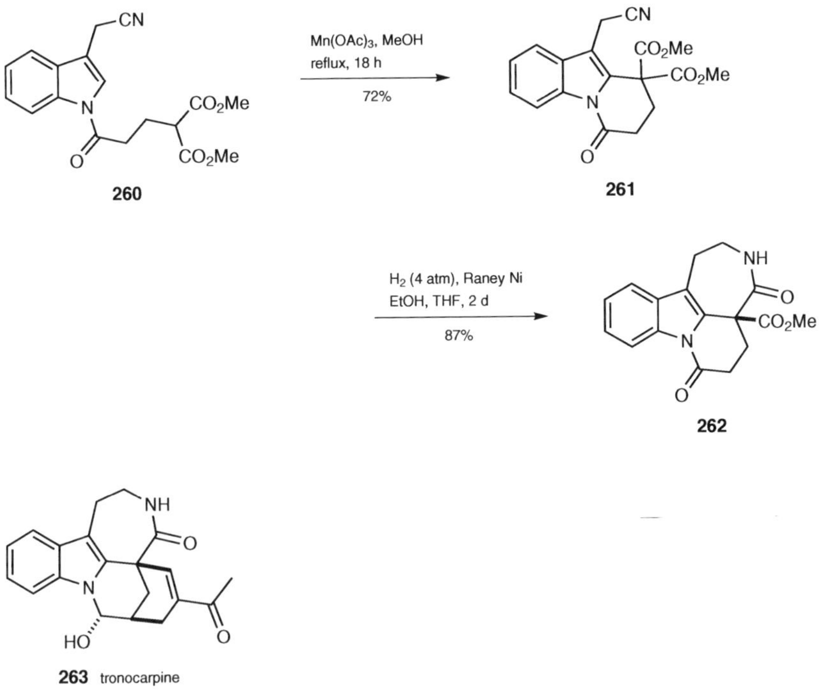

Dimethyl 10-(Cyanomethyl)-6-oxo-7,8-dihydropyrido[1,2-a]indole-9,9(6H)-dicarboxylate (261); Typical Procedure:[393]

Mn(OAc)$_3$·2H$_2$O (1.71 g, 6.39 mmol) was added to a round-bottomed flask charged with a soln of indole **260** (729 mg, 2.13 mmol) in MeOH (25 mL) and equipped with a magnetic stirrer bar. The dark brown mixture was stirred at reflux under an argon atmosphere. After 16–24 h, TLC analysis showed complete consumption of the starting material. The solvent was removed under reduced pressure, and the brown solid was partitioned between EtOAc and brine. The combined aqueous layer was extracted with EtOAc (3 ×). The organic extract was washed with H$_2$O (3 ×) and then dried (MgSO$_4$), and the solvent was removed under reduced pressure. The crude residue was purified by flash chromatography (silica gel, EtOAc/hexanes gradient) to afford an orange solid; yield: 519 mg (72%); mp 125–130 °C; R_f 0.75 (EtOAc/hexanes 7:3).

2.9.6.1.9 Synthesis of the ABC Ring of Hexacyclinic Acid

Hexacyclinic acid (**266**) with six fused rings can be isolated from *Streptomyces cellulosae* subspecies *griseorubiginosus* (strain S1013), and it displays significant cytotoxicity.[394] Toueg and Prunet have found that treatment of a mixture of β-keto esters **264A** and **264B** with manganese(III) acetate and copper(II) acetate in the presence of 2,2,2,-trifluoroethanol/acetic acid (10:1) affords hexacyclinic acid ABC ring **265** as a single diastereomer having suitable functionality so that it can be converted into the natural acid (Scheme 75).[395]

Scheme 75 Synthesis of the ABC Ring of Hexacyclinic Acid[395]

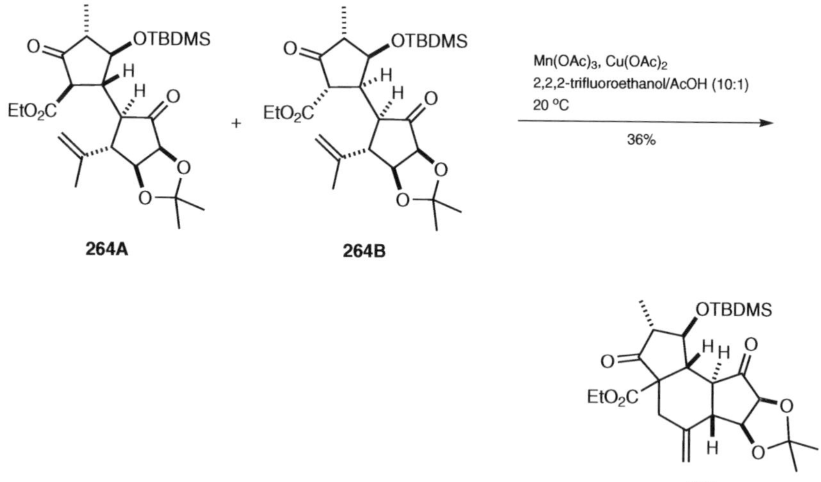

264A + **264B**

Mn(OAc)₃, Cu(OAc)₂
2,2,2-trifluoroethanol/AcOH (10:1)
20 °C
36%

265

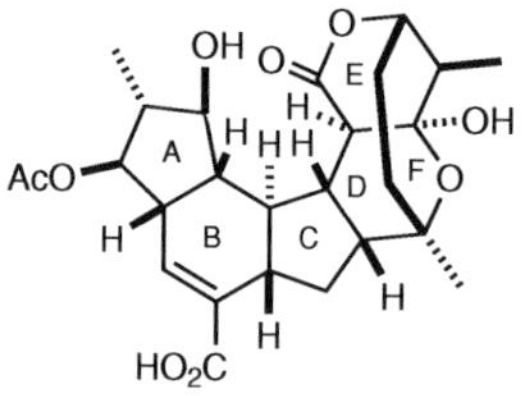

266 hexacyclinic acid

Ethyl (1S,2R,3aR,5aS,5bS,8aS,9aS,9bR)-1-(tert-Butyldimethylsiloxy)-2,7,7-trimethyl-5-methylene-3,9-dioxodecahydro-6,8-dioxacyclopenta[b]-as-indacene-3a-carboxylate (265):[395]

Cu(OAc)₂·H₂O (1.15 g, 5.78 mmol, 2 equiv) was added to a soln of β-keto esters **264A/264B** (2.5:1; 1.429 g, 2.9 mmol) in degassed 2,2,2-trifluoroethanol/AcOH (10:1; 30 mL) at rt under argon. The mixture was stirred for 30 min to dissolve Cu(OAc)₂ partially, and then the Mn(OAc)₃·4H₂O (1.55 g, 5.8 mmol, 2 equiv) was added. The mixture was stirred for another 1.5 h. Then, the mixture was dissolved in H₂O and Et₂O, and K₂CO₃ was carefully added to neutralize AcOH. After neutralization, the mixture was extracted with Et₂O (3 ×). The combined organic extract was dried (MgSO₄) and concentrated under reduced pressure. The residue was purified by flash column chromatography (petroleum ether/Et₂O 4:1) to yield a white foam; yield: 509 mg (36%).

2.9.6.1.10 Synthesis of the Tetracyclic Framework of Azadiradione

The unique ability of manganese(III) acetate to mediate a tandem cyclization reaction in a concise manner[52] has been exploited on many occasions for the synthesis of biologically active molecules. In this regard, the protocol developed by the group of Yamashita for the synthesis of **269** as the tetracyclic framework of azadiradione (**270**) is highly recognized. Azadiradione is a highly oxygenated limonoid triterpene present in the neem plant and known to possess a unique stereostructure along with a wide range of biological activities.[80] Manganese(III) acetate mediated tandem radical cyclization of acyclic β-keto ester **267** produces four C—C bonds and seven stereogenic centers in **268A/268B**, with the major product featuring a cis configuration of C13 and C14, in one single step. Subsequent reaction steps lead to the formation of azadiradione framework **269** (Scheme 76).

2.9.6 Synthesis of the Core Frameworks of Biologically Active Molecules

Scheme 76 Synthesis of the Tetracyclic Framework of Azadiradione[80]

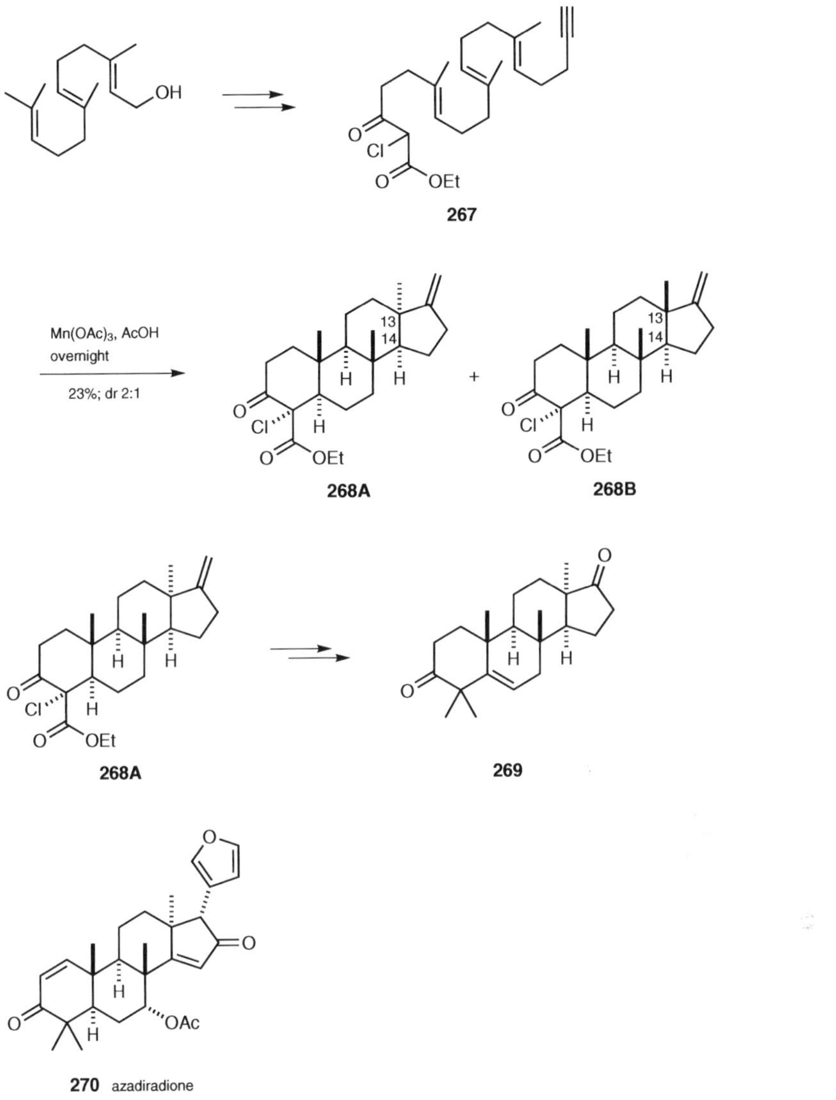

Tetracyclic Azadiradione Precursors 268A and 268B:[80]

Mn(OAc)$_3$·2H$_2$O (3.07 g, 7.06 mmol) was added to a soln of β-keto ester **267** (2.33 g, 5.72 mmol) in degassed glacial AcOH (57 mL). The mixture was stirred at rt overnight. Upon completion of the reaction, aq Na$_2$S$_2$O$_3$ was added, and the mixture was extracted with EtOAc. The organic extract was washed with brine and dried (Na$_2$SO$_4$). The combined organic extract was concentrated under reduced pressure and purified by column chromatography (silica gel, hexane/EtOAc 50:1 to 25:1); yield: 512 mg (23%); ratio (**268A/268B**) 2:1. The mixture of isomers was further purified by recrystallization (hexane) to give **268A** as colorless crystals; mp 173–178 °C.

2.9.6.1.11 Biomimetic Total Synthesis of (±)-Yezo'otogirin A

Yezo'otogirins A–C[2] are an unusual family of rearranged polycyclic polyprenylated acylphloroglucinol natural products isolated from *Hypericum yezoense* possessing a compact, tricyclic ring system with an endocyclic enol ether.[396–398] Recently, the George group reported the concise diastereoselective synthesis of yezo'otogirin A (**273**) from a commercially available starting material employing an intramolecular radical cyclization strategy.[399] This is the first report on the formation of a polycyclic enol using manganese(III) acetate/copper(II) trifluoromethanesulfonate. Pre-yezo'otogirin A (**272**), the biosynthetic precursor of yezo'otogirin A, can be synthesized in ten steps starting from 3-ethoxycyclohex-2-enone (**271**). Then, biomimetic oxidative radical cyclization allows construction of the unique 6,5,5-tricyclic ring system of yezo'otogirin A (**273**) (Scheme 77).

Scheme 77 Biomimetic Total Synthesis of (±)-Yezo'otogirin A[399]

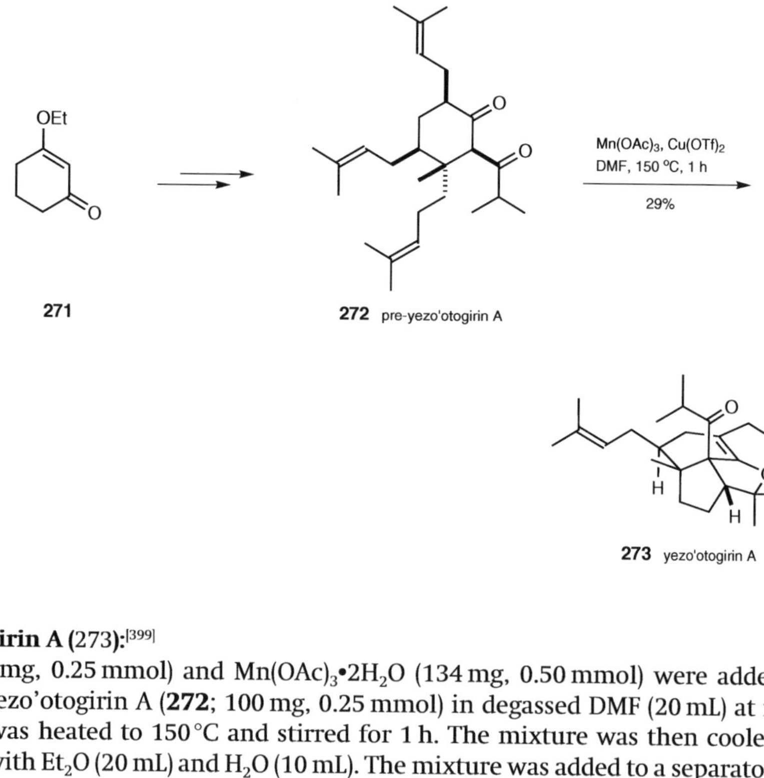

271

272 pre-yezo'otogirin A

Mn(OAc)$_3$, Cu(OTf)$_2$
DMF, 150 °C, 1 h

29%

273 yezo'otogirin A

(±)-Yezo'otogirin A (273):[399]
Cu(OTf)$_2$ (90 mg, 0.25 mmol) and Mn(OAc)$_3$·2H$_2$O (134 mg, 0.50 mmol) were added to a soln of pre-yezo'otogirin A (**272**; 100 mg, 0.25 mmol) in degassed DMF (20 mL) at rt. The suspension was heated to 150 °C and stirred for 1 h. The mixture was then cooled to rt and diluted with Et$_2$O (20 mL) and H$_2$O (10 mL). The mixture was added to a separatory funnel, and the organic layer was separated. Thereafter, the aqueous layer was extracted with Et$_2$O (2 × 20 mL). The combined organic extract was washed with H$_2$O (2 × 30 mL) and brine (30 mL), dried (MgSO$_4$), filtered, and concentrated under reduced pressure. The residue was purified by flash chromatography (silica gel, petroleum ether/EtOAc 50:1) to give a colorless oil; yield: 29 mg (29%); overall yield from **271**: 3%.

2.9.6.2 Titanium-Catalyzed Total Synthesis of Biologically Active Molecules

The remarkable chemoselectivity of titanocene(III) reagents has been used to advantage in the synthesis of various complex natural products, and these syntheses have showcased the potential of these reagents. Specific examples of the synthesis of natural and biologically active molecules using titanium-mediated oxidative cyclization as a key step include the preparation of 2-*epi*-rosmarinecine,[400] gabapentin,[401] pulchellalactam,[401]

(−)-methylenolactocin,[402] (−)-protolichesterinic acid,[402] the Taxol framework,[403] the A- and C-ring synthons of paclitaxel,[404] rostratone,[405] (±)-smenospondiol,[406] and the AB ring moiety of the fomitellic acids.[407] Significant total syntheses of naturally active compounds using titanium-mediated radical cyclization are summarized below.

2.9.6.2.1 Synthesis of Magnofargesin and 7′-Epimagnofargesin

The Roy group has reported the formal synthesis of two bioactive lignans, magnofargesin (**276A**) and 7′-epimagnofargesin (**276B**), in both racemic and optically active forms utilizing the chlorobis(η^5-cyclopentadienyl)titanium(III)-mediated radical-induced cyclization reaction.[408] Commercially available 3,4,5-trimethoxybenzene derivatives **274** and **277** can be used as the starting materials for the synthesis of corresponding epoxides *rac*-**275** and (*S*,*R*)-**275**. These epoxides are then subjected to chlorobis(η^5-cyclopentadienyl)titanium(III)-induced radical cyclization to obtain desired bioactive lignans **276A** and **276B**, in both racemic (Scheme 78) and optically active (Scheme 79) forms.

Scheme 78 Synthesis of Racemic Magnofargesin and 7′-Epimagnofargesin[408]

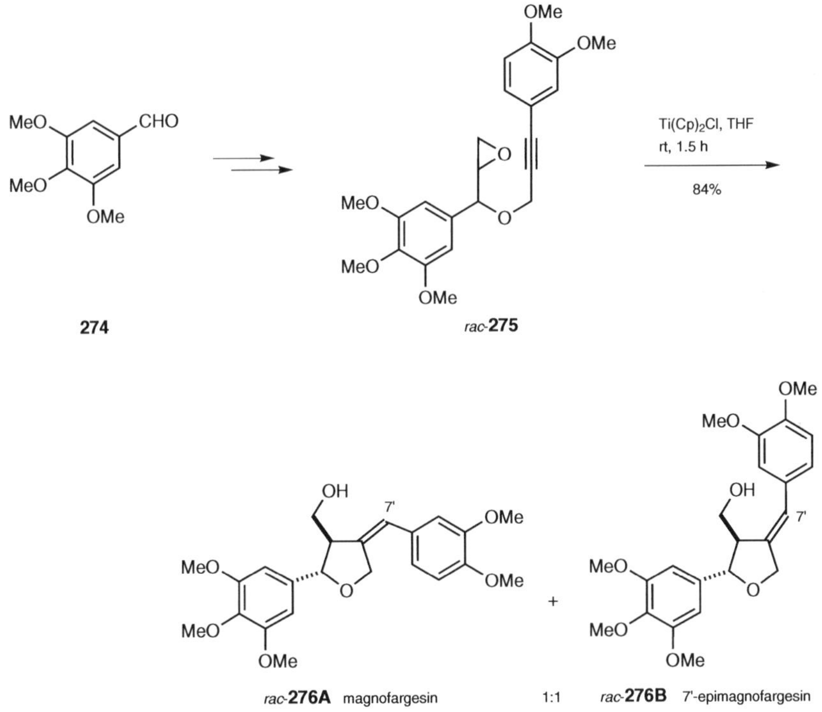

Scheme 79 Synthesis of Optically Active Magnofargesin and 7′-Epimagnofargesin[408]

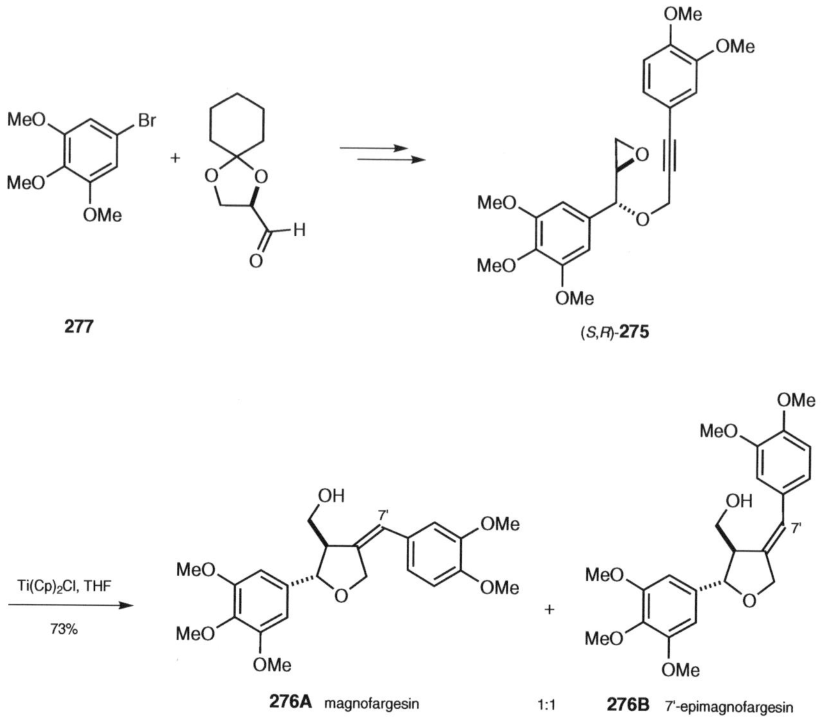

Magnofargesin (276A) and 7′-Epimagnofargesin (276B):[408]

The Ti(Cp)$_2$Cl catalyst was prepared in situ by treating commercially available Ti(Cp)$_2$Cl$_2$ (564 mg, 2.28 mmol) in dry THF (25 mL, not deoxygenated) with activated Zn dust (360 mg, 5.5 mmol) for 1 h under argon [activated Zn dust was prepared by washing commercial Zn dust (20 g) with 4 M HCl (60 mL), H$_2$O, and dry acetone, and then drying under reduced pressure]. The resulting green soln of the Ti(Cp)$_2$Cl catalyst was then added dropwise to a stirred soln of (S,R)-**275** (290 mg, 0.70 mmol) in dry THF (25 mL) at rt under argon over 1 h. The suspension was then stirred overnight and was decomposed with 10% H$_2$SO$_4$ (10 mL). The suspension was dried under reduced pressure, and the residue was extracted with Et$_2$O (4 × 30 mL). The combined Et$_2$O extract was washed with sat. NaHCO$_3$ (2 × 25 mL) and finally dried (Na$_2$SO$_4$). After removal of the solvent under reduced pressure, the crude residue was purified by column chromatography (silica gel, EtOAc/petroleum ether 1:4) to afford a mixture of isomers; yield: 213 mg (73%); ratio (**276A/276B**) 1:1.

2.9.6.2.2 Short and Stereoselective Total Synthesis of (±)-Sesamin

A short, efficient, and stereoselective synthesis of (±)-sesamin has been achieved through the radical cyclization of epoxide **280** using chlorobis(η^5-cyclopentadienyl)titanium(III) as the radical initiator.[409] Sesamin is a furofuran lignan isolated from the bark of *Fagara* plants and from sesame oil. Treatment of the known[410] isomeric mixture of epoxides **278** with 5-[(1E)-3-bromoprop-1-en-1-yl]-2H-1,3-benzodioxole (**279**) affords epoxides **280** (Scheme 80) in 78% yield as a 1:1 isomeric mixture. The isomers are not easily separable by common chromatography techniques, but treatment of crude epoxides **280** with chlorobis(η^5-cyclopentadienyl)titanium(III) and then iodine furnishes (±)-sesamin (**281**) as the only product.

2.9.6 Synthesis of the Core Frameworks of Biologically Active Molecules 513

Scheme 80 Stereoselective Synthesis of (±)-Sesamin[409]

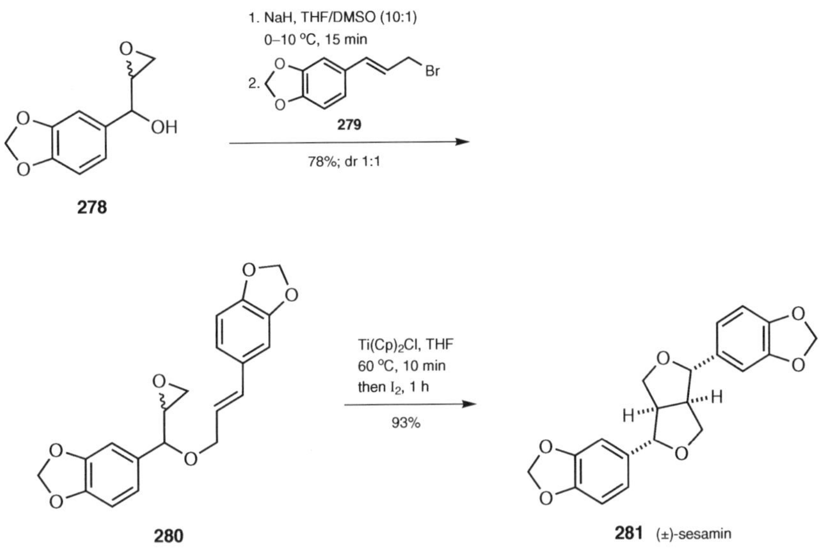

5-{(1E)-3-[(2H-1,3-Benzodioxol-5-yl)(oxiran-2-yl)methoxy]prop-1-en-1-yl}-2H-1,3-benzodioxole (280):[409]

A soln of epoxy alcohol **278** (2.0 mmol) in THF (5 mL) was added dropwise to a magnetically stirred suspension of NaH (60% dispersion; 6.0 mmol) in THF/DMSO (10:1; 20 mL) at rt under N_2. After evolution of H_2 gas ceased, a soln of 5-[(1E)-3-bromoprop-1-en-1-yl]-2H-1,3-benzodioxole (**279**; 2.6 mmol) in THF (5 mL) was added dropwise at 0 °C over 15 min up to 10 °C. The mixture was then stirred at rt for 6 h and then carefully decomposed with ice/H_2O. After removal of most of the solvent under reduced pressure, the resulting residue was extracted with Et_2O (3 × 25 mL). The combined organic extract was washed with brine and dried (Na_2SO_4). The solvent was then removed under reduced pressure, and the residue was further purified by column chromatography (silica gel, EtOAc/petroleum ether) to afford a 1:1 mixture of isomers; yield: 78%.

(±)-Sesamin (281):[409]

A green soln of Ti(Cp)$_2$Cl (4 mmol) in THF (25 mL) was transferred by cannula to an addition funnel and was added dropwise to a magnetically stirred soln of crude epoxy ether **280** in THF (25 mL) at 60 °C. After stirring for another 10 min, an excess amount of I_2 was added, and the mixture was stirred for another 1 h at that temperature. Upon completion of the reaction, the mixture was quenched with 10% aq H_2SO_4 and was extracted with Et_2O (3 × 25 mL). The organic extract was washed with sat. aq $NaHCO_3$ and brine and was then dried (Na_2SO_4). Et_2O was removed under reduced pressure, and the residue was purified by column chromatography (silica gel, EtOAc/petroleum ether); yield: 93%; mp 123–124 °C.

2.9.6.2.3 Total Synthesis of (±)-Dihydroprotolichesterinic Acid and Formal Synthesis of (±)-Roccellaric Acid

The Roy group has synthesized many polysubstituted tetrahydrofuran antitumor antibiotics, including (±)-dihydroprotolichesterinic acid (**285A**) and (±)-roccellaric acid (**285B**), using chlorobis(η^5-cyclopentadienyl)titanium(III)-mediated free-radical chemistry (Scheme 81).[411] Their short and efficient method relies on the use of epoxide **282** as the starting material, which is converted into allyl ether **283**. Radical cyclization of **283** produces tetrahydrofuran **284** as the key intermediate. The crude alcohol **284** is then oxidized to yield a diastereomeric mixture of acids **285** (dr 5:1) as a light-yellow solid. Other polysubstituted tetrahydrofuran antitumor antibiotics such as methylenolactocin and protolichesterinic acid[410] have also been synthesized by the same group.

Scheme 81 Synthesis of (±)-Dihydroprotolichesterinic Acid and (±)-Roccellaric Acid[411]

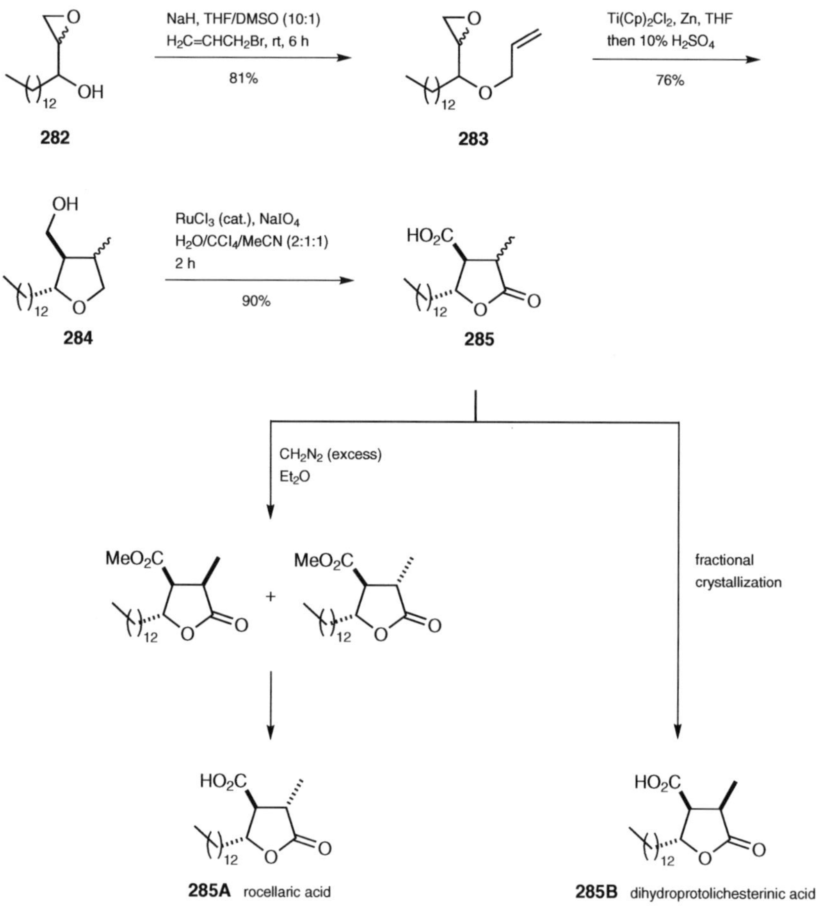

[(2R,3R)-2-Tridecyl-4-methyloxolan-3-yl]methanol (284); Typical Procedure:[411]
A green soln of Ti(Cp)$_2$Cl (4 mmol) in THF (25 mL) was added dropwise to a magnetically stirred soln of epoxide **283** (590 mg, 2 mmol) in dry THF (50 mL) under argon at rt. The suspension was stirred for 1 h and then carefully quenched with 10% H$_2$SO$_4$ (100 mL). After stirring for 30 min, THF was removed under reduced pressure, and the residue was purified by column chromatography (silica gel, EtOAc/petroleum ether 1:4) to afford an inseparable mixture of two isomers (5:1) as a viscous oil; yield: 450 mg (76%).

2.9.6.2.4 Stereoselective Total Synthesis of Furano and Furofuran Lignans

The control of stereochemistry in the polysubstituted furan moiety of furano and furofuran lignans is considered a challenging and interesting part of their synthesis. The Roy group has devised a general method to synthesize these biologically active lignans by an intramolecular radical cyclization strategy. The method generally proceeds through the generation of radicals from epoxides in the presence of a titanocene(III) radical source.[163] Substituted epoxy ethers **287** are first prepared from epoxides **286** and aryl-substituted

2.9.6 Synthesis of the Core Frameworks of Biologically Active Molecules

allyl bromides, and intramolecular radical cyclization using chlorobis(η^5-cyclopentadienyl)titanium(III) as the radical source results in the formation of trisubstituted tetrahydrofurano lignans **288** or 2,6-diaryl-3,7-dioxabicyclo[3.3.0]octane lignans **290** depending on the reaction conditions (Scheme 82). Upon radical cyclization followed by acidic workup (10% H_2SO_4), epoxy alkene ethers **287** yield furano lignans **288** (with other isomers; 5:1 ratio), that is, (±)-dihydrosesamin, (±)-lariciresinol dimethyl ether, (±)-acuminatin methyl ether, and (±)-sanshodiol methyl ether, directly, or lignans **289**, that is, (±)-lariciresinol, (±)-acuminatin, and (±)-lariciresinol monomethyl ether, after removal of the benzyl protecting group by controlled hydrogenolysis (palladium on carbon) of the corresponding cyclized products. Furofuran lignans **290**, that is, (±)-sesamin, (±)-eudesmin, and (±)-piperitol methyl ether, can also be prepared directly through radical cyclization of epoxy ethers **287** followed by treatment with iodine. (±)-Pinoresinol, (±)-piperitol, and (±)-pinoresinol monomethyl ether are prepared by controlled hydrogenolysis of the benzyl protecting group.

Scheme 82 Short and Stereoselective Total Synthesis of Furano and Furofuran Lignans[163]

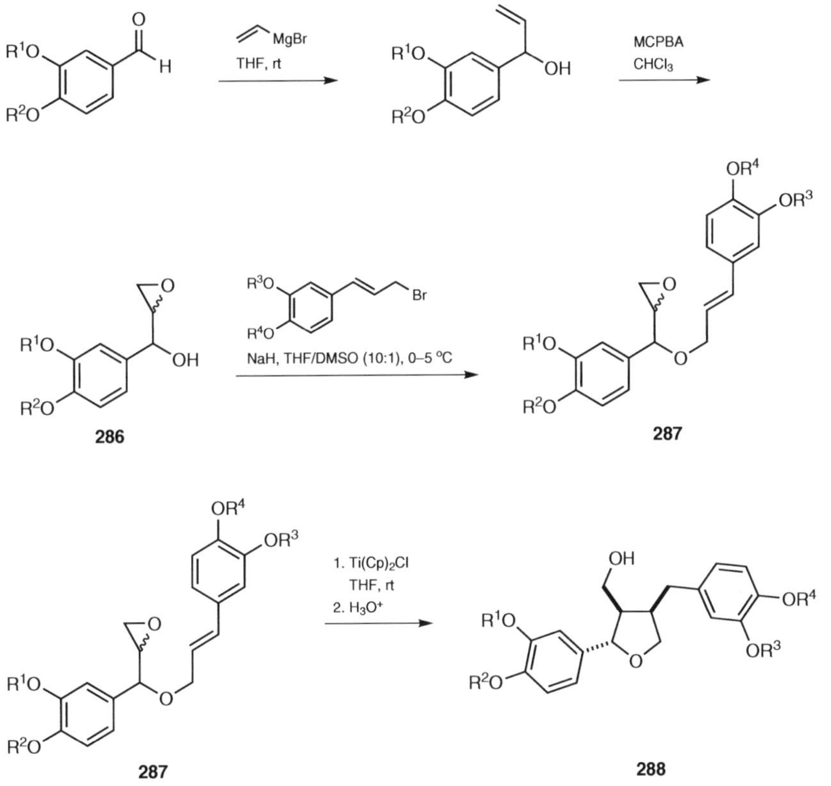

R^1	R^2	R^3	R^4	Yield (%)	Natural Product Name	Ref
CH$_2$		CH$_2$		64	(±)-dihydrosesamin	[163]
Me	Me	Me	Me	63	(±)-lariciresinol dimethyl ether	[163]
CH$_2$		Me	Me	63	(±)-acuminatin methyl ether	[163]
Me	Me	CH$_2$		62	(±)-sanshodiol methyl ether	[163]

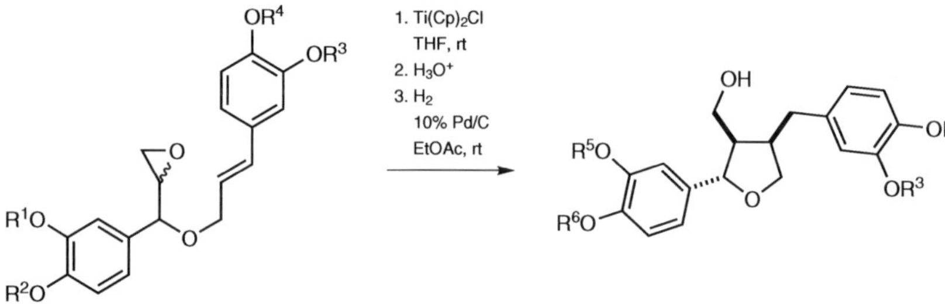

R¹	R²	R³	R⁴	R⁵	R⁶	Yield (%)	Natural Product Name	Ref
Me	Bn	Me	Bn	Me	H	58	(±)-lariciresinol	[163]
CH₂		Me	Bn	CH₂		57	(±)-acuminatin	[163]
Me	Me	Me	Bn	Me	Me	57	(±)-lariciresinol monomethyl ether	[163]

R¹	R²	R³	R⁴	Yield (%)	Natural Product Name	Ref
CH₂		CH₂		93	(±)-sesamin	[163]
Me	Me	Me	Me	90	(±)-eudesmin	[163]
CH₂		Me	Me	88	(±)-piperitol methyl ether	[163]

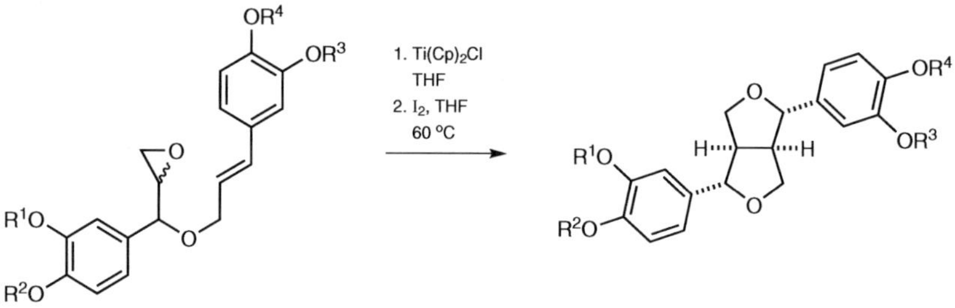

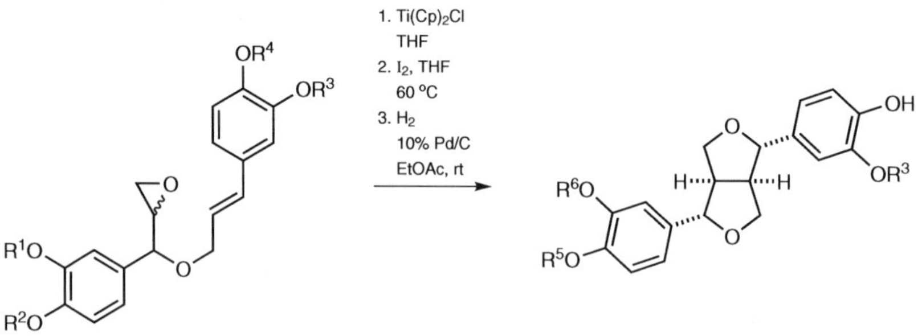

R¹	R²	R³	R⁴	R⁵	R⁶	Yield (%)	Natural Product Name	Ref
Me	Bn	Me	Bn	Me	H	84	(±)-pinoresinol	[163]
CH₂		Me	Bn	CH₂		80	piperitol	[163]
Me	Me	Me	Bn	Me	Me	81	pinoresinol monomethyl ether	[163]

Trisubstituted Tetrahydrofurano Lignans 288; General Procedure:[163]

A fresh soln of Ti(Cp)$_2$Cl [prepared from Ti(Cp)$_2$Cl$_2$ (0.292 mmol) in THF] was added dropwise to a stirred soln of epoxide **287** (0.127 mmol) in dry THF (4 mL) at rt over 30 min under an argon atmosphere. The mixture was then stirred for 1 h and decomposed with 10% H$_2$SO$_4$ (5 mL). After removal of most of the solvent under reduced pressure, the resulting residue was extracted with Et$_2$O (4 × 25 mL). The combined organic extract was washed successively with sat. aq NaHCO$_3$ (2 × 10 mL) and brine (1 × 10 mL) and was then dried (Na$_2$SO$_4$). Et$_2$O was removed under reduced pressure, and the residue was purified by column chromatography (silica gel, EtOAc/petroleum ether) to furnish a mixture of two isomers in a 5:1 ratio. The major isomer was then separated by preparative TLC.

Trisubstituted Tetrahydrofurano Lignans 289; General Procedure:[163]

The benzyl-protected epoxy ether **287** was subjected to titanium-mediated cyclization to give **288**, as described in the procedure above. A diastereomeric mixture of compound **288** (0.05 mmol), EtOAc (5 mL), and 10% Pd/C (15 mg) was subjected to hydrogenolysis at rt for 1.5 h. Upon completion of the reaction, the catalyst was filtered off, and the filtrate was concentrated under reduced pressure.

Furofuran Lignans 290; Typical Procedure:[163]

A fresh soln of Ti(Cp)$_2$Cl [prepared from Ti(Cp)$_2$Cl$_2$ (0.292 mmol) in THF] was added dropwise to a stirred soln of epoxide **287** (0.127 mmol) in dry THF (4 mL) at 60 °C over 10 min under argon. Thereafter, a soln of I$_2$ (42 mg, 0.165 mmol) in THF (1 mL) was added by syringe. The mixture was kept at 60 °C with constant stirring for another 1 h and was then decomposed by adding sat. aq NH$_4$Cl (10 mL). Most of the THF solvent was removed under reduced pressure, and the residue obtained was extracted with Et$_2$O (4 × 50 mL). The combined organic extract was thoroughly washed with 10% aq Na$_2$S$_2$O$_3$ (3 × 25 mL) and brine (1 × 20 mL) and was then dried (Na$_2$SO$_4$). The solvent was removed under reduced pressure, and the residue was purified by column chromatography (silica gel, EtOAc/petroleum ether 1:4).

2.9.6.2.5 Formal Total Synthesis of (±)-Fragranol by Template-Catalyzed 4-exo Cyclization

Titanocene(III)-catalyzed radical chemistry has been used for efficient access to cyclobutanes by employing either α,β-unsaturated carbonyl compounds or nitriles as radical acceptors.[346,412,413] The Gansäuer group has reported a novel approach for the synthesis of (±)-fragranol (**294**) that relies on a radical 4-*exo* cyclization reaction using the concept of template catalysis.[414] In template catalysis, the radical and the radical acceptor are both bound to a bis(η^5-cyclopentadienyl)titanium template [cationic functional bis(η^5-cyclopentadienyl)titanium species **291**].[415–419] The key 4-*exo* cyclization step is mediated by a cationic bis(η^5-cyclopentadienyl)titanium complex bearing a pendent amide ligand. The bis(η^5-cyclopentadienyl)titanium template can bind to the pendent amide ligand reversibly. Subsequently, a vacant coordination site is generated in situ for interaction with a substrate molecule possessing a polar functionality, and thus, the radical and its acceptor are bound to the bis(η^5-cyclopentadienyl)titanium center in a two-point mode. This transforms the usually unfavorable 4-*exo* cyclization of substrate **292** bearing *gem*-dialkyl substituents into a thermodynamically and kinetically favorable process. This method results in the formation of cyclobutane **293** with good diastereoselectivity. Finally, with the key intermediate in hand, the formal total synthesis of (±)-fragranol (**294**) can be accomplished in a few steps (Scheme 83).

Scheme 83 Formal Total Synthesis of (±)-Fragranol by Template-Catalyzed 4-*exo* Cyclization[414]

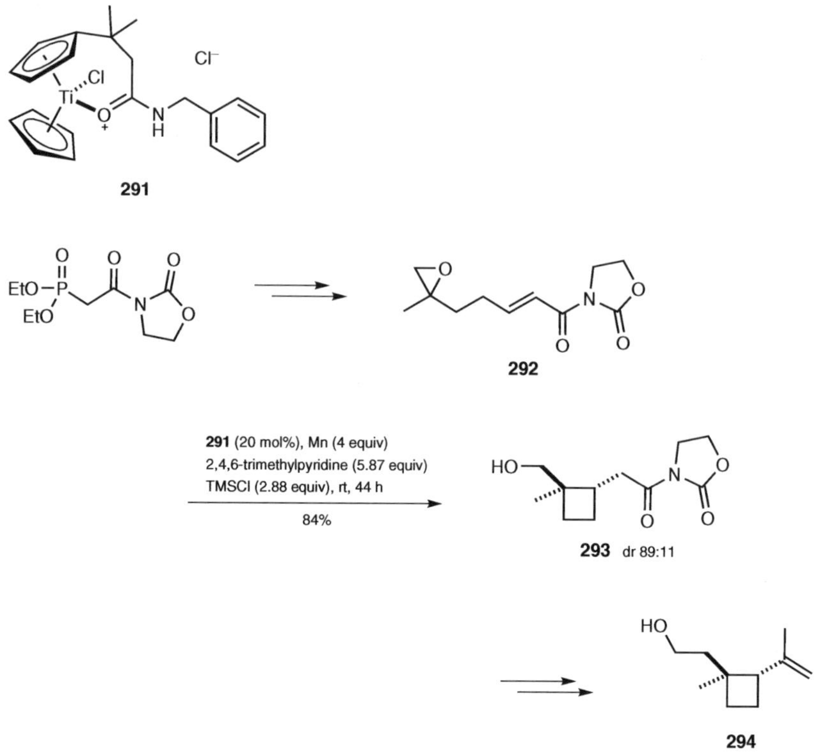

291

292

291 (20 mol%), Mn (4 equiv)
2,4,6-trimethylpyridine (5.87 equiv)
TMSCl (2.88 equiv), rt, 44 h
84%

293 dr 89:11

294

3-{[2-(Hydroxymethyl)-2-methylcyclobutyl]acetyl}oxazolidin-2-one (293):[414]
Strictly deoxygenated THF (3 mL) was added to a mixture of bis(η^5-cyclopentadienyl)titanium species **291** (48.0 mg, 100 µmol, 0.2 equiv) and Mn dust (110 mg, 2 mmol, 4 equiv) under an argon atmosphere, and the mixture was stirred at rt until it turned green (about 15 min). A soln of 2,4,6-trimethylpyridine (355 mg, 2.93 mmol, 5.87 equiv) and epoxide **292** (113 mg, 0.500 mmol, 1.00 equiv) in THF (2 mL) was added, and the mixture was stirred for 5 min. TMSCl (156 mg, 1.44 mmol, 2.88 equiv) was added, and the mixture was stirred at rt for 44 h. Phosphate buffer, KF (290 mg, 5.00 mmol, 10 equiv), TBAF (131 mg, 0.5 mmol, 1 equiv), and *t*-BuOMe were added, and the suspension was stirred for another 16 h. The solvent was removed, and the residue was extracted with Et$_2$O, EtOAc, and CH$_2$Cl$_2$. The organic extract was concentrated, and the residue was purified by flash chromatography (cyclohexane/EtOAc 5:95) to afford *cis*-**293** as a colorless oil; yield: 10.3 mg (13%, as reported); and *trans*-**293** as a colorless oil; yield: 85.5 mg (71%, as reported); dr 89:11. In the same manner, *trans*-**293** (405 mg, 69%) and *cis*-**293** (46 mg, 8%) were obtained from **292** (584 mg, 2.6 mmol).

2.9.6.2.6 Total Synthesis of Entecavir

En route to the synthesis of the hepatitis B drug entecavir (**297**) from 4-(trimethylsilyl)but-3-yn-2-one and acrolein, the titanocene(III)-catalyzed intramolecular radical addition of an epoxide to an alkyne is employed as the key step (Scheme 84).[420] Among various other antiviral agents against hepatitis B virus,[421] entecavir (BMS-200 475) is considered one of the best choices owing to its lack of significant adverse effects and the low risk of inducing long-term resistance to the drug.[422] The cyclization of **295** to **296** can be achieved using dichlorobis(η^5-cyclopentadienyl)titanium(IV) as the catalyst in the presence of

2.9.6 Synthesis of the Core Frameworks of Biologically Active Molecules 519

2,4,6-trimethylpyridine, an iridium catalyst [IrCl(CO)(PPh$_3$)$_2$], and chlorotrimethylsilane, with dihydrogen as the proton source. The method shows excellent selectivity and gives excellent yields relative to those obtained with the stoichiometric version of this reaction.

Scheme 84 Synthesis of Entecavir[420]

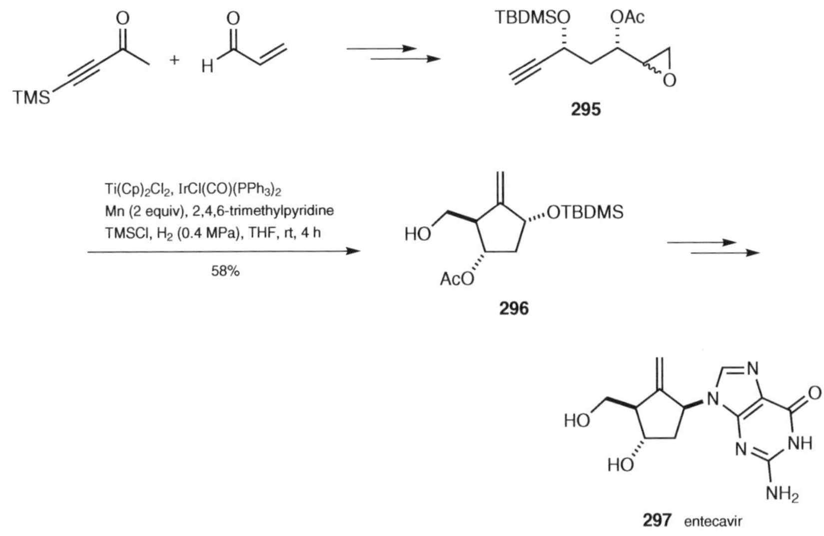

(1S,2R,4R)-4-(*tert*-Butyldimethylsiloxy)-2-(hydroxymethyl)-3-methylenecyclopentyl Acetate (296); Typical Procedure:[420]

Strictly deoxygenated THF (15 mL) was added to a mixture of Mn powder (0.368 g, 6.70 mmol) and IrCl(CO)(PPh$_3$)$_2$ (0.26 g, 0.34 mmol). A soln of **295** (1 g, 3.35 mmol) and 2,4,6-trimethylpyridine (3.5 mL, 26.8 mmol) in strictly deoxygenated anhyd THF (22 mL) was added at rt. TMSCl (1.7 mL, 13.4 mmol) was then added, followed by a soln of Ti(Cp)$_2$Cl$_2$ (0.167 g, 0.67 mmol) in strictly deoxygenated anhyd THF (12 mL). The mixture was stirred under an atmosphere of H$_2$ (0.4 MPa) for 4 h at the same temperature. Upon completion of the reaction, H$_2$O (5 mL) was added, and the mixture was stirred for 10 min and then filtered through a pad of Celite. The pad was washed with *t*-BuOMe (20 mL), and the combined organic phase was acidified to pH 2 with 2 M HCl. The mixture was stirred for another 15 min, and the phases were separated. The organic phase was washed with H$_2$O (20 mL) and dried (Na$_2$SO$_4$). The solvent was removed under reduced pressure, and the resulting oily residue was purified by flash chromatography (silica gel, hexane/EtOAc from 9:1 to 3:2) to afford a pale yellow oil; yield: 0.582 g (58%).

2.9.6.2.7 Enantioselective Synthesis of α-Ambrinol

The first enantioselective synthesis of (−)-α-ambrinol (**302**, 96% ee) was achieved from commercial (5E)-6,10-dimethylundeca-5,9-dien-2-one by combining Jacobsen's asymmetric epoxidation with titanium-catalyzed stereoselective cyclization reactions (Scheme 85).[423] The combined approach of enantioselective epoxidation of polyprene **299** using Jacobsen's catalyst **298**, followed by titanium(III)-catalyzed radical cyclization of the corresponding epoxy derivative **300** simplifies the enantioselective synthesis of cyclic odorant terpenoids. The titanocene(III)-catalyzed cyclization reaction initiates the radical opening of the epoxide ring, which proceeds with retention of configuration at the epoxide chiral center to yield secondary alcohol (−)-**301**. Alcohol (−)-**301** bears all the functionalities that are required to develop (−)-α-ambrinol in enantiomerically enriched form. Finally, after a series of reactions, alcohol (−)-**301** can be converted into (−)-α-ambrinol (**302**) in an overall yield of 18%. Secondary alcohol (−)-**301** can also be employed in the total synthesis of achilleol A (**303**) (Scheme 86).[424]

Scheme 85 Enantioselective Synthesis of (−)-α-Ambrinol[423]

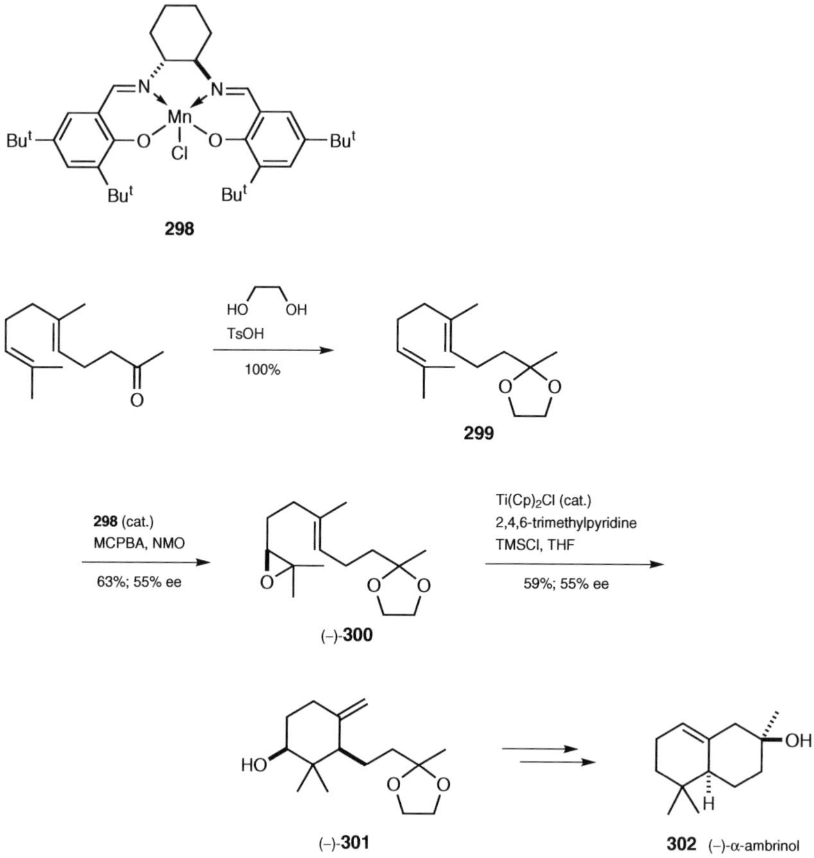

Scheme 86 Achilleol A[424]

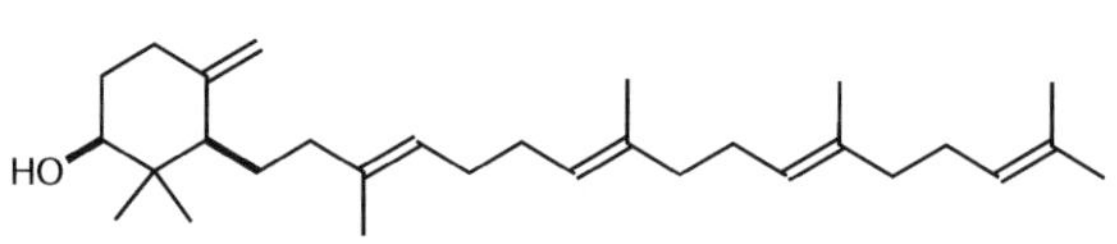

2,2-Dimethyl-3-[2-(2-methyl-1,3-dioxolan-2-yl)ethyl]-4-methylenecyclohexan-1-ol (301):[423]
A soln of epoxy alkene **300** (500 mg, 1.96 mmol) and 2,4,6-trimethylpyridine (1.8 mL) in THF (5 mL) and TMSCl (1.0 mL) were simultaneously added to a fresh soln of Ti(Cp)$_2$Cl [prepared from Ti(Cp)$_2$Cl$_2$ (98 mg, 0.39 mmol) and Mn dust (860 mg, 15.7 mmol) in THF (20 mL)]. The mixture was stirred at rt for 1.5 h. Upon completion of the reaction, sat. brine was added and the aqueous mixture was extracted with EtOAc. The organic layer was dried (Na$_2$SO$_4$) and the solvent was removed under reduced pressure. The residue obtained was purified by flash chromatography (hexane/EtOAc 85:15) to afford a colorless oil; yield: 305 mg (61%). The reaction performed on nonracemic (−)-**300** gave (−)-**301**; yield: 59%; 55% ee.

2.9.6.2.8 Total Synthesis of (±)-Platencin

The titanium(III)-mediated, radical-based carbon–carbon bond forming reaction of substrate **304** to afford tricyclic system **305** has been utilized in the total synthesis of (±)-platencin (**306**) (Scheme 87).[425] Platencin,[426–429] bearing a 3-amino-2,4-dihydroxybenzoic acid linked via an amide bond, is a highly potent antibiotic that can be isolated from the strains of *Streptomyces platensis*.

Scheme 87 Total Synthesis of (±)-Platencin[425]

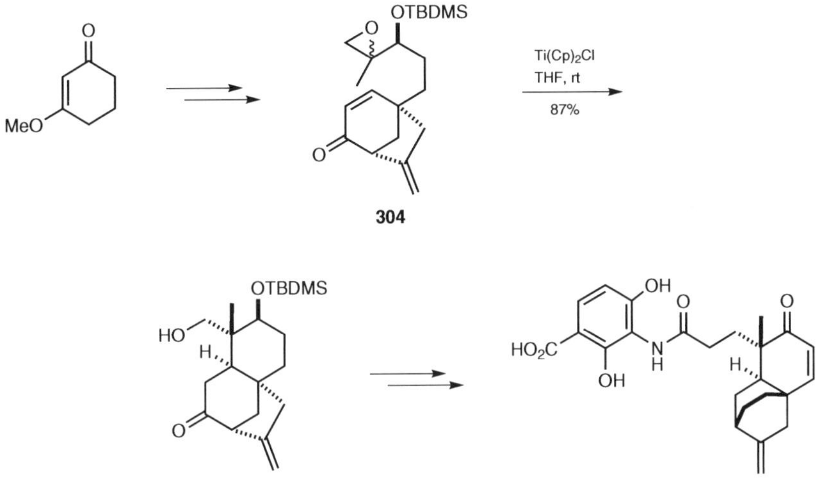

(1R*,4R*,5S*,6S*,9S*)-4-(tert-Butyldimethylsilyloxy)-5-(hydroxymethyl)-5-methyl-10-methylenetricyclo[7.2.1.0¹,⁶]dodecan-8-one (305):[425]

A freshly prepared 0.2 M soln of Ti(Cp)$_2$Cl in THF (2.9 mL, 0.580 mmol) was added dropwise to a stirred soln of epoxide **304** (70.0 mg, 0.193 mmol) in THF (0.5 mL) at rt, and the mixture was stirred for 2 h. Sat. aq NaHCO$_3$ and Celite were added, and the mixture was stirred for another 4 h, which was followed by filtration through a pad of Celite. The filtrate was diluted with EtOAc and H$_2$O and poured into a separatory funnel. The phases were separated, and the organic extract was washed with brine. The organic phase was separated, combined, dried (Na$_2$SO$_4$), filtered, and concentrated. The residue was purified by flash column chromatography (silica gel, EtOAc/hexane 1:5) to afford a colorless solid; yield: 61.0 mg (87%). Recrystallization (Et$_2$O/hexane) afforded colorless plates; mp 134–135 °C.

2.9.6.2.9 Total Synthesis of (±)-Smenospondiol

Takahashi and co-workers have described the total synthesis of (±)-smenospondiol (**311**) using a free-radical cyclization strategy (Scheme 88).[406] (±)-Smenospondiol is a member of the sesquiterpene class of compounds,[430] and it exhibits a variety of promising biological effects, including antimicrobial, cytotoxic, antiviral, and immunomodulatory activities.[431–433] The stereoselective construction of the drimane skeleton of smenospondiol is accomplished by titanium(III)-mediated tandem radical cyclization of epoxide **307**. Upon treatment with two equivalents of chlorobis(η⁵-cyclopentadienyl)titanium(III) in tetrahydrofuran, acyclic epoxide **307** provides the desired decalin as the major product in 82% yield {ratio [**308**/(**309A**/**309B**)/**310**] 88:4:7}. The use of a nonpolar solvent and heating at 60 °C are essential for this tandem cyclization.

Scheme 88 Total Synthesis of (±)-Smenospondiol by Titanium(III)-Catalyzed Tandem Radical Cyclization[406]

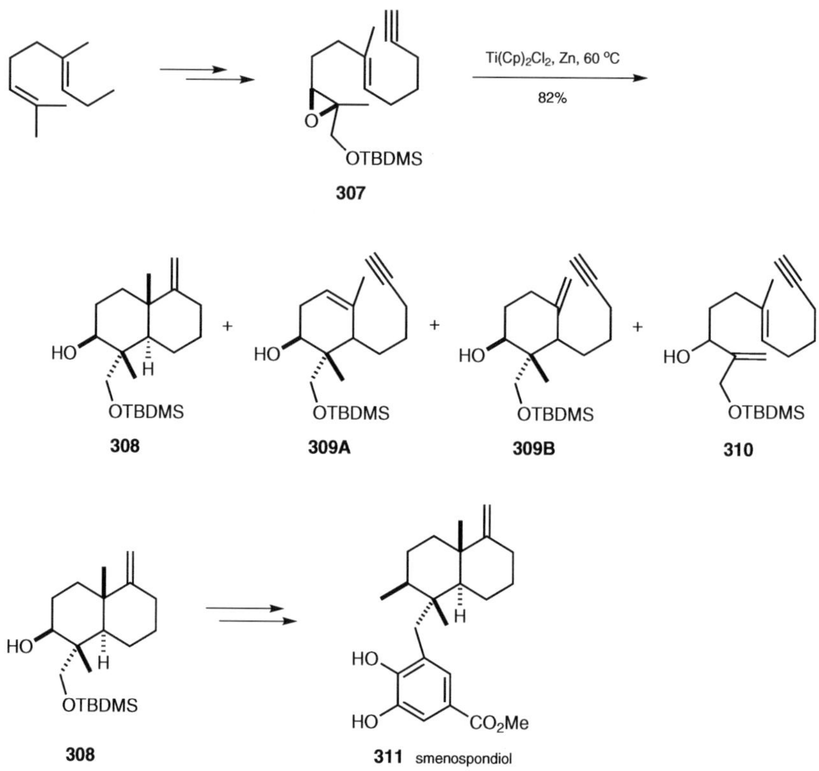

(1R*,2S*,4aS*,8aR*)-1-[(tert-Butyldimethylsiloxy)methyl]-1,4a-dimethyl-5-methylenedecahydronaphthalen-2-ol (308):[406]

Activated Zn dust (29.2 mg, 0.446 mmol) was added to a stirred soln of Ti(Cp)$_2$Cl$_2$ (73.9 mg, 0.297 mol) in dry THF (0.5 mL) and benzene (1.2 mL) (**CAUTION:** *carcinogen*) under argon at rt. The mixture was stirred for 1 h and was then added to a soln of epoxide **307** (50 mg, 0.149 mmol) in dry benzene (1.2 mL) at 60 °C. After stirring for 10 min at 60 °C, the mixture was diluted with Et$_2$O and poured into 1 M HCl (30 mL) at 0 °C. The aqueous layer was extracted with Et$_2$O (3 × 50 mL). The combined organic layer was washed with sat. brine, dried (MgSO$_4$), filtered, and concentrated. The residue was purified by column chromatography (silica gel, hexane/EtOAc 95:5) to afford a mixture; yield: 82%; ratio [**308**/(**309A**/**309B**)/**310**] 88:4:7. The mixture was further purified by HPLC (Lichrosorb Si60–5 7.5 × 300 mm column, EtOAc/hexane 6:94; flow rate 3.0 mL·min^{-1}) to yield pure **308**, which elutes first; t_R 25–27 min.

2.9.6.2.10 Total Synthesis of Fomitellic Acid B

The Kobayashi group described the first radical-based total synthesis of fomitellic acid B (**314**),[434] a potent inhibitor of DNA polymerases α and β.[435–437] The key feature of this method comprises the stereoselective synthesis of tetracyclic AB ring system **313** bearing all the requisite chiral centers in one step by means of titanium(III)-mediated radical cascade cyclization of epoxypolyene **312** (Scheme 89).

2.9.6 Synthesis of the Core Frameworks of Biologically Active Molecules

Scheme 89 Titanium(III)-Mediated Radical Cascade Cyclization of an Epoxypolyene[434]

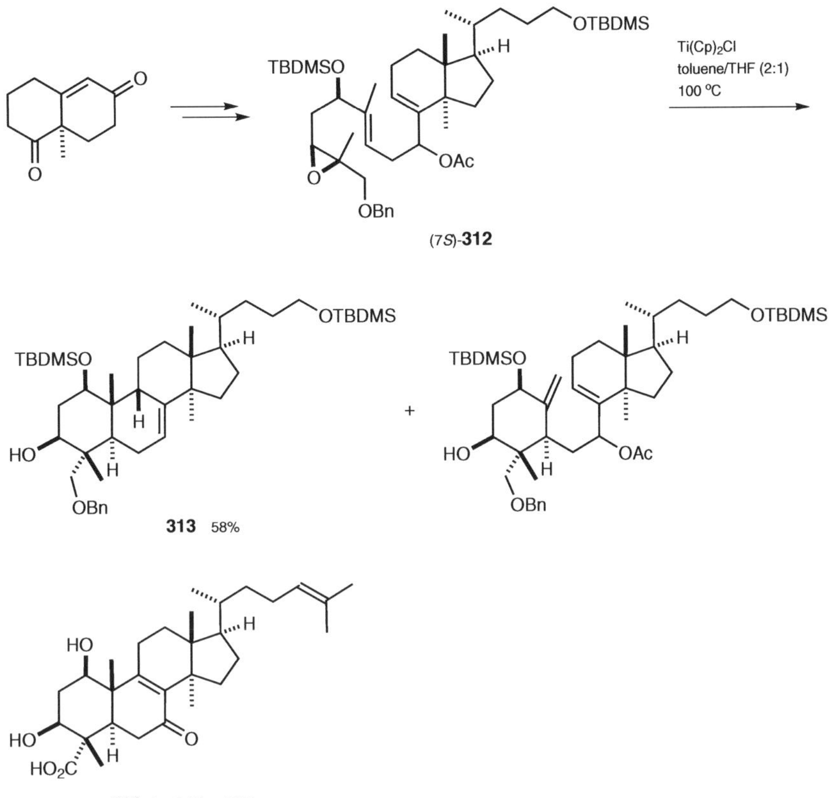

(7S)-**312**

313 58%

314 fomitellic acid B

Tetracycle 313:[434]
A fresh soln of Ti(Cp)$_2$Cl [prepared from Ti(Cp)$_2$Cl$_2$ (479 mg, 1.9 mmol) and Zn (252 mg, 3.9 mmol) in THF (6.4 mL)] was added to a soln of epoxide **312** (520 mg, 0.64 mmol) in toluene (21 mL) at 100 °C. After stirring for 10 min at this temperature, the mixture was cooled to 0 °C, diluted with Et$_2$O (50 mL), and quenched with 1 M HCl (25 mL). The organic layer was washed with brine, dried (Na$_2$SO$_4$), and concentrated under reduced pressure. The residue was purified by column chromatography (silica gel, hexane/EtOAc 4:1) to afford a white amorphous powder; yield: 285 mg (58%); R_f 0.40 (hexane/EtOAc 4:1).

2.9.6.2.11 Synthesis of the BCDE Molecular Fragment of Azadiradione

Azadiradione, a limonoid isolated from the neem tree *Azadirachta indica* (A. Juss), possesses a unique stereostructure distinct from that of the steroids and is known for its insect antifeedant activity.[438] In Section 2.9.6.1.10, the efficiency of a radical-catalyzed reaction in the synthesis of the tetracyclic framework of azadiradione was described. Here, an effective, short, and diastereoselective synthesis of the BCDE fragment (±)-**317**, the core of azadiradione, is described on the basis of titanocene(III)-promoted tandem cyclization of unsaturated epoxynitrile (±)-**315**.[439] The titanocene(III)-catalyzed stereoselective construction of the core CD rings by a 6-*endo*/4-*exo* process allows the relative configurations of the C8, C9, and C13 stereocenters of hydroxy ketone (±)-**316** to be identical to those of natural azadiradione (Scheme 90).

Scheme 90 Synthesis of the BCDE Fragment of Azadiradione[439]

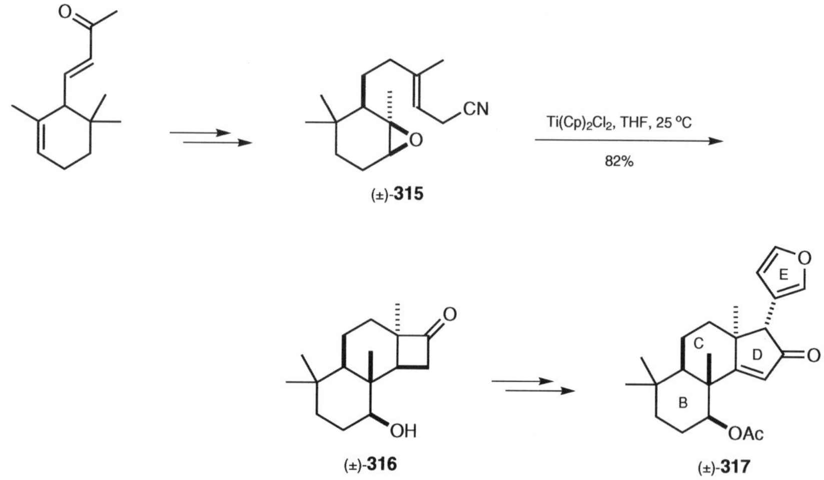

(±)-**315**

(±)-**316**

(±)-**317**

(2aR^*,4aR^*,8S^*,8aR^*,8bS^*)-8-Hydroxy-2a,5,5,8a-tetramethyldecahydrocyclobuta[a]naphthalen-2(1H)-one [(±)-316]:[439]

Under argon, a soln of (±)-**315** (1.00 g, 4.05 mmol) in THF (40.0 mL) was added dropwise by cannula to a soln of Ti(Cp)$_2$Cl [prepared from Ti(Cp)$_2$Cl$_2$ (3.34 g, 13.36 mmol) and Zn (1.71 g, 26.72 mmol) in THF (27.0 mL)]. The mixture was stirred for 15 min and then quenched by adding sat. aq NaH$_2$PO$_4$. After stirring for 30 min, the resulting mixture was filtered through a fritted glass funnel. The aqueous layer was extracted with Et$_2$O, and the combined organic extract was washed with H$_2$O and brine and then dried (Na$_2$SO$_4$). Et$_2$O was removed under reduced pressure, and the residue was purified by flash chromatography (silica gel, hexane/EtOAc 4:1) to afford a white solid; yield: 830 mg (82%); mp 72–73 °C.

2.9.6.2.12 Approach to Bis(lactone) Skeletons: Total Synthesis of (±)-Penifulvin A

An efficient total synthesis of the interesting and structurally complex tetracyclic sesquiterpenoid (±)-penifulvin A (**320**) has been accomplished through a 12-step sequence (Scheme 91).[440] Penifulvin A, a novel sesquiterpenoid having significant activity against *Spodoptera frugiperda*, has been isolated from cultures of an isolate of *Penicillium griseofulvum* (NRRL 35 584).[441,442] For the construction of this architecturally challenging molecule, the key step includes the titanium(III)-mediated reductive epoxide opening of epoxy ester **318**, followed by 5-*exo-dig* cyclization onto the alkyne. The formation of the δ-lactone ring during the radical cyclization process is achieved by the addition of 1 M aqueous hydrochloric acid to the reaction mixture. A series of additional steps starting from **319** produces (±)-penifulvin A (**320**) in an overall yield of 16%.

2.9.6 Synthesis of the Core Frameworks of Biologically Active Molecules

Scheme 91 Titanium(III)-Mediated Epoxide Opening/Cyclization[440]

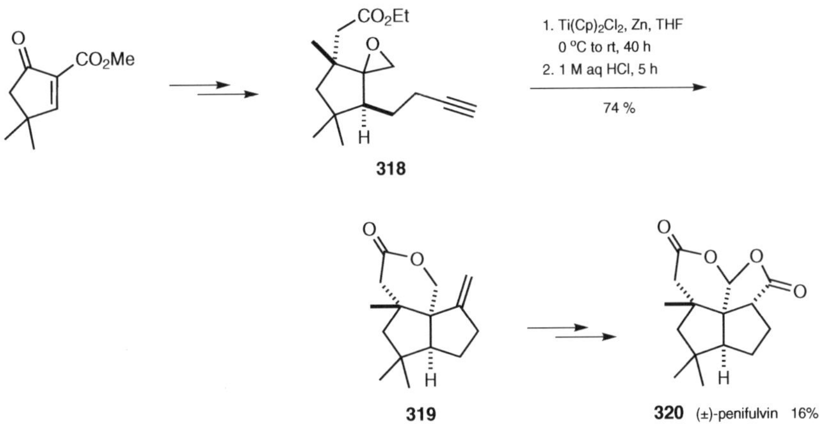

320 (±)-penifulvin 16%

(4aR*,6aS*,9aS*)-4a,6,6-Trimethyl-9-methyleneoctahydro-1H,3H-pentaleno[6a,1-c]pyran-3-one (319):[440]

Under N_2, a soln of **318** (815 mg, 2.93 mmol, 1.0 equiv) in THF (140 mL) was added dropwise by cannula to a soln of Ti(Cp)$_2$Cl [prepared from Ti(Cp)$_2$Cl$_2$ (2.18 g, 8.79 mmol, 3 equiv) and Zn (1.72 g, 26.37 mmol) in THF (70 mL)] at 0 °C. The mixture was brought to rt and stirred for 40 h. The mixture was then diluted with 1 M HCl (30 mL) and stirred for another 5 h. Most of the solvent was evaporated under reduced pressure, and the residue was extracted with EtOAc, washed with H$_2$O and brine, dried (Na$_2$SO$_4$), filtered, and concentrated. The residue was purified by column chromatography (silica gel, EtOAc/hexane 5:95) to afford a colorless dense mass; yield: 508 mg (74%).

2.9.6.2.13 Synthesis of Eudesmanolides

A general procedure for the synthesis of both 12,6- and 12,8-eudesmanolides has been developed through the titanocene-catalyzed radical cyclization of easily accessible epoxy-germacrolides **322** and **326** as the key step (Scheme 92).[158] Eudesmanolides make up the main group of natural sesquiterpenoids, which include more than 500 members.[443–445] They are known to exhibit a variety of promising biological effects, including antifungal, anti-inflammatory, and antitumor activities, and are potent inhibitors of human topoisomerase II.[446–448] The synthesis involves inexpensive reagents, works at room temperature with conditions compatible with several functional groups, and employs a catalytic quantity of titanium. As germacrolides are the biogenetic precursors of eudesmanolides, some of them such as (+)-dihydrocostunolide, (+)-costunolide, (+)-stenophyllolide (**321**), and (+)-salonitenolide (**325**) are easily accessible in (multi)gram quantities.[166,449,450] (+)-Stenophyllolide (**321**) and (+)-salonitenolide (**325**) are isolable from *Centaurea calcitrapa* and *C. malacitana* simply by immersing these weeds in chloroform. (+)-Salonitenolide (**321**), after selective hydrogenation and treatment with 3-chloroperoxybenzoic acid, affords oxirane **322**, which upon treatment with titanocene(III) undergoes cyclization to give alkene **323**. Finally, the conjugated double bond can be restored in a single-pot reaction to afford (+)-9β-hydroxyreynosin (**324**). In a similar manner, the titanocene(III)-catalyzed cyclization of **326** gives (+)-β-cyclopyrethrosin (**327**) in an overall yield of 13%.

Scheme 92 Synthesis of Eudesmanolides[158]

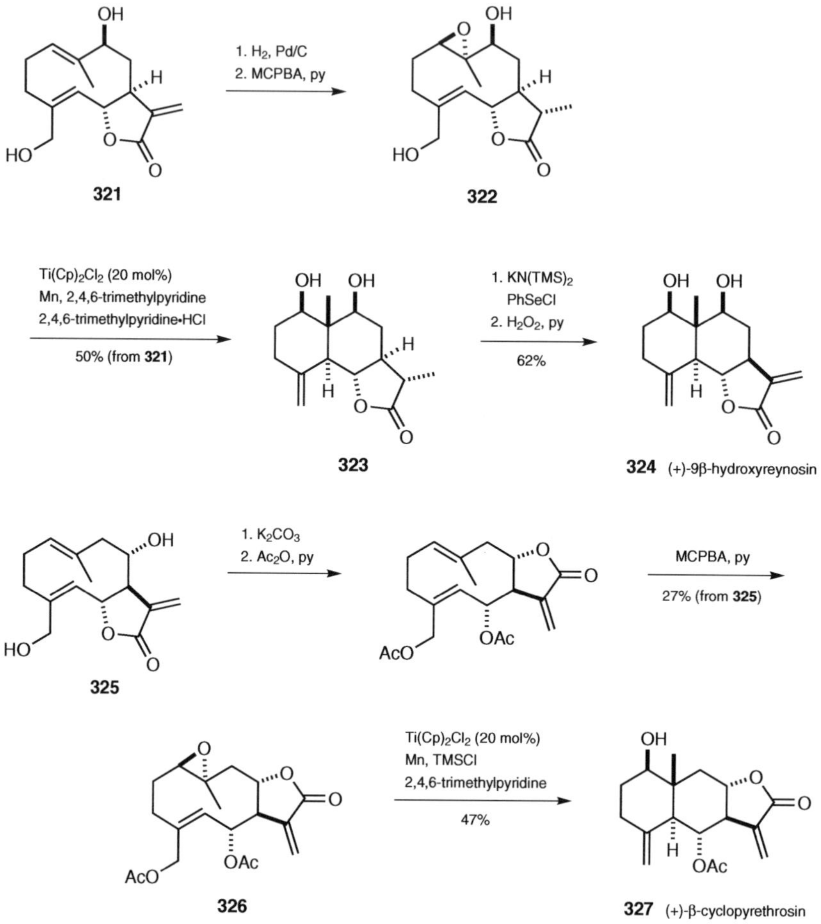

(3aS,5S,5aR,6R,9aS,9bS)-5,6-Dihydroxy-3,5a-dimethyl-9-methylenedecahydronaphtho-[1,2-b]furan-2(3H)-one (323); Typical Procedure:[158]

A mixture of Ti(Cp)$_2$Cl$_2$ (11 mg, 0.044 mmol) and Mn dust (95 mg) was added to a Schlenk-type flask containing deoxygenated THF (15 mL) under argon and the mixture was stirred at 25 °C until it turned lime green. Then, a mixture of **322** (55 mg, 0.22 mmol), H$_2$O (20 µL, 1.1 mmol), and 2,4,6-trimethylpyridine (0.13 mL, 1.1 mmol) in THF (5 mL) was added to the green suspension, followed by 2,4,6-trimethylpyridine•HCl (158 mg, 1.0 mmol). The deep-blue mixture was then stirred at 25 °C for 7 h. Then, the solvent was removed, and t-BuOMe (30 mL) was added to the residue. The ethereal soln was washed with 2 M HCl and brine. The organic extract was dried (Na$_2$SO$_4$) and concentrated. The residue was purified by flash chromatography (hexane/t-BuOMe 1:9).

2.9.6.3 Samarium(II)-Based Total Synthesis of Biologically Active Molecules

Samarium(II) iodide exhibits exceptional properties in sequential reactions[170,451–453] and with remarkable selectivity promotes a large number of key radical cyclizations that are useful in organic synthesis. The reactivity and selectivity of samarium(II) iodide can be modified by varying the catalyst, solvent, or the reaction conditions.[454,455] With the ability to promote both one- and two-electron processes, samarium(II) iodide has achieved significant status in tandem reactions. Specific examples of the synthesis of natural and bio-

2.9.6 Synthesis of the Core Frameworks of Biologically Active Molecules

logically active molecules using samarium(II) iodide mediated oxidative cyclization as a key step include the preparation of (−)-grayanotoxin III,[456] platensimycin,[457] and 3′-β-branched uridine derivatives.[458] A few significant total syntheses of naturally active compounds mediated by samarium are summarized below.

2.9.6.3.1 Total Synthesis of (±)-Lundurines A and B

The stereospecific construction of the pentasubstituted cyclopropane core of (±)-lundurine A (**330**) and (±)-lundurine B (**331**) (kopsia alkaloids) has been achieved by radical cyclization of **328** using samarium(II) iodide to give cyclopropane **329** with perfect stereoselectivity (Scheme 93).[459] These kopsia alkaloids are the only natural products having a unique hexacyclic skeleton with a cyclopropane-fused indoline and most of the stereogenic centers, including two quaternary carbon atoms, in the cyclopropane ring.

Scheme 93 Samarium(II) Iodide Mediated Cyclopropanation in the Total Synthesis of (±)-Lundurines A and B[459]

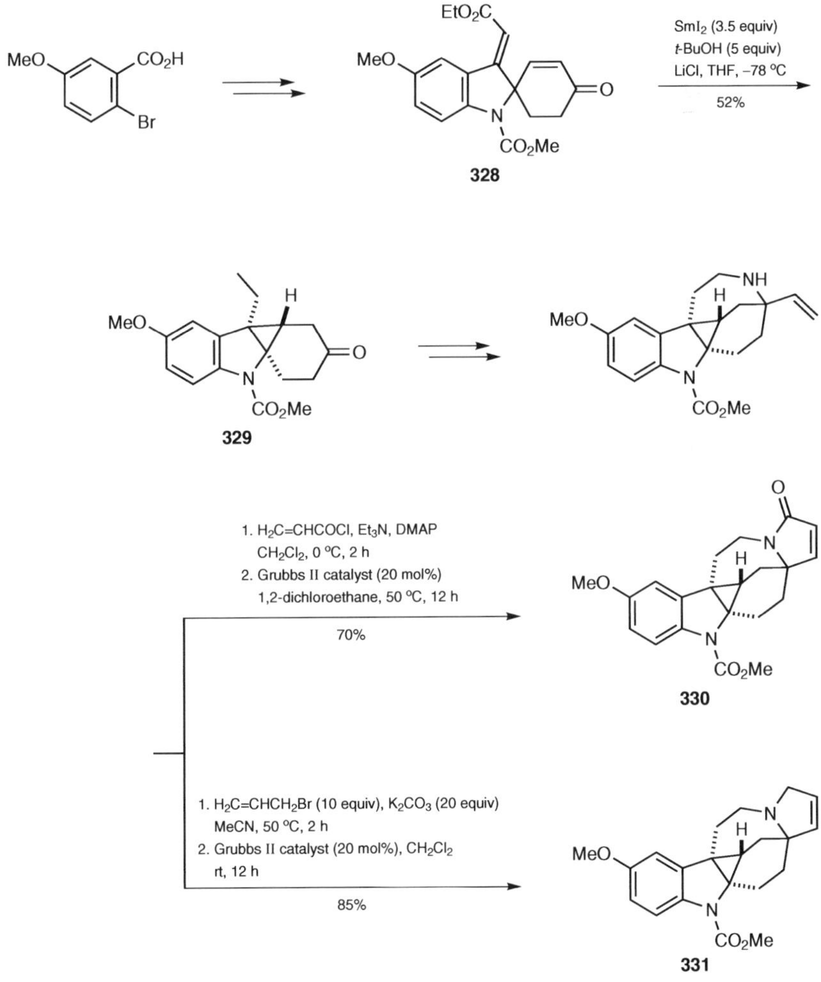

for references see p 533

Methyl (4aS*,4bR*,9aR*)-4b-(2-Ethoxy-2-oxoethyl)-6-methoxy-3-oxo-1,2,3,4,4a,4b-hexahydro-9H-benzo[1,3]cyclopropa[1,2-b]indole-9-carboxylate (329); Typical Procedure:[459]
A soln of SmI_2 in THF (4.2 mL, 0.42 mmol) was added to a suspension of LiCl (101 mg, 2.4 mmol) in THF (4.0 mL) at −78 °C, followed by the addition of a soln of spiroenone **328** (44.6 mg, 0.12 mmol) and t-BuOH (57 µL, 0.6 mmol) in THF (4.0 mL) at −78 °C. The reaction was quenched with sat. aq NH_4Cl, and the resulting mixture was stirred for 12 h at rt. Upon completion of the reaction, the mixture was extracted with EtOAc (3 × 10 mL). The resulting organic extract was washed with brine, dried (Na_2SO_4), and concentrated under reduced pressure. The residue was purified by flash column chromatography (hexane/EtOAc 5:1) to give a yellow oil; yield: 23.0 mg (52%).

2.9.6.3.2 Synthesis of Trehazolin from D-Glucose

Trehazolin (**336**) is a specific inhibitor of the enzyme trehalase, which is responsible for the breakdown of reserve carbohydrates of many insects. The Giese group described a short and efficient synthesis of trehazolin (**336**) and trehazolamine (**334**) by mimicking their hypothetical biosynthesis (Scheme 94).[460] They employed glucose as the starting molecule for the synthesis of trehazolamine, the result of which is conservation of three chiral centers during the reaction sequence. Moreover, the chirality of D-glucose allows inversion of a stereocenter carrying an amino functionality adjacent to a hydroxy group. Treatment of ketooxime ether **332** with samarium(II) iodide gives exclusively diastereomer **333** in 84% yield. Note, however, that treatment of **332** with tributylstannane/2,2′-azobisisobutyronitrile in refluxing benzene produces a mixture of product **333** and its epimer at the quaternary center. The synthesis of trehazolin (**336**) from trehazolamine (**334**) and α-D-glucopyranosyl isothiocyanate **335** follows a common procedure and is achieved in a yield of 63% over three steps.

Scheme 94 Samarium(II) Iodide Mediated Intramolecular Cyclization of a Ketooxime Ether in the Synthesis of Trehazolamine and Trehazolin[460]

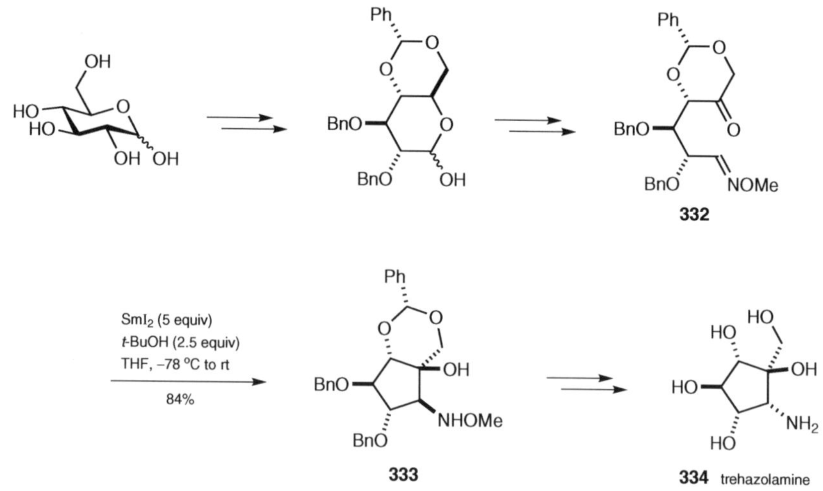

2.9.6 Synthesis of the Core Frameworks of Biologically Active Molecules

334 + **335** → **336** trehazolin 63%

(2R,4aR,5S,6S,7R,7aS)-6,7-Bis(benzyloxy)-5-(methoxyamino)-2-phenyltetrahydrocyclopenta[d][1,3]dioxin-4a(4H)-ol (333):[460]
A soln of ca. 0.1 M SmI_2 (186 mL, 18.6 mmol, 4.6 equiv) in THF was added to a stirred and strictly deoxygenated soln of dried, crude ketooxime ether **332** [(E/Z) 5:1; 1.91 g, 4.02 mmol] and t-BuOH (1.10 mL, 11.7 mmol, 2.9 equiv) in dry THF (220 mL) at −78 °C. After stirring for 2 h at this temperature, the temperature was slowly raised to rt overnight. The yellow soln was extracted with Et_2O and the extracts were washed with $NaHCO_3$, 10% $Na_2S_2O_3$, and brine, and then dried (Na_2SO_4). Removal of the solvent under reduced pressure gave an almost-pure product. Filtration over a bed of silica gel (pentane/CH_2Cl_2/Et_2O 2:1:1) afforded the title compound; yield: 1.63 g (84%).

2.9.6.3.3 Synthesis of (±)-Cryptotanshinone

The Cai group has reported an effective synthesis of (±)-cryptotanshinone and its simplified analogues by employing the samarium(II) iodide promoted radical cyclization strategy for the construction of the furan ring.[461] Cryptotanshinone bears a benzo-1,2-quinone skeleton and has been reported to be an effective inhibitor of topoisomerase I[462] and to exhibit intense cytotoxicity against a number of cultured human tumor cell lines.[463] Moreover, cryptotanshinone is a tetracyclic compound having a basic B and C naphthalene moiety, and thus, its synthesis has been attained using accessible 5-methoxynaphthalen-1-ol (**337**)[464] as the starting material (Scheme 95). The conversion of **338** into naphthofuran **339** through samarium(II) iodide promoted intramolecular cyclization can be executed in good yield. Desired (±)-cryptotanshinone (**340**) is then obtained as orange needles after a series of steps. By employing this novel method, simplified analogues **341** and **342** have also been synthesized and their structure–activity relationship as CDC25 inhibitors has been investigated. These compounds display powerful inhibitory activity against protein phosphatase CDC25B and cytotoxic activity against A-549 tumor cell lines.

Scheme 95 Synthesis of (±)-Cryptotanshinone and Analogues[461]

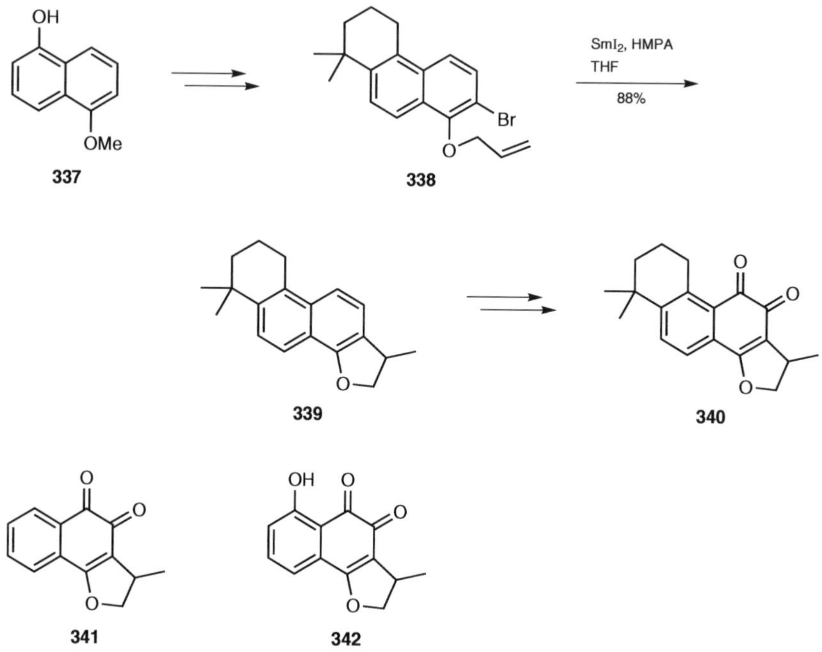

337 — **338** — SmI$_2$, HMPA, THF, 88% — **339**

340

341 **342**

1,6,6-Trimethyl-1,2,6,7,8,9-hexahydrophenanthro[1,2-b]furan (339); Typical Procedure:[461]

> **CAUTION:** *Hexamethylphosphoric triamide is a possible human carcinogen and an eye and skin irritant.*

A soln of bromide **338** (0.317 g, 0.92 mmol) in dry THF was added to a mixture of a 0.1 M soln of SmI$_2$ in THF (40 mL, 2 mmol) and HMPA (2.4 mL, 14 mmol) at 25 °C. The mixture was stirred under N$_2$ for 3 h. The reaction was quenched by adding sat. aq NH$_4$Cl and was extracted with EtOAc. The organic extract was washed with H$_2$O, 3% Na$_2$S$_2$O$_3$, and brine, and it was then dried (MgSO$_4$) and concentrated under reduced pressure. The residue was purified by column chromatography (silica gel, petroleum ether/EtOAc 20:1) to afford a colorless oil; yield: 0.215 g (88%).

2.9.6.3.4   Total Synthesis of Pradimicinone

Pradimicinone (**345**) is a common aglycon of the pradimicin–benanomicin antibiotics with significant antifungal and anti-HIV activities. The strong inhibitory action of this compound is attributed to its potential specific binding to the oligosaccharides of fungi or viral surfaces.[465–469] The key step in the synthesis of **345** is a samarium(II) iodide mediated radical cyclization of enantiopure dialdehyde **343** to *trans*-diol **344** as the sole product,[470] which is produced in an enantiomerically pure form (Scheme 96).[471]

2.9.7 Conclusions and Future Perspectives

Scheme 96 Total Synthesis of Pradimicinone[471]

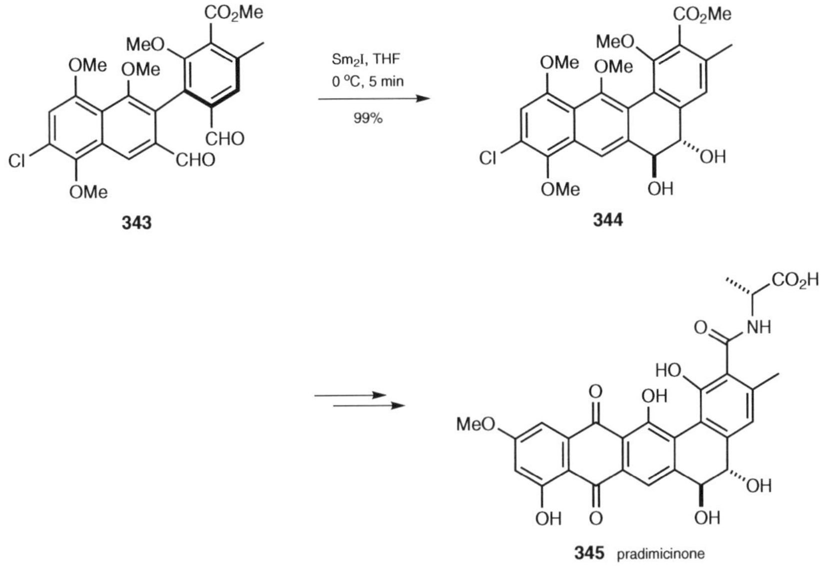

345 pradimicinone

Tetracycle 344:[471]
A 0.1 M soln of SmI_2 in THF (5.0 mL, 0.50 mmol) was added to a soln of **343** (50.0 mg, 0.109 mmol) in THF (2 mL) at 0 °C. After stirring for 5 min at 0 °C, the mixture was quenched by the addition of 1 M HCl and was then extracted with EtOAc (3 × 10 mL). The combined organic extract was washed sequentially with H_2O and brine, dried (Na_2SO_4), and concentrated under reduced pressure. The crude residue was purified by preparative TLC (hexane/EtOAc 1:1) to afford a white solid; yield: 49.9 mg (99%).

2.9.7 Conclusions and Future Perspectives

Although tremendous progress has been achieved in the field of radical cyclization reactions for selective organic synthesis, numerous areas remain unexplored. In recent years, radical reactions have evolved into many general areas in organic, inorganic, materials, supramolecular, and polymer chemistry. The present status of radical processes may be attributed to the groups of Julia, Walling, Ingold, Beckwith, Hart, Stork, Houk, RajanBabu, Mulliken,[472] Kagan,[175] Taube,[473] Marcus,[474] Huber,[475] Kochi,[476] Ashby,[477] Pross,[478] Evans,[479] Giese,[36] Curran,[37] Roy, Chakraborty, Snider, Demir, Molander, Gansäuer, Nishino, and many more. The strategies developed so far have already begun to display their utility in a wide range of applications ranging from the preparation of significant biologically active compounds or their precursors to agrochemicals and pharmaceuticals. With the advantages of higher yields and higher stereoselectivities relative to those obtained in the equivalent cationic reactions, radical cyclizations have been applied in the efficient synthesis of many (poly)cyclic skeletons that are valuable synthons in numerous synthetic approaches. Moreover, relative to their cationic counterparts, radical-bioinspired protocols are often characterized by their unique chemo- and regioselectivities. Nevertheless, some methods still suffer from certain drawbacks and need further research. For instance, the development of general methods for the synthesis of large families of organic products might be more productive than specific procedures restricted to the preparation of only one or a few compounds. Furthermore, titanium(III)-based radical cyclizations do not allow the synthesis of *cis*-fused carbocycles, which are essential elements of many natural products.

for references see p 533

As discussed in this chapter, manganese(III)-catalyzed oxidative cyclization strategies display the exceptional ability to generate radicals from carbonyl groups, unsaturated hydrocarbons, heterocycles, organophosphates, and carbohydrates, which facilitates intramolecular reactions with other radicals or neutral moieties. This chapter has presented numerous applications of radical chemistry relying on manganese(III)-based strategies.

Similarly, samarium(II) iodide has evolved as an extremely powerful reagent for the chemoselective reduction of a variety of functional groups through open-shell reaction pathways. The efficient reduction of halides, carbonyl groups, carboxylic acid derivatives, aromatics, alkenes, and heteroatom–heteroatom bonds is today a well-developed tool. However, research to develop samarium(II)-based catalytic systems with more general and effective reduction conditions will need to be undertaken in the near future.

Titanocene(III)-mediated radical processes have also led to significant advances in the synthesis of natural products of diverse nature and biological relevance. This chemistry has excelled in functional-group interconversions, mainly from epoxides to (poly)cyclic natural skeletons. In particular, the bioinspired synthesis of stereodefined terpene skeletons at room temperature is another remarkable feature of titanium(III) chemistry. Moreover, radical-based bioinspired protocols are also characterized by their distinctive chemo- and regioselectivities, which are often higher than those obtained in cationic cyclization processes. With the introduction of template catalysis[415–418] in titanium chemistry, the advancement of radical chemistry into complex natural product synthesis has gained additional strength. Thus, these reagents will greatly increase the options in the arsenal of synthetic organic chemists in the near future.

References

[1] Barton, D. H.; Beaton, J. M.; Geller, L. E.; Pechet, M. M., *J. Am. Chem. Soc.*, (1960) **82**, 2640.
[2] Parsons, A. F., *An Introduction to Free Radical Chemistry*, Wiley-Blackwell: Oxford, (2000).
[3] *Radicals in Organic Synthesis*, Renaud, P.; Sibi, M. P., Eds.; Wiley-VCH: Weinheim, Germany, (2001).
[4] *General Aspects of the Chemistry of Radicals*, Alfassi, Z. B., Ed.; Wiley: New York, (1999).
[5] *Stereochemistry of Radical Reactions: Concepts, Guidelines, and Synthetic Applications*, Curran, D. P.; Porter, N. A.; Giese, B., Eds.; VCH: Weinheim, Germany, (1996).
[6] Fossey, J.; Lefort, D.; Sorba, J., *Free Radicals in Organic Chemistry*, Wiley: New York, (1995).
[7] Zard, S. Z., *Radical Reactions in Organic Synthesis*, Oxford University Press: Oxford, (2003).
[8] Togo, H., *Advanced Free Radical Reactions for Organic Synthesis*, Elsevier: Oxford, (2004).
[9] Breslow, R.; Barrett, E.; Mohacsi, E., *Tetrahedron Lett.*, (1962) **3**, 1207.
[10] Breslow, R.; Groves, J. T.; Olin, S. S., *Tetrahedron Lett.*, (1966) **7**, 4717.
[11] Lallemand, J. Y.; Julia, M.; Mansuy, D., *Tetrahedron Lett.*, (1973) **14**, 4461.
[12] Julia, M., *Acc. Chem. Res.*, (1971) **4**, 386.
[13] Walling, C., *Tetrahedron*, (1985) **41**, 3887.
[14] Beckwith, A. L. J.; Schiesser, C. H., *Tetrahedron*, (1985) **41**, 3925.
[15] Beckwith, A. L. J.; Schiesser, C. H., *Tetrahedron Lett.*, (1985) **26**, 373.
[16] Beckwith, A. L. J.; Phillipou, G.; Serelis, A. K., *Tetrahedron Lett.*, (1981) **22**, 2811.
[17] Beckwith, A. L. J.; Boate, D. R., *Tetrahedron Lett.*, (1985) **26**, 1761.
[18] Johnson, L. J.; Lusztyk, J.; Wayner, D. D. M.; Abeywickreyma, A. N.; Beckwith, A. L. J.; Scaiano, J. C.; Ingold, K. U., *J. Am. Chem. Soc.*, (1985) **107**, 4594.
[19] Beckwith, A. L. J.; Roberts, D. H., *J. Am. Chem. Soc.*, (1986) **108**, 5893.
[20] Hart, D. J., *Science (Washington, D. C.)*, (1984) **223**, 883.
[21] Stork, G.; Baine, N. H., *J. Am. Chem. Soc.*, (1982) **104**, 2321.
[22] Stork, G.; Mook, R., Jr., *Tetrahedron Lett.*, (1986) **27**, 4529.
[23] Curran, D. P., *Synthesis*, (1988), 489.
[24] Jasperse, C. P.; Curran, D. P.; Fevig, T. L., *Chem. Rev.*, (1991) **91**, 1237.
[25] Spellmeyer, D. C.; Houk, K. N., *J. Org. Chem.*, (1987) **52**, 959.
[26] RajanBabu, T. V., *J. Am. Chem. Soc.*, (1987) **109**, 609.
[27] RajanBabu, T. V., *J. Org. Chem.*, (1988) **53**, 4522.
[28] RajanBabu, T. V.; Fukanaga, T., *J. Am. Chem. Soc.*, (1989) **111**, 296.
[29] RajanBabu, T. V.; Fukanaga, T.; Reddy, G. S., *J. Am. Chem. Soc.*, (1989) **111**, 1759.
[30] Kharasch, M. S.; Arimato, F. S.; Nudenberg, W., *J. Org. Chem.*, (1951) **16**, 1556.
[31] Kharasch, M. S.; Kawahara, F.; Nudenberg, W., *J. Org. Chem.*, (1954) **19**, 1977.
[32] *Free Radicals*, Kochi, J. K., Ed.; Wiley: New York, (1973); Vol. 1.
[33] Minisci, F., *Top. Curr. Chem.*, (1976) **62**, 1.
[34] Jahn, U., *Top. Curr. Chem.*, (2012) **320**, 121.
[35] Yorimitsu, H., In *Encyclopedia of Radicals in Chemistry, Biology and Materials*, Chatgilialoglu, C.; Studer, A., Eds.; Wiley: Chichester, UK, (2012); Vol. 2, p 1003.
[36] Giese, B., *Radicals in Organic Synthesis: Formation of Carbon-Carbon Bonds*, Pergamon: Oxford, (1986).
[37] Curran, D. P., In *Comprehensive Organic Synthesis*, Trost, B. M.; Fleming, I., Eds.; Pergamon: Oxford, (1991); Vol. 4, p 779.
[38] Curran, D. P.; Porter, N. A.; Giese, B., *Stereochemistry of Radical Reactions*, VCH: Weinheim, (1996).
[39] Parsons, A. F., *An Introduction to Free Radical Chemistry*, Wiley-Blackwell: Oxford, (2000); Chapter 7, p 139.
[40] Lee, C.-S.; Yu, T.-C.; Luo, J.-W.; Cheng, Y.-Y.; Chuang, C.-P., *Tetrahedron Lett.*, (2009) **50**, 4558.
[41] Dhimane, A.-L.; Fensterbank, L.; Malacria, M., In *Radicals in Organic Synthesis*, Renaud, P.; Sibi, M. P., Eds.; Wiley-VCH: Weinheim, Germany, (2001); Vol. 2, p 350.
[42] Srikrishna, A., In *Radicals in Organic Synthesis*, Renaud, P.; Sibi, M. P., Eds.; Wiley-VCH: Weinheim, Germany, (2001); Vol. 2, p 151.
[43] Lee, E., In *Radicals in Organic Synthesis*, Renaud, P.; Sibi, M. P., Eds.; Wiley-VCH: Weinheim, Germany, (2001); Vol. 2, pp 303.
[44] Bunce, R. A., *Tetrahedron*, (1995) **51**, 13103.
[45] McCarroll, A. J.; Walton, J., *Angew. Chem. Int. Ed.*, (2001) **40**, 2224.
[46] McCarroll, A. J.; Walton, J., *J. Chem. Soc., Perkin Trans. 1*, (2001), 3215.
[47] Demir, A. S.; Emrullahoglu, M., *Curr. Org. Synth.*, (2007) **4**, 321.

[48] Snider, B. B., *Chem. Rev.*, (1996) **96**, 339.
[49] Melikyan, G. G., *Org. React. (N. Y.)*, (1997) **49**, 427.
[50] Pan, X.-Q.; Zou, J.-P.; Zhang, W., *Mol. Diversity*, (2009) **13**, 421.
[51] Nishino, H., *Top. Heterocycl. Chem.*, (2006) **6**, 39.
[52] Mondal, M., *Synlett*, (2013) **24**, 137.
[53] Mondal, M.; Bora, U., *RSC Adv.*, (2013) **3**, 18716.
[54] Demir, A. S.; Reis, O.; Emrullahoglu, M., *J. Org. Chem.*, (2003) **68**, 10130.
[55] Wang, G.-W.; Li, F.-B., *Org. Biomol. Chem.*, (2005) **3**, 794.
[56] Snider, B. B.; Kwon, T., *J. Org. Chem.*, (1990) **55**, 1965.
[57] Kochi, J. K., *J. Am. Chem. Soc.*, (1965) **87**, 1811.
[58] Li, F.-B.; Liu, T.-X.; Huang, Y.-S.; Wang, G.-W., *J. Org. Chem.*, (2009) **74**, 7743.
[59] Zengin, M.; Sonmez, F.; Arslan, M.; Kucukislamoglu, M., *Maced. J. Chem. Chem. Eng.*, (2012) **31**, 55.
[60] Nair, V.; Deepthi, A., *Chem. Rev.*, (2007) **107**, 1862.
[61] Dhakshinamoorthy, A., *Synlett*, (2005), 3014.
[62] Li, F.-B.; Liu, T.-X.; You, X.; Wang, G.-W., *Org. Lett.*, (2010) **12**, 3258.
[63] Li, F.-B.; You, X.; Wang, G.-W., *Org. Lett.*, (2010) **12**, 4896.
[64] Li, F.-B.; Liu, T.-X.; Wang, G.-W., *J. Org. Chem.*, (2008) **73**, 6417.
[65] Molander, G. A.; Harris, C. R., *Chem. Rev.*, (1996) **96**, 307.
[66] Krief, A.; Laval, A.-M., *Chem. Rev.*, (1999) **99**, 745.
[67] Edmonds, D. J.; Johnston, D.; Procter, D. J., *Chem. Rev.*, (2004) **104**, 3371.
[68] Szostak, M.; Spain, M.; Procter, D. J., *Chem. Soc. Rev.*, (2013) **42**, 9155.
[69] Justicia, J.; Álvarez de Cienfuegos, L.; Campaña, A. G.; Miguel, D.; Jakoby, V.; Gansäuer, A.; Cuerva, J. M., *Chem. Soc. Rev.*, (2011) **40**, 3525.
[70] Márquez, I. R.; Millán, A.; Campaña, A. G.; Cuerva, J. M., *Org. Chem. Front.*, (2014) **1**, 373.
[71] Barrero, A. F.; Quílez del Moral, J. F.; Sánchez, E. M.; Arteaga, J. F., *Eur. J. Org. Chem.*, (2006), 1627.
[72] Morcillo, S. P.; Miguel, D.; Campaña, A. G.; Álvarez de Cienfuegos, L.; Justicia, J.; Cuerva, J. M., *Org. Chem. Front.*, (2014) **1**, 15.
[73] Cahiez, G.; Moyeux, A., *Chem. Rev.*, (2010) **110**, 1435.
[74] Xi, Y.; Yia, H.; Lei, A., *Org. Biomol. Chem.*, (2013) **11**, 2387.
[75] Pepper, H. P.; Tulip, S. J.; Nakano, Y.; George, J. H., *J. Org. Chem.*, (2014) **79**, 2564.
[76] Pepper, H. P.; Lam, H. C.; Bloch, W. M.; George, J. H., *Org. Lett.*, (2012) **14**, 5162.
[77] Wang, X.; Ma, Z.; Lu, J.; Tan, X.; Chen, C., *J. Am. Chem. Soc.*, (2011) **133**, 15350.
[78] Wang, X.; Wang, X.; Tan, X.; Lu, J.; Cormier, K. W.; Ma, Z.; Chen, C., *J. Am. Chem. Soc.*, (2012) **134**, 18834.
[79] Ma, Z.; Lu, J.; Wang, X.; Chen, C., *Chem. Commun. (Cambridge)*, (2011) **47**, 427.
[80] Yamashita, S.; Naruko, A.; Yamada, T.; Hayashi, Y.; Hirama, M., *Chem. Lett.*, (2013) **42**, 220.
[81] Cong, Z.; Miki, T.; Urakawa, O.; Nishino, H., *J. Org. Chem.*, (2009) **74**, 3978.
[82] Yin, J.; Wang, C.; Kong, L.; Cai, S.; Gao, S., *Angew. Chem. Int. Ed.*, (2012) **51**, 7786.
[83] Magolan, J.; Carson, C. A.; Kerr, M. A., *Org. Lett.*, (2008) **10**, 1437.
[84] Zoretic, P. A.; Fang, H.; Ribeiro, A. A., *J. Org. Chem.*, (1998) **63**, 4779.
[85] Snider, B. B.; Kiselgof, J. Y.; Foxman, B. M., *J. Org. Chem.*, (1998) **63**, 7945.
[86] Barrero, A. F.; Herrador, M. M.; Quílez del Moral, J. F.; Valdivia, M. V., *Org. Lett.*, (2002) **4**, 1379.
[87] Yang, D.; Ye, X.-Y.; Gu, S.; Xu, M., *J. Am. Chem. Soc.*, (1999) **121**, 5579.
[88] Yang, D.; Ye, X.-Y.; Xu, M., *J. Org. Chem.*, (2000) **65**, 2208.
[89] Christensen, O. T., *J. Prakt. Chem.*, (1883) **28**, 1.
[90] Christensen, O. T., *Z. Anorg. Chem.*, (1901) **27**, 321.
[91] Bush, J. B., Jr.; Finkbeiner, H., *J. Am. Chem. Soc.*, (1968) **90**, 5903.
[92] Iwahama, T.; Sakaguchi, S.; Ishii, Y., *Chem. Commun. (Cambridge)*, (2000), 2317.
[93] Kagayama, T.; Nakano, A.; Sakaguchi, S.; Ishii, Y., *Org. Lett.*, (2006) **8**, 407.
[94] Ishii, Y.; Sakaguchi, S., In *Modern Oxidation Methods*, Bäckvall, J.-E., Ed.; Wiley-VCH: Weinheim, Germany, (2004), p 119.
[95] Weinland, R. F.; Fischer, G., *Z. Anorg. Allg. Chem.*, (1921) **120**, 161.
[96] Hessel, L. W.; Romers, C., *Recl. Trav. Chim. Pays-Bas*, (1969) **88**, 545.
[97] Heiba, E. I.; Dessau, R. M.; Koehl, W. J., Jr., *J. Am. Chem. Soc.*, (1968) **90**, 5905.
[98] Vinogradov, M. G.; Verenchikov, S. P.; Nikishin, G. I., *Bull. Acad. Sci. USSR, Div. Chem. Sci. (Engl. Transl.)*, (1972) **21**, 1626.
[99] Heiba, E. I.; Dessau, R. M., *J. Org. Chem.*, (1974) **39**, 3456.
[100] Fristad, W. E.; Peterson, J. R., *J. Org. Chem.*, (1985) **50**, 10.

[101] Fristad, W. E.; Peterson, J. R.; Ernst, A. B., *J. Org. Chem.*, (1985) **50**, 3143.
[102] Fristad, W. E.; Hershberger, S. S., *J. Org. Chem.*, (1985) **50**, 1026.
[103] Huang, J.-W.; Shi, M., *J. Org. Chem.*, (2005) **70**, 3859.
[104] Corey, E. J.; Kang, M.-C., *J. Am. Chem. Soc.*, (1984) **106**, 5384.
[105] Snider, B. B.; Mohan, R.; Kates, S. A., *J. Org. Chem.*, (1985) **50**, 3659.
[106] Ernst, A. B.; Fristad, W. E., *Tetrahedron Lett.*, (1985) **26**, 3761.
[107] Çalişkan, R.; Pekel, T.; Watson, W. H.; Balci, M., *Tetrahedron Lett.*, (2005) **46**, 6227.
[108] Dengiz, C.; Çalişkan, R.; Balci, M., *Tetrahedron Lett.*, (2012) **53**, 550.
[109] Anderson, J. M.; Kochi, J. K., *J. Am. Chem. Soc.*, (1970) **92**, 2450.
[110] de Klein, W. J., In *Organic Syntheses by Oxidation with Metal Compounds*, Mijs, W. J.; de Jonge, C. R. H. I., Eds.; Plenum, New York, (1986), 261.
[111] Asahi, K.; Nishino, H., *Synthesis*, (2009), 409.
[112] Demir, A. S.; Findik, H.; Saygili, N.; Subasi, N. T., *Tetrahedron*, (2010) **66**, 1308.
[113] Demir, A. S.; Reis, O.; Karaaslan, E. O., *J. Chem. Soc., Perkin Trans. 1*, (2001), 3042.
[114] Demir, A. S.; Tanyeli, C.; Altinel, E., *Tetrahedron Lett.*, (1997) **38**, 7267.
[115] Gansäuer, A.; Justicia, J.; Fan, C.-A.; Worgull, D.; Piestert, F., *Top. Curr. Chem.*, (2007) **279**, 25.
[116] Cuerva, J. M.; Justicia, J.; Oller-López, J. L.; Oltra, J. E., *Top. Curr. Chem.*, (2006) **264**, 63.
[117] Rossi, B.; Prosperini, S.; Pastori, N.; Clerici, A.; Punta, C., *Molecules*, (2012) **17**, 14700.
[118] Gansäuer, A.; Fleckhaus, A., In *Encyclopedia of Radicals in Chemistry, Biology and Materials*, Chatgilialoglu, C.; Studer, A., Eds.; Wiley: Chichester, UK, (2012); Vol. 2, p 989.
[119] Gansäuer, A.; Ndene, N.; Lauterbach, T.; Justicia, J.; Winkler, I.; Mück-Lichtenfeld, C.; Grimme, S., *Tetrahedron*, (2008) **64**, 11839.
[120] Barrero, A. F.; Herrador, M. M.; Quílez del Moral, J. F.; Arteaga, P.; Arteaga, J. F.; Piedra, M.; Sánchez, E. M., *Org. Lett.*, (2005) **7**, 2301.
[121] Estévez, R. E.; Justicia, J.; Bazdi, B.; Fuentes, N.; Paradas, M.; Choquesillo-Lazarte, D.; García-Ruiz, J. M.; Robles, R.; Gansäuer, A.; Cuerva, J. M.; Oltra, J. E., *Chem.–Eur. J.*, (2009) **15**, 2774.
[122] Parrish, J. D.; Shelton, D. R.; Little, R. D., *Org. Lett.*, (2003) **5**, 3615.
[123] Estévez, R. E.; Paradas, M.; Millán, A.; Jiménez, T.; Robles, R.; Cuerva, J. M.; Oltra, J. E., *J. Org. Chem.*, (2008) **73**, 1616.
[124] Cuerva, J. M.; Campaña, A. G.; Justicia, J.; Rosales, A.; Oller-López, J. L.; Robles, R.; Cárdenas, D. J.; Buñuel, E.; Oltra, J. E., *Angew. Chem. Int. Ed.*, (2006) **45**, 5522.
[125] Paradas, M.; Campaña, A. G.; Marcos, M. L.; Justicia, J.; Haidour, A.; Robles, R.; Cárdenas, D. J.; Oltra, J. E.; Cuerva, J. M., *Dalton Trans.*, (2010) **39**, 8796.
[126] Diéguez, H. R.; López, A.; Domingo, V.; Arteaga, J. F.; Dobado, J. A.; Herrador, M. M.; Quílez del Moral, J. F.; Barrero, A. F., *J. Am. Chem. Soc.*, (2010) **132**, 254.
[127] Gansäuer, A., *Chem. Commun. (Cambridge)*, (1997), 457.
[128] Gansäuer, A.; Moschioni, M.; Bauer, D., *Eur. J. Org. Chem.*, (1998), 1923.
[129] Gansäuer, A.; Bauer, D., *Eur. J. Org. Chem.*, (1998), 2673.
[130] Gansäuer, A.; Bauer, D., *J. Org. Chem.*, (1998) **63**, 2070.
[131] Paradas, M.; Campaña, A. G.; Estévez, R. E.; Álvarez de Cienfuegos, L.; Jiménez, T.; Robles, R.; Cuerva, J. M.; Oltra, J. E., *J. Org. Chem.*, (2009) **74**, 3616.
[132] Streuff, J., *Synthesis*, (2013) **45**, 281.
[133] Coutts, R. S. P.; Wailes, P. C.; Martin, R. L., *J. Organomet. Chem.*, (1973) **50**, 145.
[134] Green, M. L. H.; Lucas, C. R., *J. Chem. Soc., Dalton Trans.*, (1972), 1000.
[135] Gansäuer, A.; Pierobon, M.; Bluhm, H., *J. Am. Chem. Soc.*, (1998) **120**, 12849.
[136] Gansäuer, A.; Pierobon, M.; Bluhm, H., *Synthesis*, (2001), 2500.
[137] Enemærke, R. J.; Hjøllund, G. H.; Daasbjerg, K.; Skrydstrup, T., *C. R. Acad. Sci., Ser. IIc*, (2001) **4**, 435.
[138] Rickborn, B., In *Comprehensive Organic Synthesis*, Trost, B. M.; Fleming, I., Eds.; Pergamon: Oxford, (1991); Vol. 3, p 733.
[139] Mitsunobu, O., In *Comprehensive Organic Synthesis*, Trost, B. M.; Fleming, I., Eds.; Pergamon: Oxford, (1991); Vol. 6, p 1.
[140] Nugent, W. A.; RajanBabu, T. V., *J. Am. Chem. Soc.*, (1988) **110**, 8561.
[141] RajanBabu, T. V.; Nugent, W. A., *J. Am. Chem. Soc.*, (1989) **111**, 4525.
[142] RajanBabu, T. V.; Nugent, W. A.; Beattie, M. S., *J. Am. Chem. Soc.*, (1990) **112**, 6408.
[143] RajanBabu, T. V.; Nugent, W. A., *J. Am. Chem. Soc.*, (1994) **116**, 986.
[144] Weigand, S.; Brückner, R., *Synlett*, (1997), 225.
[145] Yadav, J. S.; Srinivas, D., *Chem. Lett.*, (1997) **26**, 905.

[146] Jørgensen, K. B.; Suenaga, T.; Nakata, T., *Tetrahedron Lett.*, (1999) **40**, 8855.
[147] Chakraborty, T. K.; Das, S., *Tetrahedron Lett.*, (2002) **43**, 2313.
[148] Chakraborty, T. K.; Goswami, R. J., *Tetrahedron Lett.*, (2004) **45**, 7637.
[149] Hardouin, C.; Chevallier, F.; Rousseau, B.; Doris, E., *J. Org. Chem.*, (2001) **66**, 1046.
[150] Bermejo, F.; Sandoval, C., *J. Org. Chem.*, (2004) **69**, 5275.
[151] Gansäuer, A.; Bluhm, H.; Rinker, B.; Narayan, S.; Schick, M.; Lauterbach, T.; Pierobon, M., *Chem.–Eur. J.*, (2003) **9**, 531.
[152] Gansäuer, A.; Bluhm, H., *Chem. Commun. (Cambridge)*, (1998), 2143.
[153] Merlic, C. A.; Xu, D., *J. Am. Chem. Soc.*, (1991) **113**, 9855.
[154] Merlic, C. A.; Xu, D.; Nguyen, M. C., *Tetrahedron Lett.*, (1993) **34**, 227.
[155] Dötz, K. H.; da Silva, E. G., *Tetrahedron*, (2000) **56**, 8291.
[156] Gansäuer, A.; Rinker, B., *Tetrahedron*, (2002) **58**, 7017.
[157] Manxzer, L. E.; Deaton, J.; Sharp, P.; Schrock, R. R., *Inorg. Synth.*, (1982) **21**, 135.
[158] Barrero, A. F.; Rosales, A.; Cuerva, J. M.; Oltra, J. E., *Org. Lett.*, (2003) **5**, 1935.
[159] Fuse, S.; Hanochi, M.; Doi, T.; Takahashi, T., *Tetrahedron Lett.*, (2004) **45**, 1961.
[160] Gansäuer, A.; Fan, C.-A.; Piestert, F., *J. Am. Chem. Soc.*, (2008) **130**, 6916.
[161] Banerjee, B.; Roy, S. C., *Eur. J. Org. Chem.*, (2006), 489.
[162] Banerjee, B.; Roy, S. C., *Synthesis*, (2005), 2913.
[163] Roy, S. C.; Rana, K. K.; Guin, C., *J. Org. Chem.*, (2002) **67**, 3242.
[164] *Multiphase Homogeneous Catalysis*, Cornils, B.; Herrmann, W. A.; Horváth, I. T.; Leitner, W.; Mecking, S.; Olivier-Bourbigou, H.; Vogt, D., Eds.; Wiley-VCH: Weinheim, Germany, (2008); p 2.
[165] Spiegel, D. A.; Wiberg, K. B.; Schacherer, L. N.; Medeiros, M. R.; Wood, J. L., *J. Am. Chem. Soc.*, (2005) **127**, 12513.
[166] Barrero, A. F.; Oltra, J. E.; Cuerva, J. M.; Rosales, A., *J. Org. Chem.*, (2002) **67**, 2566.
[167] Mandal, S. K.; Jana, S.; Roy, S. C., *Tetrahedron Lett.*, (2005) **46**, 6115.
[168] Jana, S.; Guin, C.; Roy, S. C., *Tetrahedron Lett.*, (2005) **46**, 1155.
[169] Namy, J.-L.; Girard, P.; Kagan, H. B., *New J. Chem.*, (1977) **1**, 5.
[170] Girard, P.; Namy, J.-L.; Kagan, H. B., *J. Am. Chem. Soc.*, (1980) **102**, 2693.
[171] Kagan, H. B.; Namy, J.-L.; Girard, P., *Tetrahedron*, (1981) **37**, Suppl.1, 175.
[172] Molander, G. A., *Chem. Rev.*, (1992) **92**, 29.
[173] Concellón, J. M.; Rodríguez-Solla, H., *Chem. Soc. Rev.*, (2004) **33**, 599.
[174] Dahlén, A.; Hilmersson, G., *Eur. J. Inorg. Chem.*, (2004), 3393.
[175] Gopalaiah, K.; Kagan, H. B., *New J. Chem.*, (2008) **32**, 607.
[176] Nicolaou, K. C.; Ellery, S. P.; Chen, J. S., *Angew. Chem. Int. Ed.*, (2009) **48**, 7140.
[177] Fevig, T. L.; Elliott, R. L.; Curran, D. P., *J. Am. Chem. Soc.*, (1988) **110**, 5064.
[178] Matignon, C. A.; Caze, E., *Ann. Chim. Phys.*, (1906) **8**, 417.
[179] Perkins, M. J., *Radical Chemistry*, Ellis Horwood: New York, (1994); p 182.
[180] Curran, D. P., *Aldrichimica Acta*, (2000) **33**, 104.
[181] Rozantsev, E. G.; Loshadkin, D. V., *Des. Monomers Polym.*, (2001) **4**, 281.
[182] Gansäuer, A.; Lauterbach, T.; Narayan, S., *Angew. Chem. Int. Ed.*, (2003) **42**, 5556.
[183] Hicks, R. G., *Org. Biomol. Chem.*, (2007) **5**, 1321.
[184] Bashir, N.; Patro, B.; Murphy, J. A., In *Advances in Free Radical Chemistry*, Zard, S. Z., Ed.; JAI: Stamford, CT, (1999); p 123.
[185] Godineau, E.; Schenk, K.; Landais, Y., *J. Org. Chem.*, (2008) **73**, 6983.
[186] Molander, G. A.; Harris, C. R., *Tetrahedron*, (1998) **54**, 3321.
[187] Molander, G. A.; Kenny, C., *J. Org. Chem.*, (1991) **56**, 1439.
[188] Imamoto, T.; Ono, M., *Chem. Lett.*, (1987) **16**, 501.
[189] Concellón, J. M.; Rodríguez-Solla, H.; Bardales, E.; Huerta, M., *Eur. J. Org. Chem.*, (2003), 1775.
[190] Akane, N.; Kanagawa, Y.; Nishiyama, Y.; Ishii, Y., *Chem. Lett.*, (1992) **21**, 2431.
[191] Hamann, B.; Namy, J.-L.; Kagan, H. B., *Tetrahedron*, (1996) **52**, 14225.
[192] Namy, J.-L.; Colomb, M.; Kagan, H. B., *Tetrahedron Lett.*, (1994) **35**, 1723.
[193] Evans, W. J.; Gummersheimer, T. S.; Ziller, J. W., *J. Am. Chem. Soc.*, (1995) **117**, 8999.
[194] Concellón, J. M.; Bernad, P. L.; Huerta, M.; García-Granda, S.; Díaz, M. R., *Chem.–Eur. J.*, (2003) **9**, 5343.
[195] Kagan, H. B.; Namy, J.-L., *Top. Organomet. Chem.*, (1999) **2**, 155.
[196] Steel, P. G., *J. Chem. Soc., Perkin Trans. 1*, (2001), 2727.
[197] Corey, E. J.; Zheng, G. Z., *Tetrahedron Lett.*, (1997) **38**, 2048.
[198] Nomura, R.; Matsuno, T.; Endo, T., *J. Am. Chem. Soc.*, (1996) **118**, 11666.

[199] Helion, F.; Namy, J.-L., *J. Org. Chem.*, (1999) **64**, 2944.
[200] Inanaga, J.; Sakai, S.; Handa, Y.; Yamaguchi, M.; Yokoyama, Y., *Chem. Lett.*, (1991) **20**, 2117.
[201] Chopade, P. R.; Prasad, E.; Flowers, R. A., II, *J. Am. Chem. Soc.*, (2004) **126**, 44.
[202] Inanaga, J.; Ishikawa, M.; Yamaguchi, M., *Chem. Lett.*, (1987) **16**, 1485.
[203] Shabangi, M.; Flowers, R. A., II, *Tetrahedron Lett.*, (1997) **38**, 1137.
[204] Enemærke, R. J.; Daasbjerg, K.; Skrydstrup, T., *Chem. Commun. (Cambridge)*, (1999), 343.
[205] Enemærke, R. J.; Hertz, T.; Skrydstrup, T.; Daasbjerg, K., *Chem.–Eur. J.*, (2000) **6**, 3747.
[206] Shotwell, J. B.; Sealy, J. M.; Flowers, R. A., II, *J. Org. Chem.*, (1999) **64**, 5251.
[207] Fuchs, J. R.; Mitchell, M. M.; Shabangi, M.; Flowers, R. A., II, *Tetrahedron Lett.*, (1997) **38**, 8157.
[208] Miller, R. S.; Sealy, J. M.; Shabangi, M.; Kuhlman, M. L.; Fuchs, J. R.; Flowers, R. A., II, *J. Am. Chem. Soc.*, (2000) **122**, 7718.
[209] Machrouhi, F.; Hamann, B.; Namy, J.-L.; Kagan, H. B., *Synlett*, (1996), 633.
[210] Ogawa, A.; Sumino, Y.; Nanke, T.; Ohya, S.; Sonoda, N.; Hirao, T., *J. Am. Chem. Soc.*, (1997) **119**, 2745.
[211] Procter, D. J.; Flowers, R. A., II; Skrydstrup, T., *Organic Synthesis using Samarium Diiodide: A Practical Guide*, RSC: Cambridge, UK, (2009).
[212] van Wijngaarden, I.; Kruse, C. G.; van der Heyden, J. A. M.; Tulp, M. T. M., *J. Med. Chem.*, (1988) **31**, 1934.
[213] Alešković, M.; Basarić, N.; Halasz, I.; Liang, X.; Qin, W.; Mlinarić-Majerski, K., *Tetrahedron*, (2013) **69**, 1725.
[214] Sobenina, L. N.; Vasil'tsov, A. M.; Petrova, O. V.; Petrushenko, K. B.; Ushakov, I. A.; Clavier, G.; Meallet-Renault, R.; Mikhaleva, A. I.; Trofimov, B. A., *Org. Lett.*, (2011) **13**, 2524.
[215] Harman, W. H.; Chang, C. J., *J. Am. Chem. Soc.*, (2007) **129**, 15128.
[216] Gulevich, A. V.; Dudnik, A. S.; Chernyak, N.; Gevorgyan, V., *Chem. Rev.*, (2013) **113**, 3084.
[217] Estévez, V.; Villacampa, M.; Menéndez, J. C., *Chem. Soc. Rev.*, (2010) **39**, 4402.
[218] Zhao, M.-N.; Ren, Z.-H.; Wang, Y.-Y.; Guan, Z.-H., *Org. Lett.*, (2014) **16**, 608.
[219] Wen, J.; Zhang, R.-Y.; Chen, S.-Y.; Zhang, J.; Yu, X.-Q., *J. Org. Chem.*, (2012) **77**, 766.
[220] Wen, J.; Qin, S.; Ma, L.-F.; Dong, L.; Zhang, J.; Liu, S.-S.; Duan, Y.-S.; Chen, S.-Y.; Hu, C.-W.; Yu, X.-Q., *Org. Lett.*, (2010) **12**, 2694.
[221] Yamamoto, H.; Sasaki, I.; Mitsutake, M.; Karasudani, A.; Imagawa, H.; Nishizawa, M., *Synlett*, (2011), 2815.
[222] Rieth, R. D.; Mankad, N. P.; Calimano, E.; Sadighi, J. P., *Org. Lett.*, (2004) **6**, 3981.
[223] Du, W.; Zhao, M.-N.; Ren, Z.-H.; Wang, Y.-Y.; Guan, Z.-H., *Chem. Commun. (Cambridge)*, (2014) **50**, 7437.
[224] Tucker, J. W.; Narayanam, J. M. R.; Krabbe, S. W.; Stephenson, C. R. J., *Org. Lett.*, (2010) **12**, 368.
[225] Fodor, G. B.; Colasanti, B., In *Alkaloids: Chemical and Biological Perspectives*, Pelletier, S. W., Ed.; Wiley: New York, (1985); Vol. 3, p 1.
[226] Strunz, G. M.; Findlay, J. A., In *The Alkaloids: Chemistry and Pharmacology*, Brossi, A., Ed.; Academic: London, (1985); Vol. 26, p 89.
[227] Schneider, M. J., In *Alkaloids: Chemical and Biological Perspectives*, Pelletier, S. W., Ed.; Pergamon: Oxford, UK, (1996); Vol. 10, p 155.
[228] Pinder, A. R., *Nat. Prod. Rep.*, (1990) **7**, 447.
[229] Kinghorn, A. D.; Balandrin, M. F., In *Alkaloids: Chemical and Biological Perspectives*, Pelletier, S. W., Ed.; Wiley: New York, (1984); Vol. 2, p 105.
[230] Michael, J. P., *Nat. Prod. Rep.*, (2001) **18**, 520.
[231] Michael, J. P., *Nat. Prod. Rep.*, (2008) **25**, 139.
[232] Podichetty, A. K.; Wagner, S.; Schröer, S.; Faust, A.; Schäfers, M.; Schober, O.; Kopka, K.; Haufe, G., *J. Med. Chem.*, (2009) **52**, 3484.
[233] He, X.; Alian, A.; Stroud, R.; Ortiz de Montellano, P. R., *J. Med. Chem.*, (2006) **49**, 6308.
[234] Zhang, J.; Wang, Q.; Fang, H.; Xu, W. F.; Liu, A. L.; Du, G. H., *Bioorg. Med. Chem.*, (2007) **15**, 2749.
[235] Lim, H. J.; RajanBabu, T. V., *Org. Lett.*, (2009) **11**, 2924.
[236] Rao, W. D.; Chan, P. W. H., *Chem.–Eur. J.*, (2008) **14**, 10486.
[237] Bertozzi, F.; Gustafsson, M.; Olsson, R., *Org. Lett.*, (2002) **4**, 3147.
[238] Häberli, A.; Leumann, C. J., *Org. Lett.*, (2001) **3**, 489.
[239] Zhang, S. H.; Xu, L.; Miao, L.; Shu, H.; Trudell, M. L., *J. Org. Chem.*, (2007) **72**, 3133.
[240] Zhang, J. L.; Yang, C. G.; He, C., *J. Am. Chem. Soc.*, (2006) **128**, 1798.
[241] Honda, T.; Aranishi, E.; Kaneda, K., *Org. Lett.*, (2009) **11**, 1857.
[242] Tanaka, K.; Kitamura, M.; Narasaka, K., *Bull. Chem. Soc. Jpn.*, (2005) **78**, 1659.

[243] Tsuritani, T.; Shinokubo, H.; Oshima, K., *J. Org. Chem.*, (2003) **68**, 3246.
[244] Heuger, G.; Kalsow, S.; Göttlich, R., *Eur. J. Org. Chem.*, (2002), 1848.
[245] Senboku, H.; Hasegawa, H.; Orito, K.; Tokuda, M., *Heterocycles*, (1999) **50**, 333.
[246] Tsuritani, T.; Shinokubo, H.; Oshima, K., *Org. Lett.*, (2001) **3**, 2709.
[247] Inoue, K.; Koga, N.; Iwamura, H., *J. Am. Chem. Soc.*, (1991) **113**, 9803.
[248] Ciriano, M. V.; Korth, H. G.; Scheppingen, W. B.; Mulder, P., *J. Am. Chem. Soc.*, (1999) **121**, 6375.
[249] Qu, J. Q.; Katsumata, T.; Satoh, M.; Wada, J.; Masuda, T., *Macromolecules*, (2007) **40**, 3136.
[250] Aggarwal, V. K.; Lopin, C.; Sandrinelli, F., *J. Am. Chem. Soc.*, (2003) **125**, 7596.
[251] Logan, A. W. J.; Sprague, S. J.; Foster, R. W.; Marx, L. B.; Garzya, V.; Hallside, M. S.; Thompson, A. L.; Burton, J. W., *Org. Lett.*, (2014) **16**, 4078.
[252] Keane, H. A.; Hess, W.; Burton, J. W., *Chem. Commun. (Cambridge)*, (2012) **48**, 6496.
[253] Liu, W.-M.; Liu, Z.-H.; Cheong, W.-W.; Priscilla, L.-Y. T.; Li, Y.; Narasaka, K., *Bull. Korean Chem. Soc.*, (2010) **31**, 563.
[254] Bytschkov, I.; Doye, S., *Tetrahedron Lett.*, (2002) **43**, 3715.
[255] Payack, J. F.; Hughes, D. L.; Cai, D.; Cottrell, I. F.; Verhoeven, T. R., *Org. Synth.*, (2004) **10**, 355.
[256] Payack, J. F.; Hughes, D. L.; Cai, D.; Cottrell, I. F.; Verhoeven, T. R., *Org. Synth.*, (2002) **79**, 19.
[257] Takahashi, K.; Midori, M.; Kawano, K.; Ishihara, J.; Hatakeyama, S., *Angew. Chem. Int. Ed.*, (2008) **47**, 6244.
[258] Yoon, T. P.; Ischay, M. A.; Du, J., *Nat. Chem.*, (2010) **2**, 527.
[259] Narayanam, J. M. R.; Stephenson, C. R. J., *Chem. Soc. Rev.*, (2011) **40**, 102.
[260] Teplý, F., *Collect. Czech. Chem. Commun.*, (2011) **76**, 859.
[261] Andrews, R. S.; Becker, J. J.; Gagné, M. R., *Angew. Chem. Int. Ed.*, (2010) **49**, 7274.
[262] Andrews, R. S.; Becker, J. J.; Gagné, M. R., *Org. Lett.*, (2011) **13**, 2406.
[263] Rueping, M.; Vila, C.; Koenigs, R. M.; Poscharny, K.; Fabry, D. C., *Chem. Commun. (Cambridge)*, (2011) **47**, 2360.
[264] Zou, Y.-Q.; Lu, L.-Q.; Fu, L.; Chang, N.-J.; Rong, J.; Chen, J.-R.; Xiao, W.-J., *Angew. Chem. Int. Ed.*, (2011) **50**, 7171.
[265] Xuan, J.; Cheng, Y.; An, J.; Lu, L.-Q.; Zhang, X.-X.; Xiao, W.-J., *Chem. Commun. (Cambridge)*, (2011) **47**, 8337.
[266] Rueping, M.; Zhu, S.; Koenigs, R. M., *Chem. Commun. (Cambridge)*, (2011) **47**, 8679.
[267] Quiclet-Sire, B.; Zard, S. Z., *Pure Appl. Chem.*, (2011) **83**, 519.
[268] Beemelmanns, C.; Reissig, H.-U., *Chem. Soc. Rev.*, (2011) **40**, 2199.
[269] Tucker, J. W.; Stephenson, C. R. J., *Org. Lett.*, (2011) **13**, 5468.
[270] Scheiner, P., *J. Org. Chem.*, (1967) **32**, 2628.
[271] Eckelbarger, J. D.; Wilmot, J. T.; Gin, D. Y., *J. Am. Chem. Soc.*, (2006) **128**, 10370.
[272] Lu, Z.; Shen, M.; Yoon, T. P., *J. Am. Chem. Soc.*, (2011) **133**, 1162.
[273] Curran, D. P.; Liu, H., *J. Am. Chem. Soc.*, (1992) **114**, 5863.
[274] Liard, A.; Quiclet-Sire, B.; Saicic, R. N.; Zard, S. Z., *Tetrahedron Lett.*, (1997) **38**, 1759.
[275] Sortais, B.; Zard, S. Z., *Tetrahedron Lett.*, (1999) **40**, 2533.
[276] Fischer, E.; Jourdan, F., *Chem. Ber.*, (1883), 2241.
[277] Fischer, E.; Hess, O., *Chem. Ber.*, (1884), 559.
[278] Wagaw, S.; Yang, B. H.; Buchwald, S. L., *J. Am. Chem. Soc.*, (1999) **121**, 10251.
[279] Simoneau, C. A.; Ganem, B., *Tetrahedron*, (2005) **61**, 11374.
[280] Christoffers, J., *Synlett*, (2006), 318.
[281] Diedrich, C. L.; Frey, W.; Christoffers, J., *Eur. J. Org. Chem.*, (2007), 4731.
[282] Hisler, K.; Commeureuc, A. G. J.; Zhou, S.; Murphy, J. A., *Tetrahedron Lett.*, (2009) **50**, 3290.
[283] Zhu, S.; Das, A.; Bui, L.; Zhou, H.; Curran, D. P.; Rueping, M., *J. Am. Chem. Soc.*, (2013) **135**, 1823.
[284] Wipf, P.; Maciejewski, J. P., *Org. Lett.*, (2008) **10**, 4383.
[285] Gansäuer, A.; Pierobon, M.; Bluhm, H., *Angew. Chem. Int. Ed.*, (1998) **37**, 101.
[286] Gansäuer, A.; Barchuk, A.; Keller, F.; Schmitt, M.; Grimme, S.; Gerenkamp, M.; Mück-Lichtenfeld, C.; Daasbjerg, K.; Svith, H., *J. Am. Chem. Soc.*, (2007) **129**, 1359.
[287] Curran, D. P.; Liu, H.; Josien, H.; Ko, S.-B., *Tetrahedron*, (1996) **52**, 11385.
[288] Clyne, M. A.; Aldabbagh, F., *Org. Biomol. Chem.*, (2006) **4**, 268.
[289] Clive, D. L. J.; Peng, J.; Fletcher, S. P.; Ziffle, V. E.; Wingert, D., *J. Org. Chem.*, (2008) **73**, 2330.
[290] Bowman, W. R.; Heaney, H.; Jordan, B. M., *Tetrahedron*, (1991) **47**, 10119.
[291] Beckwith, A. L. J.; Bowry, V. W.; Bowman, W. R.; Mann, E.; Parr, J.; Storey, J. M. D., *Angew. Chem. Int. Ed.*, (2004) **43**, 95.
[292] Curran, D. P.; Keller, A. I., *J. Am. Chem. Soc.*, (2006) **128**, 13706.

[293] Liao, Y.-J.; Wu, Y.-L.; Chuang, C.-P., *Tetrahedron*, (2003) **59**, 3511.
[294] Marti, C.; Carreira, E. M., *Eur. J. Org. Chem.*, (2003), 2209.
[295] Galliford, C. V.; Scheidt, K. A., *Angew. Chem. Int. Ed.*, (2007) **46**, 8748.
[296] Zhou, F.; Liu, Y.-L.; Zhou, J., *Adv. Synth. Catal.*, (2010) **352**, 1381.
[297] Bowman, W. R.; Storey, J. M. D., *Chem. Soc. Rev.*, (2007) **36**, 1803.
[298] Rowlands, G. J., *Tetrahedron*, (2009) **65**, 8603.
[299] Rowlands, G. J., *Tetrahedron*, (2010) **66**, 1593.
[300] Ju, X.; Liang, Y.; Jia, P.; Li, W.; Yu, W., *Org. Biomol. Chem.*, (2012) **10**, 498.
[301] D'Annibale, A.; Nanni, D.; Trogolo, C.; Umani, F., *Org. Lett.*, (2000) **2**, 401.
[302] Tsubusaki, T.; Nishino, H., *Tetrahedron*, (2009) **65**, 9448.
[303] Bellesia, F.; Clark, A. J.; Felluga, F.; Gennaro, A.; Isse, A. A.; Roncaglia, F.; Ghelfi, F., *Adv. Synth. Catal.*, (2013) **355**, 1649.
[304] Tsai, P.-J.; Kao, C.-B.; Chiow, W.-R.; Chuang, C.-P., *Synthesis*, (2014) **46**, 175.
[305] Hirase, K.; Iwahama, T.; Sakaguchi, S.; Ishii, Y., *J. Org. Chem.*, (2002) **67**, 970.
[306] Wu, W.; Xu, J.; Huang, S.; Su, W., *Chem. Commun. (Cambridge)*, (2011) **47**, 9660.
[307] Wang, C.; Li, Z.; Ju, Y.; Koo, S., *Eur. J. Org. Chem.*, (2012), 6976.
[308] Snider, B. B.; Patricia, J. J.; Kates, S. A., *J. Org. Chem.*, (1988) **53**, 2137.
[309] Rahman, M. T.; Nishino, H., *Org. Lett.*, (2003) **5**, 2887.
[310] Dulcere, J. P.; Mihoubi, M. N.; Rodriguez, J., *J. Org. Chem.*, (1993) **58**, 5709.
[311] Okabe, M.; Abe, M.; Tada, M., *J. Org. Chem.*, (1982) **47**, 1775.
[312] Dulcere, J. P.; Rodriguez, J.; Santelli, M.; Zahra, J. P., *Tetrahedron Lett.*, (1987) **28**, 2009.
[313] Zhou, L.; Hirao, T., *J. Org. Chem.*, (2003) **68**, 1633.
[314] Ischay, M. A.; Lu, Z.; Yoon, T. P., *J. Am. Chem. Soc.*, (2010) **132**, 8572.
[315] Mahmoodi, N. O.; Salehpour, M., *J. Heterocycl. Chem.*, (2003) **40**, 875.
[316] Mahmoodi, N. O.; Tabatabaeian, K.; Kosari, M.; Zarrabi, S., *Chin. Chem. Lett.*, (2008) **19**, 1431.
[317] De Campo, F.; Lastécouères, D.; Verlhac, J.-B., *J. Chem. Soc., Perkin Trans. 1*, (2000), 575.
[318] Hulcoop, D. G.; Burton, J. W., *Chem. Commun. (Cambridge)*, (2005), 4687.
[319] Powell, L. H.; Docherty, P. H.; Hulcoop, D. G.; Kemmitt, P. D.; Burton, J. W., *Chem. Commun. (Cambridge)*, (2008), 2559.
[320] Logan, A. W. J.; Parker, J. S.; Hallside, M. S.; Burton, J. W., *Org. Lett.*, (2012) **14**, 2940.
[321] Kishida, A.; Nagaoka, H., *Tetrahedron Lett.*, (2008) **49**, 6393.
[322] Li, Z.; Jung, H.; Park, M.; Lah, M. S.; Koo, S., *Adv. Synth. Catal.*, (2011) **353**, 1913.
[323] Jana, S.; Roy, S. C., *Tetrahedron Lett.*, (2006) **47**, 5949.
[324] Chatgilialoglu, C.; Ferreri, C., In *Triple Bonded Functional Groups*, Patai, S., Ed.; Wiley: Chichester, UK, (1994); p 917.
[325] Zhan, Z.-P.; Lang, K., *Org. Biomol. Chem.*, (2005) **3**, 727.
[326] Ohkita, T.; Tsuchiya, Y.; Togo, H., *Tetrahedron*, (2008) **64**, 7247.
[327] Tucker, J. W.; Nguyen, J. D.; Narayanam, J. M. R.; Krabbe, S. W.; Stephenson, C. R. J., *Chem. Commun. (Cambridge)*, (2010) **46**, 4985.
[328] Villar, H.; Guibé, F., *Tetrahedron Lett.*, (2002) **43**, 9517.
[329] Johnston, D.; McCusker, C. F.; Muir, K.; Procter, D. J., *J. Chem. Soc., Perkin Trans. 1*, (2000), 681.
[330] Yamamoto, Y.; Matsumi, D.; Itoh, K., *Chem. Commun. (Cambridge)*, (1998), 875.
[331] Shono, T.; Kise, N.; Fujimoto, T.; Tominaga, N.; Morita, H., *J. Org. Chem.*, (1992) **57**, 7175.
[332] Corey, E. J.; Pyne, S. G., *Tetrahedron Lett.*, (1983) **24**, 2821.
[333] Miyazaki, T.; Maekawa, H.; Yonemura, K.; Yamamoto, Y.; Yamanaka, Y.; Nishiguchi, I., *Tetrahedron*, (2011) **67**, 1598.
[334] Hasegawa, E.; Okamoto, K.; Tanikawa, N.; Nakamura, M.; Iwaya, K.; Hoshi, T.; Suzuki, T., *Tetrahedron Lett.*, (2006) **47**, 7715.
[335] Kise, N.; Agui, S.; Morimoto, S.; Ueda, N., *J. Org. Chem.*, (2005) **70**, 9407.
[336] Maeda, H.; Ashie, H.; Maki, T.; Ohmori, H., *Chem. Pharm. Bull.*, (1997) **45**, 1729.
[337] Kagan, H. B., *Tetrahedron*, (2003) **59**, 10351.
[338] Zimmerman, J.; Sibi, M. P., *Top. Curr. Chem.*, (2006) **263**, 107.
[339] Sibi, M. P.; Manyem, S.; Zimmerman, J., *Chem. Rev.*, (2003) **103**, 3263.
[340] Streuff, J.; Feurer, M.; Bichovski, P.; Frey, G.; Gellrich, U., *Angew. Chem. Int. Ed.*, (2012) **51**, 8661.
[341] Molander, G. A.; Shakya, S. R., *J. Org. Chem.*, (1996) **61**, 5885.
[342] Otsubo, K.; Inanaga, J.; Yamaguchi, M., *Tetrahedron Lett.*, (1986) **27**, 5763.
[343] Fukuzawa, S.; Nakanishi, A.; Fujinami, T.; Sakai, S., *J. Chem. Soc., Chem. Commun.*, (1988), 1669.
[344] Ujikawa, O.; Inanaga, J.; Yamaguchi, M., *Tetrahedron Lett.*, (1989) **30**, 2873.

[345] Molander, G. A.; McKie, J. A., *J. Org. Chem.*, (1992) **57**, 3132.
[346] Gansäuer, A.; Lauterbach, T.; Geich-Gimbel, D., *Chem.–Eur. J.*, (2004) **10**, 4983.
[347] Kagan, H. B.; Namy, J.-L., *Tetrahedron*, (1986) **42**, 6573.
[348] Skrydstrup, T., *Angew. Chem. Int. Ed. Engl.*, (1997) **36**, 345.
[349] Wang, B.; Wang, Y.-J., *Org. Lett.*, (2009) **11**, 3410.
[350] Molander, G. A.; McKie, J. A., *J. Org. Chem.*, (1994) **59**, 3186.
[351] Frey, G.; Luu, H.-T.; Bichovski, P.; Feurer, M.; Streuff, J., *Angew. Chem. Int. Ed.*, (2013) **52**, 7131.
[352] Shinohara, I.; Okue, M.; Yamada, Y.; Nagaoka, H., *Tetrahedron Lett.*, (2003) **44**, 4649.
[353] Fu, W.; Zhu, M.; Xu, F.; Fu, Y.; Xu, C.; Zou, D., *RSC Adv.*, (2014) **4**, 17 226.
[354] Tamura, K. J.; Mizukami, H.; Maeda, K.; Watanabe, H.; Uneyama, K. J., *J. Org. Chem.*, (1993) **58**, 32.
[355] Shi, D.-Q.; Rong, S.-F.; Dou, G.-L., *Synth. Commun.*, (2010) **40**, 2302.
[356] Bogert, M. T.; Naiman, B., *J. Am. Chem. Soc.*, (1935) **57**, 1529.
[357] Mu, X.-J.; Zou, J.-P.; Zeng, R.-S.; Wu, J.-C., *Tetrahedron Lett.*, (2005) **46**, 4345.
[358] Rahim, M. A.; Fujiwara, T.; Takeda, T., *Synlett*, (1999), 1029.
[359] Snider, B. B.; Zhang, Q., *J. Org. Chem.*, (1993) **58**, 3185.
[360] Snider, B. B.; McCarthy, B. A., *J. Org. Chem.*, (1993) **58**, 6217.
[361] Paquette, L. A.; Schaefer, A. G.; Springer, J. P., *Tetrahedron*, (1987) **43**, 5567.
[362] Rama Rao, A. V.; Rao, B. V.; Reddy, D. R.; Singh, A. K., *J. Chem. Soc., Chem. Commun.*, (1989), 400.
[363] Colombo, M. I.; Signorella, S.; Mischne, M. P.; Gonzalez-Sierra, M.; Ruveda, E. A., *Tetrahedron*, (1990) **46**, 4149.
[364] Zoretic, P. A.; Shen, Z.; Wang, M.; Ribeiro, A. A., *Tetrahedron Lett.*, (1995) **36**, 2925.
[365] Zoretic, P. A.; Zhang, Y.; Ribeiro, A. A., *Tetrahedron Lett.*, (1995) **36**, 2929.
[366] Snider, B. B.; Merritt, J. E.; Domboski, M. A.; Buckman, B. O., *J. Org. Chem.*, (1991) **56**, 5544.
[367] Zhang, Q.; Mohan, R. M.; Cook, L.; Kazanis, S.; Peisach, D.; Foxman, B. M.; Snider, B. B., *J. Org. Chem.*, (1993) **58**, 7640.
[368] Burnell, R. H.; Girard, M., *Synth. Commun.*, (1990) **20**, 2469.
[369] Burnell, R. H.; Côté, C.; Girard, M., *J. Nat. Prod.*, (1993) **56**, 461.
[370] Shishido, K.; Goto, K.; Tsuda, A.; Takaishi, Y.; Shibuya, M., *J. Chem. Soc., Chem. Commun.*, (1993), 793.
[371] Snider, B. B.; O'Neil, S., *Tetrahedron*, (1995) **51**, 12 983.
[372] Yang, F. Z.; Trost, M. K.; Fristad, W. E., *Tetrahedron Lett.*, (1987) **28**, 1493.
[373] Demir, A. S.; Tanyeli, C.; Akgün, H.; Çalişkan, Z.; Özgül, E., *Bull. Soc. Chim. Fr.*, (1995) **132**, 423.
[374] Demir, A. S.; Gercek, Z.; Duygu, N.; Igdir, A. C.; Reis, O., *Can. J. Chem.*, (1999) **77**, 1336.
[375] Yang, D.; Ye, X.-Y.; Xu, M.; Pang, K.-W.; Cheung, K.-K., *J. Am. Chem. Soc.*, (2000) **122**, 1658.
[376] González, M. A.; Molina-Navarro, S., *J. Org. Chem.*, (2007) **72**, 7462.
[377] Brasholz, M.; Reissig, H.-U., *Synlett*, (2007), 1294.
[378] Bar, G.; Parsons, A. F.; Thomas, C. B., *Tetrahedron*, (2001) **57**, 4719.
[379] Lee, E.; Lim, J. W.; Yoon, C. H.; Sung, Y.; Kim, Y. K.; Yun, M.; Kim, S., *J. Am. Chem. Soc.*, (1997) **119**, 8391.
[380] Thoison, O.; Cuong, D. D.; Gramain, A.; Chiaroni, A.; Hung, N. V.; Sevenet, T., *Tetrahedron*, (2005) **61**, 8529.
[381] Davies, J. J.; Krulle, T. M.; Burton, J. W., *Org. Lett.*, (2010) **12**, 2738.
[382] Phillips, E. M.; Roberts, J. M.; Scheidt, K. A., *Org. Lett.*, (2010) **12**, 2830.
[383] Jamieson, G. R.; Reid, E. H.; Turner, B. P.; Jamieson, A. T., *Phytochemistry*, (1976) **15**, 1713.
[384] Kano, K.; Hayashi, K.; Mitsuhashi, H., *Chem. Pharm. Bull.*, (1982) **30**, 1198.
[385] Silva, L. F., *Synthesis*, (2001), 671.
[386] Brocksom, T. J.; Brocksom, U.; Constantino, M. G., *Quim. Nova*, (2008) **31**, 937.
[387] Bhat, V.; MacKay, J. A.; Rawal, V. H., *Org. Lett.*, (2011) **13**, 3214.
[388] Tanaka, M.; Ubukata, M.; Matsuo, T.; Yasue, K.; Matsumoto, K.; Kajimoto, Y.; Ogo, T.; Inaba, T., *Org. Lett.*, (2007) **9**, 3331.
[389] Rinehart, K. L., *Pure Appl. Chem.*, (1989) **61**, 525.
[390] Kobayashi, J.; Tsuda, H.; Murayama, T.; Nakamura, H.; Ohizumi, Y.; Ishibashi, M.; Iwamura, M.; Ohta, T.; Nozoe, S., *Tetrahedron*, (1990) **46**, 5579.
[391] Simpkins, N. S.; Weller, M. D., *Tetrahedron Lett.*, (2010) **51**, 4823.
[392] Yamashita, S.; Suda, N.; Hayashi, Y.; Hirama, M., *Tetrahedron Lett.*, (2013) **54**, 1389.
[393] Magolan, J.; Kerr, M. A., *Org. Lett.*, (2006) **8**, 4561.
[394] Höfs, R.; Walker, M.; Zeeck, A., *Angew. Chem. Int. Ed.*, (2000) **39**, 3258.
[395] Toueg, J.; Prunet, J., *Org. Lett.*, (2008) **10**, 45.

[396] Ciochina, R.; Grossman, R. B., *Chem. Rev.*, (2006) **106**, 3963.
[397] Njardarson, J. T., *Tetrahedron*, (2011) **67**, 7631.
[398] Richard, J.-A.; Pouwer, R. H.; Chen, D. Y.-K., *Angew. Chem. Int. Ed.*, (2012) **51**, 4536.
[399] Lam, H. C.; Kuan, K. K. W.; George, J. H., *Org. Biomol. Chem.*, (2014) **12**, 2519.
[400] Basu, S.; Kandiyal, P. S.; Ampapathi, R. S.; Chakraborty, T. K., *RSC Adv.*, (2013) **3**, 13630.
[401] Bryans, J. S.; Chessum, N. E. A.; Huther, N.; Parsons, A. F.; Ghelfi, F., *Tetrahedron*, (2003) **59**, 6221.
[402] Saha, S.; Roy, S. C., *Tetrahedron*, (2010) **66**, 4278.
[403] Doi, T.; Fuse, S.; Miyamoto, S.; Nakai, K.; Sasuga, D.; Takahashi, T., *Chem.–Asian J.*, (2006) **1**, 370.
[404] Nakai, K.; Kamoshita, M.; Doi, T.; Yamada, H.; Takahashi, T., *Tetrahedron Lett.*, (2001) **42**, 7855.
[405] Justicia, J.; Oltra, J. E.; Cuerva, J. M., *Tetrahedron Lett.*, (2004) **45**, 4293.
[406] Haruo, Y.; Hasegawa, T.; Tanaka, H.; Takahashi, T., *Synlett*, (2001), 1935.
[407] Yamaoka, M.; Fukatsu, Y.; Nakazaki, A.; Kobayashi, S., *Tetrahedron Lett.*, (2009) **50**, 3849.
[408] Chakraborty, P.; Jana, S.; Saha, S.; Roy, S. C., *Tetrahedron Lett.*, (2012) **53**, 6584.
[409] Rana, K. K.; Guin, C.; Roy, S. C., *Tetrahedron Lett.*, (2000) **41**, 9337.
[410] Mandal, P. K.; Maiti, G.; Roy, S. C., *J. Org. Chem.*, (1998) **63**, 2829.
[411] Mandal, P. K.; Roy, S. C., *Tetrahedron*, (1999) **55**, 11395.
[412] Friedrich, J.; Walczak, K.; Dolg, M.; Piestert, F.; Lauterbach, T.; Worgull, D.; Gansäuer, A., *J. Am. Chem. Soc.*, (2008) **130**, 1788.
[413] Gansäuer, A.; Piestert, F.; Huth, I.; Lauterbach, T., *Synthesis*, (2008), 3509.
[414] Gansäuer, A.; Greb, A.; Huth, I.; Worgull, D.; Knebel, K., *Tetrahedron*, (2009) **65**, 10791.
[415] Gansäuer, A.; Franke, D.; Lauterbach, T.; Nieger, M., *J. Am. Chem. Soc.*, (2005) **127**, 11622.
[416] Klawonn, T.; Gansäuer, A.; Winkler, I.; Lauterbach, T.; Franke, D.; Nolte, R. J. M.; Feiters, M. C.; Börner, H.; Hentschel, J.; Dötz, K. H., *Chem. Commun. (Cambridge)*, (2007), 1894.
[417] Gansäuer, A.; Winkler, I.; Worgull, D.; Lauterbach, T.; Franke, D.; Selig, A.; Wagner, L.; Prokop, A., *Chem.–Eur. J.*, (2008) **14**, 4160.
[418] Gansäuer, A.; Winkler, I.; Klawonn, T.; Nolte, R. J. M.; Feiters, M. C.; Börner, H. G.; Hentschel, J.; Dötz, K. H., *Organometallics*, (2009) **28**, 1377.
[419] Gansäuer, A.; Winkler, I.; Worgull, D.; Franke, D.; Lauterbach, T.; Okkel, A.; Nieger, M., *Organometallics*, (2008) **27**, 5699.
[420] Velasco, J.; Ariza, X.; Badía, L.; Bartra, M.; Berenguer, R.; Farràs, J.; Gallardo, J.; Garcia, J.; Gasanz, Y., *J. Org. Chem.*, (2013) **78**, 5482.
[421] Dienstag, J. L., *N. Engl. J. Med.*, (2008) **359**, 1486.
[422] Scott, L. J.; Keating, G. M., *Drugs*, (2009) **69**, 1003.
[423] Justicia, J.; Campaña, A. G.; Bazdi, B.; Robles, R.; Cuerva, J. M.; Oltra, J. E., *Adv. Synth. Catal.*, (2008) **350**, 571.
[424] Barrero, A. F.; Cuerva, J. M.; Alvarez-Manzaneda, E. J.; Oltra, J. E.; Chahboun, R., *Tetrahedron Lett.*, (2002) **43**, 2793.
[425] Yoshimitsu, T.; Nojima, S.; Hashimoto, M.; Tanaka, T., *Org. Lett.*, (2011) **13**, 3698.
[426] Palanichamy, K.; Kaliappan, K. P., *Chem.–Asian J.*, (2010) **5**, 668.
[427] Lu, X.; You, Q., *Curr. Med. Chem.*, (2010) **17**, 1139.
[428] Harsh, P.; O'Doherty, G. A., *Chemtracts*, (2009) **22**, 31.
[429] Tiefenbacher, K.; Mulzer, J., *Angew. Chem. Int. Ed.*, (2008) **47**, 2548.
[430] Kondracki, M. L.; Guyot, M., *Tetrahedron*, (1989) **45**, 1995.
[431] Kondracki, M. L.; Longeon, A.; Morel, E.; Guyot, M., *Int. J. Immunopharmacol.*, (1991) **13**, 393.
[432] Sarin, P. S.; Sun, D.; Thornton, A.; Müller, W. E. G., *J. Natl. Cancer Inst.*, (1987) **78**, 663.
[433] Schröder, H. C.; Wenger, R.; Gerner, H.; Reuter, P.; Kuchino, Y.; Sladić, D.; Müller, W. E. G., *Cancer Res.*, (1989) **49**, 2069.
[434] Yamaoka, M.; Nakazaki, A.; Kobayashi, S., *Tetrahedron Lett.*, (2009) **50**, 6764.
[435] Mizushina, Y.; Tanaka, N.; Kitamura, A.; Tamai, K.; Ikeda, M.; Takemura, M.; Sugawara, F.; Arai, T.; Matsukage, A.; Yoshida, S.; Sakaguchi, K., *Biochem. J.*, (1998) **330**, 1325.
[436] Mizushina, Y.; Iida, A.; Ohta, K.; Sugawara, F.; Sakaguchi, K., *Biochem. J.*, (2000) **350**, 757.
[437] Obara, Y.; Nakahata, N.; Mizushina, Y.; Sugawara, F.; Sakaguchi, K.; Ohizumi, Y., *Life Sci.*, (2000) **67**, 1659.
[438] Brahmachari, G., *ChemBioChem*, (2004) **5**, 408.
[439] Fernández-Mateos, A.; Madrazo, S. E.; Teijón, P. H.; Clemente, R. R.; González, R. R.; González, F. S., *J. Org. Chem.*, (2013) **78**, 9571.
[440] Das, D.; Kant, R.; Chakraborty, T. K., *Org. Lett.*, (2014) **16**, 2618.
[441] Shim, S. H.; Swenson, D. C.; Gloer, J. B.; Dowd, P. F.; Wicklow, D. T., *Org. Lett.*, (2006) **8**, 1225.

[442] Shim, S. H.; Gloer, J. B.; Wicklow, D. T., *J. Nat. Prod.*, (2006) **69**, 1601.
[443] Fischer, N. H.; Olivier, E. J.; Fischer, H. D., *Prog. Chem. Org. Nat. Prod.*, (1979) **38**, 134.
[444] Connolly, J. D.; Hill, R. A., *Dictionary of Terpenoids*, Chapman and Hall: London, (1991); Vol. 1, p 340.
[445] Fraga, B. M., *Nat. Prod. Rep.*, (2002) **19**, 650.
[446] Hehner, S. P.; Heinrich, M.; Bork, P. M.; Vogt, M.; Ratter, F.; Lehmann, V.; Schulze-Osthoff, K.; Dröge, W.; Schmitz, M. L., *J. Biol. Chem.*, (1998) **273**, 1288.
[447] Dirsch, V. M.; Stuppner, H.; Ellmerer-Müller, E. P.; Vollmar, A. M., *Bioorg. Med. Chem.*, (2000) **8**, 2747.
[448] Skaltsa, H.; Lazari, D.; Panagouleas, C.; Georgiadou, E.; García, B.; Sokovic, M., *Phytochemistry*, (2000) **55**, 903.
[449] Molander, G. A., *Org. React. (N. Y.)*, (1994) **46**, 211.
[450] Barrero, A. F.; Oltra, J. E.; Barragán, A.; Álvarez, M., *J. Chem. Soc., Perkin Trans. 1*, (1998), 4107.
[451] Barrero, A. F.; Oltra, J. E.; Álvarez, M.; Rosales, A., *J. Org. Chem.*, (2002) **67**, 5461.
[452] Sasaki, M.; Collin, J.; Kagan, H. B., *New J. Chem.*, (1992) **16**, 89.
[453] Soderquist, J. A., *Aldrichimica Acta*, (1991) **24**, 15.
[454] Ogawa, A.; Takami, N.; Sekiguchi, M.; Ryu, I.; Kambe, N.; Sonoda, N., *J. Am. Chem. Soc.*, (1992) **114**, 8729.
[455] Ogawa, A.; Nanke, T.; Takami, N.; Sumino, Y.; Ryu, I.; Sonoda, N., *Chem. Lett.*, (1994) **23**, 379.
[456] Kan, T.; Hosokawa, S.; Nara, S.; Oikawa, M.; Ito, S.; Matsuda, F.; Shirahama, H., *J. Org. Chem.*, (1994) **59**, 5532.
[457] Nicolaou, K. C.; Pappo, D.; Tsang, K. Y.; Gibe, R.; Chen, D. Y.-K., *Angew. Chem. Int. Ed.*, (2008) **47**, 944.
[458] Ichikawa, S.; Shuto, S.; Minakawa, N.; Matsuda, A., *J. Org. Chem.*, (1997) **62**, 1368.
[459] Arai, S.; Nakajima, M.; Nishida, A., *Angew. Chem. Int. Ed.*, (2014) **53**, 5569.
[460] Boiron, A.; Zillig, P.; Faber, D.; Giese, B., *J. Org. Chem.*, (1998) **63**, 5877.
[461] Huang, W. G.; Jiang, Y. Y.; Li, Q.; Li, J.; Li, J. Y.; Lu, W.; Cai, J. C., *Tetrahedron*, (2005) **61**, 1863.
[462] Lee, D. S.; Hong, S. D., *J. Microbiol. Biotechnol.*, (1998) **8**, 89.
[463] Ryu, S. Y.; Lee, C. O.; Choi, A. U., *Planta Med.*, (1997) **63**, 339.
[464] Hannan, R. L.; Barber, R. B.; Rapoport, H., *J. Org. Chem.*, (1979) **44**, 2153.
[465] Oki, T.; Konishi, M.; Tomatsu, K.; Tomita, K.; Saitoh, K.; Tsunakawa, M.; Nishio, M.; Miyaki, T.; Kawaguchi, H., *J. Antibiot.*, (1988) **41**, 1701.
[466] Takeuchi, T.; Hara, T.; Naganawa, H.; Hamada, M.; Umezawa, H.; Gomi, S.; Sezaki, M.; Kondo, S., *J. Antibiot.*, (1988) **41**, 807.
[467] Watanabe, M.; Gomi, S.; Tohyama, H.; Ohtsuka, K.; Shibahara, S.; Inouye, S.; Kobayashi, H.; Suzuki, S.; Kondo, S.; Takeuchi, T.; Yamaguchi, H., *J. Antibiot.*, (1996) **49**, 366.
[468] Ueki, T.; Numata, K.; Sawada, Y.; Nakajima, T.; Fukagawa, Y.; Oki, T., *J. Antibiot.*, (1993) **46**, 149.
[469] Mizuochi, T.; Nakata, M., *Jpn. J. Clin. Med.*, (1995) **53**, 2340.
[470] Ohmori, K.; Kitamura, M.; Suzuki, K., *Angew. Chem. Int. Ed.*, (1999) **38**, 1226.
[471] Kitamura, M.; Ohmori, K.; Kawase, T.; Suzuki, K., *Angew. Chem. Int. Ed.*, (1999) **38**, 1229.
[472] Mulliken, R. S.; Person, W. B., *Annu. Rev. Phys. Chem.*, (1962) **13**, 107.
[473] Taube, H., *Angew. Chem. Int. Ed. Engl.*, (1984) **23**, 329.
[474] Marcus, R. A., *Angew. Chem. Int. Ed. Engl.*, (1993) **32**, 1111.
[475] Huber, R., *Angew. Chem. Int. Ed. Engl.*, (1989) **28**, 848.
[476] Rosokha, S. V.; Kochi, J. K., *Acc. Chem. Res.*, (2008) **41**, 641.
[477] Ashby, E. C., *Acc. Chem. Res.*, (1988) **21**, 414.
[478] Pross, A., *Acc. Chem. Res.*, (1985) **18**, 212.
[479] Evans, D. H., *Chem. Rev.*, (2008) **108**, 2113.

2.10 Ring-Closing Metathesis

D. Lee and V. Reddy Sabbasani

General Introduction

Ring-closing metathesis (RCM) has emerged as a powerful tool in synthetic organic chemistry for the construction of cyclic compounds ranging from small and medium rings to macrocycles. Compared to other conventional ring-closing reactions, ring-closing metathesis based approaches have significant merits in terms of their catalytic activation of alkene and alkyne substrates with excellent spatial and temporal control. Over the past two decades, a number of ring-closing metathesis strategies along with the development of effective catalysts conferring high E/Z selectivity have evolved. Depending on the types of unsaturated π-systems involved in the ring-closure event, these metathesis processes are categorized into diene, enyne, and diyne ring-closing metathesis (Scheme 1). Whereas diene and diyne metatheses produce products containing the same kind of π-systems present in the starting materials, enyne metathesis generates 1,3-diene products as a consequence of a formal addition between an alkene and an alkyne, instead of an exchange reaction, without loss of any carbon elements.

Scheme 1 Typical Ring-Closing Metatheses: Diene, Enyne, and Diyne Ring-Closing Metathesis

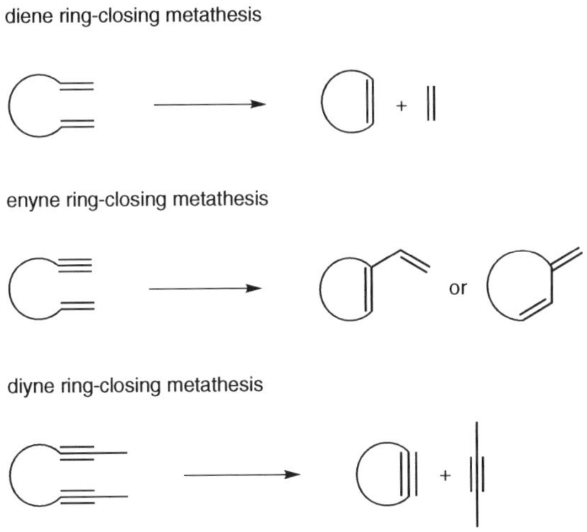

diene ring-closing metathesis

enyne ring-closing metathesis

diyne ring-closing metathesis

Ring-closing metathesis reactions of dienes and enynes are mediated by metal–alkylidene complexes whereas those of diynes are mediated only by metal–alkylidynes. Although each of these three metathesis processes involves different combinations of unsaturated functionalities, they all share two identical elementary steps. These two steps are (1) (2+2) cycloaddition between a metal–alkylidene or –alkylidyne complex with an appropriate π-system to form a metallacyclobutane (in diene metathesis), metallacyclobutene (in enyne metathesis), or metallacyclobutadiene (in diyne metathesis) intermediate, and (2) their subsequent cycloreversion (Scheme 2). Compared to diene metathesis, enyne me-

for references see p 672

tathesis involves a more complex mechanistic regime and reaction profile, which results in unpredictable substrate reactivity and poor regio- and stereoselectivity, so that fewer applications to the synthesis of complex molecules have been developed. For diyne metathesis, one of the main impediments has been the limited availability of versatile catalyst systems. Recently, however, various improvements and solutions to these problems have slowly evolved, including the development of more efficient catalysts, thus leading to an increased number of applications of the inherent capacity of enyne and diyne metathesis to the synthesis of complex molecular frameworks.

Scheme 2 Elementary Steps in Diene, Enyne, and Diyne Metathesis

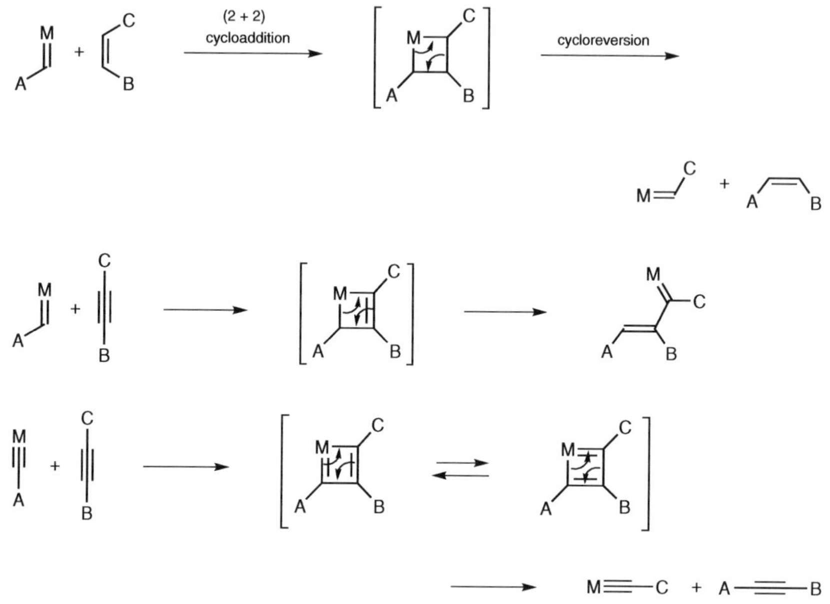

2.10.1 Brief Historical Background

In 1980, Villemin[1] and the Tsuji group[2] independently disclosed the first diene ring-closing metathesis using tungsten-based catalysts for the preparation of macrocycles from dialkenyl oxo esters and dialkenyl ketones. In 1993, Grubbs and coworkers reported a more efficient carbocyclization of 1,6-diene **3** with molybdenum-based carbene complex **1** as catalyst to generate cyclopentene **4** in high yield,[3] and subsequently with relatively air-stable ruthenium–alkylidene complex **2** (Scheme 3).[4] After this breakthrough, the development of more effective and user-friendly catalyst systems as well as their application to a variety of ring-closing metathesis reactions have been realized.

2.10.1 Brief Historical Background

Scheme 3 First Catalytic Molybdenum and Ruthenium Carbene Based Diene Ring-Closing Metathesis[3,4]

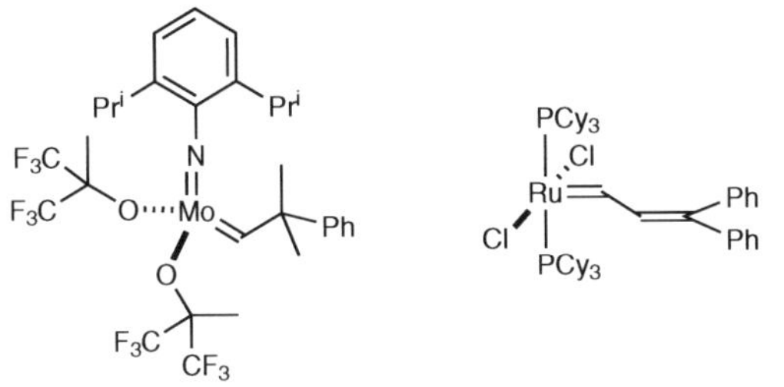

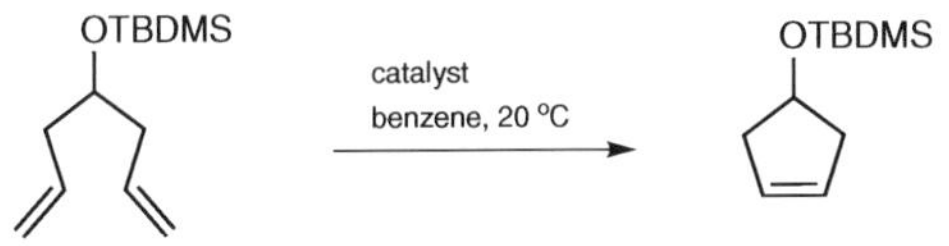

Catalyst (mol%)	Yield (%)	Ref
1 (2)	91	[3]
2 (2–4)	85	[4]

In 1985, Katz and coworkers reported the first example of enyne ring-closing metathesis by using a stoichiometric amount of a tungsten Fischer carbene complex.[5,6] The groups of Hoye and Mori independently extended enyne ring-closing metathesis with chromium variants of Fischer carbene complexes,[7–13] although these catalytic systems suffered from not only low yields and significant byproduct formation but also limited functional group tolerance and high catalyst loading. A significant improvement of catalytic enyne ring-closing metathesis was realized using ruthenium–alkylidene complex **2**; with a catalytic amount of **2** (2 mol%), enyne **5** was efficiently converted into diene **6** (Scheme 4).[4] Subsequently, Grubbs and coworkers described the tandem enyne ring-closing metathesis of dienynes to construct fused bicyclic frameworks using the same catalyst.[14] Based on these initial breakthroughs, more sophisticated enyne metathesis strategies have evolved.

Scheme 4 First Catalytic Ruthenium Carbene Based Enyne Ring-Closing Metathesis[4]

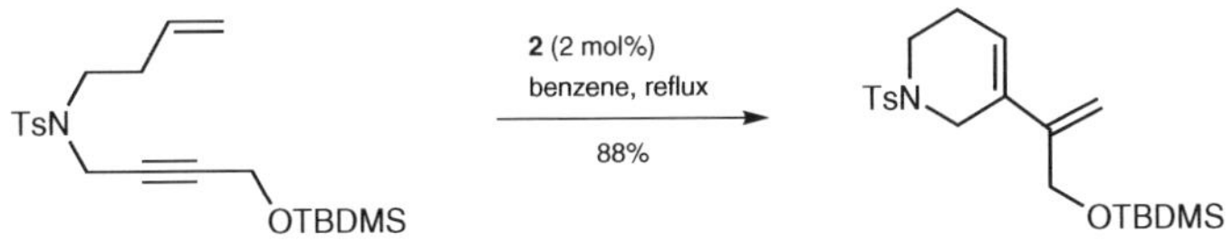

In 1998, the Fürstner group demonstrated an efficient ring-closing alkyne metathesis (RCAM) of diyne **9** to generate macrocycle **10** using tungsten–alkylidyne complex **8** (Schrock alkylidyne),[15] readily accessed by treating precursor complex **7**[16] with *tert*-butylacetylene (Scheme 5). In 1999, the same group reported the ring-closing alkyne metathesis with in situ generated molybdenum–alkylidyne complex from molybdenum(IV) chloride.[17] These two catalyst systems played a crucial role in the subsequent development of ring-closing alkyne metathesis, which has been extensively employed to construct cycloalkene-based natural products via the initial formation of macrocyclic alkynes followed by E/Z-selective semireduction.

Scheme 5 First Catalytic Tungsten–Alkylidyne Based Ring-Closing Alkyne Metathesis[15,16]

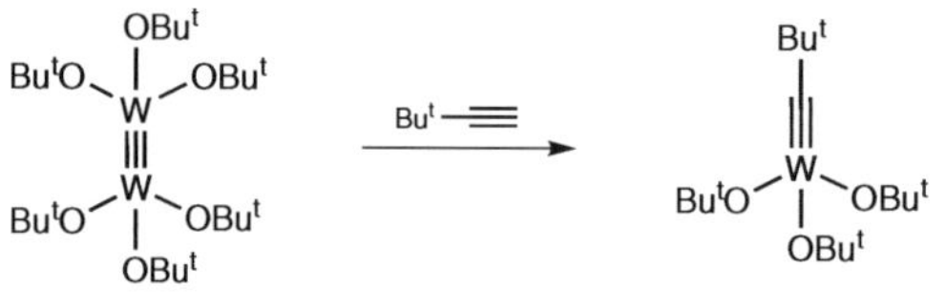

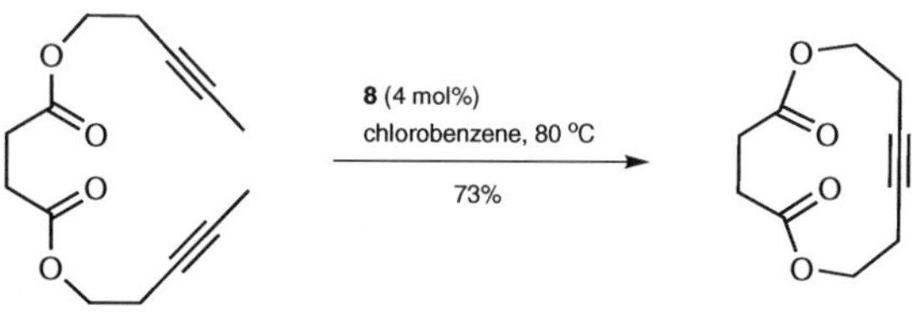

2.10.2 Diene Ring-Closing Metathesis

2.10.2.1 Catalysts and Mechanism

Based on availability, functional group tolerance, and their relatively air- and moisture-stable nature, Grubbs ruthenium–alkylidene complexes **2** and **11–16** are extensively used as common initiators of diene ring-closing metathesis (Scheme 6). Highly reactive, and thus air- and moisture-sensitive, Schrock complex **1** has also been successfully employed in the ring-closing metathesis of 1,n-dienes to form macrocycles.

Scheme 6 Commonly Used Catalysts for Diene Ring-Closing Metathesis

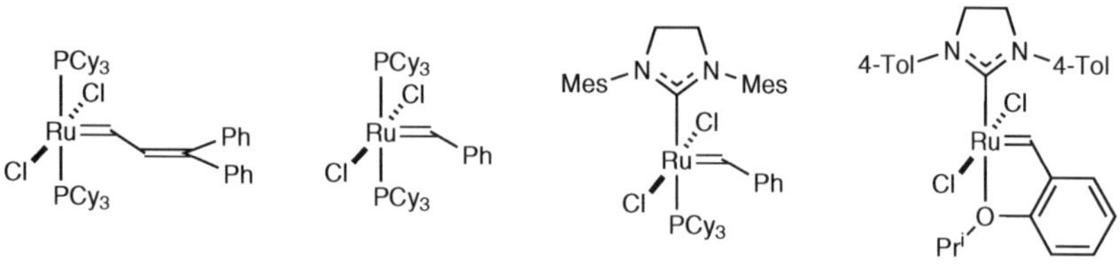

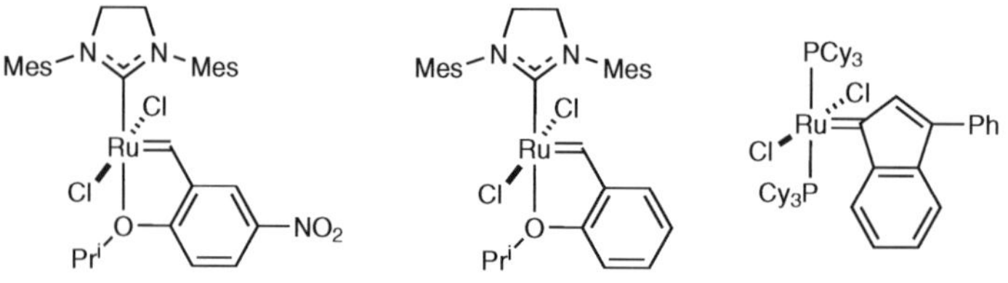

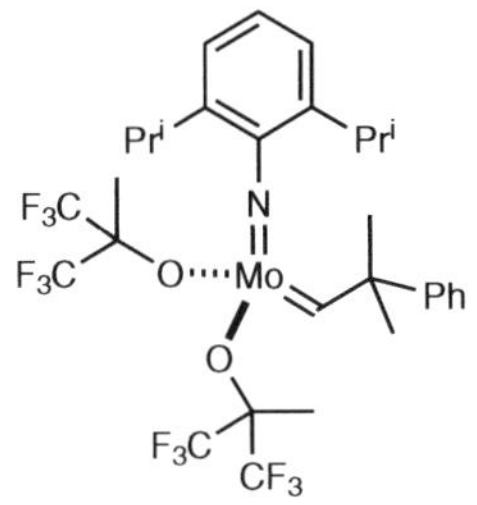

2.10.2 Diene Ring-Closing Metathesis

1

In the traditional mechanism of ring-closing metathesis of 1,n-dienes (Scheme 7),[18] the initiation step involves the loss of a phosphine ligand from an initiator ruthenium complex to form a 14-electron species **17**,[19] which then coordinates to an alkene moiety of a 1,n-diene to generate **18**. The alkylidene exchange occurs via the formation of metallacyclobutene **19**[20] followed by extrusion of styrene, which generates the first propagating species **20** in the catalytic cycle. Repetition of a similar sequence of elementary steps from **20** leads to new metallacyclobutane **21**. Cycloreversion of **21** releases a ring-closing metathesis product and a new 14-electron ruthenium–methylidene complex **22**, which reenters the catalytic cycle after releasing ethene via alkylidene exchange with another 1,n-diene. If complex **22** reacts with a previously released phosphine, the catalyst becomes a resting state in the form of ruthenium–methylidene complex **23**, which is catalytically less competent if not unreactive.

Scheme 7 Mechanism of Diene Ring-Closing Metathesis[18]

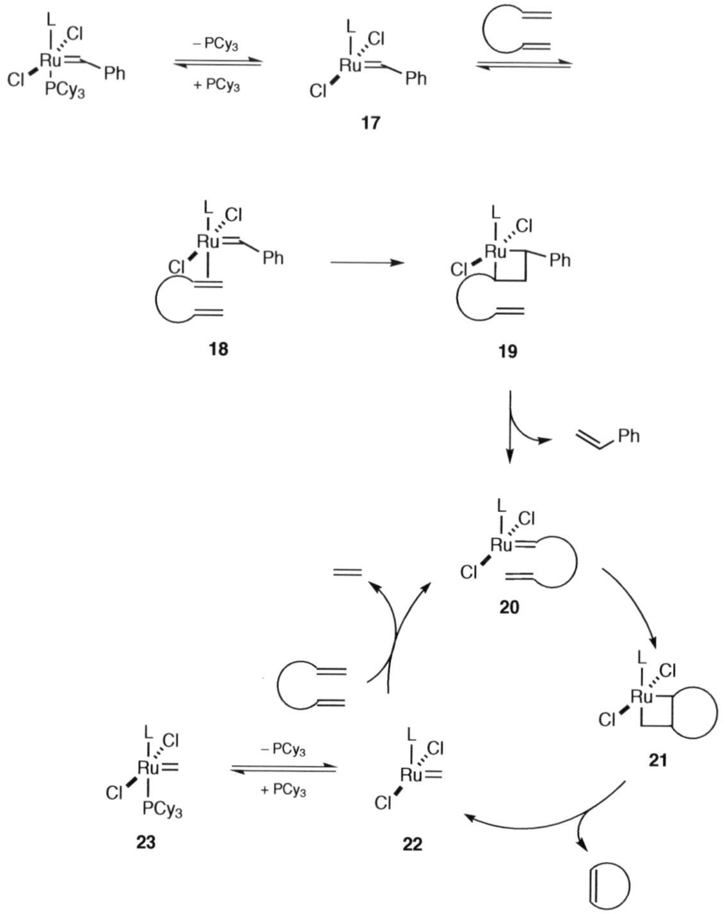

2.10.2.2 Selectivity and Ring Size

2.10.2.2.1 E/Z Selectivity for Small and Medium Rings

The applications of diene ring-closing metathesis to construct small and medium rings are most common for five- to seven-membered rings, where the newly formed alkenes invariably have Z configuration. On the other hand, due to the strain energy caused by transannular interactions, eight- and nine-membered carbocycles start to form with *E* configuration.[21] The transannular interactions in medium rings can be diminished by replacing sp³-hybridized carbons in these rings with sp²-hybridized carbons, nitrogen atoms, or other heteroatoms with large polarizable orbitals instead of C—H bonds. Hence, medium-sized cyclic silyl ethers, silaketals, and azacycles are more readily constructed by ring-closing metathesis than medium-sized carbocycles.

Although rare, an example of *E*-selective ring-closing metathesis to form an eight-membered ring is known that shows how a subtle conformational bias in the substrate plays a critical role in the outcome of the reaction. In the construction of the BC ring system of paclitaxel, ring-closing metathesis of diene **24** fails to deliver the cyclooctene product with Grubbs' first-generation catalyst **11** (0.02 M in benzene, 80 °C, 8 d), but is successful with Schrock complex **1** (0.02 M in benzene, 80 °C, 3 d), delivering the Z-cyclooctene product **25** in 56% yield. In stark contrast, the ring-closing metathesis of a diastereomeric

2.10.2 Diene Ring-Closing Metathesis

mixture of diene **26** with a different diol protecting group provides only one diastereomer of the *E*-cyclooctene product **27** along with the recovered substrate diastereomer **26A** upon treatment with Grubbs' first-generation catalyst **11** or Schrock complex **1** (Scheme 8).[22] Considering the equilibrating nature of this ring-closing metathesis, this unusual reactivity and *E*/*Z* selectivity is assumed to be the consequence of thermodynamic rather than kinetic control.

Scheme 8 Conformational Control in Ring-Closing Metathesis for the Synthesis of *E*/*Z*-Cyclooctenes[22]

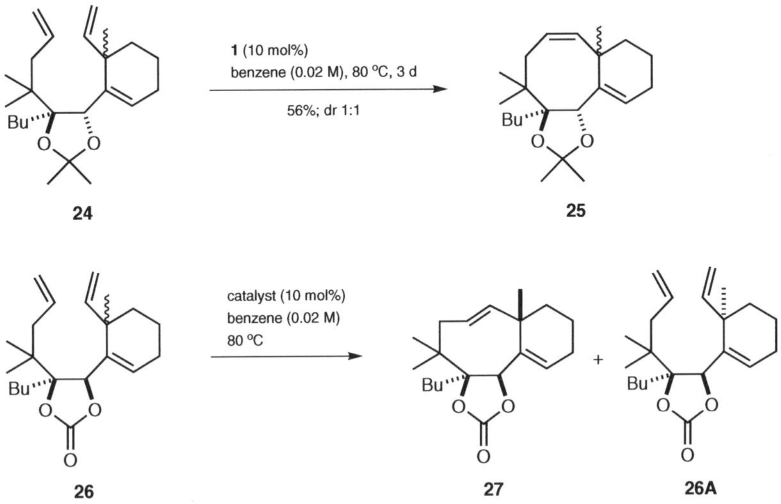

Catalyst	Yield (%)		Ref
	27	26A	
11	34	46	[22]
1	41	34	[22]

An *E*-selective ring-closing metathesis is also achieved from silicon-tethered dienes (Scheme 9).[23] When diene substrate **28A** is treated with the Grubbs–Hoveyda second-generation catalyst **15** in the presence of benzo-1,4-quinone, an eight-membered silaketal **29A** (*E*/*Z* >20:1) is obtained in 93% yield, whereas the C3 epimeric diene substrate **28B** affords the eight-membered silaketal **29B** with only *Z* configuration under otherwise identical conditions.

Scheme 9 E/Z-Selective Ring-Closing Metathesis To Form Medium Rings[23]

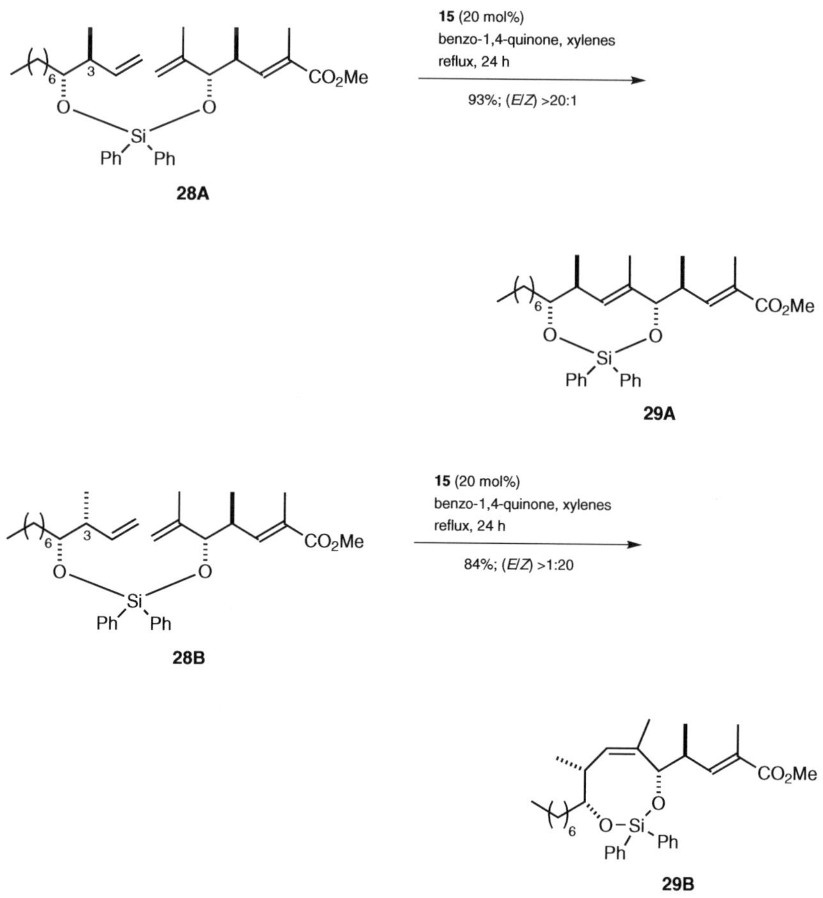

28A → 29A
15 (20 mol%)
benzo-1,4-quinone, xylenes
reflux, 24 h
93%; (E/Z) >20:1

28B → 29B
15 (20 mol%)
benzo-1,4-quinone, xylenes
reflux, 24 h
84%; (E/Z) >1:20

2.10.2.2.2 E/Z Selectivity for Macrocycles

Ring-closing diene metathesis is one of the most effective methods for the synthesis of functionalized macrocycles, yet the E/Z selectivity for the newly formed alkene is not always controllable. Predicting a major isomer of the macrocyclic product formed from a particular substrate is often not straightforward due to the dependency of the reaction on multiple factors. In some cases, the catalyst may also affect the E/Z selectivity under thermodynamically controlled conditions because it may not be active enough to equilibrate the kinetic product.[24,25] Sometimes particular substitution patterns in the substrate are required to get high E/Z selectivity.[26,27] The reaction temperature and solvents also significantly affect the E/Z selectivity in macrocyclic ring-closing metathesis reactions.[28,29]

Due to the inherent thermodynamic preference, establishing an effective equilibration generally results in E-alkene products with good selectivity. On the other hand, the preparation of Z-alkenes requires specific catalyst systems that favor the formation of metallacyclobutene intermediates with a *cis* arrangement of vicinal alkyl groups from which cycloreversion occurs to generate Z-alkene products. Specific examples of these two different regimes are described in the following sections.

2.10.2.2.2.1 E-Selective Ring-Closing Metathesis

Ring-closing metathesis of 1,n-dienes often favors the formation of macrocycles with a thermodynamically more stable E-configured double bond. This is due to the reversible nature of metathesis via the repetitive ring-opening and ring-closing events; alternatively, the double-bond isomerization may occur using the metal hydride species derived from catalyst decomposition. For example, the macrocyclic ring-closing metathesis of ester-tethered diene **30** in the presence of Grubbs' second-generation catalyst **12** provides the 14-membered lactone **31** quantitatively with high E selectivity under thermodynamic conditions (Scheme 10).[30]

Scheme 10 High E Selectivity in Ring-Closing Metathesis via Equilibration[30]

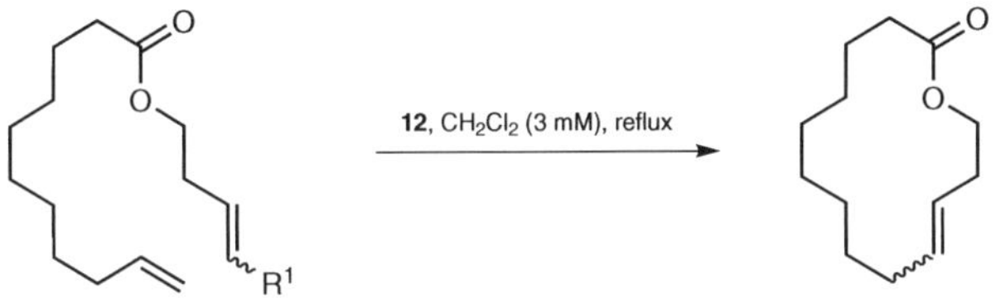

R¹	Config of 30	mol% of Catalyst	Time (min)	Ratio (E/Z)	Yield (%)	Ref
H	–	1.0	40	11.5:1	100	[30]
Et	Z	0.5	30	9.7:1	100	[30]
(CH$_2$)$_4$Me	Z	0.5	40	10.8:1	100	[30]
CH$_2$OAc	E	2.0	180	9.7:1	80	[30]

2.10.2.2.2.2 Z-Selective Ring-Closing Metathesis

The selective formation of Z-alkenes by metathesis has been a long-standing challenge because of the inherently equilibrating nature of the alkene metathesis mechanism. Thus, the formation of macrocycles via ring-closing metathesis in a substrate-controlled manner generally displays low E/Z selectivity and also frequently low yields mainly because of an entropic disadvantage. However, on some occasions, although not predictable, high Z selectivity can be realized in macrocyclic ring-closing metathesis. This is probably the result of a convoluted influence of the existing functional groups and their stereochemical array in combination with a particular ring size. To address this problem, catalyst-based E/Z selectivity control would be highly desirable, for which development of catalyst systems is crucial.

An efficient catalyst-based control strategy for the synthesis of macrocyclic natural products with high Z selectivity and high yields uses tungsten catalyst **32** (Scheme 11).[31] For example, the ring-closing metathesis of highly functionalized 1,17-diene **33** with this catalyst provides the bis(*tert*-butyldimethylsilyl)-protected form of epothilone C **34** with 96% Z selectivity at 97% conversion, from which epothilone C (**35**) is obtained upon removal of the silyl protecting groups.[31]

Scheme 11 Catalyst-Controlled Z-Selective Ring-Closing Metathesis in the Synthesis of Epothilone C[31]

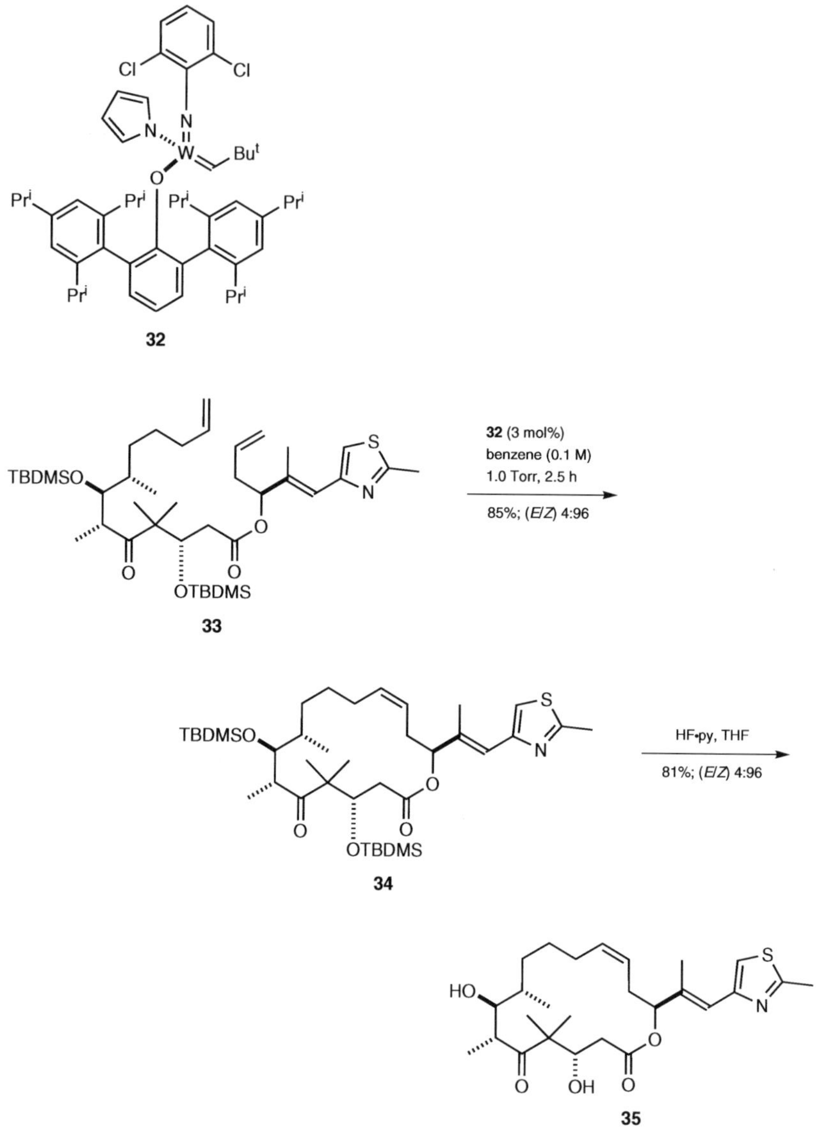

A two-step protocol for the Z-selective formation of macrocycles relies on a removable bulky silyl group, where the Z configuration of the initially formed alkene is masked. In the first step, ring-closing metathesis of a siloxy-functionalized 1,15-diene (e.g., **37**) with ruthenium complex **36** affords a silyl-functionalized E-macrocycle (e.g., **38**) with high selectivity (>95%). In the second step protodesilylation gives the Z-configured macrocycle (e.g., **39**) (Scheme 12).[32]

2.10.2 Diene Ring-Closing Metathesis

Scheme 12 Two-Step Diene Ring-Closing Metathesis Protocol for Z Selectivity in the Synthesis of Macrocycles[32]

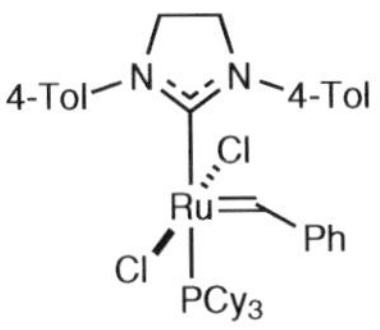

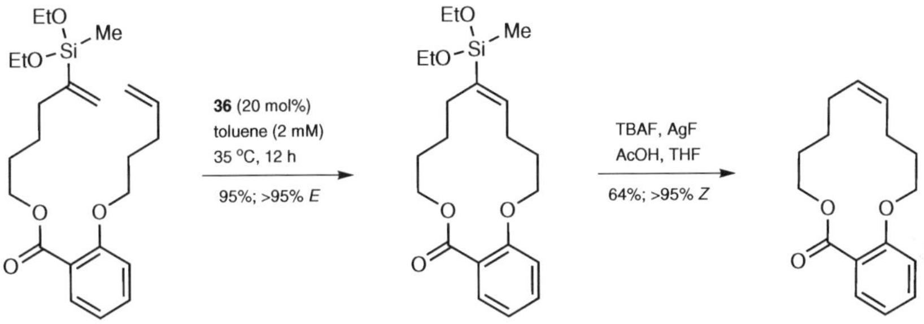

The ring-closing metathesis of dienes **42** containing ester functionality (e.g., Z=CO$_2$CH$_2$) catalyzed by adamantyl-derived ruthenium catalyst **40** produces macrolactones **43** with good Z selectivity, but the ring-closing metathesis to form macrocycles containing ketone or alcohol functionality [e.g., Z=CH$_2$C(O)CH$_2$, (CH$_2$)$_2$CHOH] provides lower selectivity (Scheme 13).[33] By replacing the N-mesityl group of ruthenium complex **40** with a sterically bulkier N-(2,6-diisopropylphenyl) group in ruthenium complex **41**, the catalyst further destabilizes the transition state leading to the E-alkene, thereby enhancing the Z selectivity of the macrocyclic ring-closing metathesis to greater than 95% (Scheme 13).[34]

Scheme 13 Ruthenium-Catalyzed Z-Selective Ring-Closing Metathesis[33,34]

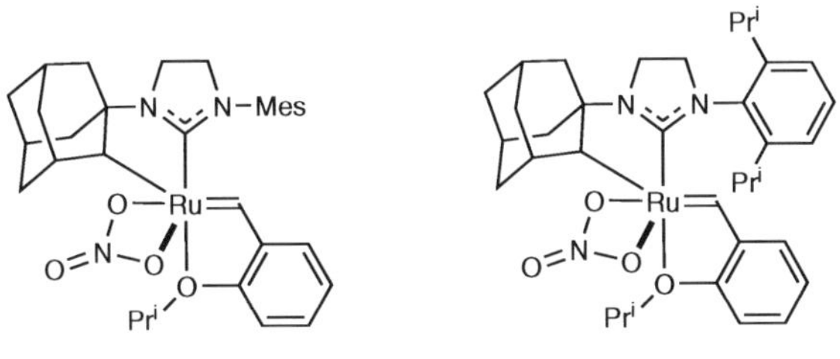

42 → **43**

catalyst (7.5 mol%)
ca. 0.02 Torr
1,2-dichloroethane (3 mM)
60 °C, 24 h

Z	Catalyst	Ratio (Z/E)	Yield (%)	Ref
CO_2CH_2	40	8.1:1	71	[33]
CO_2CH_2	41	>19:1	64	[34]
CH_2COCH_2	40	2.1:1	50[a]	[33]
CH_2COCH_2	41	>19:1	36	[34]
$(CH_2)_2CHOH$	40	1.6:1	56	[33]
$(CH_2)_2CHOH$	41	>19:1	45	[34]

[a] The reaction was quenched after 8 h.

Macrocycles 43; General Procedure:[33,34]
In a glovebox, a 500-mL Strauss flask was charged with a soln of diene **42** (0.45 mmol) in 1,2-dichloroethane (90 mL, 3 mM), and a soln of Ru complex **40** or **41** (0.034 mmol, 0.075 equiv) in 1,2-dichloroethane (1 mL) was added. The flask was sealed, brought out of the glovebox, and subjected to a single freeze–pump–thaw cycle. The flask was kept under a static vacuum (ca. 0.02 Torr), and heated at 60 °C. After 24 h, the mixture was cooled, quenched with excess ethyl vinyl ether, and concentrated. The residue was purified by flash chromatography (silica gel, Et_2O/pentanes gradient).

2.10.2.2.3 Chemoselectivity with Multiple Double Bonds

Ring-closing metathesis of substrates containing more than one pair of reactive double bonds could potentially lead to sequential ring-closing metathesis events depending on their nature and connectivity. Although this multiple ring-closing metathesis strategy may generate more complex molecular structures by a single operation, chemoselectivity between the pairs of participating alkenes to generate the desired connectivity is required to prevent the formation of other constitutional isomers or stereoisomers. There are many elements that contribute to the chemoselectivity in ring-closing metathesis reactions including steric hindrance, electronic activation/deactivation, ring size, stereochemistry of substituents, hybridization and the kind of tethering group, catalyst, concentration, and reaction temperature. Representative examples addressing various chemoselectivity problems are described in this section.

2.10.2.2.3.1 Control Based on Ring Size

The chemoselective ring-closing metathesis of benzyl-protected triene **44** with Grubbs' second-generation catalyst **12** (10 mol%) results in the exclusive formation of butenolide **45** in 58% (in CH_2Cl_2) and 52% (in benzene) yield, respectively, without formation of the six-membered lactone **46** (Scheme 14).[35,36] In the presence of dodec-1-ene, with slow addition of Grubbs' second-generation catalyst **12** (syringe pump, 0.01 M, 8 h), a ring-closing metathesis/cross-metathesis sequence generates butenolide **47**, which is hydrogenated in situ to give (−)-muricatacin (**48**).[35] The sequential ring-closing metathesis/cross metathesis with tert-butyldimethylsilyl-protected triene **49** in the presence of an external alkene **50** affords under identical conditions five-membered lactone **51** in 64% yield, which is

2.10.2 Diene Ring-Closing Metathesis

then transformed into rollicosin (**52**).[36] In the total synthesis of (+)-phomopsolide C (**55**), a chemoselective ring-closing metathesis of triene **53** with Grubbs–Hoveyda second-generation catalyst **15** selectively generates the six-membered lactone **54**.[37]

Scheme 14 Chemoselective Ring-Closing Metathesis of Trienes[35–37]

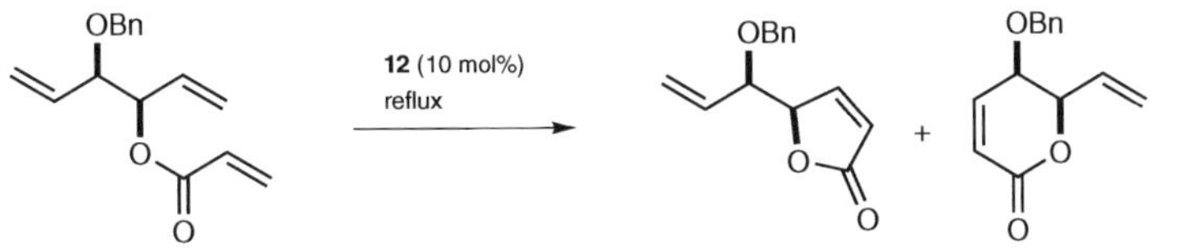

Solvent	Yield (%)	Ref
CH$_2$Cl$_2$	58	[35]
benzene	52	[35]

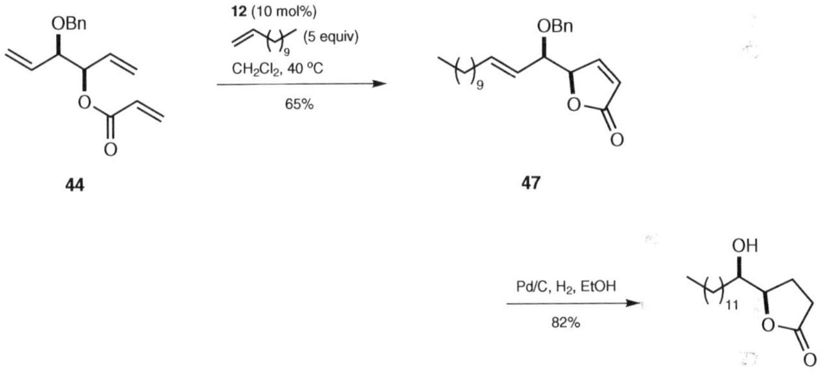

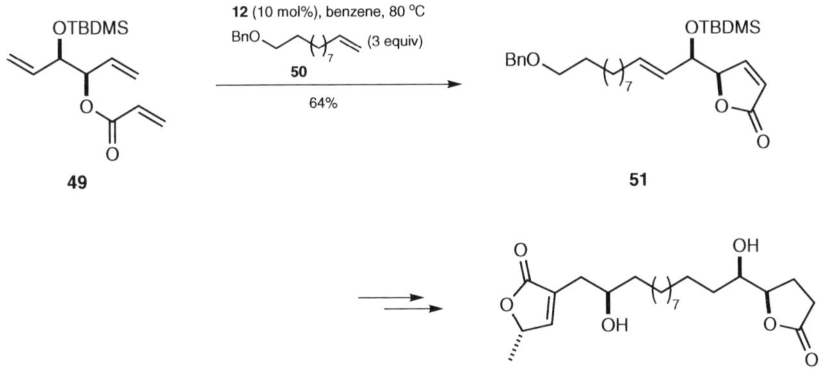

for references see p 672

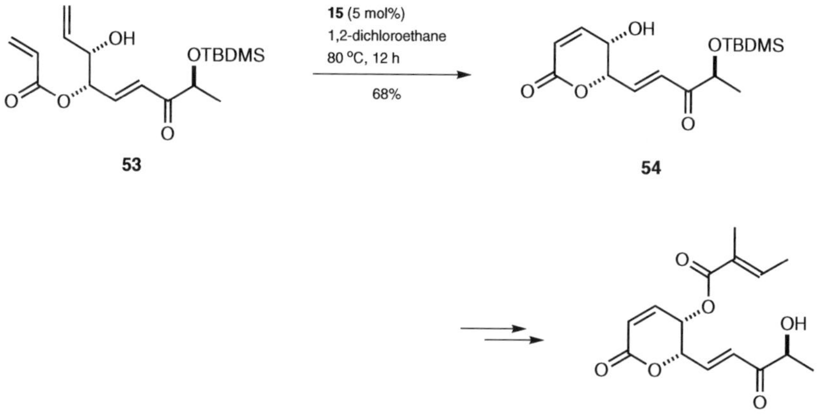

The double ring-closing metathesis of tetraene **56** with Grubbs' first-generation catalyst **11** favors the formation of the six-membered oxacycle **57** over the medium-sized carbocycle **58**, which is not observed (Scheme 15).[38] This chemoselective ring-closing metathesis has been further extended to a triple ring-closing metathesis reaction of hexaene **59** where, upon treatment with Grubbs' first-generation catalyst **11**, only the double ring-closing metathesis product **60** is obtained, and even with prolonged reaction times and freshly added catalyst the triple ring-closing metathesis product **61** is not observed. However, resubmission of the purified product **60** to the ring-closing metathesis conditions with successive additions of Grubbs' first-generation catalyst **11** affords product **61** in 59% yield after 8 days. On the other hand, the ring-closing metathesis of hexaene **59** with Grubbs' second-generation catalyst **12** directly provides the expected triple ring-closing metathesis product **61** in 75% yield in 4 hours, and none of the intermediates on the way to **61** including **60** are observed in this reaction.[39] A similar double ring-closing metathesis of tetraene **62** with Grubbs' first-generation catalyst **11** generates the medium-sized *trans*-fused tricyclic ether **63** in high yield without forming bridged cycles such as **64**.[40] The ring-closing metathesis of tetraene **62** has also been examined with the more reactive Schrock complex **1** as catalyst, but the reaction gives lower yields even at higher catalyst loadings of up to 50 mol%.

Scheme 15 Double and Triple Ring-Closing Metathesis of Tetra- and Hexaenes[38–40]

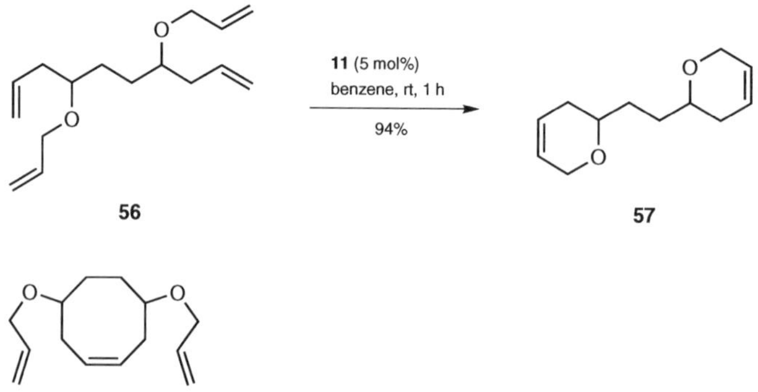

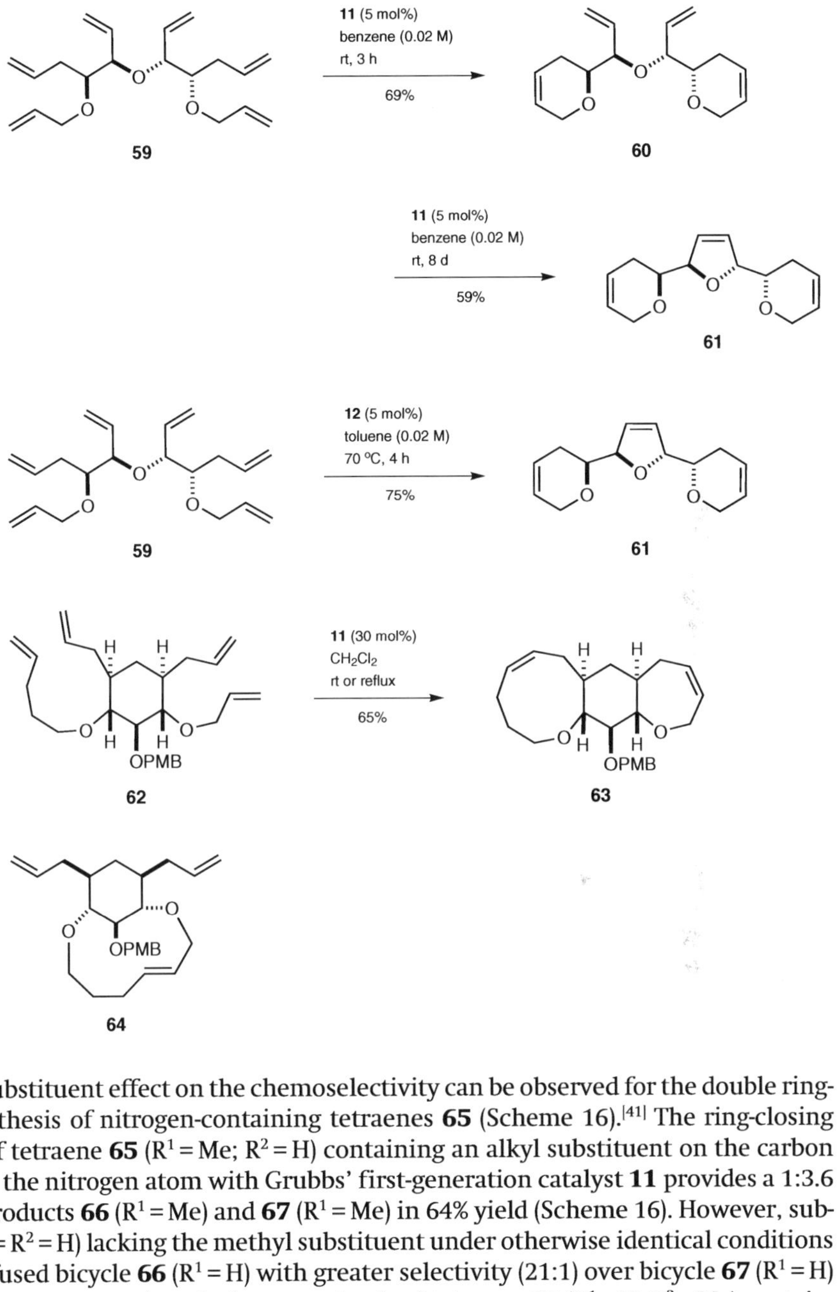

2.10.2 Diene Ring-Closing Metathesis

A dramatic substituent effect on the chemoselectivity can be observed for the double ring-closing metathesis of nitrogen-containing tetraenes **65** (Scheme 16).[41] The ring-closing metathesis of tetraene **65** (R^1 = Me; R^2 = H) containing an alkyl substituent on the carbon connected to the nitrogen atom with Grubbs' first-generation catalyst **11** provides a 1:3.6 mixture of products **66** (R^1 = Me) and **67** (R^1 = Me) in 64% yield (Scheme 16). However, substrate **65** (R^1 = R^2 = H) lacking the methyl substituent under otherwise identical conditions delivers the fused bicycle **66** (R^1 = H) with greater selectivity (21:1) over bicycle **67** (R^1 = H) (Scheme 16). In contrast, ring-closing metathesis of tetraene **65** (R^1 = H; R^2 = Me) containing a methyl substituent on the terminal position of the C=C bond completely suppresses the formation of **67** (R^1 = H), providing **66** (R^1 = H) as the sole product in 86% yield (Scheme 16).

Scheme 16 Substituent Effect in Double Ring-Closing Metathesis of Tetraenes[41]

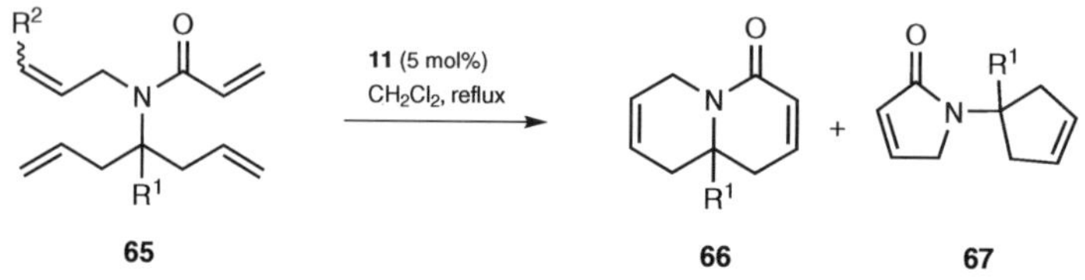

R¹	R²	Ratio (66/67)	Yield (%)	Ref
Me	H	1:3.6	64	[41]
H	H	21:1	88	[41]
H	Me	1:0	86	[41]

The chemoselective ring-closing metathesis strategy described above has been applied to a short synthesis of the bicyclic alkaloid (+)-lupinine (**71**) (Scheme 17).[42] The double ring-closing metathesis of tetraene **68** with Grubbs' second-generation catalyst **12** affords a mixture of bicyclic compounds **69** and **70** with moderate selectivity. However, the low selectivity is inconsequential because both compounds can be converted into (+)-lupinine (**71**) by hydrogenation/benzyl deprotection and reductive removal of the carbonyl group. On the other hand, the ring-closing metathesis of tetraene **68** with Grubbs' first-generation catalyst **11** delivers mono ring-closing metathesis product **72** along with only a small amount of the double ring-closing metathesis product **70**. As expected, the mono ring-closing metathesis product **72** can be converted into the bicyclic compound **69** with the more reactive Grubbs' second-generation catalyst **12**.

Scheme 17 Double Ring-Closing Metathesis for the Synthesis of (+)-Lupinine[42]

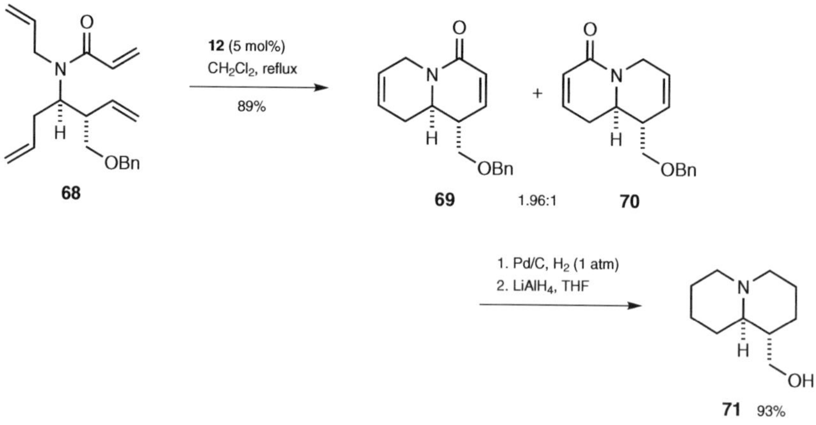

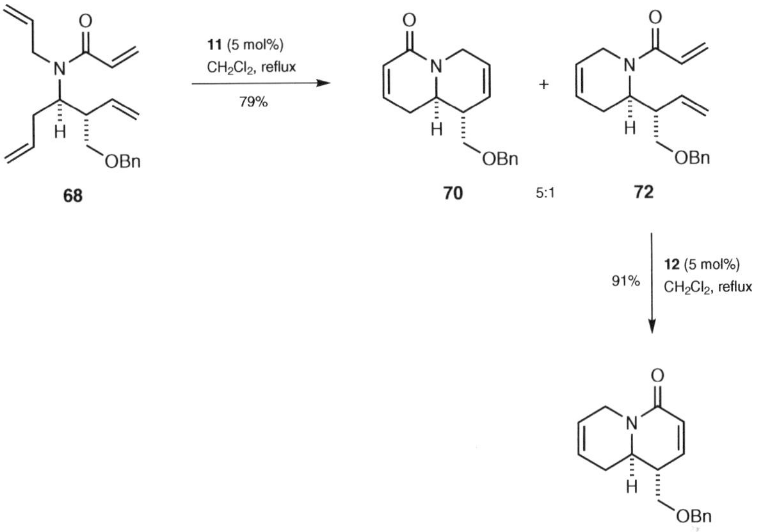

2.10.2 Diene Ring-Closing Metathesis

1,6,9,9a-Tetrahydro-4H-quinolizin-4-one (66, R^1 = H):[41]
A portion of Grubbs' first-generation catalyst **11** (37 mg, 0.045 mmol, 0.05 equiv) was added to a soln of N-allyl-N-(hepta-1,6-dien-4-yl)acrylamide (**65**, R^1 = R^2 = H; 170 mg, 0.83 mmol) in CH$_2$Cl$_2$ (28 mL) under argon, and the mixture was stirred under reflux for 2 h. The resulting soln was exposed to air with stirring, and concentrated. The residue was purified by flash column chromatography (silica gel, petroleum ether/EtOAc 3:1); yield: 104 mg (84%). 1-(Cyclopent-3-enyl)-1,5-dihydro-2H-pyrrol-2-one (**67**, R^1 = H) was also obtained; yield: 5 mg (4%).

2.10.2.2.3.2 Stereochemistry-Based Control

An interesting chemoselective ring-closing metathesis for the generation of skeletal diversity relies on the stereochemistry of the substituents (Scheme 18).[43] The ring-closing metathesis sequence of the diastereomeric substrates **73A** and **73B** containing a pseudoephedrine-derived side chain and an oxanorbornene core structure takes very different courses. With Grubbs' second-generation catalyst **12**, triene **73A** undergoes tandem ring-closing and ring-opening metathesis followed by another ring-closing process to generate fused tetracyclic compound **74**. On the other hand, the ring-closing metathesis of diastereomer **73B**, under identical conditions, affords bridged tetracyclic product **75** derived from a single macrocyclic diene ring-closing metathesis process. This dichotomy in ring-closing metathesis modes clearly illustrates the importance of conformational bias in these substrates, which is caused by the stereochemistry of substituents on flexible systems.

Scheme 18 Stereochemistry-Based Control of Ring-Closing Metathesis[43]

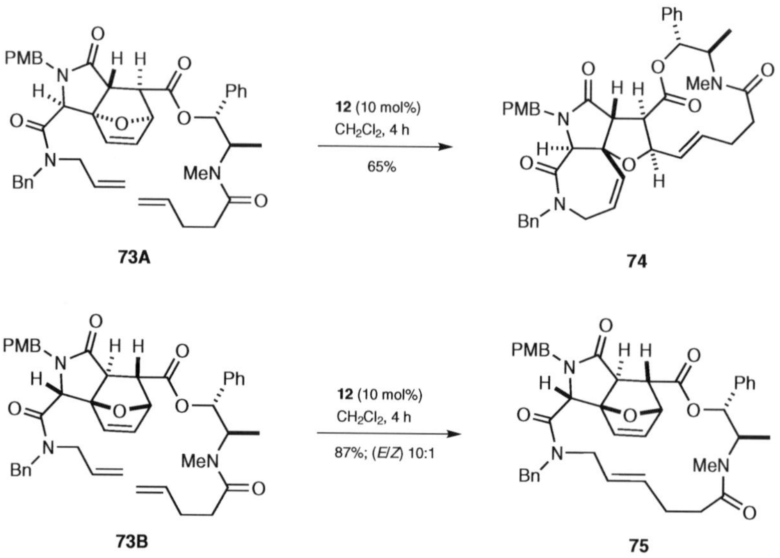

2.10.2.2.3.3 Relay Metathesis Based Control

Diene metathesis generally works well with simple alkenes, yet the presence of sterically hindered substituents or electronically deactivating groups on or near the reacting alkenes slows down the metathesis process if not completely abrogating it. This deactivating effect of the substituents on the alkene can be accommodated as a beneficial handle for realizing chemoselectivity in multiple metathesis events. In 2004, the Hoye and Lee groups started independently using the "relay metathesis" approach to control reaction pathways in ring-closing and cross metathesis, respectively (Scheme 19).[44–46]

As the Grubbs' first-generation catalyst **11** has limited reactivity toward substituted double bonds, ring-closing metathesis of dienes such as **76** would not undergo initiation at either end of the diene to generate intermediate **77**. However, by introducing a relay tether bearing a terminal double bond in triene **79**, Grubbs' first-generation catalyst **11** can initiate the ring-closing metathesis reaction by forming intermediate **80** followed by an intramolecular transfer of a propagating alkylidene species to a sterically hindered internal double bond to generate intermediate **77**, which then generates tetrasubstituted ring-closing metathesis product **78** (Scheme 19).[44] The electronically deactivated enyne **81** (R^1 = H) does not readily participate in the Grubbs' second-generation catalyst **12** catalyzed cross-metathesis reaction with propagating alkylidene **83** derived from (Z)-1,4-bis(benzyloxy)but-2-ene, delivering only a low yield of the cross metathesis product **82** (Scheme 19). By tethering a relay containing a more reactive terminal double bond in dienyne **81** (R^1 = CH_2OCH_2CH=CH_2), the initiation with Grubbs' second-generation catalyst **12** readily occurs to generate an alternative propagating alkylidene **84**, which participates in the cross metathesis with (Z)-1,4-bis(benzyloxy)but-2-ene more efficiently, affording product **82** in much higher yield.[45]

2.10.2 Diene Ring-Closing Metathesis

Scheme 19 First Relay Ring-Closing and Cross Metathesis Reactions[44–47]

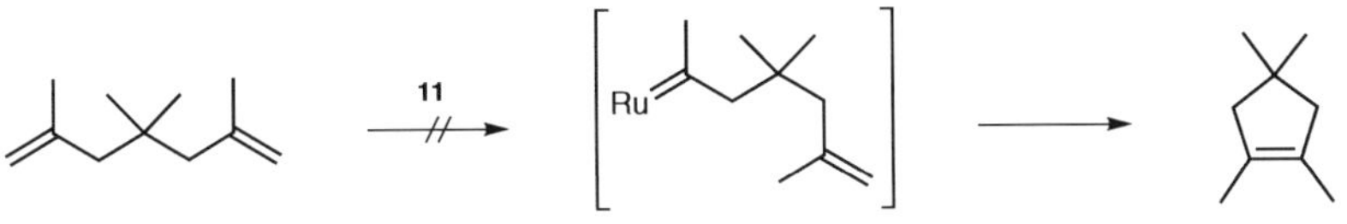

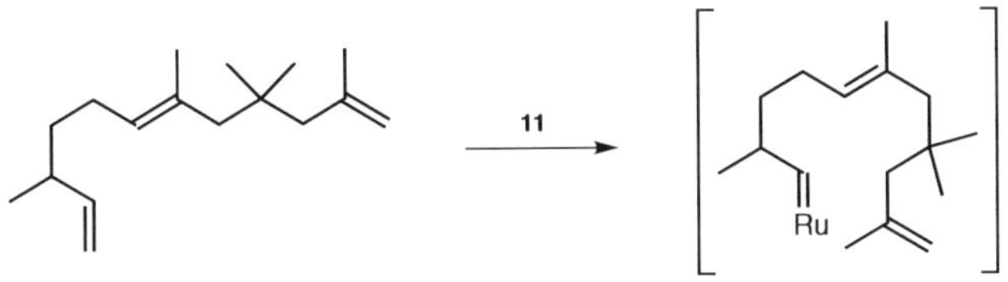

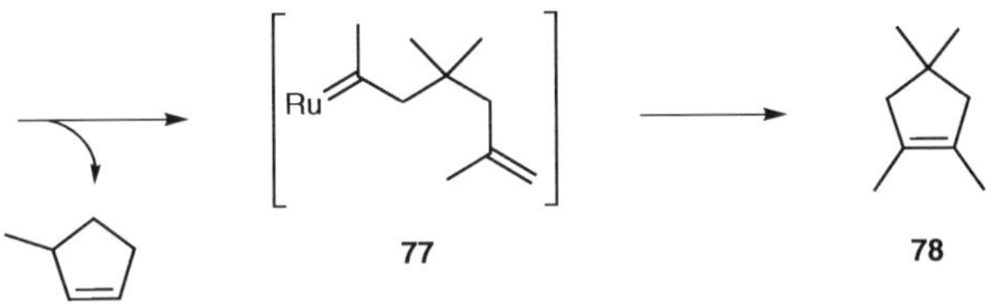

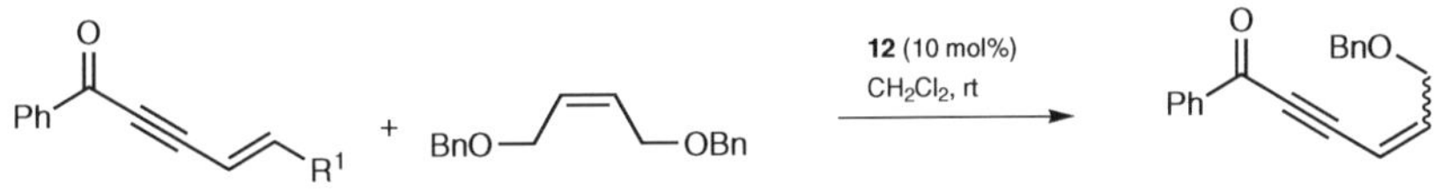

R^1	Ratio (E/Z)	Yield (%)	Ref
H	1:7	34	[45]
$CH_2OCH_2CH=CH_2$	1:4	63	[45]

The earliest relay ring-closing metathesis (RRCM) application in complex natural product synthesis was reported for the preparation of the 12-membered-ring intermediate **86** in the total synthesis of oximidine III (**87**). The initially tried ring-closing metathesis of triene **85** (R^1 = H) with a terminal double bond next to an epoxide moiety results in low conversion and low yield with Grubbs' second-generation catalyst **12** or Grubbs–Hoveyda second-generation catalyst **15**. On the other hand, substrate **85** [R^1 = $(CH_2)_4CH=CH_2$] containing a relay tether provides ring-closing metathesis product **86** in good yield with complete Z selectivity for the newly formed double bond, which is elaborated to target natural product oximidine III (**87**) (Scheme 20).[47]

for references see p 672

Scheme 20 Relay Ring-Closing Metathesis in the Synthesis of Oximidine III[47]

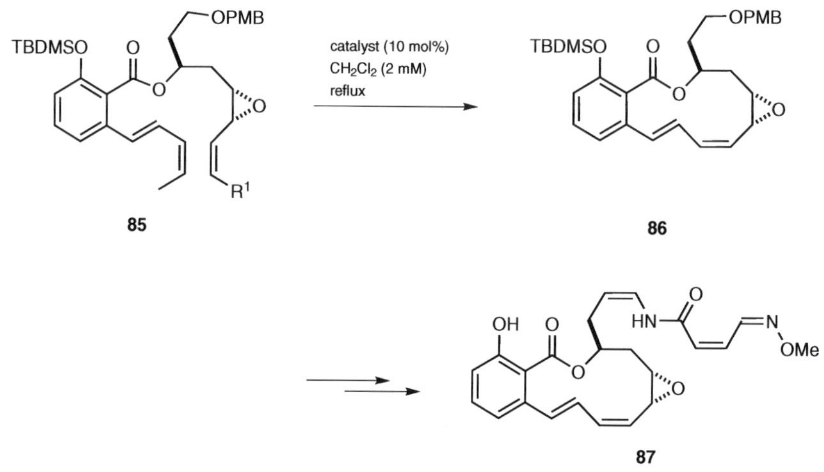

R¹	Catalyst	Conversion (%)	Yield (%)	Ref
H	12	38	15	[47]
H	15	38	15	[47]
(CH$_2$)$_4$CH=CH$_2$	12	>90	71	[47]

2.10.2.2.4 Diastereo- and Enantioselective Ring-Closing Metathesis

A diastereoselective ring-closing metathesis of a triene bearing a stereogenic center as stereochemical handle at an allylic or a homoallylic position creates a new stereogenic center at a carbon atom containing two diastereotopic alkene moieties (Scheme 21).[48] High diastereoselectivity can be obtained even if the ring-closing metathesis is carried out with an achiral catalyst as long as the initiation occurs at the double bond near the stereogenic center, which would differentiate the two diastereotopic alkene moieties. As expected, the ring-closing metathesis of triene **88** with Grubbs' first-generation catalyst **11** produces N-protected 2,5-dihydro-1H-pyrrole **89** with high diastereoselectivity (**89A/89B** = 4:96). On the other hand, the ring-closing metathesis of the same substrate with the more reactive Schrock catalyst **1** leads to the favorable formation of *syn*-product **89A** (**89A/89B** = 86:14). The reversal of selectivity is tentatively attributed to catalyst specificity for the different spatial arrangements of the respective ligands during the cyclization. The *anti*-isomer **89B** has been elaborated to (−)-azasugar **90**.[48,49]

2.10.2 Diene Ring-Closing Metathesis

Scheme 21 Diastereoselective Ring-Closing Metathesis of a Triene[48,49]

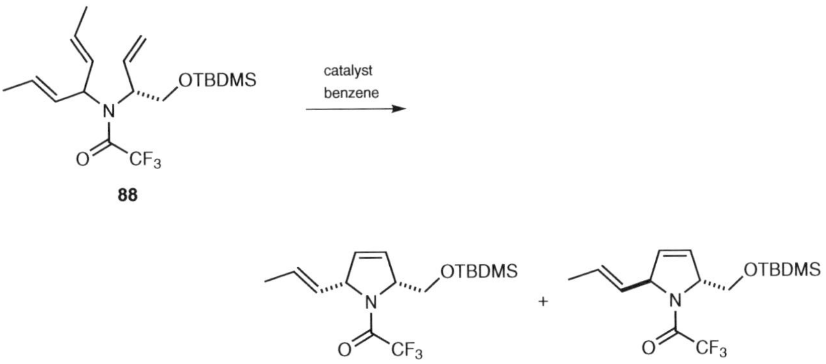

Catalyst	Time (d)	Ratio (89A/89B)	Yield (%)	Ref
11	2	4:96	88	[48]
1	3	86:14	97	[48]

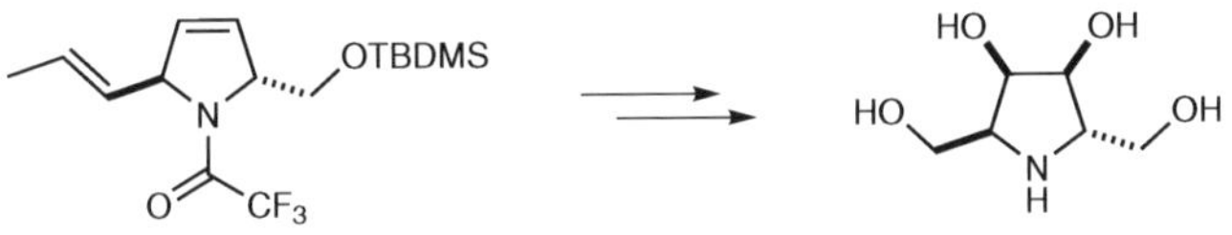

An elegant diastereoselective ring-closing metathesis reaction of trienes **91** is achieved by a long range asymmetric induction relying on a temporary silicon tether (Scheme 22).[50] It was envisioned that the group-selective ring-closing metathesis of triene **91** containing a diastereotopic pair of double bonds and different substituents (R^1 and R^2) would preferentially provide the *syn*-diastereomer **92A** in the presence of a sterically demanding R^2 substituent on the silicon atom (via transition state **93A**). The axial R^2 substituent on the silicon atom increases the unfavorable nonbonded interaction of the axially oriented propenyl group in the disfavored transition state **93B**, thus disfavoring the formation of the *anti*-diastereomer **92B**. The yield and diastereoselectivity of this transformation show significant catalyst dependency. For example, the ring-closing metathesis with the least active Grubbs' first-generation catalyst **11** affords high selectivity and yields, whereas Grubbs' second-generation catalyst **12** as the most active ruthenium complex or Schrock complex **1** show a poor reaction profile, affording lower selectivity and yields.

Scheme 22 Diastereoselective Ring-Closing Metathesis of Temporary Silicon-Tethered Trienes[50]

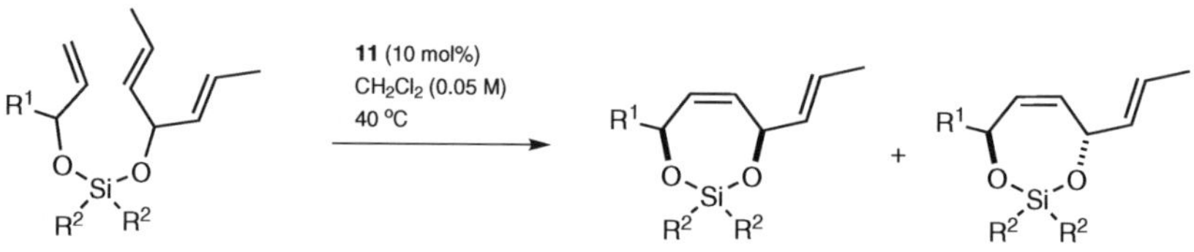

R¹	R²	Ratio (**92A/92B**)	Yield (%)	Ref
2-naphthyl	Me	23:1	41	[50]
2-naphthyl	iPr	99:1	75	[50]
Ph	iPr	99:1	90	[50]
CH₂OBn	iPr	41:1	61	[50]

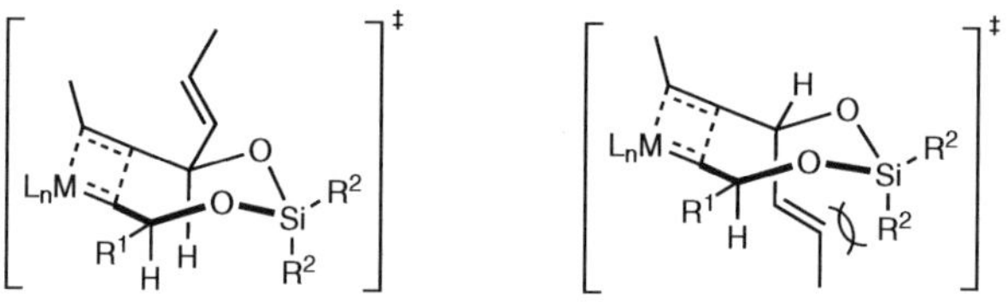

The enantioselective ring-closing metathesis of trienes involves the formation of a chiral propagating alkylidene species followed by its stereomutation with one of the two enantiotopic double bond moieties. The chiral information on the propagating alkylidene will ultimately be conferred by the chiral ligand on the metal. Over the past decade, Hoveyda and Schrock and their coworkers have developed various chiral molybdenum complexes, whereas Grubbs and coworkers have mainly focused on the ruthenium-based chiral complexes effective for the enantioselective ring-closing metathesis of trienes.[51–57] Some of the most efficient catalysts such as molybdenum catalysts **94–97** and ruthenium complexes **98** and **99** are shown in Scheme 23.

Scheme 23 Chiral Molybdenum and Ruthenium Complexes Commonly Used in the Enantioselective Ring-Closing Metathesis of Trienes[51–57]

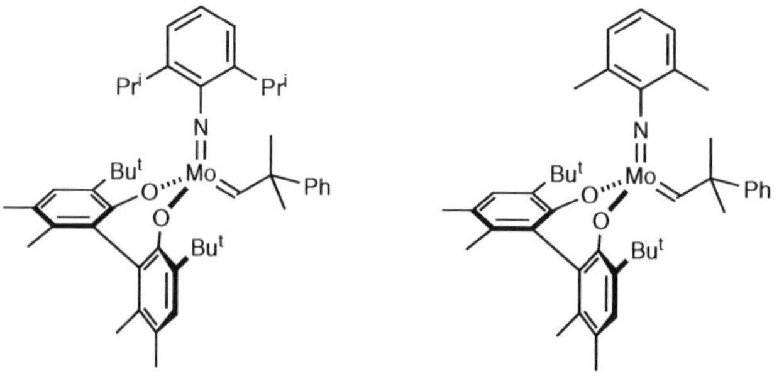

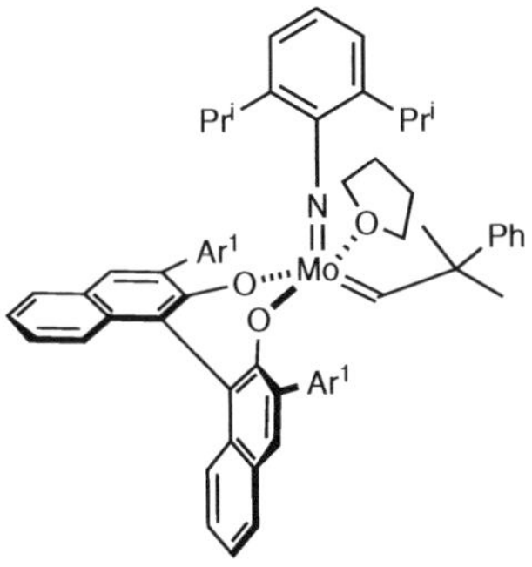

2.10.2 Diene Ring-Closing Metathesis

96 Ar¹ = 2,4,6-(iPr)₃C₆H₂

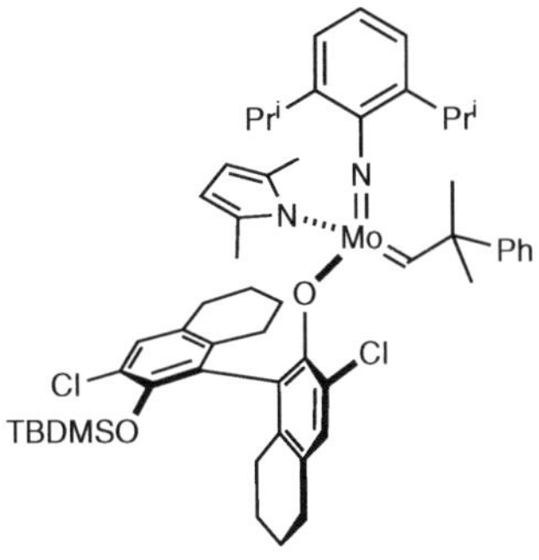

97

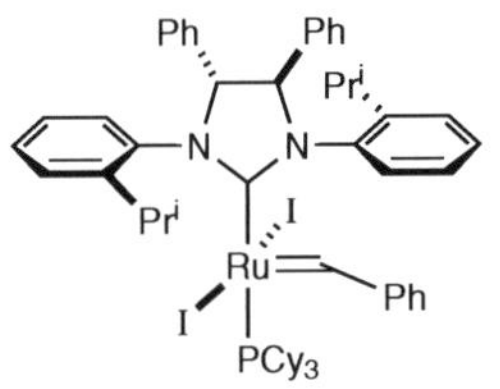

98

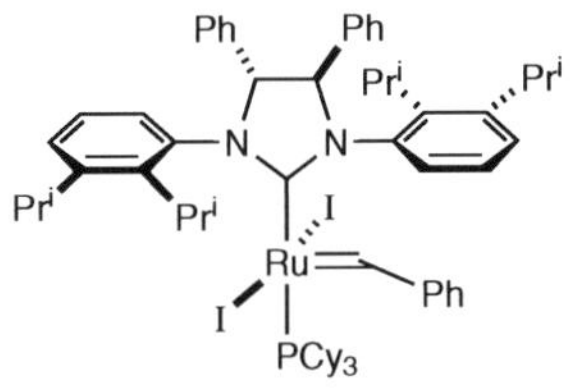

99

Diene **100** undergoes an efficient kinetic resolution through ring-closing metathesis by using molybdenum catalyst **94** (5 mol%). The cyclized product **101** (43%, 93% ee) and the dimeric product **102** (38%) are formed in 10 minutes (Scheme 24). In addition, the unreacted diene **100** can be isolated in 19% yield in nearly optically pure form (>99% ee). With molybdenum catalyst **94**, the enantioselective desymmetrization of triene **103** generates chiral dihydrofuran derivative (R)-**104** with 99% enantiomeric excess (Scheme 24).[52] The same ring-closing metathesis based desymmetrization using ruthenium complexes **98** and **99** generates the S-enantiomer with slightly lower yield and enantioselectivity (Scheme 24).[55] Desymmetrization of silyl ether based triene **105** (R¹ = H) with chiral binaphthyl ligand containing molybdenum catalyst **96** furnishes six-membered cyclic silyl ether (R)-**106** in near quantitative yield and excellent enantioselectivity (Scheme 24).[54] Similarly, chiral ruthenium complexes **98** and **99** are employed for the desymmetrization of triene **105** (R¹ = Me), providing cyclic silyl ether (S)-**106** with slightly lower yield and enantiomeric excess (Scheme 24).[55] Ring-closing metathesis based desymmetrization of amine-containing triene **107** is achieved in the presence of molybdenum catalyst **95** to form six-membered azacycle **108**, which provides an efficient entry for the synthesis of 2-propylpiperidine (**109**) (Scheme 24).[56] An effective desymmetrization of the enantiotopic double bonds of triene **110** in the presence of molybdenum catalyst **97** generates tetracyclic compound **111** in excellent yield and enantiomeric excess, which is readily converted into (+)-quebrachamine (**112**).[57]

Scheme 24 Enantioselective Ring-Closing Metathesis of Trienes[52–57]

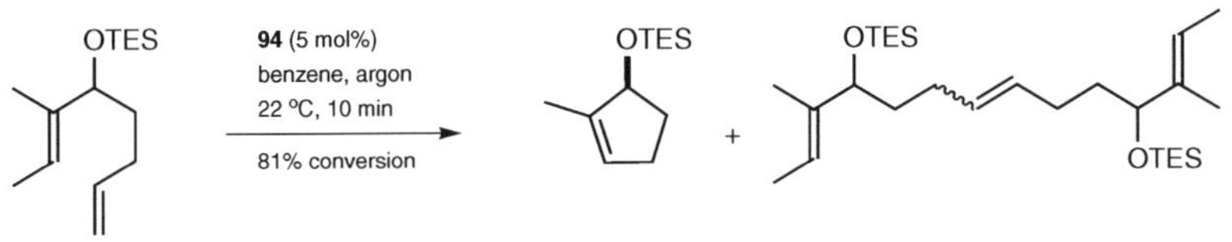

100

94 (5 mol%)
benzene, argon
22 °C, 10 min
81% conversion

101 43%; 93% ee

102 38%

for references see p 672

103 104

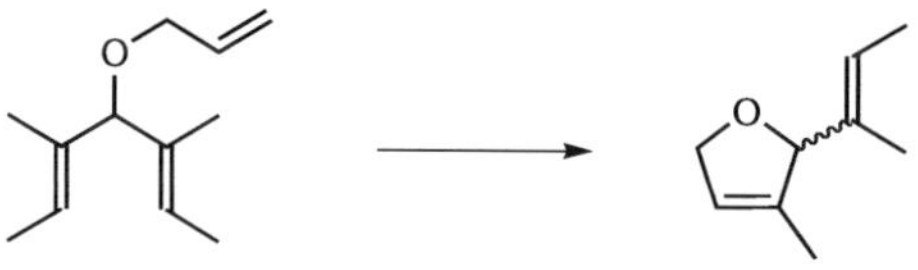

Conditions	Config	ee (%)	Yield (%)	Ref
94 (2 mol%), neat, 22 °C, 5 min	*R*	99	93	[52]
98 (4 mol%), THF, 40 °C, 4 h	*S*	90	98[a]	[55]
99 (4 mol%), THF, 40 °C, 4 h	*S*	90	98[a]	[55]

[a] Conversion.

105 106

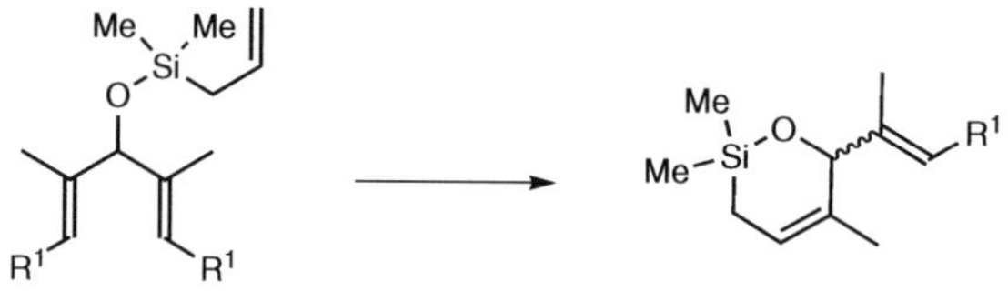

R^1	Conditions	Config	ee (%)	Yield (%)	Ref
H	**96** (2 mol%), neat, 60 °C, 3 h	*R*	99	98	[54]
Me	**98** (4 mol%), THF, 40 °C, 4 h	*S*	86	68[a]	[55]
Me	**99** (4 mol%), THF, 40 °C, 4 h	*S*	92	58[a]	[55]

[a] Conversion.

107 108 109

110 111

112

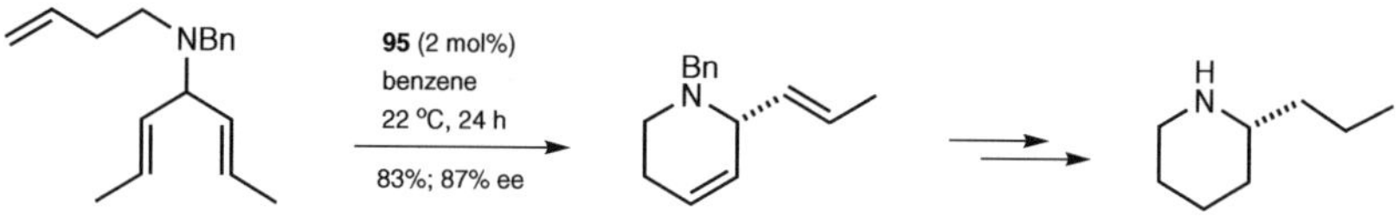

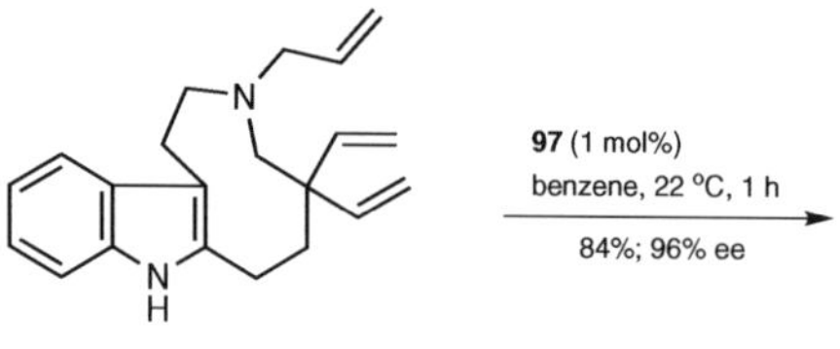

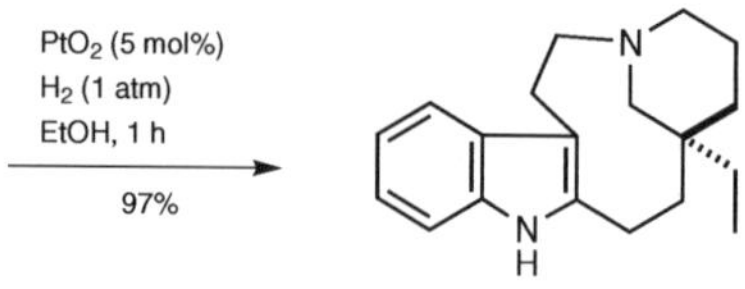

2.10.2 Diene Ring-Closing Metathesis

(4R*,7R*)-2,2-Diisopropyl-4-(2-naphthyl)-7-[(E)-prop-1-enyl]-4,7-dihydro-1,3,2-dioxasilepin (92A, R¹ = 2-Naphthyl; R² = iPr):[50]

A 10-mL, flame-dried, round-bottomed flask equipped with a magnetic stirrer bar and reflux condenser was charged with triene **91** (R¹ = 2-naphthyl; R² = iPr; 40.9 mg, 0.1 mmol) and dissolved in anhyd CH_2Cl_2 (2 mL) under argon. The soln was degassed for 5 min and Grubbs' first-generation catalyst **11** (4.1 mg, 5 µmol, 0.05 equiv) was added in a single portion resulting in a brown soln, which was heated at 40 °C for ca. 6 h. Another portion of Grubbs' first-generation catalyst **11** (4.1 mg, 5 µmol, 0.05 equiv) was added and the mixture was then heated at 40 °C for an additional ca. 6 h (monitored by TLC). The mixture was then allowed to cool to rt, and concentrated under reduced pressure. The obtained dark brown oil was purified by flash chromatography [benzene (**CAUTION:** *carcinogen*)/hexanes 5:95]; yield: 27.5 mg (75%); ratio (**92A/92B**) 99:1.

2.10.2.3 Applications of Diene Ring-Closing Metathesis to Natural Product Synthesis

The utility of diene ring-closing metathesis has been demonstrated by the synthesis of a variety of natural products containing carbo- and heterocycles ranging from small and medium rings to macrocycles. Through these examples the merits of ring-closing metathesis based strategies are clearly illustrated where the steps required for protecting group manipulations are avoided or minimized. The compatibility of alkene functionality with polar functional groups in many transformations allows more flexible strategic development. In addition, the use of only a small amount of metal–alkylidene initiator can minimize the formation of byproducts and reduces the solvents and other material involved in workup and purification because of the relatively easy removal of the catalyst through filtration. Most of all, the excellent chemo-, regio-, and stereoselectivity together with spatiotemporal control of the ring closure are unique traits that no other synthetic transformation can compete with. This section is organized in order of increasing ring size.

2.10.2.3.1 Ring-Closing Metathesis with 1,n-Dienes

The regio- and stereo-controlled total synthesis of chitinase inhibitor (−)-allosamizoline (**115**) uses a ring-closing metathesis as key step for the formation of the cyclopentene moiety (Scheme 25). The synthesis starts with commercially available D-glucosamine, which is converted into 1,6-dienes **113** (R¹ = Ac or CO_2Me). Ring-closing metathesis of these dienes with Grubbs' second-generation catalyst **12** affords the same cyclopentene derivative **114** in high yield (85 and 88%), which is then elaborated to (−)-allosamizoline (**115**).[58]

Scheme 25 Diene Ring-Closing Metathesis Based Synthesis of (−)-Allosamizoline[58]

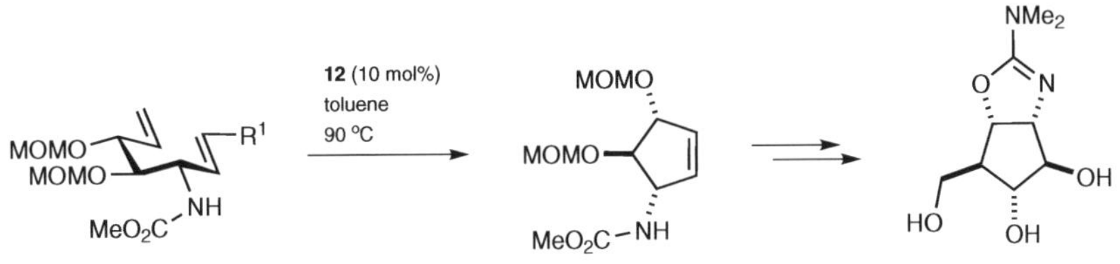

R¹	Yield (%) of **114**	Ref
Ac	85	[58]
CO_2Me	88	[58]

In the total synthesis of (−)-heptemerone B (**118**, R^1 = Ac) and (−)-guanacastepene E (**118**, R^1 = H), ring-closing metathesis of diene **116** with Grubbs' second-generation catalyst **12** forms an appropriately functionalized cyclopentene derivative **117**, from which the tetracyclic framework of the natural products was obtained (Scheme 26).[59]

Scheme 26 Ring-Closing Metathesis for the Formation of a Functionalized Cyclopentene Derivative toward the Synthesis of (−)-Heptemerone B and (−)-Guanacastepene E[59]

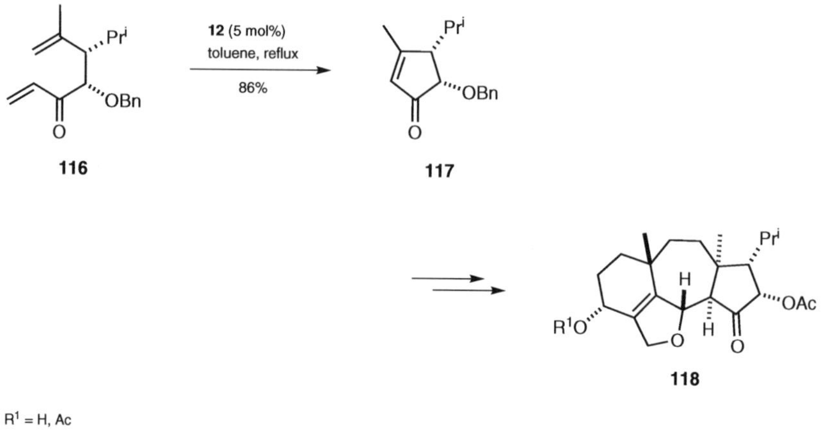

R^1 = H, Ac

In the total synthesis of spirotenuipesines A (**121**) and B (**122**), a ring-closing metathesis is employed to synthesize cyclic alcohol **120**, which had previously been synthesized in nine steps,[62] in three steps (Scheme 27).[60,61] The synthesis of tetrasubstituted cyclopentene derivatives with Grubbs-type ruthenium–alkylidenes is often challenging, and requires relay ring-closing metathesis to deliver the catalyst at a sterically hindered double bond. However, the ring-closing metathesis of 1,6-diene **119** containing two 1,1-disubstituted double bonds, in the presence of Grubbs' second-generation catalyst **12**, affords tetrasubstituted double bond containing cyclopentene derivative **120** in excellent yield (82%). This intermediate is used for the synthesis of the spirocyclic natural products **121** and **122**.

Scheme 27 Ring-Closing Metathesis for the Formation of a Tetrasubstituted Alkene for Use in the Synthesis of Spirotenuipesines A and B[60,61]

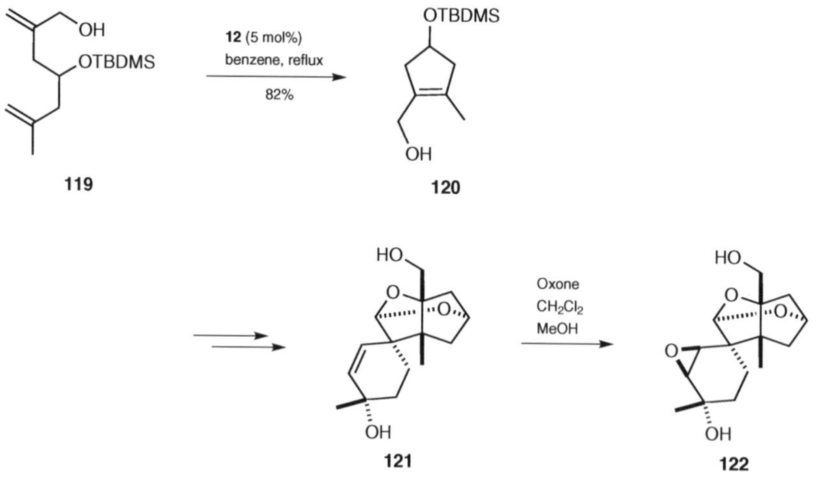

2.10.2 Diene Ring-Closing Metathesis

The ring-closing metathesis of diene **123** in the presence of Grubbs' first-generation catalyst **11** provides pyrrole derivative **124** in good yield (90%). Under similar conditions, ring-closing metathesis of 1,7-diene **125** quantitatively delivers carboxylate **126**, where a newly formed six-membered carbocycle is fused to a tetrahydropyrrole moiety (Scheme 28).[63,64] In subsequent steps, these two precursors are elaborated and merged to realize the synthesis of dysinosin A (**127**).

Scheme 28 Diene Ring-Closing Metathesis for the Synthesis of Dysinosin A[63,64]

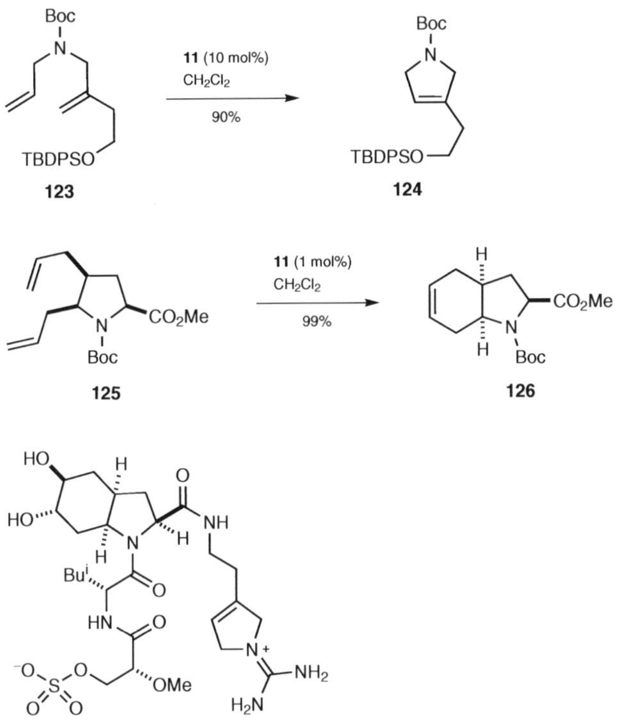

In studies toward the synthesis of azasugars 2,5-dideoxy-2,5-imino-D-mannitol (**130**), (−)-bulgecinine (**131**), and (−)-broussonetine G (**132**), a ring-closing metathesis constructs the core dihydropyrrole system of these natural products (Scheme 29). The ring-closing metathesis of 1,6-diene **128** in the presence of Grubbs' second-generation catalyst **12** delivers bicyclic dihydropyrrole derivative **129** in good yield as a common precursor.[65,66]

Scheme 29 Diene Ring-Closing Metathesis Strategy for the Synthesis of Azasugars[65,66]

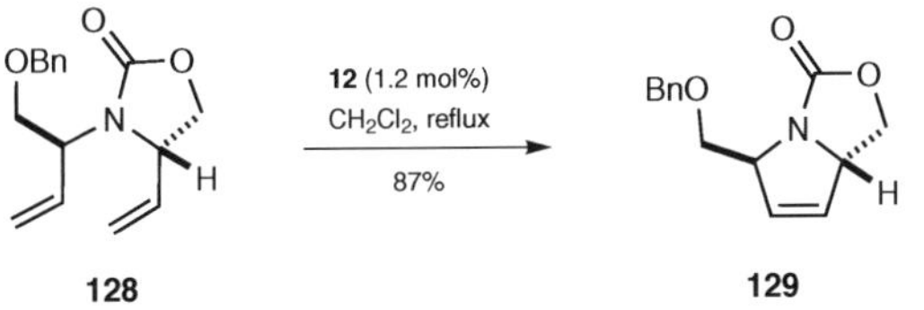

for references see p 672

130   131   132

In the total synthesis of (−)-mucocin (**138**), ring-closing metathesis constructs the heterocyclic tetrahydrofuran and tetrahydropyran moieties (Scheme 30).[67] The relay ring-closing metathesis of tetraene **133** in the presence of Grubbs' second-generation catalyst **12** affords dihydrofuran **134** in good yield. In this transformation, the relay metathesis is crucial to prevent the six-membered ring closure (between double bonds a and b).[68] On the other hand, the group-selective ring-closing metathesis of triene **135** under the same conditions affords dihydropyran derivative **136** without formation of the seven-membered ring. The cross metathesis of **134** and **135** with Grubbs–Hoveyda second-generation catalyst **15** delivers dienyne **137** in moderate yield (68%) along with the homodimerized product of **136** (23%) and recovered starting material **134** (13%). The remaining butenolide part of (−)-mucocin (**138**) is introduced by Sonogashira coupling on the terminal alkyne.

Scheme 30 Diene Ring-Closing Metathesis Based Approach to (−)-Mucocin[67]

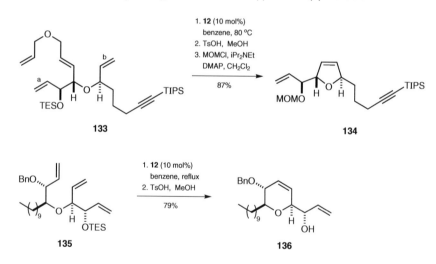

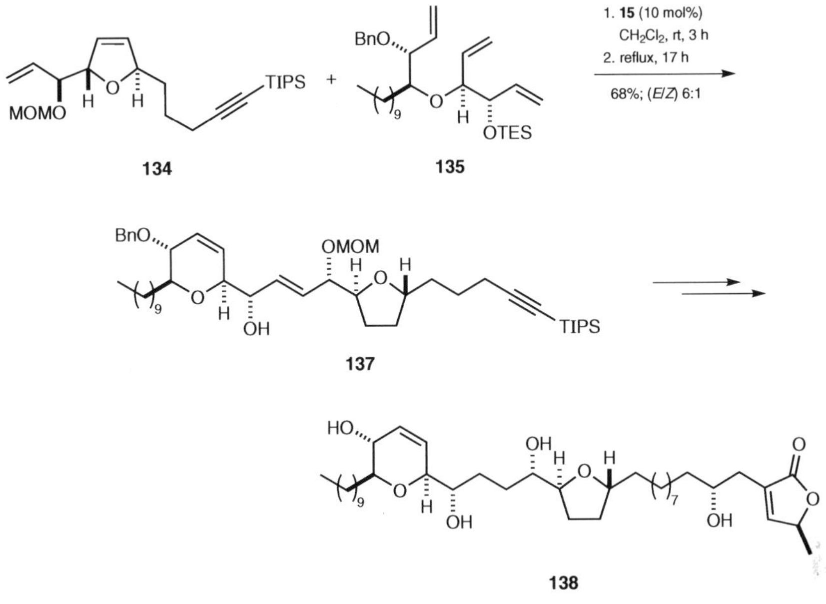

2.10.2 Diene Ring-Closing Metathesis

In the total synthesis of (±)-garsubellin A (**141**), the diene ring-closing metathesis of diene **139** in the presence of Grubbs' second-generation catalyst **12** gives a sterically congested six-membered ring as part of the highly functionalized bridged bicycle **140** in excellent yield (Scheme 31).[69]

Scheme 31 Diene Ring-Closing Metathesis in the Synthesis of Garsubellin A[69]

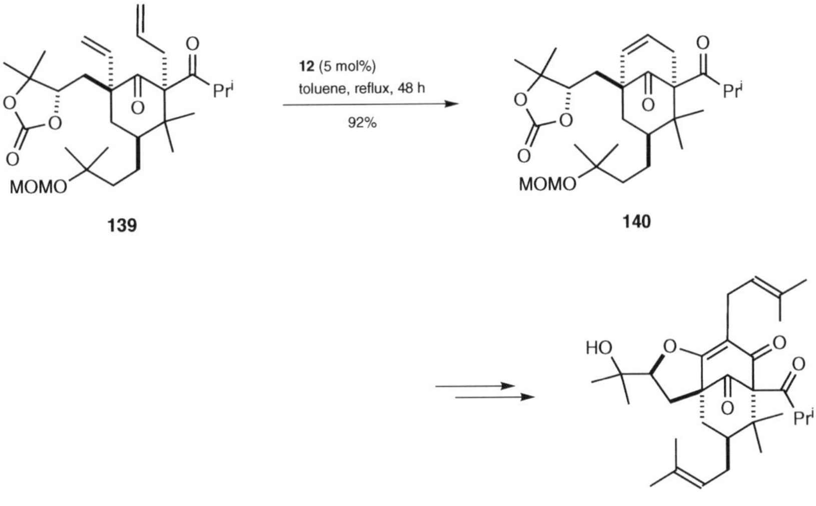

Ring-closing metathesis reactions are used for the construction of key intermediates in two formal total syntheses of platencin (**147**).[70,71] In the first, a fused six-membered ring is installed onto the existing bicyclo[2.2.2]octane moiety of triene **142** by ring-closing metathesis with Grubbs' second-generation catalyst **12** to deliver tricycle **143** in excellent yield (Scheme 32).[72] This advanced intermediate is oxidized to furnish known compound **144**. In the second, a ring-closing metathesis reaction of bicyclo[4.4.0]decenyl triene **145**

with Grubbs' second-generation catalyst **12** affords tricycle **146**. The trisubstituted endocyclic double bond of **146** is migrated to an *exo* position to generate the known intermediate **144** (Scheme 32).[73]

Scheme 32 Examples of Diene Ring-Closing Metathesis in the Formal Synthesis of Platencin[70–73]

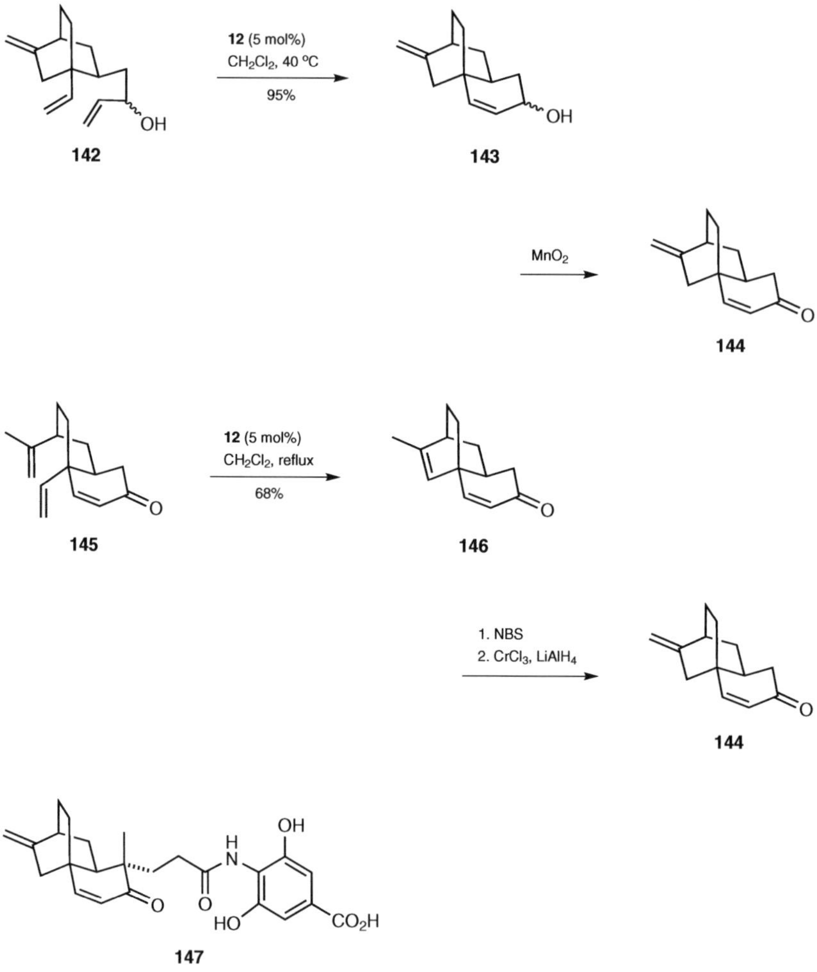

The challenging preparation of a cyclohexene moiety containing a vinylic chloride has been achieved by ring-closing metathesis in the first asymmetric total synthesis of (+)-elatol (**150**) (Scheme 33). In the key step, a ring-closing metathesis of dimedone-derived diene **148** in the presence of Grubbs–Hoveyda second-generation catalyst **15** smoothly delivers the fully substituted chlorinated spirocyclic product **149** in excellent yield (97%). This spirocyclic intermediate is converted into the target natural product **150** via a three-step sequence.[74]

2.10.2 Diene Ring-Closing Metathesis

Scheme 33 Diene Ring-Closing Metathesis in the Synthesis of (+)-Elatol[74]

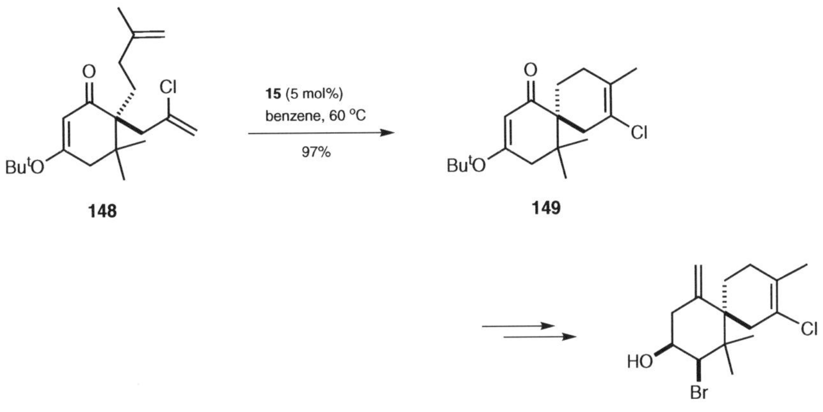

Ring-closing metathesis reactions are employed in the total synthesis of (±)-lundurine B (**155**) to construct a six-membered carbocycle at an early stage and a five-membered heterocycle at a later stage of the synthesis (Scheme 34).[75] The ring-closing metathesis of siloxy diene **151** using Grubbs' second-generation catalyst **12** delivers a cyclic silyl enol ether, which is converted into ketone **152** in 92% yield by mild acid treatment. Elaboration of **152** into diene **153** followed by ring-closing metathesis affords tetracycle **154**, from which (±)-lundurine B (**155**) is obtained by converting the *tert*-butoxycarbonyl group into a methyl carbamate moiety.

Scheme 34 Diene Ring-Closing Metathesis in the Synthesis of (±)-Lundurine B[75]

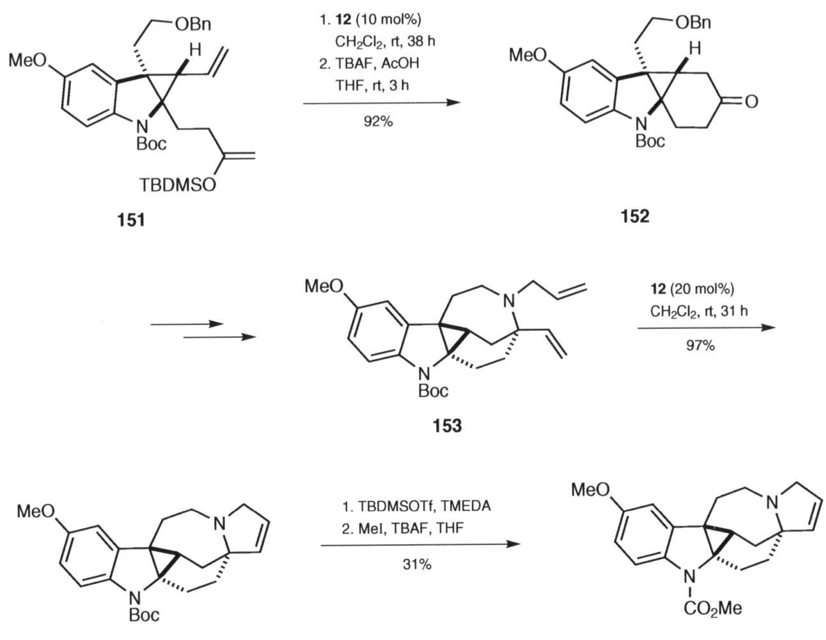

for references see p 672

In a recent total synthesis of (±)-ryanodol (**158**), a ring-closing metathesis constructs a six-membered carbocyclic substructure (Scheme 35).[76] The ring-closing metathesis between two substituted double bonds in diene **156** proceeds smoothly in the presence of Grubbs–Hoveyda second-generation catalyst **15** to provide trisubstituted cyclohexene **157** in good yield.

Scheme 35 Diene Ring-Closing Metathesis in the Synthesis of (±)-Ryanodol[76]

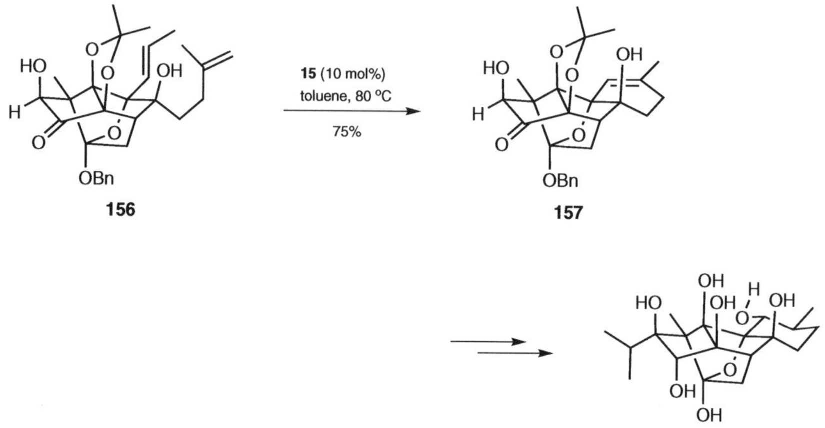

Ring strain based control of group-selective ring-closing metathesis is utilized in a short total synthesis of (±)-epimeloscine (**160**) and (±)-meloscine (**161**) (Scheme 36).[77] In the presence of Grubbs–Hoveyda second-generation catalyst **15**, triene **159** containing two diastereotopic vinyl groups smoothly delivers pentacyclic natural product **160** in excellent yield (89%). The epimerization of the proton at the α-carbon of the amide moiety in **160** by treatment with potassium *tert*-butoxide provides (±)-meloscine (**161**) in 83% yield.

Scheme 36 Group-Selective Ring-Closing Metathesis in the Synthesis of (±)-Epimeloscine and (±)-Meloscine[77]

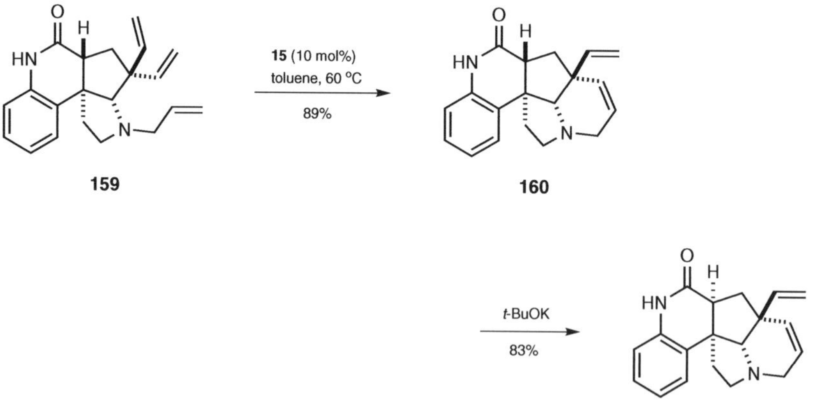

A total synthesis of lysergic acid (**165**) relies on the construction of a tetrahydropyridine moiety by ring-closing metathesis (Scheme 37).[78] The initially used diene **162** (R¹ = Me) provides an inseparable mixture of azacycles **163** and **164** by ring-closing metathesis in the presence of Grubbs–Hoveyda second-generation catalyst **15** under prolonged heating

2.10.2 Diene Ring-Closing Metathesis

in cyclopentyl methyl ether. The formation of the seven-membered azacycle **164** is the consequence of an initial double-bond isomerization to the terminal position induced by a ruthenium hydride species generated from the catalyst. The modified diene **162** (R^1 = H) containing two terminal double bonds affords the expected ring-closing metathesis product **163** in quantitative yield in the presence of Grubbs' second-generation catalyst **12**.

Scheme 37 Ring-Closing Metathesis in the Synthesis of Lysergic Acid[78]

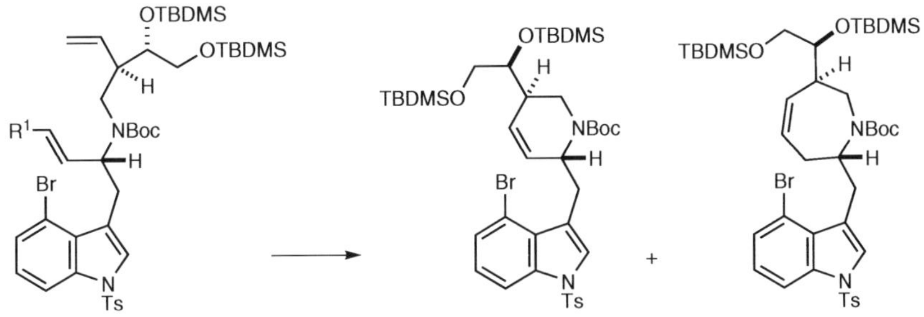

R^1	Conditions	Ratio^a (**163/164**)	Yield^a (%)	Ref
Me	**15**, cyclopentyl methyl ether, reflux	n.r.	n.r.	[78]
H	**12** (3 mol%), toluene, 80 °C	1:0	99	[78]

^{a} n.r. = not reported.

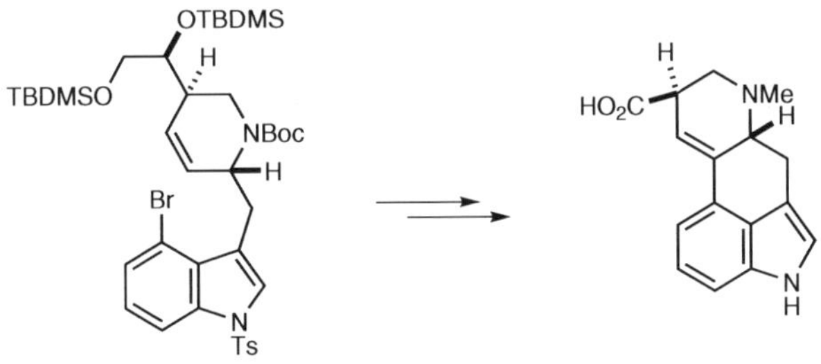

A scalable total synthesis of (+)-omphadiol (**168B**) utilizes a tandem double-bond isomerization–ring-closing metathesis strategy to construct a seven-membered ring (Scheme 38).[79] epi-Omphadiol (**168A**) was synthesized first by employing (R)-carvone-derived 1,8-diene **166** in the ring-closing metathesis process, which afforded bicycle **167** quantitatively in the presence of Grubbs' second-generation catalyst **12**. Under the same conditions, β,γ-enone **169** smoothly undergoes double-bond isomerization followed by ring-closing metathesis to deliver bicyclic enone **170** in excellent yield (95%). Carbonyl reduction of **170** to the corresponding α-allylic alcohol followed by cyclopropanation gives (+)-omphadiol (**168B**).

Scheme 38 Ring-Closing Metathesis in the Synthesis of (+)-Omphadiol[79]

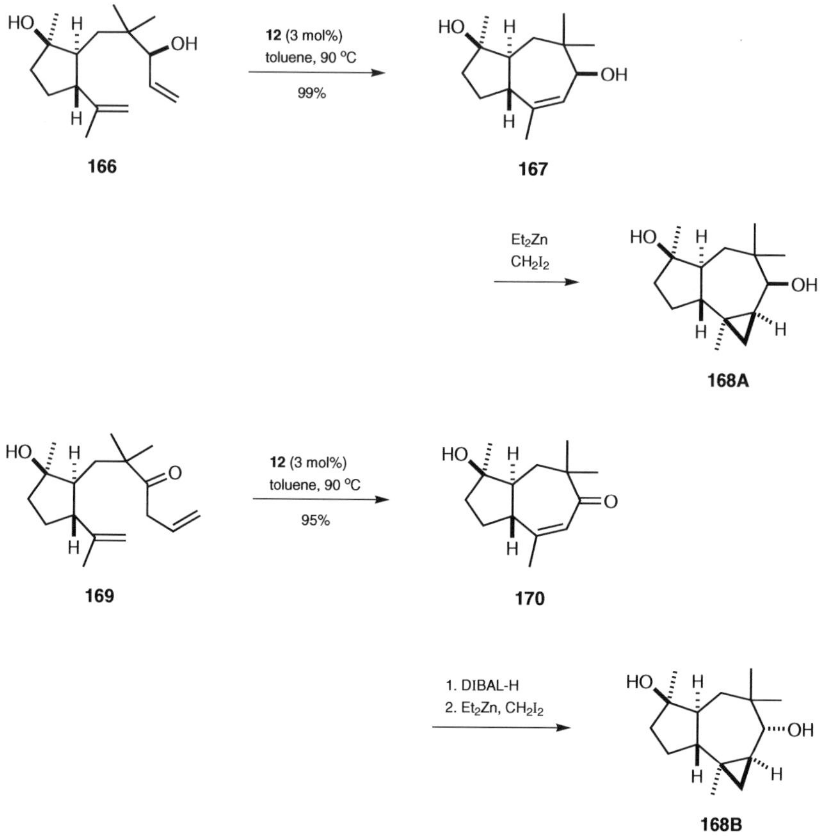

Ingenane diterpenoids such as ingenol (**176**), 13-oxyingenol (**184**), and its derivative **185** possess a synthetically challenging bicyclo[4.4.1]undecane skeleton with a highly strained intrabridgehead topology. Ring-closing metathesis has been used for the construction of this strained framework.[80–83] In a total synthesis of ingenol (**176**), the diene ring-closing metathesis precursor **172** is synthesized from the corresponding oxo ester **171** in 11 steps (Scheme 39). The ring-closing metathesis of diene **172** in the presence of Grubbs' first-generation catalyst **11** (80 mol%) provides tetracyclic compound **173** with incomplete conversion (45% yield), probably due to a competing ring-opening process of the relatively strained product. By introducing an additional substituent on one of the terminal double bonds, diene **174** provides a much higher yield for the ring-closing metathesis product **175** in the presence of Grubbs–Hoveyda second-generation catalyst **15** (25 mol%).[80] This might be the consequence of a reduced rate for the ring-opening process from ring-closing metathesis product **175**, which agrees with the result from the less-substituted substrate **172**.

2.10.2 Diene Ring-Closing Metathesis

Scheme 39 Ring-Closing Metathesis in the Synthesis of Ingenol[80]

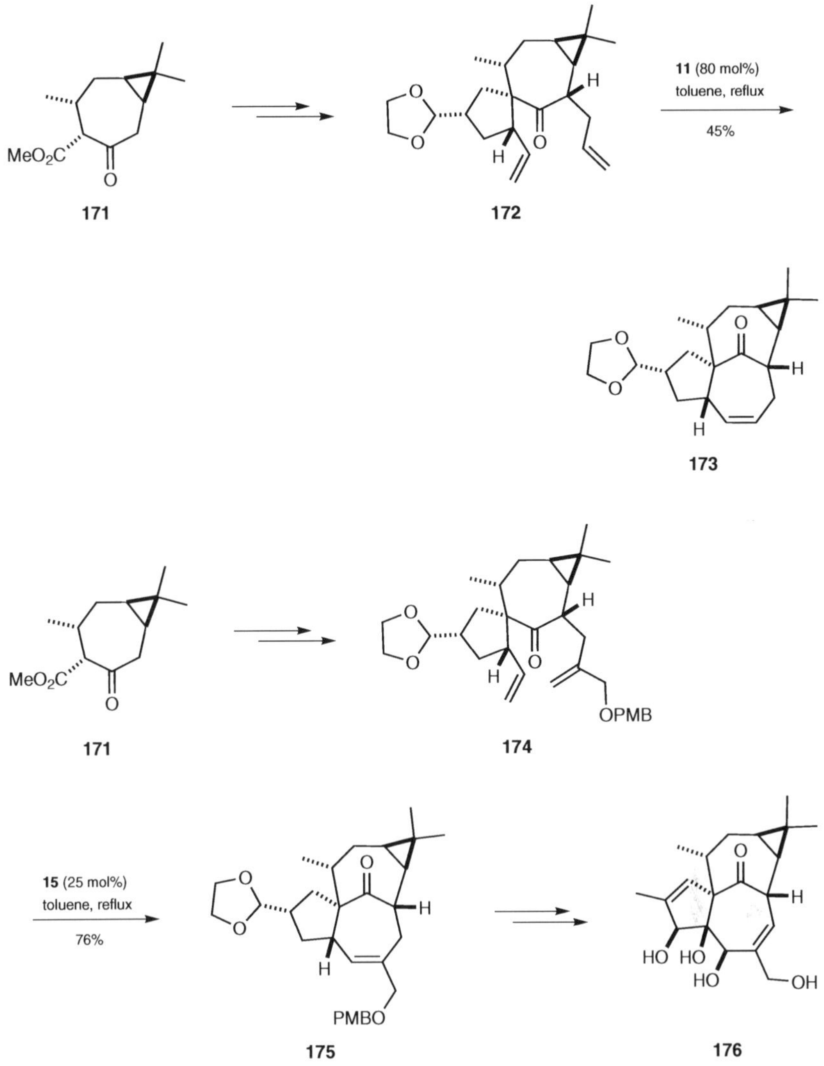

In a formal synthesis of ingenol (**176**), the ring-closing metathesis reaction of diene **177** shows dramatic dependence on solvent and temperature; no reaction is observed in dichloromethane, yet in refluxing toluene with Grubbs' second-generation catalyst **12** (80 mol%) the ring-closing metathesis product **178** is obtained in excellent yield (Scheme 40).[81] Allylic oxidation gives the known intermediate **179**.[83] A total synthesis of (−)-13-oxyingenol (**184**) and its derivative **185** is based on a similar ring-closing metathesis strategy. The key ring-closing metathesis of sterically hindered dienes **180** (R^1 = H, Ac) with Grubbs–Hoveyda second-generation catalyst **15** in the presence of a free allylic alcohol or its acetate affords the products **181** (R^1 = H, Ac) in moderate yields. In comparison, the ring-closing metathesis of diene **182** containing an α,β-unsaturated enone moiety leads to the ring-closure product **183** with significantly increased yield, which is most likely the consequence of a reduced rate of ring-opening in **183** at the newly formed double bond due to its electronically deactivated nature toward metathesis. Via simple functional group manipulations from tricycle **183**, the total synthesis of naturally occurring (−)-13-oxyingenol (**184**) and its derivative **185** is achieved (Scheme 40).[82]

for references see p 672

Scheme 40 Ring-Closing Metathesis Approach to the Synthesis of Ingenols[81,82]

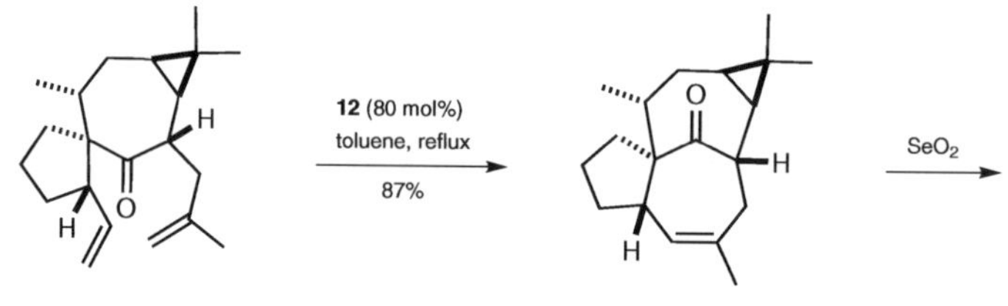

177 178

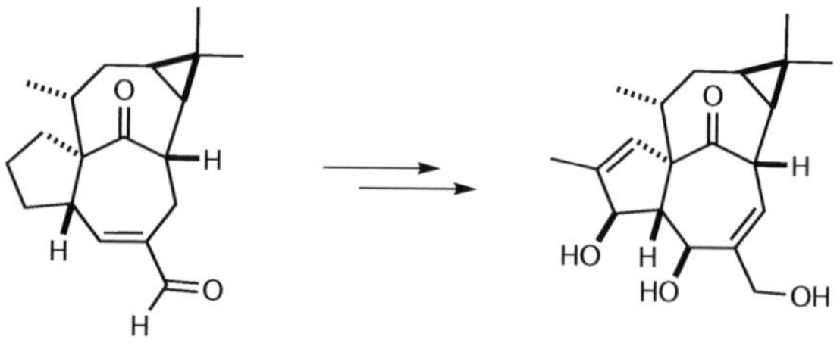

179 176

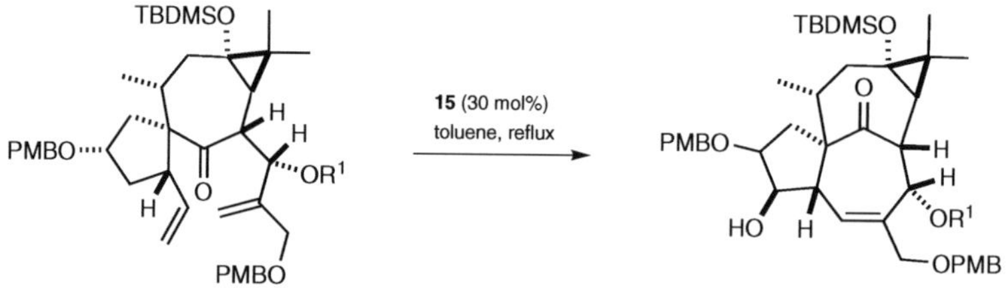

180 181

R^1	Yield (%)	Ref
H	54	[82]
Ac	64	[82]

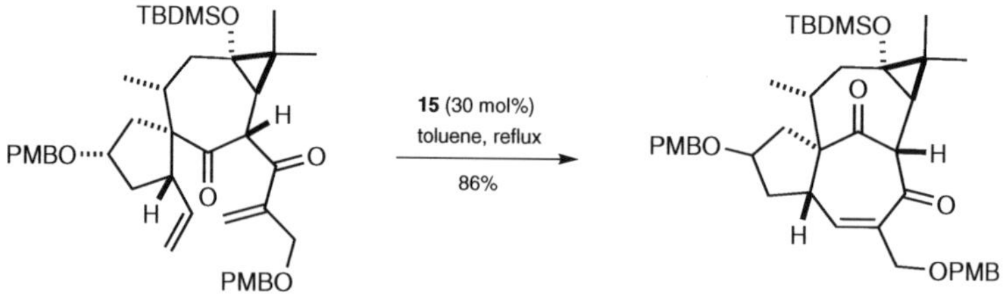

182 183

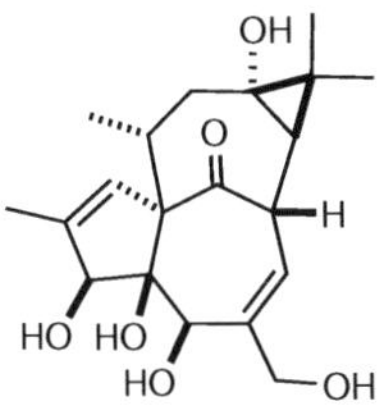

2.10.2 Diene Ring-Closing Metathesis

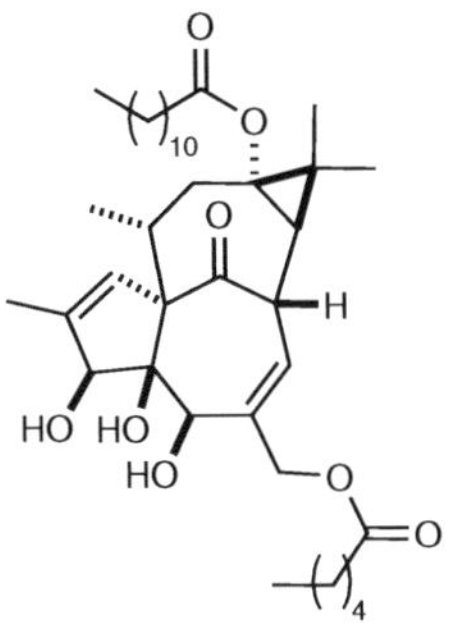

An enantioselective total synthesis of arglabin (**189**) employs a ring-closing metathesis reaction to form a challenging tetrasubstituted double bond (Scheme 41).[84] The seven-membered carbocycle in the advanced synthetic intermediate **188** is established by the ring-closing metathesis of 1,8-diene **187**, which is readily obtained from allylsilane derivative **186**. Although challenging, the ring-closing metathesis of 1,8-diene **187** under argon, purging with three batchwise additions of Grubbs' second-generation catalyst **12** (5 mol%), leads to the formation of tricycle **188** containing a newly formed tetrasubstituted double bond in remarkable 86% yield.

Scheme 41 Ring-Closing Metathesis in the Synthesis of Arglabin[84]

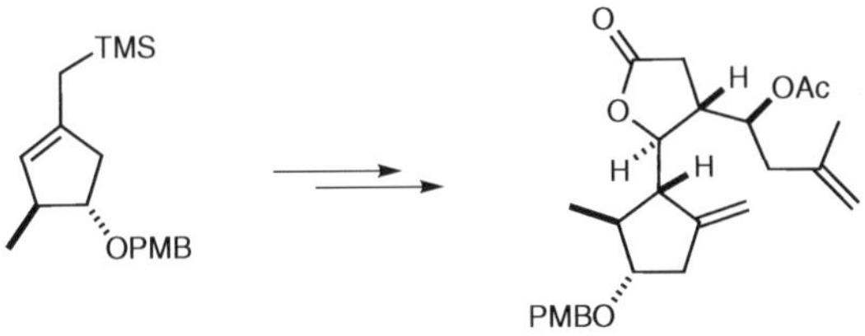

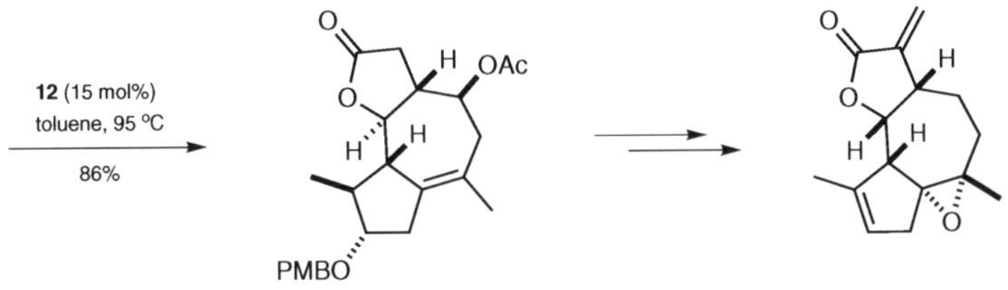

Englerin A (**194**), showing a highly selective activity against renal cancer cell lines, is a synthetically challenging target that contains an oxygen-bridged bicyclic hydroazulene framework (Scheme 42). Ring-closing metathesis based strategies for the construction of seven-membered carbocyclic ring structures have been pursued. The ring-closing metathesis of 1,8-diene **190** with Grubbs' second-generation catalyst **12** (20 mol%) affords bicycle **191** in excellent yield.[85] Another enantioselective approach relies on a similar ring-closing metathesis of 1,8-diene **192**, which affords the bicyclic system **193** in excellent yield in the presence of Grubbs' second-generation catalyst **12** (1 mol%).[86] The advanced intermediates **191** and **193** were elaborated to (+)-englerin A (**194A**) and (−)-englerin A (**194B**), respectively.

for references see p 672

Scheme 42 Ring-Closing Metathesis Reactions in the Synthesis of Englerin A[85,86]

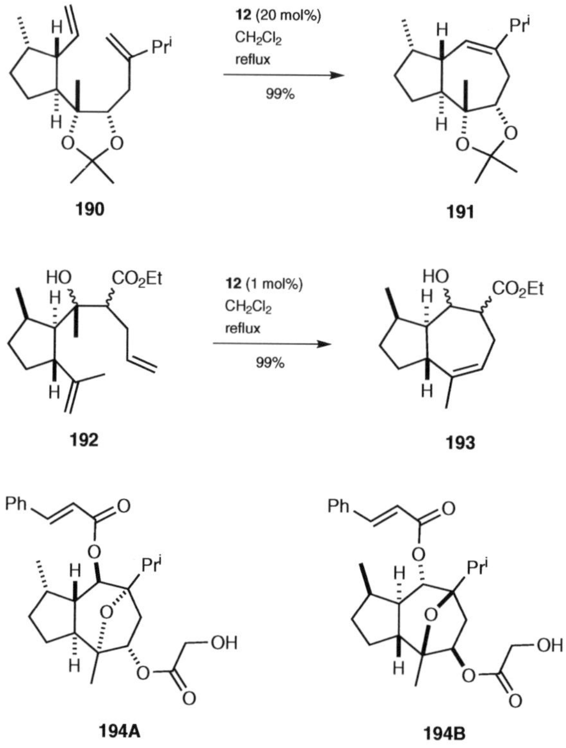

190 → (12 (20 mol%), CH$_2$Cl$_2$, reflux, 99%) → **191**

192 → (12 (1 mol%), CH$_2$Cl$_2$, reflux, 99%) → **193**

194A   **194B**

Two formal syntheses of balanol (**200**) rely on ring-closing metathesis to construct the seven-membered azacycle of the natural product. For the preparation of the known hexahydroazepine intermediate **197**,[87] the seven-membered azacycle **196** is generated in 77% yield by treatment of 1,8-diene **195** with Grubbs' first-generation catalyst **11** (Scheme 43).[88] To reduce the number of operations, a ring-closing metathesis with 1,8-diene **198** containing a free hydroxy group has also been performed (Scheme 43).[89] However, the ring-closing metathesis of diene **198** with various Grubbs-type ruthenium complexes provides only low yields and/or requires longer reaction times. On the other hand, the same ring-closing metathesis with Schrock complex **1** provides the expected ring-closure product **199** in excellent yield (94%).

2.10.2 Diene Ring-Closing Metathesis

Scheme 43 Ring-Closing Metathesis Reactions in a Formal Synthesis of Balanol[88,89]

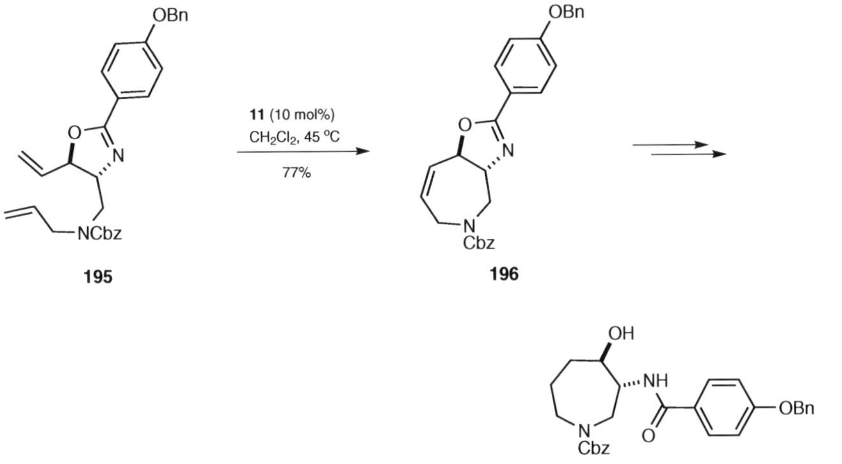

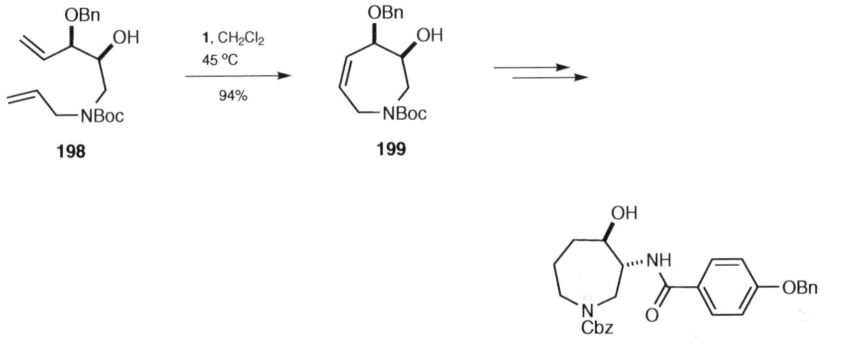

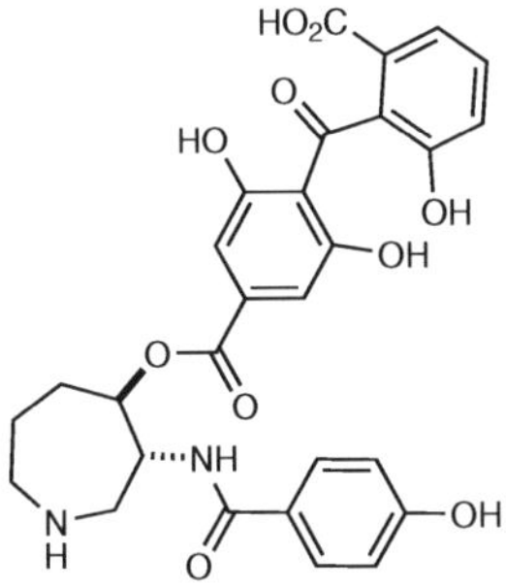

A diene ring-closing metathesis constructs the eight-membered carbocycle in an enantioselective total synthesis of asteriscanolide (**204**) (Scheme 44). Chiral ring-closing metathesis substrate **202** is readily prepared from cyclopentenyl sulfoxide **201**. Due to a relatively rigid conformation of triene **202**, its ring-closing metathesis with Grubbs' first-generation catalyst **11** delivers tricycle **203** in excellent yield, which is subsequently elaborated to (+)-asteriscanolide [(+)-**204**].[90] In a related approach, ring-closing metathesis of the more functionalized substrate **206** derived from precursor **205** forms the tricyclic lactone intermediate **207** in good yield (92%) with Grubbs' first-generation catalyst **11** (50 mol%)

(Scheme 44).[91,92] Hydrogenation of the double bond in ring-closing metathesis product **207** followed by conversion of the silyl ether into the corresponding ketone leads to the total synthesis of racemic asteriscanolide (**204**).

Scheme 44 Ring-Closing Metathesis in the Synthesis of Asteriscanolide[90–92]

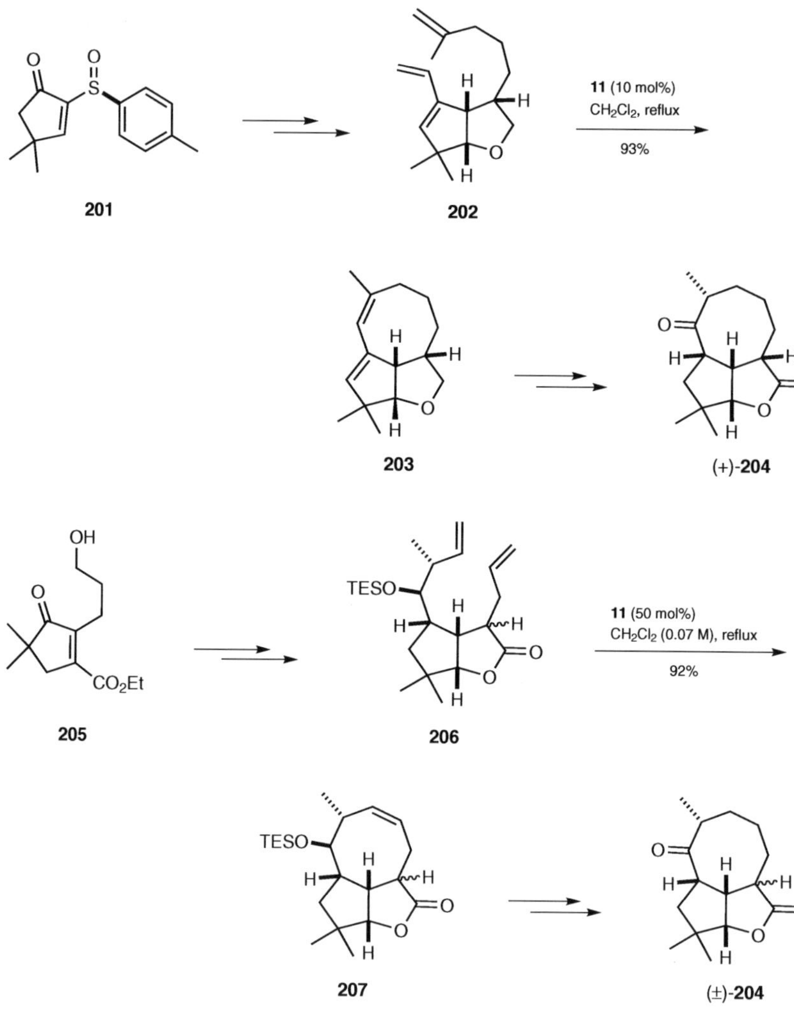

In a total synthesis of (+)-ophiobolin A (**211**), a ring-closing metathesis constructs the central eight-membered carbocyclic B ring (Scheme 45). The ring-closing metathesis of 1,9-diene **208** possessing a bulky silyl protecting group on the A ring leads only to the recovery of starting material along with small amounts of styrene derivatives of **208** under various reaction conditions with Grubbs' second-generation catalyst **12**, ruthenium complex **13**, and Grubbs–Hoveyda second-generation catalyst **15**. The unexpected failure is assumed to be the consequence of the bulky *tert*-butyldiphenylsilyl group at the secondary alcohol, and it is therefore replaced with a benzyl group. Gratifyingly, the ring-closing metathesis of substrate **209** with Grubbs–Hoveyda second-generation catalyst **15** and benzo-1,4-quinone in toluene at 110°C affords metathesis product **210** (not isolated), from which a total synthesis of (+)-ophiobolin A (**211**) is completed.[93]

2.10.2 Diene Ring-Closing Metathesis

Scheme 45 Ring-Closing Metathesis in the Synthesis of (+)-Ophiobolin A[93]

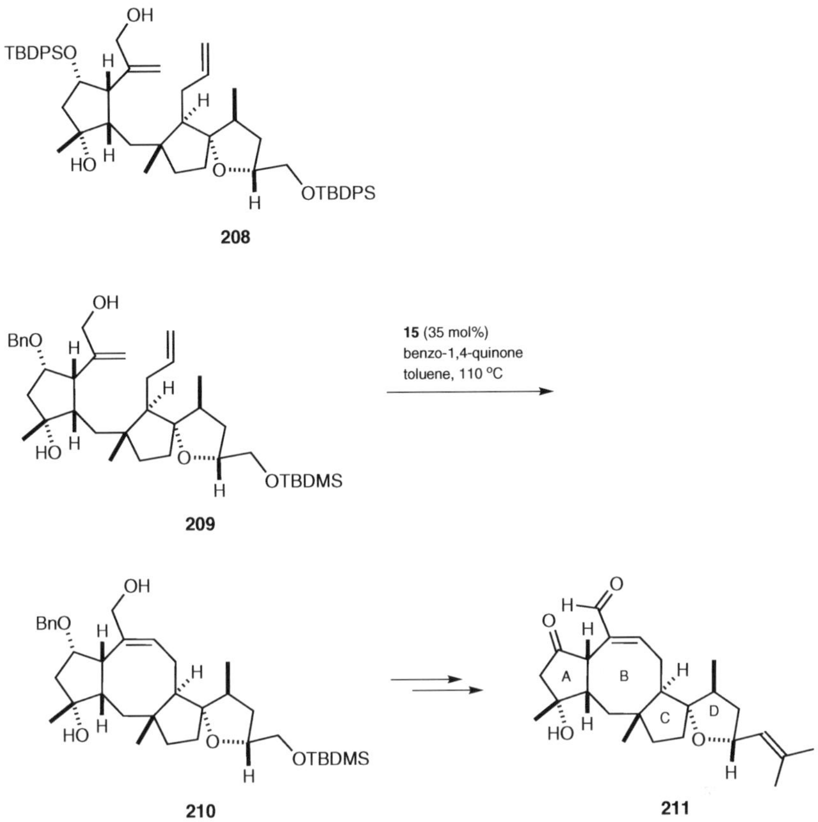

For the ring-closing metathesis of 1,9-dienes to furnish eight-membered-ring products, introducing conformational constraint or pre-organization of the reacting alkenes is a useful strategy. In a recent total synthesis of (−)-nitidasin (**218**), the unexpected participation of the tetrasubstituted double bond of triene **212** in ring-closing metathesis results in the formation of cyclopentene **213** (Scheme 46). On the other hand, the ring-closing metathesis of 1,9-diene **214**, containing an epoxide functionality, provides **215** in excellent yield. Relying on this model study, a more elaborate 1,9-diene **216** is treated under the same conditions, which affords tetracycle **217** in excellent yield, which is elaborated to the target natural product **218**.[94]

Scheme 46 Ring-Closing Metathesis in the Synthesis of (−)-Nitidasin[94]

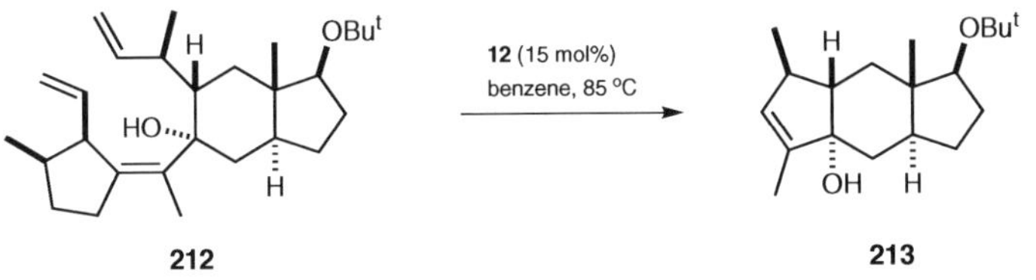

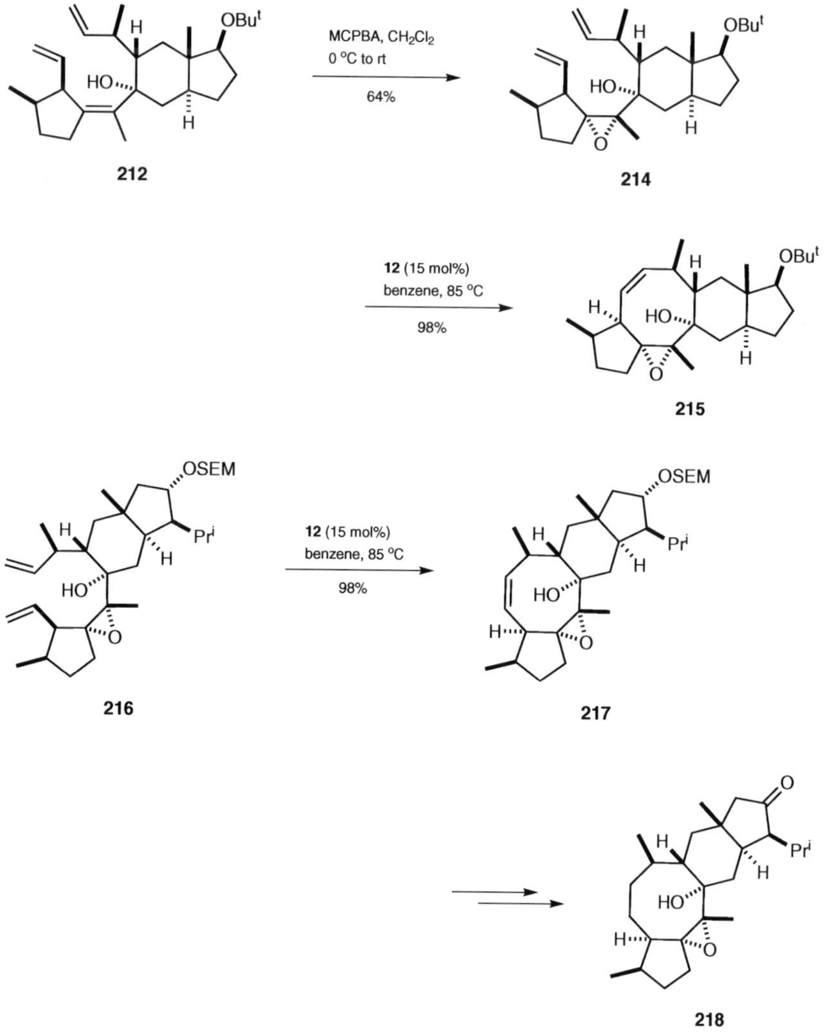

	MCPBA, CH₂Cl₂	
212	0 °C to rt, 64%	**214**

12 (15 mol%)
benzene, 85 °C
98%
→ **215**

216 — 12 (15 mol%), benzene, 85 °C, 98% → **217**

→ **218**

The conformational constraint in the ring-closing metathesis of a 1,9-diene is used for a total synthesis of mycoepoxydiene (**222**) (Scheme 47). The ring-closing metathesis of diene **220** derived from anhydride **219** in the presence of Grubbs' first-generation catalyst **11** delivers oxa-bridged cyclooctene derivative **221** in good yield. A dramatic solvent effect is observed in this ring-closing metathesis. The reaction in dichloromethane results in decomposition of the starting material.[95]

2.10.2 Diene Ring-Closing Metathesis

Scheme 47 Ring-Closing Metathesis in the Synthesis of Mycoepoxydiene[95]

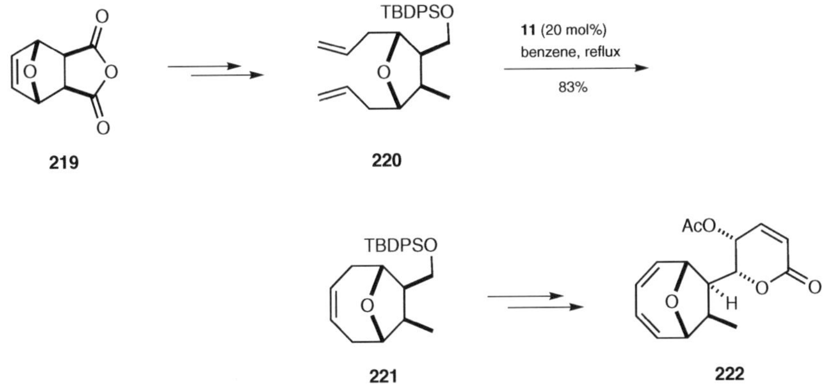

A sequential asymmetric aldol–ring-closing metathesis strategy has been used for the enantioselective construction of seven-, eight-, and nine-membered oxacycles.[96] The ring closures are achieved without conformational constraints except the gauche effect of vicinal dioxy substituents. Relying on this strategy, a formal synthesis of (+)-laurencin (**226**) is achieved via the known intermediate **225** (Scheme 48).[96,97] The ring-closing metathesis of diene **223** using Grubbs' first-generation catalyst **11** delivers oxocin **224** in excellent yield (97%). Subsequent protecting group modifications of **224** give intermediate **225**.[97] A related strategy has also been utilized in a total synthesis of (+)-laurencin (**226**).[98] In another variation of this strategy, the ring-closing metathesis of 1,9-diene **227** derived from (1R,2S)-ephedrine in the presence of Grubbs' second-generation catalyst **12** delivers oxocin **228** in good yield (85%), which is subsequently converted into the known intermediate **229** (Scheme 48).[99,100]

Scheme 48 Ring-Closing Metathesis in the Synthesis of (+)-Laurencin[96,97,99,100]

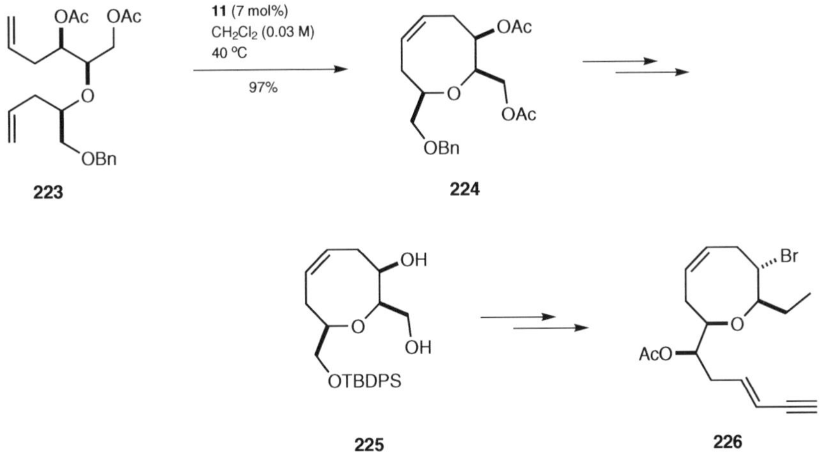

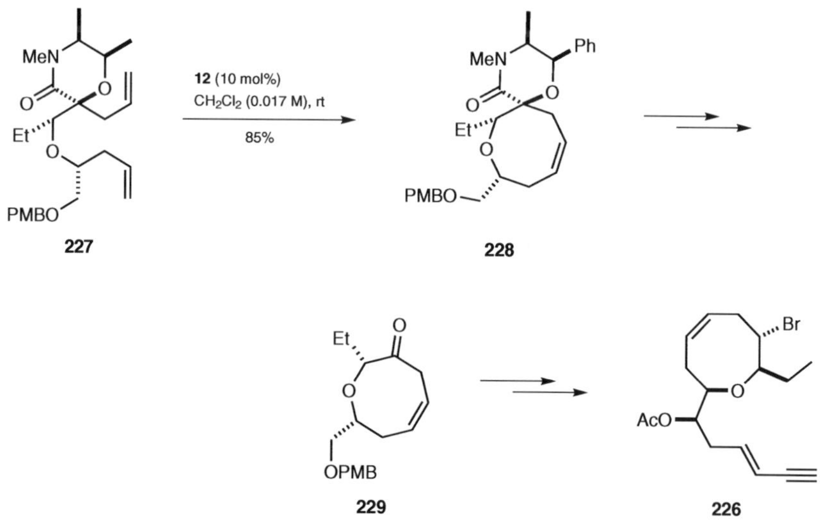

The conformational bias for *E*-selective ring-closing metathesis to form eight-membered silaketals (see also Scheme 9, Section 2.10.2.2.1) is used in a convergent total synthesis of (+)-TMC-151 C (**232**) for polyketide chain elongation. The ring-closing metathesis of silicon-tethered 1,9-diene **230** with Grubbs–Hoveyda second-generation catalyst **15** in the presence of benzo-1,4-quinone provides cyclic product **231** in excellent yield and selectivity (87%, *E/Z* >20:1). Subsequent removal of all silyl groups with hydrogen fluoride–pyridine complex delivers (+)-TMC-151 C (**232**) in moderate yield (Scheme 49).[101]

2.10.2 Diene Ring-Closing Metathesis

Scheme 49 Ring-Closing Metathesis in the Synthesis of (+)-TMC-151 C[101]

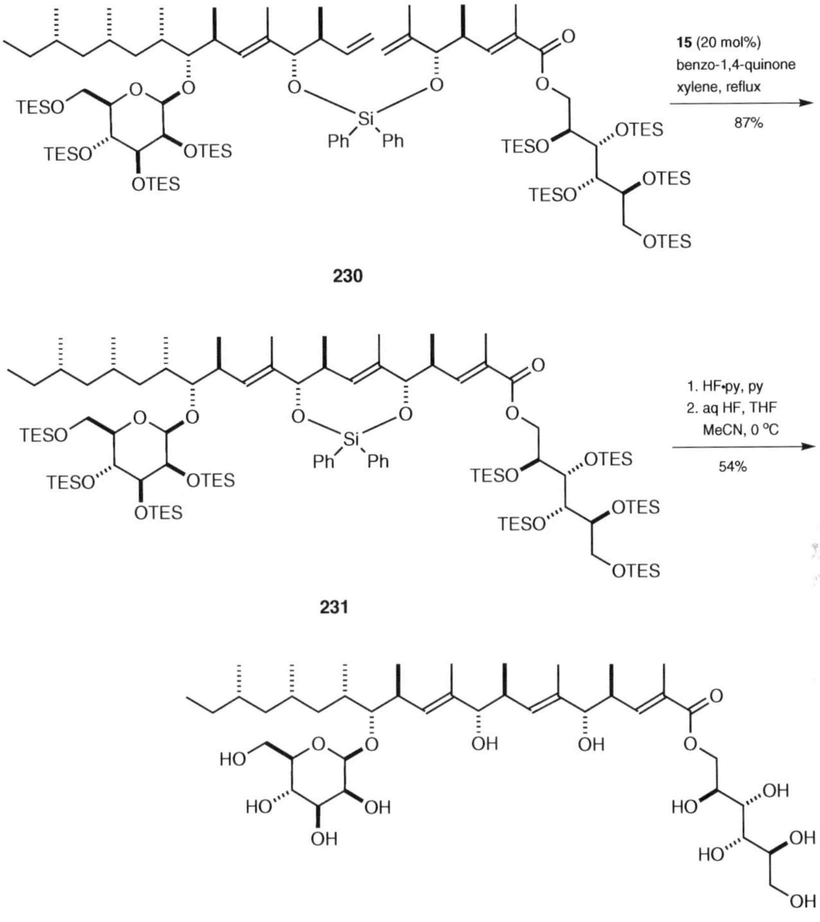

An asymmetric total synthesis of (−)-amphidinolide V (**235**) is accomplished through effective combination of various catalytic transformations (Scheme 50). The ring-closing metathesis of silyl ether **233** in the presence of Grubbs' second-generation catalyst **12** and benzo-1,4-quinone provides eight-membered silyl ether **234** in excellent yield (96%). The group-selective ring-closing metathesis is achieved due to the presence of a trimethylsilyl substituent on the 1,3-diene moiety, which prohibits the participation of this alkene in the metathesis reaction. Further elaboration of silyl ether **234** via rhenium(VII) oxide catalyzed ring-contractive allylic transposition establishes the required 1,5-diene subunit.[102]

Scheme 50 Ring-Closing Metathesis in the Synthesis of (−)-Amphidinolide V[102]

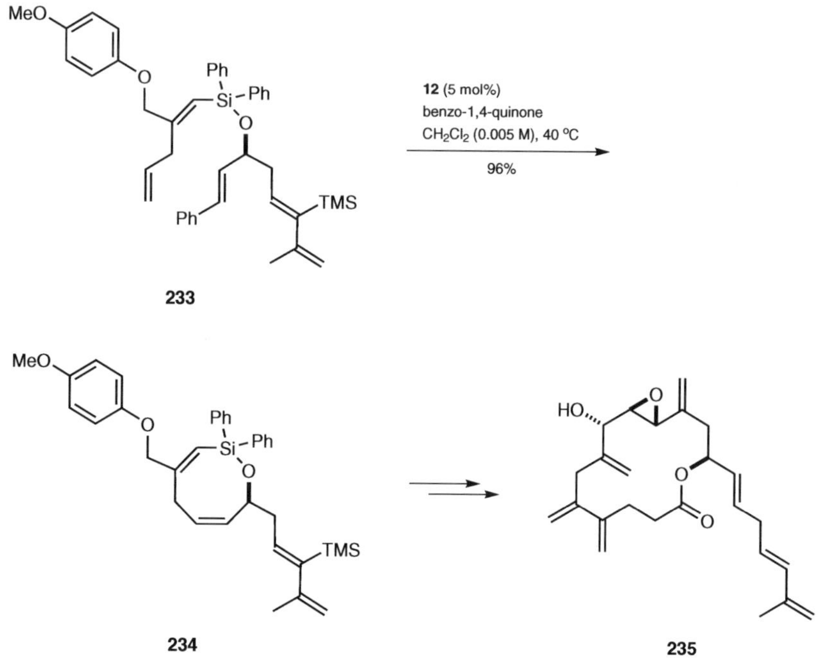

A ring-closing metathesis based strategy to construct a nine-membered oxacycle is employed in the synthesis of ophirin B (**241**) (Scheme 51). The ring-closing metathesis of chiral auxiliary containing diene **236** using Grubbs' first-generation catalyst **11** or Grubbs' second-generation catalyst **12** provides only the dimeric product **237**. Molecular modeling suggests that the dipole-stabilized conformation of the N-acyloxazolidinone moiety of diene **236** should place the two double bonds in unfavorable orientations for ring closure. Reductive removal of the auxiliary leads to alcohol **238**, and its ring-closing metathesis in the presence of Grubbs' second-generation catalyst **12** affords nine-membered heterocycle **240** and the dimer **239** in a 3:1 ratio. Further improvement of the ring-closing metathesis by increasing the reaction temperature to 80 °C in benzene gives a 15:1 ratio of the products **240** and **239** in 89% yield. Under these conditions, it is believed that the initially formed dimer **239** reenters a catalytic cycle involving a cross metathesis–ring-closing metathesis sequence to generate the nine-membered oxacycle **240**.[103]

Scheme 51 Ring-Closing Metathesis in the Synthesis of Ophirin B[103]

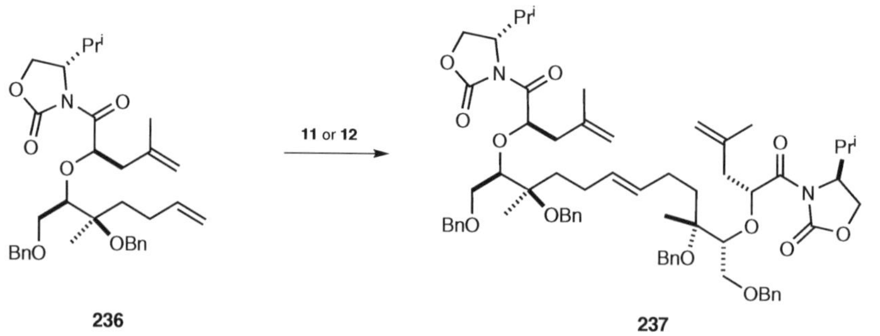

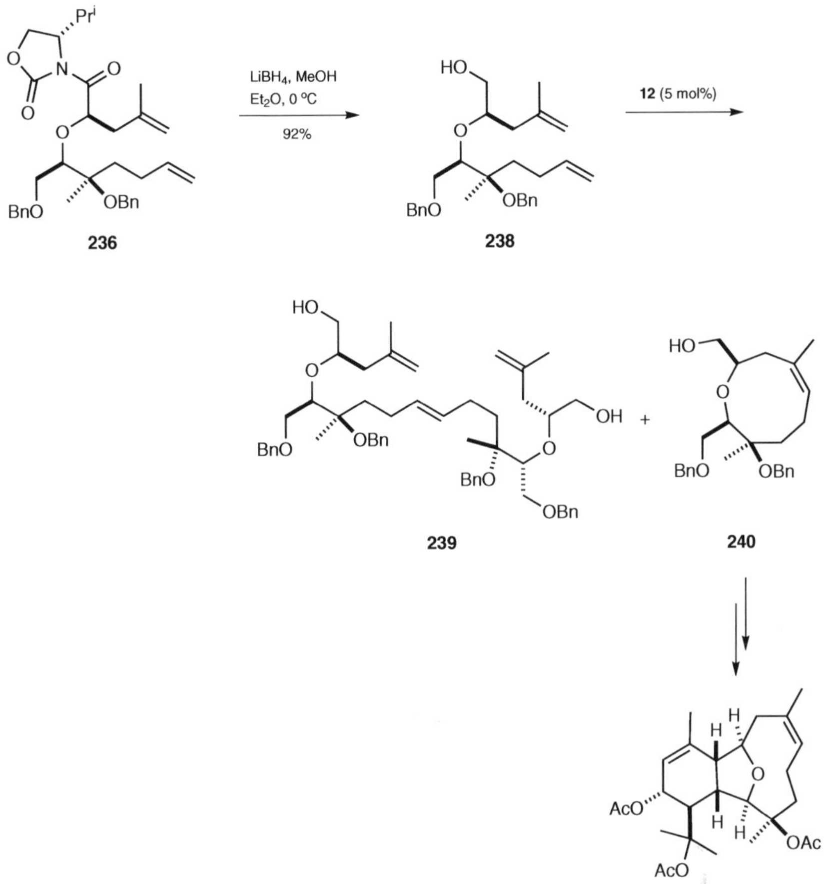

2.10.2 Diene Ring-Closing Metathesis

Conditions	Ratio (239/240)	Yield (%)	Ref
CH$_2$Cl$_2$, 40 °C	1:3	75	[103]
benzene, 80 °C	1:15	89	[103]

In the total synthesis of the enantiomer **244** of clavilactone B, a ring-closing metathesis constructs the ten-membered carbocycle (Scheme 52).[104] By slow addition of Grubbs' second-generation catalyst **12** (40 mol%) over 8 hours to a toluene solution of 1,11-diene **242** and tetrafluorobenzo-1,4-quinone (80 mol%), the ring-closing metathesis product **243** is obtained in moderate yield (65%), which is subsequently oxidized with ammonium cerium(IV) nitrate to provide the enantiomer **244** of clavilactone B. This ring-closing metathesis protocol is applied in a total synthesis of a proposed structure of clavilactone D (Scheme 52).[105] In this synthesis, ring-closing metathesis substrate **247** is prepared by converting the highly substituted benzaldehyde derivative **245** into the epoxy lactone containing alkene **246** followed by Stille coupling with tributyl(2-methylallyl)stannane. The ring-closing metathesis of 1,11 diene **247** generates the ten-membered carbocycle **248**, which is converted into the revised structure of clavilactone D (**249**).

Scheme 52 Ring-Closing Metathesis in the Synthesis of Clavilactones[104,105]

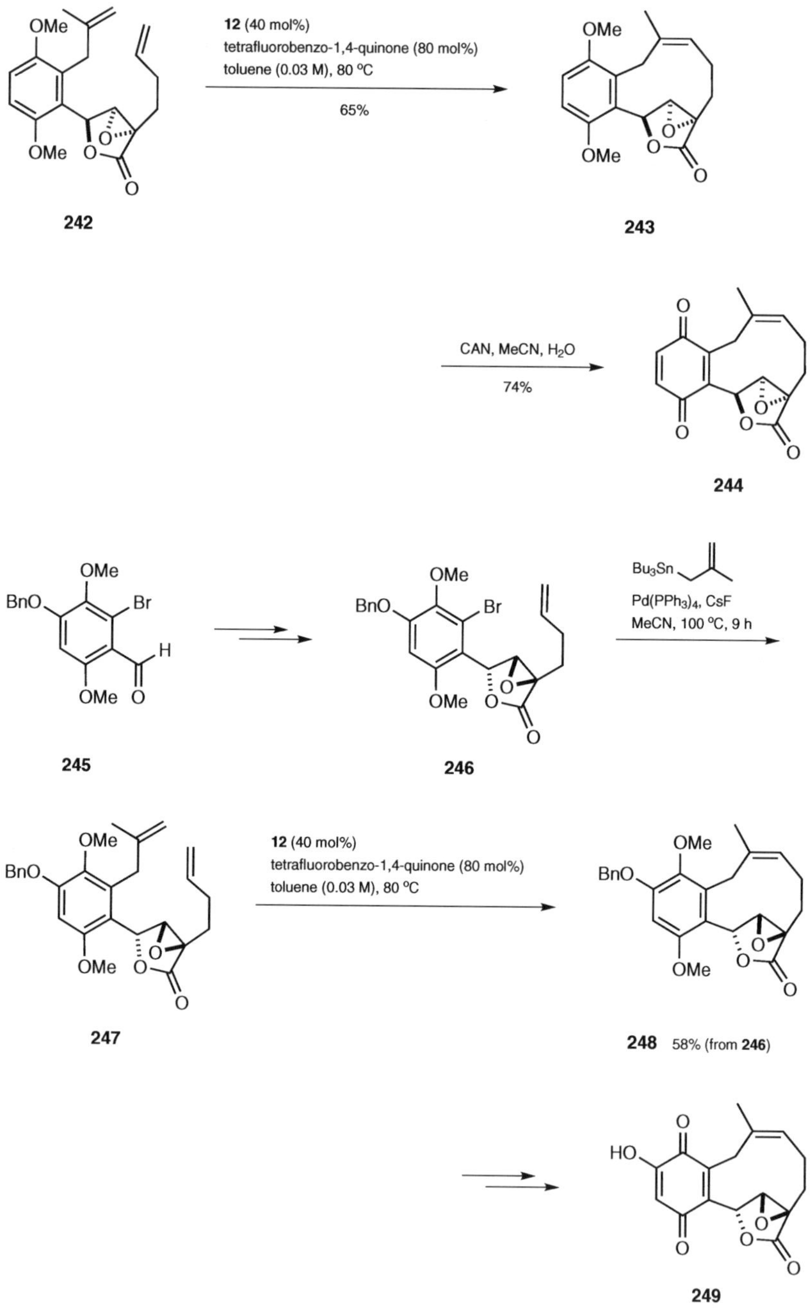

An enantioselective synthesis of clavirolide C (**253**) features a copper-catalyzed asymmetric conjugate addition and a ring-closing metathesis catalyzed by Grubbs–Hoveyda second-generation catalyst **15** as key steps (Scheme 53).[106] The ring-closing metathesis substrate **251** (with 3:2 dr at the OTES group) is prepared via copper-catalyzed asymmetric 1,4-addition with enone **250** followed by an aldol reaction. The ring-closing metathesis of the conformationally biased diene **251** with Grubbs–Hoveyda second-generation catalyst **15** under high-dilution conditions (0.001 M) provides the 11-membered cycloalkene

2.10.2 Diene Ring-Closing Metathesis

252 in good yield (75%) and high *E* selectivity (>95%). Subsequent elaboration of **252** via an eight-step sequence leads to a total synthesis of clavirolide C (**253**).

Scheme 53 Ring-Closing Metathesis in the Synthesis of Clavirolide C[106]

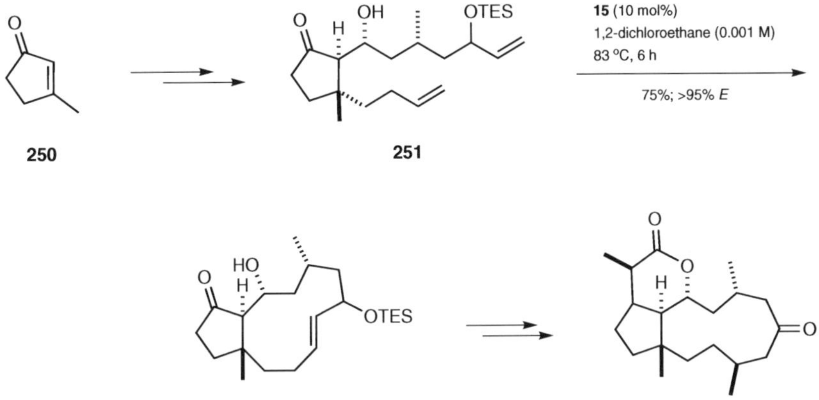

An interesting competition between cross and ring-closing metathesis as well as double-bond migration is observed in a total synthesis of the 11-membered carbocyclic natural product buddledone A (**258**). The ring-closing metathesis of oxo triene **254** delivers only dimer **255** under various conditions (Scheme 54).[107] On the other hand, the ring-closing metathesis of cyanohydrin **256** derived from oxo triene **254** with Grubbs–Hoveyda second-generation catalyst **15** affords only the double bond migrated enone **257** upon removal of the cyanohydrin moiety (Scheme 54). However, under otherwise identical conditions, the addition of benzo-1,4-quinone (20 mol%) to the reaction promotes ring closure to afford buddledone A (**258**) in moderate yield after removal of the cyanohydrin moiety.

Scheme 54 Ring-Closing Metathesis in the Synthesis of Buddledone A[107]

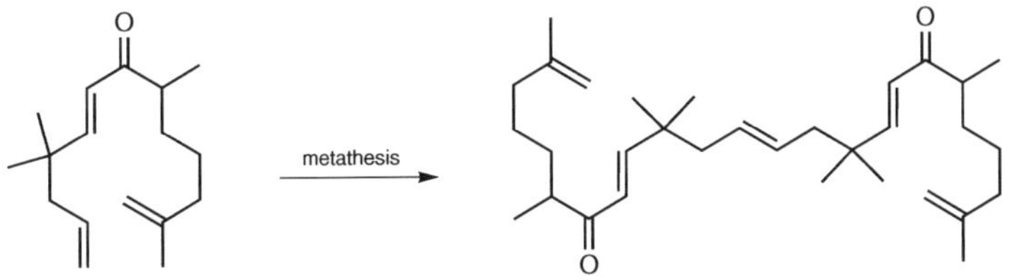

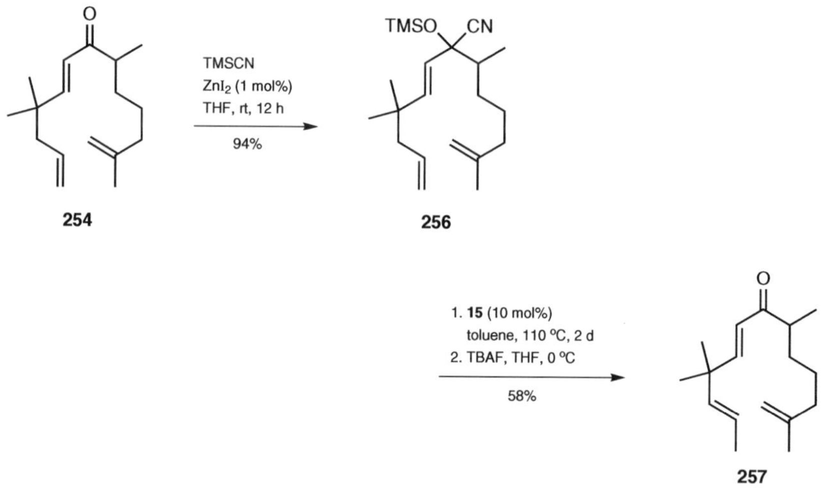

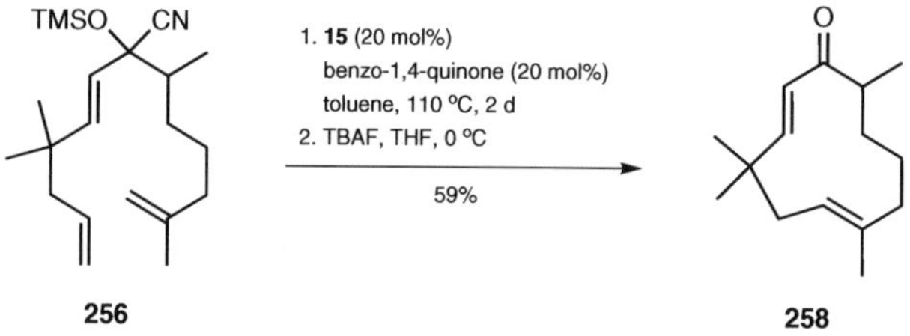

The ring-closing metathesis of triene **259** in the presence of Grubbs' second-generation catalyst **12** constructs the 12-membered macrolide **260** in good yield; however, *cis*/*trans* isomerization is observed on the side chain (Scheme 55). Among several solvents tested, including tetrahydrofuran, benzene, and methanol, high dilution (0.5 mM) dichloromethane at low temperature (0 °C to rt) provides minimal isomerization [ratio (*E*/*Z*) 1:4]. Through a seven-step sequence, macrolide **260** is converted into the target natural product cruentaren A (**261**).[108]

Scheme 55 Ring-Closing Metathesis in the Synthesis of Cruentaren A[108]

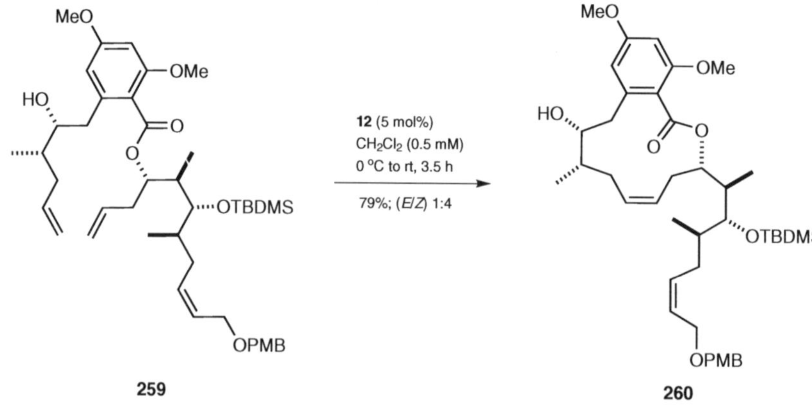

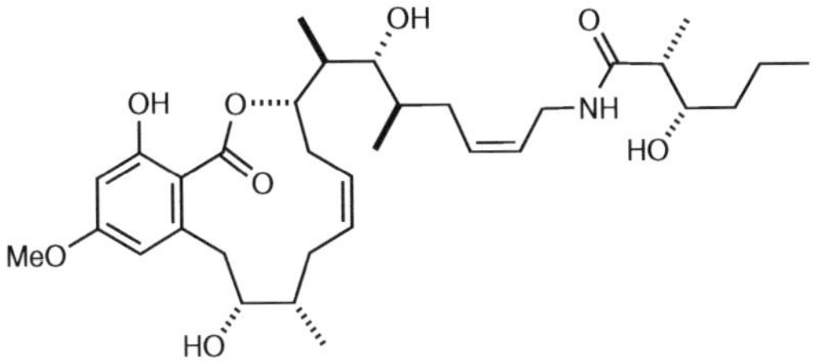

2.10.2 Diene Ring-Closing Metathesis

261

In a total synthesis of the Z-configured natural product (−)-ecklonialactone B (**267**),[109] a Gosteli-type asymmetric Claisen rearrangement followed by diastereoselective ring-closing metathesis forms a five-membered carbocycle (Scheme 56). α-Hydroxy ester **263** is prepared via a copper-catalyzed asymmetric Claisen rearrangement of the Gosteli-type allyl vinyl ether **262** followed by potassium tri-sec-butylborohydride (K-Selectride) reduction. Diastereoselective ring-closing metathesis of triene **263** with Grubbs–Hoveyda second-generation catalyst **15** delivers trans-1,2-disubstituted cyclopentene **264** with excellent diastereoselectivity (dr ≥ 95:5). Multi-step conversion of **264** into diene **265** followed by its ring-closing metathesis with ruthenium complex **13** affords macrolide **266** in good yield but low E/Z selectivity (2:1). Subsequent chemoselective epoxidation of **266** gives the Z-configured (−)-ecklonialactone B (**267**) and further hydrogenation of **267** delivers the nonnatural (+)-9,10-dihydroecklonialactone B (**268**).[110]

Scheme 56 Ring-Closing Metathesis in the Synthesis of (−)-Ecklonialactone B[109,110]

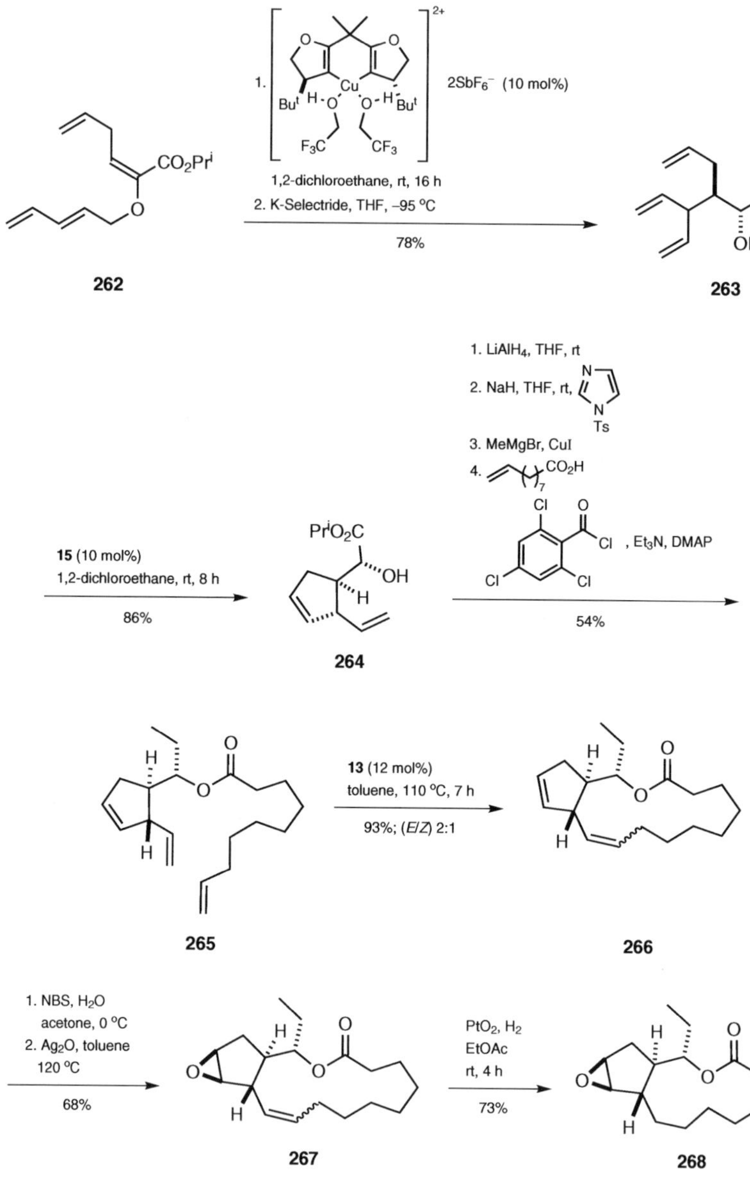

A stereoselective synthesis of E,Z-1,3-dienes by ring-closing metathesis relies on strategic positioning of a silyl group on a diene moiety (Scheme 57). Ring-closing metathesis of ester-tethered triene **269** having a silyl substituent on a disubstituted double bond provides a single regio- and stereoisomer **270** in excellent yield (91%) with ruthenium complex **14** and tricyclohexylphosphine oxide. The trialkyl silyl group in diene **270** functions as a protecting group for the internal alkene against metathesis and, at the same time, it plays an active role as a stereodirecting element in ring-closing metathesis. Treatment of macrolide **270** with tetrabutylammonium fluoride affords E,Z-1,3-diene-containing macrocycle **271** in good yield.[111]

2.10.2 Diene Ring-Closing Metathesis

Scheme 57 Two-Step Ring-Closing Metathesis for the Preparation of E,Z-1,3-Dienes[111]

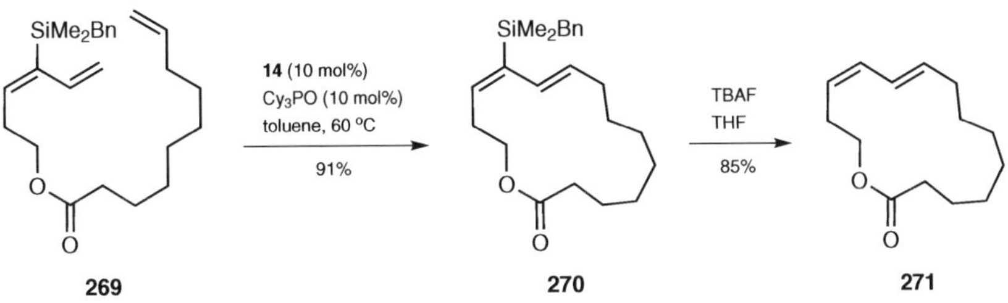

This two-step protocol has been further applied to a formal synthesis of lactimidomycin (**276**) containing an E,Z-1,3-diene moiety (Scheme 58).[111] The ring-closing metathesis of silyl-masked tetraene **272** under the same conditions delivers a mixture of 11-membered macrocycle **273** and 12-membered macrocycle **274** in a 1:1.7 ratio. The formation of 11-membered macrolide **273** is caused by the isomerization of a terminal double bond to an internal one followed by ring-closing metathesis. By increasing the reaction temperature to 120 °C, the formation of **273** decreases [ratio (**273**/**274**) 1:4]. Removal of the silyl and hydroxy protecting groups gives the known synthetic intermediate **275**.[112]

Scheme 58 Two-Step Ring-Closing Metathesis for the Synthesis of E,Z-Dienes and Its Application to the Synthesis of Lactimidomycin[111,112]

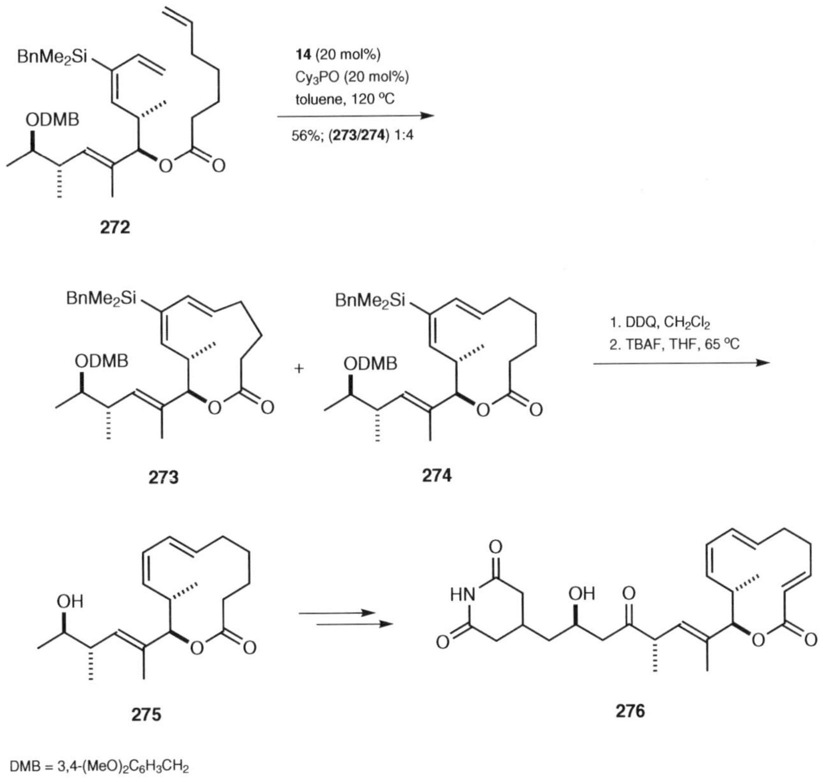

DMB = 3,4-(MeO)$_2$C$_6$H$_3$CH$_2$

A total synthesis of the macrolactone Sch 725 674 (**279**) uses ring-closing metathesis as a key step (Scheme 59). The ring-closing metathesis of acryloyl ester diene **278**, derived from bis Weinreb amide **277**, affords macrolactone **279** upon treatment with Grubbs' second-generation catalyst **12** in 36% yield. The yield of the reaction is not improved by using different catalysts. Furthermore, the ring-closing metathesis with Grubbs–Hoveyda second-generation catalyst **15** gives only an undesired mixture of products.[113]

Scheme 59 Ring-Closing Metathesis in the Synthesis of Sch 725 674 Starting from a Weinreb Amide[113]

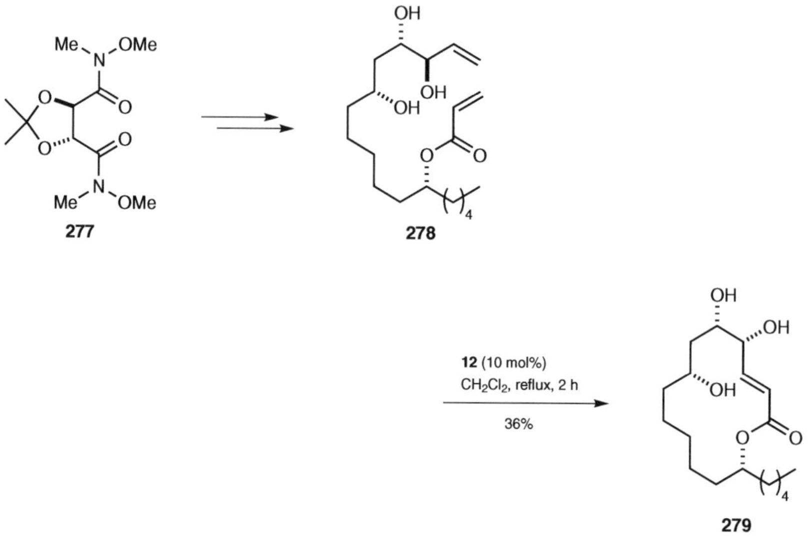

In the total synthesis of 17-deoxyprovidencin (**283**), a ring-closing metathesis forms a macrocyclic Z-alkene which is later isomerized to the corresponding E-alkene by UV light (Scheme 60).[114] The ring-closing metathesis of diene **280** with Grubbs' second-generation catalyst **12** provides Z-macrocycle **281** as a single product in good yield (76%). Acetylation of the hydroxy group, chemoselective epoxidation of the butenolide, and E/Z isomerization of the trisubstituted double bond leads to the 13-membered E-configured macrocycle **282**. Removal of the triisopropylsilyl group and oxidation of the resulting secondary alcohol, epoxidation of the E-double bond, and final conversion of the cyclobutanone moiety into the corresponding *exo*-methylene group affords the natural product **283**.

2.10.2 Diene Ring-Closing Metathesis

Scheme 60 Ring-Closing Metathesis in the Synthesis of 17-Deoxyprovidencin[114]

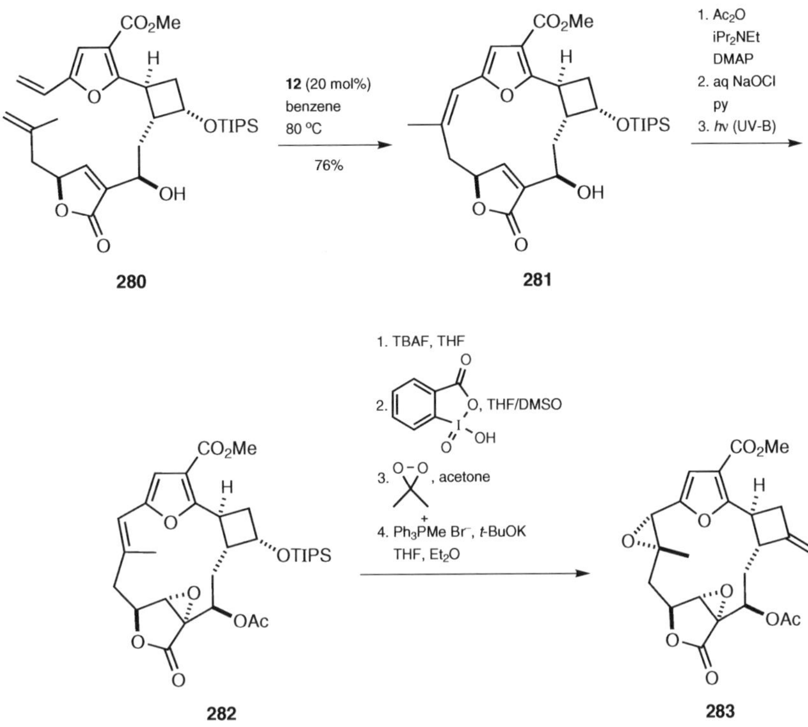

In a total synthesis of (−)-dactylolide (**286**), a key ring-closing metathesis step forms an 18-membered macrocycle (Scheme 61). Ring-closing metathesis of substrate **284** (R^1 = H) is explored under various reaction conditions using Grubbs' first-generation catalyst **11**, Grubbs' second-generation catalyst **12**, and Grubbs–Hoveyda second-generation catalyst **15** to maximize the yield of macrolactone **285**. The best result is obtained with Grubbs–Hoveyda second-generation catalyst **12** in the presence of benzo-1,4-quinone, providing lactone **285** in 45% yield. The substrate **284** (R^1 = CH$_2$OCH$_2$CH=CH$_2$) containing a relay tether does not directly cyclize to form **285**; instead, it is converted rapidly into substrate **284** (R^1 = H) via cleavage of the relay tether.[115] Starting from **285**, (−)-dactylolide (**286**) is obtained via silyl deprotection and oxidation.

Scheme 61 Ring-Closing Metathesis in the Synthesis of (−)-Dactylolide[115]

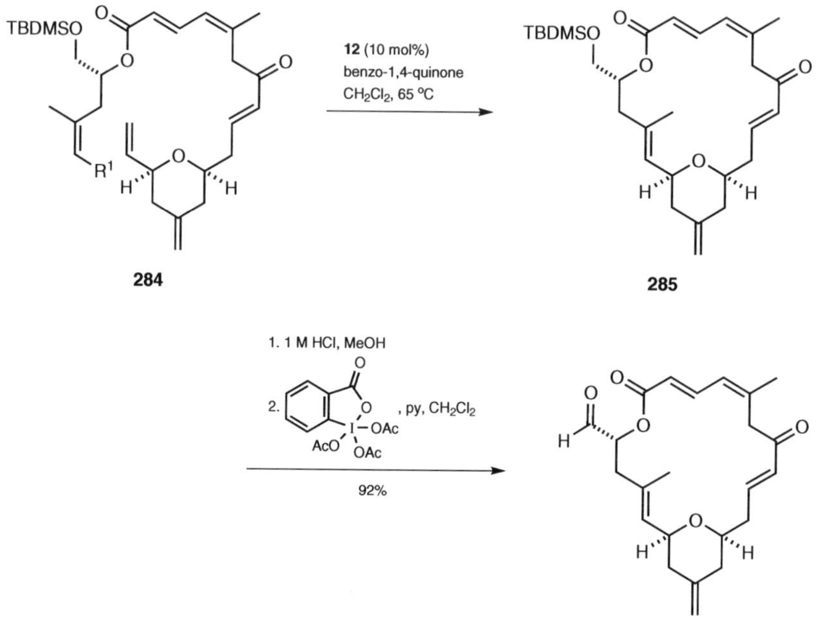

R^1	Yield (%) of **285**	Ref
H	45	[115]
$CH_2OCH_2CH=CH_2$	45	[115]

(+)-Cytotrienin A (**288**), an ansamycin-type antitumor drug containing a 21-membered-ring lactam, is synthesized via ring-closing metathesis to construct the macrolactam (Scheme 62).[116] Similar to the observed ring-closing metathesis of the ansamycin core,[117] the ring-closing metathesis of substrate **287** with ruthenium catalyst **2** delivers the metathesis product in 51% yield based on recovered starting material. Final removal of the triethylsilyl protecting groups with Amberlyst 15 provides (+)-cytotrienin A (**288**) in quantitative yield.

2.10.2 Diene Ring-Closing Metathesis

Scheme 62 Ring-Closing Metathesis in the Synthesis of (+)-Cytotrienin A[116]

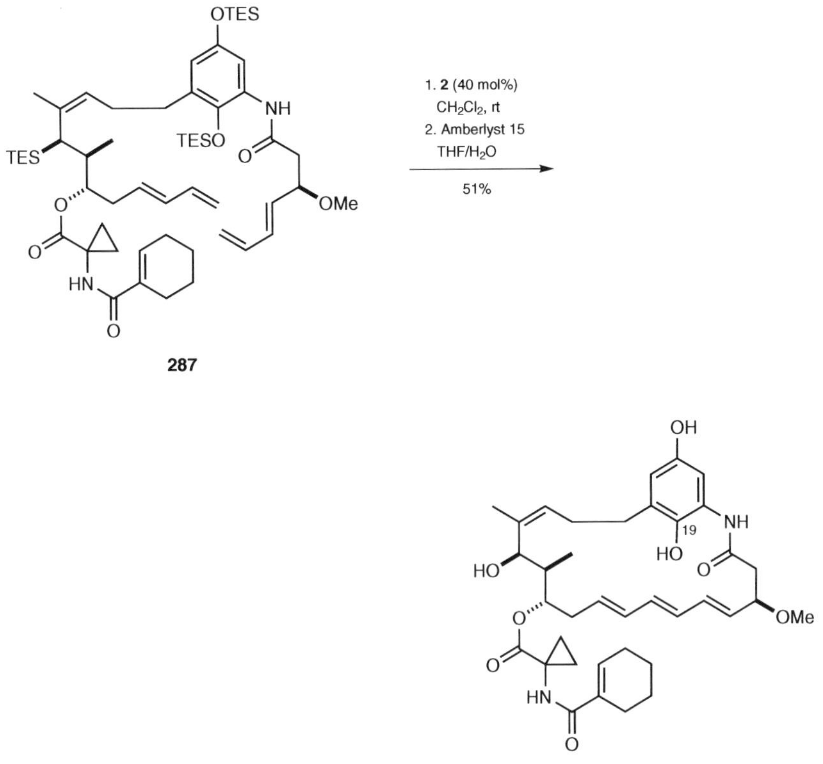

In the ring-closing metathesis based approach to the synthesis of (+)-trienomycin A (**290**, R^1 = Cy) and (+)-trienomycin F [**290**, R^1 = (E)-CMe=CHMe], the ring-closing metathesis of the corresponding substrates **289** with Grubbs' first-generation catalyst **11**, Grubbs' second-generation catalyst **12**, and Grubbs–Hoveyda second-generation catalyst **15** provides only undesired products. On the other hand, the ring-closing metathesis with ruthenium–indenylidene complex **16** selectively produces the E,E,E-triene motif of the macrolactams **290**. Removal of the protecting group from the ring-closing metathesis products provides trienomycins **290** in relatively low overall yields (Scheme 63).[118] The inferior outcome of the ring-closing metathesis is probably the result of the lack of a substituent at C19 as the synthesis of cytotrienins with a C19 substituent suggests that the C19 substituent is a conformation biasing element for an effective ring-closing metathesis.[116,117]

Scheme 63 Ring-Closing Metathesis in the Synthesis of (+)-Trienomycins A and F[118]

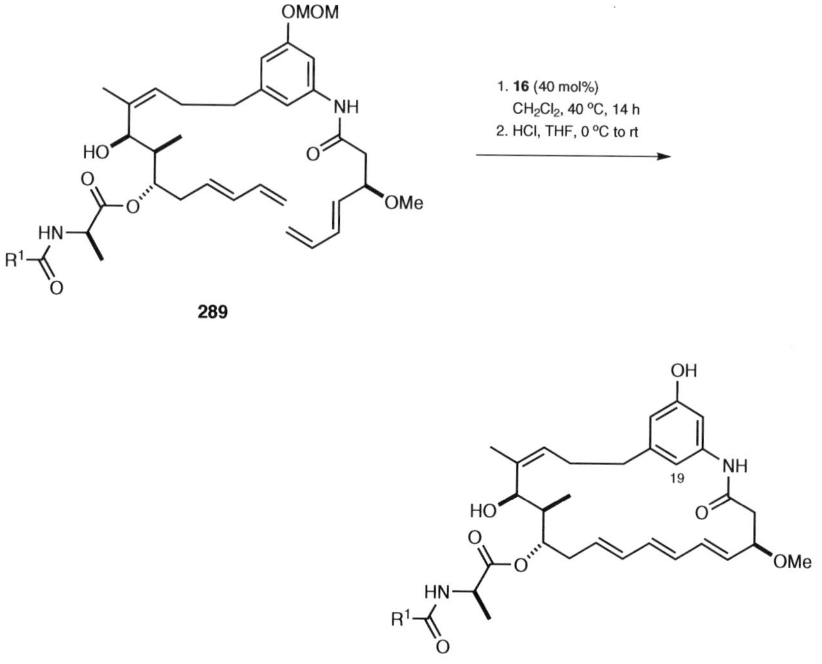

R¹	Yield (%)	Ref
Cy	21	[118]
(E)-CMe=CHMe	24	[118]

In the synthesis of pectenotoxin-2 (**293**), a ring-closing metathesis forms a macrocyclic diene (Scheme 64). The ring-closing metathesis of triene **291** with Grubbs' second-generation catalyst **12** delivers macrocycle **292** in good yield (64%) without isomerization of the spiroacetal moieties. Treatment of macrocycle **292** with trifluoroacetic acid converts the anomeric 1,6-dioxaspirodecane moiety into the nonanomeric spiroketal-containing natural product **293**.[119]

2.10.2 Diene Ring-Closing Metathesis

Scheme 64 Ring-Closing Metathesis in the Synthesis of Pectenotoxin-2[119]

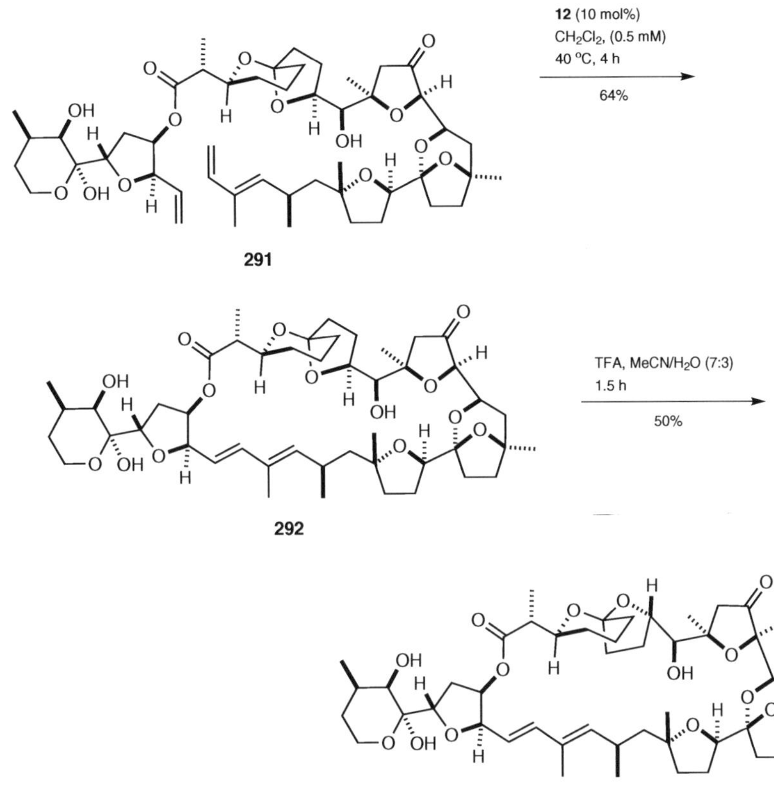

A macrocyclic ring-closing metathesis of pentaene **294** with Grubbs' second-generation catalyst **12** selectively forms a new double bond between the two terminal alkene double bonds to give 26-membered macrocycle **295** as a single *E*-isomer without forming any of the undesired *Z* double bond isomer (Scheme 65). The cleavage of all the *tert*-butyldimethylsilyl groups from **295** with tris(dimethylamino)sulfonium difluorotrimethylsilicate (TASF) completes the synthesis of the natural product amphidinolide B (**296**).[120,121]

Scheme 65 Ring-Closing Metathesis in the Synthesis of Amphidinolide B[120]

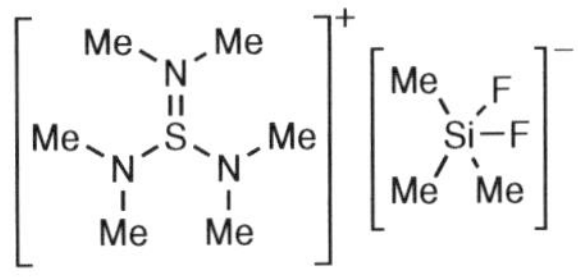

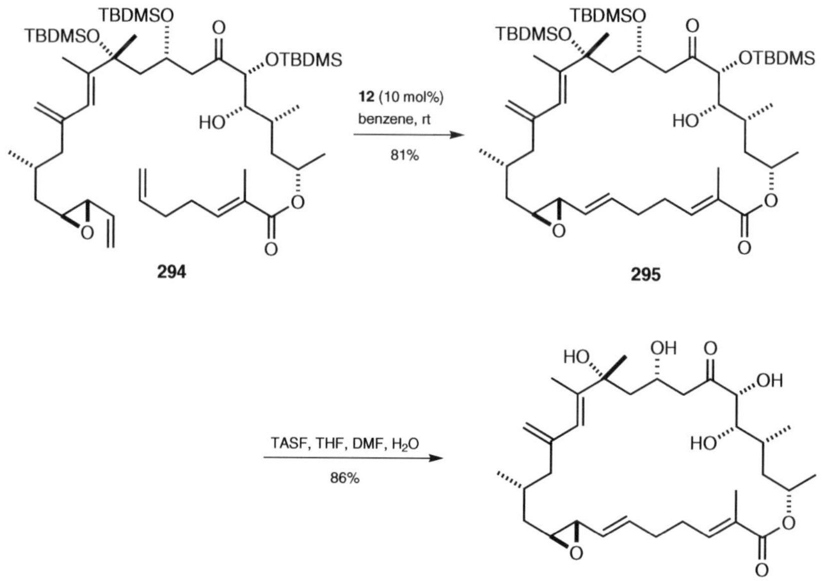

2.10.2.3.2 Ring-Closing Metathesis with Multiple Sequences (Ring Rearrangement)

Ring-rearrangement metathesis (RRM) involves the combination of a sequential ring-opening–ring-closing metathesis process, where the strained cycloalkene moiety transforms into a new cycle (Scheme 66). The ring-rearrangement metathesis process is driven by thermodynamic and kinetic factors such as loss of ring strain, release of a volatile alkene component, and formation of a less-reactive double bond or propagating alkylidene carbene species. The first ring-rearrangement reactions for trienes were shown by Grubbs and coworkers in 1996,[122] and Hoveyda and coworkers in 1997.[123]

Scheme 66 Ring-Rearrangement Metathesis via a Ring-Opening–Ring-Closing Metathesis Sequence

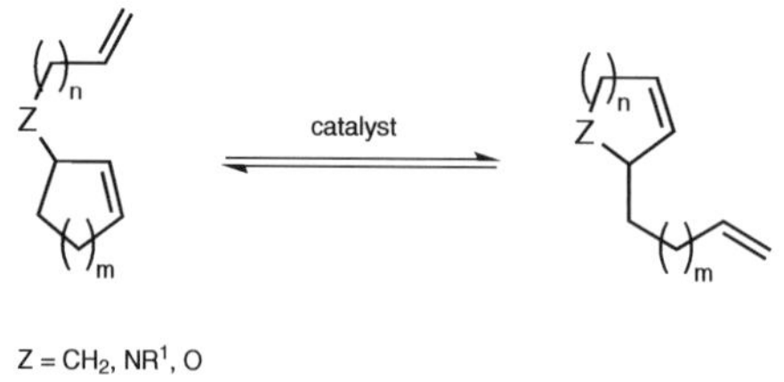

$Z = CH_2, NR^1, O$

An early application of ring-rearrangement metathesis was reported for an enantioselective synthesis of tetraponerines 4, 6, 7, and 8. Tetraponerine 7 (**299**) and tetraponerine 6 (**302**) are synthesized via ring-rearrangement metathesis of cyclopentenyl diamines **297** and **300**, respectively, bearing different side chains in the form of N-homoallyl or N-allyl substituents (Scheme 67). The ring-opening–ring-closing metathesis of dienes **297** and **300** using Grubbs' first-generation catalyst **11** provides ring rearranged diastereomers **298** and **301**, respectively, in excellent yields. The reaction rate is significantly affected by the protecting groups on nitrogen; diamine **297** with benzyloxycarbonyl protecting groups gives 100% conversion whereas the reaction with 2-nitrobenzenesulfonyl protecting groups gives only 66% conversion. One plausible explanation is that the catalyst complexes with the heteroatom functionality to a different extent.[124]

2.10.2 Diene Ring-Closing Metathesis

Scheme 67 Ring-Rearrangement Metathesis in the Synthesis of Tetraponerines[124]

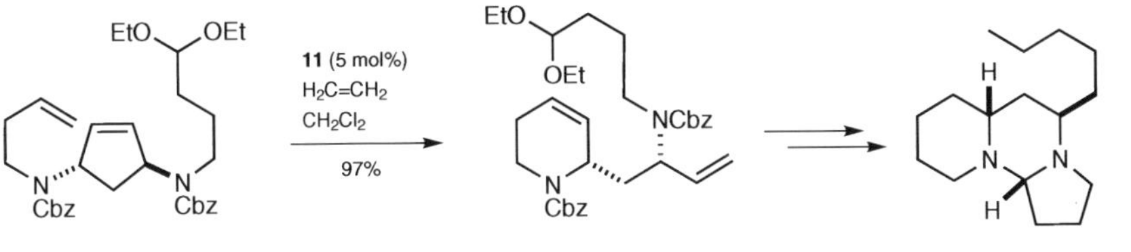

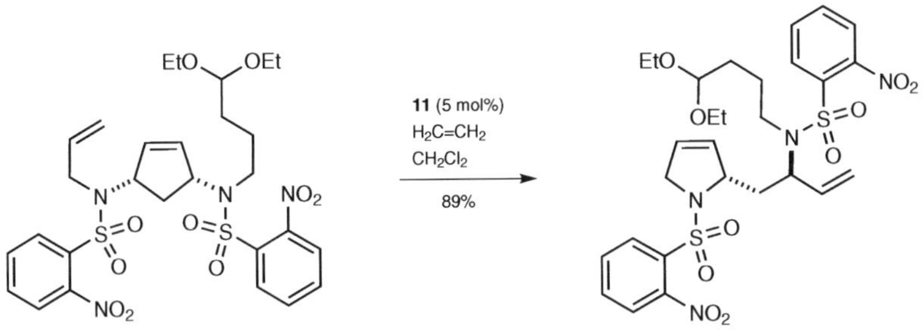

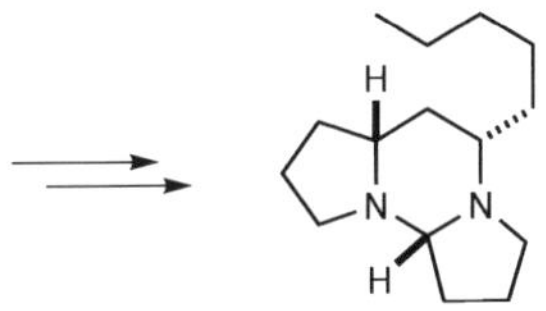

In the synthesis of (−)-swainsonine (**305**), a ring-rearrangement metathesis completely converts diene **303** containing a bulky silyl ether group into the 2,5-dihydro-1H-pyrrole **304** in excellent yield (98%) in the presence of Grubbs' first-generation catalyst **11** (Scheme 68). The rearrangement of the O-benzyl-protected diene gives only an 18:1 mixture of starting material and product. The observed reactivity difference is assumed to be the consequence of the increased population of reactive conformations induced by the bulky *tert*-butyldimethylsilyl group, which thus facilitates the ring opening and also prevents product dimerization.[125]

Scheme 68 Ring-Rearrangement Metathesis in the Synthesis of (−)-Swainsonine[125]

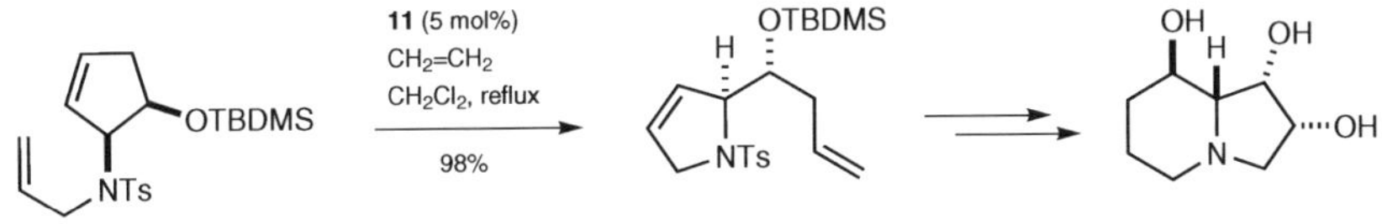

Ring-rearrangement metathesis can also be used to prepare six-membered heterocycles by ring-opening of a cyclohexene moiety (Scheme 69). The ring rearrangement of cyclohexene derivative **306** using Grubbs' first-generation catalyst **11** provides the corresponding heterocycle **307** in excellent yield (96%). Subsequent deprotection of the 2-nitrobenzenesulfonyl group followed by a Negishi coupling delivers the natural product (+)-*trans*-195A (**308**).[126]

for references see p 672

Scheme 69 Ring-Rearrangement Metathesis in the Synthesis of (+)-trans-195A[126]

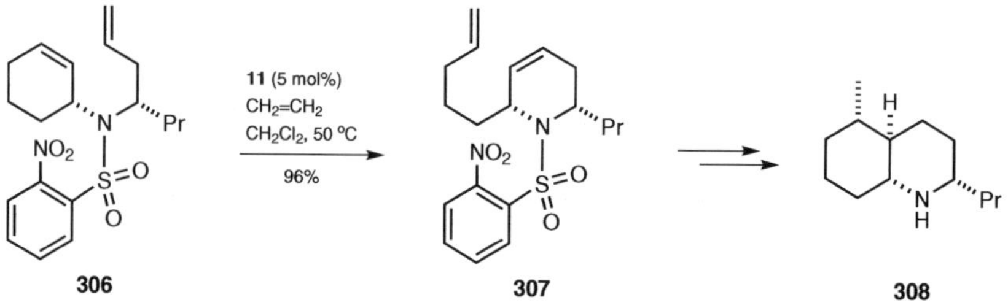

A diastereoselective ring-rearrangement metathesis[127] is utilized in the synthesis of (−)-centrolobine (**312**). Treatment of the cyclopentenol-derived ether **309** with Grubbs' second-generation catalyst **12** leads to the formation of 3,6-dihydropyran **310** as a 4:1 mixture of diastereomers, and the terminal double bond of **310** is isomerized in situ using sodium borohydride to provide 3,6-dihydropyran **311**. Cross metathesis of **311** with 4-hydroxystyrene and subsequent hydrogenation delivers the natural product **312** (Scheme 70).[128]

Scheme 70 Diastereoselective Ring-Rearrangement Metathesis in the Synthesis of (−)-Centrolobine[128]

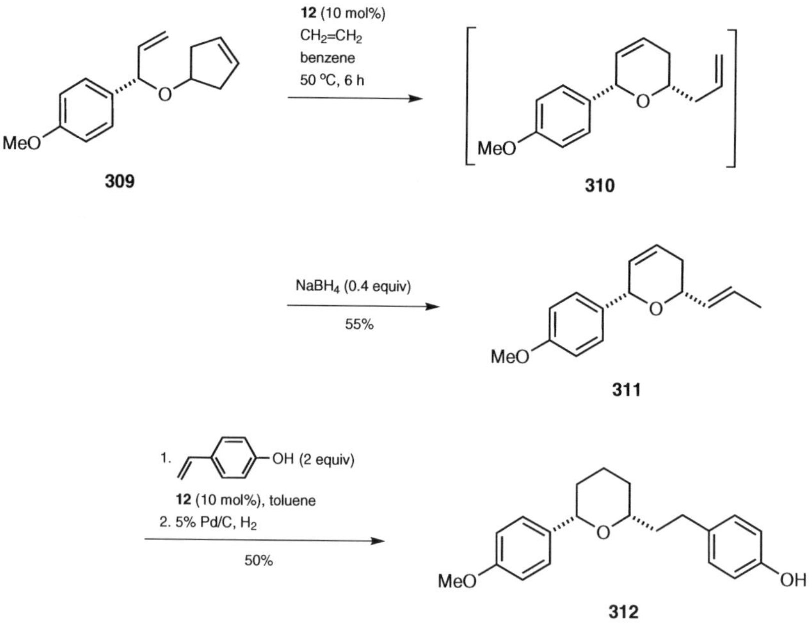

2.10.2 Diene Ring-Closing Metathesis

Ring-rearrangement metathesis of bicyclic compounds, particularly norbornene derivatives, is commonly used for sequential ring-opening–ring-closing metathesis followed by another streamlined metathesis. This ring-opening–ring-closing–cross metathesis based ring-rearrangement metathesis of norbornene derivatives[129] has been amply exploited in natural product synthesis. In a total synthesis of an indolizidine alkaloid, Grubbs' first-generation catalyst **11** effectively catalyzes the ring-rearrangement metathesis of substrate **313**, delivering bicyclo[3.3.0]octene derivative **314** in excellent yield without formation of bicyclo[3.2.1]octene derivative **323** (Scheme 71).[130] The ring-rearrangement metathesis product **314** serves as a pivotal intermediate for the synthesis of indolizidine 251F (**315**).

Deduced from the observed product, the mechanism of this ring-rearrangement metathesis involves an initial attack of the ruthenium–carbene complex on the more strained double bond within the norbornene skeleton in two different orientations to form metallacyclobutanes **316** and **320**. Their cycloreversion would lead to intermediates **317** and **321**, and they may undergo alkylidene exchange through the intermediacy of **319**. Finally, the ring-closing metathesis of **317** with the enone moiety through metallacyclobutane intermediate **318** would deliver bicycle **314**. Similarly, the regioisomeric alkylidene **321** would provide bicycle **323** via the intermediacy of metallacyclobutane **322**. The selective formation of bicyclo[3.3.0]octene derivative **314** is most likely the consequence of a kinetically favorable ring closure of **317** to **318** over that of **321** to **322** while **317** and **321** undergo rapid alkylidene exchange.

Scheme 71 Mechanism of Ring-Rearrangement Metathesis of the Norbornene Skeleton[130]

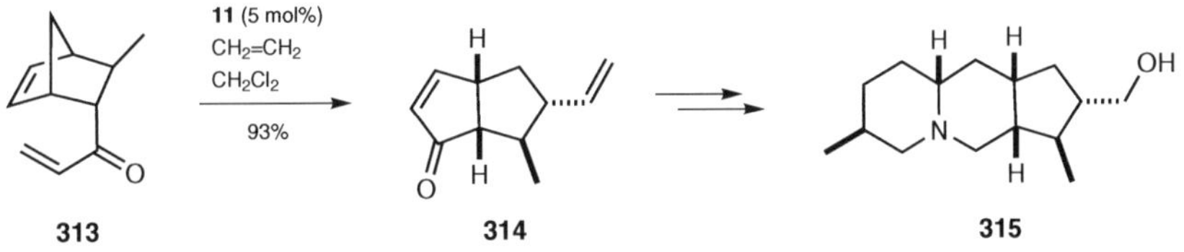

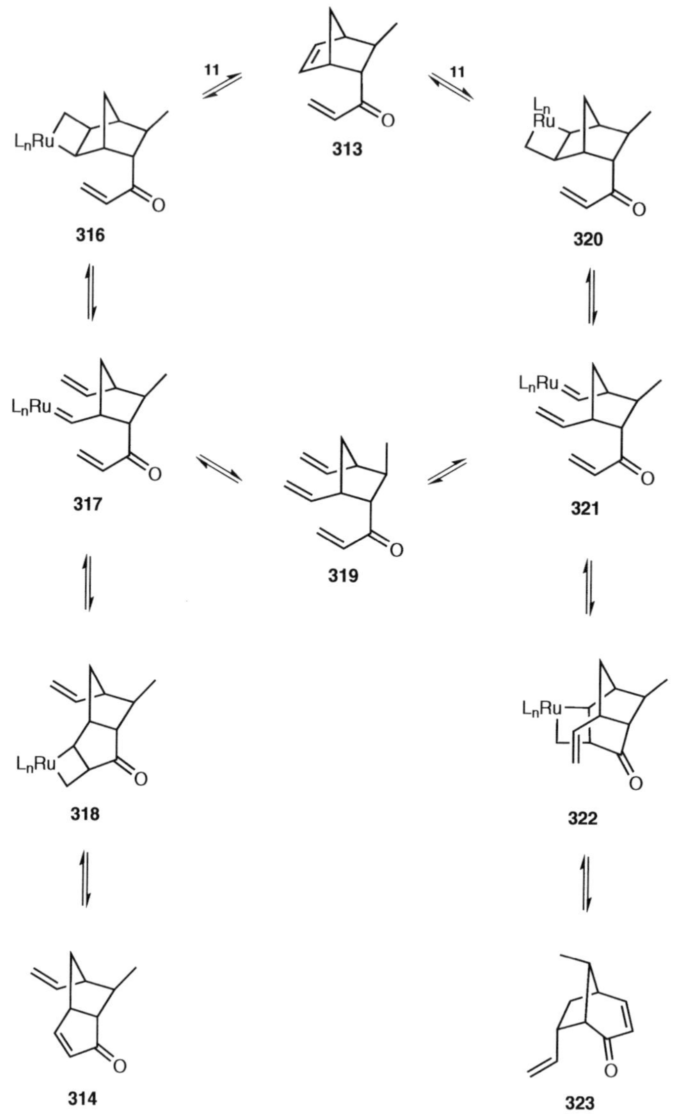

In the total synthesis of aburatubolactam A (**326**), a ring-rearrangement metathesis constructs the bicyclo[3.3.0]octane skeleton (Scheme 72).[131] Ring-rearrangement metathesis of norbornene **324** with Grubbs' first-generation catalyst **11** efficiently produces bicyclic enone **325**. This ring-rearrangement metathesis strategy has been further extended to a total synthesis of cyanthiwigin U (**329**). The ring-rearrangement metathesis of triene **327** with Grubbs' second-generation catalyst **12** provides tricyclic product **328** (45% yield over three steps from the corresponding dialdehyde precursor), which is further converted into the natural product **329** with 12 steps overall and 17% total yield.[132]

2.10.2 Diene Ring-Closing Metathesis

Scheme 72 Ring-Rearrangement Metathesis with Norbornene and Related Structures toward the Synthesis of Aburatubolactam A and Cyanthiwigin U[131,132]

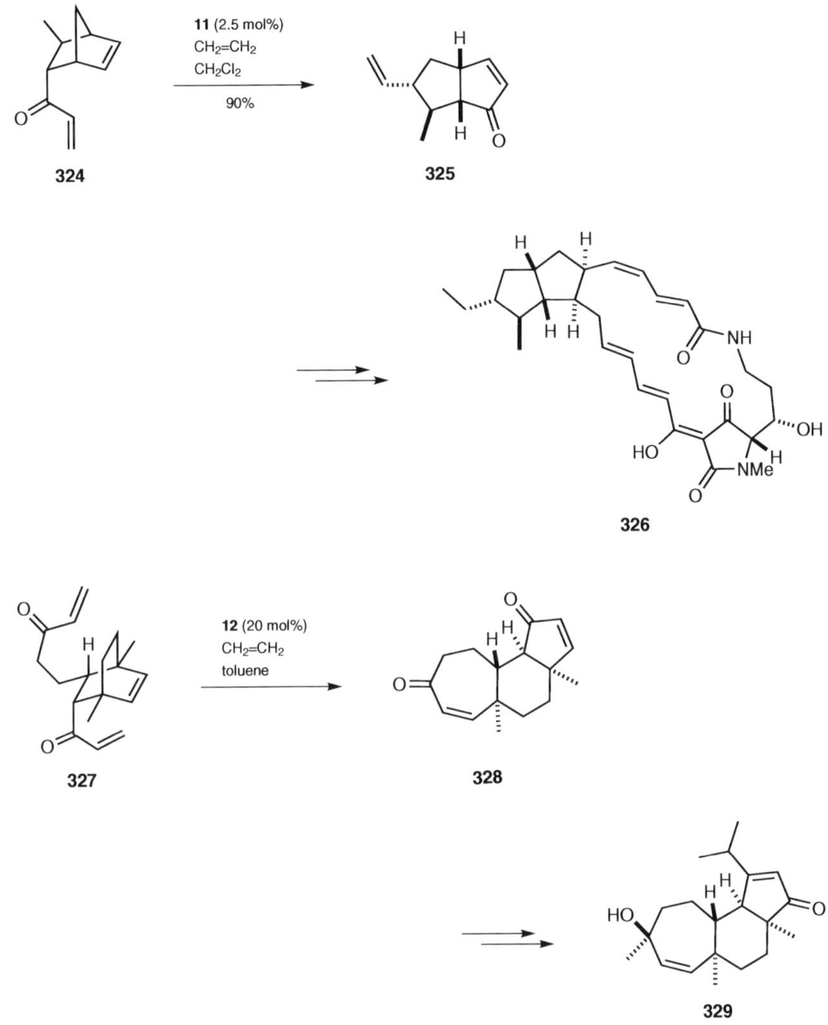

In a total synthesis of (−)-isoschizogamine (**332**), a tandem ring-rearrangement metathesis is used as a key step (Scheme 73).[133] With Grubbs–Hoveyda second-generation catalyst **15** in benzene, the ring-rearrangement metathesis of norbornene derivative **330** delivers bicyclic lactone **331** in only 24% yield. However, by using the more reactive Grubbs' third-generation catalyst **13** in the presence of hepta-1,6-diene, the efficiency of the ring-rearrangement metathesis is significantly increased, delivering the product **331** in up to 73%.

Scheme 73 Ring-Rearrangement Metathesis of a Norborene Derivative To Give a Bicyclic Lactone for the Synthesis of (−)-Isoschizogamine[133]

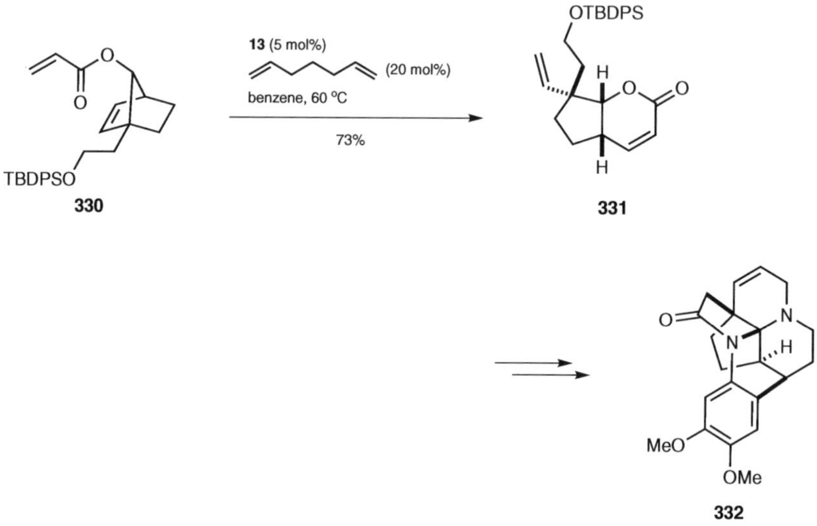

A diastereoselective ring-rearrangement metathesis protocol sets the stereochemistry of an all-carbon quaternary center on a cyclohexene skeleton (Scheme 74).[134] In the presence of Grubbs' second-generation catalyst **12**, cyclopentene-based diene **333** possessing a secondary silyl ether moiety next to a quaternary center rearranges efficiently to give cyclohexene derivative **336** with excellent diastereoselectivity. The stereochemical outcome of the ring-rearrangement metathesis process can be justified by favorable formation of *exo*-intermediate **334** over *endo*-intermediate **335**, where the role of *tert*-butyldimethylsilyl protection is critical to achieve high diastereoselectivity. Through a seven-step manipulation, cyclohexene derivative **336** is converted into the natural product (±)-nitramine (**337**).

Scheme 74 Ring-Rearrangement Metathesis To Set the Stereochemistry of an all-Carbon Quaternary Center in the Synthesis of (±)-Nitramine[134]

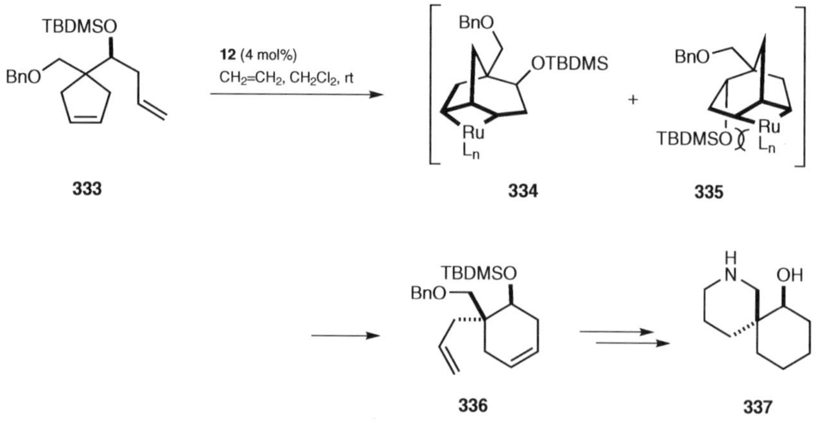

The ring-closing metathesis approach to clavilactones (see Scheme 52) has been further elaborated for the synthesis of clavilactones A (**341**) and B (**244**), where the butenolide moiety is constructed via a ring-rearrangement metathesis, setting the stage for a diene

2.10.2 Diene Ring-Closing Metathesis

ring-closing metathesis to form the required macrocycle at a later stage (Scheme 75).[135] The ring-rearrangement metathesis of cyclobutene carboxylate substrate **338** under various conditions with Grubbs' first-generation catalyst **11** or Grubbs' second-generation catalyst **12** leads to the formation of furanone **340** along with a significant amount of dimer **339**.[135] However, running the metathesis with Grubbs' first-generation catalyst **11** in the presence of 2,6-dichlorobenzo-1,4-quinone followed by treating the reaction mixture with Grubbs' second-generation catalyst **12** under ethene produces furanone **340** in good yield. This product is further converted into natural products **244** and **341** via Stille coupling followed by diene ring-closing metathesis.[135]

Scheme 75 Synthesis of the Butenolide Moiety by Ring-Rearrangement Metathesis in the Synthesis of Clavilactones A and B[135]

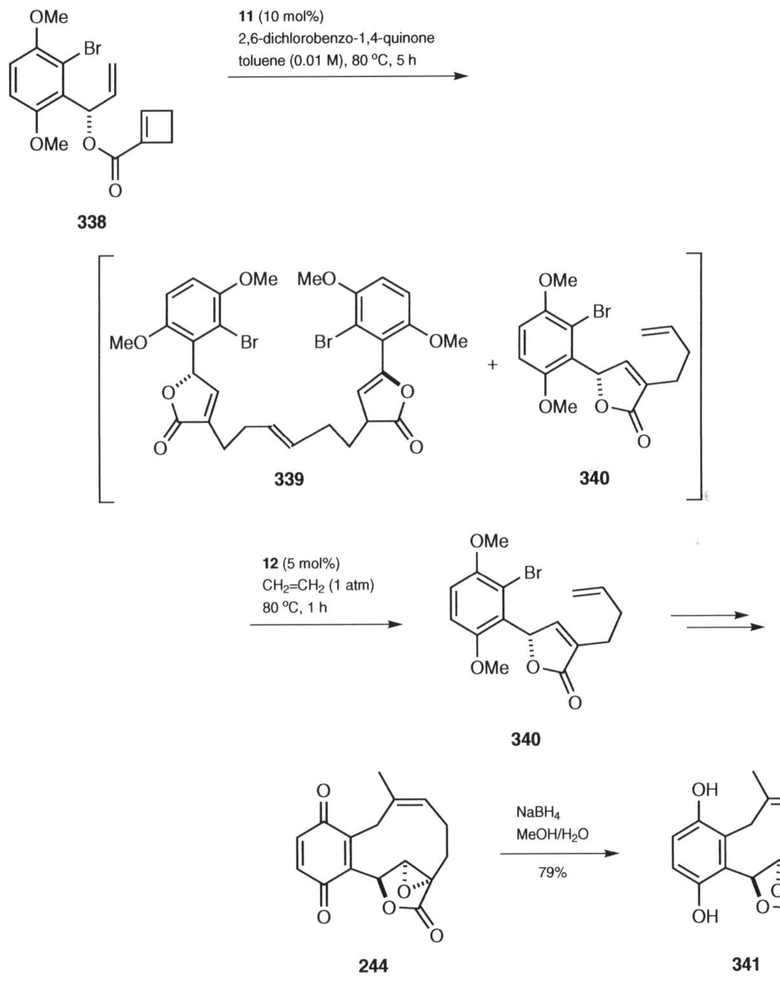

Ring-rearrangement metathesis has been exploited in the collective synthesis of the humulanolides (−)-asteriscunolide D (**345**), (−)-6,7,9,10-tetradehydroasteriscunolide (**347**), (+)-asteriscanolide (**348**), 6,7,9,10-tetradehydroasteriscanolide (**349**), (−)-asteriscunolide A (**350**), (−)-asteriscunolide B (**351**), and (−)-asteriscunolide C (**352**) (Scheme 76). The ring-rearrangement metathesis of cyclobutene carboxylate derivative **342**, synthesized from D-(−)-pantolactone, with Grubbs–Hoveyda second-generation complex **15** in toluene under reflux affords (−)-asteriscunolide D (**345**) in 36% yield (1 g), along with dimer **346**

and several unidentified byproducts. The formed ring-closing metathesis product **345** is further converted into other natural products of the same family in one- or two-step processes. In terms of mechanism, the ring-rearrangement metathesis proceeds through the formation of propagating ruthenium–alkylidene species **343** derived from an alkylidene exchange with the terminal double bond of substrate **342**. Subsequent intramolecular ring-opening metathesis of the more strained cyclobutene double bond leads to a new propagating ruthenium–alkylidene species **344**, and ring-closing metathesis with the enone moiety provides 11-membered-ring natural product (−)-asteriscunolide D (**345**).[136]

Scheme 76 Ring-Rearrangement Metathesis of a Cyclobutene Carboxylate in the Synthesis of Humulanolides[136]

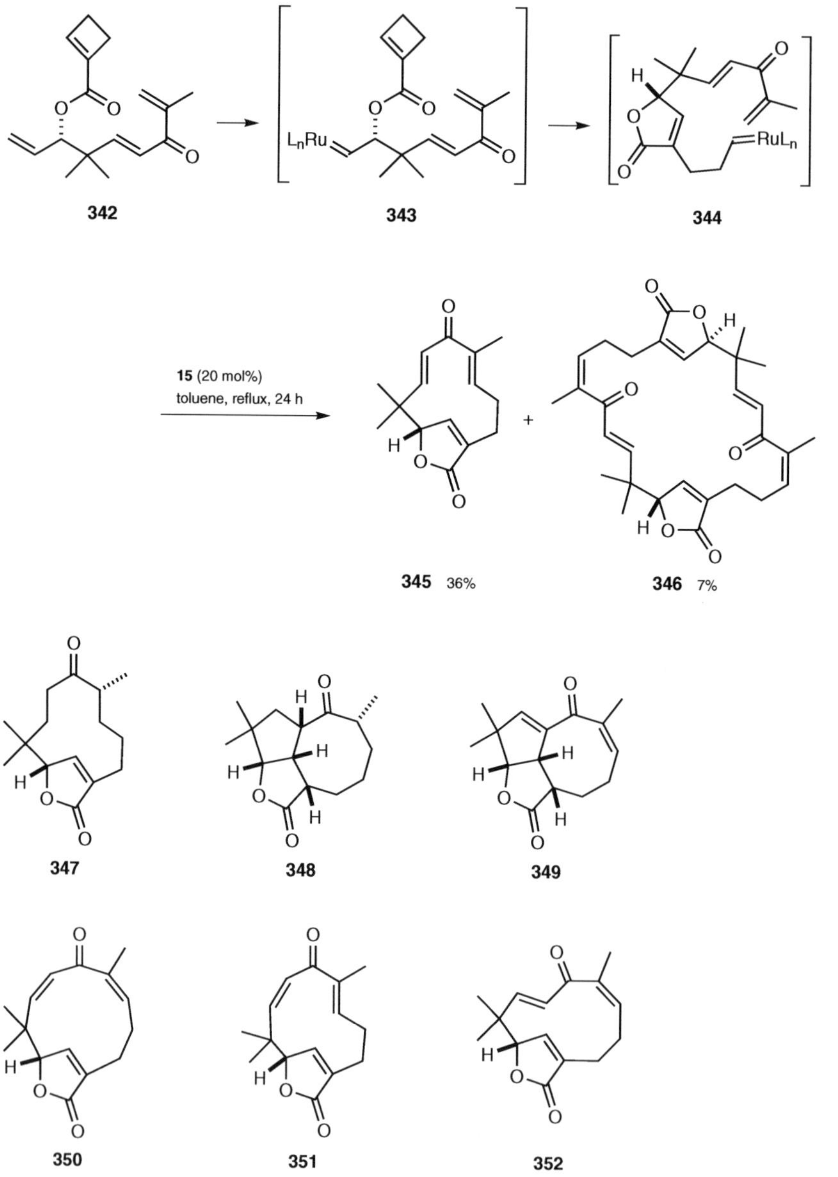

2.10.2.4 Scope and Limitations

Diene ring-closing metathesis has a broad scope for the construction of carbo- and heterocyclic frameworks ranging from small-sized rings to macrocycles. In general, strategic placement of certain functionality relative to the reacting double bonds on the ring-closing metathesis substrates is necessary to achieve high efficiency, which will be further improved by choosing a particular type of catalyst together with assorted reaction conditions. Even with optimal substrate structures and reaction conditions with a selected catalyst, the construction of certain types of cyclic alkenes are not, with some exceptions, generally achievable by ring-closing metathesis, these include: (1) E-alkenyl medium-sized rings, (2) Z-alkenyl macrocycles, (3) trisubstituted alkenyl macrocycles, and (4) tetrasubstituted alkenyl medium-sized cycles and macrocycles. Also, diastereo- and enantioselective macrocyclic ring-closing-metathesis-based desymmetrization of symmetrical triynes, using either an existing stereogenic center near the prostereogenic center or chiral catalysts, are yet to be demonstrated.

2.10.3 Enyne Ring-Closing Metathesis

2.10.3.1 Catalysts and Mechanism

Common catalysts for enyne ring-closing metathesis are the same or similar to those used for diene ring-closing metathesis. Ruthenium (e.g., **2**, **11**, **12**, **14**, **15**, and **353**), molybdenum (e.g., **1** and **354**), and tungsten complexes have been in use most frequently in enyne ring-closing metathesis (Scheme 77).

Scheme 77 Common Catalysts for Enyne Ring-Closing Metathesis

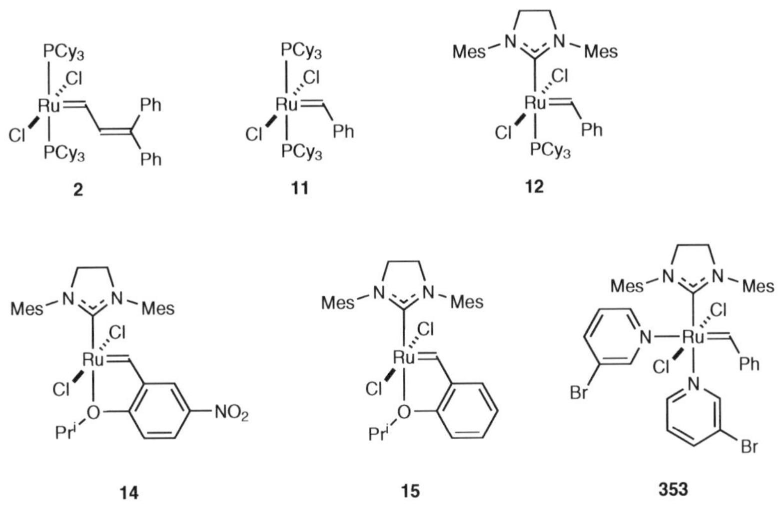

In terms of mechanisms, diene and diyne ring-closing metathesis have two initiation events from each participating alkene or alkyne resulting in two different reaction pathways, which ultimately generate the same ring-closing metathesis product regardless of the initiation events. On the other hand, enyne ring-closing metathesis has potentially three different productive initiation events, which would result in two different ene-first pathways and two different yne-first pathways.[137,138] Each of the four pathways involves unique propagating alkylidene species, yet only two distinct ring-closing metathesis products are possibly generated (Scheme 78). More specifically, in the ene-first mechanism, the *exo/endo*-mode selectivity arises from the two possible orientations of (2+2) cycloaddition between a propagating metal–alkylidene species and the tethered alkyne. The resulting different metallacyclobutenes **355** and **356** then lead to penultimate intermediates **357** and **358**, respectively, from which the *exo-* and *endo*-products **359** and **361** are formed, respectively. Under an ethene atmosphere, the enyne cross metathesis between ethene and the alkyne moiety will generate triene **360**, which re-enters the catalytic cycle via the formation of new ruthenium–alkylidene **363**. In the yne-first mechanism, the *exo/endo*-mode selectivity is the consequence of the formation of two distinct propagating species **365** and **366** via metallacyclobutenes **367** and **368**. Subsequent ring-closing metathesis of alkylidenes **365** and **366** through the formation of metallacyclobutenes **362** and **364**, respectively, ultimately produces the *exo-* and *endo*-products **359** and **361**, respectively.

2.10.3 Enyne Ring-Closing Metathesis

Scheme 78 Possible Mechanisms for Enyne Ring-Closing Metathesis[137,138]

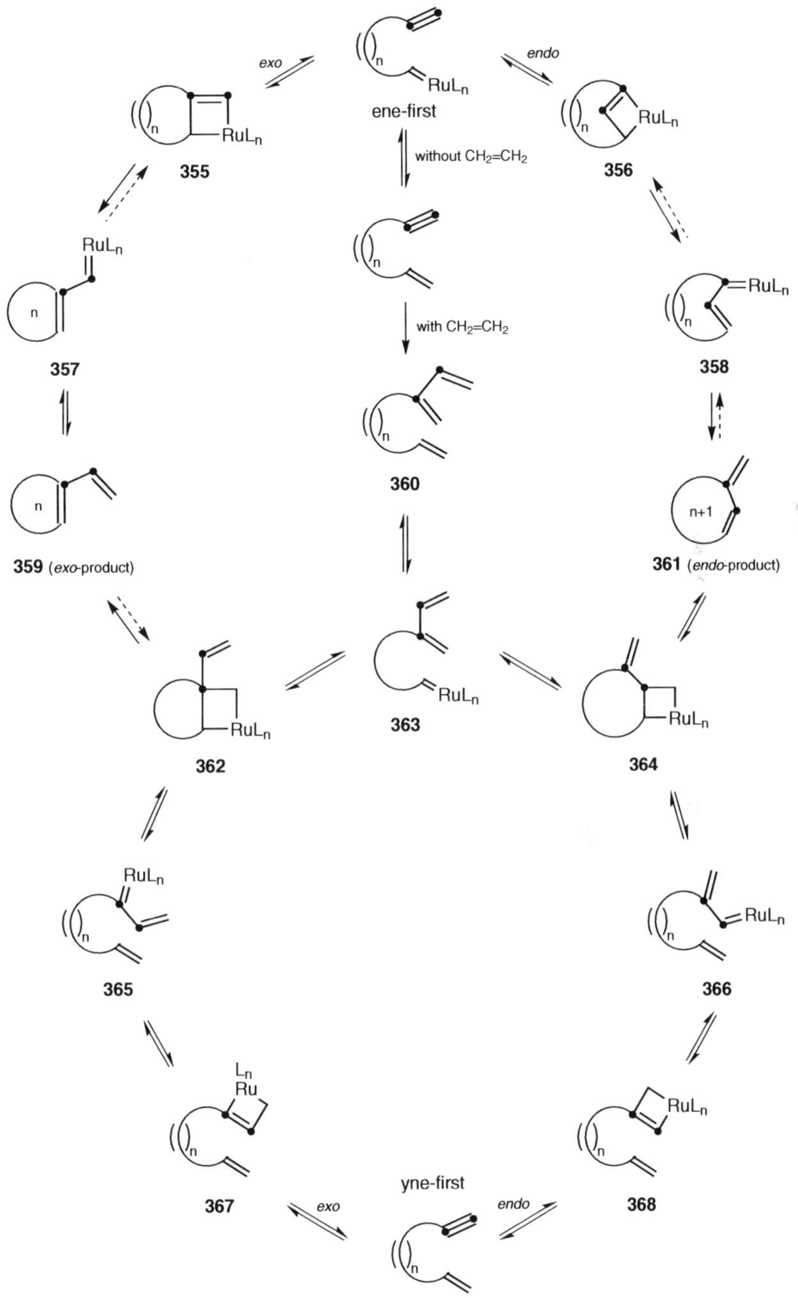

The enyne metathesis catalyzed by group VI metals (W, Cr, Mo) preferentially initiates at an alkyne counterpart. Therefore, it was at first believed that the ruthenium–alkylidene catalyzed enyne metathesis also proceeded via an yne-first mechanism although there was no supporting experimental evidence. In the early examples of enyne ring-closing metathesis using ruthenium–alkylidene complex **2**, a competition experiment to gain insight into the mechanism using dienyne substrate **369** gave an *exo*-mode enyne ring-closing metathesis product **370** (19%), a diene ring-closing metathesis product (5%), and a cyclopropanation product (3%) (Scheme 79). The observed product distribution, however, does not indicate whether an ene-first or yne-first mechanism is involved in the formation of **370**.[13]

Scheme 79 Competition Experiment for Enyne Ring-Closing Metathesis[13]

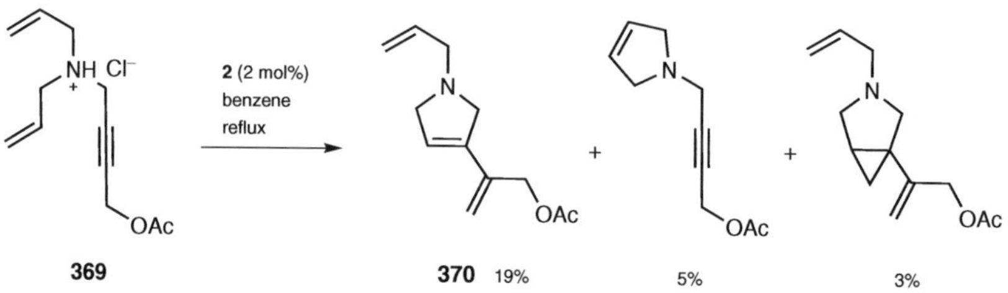

The ene-first mechanism gained increasing support based especially on the results of ring-closing metathesis of systems containing a terminal double bond. In 1998, Grubbs and Ulman described the kinetic product of metathesis reactions involving a terminal double bond, which was derived from a propagating alkylidene, one of the characteristics of ene-first mechanism, rather than a methylidene species.[139] NMR experiments carried out independently also confirmed the existence of alkylidene intermediates,[140,141] which implied that the initiation in these ring-closing metathesis occurred favorably from the alkene counterpart in the presence of either electron-deficient or electron-rich triple bonds. However, extrapolation of these observations for different systems needs to be cautious.[140,141] The mechanism of enyne ring-closing metathesis has also been investigated with deuterium-labeled enynes (Scheme 80).[142] In the predicted yne-first pathway, the methylidene **371**, generated from the conversion of **372** into **373**, may add to the alkyne moiety of substrate **374** to form a metallacyclobutene **372**, the cycloreversion of which would give (E/Z)-1,3-diene **373** nonstereoselectively. In contrast, in the "ene-first" pathway, a putative alkylidene **375**, generated from **376** via cycloreversion and ring closure, may form diastereomeric metallacyclobutanes cis- and trans-**376**, which will lead to exo-products (E)-and (Z)-**377**, respectively. However, due to the unfavorable steric interactions in cis-**376**, it is expected that the ring-closing metathesis would proceed favorably through trans-**376** to produce (Z)-**377** predominantly. When deuterium-labeled enynes (E)-**378** and (Z)-**380** are treated with Grubbs' first-generation catalyst **11** or Grubbs' second-generation catalyst **12** ring-closing metathesis products (Z)-**379** and (E)-**381** are generated with high stereoselectivity, respectively. These results strongly support the ene-first mechanism at least for simple enynes containing terminal double and triple bonds.

2.10.3 Enyne Ring-Closing Metathesis

Scheme 80 Deuterium Labeling Studies in Enyne Ring-Closing Metathesis[142]

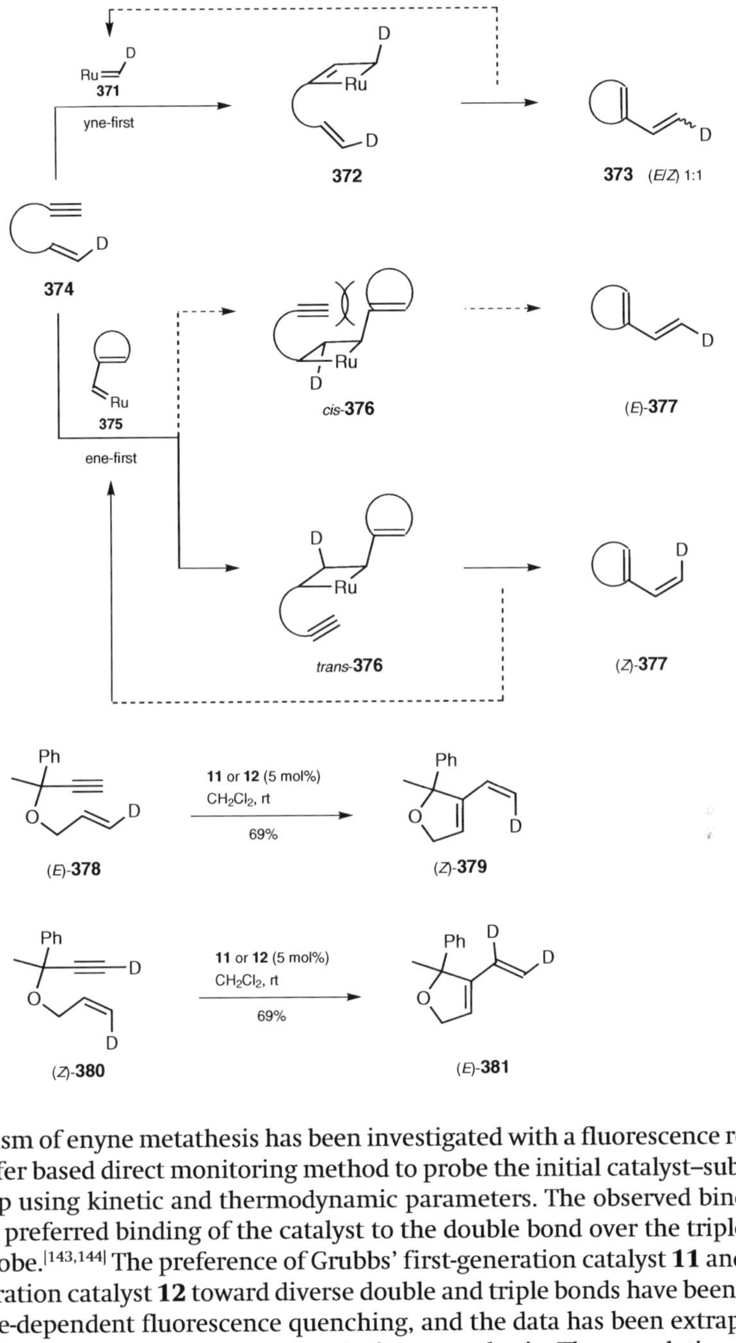

The mechanism of enyne metathesis has been investigated with a fluorescence resonance energy transfer based direct monitoring method to probe the initial catalyst–substrate association step using kinetic and thermodynamic parameters. The observed binding data supports the preferred binding of the catalyst to the double bond over the triple bond of the enyne probe.[143,144] The preference of Grubbs' first-generation catalyst **11** and Grubbs' second-generation catalyst **12** toward diverse double and triple bonds have been explored by using time-dependent fluorescence quenching, and the data has been extrapolated to the *exo/endo* selectivity of the enyne ring-closing metathesis. The correlation indicates that the reaction proceeds through either an ene-first or yne-first initiated pathway, depending on the nature of the substrate and the catalyst. For the ring-closing metathesis of enynes containing a terminal double bond and a triple bond of similar steric and electronic environment, no unique selectivity in the initiation step either at the double or triple bond is observed when Grubbs' first-generation catalyst **11** and Grubbs' second-generation catalyst **12** are used.[145]

The observed preference for the ene-first pathway with ruthenium-based catalysts in enyne ring-closing metathesis can be easily switched over to the yne-first pathway with other catalysts (Scheme 81). For example, the ring-closing metathesis of enyne **382** with molybdenum catalyst **354** provides predominantly the *endo*-product **383** and only a trace amount of the *exo*-product **384**.[146] Switching the steric environment of the enyne substrate does not change the *endo/exo* preference, as the ring-closing metathesis of enyne **385** provides the *endo*-product **386** with only traces of the *exo*-product **387**. The nearly exclusive formation of the *endo*-products **383** and **386** clearly indicates that the yne-first mechanism should be involved in these ring-closing metathesis reactions.

Scheme 81 Enyne Ring-Closing Metathesis Involving an Yne-First Mechanism[146]

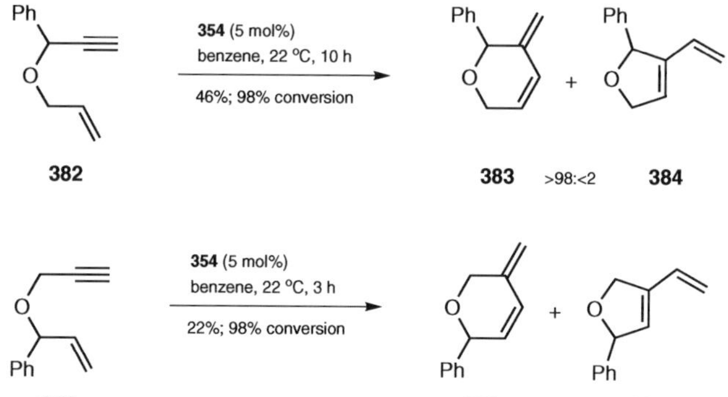

The ligands on the ruthenium catalyst can also control the preference for the ene-first or yne-first mechanism. Ring-closing metathesis with ruthenium–indenylidene complexes containing a phosphine ligand proceeds exclusively via the ene-first mechanism, whereas the yne-first mechanism is in operation when the phosphine ligand is replaced with an N-heterocyclic carbene ligand.[147]

Interestingly, a computational study shows that the vinylalkylidene formation happens via a metallacyclobutene transition state rather than a discreet intermediate. Furthermore, the ruthenium–carbene binding strength is higher with an alkyne than an alkene. However, the insertion of a ruthenium–carbene is kinetically disfavored with an alkyne compared to that with an alkene (5–7 kcal/mol), although the alkyne insertion is thermodynamically more favorable by 33 kcal/mol than that of the alkene, which thus becomes an irreversible step in the catalytic cycle of enyne ring-closing metathesis.[148] The theoretical calculations for enyne ring-closing metathesis show, however, that the ene-first and yne-first pathways have no distinctive energetic preference, and thus both should be operative when unsubstituted enynes are involved. In addition, the *exo/endo* selectivity is strongly influenced by the presence of substituents in the system.[149]

Even with all these experimental and theoretical studies, the exact operating mechanism for each enyne ring-closing metathesis is still elusive due to the many factors affecting the reaction at multiple stages. Therefore, for a more complete understanding of the enyne ring-closing metathesis mechanism, more systematic studies are required.

2.10.3.2 Selectivity, Ring Size, and Substrates

In enyne metathesis, the selectivity among possible reaction pathways depends on multiple factors including the type of metal catalyst and its ligands, the substituent patterns of the double or triple bond, and the electron-withdrawing or -releasing nature of the substituents. Due to the subtle and convoluted nature of these factors along with the two pos-

2.10.3 Enyne Ring-Closing Metathesis

sible addition modes of the metal–alkylidene catalyst to the triple bond either in the initiation step (yne-first mechanism) or in the propagating step (ene-first mechanism), enyne ring-closing metathesis has inherent selectivity problems in terms of the formation of *exo/endo*-mode 1,3-diene products (Scheme 82).

Scheme 82 Regioselectivity in Enyne Ring-Closing Metathesis

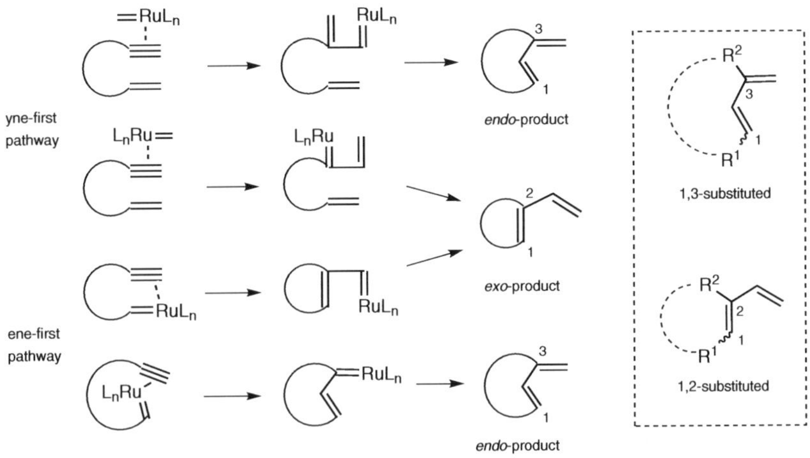

2.10.3.2.1 Chemoselectivity with Multiple Double and Triple Bonds

An outstanding chemoselectivity problem exists between diene ring-closing metathesis and enyne ring-closing metathesis when ring-closing metathesis substrates contain two double bonds and a triple bond connected by tethers in different topologies (Scheme 83). From these dienyne substrates, enyne ring-closing metathesis is known to be more favorable than diene ring-closing metathesis due to both thermodynamic (reversible nature of diene ring-closing metathesis versus irreversibility for enyne ring-closing metathesis) and kinetic (smaller ring by enyne ring-closing metathesis versus larger ring by diene ring-closing metathesis) reasons. If the triple bond is part of the main tether that connects the two double bonds such as in **388**, a cascade enyne ring-closing metathesis would favorably generate a dumbbell-shaped 1,3-diene **389** (*exo*-product) rather than a fused bicyclic 1,3-diene product **390** (*endo*-product), especially for small- and medium-sized rings. If the triple bond is directly connected to the main chain such as in **391**, two different fused bicyclic 1,3-dienes **392** and **393** would be formed depending on the mode of ring-closure (*endo* versus *exo*) or the initiation on different double bonds with a preferred *exo*-mode ring closure (sequence selectivity). Similarly, if the triple bond is connected to the main chain with a methylene group such as in **394**, bridged bicyclic 1,3-dienes **395** or **396** could be formed if the ring containing a *cis* double bond is large enough.

Over the years, various strategies have evolved to address the chemoselectivity issues associated with various systems containing multiple double and triple bonds. Due to the sensitivity of metathesis catalysts toward steric and electronic environments on or around the engaged double and triple bonds in the substrates, these selectivity-controlling strategies rely on the combined influence of these steric and electronic factors to control the overall reaction profiles.

Scheme 83 Chemoselectivity in the Ring-Closing Metathesis of Dienynes

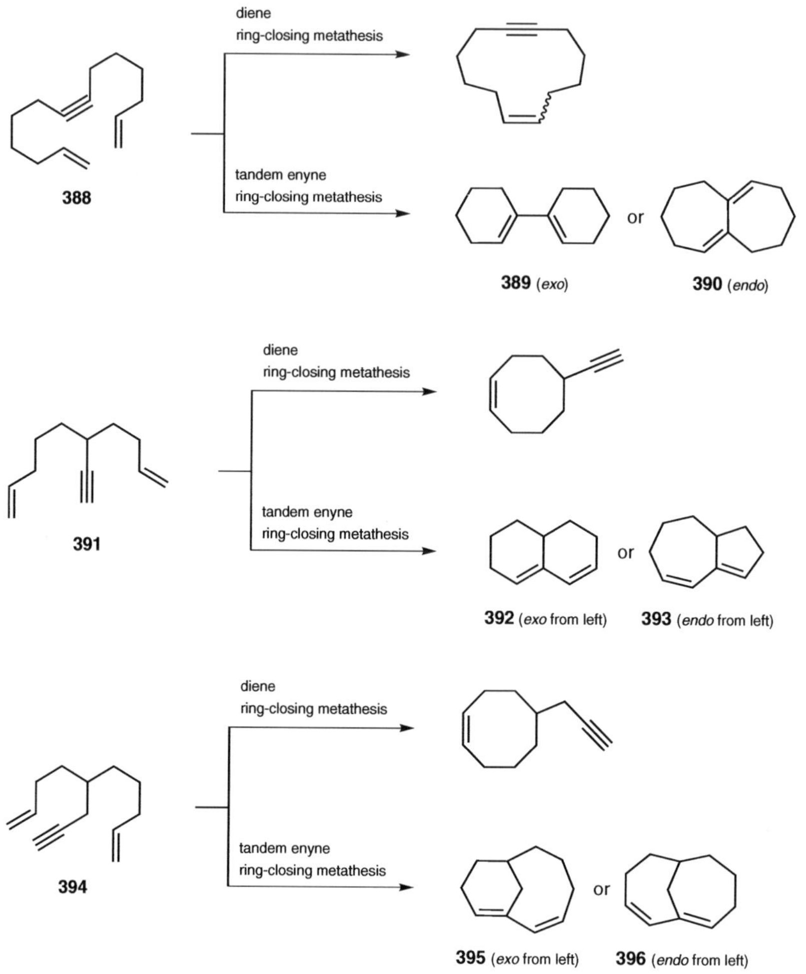

2.10.3.2.1.1 Substituent-Based Control

The sequence-selective tandem enyne ring-closing metathesis of dienynes **397** relies on a steric factor. The ring-closing metathesis of dienyne **397** (R^1 = H) in the presence of ruthenium complex **2** in benzene delivers a 1:1 mixture of [4.4.0] bicycle **398** and [3.5.0] bicycle **399** in 86% yield. In contrast, upon installing an ethyl substituent on one double bond of dienyne **397** (R^1 = Et), the same enyne ring-closing metathesis affords the [4.4.0] bicycle **398** as a single product (Scheme 84).[14] This enyne ring-closing metathesis selectivity has also been further investigated in the synthesis of polycyclic compounds using Grubbs' first-generation catalyst **11**.[150]

2.10.3 Enyne Ring-Closing Metathesis

Scheme 84 Steric Control in Enyne Ring-Closing Metathesis of Dienynes[14]

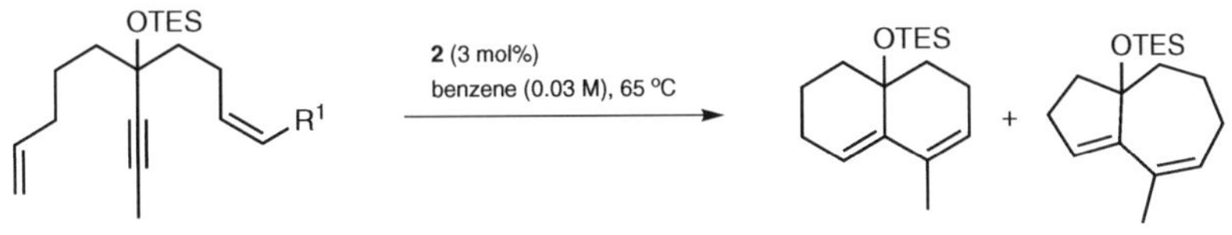

R¹	Time (h)	Ratio (398/399)	Yield (%)	Ref
H	15	1:1	86	[14]
Et	6	1:0	83	[14]

Similarly, tandem enyne ring-closing metathesis of ynamide-based dienyne **400** (R¹ = H) with Grubbs' second-generation catalyst **12** affords a 1:1 mixture of azabicycles **401** and **402**. On the other hand, the enyne ring-closing metathesis with dienyne **400** (R¹ = Me) having sterically differentiated double bonds provides a significantly increased selectivity for azabicycle **401**, which is a consequence of the preferred initiation on the sterically less-hindered terminal double bond (Scheme 85).[151]

Scheme 85 Steric Control in Enyne Ring-Closing Metathesis of Ynamide-Based Dienynes[151]

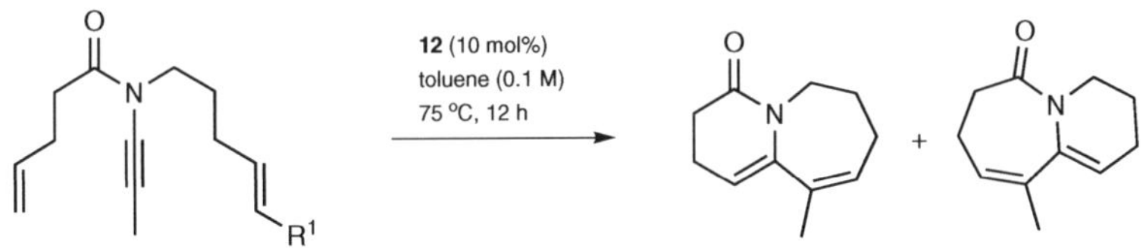

R¹	Ratio (401/402)	Yield (%)	Ref
H	1:1	77	[151]
Me	6:1	70	[151]

Grubbs and Choi have described the sequence-selective tandem ring-closing metathesis of dienynes relying on electronic differentiation in the substrates. The tandem enyne ring-closing metathesis of dienynes **403** containing terminal monosubstituted double bonds in an electronically differentiated environment provides the bicyclic lactones **405** as sole products regardless of the substitution on the triple bond. The observed selectivity can be explained by the site-selective initiation of the ring-closing metathesis with the more electron-rich double bond to form a propagating ruthenium–alkylidene intermediate **404** followed by two consecutive ring-closure processes (Scheme 86).[152]

Scheme 86 Effect of Electronic Differentiation in Selective Enyne Ring-Closing Metathesis[152]

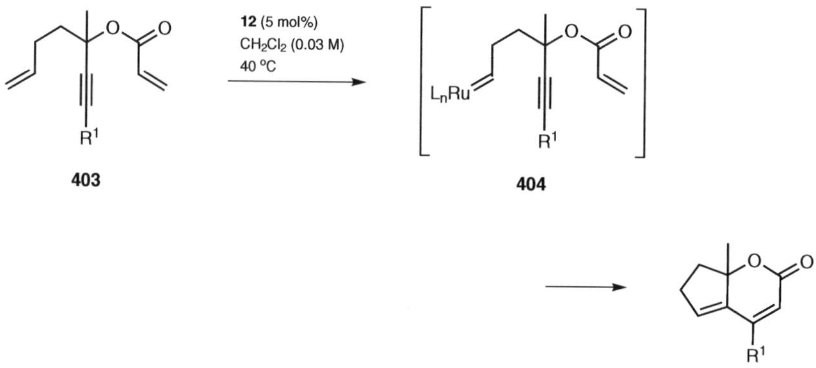

R¹	Yield (%)	Ref
H	95	[152]
Me	72	[152]

Although demonstrated with only a limited number of examples, substituent-based control for sequence-selective tandem enyne ring-closing metathesis can be expected as long as enough bias is introduced either by steric or electronic factors on one of the double bonds. This strategy has been applied to a number of natural product syntheses including examples described in Section 2.10.3.3.2.

8-Methyl-4a-(triethylsiloxy)-3,4,4a,5,6,7-hexahydronaphthalene (398):[14]
A soln of Ru catalyst **2** (5.6 mg, 6 µmol, 0.03 equiv) in benzene (1.0 mL) (**CAUTION:** *carcinogen*) was added through a cannula to a soln of dienyne **397** (R¹ = Et; 67 mg, 0.2 mmol) in benzene (5.7 mL, 0.03 M). The resulting light brown soln was placed in an oil bath at 65 °C for 6.5 h (TLC). The soln was concentrated under reduced pressure and the residue was purified by flash chromatography (petroleum ether); yield: 46 mg (83%).

2.10.3.2.1.2   Control Based on Ring Size

The ring-closing metathesis of dienyl ethynylphosphonates **406** possessing two terminal double bonds with similar steric and electronic environments using Grubbs' second-generation catalyst **12** shows ring-size-based sequence control (Scheme 87). The ring-closing metathesis of dienyne **406** (n = 1) provides a mixture of bicycle **407** (n = 1) and monocycle **408** (n = 1) in a 2.9:1 ratio, but no six-membered ring containing product is observed. The symmetrical system **406** (n = 2) gives a mixture of **407** (n = 2) and **408** (n = 2) in a 4.2:1 ratio. In contrast, nonsymmetrical substrate **406** (n = 3) exclusively affords product **407** (n = 3) by initiation at the terminal double bond with the longer tether. Unless unavoidable, the formation of the six-membered ring phosphonate seems to be disfavored.[153] In contrast, in a related study of ring-size-based control of group-selective tandem enyne ring-closing metathesis of 1,2-diol-derived dienynes, selective formation of six-membered rings over the five-membered ones is observed.[154]

2.10.3 Enyne Ring-Closing Metathesis

Scheme 87 Control Based on Ring Size in the Enyne Ring-Closing Metathesis of Dienyl Ethynylphosphonates[153]

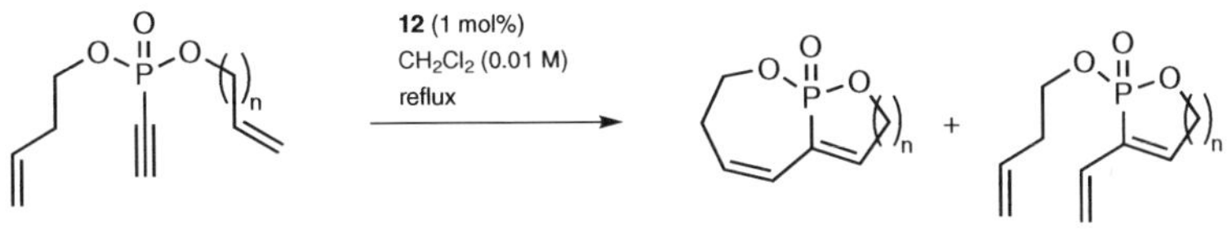

n	Ratio (**407**/**408**)	Yield (%)	Ref
1	2.9:1	89	[153]
2	4.2:1	94	[153]
3	1:0	83	[153]

The dienyne **409** containing two tethered terminal double bonds of only one methylene unit difference undergoes a highly group-selective ring-closing metathesis with Grubbs' first-generation catalyst **11**, relying on different rates for the formation of rings of different sizes (Scheme 88).[155] The reaction at higher concentrations provides higher selectivity between the two ring-closing metathesis products **411** and **413**, which is due to an increased bimolecular alkylidene exchange process to establish a pre-equilibrium between the two propagating ruthenium–alkylidene intermediates **410** and **412** at higher concentration.

Scheme 88 Concentration-Dependent Group-Selective Enyne Ring-Closing Metathesis[155]

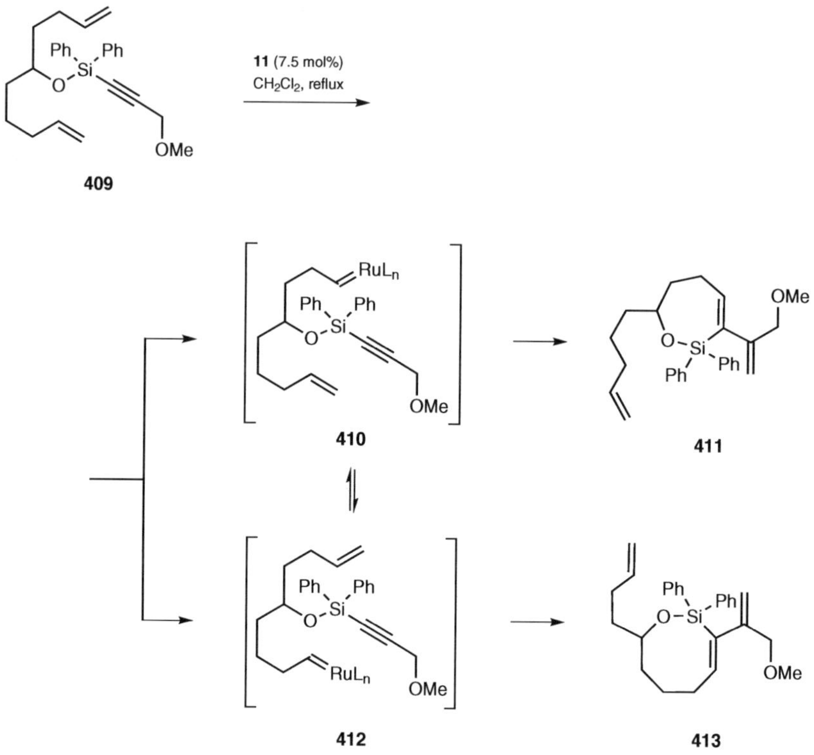

Concentration (M)	Ratio (411/413)	Yield (%)	Ref
0.01	6.4:1	80	[155]
0.03	9.0:1	67	[155]
0.07	25.0:1	74	[155]
0.1	30.0:1	74	[155]
neat	>50.0:1	79	[155]

2.10.3.2.1.3 Relay Metathesis Based Control

Relay metathesis based control can also be used in the group-selective tandem ring-closing metathesis of dienynes.[44] In this strategy, the most reactive initiating double bond on the relay tether can intramolecularly deliver the propagating alkylidene to a less reactive double bond even in the presence of a more reactive one (see Section 2.10.2.2.3.3, Scheme 19).

The effectiveness of relay tether based control of dienyne ring-closing metathesis is clearly demonstrated with dienynes **414** (Scheme 89). Dienyne **414** ($R^1 = R^2 = H$) affords the bicyclic products **415** and **416** in a 1:2 ratio in the presence of Grubbs' second-generation catalyst **12**. However, by introducing a terminal double bond based relay tether at either end of dienynes **414**, the selectivity can be improved and the highest selectivity is achieved by using the less reactive Grubbs' first-generation catalyst **11**.[44]

2.10.3 Enyne Ring-Closing Metathesis

Scheme 89 Group-Selective Dienyne Ring-Closing Metathesis by Relay Metathesis[44]

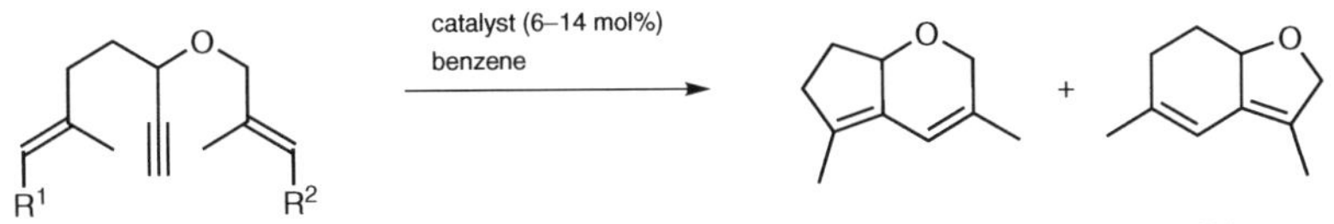

R^1	R^2	Catalyst	Temp	Ratio (**415/416**)	Yield[a] (%)	Ref
H	H	12	rt	1:2	–	[44]
$CH_2OCH_2CH=CH_2$	H	12	rt	4.7:1	–	[44]
$CH_2OCH_2CH=CH_2$	H	11	50 °C	26:1	77	[44]
H	$CH_2OCH_2CH=CH_2$	12	rt	1:7	–	[44]
H	$CH_2OCH_2CH=CH_2$	11	50 °C	1:45	78	[44]

[a] Yields determined by ^{1}H NMR spectroscopy with an added internal standard.

2.10.3.2.2 endo/exo Mode and E/Z Selectivity

The mechanistic rationale for the formation of *endo/exo*-mode products in enyne ring-closing metathesis is illustrated in Schemes 78 and 82 (see Sections 2.10.3.1 and 2.10.3.2). Based on these mechanisms, the *endo/exo*-mode selectivity should be the consequence of either a preferred ring-closure mode of a propagating alkylidene (*endo* versus *exo*) or a preferred mode of addition of a propagating alkylidene (typically methylidene) in the initiation step. These two crucial selectivity-determining events are strongly influenced by the reacting double and triple bonds along with the catalysts employed (Ru versus Mo). Several representative examples are described in the following sections.

2.10.3.2.2.1 Small and Medium Rings

Generally, the formation of small- and medium-sized rings in enyne ring-closing metathesis preferentially forms the *exo*-products, yet, the preference can be reversed by using a different catalyst (see Section 2.10.3.1, Scheme 81). For example, the reaction of enyne **417** with Grubbs–Hoveyda second-generation catalyst **15** selectively generates the *exo*-product **418** with traces of the *endo*-product **419**. However, by employing molybdenum catalyst **354** the selectivity is completely reversed, providing the *endo*-product **419** (Scheme 90).[146] Also, the same *endo*-selective ring-closing metathesis is obtained with Schrock-type monopyrrolide tungsten complexes.[156]

Scheme 90 endo/exo Selectivity in the Formation of Small Rings[146]

Enyne ring-closing metathesis of 1,6-enynes **420** and **422** using Grubbs' second-generation catalyst **12** under ethene generates *endo*-products **421** and **423**, respectively (Scheme 91).[157] Mechanistically, these metatheses proceed via cross metathesis between the alkyne moiety of the enynes and ethene followed by ring-closing metathesis between the two monosubstituted double bonds. The ring-closing metathesis products can serve as advanced intermediates for the synthesis of otteliones (e.g., **424**) and loloanolides (e.g., **425**).

Scheme 91 endo Selectivity in the Formation of Cyclohexenes for Use in the Synthesis of Otteliones and Loloanolides[157]

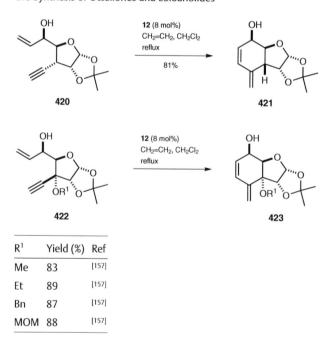

R¹	Yield (%)	Ref
Me	83	[157]
Et	89	[157]
Bn	87	[157]
MOM	88	[157]

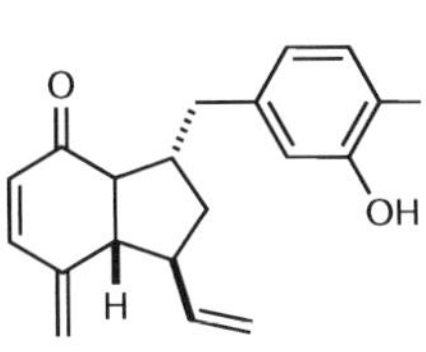

2.10.3 Enyne Ring-Closing Metathesis

424

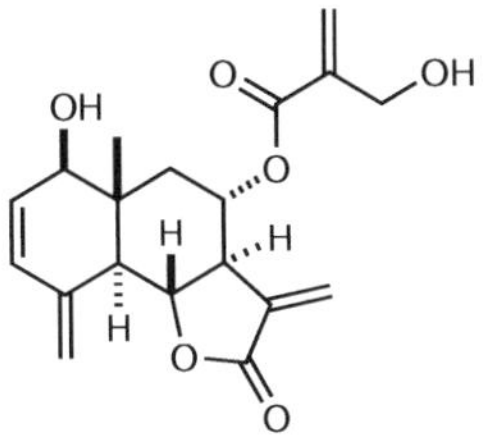

425

The *exo*-mode selective formation of small and medium rings by enyne ring-closing metathesis of silicon-tethered enynes **426** affords silacycles **428** via intermediates **427** with Z configuration of the endocyclic double bond regardless of the tether size. The exclusive formation of intermediate **427** is due to the steric effect of the silyl substituent, which prevents the formation of *endo*-intermediate **429** (Scheme 92).[158] On the other hand, enyne ring-closing metathesis of sterically less demanding silyl-based enyne **430** provides a 1:1 mixture of the products **431** and **432** (Scheme 92).[159]

Scheme 92 Selectivity in Silicon-Tethered Enynes[158,159]

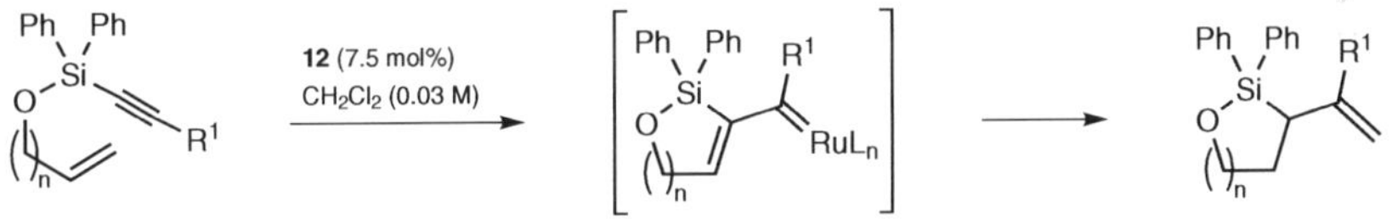

426

12 (7.5 mol%)
CH_2Cl_2 (0.03 M)

427

428 30–85%

R^1 = H, CH_2OMe, Bu; n = 1–9

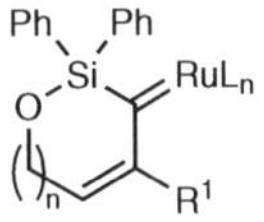

429

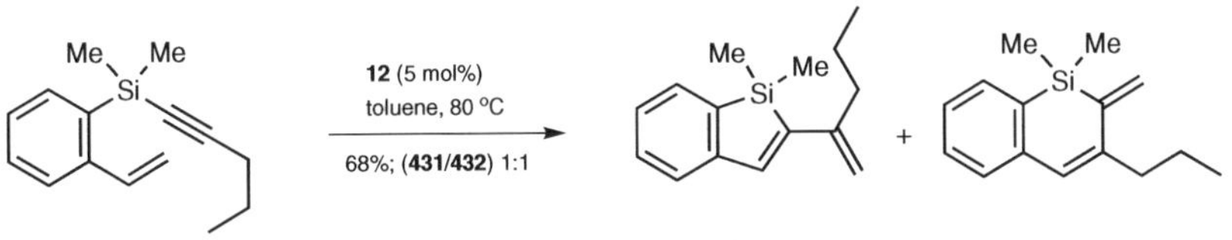

430

12 (5 mol%)
toluene, 80 °C

68%; (**431**/**432**) 1:1

431

432

2.10.3.2.2.2 Macrocycles

In the synthesis of (−)-longithorone A (**437**), a formal *endo*-selective enyne ring-closing metathesis constructs the 16-membered carbocycle (Scheme 93). The enynes **433** and **435** provide exclusively *endo*-products **434** and **436** in moderate yields with Grubbs' first-generation catalyst **11** (50 mol%) under ethene. Mechanistically, this ring-closing metathesis proceeds via cross metathesis of the terminal alkyne moiety with ethene followed by ring-closing metathesis of the monosubstituted terminal double bond. Subsequent elaborations of **434** and **436** including two Diels–Alder reactions leads to a total synthesis of (−)-longithorone A (**437**).[160,161]

for references see p 672

Scheme 93 endo-Selective Enyne Ring-Closing Metathesis in the Formation of Macrocycles Used in the Synthesis of (−)-Longithorone A[160,161]

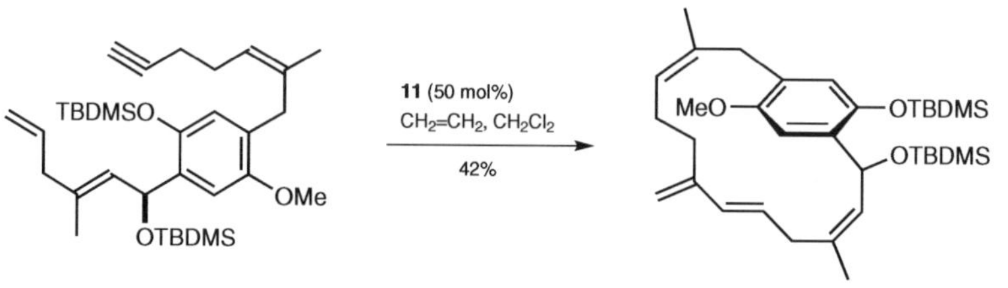

The *exo/endo* selectivity of enyne ring-closing metathesis depends on the ring size. The formation of small and medium rings generally follows the *exo*-mode ring closure. The ring-closing metathesis of **438** (m = n = 1) provides only ten-membered *exo*-product **439** (m = n = 1) (Scheme 94). However, the ring-closing metathesis of **438** (m = 1; n = 2) to form the 11-membered ring generates a 1:1 mixture of *exo*- and *endo*-products **439** (m = 1; n = 2) and **440** (m = 1; n = 2), respectively (Scheme 94).[162] In contrast, the ring-closing metathesis to form macrocyclic rings (ring size 12–15) follows *endo*-mode ring closure to provide products **440** exclusively. This outcome seems to be the result of the ene-first pathway, where the propagating alkylidene favors addition to the tethered alkyne to generate the more alkyl-substituted alkylidene intermediate through (2+2) cycloaddition followed by cycloreversion.[162,163] The ring-closing metathesis of **438** under ethene provides 1,3-diene **441** as the initially formed product, which then undergoes a diene ring-closing metathesis to afford exclusively a formal *endo*-mode product **440**, generally with high E selectivity (Scheme 94).[163] A similar trend of macrocyclic *endo*-product formation is observed in the enyne ring-closing metathesis of glycal-derived substrates.[164]

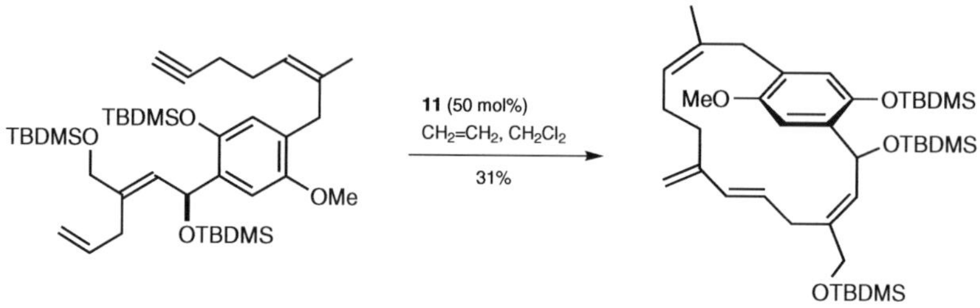

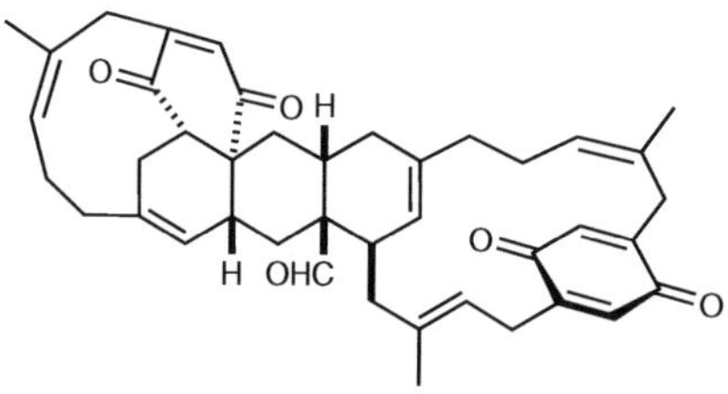

2.10.3 Enyne Ring-Closing Metathesis

Scheme 94 endo/exo Selectivity in Macrocyclic Enyne Ring-Closing Metathesis[162,163]

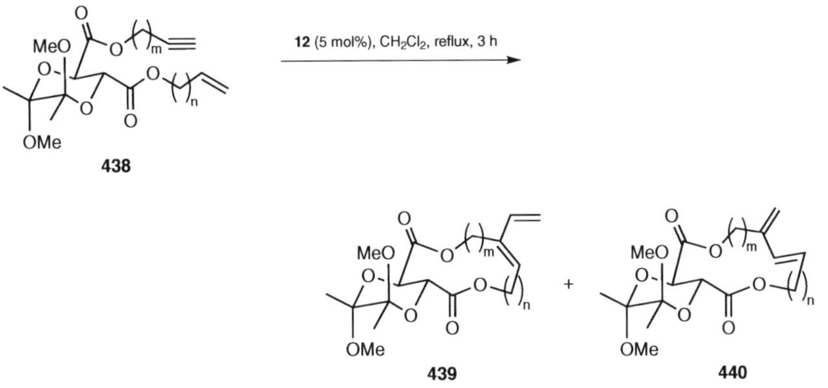

m	n	Ratio (**439**/**440**)	Ratio (E/Z)	Yield (%)	Ref
1	1	1:0	0:1	52	[162]
1	2	1:1	1:1 (**439**) and 2:1 (**440**)	50	[162]
2	1	0:1	1:1	70	[162]
2	2	0:1	0:1	61	[162]
3	2	0:1	2:1	54	[162]
3	3	0:1	1:1	55	[162]

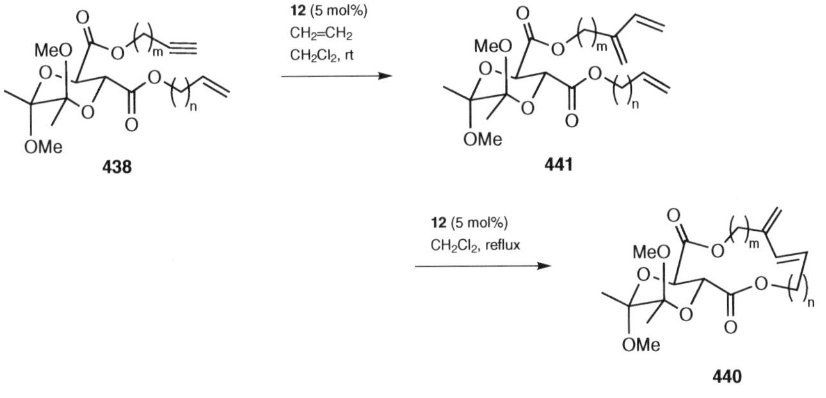

m	n	Yield (%) of **441**	Ratio (E/Z) of **440**	Yield (%) of **440**	Ref
1	1	96	0:1	51	[163]
1	2	65	1:1	91	[163]
2	2	77	1:0	65	[163]
3	2	92	1:0	85	[163]
3	3	88	15:1	90	[163]

(2R*,3R*,4aR*,12aR*,Z)-2,3-Dimethoxy-2,3-dimethyl-8-vinyl-2,3,4a,7,10,12a-hexahydro-[1,4]dioxino[2,3-c][1,6]dioxecine-5,12-dione (439, m = n = 1); Typical Procedure:[162]

Enyne **438** (m = n = 1; 30 mg, 0.09 mmol) was dissolved in CH$_2$Cl$_2$ (100 mL) in a 250-mL round-bottomed flask and N$_2$ was bubbled through the soln for 10 min. The flask was fitted with a reflux condenser and a soln of Grubbs' second-generation catalyst **12** (3.8 mg,

4.5 μmol, 0.05 equiv) in CH_2Cl_2 (1 mL) was added. The mixture was heated at reflux for 3 h under N_2. The solvent was then removed under reduced pressure and the crude material was purified by chromatography (hexanes/EtOAc 5:1); yield: 15.6 mg (52%).

2-Allyl 3-(2-Methylenebut-3-enyl) (2R*,3R*,5R*,6R*)-5,6-Dimethoxy-5,6-dimethyl-1,4-dioxane-2,3-dicarboxylate (441, m = n = 1); Typical Procedure:[163]

Enyne **438** (m = n = 1; 30 mg, 0.09 mmol) was dissolved in CH_2Cl_2 (70 mL) and ethene was bubbled through the soln for 10 min using a long syringe connected to a balloon. A soln of Grubbs' second-generation catalyst **12** (3.8 mg, 6.3 μmol, 0.07 equiv) in CH_2Cl_2 (1 mL) was then added, and the reaction was stirred at rt. After complete consumption of starting material, the solvent was removed under reduced pressure and the product was purified by chromatography (silica gel, hexanes/EtOAc 5:1); yield: 32 mg (96%).

(2R*,3R*,4aR*,13aR*,Z)-2,3-Dimethoxy-2,3-dimethyl-8-methylene-2,3,4a,8,11,13a-hexahydro-7H-[1,4]dioxino[2,3-c][1,6]dioxacycloundecine-5,13-dione (440, m = n = 1); Typical Procedure:[163]

The cross metathesis product **441** (m = n = 1; 32 mg, 0.09 mmol) was dissolved in CH_2Cl_2 (75 mL) and the soln was flushed with N_2. A soln of Grubbs' second-generation catalyst **12** (3.8 mg, 6.3 μmol, 0.07 equiv) in CH_2Cl_2 (1 mL) was added and the reaction was heated at reflux. After consumption of starting material, the mixture was cooled to rt and concentrated under reduced pressure. The residue was purified by chromatography (silica gel, hexanes/EtOAc 5:1); yield: 15.7 mg (51%).

2.10.3.2.3 Regioselectivity in Enyne Ring-Closing Metathesis–Metallotropic [1,3]-Shift

Metal complexes of alkynyl alkylidenes undergo a [1,3]-transposition known as the metallotropic [1,3]-shift. From a mechanistic point of view, metallotropic [1,3]-shift can be considered as a special case of enyne ring-closing metathesis. For the metallotropic shift to be initiated by enyne ring-closing metathesis, the *exo*-mode ring-closure by the propagating alkylidene species and an appropriate substituent on the alkyne moiety for regioselective termination seem to be necessary (Scheme 95).

Scheme 95 Possibility of Metallotropic [1,3]-Shift in Enyne Ring-Closing Metathesis

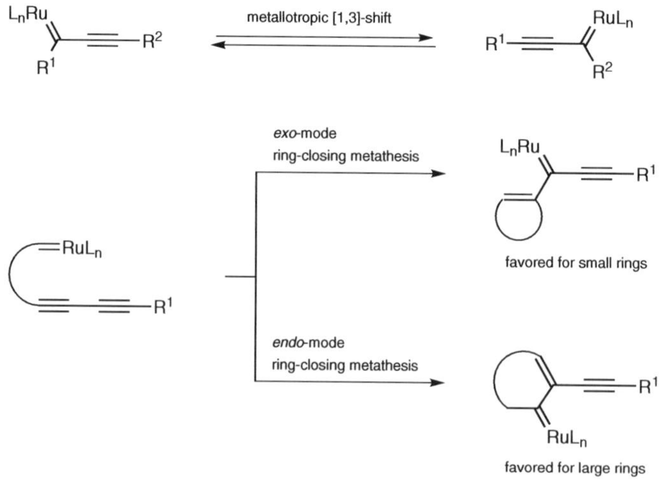

2.10.3 Enyne Ring-Closing Metathesis

The regioselective ring-closing metathesis of ene-1,3-diynes **442** followed by metallotropic [1,3]-shift is controlled by the R^1 substituent at the terminus of the 1,3-diyne moiety. If the R^1 substituent is hydrogen or an unhindered alkyl group, metallotropic [1,3]-shift occurs to generate 1,5-dien-3-ynes **443**, but if the R^1 substituent is a sterically hindered substituent, such as a triethylsilyl group, a 1,3-dien-2-yne **444** ($R^1 = SiEt_3$) is generated without involving a metallotropic shift process (Scheme 96).[165]

Scheme 96 Metallotropic [1,3]-Shift Induced by Enyne Ring-Closing Metathesis of Ene-1,3-diynes To Give Dienynes[165]

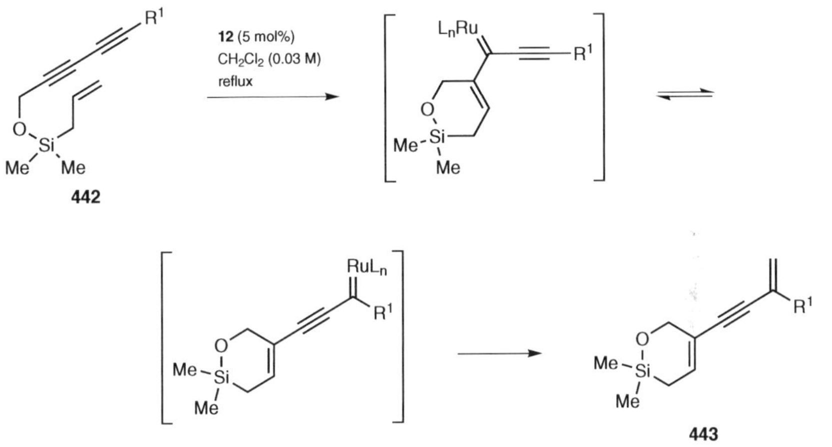

R^1	Yield (%)	Ref
H	69	[165]
Bu	96	[165]
MOM	quant	[165]

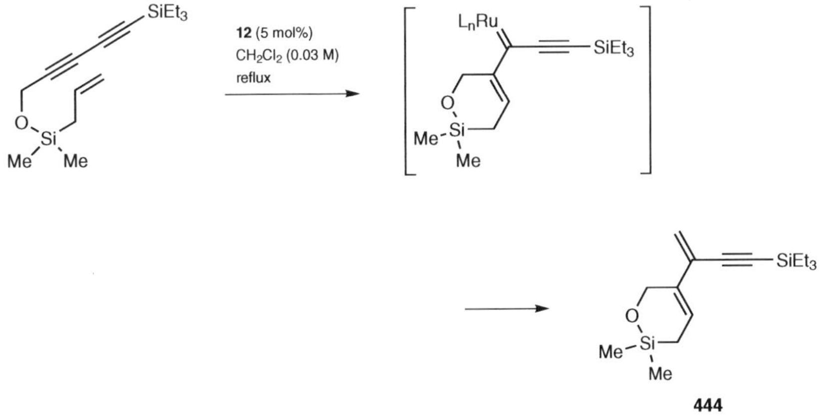

Further investigation has revealed the substituent effect on the metallotropic [1,3]-shift in enyne ring-closing metathesis. In the presence of a catalytic amount of Grubbs' second-generation catalyst **12**, triyne **445** provides dienediyne **446** as sole product, whereas the corresponding unsubstituted triyne **447** affords alkynyl alkylidene complex **448** with a full equivalent of Grubbs' second-generation catalyst **12** (Scheme 97). More examples of contrasting behaviors can be observed with triynes **449** and **451**. With a catalytic amount

(10 mol%) of Grubbs' second-generation catalyst **12**, the ring-closing metathesis of triyne **449** provides the expected ring-closing metathesis product **450** in good yield, whereas with the similar substrate **451**, no catalytic turnover is observed. However, with 100 mol% of Grubbs' second-generation catalyst **12**, alkyne-chelated ruthenium complex **452** is isolated in 79% yield.[166,167]

Scheme 97 Reactivity of Alkynyl Ruthenium–Alkylidenes[166,167]

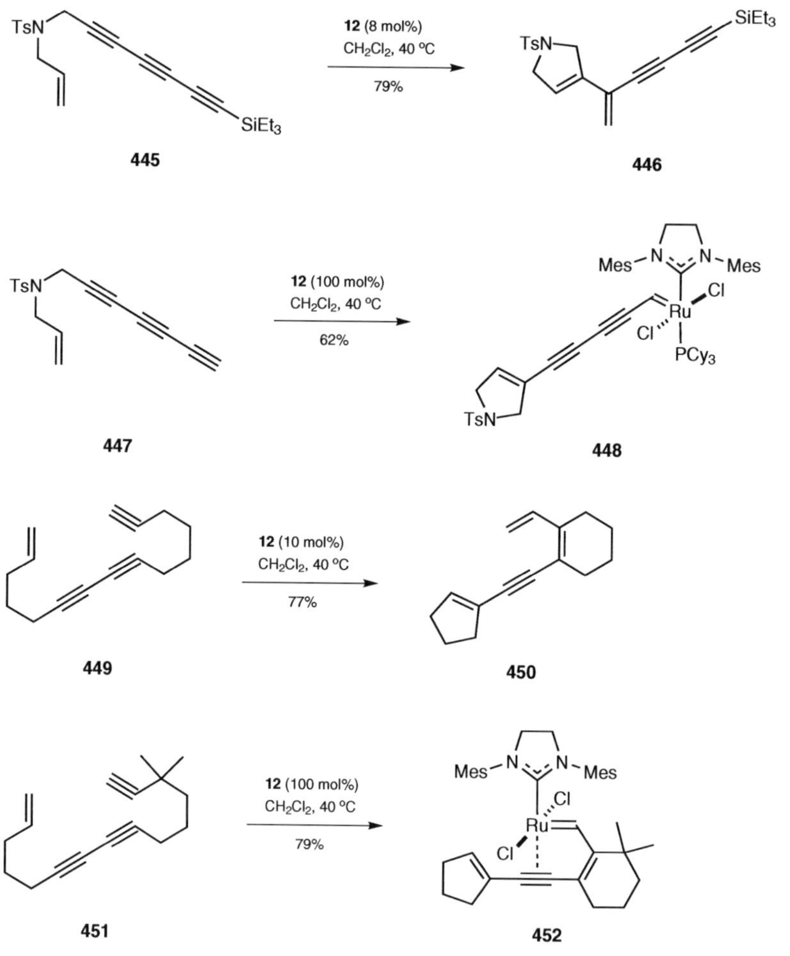

The symmetrical diynes **453** and **455** are converted into symmetrical 1,5-dien-3-ynes **454** and **456**, respectively, via enyne ring-closing metathesis in the presence of Grubbs' second-generation catalyst **12** under ethene (Scheme 98). It is proposed that the double ring-closing metathesis product (e.g., **457**) is formed first followed by thermodynamically driven ethene extrusion.[168]

2.10.3 Enyne Ring-Closing Metathesis

Scheme 98 Sequential Ring-Closing Metathesis of Symmetrical Diynes To Give 1,5-Dien-3-ynes[168]

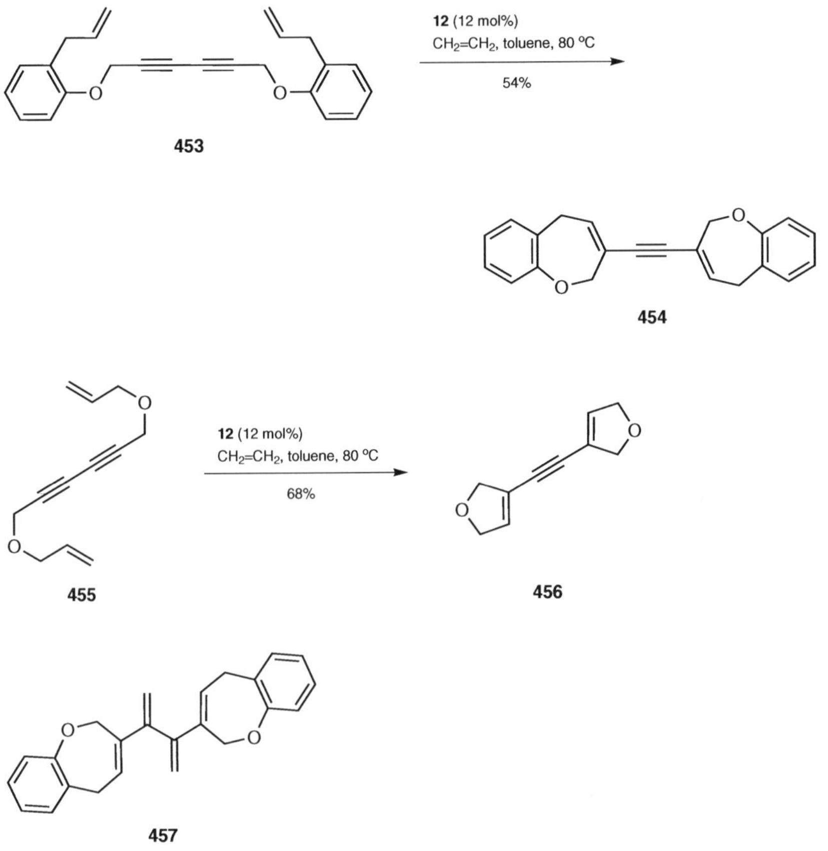

2.10.3.3 Applications of Enyne Ring-Closing Metathesis to Natural Product Synthesis

The tandem multiple bond- and ring-forming capability of enyne ring-closing metathesis, as well as its atom-economical nature, has significant synthetic potential for the preparation of complex carbocycles and heterocycles with 1,3-diene moieties. These characteristics have been exploited extensively in natural product total synthesis.

2.10.3.3.1 Ring-Closing Metathesis with 1,n-Enynes

One of the early examples of the application of enyne ring-closing metathesis to natural product synthesis is the smooth conversion of pyrrolidine-containing 1,8-enyne **458** into bicyclic lactam **459** in good yield upon treatment with Grubbs' first-generation catalyst **11** (Scheme 99). The installation of a lactone moiety on the ring-closing metathesis product **459** via a five-step sequence leads to the completion of (−)-stemoamide (**460**).[169,170]

Scheme 99 Enyne Ring-Closing Metathesis in the Synthesis of (−)-Stemoamide[169,170]

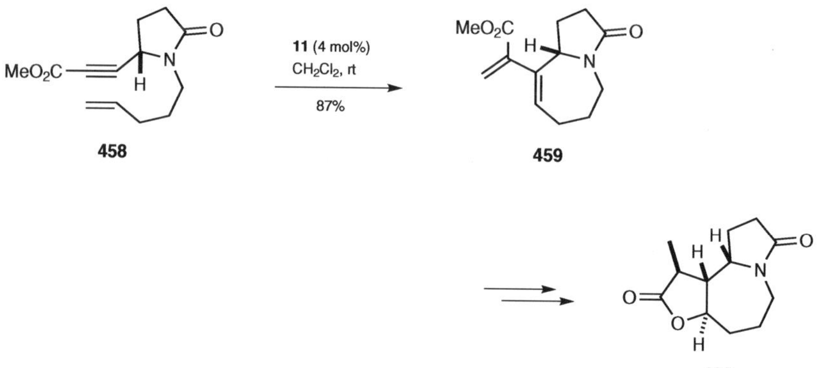

The formation of small-membered cycloalkenes (three- and four-membered rings) using ring-closing metathesis is quite challenging if not impossible because of their inherently strained nature. Even though strained rings can be constructed under metathesis conditions, they prefer ring-opening metathesis. Contrary to this general trend, a rare example of enyne ring-closing metathesis forms cyclobutenes (Scheme 100). The 1,5-enyne **461** provides cyclobutene derivative **462** in 58% yield upon exposure to Grubbs' second-generation catalyst **12** and microwave heating at 75 °C.[171] This ring-closing metathesis process is exploited in the synthesis of (±)-grandisol (**465**) from 1,5-enyne **463**. The complete carbon framework of the target natural product is established in cyclobutene **464** by enyne ring-closing metathesis, and the more strained double bond is chemoselectively hydrogenated to provide (±)-grandisol (**465**) after desilylation.[172]

Scheme 100 Enyne Ring-Closing Metathesis Based Synthesis of (±)-Grandisol[171,172]

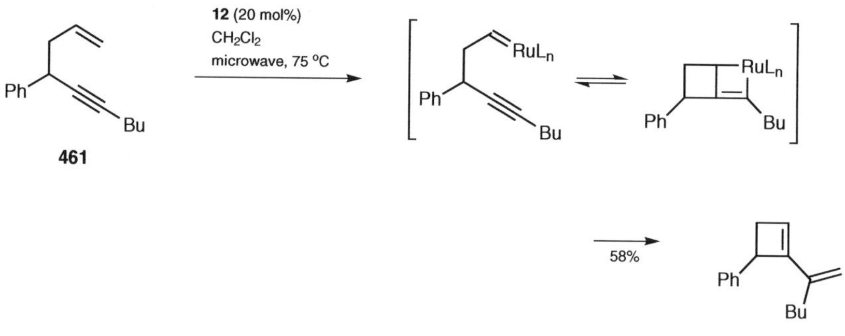

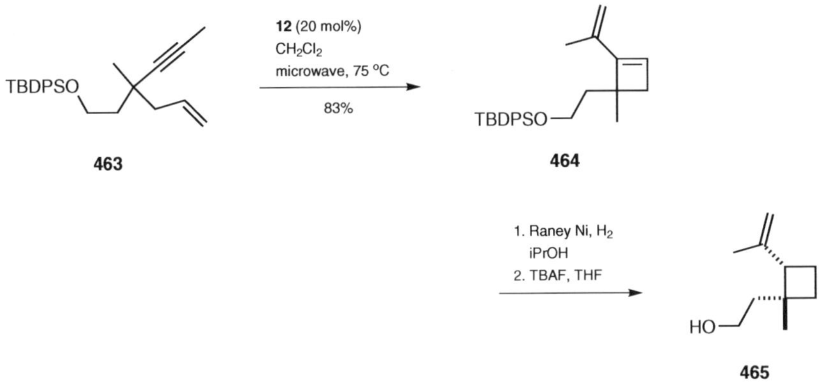

2.10.3 Enyne Ring-Closing Metathesis

Enyne ring-closing metathesis has also been successfully implemented for the synthesis of neurochemically active alkaloids **468**, **471**, and **474** (Scheme 101).[173–175] The ring-closing metathesis of 1,7-enyne **466** provides azabridged bicycle **467** in good yield by using Grubbs' first-generation ruthenium catalyst **11**. The 1,3-diene moiety of **467** is further converted into the corresponding α,β-unsaturated enone to complete the total synthesis of (+)-ferruginine (**468**) in three steps.[173] Treatment of 1,8-enyne **469** with Grubbs' second-generation catalyst **12** affords azabicycle **470**, which serves as an advanced intermediate for the synthesis of (+)-anatoxin-a (**471**).[174] In a similar manner, ring-closing metathesis of silylated alkyne **472** with Grubbs' second-generation catalyst **12** affords azabicycle **473**, which is elaborated to N-tosylanatoxin-a (**474**).[175]

Scheme 101 Enyne Ring-Closing Metathesis in the Synthesis of Azabridged Bicyclic Natural Products[173–175]

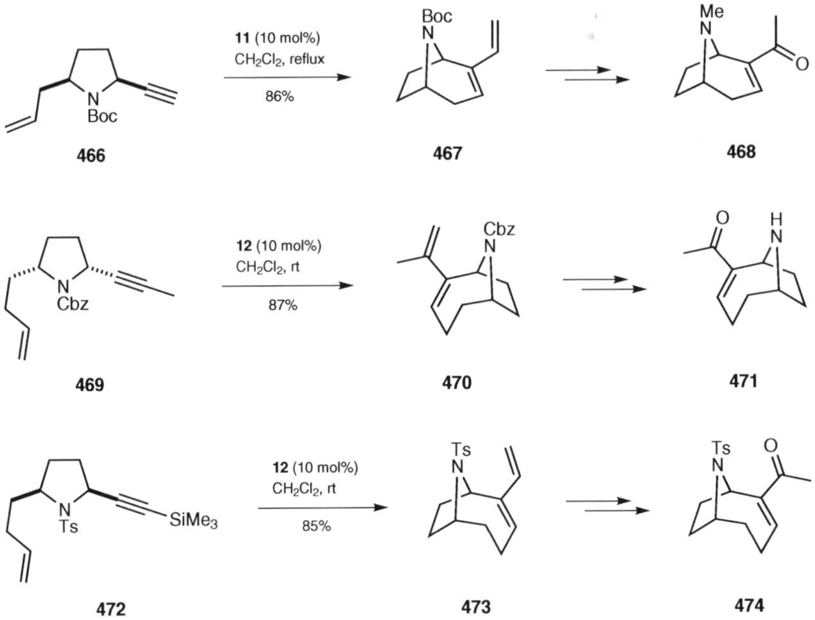

In the total synthesis of β-eremophilane (**477**) the ring-closing metathesis of enynes containing a silyl ynol ether has been applied. Treatment of silyl ynol ether **475** with Grubbs' second-generation catalyst **12** followed by in situ desilylation provides α,β-unsaturated enone **476** in high yield (85%) (Scheme 102) which is further elaborated to β-eremophilane (**477**).[176]

Scheme 102 Enyne Ring-Closing Metathesis in the Synthesis of β-Eremophilane[176]

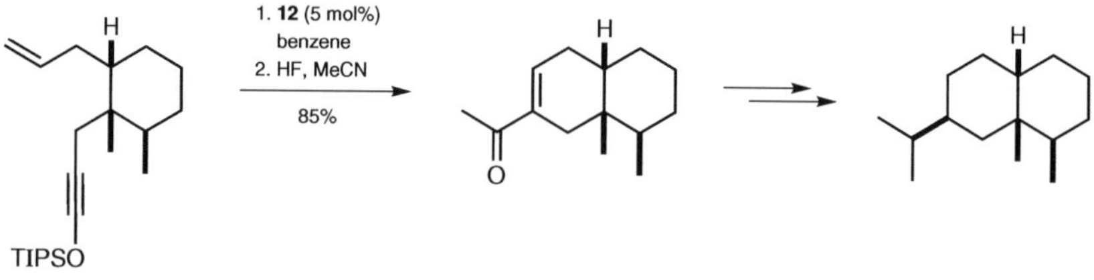

In the synthesis of 6-deoxyerythronolide B (**480**), the 14-membered ring is formed by enyne ring-closing metathesis (Scheme 103).[177] The initial attempts at ring-closing metathesis with enyne **478** in the absence of ethene, using Grubbs–Hoveyda second-generation catalyst **15**, led only to isomerization of the terminal double bond. Even under ethene at 80 °C, macrolide **479** is not formed, but the terminal triple bond is converted into the corresponding 1,3-diene in quantitative yield. However, running the reaction at 80 °C under ethene followed by degassing and further heating at 110 °C affords macrolide **479** in excellent yield, and further functionalization provides 6-deoxyerythrolonide B (**480**).

Scheme 103 Enyne Ring-Closing Metathesis in the Synthesis of 6-Deoxyerythronolide B[177]

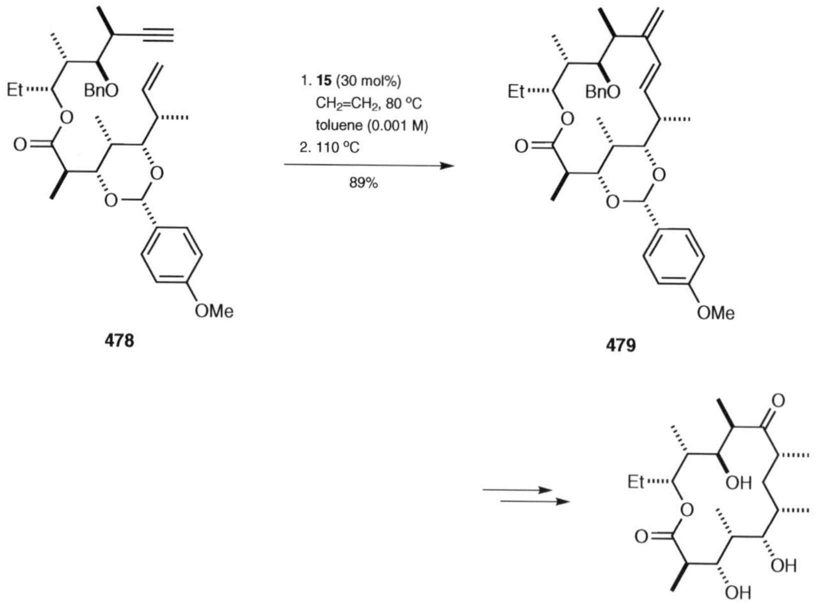

A successful synthesis of (+)-panepophenanthrin (**483**) relies on an enyne ring-closing–cross metathesis (or cross–ring-closing metathesis) sequence (Scheme 104). Upon treatment of enyne **481** (R¹ = H) and but-3-en-2-ol with Grubbs' second-generation catalyst **12**

2.10.3 Enyne Ring-Closing Metathesis

the desired product **482** is obtained in 28% yield. On the other hand, relay metathesis with enyne **481** ($R^1 = CH_2OCH_2CH=CH_2$) improves the yield of **482** to 51%, which is most likely due to a change of the metathesis sequence from cross–ring-closing metathesis to ring-closing–cross metathesis. Simple functional group manipulations complete the synthesis of the natural product **483**.[178]

Scheme 104 Enyne Ring-Closing–Cross Metathesis (or Cross–Ring-Closing Metathesis) in the Synthesis of (+)-Panepophenanthrin[178]

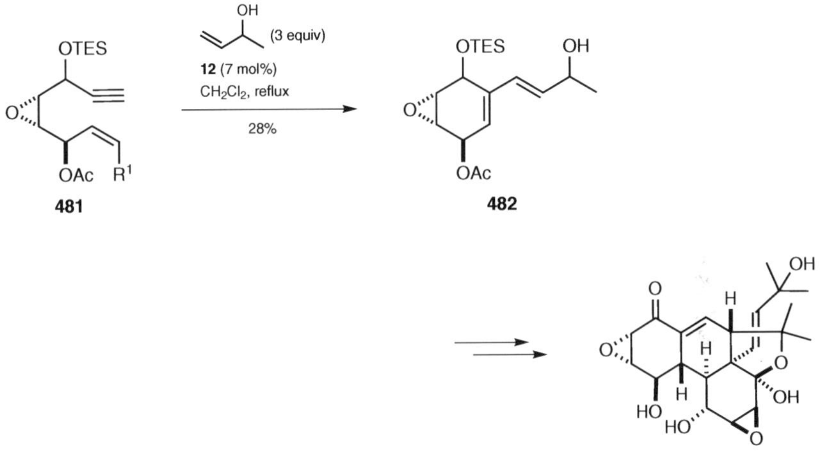

R^1	Yield (%)	Ref
H	28	[178]
$CH_2OCH_2CH=CH_2$	51	[178]

2.10.3.3.2 Double Ring-Closing Metathesis with Dienynes

The total synthesis of (−)-securinine (**486**) relies on a sequential ring-closing metathesis of a dienyne (Scheme 105). The group-selective ring-closing metathesis of dienyne **484** provides the bicyclic diene **485** in the presence of ruthenium complex **14**.[179] To control the sequence of the double ring-closing metathesis, the two double bonds are differentiated by an extra alkyl substituent such that the initiation occurs at the sterically less-hindered allyl ether group of dienyne **484**.

Scheme 105 Dienyne Ring-Closing Metathesis in the Synthesis of (−)-Securinine[179]

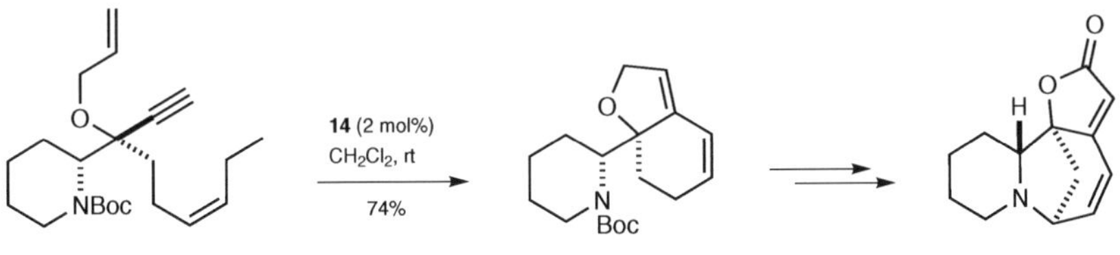

A similar dienyne ring-closing metathesis strategy is used in the total synthesis of guanacastepene A (**489**) (Scheme 106). Sterically differentiated double bonds on trienyne **487** allow the ring-closing metathesis to occur and deliver tricycle **488** in good yield (82%) with Grubbs' second-generation catalyst **12**. Installation of the required oxygen functional groups on the tricyclic ring-closing metathesis product **488** provides the natural product **489**.[180]

Scheme 106 Dienyne Ring-Closing Metathesis in the Synthesis of Guanacastepene A[180]

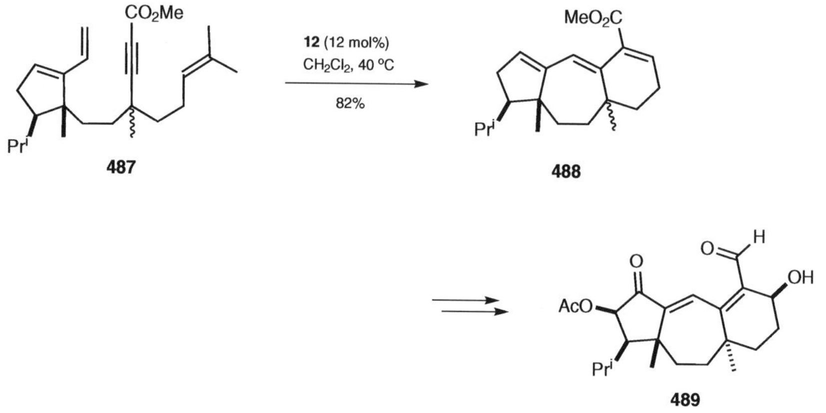

Another sequence-selective dienyne ring-closing metathesis relying on steric differentiation is illustrated in the synthesis of kempene-2 (**492**), kempene-1 (**493**), and 3-*epi*-kempene-1 (**494**) (Scheme 107). The pivotal tetracyclic core of the kempenes is constructed via ring-closing metathesis of dienyne **490** with Grubbs' second-generation catalyst **12** in refluxing dichloromethane giving the ring-closing metathesis product **491** in 92% yield (Scheme 107).[181]

Scheme 107 Dienyne Ring-Closing Metathesis in the Synthesis of Kempenes[181]

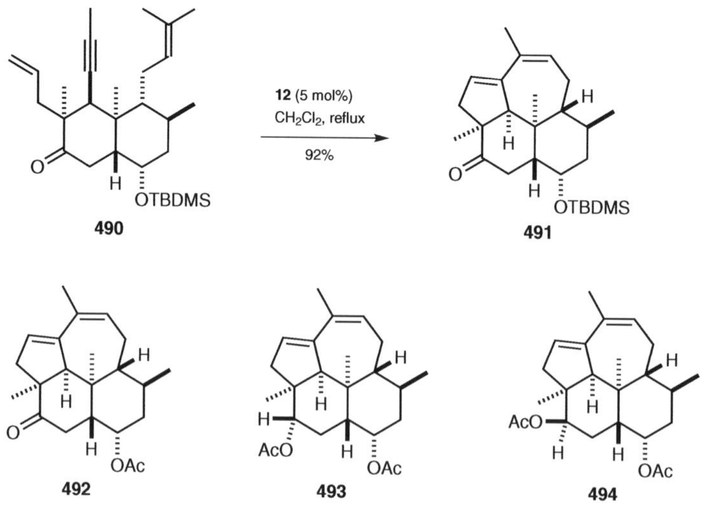

2.10.3 Enyne Ring-Closing Metathesis

A formal synthesis of (−)-cochleamycin A (**498**) makes use of a tandem dienyne ring-closing metathesis approach (Scheme 108).[182] The sequence-selective tandem enyne ring-closing metathesis of silaketal **495** promoted by Grubbs' second-generation catalyst **12** provides a bicyclic siloxane, which is transformed into an E,Z-1,3-diene **496** with tetrabutylammonium fluoride. Elaboration of **496** via a 12-step sequence gives the known intermediate **497**[183] and then completes a formal synthesis of (−)-cochleamycin A (**498**).

Scheme 108 Dienyne Ring-Closing Metathesis in the Synthesis (−)-Cochleamycin A[182,183]

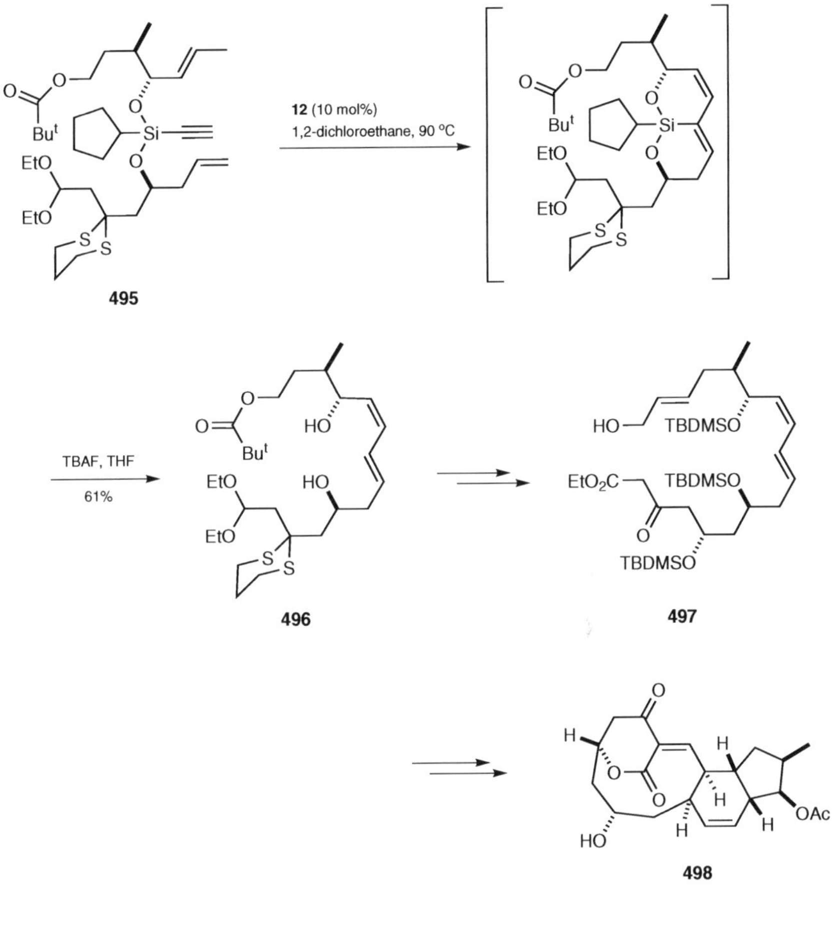

A total synthesis of (−)-acylfulvene and (−)-irofulven uses dienyne ring-closing metathesis as a key step to form a six-membered carbocycle (Scheme 109). The reaction of silicon-tethered trienyne **499** provides ring-closing metathesis product **500** accompanied by a significant amount of alternative ring-closing metathesis product **501** after in situ desilylation. By masking the trisubstituted alkene with a phenyl group in trienyne **502**, only the desired ring-closing metathesis product **503** is obtained in good yield (79%) after in situ desilylation.[184]

Scheme 109 Dienyne Ring-Closing Metathesis in the Synthesis of (−)-Acylfulvene and (−)-Irofulven[184]

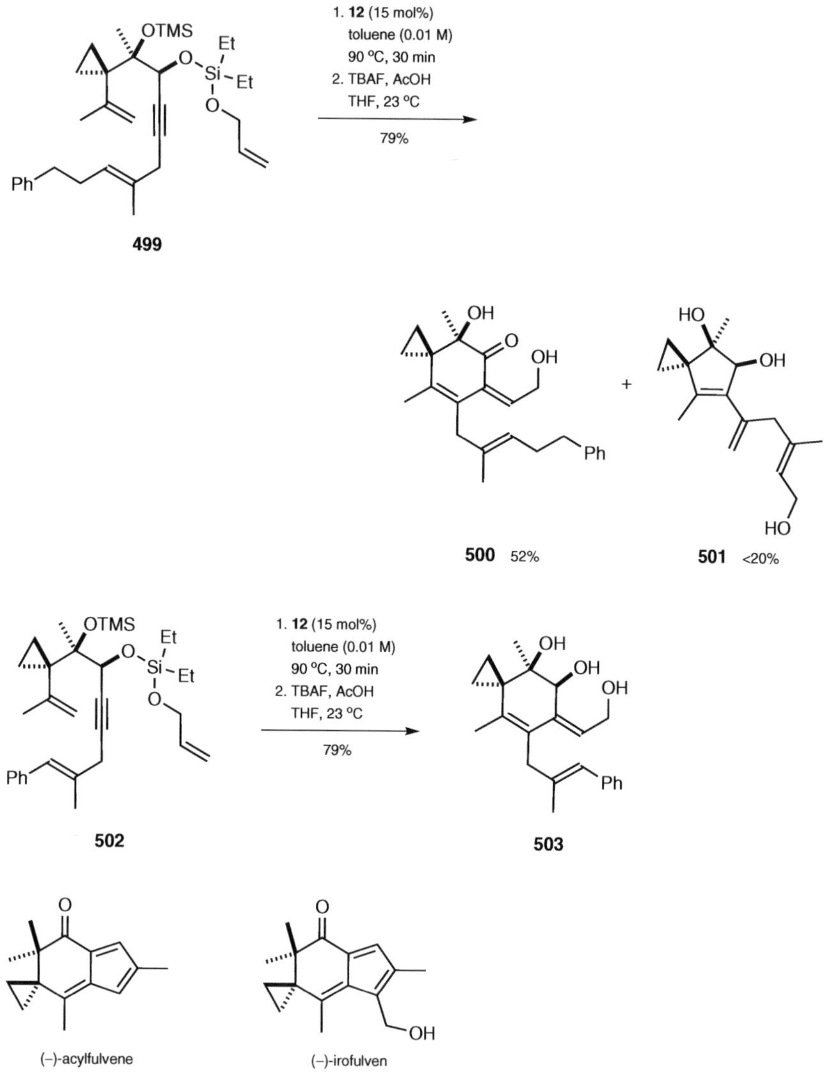

A domino enyne ring-closing metathesis–intramolecular Diels–Alder reaction sequence is used for the construction of the core structure of (−)-vinigrol (**510**) (Scheme 110). The tandem ring-closing metathesis of dienyne **504** with Grubbs' second-generation catalyst **12** in toluene at 110 °C delivers the dienyne ring-closing metathesis product **506** via a putative intermediate **505**. However, under ethene at room temperature, the propagating intermediate **505** is intercepted to generate mono ring-closing metathesis product **507**, which undergoes intramolecular Diels–Alder reaction upon treatment with tin(IV) chloride to afford tricycle **508** predominantly. Wittig alkenation of **508** followed by chemoselective hydrogenation provides known intermediate **509**[185] and further elaboration completes a formal synthesis of (−)-vinigrol (**510**).[185,186]

2.10.3 Enyne Ring-Closing Metathesis

Scheme 110 Enyne Ring-Closing Metathesis/Intramolecular Diels–Alder Sequence in the Synthesis of (−)-Vinigrol[186]

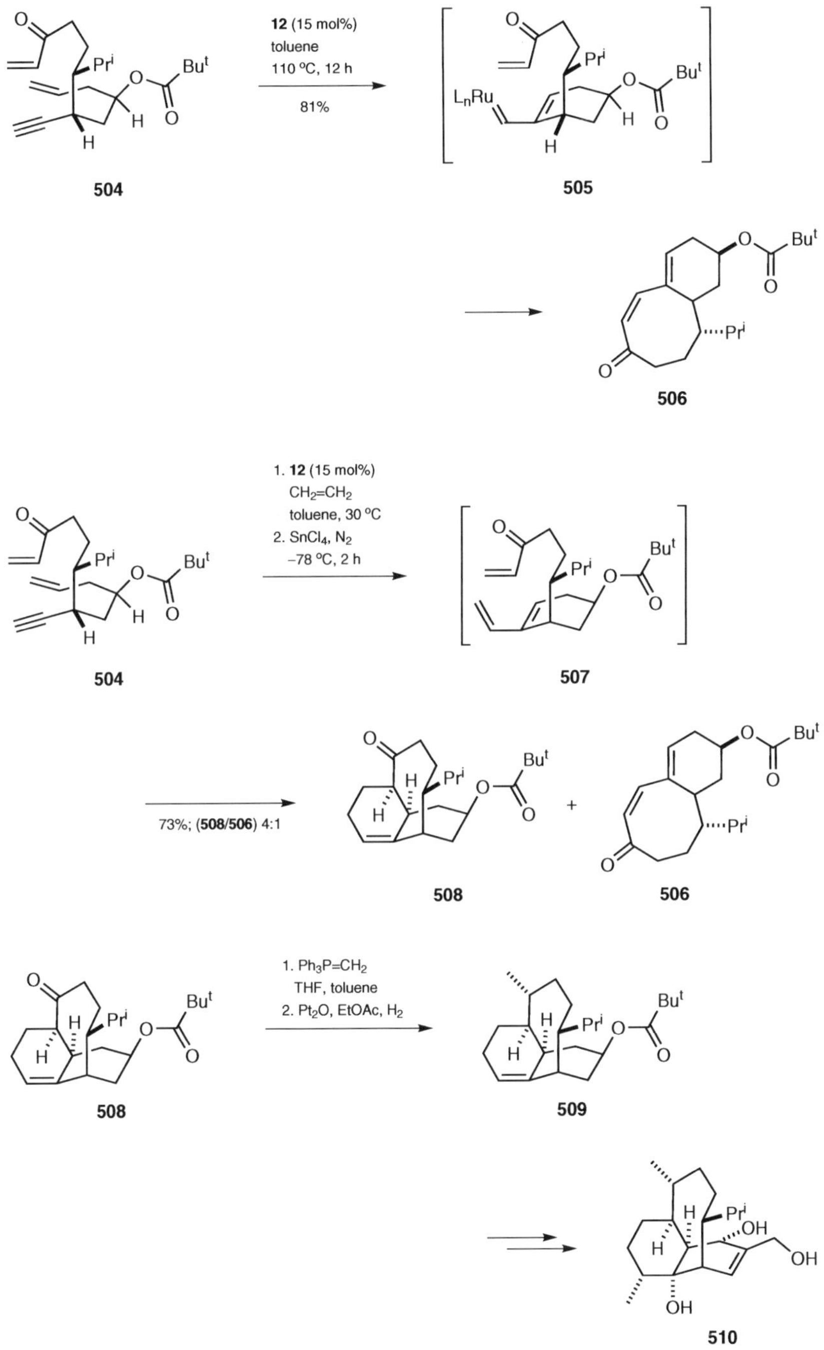

A relay metathesis initiated dienyne ring-closing metathesis is employed in the total synthesis of (−)-englerin A [(−)-**195**] (Scheme 111). To enforce chemoselective initiation at a 1,1-disubstituted double bond in the presence of a terminal triple bond, a relay tether is introduced to one of the two 1,1-disubstituted double bonds, which initiates a tandem ring-closing metathesis process of trienyne **511** with ruthenium complex **13** to generate 1,3-diene **512** in good yield. Further elaboration of **512** to the known intermediate **513**[187] constitutes a formal synthesis of (−)-englerin A [(−)-**195**].[188]

Scheme 111 Dienyne Ring-Closing Metathesis in the Synthesis of (−)-Englerin A[187,188]

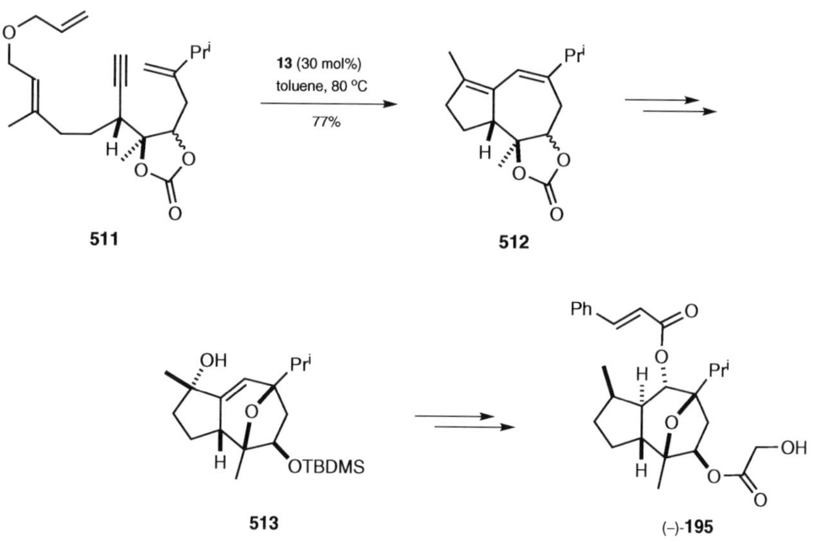

Another relay metathesis based tandem enyne ring-closing metathesis approach is employed for the synthesis of securinega alkaloids **517–519** (Scheme 112).[189] To initiate the sequential ring-closing metathesis in the right order, the initiation is required to occur at the electronically deactivated acrylate moiety. A relay tether installed at the acrylate moiety in dienyne substrates **515** serves nicely for this purpose and, upon treatment with ruthenium complex **514** in refluxing toluene, the ring-closing metathesis of the diastereomers **515A** and **515B** affords dihydrobenzofuranones **516A** and **516B** in 64 and 67% yield, respectively. These ring-closing metathesis products are separately elaborated to (−)-norsecurinine (**517A**) and (+)-allonorsecurinine (**517B**), from which the synthesis of the more complex natural products (−)-flueggine A (**518**) and (+)-virosaine B (**519**), respectively, is achieved.

Scheme 112 Dienyne Ring-Closing Metathesis in the Synthesis of Securinega Alkaloids[189]

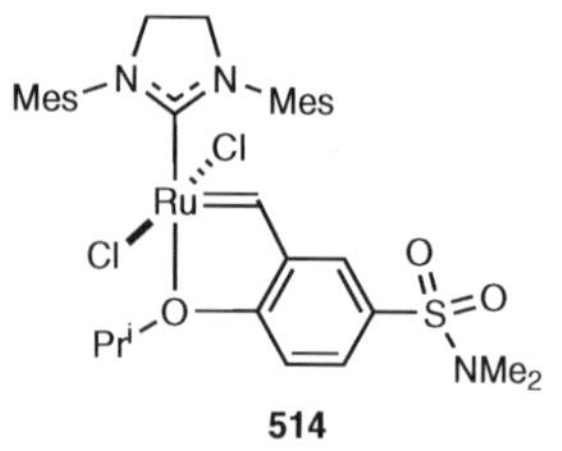

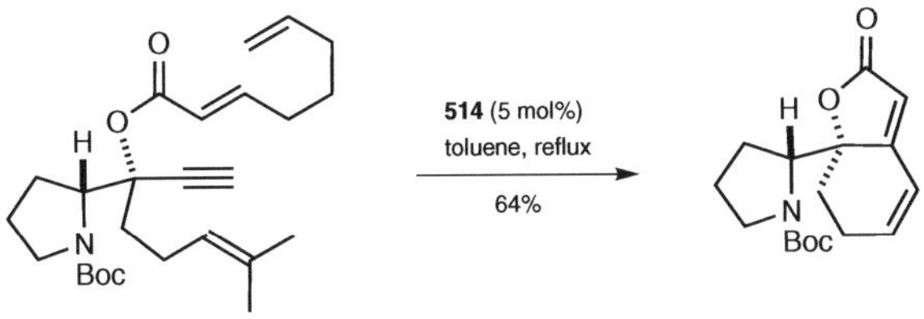

2.10.3 Enyne Ring-Closing Metathesis

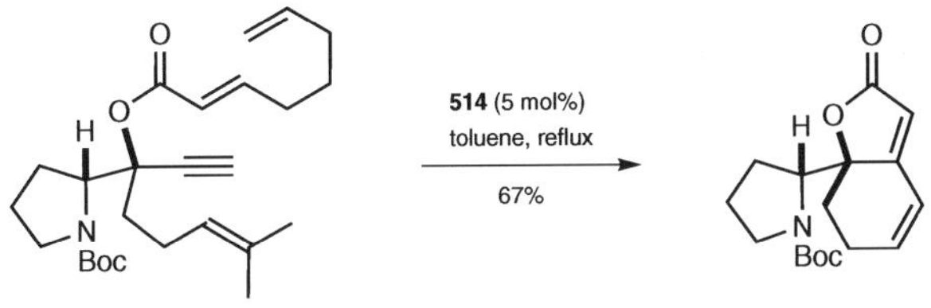

515A → (514 (5 mol%), toluene, reflux, 64%) → **516A**

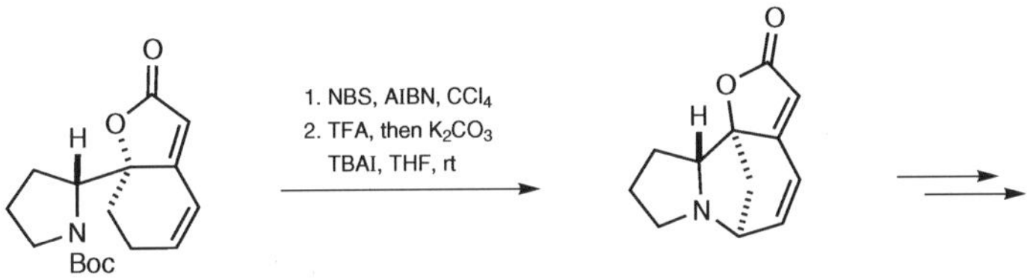

515B → (514 (5 mol%), toluene, reflux, 67%) → **516B**

516A → (1. NBS, AIBN, CCl₄; 2. TFA, then K₂CO₃, TBAI, THF, rt) → **517A** (−)-norsecurinine 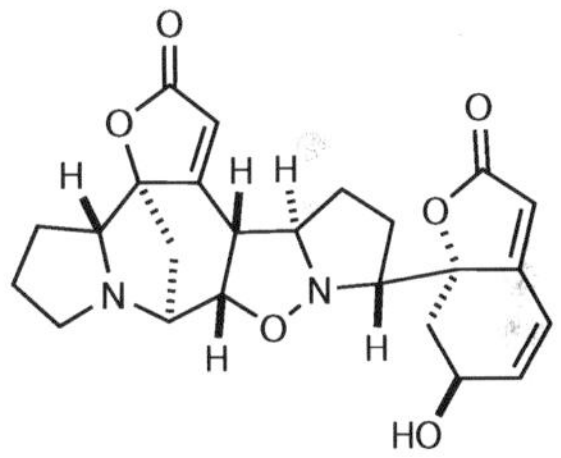 → ⟶ **518** (−)-flueggine A

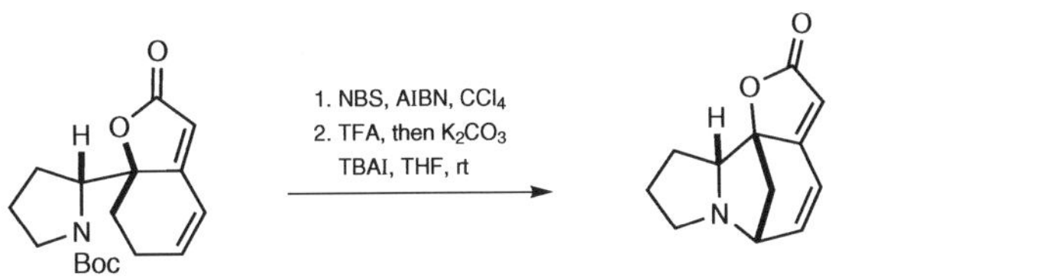

516B → (1. NBS, AIBN, CCl₄; 2. TFA, then K₂CO₃, TBAI, THF, rt) → **517B** (+)-allonorsecurinine 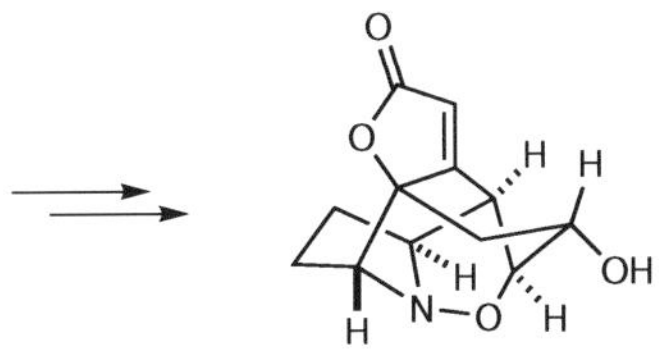 → ⟶ **519** (+)-virosaine B

2.10.3.4 Diverse Applications of Enyne Ring-Closing Metathesis in Synthesis

2.10.3.4.1 Ring-Rearrangement by Enyne Ring-Closing Metathesis

The ring-rearrangement metathesis (RRM) of enynes **520** catalyzed by Grubbs' first-generation catalyst **11** under ethene provides the skeletal rearrangement products **522** (Scheme 113).[190,191] Under ethene, it is proposed that the propagating methylidene adds onto the alkyne moiety to form vinyl alkylidene intermediate **521**, which then undergoes ring-opening metathesis to provide the product **522**. No reaction is observed under an argon atmosphere.[190]

Scheme 113 Ring-Rearrangement Metathesis of Enynes To Give 2,5-Dihydropyrroles and 2,5-Dihydrofurans[190,191]

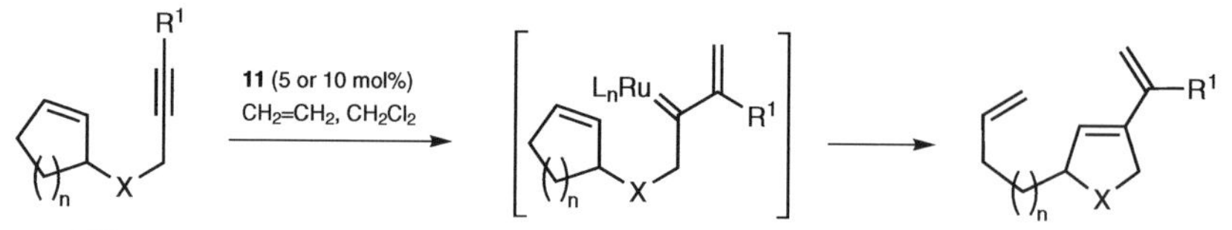

R^1	X	n	Yield (%)	Ref
Me	NTs	2	15	[190]
Me	NTs	3	56	[190]
H	NTs	2	78	[190]
H	O	1	67	[191]
H	NTs	1	75	[191]
H	NTs	6	47	[191]

Isoquinoline derivatives can be obtained via ring-rearrangement metathesis of cyclobutene-containing enynes **523** (Scheme 114). The initiation of this ring-rearrangement metathesis is proposed to occur on the internal alkyne moiety of enyne **523**, which generates alkylidene **525** via ring opening of the cyclobutene moiety by a propagating ruthenium–methylidene in intermediate **524**. The final ring-closure of intermediate **525** provides the ring-rearrangement metathesis product **526**. By varying the substituents on the alkyne (R^2), differently substituted isoquinoline derivatives **526** are obtained in good to moderate yields (Scheme 114).[192]

2.10.3 Enyne Ring-Closing Metathesis

Scheme 114 Ring-Rearrangement Metathesis of Cyclobutene-Containing Enynes To Give Isoquinoline Derivatives[192]

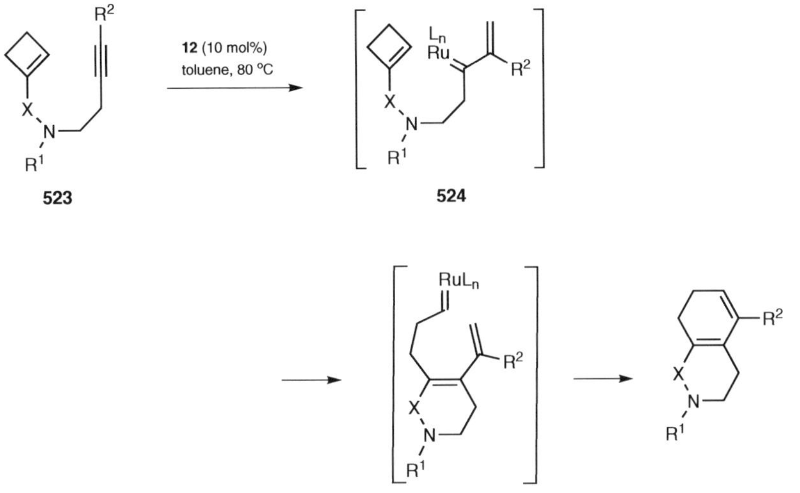

R¹	R²	X	Yield (%)	Ref
Ts	CH$_2$OAc	CH$_2$	39[a]	[192]
Bn	Me	CO	56	[192]
Bn	Ph	CO	82	[192]
Bn	naphthyl	CO	71	[192]

[a] 36% yield obtained in CH$_2$Cl$_2$ at 40 °C.

The ring-rearrangement metathesis of cyclopropene-containing 1,6-enynes gives 2,5-dihydropyrrole skeletons. Treatment of substrates **527** with Grubbs' first-generation catalyst **11** along with a monosubstituted external alkene provides ring-rearrangement metathesis products **530** (Scheme 115). Although the mechanism of this reaction is not clear, the initial ring-opening metathesis of the cyclopropenyl moiety mediated by a propagating methylidene to form intermediate **528** has been suggested. After generation of intermediate **529** from **528**, a subsequent cross metathesis between **529** and an external alkene provides the final product **530**. Although the yields are good regardless of the alkyl groups on the ester or the external alkene, the E/Z selectivity is low.[193]

Scheme 115 2,5-Dihydropyrroles and 2,5-Dihydrofurans by Ring-Rearrangement Metathesis of Cyclopropene-Containing Enynes[193]

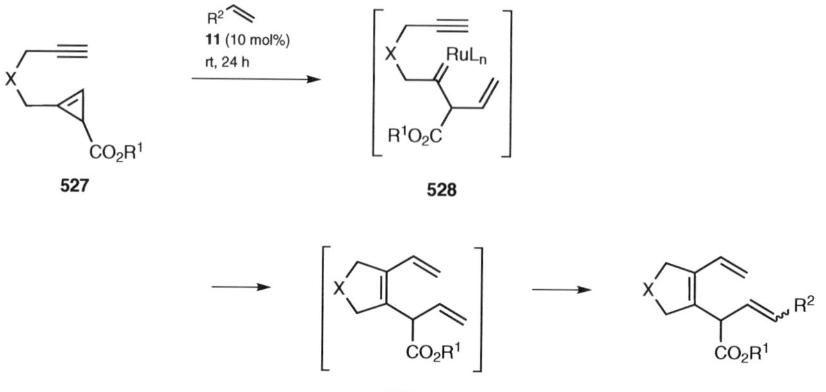

R¹	R²	X	Ratio (E/Z)	Yield (%)	Ref
Et	Ph	NTs	1:1.4	77	[193]
Me	(CH$_2$)$_4$Me	NTs	1:2.3	71	[193]
Et	Ph	4-O$_2$NC$_6$H$_4$SO$_2$N	1:0.8	74	[193]
Et	Ph	O	1:1.4	55	[193]

A concise synthesis of virgidivarine (**535**) and virgiboidine (**536**) relies on ring-rearrangement metathesis for the construction of the dipiperidine moiety (Scheme 116).[194] The ring-rearrangement metathesis of dienyne **531** with Grubbs–Hoveyda second-generation catalyst **15** under ethene provides dipiperidine derivative **534** in 83% yield. Mechanistically, it is proposed that the initiation occurs on the terminal double bond followed by ring-closing metathesis with the cyclopentene moiety to form intermediate **532**, which then undergoes enyne ring-closing metathesis to generate **533** as the penultimate intermediate. However, considering the presence of ethene, cross metathesis between a propagating methylidene and the double bond may be the equally or even more preferred initiation event of this ring-rearrangement metathesis process. Dipiperidine **534** is further elaborated to virgidivarine (**535**), which is subjected to lactam formation with 1-ethyl-3-(3-dimethylaminopropyl)carbodiimide to complete the synthesis of virgiboidine (**536**).

2.10.3 Enyne Ring-Closing Metathesis

Scheme 116 Ring-Rearrangement Metathesis of a Dienyne for the Construction of the Dipiperidine Moiety in the Synthesis of Virgidivarine and Virgiboidine[194]

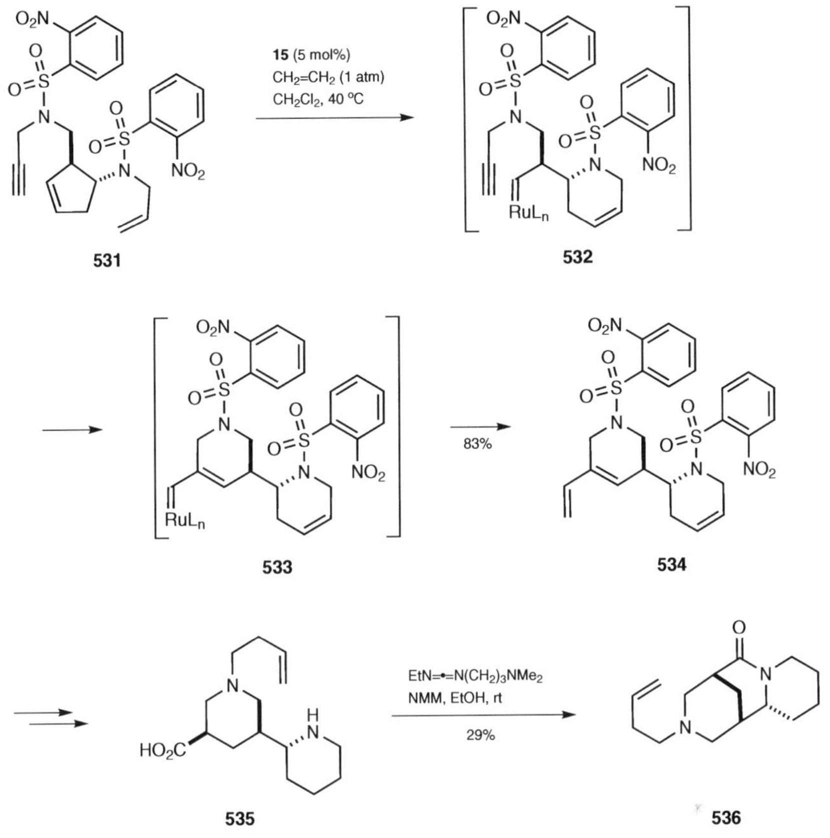

In the ring-rearrangement metathesis of norbornene derivatives, the enyne **537** provides *exo*-bicyclic product **538** upon treatment with Grubbs' first-generation catalyst **11** under ethene (Scheme 117).[195] In a similar ring-rearrangement metathesis, enediyne **539** rearranges to the tricyclic compound **540**.[196] Norbornene-based dienyne **541** is converted into tricyclic compound **542** in high yield by consecutive use of Grubbs' first-generation catalyst **11** and Grubbs' second-generation catalyst **12** under ethene and microwave irradiation at 60 °C.[197]

Scheme 117 Ring-Rearrangement Metathesis of Norbornene Derivatives[195–197]

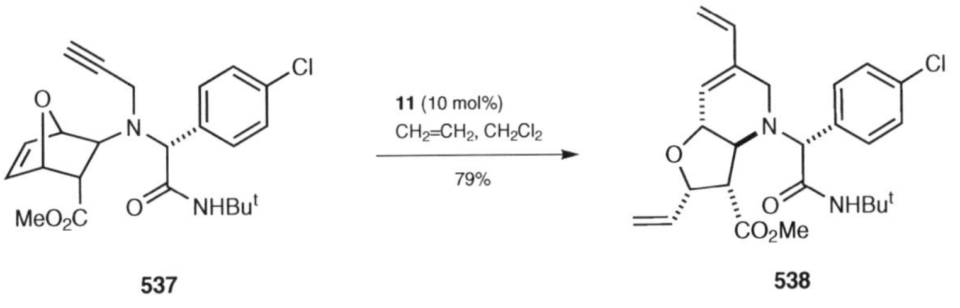

for references see p 672

539 → **540**
11 (10 mol%)
CH₂=CH₂, CH₂Cl₂
43%

541 → **542**
1. 11 (10 mol%), CH₂=CH₂, microwave, 60 °C
2. 11 (10 mol%), **12** (10 mol%)
CH₂=CH₂, microwave, 60 °C
87%

Isoquinolines 526; General Procedure:[192]
A cyclobutene-containing enyne **523** (0.12 mmol) and Grubbs' second-generation catalyst **12** (0.012 mmol, 0.1 equiv) were dissolved in anhyd toluene (0.02–0.03 M). The soln was degassed and subsequently argon was introduced into the reaction vessel. The soln was heated at 80 °C. After the starting material had disappeared (monitored by TLC), a few drops of ethyl vinyl ether were added to the soln. The volatiles were removed under reduced pressure, and the residue was purified by column chromatography (silica gel, hexanes/EtOAc 2:1).

Ethyl 4-Phenyl-2-(1-tosyl-4-vinyl-2,5-dihydro-1H-pyrrol-3-yl)but-3-enoate (530, R^1 = Et; R^2 = Ph; X = NTs); General Procedure:[193]
Under argon, ethyl 2-({[4-methyl-N-(prop-2-ynyl)phenyl]sulfonamido}methyl)cycloprop-2-ene-1-carboxylate (**527**, R^1 = Et; X = NTs; 33 mg, 0.1 mmol), Grubbs' first-generation catalyst **11** (8 mg, 0.01 mmol, 0.1 equiv), and distilled styrene (1.0 mL) were added into a flame-dried Schlenk tube. The mixture was stirred at rt for 24 h. The excess styrene was recovered by vacuum distillation and the residue was purified by a flash column chromatography (silica gel, petroleum ether/EtOAc 12:1); yield: 33 mg (77%); ratio (E/Z) 1:1.4.

2.10.3.4.2 Multiple Combinations of Metathesis

For multiple elementary steps to be streamlined in a tandem metathesis process, strategic positioning of metathesis-active π-systems at the right places is required such that among many potential pathways, only the most preferred planned pathway is to operate. The combination of enyne ring-closing metathesis and metallotropic [1,3]-shift can operate in a sequence-selective manner to generate oligoenynes (Scheme 118).[198] Enyne system **543** containing multiple copies of internal 1,3-diyne moieties provide oligoenynes **544** when treated with Grubbs' second-generation catalyst **12**. This outcome is the consequence of a chemoselective initiation at the terminal double bond followed by a series of self-regiocontrolled ring-closing metathesis events while the propagating propargylic alkylidene species are in equilibrium through metallotropic [1,3]-shifts. This multiple sequence of metathesis can be extended to other substrates to deliver the related products **545** and **546**.

2.10.3 Enyne Ring-Closing Metathesis

Scheme 118 Combination of Multiple Metathesis Reactions To Form Oligoenynes[198]

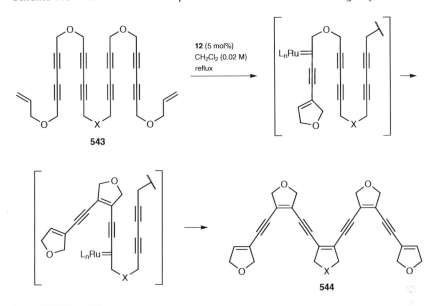

X	Yield (%)	Ref
O	51	[198]
NTs	45	[198]

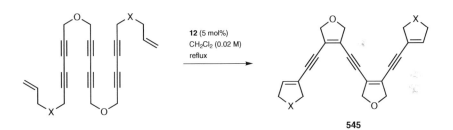

X	Yield (%)	Ref
O	63	[198]
NTs	47	[198]

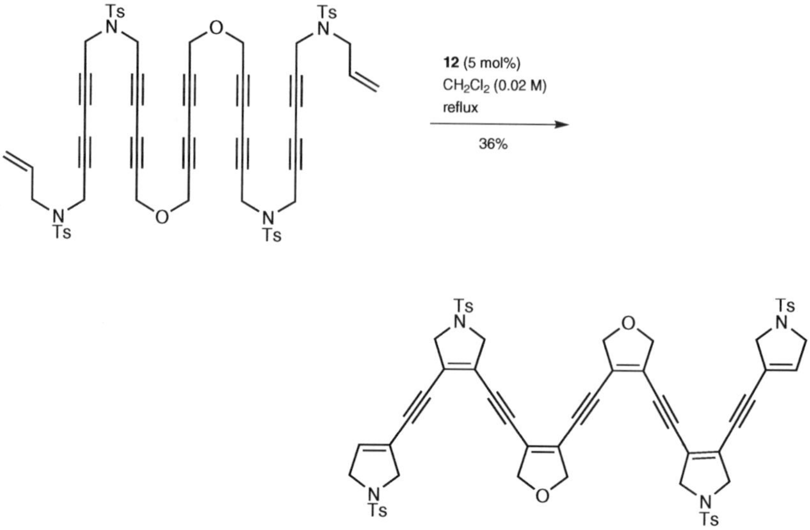

A relay metathesis-initiated enyne ring-closing metathesis and metallotropic [1,3]-shift is successfully implemented in the total synthesis of epoxyquinoids **551**–**553** (Scheme 119).[199] Enediyne **547** reacts with Grubbs' second-generation catalyst **12** to generate ruthenium–alkylidene **548**, which forms a new ruthenium–alkylidene **549** by ring closure. The preferred [1,3]-shift and final turnover of **549** affords 1,5-dien-3-ynes **550**. Subsequent functional group manipulations lead to the total synthesis of (+)-asperpentyn (**551**), (−)-harveynone (**552**), and (−)-tricholomenyn A (**553**).[199]

2.10.4 Diyne Ring-Closing Metathesis

Scheme 119 Combination of Multiple Metathesis Events in the Synthesis of Epoxyquinoids[199]

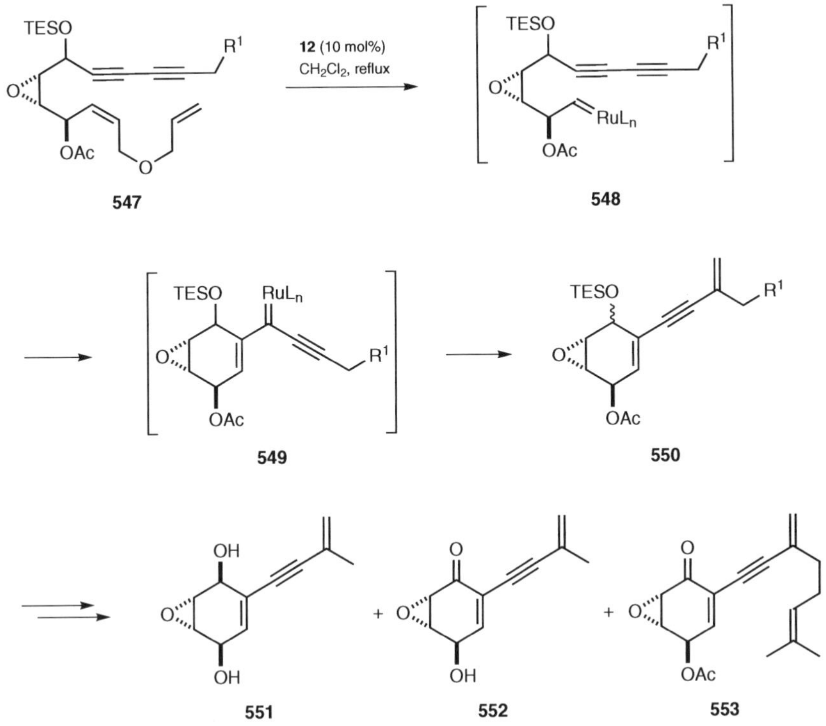

R^1	Yield (%) of **550**	Ref
H	62	[199]
$CH_2CH=CMe_2$	58	[199]

2.10.3.5 Scope and Limitations

Due to the possibility of metathesis initiation either at the alkene or alkyne part of enyne substrates, the ring closure mode in enyne ring-closing metathesis is not always predictable. This problem can usually be resolved by installing steric or electronic modulating groups on the alkene or alkyne moiety although this requires more steps in the preparation of the ring-closing metathesis substrates. These unnecessary steps can be eliminated if catalysts have predictable preferences for different initiation modes, but only a few such enyne metathesis catalysts are available. In addition, due to the higher activation barrier for the formation of the (2+2) cycloadduct of a metal–alkylidene with an alkyne than that with an alkene, the formation of medium rings and macrocycles by enyne ring-closing metathesis is challenging and thus ring closure with enyne substrates devoid of elaborated conformation-rigidifying elements will not generally be attainable for these ring systems.

2.10.4 Diyne Ring-Closing Metathesis

Alkyne metathesis is an exchange process of alkylidyne units between a pair of alkynes catalyzed by metal–carbyne complexes. Depending on the spatial relationship of the involved reacting alkyne counterparts and the products thereof, alkyne metathesis can be classified into cross alkyne metathesis, ring-closing alkyne metathesis (RCAM), ring-opening alkyne metathesis polymerization, and acyclic diyne metathesis polymerization (ADIMET). Among the different types of alkyne metathesis, ring-closing alkyne metathesis is particularly useful for the formation of macrocycles containing a *cis* or *trans* double bond via the corresponding macrocyclic alkynes. Compared to the significant impact of diene metathesis in organic synthesis, the corresponding alkyne metathesis is much less-developed, probably due to the limited availability of user-friendly catalysts of good reactivity and functional group tolerance. However, with the development of various tungsten– and molybdenum–alkylidyne complexes, the prowess and utility of alkyne metathesis has been exploited in the synthesis of various natural products, especially containing macrocyclic structures. The unique merits of alkyne metathesis are the orthogonal chemoselectivity of alkylidynes for alkyne–alkyne coupling devoid of alkene–alkene or alkene–alkyne coupling and the versatility of the alkyne products in post-metathesis elaboration.

2.10.4.1 Catalyst and Mechanism

The availability of catalysts with desirable features is one of the key requirements for their practical use. All alkyne metathesis catalysts known to date belong to either ill-defined [the so-called Mortreux system consisting of hexacarbonylmolybdenum(0) and various phenols] or well-defined single-component catalyst systems (Schrock-type complexes). The latter have played a major role in the development of new air-stable yet alkyne-metathesis-active catalysts. Well-defined single-component catalysts are generally viewed as variants of Schrock-type alkylidyne complexes of high oxidation state molybdenum (e.g., **1** and **555–560**) or tungsten (e.g., **8** and **554**) (Scheme 120). Except for the Mortreux system, all classical Schrock-type alkylidyne complexes are highly moisture- and oxygen-sensitive, and molybdenum complex **555** is even sensitive to molecular nitrogen. The first well-defined and widely used alkyne metathesis catalyst is Schrock-type tungsten complex **8** containing bulky alkoxide ligands that prevent or slow down bimolecular decomposition of the alkylidyne intermediates in the catalytic cycle.[200] Other bulky ligands having alkyl, thiolate, and amide groups do not show much improvement in catalyst activity.[201] Significant improvements of the performance of Schrock-type complexes have emerged mainly due to the systematic studies of Fürstner and coworkers, and their broad range of applications have included complex natural product synthesis.[202]

Scheme 120 Representative Catalysts for Alkyne Metathesis

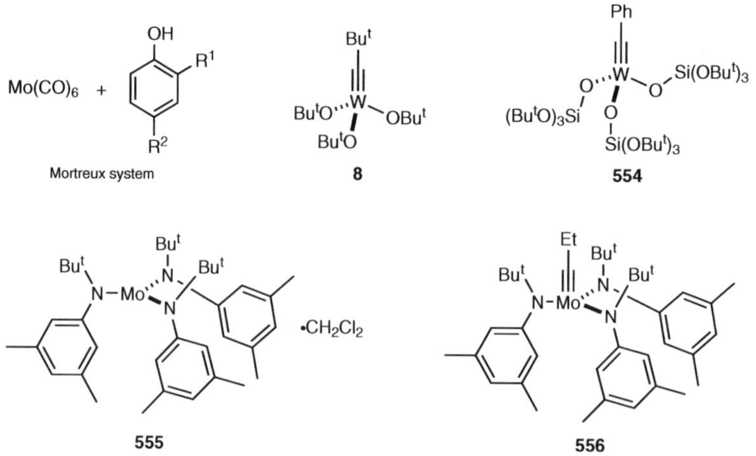

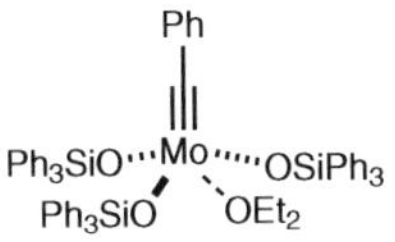

2.10.4 Diyne Ring-Closing Metathesis

557 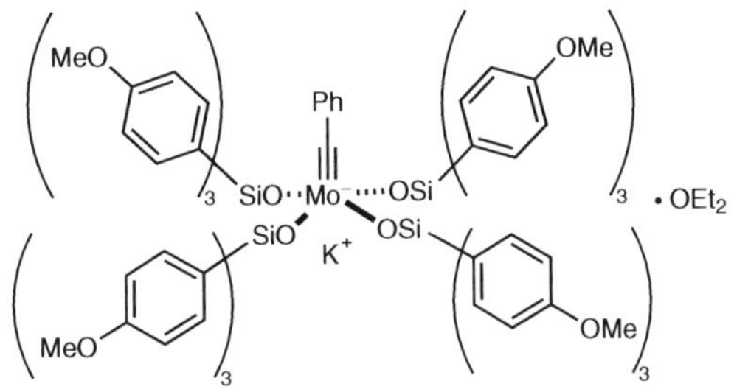 **558** 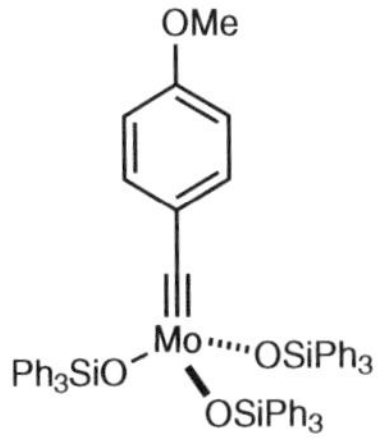 **559**  **560** 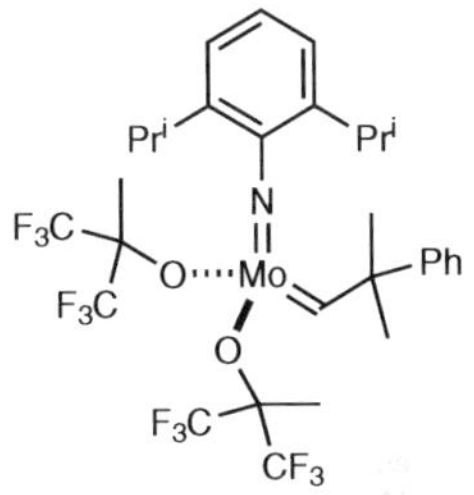 **1**

In 1975, Katz and McGinnis proposed a mechanism for alkyne metathesis which is similar to the Chauvin mechanism[203] for alkene metathesis.[204] In the catalytic cycle, a Schrock-type alkylidyne forms a metallacyclobutadiene intermediate by a formal (2+2) cycloaddition (Scheme 121). The metallacyclobutadiene undergoes consecutive double-bond migration and cycloreversion (an equivalent of two formal cycloreversions) provides a new alkyne product **561** and a propagating metal–alkylidyne species **562**.

Scheme 121 Plausible Mechanism of Diyne Metathesis[204]

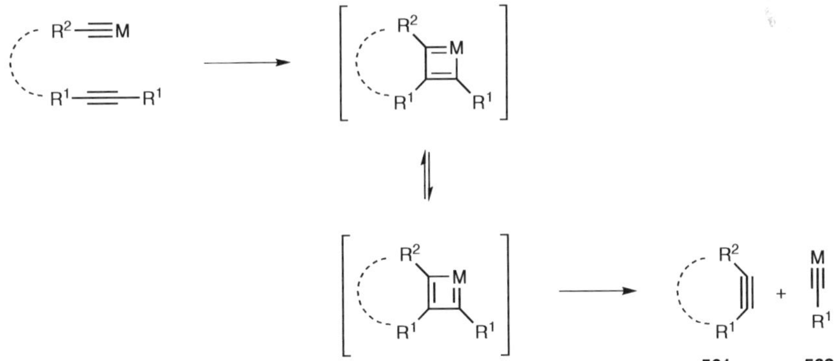

One reason for the particular problem associated with alkyne metathesis of terminal alkynes when catalyzed by a well-defined tungsten–neopentylidyne complex has been elucidated through the isolation of a deprotonated metallacyclobutadiene intermediate and its X-ray diffraction analysis.[205] Recently, theoretical investigations of tungsten– and molybdenum–alkylidynes for alkyne metathesis have been performed to elucidate their relative reactivity order.[206]

2.10.4.2 Ring Size and Substrates

The formation of cycloalkynes via diyne ring-closing metathesis should employ diyne substrates that form large enough cycles to accommodate an alkyne moiety without high ring strain. This structural characteristic of cycloalkynes limits the use of ring-closing alkyne metathesis to the formation of macrocycles larger than 12-membered rings. If alkyne-containing cycles have large enough ring strain then ring-opening metathesis would promote polymerization of the cycloalkynes.[207] Various polar functional groups on the diyne substrates can be used in ring-closing alkyne metathesis depending on the choice of catalyst.

2.10.4.2.1 1,n-Diynes

Since the first efficient ring-closing alkyne metathesis of 1,n-diynes catalyzed by tungsten–alkylidyne complex **8** described in Scheme 5 (Section 2.10.1), more examples of ring-closing alkyne metathesis with this catalyst have emerged. The ring-closing alkyne metathesis of diyne **563** with tungsten complex **8** (6 mol%) at 80 °C provides 12-membered macrocycle **564** in moderate yield (Scheme 122). This macrocyclic alkyne is converted into conformationally restricted cystine isosteres **565** and **566**.[208] A similar protocol is employed for the synthesis of alkyne-containing 12-membered macrocyclic dipeptides.[209]

Scheme 122 Ring-Closing Alkyne Metathesis in the Synthesis of Cystine Isosteres[208]

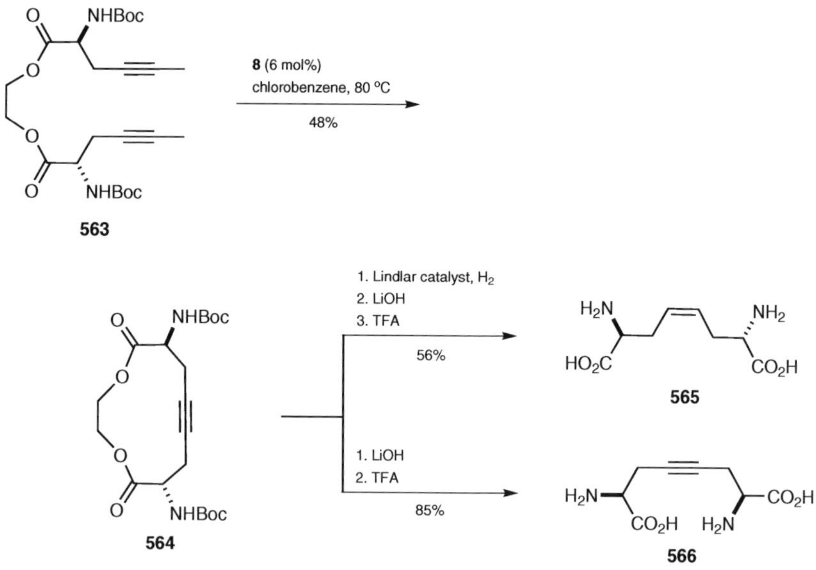

For the synthesis of cyclic β-turn mimetics, a ring-closing alkyne metathesis catalyzed by Schrock catalyst **8** converts 1,15-diynes **567** into the corresponding 14-membered macrocyclic tetrapeptides **568** in good yields (Scheme 123).[210,211]

2.10.4 Diyne Ring-Closing Metathesis

Scheme 123 Ring-Closing Alkyne Metathesis in the Synthesis of Cyclic β-Turn Mimetics[210,211]

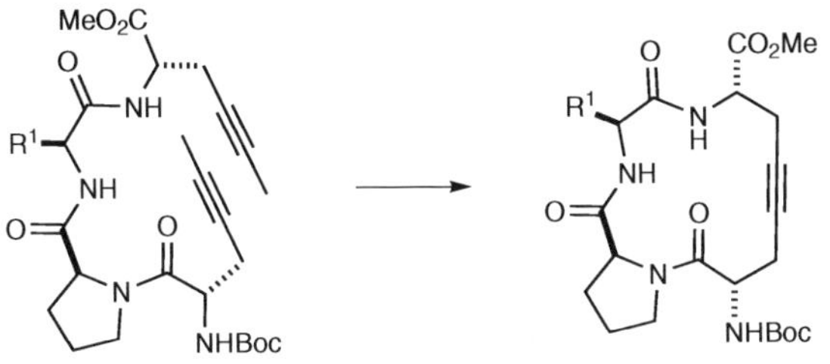

R¹	Conditions	Yield (%)	Ref
H	**8** (8 mol%), toluene (2 mM), 80 °C	82	[211]
Me	**8** (6 mol%), chlorobenzene (3.6 mM), 80 °C, 3 h	62	[210]

A ring-closing alkyne metathesis based synthesis of glycophanes uses the Mortreux catalytic system consisting of hexacarbonylmolybdenum(0) in 2-fluorophenol. The macrocyclization of diyne **569** provides phenylene-1,4-diamide **570** in 27% yield. The low yield of this cyclization seems to be a consequence of extensive decomposition of either substrate diyne **569** or product **570** under the relatively harsh reaction conditions (Scheme 124).[212]

Scheme 124 Ring-Closing Alkyne Metathesis in the Synthesis of Glycophanes[212]

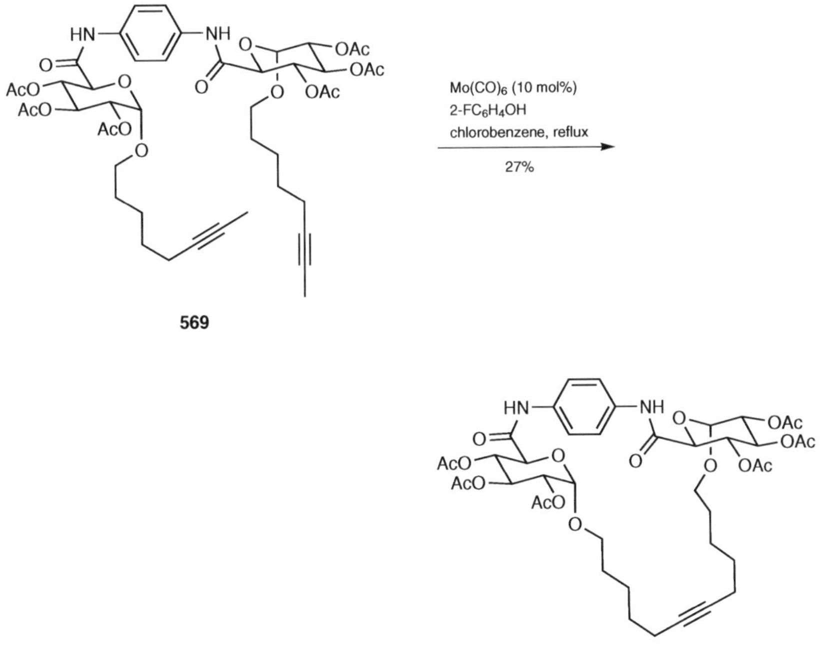

The utility of ring-closing alkyne metathesis is also demonstrated for the preparation of novel metallamacrocyclic alkynes. The ligand geometry of the produced metallamacrocycles depends on the nature of the catalyst and the reaction conditions. The ring-closing alkyne metathesis of *cis*-diyne–rhenium complex **571** with Schrock catalyst **8** delivers cis-chelated product **572** without alteration of the ligand geometry of the starting rhenium complex. On the other hand, the same rhenium complex **571** provides *trans*-chelated complex **573** in 44% yield accompanied by a small amount of *cis*-chelated product **572** upon treatment with the Mortreux system consisting of hexacarbonylmolybdenum(0) and 2-chlorophenol (Scheme 125).[213,214]

Scheme 125 Ring-Closing Alkyne Metathesis in the Synthesis of Macrocycles Containing a Transition Metal[213,214]

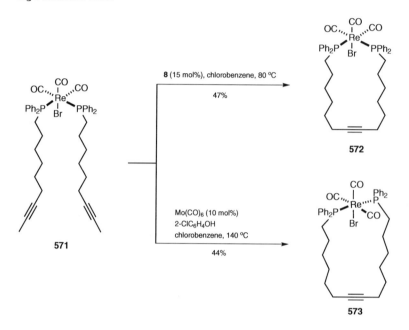

2.10.4.2.2 1,n-Bis-1,3-diynes

The ring-closing alkyne metathesis of bis-1,3-diyne systems provides an unexpected outcome. Although the formation of three different cyclic products containing one, two, or three triple bonds is possible, the ring-closing alkyne metathesis of bis-1,3-diyne **574** upon treatment with tungsten–benzylidyne complex **554** bearing siloxide ligands selectively delivers 16-membered macrocyclic 1,3-diyne **575** in 90% yield along with hexa-2,4-diyne as byproduct (Scheme 126).[215] The generality of this selectivity is yet to be confirmed as this is the only example reported to date for ring-closing alkyne metathesis of bis-1,3-diynes (except for that shown in Section 2.10.4.3.4).

Scheme 126 Macrocyclic 1,3-Diyne by Ring-Closing Alkyne Metathesis of a 1,n-Bis-1,3-diyne[215]

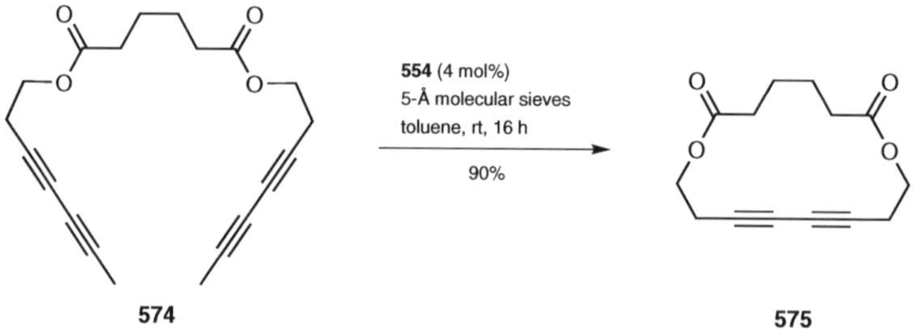

2.10.4.3 Applications of Diyne Ring-Closing Metathesis to Natural Product Synthesis

2.10.4.3.1 Ring-Closing Metathesis–Semireduction for the Construction of Z-Alkenes

Ring-closing metathesis is a powerful tool to form various cyclic frameworks, yet diene ring-closing metathesis provides a mixture of E- and Z-isomers with unpredictable ratios except for small- and medium-sized rings where Z-isomers are preferred. In macrocyclic ring-closing metathesis, usually E-isomers form more favorably due to the equilibrium nature of the metathesis process although complete stereocontrol is yet to be realized. In contrast, achieving Z-selective metathesis has been an outstanding problem in macrocyclic ring-closing metathesis. However, several successful processes including catalyst-based control have been developed recently (see Section 2.10.2.2.2.2). In this regard, ring-closing alkyne metathesis to construct macrocycles containing an alkyne moiety followed by semireduction would be an attractive solution for the stereoselective synthesis of E- or Z-alkene-containing macrocycles (Scheme 127). From ring-closing alkyne metathesis products, Z-configured cycloalkenes can be synthesized by semihydrogenation with Lindlar catalyst. Alternatively, Birch reduction, *trans* hydrosilylation–desilylation, and *trans* hydroboration–protodeborylation lead to the corresponding E-alkene-containing macrocycles.

Scheme 127 Ring-Closing Alkyne Metathesis–Semireduction for the Synthesis of E- or Z-Alkene-Containing Macrocycles[215]

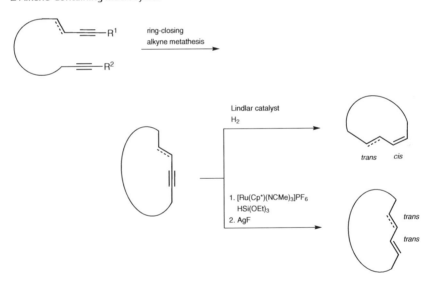

The total synthesis of olfactory macrolides **578**, **581**, and **584** relies on a ring-closing alkyne metathesis/Z-selective semihydrogenation sequence (Scheme 128). The ring-closing alkyne metathesis with the Mortreux system consisting of hexacarbonylmolybdenum(0) and 4-chlorophenol smoothly converts diynes **576** and **579** into the alkyne macrolides **577** and **580** in good yields, respectively. Semihydrogenation with the Lindlar catalyst provides yuzu lactone (**578**) and ambrettolide (**581**) in excellent yields.[17] Similarly, a Z-stereoselective synthesis of civetone (**584**) is achieved from the corresponding diyne **582** through ring-closing alkyne metathesis with tungsten–alkylidene catalyst **8** followed by Z-selective semihydrogenation of the resulting cycloalkyne **583**.[216]

Scheme 128 Ring-Closing Alkyne Metathesis–Semihydrogenation for the Preparation of Z-Alkene-Containing Macrocycles[17,216]

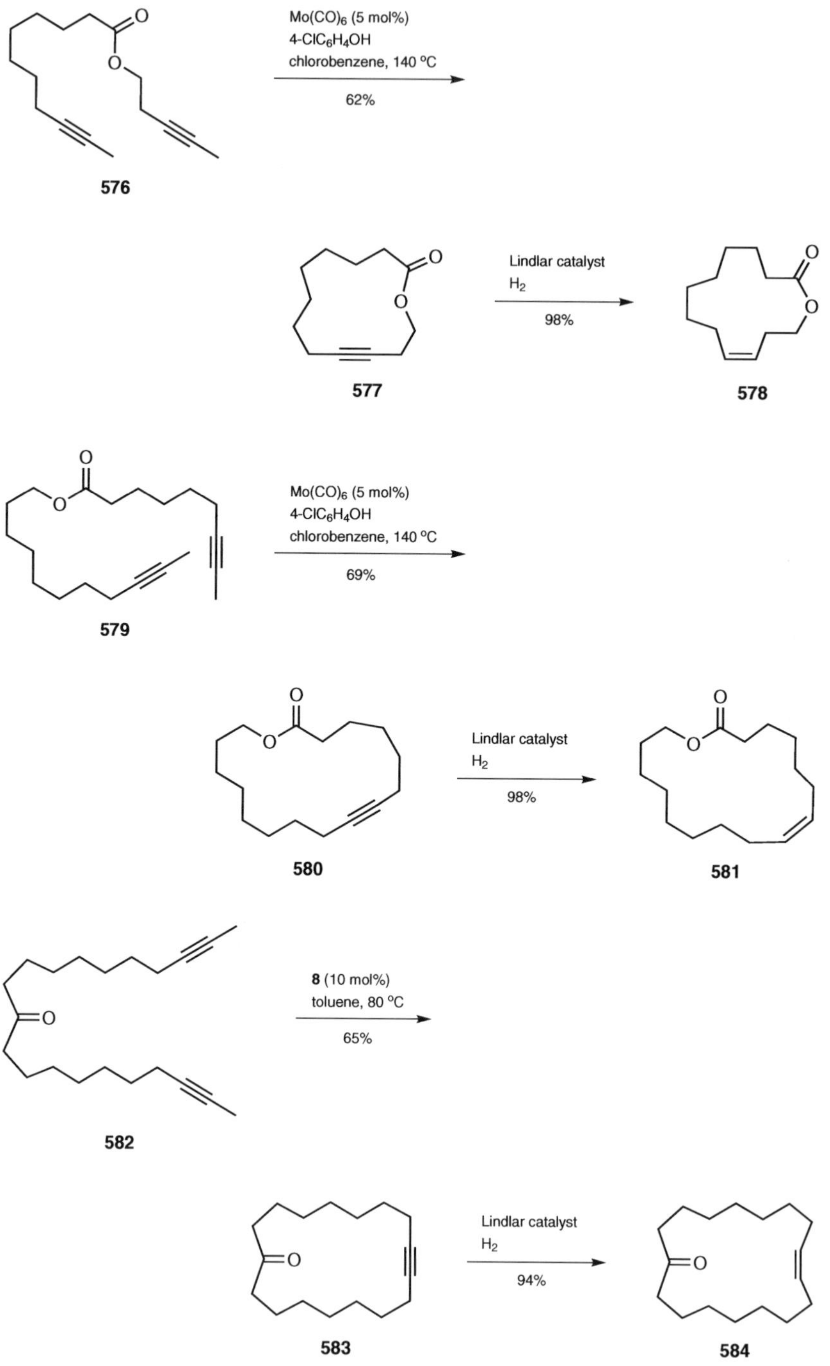

The ring-closing alkyne metathesis of diyne **585** containing a protected nitrogen moiety either with the Mortreux catalytic system consisting of hexacarbonylmolybdenum(0) and 4-chlorophenol or Schrock catalyst **8** provides macrocyclic alkyne **586** in good yields (Scheme 129). Semihydrogenation of **586** gives Z-configured macrocycle **587**, which is elaborated to motuporamine C (**588**).[217]

2.10.4 Diyne Ring-Closing Metathesis

Scheme 129 Ring-Closing Alkyne Metathesis–Semihydrogenation Strategy in the Synthesis of Motuporamine C[217]

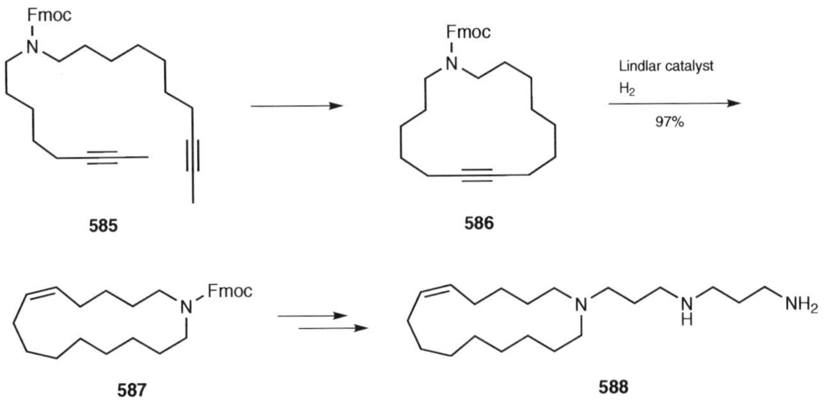

Conditions	Yield (%) of **586**	Ref
8 (10 mol%), toluene, 80 °C	62	[217]
Mo(CO)$_6$ (5 mol%), 4-ClC$_6$H$_4$OH, chlorobenzene, 140 °C	68	[217]

A ring-closing alkyne metathesis–semihydrogenation strategy is used for a stereoselective synthesis of PGE2–1,15-lactone (**591**) (Scheme 130).[218] In this synthesis, an in situ generated molybdenum chloride species, formed from the molybdenum(III) trisamide **555** in dichloromethane, smoothly converts diyne **589** into macrocyclic alkyne **590** in toluene at 80 °C. Semihydrogenation of **590** followed by protodesilylation affords PGE2–1,15-lactone (**591**).

Scheme 130 Ring-Closing Alkyne Metathesis–Semihydrogenation in the Synthesis of PGE2–1,15-Lactone[218]

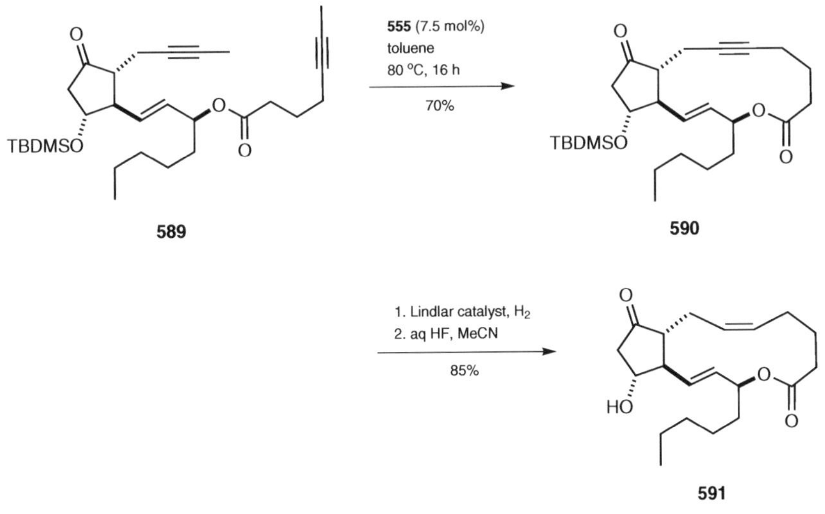

A similar strategy is employed in the synthesis of epothilone C (**35**) (Scheme 131), where the macrocyclic alkyne **593** is obtained by ring-closing alkyne metathesis of diyne **592** with molybdenum complex **555**. Semihydrogenation of the macrocyclic alkyne **593** and protodesilylation completes the synthesis of epothilone C (**35**).[219] This diyne ring-closing metathesis based strategy for epothilone C (**35**) compares favorably with that of the catalyst-controlled Z-selective diene ring-closing metathesis described in Scheme 11 (Section 2.10.2.2.2.2).

Scheme 131 Ring-Closing Alkyne Metathesis–Semihydrogenation in the Synthesis of Epothilone C[219]

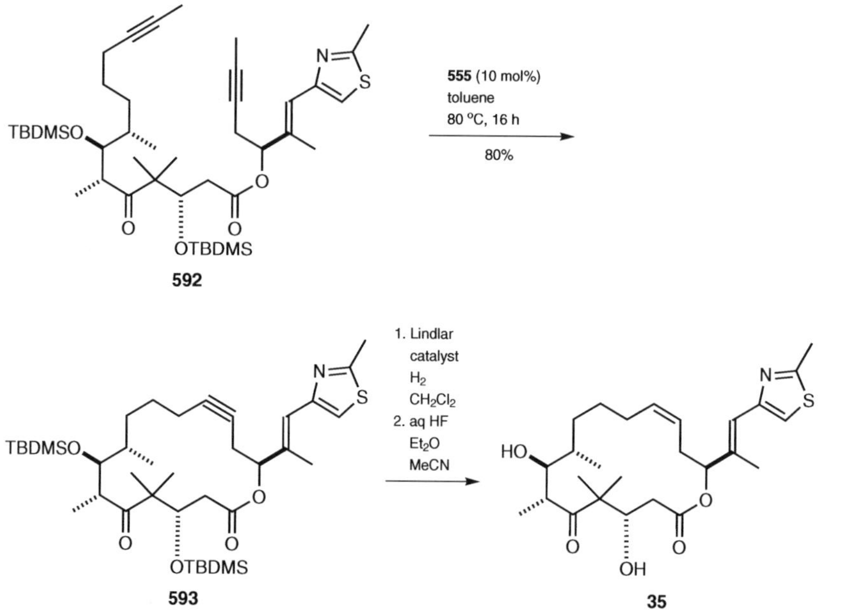

A synthesis of cruentaren A (**261**) relies on a ring-closing alkyne metathesis–semireduction sequence for the stereoselective installation of the Z-configured double bond of the macrolide. The ring-closing alkyne metathesis of diyne **594** catalyzed by tungsten–alkylidene complex **8** in toluene at 85 °C provides macrocycle **595** in 91% yield (Scheme 132).[220,221] Macrocyclic alkyne **595** is further elaborated to penultimate intermediate **596**, which is stereoselectively hydrogenated to the corresponding Z-configured macrocyclic natural product cruentaren A (**261**). Another synthesis of cruentaren A (**261**) also employs a ring-closing alkyne metathesis–semihydrogenation approach to install the macrocyclic Z-alkene using molybdenum complex **555** as catalyst,[222] and a third one uses a molybdenum nitride catalyst for the ring-closing alkyne metathesis.[223] These three ring-closing alkyne metathesis based syntheses of cruentaren A (**261**) favorably compare with the diene ring-closing metathesis based approach described in Scheme 55 (Section 2.10.2.3.1).

2.10.4 Diyne Ring-Closing Metathesis

Scheme 132 Ring-Closing Alkyne Metathesis–Semihydrogenation in the Synthesis of Cruentaren A[220]

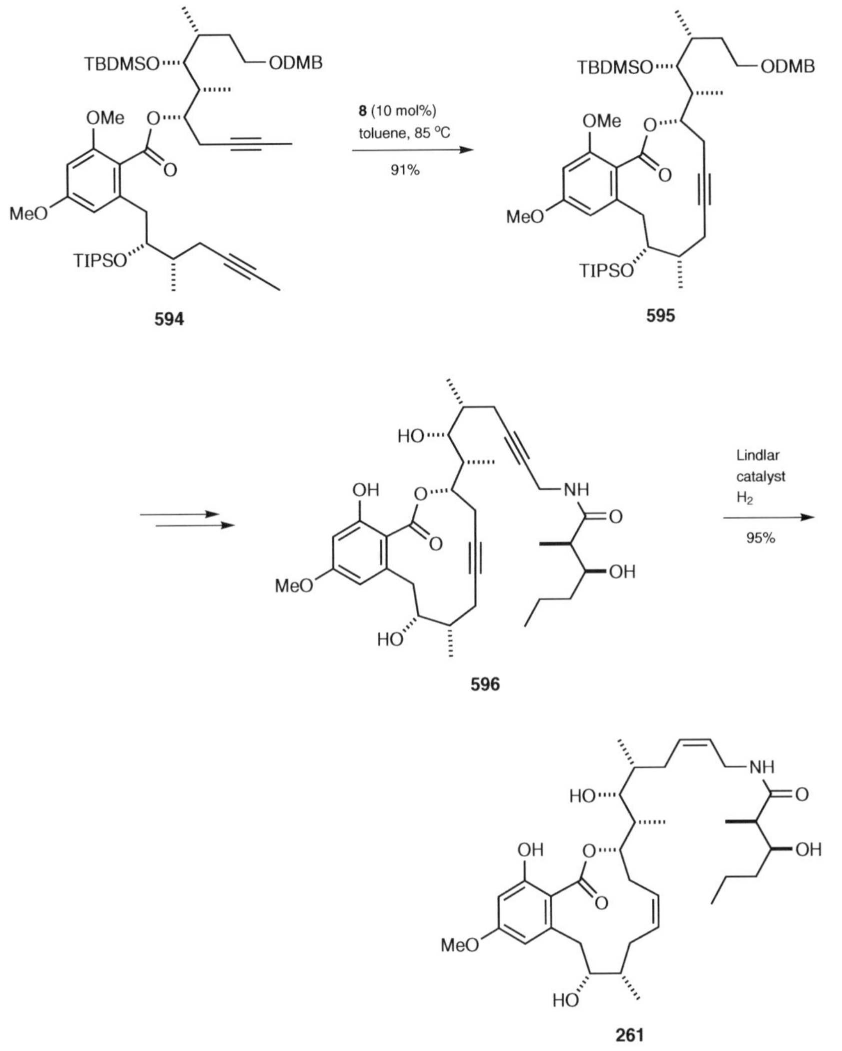

DMB = 3,4-(MeO)$_2$C$_6$H$_3$CH$_2$

A total synthesis of ecklonialactone A (**601**) and B (**267**) is based on the ring-closing alkyne metathesis–semihydrogenation sequence as the key step (Scheme 133).[224] Initial trials of the ring-closing alkyne metathesis of diynes **597** and **599** using the in situ generated catalyst from dichloromethane and molybdenum(III) trisamide **555** afforded unsatisfactory results even at high catalyst loadings.[225] Furthermore, the reaction with tungsten–alkylidyne complex **8** fails completely probably due to its Lewis acidity.[226] A triarylsilanolate pyridine molybdenum nitride precatalyst is also unsuitable due to the known propensity of the nitride function to react with epoxides.[227] On the other hand, the reaction with molybdenum complex **557** smoothly delivers the macrocyclic alkynes **598** and **600**, which are further semihydrogenated to form Z-alkenyl macrocyclic natural products ecklonialactone B (**267**) and A (**601**), respectively.

for references see p 672

Scheme 133 Ring-Closing Alkyne Metathesis–Semihydrogenation for the Synthesis of Ecklonialactone A and B[224]

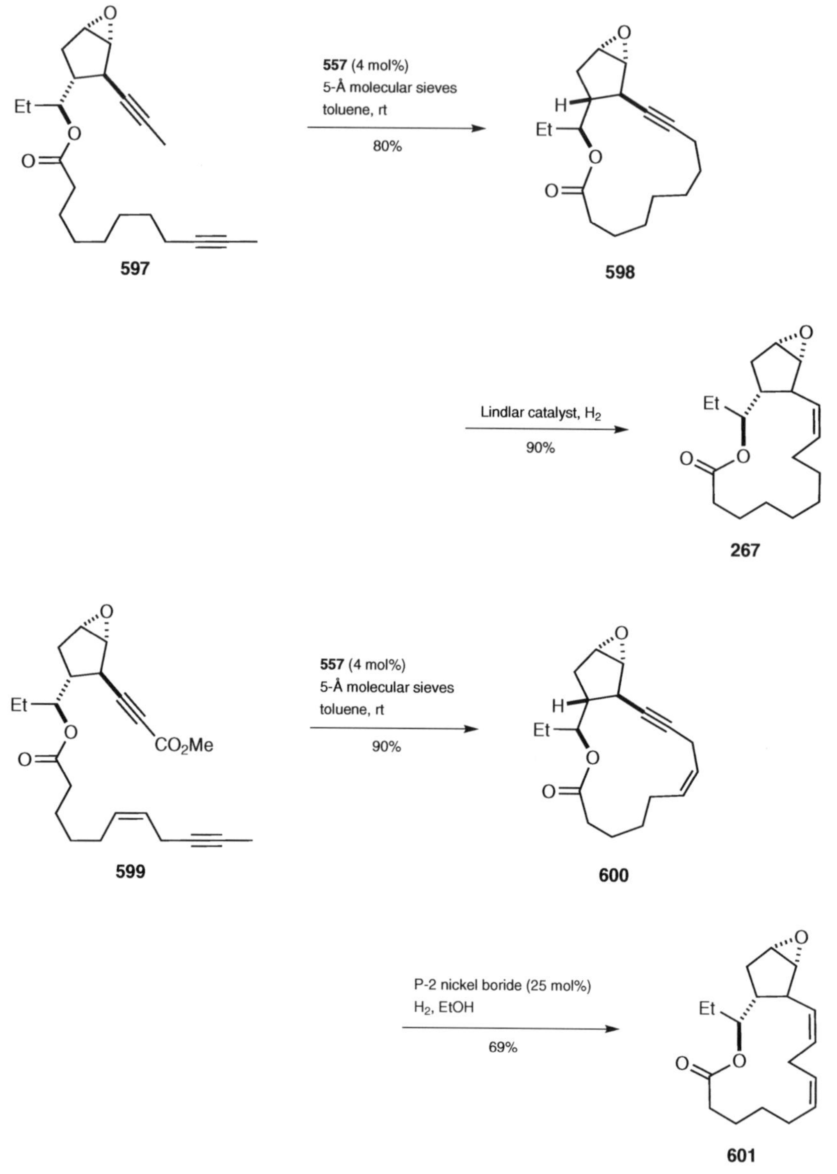

The prowess of the ring-closing alkyne metathesis–semihydrogenation sequence is clearly demonstrated in the total synthesis of nakadomarin A (**605**) (Scheme 134).[31,228] Macrocyclic ring-closing alkyne metathesis of elaborated tetracyclic diyne **602** occurs smoothly with Schrock alkylidyne **8** (30 mol%, chlorobenzene) to afford cycloalkyne **603** in 69% yield. The *cis* double bond on the 15-membered ring is then installed by semihydrogenation to generate **604**. Subsequent reduction of both amides followed by eight-membered-ring formation (diene ring-closing metathesis with Grubbs' first-generation catalyst **11**) leads to a total synthesis of nakadomarin A (**605**).[228] In comparison, the 15-membered cyclic Z-alkene of nakadomarin A (**605**) can also be installed by diene ring-closing metathesis. High Z selectivity is only realized using recently developed tungsten–alkylidene complexes such as **34**.[31] Thus, treatment of substrate **606** with tungsten catalyst **34** (5 mol%)

2.10.4 Diyne Ring-Closing Metathesis

affords nakadomarin A (**605**) in 63% yield with a 94:6 Z/E ratio.[31] Reduced pressure is required for efficient removal of ethene from the reaction medium to avoid a double-bond isomerization mediated by the tungsten–methylidene species. Although this example demonstrates the significant merit of the Z-selective ring-closing metathesis approach, extensive optimization of reaction conditions for each metathesis is necessary to achieve high substrate conversion while maintaining high Z selectivity.

Scheme 134 Comparison of Ring-Closing Alkyne Metathesis–Semihydrogenation with Diene Ring-Closing Metathesis for the Synthesis of Nakadomarin A[31,228]

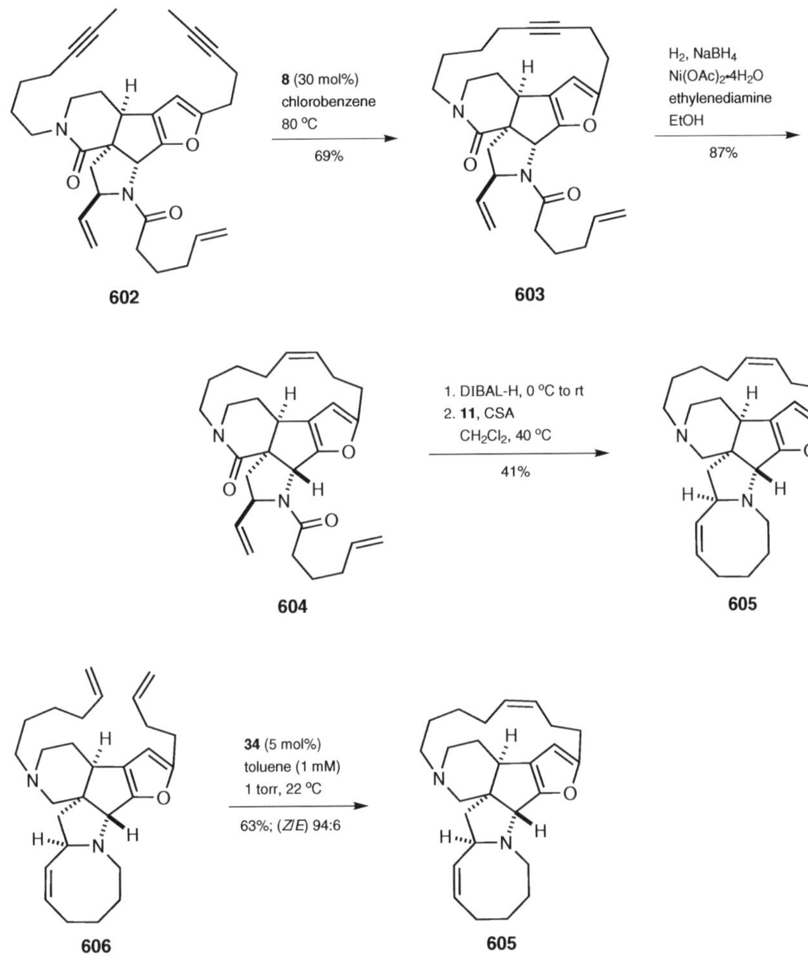

2.10.4.3.2 Ring-Closing Metathesis–Semireduction for the Construction of E-Alkenes

In addition to the successful implementation of Z-selective macrocycle synthesis in various natural product syntheses via ring-closing alkyne metathesis–semihydrogenation, procedures for the construction of the complementary macrocyclic E-alkenes have been developed, although exploited to a lesser extent by far.[229,230]

To explore the preparation of macrocyclic E-alkenes, simple diester-tethered diyne **607** is cyclized by treatment with tungsten–alkylidene complex **8**[15] or more active molybdenum complex **557**[231] to afford macrocyclic alkyne **608** (Scheme 135). Hydrosilylation of the alkyne moiety with triethoxysilane and cationic tris(acetonitrile)(η^5-pentamethylcyclopentadienyl)ruthenium(II) hexafluorophosphate {[Ru(Cp*)(NCMe)$_3$]PF$_6$} generates the corresponding vinylsilane via *anti* addition with high chemo- and stereoselectivity, which is then protodesilylated with silver(I) fluoride to yield the macrocyclic *E*-alkene **609** in good yield.[229,230] A similar two-step protocol has been developed with pinacolborane (4,4,5,5-tetramethyl-1,3,2-oxaborolidine) via *trans*-hydroboration catalyzed by the same cationic ruthenium complex (Scheme 135).[232] These catalytic hydrosilylation and hydroboration protocols have good compatibility with a variety of functional groups. However, silyl deprotection requires careful choice of desilylating agent, such as silver(I) fluoride, to achieve high selectivity and yield. Typical fluoride-based desilylation conditions suffer from double-bond isomerization and poor functional group tolerance. Further improvement in *E* selectivity is realized via one-step *trans*-hydrogenation.[233] For example, upon treatment of macrocyclic alkyne **608** with chloro(cycloocta-1,5-diene)(η^5-pentamethylcyclopentadienyl)ruthenium(II) {RuCl(Cp*)(cod)} and silver(I) trifluoromethanesulfonate under a hydrogen atmosphere, *trans*-hydrogenated product **609** is obtained in high yield and selectivity [96%; ratio (*E/Z*) 98:2]. Overall, these protocols involving ring-closing alkyne metathesis–semireduction can effectively construct macrocyclic *E*-alkenes starting from acyclic 1,n-diynes.

Scheme 135 Ring-Closing Alkyne Metathesis–Semireduction for the Synthesis of Macrocyclic *E*-Alkenes[15,229–232]

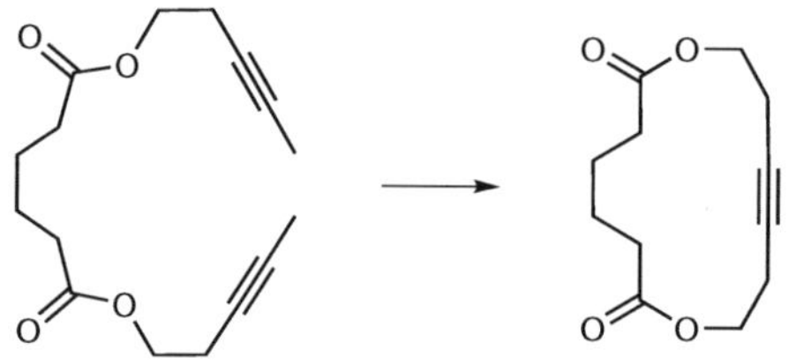

Conditions	Yield (%)	Ref
8 (10 mol%), toluene, 80 °C	79	[15,229,230]
557 (2 mol%), 5-Å molecular sieves, toluene, rt	92	[231,232]

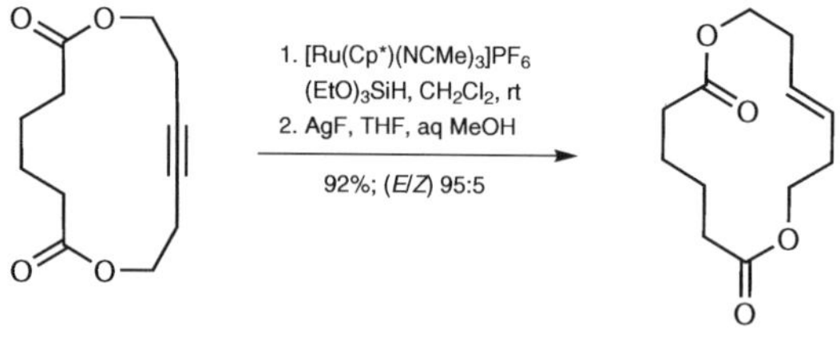

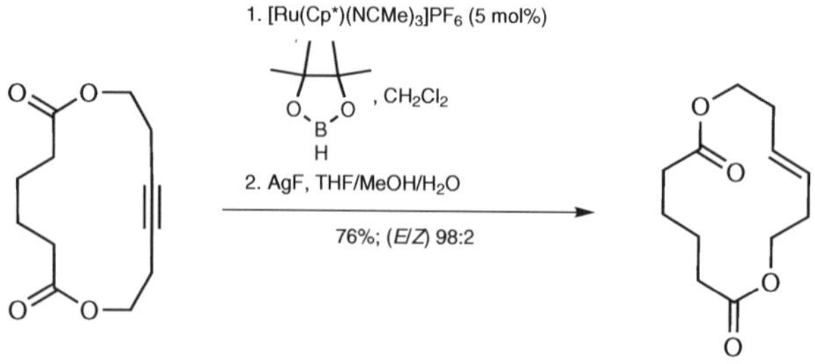

2.10.4 Diyne Ring-Closing Metathesis

In the efficient total synthesis of tulearin C (**613**), a ring-closing alkyne metathesis of diyne **610** proceeds smoothly with a remarkable rate when catalyzed by molybdenum–alkylidyne complex **558** at relatively high temperature (50 °C). Subsequent conversion of the macrocyclic alkyne **611** with tris(acetonitrile)(η^5-pentamethylcyclopentadienyl)ruthenium(II) hexafluorophosphate and triethoxysilane provides *trans*-hydrosilylated product **612**, which is directly treated with silver(I) fluoride and tetrabutylammonium fluoride for global desilylation to generate tulearin C (**613**) (Scheme 136).[234]

Scheme 136 Ring-Closing Alkyne Metathesis–Semireduction in the Synthesis of Tulearin C[234]

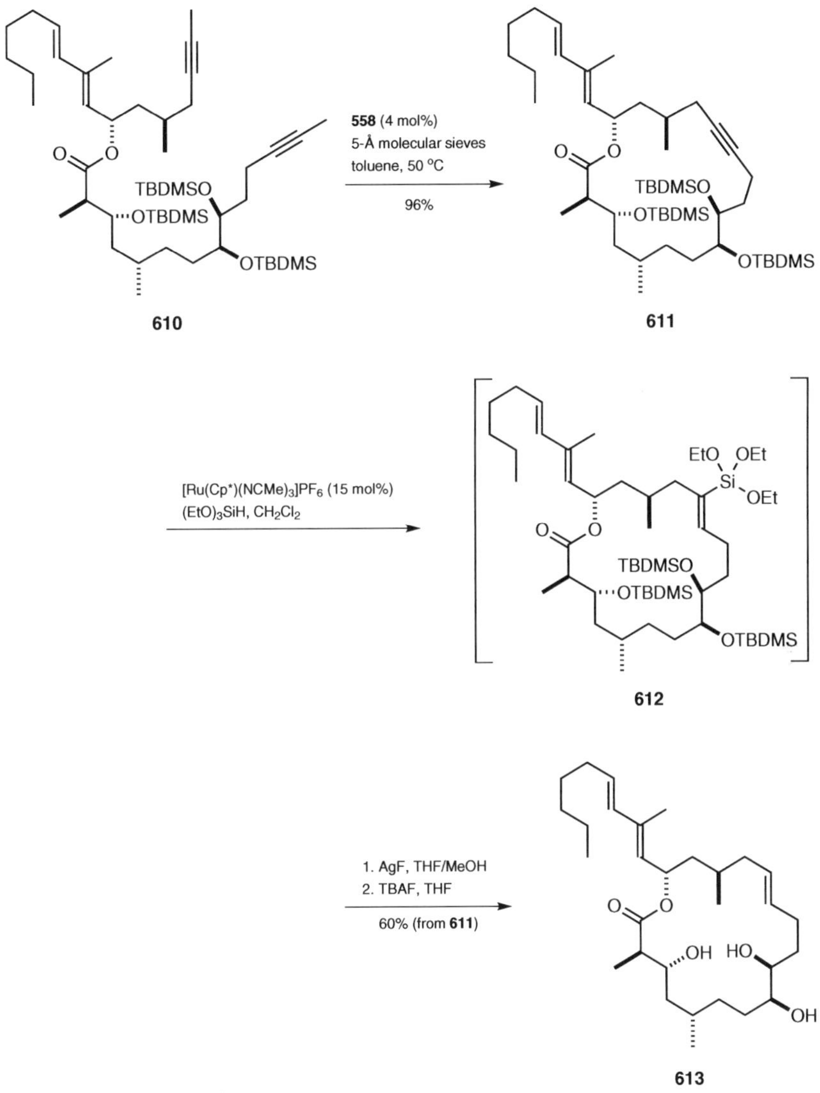

2.10.4.3.3 Ring-Closing Alkyne Metathesis–Semireduction for the Synthesis of Cyclic Conjugated 1,3-Dienes

The ring-closing alkyne metathesis–semireduction protocol can be used for the synthesis of macrocyclic 1,3-dienes by employing conjugated E- or Z-1,3-enynes as reacting counterparts in the substrate for ring-closing alkyne metathesis. This extension is made possible by the orthogonal nature of alkylidyne catalysts, which are reactive toward only alkynes but not alkenes. This strategy has been applied to effective syntheses of macrocyclic natural products containing macrocyclic E,Z- or E,E-1,3-diene moieties.

In a total synthesis of lactimidomycin (**276**), the Z-enediyne **614** smoothly undergoes ring closure with molybdenum complex **557** to provide macrocyclic Z-enyne **615** in 95% yield (Scheme 137),[112,235] The *trans* addition of benzyldimethylsilane to the Z-enyne catalyzed by cationic ruthenium catalyst tris(acetonitrile)(η^5-pentamethylcyclopentadienyl)-ruthenium(II) hexafluorophosphate provides alkenylsilane **616**, which is desilylated with tetrabutylammonium fluoride to deliver conjugated Z,E-1,3-dienyl 12-membered lactone **275** in 64% overall yield. This ring-closing alkyne metathesis based approach compares favorably with the diene ring-closing metathesis strategy to install the conjugated 1,3-diene moiety of lactimidomycin (**276**) described in Scheme 58 (Section 2.10.2.3.1).

Scheme 137 Ring-Closing Alkyne Metathesis–Semireduction in the Synthesis of Lactimidomycin[112,235]

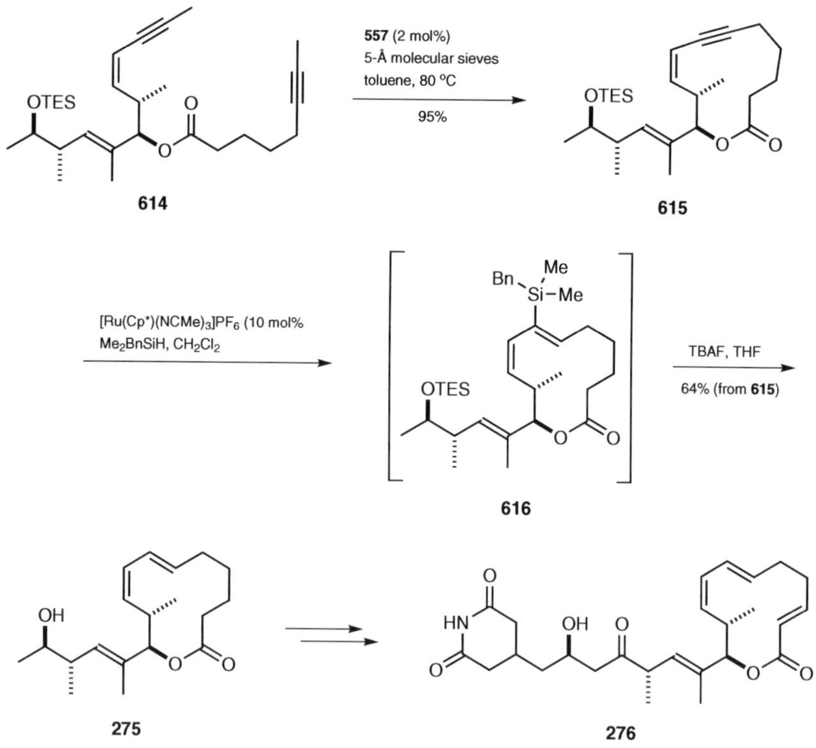

Similarly, in the recent synthesis of the proposed structure **620** of mandelalide A, a ring-closing alkyne metathesis–semireduction strategy installs the Z-double bond of the conjugated E,Z-diene on the 21-membered macrolide **619** (Scheme 138).[236] The ring-closing alkyne metathesis of E-enediyne **617** with molybdenum complex **559** provides macrocyclic E-enyne **618** in 72% yield. The triple bond of this macrocyclic enyne is converted into the corresponding Z-alkene [Zn(Cu/Ag), THF/MeOH/H$_2$O] to afford macrocyclic Z,E-1,3-

2.10.4 Diyne Ring-Closing Metathesis

diene **619** in 88% yield, which was elaborated to the proposed structure **620** of mandelalide A.

Scheme 138 Ring-Closing Alkyne Metathesis–Semireduction in the Synthesis of a Putative Structure of Mandelalide A[236]

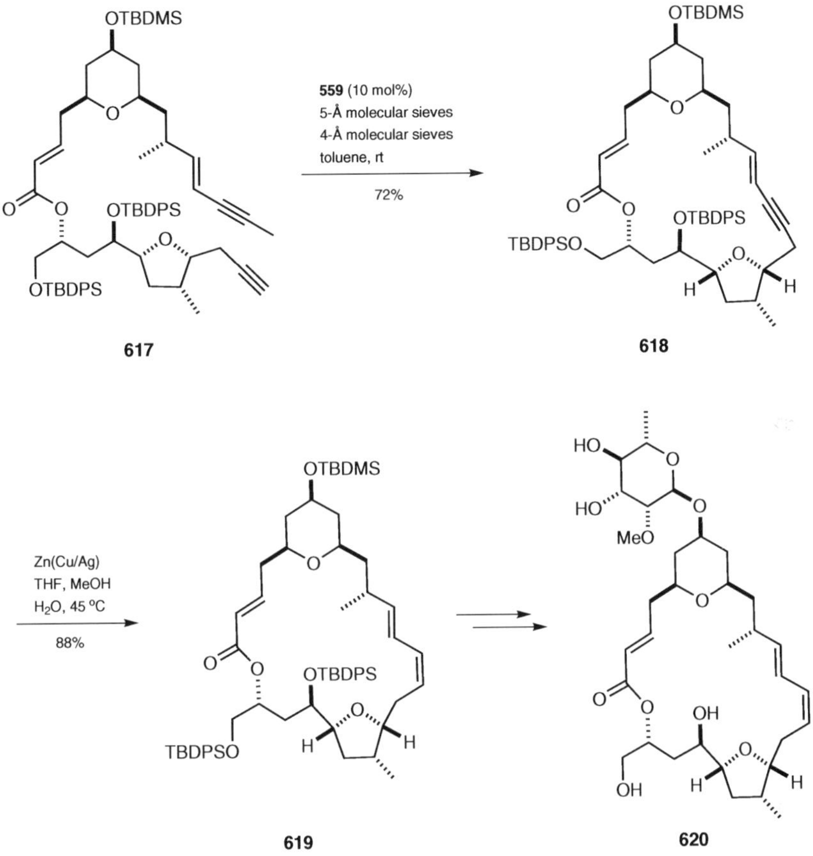

The total synthesis of WF-1360F (**625**) relies on a ring-closing alkyne metathesis–semireduction sequence to set up the conjugated E,E-diene of the macrolide (Scheme 139).[237] In this approach, E-enediyne **621** is treated with a catalytic amount of molybdenum complex **560** and manganese(II) chloride (toluene, 120 °C) to provide macrocycle **622** in 69% yield. A relatively high temperature and long reaction time (3 to 27 h) are required for complete conversion. Unexpectedly, the semireduction of the triple bond on enyne **622** under transition-metal-catalyzed hydrogenation, hydrostannylation, hydroboration, and hydrozirconation fails. An alternative method involving an acetylenehexacarbonyldicobalt complex and its reductive decomplexation with 1-ethylpiperidine hypophosphite successfully generates macrocyclic E,Z-diene **623** in 74% yield. The subsequent isomerization of the E,Z-diene to the corresponding E,E-diene with 2,2′-azobisisobutyronitrile/benzenethiol in refluxing benzene ultimately provides the desired conjugated E,E-diene-containing macrocycle **624** in 88% yield [ratio (E/Z) 20:1], which is elaborated to natural product **625**.

Scheme 139 Ring-Closing Alkyne Metathesis–Semireduction in the Synthesis of WF-1360F[237]

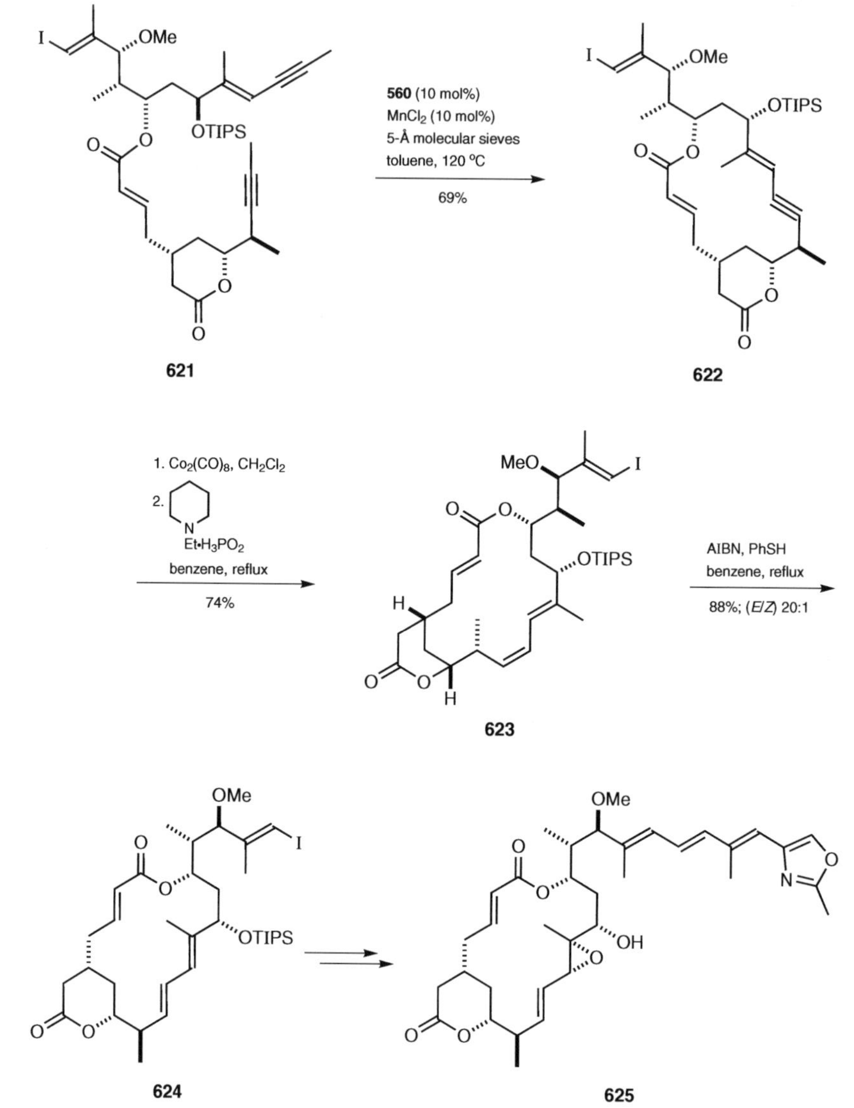

Alternatively, the macrocyclic 1,3-enyne moiety generated by ring-closing alkyne metathesis is converted into to the Z,Z-dienyl moiety in a total synthesis of macrolide leiodermatolide (**631**) (Scheme 140).[238] Ring-closing alkyne metathesis of Z-enediyne **626** catalyzed by the in situ generated molybdenum complex **555** (40 mol%, toluene, 100 °C) affords macrocycle **627** in 72% yield whereas the reaction with molybdenum complex **559** results in the formation of an acyclic dimer. Suzuki coupling of the vinyl iodide moiety of **627** with vinylboronic acid derivative **628** provides **629** in moderate yield (55%). Silyl deprotection and semireduction of the newly formed alkyne using activated zinc [Zn(Cu/Ag)] in tetrahydrofuran/water/methanol affords conjugated macrocyclic Z,Z-diene **630**, which is converted into natural product **631**.

2.10.4 Diyne Ring-Closing Metathesis

Scheme 140 Ring-Closing Alkyne Metathesis–Semireduction in the Synthesis of Leiodermatolide[238]

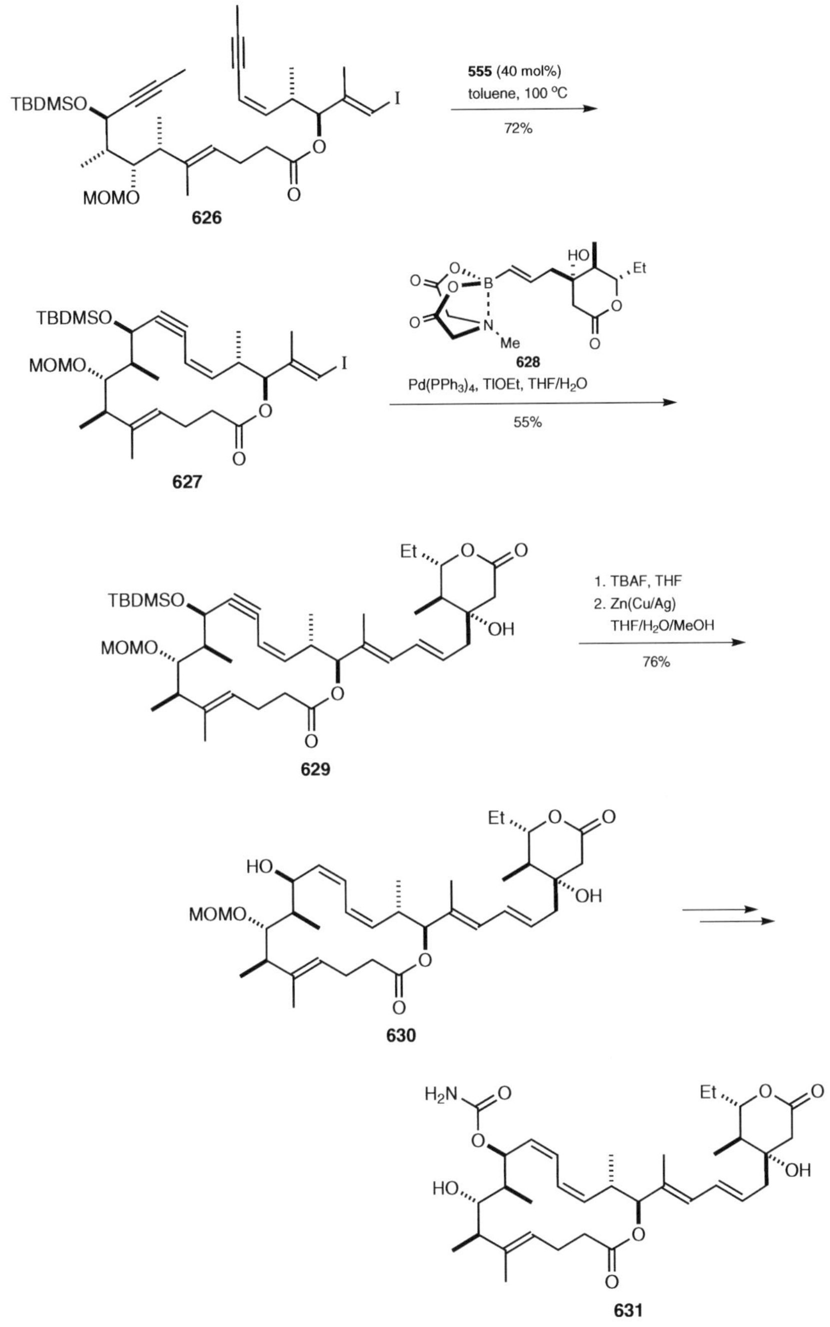

2.10.4.3.4 Ring-Closing Alkyne Metathesis for the Synthesis of Cyclic Conjugated 1,3-Diynes

The unique selectivity of ring-closing alkyne metathesis of bis-1,3-diyne systems to form only 1,3-diynes (see Scheme 126, Section 2.10.4.2.2) is used in the synthesis of 1,3-diyne-containing macrolactone **634** (Scheme 141). To directly construct the 1,3-diyne moiety embedded in the 17-membered macrolactone, bis-1,3-diyne **632** is treated with molybdenum complex **559** (40 mol%, 4-Å/5-Å molecular sieves, toluene, rt) to provide 1,3-diyne macrolactone **633** in high yield (82%). Stereoselective epoxidation followed by deprotection of the *tert*-butyldimethylsilyl group leads to the total synthesis of ivorenolide B (**634**).[239]

Scheme 141 Ring-Closing Alkyne Metathesis of a Bis-1,3-diyne in the Synthesis of Ivorenolide B[239]

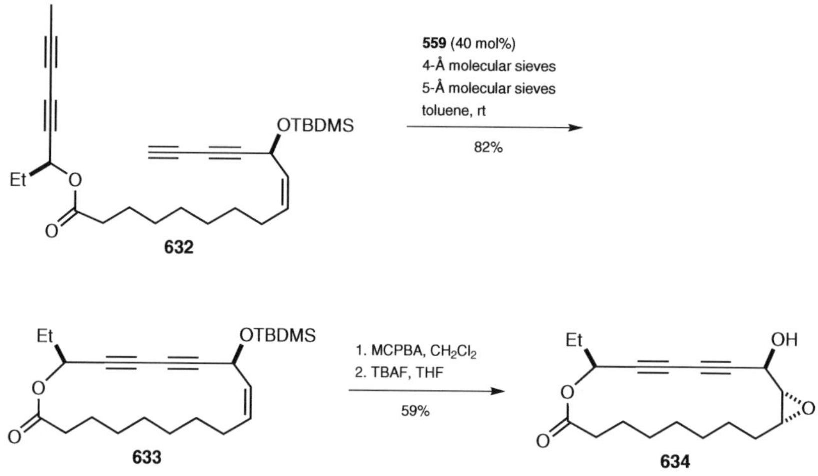

2.10.4.4 Applications of Ring-Closing Alkyne Metathesis to the Synthesis of Cyclic Oligomers

Ring-closing alkyne metathesis (RCAM) can be used for the synthesis of compounds beyond natural products. The inherently reversible nature of the metathesis-based ring closure and the linear geometry of alkynes would have a self-correction mechanism to form cyclic oligomers of certain sizes with good selectivity. This characteristic of reversible bond formation has significant merit compared to other nonreversible methods, such as Sonogashira or Glaser-type couplings, for the synthesis of alkyne-based cyclooligomers. With alkyne metathesis, appropriate diyne-containing monomers will form oligomers of sufficient length, which then cyclize into the corresponding cyclic oligomers of the most favorable ring size through ring-closing alkyne metathesis. Due to the reversibility in alkyne metathesis, cyclic oligomers can equilibrate with open-chain oligomers until they reach the thermodynamically favorable cyclic oligomers.[240,241]

Relying on this concept, cyclic 1,3-phenyleneethynylene hexamers **636** are synthesized from 1,3-dipropynylated benzenes **635** (Scheme 142).[242] By treating differently substituted dipropynylated benzenes **635** with the Mortreux catalytic system consisting of hexacarbonylmolybdenum(0) and 4-chlorophenol at 150 °C, low yields of macrocycles **636** are obtained (0.5–6%). Polymers and other cycloisomers are produced as major products and do not undergo self-reorganization under these harsh reaction conditions.

2.10.4 Diyne Ring-Closing Metathesis

Scheme 142 Alkyne Metathesis of Dipropynylated Benzenes[242]

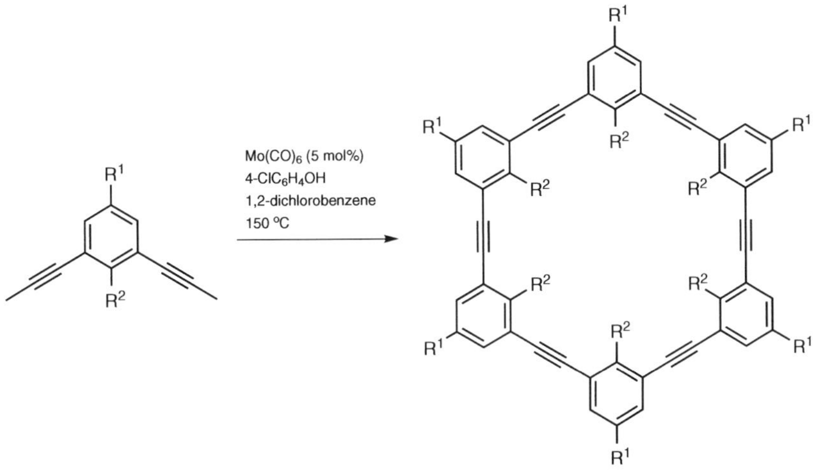

Reagents/conditions: Mo(CO)$_6$ (5 mol%), 4-ClC$_6$H$_4$OH, 1,2-dichlorobenzene, 150 °C

635 → **636**

R^1	R^2	Yield (%)	Ref
t-Bu	H	6	[242]
(CH$_2$)$_5$Me	H	1.2	[242]
t-Bu	Me	0.5	[242]

The synthesis of 1,3-phenyleneethynylene-based cyclic oligomers is significantly improved by using molybdenum complex **556** in the presence of 4-nitrophenol (Scheme 143).[243] A small-scale (15 to 33 mg) metathesis reaction of diynes **637** under vacuum (1 Torr) for the removal of but-2-yne byproduct provides the macrocycles **638** in good yield, however, the corresponding large-scale reaction is not effective. The failure is rationalized by catalyst decomposition and the consumption of the catalyst by polymerization of but-2-yne that cannot be removed effectively in large-scale reactions. This problem is solved by attaching a specific functional group on the monomers by which the resultant alkyne byproducts are readily precipitated. A benzoylbiphenyl group is the best substituent for R^2. Thus, under otherwise identical conditions, diyne substrate **637** [R^1 = CO$_2$(CH$_2$CH$_2$O)$_3$Me; R^2 = 4-(4-BzC$_6$H$_4$)C$_6$H$_4$] provides cyclic oligomer **638** [R^1 = CO$_2$(CH$_2$CH$_2$O)$_3$Me; R^2 = 4-(4-BzC$_6$H$_4$)C$_6$H$_4$] in up to 77% yield even on larger scales.

Scheme 143 Precipitation-Driven Alkyne Metathesis of Cyclic 1,3-Phenyleneethynylene Hexamers[243]

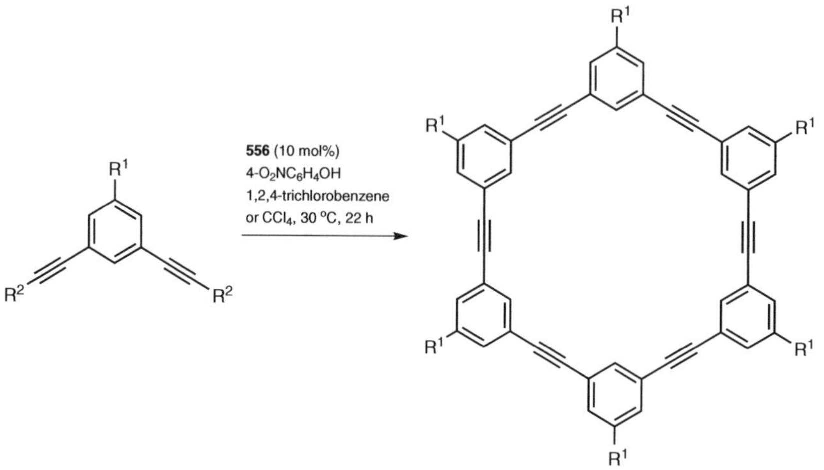

R^1	R^2	Yield (%)		Ref
		mg scale	gram scale	
t-Bu	Me	61	–	[243]
$CO_2(CH_2CH_2O)_3Me$	Me	76	<10	[243]
$CO_2(CH_2CH_2O)_3Me$	4-(4-$BzC_6H_4)C_6H_4$	81	77	[243]
CO_2t-Bu	4-(4-$BzC_6H_4)C_6H_4$	79	–	[243]

Cyclic 1,3-Phenyleneethynylene Hexamer 638 (R^1 = t-Bu); Typical Procedure for Precipitation-Driven Small-Scale Synthesis:[243]

In a glovebox under argon, a soln of molybdenum complex **556** (10 mg, 0.015 mmol, 0.1 equiv) and 4-nitrophenol (6.3 mg, 0.045 mmol, 0.3 equiv) in anhyd 1,2,4-trichlorobenzene (0.8 mL) was added to a soln of 1-(tert-butyl)-3,5-di(prop-1-ynyl)benzene (**637**, R^1 = t-Bu; R^2 = Me; 31.5 mg, 0.15 mmol) in 1,2,4-trichlorobenzene (0.4 mL). The flask was sealed and removed from the glovebox and stirred for 22 h at 30 °C under reduced pressure (1 Torr). Upon connecting to vacuum, vigorous gas evolution was observed. After removal of the solvent under high vacuum, the residue was purified by column chromatography (CH_2Cl_2/hexanes 1:9) giving the product as a white solid; yield: 14.3 mg (61%).

Cyclic 1,3-Phenyleneethynylene Hexamer 638 [R^1 = $CO_2(CH_2CH_2O)_3Me$]; Typical Procedure for Precipitation-Driven Large-Scale Synthesis:[243]

In a glovebox under argon, a soln of molybdenum complex **556** (1.69 g, 2.54 mmol, 0.1 equiv) and 4-nitrophenol (1.06 g, 7.61 mmol, 0.3 equiv) in CCl_4 (200 mL) (**CAUTION: toxic**) was added to a soln of diyne **637** [R^1 = $CO_2(CH_2CH_2O)_3Me$; R^2 = 4-(4-$BzC_6H_4)C_6H_4$; 21.0 g, 25.4 mmol] in CCl_4 (500 mL) in a 1-L round-bottomed flask. The flask was sealed and removed from the glovebox. The resulting mixture was stirred for 22 h at 30 °C, and the precipitated byproduct formed almost immediately upon warming to 30 °C. After filtration to remove the precipitate, the filtrate was concentrated under reduced pressure and the residue was purified by column chromatography (iPrOH/CH_2Cl_2 1:9); yield: 5.68 g (77%).

2.10.4.5 Scope and Limitations

One of the greatest advantages of alkyne metathesis is the orthogonal reactivity of the alkylidyne catalyst toward alkynes in the presence of alkenes, thus ring-closing metathesis can be achieved with substrates containing an alkene or alkenes as long as the ring strain of the cyclic products is tolerable. Also, a broad range of polar functional groups including ethers, esters, and amides can be accommodated in the tether of ring-closing alkyne metathesis substrates. On the other hand, due to the ring strain caused by the linear geometry of an alkyne moiety, the ring-closing alkyne metathesis does not allow for the formation of cyclic alkynes smaller than 12-membered rings, which is a unique and significant limitation of ring-closing alkyne metathesis. Another constraint in the substrates is the incompatibility of terminal diynes with most alkylidyne complexes except for with a few newly developed catalysts. From a practical perspective, contrary to a vast array of alkene and enyne ring-closing metathesis active alkylidene complexes, only a limited number of alkylidyne complexes are commercially available, and handling of most of these complexes requires a strictly inert environment free of moisture and oxygen. This will limit the use of alkyne ring-closing metathesis and alkyne metathesis in general for large-scale operation.

2.10.5 Conclusions

Ring-closing metathesis (RCM) has emerged as one of the most powerful synthetic methods to form cyclic molecular frameworks including small, medium, and large carbo- and heterocycles as illustrated by the selected examples above. Due to the compatibility and differential reactivity of alkenes and alkynes relative to other polar functional groups in the substrate, the ring-closure step can be introduced flexibly in the synthetic sequence. Over the past two decades, significant conceptual and technological advances have been achieved in all three metathesis classes (alkene, enyne, and alkyne metathesis). The limitations of ring-closing metathesis, such as Z-selective diene ring-closing metathesis, chemo- and sequence-selective enyne ring-closing metathesis, and the availability of reactive yet user-friendly and functional group tolerant catalysts for diyne ring-closing metathesis, have been addressed in remarkable fashion although further improvements are yet to be realized. Continued development of fundamentally new concepts and investigations to address the shortcomings of current ring-closing metathesis technology will certainly make ring-closing metathesis part of the state-of-the-art synthetic arsenal for the preparation of a variety of cyclic molecules including natural products and pharmaceuticals.

References

[1] Villemin, D., *Tetrahedron Lett.*, (1980) **21**, 1715.
[2] Tsuji, J.; Hashiguchi, S., *Tetrahedron Lett.*, (1980) **21**, 2955.
[3] Fu, G. C.; Grubbs, R. H., *J. Am. Chem. Soc.*, (1993) **115**, 3800.
[4] Fu, G. C.; Nguyen, S. T.; Grubbs, R. H., *J. Am. Chem. Soc.*, (1993) **115**, 9856.
[5] Katz, T. J.; Sivavec, T. M., *Tetrahedron Lett.*, (1985) **26**, 2159.
[6] Katz, T. J.; Sivavec, T. M., *J. Am. Chem. Soc.*, (1985) **107**, 737.
[7] Korkowski, P. F.; Hoye, T. R.; Rydberg, D. B., *J. Am. Chem. Soc.*, (1988) **110**, 2676.
[8] Hoye, T. R.; Rehberg, G. M., *Organometallics*, (1989) **8**, 2070.
[9] Hoye, T. R.; Rehberg, G. M., *J. Am. Chem. Soc.*, (1990) **112**, 2841.
[10] Watanuki, S.; Ochifuji, N.; Mori, M., *Organometallics*, (1994) **13**, 4129.
[11] Watanuki, S.; Mori, M., *Organometallics*, (1995) **14**, 5054.
[12] Watanuki, S.; Ochifuji, N.; Mori, M., *Organometallics*, (1995) **14**, 5062.
[13] Kinoshita, A.; Mori, M., *Synlett*, (1994), 1020.
[14] Kim, S.-H.; Bowden, N.; Grubbs, R. H., *J. Am. Chem. Soc.*, (1994) **116**, 10801.
[15] Fürstner, A.; Seidel, G., *Angew. Chem. Int. Ed.*, (1998) **37**, 1734.
[16] Akiyama, M.; Chisholm, M. H.; Cotton, F. A.; Extine, M. W.; Haitko, D. A.; Little, D.; Fanwick, P. E., *Inorg. Chem.*, (1979) **18**, 2266.
[17] Fürstner, A.; Guth, O.; Rumbo, A.; Seidel, G., *J. Am. Chem. Soc.*, (1999) **121**, 11108.
[18] Urbina-Blanco, C. A.; Poater, A.; Lebl, T.; Manzini, S.; Slawin, A. M. Z.; Cavallo, L.; Nolan, S. P., *J. Am. Chem. Soc.*, (2013) **135**, 7073.
[19] Romero, P. E.; Piers, W. E.; McDonald, R., *Angew. Chem. Int. Ed.*, (2004) **43**, 6161.
[20] van der Eide, E. F.; Romero, P. E.; Piers, W. E., *J. Am. Chem. Soc.*, (2008) **130**, 4485.
[21] Schleyer, P. V. R.; Williams, J. E.; Blanchard, K. R., *J. Am. Chem. Soc.*, (1970) **92**, 2378.
[22] Bourgeois, D.; Pancrazi, A.; Ricard, L.; Prunet, J., *Angew. Chem. Int. Ed.*, (2000) **39**, 725.
[23] Matsui, R.; Seto, K.; Fujita, K.; Suzuki, T.; Nakazaki, A.; Kobayashi, S., *Angew. Chem. Int. Ed.*, (2010) **49**, 10068.
[24] Fürstner, A.; Radkowski, K.; Wirtz, C.; Goddard, R.; Lehmann, C. W.; Mynott, R., *J. Am. Chem. Soc.*, (2002) **124**, 7061.
[25] Murga, J.; Falomir, E.; García-Fortanet, J.; Carda, M.; Marco, J. A., *Org. Lett.*, (2002) **4**, 3447.
[26] Meng, D.; Su, D.-S.; Balog, A.; Bertinato, P.; Sorensen, A. J.; Danishefsky, S. J.; Zheng, Y.-H.; Chou, T.-C.; He, L.; Horwitz, S. B., *J. Am. Chem. Soc.*, (1997) **119**, 2733.
[27] Fürstner, A.; Thiel, O. R.; Blanda, G., *Org. Lett.*, (2000) **2**, 3731.
[28] Arisawa, M.; Kato, C.; Kaneko, H.; Nishida, A.; Nakagawa, M., *J. Chem. Soc., Perkin Trans. 1*, (2000), 1873.
[29] Nakashima, K.; Ito, R.; Sono, M.; Tori, M., *Heterocycles*, (2000) **53**, 301.
[30] Lee, C. W.; Grubbs, R. H., *Org. Lett.*, (2000) **2**, 2145.
[31] Yu, M.; Wang, C.; Kyle, A. F.; Jakubec, P.; Dixon, D. J.; Schrock, R. R.; Hoveyda, A. H., *Nature (London)*, (2011) **479**, 88.
[32] Wang, Y.; Jimenez, M.; Hansen, A. S.; Raiber, E.-A.; Schreiber, S. L.; Young, D. W., *J. Am. Chem. Soc.*, (2011) **133**, 9196.
[33] Marx, V. M.; Herbert, M. B.; Keitz, B. K.; Grubbs, R. H., *J. Am. Chem. Soc.*, (2013) **135**, 94.
[34] Rosebrugh, L. E.; Herbert, M. B.; Marx, V. M.; Keitz, B. K.; Grubbs, R. H., *J. Am. Chem. Soc.*, (2013) **135**, 1276.
[35] Quinn, K. J.; Isaacs, A. K.; Arvary, R. A., *Org. Lett.*, (2004) **6**, 4143.
[36] Quinn, K. J.; Isaacs, A. K.; DeChristopher, B. A.; Szklarz, S. C.; Arvary, R. A., *Org. Lett.*, (2005) **7**, 1243.
[37] Michaelis, S.; Blechert, S., *Org. Lett.*, (2005) **7**, 5513.
[38] Baylon, C.; Heck, M.-P.; Misoskowski, C., *J. Org. Chem.*, (1999) **64**, 3354.
[39] Heck, M.-P.; Baylon, C.; Nolan, S. P.; Misoskowski, C., *Org. Lett.*, (2001) **3**, 1989.
[40] Clark, J. S.; Hamelin, O., *Angew. Chem. Int. Ed.*, (2000) **39**, 372.
[41] Ma, S.; Ni, B., *Org. Lett.*, (2002) **4**, 639.
[42] Ma, S.; Ni, B., *Chem.–Eur. J.*, (2004) **10**, 3286.
[43] Sello, J. K.; Andreana, P. R.; Lee, D.; Schreiber, S. L., *Org. Lett.*, (2003) **5**, 4125.
[44] Hoye, T. R.; Jeffrey, C. S.; Tennakoon, M. A.; Wang, J.; Zhao, H., *J. Am. Chem. Soc.*, (2004) **126**, 10210.
[45] Hansen, E. C.; Lee, D., *Org. Lett.*, (2004) **6**, 2035.

[46] Hansen, E. C., Ph.D. Thesis, University of Wisconsin, Madison, WI, (2006).
[47] Wang, X.; Bowman, E. J.; Bowman, B. J.; Porco, J. A., Jr., *Angew. Chem. Int. Ed.*, (2004) **43**, 3601.
[48] Huwe, C. M.; Velder, J.; Blechert, S., *Angew. Chem. Int. Ed.*, (1996) **35**, 2376.
[49] Huwe, C. M.; Blechert, S., *Synthesis*, (1997), 61.
[50] Evans, P. A.; Cui, J.; Buffone, G. P., *Angew. Chem. Int. Ed.*, (2003) **42**, 1734.
[51] Alexander, J. B.; La, D. S.; Cefalo, D. R.; Hoveyda, A. H.; Schrock, R. R., *J. Am. Chem. Soc.*, (1998) **120**, 4041.
[52] La, D. S.; Alexander, J. B.; Cefalo, D. R.; Graf, D. D.; Hoveyda, A. H.; Schrock, R. R., *J. Am. Chem. Soc.*, (1998) **120**, 9720.
[53] Seiders, T. J.; Ward, D. W.; Grubbs, R. H., *Org. Lett.*, (2001) **3**, 3225.
[54] Zhu, S. S.; Cefalo, D. R.; La, D. S.; Jamieson, J. Y.; Davis, W. M.; Hoveyda, A. H.; Schrock, R. R., *J. Am. Chem. Soc.*, (1999) **121**, 8251.
[55] Funk, T. W.; Berlin, J. M.; Grubbs, R. H., *J. Am. Chem. Soc.*, (2006) **128**, 1840.
[56] Sattely, E. S.; Cortez, G. A.; Moebius, D. C.; Schrock, R. R.; Hoveyda, A. H., *J. Am. Chem. Soc.*, (2005) **127**, 8526.
[57] Malcolmson, S. J.; Meek, S. J.; Sattely, E. S.; Schrock, R. R.; Hoveyda, A. H., *Nature (London)*, (2008) **456**, 933.
[58] Donohoe, T. J.; Rosa, C. P., *Org. Lett.*, (2007) **9**, 5509.
[59] Miller, A. K.; Hughes, C. C.; Kennedy-Smith, J. J.; Gradl, S. N.; Trauner, D., *J. Am. Chem. Soc.*, (2006) **128**, 17057.
[60] Dai, M.; Danishefsky, S. J., *J. Am. Chem. Soc.*, (2007) **129**, 3498.
[61] Dai, M.; Krauss, I. J.; Danishefsky, S. J., *J. Org. Chem.*, (2008) **73**, 9576.
[62] Ziegler, F. E.; Metcalf, C. A., III; Nangia, A.; Schulte, G., *J. Am. Chem. Soc.*, (1993) **115**, 2581.
[63] Hanessian, S.; Margarita, R.; Hall, A.; Johnstone, S.; Tremblay, M.; Parlanti, L., *J. Am. Chem. Soc.*, (2002) **124**, 13342.
[64] Hanessian, S.; Tremblay, M.; Petersen, J. F. W., *J. Am. Chem. Soc.*, (2004) **126**, 6064.
[65] Trost, B. M.; Horne, D. B.; Woltering, M. J., *Angew. Chem. Int. Ed.*, (2003) **42**, 5987.
[66] Trost, B. M.; Horne, D. B.; Woltering, M. J., *Chem.–Eur. J.*, (2006) **12**, 6607.
[67] Crimmins, M. T.; Zhang, Y.; Diaz, F. A., *Org. Lett.*, (2006) **8**, 2369.
[68] She, J., Ph.D. Dissertation, University of North Carolina at Chapel Hill, Chapel Hill, NC, (2004).
[69] Kuramochi, A.; Usuda, H.; Yamatsugu, K.; Kanai, M.; Shibasaki, M., *J. Am. Chem. Soc.*, (2005) **127**, 14200.
[70] Nicolaou, K. C.; Tria, G. S.; Edmonds, D. J., *Angew. Chem. Int. Ed.*, (2008) **47**, 1780.
[71] Hayashida, J.; Rawal, V. H., *Angew. Chem. Int. Ed.*, (2008) **47**, 4373.
[72] Yun, S. Y.; Zheng, J.-C.; Lee, D., *Angew. Chem. Int. Ed.*, (2008) **47**, 6201.
[73] Tiefenbacher, K.; Mulzer, J., *J. Org. Chem.*, (2009) **74**, 2937.
[74] White, D. E.; Stewart, I. C.; Grubbs, R. H.; Stoltz, B. M., *J. Am. Chem. Soc.*, (2008) **130**, 810.
[75] Hoshi, M.; Kaneko, O.; Nakajima, M.; Arai, S.; Nishida, A., *Org. Lett.*, (2014) **16**, 768.
[76] Nagatomo, M.; Koshimizu, M.; Masuda, K.; Tabuchi, T.; Urabe, D.; Inoue, M., *J. Am. Chem. Soc.*, (2014) **136**, 5916.
[77] Zhang, H.; Curran, D. P., *J. Am. Chem. Soc.*, (2011) **133**, 10376.
[78] Umezaki, S.; Yokoshima, S.; Fukuyama, T., *Org. Lett.*, (2013) **15**, 4230.
[79] Liu, G.; Romo, D., *Angew. Chem. Int. Ed.*, (2011) **50**, 7537.
[80] Nickel, A.; Maruyama, T.; Tang, H.; Murphy, P. D.; Greene, B.; Yusuff, N.; Wood, J. L., *J. Am. Chem. Soc.*, (2004) **126**, 16300.
[81] Watanabe, K.; Suzuki, Y.; Aoki, K.; Sakakura, A.; Suenaga, K.; Kigoshi, H., *J. Org. Chem.*, (2004) **69**, 7802.
[82] Ohyoshi, T.; Funakubo, S.; Miyazawa, Y.; Niida, K.; Hayakawa, I.; Kigoshi, H., *Angew. Chem. Int. Ed.*, (2012) **51**, 4972.
[83] Winkler, J. D.; Rouse, M. B.; Greaney, M. F.; Harrison, S. J.; Jeon, Y. T., *J. Am. Chem. Soc.*, (2002) **124**, 9726.
[84] Kalidindi, S.; Jeong, W. B.; Schall, A.; Bandichhor, R.; Nosse, B.; Reiser, O., *Angew. Chem. Int. Ed.*, (2007) **46**, 6361.
[85] Willot, M.; Radtke, L.; Könning, D.; Fröhlich, R.; Gessner, V. H.; Strohmann, C.; Christmann, M., *Angew. Chem. Int. Ed.*, (2009) **48**, 9105.
[86] Zahel, M.; Keßberg, A.; Metz, P., *Angew. Chem. Int. Ed.*, (2013) **52**, 5390.
[87] Nicolaou, K. C.; Bunnage, M. E.; Koide, K., *J. Am. Chem. Soc.*, (1994) **116**, 8402.
[88] Cook, G. R.; Shanker, P. S.; Peterson, S. L., *Org. Lett.*, (1999) **1**, 615.

[89] Fürstner, A.; Thiel, O. R., *J. Org. Chem.*, (2000) **65**, 1738.
[90] Paquette, L. A.; Tae, J.; Arrington, M. P.; Sadoun, A. H., *J. Am. Chem. Soc.*, (2000) **122**, 2742.
[91] Krafft, M. E.; Cheung, Y.-Y.; Juliano-Capucao, C. A., *Synthesis*, (2000), 1020.
[92] Krafft, M. E.; Cheung, Y. Y.; Abboud, K. A., *J. Org. Chem.*, (2001) **66**, 7443.
[93] Tsuna, K.; Noguchi, N.; Nakada, M., *Angew. Chem. Int. Ed.*, (2011) **50**, 9452.
[94] Hog, D. T.; Huber, F. M. E.; Mayer, P.; Trauner, D., *Angew. Chem. Int. Ed.*, (2014) **53**, 8513.
[95] Takao, K.-i.; Watanabe, G.; Yasui, H.; Tadano, K.-i., *Org. Lett.*, (2002) **4**, 2941.
[96] Crimmins, M. T.; Choy, A. L., *J. Am. Chem. Soc.*, (1999) **121**, 5653.
[97] Burton, J. W.; Clark, J. S.; Derrer, S.; Stork, T. C.; Bendall, J. G.; Holmes, A. B., *J. Am. Chem. Soc.*, (1997) **119**, 7483.
[98] Crimmins, M. T.; Emmitte, K. A., *Org. Lett.*, (1999) **1**, 2029.
[99] Adsool, V. A.; Pansare, S. V., *Org. Biomol. Chem.*, (2008) **6**, 2011.
[100] Bratz, M.; Bullock, W. H.; Overman, L. E.; Takemoto, T., *J. Am. Chem. Soc.*, (1995) **117**, 5958.
[101] Matsui, R.; Seto, K.; Sato, Y.; Suzuki, T.; Nakazaki, A.; Kobayashi, S., *Angew. Chem. Int. Ed.*, (2011) **50**, 680.
[102] Volchkov, I.; Lee, D., *J. Am. Chem. Soc.*, (2013) **135**, 5324.
[103] Crimmins, M. T.; Brown, B. H., *J. Am. Chem. Soc.*, (2004) **126**, 10264.
[104] Larrosa, I.; Da Silva, M. I.; Gómez, P. M.; Hannen, P.; Ko, E.; Lenger, S. R.; Linke, S. R.; White, A. J. P.; Wilton, D.; Barrett, A. G. M., *J. Am. Chem. Soc.*, (2006) **128**, 14042.
[105] Lv, L.; Shen, B.; Li, Z., *Angew. Chem. Int. Ed.*, (2014) **53**, 4164.
[106] Brown, M. K.; Hoveyda, A. H., *J. Am. Chem. Soc.*, (2008) **130**, 12904.
[107] Cai, Z.; Yongpruksa, N.; Harmata, M., *Org. Lett.*, (2012) **14**, 1661.
[108] Kusuma, B. R.; Brandt, G. E. L.; Blagg, S. J., *Org. Lett.*, (2012) **14**, 6242.
[109] Kurata, K.; Taniguchi, K.; Shiraishi, K.; Hayama, N.; Tanaka, I.; Suzuki, M., *Chem. Lett.*, (1989), 267.
[110] Becker, J.; Butt, L.; von Kiedrowski, V.; Mischler, E.; Quentin, F.; Hiersemann, M., *Org. Lett.*, (2013) **15**, 5982.
[111] Gallenkamp, D.; Fürstner, A., *J. Am. Chem. Soc.*, (2011) **133**, 9232.
[112] Micoine, K.; Fürstner, A., *J. Am. Chem. Soc.*, (2010) **132**, 14064.
[113] Bali, A. K.; Sunnam, S. K.; Prasad, K. R., *Org. Lett.*, (2014) **16**, 4001.
[114] Toelle, N.; Weinstabl, H.; Gaich, T.; Mulzer, J., *Angew. Chem. Int. Ed.*, (2014) **53**, 3859.
[115] Yun, S. Y.; Hansen, E. C.; Volchkov, I.; Cho, E. J.; Lo, W. Y.; Lee, D., *Angew. Chem. Int. Ed.*, (2010) **49**, 4261.
[116] Hayashi, Y.; Shoji, M.; Ishikawa, H.; Yamaguchi, J.; Tamaru, T.; Imai, H.; Nishigaya, Y.; Takabe, K.; Kakeya, H.; Osada, H., *Angew. Chem. Int. Ed.*, (2008) **47**, 6657.
[117] Evano, G.; Schaus, J. V.; Panek, J. S., *Org. Lett.*, (2004) **6**, 525.
[118] Del Valle, D. J.; Krische, M. J., *J. Am. Chem. Soc.*, (2013) **135**, 10986.
[119] Fujiwara, K.; Suzuki, Y.; Koseki, N.; Aki, Y.-i.; Kikuchi, Y.; Murata, F.; Yamamoto, F.; Kawamura, M.; Norikura, T.; Matsue, H.; Murai, A.; Katoono, R.; Kawai, H.; Suzuki, T., *Angew. Chem. Int. Ed.*, (2014) **53**, 780.
[120] Hara, A.; Morimoto, R.; Iwasaki, Y.; Saitoh, T.; Ishikawa, Y.; Nishiyama, S., *Angew. Chem. Int. Ed.*, (2012) **51**, 9877.
[121] Kobayashi, J.; Shimbo, K.; Sato, M.; Shiro, M.; Tsuda, M., *Org. Lett.*, (2000) **2**, 2805.
[122] Zuercher, W. J.; Hashimoto, M.; Grubbs, R. H., *J. Am. Chem. Soc.*, (1996) **118**, 6634.
[123] Harrity, J. P. A.; Visser, M. S.; Gleason, J. D.; Hoveyda, A. H., *J. Am. Chem. Soc.*, (1997) **119**, 1488.
[124] Stragies, R.; Blechert, S., *J. Am. Chem. Soc.*, (2000) **122**, 9584.
[125] Buschmann, N.; Rückert, A.; Blechert, S., *J. Org. Chem.*, (2002) **67**, 4325.
[126] Holub, N.; Neidhöfer, J.; Blechert, S., *Org. Lett.*, (2005) **7**, 1227.
[127] Böhrsch, V.; Neidhöfer, J.; Blechert, S., *Angew. Chem. Int. Ed.*, (2006) **45**, 1302.
[128] Böhrsch, V.; Blechert, S., *Chem. Commun. (Cambridge)*, (2006), 1968.
[129] Stragies, R.; Blechert, S., *Synlett*, (1998), 169.
[130] Wrobleski, A.; Sahasrabudhe, K.; Aubé, J., *J. Am. Chem. Soc.*, (2004) **126**, 5475.
[131] Henderson, J. A.; Phillips, A. J., *Angew. Chem. Int. Ed.*, (2008) **47**, 8499.
[132] Pfeiffer, M. W. B.; Phillips, A. J., *J. Am. Chem. Soc.*, (2005) **127**, 5334.
[133] Miura, Y.; Hayashi, N.; Yokoshima, S.; Fukuyama, T., *J. Am. Chem. Soc.*, (2012) **134**, 11995.
[134] Li, J.; Lee, D., *Chem. Sci.*, (2012) **3**, 3296.
[135] Takao, K.-i.; Nanamiya, R.; Fukushima, Y.; Namba, A.; Yoshida, K.; Tadano, K.-i., *Org. Lett.*, (2013) **15**, 5582.
[136] Han, J.-c.; Li, F.; Li, C.-c., *J. Am. Chem. Soc.*, (2014) **136**, 13610.

[137] Hansen, E. C.; Lee, D., *Acc. Chem. Res.*, (2006) **39**, 509.
[138] Villar, H.; Frings, M.; Bolm, C., *Chem. Soc. Rev.*, (2007) **36**, 55.
[139] Ulman, M.; Grubbs, R. H., *Organometallics*, (1998) **17**, 2484.
[140] Hoye, T. R.; Donaldson, S. M.; Vos, T. J., *Org. Lett.*, (1999) **1**, 277.
[141] Schramm, M. P.; Reddy, D. S.; Kozmin, S. A., *Angew. Chem. Int. Ed.*, (2001) **40**, 4274.
[142] Lloyd-Jones, G. C.; Margue, R. G.; de Vries, J. G., *Angew. Chem. Int. Ed.*, (2005) **44**, 7442.
[143] Sohn, J.-H.; Kim, K. H.; Lee, H.-Y.; No, Z. S.; Ihee, H., *J. Am. Chem. Soc.*, (2008) **130**, 16506.
[144] Kim, K. H.; Ok, T.; Lee, K.; Lee, H.-Y.; Chang, K. T.; Ihee, H.; Sohn, J.-H., *J. Am. Chem. Soc.*, (2010) **132**, 12027.
[145] Lee, O. S.; Kim, K. H.; Kim, J.; Kwon, K.; Ok, T.; Ihee, H.; Lee, H.-Y.; Sohn, J.-H., *J. Org. Chem.*, (2013) **78**, 8242.
[146] Lee, Y.-J.; Schrock, R. R.; Hoveyda, A. H., *J. Am. Chem. Soc.*, (2009) **131**, 10652.
[147] Clavier, H.; Correa, A.; Escudero-Adán, E. C.; Benet-Buchholz, J.; Cavallo, L.; Nolan, S. P., *Chem.–Eur. J.*, (2009) **15**, 10244.
[148] Lippstreu, J. J.; Straub, B. F., *J. Am. Chem. Soc.*, (2005) **127**, 7444.
[149] Nuñez-Zarur, F.; Solans-Monfort, X.; Rodríguez-Santiago, L.; Pleixats, R.; Sodupe, M., *Chem.–Eur. J.*, (2011) **17**, 7506.
[150] Zuercher, W. J.; Scholl, M.; Grubbs, R. H., *J. Org. Chem.*, (1998) **63**, 4291.
[151] Huang, J.; Xiong, H.; Hsung, R. P.; Rameshkumar, C.; Mulder, J. A.; Grebe, T. P., *Org. Lett.*, (2002) **4**, 2417.
[152] Choi, T.-L.; Grubbs, R. H., *Chem. Commun. (Cambridge)*, (2001), 2648.
[153] Timmer, M. S. M.; Ovaa, H.; Filippov, D. V.; van der Marel, G. A.; van Boom, J. H., *Tetrahedron Lett.*, (2001) **42**, 8231.
[154] Wu, C.-J.; Madhushaw, R. J.; Liu, R.-S., *J. Org. Chem.*, (2003) **68**, 7889.
[155] Maifeld, S. V.; Miller, R. L.; Lee, D., *J. Am. Chem. Soc.*, (2004) **126**, 12228.
[156] Zhao, Y.; Hoveyda, A. H.; Schrock, R. R., *Org. Lett.*, (2011) **13**, 784.
[157] Betkekar, V. V.; Panda, S.; Kaliappan, K. P., *Org. Lett.*, (2012) **14**, 198.
[158] Miller, R. L.; Maifeld, S. V.; Lee, D., *Org. Lett.*, (2004) **6**, 2773.
[159] Matsuda, T.; Yamaguchi, Y.; Murakami, M., *Synlett*, (2008), 561.
[160] Layton, M. E.; Morales, C. A.; Shair, M. D., *J. Am. Chem. Soc.*, (2002) **124**, 773.
[161] Morales, C. A.; Layton, M. E.; Shair, M. D., *Proc. Natl. Acad. Sci. U. S. A.*, (2004) **101**, 12036.
[162] Hansen, E. C.; Lee, D., *J. Am. Chem. Soc.*, (2003) **125**, 9582.
[163] Hansen, E. C.; Lee, D., *J. Am. Chem. Soc.*, (2004) **126**, 15074.
[164] Grimwood, M. E.; Hansen, H. C., *Tetrahedron*, (2009) **65**, 8132.
[165] Kim, M.; Miller, R. L.; Lee, D., *J. Am. Chem. Soc.*, (2005) **127**, 12818.
[166] Yun, S. Y.; Kim, M.; Lee, D.; Wink, D. J., *J. Am. Chem. Soc.*, (2009) **131**, 24.
[167] Wang, K.-P.; Yun, S. Y.; Lee, D., *J. Am. Chem. Soc.*, (2009) **131**, 15114.
[168] van Otterlo, W. A. L.; Ngidi, E. L.; de Koning, C. B.; Fernandes, M. A., *Tetrahedron Lett.*, (2004) **45**, 659.
[169] Kinoshita, A.; Mori, M., *J. Org. Chem.*, (1996) **61**, 8356.
[170] Kinoshita, A.; Mori, M., *Heterocycles*, (1997) **46**, 287.
[171] Debleds, O.; Campagne, J.-M., *J. Am. Chem. Soc.*, (2008) **130**, 1562.
[172] Graham, T. J. A.; Gray, E. E.; Burgess, J. M.; Goess, B. C., *J. Org. Chem.*, (2010) **75**, 226.
[173] Aggarwal, V. K.; Astle, C. J.; Rogers-Evans, M., *Org. Lett.*, (2004) **6**, 1469.
[174] Brenneman, J. B.; Martin, S. F., *Org. Lett.*, (2004) **6**, 1329.
[175] Mori, M.; Tomita, T.; Kita, Y.; Kitamura, T., *Tetrahedron Lett.*, (2004) **45**, 4397.
[176] Reddy, D. S.; Kozmin, S. A., *J. Org. Chem.*, (2004) **69**, 4860.
[177] Gao, X.; Woo, S. K.; Krische, M. J., *J. Am. Chem. Soc.*, (2013) **135**, 4223.
[178] Li, J.; Lee, D., *Chem.–Asian J.*, (2010) **5**, 1298.
[179] Honda, T.; Namiki, H.; Kaneda, K.; Mizutani, H., *Org. Lett.*, (2004) **6**, 87.
[180] Boyer, F.-D.; Hanna, I.; Ricard, L., *Org. Lett.*, (2004) **6**, 1817.
[181] Schubert, M.; Metz, P., *Angew. Chem. Int. Ed.*, (2011) **50**, 2954.
[182] Mukherjee, S.; Lee, D., *Org. Lett.*, (2009) **11**, 2916.
[183] Dineen, T. A.; Roush, W. R., *Org. Lett.*, (2004) **6**, 2043.
[184] Movassaghi, M.; Piizzi, G.; Siegel, D. S.; Piersanti, G., *Angew. Chem. Int. Ed.*, (2006) **45**, 5859.
[185] Poulin, J.; Grise-Bard, C. M.; Barriault, L., *Angew. Chem. Int. Ed.*, (2012) **51**, 2111.
[186] Betkekar, V. V.; Sayyad, A. A.; Kaliappan, K. P., *Org. Lett.*, (2014) **16**, 5540.
[187] Molawi, K.; Delpont, N.; Echavarren, A. M., *Angew. Chem. Int. Ed.*, (2010) **49**, 3517.

[188] Lee, J.; Parker, K. A., *Org. Lett.*, (2012) **14**, 2682.
[189] Wei, H.; Qiao, C.; Liu, G.; Yang, Z.; Li, C.-c., *Angew. Chem. Int. Ed.*, (2013) **52**, 620.
[190] Kitamura, T.; Mori, M., *Org. Lett.*, (2001) **3**, 1161.
[191] Rückert, A.; Eisele, D.; Blechert, S., *Tetrahedron Lett.*, (2001) **42**, 5245.
[192] Mori, M.; Wakamatsu, H.; Tonogaki, K.; Fujita, R.; Kitamura, T.; Sato, Y., *J. Org. Chem.*, (2005) **70**, 1066.
[193] Zhu, Z.-B.; Shi, M., *Org. Lett.*, (2010) **12**, 4462.
[194] Kress, S.; Weckesser, J.; Schulz, S. R.; Blechert, S., *Eur. J. Org. Chem.*, (2013), 1346.
[195] Basso, A.; Banfi, L.; Riva, R.; Guanti, G., *Tetrahedron*, (2006) **62**, 8830.
[196] Groaz, E.; Banti, D.; North, M., *Tetrahedron Lett.*, (2007) **48**, 1927.
[197] Spandl, R. J.; Rudyk, H.; Spring, D. R., *Chem. Commun. (Cambridge)*, (2008), 3001.
[198] Kim, M.; Lee, D., *J. Am. Chem. Soc.*, (2005) **127**, 18024.
[199] Li, J.; Park, S.; Miller, R. L.; Lee, D., *Org. Lett.*, (2009) **11**, 571.
[200] Wengrovius, J. H.; Sancho, J.; Schrock, R. R., *J. Am. Chem. Soc.*, (1981) **103**, 3932.
[201] Schrock, R. R., *Polyhedron*, (1995) **14**, 3177.
[202] Fürstner, A.; Davies, P. W., *Chem. Commun. (Cambridge)*, (2005), 2307.
[203] Hérisson, J.-L.; Chauvin, Y., *Makromol. Chem.*, (1971) **141**, 161.
[204] Katz, T. J.; McGinnis, J., *J. Am. Chem. Soc.*, (1975) **97**, 1592.
[205] McCullough, L. G.; Listemann, M. L.; Schrock, R. R.; Churchill, M. R.; Ziller, J. W., *J. Am. Chem. Soc.*, (1983) **105**, 6729.
[206] Zhu, J.; Jiu, G.; Lin, Z., *Organometallics*, (2006) **25**, 1812.
[207] Krouse, S. A.; Schrock, R. R., *Macromolecules*, (1989) **22**, 2569.
[208] Aguilera, B.; Wolf, L. B.; Nieczypor, P.; Rutjes, F. P. J. T.; Overkleeft, H. S.; van Hest, J. C. M.; Schoemaker, H. E.; Wang, B.; Mol, J. C.; Fürstner, A.; Overhand, M.; van der Marel, G. A.; van Boom, J. H., *J. Org. Chem.*, (2001) **66**, 3584.
[209] Jsselstijn, M. I.; Kaiser, J.; van Delft, F. L.; Schoemaker, H. E.; Rutjes, F. P. J. T., *Amino Acids*, (2003) **24**, 263.
[210] IJsselstijn, M.; Aguilera, B.; van der Marel, G. A.; van Boom, J. H.; van Delft, F. L.; Schoemaker, H. E.; Overkleeft, H. S.; Rutjes, F. P. J. T.; Overhand, M., *Tetrahedron Lett.*, (2004) **45**, 4379.
[211] Ghalit, N.; Poot, A. J.; Fürstner, A.; Rijkers, D. T. S.; Liskamp, R. M., *Org. Lett.*, (2005) **7**, 2961.
[212] Doyle, D.; Murphy, P. V., *Carbohydr. Res.*, (2008) **343**, 2535.
[213] Bauer, E. B.; Szafert, S.; Hampel, F.; Gladysz, J. A., *Organometallics*, (2003) **22**, 2184.
[214] Bauer, E. B.; Hampel, F.; Gladysz, J. A., *Adv. Synth. Catal.*, (2004) **346**, 812.
[215] Lysenko, S.; Volbeda, J.; Jones, P. G.; Tamm, M., *Angew. Chem. Int. Ed.*, (2012) **51**, 6757.
[216] Fürstner, A.; Seidel, G., *J. Organomet. Chem.*, (2000) **606**, 75.
[217] Fürstner, A.; Rumbo, A., *J. Org. Chem.*, (2000) **65**, 2608.
[218] Fürstner, A.; Grela, K., *Angew. Chem. Int. Ed.*, (2000) **39**, 1234.
[219] Fürstner, A.; Mathes, C.; Grela, K., *Chem. Commun. (Cambridge)*, (2001), 1057.
[220] Vintonyak, V. V.; Maier, M., *Angew. Chem. Int. Ed.*, (2007) **46**, 5209.
[221] Vintonyak, V. V.; Calà, M.; Lay, F.; Kunze, B.; Sasse, F.; Maier, M., *Chem.–Eur. J.*, (2008) **14**, 3709.
[222] Fürstner, A.; Bindl, M.; Jean, L., *Angew. Chem. Int. Ed.*, (2007) **46**, 9275.
[223] Fouché, M.; Rooney, L.; Barrett, A. G. M., *J. Org. Chem.*, (2012) **77**, 3060.
[224] Hickmann, V.; Alcarazo, M.; Fürstner, A., *J. Am. Chem. Soc.*, (2010) **132**, 11042.
[225] Fürstner, A.; Larionov, O.; Flügge, S., *Angew. Chem. Int. Ed.*, (2007) **46**, 5545.
[226] Freudenberger, J. H.; Schrock, R. R.; Churchill, M. R.; Rheingold, A. L.; Ziller, J. W., *Organometallics*, (1984) **3**, 1563.
[227] Bindl, M.; Stade, R.; Heilmann, E. K.; Picot, A.; Goddard, R.; Fürstner, A., *J. Am. Chem. Soc.*, (2009) **131**, 9468.
[228] Kyle, A. F.; Jakubec, P.; Cockfield, D. M.; Cleator, E.; Skidmore, J.; Dixon, D. J., *Chem. Commun. (Cambridge)*, (2011) **47**, 10037.
[229] Fürstner, A.; Radkowski, K., *Chem. Commun. (Cambridge)*, (2002), 2182.
[230] Lacombe, F.; Radkowski, K.; Seidel, G.; Fürstner, A., *Tetrahedron*, (2004) **60**, 7315.
[231] Heppekausen, J.; Stade, R.; Goddard, R.; Fürstner, A., *J. Am. Chem. Soc.*, (2010) **132**, 11045.
[232] Sundararaju, B.; Fürstner, A., *Angew. Chem. Int. Ed.*, (2013) **52**, 14050.
[233] Radkowski, K.; Sundararaju, B.; Fürstner, A., *Angew. Chem. Int. Ed.*, (2013) **52**, 355.
[234] Lehr, K.; Mariz, R.; Leseurre, L.; Gabor, B.; Fürstner, A., *Angew. Chem. Int. Ed.*, (2011) **50**, 11373.

[235] Micoine, K.; Persich, P.; Llaveria, J.; Lam, M.-H.; Maderna, A.; Loganzo, F.; Fürstner, A., *Chem.–Eur. J.*, (2013) **19**, 7370.
[236] Willwacher, J.; Fürstner, A., *Angew. Chem. Int. Ed.*, (2014) **53**, 4217.
[237] Neuhaus, C. M.; Liniger, M.; Stieger, M.; Altmann, K.-H., *Angew. Chem. Int. Ed.*, (2013) **52**, 5866.
[238] Willwacher, J.; Kausch-Busies, N.; Fürstner, A., *Angew. Chem. Int. Ed.*, (2012) **51**, 12041.
[239] Ungeheuer, F.; Fürstner, A., *Chem.–Eur. J.*, (2015) **21**, 11387.
[240] Zhang, W.; Moore, J. S., *J. Am. Chem. Soc.*, (2005) **127**, 11863.
[241] Zhang, W.; Moore, J. S., *Angew. Chem. Int. Ed.*, (2006) **45**, 4416.
[242] Ge, P.-H.; Fu, W.; Herrmann, W. A.; Herdtweck, E.; Campana, C.; Adams, R. D.; Bunz, U. H. F., *Angew. Chem. Int. Ed.*, (2000) **39**, 3607.
[243] Zhang, W.; Moore, J. S., *J. Am. Chem. Soc.*, (2004) **126**, 12796.

Keyword Index

A

Aburatubolactam A 606, 607
Acetamides, 2-bromo-N-phenyl-, iridium-catalyzed cyclization to give 3,3-disubstituted oxindoles 420, 421
Acetamides, 2-(methylsulfanyl)-N-phenyl-, cyclization to give oxindoles 418, 419
Acetylenecarboxylates, rhodium-catalyzed (2+2+2) cycloaddition with silylacetylenes and acrylamides to give cyclohexadienylcarboxamides 304, 305
Acetylenedicarboxylates, rhodium-catalyzed (2+2+2) cycloaddition with terminal alkynes to give phthalates 267
(+)-Achalensolide 159
Achilleol A 520
Acrylamides, rhodium-catalyzed (2+2+2) cycloaddition with silylacetylenes and acetylenecarboxylates to give cyclohexadienylcarboxamides 304, 305
Acrylates, asymmetric aziridination catalyzed by a copper–1,4-diimine complex 24, 25
(±)-Acuminatin 514–517
Acylaziridines, from chalcones by asymmetric aziridination catalyzed by a copper–1,8-bis(dihydrooxazolyl)anthracene complex 24
Acylcyclopropanes, bicyclic, from alkyne-substituted vinyloxiranes by rhodium-catalyzed intermolecular (5+2) cycloaddition 366
Acylcyclopropanes, from styrenes by palladium-catalyzed cross coupling with aryl methyl ketones 74
Acylcyclopropanes, from α,β-unsaturated carbonyl compounds by palladium(II)-catalyzed cyclopropanation using diazomethane 71
3-Acylcyclopropenes, rhodium-catalyzed (5+1) carbonylation with carbon monoxide to give α-pyrones 354, 355
(−)-Acylfulvene 637, 638
Acyloxiranes, from α,β-unsaturated ketones by asymmetric epoxidation catalyzed by iron(II) with a phenanthroline ligand 11, 12
Acyloxiranes, from α,β-unsaturated ketones by asymmetric epoxidation using a chiral N,N-dioxide–scandium(III) complex 21, 22
Acyloxiranes, from α,β-unsaturated ketones by manganese-catalyzed asymmetric epoxidation 8, 9
(Acyloxy)borane catalysts, chiral, in Diels–Alder reactions 206
(+)-Agelastatin A 43

Ageliferin 503
Aldehydes, aryl-, enantioselective intramolecular (3+2) cycloaddition with a carbonyl ylide to give 6,8-dioxabicyclo[3.2.1]octan-2-ones 187, 188
Aldehydes, chiral aluminum complex catalyzed oxa-Diels–Alder reactions with Danishefsky-type dienes to give dihydropyran-4-ones 225, 226
Aldehydes, chiral 1,1′-binaphthalene-2,2′-diol/titanium catalyzed oxa-Diels–Alder reactions with a Brassard-type diene to give 5,6-dihydropyran-2-ones 228, 229
Aldehydes, chiral 1,1′-binaphthalene-2,2′-diol/titanium catalyzed oxa-Diels–Alder reactions with Danishefsky-type dienes to give dihydropyran-4-ones 228–230
Aldehydes, chiral chromium complex catalyzed oxa-Diels–Alder reactions with Danishefsky-type dienes to give dihydropyran-4-ones 233
Aldehydes, chiral chromium complex catalyzed oxa-Diels–Alder reactions with 3-siloxy-substituted hexa-2,4-dienes to give tetrahydropyran-4-ones 234
Aldehydes, chiral indium complex catalyzed oxa-Diels–Alder reactions with Danishefsky-type dienes to give dihydropyran-4-ones 226, 227
Aldehydes, chiral rhodium complex catalyzed oxa-Diels–Alder reaction with Danishefsky-type dienes to give dihydropyran-4-ones 241, 242
Aldehydes, chiral substituted 1,1′-binaphthalene-2,2′-diol/zinc catalyzed oxa-Diels–Alder reaction with Danishefsky-type dienes to give dihydropyran-4-ones 238, 239
Aldehydes, chiral titanium–Schiff base complex catalyzed oxa-Diels–Alder reactions with Danishefsky-type dienes to give dihydropyran-4-ones 230, 231
Aldehydes, chiral zirconium complex catalyzed oxa-Diels–Alder reactions with Danishefsky-type dienes to give dihydropyran-4-ones 232
Aldehydes, gallium-catalyzed three-component (4+3)-cycloaddition reactions to give 5,6,9,10-tetrahydro-6,9-methanocyclohepta[b]indoles 338
Aldehydes, rhodium(II)-catalyzed diastereoselective intermolecular (3+2) cycloaddition with carbonyl ylides to give 1,3-dioxolanes 186, 187

Aldehydes, scandium-catalyzed three-component asymmetric aza-Diels–Alder reactions with 2-aminophenols and cyclopentadiene to give 3a,4,5,9b-tetrahydro-3H-cyclopenta[c]-quinolin-6-ols 255

Aldehydes, γ,δ-unsaturated, 4-*exo-trig* cyclization to give cyclobutanols 471–473

Alkenes, aliphatic, aziridination catalyzed by a copper–N-heterocyclic carbene complex 26, 27

Alkenes, asymmetric epoxidation catalyzed by an iron(III) complex 12, 13

Alkenes, asymmetric epoxidation catalyzed by a ruthenium–aqua–salen complex 15, 16

Alkenes, asymmetric epoxidation catalyzed by a ruthenium–pyboxazine complex 14, 15

Alkenes, asymmetric epoxidation catalyzed by manganese(II) and a porphyrin-inspired chiral ligand 9, 10

Alkenes, asymmetric epoxidation catalyzed by molybdenum(VI) and a chiral bishydroxamic acid ligand 16, 17

Alkenes, asymmetric epoxidation using titanium–salalen/salan complexes 1–3

Alkenes, aziridination catalyzed by a dirhodium carboxylate complex 31, 32

Alkenes, aziridination catalyzed by a silver–terpyridine complex 39, 40

Alkenes, cobalt(II)-catalyzed cyclopropanation using diazo sulfones to give cyclopropyl sulfones 60, 61

Alkenes, cobalt–porphyrin complex catalyzed aziridination to give aziridin-1-ylphosphonates 36, 37

Alkenes, cobalt–porphyrin complex catalyzed aziridination to give aziridin-1-ylsulfonates 37, 38

Alkenes, cobalt–porphyrin complex catalyzed aziridination to give N-tosylaziridines 35, 36

Alkenes, copper-catalyzed asymmetric cyclopropanation using in situ generated iodonium ylides to give 1-nitrocyclopropane-1-carboxylates 53

Alkenes, cyclopropanation with in situ generated ethyl 2-diazoacetate 69

Alkenes, electron-deficient, enantioselective rhodium-catalyzed cyclopropanation to give cyclopropanecarboxylates 59

Alkenes, epoxidation catalyzed by a Keggin-type silicodecatungstate salt 18

Alkenes, intermolecular aziridination catalyzed by a dirhodium caprolactamate complex 29, 30

Alkenes, iron-catalyzed cyclopropanation with in situ generated 2-diazo-1,1,1-trifluoroethane to give (trifluoromethyl)cyclopropanes 68

Alkenes, rhodium(II)-catalyzed aziridination to give aziridine-1-sulfonates 28, 29

Alkenes, rhodium(II)-catalyzed cyclopropanation using alkyl diazoacetates to give cyclopropanecarboxylates 55, 56

Alkenes, rhodium-catalyzed enantioselective intermolecular (3+2) cycloaddition with carbonyl ylides to give 8-oxabicyclo[3.2.1]octan-2-ones 176, 177

Alkenes, rhodium-catalyzed stereoselective cyclopropanation using 1-sulfonyl-1,2,3-triazoles to give cyclopropanecarbaldehydes 56, 57

Alkenes, ruthenium(II)-catalyzed enantioselective cyclopropanation to give cyclopropylmethanols 64

Alkenes, ruthenium–salen catalyzed aziridination to give aziridin-1-yl sulfones 32–34

Alkenes, *cis*-selective cyclopropanation using aryliridium–salen complexes to give cyclopropanecarboxylates 62, 63

Alkenes, stereoselective cyclopropanation with 2-cyano-2-diazoacetates to give 1-cyanocyclopropane-1-carboxylates 61, 62

Alkenes, stereoselective rhodium-catalyzed cyclopropanation to give cyclopropanecarbonitriles 58

Alkenes, stereoselective rhodium-catalyzed cyclopropanation to give nitrocyclopropanes 57, 58

Alkenes, terpyridine–iron complex catalyzed aziridination to give N-tosylaziridines 34, 35

Alkenyl bromides, photoredox-catalyzed radical cyclization to give five- and six-membered carbocycles 465–467

Alkenyl carbonyl ylides, rhodium-catalyzed diastereoselective intramolecular (3+2) cycloaddition to give an octahydro-6H-3a,7-epoxyazulen-6-one 170, 171

Alkenyl carbonyl ylides, rhodium-catalyzed enantioselective intramolecular (3+2) cycloaddition to give 8-oxabicyclo[3.2.1]octan-2-ones 176, 177

Alkenyl epoxides, ketyl coupling to give alkenyl cycloalkanols 476–478

Alkenyl epoxides, titanocene(III)-catalyzed reductive cyclization to give cyclopentanes 467, 468

Alkenyl epoxides, titanocene(III)-mediated intramolecular cyclization to give cyclopropanes and -butanes 480–482

Alkenyl ethers, iodo, titanium(IV)/chlorotrimethylsilane/manganese mediated cyclization to give tetrahydrofurans 434, 435

Alkenyl isocyanates, enantioselective rhodium-catalyzed (2+2+2) cycloaddition with alkynes to give piperidine derivatives 296, 297

Alkenyl isocyanates, enantioselective rhodium-catalyzed (2+2+2) cycloaddition with alkynes to give tetrahydroindolizinones 295–297

Alkenyl ketones, 8-*endo*-radical cyclization to give substituted cyclooctanols 484–486

Alkenyl ketones, samarium(II)-mediated cyclization to give cycloalkanols 478–480

2-Alkenylmalonates, manganese(III)-catalyzed cyclization to give bicyclo[3.3.0] γ-lactones 444, 445

2-(Alk-4-enyl)malonates, manganese(III)-mediated oxidative radical cyclization to give cyclopentane-fused lactones 448, 449

(Z)-Alkenylsilanes, asymmetric epoxidation with a titanium–salalen complex 2

N-Alkenylsulfonamides, copper(I)-catalyzed cyclization to give pyrrolidines 401, 402

Alkenyl trichloroacetates, cyclization to give α,α-dichloro lactones 440–442

3-Alkylidenecyclohepta-1,4-dienes, bicyclic, from alkyne-substituted allenylcyclopropanes by rhodium-catalyzed intramolecular (5+2) cycloaddition 365

Alkylidenecyclopropanes, from norbornadienes by platinum(II)-catalyzed reaction with phenylacetylene 87

2-(Alkynamido)malonates, manganese(III) acetate mediated cyclization to give 4-alkylidenepyrrolidin-2-ones 404, 405

2-(Alkynamido)malonates, manganese(III)-mediated cyclization to give piperidinones 427

Alk-4-ynamines, cyclization catalyzed by the Petasis reagent to give pyrrolidine derivatives 403, 404

Alk-3-ynamines, manganese(III)-mediated cyclization to give 3-alkylidenepyrrolidines 401

Alkynes, rhodium-catalyzed enantioselective intermolecular (3+2) cycloaddition with carbonyl ylides to give 8-oxabicyclo[3.2.1]oct-6-en-2-ones 177, 178

Alkynes, ruthenium-catalyzed diastereoselective intramolecular (3+2) cycloaddition with carbonyl ylides to give 8-oxabicyclo[3.2.1]oct-6-en-2-ones 171

Alkynes, ruthenium-catalyzed three-component cycloaddition with azomethine ylides to give 2,5-dihydropyrroles 196–198

Alkynoates, palladium-catalyzed intermolecular Pauson–Khand-type cyclization with norbornene to give 1-oxohexahydro-4,7-methanoindenylcarboxylates 134, 135

Alkynyl aldehydes, hetero-Pauson–Khand reaction to give lactones 156

Alkynyl bromides, photoredox-catalyzed radical cyclization to give five- and six-membered carbocycles 465–467

Alkynyl carbonyl ylides, rhodium-catalyzed enantioselective intramolecular (3+2) cycloaddition to give 6-oxotetrahydro-1*H*-3a,7-epoxyazulene-7(4*H*)-carboxylates 176

Alkynylcyclopropanes, rhodium-catalyzed (5+1) cycloadditions to give methylenecyclohexenones 356, 357

Alkynyl epoxides, titanocene(III)-catalyzed reductive cyclization to give cyclopentanes 467, 468

Alkynyl imines, hetero-Pauson–Khand reaction to give lactams 156

1-Alkynylnaphthalenes, rhodium-catalyzed enantioselective cycloaddition with diynes to give axially chiral biaryls 272, 273

N-Alkynylsulfonamides, copper(I)-catalyzed cyclization to give 2-alkylidene- and 2-alkynylpyrrolidines 402, 403

Allenamides, zinc-catalyzed cycloaddition with substituted furans to give 2-amino-8-oxabicyclo[3.2.1]oct-6-en-3-ones 334–337

Allene carbamates, silver(I)/1,10-phenanthroline catalyzed intramolecular aziridination to give 7-methylene-3-oxa-1-azabicyclo[4.1.0]-heptan-2-ones 40

Allenedienes, platinum(II)- and gold(I)-catalyzed intramolecular cycloaddition to give fused methylene cyclohexenes or cyclohepta-1,4-dienes 330, 331

Allenes, activation to give allylic cations 330–332

Allenes, axially chiral, rhodium-catalyzed intramolecular Pauson–Khand reaction to give bicyclic 4-alkylidenecyclopent-2-enones 139

Allenes, rhodium-catalyzed enantioselective intermolecular (3+2) cycloaddition with carbonyl ylides to give 6-methylene-8-oxabicyclo[3.2.1]octan-2-ones 179, 180

Allenoates, rhodium-catalyzed enantioselective (2+2+2) cycloaddition with ene–allenes to give fused 1,2-dimethylenecyclohexanes 307

Allenyl carbonitriles, rhodium-catalyzed Pauson–Khand-type reaction to give fused pyrrolidin-2-ones 158

Allenylcyclopropanes, alkynyl, rhodium-catalyzed intramolecular (5+2) cycloaddition to give bicyclic 3-alkylidenecyclohepta-1,4-dienes 365

Allenylcyclopropanes, iridium-catalyzed (5+1) cycloaddition with carbon monoxide to give methylenecyclohexenones 353

1-Allenylcyclopropan-1-ols, cobalt-catalyzed (5+1) cycloaddition to give hydroquinone derivatives 351, 352

Allenynes, intramolecular Pauson–Khand reaction to give fused cyclopentenones 110, 111

Allenynes, rhodium-catalyzed intramolecular Pauson–Khand reaction to give 4-alkylidenecyclopenten-2-ones 125–127

(+)-Allocyathin B2 373, 374

(+)-Allonorsecurinine 640, 641

(−)-Allosamizoline 567

N-Allylamides, α-polychloro-, reverse atom-transfer radical cyclization to give γ-lactams 423–425

O-Allyl carbamates, N-tosyloxy-, rhodium(II)-catalyzed intramolecular aziridination to give 3-oxa-1-azabicyclo[3.1.0]hexan-2-ones 29

Allyl carbonates, combined allylation/Pauson–Khand reaction sequence to give 2,3,3a,4-tetrahydrocyclopenta[c]pyrrol-5(1H)-ones 150, 151

Allyl carbonates, palladium-catalyzed cyclopropanation with acetamides to give cyclopropylacetamides 73, 74

Allyl α-diazocarboxylates, enantioselective iron-catalyzed intramolecular cyclopropanation to give 3-oxabicyclo[3.1.0]hexan-2-one 70

Allylic acetals, sulfur substituted, titanium-catalyzed (4+3) cycloaddition with furan to give 8-oxabicyclo[3.2.1]oct-6-en-3-ones 342

Allylic acetals, titanium-mediated (4+3) cycloadditions with furan or cyclopentadiene to give 2-alkoxy-3-methylenebicyclo[3.2.1]oct-6-enes 341

Allylic alcohols, asymmetric epoxidation catalyzed by a tungsten–bishydroxamic acid complex 19

Allylic alcohols, asymmetric epoxidation using vanadium(V) and a hydroxamic acid 3, 4

Allylic alcohols, asymmetric epoxidation with a niobium–salan complex 20

Allylic alcohols, asymmetric epoxidation with a vanadium–bishydroxamic acid complex 4, 5

Allylic alcohols, epoxidation catalyzed by a selenium-containing dinuclear peroxotungstate complex 18, 19

Allylic alcohols, meso, desymmetrization by epoxidation using vanadium(V) and a bis-hydroxamic acid ligand 6

Allylic cations, carbon-substituted, in cycloaddition reactions 321–333

Allylic cations, (4+3) cycloaddition with dienes 319

Allylic cations, heteroatom-substituted, in cycloaddition reactions 333–343

Allyllic chlorides, trimethylsiloxy-substituted, silver(I)-mediated cycloaddition with furans or cyclopentadiene to give bicyclo[3.2.1]oct-6-en-3-ones 325

Allylic sulfones, alkoxy, titanium-mediated intramolecular cycloaddition to give 1,2,3,6,7,8a-hexahydro-8H-3a,6-epoxyazulen-8-ones 329, 330

Allylic sulfones, sulfur substituted, titanium-catalyzed intramolecular (4+3) cycloaddition to give 7-(phenylsulfanyl)-1,2,3,6,7,8a-hexahydro-8H-3a,6-epoxyazulen-8-ones 342

Allyl oxiranylmethyl ethers, titanium(III)-mediated cyclization to give furans and furofurans 390, 391

[2-(Allyloxy)ethyl]oxiranes, titanium(III)-mediated radical cyclization to give tetrahydropyrans 456, 457

Allyl(propargyl)amines, gold(I)-catalyzed intramolecular cyclopropanation to give 3-azabicyclo[4.1.0]hept-4-enes 77, 78

Allyl(propargyl)amines, ruthenium-catalyzed asymmetric construction of 3-azabicyclo[3.1.0]hexanes 65, 66

Allyl propargyl ethers, gold(I)-catalyzed intramolecular cyclopropanation to give 3-oxabicyclo[4.1.0]hept-4-enes 77, 78

Allyl propargyl ethers, platinum-catalyzed enyne cyclization and acid-catalyzed ring-opening reaction to give dihydrobenzofurans and -benzopyrans 86

N-Allyl sulfonamides, asymmetric epoxidation catalyzed by a hafnium(IV)–bishydroxamic acid complex 23

(−)-α-Ambrinol 519, 520

β-Amino esters, α-hydroxy-, from imines and carbonyl ylides in a one-pot process 190

α-Amino lactones, from (imino)alkanenitriles by intramolecular radical cyclization 487

2-Amino-8-oxabicyclo[3.2.1]oct-6-en-3-ones, from allenamides by zinc-catalyzed cycloaddition with substituted furans 334–337

2-Aminophenols, scandium-catalyzed three-component asymmetric aza-Diels–Alder reactions with aldehydes and cyclopentadiene to give 3a,4,5,9b-tetrahydro-3H-cyclopenta[c]-quinolin-6-ols 255

Amphidinolide B 601, 602

(−)-Amphidinolide V 587, 588

(+)-Anabasine 260, 261

(+)-Anatoxin-a 633

Anguinomycin C 259, 260

Anilines, N-benzyl-2-vinyl-, iridium-mediated photoredox-catalyzed cascade reaction to give indole-3-carbaldehydes 412–414

2-(2-Anilino-2-oxoethyl)malonates, manganese(III) acetate mediated 6-endo-trig cyclization to give 3,4-dihydroquinolin-2(1H)-ones 427–429

Antiostatin A1 312

(+)-Aphanamol I 373

Arglabin 579

Aryl cyclopropyl ketones, photocatalyzed intramolecular (3+2) cycloadditions to give cyclopentanes 469, 470

N-Arylthioamides, benzoyl-containing, manganese(III)-promoted cyclization to give 2-benzoylbenzothiazoles 496

(+)-Asperpentyn 648, 649

Aspidophytine 200, 201

Asteriscanolide 379, 380, 581, 582

3-Azabicyclo[4.1.0]hept-4-enes, from allyl(propargyl)amines by gold(I)-catalyzed intramolecular cyclopropanation 77, 78

3-Azabicyclo[3.1.0]hexanes, from allyl(propargyl)amines by ruthenium-catalyzed asymmetric cyclization 65, 66
3-Azabicyclo[3.1.0]hexan-2-ones, from N-propenyl-3-oxobutanamides by manganese(III)-mediated cyclization 490, 491
Aza-Diels–Alder reactions, asymmetric 245–256
Aza-Diels–Alder reactions, asymmetric, using chiral copper complexes 245–247
Aza-Diels–Alder reactions, asymmetric, using chiral nickel complexes 252
Aza-Diels–Alder reactions, asymmetric, using chiral niobium complexes 251, 252
Aza-Diels–Alder reactions, asymmetric, using chiral rare earth metal complexes 254–256
Aza-Diels–Alder reactions, asymmetric, using chiral rhodium complexes 253
Aza-Diels–Alder reactions, asymmetric, using chiral silver complexes 249
Aza-Diels–Alder reactions, asymmetric, using chiral zinc complexes 247, 248
Aza-Diels–Alder reactions, asymmetric, using chiral zirconium complexes 250
Aza-Diels–Alder reactions, mechanistic pathways 245
Azadiradione 508, 509, 523, 524
Azaindolines, by radical cyclization reactions 415–417
Azasugars, from dienes by a ring-closing metathesis strategy 569, 570
Azetidinones, from chiral enamides, by manganese(III)-mediated cyclization 422
Aziridination, cobalt catalyzed 35–39
Aziridination, copper catalyzed 24–28
Aziridination, iron catalyzed 34, 35
Aziridination, rhodium catalyzed 28–32
Aziridination, ruthenium catalyzed 32–34
Aziridination, silver catalyzed 39–41
Aziridine-2-carboxylates, 2-acyl-, from β-keto esters by copper(II)-catalyzed aziridination 27
Aziridine-1-carboxylates, 2-aryl-, from styrenes by intermolecular aziridination catalyzed by a dirhodium carboxylate complex 30, 31
Aziridine-1-carboxylates, 2-aryl-, from styrenes by intermolecular aziridination with a copper–pyridine complex 25, 26
Aziridine-2-carboxylates, from acrylates by asymmetric aziridination catalyzed by a copper–1,4-diimine complex 24, 25
Aziridines, 1,2-diaryl-, from styrenes by cobalt–porphyrin complex catalyzed aziridination 38, 39
Aziridines, from alkenes by aziridination catalyzed by a dirhodium carboxylate complex 31, 32

Aziridines, N-tosyl-, from alkenes by aziridination catalyzed by a silver–terpyridine complex 39, 40
Aziridines, N-tosyl-, from alkenes by cobalt–porphyrin complex catalyzed aziridination 35, 36
Aziridines, N-tosyl-, from alkenes by intermolecular aziridination catalyzed by a dirhodium caprolactamate complex 29, 30
Aziridines, N-tosyl-, from alkenes by terpyridine–iron complex catalyzed aziridination 34, 35
Aziridine-1-sulfonates, from aliphatic alkenes by aziridination catalyzed by a copper–N-heterocyclic carbene complex 26, 27
Aziridine-1-sulfonates, from alkenes by rhodium(II)-catalyzed aziridination 28, 29
Aziridin-1-ylphosphonates, from alkenes by cobalt–porphyrin complex catalyzed aziridination 36, 37
Aziridin-1-ylsulfonates, from alkenes by cobalt–porphyrin complex catalyzed aziridination 37, 38
Aziridin-1-yl sulfones, from alkenes by ruthenium–salen catalyzed aziridination 32–34
Azodicarboxylates, ruthenium-catalyzed intermolecular cyclization with azomethine ylides to give 1,2,4-triazolidines 198, 199

B

Bakkenolides I, J, and S 500, 501
Balanol 580, 581
Benzaldehydes, 2-(allyloxy)- or (propargyloxy)-, titanium(III)-mediated cyclization to give dihydrobenzopyranols 458, 459
Benzamides, axially chiral, from ynamides by rhodium-catalyzed enantioselective cycloaddition with 1,6-diynes 275, 276
Benzene, manganese(III) acetate mediated alkylation with acetone to give phenylacetone 385
Benzo[c]furan-1(3H)-ones, from 2-substituted benzoic acids by copper(II)-mediated cyclization 439
Benzoic acids, 2-substituted, copper(II)-mediated cyclization to give benzo[c]furan-1(3H)-ones 439
Benzonorcaradienes, from 1,3-diynes by gold(I)-catalyzed intermolecular cyclopropanation with alkenes 82, 83
Benzopyran-4-ones, 3-vinyl-, chiral zinc complex catalyzed aza-Diels–Alder reaction with imines to give dihydropyridine derivatives 247, 248
Benzopyrans, asymmetric epoxidation catalyzed by a manganese–salen complex 7, 8
Benzopyrans, by radical cyclization reactions 458–460

Benzothiazoles, 2-aryl-, from N-arylbenzothioamides by manganese(III)-promoted cyclization 496

Benzothiazoles, 2-aryl-, from bis[(2-benzylideneamino)phenyl] disulfides by titanium/samarium catalyzed cyclization 495

Benzothiazoles, 2-benzoyl-, from benzoyl-containing N-arylthioamides by manganese(III)-promoted cyclization 496

Benzothiazoles, by intramolecular radical cyclization reactions 495–497

Benzothioamides, N-aryl-, manganese(III)-promoted cyclization to give 2-arylbenzothiazoles 496

Benz[b,f]oxepins, from 2-(9-xanthenyl)malonates by manganese(III)-mediated oxidation 461, 462

Biaryls, axially chiral, from buta-1,3-diynes by rhodium-catalyzed enantioselective cycloadditions with 1,6-diynes 273, 274

Biaryls, axially chiral, from diynes by rhodium-catalyzed enantioselective cycloaddition with 1-alkynylnaphthalenes 272, 273

N-Biaryltrifluoroacetimidoyl chlorides, cyclization to give 6-(trifluoromethyl)phenanthridines 491–494

Bicyclo[3.2.0]hepta-3,6-dien-2-ones, from cyclobutadiene surrogates by Pauson–Khand reaction 108, 109

Bicyclo[3.1.0]hexan-2-ones, from 5-en-1-yn-3-ols by platinum(II) chloride catalyzed cycloisomerization/hydrolysis 88

Bicyclo[3.1.0]hexenes, from 1,5-enynes by intramolecular gold(I)-catalyzed cyclopropanation 76

Bicyclo[3.1.0]hex-3-en-2-ones, from cyclopropenes by intramolecular Pauson–Khand reaction 108

Bicyclo[3.3.0] γ-lactone, from an alkenyl γ-lactone by manganese(III)/copper(II) mediated cyclization/methanolysis 447

Bicyclo[3.3.0] γ-lactones, from 2-alkenylmalonates by manganese(III)-catalyzed cyclization 444, 445

Bicyclo[4.3.0]nonan-8-ones, from bis(α,β-unsaturated esters) by tandem reductive coupling/Dieckmann condensation 488, 489

Bicyclo[4.2.1]non-3-en-9-ones, annulated, from dienones by iron(III)-mediated intramolecular cycloaddition 328, 329

Bicyclo[3.3.0]octan-3-ones, from bis(α,β-unsaturated esters) by tandem reductive coupling/Dieckmann condensation 488, 489

Bicyclo[3.3.0]octenes, from 5-(1-oxoprop-2-enyl)-norbornenes by ring-rearrangement metathesis 605, 606

Bicyclo[3.2.1]oct-6-en-3-ones, from trimethylsilyloxy-substituted allyl chlorides by silver(I)-mediated cycloaddition with cyclopentadiene 325

Bicyclo[3.3.0]octenones, from vinylcyclopropanes by rhodium-catalyzed intermolecular (5+2+1) cycloaddition with alkynes and carbon monoxide 376, 377

(−)-α-Bisabolol 42

Bis[(2-benzylideneamino)phenyl] disulfides, titanium/samarium catalyzed cyclization to give 2-arylbenzothiazoles 495

Bis(oxazoline)–copper complexes, catalysts for asymmetric Diels–Alder reactions 215

Bis(sulfinyl)imidoamidine–copper complexes, catalysts for Diels–Alder reactions 216, 217

Bromoageliferin 503

Brønsted acid activated borane catalysts, chiral, in Diels–Alder reactions 208–211

Buddledone A 591, 592

Buta-1,3-diene, 2,3-dimethyl-, rhodium(I)-catalyzed intermolecular Pauson–Khand reaction with alkynes to give 5-vinylcyclopent-2-enones 124, 125

Buta-2,3-dienoates, rhodium-catalyzed diastereoselective intramolecular (3+2) cycloaddition with carbonyl ylides to give 7-methylene-8-oxabicyclo[3.2.1]octan-2-ones 171, 172

Buta-1,3-diynes, rhodium-catalyzed enantioselective cycloadditions with 1,6-diynes to give axially chiral biaryls 273, 274

Butenolides, from allyl cyclobutene-1-carboxylates by ring-rearrangement metathesis 608–610

C

Carbazoles, from nitrogen-linked diynes by (2+2+2) cycloaddition with monoynes 268, 269

Carbodiimides, hetero-Pauson–Khand reaction to give fused lactams 156, 157

Carbodiimides, use in (2+2+2)-cycloaddition reactions 299–301

Carbonyl ylide generation, general approaches 167

Carbonyl ylides, catalytic models for selective (3+2)-cycloadditions 169, 170

Carbonyl ylides, competition between epoxidation and cycloaddition 168

Carboxylic esters, titanium-catalyzed reaction to give cyclopropanols 88–90

(−)-Centrolobine 604

Chalcones, asymmetric aziridination catalyzed by a copper–1,8-bis(dihydrooxazolyl)anthracene complex 24

N-Chloroamines, copper(I)-catalyzed radical cyclization to give 3-chloropiperidines 425, 426

Chlorobis(η^5-cyclopentadienyl)titanium(III), equilibrium between monomeric and dimeric species 387, 388
α-Chloroenamines, silver-mediated ionization to give allylic cations 326, 327
3-Chloropiperidines, from N-chloroamines by copper(I)-catalyzed radical cyclization 425, 426
Clavilactones 589, 590, 608, 609
Clavirolide C 590, 591
(S)-Cleonin 94
(−)-Cochleamycin A 637
Cortisone 257
Cruentaren A 592, 658, 659
(±)-Cryptotanshinone 529, 530
Cyanthiwigin U 606, 607
(5+1)-Cycloaddition reactions, chromium- or molybdenum-catalyzed 350, 351
(5+1)-Cycloaddition reactions, cobalt-catalyzed 351–353
(4+3)-Cycloaddition reactions, general mechanisms 320, 321
(2+2+2)-Cycloaddition reactions, involving alkenes 301–309
(4+3)-Cycloaddition reactions, involving allylic cations, metal-catalyzed 319–346
(2+2+2)-Cycloaddition reactions, involving carbodiimides and carbon dioxide 299–301
(2+2+2)-Cycloaddition reactions, involving carbonyl compounds 309–312
(2+2+2)-Cycloaddition reactions, involving isocyanates 292–299
(5+1)-Cycloaddition reactions, iridium catalyzed 353
(5+1)-Cycloaddition reactions, iron catalyzed 349, 350
(5+2)-Cycloaddition reactions, iron catalyzed 372
(5+2)-Cycloaddition reactions, mechanistic studies 370
(5+2)-Cycloaddition reactions, nickel catalyzed 371, 372
(2+2+2)-Cycloaddition reactions, of alkynes 266–282
(2+2+2)-Cycloaddition reactions, of alkynes with nitriles 282–291
(3+2)-Cycloaddition reactions, of carbonyl ylides with aldehydes 184–187
(3+2)-Cycloaddition reactions, of carbonyl ylides with C—C π-bonds 170–184
(3+2)-Cycloaddition reactions, of carbonyl ylides with imines 188–191
(5+1)-Cycloaddition reactions, rhodium catalyzed 354–358
(5+2)-Cycloaddition reactions, rhodium catalyzed 359–370
(5+2+1)-Cycloaddition reactions, rhodium catalyzed 376–378
(5+1)-Cycloaddition reactions, ruthenium catalyzed 353, 354
(5+2)-Cycloaddition reactions, ruthenium catalyzed 370, 371
(2+2+2)-Cycloaddition reactions, transition-metal catalyzed, mechanism 265, 266
Cycloalkadienes, macrocyclic, by E/Z-selective ring-closing alkyne metathesis–semireduction 655
Cycloalkanes, by intramolecular radical cyclization reactions 463–470
Cycloalkanols, alkenyl, from alkenyl epoxides by ketyl coupling 476–478
Cycloalkanols, trans-2-amino, from keto imines by intramolecular reductive coupling 482, 483
Cycloalkanols, from alkenyl ketones by samarium(II)-mediated cyclization 478–480
Cycloalkanones, 2-hydroxy-, from ketonitriles by enantioselective cyclization 473–476
Cycloalkenes, macrocyclic, from dienes by ruthenium-catalyzed Z-selective ring-closing metathesis 552, 553
Z-Cycloalkenes, macrocyclic, from diynes by ring-closing alkyne metathesis–semihydrogenation 655–657
E-Cycloalkenes, macrocyclic, from diynes by ring-closing alkyne metathesis–semireduction 661–663
Cycloalkenes, malonate and carboxy substituted, manganese(III)/copper(II) promoted cyclization to give fused tricyclic γ-lactones 442, 443
7,11-Cyclobotryococca-5,12,26-triene, precursor synthesis 499, 500
Cyclobutadiene surrogates, Pauson–Khand reaction to give bicyclo[3.2.0]hepta-3,6-dien-2-ones 108, 109
Cyclobutanes, from alkenyl epoxides by titanocene(III)-mediated intramolecular cyclization 480–482
Cyclobutanols, from γ,δ-unsaturated aldehydes by 4-exo-trig cyclization 471–473
Cyclobutene-1-carboxylates, allyl, ring-rearrangement metathesis to give butenolides 608–610
Cyclobutenes, 1-alk-5-ynyl-, ring-rearrangement metathesis to give isoquinoline derivatives 642, 643
Cyclodeca-1,5-diene oxide, titanium(III)-mediated reductive transannular cyclization to give decahydronaphthalen-1-ol derivatives 391
Cycloheptadienecarboxylates, from 1,3-dienes by rhodium-catalyzed (4+3) cycloaddition with vinyl diazoacetates 332, 333
Cycloheptadienes, bicyclic, from alkyne-substituted vinylcyclopropanes by iron-catalyzed (5+2) intramolecular cycloaddition 372

Cycloheptadienes, from vinylcyclopropanes by rhodium-catalyzed intermolecular (5+2) cycloaddition with alkynes 362

Cycloheptadienes, fused, from alkynyl vinylcyclopropanes by rhodium-catalyzed intramolecular (5+2) cycloaddition 359, 360

Cycloheptadienes, fused, from alkynyl vinylcyclopropanes by ruthenium-catalyzed intramolecular (5+2) cycloaddition 370, 371

Cyclohepta-1,4-dienes, fused, from allenedienes by platinum(II)- and gold(I)-catalyzed intramolecular cycloaddition 330, 331

Cycloheptatrienes, from 3-acyloxy-1,4-enynes by rhodium-catalyzed intermolecular (5+2) cycloaddition with alkynes 368, 369

Cycloheptatrienes, fused, from alkyne-substituted 3-acyloxy-1,4-enynes by rhodium-catalyzed intramolecular (5+2) cycloaddition 368

Cycloheptatrienes, intermolecular cyclopropanation with alkenes by retro-Buchner reaction using gold(I) carbenes 83, 84

Cycloheptenes, fused, from alkenyl vinylcyclopropanes by rhodium-catalyzed intramolecular (5+2) cycloaddition 360, 361

Cyclohept-4-en-1-ones, from α,α′-dihalo ketones by reductive cycloaddition with 1,3-dienes mediated by iron carbonyls 323

Cyclohept-4-en-1-ones, from 2-siloxyacroleins by metal-catalyzed (4+3) cycloadditions with 1,3-dienes 339, 340

Cycloheptenones, from vinylcyclopropanes by rhodium-catalyzed intermolecular (5+2) cycloaddition with alkynes 361, 362

Cyclohexa-1,3-diene, titanium-mediated cycloaddition with furfuryl alcohols to give 7,8-dihydro-4H-4,7-ethanocyclohepta[b]furans 327, 328

Cyclohexadienes, bicyclic, from 1,6-enynes by rhodium-catalyzed enantioselective (2+2+2) cycloaddition with alkynes 302–304

Cyclohexadienes, fused, from 1,6-diynes by rhodium-catalyzed enantioselective (2+2+2) cycloaddition with exo-methylene lactones 301, 302

Cyclohexadienylcarboxamides, from acrylamides by rhodium-catalyzed (2+2+2) cycloaddition with silylacetylenes and acetylenecarboxylates 304, 305

Cyclohexanes, 1,2-dimethylene, fused, from allenoates by rhodium-catalyzed enantioselective (2+2+2) cycloaddition with ene-allenes 307

Cyclohexanes, from alkenyl or alkynyl bromides by photoredox-catalyzed radical cyclization 465–467

Cyclohexenes, bicyclic, by rhodium-catalyzed enantioselective (2+2+2) cycloaddition of 1,6-enynes with acrylamides 305, 306

Cyclohexenes, fused, from dienynes by rhodium-catalyzed enantioselective (2+2+2) cycloaddition 306, 307

Cyclohexenes, 3-methylene-, from 1,6-enynes by endo selective ring-closing metathesis 623, 624

Cyclohexenones, from vinylcyclopropanes by cobalt-catalyzed (5+1) cycloaddition 351, 352

Cyclohexenones, from vinylcyclopropanes, by iron-mediated (5+1) cycloaddition 349, 350

Cyclohexenones, from vinylcyclopropanes by rhodium-catalyzed (5+1) cycloaddition with carbon monoxide 357, 358

Cyclooctanols, from alkenyl ketones by an 8-endo-radical cyclization process 484–486

(E)-Cyclooctene, 1-methyl-, intermolecular Pauson–Khand reaction to give octahydro-1H-cyclopenta[8]annulen-1-ones 106

(E)-Cyclooctenes, fused, from dienes by ring-closing metathesis with conformational control 548, 549

Cyclooctenones, fused, from alkenyl-tethered vinylcyclopropanes by rhodium-catalyzed intramolecular (5+2+1) cycloaddition with carbon monoxide 377, 378

Cyclopentadiene, silver(I)-mediated cycloaddition with trimethylsiloxy-substituted allyl chlorides to give bicyclo[3.2.1]oct-6-en-3-ones 325

Cyclopentanes, from alkenyl or alkynyl bromides by photoredox-catalyzed radical cyclization 465–467

Cyclopentanes, from aryl cyclopropyl ketones by photocatalyzed intramolecular (3+2) cycloadditions 469, 470

Cyclopentanes, from unsaturated epoxides by titanocene(III)-catalyzed reductive cyclization 467, 468

Cyclopentanones, by intramolecular radical cyclization reactions 487–491

Cyclopentenes, 1-vinyl-, from 1,6-enynes by exo selective ring-closing metathesis 623, 624

Cyclopenten-2-ones, 4-alkylidene-, from allenynes by rhodium-catalyzed intramolecular Pauson–Khand reaction 125–127

Cyclopent-2-enones, 4-alkylidene-, from axially chiral allenes by rhodium-catalyzed intramolecular Pauson–Khand reaction 139

Cyclopentenones, by chemically promoted intermolecular Pauson–Khand reactions 102, 103

Cyclopentenones, by Lewis base promoted intermolecular Pauson–Khand reactions 103, 104

Cyclopentenones, by Pauson–Khand reactions using chiral auxiliaries 139–142

Cyclopentenones, from enynes by rhodium-catalyzed intramolecular Pauson–Khand reactions using aldehydes or alcohols as the carbon monoxide source 127–129

Cyclopentenones, from enynes by titanium-catalyzed intramolecular Pauson–Khand reaction 130–132

Cyclopentenones, from ethene by intermolecular Pauson–Khand reaction 109, 110

Cyclopentenones, from norbornadienes by a cobalt-catalyzed intermolecular Pauson–Khand reaction/conjugate addition/retro-Diels–Alder sequence 107

Cyclopentenones, from vinyl(2-pyridyl)silanes, ruthenium-catalyzed intermolecular Pauson–Khand reaction 132, 133

Cyclopentenones, fused, by intramolecular Pauson–Khand reactions mediated by molybdenum complexes 104, 105

Cyclopentenones, fused, from enynes by intramolecular cobalt-catalyzed Pauson–Khand reactions using additives 115, 116

Cyclopentenones, fused, from enynes by intramolecular Pauson–Khand reactions catalyzed by preformed alkyne–hexacarbonyldicobalt complexes 117, 118

Cyclopentenones, fused, from enynes by rhodium-catalyzed enantioselective intramolecular Pauson–Khand reaction 120–122

Cyclopent-2-enones, 5-vinyl-, from 2,3-dimethylbuta-1,3-diene by rhodium(I)-catalyzed intermolecular Pauson–Khand reaction 124, 125

Cyclophanes, by rhodium-catalyzed (2+2+2) cycloaddition of terminal diynes with acetylenedicarboxylates 267, 268

Cyclophanes, planar chiral, from cyclic diynes by rhodium-catalyzed enantioselective cycloaddition with terminal monoynes 276, 277

Cyclopropanation, cobalt catalyzed 60–62
Cyclopropanation, copper catalyzed 50–53
Cyclopropanation, gold catalyzed 75–85
Cyclopropanation, iridium catalyzed 62, 63
Cyclopropanation, iron catalyzed 66–70
Cyclopropanation, palladium catalyzed 70–74
Cyclopropanation, platinum catalyzed 85–88
Cyclopropanation, rhodium catalyzed 54–59
Cyclopropanation, ruthenium catalyzed 63–66
Cyclopropanation, titanium catalyzed 88–90
Cyclopropanation, using diazo compounds, mechanism 49, 50

Cyclopropanecarbaldehydes, from alkenes by rhodium-catalyzed stereoselective cyclopropanation using 1-sulfonyl-1,2,3-triazoles 56, 57

Cyclopropanecarbonitriles, from alkenes by stereoselective rhodium-catalyzed cyclopropanation 58

Cyclopropanecarboxamides, N-alkynyl-, iridium-catalyzed light-mediated radical cyclization to give pyrrolidin-2-one derivatives 406–409

Cyclopropane-1-carboxylates, 1-cyano-, from alkenes by stereoselective cyclopropanation with 2-cyano-2-diazoacetates 61, 62

Cyclopropanecarboxylates, from alkenes by cyclopropanation with in situ generated ethyl 2-diazoacetate 69

Cyclopropanecarboxylates, from alkenes by rhodium(II)-catalyzed cyclopropanation using alkyl diazoacetates 55, 56

Cyclopropanecarboxylates, from alkenes by cis-selective cyclopropanation using aryliridium–salen complexes 62, 63

Cyclopropanecarboxylates, from electron-deficient alkenes by enantioselective rhodium-catalyzed cyclopropanation 59

Cyclopropanecarboxylates, 2-phenyl-, from styrene by copper-catalyzed asymmetric cyclopropanation 51

Cyclopropanes, bicyclic, from dienes by platinum(II)-catalyzed cycloisomerization 85

Cyclopropanes, from alkenyl epoxides by titanocene(III)-mediated intramolecular cyclization 480–482

Cyclopropanes, from cycloheptatrienes by intermolecular cyclopropanation with alkenes by retro-Buchner reaction using gold(I) carbenes 83, 84

Cyclopropanes, from 1,3-dibromopropanes by samarium(II) iodide mediated cyclization 464, 465

Cyclopropanes, from 1,3-diiodopropanes by samarium(II) iodide mediated cyclization 463, 464

Cyclopropanols, from carboxylic esters by titanium-catalyzed reaction 88–90

Cyclopropanols, from α,β-unsaturated δ-keto esters by samarium(II) iodide mediated cyclization 470, 471

Cyclopropenes, 1-alkynyl-, by ring-rearrangement metathesis to give 2,5-dihydropyrroles and -furans 643, 644

Cyclopropenes, intermolecular Pauson–Khand reaction to give bicyclo[3.1.0]hex-3-en-2-ones 108

Cyclopropenes, phenyl-, chromium/molybdenum-catalyzed (5+1) cycloaddition to give naphthols 350, 351

Cyclopropylacetamides, from allyl carbonates by palladium-catalyzed cyclopropanation with acetamides 73

Cyclopropylmethanimines, rhodium-catalyzed intermolecular (5+2) cycloaddition with alkynes to give dihydroazepines 365, 366

Cyclopropylmethanimines, ruthenium-catalyzed (5+1) cycloaddition with carbon monoxide to give 3,4-dihydropyridin-2(1H)-ones 353, 354

Cyclopropylmethanols, from alkenes by ruthenium(II)-catalyzed enantioselective cyclopropanation 64

Cyclopropyl sulfones, from alkenes by cobalt(II)-catalyzed cyclopropanation using diazo sulfones 60

1-Cyclopropylvinyl esters, from propargyl esters by intermolecular gold(I)-catalyzed cyclopropanation with alkenes 78–80

Cystine isosteres, from 1,13-diynes by ring-closing alkyne metathesis 652

(+)-Cytotrienin A 598, 599

D

(−)-Dactylolide 597, 598

Decahydro-1H-cyclopenta[a]pentalene, from 3-[4-(but-3-ynyl)cyclopent-2-enyl]propanal by sequential samarium(II) iodide mediated reactions 392

Decahydronaphthalen-1-ols, from cyclodeca-1,5-diene oxides by titanium(III)-mediated reductive transannular cyclization 391

(−)-Delobanone 358, 359

6-Deoxyerythronolide B 634

17-Deoxyprovidencin 596, 597

1-Desoxyhypnophilin 379

Dialdehydes, chiral diimine activated zinc complex catalyzed tandem cycloaddition/alkylation to give dihydropyran-4-one products 239, 240

Diallylamines, ruthenium-mediated (2+2) cycloaddition to give cyclobutane–pyrrolidine derivatives 435–438

Diallyl ethers, ruthenium-mediated (2+2) cycloaddition to give cyclobutane–tetrahydrofuran derivatives 435–438

1,3-Dialk-1-ynylbenzenes, alkyne metathesis to give cyclic 1,3-phenyleneethynylene hexamers 669, 670

Dibenzofurans, from oxygen-linked diynes by (2+2+2) cycloaddition with monoynes 268, 269

Dibenzosiloles, from silicon-linked diynes by (2+2+2) cycloaddition with monoynes 268, 269

Dibromoageliferin 503

α,α′-Dibromo ketones, reductive cycloaddition with chiral furfuryl alcohols mediated by diethylzinc to give (1-hydroxymethyl)-8-oxabicyclo[3.2.1]oct-6-en-3-ones 324

1,3-Dibromopropanes, samarium(II) iodide meditated cyclization to give cyclopropanes 464, 465

(+)-Dictamnol 373

Diels–Alder reaction, asymmetric 205–224

Diels–Alder reaction, asymmetric, using chiral aluminum complexes 212, 213

Diels–Alder reaction, asymmetric, using chiral boron complexes 206–212

Diels–Alder reaction, asymmetric, using chiral chromium complexes 219, 220

Diels–Alder reaction, asymmetric, using chiral cobalt complexes 220, 221

Diels–Alder reaction, asymmetric, using chiral copper complexes 215, 216

Diels–Alder reaction, asymmetric, using chiral indium complexes 213, 214

Diels–Alder reaction, asymmetric, using chiral rare earth complexes 222, 223

Diels–Alder reaction, asymmetric, using chiral ruthenium complexes 221, 222

Diels–Alder reaction, asymmetric, using chiral titanium complexes 217, 218

Diene–diynes, double Pauson–Khand reaction for the synthesis of a linear [5.5.5.5] system 153, 154

Diene–polyynes, combination of multiple metathesis reactions to form oligoenynes 646–648

Diene ring-closing metathesis, chemoselectivity 554–562

Diene ring-closing metathesis, commonly used catalysts 546

Diene ring-closing metathesis, mechanism 547, 548

Diene ring-closing metathesis, E/Z selectivity 548–554

E,Z-1,3-Dienes, cyclic, from alkene-tethered 3-silyl-1,3-dienes by two-step ring-closing metathesis 594, 595

Dienes, (4+3) cycloaddition with allylic cations 319

1,3-Dienes, metal-catalyzed (4+3) cycloadditions with 2-siloxyacroleins to give cyclohept-4-en-1-ones 339, 340

Dienes, platinum(II)-catalyzed cycloisomerization to give bicyclic cyclopropanes 85

1,3-Dienes, reductive cycloaddition with α,α′-dihalo ketones mediated by iron carbonyls to give cyclohept-4-en-1-ones 323

1,3-Dienes, rhodium-catalyzed (4+3) cycloaddition with vinyl diazoacetates to give cycloheptadienecarboxylates 332, 333

Dienes, rhodium-catalyzed hetero-Pauson–Khand reaction/ring-closing metathesis to give tricyclic products 149

Dienes, ruthenium-catalyzed Z-selective ring-closing metathesis to give macrocyclic cycloalkenes 552, 553

1,3-Dienes, 3-silyl-, alkene tethered, two-step ring-closing metathesis to give cyclic E,Z-1,3-dienes 594, 595

cis,cis-Dienetriynes, nickel-catalyzed synthesis of [n]helicenes 281

Dienones, iron(III)-mediated intramolecular cycloaddition to give annulated bicyclo[4.2.1]-non-3-en-9-ones 328, 329

Dienyl ethynylphosphonates, tandem enyne ring-closing metathesis to give bicyclic products 620, 621

Dienynes, amine tethered, ring-rearrangement metathesis in the synthesis of a dipiperidine moiety 644, 645

Dienynes, double ring-closing metathesis 635–641

Dienynes, from ene-1,3-diynes by metallotropic [1,3]-shift induced by enyne ring-closing metathesis 629

1,5-Dien-5-ynes, from symmetrical diynes by sequential enyne ring-closing metathesis 630, 631

Dienynes, gold(I)-catalyzed intramolecular bis-cyclopropanation to give tetracyclic fused cyclopropanes 81, 82

Dienynes, rhodium(I)-catalyzed intramolecular Pauson–Khand reaction 123, 124

Dienynes, rhodium-catalyzed enantioselective (2+2+2) cycloaddition to give fused cyclohexenes 306

Dienynes, tandem enyne ring-closing metathesis to give 7,7a-dihydrocyclopenta[b]pyran-2(6H)-ones 619, 620

Dienynes, tandem enyne ring-closing metathesis to give 1,2,3,7,8,8a-hexahydronaphthalenes 618, 619

Dienynes, ynamide based, tandem enyne ring-closing metathesis to give 2,6,7,8-tetrahydropyrido[1,2-a]azepin-4(3H)-ones 619

[Diethoxy(methyl)silyl]dienes, two-step ring-closing metathesis protocol for Z selective synthesis of macrocycles 552, 553

α,α′-Dihalo ketones, reduction to give oxyallylic cations 321–324

α,α′-Dihalo ketones, reductive cycloaddition with 1,3-dienes mediated by iron carbonyls to give cyclohept-4-en-1-ones 323

α,α′-Dihalo ketones, reductive cycloaddition with furan mediated by copper bronze to give 8-oxabicyclo[3.2.1]oct-6-en-3-ones 321, 322

Dihydroazepines, from cyclopropylmethanimines by rhodium-catalyzed intermolecular (5+2) cycloaddition with alkynes 365, 366

6,7-Dihydro-5H-benzo[7]annulene, asymmetric epoxidation catalyzed by a manganese–salen complex 7, 8

Dihydrobenzofurans, from allyl propargyl ethers by platinum-catalyzed enyne cyclization and acid-catalyzed ring-opening reaction 86

Dihydrobenzopyranols, from 2-(allyloxy)- or 2-(propargyloxy)benzaldehydes by titanium(III)-mediated cyclization 458, 459

Dihydrobenzopyrans, from allyl propargyl ethers by platinum-catalyzed enyne cyclization and acid-catalyzed ring-opening reaction 86

7,7a-Dihydrocyclopenta[b]pyran-2(6H)-ones, from dienynes, by tandem enyne ring-closing metathesis 619, 620

Dihydro-4H-1,3,2-dioxasilocines, from dienes by E/Z-selective ring-closing metathesis 549, 550

Dihydro-1H-1,4-epoxybenzo[c]azepin-5(2H)-ones, from imines by rhodium/Lewis acid catalyzed diastereoselective (3+2) cycloaddition with a carbonyl ylide 188, 189

7,8-Dihydro-4H-4,7-ethanocyclohepta[b]furans, from furfuryl alcohols by titanium-mediated cycloaddition with cyclohexa-1,3-diene 327, 328

2,5-Dihydrofurans, from cyclopropene-containing enynes by ring-rearrangement metathesis 643, 644

2,5-Dihydrofurans, from 1,6-enynes by ring-rearrangement metathesis 642

2,3-Dihydrofurans, from styrene by manganese(III) acetate mediated coupling with β-dicarbonyl compounds 385–387

2,5-Dihydrofurans, 2-vinyl-, from trienes by enantioselective ring-closing metathesis 565, 566

2,3-Dihydro-1H-inden-1-ones, 6-hydroxy-, from cyclopropyl 1,4-enynes by rhodium-catalyzed tandem Pauson–Khand-type reaction 154

(±)-Dihydroprotolichesterinic acid 513, 514

Dihydropyran-4-ones, by chiral diimine activated zinc complex catalyzed tandem cycloaddition/alkylation 239, 240

5,6-Dihydropyran-2-ones, from a Brassard-type diene and aldehydes by chiral 1,1′-binaphthalene-2,2′-diol/titanium catalyzed oxa-Diels–Alder reactions 228, 229

Dihydropyran-4-ones, from aldehydes by chiral rhodium complex catalyzed oxa-Diels–Alder reaction with Danishefsky-type dienes 241, 242

Dihydropyran-4-ones, from aldehydes by chiral substituted 1,1′-binaphthalene-2,2′-diol/zinc catalyzed oxa-Diels–Alder reaction with Danishefsky-type dienes 238, 239

Dihydropyran-4-ones, from Danishefsky-type dienes and aldehydes by chiral aluminum complex catalyzed oxa-Diels–Alder reactions 225, 226

Dihydropyran-4-ones, from Danishefsky-type dienes and aldehydes by chiral 1,1′-binaphthalene-2,2′-diol/titanium catalyzed oxa-Diels–Alder reactions 228, 230

Dihydropyran-4-ones, from Danishefsky-type dienes and aldehydes by chiral chromium complex catalyzed oxa-Diels–Alder reactions 233

Dihydropyran-4-ones, from Danishefsky-type dienes and aldehydes by chiral indium complex catalyzed oxa-Diels–Alder reactions 226, 227

Dihydropyran-4-ones, from Danishefsky-type dienes and aldehydes by chiral titanium–Schiff base complex catalyzed oxa-Diels–Alder reactions 230, 231

Dihydropyran-4-ones, from Danishefsky-type dienes and aldehydes by chiral zirconium complex catalyzed oxa-Diels–Alder reactions 232

Dihydropyran-4-ones, from α-oxo esters or 1,2-diones by chiral copper complex catalyzed oxa-Diels–Alder reaction with Danishefsky-type dienes 236, 237

5,6-Dihydro-2H-pyran-2-ones, from α,β-unsaturated acid chlorides by chiral erbium complex catalyzed oxa-Diels–Alder reaction with aldehydes 244

5,6-Dihydropyran-2-ones, 6-vinyl-, from trienes by chemoselective ring-closing metathesis 555, 556

3,4-Dihydro-2H-pyrans, from enones by chiral chromium complex catalyzed oxa-Diels–Alder reactions with ethyl vinyl ether 234, 235

3,4-Dihydro-2H-pyrans, from β,γ-unsaturated α-oxo esters by chiral copper complex catalyzed oxa-Diels–Alder reaction with vinyl ethers 237, 238

Dihydropyrans, fused, from 1,6-enynes by rhodium-catalyzed enantioselective (2+2+2) cycloadditions with dicarbonyl compounds 311

Dihydropyrans, polycyclic, from terpenoids by manganese(III)-mediated oxidative radical cyclization 452, 453

3,6-Dihydro-2H-pyran-2-yl ketones, from α-oxo aldehydes by chiral palladium complex catalyzed oxa-Diels–Alder reactions with dienes 242, 243

Dihydropyridines, from imines by chiral zinc complex catalyzed aza-Diels–Alder reaction with 3-vinylbenzopyran-4-ones 247, 248

3,4-Dihydropyridin-2(1H)-ones, from cyclopropylmethanimines by ruthenium-catalyzed (5+1) cycloaddition with carbon monoxide 353, 354

Dihydropyridin-4-ones, from imines by chiral copper complex catalyzed aza-Diels–Alder reaction with Danishefsky-type dienes 245, 246

Dihydropyridin-4-ones, from imines by chiral niobium complex catalyzed aza-Diels–Alder reaction with Danishefsky-type dienes 251

Dihydropyridin-4-ones, from imines by chiral scandium complex catalyzed aza-Diels–Alder reaction with Danishefsky-type dienes 254

Dihydropyridin-4-ones, from imines by chiral silver complex catalyzed aza-Diels–Alder reaction with Danishefsky-type dienes 249

Dihydropyridin-4-ones, from imines by chiral zirconium complex catalyzed aza-Diels–Alder reaction with Danishefsky-type dienes 250

2,5-Dihydropyrroles, from alkynes by ruthenium-catalyzed three-component cycloaddition with azomethine ylides 196, 197

4,5-Dihydropyrroles, from azomethine ylides by intramolecular cyclization 192–194

2,5-Dihydropyrroles, from cyclopropene-containing enynes by ring-rearrangement metathesis 643, 644

2,5-Dihydropyrroles, from 1,6-enynes by ring-rearrangement metathesis 642

3,4-Dihydroquinolin-2(1H)-ones, from substituted malonates by manganese(III) acetate mediated 6-*endo-trig* cyclization 427–429

(±)-Dihydrosesamin 514, 515

2,3-Dihydrothiophenes, from S-[3,3-bis(phenylsulfanyl)propyl] alkanethioates by titanium-catalyzed cyclization 497

α,α′-Diiodo ketones, reductive cycloaddition with furan mediated by zinc/copper couple to give 8-oxabicyclo[3.2.1]oct-6-en-3-ones 322

1,3-Diiodopropanes, samarium(II) iodide mediated cyclization to give cyclopropanes 463, 464

1,2-Diones, chiral copper complex catalyzed oxa-Diels–Alder reaction with Danishefsky-type dienes to give dihydropyran-4-ones 236

6,8-Dioxabicyclo[3.2.1]octan-2-ones, from aromatic aldehydes by enantioselective intramolecular (3+2) cycloaddition with a carbonyl ylide 187, 188

1,3-Dioxolanes, from aldehydes by rhodium(II)-catalyzed diastereoselective intermolecular (3+2) cycloaddition with carbonyl ylides 184–186

Dipiperidines, from amine-tethered dienynes by ring-rearrangement metathesis 644, 645

1,3-Dipolar (3+2)-cycloaddition reactions, of azomethine ylides 192–200

Dirhodium catalysts, for selective metal carbene transformations 168, 169

Diyne ring-closing metathesis, mechanism 651

Diyne ring-closing metathesis, permitted ring size 652

Diyne ring-closing metathesis, representative catalysts 650, 651

Diyne ring-closing metathesis–semireduction, construction of cyclic Z-alkenes 655–661

Diyne ring-closing metathesis–semireduction, construction of cyclic E-alkenes 661–663

Diyne ring-closing metathesis–semireduction, construction of cyclic 1,3-dienes 664–667

Diyne ring-closing metathesis–semireduction, construction of cyclic 1,3-diynes 668

Diyne ring-closing metathesis, synthesis of cyclic oligomers 668–670

1,6-Diynes, aryl, iridium-catalyzed enantioselective cycloaddition with monoynes to give axially chiral 1,4-teraryls 274, 275

Diynes, cobalt-catalyzed (2+2+2) cycloaddition with nitriles to give fused pyridines 283, 284

Diynes, cyclic, rhodium-catalyzed enantioselective cycloaddition with terminal monoynes to give planar chiral cyclophanes 276, 277

Diynes, diboryl-, cobalt-catalyzed (2+2+2) cycloaddition with monoynes to give 1,2,3,4-tetrahydronaphthalenes 269, 270

Diynes, diiodo, ruthenium-catalyzed (2+2+2) cycloaddition of acetylene to give iodoarenes 271

1,3-Diynes, gold(I)-catalyzed intermolecular cyclopropanation with alkenes to give benzonorcaradienes 82, 83

Diynes, heteroatom-linked, (2+2+2) cycloaddition with monoynes to give heterofluorenes 269, 269

Diynes, iridium-catalyzed (2+2+2) cycloaddition with isocyanates to give bicyclic pyridin-2-ones 294

Diynes, iridium-catalyzed (2+2+2) cycloaddition with nitriles to give fused pyridines 287–290

1,3-Diynes, macrocyclic, from 1,n-bis-1,3-diynes by ring-closing alkyne metathesis 654

Diynes, nickel-catalyzed (2+2+2) cycloaddition with isocyanates to give bicyclic pyridin-2-ones 292, 293

Diynes, nickel-catalyzed (2+2+2) cycloaddition with nitriles to give fused pyridines 285

Diynes, peptide-linked, by ring-closing alkyne metathesis to give macrocyclic tetrapeptides 652, 653

Diynes, phenylene-1,4-diamine-linked, ring-closing alkyne metathesis to give glycophanes 653

Diynes, rhodium-catalyzed (2+2+2) cycloaddition with acetylenedicarboxylates to give cyclophanes 267, 268

Diynes, rhodium-catalyzed (2+2+2) cycloaddition with carbodiimides to give bicyclic pyridin-2-imines 299

Diynes, rhodium-catalyzed (2+2+2) cycloadditions with carbonyl compounds to give fused pyrans 309, 310

Diynes, rhodium-catalyzed (2+2+2) cycloaddition with isocyanates to give bicyclic pyridin-2-ones 294

Diynes, rhodium-catalyzed (2+2+2) cycloaddition with nitriles to give fused pyridines 287

1,6-Diynes, rhodium-catalyzed enantioselective (2+2+2) cycloaddition with exo-methylene lactones to give fused cyclohexadienes 301, 302

1,6-Diynes, rhodium-catalyzed enantioselective cycloaddition with ynamides to give axially chiral benzamides 276

Diynes, rhodium- or nickel-catalyzed (2+2+2) cycloaddition with carbon dioxide to give fused pyran-2-ones 300

1,13-Diynes, ring-closing alkyne metathesis in the synthesis of cystine isosteres 652

Diynes, ring-closing alkyne metathesis–semihydrogenation to give Z-alkene-containing macrocycles 655, 656

Diynes, ring-closing alkyne metathesis–semireduction to give E-alkene-containing macrocycles 661, 662

Diynes, ruthenium-catalyzed (2+2+2) cycloaddition with isocyanates to give bicyclic pyridin-2-ones 293

Diynes, ruthenium-catalyzed (2+2+2) cycloaddition with nitriles to give fused pyridines 285

Diynes, symmetrical, sequential enyne ring-closing metathesis to give 1,5-dien-3-ynes 630, 631

1,6-Diynylbenzenes, rhodium-catalyzed enantioselective cycloaddition with isocyanates to give axially chiral arylpyridin-2-ones 295

1-Diynylnaphthalenes, cobalt-catalyzed enantioselective cycloaddition with nitriles to give axially chiral 2-naphthylpyridines 289

Dysinosin A 569

E

Echinopines A and B 91, 92

Ecklonialactone A 659, 660

Ecklonialactone B 594, 659, 660

(+)-Elatol 573

Enamides, chiral, manganese(III)-mediated cyclization to give trans-azetidinones 422

Enamines, N-(2-vinylaryl)-, oxidative radical cyclization to give quinolines 430

Ene-1,3-diynes, metallotropic [1,3]-shift induced by enyne ring-closing metathesis to give dienynes 629

Enetriynes, enyne ring-closing metathesis–metallotropic [1,3]-shift 629, 630

Englerin A 580, 641

Enones, chiral chromium complex catalyzed oxa-Diels–Alder reactions with ethyl vinyl ether to give 3,4-dihydro-2H-pyrans 235

Enones, chiral oxazaborolidine catalyzed Diels–Alder reactions to give cyclohexenes 210–212

Enones, chiral ruthenium complex catalyzed Diels–Alder reactions to give cyclohexene derivatives 221, 222

Enones, yttrium-catalyzed three-component asymmetric aza-Diels–Alder reactions with cyclic ketones and anilines to give 1,4,5,6,7,8-hexahydroquinoline derivatives 256

Entecavir 518, 519
Enyne ring-closing metathesis, chemoselectivity with dienynes 618
Enyne ring-closing metathesis, commonly used catalysts 611
Enyne ring-closing metathesis, competition experiment 614
Enyne ring-closing metathesis, concentration-dependent selectivity with a dienyne 621, 622
Enyne ring-closing metathesis, control based on ring size 620, 621
Enyne ring-closing metathesis, deuterium labeling studies 614, 615
Enyne ring-closing metathesis/intramolecular Diels–Alder sequence 638, 639
Enyne ring-closing metathesis, mechanism 612–616
Enyne ring-closing metathesis–metallotropic [1,3]-shift, regioselectivity 628–631
Enyne ring-closing metathesis, *endo/exo* mode and E/Z selectivity 623–628
Enyne ring-closing metathesis, regioselectivity 617
Enyne ring-closing metathesis, *endo/exo*-selective, for the formation of macrocycles 625, 626
Enyne ring-closing metathesis, tandem, electronic differentiaton of alkenes in a dienyne 619, 620
Enyne ring-closing metathesis, tandem, steric control with dienynes 618, 619
1,4-Enynes, 3-acyloxy-, rhodium-catalyzed (5+2) cycloaddition with alkynes to give cycloheptatrienes 368
1,4-Enynes, 3-acyloxy-, rhodium-catalyzed (5+1) cycloaddition with carbon monoxide to give resorcinols 356
Enynes, cobalt-catalyzed asymmetric intramolecular Pauson–Khand reaction using sulfoxides as chiral auxiliaries 142, 143
Enynes, cobalt-catalyzed intramolecular reductive Pauson–Khand reactions 155
1,6-Enynes, cobalt–chiral ligand catalyzed enantioselective intramolecular Pauson–Khand reaction to give 4,5,6,6a-tetrahydropentalen-2(1H)-ones 146, 147
1,4-Enynes, cyclopropyl, rhodium-catalyzed tandem Pauson–Khand-type reaction to give 6-hydroxy-2,3-dihydro-1H-inden-1-ones 154
Enynes, intramolecular cobalt-catalyzed Pauson–Khand reactions using additives 116, 117
1,6-Enynes, intramolecular cycloisomerizations 75
1,5-Enynes, intramolecular 5-*endo-dig* gold(I)-catalyzed cyclopropanation 75, 76

Enynes, intramolecular Pauson–Khand reactions catalyzed by preformed alkyne–hexacarbonyldicobalt complexes 117, 118
Enynes, iridium-catalyzed asymmetric Pauson–Khand reactions 129, 130
1,6-Enynes, Pauson–Khand reaction/allylic alkylation/Pauson–Khand reaction sequence to give fenestranes 151
1,6-Enynes, Pauson–Khand-type reaction using a heterobimetallic cluster as precatalyst to give 4,5,6,6a-tetrahydropentalen-2(1H)-ones 134
1,6-Enynes, rhodium-catalyzed enantioselective (2+2+2) cycloaddition with acrylamides to give bicyclic cyclohexenes 305, 306
1,6-Enynes, rhodium-catalyzed enantioselective (2+2+2) cycloaddition with alkynes to give bicyclic cyclohexadienes 302–304
1,6-Enynes, rhodium-catalyzed enantioselective (2+2+2) cycloaddition with dicarbonyl compounds to give fused dihydropyrans 311
Enynes, rhodium-catalyzed enantioselective intramolecular Pauson–Khand reaction 120–122
Enynes, rhodium-catalyzed intramolecular Pauson–Khand reactions using aldehydes or alcohols as the carbon monoxide source 128, 129
1,6-Enynes, ring-rearrangement metathesis to give 2,5-dihydropyrroles and 2,5-dihydrofurans 642
1,6-Enynes, *endo/exo*-selective ring-closing metathesis to give 1-vinylcyclopentenes or 3-methylenecyclohexenes 623, 624
Enynes, silicon-tethered, *endo/exo*-selective ring-closing metathesis 625
Enynes, tandem Nicholas/Pauson–Khand reaction to give [5.n.5]-systems 149, 150
Enynes, titanium-catalyzed intramolecular Pauson–Khand reaction 131, 132
5-En-1-yn-3-ols, platinum(II) chloride catalyzed cycloisomerization/hydrolysis to give bicyclo[3.1.0]hexan-2-ones 88
7′-Epimagnofargesin 511, 512
Epimeloscine 574
Epothilone C 552, 658
Epoxidation, iron catalyzed 11–14
Epoxidation, manganese catalyzed 7–11
Epoxidation, molybdenum catalyzed 16, 17
Epoxidation, ruthenium catalyzed 14–16
Epoxidation, titanium catalyzed 1–3
Epoxidation, tungsten catalyzed 17–19
Epoxidation, vanadium catalyzed 3–7
Epoxides, (arylamino)methyl, intramolecular cyclization to give indoline derivatives 415–417
Epoxides, from alkenes by asymmetric epoxidation catalyzed by an iron(III) complex 12, 13

Epoxides, from alkenes by asymmetric epoxidation catalyzed by a ruthenium–aqua–salen complex 15, 16
Epoxides, from alkenes by asymmetric epoxidation catalyzed by a ruthenium–pyboxazine complex 14, 15
Epoxides, from alkenes by asymmetric epoxidation catalyzed by molybdenum(VI) and a chiral bishydroxamic acid ligand 16, 17
Epoxides, from alkenes by asymmetric epoxidation using manganese(II) and a porphyrin-inspired chiral ligand 9, 10
Epoxides, from alkenes by asymmetric epoxidation using titanium–salalen/salan complexes 1, 2
Epoxides, from alkenes by epoxidation catalyzed by a Keggin-type silicodecatungstate salt 18
Epoxides, titanocene(III)-mediated homolytic ring opening 388, 389
β-Eremophilane 634
Ethene, intermolecular Pauson–Khand reaction to give cyclopentenones 110
Eudesmanolides 525, 526
(±)-Eudesmin 514–517

F

Fenestranes, from 1,6-enynes by a Pauson–Khand reaction/allylic alkylation/Pauson–Khand reaction sequence 151
Fenestranes, from trienynes by a Pauson–Khand/Diels–Alder reaction 152
(+)-Ferruginine 633, 634
(−)-Flueggine A 641
Fluostatins C and E 258
Fomitellic acid B 522, 523
(±)-Fragranol 517, 518
(+)-Frondosin A 375
(±)-Frondosin B 343, 344
Furan-2-carboxylates, from alkenyl β-keto esters by manganese(III)-mediated oxidation/heterocyclization 431, 432
Furano lignans 514–517
Furan-2-ones, 5-allyl-, from trienes by chemoselective ring-closing metathesis 554, 555
Furan, reductive cycloaddition with α,α′-dihalo ketones mediated by copper bronze to give 8-oxabicyclo[3.2.1]oct-6-en-3-ones 321, 322
Furan, reductive cycloaddition with α,α′-diiodo ketones mediated by zinc/copper couple to give 8-oxabicyclo[3.2.1]oct-6-en-3-ones 322
Furan, silver(I)-mediated cycloaddition with 6-chloro-1-pyrrolidinocyclohex-1-ene to give 11-oxatricyclo[4.3.1.1^{2,5}]undec-3-en-10-one 326
Furan, titanium-catalyzed (4+3) cycloadditions with sulfur-substituted allylic acetals to give 8-oxabicyclo[3.2.1]oct-6-en-3-ones 342
Furan, titanium-mediated (4+3) cycloadditions with chiral allylic acetals to give 2-alkoxy-3-methylene-8-oxabicyclo[3.2.1]oct-6-enes 341
Furans, by radical cyclization reactions 431–433
Furans, copper-catalyzed cyclopropanation to give 2-oxabicyclo[3.1.0]hex-3-enes 52
Furans, from allyloxy epoxides by titanium(III)-mediated ring formation 390, 391
Furans, from β-keto carbonyl compounds by manganese(III)-mediated heterocyclization 433
Furans, silver(I)-mediated cycloaddition with trimethylsiloxy-substituted allyl chlorides to give 8-oxabicyclo[3.2.1]oct-6-en-3-ones 325, 326
Furans, zinc-catalyzed cycloaddition with allenamides to give 2-amino-8-oxabicyclo[3.2.1]oct-6-en-3-ones 334–337
Furfuryl alcohols, reductive cycloaddition with α,α′-dibromo ketones mediated by diethylzinc to give (1-hydroxymethyl)-8-oxabicyclo[3.2.1]oct-6-en-3-ones 324
Furfuryl alcohols, titanium-mediated cycloaddition with cyclohexa-1,3-diene to give 7,8-dihydro-4H-4,7-ethanocyclohepta[b]furans 327, 328
Furofuran lignans 514–517
Furofurans, from allyloxy epoxides by titanium(III)-mediated ring formation 390, 391
(+)-Fusarisetin A 159

G

(±)-Garcibracteatone 498, 499
Garsubellin A 571
Gibberellin GA$_{103}$ 91
Glycophanes, from phenylene-1,4-diamine-linked diynes by ring-closing alkyne metathesis 653
(±)-Grandisol 632
GSK 1360707 (antidepressive agent) 92
Guanacastepene A 636
(−)-Guanacastepene E 568
(−)-Gymnodimine 259

H

α-Halo enol ethers, ionization to give allylic cations 325–327
(−)-Harveynone 648
[7]Helicenes, from biaryl-linked tetraynes by rhodium-catalyzed enantioselective cycloaddition with diynes 277, 278
[n]Helicenes, from cis,cis-dienetriynes by nickel-catalyzed synthesis 281
[7]Helicenes, from triynes by rhodium-catalyzed enantioselective synthesis 282
(−)-Heptemerone B 568
Hetero-Pauson–Khand reaction 156–159

Hexacyclinic acid 507, 508
Hexaenes, triple ring-closing metathesis 556, 557
1,2,3,4,5,8-Hexahydroazulenes, from bicyclic vinylcyclopropanes by rhodium-catalyzed intermolecular (5+2) cycloaddition with but-2-yne 362
Hexahydro-2H-1-benzopyrans, from vinyl iodides by 6-(π-exo)-exo-dig radical cyclization 460
1,2,3,6,7,8a-Hexahydro-8H-3a,6-epoxyazulen-8-ones, from alkoxy allylic sulfones by titanium-mediated intramolecular cycloaddition 330
1,2,3,6,7,8a-Hexahydro-8H-3a,6-epoxyazulen-8-ones, 7-(phenylsulfanyl)-, from sulfur-substituted allylic sulfones by titanium-catalyzed intramolecular (4+3) cycloaddition 342
1,2,3,4,7,8-Hexahydroisoquinolines, from 1-alk-5-ynylcyclobutenes by ring-rearrangement metathesis 644
1,2,3,7,8,8a-Hexahydronaphthalenes, from dienynes by tandem enyne ring-closing metathesis 618, 619
1,4,5,6,7,8-Hexahydroquinolines, from enones by yttrium-catalyzed three-component asymmetric aza-Diels–Alder reactions with cyclic ketones and anilines 255, 256
Hexaphenylenes, from diynes by ruthenium-catalyzed (2+2+2) cycloadditions with acetylene 271, 272
Hirsutene 379
Homoallyl alcohols, asymmetric epoxidation catalyzed by zirconium– and hafnium–bishydroxamic acid complexes 22, 23
Homoallyl alcohols, asymmetric epoxidation using vanadium(V) and a chiral bishydroxamic acid ligand 5, 6
Homoallyl alcohols, epoxidation catalyzed by a selenium-containing dinuclear peroxotungstate complex 18, 19
Humulanolides 610
Hydrazides, acyl-, cyclization using the Petasis reagent to give indoles 410
Hydroquinones, from 1-allenylcyclopropan-1-ols by cobalt-catalyzed (5+1) cycloaddition 351, 352
(Hydroxyalkyl)oxiranes, from allyl and homoallyl alcohols by epoxidation catalyzed by a selenium-containing dinuclear peroxotungstate complex 18, 19
(2-Hydroxyethyl)oxiranes, from homoallyl alcohols by asymmetric epoxidation catalyzed by zirconium– and hafnium–bishydroxamic acid complexes 22, 23
(2-Hydroxyethyl)oxiranes, from homoallyl alcohols by asymmetric epoxidation using vanadium(V) and a chiral bishydroxamic acid ligand 5, 6
(Hydroxymethyl)oxiranes, from allyl alcohols by asymmetric epoxidation catalyzed by a tungsten–bishydroxamic acid complex 19
(Hydroxymethyl)oxiranes, from allyl alcohols by asymmetric epoxidation using vanadium 4, 5
(Hydroxymethyl)oxiranes, from allyl alcohols by asymmetric epoxidation with a niobium–salan complex 20

I

($\pm$)-Ialibinones A and B 504, 505
Imines, chiral copper complex catalyzed aza-Diels–Alder reaction with Danishefsky-type dienes to give dihydropyridin-4-ones 245, 246
Imines, chiral niobium complex catalyzed aza-Diels–Alder reaction with Danishefsky-type dienes to give dihydropyridin-4-ones 251
Imines, chiral scandium complex catalyzed aza-Diels–Alder reaction with Danishefsky-type dienes to give dihydropyridin-4-ones 254
Imines, chiral silver complex catalyzed aza-Diels–Alder reaction with Danishefsky-type dienes to give dihydropyridin-4-ones 249
Imines, chiral zinc complex catalyzed aza-Diels–Alder reaction with 3-vinylbenzopyran-4-ones to give dihydropyridine derivatives 247, 248
Imines, chiral zirconium complex catalyzed aza-Diels–Alder reaction with Danishefsky-type dienes to give dihydropyridin-4-ones 250
Imines, diastereoselective (3+2) cycloaddition with a carbonyl ylide to give tetrahydro-1H-3,9a-epoxypyrimido[2,1-c][1,4]oxazin-4(9H)-ones 190, 191
Imines, reaction with carbonyl ylides in a one-pot process to give α-hydroxy-β-amino esters 189, 190
Imines, rhodium/Lewis acid catalyzed diastereoselective (3+2) cycloaddition with a carbonyl ylide to give dihydro-1H-1,4-epoxybenzo[c]azepin-5(2H)-ones 188, 189
Imines, three-component diastereoselective intermolecular (3+2) cycloaddition with a carbonyl ylide and an aldehyde to give oxazolidines 191
Imines, N-vinyl-, chiral rhodium complex catalyzed aza-Diels–Alder reaction with aldehydes to give 1,3-oxazinan-4-ones 253
(Imino)alkanenitriles, intramolecular radical cyclization to give α-amino cyclic ketones 487
Indole alkaloids, oxidation 502
Indole-3-carbaldehydes, from N-benzyl-2-vinylanilines by an iridium-mediated photoredox-catalyzed cascade reaction 412–414

Indole-3-methanols, as allylic cation precursors 338
Indole, 1-methyl-, gallium-catalyzed three-component (4+3) cycloaddition reactions to give 5,6,9,10-tetrahydro-6,9-methanocyclohepta[b]indoles 338
Indoles, bromoalkyl-substituted, ruthenium(II)-mediated photoredox cyclization to give fused indoles 411
Indoles, by radical cyclization reactions 410–421
Indoles, chiral copper complex catalyzed aza-Diels–Alder reaction with hydrazides to give 4,4a,9,9a-tetrahydro-1H-pyridazino[3,4-b]indoles 246, 247
Indoles, from acyl hydrazides by cyclization using the Petasis reagent 410
Indoles, rhodium-catalyzed enantioselective intermolecular (3+2) cycloaddition with carbonyl ylides to give indole-fused tetrahydropyranones 180
Indolines, by radical cyclization reactions 415–417
Indolines, from 2-[(arylamino)methyl]oxiranes by intramolecular cyclization 415–417
Ingenol 159, 577
Iodoarenes, from diiododiynes by ruthenium-catalyzed (2+2+2) cycloaddition of acetylene 272
(−)-Irofulven 638
Isocyanates, ruthenium-catalyzed hetero-Pauson–Khand reaction to give maleimides 157, 158
Isocyanates, use in (2+2+2)-cycloaddition reactions 292–299
Isoschizogamine 608
Ivorenolide B 668

J
(−)-Jiadifenin 159

K
(−)-α-Kainic acid 159
Kempenes 636
β-Keto carbonyl compounds, manganese(III)-mediated heterocyclization to give furans 433
Keto diesters, reductive cyclization/Dieckmann condensation/lactonization to give tricyclic γ-lactones 450, 451
β-Keto esters, alkenyl, manganese(III)-mediated oxidation/heterocyclization to give furan-2-carboxylates 431, 432
β-Keto esters, 2-alkyl-substituted, copper(II)-catalyzed aziridination to give 2-acylaziridine-2-carboxylates 27
δ-Keto esters, α,β-unsaturated, samarium(II) iodide mediated cyclization to give anti-cyclopropanol derivatives 470, 471

Keto imines, intramolecular reductive coupling to give trans-2-amino cycloalkanols 482, 483
Ketonitriles, enantioselective cyclization to give 2-hydroxycycloalkanones 474–476
Ketoxime acetates, copper-catalyzed cyclization to give 2-arylpyrroles 395
Ketoxime carboxylates, copper-catalyzed cyclization to give 2-phenylpyrroles 396
Kulinkovich reaction 88–90, 94

L
Lactams, by radical cyclization reactions 422–425
Lactams, from alkynyl imines by hetero-Pauson–Khand reaction 156
γ-Lactams, from α-polychloro-N-allylamides by reverse atom-transfer radical cyclization 423–425
Lactams, fused, from carbodiimides by hetero-Pauson–Khand reaction 156, 157
Lactimidomycin 595, 664
γ-Lactones, bi- and tricyclic, from malonates by manganese(III)/copper(II) promoted cyclization 445, 446
Lactones, by radical cyclization reactions 439–452
Lactones, cyclopentane-fused, from 2-(alk-4-enyl)malonates by manganese(III)-mediated oxidative radical cyclization 448, 449
Lactones, α,α-dichloro, from alkenyl trichloroacetates by cyclization 441, 442
γ-Lactones, from alkenes by manganese(III) acetate mediated oxidative free-radical cyclization 385
Lactones, from alkynyl aldehydes by hetero-Pauson–Khand reaction 156
Lactones, tricyclic, from 2-alk-3-enyl-3-alk-1-ynyloxiranes by cobalt-mediated tandem (5+1)/[2+2+1] cycloaddition 153
γ-Lactones, tricyclic, from keto diesters by reductive cyclization/Dieckmann condensation/lactonization 450, 451
γ-Lactones, tricyclic, from malonate and carboxy substituted cycloalkenes by manganese(III)/copper(II) promoted cyclization 443
(±)-Lariciresinol 514–517
(±)-Lariciresinol dimethyl ether 514–517
(+)-Laurencin 585, 586
Leiodermatolide 666, 667
(S)-Levcromakalim 42, 43
Lewis acid activated borane catalysts, chiral, in Diels–Alder reactions 211, 212
Ligusticum chuanxiong 201
Loloanolides 624, 625
Longithorone A 626, 627
(±)-Lundurine A 527, 528
(±)-Lundurine B 527, 528, 573
(+)-Lupinine 558, 559
Lysergic acid 575

M

Magnofargesin 511, 512
Maleimides, from isocyanates by ruthenium-catalyzed hetero-Pauson–Khand reaction 158
Maleimides, rhodium-catalyzed diastereoselective intermolecular (3+2) cycloaddition with an indane-based carbonyl ylide to give spiro-fused tetracycles 172, 173
Malonates, N-arylamino-2-oxoethyl, manganese(III) acetate mediated 5-*exo* cyclization to give spiro-fused pyrrolidin-2-ones 423
Malonates, manganese(III)/copper(II) promoted cyclization to give bi- and tricyclic γ-lactones 446
Mandelalide A 665
Manganese(III) acetate, synthesis 385
Meloscine 574
Metallotropic [1,3]-shift, in enyne ring-closing metathesis 628, 629
(R,R)-β-Methoxytyrosine 44
3-Methylenebicyclo[3.2.1]oct-6-enes, 2-alkoxy-, from chiral allylic acetals by titanium-mediated (4+3) cycloadditions with furan or cyclopentadiene 341
Methylenecycloheptanones, from vinylcyclopropanes by rhodium-catalyzed intermolecular (5+2) cycloaddition with allenes 363, 364
Methylenecycloheptenes, fused, from allenyl vinylcyclopropanes by rhodium-catalyzed intramolecular (5+2) cycloaddition 361
Methylenecyclohexenones, from alkynylcyclopropanes by rhodium-catalyzed (5+1) cycloaddition 356, 357
Methylenecyclohexenones, from allenylcyclopropanes by iridium-catalyzed (5+1) cycloaddition with carbon monoxide 353
7-Methylene-3-oxa-1-azabicyclo[4.1.0]heptan-2-ones, from allenyl carbamates by silver(I)/1,10-phenanthroline catalyzed intramolecular aziridination 40
6-Methylene-8-oxabicyclo[3.2.1]octan-2-ones, from allenes by rhodium-catalyzed enantioselective intermolecular (3+2) cycloaddition with carbonyl ylides 180
7-Methylene-8-oxabicyclo[3.2.1]octan-2-ones, from buta-2,3-dienoates by rhodium-catalyzed diastereoselective intramolecular (3+2) cycloaddition with carbonyl ylides 171, 172
N-Methylwelwitindolinone D isonitrile 501, 502
Motuporamine C 656, 657
(−)-Mucocin 570, 571
Mycoepoxydiene 585

N

Nakadomarin A 159, 661
1-Naphthoic acids, 8-benzyl-, potassium persulfate/copper(II) chloride mediated cyclization to give 1H,3H-naphtho[1,8-*cd*]pyran-1-ones 440
Naphthols, from 3-phenylcyclopropenes by chromium/molybdenum-catalyzed (5+1) cycloaddition 350, 351
1H,3H-Naphtho[1,8-*cd*]pyran-1-ones, 3-aryl-, from 8-benzyl-1-naphthoic acids using potassium persulfate/copper(II) chloride 440
2-Naphthylpyridines, axially chiral, from 1-diynylnaphthalenes by cobalt-catalyzed enantioselective cycloaddition with nitriles 289
(−)-Nitidasin 583, 584
Nitramine 608
Nitriles, use in (2+2+2)-cycloaddition reactions 282–292
1-Nitrocyclopropane-1-carboxylates, from alkenes by copper-catalyzed asymmetric cyclopropanation using in situ generated iodonium ylides 53
Nitrocyclopropanes, from alkenes by stereoselective rhodium-catalyzed cyclopropanation 57, 58
Norbornadiene, asymmetric intermolecular Pauson–Khand reactions using chiral metal–cobalt complexes to give 3a,4,7,7a-tetrahydro-1H-4,7-methanoinden-1-ones 144, 145
Norbornadienes, cobalt-catalyzed intermolecular Pauson–Khand reaction/conjugate addition/retro-Diels–Alder sequence to give cyclopentenones 107
Norbornadienes, platinum(II)-catalyzed reaction with phenylacetylene to give bicyclic alkylidenecyclopropanes 87
Norbornene, iridium-catalyzed asymmetric Pauson–Khand reaction 129
Norbornene, palladium-catalyzed intermolecular Pauson–Khand-type cyclization with alkynoates to give 1-oxohexahydro-4,7-methanoindenylcarboxylates 134, 135
Norbornenes, alkynyl-, ring-rearrangement metathesis 645, 646
Norbornenes, cyclopropanation via domino heck-type coupling/C—H activation of benzyl bromides to give tricyclo[3.2.1.0^{2,4}]octanes 71, 72
Norbornenes, 5-(1-oxoprop-2-enyl)-, ring-rearrangement metathesis to give bicyclo[3.3.0]-octenes 605, 606
(−)-Norsecurinine 640, 641

O

1,2,3,4,5,6,8,9-Octahydro-7H-benzo[7]annulen-7-ones, from vinylcyclopropanes by a rhodium-catalyzed (5+2) cycloaddition/vinylogous Peterson elimination/[4+2] cycloaddition cascade 363

Octahydro-1H-cyclopenta[8]annulen-1-ones, from (E)-1-methylcycloctene by intermolecular Pauson–Khand reaction 106, 107

Octahydrocyclopenta[c]pyrroles, 5-vinyl-, from alkenyl vinylcyclopropanes by rhodium-catalyzed (3+2) cycloaddition 364, 365

Octahydro-6H-3a,7-epoxyazulen-6-ones, from alkenyl carbonyl ylides by rhodium-catalyzed diastereoselective intramolecular (3+2) cycloaddition 170, 171

Octahydro-6H-3a,7-epoxyazulen-6-ones, from alkenyl carbonyl ylides by rhodium-catalyzed enantioselective intramolecular (3+2) cycloaddition 176

Oligoenynes, from diene–polyynes by combination of multiple metathesis reactions 646–648

(+)-Omphadiol 576

Ophiobolin A 583

Ophirin B 588, 589

(−)-Oseltamivir 44, 45

Otteliones 624, 625

3-Oxa-1-azabicyclo[3.1.0]hexan-2-ones, from allyl carbamates by rhodium(II)-catalyzed intramolecular aziridination 29

3-Oxabicyclo[4.1.0]hept-4-enes, from allyl propargyl ethers by gold(I)-catalyzed intramolecular cyclopropanation 77, 78

3-Oxabicyclo[3.1.0]hexan-2-ones, from allyl α-diazocarboxylates by enantioselective iron-catalyzed intramolecular cyclopropanation 69, 70

3-Oxabicyclo[3.1.0]hexan-2-ones, from 3-oxobutanoates by manganese(III)-induced oxidative intramolecular cyclization 491

8-Oxabicyclo[3.2.1]octan-2-ones, from alkenes by rhodium-catalyzed enantioselective intermolecular (3+2) cycloaddition with carbonyl ylides 176, 177

8-Oxabicyclo[3.2.1]oct-6-en-2-ones, from alkynes by rhodium-catalyzed enantioselective intermolecular (3+2) cycloaddition with carbonyl ylides 177, 178

8-Oxabicyclo[3.2.1]oct-6-en-2-ones, from alkynes by ruthenium-catalyzed diastereoselective intramolecular (3+2) cycloaddition with carbonyl ylides 171

8-Oxabicyclo[3.2.1]oct-6-en-3-ones, from α,α'-dihalo ketones by reductive cycloaddition with furan mediated by copper bronze 321, 322

8-Oxabicyclo[3.2.1]oct-6-en-3-ones, from α,α'-diiodo ketones by reductive cycloaddition with furan mediated by zinc/copper couple 322

8-Oxabicyclo[3.2.1]oct-6-en-3-ones, from sulfur-substituted allylic acetals by titanium-catalyzed (4+3) cycloaddition with furan 342

8-Oxabicyclo[3.2.1]oct-6-en-3-ones, from trimethylsiloxy-substituted allyl chlorides by silver(I)-mediated cycloaddition with furans 325, 326

8-Oxabicyclo[3.2.1]oct-6-en-3-ones, (1-hydroxymethyl)-, from α,α'-dibromo ketones by reductive cycloaddition with chiral furfuryl alcohols mediated by diethylzinc 324

Oxacyclotetradec-11-en-2-one, from dienyl esters by E-selective ring-closing metathesis 551

Oxa-Diels–Alder reaction, asymmetric 224–244

Oxa-Diels–Alder reaction, asymmetric, using chiral aluminum complexes 225, 226

Oxa-Diels–Alder reaction, asymmetric, using chiral chromium complexes 233–235

Oxa-Diels–Alder reaction, asymmetric, using chiral copper complexes 236–238

Oxa-Diels–Alder reaction, asymmetric, using chiral indium complexes 226, 227

Oxa-Diels–Alder reaction, asymmetric, using chiral palladium complexes 242, 243

Oxa-Diels–Alder reaction, asymmetric, using chiral rare earth metal complexes 244

Oxa-Diels–Alder reaction, asymmetric, using chiral rhodium complexes 241, 242

Oxa-Diels–Alder reaction, asymmetric, using chiral titanium complexes 228–231

Oxa-Diels–Alder reaction, asymmetric, using chiral zinc complexes 238, 239

Oxa-Diels–Alder reaction, asymmetric, using chiral zirconium complexes 232

Oxa-Diels–Alder reaction, reaction pathways 224

11-Oxatricyclo[4.3.1.1^{2,5}]undec-3-en-10-one, from 6-chloro-1-pyrrolidinocyclohex-1-ene by silver(I)-mediated cycloaddition with furan 326, 327

Oxazaborolidines, chiral, Brønsted acid activated, catalysts for Diels–Alder reactions 209, 210

Oxazaborolidines, chiral, Lewis acid activated, catalysts for Diels–Alder reactions 211, 212

1,3-Oxazinan-4-ones, from N-vinyl imines by chiral rhodium complex catalyzed aza-Diels–Alder reaction with aldehydes 253

Oxazolidines, from imines by three-component diastereoselective intermolecular (3+2) cycloaddition with a carbonyl ylide and an aldehyde 190, 191

Oximidine III 561, 562

Oxindoles, by radical cyclization reactions 418–421

Oxindoles, 3,3-disubstituted, from 2-bromo-*N*-phenylacetamides under iridium catalysis 420, 421

Oxindoles, from 2-(methylsulfanyl)-*N*-phenylacetamides by cyclization 418, 419

Oxiranes, 2-(alk-3-enyl)-3-(alk-1-ynyl)-, cobalt-mediated tandem (5+1)/(2+2+1) cycloaddition to give tricyclic lactones 152, 153

N-(Oxiranylmethyl) sulfonamides, from *N*-allyl sulfonamides by asymmetric epoxidation of catalyzed by a hafnium(IV)–bishydroxamic acid complex 23

α-Oxo aldehydes, chiral palladium complex catalyzed oxa-Diels–Alder reactions with dienes to give 3,6-dihydro-2*H*-pyran-2-yl ketones 242, 243

3-Oxobutanoates, manganese(III)-induced oxidative intramolecular cyclization to give 3-oxabicyclo[3.1.0]hexan-2-ones 491

α-Oxo esters, chiral copper complex catalyzed oxa-Diels–Alder reaction with Danishefsky-type dienes to give dihydropyran-4-ones 236, 237

β-Oxo esters, manganese(III) acetate mediated coupling with styrene to give 2,3-dihydrofurans 385, 386

(−)-13-Oxyingenol 577–579

P

Paecilomycine A 159
Palau'amine 503, 504
(+)-Panepophenanthrin 634, 635
Pauson–Khand reaction, catalytic reaction pathway 114
Pauson–Khand reaction, chiral auxiliaries 139, 140
Pauson–Khand reaction, cobalt catalyzed 115–119
Pauson–Khand reaction, general overview 99, 100
Pauson–Khand reaction, in cascade reactions 147–151
Pauson–Khand reaction, in the absence of carbon monoxide 127–129
Pauson–Khand reaction, iridium catalyzed 129, 130
Pauson–Khand reaction, reaction promotion 102–106
Pauson–Khand reaction, regioselectivity issues 100, 101
Pauson–Khand reaction, rhodium catalyzed 120–129
Pauson–Khand reaction, scope and limitations 106–111
Pauson–Khand reaction, stoichiometric reaction pathway 112–114
Pauson–Khand reactions, tandem 153–155
Pauson–Khand reaction, ways to induce asymmetry 138–147
Pauson–Khand reaction, with heterogeneous catalysts 135–138
Pectenotoxin-2 600, 601
(±)-Penifulvin A 524, 525
(−)-Pentalenene 159
PGE2-1,15-lactone 657
Phenols, from 3-vinylcyclopropenes by rhodium-catalyzed (5+1) carbonylation with carbon monoxide 354, 355
Phenylacetone, from benzene by manganese(III) acetate mediated alkylation with acetone 385
Phenylcyclopropanes, from styrenes by diazomethane generation followed by iron-catalyzed cyclopropanation 67
1,3-Phenyleneethynylene oligomers, cyclic, from 1,3-dialk-1-ynylbenzenes by alkyne metathesis 669, 670
Phenyloxiranes, from styrenes by asymmetric epoxidation with a titanium–salan complex 3
Phthalates, from acetylenedicarboxylates by rhodium-catalyzed (2+2+2) cycloaddition with terminal alkynes 266, 267
Pinacol-type ketone–imine reductive coupling 482–484
(±)-Pinoresinol 514–517
Piperidines, by radical cyclization reactions 425–427
Piperidines, from alkenyl isocyanates by enantioselective rhodium-catalyzed (2+2+2) cycloaddition with alkynes 296, 297
Piperidinones, from 2-(alkynamido)malonates by manganese(III)-mediated cyclization 427
(±)-Piperitol 514–517
Platencin 521, 571, 572
Pradimicinone 530, 531
(+)-Prelactone C 232
(+)-Pretazettine 41
Propargyl esters, intermolecular gold(I)-catalyzed cyclopropanation with alkenes to give 1-cyclopropylvinyl esters 78–80
Propargyl ethers, α-bromo-β-oxo, titanium(III)-mediated cyclization to give tetrahydrofurans 433, 434
[2-(Propargyloxy)ethyl]oxiranes, titanium(III)-mediated radical cyclization to give tetrahydropyrans 454–456
N-Propenyl-3-oxobutanamides, manganese(III)-mediated cyclization to give 3-azabicyclo[3.1.0]hexan-2-ones 490, 491
(−)-Pseudolaric acid B 374, 375
Pyran-2-ones, from 3-acylcyclopropenes by rhodium-catalyzed (5+1) carbonylation with carbon monoxide 354, 355
Pyran-2-ones, fused, from diynes by rhodium- or nickel-catalyzed (2+2+2) cycloaddition with carbon dioxide 300

Pyrans, fused, from diynes by rhodium-catalyzed (2+2+2) cycloadditions with carbonyl compounds 309–311

Pyridines, fused, from diynes by cobalt-catalyzed (2+2+2) cycloaddition with nitriles 283, 284

Pyridines, fused, from diynes by iridium-catalyzed (2+2+2) cycloaddition with nitriles 287–289

Pyridines, fused, from diynes by nickel-catalyzed (2+2+2) cycloaddition with nitriles 284, 285

Pyridines, fused, from diynes by rhodium-catalyzed (2+2+2) cycloaddition with nitriles 287

Pyridines, fused, from diynes by ruthenium-catalyzed (2+2+2) cycloaddition with nitriles 285, 286

Pyridin-2-imines, bicyclic, from diynes by rhodium-catalyzed (2+2+2) cycloaddition with carbodiimides 299, 300

Pyridin-2-ones, 6-aryl-, axially chiral, from 1,6-diynylbenzenes by rhodium-catalyzed enantioselective cycloaddition with isocyanates 295

Pyridin-2-ones, bicyclic, from diynes by iridium-catalyzed (2+2+2) cycloaddition with isocyanates 294

Pyridin-2-ones, bicyclic, from diynes by nickel-catalyzed (2+2+2) cycloaddition with isocyanates 292, 293

Pyridin-2-ones, bicyclic, from diynes by rhodium-catalyzed (2+2+2) cycloaddition with isocyanates 294, 295

Pyridin-2-ones, bicyclic, from diynes by ruthenium-catalyzed (2+2+2) cycloaddition with isocyanates 293, 294

Pyrroles, 2-aryl-, from ketoxime acetates by copper-catalyzed cyclization 395

Pyrroles, N-bromoalkyl-, photoredox cyclization to give 5,6,7,8-tetrahydroindolizines 397–399

Pyrroles, by radical cyclization reactions 394–398

Pyrroles, copper-catalyzed cyclopropanation to give 2-azabicyclo[3.1.0]hexanes 52

Pyrroles, from alkynes by rhodium-catalyzed intermolecular cyclization of in situ generated azomethine ylides 197, 198

Pyrroles, 2-phenyl-, from ketoxime carboxylates by copper-catalyzed cyclization 395, 396

Pyrrolidines, 3-alkylidene-, from alk-3-ynylamines by manganese(III)-mediated cyclization 401

Pyrrolidines, 2-alkynyl-, from N-alkynylsulfonamides by copper(I)-catalyzed cyclization 402, 403

Pyrrolidines, 2-allyl-1-propargyl-, chiral, cobalt-catalyzed intramolecular Pauson–Khand reaction to give chiral tricyclic products 138, 139

Pyrrolidines, by radical cyclization reactions 400–404

Pyrrolidines, cyclobutane-fused, from diallyl ethers by ruthenium-mediated (2+2) cycloaddition 435–438

Pyrrolidines, from alk-4-ynamines by cyclization catalyzed by the Petasis reagent 403, 404

Pyrrolidines, from N-alkenyl-N-(benzoyloxy)sulfonamides by copper(I)-catalyzed cyclization 401, 402

Pyrrolidines, from α,β-unsaturated esters by ruthenium-catalyzed three-component cycloaddition with azomethine ylides 195

Pyrrolidin-2-ones, 4-alkylidene-, from 2-(alkynamido)malonates by manganese(III) acetate mediated cyclization 404, 405

Pyrrolidinones, by radical cyclization reactions 404–410

Pyrrolidin-2-ones, from N-alkynylcyclopropanecarboxamides by iridium-catalyzed light-mediated radical cyclization 406–409

Pyrrolidin-2-ones, fused, from allenyl carbonitriles by rhodium-catalyzed Pauson–Khand-type reaction 158

Pyrrolidin-2-ones, spiro-fused, from 2-(2-amino-2-oxoethyl)malonates by manganese(III) acetate mediated 5-exo cyclization 423

Q

Quinolines, by radical cyclization reactions 427–431

Quinolines, from N-(2-vinylaryl)enamines by oxidative radical cyclization 430

R

Radical cyclization, intramolecular, to give cycloalkanes 463–470

Radical cyclization, intramolecular, to give cycloalkanols 470–487

Radical cyclization, introduction 383, 384

Radical cyclization, manganese(III) acetate mediated 384–387

Radical cyclization, samarium(II) iodide mediated 392–394

Radical cyclization, titanocene(III) mediated 387–392

Radical cyclization, to give N-heterocycles 394–431

Radical cyclization, to give O-heterocycles 431–463

Rameswaralide 375, 376

Relay ring-closing and cross-metathesis reactions 560–562

Resorcinols, from 3-acyloxy-1,4-enynes by rhodium-catalyzed (5+1) cycloaddition with carbon monoxide 355, 356

Ring-closing alkyne metathesis, synthesis of macrocycles containing a transition metal 653, 654

Ring-closing metathesis, diastereo- and enantioselective 562–567

Ring-closing metathesis, overview and history 543–546

Ring-rearrangement metathesis 602

(±)-Roccellaric acid 513, 514

(±)-Ryanodol 574

S

Sabina ketone 93

(−)-Salinosporamide A 400

Samarium(II) iodide mediated radical reactions 392, 393

(±)-Sanshodiol methyl ether 514–517

Securinega alkaloids 640, 641

(−)-Securinine 635

(±)-Sesamin 512–517

2-Siloxyacroleins, metal-catalyzed (4+3) cycloadditions with 1,3-dienes to give cyclohept-4-en-1-ones 339, 340

Silylacetylenes, rhodium-catalyzed (2+2+2) cycloaddition with acrylamides and acetylenecarboxylates to give cyclohexadienylcarboxamides 304, 305

Silyloxiranes, from (Z)-alkenylsilanes by asymmetric epoxidation with a titanium–salalen complex 2

(±)-Smenospondiol 521, 522

Spirobipyridines, from tetraynes by rhodium-catalyzed enantioselective intramolecular cycloaddition with nitriles 291

Spirotenuipesines A and B 568

Sporolide B 313

(−)-Stemoamide 631, 632

Styrene, copper-catalyzed asymmetric cyclopropanation to give 2-phenylcyclopropanecarboxylates 51

Styrenes, asymmetric epoxidation catalyzed by a manganese–salen complex 7, 8

Styrenes, asymmetric epoxidation with a titanium–salan complex 3

Styrenes, cobalt–porphyrin complex catalyzed aziridination to give 1,2-diarylaziridines 38, 39

Styrenes, diazomethane generation followed by iron-catalyzed cyclopropanation 67

Styrenes, intermolecular aziridination catalyzed by a dirhodium carboxylate complex 30, 31

Styrenes, intermolecular aziridination with a copper–pyridine complex 25, 26

Styrenes, palladium-catalyzed cross coupling of styrenes with aryl methyl ketones to give acylcyclopropanes 74

Sulfoxides, chiral, as auxiliaries in the cobalt-catalyzed asymmetric intramolecular Pauson–Khand reaction of enynes 142, 143

Superacidic Lewis acid catalysts, chiral, in Diels–Alder reactions 207, 208

(−)-Swainsonine 603

T

Taiwanins E and C, precursor 313, 314

Tamiflu 257, 258

1,4-Teraryls, from aryl 1,6-diynes by iridium-catalyzed enantioselective cycloaddition with monoynes 274, 275

Terpenoids, manganese(III)-mediated oxidative radical cyclization to give polycyclic dihydropyrans 452, 453

Tetraenes, double ring-closing metathesis 556–558

3,4,5,6-Tetrahydro-1H-cyclohepta[c]furans, from propargyloxy vinylcyclopropanes by nickel-catalyzed intramolecular (5+2) cycloaddition 371, 372

2,3,3a,4-Tetrahydrocyclopenta[c]pyrrol-5(1H)-ones, from allyl carbonates by a combined allylation/Pauson–Khand reaction sequence 150, 151

3a,4,5,9b-Tetrahydro-3H-cyclopenta[c]quinolin-6-ols, from 2-aminophenols by scandium-catalyzed three-component asymmetric aza-Diels–Alder reactions with aldehydes and cyclopentadiene 255

Tetrahydro-1H-3a,7-epoxyazulene-7(4H)-carboxylates, 6-oxo-, from alkynyl carbonyl ylides by rhodium-catalyzed enantioselective intramolecular (3+2) cycloaddition 176

Tetrahydro-9H-5,8-epoxybenzo[7]annulen-9-ones, from vinyl ethers by rhodium/Lewis acid catalyzed enantioselective intermolecular (3+2) cycloaddition with carbonyl ylides 180, 181

Tetrahydro-1H-3,9a-epoxypyrimido[2,1-c][1,4]-oxazin-4(9H)-ones, from imines by diastereoselective (3+2) cycloaddition with a carbonyl ylide 190, 191

Tetrahydrofurancarboxylates, oxazolidine-substituted, from chiral amides by rhodium/nickel-catalyzed diastereoselective intermolecular (3+2) cycloaddition with carbonyl ylides 173, 174

Tetrahydrofurans, by radical cyclization reactions 433–439

Tetrahydrofurans, cyclobutane-fused, from diallyl ethers by ruthenium-mediated (2+2) cycloaddition 435–438

Tetrahydrofurans, from alkenyl iodo ethers by titanium(IV)/chlorotrimethylsilane/manganese mediated cyclization 434, 435

Tetrahydrofurans, from α-bromo-β-oxo propargyl ethers by titanium(III)-mediated cyclization 433, 434

5,6,7,8-Tetrahydroindolizines, from N-(bromoalkyl)pyrroles by photoredox cyclization 397–399

Tetrahydroindolizinones, from alkenyl isocyanates by enantioselective rhodium-catalyzed (2+2+2) cycloaddition with alkynes 295, 296

3a,4,7,7a-Tetrahydro-1H-4,7-methanoinden-1-ones, from norbornadiene by asymmetric intermolecular Pauson–Khand reactions using chiral cobalt complexes 144, 145

4,4a,9,9a-Tetrahydro-1H-pyridazino[3,4-b]indoles, from indoles by chiral copper complex catalyzed aza-Diels–Alder reaction with hydrazides 246, 247

5,6,9,10-Tetrahydro-6,9-methanocyclohepta[b]-indoles, from 1-methylindole by gallium-catalyzed three-component (4+3)-cycloaddition reactions 338

1,2,3,4-Tetrahydronaphthalene, from diboryldiynes by cobalt-catalyzed (2+2+2) cycloaddition with monoynes 269–271

4,5,6,6a-Tetrahydropentalen-2(1H)-ones, from 1,6-enynes by cobalt–chiral ligand catalyzed enantioselective intramolecular Pauson–Khand reaction 145, 146

4,5,6,6a-Tetrahydropentalen-2(1H)-ones, from 1,6-enynes by Pauson–Khand-type reaction using a heterobimetallic cluster as precatalyst 133, 134

Tetrahydropyran-4-ones, from 3-siloxy-substituted hexa-2,4-dienes and aldehydes by chiral chromium complex catalyzed oxa-Diels–Alder reactions 234

Tetrahydropyranones, indole-fused, from indoles by rhodium-catalyzed enantioselective intermolecular (3+2) cycloaddition with carbonyl ylides 180

Tetrahydropyrans, by radical cyclization reactions 454–458

Tetrahydropyrans, from [2-(allyloxy)ethyl]oxiranes by titanium(III)-mediated radical cyclization 456, 457

Tetrahydropyrans, from [2-(propargyloxy)ethyl]-oxiranes by titanium(III)-mediated radical cyclization 454–456

1,2,3,4-Tetrahydropyridines, from α,β-unsaturated imines by chiral nickel complex catalyzed aza-Diels–Alder reaction with vinyl ethers 252

Tetrahydropyridines, vinyl-, from trienes by enantioselective ring-closing metathesis 565, 566

2,6,7,8-Tetrahydropyrido[1,2-a]azepin-4(3H)-ones, from ynamide-based dienynes by tandem enyne ring-closing metathesis 619

Tetrapeptides, macrocyclic, from diynes by ring-closing alkyne metathesis 652, 653

1,4-Tetraphenylenes, from triynes by rhodium-catalyzed enantioselective double homo-(2+2+2) cycloaddition 275

Tetraponerines 602, 603

Tetraynes, biaryl-linked, rhodium-catalyzed enantioselective cycloaddition with diynes to give [7]helicenes 277, 278

Tetraynes, rhodium-catalyzed enantioselective intramolecular cycloaddition with nitriles to give spirobipyridines 291

Tetraynes, ring-closing alkyne metathesis to give macrocyclic 1,3-diynes 654

(+)-TMC-151 C 586, 587

Trehazolamine 528

Trehazolin 528, 529

Tremulenediol A 374

Tremulenolide A 374

1,2,4-Triazolidines, from azodicarboxylates by ruthenium-catalyzed intermolecular cyclization with azomethine ylides 198, 199

(−)-Tricholomenyn A 648, 649

Tricyclo[3.2.1.0^{2,4}]octanes, from norbornenes by cyclopropanation via domino Heck-type coupling/C—H activation of benzyl bromides 71, 72

Tricyclo[5.2.1.0^{1,5}] bis(lactone), from an alkenyl γ-lactone by manganese(III)/copper(II) mediated cyclization 447

Triene ring-closing metathesis, enantioselective, commonly used chiral molybdenum and ruthenium complexes 564–566

Trienes, chemoselective ring-closing metathesis to give 5-allylfuran-2-ones or 6-vinyl-5,6-dihydropyran-2-ones 554–556

Trienes, diastereoselective ring-closing metathesis 562, 563

Trienes, diastereoselective ring-closing metathesis employing a temporary silicon tether 563, 564

Trienes, enantioselective ring-closing metathesis 565, 566

Trienes, stereochemistry-based control of ring-closing metathesis 559, 560

(+)-Trienomycins A and F 599, 600

Trienynes, Pauson–Khand/Diels–Alder reaction to give fenestranes 151, 152

(Trifluoromethyl)cyclopropanes, from alkenes by iron-catalyzed cyclopropanation with in situ generated 2-diazo-1,1,1-trifluoroethane 68

6-(Trifluoromethyl)phenanthridines, from N-biaryltrifluoroacetimidoyl chlorides by cyclization 491–494

Triynes, rhodium-catalyzed enantioselective double homo-(2+2+2) cycloaddition to give chiral 1,4-tetraphenylenes 275

Triynes, rhodium-catalyzed enantioselective synthesis of [7]helicene-like molecules 282

Tronocarpine 506, 507

Tulearin C 663

U

α,β-Unsaturated acid chlorides, chiral erbium complex catalyzed oxa-Diels–Alder reaction with aldehydes to give 5,6-dihydro-2H-pyran-2-ones 244

α,β-Unsaturated aldehydes, Brønsted acid activated chiral borate catalyzed Diels–Alder reactions to give cyclohexene derivatives 208, 209

α,β-Unsaturated aldehydes, chiral (acyloxy)borane catalyzed Diels-Alder reaction to give norbornenes 206

α,β-Unsaturated aldehydes, chiral chromium complex catalyzed Diels–Alder reactions to give cyclohexene derivatives 219, 220

α,β-Unsaturated aldehydes, chiral cobalt complex catalyzed Diels–Alder reactions to give cyclohexene derivatives 220, 221

α,β-Unsaturated aldehydes, chiral indium complex catalyzed Diels–Alder reactions to give cyclohexene derivatives 213, 214

α,β-Unsaturated aldehydes, chiral superacidic Lewis acid catalyzed Diels–Alder reactions to give cyclohexene derivatives 207, 208

α,β-Unsaturated amides, chiral copper complex catalyzed Diels–Alder reactions to give cyclohexene derivatives 215, 216

α,β-Unsaturated amides, chiral diamide–aluminum complex catalyzed Diels–Alder reactions to give norbornenes 212, 213

α,β-Unsaturated amides, chiral, rhodium/nickel-catalyzed diastereoselective intermolecular (3+2) cycloaddition with carbonyl ylides to give tetrahydrofurancarboxylates 173, 174

α,β-Unsaturated amides, chiral ytterbium complex catalyzed Diels–Alder reactions with a Danishefsky-type diene to give cyclohexenones 223

α,β-Unsaturated amides, chiral ytterbium complex catalyzed Diels–Alder reactions with cyclopentadiene to give norbornenes 222, 223

α,β-Unsaturated amides, rhodium/Lewis acid catalyzed enantioselective intermolecular (3+2) cycloaddition with carbonyl ylides to give epoxy-bridged nitrogen heterocycles 181, 182

α,β-Unsaturated carbonyl compounds, chiral bis(sulfinyl)imidoamidine–copper catalyzed Diels–Alder reactions to give norbornenes 216, 217

α,β-Unsaturated carbonyl compounds, chiral titanium complex catalyzed Diels–Alder reactions to give cyclohexene derivatives 217, 218

α,β-Unsaturated carbonyl compounds, palladium(II)-catalyzed cyclopropanation using diazomethane 71

α,β-Unsaturated esters, ruthenium-catalyzed three-component cycloaddition with azomethine ylides to give pyrrolidines 195, 196

α,β-Unsaturated esters, tandem reductive coupling/Dieckmann condensation to give bicyclo[4.3.0]nonan-8-ones and bicyclo[3.3.0]-octan-3-ones 488, 489

α,β-Unsaturated imines, chiral nickel complex catalyzed aza-Diels–Alder reaction with vinyl ethers to give 1,2,3,4-tetrahydropyridines 252

α,β-Unsaturated ketones, asymmetric epoxidation catalyzed by a chiral manganese complex 8, 9

α,β-Unsaturated ketones, asymmetric epoxidation catalyzed by iron(II) with a phenanthroline ligand 11, 12

α,β-Unsaturated ketones, asymmetric epoxidation using a chiral N,N-dioxide–scandium(III) complex 21, 22

(±)-Urechitol A 345, 346

V

(−)-5-epi-Vibsanin E 344, 345

(−)-Vinigrol 638, 639

Vinyl diazoacetates, decomposition to give allylic cations 332, 333

Vinyl diazoacetates, rhodium-catalyzed (4+3) cycloaddition with 1,3-dienes to give cycloheptadienecarboxylates 332, 333

Vinyl ethers, chiral copper complex catalyzed oxa-Diels–Alder reaction with β,γ-unsaturated α-oxo esters to give 3,4-dihydro-2H-pyrans 237, 238

Vinyl ethers, rhodium/Lewis acid catalyzed enantioselective intermolecular (3+2) cycloaddition with carbonyl ylides to give tetrahydro-9H-5,8-epoxybenzo[7]annulen-9-ones 180, 181

Vinyl iodides, 6-(π-exo)-exo-dig radical cyclization to give hexahydro-2H-1-benzopyrans 460

Vinylcyclopropanes, alkenyl, rhodium-catalyzed (3+2) cycloaddition to give 5-vinyloctahydrocyclopenta[c]pyrroles 364, 365

Vinylcyclopropanes, alkenyl, rhodium-catalyzed intramolecular (5+2) cycloaddition to give fused cycloheptenes 360, 361

Vinylcyclopropanes, alkenyl, rhodium-catalyzed intramolecular (5+2+1) cycloaddition with carbon monoxide to give fused cyclooctenones 377, 378

Vinylcyclopropanes, alkynyl, iron-catalyzed (5+2) intramolecular cycloaddition to give bicyclic cycloheptadienes 372

Vinylcyclopropanes, alkynyl, rhodium-catalyzed intramolecular (5+2) cycloaddition to give fused cycloheptadienes 359, 360

Vinylcyclopropanes, alkynyl, ruthenium-catalyzed intramolecular (5+2) cycloaddition to give fused cycloheptadienes 370, 371

Vinylcyclopropanes, allenyl, rhodium-catalyzed intramolecular (5+2) cycloaddition to give fused methylenecycloheptenes 360, 361

Vinylcyclopropanes, bicyclic, rhodium-catalyzed intermolecular (5+2) cycloaddition with but-2-yne to give 1,2,3,4,5,8-hexahydroazulenes 361, 362

Vinylcyclopropanes, cobalt-catalyzed (5+1) cycloaddition to give cyclohexenones 351, 352

Vinylcyclopropanes, iron-mediated (5+1) cycloaddition to give cyclohexenones 349, 350

Vinylcyclopropanes, propargyloxy, nickel-catalyzed intramolecular (5+2) cycloaddition to give fused tetrahydrofurans 371, 372

Vinylcyclopropanes, rhodium-catalyzed (5+1) cycloaddition with carbon monoxide to give cyclohexenones 357, 358

Vinylcyclopropanes, rhodium-catalyzed (5+2) cycloaddition/vinylogous Peterson elimination/(4+2) cycloaddition cascade to give 1,2,3,4,5,6,8,9-octahydro-7H-benzo[7]-annulen-7-ones 363

Vinylcyclopropanes, rhodium-catalyzed intermolecular (5+2+1) cycloaddition with alkynes and carbon monoxide to give bicyclo[3.3.0]octenones 376, 377

Vinylcyclopropanes, rhodium-catalyzed intermolecular (5+2) cycloaddition with alkynes to give cycloheptenones and cycloheptadienes 362

Vinylcyclopropanes, rhodium-catalyzed intermolecular (5+2) cycloaddition with allenes to give methylenecycloheptanones 363, 364

3-Vinylcyclopropenes, rhodium-catalyzed (5+1) carbonylation with carbon monoxide to give phenols 354, 355

Vinyliminium ions, as dienophiles in (4+3)-cycloaddition reactions 333–337

Vinyloxiranes, alkyne-substituted, rhodium-catalyzed intramolecular (5+2) cycloaddition to give bicyclic acylcyclopropanes 366

Vinyloxocarbenium ions, as dienophiles in (4+3)-cycloaddition reactions 339–341

Vinyl(2-pyridyl)silanes, ruthenium-catalyzed intermolecular Pauson–Khand reaction to give cyclopentenones 132, 133

Vinylthionium ions, as dienophiles in (4+3)-cycloaddition reactions 342, 343

Virgiboidine 644, 645

Virgidivarine 644, 645

(+)-Virosaine B 640, 641

W

Weinreb amides, precursors in the synthesis of Sch 725674 by ring-closing metathesis 595, 596

Welwitindolinone alkaloids, synthesis 501–503

WF-1360F 665, 666

(±)-Widdrol 345

X

2-(9-Xanthenyl)malonates, manganese(III)-mediated oxidation to give benz[b,f]oxepins 461, 462

Y

(±)-Yezo'otogirin A 510

Ynamides, rhodium-catalyzed enantioselective cycloaddition with 1,6-diynes to give axially chiral benzamides 275, 276

Z

Zoanthenol 505, 506

Author Index

In this index the page number for that part of the text citing the reference number is given first. The number of the reference in the reference section is given in a superscript font following this.

A

Abboud, K. A. 582[92]
Abe, M. 433[311]
Abe, T. 168[14], 184[14], 187[14], 188[14]
Abeywickreyma, A. N. 383[18]
Aburano, D. 156[171], 157[171]
Achmatowiz, B. 88[96], 89[96]
Adams, R. D. 668[242], 669[242]
Adenu, G. 345[82]
Adeosun, A. A. 89[98], 90[98]
Adolfsson, H. 1[4]
Adrio, J. 105[29], 106[29], 140[138], 142[138], 156[167]
Adsool, V. A. 585[99]
Agenet, N. 265[11], 265[18], 283[41], 284[41], 289[41]
Aggarwal, V. K. 1[3], 400[250], 633[173]
Agui, S. 473[335]
Aguilera, B. 652[208], 652[210], 653[210]
Ahmar, M. 111[48]
Ahmed, G. 332[48]
Aiguabella, N. 101[11], 107[35], 107[36]
Akane, N. 394[190]
Akgün, H. 498[373]
Aki, Y.-i. 600[119], 601[119]
Akiyama, M. 545[16], 546[16]
Akizuki, M. 344[77]
Alayrac, C. 268[21], 269[21], 278[21], 312[72]
Albaugh-Robertson, P. 112[54]
Alberico, D. 53[27]
Albizati, K. F. 341[64]
Alcarazo, M. 659[224], 660[224]
Aldabbagh, F. 416[288], 417[288]
Alder, K. 205[1]
Alemán, J. 142[140], 143[140]
Alešković, M. 395[213]
Alexander, B. 564[51], 564[52], 565[52], 566[52]
Alexanian, E. J. 307[68], 307[69], 309[69]
Alfassi, Z. B. 383[4]
Alian, A. 400[233]
Al-Midfa, L. 215[23]
Alonso, I. 330[34], 330[36], 330[42], 331[34], 331[36]
Altinel, E. 387[114]
Altmann, K.-H. 665[237], 666[237]
Álvarez, M. 525[450], 526[451]
Álvarez de Cienfuegos, L. 383[69], 383[72], 387[131]
Alvarez-Larena, A. 139[131], 139[133], 140[131], 140[133], 141[133]
Alvarez-Manzaneda, E. J. 520[424]
Ampapathi, R. S. 510[400]
An, J. 406[265]
An, M. H. 108[40], 109[40]

Anada, M. 168[14], 175[24], 175[26], 176[24], 176[26], 177[24], 178[24], 179[26], 180[26], 182[24], 183[24], 183[26], 184[14], 187[14], 188[14], 200[55], 201[55], 202[55], 241[59], 242[59], 253[72]
Anciaux, A. J. 54[31], 55[31], 55[37], 56[37]
Anderson, J. M. 387[109]
Anderson, P. S. 41[84]
Andersson, B. 71[66]
Andreana, P. R. 559[43], 560[43]
Andreotti, D. 92[101]
Andrews, R. S. 406[261], 406[262]
Andrioletti, B. 11[34]
Anilkumar, G. 11[38], 11[39]
Añorbe, L. 99[4]
Antoline, J. E. 334[53], 334[54], 334[55], 335[54], 335[55], 336[54], 336[55], 337[54], 337[55]
Antonchick, A. P. 192[40]
Antonietti, M. 137[122]
Aoki, K. 576[81], 577[81], 578[81]
Aoki, M. 18[52]
Appella, D. H. 26[67], 27[67], 28[67]
Arai, K. 251[70], 260[70], 261[70]
Arai, S. 527[459], 528[459], 573[75]
Arai, T. 522[435]
Arai, Y. 302[64], 303[64], 304[64], 308[64]
Arakawa, T. 265[19]
Araki, M. 222[30], 223[30]
Araki, T. 276[34], 277[34]
Aranishi, E. 400[241]
Arban, R. 92[101]
Arcand, H. R. 333[50]
Arceo, E. 215[20]
Arends, I. W. C. E. 1[14], 2[14]
Arif, A. M. 328[29], 329[29], 329[30]
Arikawa, T. 175[28], 180[28], 181[28], 184[28]
Arimato, F. S. 383[30]
Arisawa, M. 550[28]
Ariza, X. 518[420], 519[420]
Arrayás, R. G. 245[63], 246[63], 252[71]
Arrington, M. P. 581[90], 582[90]
Arslan, M. 383[59]
Arteaga, J. F. 383[71], 387[120], 387[126]
Arteaga, P. 387[120]
Arunachalampillai, A. 345[84], 346[84]
Arvary, R. A. 554[35], 554[36], 555[35], 555[36]
Arvidsson, L.-E. 71[66]
Asahi, K. 387[111], 490[111], 491[111]
Asano, K. 104[28]
Ashby, E. C. 531[477]
Ashcroft, M. R. 321[10], 322[10]

Ashfeld, B. L. 373[80], 373[81], 374[80], 374[81]
Ashie, H. 473[336]
Askin, D. 224[33]
Aspiotis, R. 324[20]
Astle, C. J. 633[173]
Aubé, J. 605[130]
Aubert, C. 265[11], 265[18], 269[24], 270[24], 279[24], 283[41], 284[41], 289[41], 349[6], 349[11]
Audrain, H. 236[50], 237[50], 237[52], 238[52]
Aumann, R. 349[17]
Aungst, R. A., Jr. 339[63], 340[63]
Avery, T. D. 168[17], 170[17], 171[17]
Awan, S. I. 52[21]
Azimioara, M. D. 206[9]
Azzi, S. 57[41], 58[41], 59[41]

B

Baba, T. 175[28], 180[28], 181[28], 184[28], 188[35], 189[35], 191[35]
Bäckvall, J.-E. 384[94]
Badía, L. 518[420], 519[420]
Bagheri, V. 54[30]
Baik, M.-H. 120[79], 120[80], 125[89]
Bailey, L. N. 330[43]
Baine, N. H. 383[21]
Balan, D. 1[4]
Balandrin, M. F. 400[229]
Balci, M. 386[107], 386[108], 387[107], 387[108]
Bali, A. K. 596[113]
Balog, A. 550[26]
Balsells, J. 112[55]
Bandichhor, R. 579[84]
Banerjee, B. 390[161], 390[162], 391[162], 454[161], 455[161], 456[161], 457[161], 458[161]
Banfi, L. 645[195]
Banide, E. V. 112[56]
Banti, D. 645[196]
Bao, H. 239[56]
Bao, J. 213[17]
Bao, W. 71[67], 72[67]
Bar, G. 498[378]
Baran, P. S. 159[182], 160[182]
Barash, L. 205[4]
Barber, R. B. 529[464]
Barberis, M. 93[102]
Barbi, J. 345[85]
Barbosa, L. C. de A. 321[11], 322[11]
Barchuk, A. 416[286], 417[286]
Bardales, E. 394[189]
Barlan, A. U. 16[50], 17[50]
Barnes, C. L. 329[33], 330[33], 341[66], 341[67], 345[82]
Barnes, D. M. 215[19], 216[19]
Barragán, A. 525[450]

Barrero, A. F. 383[71], 384[86], 387[120], 387[126], 390[158], 391[166], 520[424], 525[158], 525[166], 525[450], 526[158], 526[451]
Barrett, A. G. M. 589[104], 590[104], 658[223]
Barrett, E. 383[9]
Barriault, L. 638[185]
Barroso, S. 215[21], 215[23]
Bartoli, G. 49[10]
Barton, D. H. 383[1]
Bartra, M. 518[420], 519[420]
Barzilay, C. M. 362[50]
Basak, A. 4[20], 5[20], 6[20], 16[50], 17[50]
Basarić, N. 395[213]
Bashir, N. 393[184]
Basso, A. 645[195]
Basu, S. 510[400]
Battiste, M. A. 321[8], 349[12]
Bauer, D. 387[128], 387[129], 387[130]
Bauer, E. B. 654[213], 654[214]
Baylon, C. 556[38], 556[39]
Bazdi, B. 387[121], 519[423], 520[423]
Beaton, J. M. 383[1]
Beattie, M. S. 388[142], 389[142]
Becker, H. 106[32]
Becker, J. 593[110], 594[110]
Becker, J. J. 406[261], 406[262]
Beckwith, A. L. J. 383[14], 383[15], 383[16], 383[17], 383[18], 383[19], 416[291], 417[291]
Beemelmanns, C. 406[268]
Bèkhazi, M. 167[8]
Belanger, D. B. 111[49], 117[73], 118[73]
Beller, M. 11[38], 11[39], 14[47], 15[47], 16[47], 99[3]
Bellesia, F. 423[303], 424[303], 425[303]
Bello, P. 107[35]
Bencivenni, G. 49[10]
Bendall, J. G. 585[97]
Benedetti, M. 92[101]
Benesova, V. 81[78]
Benet-Buchholz, J. 616[147]
Ben-Shoshan, R. 349[15]
Bentabed-Ababsa, G. 167[6]
Berenguer, R. 518[420], 519[420]
Bergman, R. G. 283[39]
Berkessel, A. 14[46]
Berlin, J. M. 564[55], 565[55], 566[55]
Bermejo, F. 389[150]
Bernad, P. L. 394[194]
Bernard, M. 88[85], 93[85]
Bernardes, V. 139[132], 140[132], 141[132]
Bernardes-Génisson, V. 139[135], 140[135], 141[135]
Bernardinelli, G. 221[29], 222[29]
Bernhardt, P. V. 332[47]
Berrisford, D. J. 3[18]
Berry, C. R. 334[52]
Berson, J. A. 205[2]
Bertani, B. 92[101]
Bertinato, P. 550[26]
Bertozzi, F. 400[237]

Bertus, P. 89[97]
Bes, M. T. 106[32]
Betancort, J. M. 149[151]
Betkekar, V. V. 624[157], 638[186], 639[186]
Bettati, M. 92[101]
Bettelini, L. 92[101]
Bhat, V. 501[387], 502[387], 503[387]
Bhor, S. 14[47], 15[47], 16[47]
Bi, F. C. 373[79], 374[79]
Bichovski, P. 473[340], 474[340], 475[340], 476[340], 487[351], 488[351]
Bigeault, J. 87[84]
Billington, D. C. 103[22], 104[22], 109[42], 110[42]
Bindl, M. 658[222], 659[227]
Binger, P. 361[48], 362[48]
Binsch, G. 50[14]
Bitterlich, B. 11[38]
Blackburn, C. 339[60]
Bladon, P. 109[42], 110[42]
Blagg, S. J. 592[108]
Blahy, O. M. 41[84]
Blanchard, K. R. 548[21]
Blanco-Urgoiti, J. 99[4], 110[46], 115[70], 116[70], 117[70], 119[70]
Blanda, G. 550[27]
Blasi, E. 107[37], 109[44], 110[44], 111[44]
Blaszykowski, C. 88[85], 93[85]
Blay, G. 215[21], 215[23]
Blechert, S. 555[37], 562[48], 562[49], 563[48], 563[49], 602[124], 603[124], 603[125], 603[126], 604[126], 604[127], 604[128], 605[129], 642[191], 644[194], 645[194]
Bloch, W. M. 384[76], 498[76], 499[76]
Bloomquist, J. R. 200[52]
Blu, J. 112[57], 113[57]
Bluhm, H. 387[135], 387[136], 389[135], 389[151], 389[152], 390[135], 416[135], 416[285], 417[135], 417[285], 467[135], 468[135]
Boate, D. R. 383[17]
Boese, R. 88[92], 89[92]
Boezio, A. A. 259[81]
Boger, D. L. 167[7]
Bogert, M. T. 495[356]
Böhm, C. 52[22]
Böhrsch, V. 604[127], 604[128]
Boiron, A. 528[460], 529[460]
Bolm, C. 3[18], 612[138], 613[138]
Boñaga, L. V. R. 102[20], 103[20], 103[23], 104[23], 117[74], 118[74], 119[74], 119[76], 155[163], 155[164]
Bonanomi, G. 92[101]
Bonazzi, S. 259[85], 260[85]
Bonchio, M. 14[42]
Bönnemann, H. 283[40]
Bora, U. 383[53], 384[53], 490[53], 503[53], 504[53]
Borders, S. S. K. 52[25]
Bork, P. M. 525[446]
Börner, H. 517[416], 532[416]
Börner, H. G. 517[418], 532[418]

Bourgeois, D. 549[22]
Bowden, N. 545[14], 618[14], 619[14], 620[14]
Bowman, B. J. 561[47], 562[47]
Bowman, E. J. 561[47], 562[47]
Bowman, W. R. 416[290], 416[291], 417[290], 417[291], 420[297]
Bowry, V. W. 416[291], 417[291]
Boyer, F.-D. 636[180]
Boyle, G. M. 332[47]
Brackley, J. A., III 341[67]
Braggio, S. 92[101]
Brahmachari, G. 523[438]
Brahmachary, E. 265[13]
Brånalt, J. 233[47], 234[47]
Brancour, C. 355[25], 355[26], 356[25]
Brand, C. 52[26]
Brandt, G. E. L. 592[108]
Brasholz, M. 498[377]
Bratz, M. 585[100]
Brauman, J. I. 11[35]
Braunstein, P. 133[109], 134[109]
Brenneman, J. B. 633[174]
Breslow, R. 383[9], 383[10]
Brinker, D. A. 54[30]
Brinkmann, R. 283[40]
Brocksom, T. J. 500[386]
Brocksom, U. 500[386]
Brodney, M. A. 373[79], 374[79]
Broere, D. L. J. 265[3]
Brookhart, M. 49[11], 50[11]
Brossi, A. 400[226]
Brown, B. H. 588[103], 589[103]
Brown, J. A. 103[25], 104[25]
Brown, M. K. 590[106], 591[106]
Brückner, A. 11[39]
Brückner, R. 389[144]
Brummond, K. M. 105[30], 111[50], 111[51], 125[91], 126[91], 127[91], 138[124], 139[127], 139[128]
Brusoe, A. T. 307[68], 307[69], 309[69]
Bruzinski, P. R. 176[31]
Bryans, J. S. 510[401]
Bryliakov, K. P. 10[30]
Bubert, C. 52[22]
Buchner, E. 167[4]
Buchwald, S. L. 130[103], 130[104], 130[105], 131[103], 131[104], 131[105], 132[105], 146[145], 410[278]
Buckman, B. O. 498[366]
Buffone, G. P. 563[50], 564[50], 567[50]
Bui, G. 350[18], 358[35], 359[35]
Bui, K. 412[283], 413[283], 415[283]
Buisine, O. 265[11], 349[6]
Bullock, W. H. 585[100]
Bunce, R. A. 383[44]
Bunnage, M. E. 580[87]
Buñuel, E. 81[75], 82[75], 387[124], 391[124]
Bunz, U. H. F. 668[242], 669[242]
Buono, G. 87[84]
Burgess, J. M. 632[172]
Burnell, R. H. 498[368], 498[369]
Burton, J. W. 400[251], 401[252], 404[252], 405[252], 427[252], 442[318],

443[318], 444[318], 444[319], 445[319], 446[319], 447[319], 447[320], 448[320], 449[320], 499[381], 500[381], 585[97]
Buschmann, N. 603[125]
Bush, J. B., Jr. 384[91], 385[91]
Butenschön, H. 349[13]
Butt, L. 593[110], 594[110]
Bye, C. A. 52[25]
Bykowski, D. 193[47]
Bytschkov, I. 403[254], 404[254]

C

Cabot, R. 114[61]
Cahiez, G. 383[73]
Cai, D. 404[255], 404[256]
Cai, J. C. 529[461], 530[461]
Cai, S. 384[82]
Cai, Y. 69[61], 70[61]
Cai, Z. 591[107]
Calà, M. 658[221]
Calimano, E. 395[222]
Çalişkan, R. 386[107], 386[108], 387[107], 387[108]
Çalişkan, Z. 498[373]
Calomeni, E. 345[85]
Campagne, J.-M. 632[171]
Campaña, A. G. 383[69], 383[70], 383[72], 387[124], 387[125], 387[131], 391[124], 519[423], 520[423]
Campana, C. 668[242], 669[242]
Campbell, A. N. 85[82], 86[82]
Campos, K. R. 215[19], 216[19]
Canales, E. 209[14], 210[14]
Cao, H. 105[31], 153[31], 153[161], 154[161]
Caple, R. 149[152]
Carda, M. 550[25]
Cárdenas, D. J. 81[75], 82[75], 387[124], 387[125], 391[124]
Cárdenas, L. 109[44], 110[44], 111[44]
Cariou, K. 88[85], 93[85]
Carletti, R. 92[101]
Carlsson, A. 71[66]
Carofiglio, T. 14[42]
Carreira, E. M. 66[55], 66[56], 66[57], 67[60], 68[55], 68[56], 69[57], 420[294]
Carretero, J. C. 105[29], 106[29], 140[138], 140[139], 142[138], 142[139], 156[167], 245[63], 246[63], 252[71]
Carson, C. A. 384[83]
Carson, M. W. 343[75]
Carte, B. 343[73]
Carter, K. W. 342[69]
Casarrubios, L. 102[18], 103[18], 110[46], 111[52], 115[70], 116[70], 117[70], 119[70]
Casper, D. 66[58]
Castedo, L. 330[34], 330[35], 330[37], 331[34], 331[35], 331[37]
Castro, J. 112[55], 139[129], 139[130], 139[131], 140[129], 140[130], 140[131]
Catak, S. 327[27], 328[27], 338[27]
Catino, A. J. 29[71], 30[71]

Cavallo, L. 547[18], 548[18], 616[147]
Cavanni, P. 92[101]
Caze, E. 392[178], 393[178]
Cazes, B. 111[48]
Cefalo, D. R. 564[51], 564[52], 564[54], 565[52], 565[54], 566[52], 566[54]
Cha, J. K. 326[24], 326[25], 327[24], 327[25]
Chahboun, R. 520[424]
Chakraborty, P. 511[408], 512[408]
Chakraborty, T. K. 389[147], 389[148], 510[400], 524[440], 525[440]
Chan, A. S. C. 127[97], 128[97], 129[97]
Chan, P. W. H. 27[68], 28[68], 400[236]
Chan-Bacab, M. J. 345[84], 346[84]
Chang, C. J. 395[215]
Chang, K.-M. 99[6], 135[6]
Chang, K. T. 615[144]
Chang, N.-J. 406[264], 413[264]
Chanthamath, S. 64[53]
Charette, A. B. 53[27], 57[40], 57[41], 58[40], 58[41], 59[41]
Chatani, N. 132[106], 156[166], 156[168], 353[23], 354[23]
Chatgilialoglu, C. 383[35], 387[118], 460[324]
Chauvin, Y. 651[203]
Chavez, D. E. 234[49], 235[49]
Che, C.-M. 34[77], 35[77], 63[51], 168[18], 170[18], 171[18], 172[18], 174[18], 192[44], 194[44], 195[48], 196[48], 196[49], 197[49], 198[51], 199[48], 199[49], 199[51], 200[51]
Cheang, J. 66[56], 68[56]
Checchia, A. 92[101]
Chellé, I. 222[31]
Chen, B. 358[35], 359[35]
Chen, C. 384[77], 384[78], 384[79], 503[78], 504[78], 504[79]
Chen, D. 73[68], 74[68]
Chen, D. Y.-K. 91[100], 92[100], 510[398], 527[457]
Chen, G.-Q. 154[162]
Chen, H. 125[91], 126[91], 127[91]
Chen, I.-W. 41[84]
Chen, J. 115[67], 115[68], 116[67], 116[68], 119[67], 134[110], 159[184], 160[184], 161[184], 195[48], 196[48], 199[48]
Chen, J.-R. 406[264], 413[264]
Chen, J. S. 392[176], 393[176]
Chen, L. 368[62]
Chen, S.-Y. 395[219], 395[220]
Chen, W. 152[158]
Chen, X. 246[64], 247[64], 255[74], 256[74]
Chen, Z. 184[34], 228[40], 229[40]
Chen, Z.-Y. 172[20], 173[20], 175[20], 184[32], 185[32], 187[32]
Cheng, B. 168[15], 200[15]
Cheng, Y. 406[265]
Cheng, Y.-Y. 383[40]

Cheong, P. H.-Y. 370[65], 370[66], 370[67]
Cheong, W.-W. 401[253], 402[253], 403[253]
Chernyak, N. 56[38], 395[216]
Chessum, N. E. A. 510[401]
Cheung, K.-K. 498[375]
Cheung, Y. Y. 582[92]
Cheung, Y.-Y. 582[91]
Chevallier, F. 389[149]
Chevance, S. 1[6]
Chiaroni, A. 498[380]
Chiba, T. 275[31], 280[31]
Childs, R. F. 339[60]
Chiow, W.-R. 429[304], 430[304], 431[304]
Chisholm, M. H. 545[16], 546[16]
Cho, E. J. 597[115], 598[115]
Cho, S. H. 354[24], 355[24]
Cho, Y. 127[98], 128[98], 129[98]
Choi, A. U. 529[463]
Choi, C. 121[85], 122[85]
Choi, J. H. 130[102]
Choi, J.-R. 326[24], 327[24]
Choi, T.-L. 619[152], 620[152]
Choi, Y. H. 345[81]
Choi, Y. K. 120[78], 120[81], 121[81], 122[81]
Chopade, P. R. 265[12], 394[201]
Choquesillo-Lazarte, D. 387[121]
Chou, T.-C. 550[26]
Chow, S. 332[47]
Choy, A. L. 585[96]
Christensen, O. T. 384[89], 384[90]
Christmann, M. 579[85], 580[85]
Christoffers, J. 410[280], 410[281]
Chu, Y. 21[59], 22[59], 23[59]
Chuang, C.-P. 383[40], 418[293], 419[293], 420[293], 429[304], 430[304], 431[304]
Chung, Y. K. 99[6], 102[16], 103[16], 115[64], 116[64], 127[98], 128[98], 129[98], 135[6], 135[113], 135[114], 135[115], 135[116], 135[118], 135[119], 135[120], 136[113], 136[114], 136[115], 136[116], 136[118], 136[119], 136[120], 137[116], 137[118], 137[119], 137[120], 151[156], 152[157]
Chuprakov, S. 56[38], 57[39]
Churchill, M. R. 332[48], 651[205], 659[226]
Ciminale, F. 74[69]
Ciochina, R. 510[396]
Ciriano, M. V. 400[248]
Claassen, R. J., II 342[70], 343[70]
Clark, A. J. 423[303], 424[303], 425[303]
Clark, J. S. 556[40], 585[97]
Clavier, G. 395[214]
Clavier, H. 616[147]
Cleator, E. 660[228], 661[228]
Clemente, R. R. 523[439], 524[439]
Clerici, A. 387[117]
Clive, D. L. J. 416[289], 417[289]

Closser, K. D. 150[154]
Clyne, M. A. 416[288], 417[288]
Cockfield, D. M. 660[228], 661[228]
Colasanti, B. 400[225]
Coleman, M. G. 54[34], 54[35]
Collado, I. G. 345[80]
Collin, J. 526[452]
Collman, J. P. 11[35]
Colomb, M. 394[192]
Colombier, D. 111[48]
Colombo, M. I. 498[363]
Colucci, J. 324[20]
Colyer, J. T. 193[47]
Comely, A. C. 115[65], 116[65]
Commeureuc, A. G. J. 410[282]
Concellón, J. M. 392[173], 393[173], 394[189], 394[194]
Condra, J. H. 41[84]
Conesa, J. 115[62]
Cong, Z. 384[81], 461[81], 462[81]
Connolly, J. D. 525[444]
Conser, K. R. 24[63]
Constantino, M. G. 500[386]
Cook, G. R. 580[88], 581[88]
Cook, J. M. 105[31], 153[31], 153[161], 154[161], 159[176]
Cook, L. 498[367]
Corbet, M. 76[73], 77[73], 78[73], 92[73], 93[73]
Corey, E. J. 206[8], 206[9], 207[10], 208[10], 209[12], 209[13], 209[14], 210[12], 210[13], 210[14], 211[12], 212[16], 213[16], 257[76], 257[77], 258[77], 386[104], 387[104], 394[197], 473[332]
Cormier, K. W. 384[78], 503[78], 504[78]
Cornils, B. 391[164]
Correa, A. 616[147]
Corsi, M. 92[101]
Cortez, G. A. 564[56], 565[56]
Cossío, F. P. 330[41]
Costa, A. V. 321[11], 322[11]
Côté, C. 498[369]
Cotton, F. A. 545[16], 546[16]
Cottrell, I. F. 404[255], 404[256]
Cotugno, P. 74[69]
Couch, M. M. 50[13]
Coutts, R. S. P. 387[133]
Cowan, D. O. 50[13]
Crabtree, R. H. 167[3], 168[3]
Craft, D. T. 330[40], 330[43]
Cranfill, D. C. 44[88], 46[88]
Crea, A. 71[64]
Crimmins, M. T. 570[67], 585[96], 585[98], 588[103], 589[103]
Croatt, M. P. 123[86]
Cross, M. J. 292[55], 293[55], 298[55]
Croudace, M. C. 100[7]
Crowe, W. E. 102[15], 150[153]
Cuerva, J. M. 383[69], 383[70], 383[72], 387[116], 387[121], 387[123], 387[124], 387[125], 387[131], 390[158], 391[124], 391[166], 511[405], 519[423], 520[423], 520[424], 525[158], 525[166], 526[158]

Cui, J. 563[50], 564[50], 567[50]
Cui, X. 38[81], 39[81], 228[37], 229[37]
Cui, Y. 39[82], 40[82]
Cuong, D. D. 498[380]
Curran, D. P. 138[124], 139[127], 383[5], 383[23], 383[24], 383[37], 383[38], 392[177], 392[180], 393[177], 393[180], 408[273], 409[273], 412[283], 413[283], 415[283], 416[287], 416[292], 417[287], 417[292], 531[37], 574[77]
Curtius, T. 167[4]

D

Daasbjerg, K. 387[137], 388[137], 389[137], 394[204], 394[205], 416[286], 417[286]
Dahan, A. 115[71], 116[71], 117[71]
Dahlén, A. 392[174], 393[174], 394[174]
Dai, M. 568[60], 568[61]
Dai, W. 9[29], 10[29], 11[29], 42[29], 45[29]
Dalpozzo, R. 49[10]
Danishefsky, S. J. 159[179], 160[179], 224[33], 258[78], 343[75], 550[26], 568[60], 568[61]
D'Annibale, A. 422[301]
Darke, P. L. 41[84]
Das, A. 412[283], 413[283], 415[283]
Das, D. 524[440], 525[440]
Das, S. 389[147]
da Silva, E. G. 389[155]
Da Silva, M. I. 589[104], 590[104]
Daviaud, R. 11[33]
Davies, H. M. L. 54[32], 54[33], 54[34], 54[35], 59[42], 60[42], 176[31], 332[44], 332[45], 332[46], 332[47], 332[48], 332[49], 333[45], 333[46], 343[76], 344[78]
Davies, J. J. 499[381], 500[381]
Davies, P. W. 650[202]
Davis, T. C. 205[4]
Davis, W. M. 564[54], 565[54], 566[54]
DeAngelis, A. 168[13], 184[13], 185[13], 186[13], 187[13]
Deaton, J. 390[157]
Debleds, O. 632[171]
de Bruin, T. J. M. 106[33], 113[59], 114[59]
De Campo, F. 440[317], 441[317], 442[317]
DeChristopher, B. A. 554[36], 555[36]
de Cózar, A. 330[41]
Deepthi, A. 383[60]
Degennaro, L. 1[10]
de Jonge, C. R. H. I. 387[110]
de Klein, W. J. 387[110]
de Koning, C. B. 630[168], 631[168]
de la Rosa, J. C. 140[139], 142[139]
Delpont, N. 640[187]
del Pozo, C. 101[11]
Del Valle, D. J. 599[118], 600[118]
de March, P. 167[5]
de Meijere, A. 49[1], 49[8], 49[9], 88[86], 88[87], 88[91], 88[92], 88[94], 89[87], 89[91], 89[92], 89[94], 106[32], 351[21], 352[21], 353[21], 357[21], 361[48], 362[48], 363[54], 364[54]
Demir, A. S. 383[47], 383[54], 387[112], 387[113], 387[114], 498[373], 498[374]
Demuner, A. J. 321[11], 322[11]
Deng, L. 115[67], 116[67], 119[67], 134[110], 134[111], 135[111], 135[112]
Deng, Y. 255[75], 256[75]
Dengiz, C. 386[108], 387[108]
DeNicola, A. 139[129], 140[129]
Denton, J. R. 54[33], 344[78]
Derdour, A. 167[6]
de Riggi, I. 87[84]
Derrer, S. 585[97]
Deschamps, J. 105[31], 153[31]
Deschamps, N. M. 123[87], 124[87], 124[88], 125[88]
Dessau, R. M. 385[97], 385[99], 386[99], 387[99]
Devarajan, D. 31[73], 32[73]
De Vos, D. E. 18[53]
de Vries, J. G. 614[142], 615[142]
Dhakshinamoorthy, A. 383[61]
Dhimane, A.-L. 88[85], 93[85], 355[25], 355[26], 356[25], 383[41]
Díaz, D. 149[151]
Diaz, F. A. 570[67]
Díaz, M. R. 394[194]
Diedrich, C. L. 410[281]
Diéguez, H. R. 387[126]
Diels, O. 205[1]
Dienstag, J. L. 518[421]
Di Fabio, R. 92[101]
Dineen, T. A. 637[183]
Ding, H. 91[100], 92[100]
Ding, K. 25[65], 27[65], 230[41], 230[42], 230[43], 231[42], 231[44], 238[53], 239[53], 239[54], 239[55], 239[56], 240[53], 241[54]
Dirsch, V. M. 525[447]
Dittrich, B. 52[20], 52[21], 52[26]
Dixon, D. J. 551[31], 552[31], 660[31], 660[228], 661[31], 661[228]
Djukic, J.-P. 66[59]
Doan, B. D. 332[45], 333[45]
Dobado, J. A. 387[126]
Döbler, C. 14[47], 15[47], 16[47]
Docherty, P. H. 444[319], 445[319], 446[319], 447[319]
Doherty, M. Q. 85[82], 86[82]
Doi, T. 390[159], 511[403], 511[404]
Dolg, M. 517[412]
Dolgopalets, V. I. 88[89], 89[89]
Dolva, A. 66[57], 69[57]
Domboski, M. A. 498[366]
Domingo, L. R. 167[6]
Domingo, V. 387[126]
Domínguez, G. 99[4], 102[18], 103[18], 110[46], 111[52], 115[70], 116[70], 117[70], 119[70], 155[165], 265[6]
Donaldson, S. M. 614[140]
Dong, G. 43[87], 46[87], 115[67], 116[67], 119[67], 134[110]
Dong, L. 395[220]
Dong, Z. 226[35], 227[35]

Donohoe, T. J. 567[58]
Donohue, A. C. 168[17], 170[17], 171[17]
Donovan, R. J. 349[2]
Dorcet, V. 167[6]
Doris, E. 389[149]
Dorsey, B. D. 41[84]
Dossetter, A. G. 234[48], 235[48]
Dötz, K. H. 389[155], 517[416], 517[418], 532[416], 532[418]
Dou, G.-L. 495[355]
Dowd, P. F. 524[441]
Doye, S. 403[254], 404[254]
Doyle, D. 653[212]
Doyle, M. P. 29[71], 30[71], 54[29], 54[30], 54[36], 55[29], 55[36], 60[43], 168[12], 172[12], 172[20], 173[20], 175[20], 184[12], 193[45], 193[46], 193[47], 194[46], 199[45], 241[57], 241[58]
Drexler, H.-J. 289[50], 291[50]
Dröge, W. 525[446]
Du, G. H. 400[234]
Du, H. 238[53], 239[53], 239[54], 239[55], 239[56], 240[53], 241[54]
Du, J. 406[258]
Du, W. 395[223], 396[223], 397[223]
Du, Y. 334[55], 335[55], 336[55], 337[55]
Duan, Y.-S. 395[220]
Dubé, P. 82[79], 83[79]
Dubois, G. 8[27]
Du Bois, J. 28[69], 29[69]
Dudnik, A. S. 395[216]
Duhme-Klair, A. K. 100[10]
du Jourdin, X. M. 41[85], 45[85]
Dulcere, J. P. 433[310], 433[312]
Duong, H. A. 292[55], 293[55], 298[55]
Duygu, N. 498[374]
Dyckman, A. J. 359[42], 362[50]

E

Eagle, C. T. 54[30]
Earl, R. A. 292[54]
Ebihara, Y. 287[49], 288[49], 290[49]
Ebiura, Y. 188[35], 189[35], 191[35]
Ebrahimian, G. R. 41[85], 45[85]
Echavarren, A. M. 75[71], 81[75], 82[75], 83[80], 84[80], 85[80], 86[83], 640[187]
Eckelbarger, J. D. 408[271], 409[271]
Edmonds, D. J. 259[84], 383[67], 571[70], 572[70]
Edwankar, R. V. 307[69], 309[69]
Egami, H. 20[57], 20[58], 21[58]
Eggert, U. 341[68]
Eisch, J. J. 89[98], 90[98]
Eisele, D. 642[191]
Ellery, S. P. 392[176], 393[176]
Elliott, G. I. 167[7]
Elliott, R. L. 392[177], 393[177]
Ellman, J. A. 216[24], 217[24]
Ellmerer-Müller, E. P. 525[447]
Emini, E. A. 41[84]
Emmitte, K. A. 585[98]
Emrullahoglu, M. 383[47], 383[54]

Enders, D. 349[14]
Endo, K. 275[31], 280[31]
Endo, T. 394[198]
Enemærke, R. J. 387[137], 388[137], 389[137], 394[204], 394[205]
Enzell, C. 345[79]
Ernst, A. B. 386[101], 386[106], 387[106]
Escalante-Erosa, F. 345[84], 346[84]
Eschenbrenner-Lux, V. 247[65], 248[65]
Escudero-Adán, E. C. 616[147]
Esposito, A. 94[104]
Esquivias, J. 252[71]
Ess, D. H. 31[73], 32[73]
Estévez, R. E. 387[121], 387[123], 387[131]
Estévez, V. 395[217]
Evano, G. 598[117], 599[117]
Evans, D. A. 24[62], 50[18], 51[18], 52[18], 215[19], 216[19]
Evans, D. H. 531[479]
Evans, G. R. 222[31]
Evans, P. 112[56]
Evans, P. A. 120[79], 120[80], 150[155], 151[155], 302[63], 303[63], 305[63], 308[63], 349[7], 349[10], 563[50], 564[50], 567[50]
Evans, W. J. 394[193]
Exon, C. 112[54]
Expósito Castro, M. Á. 175[23], 176[23]
Extine, M. W. 545[16], 546[16]

F

Faber, D. 528[460], 529[460]
Fabry, D. C. 406[263]
Fager-Jokela, E. 100[9]
Fairlamb, I. J. S. 100[10]
Falck, J. R. 31[73], 32[73]
Fall, M. J. 176[31]
Falomir, E. 550[25]
Fan, B.-M. 121[84], 122[84]
Fan, C.-A. 387[115], 390[160]
Fang, H. 384[84], 400[234]
Fang, L. 159[183], 160[183], 161[183]
Fanwick, P. E. 545[16], 546[16]
Farnworth, M. V. 300[61], 301[61]
Farràs, J. 518[420], 519[420]
Farwick, A. 159[178], 160[178]
Faul, M. M. 24[62], 50[18], 51[18], 52[18]
Faust, A. 400[232]
Faustino, H. 330[36], 331[36]
Faveri, G. D. 1[5]
Fazzolari, E. 92[101]
Feducia, J. A. 85[82], 86[82]
Feibelmann, S. 101[12]
Feiters, M. C. 517[416], 517[418], 532[416], 532[418]
Felluga, F. 423[303], 424[303], 425[303]
Feng, J. 228[39]
Feng, J.-J. 366[59], 367[59]
Feng, X. 21[59], 22[59], 23[59], 226[35], 227[35], 228[37], 228[40], 229[37], 229[40], 254[73], 255[74], 256[74]

Fensterbank, L. 81[76], 88[85], 93[85], 349[11], 355[25], 355[26], 356[25], 383[41]
Fernandes, M. A. 630[168], 631[168]
Fernández, I. 319[1], 330[1], 330[41]
Fernández-Mateos, A. 523[439], 524[439]
Ferrand, Y. 11[33]
Ferrer, C. 86[83]
Ferreri, C. 460[324]
Feurer, M. 473[340], 474[340], 475[340], 476[340], 487[351], 488[351]
Fevig, T. L. 383[24], 392[177], 393[177]
Fiedler, P. 281[36], 282[36]
Filippov, D. V. 620[153], 621[153]
Findik, H. 387[112]
Findlay, J. A. 400[226]
Finkbeiner, H. 384[91], 385[91]
Finnegan, D. F. 149[150]
Fischer, C. 289[50], 291[50]
Fischer, E. 410[276], 410[277]
Fischer, G. 385[95]
Fischer, H. D. 525[443]
Fischer, N. H. 525[443]
Fischer, S. 104[26], 138[124], 139[127]
Fisher, K. D. 125[91], 126[91], 127[91]
Fjermestad, T. 146[147], 147[148]
Fleckhaus, A. 387[118]
Fleming, I. 383[37], 388[138], 388[139], 531[37]
Fletcher, S. P. 416[289], 417[289]
Fletcher, V. R. 342[70], 343[70]
Flippen-Anderson, J. 105[31], 153[31]
Flowers, R. A., II 394[201], 394[203], 394[206], 394[207], 394[208], 394[211]
Flügge, S. 330[39], 659[225]
Fodor, G. B. 400[225]
Fokin, V. V. 56[38], 57[39]
Fonquerna, S. 139[136], 139[137], 140[136], 140[137], 141[136], 142[137]
Fontana, S. 92[101]
Forbes, D. C. 54[29], 55[29]
Ford, J. G. 102[17], 103[17], 105[17]
Forslund, R. E. 29[71], 30[71]
Fort, A. W. 319[6], 319[7], 320[6], 320[7]
Fossey, J. 383[6]
Foster, R. W. 400[251]
Fouché, M. 658[223]
Fournogerakis, D. N. 363[53]
Fox, J. M. 108[38], 159[177], 160[177], 168[13], 184[13], 185[13], 186[13], 187[13]
Foxman, B. M. 384[85], 498[367]
Fraga, B. M. 359[39], 359[40], 525[445]
Fraile, J. M. 49[12], 50[12]
Franke, D. 517[415], 517[416], 517[417], 517[419], 532[415], 532[416], 532[417]
Freudenberger, J. H. 659[226]
Freund, A. 49[4]
Frey, G. 473[340], 474[340], 475[340], 476[340], 487[351], 488[351]
Frey, W. 410[281]
Freyer, A. J. 343[73]
Friedrich, J. 517[412]

Frings, M. 612[138], 613[138]
Fristad, W. E. 386[100], 386[101], 386[102], 386[106], 387[106], 498[372]
Fritschi, H. 50[17], 51[17]
Fröhlich, R. 579[85], 580[85]
Frontier, A. J. 343[75]
Frühauf, H.-W. 349[3]
Fruit, C. 1[7]
Fu, G. C. 544[3], 544[4], 545[3], 545[4]
Fu, L. 406[264], 413[264]
Fu, W. 491[353], 492[353], 493[353], 494[353], 668[242], 669[242]
Fu, X. 159[184], 160[184], 161[184]
Fu, X.-F. 357[34], 358[34]
Fu, Y. 491[353], 492[353], 493[353], 494[353]
Fuchs, J. R. 394[207], 394[208]
Fuchs, P. L. 41[85], 45[85]
Fuentes, N. 387[121]
Fuji, K. 127[92], 127[95], 127[96], 128[92], 128[95], 128[96], 129[95], 129[96]
Fuji, M. 362[49], 373[77]
Fujimoto, T. 274[30], 275[30], 280[30], 473[331]
Fujinami, T. 477[343]
Fujioka, K. 159[185], 160[185], 161[185]
Fujita, K. 549[23], 550[23]
Fujita, R. 642[192], 643[192], 646[192]
Fujiwara, K. 600[119], 601[119]
Fujiwara, T. 497[358]
Fukagawa, Y. 530[468]
Fukanaga, T. 383[28], 383[29]
Fukatsu, Y. 511[407]
Fukumoto, Y. 132[106], 156[166], 156[168]
Fukunag, Y. 32[75], 33[75]
Fukushima, K. 188[35], 189[35], 191[35]
Fukushima, Y. 609[135]
Fukuyama, T. 355[25], 355[26], 356[25], 574[78], 575[78], 607[133], 608[133]
Fukuyama, Y. 344[77]
Fukuzawa, S. 477[343]
Funakubo, S. 576[82], 577[82], 578[82]
Funami, H. 167[11]
Funk, R. L. 339[63], 340[63]
Funk, T. W. 564[55], 565[55], 566[55]
Fürstner, A. 76[73], 77[73], 78[73], 92[73], 93[73], 330[39], 372[76], 545[15], 546[15], 546[17], 550[24], 550[27], 580[89], 581[89], 594[111], 595[111], 595[112], 650[202], 652[208], 652[211], 653[211], 655[17], 655[216], 656[17], 656[216], 656[217], 657[217], 657[218], 658[219], 658[222], 659[224], 659[225], 659[227], 660[224], 661[15], 661[229], 661[230], 661[231], 662[15], 662[229], 662[230], 662[231], 662[232], 662[233], 663[234], 664[112], 664[235], 664[236], 665[236], 666[238], 667[238], 668[239]
Furukawa, N. 156[172], 157[172]
Furumi, S. 277[35], 278[35], 281[35]
Furuta, K. 206[6], 206[7]

Fuse, S. 390[159], 511[403]
Fuster, A. 106[34], 107[34]
Fuster, G. 139[135], 140[135], 141[135]
Fustero, S. 101[11], 107[35]
Futatsugi, K. 211[15], 212[15]

G

Gabor, B. 663[234]
Gademann, K. 234[49], 235[49], 259[85], 260[85]
Gagné, M. R. 85[82], 86[82], 406[261], 406[262]
Gaich, T. 596[114], 597[114]
Gallardo, J. 518[420], 519[420]
Gallenkamp, D. 594[111], 595[111]
Galliford, C. V. 197[50], 198[50], 200[50], 420[295]
Gallo, V. 133[109], 134[109]
Galopin, C. C. 93[103]
Gamber, G. G. 123[87], 124[87], 363[51], 376[87], 377[87], 378[87]
Gamlath, C. B. 329[32], 329[33], 330[32], 330[33], 342[32]
Gan, Y. 188[38], 190[38], 191[38], 192[38]
Gandon, V. 265[8], 265[11], 265[18], 269[24], 270[24], 279[24], 283[41], 284[41], 289[41]
Ganem, B. 410[279]
Gangadhar, A. 71[65]
Gansäuer, A. 383[69], 387[115], 387[118], 387[119], 387[121], 387[127], 387[128], 387[129], 387[130], 387[135], 387[136], 389[135], 389[151], 389[152], 389[156], 390[135], 390[160], 392[182], 393[182], 416[135], 416[285], 416[286], 417[135], 417[285], 417[286], 467[135], 468[135], 480[346], 481[346], 482[346], 517[346], 517[412], 517[413], 517[414], 517[415], 517[416], 517[417], 517[418], 517[419], 518[414], 532[415], 532[416], 532[417], 532[418]
Gao, B. 255[74], 256[74]
Gao, G.-Y. 35[78], 36[78], 36[79], 37[79]
Gao, H. 31[73], 32[73]
Gao, S. 9[29], 10[29], 11[29], 42[29], 45[29], 384[82]
Gao, X. 634[177]
García, B. 525[448]
Garcia, J. 518[420], 519[420]
García, J. I. 49[12], 50[12]
García, J. M. 215[20]
Garcia, P. 349[11]
García-Fortanet, J. 550[25]
García-Granda, S. 394[194]
Garcia Ruano, J. L. 142[140], 143[140]
García-Ruiz, J. M. 387[121]
Garzya, V. 400[251]
Gasanz, Y. 518[420], 519[420]
Gazizova, V. 88[94], 89[94]
Ge, P.-H. 668[242], 669[242]
Geib, S. J. 125[91], 126[91], 127[91]
Geich-Gimbel, D. 480[346], 481[346], 482[346], 517[346]
Gelalcha, F. G. 11[138], 11[139]

Geldenhuys, W. J. 200[52]
Geller, L. E. 383[1]
Gellrich, U. 473[340], 474[340], 475[340], 476[340]
Genêt, J.-P. 121[83], 121[85], 122[83], 122[85], 123[83], 349[8]
Gennaro, A. 423[303], 424[303], 425[303]
Geny, A. 283[41], 284[41], 289[41]
George, J. H. 384[75], 384[76], 498[76], 499[76], 510[399]
Georgiadou, E. 525[448]
Gercek, Z. 498[374]
Gerenkamp, M. 416[286], 417[286]
Gerner, H. 521[433]
Gessner, V. H. 579[85], 580[85]
Gevorgyan, V. 56[38], 395[216]
Ghadiri, M. R. 329[31]
Ghalit, N. 652[211], 653[211]
Ghelfi, F. 423[303], 424[303], 425[303], 510[401]
Ghosez, L. 50[14]
Ghosh, A. K. 259[82]
Ghosh, S. K. 138[125]
Gibby, J. E. 300[61], 301[61]
Gibe, R. 527[457]
Gibson, S. E. 108[39], 109[39], 115[65], 116[65], 146[146], 147[146]
Giese, B. 383[5], 383[36], 383[38], 528[460], 529[460], 531[36]
Gimbert, Y. 87[84], 100[8], 103[21], 106[33], 112[57], 113[57], 113[59], 114[8], 114[59]
Gin, D. Y. 408[271], 409[271]
Giordano, L. 87[84]
Girard, M. 498[368], 498[369]
Girard, P. 392[169], 392[170], 392[171], 393[169], 393[170], 393[171], 394[170], 460[170], 526[170]
Gitua, J. N. 89[98], 90[98]
Gladysz, J. A. 654[213], 654[214]
Gleason, J. D. 602[123]
Gleiter, R. 114[60]
Glen, R. 168[16], 170[16], 171[16], 174[16], 175[22], 176[22], 182[22]
Gloer, J. B. 524[441], 524[442]
Glorius, F. 360[44], 361[44], 367[44]
Goddard, R. 76[73], 77[73], 78[73], 92[73], 93[73], 330[39], 372[76], 550[24], 659[227], 661[231], 662[231]
Godineau, E. 334[56], 393[185]
Goess, B. C. 632[172]
Goettmann, F. 137[122]
Gómez, P. M. 589[104], 590[104]
Gomi, S. 530[466], 530[467]
Gong, G. 259[82]
Gong, L.-Z. 192[41]
Gong, W. 338[58]
González, A. 215[20]
González, A. Z. 330[38]
González, F. S. 523[439], 524[439]
González, M. A. 498[376]
González, R. R. 523[439], 524[439]
Gonzalez-Sierra, M. 498[363]
Goodell, J. R. 167[10]

Goodman, J. M. 322[13]
Gopakumar, G. 76[73], 77[73], 78[73], 92[73], 93[73]
Gopalaiah, K. 392[175], 393[175], 531[175]
Gorin, D. J. 78[74], 79[74], 80[74], 82[79], 83[79]
Gosselin, F. 373[79], 374[79]
Goswami, A. 283[43], 284[43], 290[43]
Goswami, R. J. 389[148]
Gotlib, L. 41[84]
Goto, K. 498[370]
Göttlich, R. 400[244], 425[244], 426[244]
Goya, T. 268[23], 269[23], 279[23]
Gradl, S. N. 568[59]
Graf, D. D. 564[52], 565[52], 566[52]
Graham, P. I. 41[84]
Graham, T. J. A. 632[172]
Gramain, A. 498[380]
Grant, T. N. 328[28]
Gray, E. E. 632[172]
Greaney, M. F. 576[83], 577[83]
Greb, A. 517[414], 518[414]
Grebe, T. P. 619[151]
Green, M. L. H. 387[134]
Greene, A. E. 100[8], 103[21], 106[33], 112[57], 113[57], 113[59], 114[8], 114[59], 139[129], 139[130], 139[131], 139[132], 139[133], 139[135], 140[129], 140[130], 140[131], 140[132], 140[133], 140[135], 141[132], 141[133], 141[135]
Greene, B. 576[80], 577[80]
Grela, K. 657[218], 658[219]
Grillet, F. 139[128]
Grima, P. M. 322[15], 322[16], 322[17]
Grimme, S. 387[119], 416[286], 417[286]
Grimwood, M. E. 626[164]
Grise-Bard, C. M. 638[185]
Groaz, E. 645[196]
Gronenberg, L. S. 193[46], 193[47], 194[46]
Gross, Z. 11[36]
Grossman, R. B. 510[396]
Groth, U. 104[26]
Groves, J. T. 11[31], 11[32], 14[42], 383[10]
Grubbs, R. H. 544[3], 544[4], 545[3], 545[4], 545[14], 551[30], 553[33], 553[34], 554[33], 554[34], 564[53], 564[55], 565[53], 565[55], 566[55], 572[74], 573[74], 602[122], 614[139], 618[14], 618[150], 619[14], 619[152], 620[14], 620[152]
Gu, S. 384[87], 498[87]
Guan, Z.-H. 395[218], 395[223], 396[223], 397[223]
Guanti, G. 645[195]
Guibé, F. 470[328], 471[328]
Guin, C. 390[163], 391[168], 433[168], 434[168], 512[409], 513[409], 514[163], 515[163], 516[163], 517[163]
Gulevich, A. V. 395[216]
Gulías, M. 330[35], 331[35]
Gummersheimer, T. S. 394[193]
Gung, B. W. 330[40], 330[43]

Guo, X. 188[39], 190[39], 191[39], 192[39]
Guptill, D. M. 59[42], 60[42]
Gustafsson, M. 400[237]
Guth, O. 546[17], 655[17], 656[17]
Guthikonda, K. 28[69], 29[69]
Gutnov, A. 289[50], 291[50]
Güttinger, S. 259[85], 260[85]
Guyot, M. 521[430], 521[431]
Guzei, I. A. 356[33], 368[60], 369[63]
Guzmán, P. E. 332[46], 333[46]
Gybin, A. S. 149[152]

H

Häberli, A. 400[238]
Hachiya, I. 222[30], 223[30]
Hacksell, U. 71[66]
Haidour, A. 387[125]
Haitko, D. A. 545[16], 546[16]
Halasz, I. 395[213]
Hales, N. J. 115[65], 116[65]
Hall, A. 569[63]
Hallside, M. S. 400[251], 447[320], 448[320], 449[320]
Hamada, M. 530[466]
Hamaker, C. G. 66[59], 167[10]
Hamann, B. 394[191], 394[209]
Hamelin, O. 556[40]
Hammond, G. S. 50[13]
Hamura, S. 50[19], 51[19]
Hamza-Reguig, S. 167[6]
Han, J.-c. 610[136]
Han, X. 188[39], 190[39], 191[39], 192[39], 338[57], 339[57]
Hanaoka, M. 125[90], 126[90]
Hanari, T. 200[55], 201[55], 202[55]
Handa, Y. 394[200]
Hanessian, S. 569[63], 569[64]
Hanna, I. 636[180]
Hannan, R. L. 529[464]
Hannen, P. 589[104], 590[104]
Hanochi, M. 390[159]
Hansen, A. S. 552[32], 553[32]
Hansen, E. C. 560[45], 560[46], 561[45], 561[46], 597[115], 598[115], 612[137], 613[137], 626[162], 626[163], 627[162], 627[163], 628[163]
Hansen, H. C. 626[164]
Hansen, M. H. 52[25]
Hapke, M. 265[5], 283[42], 284[42], 290[42]
Hara, A. 601[120]
Hara, H. 287[48]
Hara, J. 304[65], 305[65], 308[65]
Hara, T. 530[466]
Hara, Y. 167[11]
Harada, E. 50[19], 51[19]
Harada, S. 223[32], 224[32]
Harden, J. D. 35[78], 36[78], 36[79], 37[79]
Hardick, D. J. 146[146], 147[146]
Hardouin, C. 389[149]
Harman, W. H. 395[215]

Harmata, M. 319[2], 319[3], 319[4], 319[5], 320[2], 320[3], 321[2], 321[3], 329[32], 329[33], 330[32], 330[33], 333[2], 339[61], 339[62], 340[62], 341[65], 341[66], 341[67], 342[32], 342[69], 342[70], 342[71], 343[70], 343[71], 343[72], 345[82], 345[83], 591[107]
Harn, N. K. 54[30]
Harned, A. M. 170[19], 171[19], 172[19], 175[19]
Harrak, Y. 88[85], 93[85]
Harris, C. R. 383[65], 392[65], 393[65], 393[186], 480[65]
Harrison, S. J. 576[83], 577[83]
Harrity, J. P. A. 602[123]
Harsh, P. 521[428]
Hart, D. J. 383[20]
Hartmann, J. M. 349[14]
Hartung, I. V. 370[67]
Haruo, Y. 511[406], 521[406], 522[406]
Harwood, L. M. 188[38], 190[38], 191[38], 192[38]
Hasaba, S. 156[170], 157[170]
Hasegawa, E. 473[334]
Hasegawa, H. 400[245]
Hasegawa, T. 511[406], 521[406], 522[406]
Hashiguchi, S. 544[2]
Hashimoto, M. 521[425], 602[122]
Hashimoto, S. 168[14], 175[24], 175[25], 175[26], 175[27], 176[24], 176[25], 176[26], 176[27], 177[25], 177[27], 178[24], 178[25], 179[26], 180[26], 180[27], 182[24], 183[24], 183[25], 183[26], 183[27], 184[14], 187[14], 188[14], 200[55], 201[55], 202[55], 241[59], 242[59], 253[72]
Hashimoto, T. 228[38]
Hashimoto, Y. 115[66], 116[66], 172[21], 173[21], 174[21], 175[21], 175[29], 181[29], 182[29], 184[29]
Hashmi, A. S. K. 355[30]
Hatakeyama, S. 405[257]
Hattori, K. 271[25], 271[26], 272[25], 272[26], 279[25]
Haufe, G. 400[232]
Haumann, T. 88[92], 89[92]
Haustedt, L. O. 360[46]
Hayakawa, I. 576[82], 577[82], 578[82]
Hayakawa, Y. 323[19]
Hayama, N. 593[109], 594[109]
Hayashi, K. 500[384]
Hayashi, M. 115[66], 116[66]
Hayashi, N. 607[133], 608[133]
Hayashi, S. 81[78]
Hayashi, T. 360[47]
Hayashi, Y. 207[10], 208[10], 384[80], 505[392], 506[392], 508[80], 509[80], 598[116], 599[116]
Hayashida, J. 571[71], 572[71]
Haycock, P. R. 146[146], 147[146]
Hazell, R. G. 236[50], 237[50], 237[52], 238[52]
He, C. 39[82], 40[82], 400[240]

He, L. 550[26]
He, X. 400[233]
Heaney, H. 416[290], 417[290]
Heck, M.-P. 556[38], 556[39]
Hedberg, C. 241[58]
Hehner, S. P. 525[446]
Heiba, E. I. 385[97], 385[99], 386[99], 387[99]
Heilmann, E. K. 659[227]
Heinrich, M. 525[446]
Helaja, J. 100[9]
Helion, F. 394[199]
Heller, B. 289[50], 291[50]
Helmchen, G. 159[178], 160[178]
Helps, I. M. 103[22], 104[22], 109[42], 110[42]
Henderson, J. A. 606[131], 607[131]
Hennessy, E. J. 63[52]
Hentschel, J. 517[416], 517[418], 532[416], 532[418]
Heppekausen, J. 661[231], 662[231]
Herbert, M. B. 553[33], 553[34], 554[33], 554[34]
Herbert, R. B. 200[56]
Herdtweck, E. 668[242], 669[242]
Hérisson, J.-L. 651[203]
Hernández-Galán, R. 345[80]
Herout, V. 81[78]
Herrador, M. M. 384[86], 387[120], 387[126]
Herrmann, W. A. 391[164], 668[242], 669[242]
Hershberger, S. S. 386[102]
Hertz, T. 394[205]
Herzog-Krimbacher, B. 52[25]
Hess, O. 410[277]
Hess, W. 265[9], 401[252], 404[252], 405[252], 427[252]
Hessel, L. W. 385[96]
Heuger, G. 400[244], 425[244], 426[244]
Hickmann, V. 659[224], 660[224]
Hicks, F. A. 130[103], 130[104], 131[103], 131[104]
Hicks, R. G. 392[183], 393[183]
Hiersemann, M. 593[110], 594[110]
Higuchi, S. 175[28], 180[28], 181[28], 184[28]
Higuchi, T. 14[43], 14[44]
Hikichi, S. 18[54]
Hill, R. A. 525[444]
Hilmersson, G. 392[174], 393[174], 394[174]
Hilt, G. 265[9]
Hinman, M. M. 24[62], 50[18], 51[18], 52[18]
Hintzer, K. 16[49]
Hirama, M. 384[80], 505[392], 506[392], 508[80], 509[80]
Hirano, M. 266[20], 267[20], 268[20], 272[27], 273[27], 276[33], 278[20], 280[27], 280[33], 282[37], 287[48], 291[51], 311[71], 312[71]
Hirano, T. 18[55], 19[55]
Hirao, T. 394[210], 434[313], 435[313]
Hirase, K. 430[305]

Hirashima, H. 275[31], 280[31]
Hirobe, M. 14[43], 14[44]
Hirosawa, C. 103[23], 104[23], 117[74], 118[74], 119[74], 155[164]
Hirose, T. 159[180], 160[180]
Hisler, K. 410[282]
Hjøllund, G. H. 387[137], 388[137], 389[137]
Hjorth, S. 71[66]
Ho, S. 350[19], 351[19]
Hoard, D. W. 52[25]
Hoberg, H. 292[53]
Hodgson, D. M. 168[16], 168[17], 170[16], 170[17], 171[16], 171[17], 174[16], 175[22], 175[23], 176[22], 176[23], 176[30], 182[22]
Hoerner, S. 159[176]
Hoffmann, H. M. R. 321[10], 322[10], 322[12], 322[13], 322[14], 341[68]
Höfs, R. 507[394]
Hog, D. T. 583[94]
Hollander, F. J. 216[24], 217[24]
Holloway, M. K. 41[84]
Holmes, A. B. 585[97]
Holmstrom, A. 241[58]
Holub, N. 603[126], 604[126]
Honda, T. 400[241], 635[179]
Hong, P. 292[52]
Hong, S. 257[77], 258[77]
Hong, S. D. 529[462]
Hong, W. 195[48], 196[48], 199[48]
Hong, X. 370[69], 370[70]
Hong, Y. K. 345[81]
Horne, D. B. 370[74], 373[82], 375[82], 569[65], 569[66]
Horneff, T. 56[38]
Horváth, I. T. 391[164]
Horwitz, S. B. 550[26]
Hoshi, M. 573[75]
Hoshi, T. 473[334]
Hoshino, Y. 4[19], 4[20], 5[20], 6[20], 42[86], 45[86]
Hosokawa, S. 527[456]
Hosomi, A. 325[23]
Hosoya, N. 7[25]
Hossain, M. M. 66[58]
Hou, X.-L. 73[68], 74[68]
Houk, K. N. 334[53], 334[54], 334[55], 335[54], 335[55], 336[54], 336[55], 337[54], 337[55], 370[64], 370[65], 370[66], 370[67], 370[68], 370[69], 370[70], 370[71], 383[25]
Houser, J. H. 332[45], 333[45]
Hoveyda, A. H. 249[66], 551[31], 552[31], 564[51], 564[52], 564[54], 564[56], 564[57], 565[52], 565[54], 565[56], 565[57], 566[52], 566[54], 590[106], 591[106], 602[123], 616[146], 623[146], 623[156], 624[146], 660[31], 661[31]
Hoye, T. R. 545[7], 545[8], 545[9], 560[44], 561[44], 614[140], 622[44], 623[44]
Hsung, R. P. 138[125], 334[51], 334[52], 334[53], 334[54], 334[55], 335[54], 335[55], 336[54], 336[55], 337[54], 337[55], 619[151]

Hu, C.-W. 395[220]
Hu, J. 230[41], 238[53], 239[53], 240[53]
Hu, Q.-Y. 257[76]
Hu, W. 168[12], 172[12], 184[12], 188[39], 190[39], 191[39], 192[39], 193[45], 193[46], 193[47], 194[46], 199[45], 241[57], 241[58]
Hu, W.-H. 172[20], 173[20], 175[20], 184[32], 185[32], 187[32]
Hu, X. 21[59], 22[59], 23[59]
Hu, Y. 373[82], 375[82]
Huang, F. 377[88], 378[88], 379[88]
Huang, H. 326[26]
Huang, J. 159[183], 160[183], 161[183], 619[151]
Huang, J.-S. 63[51], 192[44], 194[44]
Huang, J.-W. 386[103]
Huang, R. 246[64], 247[64]
Huang, S. 368[60], 368[61], 369[61], 430[306]
Huang, W. G. 529[461], 530[461]
Huang, Y. 219[27], 220[27], 220[28], 221[28]
Huang, Y.-S. 383[58]
Hubbard, R. D. 376[87], 377[87], 378[87]
Huber, F. M. E. 583[94]
Huber, R. 531[475]
Hubert, A. J. 54[31], 55[31], 55[37], 56[37]
Huerta, M. 394[189], 394[194]
Huff, J. R. 41[84]
Hughes, C. C. 568[59]
Hughes, D. L. 404[255], 404[256]
Hughes, R. P. 338[57], 339[57]
Hugl, H. 14[47], 15[47], 16[47]
Huisgen, R. 50[14], 167[5]
Hulcoop, D. G. 442[318], 443[318], 444[318], 444[319], 445[319], 446[319], 447[319]
Hung, N. V. 498[380]
Husfeld, C. O. 359[42], 360[43], 360[44], 360[45], 361[43], 361[44], 367[43], 367[44], 373[77]
Husinec, S. 192[42]
Huth, I. 517[413], 517[414], 518[414]
Huther, N. 510[401]
Huwe, C. M. 562[48], 562[49], 563[48], 563[49]
Hwang, S. H. 109[43], 110[43], 115[64], 116[64]
Hyeon, T. 135[113], 135[114], 136[113], 136[114]

I

Iannazzo, L. 269[24], 270[24], 279[24], 283[41], 284[41], 289[41]
Ichikawa, S. 527[458]
Igdir, A. C. 498[374]
Ihara, M. 200[57]
Ihee, H. 615[143], 615[144], 615[145]
Iida, A. 522[436]

IJsselstijn, M. 652[210], 653[210]
Ikeda, M. 522[435]
Ilyashenko, G. 1[5]
Imagawa, H. 395[221]
Imai, H. 598[116], 599[116]
Imamoto, T. 393[188]
Imwinkelried, R. 212[16], 213[16]
In, K. Y. 120[78]
Inaba, T. 501[388]
Inagaki, F. 110[47], 156[47], 158[175], 159[175], 159[181], 160[181], 363[53], 365[57]
Inanaga, J. 394[200], 394[202], 477[342], 477[344]
Inglesby, P. A. 349[10]
Ingold, K. U. 383[18]
Ini, S. 11[36]
Inoue, K. 400[247]
Inoue, M. 343[75], 574[76]
Inoue, Y. 242[60], 243[60]
Inouye, S. 530[467]
Irie, R. 7[23], 7[25]
Irvine, S. 103[25], 104[25]
Isaac-Márquez, A. P. 345[85]
Isaac-Márquez, R. 345[85]
Isaacs, A. K. 554[35], 554[36], 555[35], 555[36]
Ischay, M. A. 406[258], 435[314], 436[314], 437[314], 438[314]
Ishibashi, M. 503[390]
Ishibashi, Y. 339[59]
Ishida, M. 304[65], 305[65], 308[65]
Ishigami, K. 346[86]
Ishihara, J. 405[257]
Ishihara, K. 208[11], 209[11], 210[11]
Ishihara, N. 11[37]
Ishii, J. 271[26], 272[26]
Ishii, M. 299[60], 300[60], 301[60]
Ishii, Y. 384[92], 384[93], 384[94], 394[190], 430[305]
Ishikawa, H. 167[7], 598[116], 599[116]
Ishikawa, M. 394[202]
Ishikawa, Y. 601[120]
Ishimoto, D. 175[28], 180[28], 181[28], 184[28]
Ishitani, H. 222[30], 223[30], 232[45], 232[46], 250[67], 250[68], 250[69]
Isse, A. A. 423[303], 424[303], 425[303]
Itami, K. 132[108], 133[108], 353[22]
Ito, R. 550[29]
Ito, S. 527[456]
Ito, Y. 7[23], 32[75], 33[75], 353[22]
Itoh, K. 63[50], 172[21], 173[21], 174[21], 175[21], 175[29], 181[29], 182[29], 184[29], 265[19], 285[45], 285[46], 286[46], 290[45], 293[45], 298[45], 473[330]
Itoh, T. 225[34], 226[34]
Itoh, Y. 63[50]
Iwahama, T. 384[92], 430[305]
Iwama, T. 219[27], 220[27], 220[28], 221[28]
Iwamura, H. 400[247]

Iwamura, M. 503[390]
Iwanaga, K. 206[6]
Iwasa, S. 64[53]
Iwasaki, Y. 601[120]
Iwasawa, N. 167[11], 351[20], 352[20]
Iwata, T. 158[175], 159[175]
Iwaya, K. 473[334]

J

Jacobs, P. A. 18[53]
Jacobsen, E. N. 7[24], 24[63], 233[47], 234[47], 234[48], 234[49], 235[48], 235[49], 259[80], 259[81]
Jacobsen, N. 193[47]
Jahn, U. 383[34]
Jakoby, V. 383[69]
Jakubec, P. 551[31], 552[31], 660[31], 660[228], 661[31], 661[228]
Jamieson, A. T. 500[383]
Jamieson, G. R. 500[383]
Jamieson, J. Y. 564[54], 565[54], 566[54]
Jamison, T. F. 150[153], 234[48], 235[48], 259[80]
Jana, S. 391[167], 391[168], 433[168], 434[168], 458[323], 459[323], 511[408], 512[408]
Jankowski, P. 88[96], 89[96]
Janousek, Z. 222[31]
Jarvo, E. R. 259[81]
Jasperse, C. P. 215[22], 383[24]
Jat, J. L. 31[73], 32[73]
Jean, L. 658[222]
Jeffrey, C. S. 560[44], 561[44], 622[44], 623[44]
Jeffreys, M. S. 363[53]
Jeon, Y. T. 576[83], 577[83]
Jeong, N. 102[16], 103[16], 109[43], 110[43], 115[64], 116[64], 120[78], 120[81], 121[81], 121[82], 121[83], 121[85], 122[81], 122[82], 122[83], 122[85], 123[82], 123[83], 130[102], 145[144]
Jeong, W. B. 579[84]
Jeulin, S. 121[85], 122[85]
Ji, B. 230[41], 231[44]
Ji, Y. 107[37]
Jia, P. 420[300], 421[300]
Jia, Z.-J. 192[40]
Jiang, G.-J. 357[34], 358[34], 364[56]
Jiang, Q. 350[18]
Jiang, X. 379[89], 380[89]
Jiang, Y. 228[37], 229[37]
Jiang, Y. Y. 529[461], 530[461]
Jiao, L. 364[55], 364[56], 365[55], 367[55], 373[85], 377[88], 378[88], 379[85], 379[88]
Jiao, P. 24[64], 27[64]
Jiao, Y. 338[58]
Jimenez, M. 552[32], 553[32]
Jiménez, T. 387[123], 387[131]
Jiménez-Núñez, E. 75[71], 81[75], 82[75]

Jimeno, C. 104[27]
Jin, L.-M. 38[81], 39[81]
Jiu, G. 651[206]
Johannsen, M. 236[50], 237[50], 237[51], 238[51]
Johansson, A. M. 71[66]
Johansson, M. J. 78[74], 79[74], 80[74]
Johnson, J. S. 215[19], 216[19]
Johnson, L. J. 383[18]
Johnson, R. K. 343[73]
Johnston, D. 383[67], 471[329], 472[329], 473[329]
Johnstone, C. 176[30]
Johnstone, S. 569[63]
Jones, D. E. 341[65], 341[66], 343[72], 345[83]
Jones, J. E. 36[79], 37[79]
Jones, P. G. 654[215], 655[215]
Jordan, B. M. 416[290], 417[290]
Jørgensen, K. A. 236[50], 237[50], 237[51], 237[52], 238[51], 238[52]
Jørgensen, K. B. 389[146]
Jørgensen, L. 159[182], 160[182]
Josephsohn, N. S. 249[66]
Josien, H. 416[287], 417[287]
Jourdan, F. 410[276]
Jsselstijn, M. I. 652[209]
Ju, X. 420[300], 421[300]
Ju, Y. 431[307], 433[307]
Julia, M. 383[11], 383[12]
Juliano-Capucao, C. A. 582[91]
Jung, H. 452[322], 453[322], 454[322]
Jung, M. 104[26]
Jurčík, V. 251[70], 260[70], 261[70]
Jurewicz, A. J. 343[73]
Justicia, J. 383[69], 383[72], 387[115], 387[116], 387[119], 387[121], 387[124], 387[125], 391[124], 511[405], 519[423], 520[423]

K

Kablaoui, N. M. 130[103], 131[103]
Kadereit, D. 359[42]
Kadowaki, S. 268[23], 269[23], 279[23]
Kagan, H. B. 16[49], 205[5], 392[169], 392[170], 392[171], 392[175], 393[169], 393[170], 393[171], 393[175], 394[170], 394[191], 394[192], 394[195], 394[209], 460[170], 473[337], 480[347], 526[170], 526[452], 531[175]
Kagayama, T. 384[93]
Kahraman, M. 341[66], 342[71], 343[71], 345[82]
Kaiser, J. 652[209]
Kaiser, P. 14[46]
Kajimoto, Y. 501[388]
Kakehi, A. 172[21], 173[21], 174[21], 175[21], 175[28], 175[29], 180[28], 181[28], 181[29], 182[29], 184[28], 184[29], 188[35], 189[35], 191[35]
Kakeya, H. 598[116], 599[116]

Kakiuchi, K. 127[92], 127[95], 127[96], 128[92], 128[95], 128[96], 129[95], 129[96]
Kaliappan, K. P. 521[426], 624[157], 638[186], 639[186]
Kalidindi, S. 579[84]
Kalindjian, S. B. 108[39], 109[39]
Kalsow, S. 400[244], 425[244], 426[244]
Kamata, K. 18[54], 18[55], 19[55]
Kambe, N. 526[454]
Kamijo, N. 81[78]
Kamisawa, A. 268[22], 269[22], 279[22], 282[37]
Kamitani, A. 156[168], 353[23], 354[23]
Kamoshita, M. 511[404]
Kan, T. 527[456]
Kanagawa, Y. 394[190]
Kanai, K. 350[18]
Kanai, M. 571[69]
Kanchiku, S. 62[46], 62[47], 63[47]
Kandiyal, P. S. 510[400]
Kaneda, K. 400[241], 635[179]
Kaneko, H. 550[28]
Kaneko, O. 573[75]
Kanemasa, S. 50[19], 51[19]
Kang, M.-C. 386[104], 387[104]
Kano, K. 500[384]
Kant, R. 524[440], 525[440]
Kao, C.-B. 429[304], 430[304], 431[304]
Karaaslan, E. O. 387[113]
Karasudani, A. 395[221]
Kardos, N. 139[132], 140[132], 141[132]
Kaschel, J. 49[3], 52[20], 52[21], 52[23], 52[24], 53[23], 53[28], 55[23], 55[24]
Kase, K. 283[43], 284[43], 290[43]
Kates, S. A. 386[105], 387[105], 431[308], 432[308]
Kato, C. 550[28]
Kato, N. 224[33]
Kato, T. 242[60], 243[60]
Katoono, R. 600[119], 601[119]
Katsuki, T. 1[11], 1[12], 1[13], 2[15], 3[15], 3[16], 7[23], 7[25], 7[26], 8[26], 11[26], 15[48], 16[48], 20[57], 20[58], 21[58], 32[74], 32[75], 33[75], 33[76], 34[76], 62[46], 62[47], 63[47]
Katsumata, T. 400[249]
Kattnig, E. 372[76]
Katz, T. J. 545[5], 545[6], 651[204]
Kaufmann, K. A. C. 146[146], 147[146]
Kausch-Busies, N. 666[238], 667[238]
Kawabata, H. 32[74]
Kawaguchi, H. 530[465]
Kawahara, F. 383[31]
Kawai, H. 600[119], 601[119]
Kawamura, M. 600[119], 601[119]
Kawano, K. 405[257]
Kawase, T. 530[471], 531[471]
Kazanis, S. 498[367]
Keane, H. A. 401[252], 404[252], 405[252], 427[252]
Keating, G. M. 518[422]
Keck, G. E. 228[36]
Kędzia, J. L. 101[13]
Keese, R. 153[160]

Keitz, B. K. 553[33], 553[34], 554[33], 554[34]
Keller, A. I. 416[292], 417[292]
Keller, F. 416[286], 417[286]
Kemmitt, P. D. 444[319], 445[319], 446[319], 447[319]
Kende, A. S. 326[26]
Kennedy, R. A. 339[60]
Kennedy-Smith, J. J. 568[59]
Kenny, C. 393[187]
Kent, J. L. 111[50]
Kerekes, A. D. 125[91], 126[91], 127[91]
Kerr, M. A. 384[83], 506[393], 507[393]
Kerr, W. J. 101[13], 102[17], 103[17], 103[25], 104[25], 105[17], 109[45], 110[45]
Keßberg, A. 579[86], 580[86]
Khand, I. U. 99[5]
Kharasch, M. S. 383[30], 383[31]
Kienle, M. 188[36], 189[36], 190[36], 191[36]
Kigoshi, H. 576[81], 576[82], 577[81], 577[82], 578[81], 578[82]
Kii, S. 228[38]
Kikuchi, Y. 600[119], 601[119]
Killmer, L. 343[73]
Kim, B. W. 345[81]
Kim, C. 33[76], 34[76]
Kim, D. E. 121[82], 121[83], 121[85], 122[82], 122[83], 122[85], 123[82], 123[83], 145[144]
Kim, D. H. 130[102], 152[157]
Kim, H. 326[25], 327[25]
Kim, I. S. 121[83], 121[85], 122[83], 122[85], 123[83], 145[144]
Kim, J. 615[145]
Kim, J. S. 120[78]
Kim, K. H. 615[143], 615[144], 615[145]
Kim, M. 629[165], 630[166], 646[198], 647[198]
Kim, S. 498[379]
Kim, S.-H. 545[14], 618[14], 619[14], 620[14]
Kim, S.-W. 135[113], 135[114], 136[113], 136[114]
Kim, Y. K. 498[379]
Kimura, J. 287[49], 288[49], 290[49]
Kinebuchi, M. 159[181], 160[181]
King, G. R. 91[99]
Kinghorn, A. D. 345[85], 400[229]
Kinoshita, A. 545[13], 614[13], 631[169], 631[170], 632[169], 632[170]
Kinpara, K. 285[45], 285[46], 286[46], 290[45], 293[45], 298[45]
Kireev, S. L. 102[14]
Kirk, G. G. 102[17], 103[17], 105[17]
Kirschbaum, K. 330[43]
Kise, N. 473[331], 473[335]
Kisel, M. A. 88[89], 89[89]
Kiselgof, J. Y. 384[85]
Kishida, A. 449[321], 450[321], 451[321]
Kita, Y. 633[175]
Kitagaki, S. 110[47], 111[53], 156[47], 241[59], 242[59]

Kitajima, M. 159[186], 160[186], 161[186]
Kitamura, A. 522[435]
Kitamura, K. 344[77]
Kitamura, M. 400[242], 530[470], 530[471], 531[471]
Kitamura, T. 633[175], 642[190], 642[192], 643[192], 646[192]
Kizirian, J.-C. 107[35]
Klager, K. 265[14]
Klankermayer, J. 135[117], 136[117], 137[117]
Klawonn, M. 14[47], 15[47], 16[47]
Klawonn, T. 517[416], 517[418], 532[416], 532[418]
Klute, W. 349[1]
Knebel, K. 517[414], 518[414]
Knox, G. R. 99[5]
Ko, E. 589[104], 590[104]
Ko, S.-B. 416[287], 417[287]
Kobayashi, H. 530[467]
Kobayashi, J. 287[49], 288[49], 290[49], 503[390], 601[121]
Kobayashi, M. 304[65], 305[65], 308[65]
Kobayashi, S. 222[30], 223[30], 232[45], 232[46], 250[67], 250[68], 250[69], 251[70], 260[70], 261[70], 511[407], 522[434], 523[434], 549[23], 550[23], 586[101], 587[101]
Kobayashi, T. 120[77]
Kochi, J. K. 383[32], 383[57], 387[109], 531[476]
Kodadek, T. 66[59]
Kodama, M. 344[77]
Koehl, W. J., Jr. 385[97]
Koenigs, R. M. 406[263], 406[266]
Koga, N. 400[247]
Koga, Y. 120[77]
Kogure, N. 159[186], 160[186], 161[186]
Koide, K. 580[87]
Kollárovič, A. 281[36], 282[36]
Komine, Y. 268[22], 269[22], 279[22]
Komiyama, S. 250[67], 250[68]
Kondo, K. 287[49], 288[49], 290[49]
Kondo, S. 1[13], 530[466], 530[467]
Kondo, T. 132[107], 157[174], 158[174]
Kondracki, M. L. 521[430], 521[431]
Kong, K. 259[79]
Kong, L. 384[82]
Kong, N. 176[31]
Konishi, M. 530[465]
Könning, D. 579[85], 580[85]
Konya, D. 100[8], 114[8]
Koo, S. 431[307], 433[307], 452[322], 453[322], 454[322]
Kopecky, K. R. 50[13]
Kopka, K. 400[232]
Koradin, C. 370[74], 373[86], 375[86], 376[86]
Kordian, M. 52[25]
Korica, N. 332[47]
Korkowski, P. F. 545[7]
Korolev, A. 205[3]
Korth, H. G. 400[248]

Kosari, M. 440[316]
Koseki, N. 600[119], 601[119]
Koshimizu, M. 574[76]
Kosugi, Y. 4[20], 5[20], 6[20]
Kotha, S. 265[13]
Koya, S. 15[48], 16[48]
Kozhevko, A. N. 88[89], 89[89]
Kozhushkov, S. I. 49[8], 49[9], 88[87], 88[91], 88[92], 89[87], 89[91], 89[92], 106[32], 361[48], 362[48]
Kozmin, S. A. 85[81], 614[141], 634[176]
Krabbe, S. W. 397[224], 399[224], 411[224], 465[327], 466[327], 467[327], 501[224]
Krafft, M. E. 101[12], 102[20], 103[20], 103[23], 104[23], 117[74], 118[74], 119[74], 119[76], 155[163], 155[164], 582[91], 582[92]
Krainz, T. 332[47]
Kratzert, D. 52[24], 53[28], 55[24]
Krause, H. 372[76]
Krauss, I. J. 568[61]
Krenske, E. H. 334[53], 334[54], 334[55], 335[54], 335[55], 336[54], 336[55], 337[54], 337[55]
Kress, S. 644[194], 645[194]
Krief, A. 383[66], 480[66]
Krische, M. J. 349[4], 599[118], 600[118], 634[177]
Krishna, U. M. 359[36]
Krishnamurthi, J. 167[1], 168[1], 175[26], 175[27], 176[26], 176[27], 179[26], 180[26], 180[27], 183[26], 183[27], 253[72]
Krishnamurthy, D. 228[36]
Krouse, S. A. 652[207]
Krulle, T. M. 499[381], 500[381]
Kruse, C. G. 395[212]
Kuan, K. K. W. 510[399]
Kubo, M. 344[77]
Kubo, T. 2[15], 3[15]
Kuchino, Y. 521[433]
Küchler, P. 247[65], 248[65]
Kucukislamoglu, M. 383[59]
Kuhlman, M. L. 394[208]
Kulinkovich, O. 88[94], 89[94]
Kulinkovich, O. G. 88[86], 88[88], 88[89], 88[90], 88[93], 88[95], 89[88], 89[89], 89[90], 89[93], 89[95], 90[88]
Kumar, K. 247[65], 248[65]
Kumar, P. 284[44], 285[44], 290[44]
Kündig, E. P. 221[29], 222[29]
Kunze, B. 658[221]
Kurahashi, T. 351[21], 352[21], 353[21], 357[21]
Kuramochi, A. 571[69]
Kurata, K. 593[109], 594[109]
Kurosaki, Y. 200[55], 201[55], 202[55]
Kürti, L. 31[73], 32[73]
Kusakabe, K. 250[68], 250[69]
Kusama, H. 167[11]
Kushi, Y. 81[78]
Küster, W. 115[62]
Kusuma, B. R. 592[108]
Kutay, U. 259[85], 260[85]

Kutsumura, N. 156[173]
Kuttruff, C. A. 159[182], 160[182]
Kuwata, Y. 301[62], 302[62], 308[62]
Kuzuya, S. 18[55], 19[55]
Kwak, J. 145[144]
Kwok, S. W. 57[39]
Kwon, H. J. 345[81]
Kwon, K. 615[145]
Kwon, T. 383[56]
Kwong, F. Y. 127[97], 128[97], 129[97]
Kwong, H. L. 127[97], 128[97], 129[97]
Kwong, H.-L. 63[51]
Kyle, A. F. 551[31], 552[31], 660[31], 660[228], 661[31], 661[228]
Kyz'mina, L. G. 149[152]

L

La, D. S. 564[51], 564[52], 564[54], 565[52], 565[54], 566[52], 566[54]
Labahn, T. 52[22]
Labande, A. H. 175[22], 175[23], 176[22], 176[23], 182[22]
Lacombe, F. 661[230], 662[230]
Lagunas, A. 104[27]
Lah, M. S. 452[322], 453[322], 454[322]
Lahiri, K. 265[13]
Lai, K. W. 302[63], 303[63], 305[63], 308[63]
Lai, T.-S. 63[51]
Laib, F. 105[31], 153[31]
Lake, D. H. 176[31]
Lakshminarayana, G. 71[65]
Lallemand, J. Y. 383[11]
Lam, H. C. 384[76], 498[76], 499[76], 510[399]
Lam, M.-H. 664[235]
Lam, W. H. 127[97], 128[97], 129[97]
Lan, Y. 135[112]
Landais, Y. 393[185]
Lang, K. 460[325], 461[325]
Langkopf, E. 360[43], 360[44], 360[45], 361[43], 361[44], 367[43], 367[44]
Laplace, D. R. 343[74], 344[74]
Larionov, O. 659[225]
Larrosa, I. 589[104], 590[104]
Larson, E. 224[33]
Lastécouères, D. 440[317], 441[317], 442[317]
Laurila, M. E. 52[25]
Lautens, M. 324[20], 349[1]
Lauterbach, T. 387[119], 389[151], 392[182], 393[182], 480[346], 481[346], 482[346], 517[346], 517[412], 517[413], 517[415], 517[416], 517[417], 517[419], 532[415], 532[416], 532[417]
Laval, A.-M. 383[66], 480[66]
Lawrence, B. M. 259[81]
Lawrie, D. J. 333[50]
Lay, F. 658[221]
Layton, M. E. 625[160], 625[161], 626[160], 626[161]
Lazari, D. 525[448]
Lebel, H. 25[66], 26[66], 28[66], 29[70], 30[72], 31[72], 32[72]
Lebl, T. 547[18], 548[18]

Leboeuf, D. 265[8]
Lectard, S. 25[66], 26[66], 28[66], 29[70]
Lectka, T. 215[19], 216[19]
Lee, B. Y. 102[16], 103[16]
Lee, C. 259[79]
Lee, C. O. 529[463]
Lee, C.-S. 383[40]
Lee, C. W. 551[30]
Lee, D. 559[43], 560[43], 560[45], 561[45], 571[72], 572[72], 587[102], 588[102], 597[115], 598[115], 608[134], 612[137], 613[137], 621[155], 622[155], 625[158], 626[162], 626[163], 627[162], 627[163], 628[163], 629[165], 630[166], 630[167], 635[178], 637[182], 646[198], 647[198], 648[199], 649[199]
Lee, D. S. 529[462]
Lee, E. 383[43], 498[379]
Lee, E. W. 345[81]
Lee, F.-W. 63[51]
Lee, H. W. 127[97], 128[97], 129[97]
Lee, H.-Y. 615[143], 615[144], 615[145]
Lee, J. 640[188]
Lee, K. 615[144]
Lee, M. C. 350[19], 351[19]
Lee, O. S. 615[145]
Lee, S. H. 102[16], 103[16]
Lee, S. I. 135[113], 136[113]
Lee, S. S. 135[114], 136[114]
Lee, T. W. 209[12], 210[12], 211[12]
Lee, V. J. 11[35]
Lee, W.-S. 63[51]
Lee, Y. 115[64], 116[64]
Lee, Y.-J. 616[146], 623[146], 624[146]
Lefort, D. 383[6]
Legault, C. Y. 370[65]
Lehmann, C. W. 372[76], 550[24]
Lehmann, V. 525[446]
Lehr, K. 663[234]
Lei, A. 383[74]
Leitner, T. 52[25]
Leitner, W. 135[117], 135[121], 136[117], 136[121], 137[117], 137[121], 391[164]
Le Maux, P. 1[6]
Lenger, S. R. 589[104], 590[104]
Leogane, O. 30[72], 31[72], 32[72]
Lercher, L. 57[39]
Lesage, D. 112[57], 113[57]
Leseurre, L. 663[234]
Lesser, A. 370[68]
LeTourneau, M. E. 52[25]
Leumann, C. J. 400[238]
Leutenegger, U. 50[17], 51[17]
Levin, R. B. 41[84]
Lex, J. 14[46]
Leyrer, U. 16[49]
Lezama-Dávila, C. M. 345[85]
Li, C. 134[111], 135[111]
Li, C.-c. 159[183], 160[183], 161[183], 610[136], 640[189]
Li, F. 610[136]
Li, F.-B. 383[55], 383[58], 383[62], 383[63], 383[64]
Li, G. 9[29], 10[29], 11[29], 42[29], 45[29]

Li, G.-Y. 195[48], 196[48], 196[49], 197[49], 199[48], 199[49]
Li, H. 338[57], 339[57]
Li, J. 9[29], 10[29], 11[29], 42[29], 45[29], 246[64], 247[64], 529[461], 530[461], 608[134], 635[178], 648[199], 649[199]
Li, J. Y. 529[461], 530[461]
Li, Q. 357[34], 358[34], 364[56], 529[461], 530[461]
Li, S. 121[84], 122[84]
Li, W. 21[59], 22[59], 23[59], 420[300], 421[300]
Li, X. 230[43], 238[53], 239[53], 240[53], 356[32], 356[33], 357[32], 368[61], 368[62], 369[61]
Li, X.-Y. 228[36]
Li, Y. 63[51], 167[7], 338[58], 401[253], 402[253], 403[253]
Li, Y.-M. 127[97], 128[97], 129[97]
Li, Z. 6[22], 7[22], 22[60], 23[60], 23[61], 24[61], 24[63], 54[34], 431[307], 433[307], 452[322], 453[322], 454[322], 589[105], 590[105]
Li, Z. Y. 349[2]
Lian, Y. 332[46], 333[46], 344[78]
Liang, X. 395[213]
Liang, Y. 377[88], 378[88], 379[88], 379[89], 380[89], 420[300], 421[300]
Liao, Y.-J. 418[293], 419[293], 420[293]
Liard, A. 408[274], 409[274]
Liebeskind, L. S. 354[24], 355[24]
Lim, H. J. 400[235]
Lim, J. 360[46]
Lim, J. W. 498[379]
Lin, J. 41[84]
Lin, L. 21[59], 22[59], 23[59], 228[40], 229[40], 255[74], 256[74]
Lin, Z. 651[206]
Lindberg, P. 71[66]
Lindsay, D. M. 102[17], 103[17], 105[17]
Lindsay, V. N. G. 53[27], 57[40], 58[40]
Liniger, M. 665[237], 666[237]
Linke, S. R. 589[104], 590[104]
Lippstreu, J. J. 616[148]
Lipton, M. A. 44[88], 46[88]
Liskamp, R. M. 652[211], 653[211]
Listemann, M. L. 651[205]
Little, D. 545[16], 546[16]
Little, R. D. 387[122]
Liu, A. L. 400[234]
Liu, G. 575[79], 576[79], 640[189]
Liu, H. 172[20], 173[20], 175[20], 184[32], 185[32], 187[32], 228[37], 229[37], 408[273], 409[273], 416[287], 417[287]
Liu, J. 135[112], 138[125]
Liu, J.-L. 228[39]
Liu, L. 134[111], 135[111], 255[75], 256[75]
Liu, M. 215[22]
Liu, P. 34[77], 35[77], 370[65], 370[66], 370[67], 370[68], 370[69], 370[71]
Liu, Q. 134[111], 135[111]
Liu, R.-S. 152[159], 153[159], 620[154]

Liu, S.-S. 395[220]
Liu, T.-X. 383[58], 383[62], 383[64]
Liu, W. 73[68], 74[68]
Liu, W.-M. 401[253], 402[253], 403[253]
Liu, X. 21[59], 22[59], 23[59], 226[35], 227[35], 228[40], 229[40], 254[73], 255[74], 256[74]
Liu, Y. 198[51], 199[51], 200[51], 254[73], 338[58]
Liu, Y.-L. 420[296]
Liu, Z.-H. 401[253], 402[253], 403[253]
Livinghouse, T. 111[49], 117[72], 117[73], 118[73]
Llaveria, J. 664[235]
Lledó, A. 106[34], 107[34], 107[37], 114[61], 144[141], 144[143], 145[141]
Lledós, A. 330[34], 330[37], 330[41], 330[42], 331[34], 331[37]
Lloyd-Jones, G. C. 614[142], 615[142]
Lo, W.-C. 63[51]
Lo, W. Y. 597[115], 598[115]
Locatelli, C. 111[48]
Loebach, J. L. 7[24]
Logan, A. W. J. 400[251], 447[320], 448[320], 449[320]
Loganzo, F. 664[235]
Loh, K.-L. 54[30]
Loh, T.-P. 206[8], 206[9], 213[18], 214[18]
Lohse, A. G. 334[51], 334[53], 334[54], 334[55], 335[54], 335[55], 336[54], 336[55], 337[54], 337[55]
Long, J. 230[41], 230[42], 231[42], 238[53], 239[53], 240[53]
Long, R. 159[183], 160[183], 161[183]
Longeon, A. 521[431]
López, A. 387[126]
López, F. 330[34], 330[35], 330[36], 330[37], 330[42], 331[34], 331[35], 331[36], 331[37]
López, S. 81[75], 82[75]
López-Mosquera, C. 144[141], 145[141]
Lopin, C. 400[250]
Lord, R. L. 125[89]
Loshadkin, D. V. 392[181], 393[181]
Louie, J. 265[12], 284[44], 285[44], 290[44], 292[55], 293[55], 298[55], 300[61], 301[61], 371[75], 372[75]
Love, J. A. 359[42], 360[43], 360[44], 360[45], 360[46], 361[43], 361[44], 367[43], 367[44], 373[77]
Lu, C.-D. 172[20], 173[20], 175[20], 184[32], 185[32], 187[32]
Lu, H. 38[81], 39[81], 60[44], 61[44]
Lu, J. 111[51], 384[77], 384[78], 384[79], 503[78], 504[78], 504[79]
Lu, L.-Q. 406[264], 406[265], 413[264]
Lu, W. 529[461], 530[461]
Lu, X. 521[427]
Lu, Z. 408[272], 409[272], 435[314], 436[314], 437[314], 438[314], 469[272], 470[272]
Lu, Z.-L. 130[101]
Lucas, C. R. 387[134]

Luisi, R. 1[10]
Luo, J.-W. 383[40]
Lusztyk, J. 383[18]
Luu, H.-T. 487[351], 488[351]
Luzung, M. R. 76[72]
Lv, L. 589[105], 590[105]
Lyakin, O. Y. 10[30]
Lynam, J. M. 100[10]
Lynch, V. 54[36], 55[36]
Lysenko, S. 654[215], 655[215]

M

Ma, L. 24[64], 27[64]
Ma, L.-F. 395[220]
Ma, S. 265[8], 557[41], 558[41], 558[42], 559[41]
Ma, Z. 384[77], 384[78], 384[79], 503[78], 504[78], 504[79]
Maas, G. 63[48]
Maaß, C. 52[23], 53[23], 55[23]
McCarroll, A. J. 383[45], 383[46]
McCarthy, B. A. 498[360]
McCormack, M. P. 138[126], 139[126]
McCullough, L. G. 651[205]
McCusker, C. F. 471[329], 472[329], 473[329]
McDaniel, S. L. 41[84]
McDonald, R. 547[19]
McGinnis, J. 651[204]
McGlinchey, M. J. 112[56]
Machrouhi, F. 394[209]
Maciejewski, J. P. 415[284], 416[284], 417[284]
MacKay, J. A. 501[387], 502[387], 503[387]
McKerrall, S. J. 159[182], 160[182]
McKie, R. 478[345], 479[345], 480[345], 484[350], 485[350], 486[350], 487[350]
McLaughlin, M. 109[45], 110[45]
McPherson, A. R. 101[13]
Maderna, A. 664[235]
Madhushaw, R. J. 152[159], 153[159], 620[154]
Madrazo, S. E. 523[439], 524[439]
Maeda, H. 473[336]
Maeda, K. 491[354]
Maekawa, H. 473[333]
Maekawa, S. 129[100], 130[100]
Maestro, M. A. 144[141], 144[142], 144[143], 145[141]
MaGee, D. I. 334[56]
Mägerlein, W. 14[47], 15[47], 16[47]
Magnus, P. 112[54]
Magnusson, T. 71[66]
Magolan, J. 384[83], 506[393], 507[393]
Mahmoodi, N. O. 439[315], 440[316]
Maier, G. 49[5]
Maier, M. 658[220], 658[221], 659[220]
Maifeld, S. V. 621[155], 622[155], 625[158]
Mainetti, E. 81[76], 88[85], 93[85]
Mainolfi, N. 108[39], 109[39]
Mairata i Payeras, A. 104[27]
Maiti, G. 512[410], 514[410]

Majima, K. 372[76]
Maki, T. 473[336]
Makino, S. 323[19]
Makita, N. 42[86], 45[86]
Malacria, M. 81[76], 88[85], 93[85], 265[8], 265[11], 265[18], 269[24], 270[24], 279[24], 283[41], 284[41], 289[41], 349[6], 349[11], 355[25], 355[26], 356[25], 383[41]
Malan, S. F. 200[52]
Malcolmson, S. J. 564[57], 565[57]
Mancheño, O. G. 245[63], 246[63]
Mandal, P. K. 512[410], 513[411], 514[410], 514[411]
Mandal, S. K. 391[167]
Mander, L. N. 91[99]
Mankad, N. P. 395[222]
Mann, E. 416[291], 417[291]
Mann, J. 321[9]
Manning, A. R. 112[56]
Mansuy, D. 383[11]
Mantilli, L. 76[73], 77[73], 78[73], 92[73], 93[73]
Manxzer, L. E. 390[157]
Manyem, S. 473[339]
Manzini, S. 547[18], 548[18]
Mao, J. 71[67], 72[67]
Marchand, A. P. 200[52]
Marchioro, C. 92[101]
Marchueta, I. 102[19], 103[19], 112[55]
Marco, J. A. 550[25]
Marco-Contelles, J. 81[76], 355[29]
Marcos, M. L. 387[125]
Marcoux, D. 57[41], 58[41], 59[41]
Marcus, R. A. 531[474]
Margarita, R. 569[63]
Margue, R. G. 614[142], 615[142]
Marion, N. 355[28]
Mariz, R. 663[234]
Mark, C. 16[49]
Markham, J. P. 76[72]
Markó, I. E. 222[31]
Markovic, T. Z. 65[54], 66[54]
Márquez, I. R. 383[70]
Marshall, W. J. 63[52]
Marti, C. 420[294]
Martín, R. 372[76]
Martin, R. L. 387[133]
Martin, S. F. 373[80], 373[81], 374[80], 374[81], 633[174]
Martin, T. J. 296[59], 297[59], 299[59]
Martín, V. S. 149[151]
Martín-Castro, A. M. 142[140], 143[140]
Martínez-Merino, V. 49[12], 50[12]
Maruoka, K. 217[26], 218[26], 219[26], 225[34], 226[34], 228[38]
Maruyama, K. 11[37]
Maruyama, T. 576[80], 577[80]
Marx, L. B. 400[251]
Marx, V. M. 553[33], 553[34], 554[33], 554[34]
Mascareñas, J. L. 319[1], 330[1], 330[34], 330[35], 330[36], 330[37],

330[41], 330[42], 331[34], 331[35], 331[36], 331[37]
Maseras, F. 146[147], 147[148]
Mastrorilli, P. 133[109], 134[109]
Masuda, K. 574[76]
Masuda, T. 400[249]
Masutomi, K. 305[66], 306[66], 308[66]
Mathes, C. 658[219]
Matignon, C. A. 392[178], 393[178]
Matsuda, A. 527[458]
Matsuda, F. 527[456]
Matsuda, T. 268[23], 269[23], 279[23], 625[159]
Matsue, H. 600[119], 601[119]
Matsui, R. 549[23], 550[23], 586[101], 587[101]
Matsukage, A. 522[435]
Matsumi, D. 473[330]
Matsumoto, H. 63[50]
Matsumoto, K. 1[12], 1[13], 2[15], 3[15], 3[16], 32[75], 33[75], 62[46], 501[388]
Matsumoto, N. 7[23]
Matsuno, T. 394[198]
Matsuo, A. 81[78], 344[77]
Matsuo, T. 351[20], 352[20], 501[388]
Mauléon, P. 330[38]
Maux, P. L. 11[133]
Mayer, P. 583[94]
Mayoral, J. A. 49[12], 50[12]
Mazumder, S. 120[80]
Meallet-Renault, R. 395[214]
Mecking, S. 391[164]
Medeiros, M. R. 391[165]
Meek, S. J. 564[57], 565[57]
Mégard, P. 115[62]
Meis, A. R. 40[83], 41[83]
Mejía-Oneto, J. M. 200[54], 201[54]
Melikyan, G. G. 383[49], 386[49]
Memboeuf, A. 112[57], 113[57]
Mende, U. 71[63]
Menéndez, J. C. 395[217]
Meng, D. 550[26]
Meng, J. 231[44]
Merlic, C. A. 389[153], 389[154]
Merlo-Pich, E. 92[101]
Merritt, J. E. 498[366]
Metcalf, C. A., III 568[62]
Metz, P. 579[86], 580[86], 636[181]
Meyer, A. 373[83], 373[84], 374[83], 374[84], 375[83]
Mezzetti, A. 14[45]
Mi, A.-Q. 172[20], 173[20], 175[20], 184[32], 185[32], 187[32]
Miao, L. 400[239]
Michael, J. P. 400[230], 400[231]
Michaelis, S. 555[37]
Michaelson, R. C. 3[17]
Michelet, V. 349[8]
Micheli, F. 92[101]
Micoine, K. 595[112], 664[112], 664[235]
Middlemiss, D. 102[17], 103[17], 105[17]
Midori, M. 405[257]
Miguel, D. 383[69], 383[72]
Mihoubi, M. N. 433[310]

Mijs, W. J. 387[110]
Mikaelian, G. S. 149[152]
Mikami, K. 217[25], 218[25]
Mikhaleva, A. I. 395[214]
Miki, T. 384[81], 461[81], 462[81]
Milella, A. 74[69]
Milet, A. 100[8], 103[21], 106[33], 112[57], 113[57], 113[59], 114[8], 114[59]
Millán, A. 383[70], 387[123]
Miller, A. K. 568[59]
Miller, J. A. 63[52]
Miller, N. A. 259[83]
Miller, R. D. 52[25]
Miller, R. L. 621[155], 622[155], 625[158], 629[165], 648[199], 649[199]
Miller, R. S. 394[208]
Miller, S. J. 215[19], 216[19]
Min, S.-J. 159[179], 160[179]
Minakawa, N. 527[458]
Minami, H. 344[77]
Mingos, D. M. P. 167[3], 168[3]
Minisci, F. 383[33]
Mischler, E. 593[110], 594[110]
Mischne, M. P. 498[363]
Mishra, A. K. 184[33], 186[33], 187[33]
Misoskowski, C. 556[38], 556[39]
Mitasev, B. 105[30], 138[124], 139[127]
Mitchell, M. M. 394[207]
Mitchell, T. A. 167[10]
Mitsudo, K. 132[108], 133[108]
Mitsudo, T.-a. 132[107], 157[174], 158[174]
Mitsuhashi, H. 500[384]
Mitsunobu, O. 388[139]
Mitsutake, M. 395[221]
Miura, Y. 607[133], 608[133]
Miwa, Y. 206[6], 206[7]
Miyagawa, K. 355[26]
Miyaki, T. 530[465]
Miyakoshi, N. 156[171], 157[171], 159[180], 159[181], 160[180], 160[181]
Miyamoto, S. 511[403]
Miyashita, Y. 365[57]
Miyazaki, A. 146[146], 147[146]
Miyazaki, T. 473[333]
Miyazawa, Y. 576[82], 577[82], 578[82]
Mizoguchi, H. 15[48], 16[48]
Mizukami, H. 491[354]
Mizuno, N. 18[54], 18[55], 19[55]
Mizuochi, T. 530[469]
Mizushina, Y. 522[435], 522[436], 522[437]
Mizutani, H. 635[179]
Mlinarić-Majerski, K. 395[213]
Moebius, D. C. 564[56], 565[56]
Mohacsi, E. 383[9]
Mohan, R. 386[105], 387[105]
Mohan, R. M. 498[367]
Moher, E. D. 52[25]
Mol, J. C. 652[208]
Molander, G. A. 383[65], 392[65], 392[172], 393[65], 393[172], 393[186], 393[187], 477[341], 478[341], 478[345], 479[345], 480[65], 480[172], 480[345],

484[350], 485[350], 486[350], 487[350], 525[449]
Molawi, K. 640[187]
Molina-Navarro, S. 498[376]
Monck, N. J. T. 91[99]
Mondal, M. 383[52], 383[53], 384[52], 384[53], 490[53], 503[53], 504[53], 508[52]
Mongin, F. 167[6]
Monopoli, A. 74[69]
Montaña, A. M. 322[15], 322[16], 322[17]
Montserrat, S. 330[34], 330[37], 330[42], 331[34], 331[37]
Mook, R., Jr. 383[22]
Moore, J. S. 668[240], 668[241], 669[243], 670[243]
Morales, C. A. 625[160], 625[161], 626[160], 626[161]
Morandi, B. 66[55], 66[56], 66[57], 67[60], 68[55], 68[56], 69[57]
Morcillo, S. P. 383[72]
Moreau, B. 53[27]
Morel, E. 521[431]
Mori, F. 299[60], 300[60], 301[60]
Mori, M. 313[74], 314[74], 545[10], 545[11], 545[12], 545[13], 614[13], 631[169], 631[170], 632[169], 632[170], 633[175], 642[190], 642[192], 643[192], 646[192]
Morimoto, R. 601[120]
Morimoto, S. 473[335]
Morimoto, T. 127[92], 127[95], 127[96], 128[92], 128[95], 128[96], 129[95], 129[96], 132[106], 156[166], 156[168], 353[23], 354[23]
Morin, C. 139[130], 140[130]
Morita, H. 473[331]
Moriuti, S. 50[15], 50[16]
Morris, J. C. 91[99]
Morton, D. 54[32]
Moschioni, M. 387[128]
Motoyama, Y. 217[25], 218[25]
Moulton, B. E. 100[10]
Mouriès, V. 81[76], 88[85], 93[85]
Moussa, Z. 259[79]
Movassaghi, M. 637[184], 638[184]
Moyano, A. 102[19], 103[19], 112[55], 139[129], 139[130], 139[131], 139[132], 139[133], 139[134], 139[135], 139[136], 139[137], 140[129], 140[130], 140[131], 140[132], 140[133], 140[134], 140[135], 140[136], 140[137], 141[132], 141[133], 141[134], 141[135], 141[136], 142[137]
Moyeux, A. 383[73]
Mu, X.-J. 496[357], 497[357]
Mück-Lichtenfeld, C. 387[119], 416[286], 417[286]
Muir, K. 471[329], 472[329], 473[329]
Mukai, C. 110[47], 111[53], 125[90], 126[90], 156[47], 156[171], 157[171], 158[175], 159[175], 159[180], 159[181], 160[180], 160[181], 365[57]
Mukherjee, S. 637[182]
Mulder, J. A. 619[151]

Mulder, P. 400[248]
Muller, J.-L. 135[117], 135[121], 136[117], 136[121], 137[117], 137[121]
Müller, P. 1[7]
Müller, W. E. G. 521[432], 521[433]
Müller-Bunz, H. 112[56]
Mulliken, R. S. 531[472]
Müllner, M. 52[25]
Mulzer, J. 521[429], 572[73], 596[114], 597[114]
Mundla, S. R. 153[161], 154[161]
Muñoz, M. C. 215[23]
Muñoz, M. P. 81[75], 82[75]
Mur, V. 205[3]
Murai, A. 600[119], 601[119]
Murai, S. 132[106], 156[166], 156[168], 353[23], 354[23]
Murakami, M. 268[23], 269[23], 279[23], 353[22], 625[159]
Murase, N. 217[26], 218[26], 219[26]
Murata, F. 600[119], 601[119]
Murayama, T. 503[390]
Murga, J. 550[25]
Murphy, A. 8[27]
Murphy, J. A. 393[184], 410[282]
Murphy, P. D. 576[80], 577[80]
Murphy, P. V. 653[212]
Murray, D. H. 341[64]
Murry, J. A. 215[19], 216[19]
Musaev, D. G. 59[42], 60[42]
Muthusamy, S. 167[1], 168[1], 184[33], 186[33], 187[33]
Muuronen, M. 100[9]
Myers, R. S. 11[31]
Mynott, R. 550[24]

N

Nacci, A. 74[69]
Naganawa, H. 530[466]
Nagaoka, H. 449[321], 450[321], 451[321], 488[352], 489[352]
Nagatomo, M. 574[76]
Naiman, B. 495[356]
Nair, V. 383[60]
Nakada, M. 12[41], 13[41], 14[41], 582[93], 583[93]
Nakagawa, M. 550[28]
Nakahata, N. 522[437]
Nakai, K. 511[403], 511[404]
Nakajima, M. 168[14], 184[14], 187[14], 188[14], 241[59], 242[59], 527[459], 528[459], 573[75]
Nakajima, T. 530[468]
Nakamura, E. 113[58], 114[58]
Nakamura, H. 503[390]
Nakamura, M. 473[334]
Nakamura, S. 168[14], 175[24], 176[24], 177[24], 178[24], 182[24], 183[24], 184[14], 187[14], 188[14]
Nakanishi, A. 477[343]
Nakano, A. 384[93]
Nakano, Y. 384[75]
Nakashima, K. 550[29]
Nakata, M. 530[469]
Nakata, T. 389[146]

Nakatsu, H. 360[47]
Nakayama, A. 159[186], 160[186], 161[186]
Nakayama, M. 81[78]
Nakazaki, A. 511[407], 522[434], 523[434], 549[23], 550[23], 586[101], 587[101]
Namba, A. 609[135]
Nambu, H. 168[14], 175[24], 175[25], 175[26], 175[27], 176[24], 176[25], 176[26], 176[27], 177[24], 177[25], 178[24], 178[25], 179[26], 180[26], 180[27], 182[24], 183[24], 183[25], 183[26], 183[27], 184[14], 187[14], 188[14], 200[55], 201[55], 202[55]
Namiki, H. 635[179]
Namy, J.-L. 392[169], 392[170], 392[171], 393[169], 393[170], 393[171], 394[170], 394[191], 394[192], 394[195], 394[199], 394[209], 460[170], 480[347], 526[170]
Nanamiya, R. 609[135]
Nangia, A. 568[62]
Nanke, T. 394[210], 526[455]
Nanni, D. 422[301]
Nara, S. 527[456]
Narasaka, K. 120[77], 400[242], 401[253], 402[253], 403[253]
Narayan, R. 192[40]
Narayan, S. 389[151], 392[182], 393[182]
Narayanam, J. M. R. 397[224], 399[224], 406[259], 411[224], 465[327], 466[327], 467[327], 501[224]
Naruko, A. 384[80], 508[80], 509[80]
Naruta, Y. 11[37]
Nassar, E. 167[6]
Ndene, N. 387[119]
Neale, S. 40[83], 41[83]
Nefedov, O. M. 102[14]
Negishi, E.-i. 70[62]
Negri, M. 92[101]
Neidhöfer, J. 603[126], 604[126], 604[127]
Neuhaus, C. M. 665[237], 666[237]
Neumann, E. 130[101]
Nevado, C. 86[83]
Ngidi, E. L. 630[168], 631[168]
Nguyen, H. M. 373[86], 375[86], 376[86]
Nguyen, J. D. 465[327], 466[327], 467[327]
Nguyen, M. C. 389[154]
Nguyen, S. T. 63[52], 544[4], 545[4]
Nguyen, T. V. 349[14]
Ni, B. 557[41], 558[41], 558[42], 559[41]
Nichols, J. M. 29[71], 30[71]
Nickel, A. 576[80], 577[80]
Nicolaou, K. C. 91[100], 92[100], 259[84], 313[73], 392[176], 393[176], 527[457], 571[70], 572[70], 580[87]
Nicolas, C. 57[40], 58[40]
Nie, X. 215[22]
Nieczypor, P. 652[208]
Nieger, M. 517[415], 517[419], 532[415]
Nieto-Oberhuber, C. 81[75], 82[75]
Nihei, H. 156[172], 157[172]

Niida, K. 576[82], 577[82], 578[82]
Nikishin, G. I. 385[98]
Nilsson, J. L. G. 71[66]
Nishida, A. 223[32], 224[32], 527[459], 528[459], 550[28], 573[75]
Nishida, G. 272[27], 273[27], 273[29], 274[29], 280[27], 287[47], 287[48], 290[47]
Nishigaya, Y. 598[116], 599[116]
Nishiguchi, I. 473[333]
Nishikawa, Y. 11[40], 12[40], 14[40]
Nishino, H. 383[51], 384[81], 387[111], 423[302], 427[302], 428[302], 429[302], 431[309], 432[309], 461[81], 462[81], 490[111], 491[111]
Nishio, M. 530[465]
Nishioka, Y. 15[48], 16[48]
Nishiwaki, N. 343[72]
Nishiyama, H. 63[50], 271[25], 271[26], 272[25], 272[26], 279[25], 285[45], 285[46], 286[46], 290[45], 293[45], 298[45]
Nishiyama, S. 601[120]
Nishiyama, Y. 394[190]
Nishizawa, M. 103[24], 104[24], 106[24], 395[221]
Niwa, T. 12[41], 13[41], 14[41]
Njardarson, J. T. 510[397]
No, Z. S. 615[143]
Nobile, C. F. 133[109], 134[109]
Noda, K. 7[23]
Noe, M. C. 206[9]
Noels, A. F. 54[31], 55[31], 55[37], 56[37]
Noguchi, K. 272[27], 272[28], 273[27], 273[28], 273[29], 274[29], 276[32], 276[33], 276[34], 277[34], 277[35], 278[35], 280[27], 280[28], 280[33], 281[35], 282[37], 291[51], 294[56], 295[56], 298[56], 304[65], 305[65], 305[66], 306[66], 308[65], 308[66], 311[71], 312[71]
Noguchi, N. 582[93], 583[93]
Nojima, S. 521[425]
Nolan, S. P. 355[28], 547[18], 548[18], 556[39], 616[147]
Nolte, R. J. M. 517[416], 517[418], 532[416], 532[418]
Noltemeyer, M. 106[32]
Nomura, I. 111[53], 125[90], 126[90]
Nomura, M. 157[174], 158[174]
Nomura, R. 394[198]
Norcross, R. D. 215[19], 216[19]
Norikura, T. 600[119], 601[119]
North, M. 645[196]
Nosse, B. 579[84]
Noyori, R. 18[52], 50[15], 50[16], 323[19]
Nozaki, H. 50[15], 50[16], 81[77], 81[78]
Nozoe, S. 503[390]
Nudenberg, W. 383[30], 383[31]
Nugent, W. A. 388[140], 388[141], 388[142], 388[143], 389[140], 389[141], 389[142], 389[143], 392[143], 416[143], 417[143], 454[143]
Numata, K. 530[468]
Nuñez, Y. O. 345[80]

Nuñez-Zarur, F. 616[149]

O

Obara, Y. 522[437]
Ochifuji, N. 545[10], 545[12]
Odedra, A. 152[159], 153[159]
O'Doherty, G. A. 521[428]
Offen, P. 343[73]
Ogaki, S. 272[28], 273[28], 273[29], 274[29], 280[28]
Ogawa, A. 394[210], 526[454], 526[455]
Ogawa, R. 265[19], 285[46], 286[46]
Oghumu, S. 345[85]
Ogo, T. 501[388]
Oguma, T. 3[116], 20[58], 21[58]
Oh, J. 326[24], 326[25], 327[24], 327[25]
Ohizumi, Y. 503[390], 522[437]
Ohkita, T. 463[326], 464[326], 465[326]
Ohmori, H. 473[336]
Ohmori, K. 530[470], 530[471], 531[471]
Ohno, M. 339[59]
Ohta, K. 522[436]
Ohta, T. 503[390]
Ohta, Y. 355[25], 355[26], 356[25]
Ohtake, H. 14[43], 14[44]
Ohtaki, K. 283[43], 284[43], 290[43]
Ohtsuka, K. 530[467]
Ohtsuka, M. 175[28], 180[28], 181[28], 184[28]
Ohuchi, K. 242[60], 243[60]
Ohya, S. 394[210]
Ohyoshi, T. 576[82], 577[82], 578[82]
Oi, S. 242[60], 243[60]
Oiarbide, M. 215[20]
Oikawa, M. 527[456]
Ojima, I. 349[2]
Ok, T. 615[144], 615[145]
Okabe, M. 433[311]
Okada, T. 132[107]
Okamoto, K. 473[334]
Okamoto, S. 265[2], 283[43], 284[43], 290[43]
Oki, T. 530[465], 530[468]
Okkel, A. 517[419]
Okuda, S. 285[45], 290[45], 293[45], 298[45]
Okue, M. 488[352], 489[352]
Olin, S. S. 383[10]
Oliosi, B. 92[101]
Olivares-Romero, J. L. 23[61], 24[61]
Oliver, A. G. 216[24], 217[24]
Olivier, E. J. 525[443]
Olivier-Bourbigou, H. 391[164]
Oller-López, J. L. 387[116], 387[124], 391[124]
Olson, J. P. 343[76]
Olsson, R. 400[237]
Oltra, J. E. 387[116], 387[121], 387[123], 387[124], 387[125], 387[131], 390[158], 391[124], 391[166], 511[405], 519[423], 520[423], 520[424], 525[158], 525[166], 525[450], 526[158], 526[451]
Omae, I. 99[2]
O'Mahony, D. J. R. 111[49]
Omura, K. 32[74]

O'Neil, S. 498[371]
Ono, M. 393[188]
Onodera, G. 287[49], 288[49], 290[49], 294[57], 295[57], 298[57]
Oohara, T. 175[25], 175[27], 176[25], 176[27], 177[25], 178[25], 180[27], 183[25], 183[27]
Orito, K. 400[245]
Ortiz de Montellano, P. R. 400[233]
Osada, H. 156[173], 598[116], 599[116]
Oshima, K. 400[243], 400[246]
Oster, B. W. 292[53]
Ostovic, D. 41[84]
Otake, Y. 309[70], 310[70], 311[70], 311[71], 312[71]
Otani, T. 156[169], 156[170], 156[172], 156[173], 157[169], 157[170], 157[172]
Otsubo, K. 477[342]
Ottenbacher, R. V. 10[30]
Ovaa, H. 620[153], 621[153]
Overhand, M. 652[208], 652[210], 653[210]
Overkleeft, H. S. 652[208], 652[210], 653[210]
Overman, L. E. 585[100]
Owada, Y. 351[20], 352[20]
Owens, T. D. 216[24], 217[24]
Ozawa, T. 1[113]
Özgül, E. 498[373]

P

Padwa, A. 167[2], 168[2], 168[15], 200[15], 200[53], 200[54], 201[54]
Pagenkopf, B. L. 111[49], 117[72]
Palanichamy, K. 521[426]
Palermo, R. E. 3[17]
Pallerla, M. K. 108[38], 159[177], 160[177]
Palomo, C. 215[20]
Pan, L. 345[85]
Pan, X.-Q. 383[50]
Panagouleas, C. 525[448]
Pancrazi, A. 549[22]
Panda, S. 624[157]
Panek, J. S. 598[117], 599[117]
Pang, K.-W. 498[375]
Panne, P. 168[13], 184[13], 185[13], 186[13], 187[13]
Pansare, S. V. 585[99]
Paolillo, R. 133[109], 134[109]
Pappo, D. 527[457]
Paquette, L. A. 498[361], 581[90], 582[90]
Paradas, M. 387[121], 387[123], 387[125], 387[131]
Parinandi, P. 345[85]
Parisini, E. 52[22]
Park, C. 345[81]
Park, J. H. 99[6], 127[98], 128[98], 129[98], 135[6], 135[119], 136[119], 137[119]
Park, K. H. 135[115], 135[116], 135[118], 135[120], 136[115], 136[116], 136[118], 136[120], 137[116], 137[118], 137[120], 151[156]

Park, M. 452[322], 453[322], 454[322]
Park, S. 648[199], 649[199]
Park, S. B. 120[78]
Park, S.-B. 63[50]
Parker, J. S. 447[320], 448[320], 449[320]
Parker, K. A. 640[188]
Parlanti, L. 569[63]
Parmentier, M. 25[66], 26[66], 28[66], 30[72], 31[72], 32[72]
Parr, J. 416[291], 417[291]
Parra, A. 142[140], 143[140]
Parrish, J. D. 387[122]
Parsons, A. F. 383[2], 383[39], 498[378], 510[401]
Parsons, P. G. 332[47]
Pastori, N. 387[117]
Patai, S. 460[324]
Paterson, I. 259[83]
Patil, A. D. 343[73]
Patricia, J. J. 431[308], 432[308]
Patro, B. 393[184]
Patzschke, M. 100[9]
Paudyal, M. P. 31[73], 32[73]
Pauson, P. L. 99[5], 103[22], 104[22], 109[42], 109[45], 110[42], 110[45]
Pavlov, V. 88[94], 89[94]
Payack, J. F. 404[255], 404[256]
Payne, G. B. 17[51]
Pearson, C. M. 103[25], 104[25]
Pech-Dzib, M. Y. 345[85]
Pechet, M. M. 383[1]
Pech-López, M. 345[84], 346[84]
Pedersen, T. M. 365[58], 366[58]
Pedro, J. R. 215[21], 215[23]
Peisach, D. 498[367]
Pekel, T. 386[107], 387[107]
Pelletier, S. W. 400[225], 400[227], 400[229]
Pellissier, H. 1[9], 63[49], 359[37]
Pelphrey, P. M. 321[8], 349[12]
Peña-Rodríguez, L. M. 345[84], 346[84]
Peng, J. 416[289], 417[289]
Peng, S.-M. 63[51]
Pepper, H. P. 384[75], 384[76], 498[76], 499[76]
Pérez-Castells, J. 99[4], 102[18], 103[18], 110[46], 111[52], 115[70], 116[70], 117[70], 119[70], 147[149], 155[165], 265[6]
Perez del Valle, C. 103[21]
Pérez-Prieto, J. 93[102]
Pérez-Serrano, L. 99[4], 102[18], 103[18], 110[46], 111[52], 155[165]
Pericàs, M. A. 102[19], 103[19], 104[27], 112[55], 139[129], 139[130], 139[131], 139[132], 139[133], 139[134], 139[135], 139[136], 139[137], 140[129], 140[130], 140[131], 140[132], 140[133], 140[134], 140[135], 140[136], 140[137], 141[132], 141[133], 141[134], 141[135], 141[136], 142[137], 144[141], 145[141], 146[147], 147[148]
Perkins, M. J. 392[179], 393[179]

Perman, J. A. 37[80], 38[80], 39[80], 61[45], 62[45]
Persich, P. 664[235]
Person, W. B. 531[472]
Pesquer, A. 107[35], 107[36]
Peters, R. 244[61], 244[62]
Peters, W. 71[64]
Petersen, J. 111[51]
Petersen, J. F. W. 569[64]
Peterson, J. R. 386[100], 386[101]
Peterson, S. L. 580[88], 581[88]
Petiniot, N. 55[37], 56[37]
Petrova, O. V. 395[214]
Petrushenko, K. B. 395[214]
Pfaffenbach, M. 363[53]
Pfaltz, A. 50[17], 51[17], 130[101]
Pfeiffer, M. W. B. 606[132], 607[132]
Phillipou, G. 383[16]
Phillips, A. J. 606[131], 606[132], 607[131], 607[132]
Phillips, E. M. 500[382], 501[382]
Phillips, I. M. 241[57]
Phillips, M. L. 52[25]
Phomkeona, K. 64[53]
Picot, C. 659[227]
Piedra, M. 387[120]
Pierard, F. Y. T. M. 175[23], 176[23]
Pierobon, M. 387[135], 387[136], 389[135], 389[151], 390[135], 416[135], 416[285], 417[135], 417[285], 467[135], 468[135]
Piers, W. E. 547[19], 547[20]
Piersanti, G. 637[184], 638[184]
Piestert, F. 387[115], 390[160], 517[412], 517[413]
Piizzi, G. 637[184], 638[184]
Pikul, S. 212[16], 213[16]
Pinder, A. R. 400[228]
Piniella, J. F. 139[131], 139[133], 140[131], 140[133], 141[133]
Piras, P. P. 94[104]
Pitchen, P. 16[49]
Pitcock, W. H., Jr. 125[89]
Pleixats, R. 616[149]
Pleuss, N. 360[45]
Poater, A. 547[18], 548[18]
Poch, M. 139[129], 140[129]
Podichetty, A. K. 400[232]
Poot, A. J. 652[211], 653[211]
Porco, J. A., Jr. 561[47], 562[47]
Porter, N. A. 383[5], 383[38]
Portnoy, M. 115[71], 116[71], 117[71]
Poscharny, K. 406[263]
Potowski, M. 192[40]
Poulin, J. 638[185]
Pouwer, R. H. 510[398]
Powell, L. H. 444[319], 445[319], 446[319], 447[319]
Prasad, E. 394[201]
Prasad, K. R. 596[113]
Prescher, S. 284[44], 285[44], 290[44]
Priscilla, L.-Y. T. 401[253], 402[253], 403[253]
Pritytskaya, T. S. 88[90], 89[90]

Procter, D. J. 383[67], 383[68], 394[211], 471[329], 472[329], 473[329]
Prokop, A. 517[417], 532[417]
Prosperini, S. 387[117]
Pross, A. 531[478]
Prunet, J. 507[395], 508[395], 549[22]
Pu, L. 152[158]
Punta, C. 387[117]
Pyne, S. G. 473[332]

Q

Qian, D. 75[70]
Qiao, C. 640[189]
Qin, S. 395[220]
Qin, W. 395[213]
Qiu, L. 127[97], 128[97], 129[97]
Qu, J. Q. 400[249]
Quelquejeu-Ethève, M. 11[34]
Quentin, F. 593[110], 594[110]
Quiclet-Sire, B. 406[267], 408[274], 409[274]
Quílez del Moral, J. F. 383[71], 384[86], 387[120], 387[126]
Quinn, K. J. 554[35], 554[36], 555[35], 555[36]
Quintal, M. M. 150[154]
Quintero, J. C. 41[84]
Quiroz, R. V. 363[53]

R

Radkowski, K. 550[24], 661[229], 661[230], 662[229], 662[230], 662[233]
Radtke, L. 579[85], 580[85]
Radüchel, B. 71[63]
Rahim, M. A. 497[358]
Rahman, M. T. 431[309], 432[309]
Raiber, E.-A. 552[32], 553[32]
RajanBabu, T. V. 383[26], 383[27], 383[28], 383[29], 388[140], 388[141], 388[142], 388[143], 389[140], 389[141], 389[142], 389[143], 392[143], 400[235], 416[143], 417[143], 454[143]
Rama Rao, A. V. 498[362]
Ramazzotti, D. 94[104]
Rameshkumar, C. 334[52], 619[151]
Ramkumar, R. 184[33], 186[33], 187[33]
Rana, K. K. 390[163], 512[409], 513[409], 514[163], 515[163], 516[163], 517[163]
Randjelovic, J. 192[42]
Rao, B. V. 498[362]
Rao, M. 65[54], 66[54]
Rao, W. D. 400[236]
Rapoport, H. 529[464]
Rappoport, Z. 49[6], 49[7]
Rashatasakhon, P. 339[61]
Ratovelomanana-Vidal, V. 121[82], 121[83], 121[85], 122[82], 122[83], 122[85], 123[82], 123[83]
Ratter, F. 525[446]
Ratti, E. 92[101]
Rauch, G. 52[26]
Rautenstrauch, V. 115[62], 355[27]

Rawal, V. H. 219[27], 220[27], 220[28], 221[28], 501[387], 502[387], 503[387], 571[71], 572[71]
Rea, K. D. 92[101]
Reddy, D. R. 498[362]
Reddy, D. S. 614[141], 634[176]
Reddy, G. S. 383[29]
Reddy, R. P. 332[49]
Redgrave, A. J. 168[16], 170[16], 171[16], 174[16], 175[22], 176[22], 182[22]
Rehberg, G. M. 545[8], 545[9]
Reid, E. H. 500[383]
Reis, O. 383[54], 387[113], 498[374]
Reiser, O. 52[22], 70[62], 579[84]
Reissig, H.-U. 49[2], 406[268], 498[377]
Ren, Z.-H. 395[218], 395[223], 396[223], 397[223]
Renaud, P. 383[3], 383[41], 383[42], 383[43]
Reppe, W. 265[14]
Reuter, P. 521[433]
Revés, M. 106[34], 107[34], 107[37], 114[61]
Rheingold, A. L. 213[17], 659[226]
Riant, O. 205[5]
Ribeiro, A. A. 384[84], 498[364], 498[365]
Ricard, L. 549[22], 636[180]
Ricci, P. 120[80]
Richard, J.-A. 91[100], 92[100], 510[398]
Richards, S. C. 188[38], 190[38], 191[38], 192[38]
Rickards, B. 125[91], 126[91], 127[91]
Rickborn, B. 388[138]
Rickers, J. 221[29], 222[29]
Rickers, A. 135[121], 136[121], 137[121]
Rieck, H. 359[42], 362[49], 370[67]
Rieder, C. J. 328[28]
Riera, A. 101[11], 102[19], 103[19], 106[34], 107[34], 107[35], 107[36], 107[37], 109[44], 110[44], 111[44], 112[55], 114[61], 139[130], 139[131], 139[132], 139[133], 139[134], 139[135], 139[136], 139[137], 140[130], 140[131], 140[132], 140[133], 140[134], 140[135], 140[136], 140[137], 141[132], 141[133], 141[134], 141[135], 141[136], 142[137], 144[141], 144[142], 144[143], 145[141]
Rieth, R. D. 395[222]
Rigoli, J. W. 40[83], 41[83]
Rijkers, D. T. S. 652[211], 653[211]
Rinehart, K. L. 503[389]
Rinker, B. 389[151], 389[156]
Rios, R. 139[136], 140[136], 141[136]
Riva, R. 645[195]
Robert, F. 100[8], 113[59], 114[8], 114[59]
Roberts, D. H. 383[19]
Roberts, J. M. 500[382], 501[382]
Robertson, S. M. 109[45], 110[45]
Robichaux, P. J. 356[32], 356[33], 357[32], 368[61], 368[62], 369[61]
Robinson, J. E. 150[155], 151[155]

Robles, R. 387[121], 387[123], 387[124], 387[125], 387[131], 391[124], 519[423], 520[423]
Rodriguez, J. 433[310], 433[312]
Rodríguez Rivero, M. 105[29], 106[29], 140[139], 142[139]
Rodríguez-Santiago, L. 616[149]
Rodríguez-Solla, H. 392[173], 393[173], 394[189]
Rogers-Evans, M. 633[173]
Rohde, J. J. 207[10], 208[10]
Roisnel, T. 167[6]
Romero, P. E. 547[19], 547[20]
Romero, R. H. 101[12]
Romers, C. 385[96]
Rominger, F. 114[60]
Romo, D. 259[79], 575[79], 576[79]
Roncaglia, F. 423[303], 424[303], 425[303]
Rong, J. 406[264], 413[264]
Rong, S.-F. 495[355]
Rooney, L. 658[223]
Roper, T. D. 206[9]
Rosa, C. P. 567[58]
Rosales, A. 387[124], 390[158], 391[124], 391[166], 525[158], 525[166], 526[158], 526[451]
Roscic, M. 92[101]
Rose, E. 11[34]
Rosé, J. 133[109], 134[109]
Rosebrugh, L. E. 553[34], 554[34]
Rosokha, S. V. 531[476]
Rossi, B. 387[117]
Roth, E. 41[84]
Rouse, M. B. 576[83], 577[83]
Roush, W. R. 637[183]
Rousseau, B. 389[149]
Rout, L. 170[19], 171[19], 172[19], 175[19]
Rovis, T. 295[58], 296[58], 296[59], 297[59], 299[58], 299[59]
Rowlands, G. J. 420[298], 420[299]
Roy, S. C. 390[161], 390[162], 390[163], 391[162], 391[167], 391[168], 433[168], 434[168], 454[161], 455[161], 456[161], 457[161], 458[161], 458[323], 459[323], 511[402], 511[408], 512[408], 512[409], 512[410], 513[409], 513[411], 514[163], 514[410], 514[411], 515[163], 516[163], 517[163]
Rozantsev, E. G. 392[181], 393[181]
Rückert, A. 603[125], 642[191]
Rudolph, M. 355[30]
Rudyk, H. 645[197]
Rueping, M. 406[263], 406[266], 412[283], 413[283], 415[283]
Ruijter, E. 265[3]
Rulíšek, L. 281[36], 282[36]
Rumbo, A. 546[17], 655[17], 656[17], 656[217], 657[217]
Ruppel, J. V. 37[80], 38[80], 39[80], 60[44], 61[44]
Rutjes, F. P. J. T. 652[208], 652[209], 652[210], 653[210]
Ruveda, E. A. 498[363]

Ryan, M. C. 65[54], 66[54]
Rydberg, D. B. 545[7]
Ryu, D. H. 209[12], 209[13], 210[12], 210[13], 211[12]
Ryu, I. 355[25], 355[26], 356[25], 526[454], 526[455]
Ryu, S. Y. 529[463]

S

Sadighi, J. P. 395[222]
Sadoun, A. H. 581[90], 582[90]
Sáez, J. A. 167[6]
Sagae, H. 311[71], 312[71]
Saha, A. K. 66[58]
Saha, S. 511[402], 511[408], 512[408]
Sahasrabudhe, K. 605[130]
Saicic, R. N. 408[274], 409[274]
Saigo, K. 115[66], 116[66]
Saigoku, T. 285[45], 290[45], 293[45], 298[45]
Saino, N. 283[43], 284[43], 290[43]
Saito, B. 1[12], 1[13]
Saito, M. 156[172], 157[172]
Saito, S. 232[45], 232[46]
Saito, T. 156[169], 156[170], 156[172], 156[173], 157[169], 157[170], 157[172]
Saitoh, K. 530[465]
Saitoh, T. 601[120]
Sakaguchi, K. 522[435], 522[436], 522[437]
Sakaguchi, S. 384[92], 384[93], 384[94], 430[305]
Sakai, K. 1[12]
Sakai, S. 394[200], 477[343]
Sakakura, A. 576[81], 577[81], 578[81]
Sakata, K. 287[49], 288[49], 290[49]
Sakiyama, N. 305[66], 306[66], 308[66]
Sakurai, H. 325[23]
Salabarria, I. S. 345[80]
Salaün, J. 106[32]
Salehpour, M. 439[315]
Salter, M. M. 251[70], 260[70], 261[70]
Salvatella, L. 49[12], 50[12]
Šaman, D. 281[36], 282[36]
Sammakia, T. 167[9]
Sanchez, D. 71[66]
Sánchez, E. M. 383[71], 387[120]
Sancho, J. 650[200]
Sandoval, C. 389[150]
Sandrinelli, F. 400[250]
Santelli, M. 433[312]
Sardana, V. 41[84]
Sarel, S. 349[15], 349[16], 350[16]
Sarin, P. S. 521[432]
Sarkisian, R. G. 255[75], 256[75]
Sartori, I. 92[101]
Sasaki, I. 395[221]
Sasaki, M. 526[452]
Sasaki, T. 339[59]
Sasse, F. 658[221]
Sasuga, D. 511[403]
Sato, K. 18[52]
Sato, T. 601[121]
Sato, Y. 313[74], 314[74], 586[101], 587[101], 642[192], 643[192], 646[192]

Satoh, M. 400[249]
Satoskar, A. R. 345[85]
Sattely, E. S. 564[56], 564[57], 565[56], 565[57]
Savchenko, A. I. 88[87], 88[88], 89[87], 89[88], 90[88]
Savic, V. 192[42]
Sawada, Y. 1[112], 1[113], 277[35], 278[35], 281[35], 530[468]
Sawyer, J. R. 120[79], 120[80], 302[63], 303[63], 305[63], 308[63]
Saxton, J. E. 200[56]
Saygili, N. 387[112]
Sayyad, A. A. 638[186], 639[186]
Scaiano, J. C. 383[18]
Scanio, M. J. C. 363[51], 365[58], 366[58]
Schacherer, L. N. 391[165]
Schaefer, A. G. 498[361]
Schäfers, M. 400[232]
Schall, A. 579[84]
Schaus, J. V. 598[117], 599[117]
Schaus, S. E. 233[47], 234[47]
Scheidt, K. A. 197[50], 198[50], 200[50], 420[295], 500[382], 501[382]
Scheiner, P. 408[270], 409[270]
Schenk, K. 393[185]
Schenkluhn, H. 283[40]
Scheppingen, W. B. 400[248]
Schick, M. 389[151]
Schienebeck, C. M. 355[31], 368[61], 368[62], 369[61], 369[63]
Schiesser, C. H. 383[14], 383[15]
Schill, B. D. 329[30]
Schill, H. 49[9]
Schinnerl, M. 52[22]
Schirmer, P. 52[23], 53[23], 55[23]
Schlabach, A. J. 41[84]
Schleif, W. A. 41[84]
Schleyer, P. V. R. 548[21]
Schlichting, O. 265[14]
Schmidt, C. D. 53[28]
Schmitt, M. 416[286], 417[286]
Schmitz, M. L. 525[446]
Schneider, A. 104[26]
Schneider, M. J. 400[227]
Schneider, T. F. 49[3], 52[20], 52[21], 52[23], 52[24], 53[23], 53[28], 55[23], 55[24]
Schober, O. 400[232]
Schoemaker, H. E. 652[208], 652[209], 652[210], 653[210]
Scholl, M. 618[150]
Schollmeyer, D. 312[72]
Schomaker, J. M. 40[83], 41[83]
Schore, N. E. 100[7], 109[41], 110[41]
Schramm, M. P. 614[141]
Schreiber, S. L. 102[15], 150[153], 552[32], 553[32], 559[43], 560[43]
Schrock, R. R. 390[157], 551[31], 552[31], 564[51], 564[52], 564[54], 564[56], 564[57], 565[52], 565[54], 565[56], 565[57], 566[52], 566[54], 616[146], 623[156], 624[146],
650[200], 650[201], 651[205], 652[207], 659[226], 660[31], 661[31]
Schröder, H. C. 521[433]
Schröer, S. 400[232]
Schubert, M. 636[181]
Schulte, G. 568[62]
Schulte, J. H. 114[60]
Schulz, S. R. 644[194], 645[194]
Schulze-Osthoff, K. 525[446]
Schurig, V. 16[49]
Schwartz, B. D. 344[78]
Scialdone, M. A. 63[52]
Scott, I. L. 101[12]
Scott, L. J. 518[422]
Sealy, J. M. 394[206], 394[208]
Seidel, G. 545[15], 546[15], 546[17], 655[17], 655[216], 656[17], 656[216], 661[15], 661[230], 662[15], 662[230]
Seiders, T. J. 564[53], 565[53]
Seigal, B. A. 108[40], 109[40]
Seitz, W. J. 66[58]
Sekiguchi, M. 526[454]
Selig, A. 517[417], 532[417]
Sello, J. K. 559[43], 560[43]
Sels, B. F. 18[53]
Semmelhack, M. F. 350[19], 351[19]
Senboku, H. 400[245]
Seo, S. D. 120[78]
Serelis, A. K. 383[16]
Seres, P. 222[31]
Seto, K. 549[23], 550[23], 586[101], 587[101]
Sevenet, T. 498[380]
Sezaki, M. 530[466]
Shabangi, M. 394[203], 394[207], 394[208]
Shair, M. D. 625[160], 625[161], 626[160], 626[161]
Shakya, S. R. 477[341], 478[341]
Shalyaev, K. 14[42]
Shambayati, S. 102[15], 150[153]
Shang, D. 254[73]
Shanker, P. S. 580[88], 581[88]
Sharma, U. 339[62], 340[62], 341[66]
Sharp, P. 390[157]
Sharpless, K. B. 1[11], 3[17], 3[18]
Shashkov, A. S. 149[152]
Shaughnessy, E. A. 215[19], 216[19]
She, J. 570[68]
Shea, K. M. 150[154]
Shelton, D. R. 387[122]
Shen, B. 589[105], 590[105]
Shen, H. 370[72], 371[72]
Shen, H. C. 370[73], 370[74]
Shen, H.-J. 159[183], 160[183], 161[183]
Shen, J.-J. 69[61], 70[61]
Shen, M. 408[272], 409[272], 469[272], 470[272]
Shen, X. 230[41]
Shen, Z. 498[364]
Shevchuk, T. A. 88[95], 89[95]
Shi, B.-F. 71[67], 72[67]
Shi, D.-Q. 495[355]
Shi, F. 192[41]
Shi, L.-L. 159[183], 160[183], 161[183]
Shi, M. 154[162], 386[103], 643[193], 644[193], 646[193]
Shi, Y. 1[1]
Shibahara, S. 530[467]
Shibasaki, M. 571[69]
Shibata, T. 115[63], 127[93], 127[94], 128[93], 128[94], 129[99], 129[100], 130[99], 130[100], 265[10], 274[30], 275[30], 275[31], 280[30], 280[31], 301[62], 302[62], 302[64], 303[64], 304[64], 306[67], 307[67], 308[62], 308[64], 309[67]
Shibata, Y. 265[4], 272[28], 273[28], 280[28]
Shibatomi, K. 64[53]
Shibuya, M. 498[370]
Shido, M. 167[11]
Shim, S. H. 524[441], 524[442]
Shimada, N. 168[14], 175[24], 175[25], 175[27], 176[24], 176[25], 176[27], 177[24], 177[25], 178[24], 178[25], 180[27], 182[24], 183[24], 183[25], 183[27], 184[14], 187[14], 188[14], 200[55], 201[55], 202[55], 241[59], 242[59]
Shimbo, K. 601[121]
Shimizu, N. 325[21], 325[22], 326[22]
Shimizu, S. 206[7]
Shimizu, Y. 287[49], 288[49], 290[49]
Shin, J. Y. 120[78]
Shinohara, I. 488[352], 489[352]
Shinokubo, H. 400[243], 400[246]
Shintani, R. 360[47]
Shiotani, M. 156[170], 157[170]
Shirahama, H. 527[456]
Shirahata, A. 325[23]
Shiraishi, K. 593[109], 594[109]
Shirasaka, K. 266[20], 267[20], 268[20], 278[20]
Shirasaka, T. 225[34], 226[34]
Shirasaki, D. 223[32], 224[32]
Shiro, M. 172[21], 173[21], 174[21], 175[21], 200[55], 201[55], 202[55], 241[59], 242[59], 601[121]
Shishido, K. 159[185], 160[185], 161[185], 498[370]
Shitama, H. 7[26], 8[26], 11[26]
Shoji, M. 598[116], 599[116]
Shono, T. 473[331]
Shotwell, J. B. 394[206]
Shu, D. 355[31], 356[32], 356[33], 357[32], 368[60], 368[61], 369[61]
Shu, H. 400[239]
Shu, X.-z. 355[31], 368[60], 368[61], 369[61], 369[63], 370[71]
Shuto, S. 527[458]
Sibi, M. P. 215[22], 383[3], 383[41], 383[42], 383[43], 473[338], 473[339]
Siegel, D. S. 637[184], 638[184]
Signorella, S. 498[363]
Sill, P. C. 125[91], 126[91], 127[91]
Silva, A. A. 321[11], 322[11]
Silva, L. F. 500[385]
Simanis, J. A. 167[10]
Simic, M. 192[42]

Simmons, H. E. 322[18]
Simoneau, C. A. 410[279]
Simonneau, A. 349[11]
Simonneaux, G. 1[6], 11[33]
Simonsen, S. H. 54[36], 55[36]
Simpkins, N. S. 505[391]
Singh, A. K. 498[362]
Singh, V. 359[36]
Sirois, L. E. 363[52], 370[67], 370[68]
Sivavec, T. M. 545[5], 545[6]
Skaltsa, H. 525[448]
Skidmore, J. 660[228], 661[228]
Skrydstrup, T. 387[137], 388[137], 389[137], 394[204], 394[205], 394[211], 480[348]
Skuballa, V. 71[63]
Sladić, D. 521[433]
Slawin, A. M. Z. 547[18], 548[18]
Slowinski, F. 265[11]
Smit, W. A. 102[14], 149[152]
Smith, I. E. D. 188[38], 190[38], 191[38], 192[38]
Smith, R. D. 322[18]
Smith, R. S. 167[9]
Snapper, M. L. 108[40], 109[40], 149[150], 249[66]
Snider, B. B. 383[48], 383[56], 384[48], 384[85], 386[48], 386[105], 387[105], 431[308], 432[308], 498[359], 498[360], 498[366], 498[367], 498[371]
Sobenina, L. N. 395[214]
Soderquist, J. A. 526[453]
Sodupe, M. 616[149]
Sohn, J.-H. 615[143], 615[144], 615[145]
Sokovic, M. 525[448]
Solà, J. 144[142], 144[143]
Solans-Monfort, X. 616[149]
Solorio-Alvarado, C. R. 83[80], 84[80], 85[80]
Somfai, P. 188[36], 188[37], 189[36], 189[37], 190[36], 190[37], 191[36]
Son, S. U. 135[113], 135[114], 135[115], 135[116], 135[118], 136[113], 136[114], 136[115], 136[116], 136[118], 137[116], 137[118], 151[156]
Song, W. 369[63]
Song, Y.-G. 120[80]
Sonmez, F. 383[59]
Sono, M. 550[29]
Sonoda, N. 394[210], 526[454], 526[455]
Sorba, J. 383[6]
Sorensen, A. J. 550[26]
Sörensen, H. 139[130], 140[130]
Soriano, E. 355[29]
Sortais, B. 408[275], 409[275]
Spada, S. 92[101]
Spaeth, T. 88[91], 89[91]
Spain, M. 383[68]
Spandl, R. J. 645[197]
Spannenberg, A. 283[42], 284[42], 289[50], 290[42], 291[50]
Spellmeyer, D. C. 383[25]
Spiegel, D. A. 391[165]
Spitz, C. 30[72], 31[72], 32[72]
Sponholtz, D. J. 334[56]

Sprague, S. J. 400[251]
Spring, D. R. 645[197]
Springer, J. P. 498[361]
Spur, B. 71[64]
Srikrishna, A. 383[42]
Srinivas, D. 389[145]
Srour, H. 1[6]
Staben, S. T. 78[74], 79[74], 80[74]
Stack, T. D. P. 8[27]
Stade, R. 659[227], 661[231], 662[231]
Stafford, D. G. 332[45], 333[45]
Stalke, D. 52[23], 52[24], 53[23], 53[28], 55[23], 55[24]
Stanley, L. M. 215[22]
Stará, I. G. 281[36], 282[36]
Stark, C. B. W. 341[68]
Starý, I. 281[36], 282[36]
Steel, P. G. 394[196], 480[196]
Steigerwald, M. 350[19], 351[19]
Steinmetz, B. G. 370[74]
Stemmler, R. T. 363[52]
Stephenson, C. R. J. 397[224], 399[224], 406[259], 406[269], 407[269], 408[269], 409[269], 411[224], 465[327], 466[327], 467[327], 501[224]
Sterner, O. 345[84], 346[84]
Stevenazzi, A. 115[65], 116[65]
Stewart, I. C. 572[74], 573[74]
Stieger, M. 665[237], 666[237]
Stolle, A. 106[32]
Stoltz, B. M. 572[74], 573[74]
Stoop, R. M. 14[45]
Storey, J. M. D. 416[291], 417[291], 420[297]
Stork, G. 383[21], 383[22]
Stork, T. C. 585[97]
Stragies, R. 602[124], 603[124], 605[129]
Straub, B. F. 616[148]
Streuff, J. 387[132], 473[340], 474[340], 475[340], 476[340], 487[351], 488[351]
Strohmann, C. 579[85], 580[85]
Stroud, R. 400[233]
Strübing, D. 99[3]
Struchkov, Y. T. 149[152]
Strunz, G. M. 400[226]
Stryker, J. M. 359[38]
Studabaker, B. 49[11], 50[11]
Studer, A. 383[35], 387[118]
Stupple, P. A. 176[30]
Stuppner, H. 525[447]
Sturla, S. J. 130[105], 131[105], 132[105], 146[145]
Su, D.-S. 550[26]
Su, J. 377[88], 378[88], 379[88]
Su, W. 430[306]
Subasi, N. T. 387[112]
Subbarao, R. 71[65]
Subbarayan, V. 37[80], 38[80], 39[80]
Suda, N. 505[392], 506[392]
Suda, T. 276[33], 280[33], 282[37]
Sudo, Y. 223[32], 224[32]
Suematsu, H. 62[46], 62[47], 63[47]
Suenaga, K. 576[81], 577[81], 578[81]
Suenaga, T. 389[146]

Suga, H. 172[21], 173[21], 174[21], 175[21], 175[28], 175[29], 180[28], 181[28], 181[29], 182[29], 184[28], 184[29], 188[35], 189[35], 191[35]
Sugawara, F. 522[435], 522[436], 522[437]
Sugihara, T. 103[24], 104[24], 106[24], 115[69], 116[69], 117[69], 119[75]
Sugikubo, K. 365[57]
Sugiyama, Y. 265[2]
Sugizaki, K. 156[169], 156[173], 157[169]
Sullivan, K. A. 52[25]
Sumida, Y. 18[54]
Sumino, Y. 394[210], 526[455]
Sumiya, T. 346[86]
Sun, D. 521[432]
Sun, J. 85[81], 230[42], 230[43], 231[42]
Sun, W. 8[28], 9[28], 11[28]
Sundararaju, B. 662[232], 662[233]
Sundell, S. 71[66]
Sundermann, B. 289[50], 291[50]
Sundermann, C. 289[50], 291[50]
Sung, B. K. 120[78], 120[81], 121[81], 122[81]
Sung, Y. 498[379]
Sunnam, S. K. 596[113]
Suto, M. 294[57], 295[57], 298[57]
Suyama, T. 156[169], 156[172], 157[169], 157[172]
Suzuki, K. 1[13], 530[470], 530[471], 531[471]
Suzuki, M. 593[109], 594[109]
Suzuki, N. 132[107], 287[47], 290[47]
Suzuki, S. 530[467]
Suzuki, T. 473[334], 549[23], 550[23], 586[101], 587[101], 600[119], 601[119]
Suzuki, Y. 576[81], 577[81], 578[81], 600[119], 601[119]
Svensson, K. 71[66]
Sviridov, S. V. 88[88], 88[90], 88[93], 89[88], 89[90], 89[93], 90[88]
Svith, H. 416[286], 417[286]
Swanson, E. D. 149[152]
Sweeney, J. B. 1[8]
Swenson, D. C. 524[441]
Szafert, S. 654[213]
Szklarz, S. C. 554[36], 555[36]
Szostak, M. 383[68]
Szymoniak, J. 89[97]

T

Tabatabaeian, K. 440[316]
Taber, D. F. 350[18], 358[35], 359[35]
Tabet, J.-C. 112[57], 113[57]
Tabuchi, T. 574[76]
Tachibana, Y. 242[60], 243[60]
Tada, M. 433[311]
Tadano, K.-i. 584[95], 585[95], 609[135]
Taddei, M. 94[104]
Tae, J. 581[90], 582[90]
Tahara, Y. 302[64], 303[64], 304[64], 306[67], 307[67], 308[64], 309[67]
Takabe, K. 598[116], 599[116]

Takagi, K. 127[93], 127[94], 128[93], 128[94], 129[99], 129[100], 130[99], 130[100], 274[30], 275[30], 280[30]
Takagishi, H. 285[45], 290[45], 293[45], 298[45]
Takahashi, H. 359[41], 360[41], 367[41]
Takahashi, K. 405[257]
Takahashi, T. 390[159], 511[403], 511[404], 511[406], 521[406], 522[406]
Takai, A. 277[35], 278[35], 281[35]
Takaishi, Y. 498[370]
Takami, N. 526[454], 526[455]
Takao, K.-i. 584[95], 585[95], 609[135]
Takatsu, K. 360[47]
Takaya, H. 50[16], 323[19]
Takaya, J. 167[11]
Takayama, H. 159[186], 160[186], 161[186]
Takeda, K. 175[25], 175[26], 176[25], 176[26], 177[25], 178[25], 179[26], 180[26], 183[25], 183[26], 200[55], 201[55], 202[55]
Takeda, T. 497[358]
Takeishi, K. 276[32]
Takemoto, T. 585[100]
Takemura, M. 522[435]
Takeuchi, M. 277[35], 278[35], 281[35]
Takeuchi, R. 287[49], 288[49], 290[49], 294[57], 295[57], 298[57]
Takeuchi, T. 530[466], 530[467]
Takezawa, R. 175[29], 181[29], 182[29], 184[29]
Talsi, E. P. 10[30]
Tam, W. 349[1]
Tamai, K. 522[435]
Tamaru, T. 598[116], 599[116]
Tamm, M. 654[215], 655[215]
Tamura, K. J. 491[354]
Tamura, T. 313[74], 314[74]
Tan, X. 384[77], 384[78], 503[78], 504[78]
Tanabe, E. 283[43], 284[43], 290[43]
Tanaka, H. 511[406], 521[406], 522[406]
Tanaka, I. 593[109], 594[109]
Tanaka, K. 265[1], 265[4], 265[7], 266[20], 267[20], 268[20], 268[22], 269[22], 272[27], 272[28], 273[27], 273[28], 273[29], 274[29], 276[32], 276[33], 276[34], 277[34], 277[35], 278[20], 278[35], 279[22], 280[27], 280[28], 280[33], 281[35], 282[37], 287[47], 287[48], 290[47], 291[51], 294[56], 295[56], 298[56], 299[60], 300[60], 301[60], 304[65], 305[65], 305[66], 306[66], 308[65], 308[66], 309[70], 310[70], 311[70], 311[71], 312[71], 400[242]
Tanaka, M. 325[22], 326[22], 501[388]
Tanaka, N. 522[435]
Tanaka, R. 309[70], 310[70], 311[70]
Tanaka, T. 521[425]
Tang, H. 576[80], 577[80]

Tang, W. 355[31], 356[32], 356[33], 357[32], 368[60], 368[61], 368[62], 369[61], 369[63], 370[71]
Tang, Y. 115[67], 116[67], 119[67], 134[110], 313[73]
Tani, F. 11[37]
Taniguchi, K. 593[109], 594[109]
Tanikawa, N. 473[334]
Tanyeli, C. 387[114], 498[373]
Tao, H. 246[64], 247[64]
Tao, H.-Y. 192[43]
Tarasov, V. A. 102[14]
Tarsi, L. 92[101]
Tasic, G. 192[42]
Taube, H. 531[473]
Tay, J.-H. 152[158]
Tedesco, G. 92[101]
Teijón, P. H. 523[439], 524[439]
Tejo, C. 27[68], 28[68]
Tekavec, T. N. 300[61], 301[61]
Teller, H. 76[73], 77[73], 78[73], 92[73], 93[73], 330[39]
Teng, P.-F. 63[51]
Tennakoon, M. A. 560[44], 561[44], 622[44], 623[44]
Teo, Y.-C. 213[18], 214[18]
Teplý, F. 281[36], 282[36], 406[260]
Terada, E. 242[60], 243[60]
Terada, M. 217[25], 218[25]
Terrazas, C. 345[85]
Terreni, S. 92[101]
Teyssié, P. 54[31], 55[31], 55[37], 56[37]
Thiel, O. R. 550[27], 580[89], 581[89]
Thiel, W. 76[73], 77[73], 78[73], 92[73], 93[73]
Thoison, O. 498[380]
Thomas, A. 137[122]
Thomas, C. B. 498[378]
Thompson, A. L. 400[251]
Thompson, C. F. 259[80]
Thomson, W. 103[22], 104[22], 109[42], 110[42]
Thorhauge, J. 237[51], 237[52], 238[51], 238[52]
Thornton, A. 521[432]
Thornton, P. D. 334[56]
Tiefenbacher, K. 521[429], 572[73]
Timmer, M. S. M. 620[153], 621[153]
Timmons, D. J. 168[12], 172[12], 184[12], 193[45], 199[45]
Tiong, D. L. Y. 27[68], 28[68]
Tiseni, P. S. 244[61], 244[62]
Toelle, N. 596[114], 597[114]
Togo, H. 383[8], 463[326], 464[326], 465[326]
Tohyama, H. 530[467]
Tokuda, M. 400[245]
Tomatsu, K. 530[465]
Tominaga, N. 473[331]
Tomita, K. 530[465]
Tomita, T. 633[175]
Ton, T. M. U. 27[68], 28[68]
Tong, M.-C. 246[64], 247[64]

Tonogaki, K. 642[192], 643[192], 646[192]
Tori, M. 550[29]
Tormo, J. 139[134], 140[134], 141[134]
Torrente, E. 142[140], 143[140]
Torres, R. R. 99[1], 138[123], 159[176]
Torssell, S. 188[36], 188[37], 189[36], 189[37], 190[36], 190[37], 191[36]
Toshida, N. 127[93], 127[94], 128[93], 128[94], 129[100], 130[100]
Toste, F. D. 76[72], 78[74], 79[74], 80[74], 82[79], 83[79], 330[38], 370[72], 370[74], 371[72]
Toueg, J. 507[395], 508[395]
Toullec, P. Y. 349[8]
Toyoda, K. 266[20], 267[20], 268[20], 278[20]
Toyota, M. 200[57]
Tozer, M. J. 146[146], 147[146]
Trauner, D. 568[59], 583[94]
Tremblay, M. 569[63], 569[64]
Treutwein, J. 265[9]
Tria, G. S. 259[84], 571[70], 572[70]
Trillo, B. 330[34], 330[35], 330[37], 331[34], 331[35], 331[37]
Trinchera, P. 1[10]
Trivedi, G. K. 359[36]
Trofimov, B. A. 395[214]
Trogolo, C. 422[301]
Trost, B. M. 43[87], 44[89], 46[87], 46[89], 65[54], 66[54], 329[31], 349[4], 370[70], 370[72], 370[73], 370[74], 371[72], 373[82], 373[83], 373[84], 373[86], 374[83], 374[84], 375[82], 375[83], 375[86], 376[86], 383[37], 388[138], 388[139], 531[37], 569[65], 569[66]
Trost, M. K. 498[372]
Trudel, C. 30[72], 31[72], 32[72]
Trudell, M. L. 400[239]
Tsai, P.-J. 429[304], 430[304], 431[304]
Tsang, K. Y. 527[457]
Tse, M. K. 11[38], 11[39], 14[47], 15[47], 16[47]
Tsubusaki, T. 423[302], 427[302], 428[302], 429[302]
Tsuchida, T. 175[28], 180[28], 181[28], 184[28]
Tsuchikama, K. 265[10], 301[62], 302[62], 308[62]
Tsuchiya, Y. 463[326], 464[326], 465[326]
Tsuda, A. 498[370]
Tsuda, H. 503[390]
Tsuda, M. 601[121]
Tsuji, I. 353[22]
Tsuji, J. 544[2]
Tsuna, K. 582[93], 583[93]
Tsunakawa, M. 530[465]
Tsuno, Y. 325[21], 325[22], 326[22]
Tsuritani, T. 400[243], 400[246]
Tsutsui, H. 168[14], 175[24], 176[24], 177[24], 178[24], 182[24], 183[24], 184[14], 187[14], 188[14]

Author Index

Tsutsumi, K. 127[92], 127[95], 127[96], 128[92], 128[95], 128[96], 129[95], 129[96]
Tu, Y.-Q. 121[84], 122[84]
Tucker, J. W. 397[224], 399[224], 406[269], 407[269], 408[269], 409[269], 411[224], 465[327], 466[327], 467[327], 501[224]
Tulip, S. J. 384[75]
Tulp, M. T. M. 395[212]
Turner, B. P. 500[383]
Tzamarioudaki, M. 349[2]

U

Ubukata, M. 353[22], 501[388]
Uchida, T. 15[48], 16[48], 32[75], 33[75], 33[76], 34[76], 62[46], 62[47], 63[47]
Ueda, N. 473[335]
Ueki, T. 530[468]
Ueno, Y. 275[31], 280[31]
Uesugi, Y. 104[28]
Uffelman, E. S. 11[35]
Ujaque, G. 330[34], 330[37], 330[42], 331[34], 331[37]
Ujikawa, O. 477[344]
Ulman, M. 614[139]
Umani, F. 422[301]
Umezaki, S. 574[78], 575[78]
Umezawa, H. 530[466]
Uneyama, K. J. 491[354]
Ungeheuer, F. 159[182], 160[182], 668[239]
Ura, Y. 157[174], 158[174]
Urabe, D. 574[76]
Urakawa, O. 384[81], 461[81], 462[81]
Urbina-Blanco, C. A. 547[118], 548[118]
Ushakov, I. A. 395[214]
Usuda, H. 571[69]
Usuki, J. 115[66], 116[66]

V

Vacca, J. P. 41[84]
Valdivia, M. V. 384[86]
Valentí, E. 139[129], 140[129]
Valenzuela, M. 241[58]
Vallet, M. 221[29], 222[29]
van Boom, J. H. 620[153], 621[153], 652[208], 652[210], 653[210]
van Delft, F. L. 652[209], 652[210], 653[210]
van der Eide, E. F. 547[20]
van der Heyden, J. A. M. 395[212]
van der Marel, G. A. 620[153], 621[153], 652[208], 652[210], 653[210]
Van der Schyf, C. J. 200[52]
van der Waals, A. 153[160]
Van Hecke, K. 343[74], 344[74]
van Hest, J. C. M. 652[208]
Van Ornum, S. G. 105[31], 153[31], 159[176]
van Otterlo, W. A. L. 630[168], 631[168]
Van Pelt, C. E. 101[12]
Van Speybroeck, V. 327[27], 328[27], 338[27]

van Wijngaarden, I. 395[212]
Varela-Alvarez, A. 59[42], 60[42]
Vasilevski, D. A. 88[88], 88[90], 89[88], 89[90], 90[88]
Vasilevskii, D. A. 88[93], 89[93]
Vasil'tsov, A. M. 395[214]
Vastag, K. 41[84]
Vázquez, J. 112[55], 139[135], 140[135], 141[135]
Vázquez-Romero, A. 109[44], 110[44], 111[44]
Velasco, J. 518[420], 519[420]
Velcicky, J. 167[7]
Velder, J. 562[48], 563[48]
Venkatraman, L. 215[22]
Ventura, D. L. 54[34], 54[35]
Verbraeken, B. 343[74], 344[74]
Verdaguer, X. 101[11], 102[19], 103[19], 106[34], 107[34], 107[35], 107[36], 107[37], 109[44], 110[44], 111[44], 112[55], 114[61], 139[132], 139[133], 139[134], 139[135], 140[132], 140[133], 140[134], 140[135], 141[132], 141[133], 141[134], 141[135], 144[141], 144[142], 144[143], 145[141]
Verenchikov, S. P. 385[98]
Verhoeven, T. R. 404[255], 404[256]
Verlhac, J.-B. 440[317], 441[317], 442[317]
Vikrant 359[36]
Vila, C. 406[263]
Villa, A. L. 18[53]
Villacampa, M. 395[217]
Villar, H. 470[328], 471[328], 612[138], 613[138]
Villemin, D. 544[1]
Vinader, V. 188[38], 190[38], 191[38], 192[38]
Vinogradov, M. G. 385[98]
Vinter, J. G. 322[13], 322[14]
Vintonyak, V. V. 658[220], 658[221], 659[220]
Visentini, F. 92[101]
Viski, P. 11[32]
Visser, M. S. 602[123]
Viton, F. 221[29], 222[29]
Vo, B. T. 40[83], 41[83]
Vogt, D. 391[164]
Vogt, M. 525[446]
Volbeda, J. 654[215], 655[215]
Volchkov, I. 587[102], 588[102], 597[115], 598[115]
Volkov, S. M. 88[89], 89[89]
Vollhardt, K. P. C. 265[15], 265[16], 265[18], 269[24], 270[24], 279[24], 283[39], 292[54]
Vollmar, A. M. 525[447]
von Kiedrowski, V. 593[110], 594[110]
von Matt, P. 215[19], 216[19]
Vorbrüggen, H. 71[63]
Vos, T. J. 614[140]
Vyas, R. 36[79], 37[79]

W

Wada, A. 266[20], 267[20], 268[20], 278[20], 291[51], 294[56], 295[56], 298[56]
Wada, J. 400[249]
Wada, K. 157[174], 158[174]
Wagaw, S. 410[278]
Wagner, D. 322[12]
Wagner, L. 517[417], 532[417]
Wagner, S. 400[232]
Wailes, P. C. 387[133]
Wakamatsu, H. 642[192], 643[192], 646[192]
Wakatsuki, Y. 283[38]
Walborsky, H. M. 205[4]
Walczak, K. 517[412]
Waldmann, H. 192[40], 247[65], 248[65]
Walker, M. 507[394]
Walling, C. 383[13]
Walters, M. A. 333[50], 334[56]
Walton, J. 383[45], 383[46]
Wan, H. 111[50]
Wan, X.-L. 73[68], 74[68]
Wandless, T. J. 54[30]
Wang, B. 8[28], 9[28], 11[28], 228[37], 229[37], 482[349], 483[349], 484[349], 652[208]
Wang, C. 19[56], 135[112], 384[82], 431[307], 433[307], 551[31], 552[31], 660[31], 661[31]
Wang, C.-J. 192[43], 246[64], 247[64]
Wang, G.-W. 383[55], 383[58], 383[62], 383[63], 383[64]
Wang, H. 59[42], 60[42], 120[79], 120[80], 255[75], 256[75]
Wang, J. 167[3], 168[3], 313[73], 377[88], 378[88], 379[88], 560[44], 561[44], 622[44], 623[44]
Wang, K.-P. 630[167]
Wang, L. 9[29], 10[29], 11[29], 42[29], 45[29], 228[39]
Wang, M. 498[364]
Wang, M.-Z. 198[51], 199[51], 200[51]
Wang, N. 228[39]
Wang, Q. 400[234]
Wang, S. 8[28], 9[28], 11[28]
Wang, S.-L. 152[159], 153[159]
Wang, X. 25[65], 27[65], 384[77], 384[78], 384[79], 503[78], 504[78], 504[79], 561[47], 562[47]
Wang, Y. 83[80], 84[80], 85[80], 115[68], 116[68], 328[29], 329[29], 329[30], 349[9], 377[88], 378[88], 379[9], 379[88], 552[32], 553[32]
Wang, Y.-J. 482[349], 483[349], 484[349]
Wang, Y.-Y. 395[218], 395[223], 396[223], 397[223]
Wang, Z. 239[55], 239[56]
Ward, D. W. 564[53], 565[53]
Ward, J. A. 52[25]
Warkentin, J. 167[8]
Waroquier, M. 327[27], 328[27], 338[27]
Wartchow, R. 322[12]

Waser, J. 373[83], 373[84], 374[83], 374[84], 375[83]
Waser, M. 52[25]
Washio, T. 241[59], 242[59], 253[72]
Watanabe, G. 584[95], 585[95]
Watanabe, H. 1[13], 346[86], 491[354]
Watanabe, K. 576[81], 577[81], 578[81]
Watanabe, M. 530[467]
Watanabe, Y. 253[72]
Watanuki, S. 545[10], 545[11], 545[12]
Waters, S. P. 138[126], 139[126]
Watkinson, M. 1[5]
Watson, W. H. 386[107], 387[107]
Watts, W. E. 99[5]
Wayner, D. D. M. 383[18]
Weatherly, C. D. 40[83], 41[83]
Weckesser, J. 644[194], 645[194]
Wedemann, P. 361[48], 362[48]
Weding, N. 265[5], 283[42], 284[42], 290[42]
Wegner, H. A. 363[54], 364[54]
Wei, H. 640[189]
Wei, L. 184[34]
Weigand, S. 389[144]
Weinland, R. F. 385[95]
Weinstabl, H. 596[114], 597[114]
Weller, M. D. 505[391]
Wen, J. 395[219], 395[220]
Wender, P. A. 123[86], 123[87], 124[87], 124[88], 125[88], 359[41], 359[42], 360[41], 360[43], 360[44], 360[45], 360[46], 361[43], 361[44], 362[49], 362[50], 363[51], 363[52], 363[53], 363[54], 364[54], 365[58], 366[58], 367[41], 367[43], 367[44], 370[64], 370[65], 370[66], 370[67], 370[68], 373[77], 373[78], 373[79], 374[79], 376[87], 377[87], 377[88], 378[87], 378[88], 379[88]
Wendt, O. F. 345[84], 346[84]
Wenger, R. 521[433]
Wengrovius, J. H. 650[200]
Werz, D. B. 49[3], 52[20], 52[21], 52[23], 52[24], 52[26], 53[23], 53[28], 55[23], 55[24]
West, F. G. 328[28], 328[29], 329[29], 329[30]
Wheeler, K. 255[75], 256[75]
White, A. J. P. 146[146], 147[146], 589[104], 590[104]
White, D. E. 572[74], 573[74]
Whitwood, A. C. 100[10]
Wiberg, K. B. 391[165]
Wicha, J. 88[96], 89[96]
Wicklow, D. T. 524[441], 524[442]
Wiest, O. 135[112]
Wikström, H. 71[66]
Williams, C. M. 332[47], 344[78]
Williams, J. E. 548[21]
Williams, P. H. 17[51]
Williams, T. J. 124[88], 125[88], 360[46]
Willison, D. 103[22], 104[22], 109[42], 110[42]
Willot, M. 579[85], 580[85]

Willwacher, J. 664[236], 665[236], 666[238], 667[238]
Wilmot, J. T. 408[271], 409[271]
Wilson, S. R. 7[24]
Wilton, D. 589[104], 590[104]
Wingert, D. 416[289], 417[289]
Wink, D. J. 630[166]
Winkler, I. 387[119], 517[416], 517[417], 517[418], 517[419], 532[416], 532[417], 532[418]
Winkler, J. D. 576[83], 577[83]
Winn, C. L. 1[3]
Winne, J. M. 327[27], 328[27], 338[27], 343[74], 344[74]
Winsel, H. 88[94], 89[94]
Wipf, P. 415[284], 416[284], 417[284]
Wirtz, C. 550[24]
Witulski, B. 268[21], 269[21], 278[21], 312[72], 359[41], 360[41], 367[41]
Woerpel, K. A. 24[62], 50[18], 51[18], 52[18]
Wojtas, L. 38[81], 39[81], 60[44], 61[44]
Wolf, J. R. 66[59]
Wolf, L. B. 652[208]
Woltering, M. J. 569[65], 569[66]
Wong, E. L.-M. 34[77], 35[77]
Wong, M.-K. 196[49], 197[49], 198[51], 199[49], 199[51], 200[51]
Woo, L. K. 66[59]
Woo, S. K. 634[177]
Wood, J. L. 391[165], 576[80], 577[80]
Woodall, E. L. 167[10]
Worgull, D. 387[115], 517[412], 517[414], 517[417], 517[419], 518[414], 532[417]
Wright, D. L. 321[8], 349[12]
Wright, J. A. 102[20], 103[20], 119[76], 155[164]
Wright, P. T. 108[39], 109[39]
Wrobleski, A. 605[130]
Wu, C.-J. 152[159], 153[159], 620[154]
Wu, J. 338[57], 339[57]
Wu, J.-C. 496[357], 497[357]
Wu, M. 8[28], 9[28], 11[28]
Wu, N. 134[111], 135[111]
Wu, W. 430[306]
Wu, Y.-D. 135[112]
Wu, Y.-L. 418[293], 419[293], 420[293]
Wulff, W. D. 213[17]

X

Xi, Y. 383[74]
Xia, C. 8[28], 9[28], 11[28]
Xiang, Y. B. 212[16], 213[16]
Xiao, W.-J. 406[264], 406[265], 413[264]
Xie, C. 52[25]
Xie, J.-H. 121[84], 122[84]
Xie, M. 226[35], 227[35], 255[74], 256[74]
Xie, X.-L. 69[61], 70[61]
Xin, J. 254[73]
Xiong, H. 334[52], 619[151]
Xu, C. 491[353], 492[353], 493[353], 494[353]
Xu, D. 389[153], 389[154]

Xu, F. 491[353], 492[353], 493[353], 494[353]
Xu, H. 69[61], 70[61]
Xu, H.-W. 196[49], 197[49], 198[51], 199[49], 199[51], 200[51]
Xu, J. 24[64], 27[64], 430[306]
Xu, L. 115[68], 116[68], 400[239]
Xu, M. 384[87], 384[88], 498[87], 498[88], 498[375]
Xu, Q. 26[67], 27[67], 28[67]
Xu, Q.-L. 31[73], 32[73]
Xu, W. F. 400[234]
Xu, X. 61[45], 62[45], 188[39], 190[39], 191[39], 192[39], 370[68], 370[71]
Xu, Z. 255[75], 256[75]
Xuan, J. 406[265]
Xue, J. 338[58]

Y

Yadav, J. S. 389[145]
Yamabe, M. 50[15]
Yamada, H. 511[404]
Yamada, M. 103[24], 104[24], 106[24]
Yamada, T. 384[80], 508[80], 509[80]
Yamada, Y. 488[352], 489[352]
Yamaguchi, H. 530[467]
Yamaguchi, J. 598[116], 599[116]
Yamaguchi, K. 18[54]
Yamaguchi, M. 103[24], 104[24], 106[24], 115[69], 116[69], 117[69], 119[75], 394[200], 394[202], 477[342], 477[344]
Yamaguchi, Y. 625[159]
Yamamoto, F. 600[119], 601[119]
Yamamoto, H. 4[19], 4[20], 5[20], 5[21], 6[20], 6[21], 6[22], 7[21], 7[22], 11[40], 12[40], 14[40], 16[50], 17[50], 19[56], 22[60], 23[60], 23[61], 24[61], 42[86], 45[86], 50[19], 51[19], 206[6], 206[7], 208[11], 209[11], 210[11], 211[15], 212[15], 217[26], 218[26], 219[26], 225[34], 226[34], 395[221]
Yamamoto, Y. 115[66], 116[66], 265[19], 271[25], 271[26], 272[25], 272[26], 279[25], 285[45], 285[46], 286[46], 290[45], 293[45], 298[45], 473[330], 473[333]
Yamanaka, M. 113[58], 114[58]
Yamanaka, Y. 473[333]
Yamanishi, K. 125[90], 126[90]
Yamano, A. 175[26], 176[26], 179[26], 180[26], 183[26]
Yamaoka, M. 511[407], 522[434], 523[434]
Yamasaki, M. 129[100], 130[100]
Yamashita, S. 384[80], 505[392], 506[392], 508[80], 509[80]
Yamashita, Y. 232[45], 232[46], 251[70], 260[70], 261[70]
Yamatsugu, K. 571[69]
Yamazaki, H. 265[17], 283[38], 292[52]
Yam-Puc, A. 345[84], 346[84]
Yan, M. 193[46], 193[47], 194[46]
Yang, B. H. 410[278]
Yang, C. G. 400[240]

Yang, D. 1[2], 377[88], 378[88], 379[88], 384[87], 384[88], 498[87], 498[88], 498[375]
Yang, F. Z. 498[372]
Yang, H. 9[29], 10[29], 11[29], 42[29], 45[29]
Yang, L. 188[39], 190[39], 191[39], 192[39]
Yang, X. 228[40], 229[40]
Yang, X.-B. 228[39]
Yang, Y. 159[184], 160[184], 161[184]
Yang, Z. 115[67], 115[68], 116[67], 116[68], 119[67], 134[110], 134[111], 135[111], 135[112], 159[183], 160[183], 161[183], 640[189]
Yao, S. 236[50], 237[50]
Yap, G. P. A. 108[38], 168[13], 184[13], 185[13], 186[13], 187[13]
Yasue, K. 501[388]
Yasui, H. 584[95], 585[95]
Yasumura, S. 175[29], 181[29], 182[29], 184[29]
Ye, S. 364[55], 365[55], 367[55]
Ye, X.-Y. 384[87], 384[88], 498[87], 498[88], 498[375]
Yet, L. 349[5]
Yeung, Y.-Y. 257[77], 258[77]
Yia, H. 383[74]
Yin, J. 384[82]
Ying, J. 152[158]
Ylijoki, K. E. O. 359[38]
Yokoe, H. 159[185], 160[185], 161[185]
Yokoshima, S. 574[78], 575[78], 607[133], 608[133]
Yokota, K. 274[30], 275[30], 280[30]
Yokoyama, Y. 394[200]
Yokozawa, T. 273[29], 274[29]
Yonehara, K. 18[54]
Yonemura, K. 473[333]
Yongpruksa, N. 591[107]
Yoo, S.-E. 102[16], 103[16]
Yoon, C. H. 498[379]
Yoon, J.-Y. 360[46]
Yoon, T. P. 406[258], 408[272], 409[272], 435[314], 436[314], 437[314], 438[314], 469[272], 470[272]
Yorimitsu, H. 383[35]
Yoshida, J.-i. 104[28], 132[108], 133[108]
Yoshida, K. 609[135]
Yoshida, M. 159[185], 160[185], 161[185]
Yoshida, S. 522[435]
Yoshida, T. 156[171], 157[171]
Yoshimitsu, T. 521[425]
You, Q. 521[427]
You, T. 239[56]
You, X. 383[62], 383[63]
Young, D. W. 552[32], 553[32]
Yousufuddin, M. 31[73], 32[73]
Yu, J. 192[41]
Yu, M. 258[78], 551[31], 552[31], 660[31], 661[31]
Yu, R. 115[68], 116[68]
Yu, R. T. 295[58], 296[58], 299[58]
Yu, T.-C. 383[40]
Yu, W. 420[300], 421[300]
Yu, W.-Y. 168[18], 170[18], 171[18], 172[18], 174[18], 195[48], 196[48], 199[48]
Yu, X.-Q. 152[158], 228[39], 395[219], 395[220]
Yu, Z. 226[35], 227[35]
Yu, Z.-X. 349[9], 357[34], 358[34], 364[55], 364[56], 365[55], 367[55], 370[64], 370[65], 370[66], 370[67], 373[85], 377[88], 378[88], 379[9], 379[85], 379[88], 379[89], 380[89]
Yuan, C. 373[85], 379[85]
Yuan, Y. 230[42], 230[43], 231[42], 231[44]
Yudin, A. K. 1[4], 1[8]
Yuen, A. W.-H. 34[77], 35[77]
Yun, H. J. 345[81]
Yun, M. 498[379]
Yun, S. Y. 571[72], 572[72], 597[115], 598[115], 630[166], 630[167]
Yusa, Y. 273[29], 274[29]
Yusuff, N. 576[80], 577[80]

Z

Zabel, M. 52[22]
Zahel, M. 579[86], 580[86]
Zahra, J. P. 433[312]
Zanoni, M. 52[26]
Zard, S. Z. 383[7], 393[184], 406[267], 408[274], 408[275], 409[274], 409[275]
Zarrabi, S. 440[316]
Zeeck, A. 507[394]
Zefirov, N. S. 88[91], 89[91]
Zeldin, R. M. 330[38]
Zemp, I. 259[85], 260[85]
Zeng, R.-S. 496[357], 497[357]
Zengin, M. 383[59]
Zhai, H. 159[184], 160[184], 161[184]
Zhan, Z.-P. 460[325], 461[325]
Zhang, H. 91[99], 574[77]
Zhang, J. 75[70], 184[34], 228[39], 338[58], 366[59], 367[59], 395[219], 395[220], 400[234]
Zhang, J. L. 400[240]
Zhang, L. 57[39], 85[81], 373[78], 376[87], 377[87], 378[87]
Zhang, M. 356[32], 356[33], 357[32]
Zhang, Q. 498[359], 498[367]
Zhang, R.-Y. 395[219]
Zhang, S. 377[88], 378[88], 379[88]
Zhang, S. H. 400[239]
Zhang, S.-Q. 71[67], 72[67]
Zhang, T. 44[89], 46[89]
Zhang, W. 4[20], 5[20], 5[21], 6[20], 6[21], 6[22], 7[21], 7[22], 7[24], 383[50], 668[240], 668[241], 669[243], 670[243]
Zhang, X. 11[35], 239[55], 239[56]
Zhang, X. P. 35[78], 36[78], 36[79], 37[79], 37[80], 38[80], 38[81], 39[80], 39[81], 60[44], 61[44], 61[45], 62[45]
Zhang, X.-X. 406[265]
Zhang, Y. 115[67], 116[67], 119[67], 134[110], 498[365], 570[67]
Zhao, H. 560[44], 561[44], 622[44], 623[44]
Zhao, M.-N. 395[218], 395[223], 396[223], 397[223]
Zhao, Y. 623[156]
Zheng, G. Z. 394[197]
Zheng, J.-C. 571[72], 572[72]
Zheng, Y.-H. 550[26]
Zhou, C.-Y. 168[18], 170[18], 171[18], 172[18], 174[18], 192[44], 194[44]
Zhou, F. 420[296]
Zhou, G. 257[76]
Zhou, H. 412[283], 413[283], 415[283]
Zhou, J. 420[296]
Zhou, L. 434[313], 435[313]
Zhou, Q.-L. 54[36], 55[36], 69[61], 70[61], 121[84], 122[84]
Zhou, S. 410[282]
Zhou, X. 254[73], 368[61], 369[61]
Zhou, Z.-Y. 63[51]
Zhu, J. 651[206]
Zhu, M. 491[353], 492[353], 493[353], 494[353]
Zhu, S. 37[80], 38[80], 39[80], 60[44], 61[44], 61[45], 62[45], 406[266], 412[283], 413[283], 415[283]
Zhu, S.-F. 69[61], 70[61]
Zhu, S. S. 564[54], 565[54], 566[54]
Zhu, X.-Z. 73[68], 74[68]
Zhu, Y. 255[74], 256[74]
Zhu, Z.-B. 643[193], 644[193], 646[193]
Ziani-Cherif, C. 326[24], 326[25], 327[24], 327[25]
Ziegler, F. E. 568[62]
Ziegler, S. 247[65], 248[65]
Ziffle, V. E. 416[289], 417[289]
Ziller, J. W. 394[193], 651[205], 659[226]
Zillig, P. 528[460], 529[460]
Zimmer, R. 49[2]
Zimmerman, J. 473[338], 473[339]
Zocchi, A. 92[101]
Zonzini, L. 92[101]
Zoretic, P. A. 384[84], 498[364], 498[365]
Zou, D. 491[353], 492[353], 493[353], 494[353]
Zou, J.-P. 383[50], 496[357], 497[357]
Zou, Y. 168[15], 200[15]
Zou, Y.-Q. 406[264], 413[264]
Zrig, S. 11[34]
Zuercher, W. J. 602[122], 618[150]
Zugay, J. 41[84]
Zuo, G. 371[75], 372[75]

Abbreviations

Chemical

Name Used in Text	Abbreviation Used in Tables and on Arrow in Schemes	Abbreviation Used in Experimental Procedures
(R)-1-amino-2-(methoxymethyl)pyrrolidine	RAMP	RAMP
(S)-1-amino-2-(methoxymethyl)pyrrolidine	SAMP	SAMP
ammonium cerium(IV) nitrate	CAN	CAN
2,2′-azobisisobutyronitrile	AIBN	AIBN
barbituric acid	BBA	BBA
benzyltriethylammonium bromide	TEBAB	TEBAB
benzyltriethylammonium chloride	TEBAC	TEBAC
N,O-bis(trimethylsilyl)acetamide	BSA	BSA
9-borabicyclo[3.3.1]nonane	9-BBNH	9-BBNH
borane–methyl sulfide complex	BMS	BMS
N-bromosuccinimide	NBS	NBS
tert-butyldimethylsilyl chloride	TBDMSCl	TBDMSCl
tert-butyl peroxybenzoate	TBPB	tert-butyl peroxybenzoate
10-camphorsulfonic acid	CSA	CSA
chlorosulfonyl isocyanate	CSI	chlorosulfonyl isocyanate
3-chloroperoxybenzoic acid	MCPBA	MCPBA
N-chlorosuccinimide	NCS	NCS
chlorotrimethylsilane	TMSCl	TMSCl
1,4-diazabicyclo[2.2.2]octane	DABCO	DABCO
1,5-diazabicyclo[4.3.0]non-5-ene	DBN	DBN
1,8-diazabicyclo[5.4.0]undec-7-ene	DBU	DBU
dibenzoyl peroxide	DBPO	dibenzoyl peroxide
dibenzylideneacetone	dba	dba
di-tert-butyl azodicarboxylate	DBAD	di-tert-butyl azo-dicarboxylate
di-tert-butyl peroxide	DTBP	DTBP
2,3-dichloro-5,6-dicyanobenzo-1,4-quinone	DDQ	DDQ
dichloromethyl methyl ether	DCME	DCME
dicyclohexylcarbodiimide	DCC	DCC
N,N-diethylaminosulfur trifluoride	DAST	DAST
diethyl azodicarboxylate	DEAD	DEAD
diethyl tartrate	DET	DET
2,2′-dihydroxy-1,1′-binaphthyllithium aluminum hydride	BINAL-H	BINAL-H
diisobutylaluminum hydride	DIBAL-H	DIBAL-H
diisopropyl tartrate	DIPT	DIPT

Chemical (cont.)

Name Used in Text	Abbreviation Used in Tables and on Arrow in Schemes	Abbreviation Used in Experimental Procedures
1,2-dimethoxyethane	DME	DME
dimethylacetamide	DMA	DMA
dimethyl acetylenedicarboxylate	DMAD	DMAD
2-(dimethylamino)ethanol	Me$_2$N(CH$_2$)$_2$OH	2-(dimethylamino)ethanol
4-(dimethylamino)pyridine	DMAP	DMAP
dimethylformamide	DMF	DMF
dimethyl sulfide	DMS	DMS
dimethyl sulfoxide	DMSO	DMSO
1,3-dimethyl-3,4,5,6-tetrahydro-pyrimidin-2(1*H*)-one	DMPU	DMPU
ethyl diazoacetate	EDA	EDA
ethylenediaminetetraacetic acid	edta	edta
hexamethylphosphoric triamide	HMPA	HMPA
hexamethylphosphorous triamide	HMPT	HMPT
iodomethane	MeI	MeI
N-iodosuccinimide	NIS	NIS
lithium diisopropylamide	LDA	LDA
lithium hexamethyldisilazanide	LiHMDS	LiHMDS
lithium isopropylcyclohexylamide	LICA	LICA
lithium 2,2,6,6-tetramethylpiperidide	LTMP	LTMP
lutidine	lut	lut
methylaluminum bis(2,6-di-*tert*-butyl-4-methylphenoxide)	MAD	MAD
methyl ethyl ketone	MEK	methyl ethyl ketone
methylmaleimide	NMM	NMM
4-methylmorpholine *N*-oxide	NMO	NMO
1-methylpyrrolidin-2-one	NMP	NMP
methyl vinyl ketone	MVK	methyl vinyl ketone
petroleum ether	PE[a]	petroleum ether
N-phenylmaleimide	NPM	NPM
polyphosphoric acid	PPA	PPA
polyphosphate ester	PPE	polyphosphate ester
potassium hexamethyldisilazanide	KHMDS	KHMDS
pyridine	pyridine[b]	pyridine
pyridinium chlorochromate	PCC	PCC
pyridinium dichromate	PDC	PDC
pyridinium 4-toluenesulfonate	PPTS	PPTS
sodium bis(2-methoxyethoxy)aluminum hydride	Red-Al	Red-Al
tetrabutylammonium bromide	TBAB	TBAB

[a] Used to save space; abbreviation must be defined in a footnote.
[b] py used on arrow in schemes.

Chemical (cont.)

Name Used in Text	Abbreviation Used in Tables and on Arrow in Schemes	Abbreviation Used in Experimental Procedures
tetrabutylammonium chloride	TBACl	TBACl
tetrabutylammonium fluoride	TBAF	TBAF
tetrabutylammonium iodide	TBAI	TBAI
tetracyanoethene	TCNE	tetracyanoethene
tetrahydrofuran	THF	THF
tetrahydropyran	THP	THP
2,2,6,6-tetramethylpiperidine	TMP	TMP
trimethylamine N-oxide	TMANO	trimethylamine N-oxide
N,N,N′,N′-tetramethylethylenediamine	TMEDA	TMEDA
tosylmethyl isocyanide	TosMIC	TosMIC
trifluoroacetic acid	TFA	TFA
trifluoroacetic anhydride	TFAA	TFAA
trimethylsilyl cyanide	TMSCN	TMSCN

Ligands

acetylacetonato	acac
2,2′-bipyridyl	bipy
1,2-bis(dimethylphosphino)ethane	DMPE
2,3-bis(diphenylphosphino)bicyclo[2.2.1]hept-5-ene	NORPHOS
2,2′-bis(diphenylphosphino)-1,1′-binaphthyl	BINAP
1,2-bis(diphenylphosphino)ethane	dppe (not diphos)
1,1′-bis(diphenylphosphino)ferrocene	dppf
bis(diphenylphosphino)methane	dppm
1,3-bis(diphenylphosphino)propane	dppp
1,4-bis(diphenylphosphino)butane	dppb
2,3-bis(diphenylphosphino)butane	Chiraphos
bis(salicylidene)ethylenediamine	salen
cyclooctadiene	cod
cyclooctatetraene	cot
cyclooctatriene	cte
η^5-cyclopentadienyl	Cp
dibenzylideneacetone	dba
6,6-dimethylcyclohexadienyl	dmch
2,4-dimethylpentadienyl	dmpd
ethylenediaminetetraacetic acid	edta
isopinocampheyl	Ipc
2,3-O-isopropylidene-2,3-dihydroxy-1,4-bis(diphenylphosphino)butane	Diop
norbornadiene (bicyclo[2.2.1]hepta-2,5-diene)	nbd
η^5-pentamethylcyclopentadienyl	Cp*

Radicals

acetyl	Ac
aryl	Ar
benzotriazol-1-yl	Bt
benzoyl	Bz
benzyl	Bn
benzyloxycarbonyl	Cbz
benzyloxymethyl	BOM
9-borabicyclo[3.3.1]nonyl	9-BBN
tert-butoxycarbonyl	Boc
butyl	Bu
sec-butyl	*s*-Bu
tert-butyl	*t*-Bu
tert-butyldimethylsilyl	TBDMS
tert-butyldiphenylsilyl	TBDPS
cyclohexyl	Cy
3,4-dimethoxybenzyl	DMB
ethyl	Et
ferrocenyl	Fc
9-fluorenylmethoxycarbonyl	Fmoc
isobutyl	iBu
mesityl	Mes
mesyl	Ms
4-methoxybenzyl	PMB
(2-methoxyethoxy)methyl	MEM
methoxymethyl	MOM
methyl	Me
4-nitrobenzyl	PNB
phenyl	Ph
phthaloyl	Phth
phthalimido	NPhth
propyl	Pr
isopropyl	iPr
tetrahydropyranyl	THP
tolyl	Tol
tosyl	Ts
triethylsilyl	TES
triflyl, trifluoromethanesulfonyl	Tf
triisopropylsilyl	TIPS
trimethylsilyl	TMS
2-(trimethylsilyl)ethoxymethyl	SEM
trityl [triphenylmethyl]	Tr

General

absolute	abs
anhydrous	anhyd
aqueous	aq
boiling point	bp
catalyst	no abbreviation
catalytic	cat.
chemical shift	δ
circular dichroism	CD
column chromatography	no abbreviation
concentrated	concd
configuration (in tables)	Config
coupling constant	J
day	d
density	d
decomposed	dec
degrees Celsius	°C
diastereomeric ratio	dr
dilute	dil
electron-donating group	EDG
electron-withdrawing group	EWG
electrophile	E^+
enantiomeric excess	ee
enantiomeric ratio	er
equation	eq
equivalent(s)	equiv
flash-vacuum pyrolysis	FVP
gas chromatography	GC
gas chromatography-mass spectrometry	GC/MS
gas–liquid chromatography	GLC
gram	g
highest occupied molecular orbital	HOMO
high-performance liquid chromatography	HPLC
hour(s)	h
infrared	IR
in situ	in situ
in vacuo	in vacuo
lethal dosage, e.g. to 50% of animals tested	LD_{50}
liquid	liq
liter	L
lowest unoccupied molecular orbital	LUMO
mass spectrometry	MS
medium-pressure liquid chromatography	MPLC
melting point	mp
milliliter	mL
millimole(s)	mmol
millimoles per liter	mM
minute(s)	min
mole(s)	mol
nuclear magnetic resonance	NMR
nucleophile	Nu^-
optical purity	op
phase-transfer catalysis	PTC
proton NMR	^{1}H NMR

General (cont.)

quantitative	quant
reference (in tables)	Ref
retention factor (for TLC)	R_f
retention time (chromatography)	t_R
room temperature	rt
saturated	sat.
solution	soln
temperature (in tables)	Temp (°C)
thin layer chromatography	TLC
ultraviolet	UV
volume (literature)	Vol.
via	via
vide infra	*vide infra*
vide supra	*vide supra*
yield (in tables)	Yield (%)

List of All Volumes

Science of Synthesis, Houben–Weyl Methods of Molecular Transformations

Category 1: Organometallics

1. Compounds with Transition Metal—Carbon π-Bonds and Compounds of Groups 10–8 (Ni, Pd, Pt, Co, Rh, Ir, Fe, Ru, Os)
2. Compounds of Groups 7–3 (Mn···, Cr···, V···, Ti···, Sc···, La···, Ac···)
3. Compounds of Groups 12 and 11 (Zn, Cd, Hg, Cu, Ag, Au)
4. Compounds of Group 15 (As, Sb, Bi) and Silicon Compounds
5. Compounds of Group 14 (Ge, Sn, Pb)
6. Boron Compounds
7. Compounds of Groups 13 and 2 (Al, Ga, In, Tl, Be ··· Ba)
8a. Compounds of Group 1 (Li ··· Cs)
8b. Compounds of Group 1 (Li ··· Cs)

Category 2: Hetarenes and Related Ring Systems

9. Fully Unsaturated Small-Ring Heterocycles and Monocyclic Five-Membered Hetarenes with One Heteroatom
10. Fused Five-Membered Hetarenes with One Heteroatom
11. Five-Membered Hetarenes with One Chalcogen and One Additional Heteroatom
12. Five-Membered Hetarenes with Two Nitrogen or Phosphorus Atoms
13. Five-Membered Hetarenes with Three or More Heteroatoms
14. Six-Membered Hetarenes with One Chalcogen
15. Six-Membered Hetarenes with One Nitrogen or Phosphorus Atom
16. Six-Membered Hetarenes with Two Identical Heteroatoms
17. Six-Membered Hetarenes with Two Unlike or More than Two Heteroatoms and Fully Unsaturated Larger-Ring Heterocycles

Category 3: Compounds with Four and Three Carbon—Heteroatom Bonds

18. Four Carbon—Heteroatom Bonds: X—C≡X, X=C=X, X_2C=X, CX_4
19. Three Carbon—Heteroatom Bonds: Nitriles, Isocyanides, and Derivatives
20a. Three Carbon—Heteroatom Bonds: Acid Halides; Carboxylic Acids and Acid Salts
20b. Three Carbon—Heteroatom Bonds: Esters and Lactones; Peroxy Acids and R(CO)OX Compounds; R(CO)X, X = S, Se, Te
21. Three Carbon—Heteroatom Bonds: Amides and Derivatives; Peptides; Lactams
22. Three Carbon—Heteroatom Bonds: Thio-, Seleno-, and Tellurocarboxylic Acids and Derivatives; Imidic Acids and Derivatives; Ortho Acid Derivatives
23. Three Carbon—Heteroatom Bonds: Ketenes and Derivatives
24. Three Carbon—Heteroatom Bonds: Ketene Acetals and Yne—X Compounds

Category 4: Compounds with Two Carbon–Heteroatom Bonds

25 Aldehydes
26 Ketones
27 Heteroatom Analogues of Aldehydes and Ketones
28 Quinones and Heteroatom Analogues
29 Acetals: Hal/X and O/O, S, Se, Te
30 Acetals: O/N, S/S, S/N, and N/N and Higher Heteroatom Analogues
31a Arene—X (X = Hal, O, S, Se, Te)
31b Arene—X (X = N, P)
32 X—Ene—X (X = F, Cl, Br, I, O, S, Se, Te, N, P), Ene—Hal, and Ene—O Compounds
33 Ene—X Compounds (X = S, Se, Te, N, P)

Category 5: Compounds with One Saturated Carbon–Heteroatom Bond

34 Fluorine
35 Chlorine, Bromine, and Iodine
36 Alcohols
37 Ethers
38 Peroxides
39 Sulfur, Selenium, and Tellurium
40a Amines and Ammonium Salts
40b Amine *N*-Oxides, Haloamines, Hydroxylamines and Sulfur Analogues, and Hydrazines
41 Nitro, Nitroso, Azo, Azoxy, and Diazonium Compounds, Azides, Triazenes, and Tetrazenes
42 Organophosphorus Compounds (incl. RO—P and RN—P)

Category 6: Compounds with All-Carbon Functions

43 Polyynes, Arynes, Enynes, and Alkynes
44 Cumulenes and Allenes
45a Monocyclic Arenes, Quasiarenes, and Annulenes
45b Aromatic Ring Assemblies, Polycyclic Aromatic Hydrocarbons, and Conjugated Polyenes
46 1,3-Dienes
47a Alkenes
47b Alkenes
48 Alkanes